Fundamentals of Materials for Energy and Environmental Sustainability

How will we meet rising energy demands? What are our options? Are there viable long-term solutions for the future?

Learn the fundamental physical, chemical, and materials science at the heart of:

- nonrenewable energy sources
- renewable energy sources
- future transportation systems
- energy efficiency
- energy storage

Whether you are a student taking an energy course or a newcomer to the field, this book will help you understand the critical relationships of the environment, energy, and sustainability. Leading experts provide comprehensive coverage of each topic, bringing together diverse subject matter by integrating theory with engaging insights. Each chapter also includes helpful features to aid understanding, including a historical overview to provide context, references for further reading, and questions for discussion. The subject is beautifully illustrated and brought to life with full-color images and color-coded sections for easy browsing, making this a complete educational package.

Fundamentals of Materials for Energy and Environmental Sustainability
Enabling today's scientists. Educating future generations.

DAVID S. GINLEY is a Research Fellow and Group Manager in the Process Technology Group, National Center for Photovoltaics at the National Renewable Energy Laboratory (NREL). He received his B.S. in Mineral Engineering Chemistry from the Colorado School of Mines and his Ph.D. in Inorganic Chemistry from MIT. He is also a former President of the Materials Research Society (MRS). His principal areas of interest are in the application of organic/polymer materials, transition metal oxides and hybrid inorganic–organic semiconductor-based nanomaterials to energy conversion, and energy efficiency in the areas of photovoltaics, batteries, fuel cells, and OLEDs. He has over 400 publications and 30 patents.

DAVID CAHEN is a Professor at the Weizmann Institute of Science. He received his B.Sc. in Chemistry and Physics at the Hebrew University of Jerusalem (HUJI) and his Ph.D. in Materials Research and Physical Chemistry from Northwestern University. He then joined the Weizmann Institute in 1976, where he started work on solar cells. Today, his research focuses on alternative, especially solar, energy sources, aiming to understand electronic transport across (bio)molecules, investigating how molecules can control such transport, and looking for novel science in such systems and for possible relevance to solar cells.

This book fills an information gap in energy, environment, and sustainability, presenting broad overviews of energy challenges and solutions along with the materials advances needed to enable rapid progress. It is authoritative, insightful, and a thoroughly enjoyable read for the general audience, for energy analysts, and for students entering the field.

George Crabtree, Argonne National Laboratory

Fundamentals of Materials for Energy and Environmental Sustainability is essential reading for anyone who wishes to understand today's (and tomorrow's) energy landscape. The book focuses on the materials that are essential to energy production and the materials science research challenge posed by meeting the demands made on them. The articles, written by experts in their respective fields, cover a comprehensive range of subjects related to energy and environmental concerns. Titles of articles are broadly grouped into themes ranging from environmental concerns to an examination of energy sources to transportation issues to schemes for energy storage. Useful tutorials are included on basic issues. This book is an invaluable source of material for college courses on energy, environmental consequences, and sustainability. It also serves as an excellent primer for those who wish to become knowledgeable on this major challenge of the present century: meeting global energy demands while preserving our environment.

Julia Weertman, Northwestern University

This book represents one of the most integrated and inclusive texts on the topics of materials for energy and environmental sustainability. It provides an inclusive and fair picture of all the different alternatives for energy generation, energy efficiency, and covers many new approaches to clean energy use. You will find here all the information and data that cannot be found in other textbooks. The different chapters are written by very high profile scientists but at a consistent level. The book is beautifully illustrated and the reader will not find preselected solutions to the energy problem but will have all the information necessary to propose his\her own solution. The chapters are completed by a very useful section on "questions for discussion" to challenge the reader and by a list of further reading for more in-depth analysis. This book was really needed and I'm sure it will become a reference in the field of materials for energy and environmental sustainability not only for the students but also for researchers working in the field.

Francesco Priolo, MATIS CNR-IMM

This book addresses all the critical energy and environmental issues faced by the world and critically assesses the current options for mankind. Written by leading experts, it presents the most timely and comprehensive review, and is by far the best resource available for all students, educators, scientists, economists, and policy makers interested in understanding the options provided by advanced materials for solving the upcoming energy problems. A monumental work!

Yves Chabal, University of Texas at Dallas

This book explains clearly the dilemma we are facing: what are the implications of replacing roughly 1 Terawatt of energy (which presently is produced from approximately 10 billions tons of fossil fuels) by renewable resources (mainly solar). The materials and devices needed will be substantially different in the form of photovoltaic panels, smart grid, energy storage, batteries, and many others. The book discusses all these issues and should be very useful for anyone working or teaching on paths for a sustainable energy future.

José Goldemberg, University of São Paulo

Fundamentals of Materials for Energy and Environmental Sustainability

EDITED BY

DAVID S. GINLEY
National Renewable Energy Laboratory (NREL)

and

DAVID CAHEN
Weizmann Institute of Science, Israel

CAMBRIDGE UNIVERSITY PRESS
Cambridge, New York, Melbourne, Madrid, Cape Town
Singapore, São Paulo, Delhi, Tokyo, Mexico City

Cambridge University Press
The Edinburgh Building, Cambridge CB2 8RU, UK

Published in the United States of America by
Cambridge University Press, New York

www.cambridge.org
Information on this title: www.cambridge.org/9781107000230

First published 2012

Printed in the United States of America

A catalogue record for this publication is available from the British Library

Library of Congress Cataloging-in-Publication Data

Fundamentals of materials for energy and environmental sustainability / edited by
David S. Ginley, David Cahen.
     p.   cm.
  ISBN 978-1-107-00023-0 (Hardback)
  1. Energy conservation–Equipment and supplies.   2. Renewable energy
sources. 3. Power resources.   4. Fuel.   5. Sustainable engineering–Materials.
I. Ginley, D. S. (David S.)   II. Cahen, David.   III. Title.
  TJ163.3.F855 2011
  621.042–dc23

                                                                              2011027888

ISBN 978-1-107-00023-0 Hardback

Additional resources for this publication at www.cambridge.org/9781107000230

Contents

Part 1 Energy and the environment: the global landscape

Part 2 Nonrenewable energy sources

Part 3 Renewable energy sources

Part 4 Transportation

Part 5 Energy efficiency

Part 6 Energy storage, high-penetration renewables, and grid stabilization

Contributors

Wali Akande
Princeton University

Joe A. Almaguer
The Dow Chemical Company

Elisa Alonso
Massachusetts Institute of Technology

S. Massoud Amin
University of Minnesota

V. S. Arunachalam
Carnegie Mellon University

Robin G. Bennett
The Boeing Company

Sally M. Benson
Stanford University

Anshu Bharadwaj
Center for Study of Science, Technology, and Policy (CSTEP)

Benjamin Bollinger
SustainX, Inc.

Christopher E. Borroni-Bird
General Motors

James W. Bunger
James W. Bunger and Associates, Inc.

Linda A. Cadwell Stancin
The Boeing Company

David Cahen
Weizmann Institute of Science

William L. Carberry
The Boeing Company

J. William Carey
Los Alamos National Laboratory

Russell R. Chianelli
University of Texas at El Paso

Corrie E. Clark
Argonne National Laboratory

Adam Cohen
Princeton University

Ellann Cohen
Massachusetts Institute of Technology

Reuben Collins
Colorado School of Mines

Miguel A. Contreras
National Renewable Energy Laboratory (NREL)

Satyen Deb
National Renewable Energy Laboratory (NREL)

Huub J. M. de Groot
Leiden University

Paul Denholm
National Renewable Energy Laboratory (NREL)

Matthias Englert
Stanford University

Michael Epstein
Weizmann Institute of Science

Rodney C. Ewing
University of Michigan

Frank R. Field, III
Massachusetts Institute of Technology

Nathaniel J. Fisch
Princeton University

Daniel J. Friedman
National Renewable Energy Laboratory (NREL)

List of contributors

Thomas Gennett
National Renewable Energy Laboratory (NREL)

Anthony M. Giacomoni
University of Minnesota

Jerry Gibbs
U.S. Department of Energy

David Ginley
National Renewable Energy Laboratory (NREL)

Leon Glicksman
Massachusetts Institute of Technology

Robin W. Grimes
Imperial College London

Jean-François Guillemoles
Institut de Recherche et Développement sur l'Energie Photovoltaïque

Ajay K. Gupta
SUNY College of Environmental Science and Forestry

Charles A. S. Hall
SUNY College of Environmental Science and Forestry

Jiabin Han
Los Alamos National Laboratory

Siegfried S. Hecker
Stanford University

Stephen A. Holditch
Texas A&M University

Colin J. Humphreys
University of Cambridge

Christian Jooss
Georg August University Göttingen

Ron Judkoff
National Renewable Energy Laboratory (NREL)

Neil Kelley
National Renewable Energy Laboratory (NREL)

Dax Kepshire
SustainX, Inc.

Randolph Kirchain
Massachusetts Institute of Technology

Shayam Kocha
National Renewable Energy Laboratory (NREL)

Carolyn A. Koh
Colorado School of Mines

Fridolin Krausmann
Alpen Adria Universität (AAU)

Xiomara C. Kretschmer
University of Texas at El Paso

Abraham Kribus
Tel Aviv University

Lester B. Lave
Carnegie Mellon University

Stuart Licht
George Washington University

Todd A. Lloyd
Mascoma Corporation

David R. Luebke
National Energy Technology Laboratory (NETL)

Jennifer A. Nekuda Malik
Imperial College London

Stuart A. Maloy
Los Alamos National Laboratory

Laura D. Marlino
Oak Ridge National Laboratory (ORNL)

Melinda Marquis
National Oceanic and Atmospheric Administration (NOAA)

Troy McBride
SustainX, Inc.

Johannes Messinger
Umeå University

Michael C. Miller
Los Alamos National Laboratory

Patrick Moriarty
National Renewable Energy Laboratory (NREL)

Bryan D. Morreale
National Energy Technology Laboratory (NETL)

Ronny Neumann
Weizmann Institute of Science

Trent R. Northen
Lawrence Berkeley National Laboratory

Franklin M. Orr, Jr.
Stanford University

Ahmad A. Pesaran
National Renewable Energy Laboratory (NREL)

J. Luc Peterson
Princeton University

Bryan Pivovar
National Renewable Energy Laboratory (NREL)

Cynthia A. Powell
National Energy Technology Laboratory (NETL)

Qing Qing
University of California, Riverside

Timothy F. Rahmes
The Boeing Company

Bruce A. Robinson
Los Alamos National Laboratory

Michael Robinson
National Renewable Energy Laboratory (NREL)

Boris Rybtchinski
Weizmann Institute of Science

Richard Sassoon
Stanford University

Scott Schreck
National Renewable Energy Laboratory (NREL)

Sara C. Scott
Los Alamos National Laboratory

Dmitriy Shevela
University of Stavanger

Jian Shi
University of California, Riverside

David Simms
National Renewable Energy Laboratory (NREL)

Philip S. Sklad
Oak Ridge National Laboratory (ORNL)

E. Dendy Sloan
Colorado School of Mines

Winston Soboyejo
Princeton University

Christopher R. Stanek
Los Alamos National Laboratory

Amadeu K. Sum
Colorado School of Mines

Xinfeng Tang
Wuhan University of Technology

Pieter Tans
National Oceanic and Atmospheric Administration (NOAA)

Peter M. Thompson
The Boeing Company

Tiffany Tong
Princeton University

Helmut Tributsch
Free University Berlin and Helmholtz Center Berlin for Materials and Energy

Terry M. Tritt
Clemson University

Cetin Unal
Los Alamos National Laboratory

Mark. W. Verbrugge
General Motors

Michael R. Wasielewski
Northwestern University

William J. Weber
University of Tennessee

M. Stanley Whittingham
Binghamton University

Alan Wright
National Renewable Energy Laboratory (NREL)

David T. Wu
Colorado School of Mines

List of contributors

Charles E. Wyman
University of California, Riverside

Wenjie Xie
Clemson University and Wuhon University of Technology

Jeanne C. Yu
The Boeing Company

Qingjie Zhang
Wuhan University of Technology

Taiying Zhang
University of California, Riverside

Dandan Zhu
University of Cambridge

Preface

Academically and industrially there is increasing awareness that energy and the environment present society with issues that are pressing and need to be approached globally. Many of the global effects are driven by two factors: the continuing increase in population as shown in Figure 1 from the IEA World Energy Outlook report for 2009 and the increasing demand for energy, both from the new developing and from the developed countries, as shown in Figure 2, which comes from the BP Energy Outlook 2030 Report page 8, published in January 2011, which summarizes global energy use over 60 years.

This growing global energy demand alone, independent of environmental concerns, is such a problem globally that innovative and creative solutions must be found to meet this demand. This is probably more challenging in the developing countries, where resource limitations may limit the number of options available, than in the developed ones. We note this, because as this text will emphasize, there is no obvious way to achieve this. In fact, it is clear that the demand will have to be met by a mosaic of current, *as well as new, sources*. Overall, the fact that a diversity of new energy sources is needed, will create new, large-scale industries, a development that may lead to significant changes in, or even the end of, some of our current established industries. If we now include the various environmental concerns that accompany modern society's functioning, this leads to the drive to achieve new energy-generating and -storage capacity via sustainable and clean technologies. At the same time, both these concerns, and the increasing energy demand, re-emphasize the need to accelerate the adoption of more efficient technologies worldwide.

Figure 1. Population by major region.

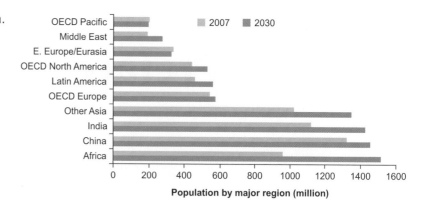

Figure 2. Historical plots of population, energy use and GDP vs time current and projected for OECD and non OECD countries.

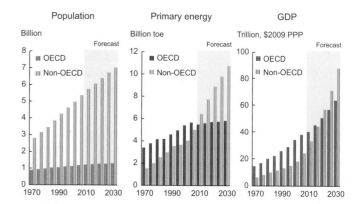

Figure 3. Shares of energy sources in world primary demand by scenario, where the 2008 line is current data, the current policy scenario is based on existing policy, the new policy scenario is a conservative view of emerging policy, and 450 is policy necessary to limit CO2 to 450 ppm. IEA World Energy Outlook 2010, International Energy Agency 9 rue de la Fédération 75739 Paris Cedex 15, France, pp. 80.

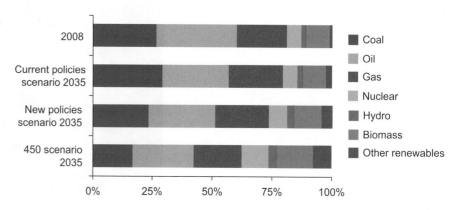

Underlying all of these changes is the basic understanding that the approaches that will be implemented must be ultimately sustainable not for tens or hundreds of years, but for thousands of years.

The purpose of this book is to serve as a college-level text that brings together the themes of environment and energy in the context of defining the issues, and subsequently focuses on the materials science and research challenges that need to be met. Part of the reason for this is illustrated in Figure 3.

These projections show that, based on current worldwide projections, while energy demand is expected to grow substantially to 2035, and use of fossil fuels will grow comparably (depending on world policy decisions), the fraction of the total energy needs supplied by fossil fuel is projected not to change substantially unless there are significant new policies enacted. This

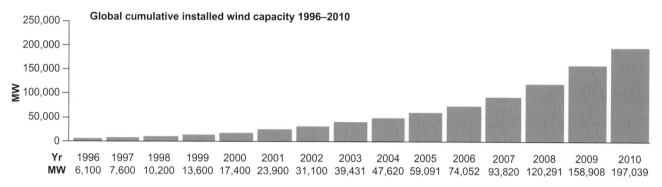

Figure 4. Global cumulative installed wind capacity to 2010 (GWEC – Global Wind 2010 Report page 14, Global Wind energy Council Headquarters Rue d'Arlon 80 1040 Brussels Belgium).

illustrates what we can term **the TW challenge** for alternative sustainable energy, because, if we want to change our dependence on fossil fuels (or have to do so, because of supply), the need is for TWs of power from new sources (where 1 TW of continuous power generation provides 8760 TW h of energy.) On a global scale, the systems that (will) provide for our energy needs are (will be) so large that for them to have any substantial effect on the production, transmission, or use of energy, is very difficult. *That is what makes this a true challenge!*

Materials are central and often critical to the potential ways to accelerate the evolution and future impact of new technologies. *This then explains the focus of the book.*

To be able to change the relative contribution of the different energy sources substantially to allow us to meet our demands in the next 20 years, rather than keep things the same as predicted (Fig. 3), we need to do things differently. Thus, we have to learn, for example, how to implement renewables on a much larger scale and how to ensure that nuclear energy will be safer (so as to decrease our innate fear of it and make it more acceptable).

Figure 4 illustrates the growth of wind power through 2009. While the amount of deployed wind is 200 GW_p (= 0.2 TW_p ≈ 0.07 TW continuous power; the substrate "p" stands for peak) worldwide (Fig. 4) and thus represents a very large industry, this is far short of the TWs of continuous power needed on a global scale.

It is increasingly clear that power generation and distribution in the future represent a much more complicated picture than the situation now with carbon(C)-based materials as the predominant source for transportation, heating, and electricity generation. As far as we can see, in the future, use of electrical energy from any primary energy source is likely to increase, and solar, geothermal, and other sustainable sources will be more important for all of these areas. For example, electric vehicles can substantially change the picture, if widely adopted. As increasingly spatially dispersed and time-dependent sources such as wind, solar, geothermal, and tidal energy are added to the grid, together with a greater use of nuclear energy for base load, the need for a smart grid with storage becomes increasingly important. This storage can be virtual, electrical, mechanical, and, last but not least, chemical (fuel).

This complex energy landscape, together with the equally complex worldwide political and economic climate, is a tremendous driving force for innovation in existing and emerging technologies. This text focuses on these challenges and has as its goals:

- To serve as a broad introduction to these challenges for the student;
- To stimulate a new generation of future scientists and engineers to focus on solving these critical problems;

- To help integrate the diverse backgrounds/disciplines that will be needed to solve these very complex challenges;
- To clearly state some of the materials challenges, as we perceive them today, that must be overcome to achieve terawatt power generating technologies.

We hope this book will both stimulate your imagination, and spike your interests in being part of the solution, as this is going to be without a doubt the greatest challenge globally for the next generation(s).

David S. Ginley
David Cahen

Acknowledgments

The editors first and foremost acknowledge the authors of the book's chapters. This distinguished group put aside their many commitments to meet the very demanding timeline and made it possible to take this book from an idea to reality. Second, we would like to thank Eileen Kiley Novak of the Materials Research Society (MRS), who was always available to solve problems and has worked many long hours to ensure that the book not only got done, but was of the quality we desired. It would not exist without her efforts. We also thank Michelle Carey of Cambridge University Press, who has been the advocate for the project from the beginning and very instrumental in helping the book take on its current form. In addition, there are many others who were instrumental in making this book possible. These include, from MRS, Todd Osman and Gail Oare, who supported this project from the outset; Kasia Bruniany, the graphic designer who set the style of the book; Andrea Peckelnicky, who redrew almost all the figures; Anita Miller for design and marketing expertise; Evan Oare and Lorraine Wolf, who coordinated the reviews and obtained final versions and permissions for the many figures; and, from Cambridge, Sarah Matthews and Caroline Mowatt, who coordinated editing and production of all final manuscripts. We thank the National Renewable Energy Laboratory (NREL) in the USA, and the Weizmann Institute of Science (WIS) in Israel, who made it possible for us to dedicate the time, and at times the resources, necessary to realize this project. We thank the WIS for a visiting professorship to Dave G., a period that led the foundation for the collaboration that culminated in this book. Dave G. thanks Diane Wiley at NREL who facilitated his involvement in the book. Finally, we would be remiss if we did not thank Lucy and Geula, our spouses, who allowed us to put much aside in the pursuit of this project.

PART 1 Energy and the environment: the global landscape

1 A primer on climate change

Melinda Marquis and Pieter Tans

NOAA Earth System Research Laboratory, Boulder, CO, USA

1.1 Focus

Despite the Kyoto Protocol and a wealth of good intentions, emissions of greenhouse gases (GHGs) – the primary cause of climate change – have continued to increase, not decrease, in recent years.

We face a global environmental crisis that is expected to include increased temperatures over land and in oceans, rising sea levels, more acidification of the oceans, increased flooding as well as drought, and extinction of many species as a result. The climate–energy crisis could cause major disruptions to ecosystems, the availability of fresh water, farming, economic activity, and global political stability on many levels.

1.2 Synopsis

Current atmospheric levels of carbon dioxide, methane, and nitrous oxide greatly exceed their pre-industrial levels, at least those that existed for hundreds of thousands of years previously. The increase in global atmospheric concentration of carbon dioxide (CO_2) is caused mostly by the burning of fossil fuels and changing land use, such as deforestation. The increase in atmospheric concentrations of methane (CH_4) and nitrous oxide (N_2O) is caused mostly by agriculture. Global climate is already changing: average surface temperatures have increased, ocean pH has decreased, precipitation patterns have changed, and sea level has risen since the industrial revolution.

Projections indicate that the global average surface air temperature will increase by at least several degrees Celsius for a range of expectations of human population growth and economic development. Further acidification of the world's oceans is expected to be caused directly by higher atmospheric CO_2 levels. The extent of sea ice at both poles is expected to shrink. The frequency of heat waves and that of heavy precipitation events are expected to increase.

To avoid the worst impacts of climate change, society would have to act quickly to transform its energy system to one that is sustainable – one that results in zero emission of CO_2. This would be no small feat. Global primary energy demand, which is inextricably tied to current and future GHG emissions, is projected to more than double from 13 TW (one terawatt equals one million megawatts (MW)) at the start of this century to 28 TW by the middle of the century [1]. This translates into obtaining 1,000 MW (the amount produced by an average nuclear or coal power plant) of new energy every single day for the next 46 years.

1.3 Historical perspective

In 1859, an Irish scientist named John Tyndall began studying the radiative properties of certain gases, including water vapor, ozone, and carbon dioxide, and recognized that they absorb and re-emit heat. He recognized also the role of water vapor in affecting Earth's surface temperature. In 1896, the Swedish scientist Svante Arrhenius recognized that burning fossil fuels and the resulting GHG emission would warm the Earth. He calculated that a doubling of the atmospheric concentration of carbon dioxide would lead to an increase in temperature of 5–6 °C [2]. A British engineer, Guy Callendar, used measurements of atmospheric carbon dioxide concentration from the years 1866 and 1956 and reported an increase over this period that correlated with the increased use of fossil fuels. Callendar recognized the difficulty of obtaining accurate measurements of carbon dioxide concentrations and documented the need for additional, more reliable observations [3].

In 1957, Roger Revelle and Hans Suess studied carbon isotopic ratios ($^{14}C/^{12}C$ and $^{13}C/^{12}C$) in tree rings and in marine material. ^{14}C is radioactive, with a half-life of 5,730 years, and is produced by cosmic rays in the upper atmosphere. The carbon dioxide created upon combustion of fossil fuels contains no ^{14}C, in contrast to carbon in the atmosphere, oceans, and living organisms, and thus a comparison of carbon isotopic ratios is a method to identify carbon dioxide that results from fossil-fuel burning. Revelle and Suess found a slight decreasing trend of the ^{14}C concentration in terrestrial plants and postulated that most of the carbon dioxide released by fossil-fuel combustion since the beginning of the industrial revolution must have been absorbed by the oceans. They wrote that "The increase of atmospheric CO_2 from this cause is at present small but may become significant during future decades if industrial fuel combustion continues to rise exponentially [4]." They documented the need for accurate measurements of atmospheric concentrations of carbon dioxide:

Thus human beings are now carrying out a large scale geophysical experiment of a kind that could not have happened in the past nor be reproduced in the future. Within a few centuries we are returning to the atmosphere and oceans the concentrated organic carbon stored in sedimentary rocks over hundreds of millions of years. This experiment, if adequately documented, may yield a far-reaching insight into the processes determining weather and climate. It therefore becomes of prime importance to attempt to determine the way in which carbon dioxide is partitioned between the atmosphere, the oceans, the biosphere and the lithosphere.

As part of the International Geophysical Year (1957–1958), Revelle hired Charles David Keeling at the Scripps Institution of Oceanography to begin obtaining these much needed, accurate measurements of carbon dioxide. Keeling began by taking measurements atop Mauna Loa, a volcanic mountain on the big island of Hawaii, and by measuring air samples obtained in Antarctica.

At the time of Dr. Keeling's initial efforts, many scientists were not certain that one could detect meaningful patterns of such a low-abundance constituent of the atmosphere. Keeling revolutionized the measurement techniques, and as a result discovered the seasonal cycle, the steady annual increase, and the difference between the hemispheres. This was the beginning of what was to become a coordinated global monitoring network involving scientists and agencies from countries around the world. Information derived from this network, which now includes concentrations of many greenhouse gases, isotopic ratios, and other tracers, has been crucial for informing national and international assessments of global climate change, not the least of which are the International Panel on Climate Change (IPCC; www.ipc.ch) assessment reports. Figure 1.1(a) shows the atmospheric concentration of carbon dioxide from 1957 to 2010 as measured at altitude 3,400 m on Mauna Loa. Its most striking feature is the accelerating increase that overwhelms the natural annual cycle caused by land plants (mostly in the Northern Hemisphere) during the growing season followed by respiration and decay during fall and winter.

International partners from all continents and many islands contribute to the global network of GHG measurements, which is now under the umbrella of the United Nations World Meteorological Organization (WMO) Global Atmosphere Watch. Measurements made as part of this network are subject to stringent quality-control procedures and are archived in globally distributed repositories. Figure 1.1(b) shows measurement sites from the current cooperative air sampling maintained by NOAA, which is a large component of the WMO network.

1.4 Greenhouse gases and radiative forcing

Earth's atmosphere is composed of multiple gases, the bulk of the (dry) atmosphere being made of two main gases: nitrogen (N_2, 78.09%) and oxygen (O_2, 20.94%). Most of the balance is composed of the inert gas argon (Ar, 0.93%). The very small remainder is composed of various gases, including *greenhouse gases* (GHGs). Greenhouse gases, which can be both natural and human-caused, are so called because they absorb and emit radiation at certain wavelengths in the infrared (IR) portion (~0.7 to ~300 μm) of the electromagnetic

Figure 1.1. (a) Atmospheric concentration at the Mauna Loa Observatory, from 1958 to 2010. (b) Carbon dioxide measurement sites from the current global network. Courtesy NOAA Earth System Research Laboratory, Global Monitoring Division.

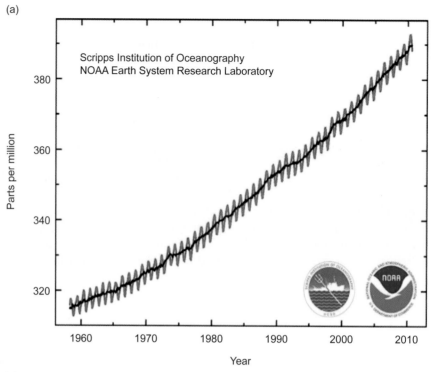

(a)

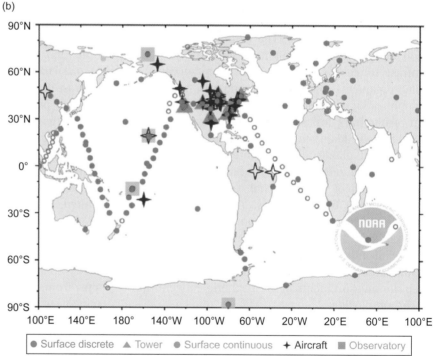

(b)

● Surface discrete ▲ Tower ● Surface continuous ✦ Aircraft ■ Observatory

Open symbol represents inactive site

spectrum. They absorb IR radiation that is emitted by Earth's surface, by gases in the atmosphere, and by clouds. The energy absorbed by the GHGs is quickly shared with the other molecules in the atmosphere. The GHGs then re-emit IR radiation at the same wavelengths as those at which they absorb in all directions, including back toward Earth, which leads to heating of the Earth and the portion of the atmosphere below where the emission takes place. This heating is called the *greenhouse effect*. See Figure 1.2.

The most important GHGs are water vapor (H_2O, 0.40%), carbon dioxide (CO_2, 0.039%), methane (CH_4, 0.00018%), and nitrous oxide (N_2O, 0.00003%). Many other important GHGs exist, for instance halocompounds, at very low concentrations; these are produced mostly by human activities.

Figure 1.2. An idealized model of the natural greenhouse effect. See the text for explanation. © IPCC AR4 WG1.

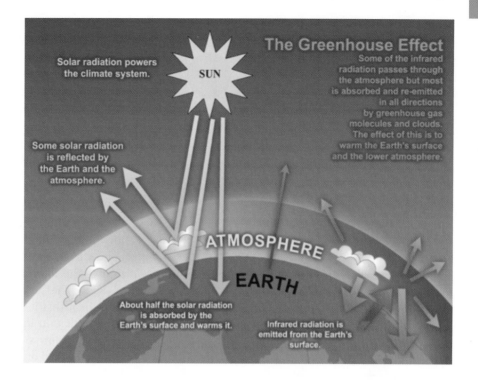

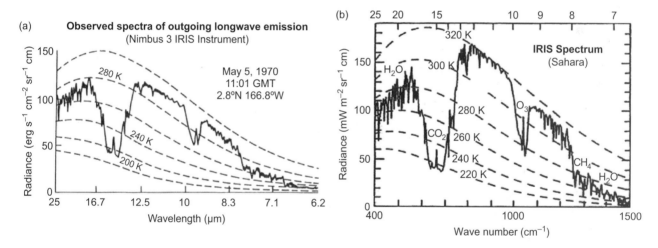

Figure 1.3. Infrared radiance as a function of (a) wavelength and (b) wave number from Earth outgoing to space as measured by satellite under cloud-free conditions, with major absorption visible for H_2O, CO_2, O_3, and CH_4. The dashed lines show what the outgoing spectrum would look like at different temperatures of the Earth's surface without the presence of GHGs (280 K corresponds to 7 °C).

Figure 1.3(a) [5] shows a spectrum of outgoing long-wave (IR) emission measured by a satellite (Nimbus 3 IRIS) over Micronesia in 1970. The dashed lines show what the spectrum would look like without the presence of an absorbing atmosphere, at different postulated temperatures of the Earth's surface. These curves are called blackbody radiation. The field of view was nearly cloud-free, and so the radiance of the atmospheric window (the spectral region where the atmosphere is transparent and lets radiation be transmitted, between 8–9 μm and 10–12 μm) almost follows a blackbody brightness temperature of ~290 K. The radiance curve differs a bit from 290 K because there are very minor atmospheric constituents that do absorb. In the presence of GHGs the satellite measures lower emission temperatures in spectral regions where GHGs absorb. In those spectral regions, the observed temperature corresponds to that of the (colder) air at the altitude from

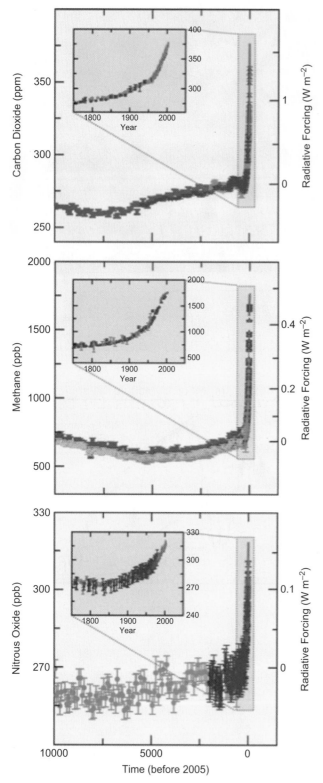

which the radiation that is able to reach space is emitted (approximately). What you cannot see in Figure 1.3(a) is that, from those altitudes, radiation is also emitted toward the surface, and, if the optical thickness (i.e., the amount of gas times the absorptive properties) in the same absorption band below the altitude of the GHG emissions is high enough, this emitted radiation will not even reach the ground. An upward-looking spectrometer on the ground will see a higher radiance in the absorption band because the radiation that reaches the ground is emitted from the lower part of the atmosphere where it is warmer. On the ground, without GHGs, one would see IR from space at a very low temperature, except for the IR radiation that comes from the Sun, which is an extremely low fraction of the total solar energy received.

Once the surface has warmed up as a result of the extra heat that is being retained by the GHGs, the outgoing radiation in the window regions should increase because a new steady-state Earth surface temperature can be reached only when the total outgoing IR radiance is the same as it was before and exactly balances incoming absorbed solar energy. Thus, with increasing GHG concentration, the outgoing radiation decreases in the absorption regions and increases in the window region. Figure 1.3(b) [5] shows a spectrum observed over North Africa (Sahara) from the same satellite, in 1970. The maximum brightness temperature measured in the atmospheric window from 830 to 970 cm^{-1} indicates that the surface temperature is about 320 K (~47 °C or 116 °F). Notice the absorption bands of the main GHGs in the atmosphere (which is not as hot as the Earth's surface). Absorption by carbon dioxide is clearly visible at ~600–750 cm^{-1}, with absorption by ozone at 1,000–1,060 cm^{-1}, by methane at ~1,240–1,300 cm^{-1}, and by water above 1,300 cm^{-1}.

The greenhouse effect existed throughout Earth's history. In fact, the natural greenhouse effect, which during the last few million years has kept the Earth about 33 °C warmer than it would otherwise be, has allowed life to evolve and to continue as we know it. However, human activities have recently led to increasing concentrations of GHGs, which have added significantly to the natural greenhouse effect since the beginning of the industrial era.[1] Figure 1.4 shows the atmospheric concentrations of carbon dioxide, methane, and nitrous oxide, which have increased

Figure 1.4. Atmospheric concentrations of carbon dioxide, methane, and nitrous oxide over the last 10,000 years (large panels) and since 1750 (inset panels). Measurements from ice cores (symbols with different colours for different studies) and atmospheric samples (red lines) are shown. The corresponding radiative forcings are shown on the right-hand axes of the large panels. © IPCC AR4 WG1.

---

[1] The Industrial Era is generally considered to have begun ~1750. The IPCC Fourth Assessment Report (Working Group I) states radiative-forcing values for 2005 relative to pre-industrial conditions defined at 1750. It is important to note, however, that before 1850 human-caused radiative forcing was quite small, e.g., GHG emissions were less than 1% of GHG emissions in the present day. As of 2010, three-quarters of all cumulative fossil-fuel emissions have taken place after 1962. Therefore, most human-caused climate change has occurred since 1960.

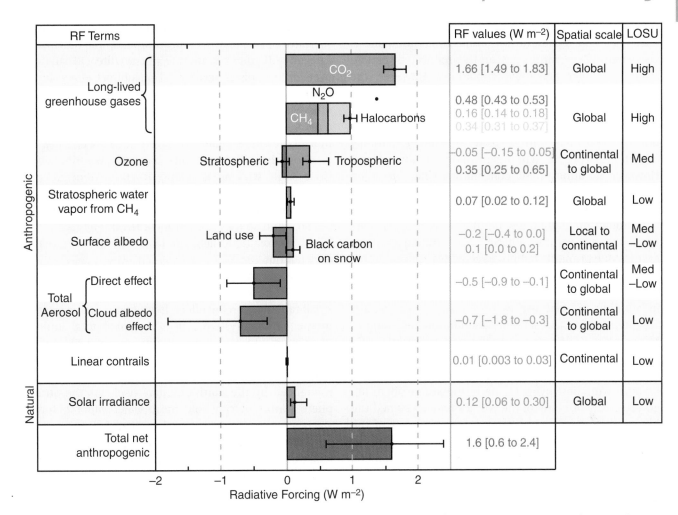

Figure 1.5. Global average radiative-forcing (RF) estimates and ranges in 2005 for anthropogenic carbon dioxide (CO_2), methane (CH_4), nitrous oxide (N_2O), and other important agents and mechanisms, together with the typical geographical extent (spatial scale) of the forcing and the assessed level of scientific understanding (LOSU). The net anthropogenic radiative forcing and its range are also shown. These require summing asymmetric uncertainty estimates from the component terms, and cannot be obtained by simple addition. Additional forcing factors not included here are considered to have a very low LOSU. Volcanic aerosols contribute an additional natural forcing but are not included in this figure due to their episodic nature. The range for linear contrails does not include other possible effects of aviation on cloudiness. © IPCC AR4 WG1.

dramatically and abruptly. Today's concentrations of carbon dioxide and methane greatly exceed their ranges measured in ancient air locked up in ice cores for the last 800,000 years [6].

When comparing warming and cooling of Earth's climate due to natural processes or human activities, a useful concept is *radiative forcing* (RF) [7], which quantifies the globally averaged *change* in the heat balance of the Earth–atmosphere system due to factors external to the climate system. Figure 1.5 gives the RF values, spatial scales, and levels of scientific understanding for important agents, relative to pre-industrial levels, in 2005.

The units of RF are watts (energyflow) per meter squared (averaged over the entire area of the globe). Positive values indicate that an agent has a warming effect on the troposphere (the lowest ~10 km of the atmosphere); negative values indicate a cooling effect. Figure 1.5 shows that carbon dioxide has an RF value of +1.66 W m^{-2}; for methane, the RF value is +0.48 W m^{-2}; and for nitrous oxide, the RF value is +0.16 W m^{-2} (all in 2005). As shown in Figures 1.1 and 1.4, there is no question that the atmospheric concentration of CO_2 has increased during the industrial era because of use of fossil fuels in the transportation, manufacturing, heating, and cooling sectors. Deforestation and cement production also release carbon dioxide to the atmosphere. Methane is also a naturally occurring gas, and is produced by certain microbes under anaerobic conditions. However, the atmospheric concentration of methane has increased in the last two and a half centuries mostly because of agricultural practices, landfills, and the mining and distribution of fossil fuels. Nitrous oxide is another naturally occurring gas produced by organisms in soils and oceans. However, its atmospheric concentration has increased since the industrial revolution mostly because of the use of nitrogen-containing fertilizers for agriculture.

Though there are a few natural sources of halocarbons, the rise in atmospheric concentration of these gases is due primarily to human activities. Chlorofluorocarbons were used as aerosol propellants (e.g., in hair sprays and deodorants) and in foam insulation, in addition to their use as refrigerants (e.g., for air-conditioning in offices, home, and automobiles) and in refrigerators. Halocarbons are not only powerful GHGs with a collective RF value of $+0.34$ W m^{-2}, but are also ozone-depleting substances. After recognizing that halocarbons were destroying ozone in the stratosphere [8], governments of many of the world's nations signed the Montreal Protocol, which requires a phase-out of the production, consumption, and emission of a number of these substances. As a result of the Montreal Protocol, the atmospheric concentrations of the most destructive of the substances have stabilized or decreased. This international agreement had an important unintended consequence, namely it slowed down the rate at which climate forcing would otherwise have increased [9].

Referring again to Figure 1.5, note that the RF value for ozone depends on which part of the atmosphere it inhabits. Up high, in the stratosphere, where ozone helps protect humans and other species from damaging ultraviolet radiation, its RF value is -0.05 W m^{-2}, which means it has a cooling effect on the Earth's radiation budget; but closer to Earth's surface where it is an air pollutant (it damages plant leaves as well as the respiratory tract of humans), it has an RF value of $+0.35$ W m^{-2}.

Changes in the surface reflectivity of the Earth resulting from changes in land cover, primarily the result of deforestation, have had a cooling effect (an RF value of -0.2 W m^{-2}). (This effect is separate from the warming effect of carbon dioxide released upon burning of forests.) Another notable change to land cover is the deposition of black carbon, from incomplete combustion of fossil fuels and biomass, on snow, which reduces the Earth's surface albedo (an RF value of $+0.1$ W m^{-2}).

Note that aerosols (small particles suspended in air) tend to produce cooling, which originates from a direct effect (an RF value of -0.5 W m^{-2}) and an indirect effect (-0.7 W m^{-2}). The direct effect comes from the fact that aerosols scatter and absorb short-wave (solar, visible) and long-wave (Earth, infrared) radiation. Though some aerosols have a positive RF (due to absorption) and others have a negative RF (due to scattering), the net direct RF summed over all aerosols is negative. The indirect effect comes from aerosols serving as cloud-condensation nuclei and otherwise affecting the amount and lifetime of clouds, with a net cooling effect (increased cloud albedo). The magnitude of the indirect effect has a very large uncertainty.

Note that solar irradiance has had a warming effect since 1750, with an RF value of $+0.12$ W m^{-2}. Large volcanic eruptions are another natural source of RF agents, but the stratosphere was free of volcanic aerosols in 2005, and so they are not included in Figure 1.4. The RF of stratospheric aerosols lasts several years before they fall out.

Water vapor is a special case: it is the most important and abundant greenhouse gas, but it has only a very small radiative forcing effect. It is a factor that is internal to the climate system [10]. The global total amount of evaporation and precipitation is ~500,000 billion metric tons per year. Direct human influence on this amount is negligible. However, water vapor is greatly influenced by the Earth's climate itself. As the atmosphere warms, it can hold more water vapor before it starts condensing into droplets that fall toward the Earth's surface. Since water vapor is a powerful GHG, it leads to more warming, which in turn leads to more atmospheric water vapor. So water vapor is a feedback agent rather than an external RF agent. See a recent publication [11] for a further description of the greenhouse effect of water and carbon dioxide. There is one exception: as a result of methane emissions, additional water vapor is created in the stratosphere, which is naturally extremely dry. Methane is oxidized in the stratosphere, and this chemical reaction produces water vapor, with an RF value of $+0.07$ W m^{-2}.

The total RF caused by human activities (in 2005) is $+1.6$ W m^{-2}, and the total forcing due to long-lived GHGs is $+2.6$ W m^{-2}. Let us put this last number into perspective. The total amount of solar energy, averaged over the Earth, absorbed by the atmosphere and the surface is close to 240 W m^{-2}. This is the energy that drives the climate system, and it has been increased by ~1.1% due to long-lived GHGs. As we will discuss below, the increase is hard to reverse. In fact, the number has increased to $+2.8$ W m^{-2} in 2010. The natural brightening of the Sun since 1750 has been only $+0.12$ W m^{-2}. As scientists we would be greatly surprised if the Earth were not warming as a result of this increase in forcing.

The RF of a species does not tell the complete story about its potential to affect global climate because the RF does not account for the lifetime of species. Another concept that helps us to compare the potential impact on climate of various species is the Global Warming Potential (GWP). The GWPs indicate the integrated RF

Figure 1.6. Integrated RF of year-2000 emissions over two time horizons (20 and 100 years). The figure gives an indication of the future climate impact of current emissions. The values for aerosols and aerosol precursors are essentially equal for the two time horizons. It should be noted that the RFs of short-lived gases and aerosols depend critically on both when and where they are emitted; the values given in the figure apply only to total global annual emissions. For organic carbon and black carbon, both fossil-fuel (FF) and biomass-burning emissions are included. The uncertainty estimates are based on the uncertainties in estimates of emission sources, lifetime, and radiative efficiency. © IPCC AR4 WG1.

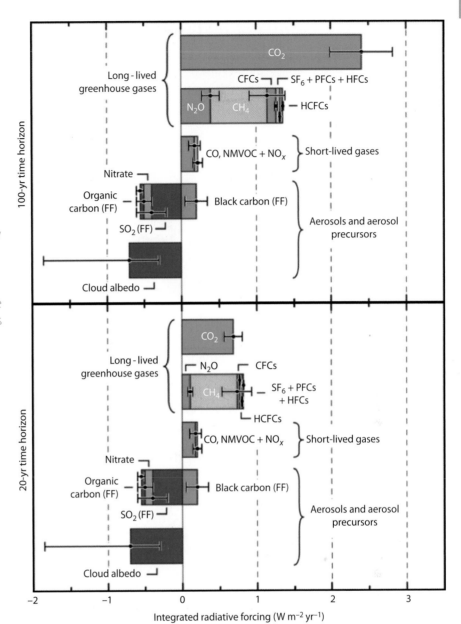

over certain time periods (e.g., 20 years or 100 years) from a unit pulse emission. A species that is removed from the atmosphere in 30 years will have ten times the impact of a species that is removed in 3 years if the two species have the same RF. Though a full discussion of this topic is beyond the scope of this chapter, it is worth considering the lifetimes of several important species. The lifetime concept is well suited to species that are destroyed by chemical processes in the atmosphere and oceans, but that concept does not describe the behavior of carbon dioxide particularly well. Carbon dioxide is not chemically destroyed in the atmosphere and oceans like other GHGs [12]. About 80% of the emissions can be expected to eventually dissolve in the oceans as carbonic acid and (bi)carbonate ions, but CO_2 keeps exchanging between the atmosphere and oceans. Carbonic acid is eventually neutralized by dissolution of calcium and magnesium carbonate minerals [13], which lower the atmospheric concentration of CO_2, but the estimated time scale is 3,000–7,000 years [14]. In the shorter term, the rate of net CO_2 uptake from the atmosphere is dominated by the oceans. It becomes ever slower as time progresses because it takes longer for deeper ocean layers to be reached. This process, and also uptake and release of carbon dioxide by the terrestrial biosphere, has been described by, among other models, a revised Bern carbon-cycle model [15], which has been used to define the integrated (100 years) climate impact of a pulse of carbon dioxide. The lifetimes of methane and nitrous oxide are more straightforward: methane's is ~9 yr and nitrous oxide's is ~110 yr.

The GWPs indicate the future climate impact of current (year 2000) emissions. Figure 1.6 shows the climate impact of various species on a 100-yr time scale

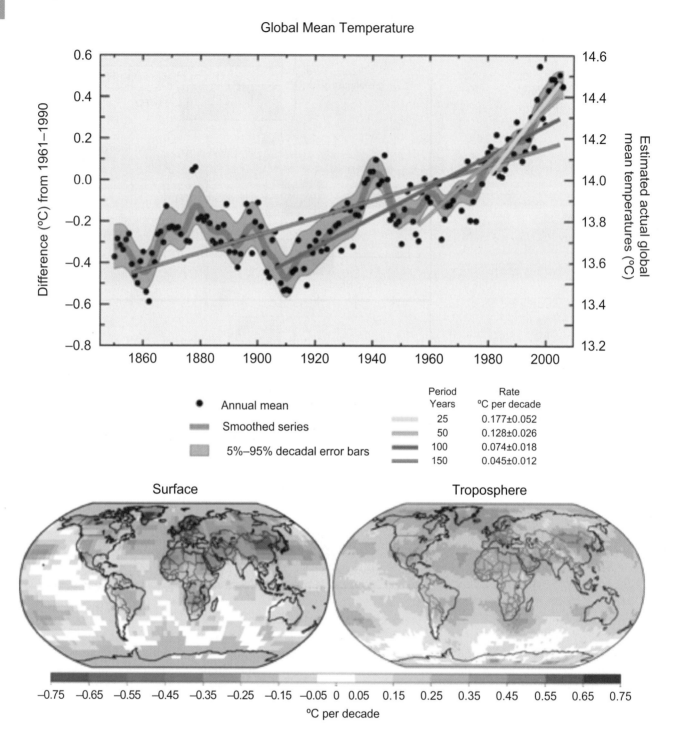

Figure 1.7. (Top) Annual global mean observed temperatures (black dots) together with simple fits to the data. The left-hand axis shows anomalies relative to the 1961 to 1990 average and the right-hand axis shows the estimated actual temperature (°C). Linear trend fits to the last 25 (yellow), 50 (orange), 100 (purple), and 150 years (red) are shown, and correspond to 1981 to 2005, 1956 to 2005, 1906 to 2005, and 1856 to 2005, respectively. Note that, for shorter recent periods, the slope is greater, indicating accelerated warming. The blue curve is a smoothed depiction to capture the decadal variations. To give an idea of whether the fluctuations are meaningful, decadal 5% to 95% (light gray) error ranges about that line are given (accordingly, annual values do exceed those limits). Results from climate models driven by estimated radiative forcings for the

(top half) and on a 20-yr time scale (bottom half). Note the larger impact of carbon dioxide on the longer (100-yr) time scale versus the shorter (20-yr) time scale. The climate impact of carbon dioxide and several other agents lasts much longer, however.

The shortcomings of GWPs have been debated [16] [17]. For example, on a 100-yr time scale, current reductions in emissions of short-lived species (e.g., methane) will yield a smaller cooling effect toward the end of the 100-yr period than the cooling effect from reducing carbon dioxide emissions. However, the GWP metric can be useful if its limitations are understood.

1.5 Observations of climate change

Throughout the Earth system, climate change has been and is continuing to be observed. Temperatures over land and in the ocean are rising, changes in precipitation are occurring, snow and ice are melting, oceans are becoming more acidic, and the global average sea level is rising. Changes in temperature extremes have occurred. All of these observations are consistent with warming of the Earth's climate [18].

1.5.1 Surface and atmospheric observations

Global mean surface temperatures increased from 1906 to 2005 by ~0.7 °C. See Figure 1.7 (top).[2]

The rate of warming over the second half of that period (0.13 °C/decade) is almost twice as fast as the rate over the entire period (0.07 °C/decade). The warming has not been steady. There was little overall change (just short periods of warming and cooling that reflect natural variability, volcanic activity, and changes in solar radiation, and possibly also lower-quality measurements) until 1915. However, from about 1915 to the early 1940s, the global mean temperature increased by 0.35 °C. For the next few decades, the temperature was approximately steady, followed by a second period of large warming through the end of 2006 (0.18 °C/decade).

The bottom panels of Figure 1.7 show global temperature trends during the satellite record era (1979–2005). The left panel shows temperature trends at the Earth's surface, and the right panel shows temperature trends in the troposphere (0 to ~10 km altitude) from satellite data.

Warming has not been uniform across the globe. Surface air temperatures over land have warmed roughly twice as much (0.27 °C/decade) as temperatures over the ocean (0.13 °C/decade) since 1979. Further, average temperatures in the Arctic have increased at almost twice the global rate in the last century. This response of the climate system has been expected by virtually all climate models, and is likely due to a feedback effect: warmer temperatures lead to less ice and snow cover, which in turn causes more solar energy to be absorbed at the surface.

There have been spurious reports in the media that global average temperatures have been decreasing recently. This is not true. Though the average temperature for 2008 was cooler than temperatures for the preceding seven years and 1998 was unusually hot, the warming trend of the last century is not changed by a few exceptional years. Climate change is understood over long time periods, and a single event is not significant. What those spurious reports omit is the fact that 2001–2010 make up all except for one (1998) of the ten warmest years on record. There is no rational way to conclude that recent global average temperatures are decreasing.

1.5.2 Changes in precipitation

Changes in precipitation have been observed and they are rather complex. Figure 1.8 shows trends in data from the Global Historical Climatology Network precipitation data set at the National Climatic Data Center.

Figure 1.7. (cont.) twentieth century suggest that there was little change prior to about 1915, and that a substantial fraction of the early-twentieth-century change was contributed by naturally occurring influences including changes in solar radiation, volcanism, and natural variability. From about 1940 to 1970 the increasing industrialization following World War II increased pollution in the Northern Hemisphere, contributing to cooling, and increases in emissions of carbon dioxide and other GHGs dominate the observed warming after the mid 1970s. (Bottom) Patterns of linear global temperature trends from 1979 to 2005 estimated at the surface (left) and for the troposphere (right) from the surface to about 10 km altitude, from satellite records. Gray areas indicate incomplete data. Note the more spatially uniform warming in the satellite tropospheric record while the surface temperature changes more clearly relate to land and ocean. © IPCC AR4 WG1.

[2] From the HadCRUT3 data set [18].

Figure 1.8. Trend of annual land precipitation amounts for 1901 to 2005 (top, % per century) and 1979 to 2005 (bottom, % per decade), using the GHCN precipitation data set from the NCDC. The percentage is based on the means for the period 1961 to 1990. For areas in gray there are insufficient data to produce reliable trends. The minimum number of years required to calculate a trend value is 66 for 1901 to 2005 and 18 for 1979 to 2005. An annual value is complete for a given year if all 12 monthly percentage anomaly values are present. Note the different color bars and units in each plot. Trends significant at the 5% level are indicated by black + marks. © IPCC AR4 WG1.

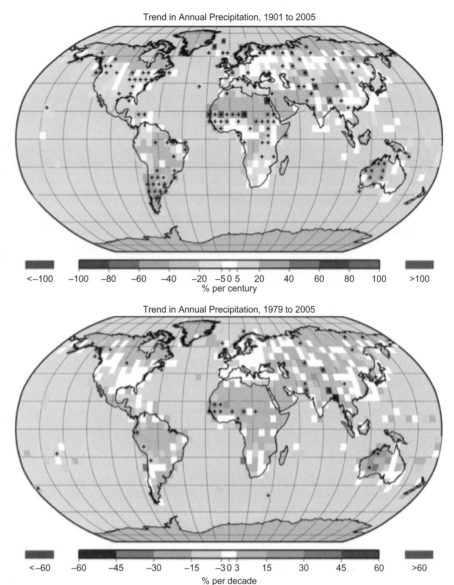

Trend in Annual Precipitation, 1901 to 2005

<−100 −100 −80 −60 −40 −20 −5 0 5 20 40 60 80 100 >100
% per century

Trend in Annual Precipitation, 1979 to 2005

<−60 −60 −45 −30 −15 −3 0 3 15 30 45 60 >60
% per decade

From 1901 to 2005, it has become wetter in northern and central Asia, in northern Europe, in southern South America, and in eastern North America. It has become drier in the Mediterranean, southern Asia, northwestern Mexico, southwestern USA, and southern Africa, and especially in the Sahel (a belt that spans Africa from its west coast to its east coast, just south of the Sahara Desert).

More severe and longer droughts have become more common since the 1970s, especially in the tropics and subtropics. Decreased precipitation over large land areas, coupled with warmer temperatures, leads to more evaporation and drying of land, resulting in droughts. The western USA, Australia, and Europe have experienced an increase in droughts since the 1970s. Drought has become widespread in much of Africa, especially in the Sahel.

1.5.3 Changes in the cryosphere (snow and ice)

Arctic sea ice

In the last three decades, there has been a decrease in Arctic sea ice extent. During the satellite record, which began in 1979, annual Arctic sea ice extent has decreased by 4.1% per decade through 2009 [19]. The decrease has been especially pronounced in the summer, when the ice extent reaches its minimum in September each year. The rate of decline of sea ice extent in the Arctic summer since 1979 has now (2010) increased to −11.5% per decade [19]. The sea ice extent in September 2007 was the lowest ever recorded, shattering the previous record for the month, set in 2005, by 23%, which was well below any of the model predictions. See Figure 1.9.

The September 2010 sea ice extent was the third lowest since satellite observations began in 1979, and

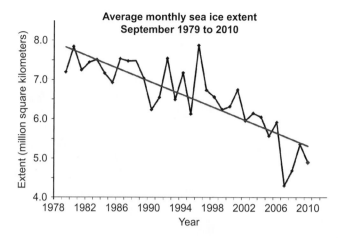

Figure 1.9. The Northern Hemisphere September sea ice extent anomaly is −11.5% per decade relative to the 1979 to 2000 average. Graphic courtesy of Walt Meier, NSIDC.

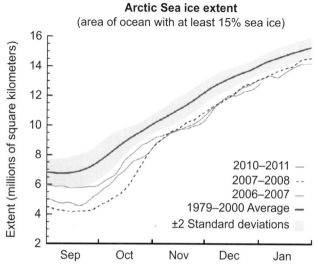

Figure 1.10. The graph above shows daily Arctic sea ice extent as of January 2, 2011, along with daily ice extents for previous low-ice-extent years in the month of November. Light blue indicates 2010–2011, pink shows 2006–2007 (the record low for the month was in 2006), green shows 2007–2008, and dark gray shows the 1979 to 2000 average. The gray area around the average line shows the 1979 to 2000 average. The gray around the average line shows the two-standard-deviation range of the data. Credit: National Snow and Ice Data Center.

the past six years (2005–2010) have seen the six lowest sea ice extents in the satellite record. See Figure 1.10.

Related to the decrease of Arctic sea ice area is the increased proportion of young (first- or second-year ice) versus old (3-year-old or older) ice. Because younger ice correlates well with thinner ice, at the end of 2009, Arctic ice was thinner and younger than the historical average, leaving the remaining ice vulnerable to melting in future summers. Scientists use satellites to measure ice age, a proxy for ice thickness. In 2009, first-year ice accounted for 49% of the ice cover at the end of summer. Second-year ice made up 32%, compared with 21% in 2007 and 9% in 2008. Only 19% of the ice cover was over 2 years old, far below the 1981–2000 average of 52% (updated from [20]. See Figure 1.11. Other methods to estimate ice age [21] [22] and directly measured thickness [23] confirm a younger and thinner ice cover.

Decreases in ice extent are important because Arctic sea ice influences global climate. In the Arctic, sea ice begins to melt in March and stops melting in September. The decrease in Arctic sea ice is not only a result of a warming climate, but also acts as a positive feedback (a cause of further warming). Ice reflects most of the incident solar radiation. Open sea water absorbs most of the solar radiation so that, as the ice melts, more solar radiation is absorbed, which leads to increased melting of ice, and so on.

Antarctic sea ice

In contrast to the Arctic, sea ice has increased slightly (+1.1% per decade) in extent around the Antarctic over the last 30 years. This increase is not inconsistent with a warming climate. Antarctic sea ice variability is dominated by winds and ocean circulations. Changes in circulation as a result of the stratospheric ozone hole above Antarctica have served to enhance sea ice formation

over much of the Antarctic [24], giving an increase of +1.1 (±0.8)% per decade (95% confidence level, data courtesy of Walt Meier, NSIDC). Sea ice changes vary regionally, and this can be seen on the western side of the Antarctic Peninsula, where the circulation changes have led to a large regional decrease of sea ice.

The ozone hole over Antarctica has shielded much of the continent from the effects of global warming experienced elsewhere [24]. The mechanism of the shielding is as follows: "The loss of stratospheric ozone has intensified the polar vortex, a ring of winds around the South Pole, altered weather patterns around the continent, and increased westerly winds by about 15% over the Southern Ocean in summer and autumn. This has resulted in the Antarctic becoming more isolated and there being little change in surface temperature across the bulk of the continent over the last 30 years." [24] However, the Antarctic Peninsula region has been one of the fastest warming regions on Earth [25], and recent studies show that Antarctica, including the interior, is beginning to warm [26].

Snow cover and snowfall

Analogously to the ice albedo–temperature feedback discussed above, snow cover has a similar albedo–temperature feedback, which depends on the depth and age of the snow, as well as on vegetation cover and

Figure 1.11. Arctic sea ice age at the end of melt season. These images compare ice age, a proxy for ice thickness, in 2007, 2008, 2009, and the 1981 to 2000 average. In 2009 there was an increase in second-year ice (in blue) over that for 2008. At the end of summer 2009, 32% of the ice cover was second-year ice. Three-year and older ice amounted to 19% of the total ice cover, the lowest in the satellite record. Credit: National Snow and Ice Data Center, courtesy C. Fowler and J. Maslanik, University of Colorado at Boulder.

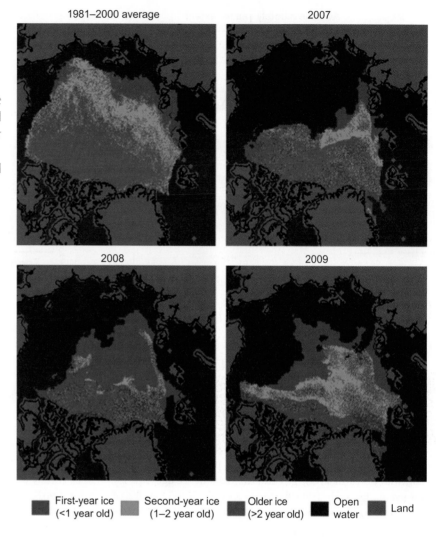

cloud cover. As noted earlier, human-caused soot is accumulating on snow [27], thereby reducing its albedo (RF value +0.1 W m^{-2}).

Snow cover has decreased in most regions globally, especially in spring and summer. From 1966 to 2005, snow cover in the Northern Hemisphere decreased in every month except for November and December. From 1922 to 2005, the Northern Hemisphere snow-covered area decreased by 7.5% (updated from [28], in [25]).

In the Antarctic, there has been no significant change in snowfall as a whole over the last 50 years, although snowfall has increased across the Antarctic Peninsula [24]. Outside of Antarctica, very little land in the Southern Hemisphere experiences snow cover.

1.5.4 Ocean temperature, acidification, and sea-level rise

The world's oceans are warming, becoming more acidic (as a direct and unavoidable result of atmospheric carbon dioxide dissolving in ocean water), and sea level is rising [29].

From 1961 to 2003, the global ocean temperature increased by an average of 0.10 °C, from the sea surface down to 700 m depth. The amount of dissolved inorganic carbon in the global oceans has been measured to have increased by ~118 billion metric tons of carbon from pre-industrial times to 1994 and is continuing to increase; the source is atmospheric carbon dioxide. The increase in inorganic carbon has led to a decrease in pH of ~0.1 units (lower pH means more acidic) near the ocean surface. Acidification of oceans is threatening the marine food chain, because corals and certain species of plankton are expected to have difficulty maintaining their calcium carbonate skeletons in ocean water with lower pH [30].

Global average sea level is rising, but there is significant decadal variability. Most of the sea-level rise is attributed to thermal expansion and to glaciers melting. From 1961 to 2003, global average sea level rose at an average rate of 1.8 mm/yr, with a ratio of thermal expansion to glacial melting of roughly one to one. From 1993 to 2003, the rate increased to 3.1 mm/yr, with a ratio of thermal expansion to glacial melting of roughly two to

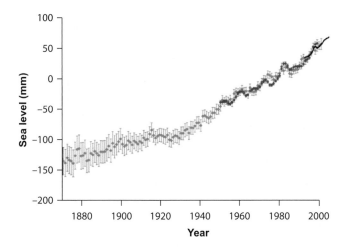

Figure 1.12. Annual averages of the global mean sea level (mm). The red curve shows reconstructed sea-level fields since 1870 (updated from [53]); the blue curve shows coastal tide gauge measurements since 1950 (from [54]) and the black curve is based on satellite altimetry ([55]). The red and blue curves are deviations from their averages for 1961 to 1990, and the black curve is the deviation from the average of the red curve for the period 1993 to 2001. Error bars show 90% confidence intervals. © IPCC AR4 WG1.

one.[3] It is not known whether the faster rate from 1993 to 2003 reflects natural decadal variability or indicates an increase in the long-term rate of sea-level rise. See Figure 1.12.

From 2000 to 2010 about 33% of carbon dioxide emitted by humans (including by deforestation) has entered the oceans, ~21% has been stored in the biosphere, and ~46% has remained in the atmosphere. The ability of the oceans to absorb additional carbon dioxide is expected to decrease because of well-understood acid–base chemistry [31].

1.6 Paleoclimate

Paleoclimate reconstructions employ climatically sensitive indicators to infer changes in global climate that occurred decades, millennia, and even millions of years ago. Some methods include direct measurements of past conditions, such as the composition of air bubbles trapped in ice cores, with high precision and relatively low uncertainty. Other methods involve the use of "proxies." These methods take advantage of changes in chemical, biological, or physical parameters that reflect changes (e.g., temperature, precipitation) in the environment in which the proxy lived or otherwise existed. Many organisms, such as trees, corals, and insects, alter their growth or population as climate changes. These alterations in growth or population are often recorded in the past growth of living and dead specimens. Tree rings, plankton, and pollen are "proxies" for climate conditions as far back as many thousands of years ago. Generally, the dating accuracy is lower and the uncertainty is higher the further back in time these proxies go. However, paleoclimatologists usually use multiple methods and proxies to reduce uncertainty in dating. Assessments of the robustness and precision of these methods are available [32] [33] [34] [35].

Paleoclimate data indicate that today's atmospheric levels of carbon dioxide and methane are far greater than previous levels, going as far back as 800,000 years ago [6] [36]. See Figure 1.13.

The atmospheric concentration of carbon dioxide has increased from pre-industrial values that ranged in the last 10,000 years from ~180 to ~280 ppm to a value of 390 ppm today. The atmospheric concentration of methane has more than doubled from its pre-industrial values that ranged from ~580 ppb to 730 ppb to a value of ~1800 ppb. The atmospheric concentration of nitrous oxide has increased by ~20% over its pre-industrial values, which ranged from ~220 to 270 ppb, to 323 ppb today.

Multiple lines of scientific evidence, such as observations of isotopic ratios of GHGs in modern air and in ice, the north–south gradients of GHGs, their correlation with the increase in global economic activity (and known processes such as mining, agricultural practices, and land use), show that that the rise in atmospheric concentrations of carbon dioxide, methane, and nitrous oxide since the industrial revolution has been caused primarily by human activities. The increases in carbon dioxide result primarily from fossil-fuel use (~65%–75%) and changes in land use (~25%–35%), whereas the increases in methane and nitrous oxide result primarily from agriculture (crops and animals).

Paleoclimate data indicate that global average sea level in the last interglacial period (~125,000 years ago) was ~4–6 m higher than it is today, mainly due to partial melting of polar ice caused by changes in the Earth's orbit around the Sun. Ice-core data indicate that global average temperatures were 3–5 °C warmer than today. This correlation of a warming of 3–5 °C with an average sea-level rise of 4–6 m could be relevant to conditions in the future, since climate models project that global average temperatures by the year 2100 will likely be 3–5 °C warmer than today [37].

It is important to note that the warmer temperatures during the last interglacial period, as well as the start and cessation of other ice ages, were most likely caused by variations in Earth's orbit around the Sun (Milankovitch cycles). These cycles affect the distribution of solar radiation hitting the Earth at various

---

[3] See Table 5.3 in [29].

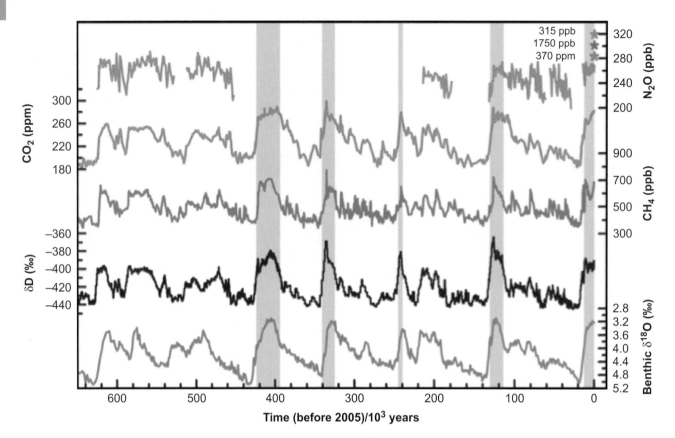

Figure 1.13. Variations of deuterium (δD) in antarctic ice, which is a proxy for local temperature, and the atmospheric concentrations of the greenhouse gases carbon dioxide (CO_2), methane (CH_4), and nitrous oxide (N_2O) in air trapped within the ice cores and from recent atmospheric measurements. Data cover 650,000 years and the shaded bands indicate current and previous interglacial warm periods. © IPCC AR4 WG1.

latitudes in each season, but not the total amount of radiation emitted by the Sun toward Earth. The amount of solar radiation hitting the northern continents in the summer season appears to be key to initiating or ending ice ages. If the solar radiation in the northern summer falls below a critical value, snow from previous winters does not melt away in the summer and an ice sheet grows, starting a positive feedback cycle. (The ice sheet reflects more solar radiation than the water, soil, or vegetation in the same location in previous years did.) On the other hand, if the amount of solar radiation hitting the northern continents in summer exceeds a certain value, glaciers and ice begin to melt quickly, triggering a period of relative warmth (interglacial). A secondary factor, which could very well be part of a natural feedback cycle, contributing to the start and maintenance of an ice age appears to be atmospheric carbon dioxide concentration. During ice ages (cold glacial times), carbon dioxide levels are low (~190 ppm), according to Antarctic ice-core data; whereas they are high (~280 ppm) during interglacial periods (warm times). Atmospheric carbon

dioxide concentration lags behind Antarctic temperature changes by hundreds of years.[4]

Further back in time, ~55 million years ago, an abrupt (over a period of several thousand years) warming occurred. This warm period of ~60,000 years is called the Paleocene–Eocene Thermal Maximum (PETM), because the rapid climate shift caused a mass extinction that defines the end of the Paleocene epoch and the start of the Eocene [38] [39]. Evidence indicates that a massive amount of carbon was released to the atmosphere and ocean, on the order of the amount that humans are likely to have emitted by ~2100 [36]. Global average temperature increases were ~5° C, and temperature increases at high latitudes were ~9° C.

Earth's climate has been warmer than it is now, and sea level has been higher than it is now; however, modern humans never lived in such conditions. The

---

[4] For an explanation of possible mechanisms that could regulate the concentration of carbon dioxide on glacial–interglacial time scales, see Box 6.2 of [36].

Table 1.1. Projected global average surface warming and sea-level rise at the end of the twenty-first century. Reprinted with permission from IPCC, *Climate Change 2007: The Physical Science Basis. Working Group I Contribution to the Fourth Assessment Report of the Intergovernmental Panel on Climate Change*, Figure 6.3. Cambridge University Press

Case	Temperature change (°C for 1980–1999)[a]		Sea-level rise (m for 2090–2099 relative to 1980–1999), model-based range excluding future rapid dynamical changes in ice flow
	Best estimate	Likely range	
Constant year-2000 concentrations[b]	0.6	0.3–0.9	NA
B1 scenario	1.8	1.1–2.9	0.18–0.38
A1 scenario	2.4	1.4–3.8	0.20–0.45
B2 scenario	2.4	1.4–3.8	0.20–0.43
A1B scenario	2.8	1.7–4.4	0.21–0.48
A2 scenario	3.4	2.0–5.4	0.23–0.51
A1F1 scenario	4.0	2.4–6.4	0.26–0.59

[a] These estimates are assessed from a hierarchy of models that encompass a simple climate model, several Earth System Models of Intermediate Complexity and a large number of atmosphere–ocean general circulation models (AOGCMs).
[b] Year-2000 constant composition is derived from AOGCMs only. © IPCC AR4 WG1.

impacts of climate change, such as the projected shortages of water, difficulty growing enough food, new ecosystems and species extinctions, and damage to life, property, and infrastructure along coastlines, are uncharted territory in human history. There are records of smaller climatic changes with massive effects on human cultures, including migrations [40].

1.7 Climate models, projections, and uncertainties

Projections of climate change are based on climate models. Different models have various strengths and weaknesses, and all have uncertainty. A chapter within the IPCC Fourth Assessment Report discusses and evaluates the models [41].

The IPCC Fourth Assessment Report projects that, if no more GHGs were released to the atmosphere, the Earth would continue to warm at about 0.1 °C per decade for the next two decades, mostly because the oceans are still absorbing heat – their temperature increase is still too small to restore the radiative balance of the Earth–atmosphere system at the already existing higher levels of GHGs. Human-caused warming and sea-level rise would continue for hundreds of years due to the time scales of environmental processes and climate feedbacks even if GHG concentrations stabilized.

The IPCC further projects that for the next two decades an increase in temperature of ~0.2 °C per decade will occur if global GHG emissions are within

the range of the *Special Report on Emissions Scenarios* (SRES, 2000). However, for most of the last decade global increases in GHG emissions have been near the worst-case (highest-emission) SRES scenario's estimates of emissions.

Table 1.1 shows the best estimates and likely (>66% probability) ranges for global average surface air warming for six SRES scenarios, as well as projections of global average sea-level rise by the year 2099.

For instance, the highest-emission scenario (A1F1) shows a best-estimate increase in temperature of 4 °C (the likely range of warming being 2.4–6.4 °C) by the end of this century compared with the average temperature for the period 1980–1999. For this A1F1 scenario, the sea-level rise by 2099 is best estimated to be an increase of 0.26–0.59 m. Note that the climate models used to make these estimations could not include much information about climate–carbon-cycle feedback or ice-sheet flow dynamics because these processes are not well enough understood to include in the models. Research to better understand these and other feedbacks is being vigorously pursued. It is well recognized that society urgently needs information about how these processes could affect projections of temperature and sea level, as well as other parameters, especially at the high end of the range.

Figure 1.14 shows projected increases in surface temperature for the remainder of this century for three SRES scenarios.

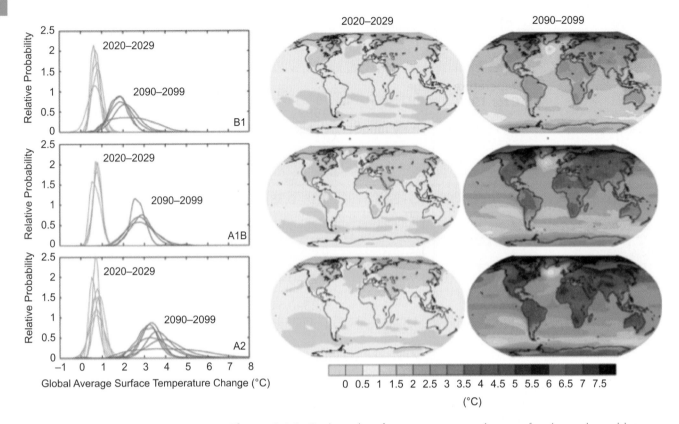

Figure 1.14. Projected surface temperature changes for the early and late twenty-first century relative to the period 1980–1999. The central and right panels show the AOGCM multi-model average projections for the B1 (top), A1B (middle), and A2 (bottom) SRES scenarios averaged over the decades 2020–2029 (center) and 2090–2099 (right). The left panels show corresponding uncertainties as the relative probabilities of estimated global average warming from several different AOGCM and Earth System Model of Intermediate Complexity studies for the same periods. Some studies present results only for a subset of the SRES scenarios, or for various model versions. Therefore the difference in the number of curves shown in the left-hand panels is due only to differences in the availability of results. © IPCC AR4 WG1.

In the left column of Figure 1.14 are probability distribution functions indicating projected increases in Earth's surface temperature for two time periods, both relative to the mean of the period 1980–1999: the first period is 2020–2029, and the second period is the last decade of this century, 2090–2099. The top panel is for the B1 SRES scenario, the middle panel is for the A1B scenario, and the bottom panel is for the A2 scenario. The middle (years 2020–2029) and right (2090–2099) columns show projected temperature increases across the globe, again for the same two periods. From top to bottom in each column are the same SRES scenarios: B1 (top), A1B (middle), and A2 (bottom).

In the section on observations, we noted that temperatures in the Arctic have increased about twice as fast as the global average temperature, and that temperature increases over land have been about double

those in the ocean. These trends are expected to continue. Figure 1.14 illustrates three SRES scenarios (B1, low emissions; and A1B and A2, moderate emissions). Note that, in the real world, emissions have been at the high end of SRES scenarios' emissions during the first decade of the twenty-first century.

As one can see in Figure 1.14, the Antarctic is expected to warm, especially in the latter half of this century. This is based partly on the expected recovery of the ozone hole [42], which has, thus far, protected Antarctica from much of the human-caused warming that the rest of the globe has experienced. If stratospheric ozone levels recover (e.g., reach levels near their pre-1969 concentration), then the positive phase of the Southern Annular Mode (SAM, an atmospheric circulation pattern, which has been shielding the Antarctic interior from warmer air masses at lower latitudes) will decrease during the Southern Hemisphere summer. As

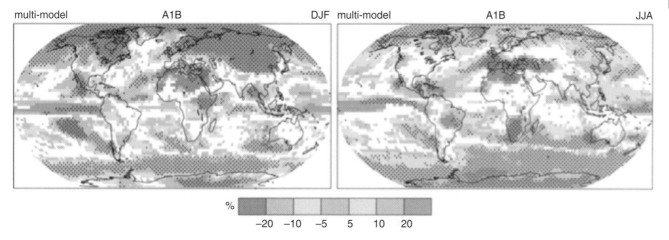

Figure 1.15. Relative changes in precipitation (in percent) for the period 2090–2099, relative to 1980–1999. Values are multi-model averages based on the SRES A1B scenario for December to February (left) and June to August (right). White areas are where less than 66% of the models agree on the sign of the change and stippled areas are where more than 90% of the models agree on the sign of the change. © IPCC AR4 WG1.

the stratospheric ozone levels recover, the lower stratosphere over Antarctica will absorb more ultraviolet radiation from the Sun, causing substantial increases in temperature and reducing the strong north–south temperature gradient that currently favors the positive phase of the SAM. Thus, it is expected that Antarctica will no longer be protected from the warming patterns affecting the rest of the world.

Additional projections include a reduction in global surface pH of between 0.1 and 0.3 units through the twenty-first century. This is acidification on top of the 0.1-unit reduction that has already occurred in the industrial era.

In the cryosphere, snow cover is expected to shrink, and widespread increases in the depth at which frozen soil thaws are expected in most areas with permafrost. Sea ice is projected to shrink at both poles. Arctic sea ice is projected to disappear during the summer by the latter part of this century, with some studies indicating that this could occur within three decades [43].

It is virtually certain that the number of heat waves will increase, very likely that heavy precipitation events will increase, and likely that future tropical cyclones (typhoons and hurricanes) will become more intense during this century [44].

Patterns of precipitation are projected to change as the Earth's climate warms. It is very likely that high latitudes will see increased precipitation, while it is likely that subtropical land regions (those adjacent to the tropics, generally from ~20 and ~40 degrees of latitude in both hemispheres) will see decreased precipitation. See Figure 1.15.

Lastly, the current partitioning of GHG emissions into the atmosphere (~46%), oceans (~33%), and biosphere (~21%) is expected to change as GHG emissions continue, with more GHGs expected to stay in the atmosphere. The rate of carbon dioxide uptake by land and ocean is predicted to decrease with further GHG emissions [45].

1.8 Causes of climate change

The vast majority of scientific evidence indicates that most of the observed warming of the Earth in the last half century has been caused by increased concentrations of GHGs due to human activities (e.g., burning of fossil fuels and cutting down forests). The IPCC Fourth Assessment report concludes that "The observed widespread warming of the atmosphere and ocean, together with ice mass loss, support the conclusion that it is extremely unlikely that global climate change of the past 50 years can be explained without external forcing, and very likely that it is not due to known natural causes alone" [46].

Climate simulations that incorporate both human-caused forcings (e.g., emissions of GHGs and aerosols, and land-use changes) and natural external forcings (e.g., volcanic eruptions, which emit aerosols, and solar variations) match the observed increase in temperature in the twentieth century, whereas simulations that include only natural forcings (and exclude human-caused forcings) do not match the observations. See Figure 1.16, which shows results of climate models that include both human-caused and natural forcings (in pink) and results of models that include only natural forcings (in blue). The observed temperature anomalies are shown in black, and match only the (pink) results

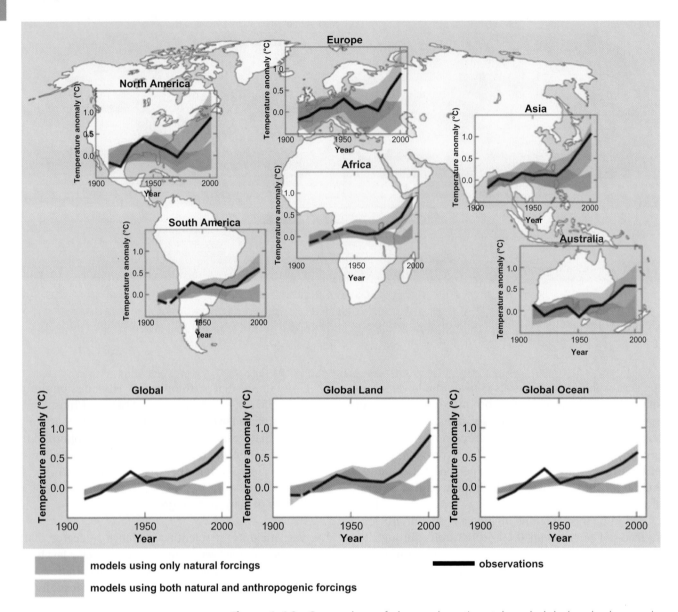

models using only natural forcings

models using both natural and anthropogenic forcings

observations

Figure 1.16. Comparison of observed continental- and global-scale changes in surface temperature with results simulated by climate models using natural and anthropogenic forcings. Decadal averages of observations are shown for the period 1906 to 2005 (black line) plotted against the center of the decade and relative to the corresponding average for 1901–1950. Lines are dashed where the spatial coverage is less than 50%. Blue shaded bands show the 5%–95% range for 19 simulations from five climate models using only the natural forcings due to solar activity and volcanoes. Red shaded bands show the 5%–95% range for 58 simulations from 14 climate models using both natural and anthropogenic forcings. © IPCC AR4 WG1.

that incorporate human-caused and natural forcings. This is shown for global temperatures, ocean temperatures, land temperatures, and temperatures on each continent except Antarctica.

After publication of the IPCC Fourth Assessment report, a research group found a human influence in the Antarctic continent [47].

In addition to these climate simulations showing that only models that include both human-caused and natural forcings are consistent with observed temperatures is the fact that the physical properties of GHGs, i.e., their radiative properties, indicate that they will, in large enough quantities, warm our planet. As described in Section 1.3 about the history of studying carbon dioxide,

RESTORING THE QUALITY
OF
OUR ENVIRONMENT

Report of The
Environmental Pollution Panel
President's Science Advisory Committee

THE WHITE HOUSE

NOVEMBER 1965

"The climatic changes that may be produced by the increased CO_2 contents could be deleterious from the point of view of human beings. The possibilities of deliberately bringing about countervailing climatic changes therefore need to be thoroughly explored. A change in the radiation balance in the opposite direction to that which might result from the increase of atmospheric CO_2 could be produced by raising the albedo, or reflectivity, of the earth (from p.127)."

APPENDIX Y4   123

ing this period more than 40% of the total CO_2 increase from fossil fuel combustion occurred. We must conclude that climatic "noise" from other processes has at least partially masked any effects on climate due to past increases in atmospheric CO_2 content.

OTHER POSSIBLE EFFECTS OF AN INCREASE IN ATMOSPHERIC CARBON DIOXIDE

Melting of the Antarctic ice cap.—It has sometimes been suggested that atmospheric warming due to an increase in the CO_2 content of the atmosphere may result in a catastrophically rapid melting of the Antarctic ice cap, with an accompanying rise in sea level. From our knowledge of events at the end of the Wisconsin period, 10 to 11 thousand years ago, we know that melting of continental ice caps can occur very rapidly on a geologic time scale. But such melting must occur relatively slowly on a human scale.

The Antarctic ice cap covers 14 million square kilometers and is about 3 kilometers thick. It contains roughly 4×10^{16} tons of ice, hence 4 x
f heat energy would be required to melt it. At the
eward heat flow across 70° latitude is 10^{22} gram
his heat is being radiated to space over Antarctica
e effect on the ice cap. Suppose that the pole-
sed by 10% through an intensification of the
ulation, and that all of this increase in the
o melt the ice. Some 4,000 years would

ting time by supposing a change in the
of which would be used to melt the
he year 2000, when the atmospheric
aps by 25%. Since the average
2×10^8 gram calories per square
centimeter per year, a 2% change would amount to 2×10^{22} calories per year. If half this energy were concentrated in Antarctica and used to melt the ice, the process would take 400 years.

Rise of sea level.—The melting of the Antarctic ice cap would raise sea level by 400 feet. If 1,000 years were required to melt the ice cap, the sea level would rise about 4 feet every 10 years, 40 feet per century. This is a hundred times greater than present worldwide rates of sea level change.

Warming of sea water.—If the average air temperature rises, the temperature of the surface ocean waters in temperate and tropical regions could be expected to rise by an equal amount. (Water temperatures in the polar regions are roughly stabilized by the melting and freezing of ice.) An oceanic warming of 1° to 2°C (about 2°F) oc-

Figure 1.17. The front page of a paper prepared by the Environmental Pollution Panel of the President's Science Advisory Committee, "Restoring the Quality of Our Environment" (November, 1965).

scientists anticipated the warming effect of carbon dioxide in the middle of the nineteenth century. About a century later, US President Lyndon Johnson recognized the danger posed by GHGs, and stated such in a "Special Message to the Congress on Conservation and Restoration of Natural Beauty" on February 8, 1965. A paper was prepared by the Environmental Pollution Panel of the President's Science Advisory Committee [48] Figure 1.17 shows the cover of this White House paper.

Despite the fact that scientific evidence for human-caused climate change continues to become stronger every year, some scientists do not accept that human activities are causing Earth to warm. It is our impression that, in most cases, the reasons have nothing to do with evidence. Many people, some of them scientists, see climate change as a threat to our existing economic system and/or our social values. Our society has come to expect a continuation of never-ending economic growth of 2% or more per year, a portion of which involves increasing

the rate of exploitation of Earth's resources. However, climate change is a clear manifestation that there are limits to exponential growth. Our collective influence on planet Earth is now a geological force large enough to make it our responsibility to protect it for future generations. There is an urgent need to develop a new energy-supply system, which is fundamental to our economic system, so that both our energy system and our economic system are sustainable.

1.9 Conclusions

The IPCC Fourth Assessment Report concluded that "Warming of the climate system is unequivocal, as is now evident from observations of increases in global average air and ocean temperatures, widespread melting of snow and ice, and rising global average sea level" and that "Global atmospheric concentrations of carbon dioxide, methane, and nitrous oxide have increased markedly as a result of human activities since 1750 and now far exceed pre-industrial values determined from ice cores spanning many thousands of years. [49]. Paleoclimatic data indicate that current levels of these three gases greatly exceed their previous levels of many hundreds of thousands of years. Our knowledge of ocean chemistry tells us that we have already committed the atmosphere to thousands of years of elevated CO_2 level unless we learn how to extract it from ambient air on a vast industrial scale or extract it from power plant exhaust stacks and permanently sequestor it, e.g. in underground formations. Carbon capture and sequestration (also called "clean coal") is technology that has been demonstrated on a small scale but not on a commercial scale, and is the topic of later chapters in this book. Our knowledge of the radiative properties of these gases and climate simulations indicates that most of the observed increase in global and continental-scale warming is caused by human activities that increase the atmospheric concentrations of these gases.

The predicted impacts of climate change are beyond the scope of this chapter, but include decreased availability of water and increased drought at mid latitudes and semi-arid low latitudes, with hundreds of millions of people exposed to increased water stress; disruption of ecosystems, including extinctions of species, changes in the carbon cycle, and further acidification of the global ocean, which endangers marine shell-forming organisms near the base of the marine food web, as well as corals; decreased food-crop productivity at atmospheric concentrations of carbon dioxide of above ~550 ppm and temperature increases above 1–3 °C (compared with pre-industrial times); increased damage from floods and storms; loss of coastal wetlands; increased incidence of malnourishment, diarrheal, cardio-respiratory, and infectious diseases, as well as increased incidence of injury, disease, and death caused by extreme weather events, such as heat waves, droughts, floods, fires, and storms.

Climate-model projections contain various levels of uncertainties, depending upon multiple factors, including which model(s) are used and what the output parameters in question are. From recent observations, it appears that the *IPCC Fourth Assessment Report (Physical Science Basis)* has not overstated any projected changes, and in some cases, e.g., Arctic sea ice loss, has understated the rate at which changes may occur. The gaps in climate models that were documented as not being fully incorporated into projections, e.g., carbon-cycle feedbacks and ice-sheet dynamics, increase the urgency of advancing our knowledge of these processes.

The degree of uncertainty in many climate projections, which grows very rapidly beyond the second half of the twenty-first century, leaves society with many questions. As noted above, permafrost degradation is being observed, but is not yet pervasive enough to be measured as enhanced methane mole fractions in air in the larger Arctic basin [50]. The total amount of carbon stored in Arctic permafrost is estimated to be ~1,400 GtC, or 1,700 GtC when northern peat-lands are included [51]. This is almost five times the cumulative fossil-fuel emissions through 2008. There is evidence of a geologically sudden release of massive amounts of carbon coinciding with the abrupt warming of the PETM, described above, which occurred ~55 million years ago. During the PETM, global average temperature increases were ~5 °C, and at high latitudes, temperature increases were ~9 °C. There is the possibility, though it is difficult to quantify, that permafrost degradation could lead to an abrupt, large release of methane to the atmosphere, which would constitute a large positive feedback that would lead to greater warming [31].

Whether we are ready to accept it or not, we have become collectively responsible for safeguarding Earth's climate on which ecosystems, millions of plant and animal species, and our civilization depend. Our understanding of the fate of the main long-lived GHGs, namely carbon dioxide, methane, and nitrous oxide, and of a host of others, tells us that we will have to bring net emissions to near zero within this century, and probably sooner. We do not know what the maximum safe atmospheric concentrations of GHGs are. What we do know is that their current levels greatly exceed the levels that existed for hundreds of thousands of years in the past. We do not know at which point the emissions may push the climate system to a state that is different enough from the present to cause us great difficulty in coping with the changes in precipitation amounts and distribution, sea level, ocean pH, temperature, food production, and extreme events, to name just a few impacts. To limit climate change to a level that, we hope, remains manageable, emissions should peak in the next 10 to 15 years and then would have to decline thereafter. The longer we wait, the more difficult it will be to reduce GHG emissions quickly so that concentrations can begin to decrease rather than continue to increase.

We hope that this book will contribute to the information that charts the path to conserving energy, increasing efficiency in the way we use energy, and transforming our energy system into one that is sustainable.

1.10 Summary

Greenhouse gases, most importantly carbon dioxide, methane, and nitrous oxide, absorb infrared radiation (heat) that is emitted by the Earth's surface, atmosphere, and clouds, thereby trapping heat in the Earth system. Radiative forcing (expressed in units of W m^{-2}) refers to the globally averaged change in the heat balance of the Earth–atmosphere system that results from various changes in the atmosphere or at the surface. The concentrations of carbon dioxide, methane, and nitrous oxide are far greater today than they were before the industrial revolution, and have contributed a positive radiative forcing (warming). Other human-caused radiative-forcing agents, such as tropospheric ozone, halocarbons, and black carbon on snow, have also contributed a positive forcing. Human changes in land use that affect Earth's surface reflectance, as well as human-produced aerosols, have contributed a negative forcing (cooling). The total radiative forcing caused by human activities (in 2005) is positive (warming), with a value of $+1.6\,\mathrm{W\,m^{-2}}$. This is likely more than ten times the forcing caused by natural brightening of the Sun since 1750.

Observed changes in the Earth's surface, atmosphere, precipitation, cryosphere, and oceans make warming of the climate system undeniable, and many of the changes appear to be accelerating. Global mean surface temperatures increased by 0.7 °C from 1906 to 2005. The rate of warming over the second half of that period (0.13 °C per year) is almost twice as fast as the rate over the entire period (0.07 °C per year). Surface temperatures over land have warmed at a faster rate than surface temperatures over oceans (~0.27 °C and 0.13 °C, respectively, per decade in the last two decades).

Significant increases in precipitation have been observed in the eastern parts of North America and South America, Northern Europe, and northern and central Asia. Significant decreases in precipitation have been observed in parts of Africa, the Mediterranean, and part of southern Asia.

The extents of snow cover, permafrost, and seasonally frozen ground have decreased. Arctic sea ice extent has decreased in the last three decades. The decrease has been especially pronounced in the summer, when the ice extent reaches its minimum in September each year. During the satellite record, which began in 1979, annual Arctic sea ice extent has decreased by 4.1% per decade through 2009. The rate of sea ice decline in the Arctic summer since 1979 has now (2009) increased to −11.2% per decade. Glaciers and ice caps have lost significant mass (melted) and contributed to sea-level rise.

The world's oceans are warming and becoming more acidic (as a result of atmospheric carbon dioxide dissolving in ocean water), and sea level is rising because of expansion of water and melting of glaciers. From 1961 to 2003, the global ocean temperature increased by 0.10 °C, from the sea surface down to 700 m depth. The global ocean near the surface is more acidic now, by ~0.1 pH units, than it was in pre-industrial times. Global average sea level is rising, but there is significant decadal variability. From 1961 to 2003, global average sea level rose at an average rate of 1.8 mm/yr. From 1993 to 2003, the rate increased to 3.1 mm/yr.

Even if GHG concentrations were, as a thought experiment, held constant at their 2000 levels, the Earth would continue to warm at a rate of ~0.1 °C per decade for the next two decades, primarily because the response of oceans and ice sheets to forcing agents is slow. The upper ocean adjusts over a period of years to decades, while the deep ocean and ice sheets adjust over periods of centuries to millennia.

Warming at the last decade of this century, compared with the average temperature from 1980–1999 and for constant year-2000 concentrations, depends on the SRES scenario. Projections range from a best estimate of 1.8 °C of warming (likely range of warming 1.1–2.9 °C) to a best estimate of 4 °C (with the likely range of warming being 2.4–6.4 °C). Projections of sea-level rise range from 0.18–0.38 m during the decade 2090–2099 relative to 1980–1999 for the SRES scenario with the least warming to a sea-level rise of 0.26–0.59 m for 2090–2099 relative to 1980–1999 for the SRES scenario with the greatest amount of warming. The models used to make these estimates of sea-level rise did not include climate–carbon-cycle feedbacks or ice-sheet dynamical flow because these processes are not well enough understood for us to address them adequately; therefore, a sea-level rise greater than these projected values is not unlikely.

We therefore conclude that our current energy system is not sustainable. A recent National Academies Report, *America's Energy Future*, concludes "... there has been a steadily growing consensus that our nation must fundamentally transform the size and complexity of the U.S. energy system and its reach into all aspects of American life, this transformation will be an enormous undertaking; it will require fundamental changes, structural as well as behavioral, among producers and consumers alike. ... *a meaningful and timely transformation to a more sustainable and secure energy system will likely entail a generation or more of sustained efforts by both the public and the private sectors.* [52].

1.11 Questions for discussion

1. Can a given extreme event, such as a hurricane or heat wave, be attributed to the greenhouse effect?
2. What would be some effective ways to rapidly decrease emissions of GHGs?
3. Can the observed global average temperature increase be explained by natural causes?
4. If society decided to reduce GHG emissions, how quickly would atmospheric concentrations stabilize and eventually decrease?
5. How could we raise the level of awareness of the predicament we are finding ourselves in?
6. What could be some potential global geo-engineering solutions, and what other consequences, not necessarily desirable, might they have?
7. How much more coal, oil, and natural gas are humans likely to burn?

1.12 Further reading

- **N. Oreskes** and **E. Conway**, 2010, *Merchants of Doubt*, New York, Bloomsbury Press. This book documents how several influential scientists have misled the public on a number of scientific matters.
- **J. Lovelock**, 2009, *The Vanishing Face of Gaia*, New York, Basic Books. This book challenges the consensus projections of the IPCC, posits that abrupt changes in Earth's climate may occur in the coming decades, and suggests that nuclear energy is the best option.
- **R. Pielke Jr.**, 2010, *The Climate Fix*, New York, Basic Books. This book promulgates the need to extricate climate policy from energy policy, and to make the goal of reducing GHG emissions secondary to efforts to provide access to all people to affordable clean energy.
- **N. Stern**, 2009, *The Global Deal: Climate Change and the Creation of a New Era of Progress and Prosperity*, New York, Public Affairs. This book argues that unchecked climate change will put all development goals out of reach, and that developing countries need to be persuaded to choose a new development path through a large amount of assistance from rich countries.

1.13 References

[1] **N. S. Lewis** and **D. G. Nocera**, 2006, "Powering the planet: chemical challenges in solar energy utilization," *Proc. Natl. Acad. Sci.*, **103**, 15729–15735.

[2] **S. Arrhenius**, 1896, "On the influence of carbonic acid in the air upon the temperature of the ground," *Phil. Mag.*, **41**, 237–276.

[3] **G. Callendar**, 1958, "On the amount of carbon dioxide in the atmosphere," *Tellus*, **10**(2), 243–248.

[4] **R. Revelle**, and **Suess, H.**, 1957, "Carbon dioxide exchange between atmosphere and ocean and the question of an increase of atmospheric CO_2 during the past decades," *Tellus*, **9**(1), 18–27.

[5] **R. A. Hanel, B. J. Conrath, V. G. Kunde** et al., 1972, "The Nimbus 4 infrared spectrometry experiment 1. Calibrated thermal emission spectra," *J. Geophys. Res.*, **77** (15), 2629–2641.

[6] **D. Lüthi, M. Le Floch, B. Bereiter** et al., 2008, "High-resolution carbon dioxide concentration record 650,000–800,000 years before present," *Nature*, **453**, 379–382.

[7] **P. Foster, V. Ramaswamy, P. Artaxo** et al., 2007, "Changes in Earth's atmospheric constituents and in radiative forcing," in *Climate Change 2007: The Physical Science Basis. Contribution to WG1 Fourth Assessment Report of the IPCC*, eds. **S. Solomon, D. Qin, M. Manning** et al., Cambridge, Cambridge University Press, pp. 131–234.

[8] **M. J. Molina** and **F. S. Rowland**, 1974, "Stratospheric sink for chlorofluoromethanes: chlorine atom-catalysed destruction of ozone," *Nature*, **249**, 810–812.

[9] **G. J. M. Velders**, 2007, "The importance of the Montreal Protocol in protecting climate change," *Proc. Natl. Acad. Sci.*, **104**(12), 4814–4819.

[10] **A. A. Lacis, G. A. Schmidt, D. Rind**, and **R. A. Ruedy**, 2010, "Atmospheric CO_2: principal control knob governing Earth's temperature," *Science*, **330**, 356–359.

[11] **R. T. Pierrehumbert**, 2011, "Principals of planetary climate," *Phys. Today, January*, pp. 33–38.

[12] **P. P. Tans** and **P. S. Bakwin**, 1995, "Climate change and carbon dioxide forever," *Ambio*, **24**(6), 376–378.

[13] **C. D. Keeling** and **Bacastow R. B.**, 1977, "Impact of industrial gases on climate," in *Energy and Climate: Studies in Geophysics*, Washington, DC, National Academy of Science, pp. 72–95.

[14] **D. Archer, M. Eby, V. Brovkin** et al., 2009, "Atmospheric lifetime of fossil-fuel carbon dioxide," *Ann. Rev. Earth Planet. Sci.*, **37**, 117–134.

[15] **F. Joos, I. C. Prentice, S. Sitch** et al., 2001, "Global warming feedbacks on terrestrial carbon uptake under the Intergovernmental Panel on Climate Change (IPCC) emission scenarios," *Global Biogeochem. Cycles*, **15**, 891–907.

[16] **B. O'Neill**, 2000, "The jury is still out on global warming potentials," *Clim. Change*, **44**, 427–443.

[17] **J. S. Fuglestvedt, T. K. Berntsen, O. Godal** et al., 2003, "Metrics of climate change: assessing radiative forcing and emission indices," *Clim. Change*, **58**, 267–331.

[18] **K. E. Trenberth, P. D. Jones, P. Ambenje** et al., 2007, "Observations: surface and atmospheric climate change," in *Climate Change 2007: The Physical Science Basis. Contribution to WG1 Fourth Assessment Report of the IPCC*, eds. **S. Solomon, D. Qin, M. Manning** et al., Cambridge, Cambridge University Press, pp. 236–5.M.3–11.

[19] **F. Fetterer, K. Knowles, W. Meier**, and **M. Savoie**, 2002, *Sea Ice Index*, Boulder, CO, National Snow and Ice Data Center (digital media, updated 2010).

[20] **J. A. Maslanik, C. Fowler, J. Stroeve** et al., 2007, "A younger, thinner Arctic ice cover: increased potential for rapid, extensive sea-ice loss," *Geophys. Res. Lett.*, **34**, L24501, doi:10.1029/2007GL032043.

[21] **S. V. Nghiem, I. G. Rigor, D. K. Perovich** et al., 2007, "Rapid reduction of Arctic perennial sea ice," *Geophys. Res. Lett.*, **34**, L19504, doi:10.1029/2007GL031138.

[22] **I. G. Rigor** and **J. M. Wallace**, 2004, "Variations in the age of Arctic sea-ice and summer sea-ice extent," *Geophys. Res. Lett.*, **31**, L09401, doi:10.1029/2004GL019492.

[23] **R. Kwok** and **D. A. Rothrock**, 2009, "Decline in Arctic sea ice thickness from submarine and ICES at records: 1958–2008," *Geophys. Res. Lett.*, **36**, L15501, doi:10.1029/2009GL039035.

[24] **J. Turner, R. A. Bindschadler, P. Convey** et al. (eds.), 2009, *Antarctic Climate Change and the Environment*, Cambridge, Scientific Committee on Antarctic Research (SCAR).

[25] **P. Lemke J. Ren, R. B. Alley** et al., 2007, "Observations: changes in snow, ice, and frozen ground," in *Climate Change 2007: The Physical Science Basis. Contribution to WG1 Fourth Assessment Report of the IPCC*, eds. **S. Solomon, D. Qin, M. Manning** et al., Cambridge, Cambridge University Press, pp. 337–383.

[26] **A. J. Monaghan, D. H. Bromwich, W. Chapman** et al., 2008, "Recent variability and trends of Antarctic near

surface temperature," *J. Geophys. Res.*, **113**, D04105, doi:10.1029/2007JD009094.

[27] **J. Hansen** and **L. Nazarenko**, 2004, "Soot climate forcing via snow and ice albedos," *Proc. Natl. Acad. Sci.*, **101**(2), 423–428.

[28] **R. D. Brown**, 2000, "Northern Hemisphere snow cover variability and change 1915–1997," *J. Clim.*, **13**, 2339–2355.

[29] **N. L. Bindoff, J. Willebrand, V. Artale** *et al.*, 2007, "Observations: oceanic climate change and sea level," in *Climate Change 2007: The Physical Science Basis. Contribution to WG1 Fourth Assessment Report of the IPCC*, eds. **S. Solomon, D. Qin, M. Manning** *et al.*, Cambridge, Cambridge University Press, pp. 387–432.

[30] **J. C. Orr, V. J. Fabry, O. Aumont**, 2005, "Anthropogenic ocean acidification over the twenty-first century and its impact on calcifying organisms," *Nature*, **437**, 681–686, doi:10.1038/nature04095.

[31] **P. Tans**, 2009, "An accounting of the observed increase in oceanic and atmospheric CO_2 and an outlook for the future," *Oceanography*, **22**(4), 26–35.

[32] **R. S. Bradley**, 1999, "Climatic variability in sixteenth-century Europe and its social dimension – preface," *Clim. Change.* **43**(1), 1–2.

[33] **T. M. Cronin**, 1999, *Principles of Paleoclimatology. Perspectives in Paleobiology and Earth History*, New York, Columbia University Press.

[34] **A. Mackay, R. W. Battarbee, J. Birks**, and **F. Oldfield** (eds.), 2003, *Global Change in the Holocene*, London, Hodder Arnold.

[35] **National Research Council**, 2006, *Surface Temperature Reconstructions for the Last 2,000 Years*, Washington, DC, National Academy Press.

[36] **E. Jansen, J. Overpeck, K. R. Briffa** *et al.*, 2007, "Paleoclimate," in *Climate Change 2007: The Physical Science Basis. Contribution to WG1 Fourth Assessment Report of the IPCC*, eds. **S. Solomon, D. Qin, M. Manning** *et al.*, Cambridge, Cambridge University Press, pp. 433–497.

[37] **J. T. Overpeck, B. L. Otto-Bliesner, G. H. Miller** *et al.*, 2006, "Paleoclimatic evidence for future ice-sheet instability and rapid sea-level rise," *Science*, **311** (5768), 1747–1750.

[38] **J. C. Zachos, U. Röhl, S. A. Schellenberg** *et al.*, 2005, "Rapid acidification of the ocean during the Paleocene–Eocene thermal maximum," *Science*, **308**, 1611–1615.

[39] **S. L. Wing, G. J. Harrington, F. A. Smith** *et al.*, 2005, "Transient floral change and rapid global warming at the Paleocene–Eocene boundary," *Science*, **310**, 993–996.

[40] **J. Diamond**, 2005, *Collapse, How Societies Choose to Fail or Succeed*, New York, Viking Press.

[41] **D. A. Randall, R. A. Wood, S. Bony** *et al.*, 2007, "Climate models and their evaluation," in *Climate Change 2007: The Physical Science Basis. Contribution to WG1 Fourth Assessment Report of the IPCC*, eds. **S. Solomon, D. Qin, M. Manning** *et al.*, Cambridge, Cambridge University Press, pp. 589–662.

[42] **J. Perlwitz, S. Pawson, R. L. Fogt, J. E. Nielsen**, and **W. D. Neff**, 2008, "Impact of stratospheric ozone hole recovery on Antarctic climate," *Geophys. Res. Lett.*, **35**, L08714, doi:10.1029/2008GL033317.

[43] **M. Wang** and **J. E. Overland**, 2009, "A sea ice free summer Arctic within 30 years," *Geophys. Res. Lett.*, **36**, L07502, doi:10.1029/2009GL037820.

[44] **G. A. Meehl, T. F. Stocker, W. D. Collins** *et al.*, 2007, "Global climate projections," in *Climate Change 2007: The Physical Science Basis. Contribution to WG1 Fourth Assessment Report of the IPCC*, eds. **S. Solomon, D. Qin, M. Manning** *et al.*, Cambridge, Cambridge University Press, pp. 747–845.

[45] **P. Friedlingstein, P. Cox, R. Betts**, *et al.*, 2006, "Climate–carbon cycle feedback analysis: results from the (CMIP)-M-4 model intercomparison," *J. Clim.*, **19**, 3337–3353.

[46] **G. C. Hegerl, F. W. Zwiers, P. Braconnot** *et al.*, 2007, "Understanding and attributing climate change," in *Climate Change 2007: The Physical Science Basis. Contribution to WG1 Fourth Assessment Report of the IPCC*, eds. **S. Solomon, D. Qin, M. Manning** *et al.*, Cambridge, Cambridge University Press, pp. 665–744.

[47] **N. P. Gillette, D. A. Stone, P. A. Stott** *et al.*, 2008, "Attribution of polar warming to human influence," *Nature Geosci.*, **1**, 750–754.

[48] **Environmental Pollution Panel of the President's Science Advisory Committee**, 1965, *Restoring the Quality of Our Environment*, http://dge.stanford.edu/labs/caldeiralab/Caldeira%20downloads/PSAC,%201965,%20Restoring%20the%20Quality%20of%20Our%20Environment.pdf.

[49] **IPCC**, 2007, "Summary for policy makers," in *Climate Change 2007: The Physical Science Basis. Contribution to WG1 Fourth Assessment Report of the IPCC*, eds. **S. Solomon, D. Qin, M. Manning** *et al.*, Cambridge, Cambridge University Press, pp. 1–18.

[50] **E. J. Dlugokencky, L. Bruhwiler, J. W. C. White** *et al.*, 2009, "Observational constraints on recent increases in the atmospheric CH_4 burden," *Geophys. Res. Lett.*, **36**, L18803, doi:10.1029/2009GL039780.

[51] **E. A. G. Schuur, J. Bockheim, J. G. Canadell** *et al.*, 2008, "Vulnerability of permafrost carbon to climate change: implications for the global carbon cycle," *BioScience*, **58**, 701–714.

[52] **Committee on America's Energy Future, Harold Shapiro, Chair**, 2009, *America's Energy Future: Technology and Transformation* Washington, DC, National Academy of Sciences.

[53] **J. A. Church** and **N. C. White**, 2006, "A 20th century acceleration in global sea level rise," *Geophys. Res. Lett.*, **33**, L01602, doi:10.1029/2005GL024826.

[54] **S. J. Holgate** and **P. L. Woodworth**, 2004, "Evidence for enhanced coastal sea level rise during the 1990s," *Geophys. Res. Lett.*, **31**, L07305, doi:10.1029/2004GL019626.

[55] **E. W. Leuliette, R. S. Nerem**, and **G. T. Mitchum**, 2004, "Calibration of TOPEX/Poseidon and Jason altimeter data to construct a continuous record of mean sea level change," *Marine Geodesy*, **27**, 79–94.

2 The global energy landscape and energy security

V. S. Arunachalam[1] and Anshu Bharadwaj[2]

[1]Carnegie Mellon University; [2]Center for Study of Science, Technology, and Policy, Bangalore, India

2.1 Focus

The global energy landscape encompasses the distribution of energy resources, as well as the related aspects of energy production, storage, transmission, use, and efficiency. In addition, energy use has been correlated with economic development. Together, these attributes define the context within which countries strive to satisfy their energy demands, in terms of both economic productivity and quality of life. With the current rapid increase in demand for energy, the question of how countries will provide their populations with access to a clean and affordable energy supply in an environmentally sustainable manner has emerged as a grand challenge for today's society.

2.2 Synopsis

This chapter summarizes the global energy resources and their availability, economic viability, and environmental consequences. Although these topics are discussed with respect to individual fuels and technologies in greater detail in other chapters of this book, they are examined here in a broader sense relating to the overall global energy landscape.

Despite the dramatically increasing worldwide demand for energy, there exist sufficient resources to ensure that the world will not 'run out of energy' in the near future. However, in addition to the theoretical availability of a resource, several practical considerations determine whether and to what extent that resource is employed. In particular, various resources differ in energy content, price, ease of resources, conversion efficiency, waste, and CO_2 emissions. Moreover, some important sources of energy – such as oil, gas, and uranium – are concentrated in just a few countries, making them more susceptible to price and supply volatility. Increasing concerns over environmental impacts of energy use have led to a global emphasis on generating and using energy efficiently, as well as developing more environmentally benign resources. All of these factors combine to influence countries' decisions on energy policy and practices and, in turn, affect the further evolution of the global energy landscape.

The interplay between energy and politics, specifically in terms of energy security, can be illustrated through two sources: biofuels and nuclear power. Because of the perceived depletion of fossil energy resources and fears of supply denials by energy-rich countries, many nations of the world have prioritized energy security in order to ensure an adequate energy supply for their citizens. Governments have intervened on the basis that new innovations in renewable energy technologies continue to be expensive and require subsidies before they can become cost-competitive with existing energy sources.

The goal of energy security is also being pursued through the transformation of the electrical grid into a "smart grid." This transformation will become important not only for the technical reasons highlighted in Chapter 42, such as robustness and efficiency, but also because of the interrelated policy issues of pricing and control. These are challenges facing many renewable sources of energy: cost ($/watt) and the ability to integrate variable (and unpredictable) sources of power.

2.3 Historical perspective

A country's energy consumption is directly related to both its economic output and the individual well-being of its citizens. Both population growth and the desire to maintain growth while raising standards of living result in increased energy consumption as a society develops. For example, in 1800, before the full-scale onset of the Industrial Revolution, the world's population was about 1 billion, with a total annual energy consumption of 0.6 TW·year. (By analogy with kilowatt-hours, a terawatt-year is a unit of energy equal to 1 TW of power expended for one year of time. In terms of other energy units, 1 TW·year $= 8.76 \times 10^{12}$ kW·h $= 3.15 \times 10^{19}$ J $= 31.5$ EJ $= 30$ Quads.) By the turn of the millennium, the global population had increased by a factor of 6, whereas the annual primary energy consumption had increased by a factor of more than 20, reaching about 14 TW·year today. The reason why growth in energy is not in step with the population are those changes that transformed the daily lives of many individuals. For example, agricultural improvements allowed larger amounts of food to be grown with less human effort, thus shifting increasing numbers of people away from employment in farming. Increased manufacturing made a greater variety of goods available, and more extensive transportation networks delivered those goods more widely and efficiently to the public. Labor-saving devices such as washing machines and power lawn mowers led to increased leisure time, which, in turn, increased demand for leisure goods, such as televisions. More recently, computers, cell phones, and video-conferencing equipment have become symbols of far-reaching transformations. Despite improvements in energy efficiency, both the manufacture and the use of such lifestyle-enhancing devices entail increased energy consumption.

However, the increased energy consumption, enhanced prosperity, and improved standard of living have not been shared equitably among all countries, nor are they shared equally among all individuals of a country or all sectors of an economy. For example, Figure 2.1 presents the trends in energy consumption of a variety of countries between 1980 and 2006. The values are given on a per capita basis because this allows an easier comparison among individuals living in the different countries. Bear in mind, however, that each country's total energy consumption is a product of the per capita value and the total population. Thus, the total energy consumption of China, with a population of 1.3 billion, far exceeds that of the Netherlands (population 16.5 million), even though the per capita consumption of the latter is over four times higher.

For comparison, the 2010 **gross domestic products (GDPs)** per capita of the same countries are presented

Table 2.1. The 2010 gross domestic product (GDP) per capita for selected countries [2]

Country	GDP per capita (USD)
USA	47,132
Netherlands	40,477
Russia	15,806
Brazil	11,289
China	7,517
India	3,290
Zimbabwe	395
Somalia	300[a]

[a] Estimated.

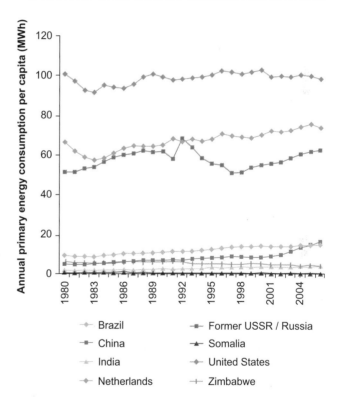

Figure 2.1. Annual primary energy consumption per capita of selected countries (1980–2006) [1].

in Table 2.1. The GDP is a measure of the total market value of all goods and services produced in a country in a given year and is considered an approximate measure of a country's prosperity. As can be seen, per capita energy usage and per capita GDP generally follow the same trends. Thus, the disparities in energy use are reflected in similar disparities in economic growth and, in turn, in individual well-being. It is only natural, then, that countries strive to satisfy their energy demands, to foster both economic development and improved quality of life. As the global energy landscape evolves over time, they must adapt the strategies they use to do so.

2.4 The global energy landscape and its implications

2.4.1 Energy resources and their availability

Energy is harnessed from a mix of sources, from the burning of fossil fuels to the fission of uranium atoms in a nuclear reactor. Various issues determine the fuel choice, such as the availability, cost, and efficiency of energy generation (and conversion). Additional factors are concerns about the environment and the reliability of supply.

In theory, there is no dearth of global energy resources (Table 2.2). The existing fossil energy and uranium reserves together are projected to be able to sustain the global energy demand for several centuries. In fact, the limits imposed by the depletion of uranium ore could be more than compensated for by reprocessing spent fuel to recover plutonium and using thorium after conversion to ^{233}U. However, such advanced fuel cycles would have associated cost and proliferation concerns, which are discussed in more detail in Chapters 13 and 14.

Although the theoretical availability of energy resources is not an issue, several considerations have shifted the global fuel mix over the years. Figure 2.2 reflects this trend between 1973 and 2008 [4]. Oil still dominates as the single largest source of energy, although its preeminence is somewhat subdued. Globally, it still remains the transportation fuel of choice because of its energy content, easy transportability, and reasonable availability. In fact, almost 55% of oil used worldwide is for transportation, and the recent growth of the automobile sector in developing countries is making further demands on this fuel. It is estimated that, at the present rate of consumption, the world's current conventional oil reserves would last for a little more than four decades [5]. In addition, although they are generally more difficult to access, costlier to develop, and more controversial, untapped resources such as deepwater reservoirs could be harnessed. Moreover, substitute resources for extracting oil, such as tar sands and oil shale, could also be developed (see Chapter 11).

Although the shifts in shares of various fuels appear modest, it is important to remember the enormous increase in total production for every type of energy source. The uses have also changed, with a much greater emphasis on the generation of electricity because of its cleanliness (from the consumer's perspective) and convenience (ability to perform multiple tasks such as

Table 2.2. Global availability of energy resources (present world primary energy consumption is about 14 TW·year) [3]. The energy resource availability of nuclear breeder reactors has been assessed by the authors

Resource	Energy potential (TW·year)
Coal	5,000
Gas and oil (conventional)	1,000
Gas and oil (unconventional)	2,000
Methane clathrates	20,000
Oil shale	30,000
Nuclear fission (conventional)	370
Nuclear fission (breeders)	7,400
Nuclear fusion	unlimited
Sunlight (on land)	30,000 per year
Wind	2,000 per year

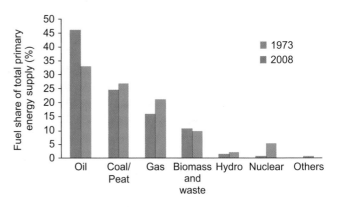

Figure 2.2. Fuel share of world total primary energy supply (1973 and 2008) [4]. Total world primary energy supply was 0.8 terawatt year equivalent in 1973 and 1.6 terawatt year in 2008.

lighting, heating, cooling, and even transportation). Essentially all nuclear power is used for electricity (naval propulsion being a negligible use), and most of the growth in coal has been for electric power production. Use of natural gas has also grown for electricity generation, where it has replaced oil in many cases because it is more efficient and more environmentally friendly. For example, natural gas combustion emits about 0.5 kg of

CO_2 per kW·h of electricity as compared with 0.85 kg per kW·h for oil and more than 1 kg per kW·h for coal. In addition, from a power-generation perspective, it is much faster to build plants for natural gas (often requiring less than half the time) than for coal. Over land, natural gas is transported under pressure through pipelines with diameters of several feet. For transport across water, natural gas is liquefied relatively easily at a reasonable cost and carried in giant tankers. As a result, several countries have established in-port facilities for liquefaction and regasification of natural gas. Further details on natural gas, as well as oil, are provided in Chapter 9.

As shown in Figure 2.2, nuclear power has grown tremendously since 1970. During this period, for example, France, Japan, and the USSR/Russia all built dozens of reactors, and India and China also embarked on major nuclear power programs. Despite this expansion, there are continuing concerns about the safety of nuclear reactors, the storage and disposal of nuclear wastes, and the proliferation of nuclear materials for use in weapons (see Chapters 13–15). Although strict international controls are in place regarding supply and use of nuclear materials, efforts to improve the safety and security of nuclear operations through both technology and policy continue. In a recent case, described later in this chapter, India was considered an exception to the non-proliferation regime because of its urgent needs for energy and also its impeccable record in non-proliferation.

Despite the emergence of natural gas and nuclear power, coal continues to hold its prominence both in developing and in developed countries. The global coal reserves are extensive and expected to last for centuries (see Table 2.2). Other attractive qualities of coal include its widespread availability, low cost of energy generation, and mature conversion technologies. However, as already mentioned, use of coal has significant environmental drawbacks, including high CO_2 emissions, as well as emissions of other pollutants such as SO_2, NO_x, mercury, and particulates. At present, there are no renewable energy technologies that could replace coal on a large scale. Thus, continuing research efforts are focusing on improving the efficiency and reducing the environmental impact of power generation from coal. These issues are discussed in detail in Chapter 10.

In terms of renewable resources, the share of hydroelectric power has remained constant (see Figure 2.2), as concerns have grown about its social and environmental consequences. In particular, hydroelectric plants generally require large land areas for their reservoirs. It is estimated, for example, that the massive Three Gorges System in China (~18,200 MW upon completion in 2009, with the scope for some expansion) displaced some 1.5 million people from the regions submerged for its upper reservoir. Nevertheless, hydroelectric power can still be an important resource under appropriate conditions. For example, in Norway, which has a low population density, hydroelectric dams contribute more than 90% of the power generation. Further details on hydroelectricity can be found in Chapter 45.

Wind energy has matured considerably over the past few decades. In 2010, the installed capacity was 0.2 terawatts [6], and it is projected to increase to 1.5 terawatts by 2020. Moreover, its cost of generation is now almost comparable to that of conventional power-generation technologies. However, wind power entails several difficulties, including that it is intermittent and location-specific. Lack of capacity for transmitting power from wind installations to load centers has hampered wind power in many parts of the world. In addition, aesthetic concerns about the location of wind generators and noise are stunting its expansion in the USA. See Chapter 30 for a detailed discussion of the current status of wind energy and the challenges it faces.

Biofuels have achieved increasing prominence because of the perceived scarcity (and distributional concentration in selected areas) of petroleum resources and the volatility in their price and supply. Biofuels are at present produced from feedstock such as sugarcane, molasses, corn, palm oil, and other oil-bearing crops. In addition, cellulosic materials can be converted into ethanol by enzymatic or thermochemical processes. These processes are discussed in detail in Chapters 25 and 26. There are increasing concerns about deforestation of tropical forests to plant palm-oil plantations, as well as diversion of food-producing lands to energy crops. Some of these issues are examined in the next section as an example of the challenges encountered in developing alternatives to the fossil-fuel-dominated energy landscape of today.

In theory, solar irradiation provides limitless potential for meeting global energy needs. One hour of sunshine falling on the Earth's surface could potentially meet the entire world's energy needs for an entire year (around 14 TW·year). Nevertheless, solar energy currently provides less than 0.1% of the world's energy supply. Unfortunately, the flux of solar radiation is low (less than 1,000 $W\,m^{-2}$ at the noon peak) and intermittent, and conversion technologies currently are not at grid parity with conventional fossil fuels. However, considerable research development and deployment is ongoing worldwide in a range of solar technologies and some are demonstrating a learning curve that will make them competitive, as discussed in detail in Chapters 17–22.

2.4.2 The quest for energy security: some examples

All nations of the world are concerned about providing an uninterrupted supply of energy at affordable prices and in the form required. Because energy resources are not equitably distributed, this can be a major challenge. In January 2009, for example, a dispute between Russia and Ukraine led to disruptions in the supply of natural gas not just in Ukraine but in several countries of the European Union as well. Similarly, a late-2010 ban by China on the export of rare-earth minerals (critical components of advanced batteries and many catalysts) illustrated the deep vulnerability of developed economies to supply disruptions. Countries have undertaken many initiatives to overcome such disruptions, including stockpiling fuel and resorting to unconventional and often environmentally contentious technologies. Another option is for governments to subsidize and support newer technologies that are not at present cost-competitive.

In this section, we discuss three responses to challenges in **energy security**. The first is the development of ethanol and biodiesel as substitutes for oil. The second concerns the implementation of a political agreement that enabled India to pursue domestic nuclear power while still addressing international concerns about proliferation. The third addresses the technical–economic and even social challenges involved in designing and implementing smart grids superimposed on the existing electricity distribution and transmission grids. These three scenarios were chosen to highlight different aspects of the challenges presented by the current global energy landscape, beyond the obvious issue of cost-effectiveness. With biofuels, the challenge is one of scale and impacts; with nuclear power, it is one of policy and global regimes; and with smart grids, it is one of managing a fundamental transformation of the existing energy infrastructure.

Biofuels

In the past few decades, biofuels have received a great deal of attention. As oil prices reached record highs in recent years, several countries announced programs in biofuels. For example, the USA has an ambitious target of producing 36 billion gal (1.4×10^{11} l = 1.4×10^8 m^3) of biofuels per year by 2020 [7]. Such responses follow a trend similar to that of the oil shocks of the 1970s, which exposed the vulnerability of the world economy to the volatile geopolitics in oil-producing countries. Then, as now, several countries came forward to invest in biofuels, as such an investment promises to provide some amount of energy security through reduced reliance on imported fossil fuels. In addition, biofuels are embraced because of their expected environmental benefits compared with fossil fuels.

In the 1970s, Brazil initiated a program for the large-scale production of ethanol from sugarcane. Large tracts of land were converted for growing sugarcane, and a sizable fraction of the harvest was diverted for producing ethanol. Cars were designed to handle any blend of ethanol, even up to 100%, and fueling stations supplying/pumping such blends became ubiquitous. The cost of producing ethanol from sugarcane is reasonable, and the Brazilian government provides no subsidies. Currently, Brazil produces 25 billion liters annually. Nevertheless, the USA is actually the world's largest ethanol producer, using mainly starch-containing crops, primarily corn. In the USA, corn is grown on an area of 10 million hectares (ha), and the yield is about 1,060 gal ha^{-1} (4,000 l ha^{-1}), for a total of 10.6 billion gal (40 billion l) in 2009. Ethanol from corn is about 30% more expensive than that from sugarcane because the corn starch must first be converted into sugar before it can be distilled into alcohol. The US government thus provides a federal tax credit of $0.51 gal^{-1} ($1.93 l^{-1}) and imposes a tariff of $0.54 gal^{-1} ($2.04 l^{-1}) on ethanol imports from Brazil, to protect its domestic ethanol industry.

Given the intended purposes of national biofuels policies, namely to increase energy security and decrease environmental impact, it is natural to ask whether they are fulfilling their promise. In fact, the life-cycle energy assessment of ethanol has been a subject of much debate. (See Chapter 41 for more information on life-cycle assessments.) Initial calculations suggested that it takes more energy, generally derived from fossil fuels, to make corn-based ethanol (as in the USA) than can be obtained from it, leading to an energy output-to-input ratio of less than 1 [8] [9]. These calculations were subsequently refuted, and now it is generally accepted that this ratio is around 1.3 [10]. In contrast, the energy balance for ethanol from sugarcane is estimated to be 8.3–10.2 [11].

The marginal energy and environmental benefits of corn ethanol notwithstanding, the implementation of the US federal incentives motivated farmers in corn-growing states to increase the area under corn cultivation and devote significant amounts of corn to making ethanol. This coincided with a period of increased global food grain prices. A World Bank study reported that large-scale production of biofuels and the related consequences of low grain stocks, speculative activity, and export bans accounted for almost 75% of the total price rise [12]. The report mentioned that the large increase in biofuel production in the USA and Europe was the main reason for the steep price rise, whereas Brazil's sugarcane-based ethanol did not have an appreciable impact on food prices. It also argued that the presence of subsidies and tariffs on imports added to the price rise and that, without such policies, the price increase would have been much lower. The World Bank study recommended that the USA and European Union remove their tariffs on ethanol imports to support efficient ethanol production in countries such as Brazil. Another study determined that the current biofuel support policies in the European Union and USA would reduce greenhouse gas emissions from transport fuel by no more than 0.8% by 2015, but that Brazilian ethanol from sugarcane would reduce greenhouse gas emissions by at least 80% compared with fossil fuels [13]. It thus called for more open markets in biofuels to improve efficiency and lower costs.

Another unanticipated impact of biofuels policy is illustrated by the experience of Europe. In 2008, the European Union announced a target for 10% of transportation fuels to come from renewable energy sources, mostly biofuels, by 2020. However, a March 2010 study reported that a biofuels level of more than 5.6% could actually harm the environment, mostly as a result of "**indirect land-use change**." Specifically, the initial EU announcement led to the large-scale clearing of forest and peat lands in Indonesia and Malaysia to support the cultivation of biofuel crops. The process of land clearing results in such a large initial release of CO_2 to the atmosphere that it could take a few decades for the CO_2 to be recovered by the annual biofuel cycle [8] [14]. In fact, deforestation significantly increased Indonesia's CO_2 emissions and made the country among the world's leading emitters (Figure 2.3). In terms of land-use change, the area under development for palm-oil plantations in Indonesia increased from less than 2,000 km^2 to more than 30,000 km^2. There was widespread deforestation (and illegal logging) during this period, and palm-oil plantations were identified as the greatest threat to forests and wildlife in southeast Asia. Of course, the decisions to cut down forests had local support because doing so contributed to job creation and economic growth. In response, the international community implemented the Reducing Emissions from Deforestation and Forest Degradation (REDD) program to help developing countries reduce emissions from forested lands and invest in low-carbon paths to sustainable development, with financial support from developed economies. For example, in May 2010, Norway pledged $1 billion to help Indonesia reduce further deforestation.

Biofuels, particularly corn ethanol, provide one instance where politics has taken precedence over sound science and economics in decision making. The production of biofuels is now largely from food crops such as sugarcane, corn, and beets. However, other plants that are not in the food chain and grow wildly in tropical and subtropical climes could be used instead. **Jatropha** is one such hardy plant and has attracted considerable attention because of its oil content and widespread growth from Africa to South America and south Asia.

Jatropha can be grown in wastelands and has minimal nutrient and care requirements. The fruits and seeds of the plant are poisonous and contain about 35% oil that has properties suitable for making biodiesel

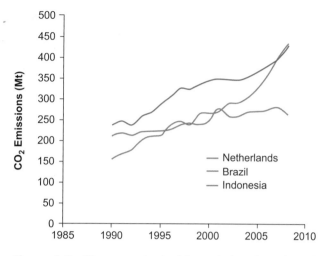

Figure 2.3. The steep rise in CO_2 emissions in Indonesia over the past two decades. Over the same period, emissions from the Netherlands started at a higher level but grew at a much lower rate. Even though emissions from Brazil started at a much higher level and grew at a substantial rate, they are now slightly surpassed by those from Indonesia.

[15]. In 2008, jatropha was planted on an estimated 900,000 ha, mainly in Indonesia. By 2015, according to some forecasts, the area should increase to 12.8 million ha.

The yield from this plant can vary significantly from 1 to 3 tonnes per hectare, and the plant starts yielding fruits even from the second year of planting [16]. A recent study concluded that jatropha cultivation in lower-quality wastelands would help small and marginal farmers, creating rural employment and making biodiesel available for rural areas [15] [17]. This study also underlined the potential of jatropha for sustainable development that helps eliminate poverty.

Despite these initiatives, it is unlikely that biofuels alone can meet even a modest fraction of global transportation requirements, which account for 30% of total primary energy use. Even for a relatively high-yield crop such as jatropha, the acreage required for a country to achieve even a modest (10%–20%) biofuel blend would make it one of the largest crops planted in the country. However, if targeted properly, biofuels can help meet a fraction of the transportation fuel needs of some countries, especially in the tropics, and also enable the alleviation of rural poverty, unemployment, and lack of energy sources.

Nuclear power

In the 1950s, nuclear power was considered the answer for the energy needs of all countries, because it is efficient, compact, and even cost-effective. In fact, the Second Geneva Conference on Peaceful Uses of Atomic Energy of 1958 recommended nuclear fission as the preferred technology until nuclear fusion became mature by the end of the twentieth century.

However, three concerns shattered this hope. First, three major nuclear reactor accidents [Windscale (1957) in the UK, Three Mile Island (1979) in the USA, and Chernobyl (1986) in the USSR] demonstrated the devastating and long-lasting damage from such events. The second concern is the possible proliferation of nuclear materials that could result in rogue states and non-state actors obtaining the material and the know-how for making nuclear weapons. Chapter 14 details the technology and policy countermeasures employed by the international community as safeguards against nuclear proliferation. The third concern, addressed in Chapter 15, regards the safe disposal of nuclear wastes from the spent fuel and other radioactive wastes. These concerns grew so intense that some countries, such as Italy, abandoned their nuclear power programs altogether. Even in the USA, no new nuclear reactors have been commissioned for over two decades.

The case of India

One country that has continued to pursue nuclear power options is India, through a program that began soon after the country gained independence (1947). However, its refusal to sign the Treaty on the Non-Proliferation of Nuclear Weapons (NPT) in 1970 and its 1974 and 1998 nuclear weapons tests resulted in India being shunned by the global nuclear community. During the subsequent decades of isolation, India was unable to obtain commercial nuclear fuel, nuclear power plant components, and services from the international market, meaning that it had to rely on its own scientists and resources for the development of its nuclear industry.

Because of limited domestic uranium resources, the Indian political and scientific leadership decided to pursue a fast breeder reactor program (see Chapter 13), in which the spent fuel of thermal nuclear reactors can be reprocessed to recover plutonium, which can then be used as a fuel for fast reactors. It was felt that plutonium is too precious a resource to be buried as nuclear waste and that it should be used for generating additional power instead. Because of the extensive availability of thorium, India went a step further to use fast breeder reactors to convert thorium, which is not fissionable, into additional ^{233}U fuel.

It is only in the past few years that there have been some landmark changes in relations between the Indian nuclear power sector and the global community. This series of steps started with the US–India Civil Nuclear Cooperation Initiative (July 2005), under which India agreed to separate its civil and military nuclear facilities and place several of its civil nuclear facilities under International Atomic Energy Agency (IAEA) safeguards in return for civil nuclear assistance from the USA. In addition, a waiver was obtained from the Nuclear Suppliers Group (NSG), the export-control organization established in response to India's 1974 nuclear test to ensure that nuclear technology is not diverted from peaceful use to weapons programs. This 2008 waiver allows India to participate in nuclear trade and commerce, even though it has nuclear weapons and has not signed the NPT. According to the Indian nuclear establishment, these arrangements will allow a significant growth in nuclear power generation in India and also enable plutonium and thorium to be harnessed successfully [18] [19]. However, it was preceded by intense national and international negotiations, which almost brought India's coalition government to the brink of collapse. The threat of proliferation and waste disposal were among the two main concerns. In any event, this example illustrates one country's efforts to ensure a reliable and efficient energy supply for its citizens.

The upcoming transformation of the power system

The power grid of today looks essentially the same as that built a century ago, unlike, say, the telecommunications network, which has benefited from rapid advances in materials and technologies, especially digital technologies. The "**smart grid**," aims to use digital communications and control technologies to make the electricity supply system more robust, efficient, cost-effective, and amenable to renewables. A detailed discussion of smart grids appears in Chapter 42.

In today's "dumb" grid, power is treated as a commodity (all kilowatt-hours are mostly treated as equal) and flows like water across the path of least resistance, with limited measurements (except for broad operations and billing) and few controls. In the envisioned smart system, one would know exactly what power was going where and when, and be able to act in response to conditions, either through direct control mechanisms or through economic signaling (changing the price). For example, today's retail consumers have mostly enjoyed flat-rate tariffs for electricity, even though power at 5 pm is typically more expensive to supply than that at 5 am. With a smart grid (including smart meters that record the time as well as the amount of usage), the billing could be done accordingly.

Under the proposed transition, something as mundane as an electricity meter has enormous policy and political implications. At base, it would mean that consumers would be paying for the electricity they use, ideally at prices that directly reflect costs. It is just such microeconomic efficiency that has proponents excited. Conversely, fears have been raised about the complexity

Table 2.3. Metering and underlying functionalities over time [20]

	Phase I, options define service, 1800s to early 1900s	Phase II, option consolidation, 1920s to 1960s	Phase III, separate options, 1970s to 2000	Phase IV, integrated options, after 2000
Pricing	End-use rates	Usage-based rates	TOU-based rates	Real-time pricing
Metering	None	Total kilowatt-hour usage	Time-period loads	Hourly loads
Load-shape objectives	Load growth	Load growth, valley filling	Peak shaving, shifting, conservation	Preserve electric reliability, customer cost management
Customer involvement	Active, fuel switching	Passive, few options	Utility command and control	Interactive participation
Demand response	Contracts for service	Water-heater time clocks	Curtailable, interruptible, direct control	Demand bidding, risk management
	⇩	⇩	⇩	⇩
	Increased choice, service tailored to customer needs	Reduced choice, increasing value to customers, declining cost	Reduced choice, increasing costs, loss of control, declining value to customers	Increasing choice, cost volatility, value of information

TOU, time of use.

of such a system, about the undue financial burden it could place on those least prepared to respond to dynamic prices (e.g., senior citizens), and about the potential it could pose for invasion of privacy and risk to consumer data.

Table 2.3 shows how metering technology has changed over time with different policy, political, and business needs.

At this point, the structure of the smart grid has many unknowns that will need to be resolved by regulators, policy makers, and consumers as technologies evolve. Beyond monitoring, pricing, and control, smart technologies could change the power grid in more fundamental ways. For example, a smart grid could more easily integrate distributed generation resources, including intermittent sources such as solar or wind plants or new storage technologies such as electric cars.

Beyond unknowns regarding what the future grid will look like, there are also challenges relating to how to get there. The benefits of smart grids will take time to be realized, but there will be a need for very large investments up front (estimated at $100–300 per consumer). It is unclear who will pay for this and what the effect on costs for consumers will be.

Nevertheless, the potential for smart grids is very large. Beyond a return on investment for utilities, compelling societal benefits could be realized, including allowing for far greater implementation of "green" power. Indeed, without a smart(er) grid, the variability of some renewables can impose unmanageable operating burdens on utilities. Most importantly, today's grid (which is considered strained in some developed countries such as the USA or fledgling in growing economies) simply cannot provide the reliability, quality, and environmental sustainability needed for the twenty-first century.

2.5 Summary

According to current projections, the available energy resources could sustain the world's energy requirements for centuries. However, merely having energy resources alone or the technologies to harness them is not adequate. These options must be cost-effective, environment – friendly, and socially and politically acceptable, in terms of both accessibility and security. It is within this context that each country must satisfy its energy needs. The examples of biofuels, nuclear power, and a smart electricity distribution grid illustrate the many factors that must be considered in addressing these challenges.

2.6 Questions for discussion

1. What are the global and country-specific trends in energy intensity? Is the global energy intensity likely to converge in the long run?

2. The longevity of world energy reserves has been a subject of considerable debate and speculation and there has been varied experience with different resources such as oil, natural gas, coal, and uranium. It is generally believed that the world will soon run out of oil and gas reserves whereas coal and uranium will remain for a much longer time. Given the growing energy demand in developing countries, what are the estimates for longevity of coal and uranium reserves? Further, oil and gas prices have shown considerable volatility, whereas coal and uranium prices have been relatively stable. What are the reasons for the differences in price behavior of these sources?

3. How do different biofuels compare in their life-cycle costs and CO_2 emissions? Which biofuels are likely to impact food security?

4. What is the potential of emerging biofuel technology options such as cellulosic ethanol and algae-based ethanol? How does their economics compare with that of conventional biofuels and also oil?

5. The spent fuel of nuclear reactors consists of plutonium, which is a fuel for nuclear power. However, most countries having nuclear power programs follow the once-through cycle and do not reprocess spent fuel. How does the economics of reprocessing compare with that of direct disposal of the spent fuel?

6. The motivation for adoption of smart electricity grids varies. In developed countries, the main drivers are load control and peak shaving, whereas in developing countries, the main driver is loss reduction. How do these differences translate into country-specific technology and business models?

2.7 Further reading

- International Energy Agency (IEA), *Energy Technology Perspectives*, provides an overview of world energy scenarios and the role of emerging technologies (http://www.iea.org/textbase/nppdf/free/2008/etp2008.pdf).
- **D. Victor** and **T. C. Heller** (eds.), 2007, *The Political Economy of Power Sector Reform: The Experiences of Five Major Developing Countries*, Cambridge, Cambridge University Press. This book provides a comparison of the electricity reforms in five developing countries.

- *The Future of Coal: An Interdisciplinary MIT Study*, Massachusetts Institute of Technology, 2007; available at http://web.mit.edu/coal/The_Future_of_Coal.pdf. This report gives an overview of coal's energy-generation potential and the techno-economics of present and future conversion technologies including carbon capture and sequestration.

- *Update of the MIT 2003 Future of Nuclear Power: An Interdisciplinary MIT Study*, Massachusetts Institute of Technology, 2009; available at http://web.mit.edu/nuclearpower/pdf/nuclearpower-update2009.pdf. This report discusses the potential, technologies, and economics of nuclear power, including the option of recycling of spent fuel.

2.8 References

[1] **Energy Information Administration (EIA)**, US Department of Energy, www.eia.doe.gov.

[2] **International Monetary Fund**, 2010, *World Economic Outlook Database*.

[3] **J. Holdren**, 2007, *Meeting the Intertwined Challenges of Energy and Environment*, 2007 Robert C. Barnard Environmental Lecture, Washington, DC, American Association for the Advancement of Science.

[4] **International Energy Agency (IEA)**, 2010, *Key World Energy Statistics 2010*; available at http://www.iea.org/textbase/nppdf/free/2010/key_stats_2010.pdf.

[5] **BP**, 2010, *Statistical Review of World Energy 2010*.

[6] **World Wind Energy Association**, 2010, *World Wind Energy Report*.

[7] **US Department of Agriculture**, 2010, *Biofuels Strategic Production Report*, Washington, DC, US Department of Agriculture.

[8] **D. Pimentel**, 2003, "Ethanol fuels: energy balance, economics, and environmental impacts are negative," *Nat. Resources Res.*, **12**, 127–134.

[9] **T. W. Patzek**, 2006, "Thermodynamics of the corn–ethanol biofuel cycle," *Crit. Rev. Plant Sci.*, **23**, 519–567.

[10] **A. E. Farrell, R. J. Plevin, B. T. Turner** et al., 2006, "Ethanol can contribute to energy and environmental goals," *Science*, **311**, 506–508.

[11] **Bioenergy Task 40**, International Energy Agency.

[12] **D. Mitchell** 2008, *A Note on Rising Food Crisis.*, The World Bank, Policy Research Working Paper No. 468.

[13] **Directorate for Trade and Agriculture, Organisation for Economic Co-operation and Development (OECD)**, 2008, *Economic Assessment of Biofuel Support Policies*.

[14] **T. Searchinger, R. Heimlich, R. A. Houghton** et al., 2008, "Use of U.S. croplands for biofuel increases greenhouse gases through emissions from land-use change," *Science*, **319**, 1238–1240.

[15] **Food and Agricultural Organization, United Nations**, 2010, *Jatropha: A Smallholder Bioenergy Crop*.

[16] **A. Bharadwaj**, **R. Tongia**, and **V. S. Arunachalam**, 2007, "Scoping technology options for India's oil security: part II – coal to liquids and bio-diesel," *Current Sci*, **92**(9), 1071–1077.

[17] **World Bank, Agriculture for Development**, 2008, *World Development Report*, Washington, DC.

[18] **A. Kakodkar**, 2008, *Evolving Indian Nuclear Power Program*, Atomic Energy Commission, Government of India, public lecture at Indian Academy of Sciences, July 4, 2008, available at www.dae.gov.in.

[19] **A. Bharadwaj**, **L. V. Krishnan**, and **S. Rajgopal**, 2008, *Nuclear Power in India: The Road Ahead*, Center for Study of Science, Technology and Policy, Bangalore, http://www.cstep.in/docs/CSTEP%20Nuclear%20Report.pdf.

[20] **Electric Power Research Institute (EPRI)**, *New Principles for Demand Response Planning*, Palo Alto, CA, EPRI.

3 Sustainability and energy conversions

Franklin M. Orr, Jr. and Sally M. Benson

Department of Energy Resources Engineering, Stanford University, Stanford, CA, USA

3.1 Focus

Energy use is inexorably woven into the fabric of modern civilization. Human well-being, economic productivity, and national security all depend on the availability of plentiful and affordable energy supplies. However, over the past half century, we have come to understand that continued growth of energy use along the lines of current energy systems will lead to unacceptable consequences for the Earth's climate and oceans. Maintaining and increasing the access to energy services to satisfy crucial societal needs requires the development of a sustainable global energy system that transitions away from energy supply options with high greenhouse gas (GHG) emissions and unhealthy air pollutants. Disparity in energy access is also not sustainable. We must provide sufficient energy for the estimated 1.6 billion people who do not have access to modern energy systems today. Fortunately, plentiful energy resources are available to meet our needs, and technology pathways for making this transition exist. Continuing to lower the cost and increase the reliability of energy from sustainable energy resources will facilitate this transition. Changing the world's energy systems to reduce GHG emissions is one of the critical challenges that humans must face in this century. The required transition can begin now with improvements in efficiency of energy conversion and use, and with continuing deployment over the coming decades of a variety of existing and innovative technologies. With continuing attention to energy conversions that minimize wastes, have low life-cycle impacts, and maximize recycling of materials, a set of sustainable energy systems can be created.

3.2 Synopsis

Feeding, clothing, and housing a growing world population will be a significant challenge in this century, as will supplying the fresh water, heat, lighting, and transportation we will need to live comfortable and productive lives. This chapter discusses energy sustainability, with emphasis on the requirement to reduce GHG emissions. A sustainable energy system is one in which energy is supplied and converted to energy services in ways that avoid unacceptable consequences for local, regional, and global natural systems that control climate and support ecosystems that provide essential services. Figure 3.1 illustrates typical conversions of a primary energy resource (solar, wind, geothermal energy, fossil or nuclear resources, etc.) into a product, such as a fuel or an energy carrier, like electricity, that then can be converted to a service like heat, light, or mechanical work. Sustainable processes and systems that convert some primary energy resource into energy services will be ones that are as efficient as possible – smaller quantities of the primary energy resource are needed and fewer waste materials are created if the conversions are efficient. Some have argued that only energy flows such as solar, wind, and wave power should be considered sustainable. Others note that any system of energy conversions has some footprint and impact, and that sustainability is necessarily a relative measure, not an absolute one. In any case, sustainable systems will have low impacts over the full life cycle of the conversions (see Chapter 41). Recycling of materials used in energy conversions will be maximized, and amounts of waste materials created in the chain of energy conversions to services will be minimized when the whole cycle from primary resource to services (such as mechanical work) to waste heat and products is considered.

A transition away from energy supply options with high GHG emissions is an essential component of a sustainable global energy system (see Chapter 1). Reducing GHG emissions will also lower emissions of unhealthy air pollutants and can increase energy security through increases in energy efficiency and diversification of energy supply. There is no shortage of energy resources available to meet our needs (see Chapter 6), both from sustainable energy flows from the Sun and from large accumulations of chemical and thermal energy in the Earth's crust – the challenge is to find cost-effective ways to convert those resources into forms that provide the energy services needed to support modern societies and to avoid adverse environmental and societal impacts. Technology pathways for making the transitions that will be required include increasing the efficiency of energy production and use, conserving energy by avoiding unneeded use, increasing the fraction of energy from sustainable sources that do not emit GHGs, understanding and minimizing life-cycle impacts of energy production and use, and mitigating unavoidable impacts through technologies such as carbon dioxide capture and storage. Reducing the cost and increasing the reliability and quality of energy from sustainable energy sources will facilitate this transition. Changing the world's energy systems is a huge challenge, and it will require a sustained effort to put the new system in place given the scale of the investments and infrastructure required. That effort can be begun now with improvements in energy efficiency and with continuing deployment of a variety of technologies now and over the decades to come.

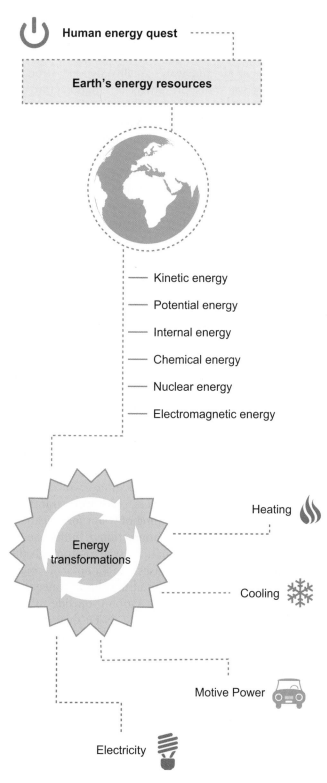

Figure 3.1. A schematic diagram illustrating the concept of conversion of primary energy resources into energy services.

3.3 Historical perspective

The development of modern, technology-based societies has required a series of energy transitions as new energy services were provided and as the scale of energy use grew. As the industrial revolution began, wood was a primary fuel for heating and cooking, candles were used for lighting by those who could afford them, and wind was used to drive ships and pump water. With the invention of the steam engine, which was used first to pump water out of coal mines, mining of coal expanded, and coal combustion began to provide heating and transportation (steam-driven ships and locomotives). For a brief period whale oil was used for lighting, though, in what may have been the first example of an energy transition driven by sustainability, it soon became clear that continued large-scale use of whale oil would be limited by the rapid decline in the number of whales. The demand for lighting led to development of petroleum resources, which were used first to provide kerosene for lighting. Refining was developed to separate high-value kerosene from petroleum. Subsequently, the development of the internal combustion engine and the automobile created markets for fuels (diesel and gasoline) made from petroleum. Electric power generation, used initially to provide lighting, shifted the balance again. Use of kerosene declined as the cleaner, brighter electric lights penetrated the lighting market, and coal became the primary fuel for electric power generation. Use of oil for transportation soared with large-scale use of automobiles and trucks for transportation. Markets for electricity expanded dramatically, first as electricity was provided in more and more rural settings and then as new services (air conditioning, electric appliances, etc.) became available. The employment of natural gas for space heating, cooking, and electric power generation made use of another fossil energy resource that was cleaner-burning than coal, even after modifications to coal-fired power plants to reduce emissions of sulfur oxide, nitrogen oxide, and particulates had been made. Throughout this period of rapid development and consequent improvements in the quality of life in developed economies, emissions from the combustion of coal, oil, and natural gas grew rapidly (see Figure 3.2 [1]).

Now, motivated by the risks of climate change and growing concerns about providing energy to billions more people, energy security, and health effects from air pollution, the world is embarking on another set of energy transitions – ones that will transform the energy systems of the planet in this century. These will be driven by the recognition that human activities now take place at a scale that is large enough to influence the planet as a whole. The complex, interrelated issues that arise from the need to supply the energy the world needs to operate complex modern societies and at the same time protect and restore the planet are the subject of this chapter.

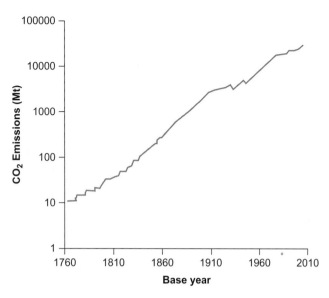

Figure 3.2. History of CO_2 emissions from use of fossil fuels from 1750 to the present (Data from [1]). One metric tonne of CO_2 is 1,000 kg CO_2.

3.4 The sustainability challenge

Nearly 7 billion people occupy our planet at present, and in this century several billion more will join us. Current estimates indicate that the population of the planet will peak in the middle of this century at about 9 billion people [2]. Feeding, clothing, and housing all of us will be a significant challenge, as will supplying the fresh water, heat, lights, and transportation we will need to live comfortable and productive lives. For example, providing the current world average per capita power use (about 2 kW, See Box 3.1) to the 1.6 billion people without commercial energy services and to 2 billion additional people would require an increase of about 7 TW, roughly half of present world power consumption; increased energy consumption in fast-growing economies is likely to add to that total.

Box 3.1. Energy and power

The standard unit of energy is the joule (J). Power is the rate of energy transfer or use, energy per unit time, measured in joules per second or watts (1 W = 1 J s^{-1}). An average power use of 2 kW for one hour is equivalent to energy use of 7.2 MJ (2,000 W × 3,600 s), though a commonly used unit for energy in the form of electricity is the kilowatt-hour (kWh), which equals 3.6 MJ. A person using average power of 2 kW on average for a full year would consume energy totaling 17,520 kWh (2 kW × 8760 h).

It is also clear that we humans are interacting at local, regional, and global scales with the natural systems that we count on to provide us with many services. Local air and water quality depend strongly on the way we transport ourselves, manufacture all manner of products, grow food, and handle the wastes we generate. Humans now use a significant fraction of the fresh water available on Earth, and fresh water is in short supply in many places around the globe. We recognize, also, that air pollution emitted in one location can affect air quality over large distances. A decades-long effort in developed economies to reduce local and regional impacts on air and water quality has been very successful, though local and indoor air pollution remain very important issues in poor economies where biomass and coal are used for heating and cooking, and in rapidly developing economies using fossil fuels without air-quality-control technology. Efforts to address the global-scale environmental impacts are just beginning. At the same time, maintaining and preserving habitats on a variety of scales for the plant and animal species with which we share the planet will require careful and complex balancing with human needs. It is now too late to avoid all ecosystem and species-diversity impacts, but recognition now that ecosystems provide essential services on which we depend and future stewardship of remaining assets is an important component of converting energy resources for human use.

Energy use, together with agriculture, which makes heavy use of energy for fertilizers, cultivation, transportation of products, and pumping of water for irrigation, is a prime component of the interaction of human activities with the global-scale natural systems through the emissions of greenhouse gases. As Chapter 1 documents, significant increases in atmospheric concentrations of key GHGs, namely carbon dioxide (CO_2), methane (CH_4), and nitrous oxide (N_2O) over the 250 years since the beginning of the industrial revolution have caused the capture of additional energy in the atmosphere over pre-industrial levels, with climate change as one result. In addition, the pH of the upper ocean has declined as the additional CO_2 in the atmosphere slowly equilibrates with seawater. Thus, it is clear that human activities now modify the large natural systems that operate at global scale, and there are also impacts at regional and local scales.

As we seek to provide and use the energy that is a fundamental underpinning of modern societies, we humans need to bring that use into balance with the large natural systems that cycle carbon and nitrogen. Even as we work to meet our energy needs, we must not disrupt the natural processes that provide other important resources such as food and water – and ecosystems that provide habitat and sustenance for the myriad of species that share the planet with us. For example, large-scale use

Figure 3.3. World primary energy consumption by fuel type (from [3]). The category "Other" includes geothermal, solar, wind, and heat. Note that one million metric tonnes of oil equivalent (Mtoe) equals 0.04476 exajoules (1 exajoule = 10^{18} J). From *Key World Energy Statistics 2010*, © OECD/IEA.

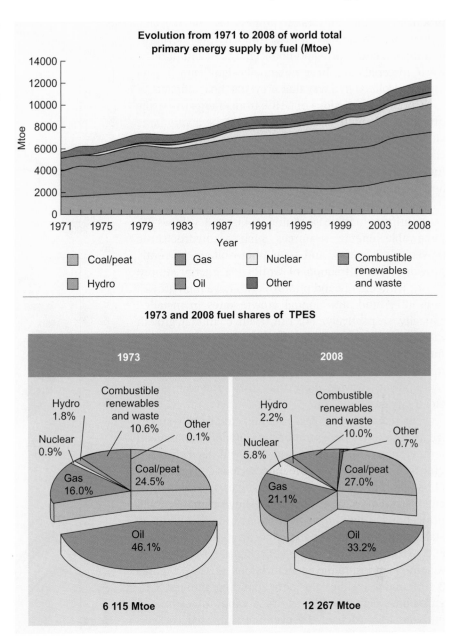

of fertilizers has also altered another important global cycle, that of nitrogen, in a significant way. Fertilizer use for agriculture in the central part of the USA has increased the availability of nitrogen in runoff to the Mississippi River drainage, supporting algal blooms in the Gulf of Mexico that decrease oxygen concentrations and create large dead zones. We humans are just beginning to see ourselves as a global biogeochemical force – a first step toward more sustainable human enterprise that will require managing our activities more effectively. Today, our growing impact on the global-scale mass and energy balances motivates some to call this geological era the beginning of the "Anthropocene."

Sustainable energy systems are those that provide sufficient energy services and also minimize long-term impacts. These systems will make use of relatively efficient energy conversions, which reduce the quantity of primary energy resources needed to provide the energy service and reduce the amount of waste heat and waste products. For example, a combined-cycle, natural-gas-fired power plant (Chapter 9) with an efficiency of 50% requires about 30% less input energy in the fuel and emits about 80% less CO_2 per kWh than does a conventional pulverized-coal-fired power plant with an efficiency of about 35%. More sustainable energy systems will minimize impacts across the full life cycle (see Chapter 41). Replacing the internal combustion engine with a battery-powered electric motor will reduce CO_2 emissions at the vehicle, but whether this reduces overall emissions will depends on how the electricity stored in the battery on the vehicle is generated. Further, sustainability is improved if materials used in conversion devices or produced in the conversion process can be recycled rather than treated as wastes, though this inevitably requires expenditure of additional energy. At a minimum, waste materials that have potentially significant impact must be isolated in a way that prevents those impacts.

Curtailing emissions of GHGs from energy use while providing adequate energy supplies will be a very large challenge for a number of reasons: our current socio-economic infrastructure is built around use of fossil fuels, the needed reductions in GHG emissions are large, and the time-frame to begin reducing emissions is short.

In fact, demand for energy from oil, coal, and natural gas continues to increase rapidly as Figure 3.3 shows [3]. Renewable energy resources such as hydroelectric power, wind energy, and solar photovoltaics currently provide a small fraction of worldwide energy supply. Today, wind energy and photovoltaics have high growth rates (27% and 40% annual growth rates in installed capacity, respectively) [4], but, even if annual growth rates of 15% (of the installed base) can be sustained over the coming decades, it could take nearly 20 years before renewable energy sources (not including large-scale hydropower) provide more than 20% of overall global electricity consumption.

The reductions in GHG emissions required to stabilize atmospheric concentrations of CO_2 and global mean temperature will be large, and they will need to begin quickly if we are to avoid the most serious consequences of global climate change and changes in ocean geochemistry. For example, Figure 3.4 shows climate-model projections of the ranges of emissions that would allow stabilization of atmospheric concentrations at 450, 550, 650, 750, and 1,000 ppm [5]. Figure 3.4(a) shows assumed atmospheric concentrations of CO_2. Figures 3.4(b) and (c) show the corresponding implied CO_2 emissions. Figure 3.4(b) shows a range of estimated emissions for each of the eventual stabilized concentrations. These ranges reflect differences in parametrizations used to represent climate feedback mechanisms in the various models and differences in assumptions within the stabilization scenarios. Figure 3.4(c) illustrates the range of estimates for stabilization at 550 ppm, for example. While there is variation in the model projections, all of them show that quite substantial reductions in emissions will be required and that stabilization at 450 or 550 ppm would require that the peak in emissions occur in the next decade or two.

Some investigators and organizations have argued that avoiding the most serious effects of climate change will require limiting the global mean temperature rise to 2 °C [6]. In March 2007 the European Council accepted this target as the basis for an aggressive plan to reduce GHG emissions. A recent assessment concluded "... limiting warming to 2 °C above pre-industrial levels with a relatively high certainty requires the equivalent

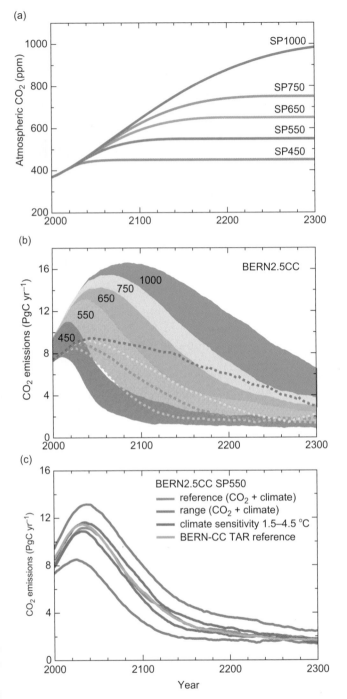

Figure 3.4. Projections of GHG emissions needed to achieve stabilization at a range of concentrations and temperatures (from [5], p. 792). (a) Assumed emission profile. (b) Corresponding estimated range of CO_2 emissions. The lines indicate the lower end of the estimated range needed to achieve stabilization at the respective values. (c) Detail of the range of emissions for stabilization at 550 ppm. The differences reflect variations in parametrizations of climate feedback in the models and differences in assumptions within scenarios. From *Climate Change 2007: The Physical Science Basis. Working Group I Contribution to the Fourth Assessment Report of the Intergovernmental Panel on Climate Change*, Figure 10.22(b), © Cambridge University Press.

Table 3.1. Estimates of global mean temperature increases, years for peak emissions and reduction in global emissions needed to achieve these stabilization levels (from [6])

Stabilization level (ppm)	Global mean temperature increase (°C)	Year global CO_2 needs to peak	Year global CO_2 emissions return to 2000 levels	Change in 2050 global CO_2 emissions compared with 2000 (%)
445–490	2.0–2.4	2000–2015	2000–2030	−85 to −50
490–535	2.4–2.8	2000–2020	2000–2040	−60 to −30
535–590	2.8–3.2	2010–2030	2020–2060	−30 to +5
590–710	3.2–4.0	2020–2060	2050–2100	+10 to +60
710–855	4.0–4.9	2050–2080		+25 to +85
855–1130	4.9–6.1	2060–2090		+90 to +140

concentration of CO_2 to stay below 400 ppm." (Chapter 1) The current CO_2 concentration alone (as of 2008) is about 385 ppm, and the CO_2-equivalent concentration is about 430 ppm, so it is no longer possible to avoid exceeding the 400 ppm concentration. If concentrations were to rise to 550 ppm CO_2 equivalent, then it is unlikely that the global mean temperature increase would stay below 2 °C. Limiting climate change to 2 °C above pre-industrial implies limiting the atmospheric concentrations of all GHGs. According to new insights into the uncertainty ranges of climate sensitivity, stabilization at 450 ppm CO_2 equivalent would imply a medium likelihood (~50%) of staying below 2 °C warming [6] [7].

The summary provided in Table 3.1 [7] indicates that, to achieve this goal, the peak emissions would need to occur within the next decade, emissions would have to return to year-2000 levels by several decades from now, and emissions in 2050 would be reduced by 50%–85% compared with emissions in 2000. Urgent and large-scale action is needed to stem the growing concentration of CO_2 in the atmosphere. Fortunately, there are steps that can be taken on a variety of time scales to meet those challenges.

The feasibility of this transition hinges on four critical factors.

1. Increased efficiency of energy use. Today's energy system is relatively inefficient. For example, only about a third of the energy in coal is converted to electricity and less than 10% of the energy in electricity is converted to light from an incandescent light bulb. The rest is converted to heat that is not useful. Increasing energy efficiency – sometimes also called reducing energy intensity – can dramatically reduce energy emissions.
2. Availability of large amounts of low-carbon emission energy supplies. Humans use an enormous amount of energy. Finding abundant supplies of energy will be needed to sustain modern civilizations.

3. Availability of efficient and low-cost energy conversion and storage technologies. Energy in resources such as oil, gas, coal, wind, and the Sun is not particularly useful until we convert it into heat, light, electricity, or mechanical work. The devices that are used to turn these so-called primary energy resources into useful energy services are called energy conversion technologies (see Figure 3.1). Advances in energy conversion technologies go hand-in-hand with the exploitation of new energy resources. While many energy conversion technologies are available today, improvements in energy conversion and energy storage technologies are crucial for achieving a sustainable energy system.
4. Effective institutions, policies, business models, and global cooperation are needed to shepherd the transition to a more sustainable energy system. Significant economic and security risks together with environmental uncertainties stand in the way of effective and rapid action. Technology alone cannot resolve these issues. Commitment to finding effective incentives to initiate and sustain this energy-system transition over the coming decades will be needed. Indeed herein may lie the biggest challenge.

3.4 Technology pathways for a sustainable energy future

Transitioning from today's energy system to a more sustainable one will require a number of technological approaches – there is no single solution to this challenge (see Pacala and Socolow [8] or the September 2006 issue of *Scientific American* [9] for discussions of many of the options available for energy systems of the future).

3.4.1 Improving the efficiency of energy conversions and end uses

There are numerous opportunities to improve significantly the efficiency of energy conversions – from improving the efficiency of engines and chemical processing to improving the efficiency of lighting, heating, cooling, and other uses of energy in buildings (see Chapters 35–39). All improvements in efficiency reduce the amount of energy supply needed and reduce the associated environmental impacts. More efficient use of energy, especially in buildings and transportation, offers significant reductions that reduce costs, substantially in the case of building insulation, and improved air conditioning and water heating, for example. About 40% of the primary energy used in the USA is used in buildings, and recent estimates [10] indicate that, if energy technologies that exist now or can reasonably be expected to be developed in the normal course of business are fully deployed (a challenge, to be sure), 25%–31% less primary energy would be used by the US economy in 2030, and the cost of deploying the technologies would be more than paid for by savings on energy costs. The component reductions in electricity use would be sufficient to eliminate the need for net new electricity generation in 2030. Globally, the most cost-effective options for reducing GHG emissions include efficiency improvements such as improving insulation, increasing the efficiency of commercial vehicles, and replacing existing lighting with high-efficiency lighting [11].

3.4.2 Conserving energy

The availability of low-cost and abundant energy in the developed world has led to considerable unneeded use of energy – such as energy consumed while electronic equipment is in standby mode; lighting, heating, and cooling in buildings that are not occupied; use of energy during peak demand periods when less efficient methods are used to produce power; and driving during periods of high traffic congestion. In addition to behavioral changes to reduce energy consumption, advanced communications and control technology can play a role in conservation by providing real-time pricing and emissions signals, optimizing heating and cooling in buildings, and providing "intelligent" transportation systems that increase transit efficiency.

3.4.3. Increasing the fraction of energy supply coming from sustainable energy flows rather than stored resources

Stored energy resources are those for which the replacement rate is lower than the rate of use. Any stored resource, therefore, has some limit on total use (even coal), though the availability of the resources might not be the factor limiting total use. A transition away from relying so heavily on stored reservoirs of energy (e.g., fossil fuels) to using sustainable energy flows such as solar and wind power that reduces GHG emissions will put us on a more sustainable energy pathway. Determining whether a particular resource is sustainable requires careful consideration both of the flow and of the ways in which it is converted. For example, whether biofuels made from the solar flux are sustainable depends on how much water and fertilizers are used and how the energy stored temporarily in chemical form in biomaterials is converted to fuel. Solar and wind resources are very large compared with current and projected energy demands. However, as Figure 3.5 shows, the costs of solar photovoltaics and solar thermal at present are high compared with those of other electricity generation sources, even with a $100 per metric ton carbon tax (the corresponding tax on CO_2 would be about $27 per metric ton). Although the

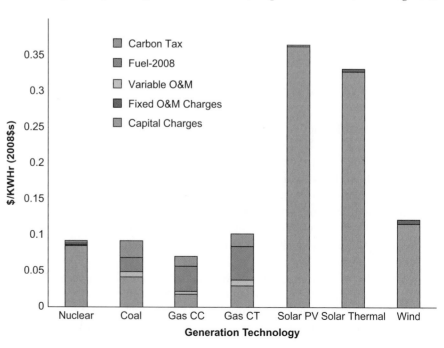

Figure 3.5. Comparison of the costs for various electricity generation options, including the cost of a $100 per ton price on carbon emissions. (Courtesy of John Weyant, Stanford University.) O&M, operations and maintenance; CC, combined cycle; CT, combustion turbine; PV, photovoltaics.

Box 3.2. An example: reducing CO$_2$ emissions in the electric power sector in the USA

Reducing CO$_2$ emissions from electric power generation in the USA will require a portfolio of changes to the large infrastructure that currently exists. Figure 3.6 illustrates one set of estimates of the changes in CO$_2$ emissions that would result if some aggressive assumptions concerning deployment of technologies were met.

The estimates shown in Figure 3.6 were constructed on the basis of an assessment of the technical capabilities of the technologies listed, analyses of technology pathways involving research, development, and demonstration for each of the technologies, and an economic analysis that attempted to select the least-cost combination of technologies that would satisfy a specified CO$_2$ emission target.

The key idea illustrated in Figure 3.6 is that a single, simple solution to reducing emissions is not likely to be what happens. Instead, the reductions will result at lower cost from deployment of a portfolio of technologies that contribute to the emissions reductions compared with estimates produced by the US Energy Information Administration – EIA (based on assumptions that are listed in the figure). Improved energy efficiency reduces the amount of electricity that would need to be generated, an important contribution to emissions reductions because less fuel needs to be burned in the remaining fossil-fuel plants. Renewables, such as wind and solar generation, play a significant role in the reductions, as does an increase in the amount of nuclear generation. Both provide power with emissions of CO$_2$ limited to that produced when the plants are built. Replacement of relatively inefficient coal-fired power plants (typically 34% efficient) with newer plants that operate at higher temperatures and pressures and have correspondingly higher efficiencies (46%–49%) also contributes, and large reductions result from deployment of carbon capture and storage (CCS) for a portion of the power plants that use fossil fuels. Plug-in hybrid vehicles (PHEVs) begin to penetrate the transportation market in this estimate. This will reduce emissions because the vehicles are more efficient than conventional automobiles, and their use moves some of the reductions in emissions from electric power generation to the transportation sector. Deployment of distributed energy resources also reduces emissions.

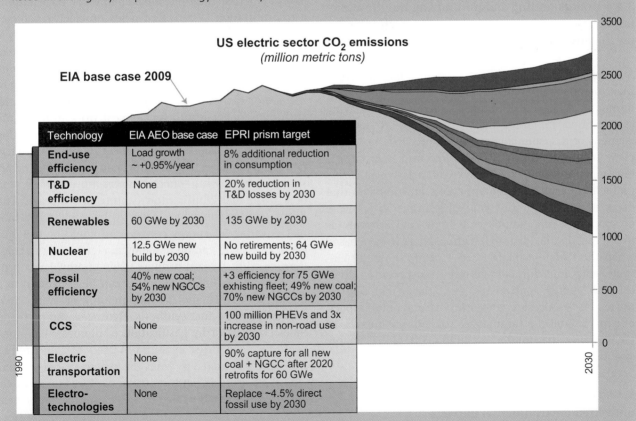

US electric sector CO$_2$ emissions
(million metric tons)

EIA base case 2009

Technology	EIA AEO base case	EPRI prism target
End-use efficiency	Load growth ~ +0.95%/year	8% additional reduction in consumption
T&D efficiency	None	20% reduction in T&D losses by 2030
Renewables	60 GWe by 2030	135 GWe by 2030
Nuclear	12.5 GWe new build by 2030	No retirements; 64 GWe new build by 2030
Fossil efficiency	40% new coal; 54% new NGCCs by 2030	+3 efficiency for 75 GWe exhisting fleet; 49% new coal; 70% new NGCCs by 2030
CCS	None	100 million PHEVs and 3x increase in non-road use by 2030
Electric transportation	None	90% capture for all new coal + NGCC after 2020 retrofits for 60 GWe
Electro-technologies	None	Replace ~4.5% direct fossil use by 2030

Figure 3.6. Emissions reductions that would result from aggressive targets for improved energy efficiency, deployment of renewables, addition of nuclear generation, replacement of less efficient coal-fired power plants with more efficient ones, deployment of significant carbon capture and storage (CCS), significant market penetration of plug-in hybrid vehicles (PHEV), and installation of significant distributed energy resources (DER), gigawatt electrical power (GWe) [13].

Box 3.2. (cont.)

It is likely, of course, that what actually does happen will differ from what is shown in Figure 3.6 as government policies are put in place and as research and development proceed. Obviously, if one technology is not available (for example, because of a policy choice not to use nuclear power), then the contributions of others would have to be expanded to meet emissions goals. Also, the remaining fleet of electric power generation assets would still provide many additional sustainability challenges. Even so, Figure 3.6 suggests that significant progress can be made toward improved sustainability using technologies that exist or can reasonably be expected to be developed.

cost of wind power is more competitive, large-scale deployment is still more costly than many other options for reducing GHG emissions [11]. Driving down the cost, increasing the reliability, and developing the infrastructure to take advantage of these sustainable resources will require significant technological advances, experience, and time – given that today, less than 1% of the current worldwide energy supply comes from these resources (see Figure 3.3). See Box 3.2 for an example of estimates of the impact of deployment of more sustainable energy technologies in the electricity sector on CO_2 emissions in the USA.

3.4.4 Understanding and minimizing the life-cycle impact of energy supply and end-use technologies on the environment

All energy systems and human activities have environmental impacts. Understanding the full life-cycle environmental impacts of energy systems and making choices that minimize their impacts is an essential element of a sustainable energy future. It is not sufficient to consider only GHG emissions and environmental impacts associated with the use of fuels; we must also understand and limit the emissions and impacts that occur during fuel production. For example, the process of manufacturing liquid fuels from coal releases as much carbon dioxide into the atmosphere as is released when the fuels are finally used – nearly doubling life-cycle emissions for coal-derived transportation fuels [12]. All energy conversions, even those that rely on renewable resources, have some environmental impact, of course. Table 3.2 summarizes some of the impacts of technologies that might be employed in global energy systems with lower GHG emissions.

3.4.5 Understanding and mitigating the undesired environmental impacts of energy production and use

The pathway toward a sustainable energy future will inevitably require economic trade-offs and societal decisions that reflect the state of the existing energy systems, the cost of advanced energy systems, the desired pace of change, and the ability to build new infrastructure to support advanced energy systems. For example, coal-fired electricity generation provides abundant and relatively low-cost (if not low-impact) energy in many parts of the world, and its use is growing rapidly in China and India (Chapter 10). Capturing and sequestering carbon dioxide from these facilities provides a means of mitigating some of the impacts of coal-fired power generation while supporting economic growth and providing electricity to improve the quality of life as people move out of poverty (see Chapter 8 for more information on carbon dioxide capture and sequestration).

The changes in the energy systems of the planet are likely to include multiple, linked pathways. For example, a transportation system that relies on electric vehicles can be imagined (see Chapters 31 and 34). Such a system would require use of one of several potential energy resources, conversion of those resources to some form that could be stored on a vehicle, transportation of the energy carrier to the vehicle, storage on the vehicle, and then conversion to mechanical work to move the vehicle. The primary resources could be solar (including photovoltaics, concentrating solar thermal electric power generation or bioconversions to chemically stored energy), wind, nuclear, waves, tides, geothermal, or a fossil fuel with carbon capture and storage. The energy carrier could be electricity, hydrogen, or a biofuel. A variety of energy conversions could be used for each. Electricity can be made by nuclear power or solar thermal with a steam turbine, while photovoltaics produce electrons directly. Biomass can be converted to a liquid fuel like ethanol by fermentation or by gasification followed by fuel synthesis. Hydrogen can be made by electrolysis or by gasification of biomass. Electrons could be transported by a grid to users or converted to hydrogen. Hydrogen in tanks (or various storage materials) or electrons in batteries could be stored on a vehicle and then converted to electricity by a fuel cell or burned directly in an internal combustion engine. Finally electric motors could convert the electricity to mechanical work. Each of the many other energy conversion pathways we use to provide energy services to humans offers similar sets of options, and reinvention of the world's energy systems will offer many opportunities to improve efficiency and reduce GHG emissions.

Table 3.2. Examples of environmental benefits and drawbacks of technologies for producing electricity

Energy supply	Benefits	Drawbacks
Coal power production	High density of stored primary energy	Large GHG emissions Mining impacts Air pollution if controls not used (e.g., SO_x, NO_x, Hg) Ash-disposal impacts
Natural-gas power production	Small to modest footprint of primary energy supply High density of stored primary energy Lowest CO_2 emissions of fossil resources Relatively efficient power plants	Moderate to high GHG emissions Microseismicity from fracturing tight gas sands and shales Disposal of fluids used for hydraulic fracturing
Fossil-fuel power production with CCS	Small footprint of primary energy supply High energy density of primary energy supply 80%–90% reduction in GHG emissions	Mining impacts Potential groundwater impacts from brine migration or CO_2 leakage
Nuclear fission	Small footprint of primary energy supply Small waste volume High density of stored energy	Potential contamination from nuclear accidents Potential for groundwater contamination from waste disposal. Potential for nuclear-weapons proliferation
Solar PV and solar thermal power production	No significant GHG emissions No air pollution Renewable energy supply	Low energy density for power supply Large footprint for solar collectors Competing uses for land. Local changes in surface albedo Potential impacts on sensitive desert habitats Use of toxic materials (e.g., cadmium) in some PV materials
Hydropower	No significant GHG emissions No air pollution Renewable energy supply	Impacts on fish habitat Loss of terrestrial habitat
Wind Turbines	No significant GHG emissions No air pollution Renewable energy supply	Large footprint needed for primary energy supply Potential impacts on birds and bats, depending on siting Noise and visual impacts
Geothermal Energy	Zero to small GHG emissions No significant air pollution	Potential impacts on water resources from brine disposal Microseismicity from water or brine injection
Biomass power production	Net GHG emissions may be reduced, depending on the life-cycle GHG emissions from biomass production, transport, and use May be sustainable dependent on agricultural practices and continued availability of water resource	Low spatial energy density for power supply Groundwater impacts from agricultural chemicals Competition with other uses for land and water resources GHG emissions from biomass production and processing Potential to increase GHG emissions from land conversion (e.g., rainforest to biofuels)

CCS, carbon capture and storage; PV, photovoltaics.

3.5 Research and development needs for sustainable energy

Each energy conversion has an efficiency, a cost, and an environmental footprint that will influence where and to what extent they are used, and for some of them considerable additional development of materials and technologies will be required. Moreover, virtually all energy resources offer interesting research opportunities for the materials science community. For example, much of the work under way now to improve the efficiency and reduce the cost of photovoltaics is aimed at the fundamental science and engineering of the materials used to convert solar photons to electrons (see Chapters 17–23), and efficient conversion of electrons to photons for lighting will impact overall energy use (Chapter 35). Advances in fundamental understanding and the ability to create ordered and disordered nanostructures (nanowires and quantum dots, for example) offer many potential paths to adjust bandgaps, increase photon absorption, enhance electron and hole transport, create photonic devices that distribute light, and reduce energy requirements and costs for making devices.

Many of the same ideas will find application in fuel cells, which offer the potential of high energy conversion efficiencies. Nanostructured materials will also play a role in electrochemical energy storage; media that are efficient in weight and volume could transform transportation, through advanced battery electrochemistry derived from abundant elements with reduced use of toxic materials (Chapter 44). The common thread that connects all of these efforts is the ability to optimize the properties of materials in new ways by controlling structure at the nanoscale and microscale and creating materials with very high surface area with the potential for high-efficiency electrochemical cycling.

More widespread use of nuclear power will require improved fuel cycles, more efficient conversions of neutron fluxes to electricity, and waste storage isolation for nuclear power, all of which imply development and selection of advanced materials with properties that can tolerate harsh environments (Chapters 14–16). Materials with high creep strength at high temperature would allow more efficient operation in a variety of power-plant settings, and materials that retain their properties in the face of high radiation fluxes will be required if fusion power is to find its way into commercial use.

Questions of biomaterials and related separations abound in the conversion of solar photons to fuels through biological processes (Chapters 26–29), and improved knowledge of genetics and the ability to design enzymes offers many opportunities to improve the efficiency of biological processes that self-assemble the molecular structures that convert sunlight to electrons and subsequently store energy in chemical bonds.

Whether hydrogen is used as an energy carrier will depend on the development of advanced materials for storage, efficient and less expensive catalysts for fuel cells, and methods to produce hydrogen with low GHG emissions (e.g., solar water splitting) (Chapters 46–49). Use of abundant elements in new catalytic structures will offer the potential cost reductions needed for commodity-scale applications of improved electrodes. Catalysts will also continue to play important roles in many chemical reactions, including those that transform biofeedstocks to fuels and other products. Improved solid-state electrolytes that allow efficient transport of ions at modest temperatures would expand the range of fuel-cell applications and might offer opportunities for separation of oxygen or CO_2 from other gases. Finally, reductions in emissions of CO_2 to the atmosphere could be achieved by use of novel materials that balance selectivity and permeance in the separation of CO_2 from product gases of combustion or gasification processes (that make hydrogen, for example) followed by geological storage of CO_2.

3.6 Summary

Observed changes in the concentrations of GHGs in the atmosphere and in the pH of the upper ocean and observed and projected impacts of those changes indicate that significant action is needed to reduce emissions of GHGs from fossil fuels. Changing the world's energy systems is a huge challenge, but it is one that can be undertaken now with improvements in energy efficiency and with continuing deployment of a variety of technologies now and over the decades to come. Abundant sustainable energy resources are available. However, there are many barriers in terms of efficiencies, impacts, and costs that will have to be overcome. Doing so will require worldwide focus on the challenge and the talents of many participants, and significant contributions from the materials science community will be essential.

3.7 Questions for discussion

1. Consider the energy technologies listed in Table 3.2. Are there others that should be listed? What would you need to consider in order to rank the various technologies in terms of sustainability? Construct and discuss a ranking of the technologies, stating the assumptions you make in order to do so.
2. Consider three potential future vehicles with potential for improving the sustainability of automobile energy use: battery electric vehicles, hydrogen-powered fuel-cell vehicles, and internal combustion engines that use methane as the fuel. Consider the sequence of energy conversions required for each type of vehicle and estimate the efficiency of each conversion. State

the assumptions you make, rank the vehicles in terms of sustainability, and discuss what technology advances will be required to permit that vehicle to penetrate the auto market at large scale.

3. Energy flows such as sunlight, wind, biomass, waves, and flowing water are termed "renewable." Stored energy resources such as fossil fuels, nuclear fuels, and even geothermal resources are often assumed to be unsustainable. Are renewable energy resources always sustainable? Can nonrenewable resources ever be sustainable (what about use of nuclear fuels in a breeder reactor, for example)?

4. In recent years, advances in horizontal well drilling and hydraulic fracturing technologies have been developed that permit recovery of methane from natural-gas shales. Those resources appear to be large, and their use has led to an increase in natural gas production in the USA. Discuss the aspects of sustainability of the use of those resources for electric power generation. Should the substitution of natural gas as a fuel for coal in electric power generation be considered as a way to reduce GHG emissions?

5. Discuss the sustainability of biofuels. Is it preferable to use biomass to make liquid fuels for transportation or to produce electricity? Why?

6. Find a recent estimate of per capita energy use in your own country and compare it with that cited in Box 3.1. Estimate your personal energy use, including fuels for transportation, electricity, and building heating and cooling. What other energy uses should be included in your personal total?

3.8 Further reading

- Up-to-date statistics on world energy production and use can be found in the annual statistical reviews produced by the International Energy Agency (http://www.iea.org/publications/free_new_Desc.asp?PUBS_ID=1199) and by BP (http://www.bp.com/productlanding.do?categoryId=6929&contentId=7044622).

- A comprehensive review of the sustainability issues involved in energy production can be found in *Sustainable Energy: Choosing Among Options* by J. W. Tester, E. M. Drake, M. J. Driscoll, M. W. Golay, and W. A. Peters, MIT Press, 2005.

- Future energy options for the USA are reviewed in *Americas's Energy Future: Technology and Transformation*, National Academies Press, 2009. Separate, more-detailed reports were also issued by the National Academies on Energy Efficiency Technologies, Alternative Liquid Transportation Fuels, and Electricity from Renewable Resources. Taken

together these reports give a comprehensive snapshot of energy technologies available to be deployed in the next two decades.

3.9 References

[1] **T. A. Boden**, **G. Marland**, and **R. J. Andres**, 2009, *Global, Regional, and National Fossil-Fuel CO_2 Emissions*, Carbon Dioxide Information Analysis Center, Oak Ridge National Laboratory, US Department of Energy, Oak Ridge, TN, doi:10.3334/CDIAC/00001.

[2] **W. Lutz**, **W. Sanderson**, and **S. Scherbov**, 2008, *IIASA's 2007 Probabilistic World Population Projections*, IIASA World Population Program Online Data Base of Results, http://www.iiasa.ac.at/Research/POP/proj07/index.html?sb=5.

[3] **IEA**, 2010, *Key World Energy Statistics*, International Energy Agency, Paris.

[4] **E. Martinot**, 2007, *Renewables 2005, Global Status Report*, World Resources Institute.

[5] **G. A. Meehl**, **T. F. Stocker**, **W. D. Collins** et al., 2007, *Climate Change 2007: The Physical Science Basis. Contribution of Working Group I to the Fourth Assessment Report of the Intergovernmental Panel on Climate Change*, eds. **S. Solomon**, **D. Qin**, **M. Manning** et al., Cambridge, Cambridge University Press, pp. 747–845.

[6] **A. J. Tirpak**, **Z. Dadi**, **L. Gylanm** et al., 2005, *Avoiding Dangerous Climate Change*, Report of the International Scientific Steering Committee, International Symposium on the Stabilisation of Greenhouse Gas, Concentrations, Hadley Center, UK.

[7] **T. Barker**, **I. Bashmakov**, **L. Bernstein** et al., 2007, "Technical Summary," in *Climate Change 2007: Mitigation. Contribution of Working group III to the Fourth Assessment Report of the Intergovernmental Panel on Climate Change*, eds. **B. Metz**, **O. R. Davidson**, **P. R. Bosch**, **R. Dave**, and **L. A. Meyer**, Cambridge, Cambridge University Press, pp. 25–93.

[8] **S. Pacala** and **R. Socolow**, 2004, "Stabilization wedges: solving the climate problem for the next 50 years with current technologies," *Science* **305**, 968–972.

[9] **Energy's Future Beyond Carbon**, *Scientific American* special issue, September 2006.

[10] *Real Prospects for Energy Efficiency in the United States*, Washington, DC, National Academies Press, 2010.

[11] **P. Enkvist**, **T. Nauclér**, **J. Rosander**, and **Q. McKinsey**, 2007, www.mckinseyquarterly.com/Energy_Resources_Materials/A_cost_curve_for_greenhouse_gas_reduction_1911.

[12] **A. E. Farrell** and **A. R. Brandt**, 2006, "Risks of the oil transition," *Environ. Res. Lett.*, **1**, 1–6.

[13] **R. James**, **R. Richels**, **G. Blanford**, and **S. Gehl**, *The Power to Reduce CO_2 Emissions*, Discussion Paper, Energy Technology Assessment Center, Electric Power Research Institute, Palo Alto, CA.

Energy cost of materials: materials for thin-film photovoltaics as an example

Ajay K. Gupta and Charles A. S. Hall

Department of Environmental and Forest Biology and Environmental Sciences, College of Environmental Science and Forestry, State University of New York, Syracuse, NY, USA

4.1 Focus

Renewable forms of energy are being sought to fulfill future needs in the face of declining fossil-fuel reserves. Much attention is being paid to energy availability, the economy, and the effects that changes in the two might have on daily life. However, little emphasis has been placed on the question of how much energy must be spent to get new sources of energy into the economy in the first place. This chapter examines this question using thin-film photovoltaics as an example.

4.2 Synopsis

Prior to the 1740s, only 13 elements in what is now called the periodic table were known to exist. By the twentieth century, all 90 naturally occurring chemical elements had been discovered and put to use in the economy. Industry has found methods of extracting, refining, and using just about every material humans have found on Earth, and this process continues to evolve today as the demands of a continually growing industrial society require ever more complex materials. As much as materials make today's industries possible, they also represent a constraint – because raw materials are needed to produce desired goods and, in turn, energy is required to develop the materials as well. Nowhere is this more important than in the energy production systems that power today's world and those that will power future societies as well.

Specifically, "low-carbon" renewable energy sources are gaining popularity in the international energy policy sphere, and more attention is being paid to the prospect of capturing solar energy in various ways (see Chapters 17–23, 27, 28, and 47–49). One particularly well-known example is through photovoltaics (PV), in which solar radiation is converted into DC electricity using semiconductors that exhibit the **photovoltaic effect**. (Chapter 18 provides a detailed discussion of PV technology.) Photovoltaic systems are rarely economically competitive with coal and currently provide less than 1% of global energy. However, PV technology is thought to have the potential not only to reduce the dependency on fossil fuels in developed countries but also to bring a source of renewable energy to the developing world, so that its use can promote economic growth and fulfill basic needs in a manner that is "carbon-neutral" [1][2][3]. To achieve this potential, a rapid increase in PV production is necessary. This chapter examines the energy cost of achieving such an increase.

4.3 Historical perspective

First-generation (1G) PV technology uses flat sheets of crystalline silicon (c-Si) or multicrystalline silicon (mc-Si) called wafers that employ single-junction circuits [4] (see Section 18.4.2 in Chapter 18). These wafers consist of very-high-quality silicon, 99.999999% pure [5], and are therefore expensive, in terms of both money and energy required for production.

The PV industry borrowed materials, production methods, and manufacturing tools from the integrated-circuits industry. The resulting screen-printing-based production line allowed 1G PV wafers to enter rapid large-scale production [6]. Such wafers are also efficient at converting sunlight into electricity. As a result, mc-Si accounts for about 63% of the world market in PV.

However, about half of the monetary cost of a 1G PV collector can be attributed to silicon production [5], which may explains some of the difficulty this technology has had in penetrating the energy market and might continue to limit its competitiveness [6]. Specifically, an energy input of about 86 GJ t^{-1} [7] is required to produce metallurgical-grade silicon, which must then be further refined to semiconductor grade. Also, high-quality silicon is used in various other high-technology industries that produce high-value products in competition with 1G PV devices [8]. To decrease costs, then, a logical step would be to create PV using lower quantities of silicon.

Indeed, second-generation (2G) PV technology employs thin-film devices, which are made using smaller volumes of materials, can be mass-produced using printing-press technology, and can be integrated into building materials [8]. The techniques used to create the layers, including vapor deposition and electroplating, cost less and are less energy intensive than those used for 1G wafers. However, the lower-quality processing gives rise to more defective materials [4].

Amorphous silicon (a-Si), $CuIn(Ga)Se_2$ (CIGS), CdTe/CdS (CdTe), and polycrystalline silicon (p-Si) comprise the thin-film PV family. The lesser quality of a-Si and p-Si compared with c-Si greatly reduces their efficiency in converting photon energy into electricity. Of all 2G PV technologies, CIGS and CdTe are currently thought to be the most economically promising because of their favorable cost versus efficiency relation [5][8]. Indeed, studies on increasing the efficiency and decreasing the costs of thin-film technologies continue (Figure 4.1) [9].

These technologies can be assessed in terms of the energy payback time (EPBT), or the time required for a system to produce an amount of energy equal to that embodied in the system, where the embodied energy refers to primary energy consumption during the life cycle of the system. For CdTe modules, the EPBT decreased by 15%–35% from 1.1 years between 2006 and 2009 [10]. This reflects an increase in both production and solar conversion efficiency. However, if the energy cost in producing the raw materials to develop modules were to increase, it could offset such gains in the future, and even increase the EPBT over time.

Given the issues with both 1G and 2G PV technologies [11], third-generation (3G) PV aims to dramatically increase efficiency while eliminating the use of large quantities of high-quality silicon, thereby realizing the same cost reductions as 2G PV (see Figure 19.1 in Chapter 19) [9].

Technologies employed to create 3G PV include multijunction concepts, optical metamaterials, quantum technology, nanotechnology, and polymer semiconductor science [6]. However, even dye-sensitized cells [12][13], which were once thought to be the next generation of PV, have yet to be commercialized. Therefore, 3G PV capacities cannot be reliably projected into the future at this time. (See Chapter 19 for a thorough discussion of 3G PV technology.)

4.4 Accounting for the costs of obtaining materials

4.4.1 Introduction

Massive global PV production faces certain material and energy constraints. Material constraints are those that exist because of the limited amounts of critical metals required for PV production [14][15][16]. This is true for both primary ores (those extracted from the ground to obtain the metal) and secondary ores (those that must be concentrated and refined from primary ores). For example, Ga is an impurity refined when the primary resource Al is obtained from bauxite ore. With increasing use of these ores, the remaining resources are reduced in grade, meaning that increasing amounts of energy are required to obtain them. In the face of increasing global energy prices, this gives rise to energy constraints for PV production [17][18].

Since 1956, when Hubbert [19] predicted a coming peak in oil production in the USA, efforts have been made to predict similar cases in the world for other nonrenewable resources. For example, studies suggest that Canada and the USA have already reached peak copper production [18][20]. Other materials that might already have passed their production peaks, or might soon do so, include Au, Ag, P (important for agriculture), and Zn (important for PV-related materials) [14][21]. However, despite the popularity of peak-production studies, it is often misunderstood that they refer not to the amount of material on Earth, but to the rate at which the material can be made available for cost-effective use. More importantly, the role of energy in the processes by which materials are obtained is rarely mentioned [22][23][24].

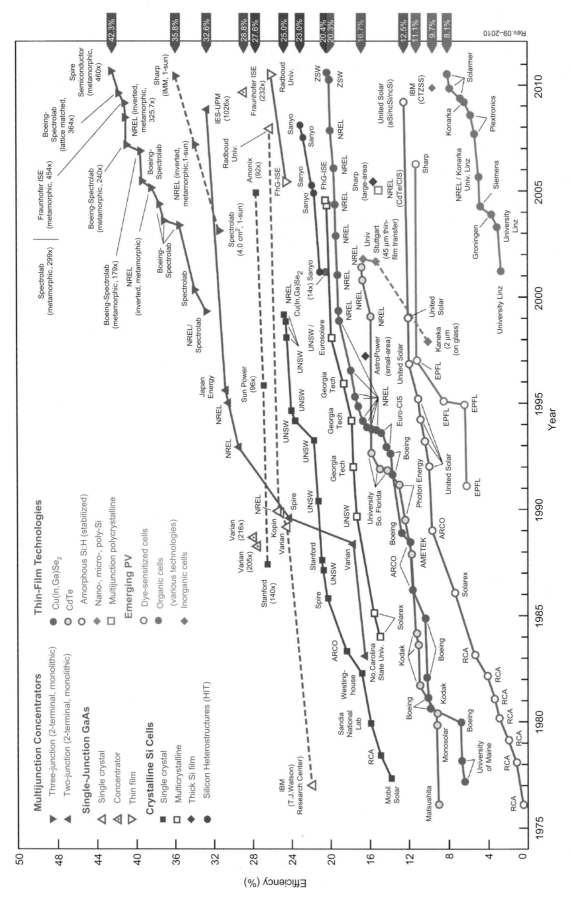

Figure 4.1. Efficiency gains over time by module type, as indicated by the best laboratory research cell efficiencies from 1975 to 2009 (from the NREL).

Rev.09-2010

For example, little work is available on the energy requirements to generate elements from primary ores, and even less for those secondary ores that are important to many of the advanced technologies used for PV systems. In 1986, Hall *et al.* provided a fairly thorough examination of resource materials and their energy requirements in the book *Energy and Resource Quality: The Ecology of the Economic Process* (for an example, see Figure 4.2) [25]. Unfortunately, their data refer mostly to the 1970s and previous years, and subsequent data are rare or inconsistent when available, but are usually withheld by private interests.

What follows is an attempt to fill the gaps left in the energy accounting for materials today, using the example of materials needed for PV production. This chapter begins with a brief description of some of the processes involved in obtaining concentrated ores and then attempts to estimate the energy costs of the following materials identified as useful to the PV industry: Cu, In, Ga, Se, Cd, Te, Ge, Al, Zn, Fe, steel, Pb, phosphate, and cement (summarized in Table 4.1).

4.4.2 Mining

Mining includes searching for, extracting, and beneficiating materials from the Earth. (**Beneficiation** is the treatment of raw ore to separate the mineral content from the surrounding commercially worthless material, called gangue.) It has been performed by humans for thousands of years. Today, it is the backbone not just for industry, but for the whole economy of many nations. There are thousands of mines across the globe. In the USA alone, mining accounts for millions of jobs and billions of dollars each year. Because of the scope and dynamics of the mining sector, it is very difficult to derive a fixed energy cost [31].

There are seven main steps in any mining operation: (1) exploration, (2) design, (3) construction, (4) extraction, (5) beneficiation, (6) processing, and (7) reclamation [32]. However, each mine differs regarding the implementation of each step. Surface mining and underground mining account for the majority of all mining operations globally (deep-sea mining is not yet a significant contributor). Solution (*in situ*) mining, which uses an acid solution to dissolve metals and then recovers the solution and separates the ore, is also gaining in popularity.

The type of mine is determined by physical factors of the deposits, including their nature, location, size, depth, and grade. In 2000, the USA obtained 92% of its metals through surface mining [33]. Underground mines are employed only when the deposits are very deep and the ore grade is high enough to make such extensive operations economically profitable. Underground mines are more energy-intensive than surface ones, because

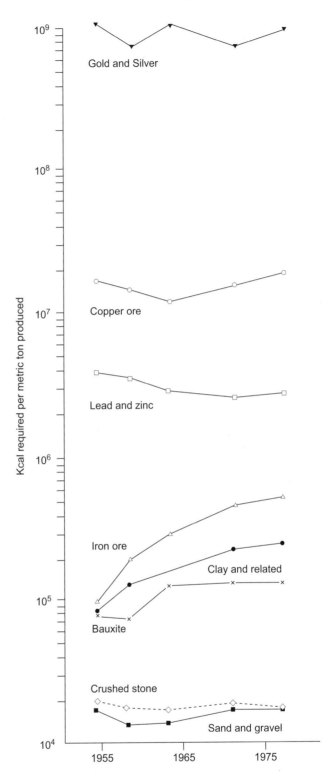

Figure 4.2. Trends over time of energy costs (i.e., direct fuel and electricity use per tonne of output) for some metal and nonmetal minerals from [25]. Note: 1 kcal = 4.184 kJ.

they require much more planning and drilling (as opposed to explosives for surface mining). However, no two mines use the same amount of energy, mostly because of geographical differences.

Table 4.1. Reserves, production, energy cost, years of production at current rate, and total energy costs for materials associated with the PV industry

	2009 Reserves (Mt)	2008 Production (Mt)	Approximate energy cost (GJ t^{-1})	Remaining production (years)	2008 Energy cost (EJ)	Energy cost to produce reserves[a] (EJ)
Primary products						
Aluminum	6,750[b]	39	188	173	7.35	1,272
Steel	77,000[c]	2,220	29	35	64	2,218
Copper	540	15	135	35	2.09	74
Cement	NA	2,840	6	NA	16	NA
Iron ore	77,000	2,220	3	35	6	223
Lead	79	3.84	31	21	0.12	2.46
Zinc	200	12	76	17	0.88	15
Phosphate	16,000	161	0.35	99	0.06	5.58
Secondary products						
Gallium	1.0×10^{-2d}	1×10^{-4}	50[e]	105	5.04×10^{-6}	5.28×10^{-4}
Germanium	2.52×10^{-3f}	1.4×10^{-4}	40[g]	18	5.60×10^{-6}	1.01×10^{-4}
Indium	1.1×10^{-2f}	5.7×10^{-4}	40[g]	19	2.3×10^{-5}	4.32×10^{-4}
Selenium	8.8×10^{-2}	1.81×10^{-3h}	116[g]	49	2.10×10^{-4}	1.0217×10^{-2}
Tellurium	2.2×10^{-2}	6.4×10^{-4i}	116[g]	34	7.4×10^{-5}	2.554×10^{-3}
Cadmium	5.90×10^{-1}	2.0×10^{-2}	4.5	30	8.8×10^{-5}	2.655×10^{-3}

Data from [26], unless stated otherwise.
[a] Energy cost to produce all 2009 reserves.
[b] Value calculated by dividing 2009 bauxite reserves by 4 [27].
[c] Because steel production consumes 98% of world Fe production, the USGS does not provide a separate accounting for steel, so the values for Fe were used instead.
[d] Value calculated from the 2009 ratio of Ga to bauxite production applied to bauxite reserves.
[e] Approximate energy cost is proprietary [28], so the cost of electrowinning for Al [29] was used.
[f] Value calculated from the ratio of Ge or In to Zn production applied to Zn reserves.
[g] Approximate energy cost is proprietary and could only be estimated from total mine operations data [28].
[h] Value calculated as the sum of 2008 production excluding the US contribution [26] plus estimated 2009 US production [30].
[i] Value calculated from the ratio of Te to Cu reserves applied to Cu production.

The amount of energy a mine uses depends on the quantity and grade of the ore. For metals in the USA, only about 4% of material handled is produced [32], although this varies depending on the metal and the ore grade. As the ore grade is reduced, more material must be handled to produce the same amount of metal, and the energy cost per tonne of product goes up. The cost of energy used directly as a percentage of that cost of supplies consumed by the overall US mining industry increased from 16% in 1992 to 17% in 1997 [34]. This appears to reflect an industry-wide decrease in ore grade.

In 2000, the USA consumed about 1.2×10^{18} J (1.2 EJ) in all mining operations, approximately 3.3% of total industrial energy consumption [35][36] and 1.5% of all energy used in the USA. Of the energy consumed, approximately 35% is in the form of electricity; 32% fuel oil; and 33% coal, gas, and gasoline [32][36]. Electricity is used mostly for ventilation, water pumping, crushing, and grinding, whereas diesel fuel serves for hauling and other transportation needs [32]. Typically, electricity is the major energy source for underground mining because of its ventilation requirements, whereas surface mining uses mostly diesel fuel for digging and hauling. Overall, about two-thirds of energy consumption in mining is due to the crushing and grinding process [32]; however, this differs among metal, mineral, and industrial mineral mines.

There is room for improvement in the energy consumption of mining. Approximately 41% of energy consumed is lost onsite to old or inadequate process equipment and poor plant configurations [36]. Granade [37] suggested that mining could become 60%–95% more efficient if different approaches were used for onsite transportation and the amount of material being transported. However, such efficiency gains are mostly theoretical and uneconomical for the near future, because the energy cost of the capital equipment required would exceed the potential energy savings.

4.4.3 Refining

Of the two general forms of refining ore, one uses physical properties, the other uses chemical properties, and both use physics and chemistry to gain higher concentrations. The more widespread approach, **pyrometallurgical processing** (physical), utilizes smelting. **Hydrometallurgical beneficiation** (chemical) involves leaching ores with strong acids and then recovering the metal by precipitation or solvent extraction/electrowinning (SX/EW), which is gaining in popularity. Both processes have many steps that might or might not be required depending on the material, ore grade, and concentrations of other desirable byproducts.

In pyrometallurgical processing (smelting), the ore is heated in a furnace of some kind so that the metals within separate. Electric, or flash, furnaces are gaining popularity because of their higher energy efficiencies (the higher cost of the electricity is balanced by its more specific application), lower material volumes, and higher byproduct recovery rates [32] compared with furnaces. The concentrate can be further refined by blowing air and other gases into the molten form to oxidize impurities. For the very pure concentrations required for electrical devices such as photovoltaics, metals must then be electrolytically refined (i.e., electrorefined).

In electrorefining, the concentrate is connected to a circuit and dissolved into an electrolyte, and direct current is passed through the cell. Upon application of the right current with the proper electrolyte, the desired metal accumulates on the cathode in almost pure concentrations. This metal is referred to as cathode-grade material. The energy cost for this process can be very low compared with those for other refinement processes [32], but it varies for each material.

Hydrometallurgical beneficiation involves crushing, grinding, washing, filtering, sorting, and sizing, followed by gravity concentration, flotation, roasting, autoclaving, chlorination, leaching (dump and *in situ*), ion exchange, solvent extraction, electrowinning, and

precipitation, depending on the metal and ore grades. In essence, the ore is put into solution using various chemical processes (leaching), and then the desired metal is refined through precipitation (cementation) or solvent extraction/electrowinning (SX/EW). If removed through precipitation, the metal is cast into an anode for electrorefining. If solvent extraction is used, the metal exists within the electrolyte used in electrorefining, and some other insoluble anode is used. The energy cost of SX/EW depends on the concentration of desirable metal in the electrolyte, which is a reflection of ore grade. Once electrorefined, the leftover electrolyte often contains other desirable metals, which can then be further refined if concentrations prove profitable.

Crushing and grinding ore typically constitute the major energy costs of processing. In the case of Cu ore, this can be as high as 45% of the energy required for refining, depending on grade [32][38]. The lower the ore grade, the more grinding is necessary to recover metals, so ore grade ultimately defines the energy cost of a particular mine. Given the numerous additional processes potentially involved in pryometallurgical processing compared with hydrometallurgical beneficiation, the former can be up to 325 times more energy intensive than the latter [32].

In 1985, 80% of Cu mined was pyrometallurgically refined, and 15% was hydrometallurgically refined [29]. However, the latter method has since gained in popularity because of costs. Ullmann [29] provides the typical energy requirements of electrorefining three primary metals as follows: Cu electrorefining at 0.9 GJ t, Cu SX/EW at 7.2 GJ t^{-1}, Zn SX/EW at 11.9 GJ t^{-1}, and Al SX/EW at 50.4 GJ t^{-1} (but 22.8 GJ t^{-1} theoretically).

4.4.4 Materials needed by an expanding PV industry

Summary information is provided in Table 4.1. The following sections provide details for each element required.

Copper
Cu is one of the oldest materials used by humans. It is currently produced by a variety of multistage processes, including mining and concentrating low-grade ores (<0.5% Cu) containing sulfide minerals and then smelting and electrorefining to produce cathode-grade materials (>99.99% pure Cu). Cu ore is also the primary ore from which Se and Te are obtained.

Because of the variety of process routes, ore grades, and other mining logistics, it is very difficult to derive a single energy cost of producing a material [28]. However, Table 4.2 lists the energy costs for some general process steps used in producing Cu. In addition, some processes are popular and old enough that some energy information is available, even though the specific steps of the

Table 4.2. Energy costs of general steps in producing 1 t of Cu[a] by pyrometallurgical breakdown

Process	Energy cost (GJ t^{-1})	Deviation (%)
Mining	5.5	35
Beneficiation	3.8	35
Smelting	8.4	20
Converting	0	NA[b]
Anode casting	0.2	10
Electrolysis	1.3	10
Cathode casting	1.7	10

From [29].
[a] For reference, a barrel of oil contains about 6.1 GJ.
[b] Not applicable.

Table 4.3. Energy costs of popular methods for refining 1 t of Cu using pyrometallurgical beneficiation

Process	Energy cost (GJ t^{-1})
INCO flash smelting	1.7
Electric furnace	6.7
Noranda process	6.7
Outokumpu process	8.4
Mitsubishi process	10.1
Reverberatory furnace with roasted concentrate	13.4
Brixlegg process	19.3
Reverberatory furnace with raw concentrate	21.8

From [29]. INCO, International Nickel Company.

process might remain proprietary. These are summarized in Table 4.3.

Although each copper mine has its own specific process route and energy balance, some generalizations can be made. The first is that mining, milling, and concentrating typically represent the largest energy consumers in Cu production. This is due to the larger volumes of material transported and processed relative to processes further down the chain. Also, electricity comprises the majority of energy consumption. Although natural gas and diesel fuel are also integral to transporting and refining ore, electricity is used throughout the operation. Table 4.4 provides a breakdown of many Cu production steps (both hydrometallurgical and pyrometallurgical) and their energy costs by fuel type.

Overall, the energy cost of producing 1 t of Cu can range from 51.7 to 200 GJ, equivalent to the energy in 10–33 barrels of oil [17][38][28][39][40][41]. However, as hydrometallurgical refinement gains in popularity, some measures are being developed to increase energy efficiency, including enhanced biological heap and stockpile leaching of chalcopyrite-dominant ores, low-solution-application-rate leaching techniques, recirculation of low-grade solutions to leach fresh ore, developments of alternatives to SX processes, and expanded applications of concentrated leaching processes to lower-grade feed materials [38].

Selenium

Se is a semiconductor obtained by the electrolytic refining of Cu. Although integral for thin-film PV production in CIGS modules (among other industrial and health applications), it does not occur in high enough

concentrations to be economically mined as a primary ore [14][28].

Se accumulates within the electrolyte or "anode slime" after Cu electrolysis has been performed. Concentrations of Se in anode slime can range from 5% to 20% [29] depending on the qualities of the primary ore, and this is what typically drives the energy cost of its production for commercial applications. In general, Se is concentrated using very strong acids or roasting and leaching, followed by hydrometallurgical processing and precipitation [29].

Although some process steps are available for Se refinement from anode slime – namely leaching, roasting, precipitation, distillation, and vaporization – the energy costs of these steps are proprietary information. Furthermore, because the process is dependent on levels of concentration that cannot be controlled by industry, energy costs are unique not only for each processing plant, but also for each batch. However, Gupta and Hall [42] provide a crude estimation of 116 GJ t^{-1} on the basis of overall mine operations data reported by Fthenakis *et al.* [28].

Tellurium

Te is relatively rare in comparison with Cu and Se. It is most relevant to the PV industry in CdTe thin-film modules. Like Se, Te is a semiconductor. It is found in the same Cu anode slime as Se, but at much lower concentrations. In general, 0.065 kg of Te is recovered per tonne of Cu. In order for its extraction to be commercially viable, Cu anode slime must contain a range of 0.5%–10% Te [29]; Te cannot be mined economically as a primary ore [14].

Table 4.4. Step-by-step energy costs by fuel type available to process routes for producing Cu

Energy source	Electricity (MJ t^{-1})	Natural gas (MJ t^{-1})	Diesel/oil (MJ t^{-1})	Wear steel energy equiv. (MJ t^{-1})	Total (MJ t^{-1})
Mining (to feed ROM leach)	3.177	0	31.064	0.044	34.285
Mining (to feed all other processes)	5.083	0	49.702	0.044	54.829
ROM leaching	3.968	0	0	0	3.968
Primary crushing and conveying	7.915	0	0	0.022	7.937
Secondary crushing	3.946	0	0	0.022	3.968
Tertiary crushing	3.946	0	0	0.022	3.968
Ball milling (after tertiary crushing)	37.781	0	0	21.621	59.402
SAG milling (SABC)	27.800	0	0	16.185	43.985
Ball milling (after SAG)	31.786	0	0	12.890	44.676
HPGR	9.530	0	0	0.042	9.572
Ball milling (after HPGR)	29.667	0	0	12.890	42.557
Floatation and regrinding	15.885	0	0	2.079	17.964
Transportation to smelter	0	0	7,158	0	7,158
Smelting	5,046	6,296	0	0	11,342
Refining	1,592	4,356	0	0	5,948
MT concentrate leaching	4,885	205	46	0	5,136
HT concentrate leaching	6,583	205	46	0	6,834
SX	4,354	0	0	0	4,354
Incremental SX (concentrate leaching)	441	0	0	0	441
EW	7,145	1,325	0	0	8,470
EW (alternative anode)	6,071	1,325	0	0	7,396
EW (ferrous/ferric)	4,786	117	0	0	4,903
Transportation to market	0	0	269	0	269
Total (MJ t^{-1})	41.08	13.83	7.60	0.07	62.58

From [38]. EW, electrowinning; HPGR, high-pressure grinding rolls; HT, high-temperature; MT, moderate-temperature; ROM, run-of-mine; SABC, semi-autogenous mill/ball mill/pebble crusher; SAG, semi-autogenous grinding, in which steel balls are used in addition to large rocks; SX, solvent extraction.

Typically, Te exists in anode slime at one-third the concentration of Se [29]. As with Se refinement, the processes for obtaining Te are proprietary information, but some general process steps can be defined – namely separation, roasting, leaching, cementation, electrolysis, vacuum distillation, and zone refinement. Also, as with Se, energy costs are unique for each batch of Te from each mine. However, Gupta and Hall [42] provide a crude estimation of 116 GJ t^{-1} on the basis of overall mine operations data reported by Fthenakis et al. [28].

Aluminum

Al is one the most abundant industrial materials on Earth; however, its application is only about 100 years old, because it is very reactive and, hence, difficult to separate from other materials with which it is combined, such as oxygen. For PV production, Al is most relevant as a primary ore for Ga, although it is also heavily used in supports and for electricity transmission. Bauxite is the primary source for alumina, from which Al and Ga are derived. The energy cost of producing primary Al ranges from 100 to 272 GJ t^{-1} [17][36][43][44], and that for secondary Al is approximately 11.7 GJ/t [44]. Because of the availability and profitability of producing the material, no energy gains are projected in the near future.

Gallium

Ga is a secondary byproduct from Al production. It is most relevant in the thin-film PV industry in its application to CIGS modules. Given the enormous availability of bauxite and thus Al production, no material shortages of Ga for PV systems are projected in the near future [14]. Unfortunately, as for all of the secondary metals discussed in this chapter, the refining methods for Ga are proprietary information. Furthermore, the USA does not produce any of the Ga it consumes, even though there are over 100 US patents for its processing. Fortunately, Ullmann [29] does provide a list of available processes, although it is likely incomplete. This list comprises fractional distillation (an average of three times), electrolysis, extracting, vacuum distillation, fractional crystallization, zone melting, and single-crystal growth.

Typically, there must be 70–150 mg of Ga available per liter of alumina liquor for refining of the Ga to be commercially viable [29]. Thus, the number of times the Ga concentrate undergoes fractional distillation is dependent on the source concentration. Unfortunately, this presents the same problem as with the previously mentioned secondary metals, in that the energy cost for refining Ga to semiconductor grade varies by plant, process, and also batch.

Zinc

The primary source of Zn is a mineral called sphalerite. Zn is also found with concentrations of Cu and Pb and usually with smaller concentrations of Au and Ag, but it is never found in nature in its elemental form [32]. Although it is not used in the PV industry at all, Zn is the primary metal for producing Cd, In, and Ge and is therefore integral to the industry.

Zn, which is almost exclusively mined underground, has an energy cost of about 76 GJ t^{-1} [32], which can be attributed mostly to electricity and fuel costs [28][32]. In 1985, around 80% of Zn was electrolytically refined, which has a process energy cost of 11.9 GJ t^{-1} [29]. Today, about 90% is electrowinned [25]. Zn distillation has an energy cost of 6.5–7.3 GJ t^{-1} [29].

Cadmium

Cd is relevant to the thin-film PV industry because of its use in CdTe modules. It is obtained as a secondary byproduct from electrolytic refinement or smelting of Zn, with minerals containing 0.2% Cd or higher considered economically profitable [29]. Cd is refined from concentrate in an SX/EW process, although it can also be further refined using vacuum distillation or zone refinement [28]. Unfortunately, the energy costs for these processes are not available. However, Ullmann provides the process steps – leaching, precipitation, electrowinning, vacuum distillation, and zone refinement – and an overall cost for producing Cd of 4.5 GJ t^{-1} [29]. Ullmann also estimates that the Zn leachate contains 30–40 mg of Cd per liter [29], which suggests that, as with all materials, the energy cost depends on the production-site logistics, processes, and batch.

Indium

In is a relatively rare metal derived from the fumes, dusts, slags, and residues of Zn smelting. It is relevant to the thin-film PV industry in its application to CIGS modules. Although the USA does not produce any In, it does have refineries that take standard-grade In from other countries and purify it to PV-grade material [28].

Typically, the residues from Zn smelting and roasting are leached and concentrated, and In is precipitated. A common process for the concentration and purification of In is called the Mitsubishi process. The process steps are known to include recovery from zinc leaching, roasting, concentration, solvent extraction, back extraction, plate immersion, anode casting, electrorefinement, vacuum distillation, and zone refinement; however, the energy costs are not available. To produce In of purity sufficient for use in CIGS, the metal can be electrowinned multiple times, vacuum distilled, zone refined, or all three [29]. However, because of varying grades of In from Zn residues, once again, the energy cost is relative to mine logistics, process steps, and batch. Although Fthenakis et al. [28] provided a crude estimate of 40 GJ t^{-1}, on the basis of the total operations of a Zn mine, they cautioned that the information is difficult to derive.

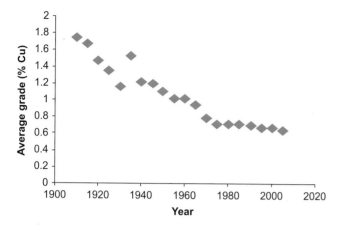

Figure 4.3. Global average Cu grade, 1910–2008, for all porphyry copper ores mined in the free-market economies (USA, Canada, Mexico, Chile, Peru, Papua New Guinea, Philippines, etc.). From [50]. Note that this plot excludes data on higher-grade strata-bound ores (Africa, Poland) and other non-porphyry ores. Porphyry deposits, consisting of large crystals in a fine-grained igneous matrix, are the principal sources of the world's copper [20].

Germanium

Ge once appeared to be very promising to the semiconductor industry. However, Ge proved difficult to purify to the required levels, and its sensitivity to temperature limited applications. Consequently, although Ge remained integral to other products such as fiber optics, very-high-quality Si replaced Ge in semiconductor devices such as transistors. As manufacturers begin to pay more attention to the very high energy intensity of producing pure Si crystals, however, Ge is once again being considered as an option. Research continues into using Ge to improve the energy and output efficiencies of PV systems, including those based on amorphous silicon (a-Si). Yet, Ge is of most interest for concentrating photovoltaics (CPV), as discussed in detail in Chapter 20.

Like In and Cd, Ge is a byproduct of Zn production, in that it is not regularly mined for the ore itself but occurs in germanite, which is associated with Zn ores [28]. Typically, it is either precipitated and electro-refined or leached and subjected to solvent extraction; it then requires zone refinement to reach semiconductor grade [29]. Other more specific processes also exist [28].

Ullmann [29] provides a breakdown of the process steps to include leaching, precipitation, solvent extraction, electrolysis, and zone refinement, but the energy costs of each step are proprietary. Again, Fthenakis *et al.* [28] provided a crude estimate of 40 GJ t^{-1}, on the basis of the total operations of a Zn mine, but cautioned that the information is difficult to derive.

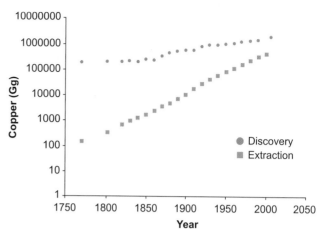

Figure 4.4. Historic trend of cumulative copper discovery versus extraction. From [20]. © 2006, National Academy of Sciences, USA.

4.4.5 Future energy costs using copper as an example

Cu is critical to the future of the PV industry, since it is required in PV systems themselves and in the infrastructure necessary to transport electricity. As is most often the case for resource extraction, as Cu is mined, the grade of the ore tends to decline over time [25][45][46][47][48][49]. This is because higher-grade, easier-to-obtain Cu is mined first and exhausted, often because companies seek maximum return on their capital investments [46]. It is therefore no surprise that the average yield of concentrated Cu ore has declined throughout recent history (Figure 4.3).

The easiest way to increase ore grade might seem to be to discover and/or mine new deposits [45]. However, although the increase in the reserve base of Cu has so far kept pace with the rate of extraction (Figure 4.4), only a small fraction of this increase is due to the discovery of new sources; the vast majority is due to the reclassification of known sources as economically retrievable [20]. Extrapolation of the trends in the data suggests that this situation cannot continue indefinitely.

Recycling from secondary Cu (scrap) is an important part of Cu availability, since grades can, in some cases, be as high as 100%. Unfortunately, although the availability and consumption of Cu scrap are growing, recycling is not. This might be because the cost of recycling increases if it is not done immediately when the scrap is discarded [51].

Another factor that could affect the grade of Cu available to the global market is constraints on international trade. For example, in the second half of 2010, China restricted the export of some of its minerals [47]. Such factors can also have an effect on the price of Cu in the global market. Already, there has been a 454% increase in the price of Cu since the year 2000 [52].

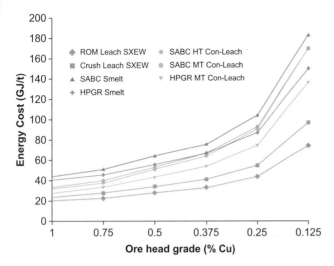

Figure 4.5. Energy cost of copper production by process route as a function of ore grade. From [38]. Con-leach, concentrate leaching; HPGR, high-pressure grinding rolls; HT, high-temperature; MT, moderate-temperature; ROM, run-of-mine; SABC, semi-autogenous mill/ball mill/pebble crusher; SX/EW, solvent extraction/electrowinning.

By extrapolating primary Cu ore grade into the future and establishing a relationship between grade and energy cost (Figure 4.5), it is possible to determine what the future costs of producing primary Cu might be. Given the issue of decreasing ore grades, it seems likely that, without a leap in technological advancement resulting in increased efficiency for mining, hauling, crushing, and concentrating ore, the energy cost of producing Cu must increase exponentially. As a result, any energy-producing system that relies on Cu being added to the economy, including PV, must suffer a decrease in net energy returned to the economy. This conclusion can apply to any of the primary materials mentioned in this chapter if their ore grade is in decline and if the relationship between ore grade and energy cost is similar. Also, because the secondary ores relevant to the PV industry are mined only as a result of the profits derived through their primary ores [14], extrapolating their future energy costs over time would be redundant.

Options to relieve this stress on the PV industry are limited. They include increasing the energy efficiency in production at a rate that at least offsets the effect of decreasing ore grades and improving the ore grades themselves. This can be achieved through greater recycling efforts, discoveries of high-quality materials, and better international collaboration. Another possible outcome could be an emphasis on switching, in the near term, to other materials that are not constrained by energy and monetary costs. For example, Al could be used for electricity transmission instead of Cu, or, in the case of thin-film PV, secondary materials that are derived from less constrained primary ores could be employed.

4.5 Summary

Materials play a critical role in the systems that provide energy to industry and society. More importantly, the energy costs of these materials can have serious effects on these integral systems. For example, although PV efficiency is increasing as research into system configurations continues, the energy cost of securing materials is increasing as well – and, using copper as an example, is likely to increase exponentially in the future. Furthermore, assessing precisely what these costs are is quite challenging, since much of the information needed is difficult to manage or is simply unavailable to the public, especially for secondary materials.

Not all materials are constrained, and there are ways to reduce the energy spent in mining operations; however, they would require heavy investments in newer infrastructure and equipment. For now, obtaining materials of high purity relies mostly on nonrenewable sources of fuel, the energy costs of which are also likely to increase. The most critical need if we are to achieve a better understanding of the energy future is better information, especially about the energy costs of obtaining materials. Indeed, energy solutions and efficiency gains cannot be fully understood without also paying attention to these energy costs.

4.6 Questions for discussion

1. What role do environmental factors play in the energy costs of mining and refining materials?
2. What effect does the geographical location of resources have on the types of fuel used to secure materials?
3. Over time, what are some factors that can both increase and decrease the energy costs of producing PV systems other than environmental and geographical factors?
4. Are there any materials integral to PV systems that were not discussed in this chapter?
5. Given the materials necessary to build CIGS and CdTe modules, what other industries could be affected by constrained stocks?

4.7 Further reading

*   **C. Wadia**, **A. P. Alivisatos** and **D. M. Kammen**, 2009, "Materials availability expands the opportunity for large-scale photovoltaics deployment," *Environ. Sci. Technol.*, **43**, 2072. (2009). Applies abundance of materials to cost analysis in considering PV technology development.

- **B. A. Andersson**, 2001, *Material Constraints on Technology Evolution: The Case of Scarce Metals and Emerging Energy Technologies*, Thesis for the Degree of Doctor of Philosophy, Department of Physical Resource Theory, Chalmers University of Technology and Göteborg University, Göteborg, available at http://bscw-app1.ethz.ch/pub/bscw.cgi/d170262/Andersson_2001.pdf. Describes material constraints as they pertain to other emerging technologies including photovoltaics.
- **C. A. S. Hall**, **S. Balogh**, and **D. J. R. Murphy**, 2009, "What is the minimum EROI that a sustainable society must have?," *Energies*, **2**, 25–47. Demonstrates an application of calculating energy costs versus energy gains.
- **K. Zweibel**, 2005, *The Terawatt Challenge for Thin-Film PV*, NREL Technical Report. Describes the challenges for scaling up of thin-film photovoltaics.

4.8 References

[1] **J. Goldemberg**, **T. B. Johansson**, **A. K. N. Reddy**, and **R. H. Williams**, 1985, "Basic needs and much more with one kilowatt per capita," *Ambio*, **14**(4–5), 190–200.

[2] **J. M. Blanco**, 1995, "Photovoltaics, wind, and other dispersed energy sources," in *UNDP: Energy as an Instrument for Socio-Economic Development: Part 2 – Removing the Obstacles: The Small-Scale Approach*, Chapter 6.

[3] **UNDP**, 2004, *Reducing Rural Poverty Through Increased Access to Energy Services: A Review of the Multifunctional Platform Project in Mali*, UNDP Mali Office.

[4] **SPREE School of Photovoltaic and Renewable Energy Engineering**, 2010, *Third Generation Photovoltaics*, http://www.pv.unsw.edu.au/Research/3gp.asp.

[5] **USDOE**, 2010, *Energy Efficiency and Renewable Energy: 2008 Solar Technologies Market Report*.

[6] **D. M. Bagnall** and **M. Boreland**, 2008, "Photovoltaic technologies," *Energy Policy*, **36**, 4390–4396.

[7] **M. A. Green**, 1982, *Solar Cells: Operating Principles, Technology, and System Applications*, Englewood Cliffs, NJ, Prentice-Hall, Inc.

[8] **D. Butler**, 2008, "Thin-films: ready for their close-up?," *Nature*, **454**, 558–559.

[9] **ARCPCE Australian Research Council Photovoltaics Centre of Excellence**, 2008, *ARC Photovoltaics Centre of Excellence Annual Report 2008*, Sidney, The University of South Wales.

[10] **V. Fthenakis**, **H. C. Kim**, **M. Held**, **M. Raugei**, and **J. Krones**, 2009, "Update of PV energy update times and life-cycle greenhouse gas emissions," in *24th European Photovoltaic Solar Energy Conference*, Hamburg.

[11] **W. Shockley** and **H. J. Queisser**, 1961, "Detailed balance limit of efficiency of p–n junction solar cells," *J. Appl. Phys.* **32**, 510.

[12] **M. Gratzel**, 2001, "Photoelectrochemical cells," *Nature*, **414**, 338.

[13] **M. A. Green**, (ed.), 2001, *Proceedings – Electrochemical Society, 2001–10*, vol.3 (*Photo-voltaics for the 21st Century II*).

[14] **B. A. Andersson**, 2001, *Material Constraints on Technology Evolution: The Case of Scarce Metals and Emerging Energy Technologies*, Thesis for the Degree of Doctor of Philosophy, Department of Physical Resource Theory, Chalmers University of Technology and Göteborg University, Göteborg.

[15] **D. A. Cranstone**, 2002, *A History of Mining and Mineral Exploration in Canada and Outlook for the Future*, Ottawa, Ontario, Department of Natural Resources.

[16] **USGS**, 2006, *Copper Statistics*, http://minerals.usgs.gov/ds/2005/140/copper.pdf.

[17] **M. Ruth**, 1995, "Thermodynamic constraints on optimal depletion of copper and aluminum in the United States: a dynamic model of substitution and technical change," *Ecol. Econ.*, **15**, 197–213.

[18] **R. N. Rosa** and **D. R. N. Rosa**, 2007, "Exergy cost of extracting mineral resources," *Proceedings of the 3rd International Energy, Exergy and Environment Symposium*, eds. **A. F. Miguel**, **A. Heitor Reis**, and **R. N. Rosa**, Évora, Portugal, CGE-Évora Geophysics Centre.

[19] **M. K. Hubbert**, 1956, *Nuclear Energy and the Fossil Fuels* (presented before the Spring Meeting of the Southern District Division of Production, American Petroleum Institute, San Antonio, Texas, March 8, 1956), Houston, TX, Shell Development Company, Exploration and Production Research Division.

[20] **R. B. Gordon**, **M. Bertram**, and **T. E. Graedel**, 2006, "Metal stocks and sustainability," *Proc. Natl. Acad. Sci.*, **103**(5), 1209–1214.

[21] **J. Laherrere**, 2010, "Peaks in Argentina, Latin America and the world," in *ASPO Conference*.

[22] **B. Hannon**, **M. Ruth**, and **E. Delucia**, 1993, "A physical view of sustainability," *Ecol. Econ.*, **8**, 253–268.

[23] **C. A. S. Hall**, **D. Lindenberger**, **R. Kummel**, **T. Kroeger**, and **W. Eichhorn**, "The need to reintegrate the natural sciences with economics," *BioScience*, 51(8), 663–673.

[24] **R. U. Ayers**, 1994, *Information, Entropy, and Progress: A New Evolutionary Paradigm*, New York, American Institute of Physics.

[25] **C. A. S. Hall**, **C. J. Cleveland**, and **R. Kaufmann**, 1986, *Energy and Resource Quality: The Ecology of the Economic Process*, New York, John Wiley & Sons, Inc., pp. 221–228.

[26] **USGS**, 2010, *Statistics and Information*, http://minerals.usgs.gov/minerals/pubs/commodity/.

[27] **ICF International**, 2007, *Energy Trends in Selected Manufacturing Sectors: Opportunities and Challenges for Environmentally Preferable Energy Outcomes*, USEPA.

[28] **V. M. Fthenakis**, **H. C. Kim**, and **W. Wang**, 2007, *Life Cycle Inventory Analysis in the Production of Metals Used in Photovoltaics*, Energy Sciences and Technology Department, Brookhaven National Laboratory, New York.

[29] **F. Ullmann**, 1985, *Ullmann's Encyclopedia of Industrial Chemistry*, 5th completely rev. edn. (various volumes), Deerfield Beach, FL, VCH Publishers.

[30] **M. George**, Mineral Commodity Specialist, National Minerals Information Center, US Geological Survey, personal communication, 2010.

[31] **N. J. Page** and **S. C. Creasy**, 1975, "Ore grade, metal production, and energy," *J. Res. U.S. Geol. Surv.*, **3**, 9–13.

[32] **BCS Inc.**, *Mining Industry of the Future: Energy and Environmental Profile of the U.S. Mining industry*, Washington, DC, US Department of Energy, Office of Energy Efficiency and Renewable Energy.

[33] **USDOI**, 2000, *Mining and Quarrying Trends*, Washington, DC, US Geological Survey.

[34] **USDOC**, 1997, Mining Industry Series, 1992–1997, US Department of Commerce, Bureau of Census.

[35] **USDOE**, 2000, *Estimated by Energy Efficiency and Renewable Energy*, Office of Industrial Technologies.

[36] **Energetics Inc. and E3M Inc.**, 2004, *Energy Use, Loss and Opportunities Analysis: U.S. Manufacturing & Mining*, Washington, DC, US Department of Energy, Office of Energy Efficiency and Renewable Energy, Industrial Technologies Program.

[37] **H. C. Granade**, 2009, *Unlocking Energy Efficiency in the U.S. Economy*, McKinsey Global Energy and Materials, McKinsey & Company.

[38] **J. O. Marsden**, 2008, "Energy efficiency and copper hydrometallurgy," in *Hydrometallurgy 2008: Proceedings of the Sixth International Symposium*, ed. **C. A. Young**, SME.

[39] **J. F. Castle**, 1989, *Energy Consumption and Costs of Smelting*, London, Institute of Mining and Metallurgy.

[40] **DOE**, 1980, *An Assesment of Energy Requirements in Proven and New Copper Processes*, Washington, DC, US Department of Energy.

[41] **K. S. Yoshiki-Gravelsins**, **J. M. Toguri**, and **R. T. C. Choo**, 1993, "Metals production, energy, and the environment, part I: energy consumption," *JOM*, **45**(5), 15.

[42] **A. K. Gupta** and **C. A. S. Hall**, 2011, *Material Constraints and Energy Costs Associated with Rapid Upscale of PV Systems*, New York, SUNY.

[43] **J. A. Demkin** and **American Institute of Architects (AIA)**, 1996, *Environmental Resource Guide*, New York, John Wiley.

[44] **University of Waterloo**, 2010, *Candian Raw Materials Database*, http://crmd.uwaterloo.ca/index.html.

[45] **T. Kenji**, **J. E. Strongman**, and **S. Maeda**, 1986, *The World Copper Industy: Its Changing Structure and Future Prospects*, Washington, DC, The World Bank.

[46] **K. E. Porter** and **G. R. Peterson**, 1992, "The availability of primary copper in market economy countries: a minerals availability appraisal," *Information Circular*, **9310**, 1–32, U.S. Department of the Interior, Bureau of Mines.

[47] **R. B. Gordon**, **M. Bertram**, and **T. E. Graedel**, 2007, "On the sustainability of metal supplies: a response to Tilton and Lagos," *Resources Policy*, **32**, 24–28.

[48] **J. E. Tilton** and **G. Lagos**, 2007, "Assessing the long-run availability of copper," *Resources Policy*, **32**, 19–23.

[49] **USGS**, 2008, *U.S. Geological Survey Mineral Yearbook: Copper, 2008*, http://minerals.usgs.gov/minerals/pubs/commodity/copper/index.html#myb.

[50] **D. Edelstein** and **K. R. Long**, US Geological Survey, personal communication, 2008.

[51] **F. Gomez**, **J. I. Guzman**, and **J. E. Tilton**, 2007, "Copper recycling and scrap availability," *Resources Policy*, **32**, 183–190.

[52] **V. Matthews**, 2007, "China and India's ravenous appetite for natural resources – their potential impact on the United States," in *ASPO-USA 2007 Houston World Oil Conference*.

5 Economics of materials

Lester B. Lave[1] (died May 2011)
and Frank R. Field III[2]

[1]*Tepper School of Business, Carnegie Mellon University, Pittsburgh, PA, USA*
[2]*Engineering Systems Division, Massachusetts Institute of Technology, Cambridge, MA, USA*

5.1 Focus

Improvements in materials have played a primary role in the transition from a bare subsistence economy to current high living standards in the developed world. They will be even more important in meeting the challenges of increasing pollution, changing global climate, growing population, and increasing resource demands that humankind will face in the twenty-first century. However, the marvelous structures created by materials science are only scientific curiosities unless they can compete with (and, ultimately, supplant) existing materials and technologies by being cheaper and more useful. This chapter explores the close connection between materials and economics.

5.2 Synopsis

The fierce competition among materials for markets drives innovation in materials, cost reductions, and new designs both in terms of materials and in terms of the products and processes that compose US and world economic activity. The invention and discovery of new materials in the laboratory drive innovation, making possible products and processes that were only engineering dreams in the past. New products and designs require new materials, spurring innovation. Indeed, some materials have the potential to lead to more desirable products and services, decreased energy use, reduced risks to the environment and human health, and reduced consumption of scarce resources.

The thousands of materials that a designer/engineer might choose, together with the millions of applications for materials, create an embarrassment of riches. Different materials offer a vast range of properties: strength, stiffness, melting point, cost, and appearance. Finding the best material for each application is therefore an extremely challenging task, particularly because there are alternative product designs, each favoring a different material.

Further, once a material has been selected for a product, dynamic factors work to impede the substitution of one material for another. Because engineers and workers become comfortable with a particular material, they are reluctant to pay the startup costs of substituting a new material, even when it offers superior characteristics in some respects.

In particular, a material must offer a combination of performance and cost that is more valuable than that provided by incumbent materials and processes in order for it to be useful [1]. For example, high-temperature superconducting materials are a scientific marvel, but they will not be useful unless they can compete with more conventional ways of transmitting electricity, accounting for the life-cycle costs of the materials, fabrication, and operation. At the same time, there are niche applications where the advantageous properties of such materials justify the expense of using and maintaining them, e.g., in this case, some motors and transformers and medical devices such as MRI machines.

New materials are engines driving the economy, making new or cheaper products possible. At the same time, innovation, productivity and income increases, and job creation require the development of new, or cheaper, materials that can solve important economic, environmental, or social problems. In effect, it is important to recognize that, as Moretti [2] remarked, "Competition among materials is, in concrete terms, competition among social entities."

5.3 Historical perspective

Improving materials has been a major concern since early humans learned to use tools. Early humans found that, by striking stones in certain ways, they could produce axes, arrowheads, or fire. The ability to extract and fashion weapons made first of stone, then copper, bronze, iron, and steel often was the difference between winners and losers in battle.

More fundamental than the waging of war is the role of materials in growing the economy. For example, the transition from wooden to metal plows increased productivity and opened new land to farming. Likewise, among the first modern materials to arrive in remote parts of Europe in the 1960s were cherished plastic gallon milk jugs. Compared with carrying water in the traditional heavy clay pot, the plastic jug is a miracle that greatly reduced the labor involved in transporting water.

Conversely, inventors have been stymied by not having a material that can perform a particular function. For example, the village of Vinci, in Italy, has a display of small wooden versions of machines that Leonardo sketched in the late 1400s and early 1500s but could not make practical. Because the machines are small, wood has sufficient strength and stiffness to make the designs functional. However, scaled-up models would quickly go beyond the ability of wood to handle the stresses. If Leonardo had had steel or aluminum in which to fashion his machines, they could have truly transformed the world.

Before the 1970s, the choice among materials typically focused on the accumulated body of working knowledge on which materials had given satisfactory performance in that function. Both design and materials selections were largely undertaken using empirical techniques, relying on past experience and incremental changes in existing designs. When major changes in a design or material were required, extensive testing protocols centering on product and process prototyping were required – slow, tedious, and expensive efforts.

Worse, product failures attributable to materials innovations have been painful reminders to designers of the risks that accompany the deployment of material innovations. Two classic examples of this issue are the de Havilland Comet (see Case Study, Box 5.5, later in this chapter) and the Grumman Flxible 870 bus. In 1980, the Grumman Flxible 870 bus was sold to New York City's metro services as a fuel-efficient and durable bus, owing in part to its unique composite material construction, which would arguably be more corrosion-resistant and more lightweight than conventional aluminum bus construction. However, it instead had to be taken out of service when structures failed in use, leading to a black eye for the firm and a long history of litigation [3].

Even today, incremental, experience-based approaches to materials selection are common [4]. However, beginning in the 1970s, improvements in materials science and in finite-element analysis together with the growing availability of computers gave engineers an increasing ability to model, analyze, and evaluate new designs and substitute materials [5]. The ultimate goal is an integrated design–materials process based on a computer model that can evaluate the performance, cost, safety, durability, and manufacturability of alternative designs constructed with a variety of materials. Although this goal is far from realization, great progress has been made in being able to explore alternative designs and materials substitution without having to build and test prototypes. Indeed, codes now exist to examine the manufacturability of a design/material and predict its cost. For example, Ashby's work is replete with examples that show how to apply analytical methods to refine the materials-selection process [5][6][7]. These developments are helping to turn materials selection into more of a science and less of an art.

5.4 Value versus cost: competition among materials

5.4.1 Determining material value

The choice of materials can be characterized as being dependent on (a) each material's intrinsic properties, (b) its resultant consequences/properties, (c) its emergent consequences/properties, and (d) myriad "soft" characteristics [8]. The intrinsic properties of a material include strength, stiffness, and unit weight. The resultant/consequent properties depend on the design of the product, rather than only on a component. The consequent properties of using a particular material for a particular component of a product include the loadings, number of cycles, and failure modes. For example, the reinforced plastic used as a step assist in a truck is less challenged than that used as a component of the truck's fuel pump. The emergent properties are those that can be seen only by examining the resultant product in use and include the product's resultant stylishness, reliability, safety, and cost. For example, the appearance of a car is important to American buyers, making the stylishness of the plastic important in any part that is visible. Current design-support software is good at handling intrinsic materials characteristics, increasingly good at resultant properties, less so for emergent properties, and not useful after that.

The soft characteristics of a material, such as prior experience in working with it, are often of greatest importance in materials selection, largely because of the extraordinary effort required to undertake a comprehensive materials-selection decision [8]. For example, a reinforced plastic might be lighter and cheaper and satisfy stiffness, strength, and other requirements but still not be chosen because of extensive experience in working with steel. It is one thing to demonstrate feasibility in a laboratory and another thing altogether to achieve the expected performance in the production of a thousand parts each day.

Many different materials could be selected for the construction of a given product. In each case, a designer would scrutinize the qualifying materials to explore which were superior for particular designs and objectives. For example, a desk could be made out of solid wood, veneer, composition board, plastic, steel, aluminum, brass, iron, or even gold. All of these materials have intrinsic properties that make them suitable for use in a desk – a not very demanding application. At least one of the materials, gold, might be ruled out by its resultant properties in this application: a solid gold desk would be so heavy that many floors would not be able to support it. In addition, gold's weight and softness would be unsuitable for drawers. Still more of the material alternatives would be excluded on the basis of their emergent properties. The appearance of particle board would presumably rule it out on the basis of aesthetics. The cost of an aluminum, brass, or gold desk would rule these materials out for a mass market. However, several materials remain that are competitive: solid wood, veneer, plastic, and steel. Desks made of each of these materials are offered in the marketplace, and customers can choose among their attributes. No one material/design is dominant among all customers.

5.4.2 Static competition: selecting the best material for today

The economics of materials encompasses both the difficulty/cost in obtaining the materials and the utility/value in being able to use them. For example, while platinum is difficult and expensive to extract and the supply of ore is limited, when employed as a catalyst in certain operations, its use generates considerably more economic value than the cost of obtaining it. Similarly, although their supply is limited, the rare-earth elements are essential components of nickel metal-hydride batteries and neodymium–boron magnets. The point is that the cost of a material is important, but it is the ratio of performance in a given task to the cost that is the determining factor in selecting among materials. See Box 5.1.

In some applications, costs are secondary to performance, such as in the space program or in defense, although, even in those cases, there are cost constraints. For commercial products, the material with the best physical properties (i.e., strength, density, etc.) often loses to lower-priced materials. For example, the 1980s were an important era for the introduction of polymeric

Box 5.1.

A component or product can be made from a number of materials, with resulting differences in performance, cost, and other attributes. While one material, such as gold, might be uniquely superior in terms of a task (electrical contacts), it is rarely used because lower-performing alternative materials can achieve close to the desired performance at a fraction of the cost. For example, copper and aluminum compete for the markets in electrical wires and transmission lines because, even though aluminum is an inferior conductor, it is cheaper – particularly once one takes into account the cost of the transmission towers that the less dense aluminum cables require.

materials into the automobile. Major plastics firms spent millions of dollars to demonstrate that their engineering polymers could achieve the performance that automakers required as they strived to meet fuel-economy regulations while satisfying customer preferences. However, once the automakers were convinced of the feasibility of advanced engineering polymers, second-tier polymer companies were able to show that their less technically advanced resins would work equally well – and at much lower costs. Suppliers of acrylonitrile butadiene styrene (ABS), a tough, inexpensive blend of resins, were particularly successful in using this strategy. Even though ABS was an outstanding material, it did not have the engineering cachet of the newer resins. (This is surprising, in retrospect, given that every Western Electric telephone, a product renowned for its ability to endure all sorts of abuse, was made of this material.) However, once the automakers were willing to consider plastics, the cost–performance features of ABS made it a formidable option.

Similarly, in some applications, appearance is of greatest importance. Thermoplastic olefins were proposed for soft automobile bumper covers to improve aerodynamics and reduce collision costs. Although the thermoplastic olefins were molded well and satisfied almost all goals, they could not be painted to attain the desired appearance. To avoid retooling costs, expensive thermoplastic elastomers that could be satisfactorily painted were used in the short run. But, because of that bad experience, the US auto industry abandoned thermoplastic olefins in favor of thermoset urethanes for soft bumper covers for many years. Appearance was regarded as a less dominant attribute in Europe, so European automakers just went with the olefin with carbon-black filler or, later, more colorful additions.

Selection of a material for a particular application is complicated by the fact that each material is a discrete

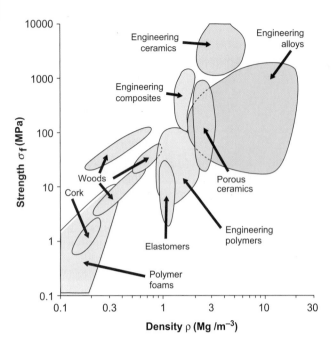

Figure 5.1. Material classes plotted as a function of their strength and density (after "Materials Selection Chart 2" in [5]).

"package" of characteristics. For example, while there are several materials that have the same yield strength as steel, the only material that has the same yield strength, Young's modulus, and fatigue performance as steel is ... steel. Materials science is less about finding materials that outperform other materials in one characteristic than about finding materials offering combinations of properties that can "fill the gaps," in terms of performance, between existing materials (e.g., provide the strength and stiffness of steel, the density of aluminum, and the corrosion resistance of titanium, all at the cost of scrap iron).

Perhaps the easiest way to see an example of this discontinuous nature of materials properties is to examine Figure 5.1, which plots the strength of material classes against their density. Note that, although all of these materials lie roughly along a line, each class of materials occupies a well-defined space. Although there is some overlap, note, for example, the fact that engineering polymers occupy a space between wood and engineering alloys (i.e., metals) that would otherwise be unavailable to designers. Similarly, note that engineering composites create a new space of comparable density but higher strength than engineering polymers.

Of course, there are other equally important design spaces (e.g., specific stiffness, fatigue performance, cost, and manufacturability), and materials classes tend to "bunch up" along different property dimensions in ways that afford designers the luxury of developing general design rules for materials by class. However, these "clumps of feasibility" also point to domains for

materials development as designers seek to push the performance envelope in their product development. It is to the filling of these gaps (represented by the empty regions in Figure 5.1) that materials science devotes much of its efforts. For example, the development of better batteries is constrained by the materials currently available. Applications as diverse as electric vehicles and efficient electrical grids are limited by the deficiencies in the combinations of strength, density, chemical stability, and electrochemical performance that current materials afford battery producers. (See Chapter 44 for more details on the challenges being addressed in current battery research.) See Box 5.2.

In selecting a design, engineers also consider the toxicity of the material, the manufacturability of the design materials, and the history of using a particular material in a given application. A material that is excellent in all other senses can be ruled out because of toxicity; for example, the incidence of lung cancer and mesothelioma due to exposure to asbestos led to the discontinuation of its use in insulation [9]. In terms of manufacturability, steel parts can be stamped in seconds, making them cheaper for large-scale production than plastic, which typically requires ten times as long to make a part. The effects of investment in current materials are illustrated well by the case of beverage containers. Changing the material in the container represents a substantial investment in revamping the production line, which beverage companies are reluctant to make. In the USA, an early commitment to aluminum cans was a great competitive advantage. However, dependence on a single class of material worried European can makers enough that they designed their plants to handle either steel or aluminum. Further, these producers have carefully managed their production so as to keep the market fairly balanced between the two materials. The cans look the same, but the chance of getting a steel can when you buy a beverage in Europe is roughly 50:50.

5.4.3 Dynamic competition: the role of predicted changes over time

As suggested by the example of beverage containers just discussed, the fact that manufacturers consider the history of use in an application demonstrates that there is an important dynamic to materials selection. In any application, there are difficulties to be overcome in design and manufacturing. Once a material has been chosen, the material's manufacturer has resources for research and development (R&D) to improve the material. The product manufacturer becomes comfortable with the material, its inevitable problems are ironed out, and it becomes ever more difficult for a new material to displace the incumbent, even when the new material has superior properties or lower cost. Once the designers and operating engineers have overcome these initial difficulties, they are reluctant to substitute a new material and go through the startup difficulties again. For example, motor-vehicle manufacturers have stayed with steel body panels, despite the attractive features of aluminum and plastic substitutes. In fact, in many of the examples where an incumbent material has been replaced, the reason was supply unavailability of the current materials, a change in design that the existing material could not satisfy, increasing cost, or discovery of the toxicity of the material.

Another dynamic factor affecting materials use is the difficulty in obtaining the material. For example, as extraction of a material increases, the richest (i.e., highest-quality) ores become depleted, and the mining industry is faced with making use of ever leaner ores. Indeed, the concentration of iron in the mines of Minnesota has declined over time as the richest ores have been used up, and the same has happened with copper. If extraction technology had not changed, these materials would be much more expensive as a result. However, technological changes in mining and milling processes have lowered the prices of many raw materials, despite the use of leaner ores.

Falling prices and greater availability have also characterized specialty materials as production technology has improved. For example, after accounting for inflation, specialty steels have declined in price, as have carbon fiber, aluminum, and other materials. The greater availability of exotic materials, from rare-earth metals to carbon fiber to lithium, has encouraged materials engineers to take advantage of their special properties in new products and processes. The falling prices and greater availability have resulted in the growth of mass-market products that were only laboratory curiosities in the past, including superconducting wires and carbon-fiber-reinforced plastics.

Further, the properties of a material change as extraneous materials are removed and as the particle size

Box 5.3.

Materials science has done an exemplary job in developing materials, such as carbon nanotubes (http://en.wikipedia.org/wiki/Carbon_nanotube) and high-temperature superconductors (http://en.wikipedia.org/wiki/Room-temperature_superconductor). However, knowing the remarkable properties of some materials is only the beginning of making them economically useful. A new material will be used only if the cost of the material is justified by its usefulness. For carbon nanotubes, the properties are so compelling that a research focus has been on producing the materials at lower cost so that they can be used in commercial products. For example, they could add strength to lightweight materials, improve electrical conductivity, and offer other benefits due to their unique characteristics.

changes. Ultrapure materials have distinct properties that make them more valuable in some applications than the same materials with some impurities. The classic examples are conductors such as copper or aluminum, whose conductivity increases with purity. Most metals become more ductile and malleable with increased purity, making them easier to form, at the expense of reduced strength. Similarly, because of its improved conductivity with increasing purity, silicon's use for electronics applications is dependent on the purity of the material employed.

Nanotechnology is a burgeoning field where the properties of materials can be changed in important ways by reducing the particle size. Depending on their configuration, carbon nanotubes can become extraordinary thermal and electrical conductors, can form ultra-low-friction surfaces, or can exhibit other properties not displayed by conventional forms of carbon. At these nanometer scales, the physical characteristics of materials can differ dramatically from those experienced at the macroscopic scale. This is partly a consequence of the change in the ratio of surface area to volume, leading to an increasing influence of surface effects (e.g., catalysis, chemical activity, thermal conductivity, etc.). Many conventional materials take on strikingly different characteristics at the nanoscale, and there has been an explosion of research into the potential applicability of these materials. See Box 5.3.

New materials properties, as well as lower costs, bring about possibilities for important new uses. Examples go back to the making of steel from iron, the formation of glass into fibers, and the explosion of aluminum applications following the development of the Hall–Héroult process. New industries have been created

Box 5.4.

Before 1886, aluminum was more valuable than gold, despite the fact that it is the third most abundant element in Earth's crust (after oxygen and silica). The Hall–Héroult process made aluminum into one of the most commonly used metals. Aluminum is the key to strong, lightweight structures, such as are needed for aircraft. Increasingly, subway cars, buses, and even freight cars have been constructed of aluminum rather than steel (see http://www.aluminum.org/Content/NavigationMenu/TheIndustry/TransportationMarket/Rail). Although the aluminum railcar is somewhat more expensive, it saves energy for subways and increases the payload for freight cars.

Box 5.5. Case study

The classic aerospace story of unanticipated materials failures is the de Havilland Comet I, an aluminum aircraft with square windows developed in Britain in the 1940s [12]. The Comet I was supposed to make de Havilland the leader in commercial jet transportation. It made use of innovative aluminum alloys and clad aluminum structures to achieve a light weight. Unfortunately, the properties of the material, combined with the design and the manufacturing processes, made the Comet susceptible to fatigue cracking, or failure at a stress level below the nominal strength of the material. Specifically, the square design of the windows served to concentrate stress, leading to fatigue cracking as a result of repeated pressurization/depressurization. The initial commercial success of the Comet was cut short by catastrophic failures in the early 1950s, including the first-ever fatalities in a passenger-jet crash and the disintegration of the Comet in midair twice in 1954 [13]. By the time de Havilland had resolved the Comet I's problems, Boeing had delivered its commercial jet, the 707, and proceeded to dominate the market. The experience with the Comet I was so dramatic that it led to development of the field of fracture mechanics, and any modern text on fracture mechanics will allude to, if not discuss in detail, the saga of the Comet.

by the "new" materials, even though the basis material, iron or glass, is the same. See Box 5.4.

5.4.4 Unforeseen design/materials failures

Whenever a material or design is implemented for the first time, there is a chance that some unforeseen aspect of the design/material will lead to unsatisfactory performance or even catastrophic failure. Changing to a new material design reopens the possibility of failure (e.g., fires caused by lithium batteries in laptop computers). These aspects also lead to a strong preference for keeping the current design/materials or making only minor changes in design/materials.

For example, in the 1960s the US navy built the superstructure of some warships out of aluminum rather than steel in order to get lighter, faster, less top-heavy vessels, since a lower center of gravity and a lower center of buoyancy add to stability. Unfortunately, a collision between the USS Belknap and the USS John F. Kennedy resulted in a superstructure fire on the Belknap whose severity was so great that the US Navy sought to avoid aluminum superstructures in war vessels when possible. Navies have thus had to balance the advantages of a lightweight warship superstructure against the increased flammability of the material employed to achieve it, as well as the corrosion cracking that can accompany the use of this material [10]. Aluminum continues to be used because its low density makes for nimbler boats with shallower drafts, while advances in electronics have reduced the need for tall boat superstructures, making them smaller targets [11]. See Box 5.5.

5.4.5 The role of supply shortages

However comfortable and committed to a material they may be, designers sometimes have no alternative to

seeking a replacement. For example, miners' strikes have led to shortages of tin for the canned-food industry and cobalt for aerospace applications. The periodic shortages of tin, when Bolivian miners' strikes interrupted the supply, forced the tin-plate industry to find ways to use less tin. Eventually, materials scientists were able to eliminate the tin in food cans [14]. Similarly, cobalt superalloys dominated aerospace turbine applications until problems in the mines in Zaire disrupted supply, starting in 1978. With the prospect of a prolonged shortage of cobalt and continuing supply problems, designers switched to nickel-based superalloys [15] [16]. When the supply difficulties were solved and cobalt became available again, the market did not return to cobalt, but continued to use nickel – an irreversible change. These examples illustrate that the trajectory of material choice typically is not an inevitable trajectory from one material to a "better" one, but is instead a path marked by episodes of upset, largely socially inspired, that can produce new trajectories.

In selecting materials, designers/materials engineers also need to be conscious that there are sometimes discontinuities as a result of supply issues. For example, extracting gold is the primary reason for much mining. The platinum and other metals that are extracted are

Box 5.6.

During World Wars I and II, the USA and its adversaries found that their usual supplies were disrupted, including petroleum (Germany) and latex (USA). The shortages precipitated a race to find substitutes for the unavailable materials. As a result, Germany perfected the Fischer–Tropsch process for gasifying coal and then producing diesel fuel from the synthesis gas. Likewise, the USA learned to make synthetic rubber from petroleum (see http://acswebcontent.acs.org/landmarks/landmarks/rbb/). Following the war, Germany returned to petroleum because it was less expensive, but latex never regained its market dominance. Forty years later, when the world embargoed South Africa because of apartheid, that nation turned to its coal reserves and improved the Ficher–Tropsch process.

byproducts of the gold extraction that are sold for whatever price they can fetch in the market. However, if the demand for platinum, rare-earth metals, or another "co-product" rose to the point where it became the primary reason for mining and the material that was originally the primary product became a co-product, then the price of that material would rise discontinuously. See Box 5.6.

5.4.6 The growing concern about toxic materials

The toxicity of lead and mercury led to stringent regulations concerning worker exposure, environmental discharge, and customer exposure [17]. Removing lead from ammunition was not particularly difficult, but developing lead-free solder has posed a serious challenge for materials scientists. The formation of intermetallics has limited joint reliability, while wettability has challenged manufacturing engineers trying to make use of the over 70 lead-free solder alloy compositions that have been developed [18]. Mercury has been all but banned in consumer products and is subject to stringent rules for worker safety and environmental protection. In fact, fluorescent lamps are perhaps the only major consumer product with mercury, which relies upon the emission of ultraviolet radiation from mercury vapor to cause the phospors on the tube surface to glow [19]. However, as the technology for light-emitting diodes advances to the point where they are more efficient and of comparable lifetime cost, regulators could ban the use of mercury in fluorescent lamps as well.

Conversely, vinyl chloride monomer is an example of the usefulness of a material preventing the banning of a highly toxic precursor [9]. Vinyl chloride monomer was found to cause a unique disease, angiosarcoma of the liver. One alternative was to ban the material, but this would mean that vinyl plastics would no longer be available. Instead, regulators in the USA, the UK, and Germany imposed a stringent standard for worker exposure. Fearing the loss of a major product, manufacturers in the three nations focused intense efforts on meeting the standard. Although many experts believed that meeting the standard was impossible, manufacturers were able to meet the standard within two years at an acceptable cost.

5.4.7 Recycling

The USA and Western societies more generally have been characterized as "throw-away" societies that pollute the environment with discarded items. Thus, concerns about environmental quality and depleting the supply of nonrenewable resources have led to social policies to promote or require recycling. Unfortunately, reclaiming materials from used products is more contentious than might seem.

Closed-loop recycling (Chapter 40), where a material is returned to its original state, is often held up as the most desirable goal [16]. Where the material is extremely valuable or easy to reclaim, closed-loop recycling can be the best goal. However, it is important to recognize that closed-loop recycling uses energy, consumes other materials, and results in some material loss. One way of thinking of this is to estimate the amount of energy, materials, environmental pollution, and labor associated with extracting a certain amount of a material. If the same amount of material can be recovered for less than that amount of energy, materials, environmental pollution, and labor, then recycling is worthwhile.

For many materials, closed-loop recycling is not a reasonable option. For example, the rare-earth metals in a nickel metal-hydride battery are not currently recovered; instead, they become innocuous contaminants in recycled nickel going into stainless steel. When various aluminum alloys are mixed together, the quality of the result is downgraded. However, expensive testing would be required to determine the specific alloy and to separate and store it so that it was not mixed with other alloys. The same is true of plastics, where mixing different resins lowers quality. Closed-loop recycling in this case would again require extensive testing and storage.

Likewise, for products that are not designed to be recycled, recycling can be difficult and expensive and might not even contribute to sustainability. For example, seeking to recycle conventional nylon carpet might not make sense, because carpet is a combination of different materials that need to be identified and separated before the carpet can be recycled [20]. In order for recycling to make sense, the product typically needs to be designed and manufactured with that specific goal in mind, without detracting from other qualities preferred by consumers. Even in that case, whether recycling is an appropriate goal depends on a life-cycle analysis (Chapter 41) of the alternatives.

In some cases, reclamation of a material from a product at the end of its life might be required because of toxicity (e.g., mercury, lead, radioactive material). For example, the toxicity of lead convinced law makers to impose a deposit-refund of $5 per automobile battery. Even this relatively low price leads auto recyclers to take out the battery before the vehicle is turned into scrap.

Social policies have also been used in other cases to encourage recycling (i.e., tilt the life-cycle analysis in favor of recycling). Placing a deposit on a beverage container increases the value of returning that container by a large amount. For example, an aluminum can is worth less than 1 cent for recycling, but a deposit of 5 or 10 cents significantly increases the incentive to return the can rather than discard it into the environment. Placing a deposit on the can serves an environmental purpose, rather than strictly serving the goals of lowering energy use and using fewer resources.

As another example, European automobile manufacturers have entered into an agreement to recycle up to 95% of the weight of a vehicle at the end of its life by 2015 [21]. The lead–acid battery is easily recycled. The ferrous metals are easy to separate and recycle. Although the nonferrous metals are more difficult to separate, they can then be recycled. However, plastics and composites are difficult to recycle in a meaningful way. Thus, the voluntary agreement limits the use of plastics and composites in new vehicles in Europe. This somewhat arbitrary restriction can lead to greater fuel use, less safety, and higher cost. On a life-cycle basis, it is far from evident that excluding the use of these materials brings society closer to the goal of having a product that pleases consumers, is safe, uses less energy and raw materials, and results in fewer environmental discharges. See, for example, [22].

Thus, the trade-off between closed-loop recycling and use of new materials or alternative materials better suited to the task should be evaluated carefully. Recycling in itself is not the social goal. Rather the goal is to lower the use of both depletable and non-depletable resources as well as energy, to improve environmental quality, and to provide consumers with the products and services they desire.

5.4.8 Limits of materials availability

As noted above, technology improvements have more than kept pace with degradation in ore quality and difficulty of extraction. Whether that can continue is more of a philosophical question. In a fundamental sense, going

to leaner ores and more difficult extraction requires more energy. According to Goeller and Weinberg [23], there is a virtually unlimited quantity of mineral resources in seawater, albeit at low concentrations. They argue (echoing the classic argument of nineteenth-century economist David Ricardo) that raw materials will not run out, but that, at some point, it will no longer be worthwhile to extract the materials from lean ores or seawater. They regard energy as the limiting resource. Their energy "theory of value" equates, in the limit, the standard of living to the availability and cost of energy.

In the Goeller–Weinberg view, closed-loop recycling is good insofar as it saves energy. Recycling that does not save energy is not useful. If products are designed to be recycled and the resulting recycling saves energy on a life-cycle basis, that is good. However, this framework neglects the fact that the economy is not designed to save energy or minimize entropy. Rather, it is designed to satisfy the desires of customers. No matter how energy-efficient a product is, if it fails to please consumers, it will not be purchased.

How can the Goeller–Weinberg view be reconciled with a free-market economy? Imposing regulations that require products that impose less of a burden on the environment and use less energy might be good for sustainability, but might not be good for the economy. In a free-market economy, the consumer is the focus. However, for consumers to make informed choices, they have to know the environmental-sustainability burden of each product. Focusing solely on designing and making products that have low environmental-sustainability burdens is unlikely to be successful, since there are many ways to lower the environmental-sustainability burden of a product but customers might not find the resulting product desirable. Rather, the goal of environmental sustainability must be considered within the context of the entire package of properties affecting the design and materials selection for a product.

The best answer is to give customers free choice among products, but have the price of each product reflect the externality costs, including the full energy use, depletion of natural resources, and environmental discharges [24]. Using such a price system would similarly send signals to materials engineers. If the price of each material and of each process reflected the full energy and resources used and the environmental consequences, product designers and materials engineers would be led to select materials and designs that were cheaper and more friendly to the environment and sustainability.

5.5 Summary

A component or product can be made from a number of materials, with resulting differences in performance, cost, and other attributes. Selecting a material involves balancing the usefulness, longevity, safety, cost, and style of a product, as well as its life-cycle energy and materials use, environmental discharges, recyclability, and depletion of scarce resources. In choosing a material for a particular application, there are always unexpected outcomes, some beneficial and some detrimental. As a consequence, most materials/design choices are incremental, tiny modifications of a design known to work.

The selection of design/materials is becoming more science and less art as computer modeling is able to encompass more of the dimensions of the selection process. These improvements have important implications for the quality and price of products because non-incremental changes in designs and materials aided by computer analysis will lead more quickly to better, cheaper, safer products.

5.6 Questions for discussion

1. When a material is introduced into a new application, the manufacturer often limits production for a year or so. Can you explain why?
2. In considering a material for a particular application, which is more important, the cost of the material or its particular properties for this application?
3. What are the comparative properties of a material that are sufficient to get it to replace an existing material in a product?
4. Why does a material that is replaced because of supply shortages not necessarily come back to dominate that application when the supply shortage passes?
5. Under what conditions is the search for a new material dominated by the need to satisfy a particular application? When is new material development dominated by materials scientists exploring the development of new molecules?

5.7 Further reading

- **M. F. Ashby**, 2005, *Materials Selection in Mechanical Design*, 3rd edn., New York, Butterworth-Heinemann.
- **F. Field**, **R. Kirchain**, and **R. Roth**, 2007, "Process cost modeling: strategic engineering and economic evaluation of materials technologies," *J. Minerals, Metals Mater. Soc.*, **49**, 1543–1851.
- **M. Kutz**, 2002, *Handbook of Materials Selection*, New York, Wiley.
- **E. Moretti**, 1986, *The Material of Invention*, Cambridge, MA, MIT Press.

5.8 References

[1] **F. R. Field**, **J. P. Clark**, and **M. F. Ashby**, 2001, "Market drivers for materials and process development in the 21st century," *MRS Bull.*, **26**, 716–725.

[2] **E. Moretti**, 1986, *The Material of Invention*, Cambridge, MA, MIT Press.

[3] 748 F2d 729, 1984, Grumman versus Rohr.

[4] **M. Kutz**, 2002, *Handbook of Materials Selection*, New York, Wiley.

[5] **M. F. Ashby**, 2005, *Materials Selection in Mechanical Design*, 3rd edn., New York, Butterworth-Heinemann.

[6] **M. Ashby**, 2000, "Multi-objective optimization in materials selection and design," *Acta Mater.*, **48**, 359–369.

[7] *The Ashby Methods – The Power behind CES Selector*, http://www.grantadesign.com/products/ces/ashby.htm.

[8] **F. Field**, **R. Kirchain**, and **R. Roth**, 2007, "Process cost modeling: strategic engineering and economic evaluation of materials technologies," *J. Minerals, Metals Mater. Soc.*, **59**, 1543–1851.

[9] **D. D. Doniger**, 1979, *The Law and Policy of Toxic Substances Control: A Case Study of Vinyl Chloride*, Baltimore, MA, Johns Hopkins University Press.

[10] **AP Newswire**, 1987, "Navy Reverting to Steel in Shipbuilding after Cracks in Aluminum," *The New York Times*, August 10.

[11] **G. R. Lamb** (ed.), 2003, *High-Speed, Small Naval Vessel Technology Development Plan*, Bethesda, MD, Naval Surface Warfare Center.

[12] **R. J. H. Wanhill**, 2002, "Milestone case histories in aircraft structural integrity," in *Comprehensive Structural Integrity*, vol. 1, 61–72, doi:10.1016/B0-08-043749-4/01002-8.

[13] **P. A. Withey**, 2007, "Fatigue failure of the de Havilland Comet I," *Eng. Failure Analysis*, **4**, 147–154.

[14] **J. E. Tilton**, 1983, *Material Substitution*, Sterling, VA, RFF Press.

[15] **E. Alonso**, 2010, *Materials Scarcity from the Perspective of Manufacturing Firms: Case Studies of Cobalt and Platinum*, Cambridge, MA, Department of Materials Science and Engineering, MIT.

[16] US Government, Office of Technology Assessment, 1985, *"Strategic Materials: Technologies to Reduce US Import Vulnerability,"* http://www.fas.org/ota/reports/8525.pdf.

[17] **T. Gayer** and **R. Hahn**, 2005, "The political economy of mercury regulation," *Regulation*, **28**(2), 26–33.

[18] **M. Abtew** and **G. Selvadury**, 2000, "Lead-free solders in microelectronics," *Mater. Sci. Eng.*, **27**, 95–141, doi: 10.1016/50927-796X(00)00010-3.

[19] **T. G. Goonan**, 2006, *Mercury Flow Through the Mercury-Containing Lamp Sector of the Economy of the United States*, Reston, VA, United States Geological Survey.

[20] **L. Lave**, **N. Conway-Schempf**, **J. Harvey** et al., 1998, "Recycling postconsumer nylon carpet: a case study of the economics and engineering issues associated with recycling postconsumer goods," *J. Industrial Ecol.*, **2**: 117–126.

[21] **N. Kanari**, **J. L. Pineau**, and **S. Shallari**, 2003, "End-of-life vehicle recycling in the European Union," *JOM*, **55**, 15–19.

[22] **J. Gerrard** and **M. Kandlikar**, 2007, "Is European end-of-life vehicle legislation living up to expectations? Assessing the impact of the ELV Directive on 'green' innovation and vehicle recovery," *J. Cleaner Production*, **15**, 17–27.

[23] **H. E. Goeller** and **A. M. Weinberg**, 1978, "The age of substitutability," *American Econ. Rev.*, **68**, 1–11.

[24] National Research Council, Committee on Health, Environmental, and Other External Costs and Benefits of Energy Production and Consumption, Hidden Costs of Energy, 2010, *Unpriced Consequences of Energy Production and Use*, Washington, DC, National Academy Press.

6 Global energy flows

Richard Sassoon

Global Climate & Energy Project, Stanford University, Stanford, CA, USA

6.1 Focus

As we search for solutions to providing energy in a sustainable manner to a growing world population that demands a higher quality of life, it is instructive to understand the distribution of energy available on our planet and how it is used. This chapter provides a description of the energy resources on Earth and the transformations that they undergo in both natural and human-driven processes.

6.2 Synopsis

Earth is continuously exposed to large quantities of energy, primarily in the form of solar radiation, a small fraction of which is stored through a series of biochemical and chemical transformations, whereas most of it is dissipated through natural processes. Additionally, huge reservoirs of energy exist on the planet from the time of its creation several billion years ago, mainly as radionuclides and thermal energy embedded in the Earth's crust and interior.

This chapter explores the flows and stores of energy that occur on the planet, describes the current exploitation of these resources by humans, and discusses the impacts of human intervention in these energy flows on the natural carbon cycle. It will introduce the concept of *exergy*, that is the fraction of energy that is available to do work, as a means for describing and quantifying the energy resources and transformations that take place. The concepts and data discussed in this chapter should provide a broad context for the many energy technologies and systems described in this book.

6.3 Historical perspective

The energy balance of the Earth is driven primarily by a steady flux of solar radiation striking its exterior, release of thermal energy from radioactive decay and geological events within the Earth, and reflection and re-radiation of heat back into space. These fluxes trigger many other transformations involving energy changes as physical and chemical changes take place and life forms pass through their natural life cycles and interact with the natural environment in the biosphere.

Beginning with the Industrial Revolution in the early 1800s, humans have been intercepting these energy flows and exploiting the naturally available stores of energy to develop a complex anthropogenic flow of energy. This flow continuously changes as the demand for energy grows and new technologies for supplying this demand are introduced. Examples of technology changes that have had an impact on the human energy system include the development of the steam engine, the transition from wood to coal, the use of oil and natural gas as primary fuels, the wide-scale use of nuclear power for the generation of electric power, and the predominant use of gasoline for transportation. As we look to a future in which energy demand is expected to continue to significantly increase while at the same time more energy-efficient practices and technologies are introduced, the global human energy system will, of course, change. New energy technologies involving, for example, an increasing use of renewable energy resources and a potential transition to an electric personal transportation system are expected to have a significant impact on global use of energy in the coming decades.

The exploitation of energy resources by humans during the past two centuries has led to significant environmental impacts. For example, the average temperature on Earth has risen by around 0.74 °C during the past century [1], due to the accumulation of greenhouse gases in Earth's atmosphere resulting primarily from the burning of carbon-based fossil fuels. The changes to Earth's climate as a result of this energy use are described in Chapter 1. The challenge for the future is to design global energy systems that meet the energy needs of the human race in a sustainable fashion such that the ecosystems of the planet remain in balance. This challenge is discussed in Chapter 3. The data in this chapter are intended to provide a quantitative snapshot of today's global energy system.

6.4 Global energy flows

The largest energy flow on the planet is the 162,000 TW of solar radiation that strikes the Earth. (One terawatt equals one trillion joules per second, i.e., 1 TW = 10^{12}

J s^{-1}.) This compares with a rate of exploitation of around 18 TW by all of humankind to meet its energy needs. Therefore, in terms of the solar radiation flux alone, there is clearly no shortage of energy on the planet to supply the needs of a global human population equal to about 6.5 billion people today and expected to grow to close to 10 billion by the middle of this century. The questions this chapter attempts to address are essentially what happens to all of the energy that is received by the planet, and how can these energy resources be exploited? Having a good understanding of the answers to these questions can then allow us to consider how the available energy resources can more effectively be utilized in a manner that is both more efficient and more sustainable.

Figure 6.1 provides a simple schematic representation of the key components of the energy flow in natural systems and the primary intersection of these flows with the anthropogenic energy flow, corresponding to how humans intersect with Earth's natural systems. The Sun, Moon, and geothermal heat from the Earth provide constant flows of energy into the natural energy system of the planet. This energy undergoes a series of transformations leading to, for example, photosynthesis to grow plants, movement of air to create winds, and movement of water to create waves and tides. Some of the energy is eventually stored in the form of plants and fossil fuels. It is through these stores that humans primarily exploit the natural energy system by mining these energy reservoirs. The energy then undergoes another set of transformations in the human energy system as it is converted to forms required for final end use. Not shown in Figure 6.1 is what happens to all of the excess and wasted energy. The **first law of thermodynamics** (Appendix A) states that energy is neither created nor destroyed. Waste heat from all the processes in the energy systems of the Earth is dissipated out of the atmosphere at the same rates as that at which energy enters the systems to achieve an energy balance on the planet.

To quantitatively analyze the energy flows on our planet, it is useful to consider the concept of **exergy** [2]. Exergy is the portion of energy that allows useful thermodynamic work to be done. As energy flows through natural and anthropogenic processes, it is transformed from one form to another. Whereas energy is not created or destroyed in a transformation, the exergy content of the energy decreases as it is converted from one form to another because of the inherent inefficiencies of the process. Figure 6.2 provides an illustration of this concept, in which fuel is burned in the engine of a vehicle to move it. The fuel added has a high exergy content. The energy in the chemical bonds of the fuel is transformed through the combustion process in the engine into propulsive energy, which turns the wheels and moves the vehicle forward, and also into waste heat, which is mostly carried away by the

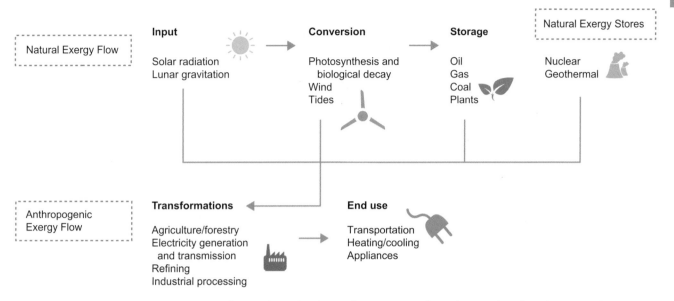

Figure 6.1. A schematic representation of natural and anthropogenic energy flows on Earth.

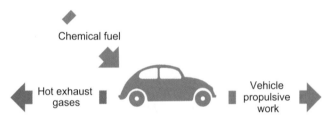

Figure 6.2. An example of exergy destruction. The energy of the fuel is conserved as it is burned to propel the car. However, because some of the energy is transformed into waste heat, much of the exergy content of the fuel is destroyed.

gaseous combustion products leaving the vehicle. Overall, energy is conserved, but much of the exergy content of the fuel is destroyed, insofar as the waste heat is generally dissipated into the environment and cannot be used again.

6.4.1 Exergy and energy relationships

Exergy (X) is calculated from numerical properties of a substance relative to the properties of a reference environment and is related to the energy content (E) of a substance. For terrestrial energy conversion, this reference state is chosen to be the environment on the surface of the Earth, that is to say the current state of the Earth's atmosphere, oceans, and crust. It should be noted, of course, that human activities are continuously affecting the state of this equilibrium. A summary of this relationship for various forms of exergy is provided below.

Gravitational and kinetic exergy
The energy associated with a moving object or the potential energy of an object above or below the reference height in a gravitational field is fully convertible to useful work and therefore numerically identical to its exergy:

$$X_{\text{Potential}} = E_{\text{Potential}} = mgh \qquad (6.1)$$

where m is the mass of the object, g is the acceleration due to gravity, and h is its height. Precipitation due to rainfall and snowfall, for example, includes gravitational exergy as a key component, together with a diffusive exergy component due to the difference in chemical potential which applies when freshwater mixes with seawater.

Similarly kinetic exergy is the same as kinetic energy:

$$X_{\text{Kinetic}} = E_{\text{Kinetic}} = \frac{1}{2}mv^2 \qquad (6.2)$$

where v is the velocity of the object. This equality is applied in calculations of exergy values for wind, waves, and transportation, and in all other energy systems involving motion.

Thermal and radiation exergy
The exergy content of thermal and radiation flows can be very different from their energy contents. While thermal and radiation flows are common in the environment, they are useful for performing work only when their temperature is much warmer or cooler than ambient. At the temperature of the environment, the quality of energy flow is zero and its exergy content is zero. Non-radiative thermal energy flows can be heat flowing through a substance, heat flowing from one substance to another, or thermal energy associated with a stream of matter. The ratio of exergy to energy associated with heat flow is shown in Figure 6.3(a) at the ambient temperatures on the planet.

Table 6.1. Adding up the chemical exergy of various substances

Chemical name	Bond potential energy	+	Degree of freedom potential	+	Diffusive potential energy	=	Chemical exergy (MJ kg^{-1})	Energy quality[a]
Methane	50.0		−0.1		2.0		51.9	1.037
Octane	44.7		1.3		1.5		47.5	1.063
Ethanol	27.7		0.8		1.1		29.6	1.067
Acetylene	48.3		−1.2		1.6		48.7	1.008
Hydrogen	120.0		−8.4		5.6		117.2	0.977
Aluminum	31.0		−0.8		2.7		32.9	1.061
Fresh water	0		0		0.0049		0.0049	NA

[a] The energy quality is defined as the ratio of exergy to energy. NA, not applicable.

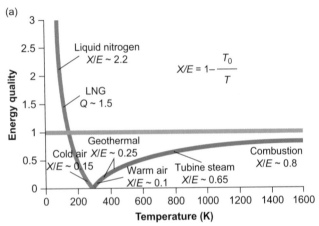

(a)

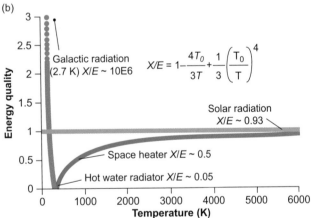

(b)

Figure 6.3. The exergy content, of (a) thermal and (b) radiation energy transfers. (See Table 6.1, note a.)

The quality of radiation energy transfer also depends on its effective temperature and the difference from the environmental temperature, and the relation is illustrated in Figure 6.3(b), together with some examples [3]. Sunlight has a temperature of about 5,800 K and consists of 93% exergy when direct, although scattered sunlight has a lower effective temperature and lower energy content.

Chemical exergy

Chemical exergy is a combination of the following three components: useful energy stored in chemical bonds; changes in the number of ways in which energy can be stored in the compounds; and useful energy extractable by diffusing relatively pure substances such as chemical reaction products to concentrations found in the environment:

$$X_{ch_i} = -\Delta_R h + T_0 \Delta_R s + R T_0 \sum_i \ln\left(\frac{y_i}{y_0}\right) \quad (6.3)$$

where $\Delta_R h$ and $\Delta_R s$ are the differences in enthalpy and entropy, respectively, between the products and reactants, T_0 is the reference temperature, R is the universal gas constant, y_i is the concentration of a substance, i, and y_0 is its environmental concentration. The sum of the bond potential energy, often referred to as the heating value, in the first term plus the useful work change associated with the degree of freedom potential in the second term is often called the Gibbs free energy. The components of the chemical exergy of several common compounds are listed in Table 6.1.

The chemical exergies of many substances we consider fuels are dominated by the bond potential energy, or heating value. Calculating the chemical exergy of a given substance is straightforward if the properties of the reactants and products of a chemical reaction with a reference-state substance, such as oxygen, are known. However, many exergy-carrying substances found in nature are non-uniform and complex. Petroleum, coal, and biomass contain a wide range of elements, compounds, and structures for which the terms of Equation (6.3) are not defined. In such cases formulas that permit calculation of an approximate chemical exergy of a fuel solely on the basis of the chemical exergies of its constituent elements have been developed [4][5].

Nuclear exergy

Like chemical fuels, nuclear fuels also have energy stored in bonds, but these bonds are between protons and neutrons in the nucleus. As nuclear fuels undergo both fission, in which nuclei break into lighter nuclei, and fusion, in which nuclei combine to form heavier nuclei, reactions the potential energy stored in these bonds is released and the mass of the substances involved is reduced according to Einstein's mass–energy equivalence. Using this relationship and assuming that the energy released has a quality equal to unity, nuclear exergy can be calculated from the difference in mass before and after the reaction. Nuclear reactions also experience a change in number of degrees of freedom and have useful work associated with the concentration of reaction products relative to the environment. However, these quantities are generally negligible for the reactions in our energy systems, so the energy and exergy values for nuclear fuel are essentially the same and governed by

$$X_{\text{nuclear}} = \frac{(\sum m_{i,\text{p}} - \sum m_{i,\text{r}})c^2}{\sum m_{i,\text{r}}} \tag{6.4}$$

where c represents the speed of light in a vacuum. Also unaccounted for is the energy carried away by neutrinos since the interaction between neutrinos and our environment is very mild.

6.4.2 Exergy resources and human use

A detailed analysis of the global energy system has been conducted on the basis of data collected from numerous literature sources on global energy resources and use. The exergy contents of the principal components of the global energy system were calculated using the principles described above adapted to each energy resource and transformation. These data have been collected and mapped as a set of exergy flows [6]. These maps are presented in Figures 6.4 and 6.5.

Figures 6.4 and Figure 6.5(a) trace the flow of exergy through the biosphere and the human energy system together with its associated accumulations and eventual destruction. Figure 6.5(b) tracks the carbon cycle associated with these flows as carbon changes its chemical form. Flows of exergy are given in units of TW (10^{12} watts or 10^{12} J s^{-1}) and flows of carbon are given in Mg carbon per second (*Carbon flow = 1 MgC s^{-1} = 31.54 MtC per year = 116 MtCO$_2$ per year*). The different forms of exergy and carbon are represented by different colors as explained in the keys to the figures. The data are presented in the form of Sankey diagrams in which the widths of the lines correspond to size of the flux. Also shown in the figures are stores of exergy and carbon that occur within the natural and

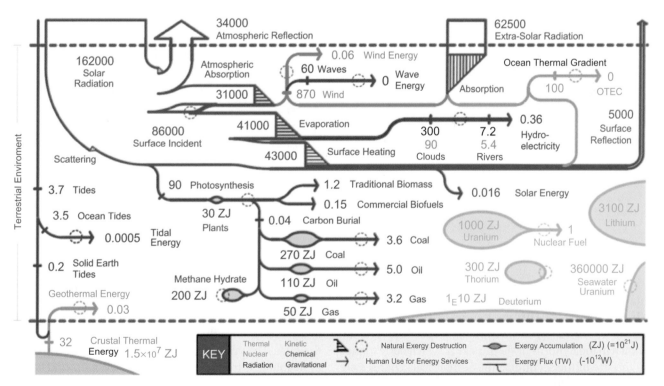

Figure 6.4. Exergy resources of the planet (from [6]). Note: the use of ZJ is for the total accumulation of that resource.

(a) Global Exergy Accumulation, Flow, and Destruction

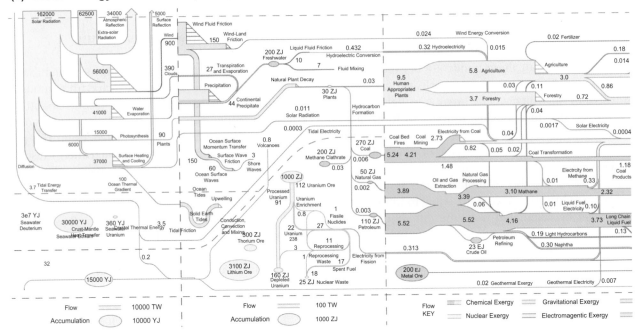

(b) The Natural and Anthropogenic Carbon Cycle

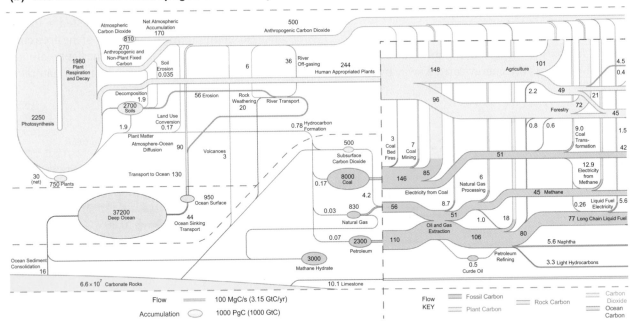

Figure 6.5. Exergy and carbon maps of the global natural and human energy systems (from [6]).

human energy systems. These are measured in units of YJ (10^{24} joules), ZJ (10^{21} joules), or EJ (10^{18} joules) and PgC (1 GtC or 10^{15} grams of carbon), respectively, and are shown in the figure as ovals, the size of which corresponds to the size of the accumulation. Several scale changes are included in Figure 6.5 and these occur at the dashed lines.

Figure 6.4 represents the exergy resources on the planet. These come in two forms, renewable exergy flows

such as solar that will be sustained for the foreseeable future, and depletable exergy resources such as fossil fuels and radionuclides that will be depleted in the course of their use. The primary exergy flux arriving at the top of the atmosphere is from the Sun (162,000 TW). Around half of this flux is dissipated through atmospheric reflection and absorption. Some of this absorbed exergy is converted into wind exergy (around 870 TW) and wave exergy

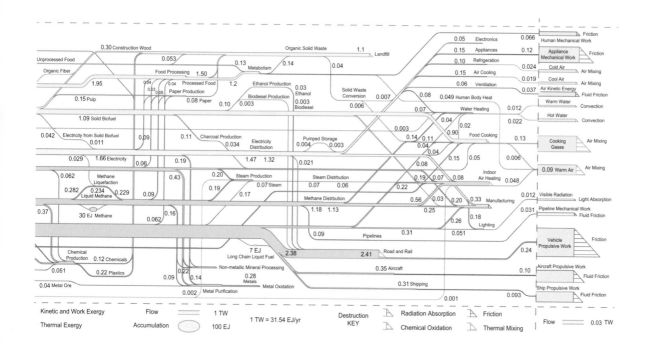

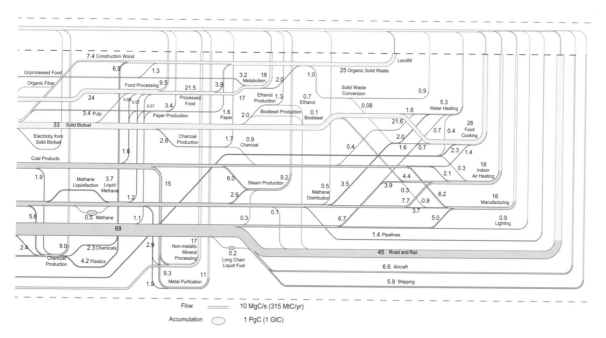

(around 60 TW). The remaining half of the primary exergy flux (around 86,000 TW) reaches the surface of the Earth and most of it is used either to evaporate water, creating the potential exergy that forms the hydropower resource in rivers, or to warm the surface, creating the exergy available in the thermal gradients of the oceans. Only a tiny fraction of this flux (90 TW) is converted by photosynthesis into plant material. Some of this (about 1.2 TW) is used directly for cooking and heating, and an even smaller fraction (around 0.04 TW) leads to the accumulation over millions of years of the fossil-fuel resources (coal, natural gas, and oil) that we use extensively at present.

There is also an enormous nuclear resource, though much of it is widely dispersed in the form of uranium or deuterium in seawater, for example. There is also a very large geothermal resource (1.5×10^7 ZJ) stored in hot rock in the upper 40 km of the Earth's crust. The flux of geothermal exergy from this resource, which can be observed in forms such as earthquakes, volcanoes, and hot springs, is about 32 TW, roughly equivalent to projected human use of energy by the middle of this century.

Figure 6.6 provides an alternative graphical description of the relative magnitudes of the exergy resources on the planet. Among the renewable exergy flows, solar

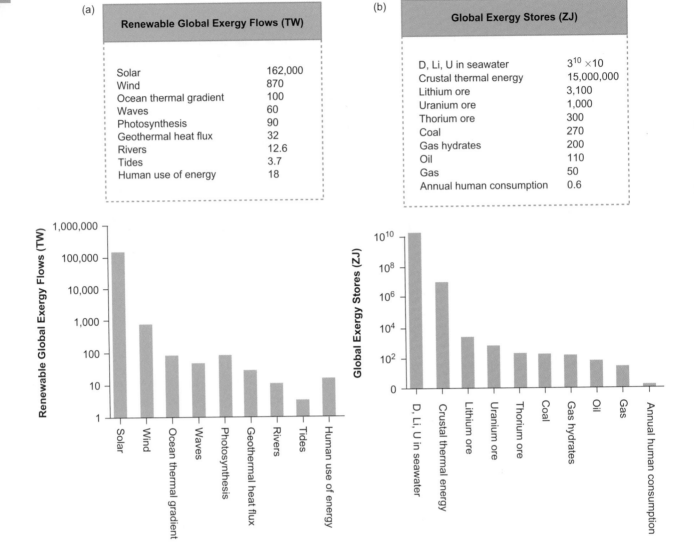

Figure 6.6. Major exergy flows and stores on Earth: (a) renewable exergy flows and (b) depletable exergy stores.

is by far the largest resource, with an exergy flow at Earth's surface equal to around 5,000 times current human energy use of about 18 TW, as can be seen in Figure 6.6(a). Other renewable exergy resources, including wind, waves, tides, and hydropower, all represent much smaller exergy flows. Direct utilization of these resources is still very low since many issues, such as geographical distribution, intermittency, and economics all need to be addressed. The flow of exergy in photosynthesis (Chapter 29) creates biomass, which has the advantage of being able to store exergy. However, again its use as an exergy resource has been limited, since most human exploitation of plant materials has been associated with agriculture and forestry rather than energy.

A graphical description of stored exergy resources that cannot be replenished is provided in Figure 6.6(b). These stores, which include stored geothermal energy and various forms of nuclear and fossil fuels, all exceed

the present annual human exergy consumption of around 0.57 ZJ by several orders of magnitude, in some cases by many orders of magnitude, although it should be noted that only a fraction of these exergy reserves may be easily recoverable with today's technologies. A discussion of the sustainability issues associated with exploiting these resources is provided in Chapter 3.

The flow maps in Figure 6.5 show the global totals for various carriers of exergy and carbon as they undergo a set of transformations in which exergy and carbon are converted from one carrier to another. The top left-hand corner of Figure 6.5(a) reproduces the exergy resources shown in Figure 6.4. Exploitation of these exergy resources takes place primarily with oil, coal, and gas, and occurs at rates of 5.5, 5.2, and 3.9 TW, respectively. Associated with this use is transfer of carbon from geological reservoirs to the active atmospheric/biological carbon cycle, which occurs at rates of 110, 146, and

Box 6.1 Energy analysis example – electricity generation from coal

The process of electricity generation from coal represents one of the largest destructions of exergy in the human energy system and is accompanied by the emission of large quantities of greenhouse gases. Table 6.2 shows the global values for exergy and carbon fluxes for this transformation derived from the sources listed in [6].

Table 6.2. Global exergy and carbon flux in the generation of electricity from coal

	Energy flux (TW)	Exergy flux (TW)	Carbon flux (MgC s^{-1})
Input			
Mined coal	2.60	2.73	85
Output			
Electricity	0.82	0.82	–
Atmospheric CO_2	–	–	85

Coal exists in a wide variety of different forms with varying amounts of carbon content. The exergy content of coal is calculated as chemical exergy as described in Section 6.4.1 and it increases within the general range of 20–30 MJ kg^{-1} with increasing carbon content. The energy quality, or ratio of exergy to energy, for coal is approximately 1.05. When coal is burned to produce electricity, the carbon content of coal is transformed primarily into CO_2. The global exergy production of electricity from coal is 0.82 TW compared with a global exergy input for coal of 2.60 TW, representing an overall process exergy efficiency of 31.5%.

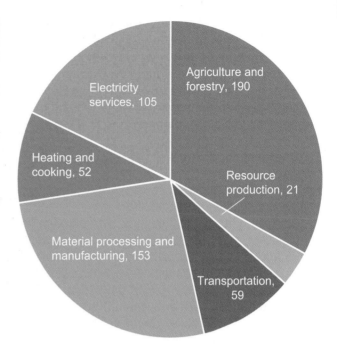

Figure 6.7. Global carbon flows from service sectors of the human energy system (MgC s^{-1}) (from [7]).

56 MgC s^{-1} for these three fossil fuels, respectively, as may be determined from Figure 6.5(b). See Box 6.1.

Analysis of human exploitation of our exergy resources reveals that agriculture and forestry represent the industries that correspond to the largest destruction of exergy. Maximizing exergy efficiency is generally not the primary goal of these transformations, and much of the exergy value of the plant materials is not maintained in the finished products of food and wood products. Transportation, generation of electricity from coal, cooking, air heating, manufacturing, petroleum refining, and landfills represent the next largest exergy consumptions. The efficiencies of these exergy-destruction

processes in the human energy system can, however, vary greatly, and an analysis of the exergy destruction in these processes can provide useful data for identifying opportunities for improving overall exergy efficiencies of these processes. [7]

Figure 6.5(b) traces the natural and anthropogenic carbon cycle in the biosphere and the human energy system and allows us to identify the energy conversions that are sources of anthropogenic CO_2. The largest carbon flow on the planet is the fixation and release of carbon in plant matter through photosynthesis and natural plant decay which occurs in the natural carbon cycle. The next largest flow is the approximately 580 MgC s^{-1} of carbon dioxide emitted from human systems. Figure 6.7 shows how each of the major sectors of the human energy system contributes to that flow. While much of the agriculture and forestry emissions comes from the decay of plants, the remaining emissions are derived almost exclusively from geological carbon. One of the largest challenges we face for the human energy system is to limit the emissions of CO_2 while still providing the energy services required by humankind.

6.5 Summary

In this chapter, we have described the flow of energy on the planet both in the natural and in human energy systems. We have demonstrated how we can apply the concept of exergy at the global level to obtain a quantitative picture of the reservoirs of energy available and the energy flows in these systems. From this analysis, it is possible to identify the efficiencies of energy conversions and the levels of CO_2 emissions that result from these transformations. It is clear that there is sufficient exergy to easily meet human needs into the future. The challenge, however, is to transform the energy resources on the planet into the required energy forms through pathways that are both sustainable and cost-effective.

6.6 Questions for discussion

1. What is the difference between energy and exergy? Discuss this difference with examples.

2. Oil, coal, natural gas, biomass, and uranium ore represent some of the largest stores of exergy resources on the planet that are currently exploited by humans. Using the data provided in Figures 6.4 and 6.5, determine the sizes of these reservoirs and the rate at which we are using up these resources. Assuming that all these resources are recoverable, calculate how many years of each resource remain. Much of these resources is not recoverable with today's technologies. Discuss the reasons why we may be able to recover only a fraction of these resources.

3. Solar radiation, lunar gravitation, and geothermal exergy represent significant exergy flows reaching the surface of our planet, yet only a tiny fraction of these resources is currently exploited by humans either directly (e.g., via solar photovoltaics) or indirectly (e.g., via wind or tidal power). Using the data provided in Figures 6.4 and 6.5, calculate the fractions of these exergy flows that are exploited by humans. Assuming a total global rate of energy use in 2050 of 30 TW, calculate the annual rate of increase of utilization of these resources that is necessary such that in total they will meet 50% of overall world energy needs by then.

4. The overall exergy process efficiency for road and rail transportation in which the exergy of liquid hydrocarbon fuels is converted to propulsive work is around 10%. What are the factors that contribute to such a low efficiency? Estimate what this efficiency would be if all road and rail transportation were electric.

5. The emissions associated with global road and rail transportation are around $45\,\mathrm{MgC\,s^{-1}}$. Yet additional emissions accompany the production of the fuels necessary to operate the vehicles. Using the data in Figure 6.5, calculate the total emissions for both fuel production and transportation in the road and rail sector. If all road and rail transportation were electric, calculate the total emissions for both the electricity production and vehicle transportation in this sector, assuming that the electricity is generated (a) using the current mix of energy resources and (b) using a mix that includes 50% solar and wind.

6.7 Further reading

- **V. Smil**, 2008, *Energy in Nature and Society: General Energetics of Complex Systems*, Cambridge, MA, MIT Press. This book provides a broad, yet comprehensive, description of global energy flows.
- **W. A. Hermann**, 2006, "Quantifying global exergy resources," *Energy*, **31**, 1685–1702. This article describes the calculations used in deriving the exergy values of the resources on the planet.
- **G. V. Kaiper**, 2002, *U.S. Energy Flow Trends – 2002*, Lawrence Livermore Technical Report UCRL-TR-129990–02. This report provides models of US and world energy flow together with their associated carbon emissions.
- Global Climate and Energy Project, Stanford University, exergy flow charts, http://www.gcep.stanford.edu/research/exergycharts.html. This website is a valuable resource for locating the source data behind the global exergy and carbon flow analysis provided here.
- **D. J. C. MacKay**, 2008, *Sustainable Energy – Without the Hot Air*, Cambridge, UIT. A refreshingly clear and readable book that provides a straightforward discussion on current and possible future energy use at the national and global level.

6.8 References

[1] **R. K. Pachauri** and **A. Reisinger** (eds.), 2007, *Fourth Assessment Report (AR4) of the United Nations Intergovernmental Panel on Climate Change (IPCC)*, Geneva, IPCC.

[2] **J. H. Keenan**, 1951, "Availability and irreversibility in thermodynamics," *Br. J. Appl. Phys.* **2**, 183–192.

[3] **J. T. Szargut**, 2003, "Anthropogenic and natural exergy losses (exergy balance of the Earth's surface and atmosphere)," *Energy*, **28**, 1047–1054.

[4] **V. S. Stepanov**, 1995, "Chemical energies and exergies of fuels," *Energy*, **20**, 235–242.

[5] **J. H. Sheih** and **L. T. Fan**, 1982, "Estimation of energy (enthalpy) and exergy (availability) contents in structurally complicated materials," *Energy Sources*, **6**, 1–46.

[6] **Global Climate and Energy Project**, Stanford University, exergy flow charts, http://www.gcep.stanford.edu/research/exergycharts.html.

[7] **R. E. Sassoon, W. A. Hermann, I. Hsiao** *et al.*, 2009, "Quantifying the flow of exergy and carbon through the natural and human systems," in *Materials for Renewable Energy at the Society and Technology Nexus*, eds. **R. T. Collins**, Warrendale, PA, Materials Research Society, pp. 1170-R01–03.

7 Global materials flows

Fridolin Krausmann

Institute of Social Ecology, Alpen Adria Universität Klagenfurt–Graz–Wien, Vienna, Austria

7.1 Focus

Human society requires increasing material inputs to sustain its production and reproduction. Many of the environmental problems we are currently facing can be attributed to this metabolism of society. This chapter explores the evolution of material use during human history and discusses current trends and patterns of global material consumption.

7.2 Synopsis

This chapter introduces the concept of social metabolism and the corresponding methodology of material flow accounting, which can be used to investigate exchange processes of materials and energy between societies and their natural environment and to address corresponding environmental problems. Against this background, the evolution of material use during human history from hunter-gatherers to current industrial societies is explored. The multiplication of global material extraction in the twentieth century, the shift from renewable biomass toward mineral and fossil materials and the growing share of materials used for non-energy applications are discussed in more detail. The chapter analyzes growth in material flows during industrialization and explores the interrelationships of material and energy use, population growth, economic development, and technological change. Inequalities in the global use of materials across countries and world regions are addressed. The chapter ends with a brief account of the challenges arising from the expected future growth of materials use.

Table 7.1. Key characteristics of material use in different societies. Material use refers to domestic material consumption (DMC) as defined in Section 7.4.3

	Hunter-gatherer	Agrarian	Current industrial
Material use (DMC) (tons per capita per year)	0.5–1	3–6	15–40
Share of biomass in DMC (% of total)	>99	>95	20–30
Physical stocks (artifacts) (t per capita)	<0.01	<10	100–1000
Share of materials for non-energy use (% of total)	<1	<20	>50
Peak global extraction (Gt)	0.005–0.01	1.8–3.6	>60

Values have been calculated by the author using data derived from [1] [2] [3].

7.3 Historical perspective

Prehistoric hunter-gatherers did not extract much more from their natural environment than what they needed for their daily nutrition. Only with the spreading use of fire some 800,000 years ago were their material needs expanded to encompass the use of wood as a fuel. Taken together, these two main components of prehistoric material use probably averaged less than a ton of biomass per person per year. The nomadic lifestyle of hunter-gatherers did not permit the accumulation of large amounts of artifacts, hence the quantity of materials used for shelter, tools, or clothing was minimal and rarely exceeded several kg per person per year. In contrast to what the commonly used notion of a "stone-age" suggests, mineral materials were of very minor significance in terms of mass, while biomass accounted for more than 99% of all material inputs. Assuming a per-capita materials use of 0.5–1 t per year and a maximum population of 9 million people at the onset of the Neolithic revolution, global material extraction, practically all of it biomass, may have amounted to between 5 and 10 million tons per year (Table 7.1).

The emergence of agriculture and a sedentary lifestyle in the Neolithic also changed the patterns of material use. Feeding a growing number of domesticated livestock boosted biomass extraction. Additionally, a large amount of biomass and mineral materials were used in buildings in rural and urban settlements and to manufacture tools and durable goods. The variability of materials use over time and space in different agrarian societies has been considerable, the prevailing patterns depending to a large extent on the significance of animal husbandry and housing and settlement patterns. Reliable estimates hardly exist, but it can be assumed that the material demand of an average agriculturalist exceeded that of a hunter-gatherer by up to one order of magnitude, averaging at larger scales between 3 and 6 t per capita per year (Table 7.1). Although the extraction and processing of ores such as iron, copper, and lead, and nonmetallic minerals (clay, dimension stone) gained significance, the overall mass of extracted mineral materials remained comparatively small. Biomass still accounted for more than 95% of all extraction and continued to dominate materials use.

Agriculture not only brought with it a higher rate of material extraction per person, but also triggered global population growth. Together these two factors resulted in a slow but steady growth of global material use after the Neolithic revolution. Assuming a global average rate of materials extraction of 3–6 t per capita per year and a global population of 600 million at the dawn of the industrial revolution, this amounted to a global material extraction of roughly 1.8–3.6 gigatons (Gt) around 1700.

The industrial revolution and the transformation of the socioeconomic energy system which started in Great Britain in the mid eighteenth century had a major impact on material use [4] [5]. It began with the rapidly spreading use of coal and the emerging technology complex comprising the steam engine, iron and steel production, and the railroad system in the nineteenth century and continued in the twentieth century, when oil, natural gas, and electricity replaced coal as the key energy carrier. This energy transition [6] fundamentally changed the possibilities for extracting, transporting, and processing materials. It was the physical basis of mass production and mass consumption, and unprecedented economic growth and new patterns of housing, mobility, and communication led to a fast rise in material use, in particular in the second half of the twentieth century [7]. In industrializing countries in Western Europe, North America, and East Asia annual per capita material use grew as much as five-fold and reached a level of 15–40 t in a period of not even

150 years. The share of biomass declined to less than a third of total domestic material consumption (DMC) and material stocks in buildings, infrastructures, and durable goods piled up to several hundred tons per capita (Table 7.1). In modern industrial economies typically 50% to 70% of all annual material input is used for non-energy purposes. During the last few decades and in particular since the oil-price shocks in the 1970s, growth in material use has slowed down markedly in industrial countries. In most high-income countries material use is now growing at a slower pace than the economy, and material intensity (material use per unit of GDP) is declining. This phenomenon has been termed relative dematerialization.[1] Absolute dematerialization, that is an actual reduction of material use while GDP continues to grow, is a rare exception. An enduring decline in material use has been observed for only a few countries, for example the UK, Germany, and Japan. In most cases this was a result of significant deindustrialization and the externalization of resource-intensive production processes to foreign countries, rather than of changes in actual consumption patterns.

7.4 Global material flows

7.4.1 Social metabolism

Human beings, like all other organisms, depend on a continuous throughput of materials. Water and food are the most essential of these material requirements. Food, a specific mix of digestible biomass, provides humans with nutritional energy and with the organic and mineral components necessary to build up and maintain the body and its functioning. After being used or stored in the human metabolic system, all intakes leave the body in the form of excrements or through the respiratory system. The same applies to human society. It requires a permanent input of materials and energy for the production and reproduction of its physical stocks; but the material needs of a particular society are generally much larger and more diverse than the basic metabolic needs of the sum of the individuals forming its population. Next to population, the physical stock of any socioeconomic system comprises domesticated livestock and a vast number of artifacts, from simple tools and shelters in traditional agrarian communities to machinery, durable goods, infrastructure, and buildings in modern industrial societies. To build up these stocks, and to maintain, renew, and operate them, industrial societies extract a broad range of biotic and mineral materials from their natural environment. These materials are processed and transformed; some are consumed within a short period of time, while others are stored for

centuries; some are discarded after use and some recycled. At some point, however, all materials leave the socioeconomic system in the form of wastes or emissions. In analogy to the metabolism of biological organisms the term social or industrial metabolism has been coined [8] [9] [10]. Many of the regional and global environmental problems that human society faces at the beginning of the twenty-first century are directly related to its metabolism: from the extraction of raw materials in agriculture and mining and their processing in industry to the consumption of the supplied goods and services and the inevitable formation of wastes and emissions, our metabolism causes a huge variety of environmental pressures, including the most serious threats to global sustainability.

In this context, the concept of social metabolism has not only proven a useful metaphor to stress the biophysical foundation of social systems and their economy, but also has emerged as a key analytical concept in sustainability science. In new interdisciplinary fields like industrial ecology and ecological economics sophisticated methods and tools have been developed to study material and energy flows in socioeconomic systems in order to contribute to the design and implementation of more sustainable types of industrial metabolism [11].

7.4.2 Materials and energy – two faces of the same coin

Material and energy flows are closely intertwined in social metabolism. First of all, materials, and in particular combustible materials, are important energy carriers. Materials form by far the largest fraction of all primary energy inputs: biomass in the form of food and feed is the primary energy source for human and animal work; wood, coal, oil, and natural gas are fueling combustion processes and uranium is used as fuel in nuclear power plants. Vice versa, this means that a significant fraction of all the materials extracted by humans is converted into energy. In hunter-gatherer and even in historical agrarian societies material and energy supply were practically identical; more or less all materials used (food, feed, and wood) also served the purpose of energy provision (see Section 7.3). This changed with the industrial revolution and the new possibilities of energy conversion, which boosted not only the overall amount but also the variety of materials used in the economy. Meanwhile, more than half of all extracted materials (roughly 32 Gt) are not converted into energy but serve as the material basis for stocks of artifacts.

Moreover, using materials inevitably involves considerable quantities of energy: energy (or, more precisely, exergy or useful work) is needed for all activities involving materials. Energy has to be invested to extract

---

[1] The terms relative and absolute decoupling (of economic growth and material use) are used synonymously to dematerialization.

raw materials, to transport them, and to mechanically work or chemically transform them. It is obvious that the availability of energy and corresponding conversion technologies determines the basic range of which types and quantities of materials can be extracted and for which applications they can be used [12].

Last but not least, all technologies to transport, convert, and store energy and to provide the requested energy services involve materials and material artifacts – from the early steam engine to the turbine in a power plant, from the horse cart to the airplane and from flint to light a fire to the lithium used to store electricity in batteries or the silicon of solar cells. Energy and materials simply are two sides of the same coin. This chapter takes a closer look at the material aspects of social metabolism.

7.4.3 How should one account for material flows?

While the accounting of energy flows and the analysis of socioeconomic energy systems has a comparatively long tradition and ample knowledge about the long- and short-term development of energy provision and use exists, much less is known about the material component of social metabolism. After pioneering conceptual and empirical studies conducted in the 1970s (e.g., [13]) material flow accounting (MFA) emerged as an important methodological framework in industrial ecology and ecological economics only in the 1990s (e.g., [14] [15]). Since then, a variety of tools for studying material flows at different temporal and spatial scales and for different applications has been developed [11]. At the national level, economy-wide MFA is now an agreed-upon and standardized tool for investigating the size and structure of socioeconomic material flows. It provides reliable and comparable data and aggregate headline indicators for material use and is widely used by statistical offices in their environmental reporting systems (e.g., [16] [17] [18]). Full-scale economy-wide material flow accounts are consistent compilations of the overall material inputs (except for water and air)[2] into national economies, the changes of material stock within the economic system, and the material outputs to other economies or to the environment (Figure 7.1). An important MFA-derived indicator of material use is the DMC. It is calculated as the sum total of all materials extracted from the domestic environment of the observed economy (domestic extraction, DE) and all materials imported from other socioeconomic systems minus all exports. It is a measure of apparent consumption similar to corresponding indicators used in energy

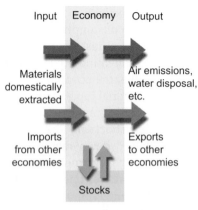

Figure 7.1. Material flow accounting scheme, based on [16].

statistics (e.g., TPES Total Primary Energy Supply). It has been argued that the DMC also indicates the waste and emissions potential of an economy [19].

The variety of materials used is huge. Some, such as the platinum group metals, are used in small quantities only, while others, such as limestone, are extracted in vast amounts. Material flow accounts distinguish several hundred different materials or groups of materials. For presentation these are usually aggregated to only a few groups of materials with similar properties. Most commonly these are biomass (including crops, used crop residues, grazed biomass and roughage, wood, and fish), fossil energy carriers (coal, petroleum, natural gas, and peat), into the strategically important group of non-metallic minerals for industrial use, and minerals. Minerals are often further separated into strategically important group of ores and nonmetallic minerals for industrial use and the much larger flow of bulk minerals used predominantly for construction (sand, gravel, crushed stone).

During the last few years an increasing body of empirical evidence on material flows in socioeconomic systems at various spatial and temporal scales has been collected, and databases with global country by country coverage of materials extraction, trade and use have been made available [3] [20] [21].[3] From these data it is possible to provide a reliable picture of the pattern of material use across the globe and of its development over time.

7.4.4 Growth of global material extraction in the twentieth century

During the last century global materials extraction (which equals material use at the global scale) has been

---

[2] Although water and air are important components of social metabolism, these flows are typically not accounted for in material flow accounts. The enormous size of these flows would simply dwarf all other material flows. Likewise, water and air are not included in the discussion of materials use in this chapter.

[3] Global material flow data are available for public download at http://www.uni-klu.ac.at/socec/inhalt/1088.htm and http://www.materialflows.net.

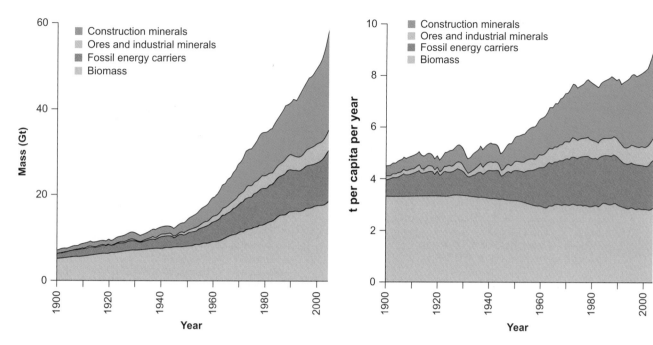

Figure 7.2. Global materials extraction by main material groups since 1900 in Gt, based on data from [21].

Figure 7.3. Development of per capita material use since 1900, based on data from [21].

growing at exponential rates. By 1900, total material extraction had risen to roughly 7 Gt, of which three-quarters were still renewable biomass. A century later, in 2005, annual extraction amounted to almost 60 Gt (Figure 7.2). Physical growth in the twentieth century was mainly driven by mineral and fossil materials: the use of these key materials of industrialization multiplied 10–30-fold. The use of biomass, in contrast, grew much slower, at roughly the same pace as global population. Nonetheless, for most of the twentieth century biomass was still the most significant of the four material types in terms of mass and only in the 1990s was it overtaken by construction minerals. By 2000 its share had declined to only one-third of total material input. In particular, the period between World War II and the oil-price peaks in the 1970s saw a rapid shift from renewable biomass toward mineral and fossil materials. Throughout the twentieth century, material use increased continuously, with annual growth rates between 1% and 4%. Periods with declining or stagnating DMC were rare and never lasted for more than a few years. All periods of absolute dematerialization coincided with economic stagnation or recession (e.g., the world economic crises in the 1930s and World War II).

The development of per capita materials use (Figure 7.3) shows three periods with distinct growth dynamics. In the first half of the twentieth century, a period characterized by two World Wars and the World Economic Crises, the DMC grew very modestly, at only half the pace of economic growth and only slightly faster than population. After World War II, physical growth accelerated and kicked off a period of uninterrupted

and rapid increase of material use that lasted for three decades. In particular, the use of fossil energy carriers, ores and construction minerals, which are key materials for industrial development, showed annual growth rates of up to 6% and periodically grew faster than the global economy. Even biomass use rose faster (1.5% per year) than ever before or since. In this period, growth rates of materials use by far exceeded population growth and led to an unprecedented increase in the rate of materials used per capita. This *great acceleration* [22] has been driven by the availability of cheap energy and was related to the emergence of a society of mass production and consumption in the industrializing world, the rapid expansion of built infrastructures, and the green revolution. It led to drastic increases in materials use in a few industrialized countries, while the metabolic pattern of the majority of the world population hardly changed. The oil-price peaks of the 1970s imposed an abrupt end to these heydays and growth slowed down markedly. With the exception of biomass use, which continued to rise at a moderate pace together with global population, average annual growth rates declined by 50% or more. Material use per capita almost stabilized.

During the whole of the twentieth century, global material use grew significantly faster than population but much less than GDP. Consequently, the per-capita DMC doubled, while material intensity declined continuously: in 2005 roughly 1.3 kg of materials were used per dollar GDP, compared with 3.6 kg per dollar in 1900. Despite these significant increases in material

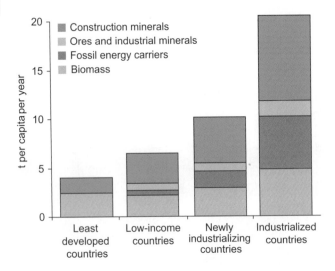

Figure 7.4. Average per capita DMC in groups of countries at different levels of economic development in 2000, based on data provided in [3].

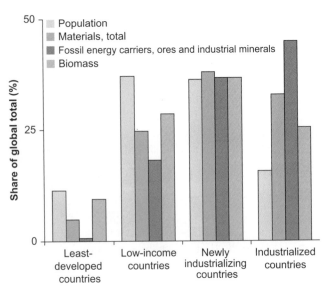

Figure 7.5. Share of global population, total material use, biomass and fossil fuels, ores and industrial minerals by country groups, based on data provided in [3].

productivity, there is no sign that the global economy is dematerializing. On the contrary, toward the turn of the new millennium, growth of materials use accelerated again and per-capita DMC has increased markedly since the year 2000 (Figure 7.3).

7.4.5 Global patterns of material use at the turn of the millennium

At the beginning of the twenty-first century, humans use an enormous amount and variety of materials. Global extraction of materials has been estimated to range between 50 and 60 Gt [21] [23]. This means that for every human being on Earth between 8 and 10 t of new raw materials are extracted every year. Nonmetallic minerals used in construction account for almost half of the annual material inputs. Renewable biomass is the second largest flow, accounting for one-third of the global total; over 60% of all biomass is used to feed domesticated livestock. The rest of global DE is composed of fossil energy carriers (20%) and the diverse group of ores and nonmetallic minerals for industrial use (5%). About half of all materials, mostly biomass and fossil energy carriers, are "consumed" within one year and then returned to the natural environment in the form of wastes and emissions. The other half are accumulated and pile up in enormous stocks of artifacts such as buildings, transport networks, sewer systems, and machinery. In industrialized countries these stocks amount to several hundred tons per capita and contain large amounts of important resources [24] [25]. For the USA it has been calculated that the in-use stocks of iron roughly equal the remaining US iron stocks in identified ores and that landfills are now the third largest iron

reservoir [26]. However, irrespective of the duration of use, sooner or later all material inputs leave the economic systems to be disposed to the natural environment.

Material consumption is distributed more equally around the globe than, for instance, energy use or income [27]. Still, the differences in the amount of materials used per capita across countries are huge: globally the average per-capita DMC varies by an order of magnitude between individual countries. In general per-capita resource consumption and the share of mineral and fossil materials in the DMC increase with economic development (Figure 7.4): among the countries with the lowest level of material use are many African and South Asian countries where the DMC averages less than three tons per capita per year and the share of biomass is very high. In contrast, industrialized countries like the USA, Canada, and Australia, which are among the nations with the highest level of material use, show a per-capita DMC of 27–44 t. The metabolism of these high-income countries is dominated by mineral resources, which often account for more than three-quarters of total material use. Figure 7.5 shows the distribution of material use across groups of countries at different levels of economic development. Less than one billion people in the most industrialized countries in the world, which account for 16% of world population, use one-third of all extracted materials and almost half of all fossil fuels, ores, and industrial minerals. In contrast, the world's least developed countries, which are inhabited by 11% of the global population, harness only 4.8% of all resources and less than 1% of the strategically important energy carriers and minerals.

A good deal of the observed differences in material use across countries can be explained by differences in income (measured in GDP per capita), but income does not explain it all [27]. Even within groups of countries with a similar level of economic development or income variations can be large. Weisz and colleagues [28] have shown that within the 15 European Union countries of 2000 per-capita DMC ranged between 12 and 37 t. Next to income a host of bio-geographic and socioeconomic factors influence the level and composition of a country's material use, such as endowment with natural resources, climate, life-style, and population density. Also trade relations are important: countries that import a large share of their material needs tend to have a lower DMC than exporting countries. Upstream material requirements associated with imports are on average 2–3 times larger than the actually imported materials. Industrial countries, where the mass of imports can reach a similar size to that of domestic extraction, thus externalize material-intensive production processes and waste generation [29] [30].

7.4.6 Outlook

Human society is in the midst of a global metabolic transition [17]. The last century witnessed an eight-fold multiplication of the size of the global social metabolism and a transition from the dominance of renewable biomass toward mineral and fossil materials. In the most recent past, per-capita material use in newly industrializing countries like China, India, Mexico, and Brazil started to rise, while the world's least developed countries are only now beginning the transition toward an industrial-type metabolism. With global economic development continuing in a business-as-usual mode and a projected population growth of 40%–50% by 2050, we should expect another sharp rise in global materials extraction. The few existing scenarios assume that, if the current development continues, this might lead to a doubling of global material extraction during the next decades [3] [31]. In the light of already existing environmental pressures associated with the human use of materials and increasing scarcity of a number of key resources, such a development does not seem very likely. Instead, industrial countries will have to significantly reduce their throughput of materials and need to find ways to decouple human well-being and quality of life from material and energy use. Materials and energy are closely intertwined in social metabolism. Any attempt to reduce the input of one will inevitably affect the other. Strategies toward a more sustainable use of natural resources and the shift toward a new, environmentally sounder type of industrial metabolism must equally consider both sides of the metabolic coin.

7.5 Summary

We have used the concept of social metabolism to explore the global patterns of society's material use and their development over time. All socioeconomic systems require a permanent throughput of materials and energy to maintain their functioning and to grow. This metabolism is causing major sustainability problems, with regard to both the provision of sufficient materials and the disposal of outflows. During human history, the composition of material use and the size of material flows have fundamentally changed. For most of human history renewable biomass was the dominating material; only with industrialization has the share of mineral materials increased, and it now constitutes the major part. Also the relation between material and energy use has changed. Two hundred years ago, more than 90% of all extracted materials were still used as primary energy sources to provide food, animal feed, and fuel. The increasing availability of energy in combination with new technologies allowed the extraction and processing of an ever increasing amount of materials also for non-energy use. In modern industrial societies, the share of non-energy materials in the DMC has surged to over 60%. The largest part of all extracted materials is accumulated in infrastructure networks, buildings, and durable artifacts. Overall, the amount of raw materials extracted annually worldwide multiplied during industrialization, and had increased to roughly 60 Gt in 2005. This growth was only partly due to a growing world population, but increases in per-capita material use in the industrial world associated with the emergence of a society of mass production and mass consumption have been a major driver of growth. Currently material use in industrial countries has stabilized at a very high level of up to 40 t per capita per year. While industrial countries are still consuming a disproportionally large share of all extracted materials, in particular of ores, industrial minerals and fossil energy carriers, newly industrializing countries are rapidly expanding their material requirements. As a consequence, growth in global material use has accelerated during the last decade and no sign of an absolute dematerialization of the global economy is in sight.

7.6 Questions for discussion

1. Which sustainability problems does the concept of social metabolism allow one to address, and which cannot be captured?
2. What is the size of your personal metabolism? Try to estimate the annual material input of your household.
3. How are material and energy use related?
4. Which major transitions in material use can be observed during human history?

5. What are the factors that drive changes in material use?

6. Why are socioeconomic stocks of materials important for annual flows?

7. What is dematerialization and why is it important for sustainable development?

8. Which factors could contribute to a decoupling of material use from economic growth?

7.7 Further reading

- **P. Baccini** and **P. H. Brunner** 1991, *The Metabolism of the Anthroposphere*, Berlin, Springer. This book provides a concise introduction to the concept of social metabolism.

- **M. Fischer-Kowalski** and **W. Hüttler**, 1998, "Society's metabolism. The intellectual history of material flow analysis, part II: 1970–1998," *J. Indust. Ecol.*, **2**, 107–137. This contribution explores the origins of material flow accounting and its application to environmental science.

- **A. Adriaanse, S. Bringezu, A. Hammond,** *et al.* 1997, *Resource Flows: The Material Basis of Industrial Economies*. Washington, DC, World Resources Institute. The first comparative and multinational study of economy-wide material flows is still a key publication in the field.

- **D. Rogich, A. Cassara, I. Wernick**, and **M. Miranda**, 2008, *Material Flows in the United States: A Physical Accounting of the U.S. Industrial Economy. WRI Report*. Another study from the World Resources Institute, which provides interesting insights into material use in the US economy. Available for download at http://www.wri.org/publication/material-flows-in-the-united-states.

- **H. Schandl** and **H. Weisz**, 2008, *Materials Use Across World Regions: Inevitable Pasts and Possible Futures*. Special issue of the *Journal of Industrial Ecology*, **12**, 629–798. This special issue compiles a number of excellent case studies and applications of economy-wide material flow accounting and provides an overview of the state of the art.

- **S. Bringezu** and **R. Bleischwitz**, 2009, *Sustainable Resource Management*, Sheffield, Greenleaf Publishing Limited. A recent edited volume with a good introduction to material flow accounting and its use for resource management from a European perspective.

- **Reports of the UNEP resource panel**: the reports of the UNEP resource panel summarize the current understanding of material use and the future challenges. Available for download at http://www.unep.fr/scp/rpanel/.

7.8 References

[1] **M. Fischer-Kowalski** and **H. Haberl**, 1997, "Stoffwechsel und Kolonisierung: Ein universalhistorischer Bogen," in *Gesellschaftlicher Stoffwechsel und Kolonisierung von Natur*, eds. **M. Fischer-Kowalski, H. Haberl, W. Hüttler** *et al.*, Amsterdam, Gordon & Breach Fakultas, pp. 25–36.

[2] **M. Fischer-Kowalski** and **H. Haberl**, 2007, *Socioecological Transitions and Global Change: Trajectories of Social Metabolism and Land Use*, Cheltenham, Edward Elgar.

[3] **F. Krausmann, M. Fischer-Kowalski, H. Schandl**, and **N. Eisenmenger**, 2008, "The global socio-metabolic transition: past and present metabolic profiles and their future trajectories," *J. Industrial Ecol.*, **12**, 637–656.

[4] **R. P. Sieferle**, 2001, *The Subterranean Forest. Energy Systems and the Industrial Revolution*, Cambridge, The White Horse Press.

[5] **J. -C. Debeir, J. -P. Deléage**, and **D. Hémery**, 1991, *In the Servitude of Power: Energy and Civilization through the Ages*, London, Zed Books.

[6] **A. Grübler**, 2004, "Transitions in energy use," in *Encyclopedia of Energy*, ed. **C. J. Cleveland**, Amsterdam, Elsevier, pp. 163–177.

[7] **A. Grübler**, 1998, *Technology and Global Change*, Cambridge, Cambridge University Press.

[8] **P. Baccini** and **P. H. Brunner**, 1991, *The Metabolism of the Anthroposphere*, Berlin, Springer.

[9] **M. Fischer-Kowalski** and **W. Hüttler**, 1998, "Society's metabolism. The intellectual history of material flow analysis, part II: 1970–1998," *J. Industrial Ecol.*, **2**, 107–137.

[10] **R. U. Ayres** and **U. E. Simonis**, 1994, *Industrial Metabolism: Restructuring for Sustainable Development*. Tokyo, United Nations University Press.

[11] **S. Bringezu, I. van der Sand, H. Schütz, R. Bleischwitz**, and **S. Moll**, 2009, "Analysing global resource use of national and regional economies across various levels," in *Sustainable Resource Management. Global Trends, Visions and Policies*, eds. **S. Bringezu** and **R. Bleischwitz**, Sheffield, Greenleaf, pp. 10–52.

[12] **C. A. S. Hall, C. J. Cleveland**, and **R. K. Kaufmann**, 1986, *Energy and Resource Quality. The Ecology of the Economic Process*, New York, Wiley Interscience.

[13] **A. V. Kneese, R. U. Ayres** and **R. d'Arge**, 1970, *Economics and the Environment. A Materials Balance Approach*, Washington, DC, Resources for the Future, Inc.

[14] **R. U. Ayres** and **L. W. Ayres**, 1998, *Accounting for Resources, 1, Economy-Wide Applications of Mass-Balance Principles to Materials and Waste*, Cheltenham, Edward Elgar.

[15] **A. Adriaanse**, 1997, *Resource Flows: The Material Basis of Industrial Economies*, Washington, DC, World Resources Institute.

[16] **Eurostat, 2007**, *Economy-Wide Material Flow Accounting. A Compilation Guide*, Luxembourg, European Statistical Office.

[17] **OECD, 2008**, *Measuring Material Flows and Resource Productivity. Volume I. The OECD Guide*, Paris, OECD.

[18] **L. A. Wagner**, 2002, "Materials in the economy – Material flows, scarcity, and the environment," *U.S. Geological Survey Circular* **1221**, 1–29.

[19] **N. Eisenmenger**, **M. Fischer-Kowalski**, and **H. Weisz**, 2007, "Indicators of natural resource use and consumption," in *Sustainability Indicators. A Scientific Assessment*, eds. **T. Hak**, **B. Moldan**, and **A. L. Dahl**, Washington, DC, SCOPE, Island Press, pp. 193–209.

[20] **SERI**, 2008, *Global Resource Extraction 1980 to 2005*, online database, Sustainable Europe Research Institute, http://www.materialflows.net/

[21] **F. Krausmann**, **S. Gingrich**, **N. Eisenmenger** *et al.*, 2009, "Growth in global materials use, GDP and population during the 20th century," *Ecol. Econ.*, **68**, 2696–2705.

[22] **K. Hibbard**, **P. J. Crutzen**, **E. F. Lambin** *et al.*, 2007, "Decadal scale interactions of humans and the environment," in *Sustainability or Collapse? An Integrated History and Future of People on Earth. Dahlem Workshop Reports*, eds. **R. Costanza**, **L. J. Graumlich**, and **W. Steffen**, Cambridge, MA, MIT Press, pp. 341–378.

[23] **A. Behrens**, **S. Giljum**, **J. Kovanda**, and **S. Niza**, 2007, "The material basis of the global economy: worldwide patterns of natural resource extraction and their implications for sustainable resource use policies," *Ecol. Econ.*, **64**, 444–453.

[24] **S. Hashimoto**, **H. Tanikawa**, and **Y. Moriguchi**, 2007, "Where will large amounts of materials accumulated within the economy go? – A material flow analysis of construction minerals for Japan," *Waste Management*. **27**, 1725–1738.

[25] **R. B. Gordon**, **M. Bertram**, and **T. E. Graedel**, 2006, "Metal stocks and sustainability," *Proc. Natl. Acad. Sci.*, **103**, 1209–1214.

[26] **USGS**, 2005, *Metal Stocks in Use in the United States*. Science for a Changing World, http://pubs.usgs.gov/fs/2005/3090/2005-3090.pdf.

[27] **J. K. Steinberger**, **F. Krausmann**, and **N. Eisenmenger**, 2010, "Global patterns of material use: a socioeconomic and geophysical analysis," *Ecol. Econ.*, **69**, 1148–1158.

[28] **H. Weisz**, **F. Krausmann**, **C. Amman** *et al.*, 2006, "The physical economy of the European Union: cross-country comparison and determinants of material consumption," *Ecol. Econ.*, **58**, 676–698.

[29] **S. Giljum**, and **N. Eisenmenger**, 2008, "North–south trade and the distribution of environmental goods and burdens: a biophysical perspective," *J. Environment Development*, **13**(1), 73–100; also in J. Martinez-Alier and I. Roepke, 2009, *Recent Developments in Ecological Economics*, Cheltenham, Edward Elgar, vol. 1, pp. 383–410.

[30] **S. Bringezu**, **H. Schütz**, **M. Saurat**, **S. Moll**, and **J. Acosta Fernandez**, 2009, "Europe's resource use. Basic trends, global and sectoral patterns and environmental and socioeconomic impacts," in *Sustainable Resource Management. Global Trends, Visions and Policies*, ed. **S. Bringezu** and **R. Bleischwitz**, Sheffield, Greenleaf, pp. 52–155.

[31] **C. Lutz**, and **S. Giljum**, 2009, "Global resource use in a business-as-usual world up to 2030: updated results from the GINFORS model," in *Sustainable Growth and Resource Productivity. Economic and Global Policy Issues*, eds. **R. Bleischwitz**, **P. J. J. Welfens**, and **Z. Zhang**, Sheffield, Greenleaf, pp. 30–42.

8 Carbon dioxide capture and sequestration

Sally M. Benson

Department of Energy Resources Engineering, Stanford University, Stanford, CA, USA

8.1 Focus

A transition to a low-carbon economy can be facilitated by CO_2 capture and sequestration. This chapter focuses on capture of carbon dioxide from industrial emission sources such as electricity generation and sequestration in deep geological formations. A detailed description of the technology is provided, including the potential scale of application, estimated costs, assessment of risks, and emerging research issues.

8.2 Synopsis

Today, 60% of global CO_2 emissions come from large point sources such as power plants, refineries, cement plants, and steel mills. Reducing emissions from these sources will require reducing demand for the services or materials they provide, finding alternative ways to provide similar services with lower carbon dioxide emissions, or directly reducing emissions by capturing and sequestering emissions. Technology to capture carbon dioxide is available today, but capturing and sequestering CO_2 will increase the cost of electricity production by an estimated 50%–100% compared with today's generating costs. Moreover, an estimated increase of 15%–30% of the primary energy supply needed to deliver these services or goods would be required. Captured carbon dioxide can be sequestered in deep geological formations, either onshore or offshore. Sedimentary basins are the preferred location for carbon dioxide sequestration, since they are known to contain both the porous and permeable sandstone formations needed to sequester CO_2 and low-permeability rocks such as shale that can trap CO_2 for geological time periods of millions of years. The estimated capacity for sequestering CO_2 is large and expected to be sufficient for at least 100 years of needed demand. However, the actual capacity for safe and environmentally benign sequestration remains uncertain, since CO_2 sequestration has been employed for little more than a decade and only on a small scale. Nevertheless, the basic technologies for sequestration and performance prediction are mature, building on nearly a century of oil and gas production, natural-gas storage, CO_2-enhanced oil recovery (CO_2-EOR), and acid gas disposal. Enhancements of these technologies will arise as geological sequestration itself matures – but they are sufficiently developed to initiate sequestration today. Regulatory and legal issues remain to be resolved, including issues such as permits for sequestration-project siting, well drilling, and completion, operational parameters such as maximum injection pressures, ownership of underground power space, supremacy of mineral or groundwater rights, and liability for long-term stewardship. Resolving these issues and gaining support for this approach from the public are likely to be the greatest challenges for implementing CO_2 capture and sequestration on a meaningful scale.

8.3 Historical perspective

Currently, all of the CO_2 produced by burning fossil fuels is emitted directly into the atmosphere. Historically, this approach was used because it had no immediate cost and, until about 50 years ago, it was believed to be harmless [1] [2]. Beginning with the pioneering data collected by Charles David Keeling of the Scripps Institute in the 1950s, however, humanity's fingerprint on the composition of the atmosphere has become clear. Since then, the impacts of air-quality pollutants such as soot, sulfur dioxide, and nitrogen oxides (NO_x) have been recognized, as have the global climate implications of CO_2 emissions. When sulfur dioxide and nitrous oxide emissions from power plants were found to cause air pollution, technologies were developed to remove these species from the smokestacks of power plants. Indeed, through advances in technology and effective policy measures, emissions of these pollutants have been reduced significantly. Many similar examples demonstrate that the management of such byproducts can be improved to reduce or eliminate their environmental impacts. In each case, treatment was initially believed to be too costly. However, once requirements to reduce discharges were put in place, technological innovations resulted in effective new technologies at acceptable costs. In some cases, reducing discharges of contaminants even created new market opportunities to recycle wastes and provide valuable products – such the production of wall-board or sheetrock from desulfurization of power-plant emissions. So today, a similar challenge faces continued use of fossil fuels: can new technologies that reduce or eliminate CO_2 emissions from the burning of fossil fuels be developed and implemented at an acceptable cost? Also, can value-added uses of CO_2 help to build infrastructure and offset the costs of early deployment?

Beginning in the early 1990s, a small group of scientists began to ask whether CO_2 emissions could be reduced through the use of pollution-abatement technology, giving rise to the new suite of technologies called **carbon capture and storage** or **carbon sequestration**. Scientists studied options for capturing CO_2 from industrial emission sources and directly from the air itself. Initially, there was a great deal of interest in sequestering CO_2 in the ocean, either by pumping it into the mid-depth ocean or by enhancing the natural carbon cycle in the ocean. If CO_2 is injected into the mid-depth ocean (1,000–3,000 m deep), it can be stored for hundreds to thousands of years before returning to the atmosphere through ocean circulation [1]. Although the potential capacity for ocean storage is large – on the order of a trillion tonnes of CO_2 – concerns about biological impacts, high costs, impermanence of ocean storage, and public acceptance have decreased interest and

investment in this technology over the past 5 years [1]. Alternatively, by fertilizing the ocean with deficient nutrients such as iron, the ocean carbon cycle can be accelerated [3]. Some fraction of increased biomass production, and hence, indirectly, CO_2, can then be exported from the near-surface environment and sequestered in the deep ocean. Results regarding carbon export (sequestration) from these experiments have been ambiguous, and uncertain biological impacts have likewise lessened interest in this approach to carbon sequestration.

There has also been considerable interest in terrestrial carbon sequestration through regrowth and new growth of forests, as well as changed tillage practices to increase soil carbon storage [4]. These options are attractive because they often provide additional ecosystem benefits such as habitat restoration, improved soil quality, and watershed improvement. Terrestrial sequestration directly captures CO_2 from the atmosphere through the process of photosynthesis. However, issues of permanence, land-use practices that emit large amounts of CO_2, accounting, and global climate policy have raised questions about the widespread implementation of these technologies. Regardless, terrestrial sequestration alone is not sufficient to compensate for the current or likely future atmospheric loadings.

New options for sequestering CO_2 are being proposed on a regular basis. Examples include making biochar for soil augmentation by pyrolysis of biomass, mineralization to form magnesium or calcium carbonate, use of CO_2 to augment growth of algae, and production of synthetic fuels. However, today, the most well-developed option for sequestering CO_2 is to inject it into deep geological formations, where it would remain permanently trapped or sequestered. Consequently, this sequestration option is the focus of the remainder of this chapter.

8.4 The case for carbon capture and sequestration

Before investigating the details of **carbon capture and sequestration (CCS)**, consider why this technology should be an option at all. Some arguments in favor of including CCS in the portfolio of options for reducing emissions are as follows.

- Fossil fuels remain abundant and are unlikely to be replaced as the primary energy supply in the near future. In fact, the International Energy Agency (IEA) expects that, despite rapid growth of renewable energy supplies, fossil fuels will provide around 75% of the global energy supply in 2030. Therefore, CCS could be of value to reduce emissions while these fossil-fuel resources continue to be used [5].

- Historical analysis of past energy transitions has demonstrated that many decades are needed for new energy technologies to achieve even a small fraction of the overall energy supply [6]. Consequently, to reduce emissions now, we cannot wait for renewable energy resources to achieve these reductions.

- Modern societies rely on 24 × 7 × 365 energy supplies. Reliable supplies of fossil fuels can provide the "dispatchable" energy supplies needed in order to meet these demands, that is, energy supplies that can readily be turned on or off on demand. Fossil fuels with CCS, combined with renewable energy supplies and demand-side management, can provide dispatchable energy services while reducing CO_2 emissions.

- Some regions of the world do not have adequate supplies of renewable energy or access to nuclear power. In such cases, fossil fuels with CCS might provide the best option for meeting energy demand, including helping to provide energy to the 1.5 billion people currently lacking access to electricity [7], while reducing emissions.

- Economic analysis indicates that the cost of reducing emissions will be lower in a portfolio where CCS is an option [1].

- The political coalition to reduce emissions will be stronger and more effective if traditional energy providers (e.g., coal, gas, and oil companies) are partners in reducing emissions. Thus CCS provides the opportunity for all sectors of the current energy suppliers to participate in providing low-carbon energy supplies.

Although many believe the case for CCS to be strong, making significant inroads into the energy supply system is fraught with technical, financial, political, and societal challenges. In the coming sections, these issues are reviewed with an eye to the prospects for CCS in an evolving energy landscape.

8.4.1 Overview of CO_2 capture and sequestration technology

Carbon dioxide capture and sequestration is expected to be most useful for large, stationary sources of CO_2 such as power plants, petroleum refineries, gas-processing facilities, steel-manufacturing plants, and cement factories. The high capital cost of separation technology favors large facilities where economies of scale can be found. Today, there are over 7,500 point sources with emissions over 0.1 Mt per year (see Box 8.1). Power plants account for more than two-thirds of the sources, with an average CO_2 emission rate of about 4 Mt per year for coal-fired power plants and about 1 Mt per year for natural-gas- and oil-fired power plants. Industrial emission sources also tend to average about 1 Mt per

Box 8.1. Sources of CO_2 amenable to CCS

Over 7,500 industrial sources of CO_2 amenable to CCS have been identified [1]. These include coal-, gas-, and oil-fired power plants, gas-cleanup facilities, cement plants, steel makers, and refineries (see Figure 8.1). About two-third of these emission sources are related to electricity production. In general, the higher the concentration of CO_2 in the gas stream, the lower the costs and energy requirements needed for separation. The concentration of CO_2 in these sources ranges from less than several percent for boilers and gas turbines to nearly 100% for H_2 production facilities in refineries. Table 8.1 summarizes the average emissions and distribution of these sources. As shown, a single CCS project can remove several million tonnes per year of emissions, comparable to 1,000 MW of installed wind capacity operating at a capacity factor of 30%.

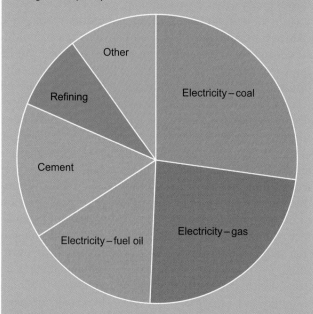

Figure 8.1. Distribution of CO_2 point sources among various industries.

Fuel	No. of sources	Average emissions per source (Mt of CO_2 per source)
Coal	2,025	3.9
Natural gas	1,728	0.8–1.0
Fuel oil	1,108	0.6–1.3
Cement	1,175	0.8
Refining	638	1.35
Other	736	0.15–3.5

Table 8.1. Emission sources and sizes amenable to CCS [1]

year. In the short term, applications of CCS are likely to focus on the electricity sector; however, over the long run, applications to the industrial and transportation sectors are also promising. Finally, the possibility of combining biomass-based energy production with CCS holds promise for even greater emission reductions. Early progress in this regard could be achieved by co-firing fossil fuels and biomass [1].

Carbon dioxide capture and sequestration is a four-step process, as shown in Figure 8.2. First, the CO_2 is separated from emissions and concentrated into a nearly pure form. For today's natural-gas- and coal-fired power plants, from 4% to 14% of the flue gas is CO_2; the rest is primarily nitrogen and oxygen. After the CO_2 has been separated from the flue gas, it is compressed to at least 100 bar, at which pressure it is in the liquid phase. Next, it is transported by pipeline to the location where it is to be stored. Pipelines transporting CO_2 for hundreds of kilometers exist today, with more than 5,000 km of such pipelines, primarily in the USA. The last step is

to pump the CO_2 into the deep geological formation in which it will be sequestered.

As illustrated in Figure 8.3, there are a number of options for geological sequestration. CO_2 can be injected into deep underground formations such as depleted oil and gas reservoirs, brine-filled formations, and deep unmineable coal beds [1][2]. Today, **sedimentary basins**, which are thick accumulations of sediments eroded from mountain ranges, are considered to have the greatest potential for sequestration. However, sequestering CO_2 in basaltic rocks and ocean-bottom sediments is also being investigated. The potential sequestration capacity in geological formations is highly uncertain, particularly for saline aquifers and coal beds. Estimates of sequestration capacity in oil and gas fields range from 996 to 1,150 Gt of CO_2; estimates for brine-filled formations range from 5,900 to 16,000 Gt [1] [2] [8]; and estimates for coal beds range from 210 to 240 Gt.

Four industrial-scale CCS projects are operating today. Three of them are sequestering CO_2 captured by separating CO_2 from natural gas. Natural gas containing more than several percent CO_2 must be "cleaned up" to the specifications of pipeline and purchase agreements. The first of these projects, the Sleipner Saline Aquifer Storage Project, began in 1996. Annually, 1 Mt of CO_2 is separated from natural gas and stored in a deep sub-sea brine-filled sandstone formation [9]. The In Salah Gas Project in Algeria began in 2004 and is storing about 1 Mt of CO_2 annually in the flanks of a depleting gas field [10]. The third industrial-scale CCS project, located in Saskatchewan, Canada, uses CO_2 produced from the Dakota Gasification Plant in Beulah, ND, USA, to simultaneously enhance oil production and store CO_2 in the Weyburn Oil Field [11], with 2–3 Mt of CO_2 injected annually. Although CCS is not the primary purpose for injecting CO_2 underground at Weyburn, a significant research and monitoring program was implemented to evaluate the efficacy of CO_2 sequestration, and now the operators intend to continue storing CO_2 in the reservoir

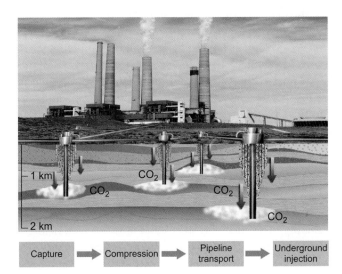

Figure 8.2. The four steps involved in CO_2 capture and sequestration.

Figure 8.3. Options for geological storage of CO_2.

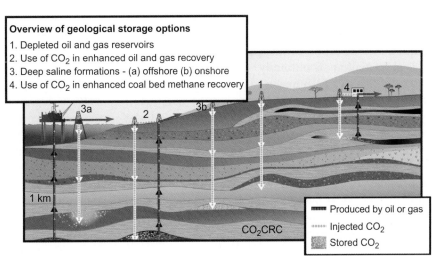

after oil-recovery operations cease. The Snøhvit Project located in the northern part of Norway began in 2008 and is expected to sequester about 1 Mt per year into a saline aquifer below the gas fields from which CO_2 is produced.

Electricity generators are expected to become the principal application for CCS. The feasibility of economically separating CO_2 from flue gas and compressing it into a liquid is crucial to its successful application. Pre- and post-combustion approaches for separating (or capturing) CO_2 from power plants are available [12]. **Integrated gasification combined cycle (IGCC)** technology holds particular promise as an efficient means of capturing CO_2 because of its lower costs for capture and higher operating efficiency. The IGCC approach requires gasification of the coal, followed by a **water–gas shift (WGS) reaction** (a chemical reaction to convert CO and H_2O to CO_2 and H_2) to produce hydrogen for running a gas turbine and CO_2 for sequestration (Chapter 39) [13]. Hydrogen could also be available for other purposes such as a feedstock for chemical processes or a fuel for transportation. Similar approaches have been proposed for natural-gas power stations [14]. Emission reductions of about 85%–90% are possible using CCS with fossil-fuel-powered electricity generating stations [1]. Full life-cycle emissions reductions, from extracting the hydrocarbon resource to sequestering the CO_2, are in the range of 65%–85%.

If CCS is combined with biomass-based energy production (e.g., energy from agricultural wastes, energy crops such as switchgrass, and trees), even deeper reductions in emissions are possible [1], since use of biomass will displace CO_2 emissions from fossil fuels and sequestration of the CO_2 generated by combusting biomass will result in negative overall emissions. This strategy might be useful to compensate for or offset emissions from the transportation sector or other sectors for which traditional CCS is too costly.

8.4.2 Technological status of CO_2 capture and geological sequestration technology

Carbon dioxide capture with sequestration in deep geological formations is currently the most advanced and the most likely CCS approach to be deployed on a large scale in the coming decades. Almost every component of this technology is used on an industrial scale today, for a variety of purposes from fertilizer production to enhanced oil recovery. Lacking is the integrated experience applied to electricity generation and cost-effective approaches to CO_2 capture. Large-scale CO_2 sequestration in geological systems will also require practical experience in selection of appropriate sequestration sites, monitoring, risk management. and regulatory oversight.

Capture options

Today there are three principal methods for capturing CO_2 from fossil-fired power plants: **post-combustion capture**, **pre-combustion capture**, and **oxygen combustion** (see Box 8.2). Worldwide, significant scientific research and engineering development has been invested in all of these approaches over the past decade. Each has its advantages and disadvantages, as detailed in Table 8.2.

Post-combustion capture removes CO_2 from the flue gas of a power plant using chemical solvents, primarily regeneratable amines or, more recently, chilled ammonia [12]. Flue gas is pumped through a low-pressure gas/liquid contactor, where CO_2 partitions into the solvent, and then the amine solvent is heated to release nearly pure CO_2. Some solvents can be regenerated with so-called pressure-swing adsorption, which can reduce the amount of steam needed to regenerate the solvent. The advantages of post-combustion capture are that existing electricity generation plants could be retrofitted with a post-combustion capture unit and the technology is well established. Disadvantages include the large energy requirements to regenerate the amine and compress the CO_2 from atmospheric pressure to pipeline pressures in excess of 100 bar. For newly constructed plants with post-combustion capture, it might be possible to reduce costs and energy requirements significantly by more efficient energy integration and pre-concentration of the CO_2 before separation [15].

Carbon dioxide also can be captured using pre-combustion technology. The most efficient such approach, IGCC, integrates gasification of the coal, electricity generation with a gas turbine, and a low-pressure steam generator [12]. Pre-combustion capture separates the CO_2 by reacting fossil fuel with steam and air or oxygen before it is combusted. The reaction creates "**synthesis gas**" (or **syngas**), a mixture primarily of CO and H_2 (Chapter 39). The CO then reacts with steam to produce CO_2 and more H_2. The mixture is then separated to produce H_2, which can be used for electricity generation or other purposes, and CO_2 for sequestration. Advantages include lower compression requirements, lower energy use for the separation process, and the ability to produce hydrogen. Disadvantages include the lack of experience of the electrical utility industry with advanced chemical processing facilities such as IGCC; highly integrated electricity generation and capture, which could reduce reliability; and lack of widespread application of IGCC for the purpose of power production. Gasification is used today to produce ammonia for fertilizers and H_2 for petroleum processing. Currently, there are four IGCC units, two in Europe and two in the USA. Two additional commercial units are also being built in the USA.

A number of promising technologies utilize oxygen, rather than air, for combusting fossil fuels, so-called

Box 8.2. Overview of approaches for CO_2 capture

Three principal approaches to capturing CO_2 from an emission source are available: post-combustion capture, pre-combustion capture, and oxy-combustion.

Post-combustion capture is the most straightforward and, in principle, requires the least amount of changes to the power-generating plant. The schematic diagram shown in Figure 8.4 illustrates how a post-combustion capture unit scrubs CO_2 from the flue gas stream leaving the power plant. The flue gas stream is treated using chemical solvents to separate CO_2 (the combustion product) from nitrogen. Some fraction of the steam produced in the boiler is used in the separation process, resulting in a significant decrease in the efficiency of the power plant.

Oxy-combustion uses an entirely different approach for capture (see Figure 8.5). Instead of burning the fossil fuel in air, which consists of about 80% nitrogen, the fossil fuel is burned in oxygen. The oxygen is obtained by separating it from air, which requires a significant amount of electricity to operate the oxygen separation unit. In this case, the flue gas is a mixture primarily of CO_2 and water with minor amounts of contaminants, making the separation process comparatively simple.

Figure 8.4. Post-combustion capture (courtesy of ZEP).

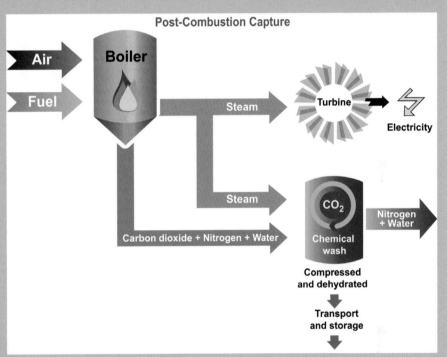

Figure 8.5. Oxy-combustion capture (courtesy of ZEP).

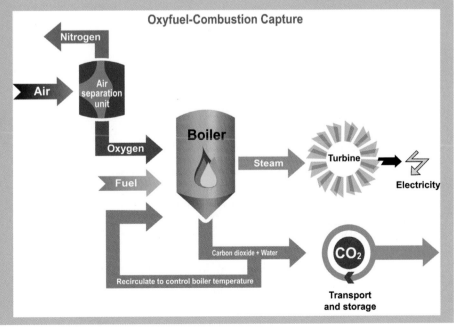

Box 8.2. (cont.)

The final and most complex approach to capture is called pre-combustion capture, which is often called integrated gasification combined cycle (IGCC) power generation (see Figure 8.6). Although complex, it is considered to be one of the most effficient approaches to CO_2 capture. Here, the fossil fuel is converted into H_2 before it is used to produce electricity. In the case of coal, the solids are converted into a gas before being combusted. This **gasification** involves partial combustion of coal in a low-oxygen environment to create syngas, a mixture of H_2 and CO. A second step involves the water–gas shift (WGS) reaction to create a mixture of CO_2 and H_2, which can be separated relatively easily with physical sorbents. The hydrogen can then be used in a gas turbine to produce electricity or sold for use in transportation or for upgrading petroleum products.

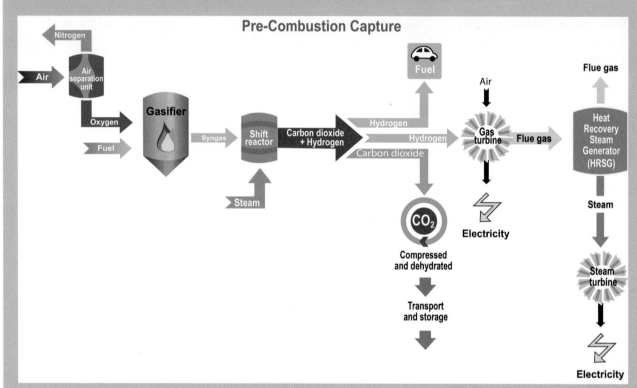

Figure 8.6. Pre-combustion capture (courtesy of ZEP).

Table 8.2. Comparative benefits of post-combustion, pre-combustion, and oxygen combustion CO_2 capture technologies

Technology	Advantages	Challenges
Post-combustion	Mature technology Standard retrofit	High energy penalty (~30%) High cost for capture
Pre-combustion (IGCC)	Lower costs than post-combustion Lower energy penalties (10%–15%) H_2 production	Complex chemical process Repowering Large capital investment
Oxygen combustion	Avoids complex post-combustion separation Potentially higher generation efficiencies	Oxygen separation Repowering

oxygen combustion [16]. The advantage of doing so is that the only gaseous emissions are water vapor, CO_2, and 5%–10% O_2 and N_2, with small amounts of SO_2 and NO_x. These are easily separated, and CO_2 can then be readily captured. The disadvantage is the energy and economic penalties associated with the production of oxygen as part of the process. However, some electricity generators favor oxygen combustion because it avoids the need for integrating complex chemical processing with electricity generation.

Large-scale demonstration projects of electricity generation with CCS are planned in many countries, namely the USA, Australia, Canada, China, the UK, Germany, Norway, and Abu Dhabi. Additional commercial and government-sponsored demonstration projects are likely to be announced worldwide over the next few years. Within less than a decade, it is expected that significant experience operating IGCC power plants with CO_2 sequestration will be available.

Over the past 5 years or so there has been a growing interest by governments, a new community of academic scientists, and industry in developing new materials and separation techniques that could dramatically improve the efficiency and lower the cost of CO_2 capture. While these new approaches are a long way from commercial implementation, in principle, they hold significant promise. Among these new methods are

- metal–organic frameworks, a new class of nano-structured hybrid materials with exceptionally high surface area that can improve the efficiency of absorption of traditional and novel organic solvents;
- ionic liquids with higher absorption rates and comparatively smaller energy penalties for regenerating the solvent;
- Si-based solvents requiring much less water for CO_2 capture;
- biologically motivated approaches that utilize nature-inspired catalysts such as carbonic anhydrase to capture and convert gaseous CO_2 to liquid or solid forms;
- membranes coated with CO_2-affinity materials to improve the selectivity and permeance to CO_2;
- hydrogen-conducting membranes for pre-combustion capture;
- catalytic membranes to simultaneously separate CO_2 and carry out the water–gas shift reaction;
- oxygen-seperation membranes to lower the cost of oxygen production; and
- solid adsorbents such as activated carbon, carbon nanotubes, and other nanostructured solids.

As these new approaches reach maturity they will need to compete on the basis of cost and performance with existing approaches for CO_2 capture as described above. These existing technologies are of course expected to improve as they become more widely deployed, For this reason, it is difficult to predict which of the existing and new technologies will emerge as market leaders.

Carbon dioxide sequestration in deep geological formations

Everywhere, under a thin veneer of soils or sediments, the Earth's surface is made up primarily of two types of rocks: those formed by cooling magma, either from volcanic eruptions or from magmatic intrusions far beneath the land surface; and those formed as thick accumulations of sand, clay, salts, and carbonates over millions of years. The latter types of rocks occur primarily in what are termed sedimentary basins. Geographical locations overlying sedimentary basins are best suited for geological sequestration of CO_2, and, fortuitously, the majority of CO_2 sources are located in or near sedimentary basins.

Sedimentary basins often contain many thousands of meters of rocks where the tiny pores in the rocks are filled with salt water (saline formations) or where oil and gas reservoirs are found. Sedimentary basins consist of many layers of sand, silt, clay, carbonate, and evaporite (rock formations composed of salt deposited from evaporating water), as illustrated in Figure 8.7. The sand layers provide sequestration space for oil, water, and natural gas. The silt, clay, and evaporite layers provide the seal that can trap these fluids underground for millions of years and longer. Geological sequestration of CO_2 would take place deep in sedimentary basins trapped below silt and clay layers, in much the same way as oil and natural gas are trapped today [17,18].

The presence of an overlying, thick, and continuous layer of silt, clay, or evaporite is the single-most important feature of a geological formation that is suitable for geological sequestration of CO_2. These fine-textured rocks physically prevent the upward migration of CO_2 by a combination of viscous and capillary forces. Oil and gas reservoirs are found under such fine-textured rocks, and the mere presence of the oil and gas demonstrates the existence of a suitable reservoir seal. In saline formations, where the pore space is initially filled with water, after the CO_2 has been underground for hundreds to thousands of years, chemical reactions will dissolve some or all of the CO_2 in the salt water, and, eventually, some fraction of the CO_2 will be converted to carbonate minerals, thus becoming a part of the rock itself [18].

One of the key questions for geological sequestration is how long will the CO_2 remain trapped underground? This question is best addressed from the perspective of how long it must remain trapped for this to be an effective method for avoiding CO_2 emissions to the atmosphere. Although there is no generally accepted answer to the latter question, most studies agree that, if

Figure 8.7. An example of a sedimentary basin suitable for sequestration, showing a cross section of the rock types, including the sandstone reservoirs and shale seals.

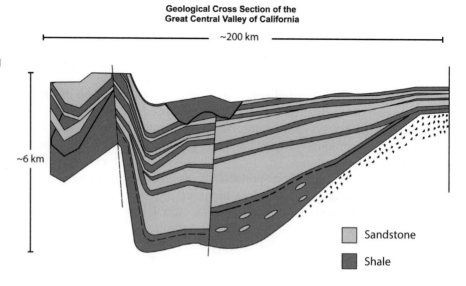

Geological Cross Section of the
Great Central Valley of California

~200 km

~6 km

Sandstone

Shale

more than 90% of the CO_2 remains underground over a 1000-year period, CCS will be a very effective method for avoiding CO_2 emissions [19]. The question then becomes will geological formations retain CO_2 over such time periods?

A number of lines of evidence suggest that, for well-selected and -managed sequestration formations, retention rates will be very high and more than sufficient for the purpose of avoiding CO_2 emissions into the atmosphere [1]. Specifically, these items of evidence are as follows.

- Natural oil, gas, and CO_2 reservoirs demonstrate that buoyant fluids such as CO_2 can be trapped underground for millions of years.
- For industrial analogs such as natural-gas sequestration, CO_2-EOR, acid-gas injection, and liquid-waste-disposal operations methods for injecting and storing fluids without compromising the integrity of the caprock or the sequestration formation have been developed.
- Multiple processes contribute to long-term retention of CO_2, including physical trapping beneath low-permeability rocks, dissolution of CO_2 in brine, capillary trapping of CO_2, adsorption on coal, and mineral trapping. Together, these trapping mechanisms increase the security of sequestration over time, thus further diminishing the possibility of potential leakage and surface release.
- Early experiences at the Sleipner Project in the North Sea, the Weyburn Project in Saskatchewan, Canada, and the In Salah Project in Algeria have shown that the sequestered CO_2 can successfully be contained with no significant safety problems.

The technology for storing CO_2 in deep underground formations is adapted from oil and gas exploration and production technology. For example, technologies for drilling and monitoring wells that can safely inject

CO_2 into the sequestration formation are available. Methods for characterizing a site are fairly well developed. Models to predict where the CO_2 moves when it is pumped underground are available, although more work is needed to further develop and test these models, particularly over the long time-frames and large spatial scales envisioned for CO_2 sequestration. The subsurface movement of CO_2 is currently being successfully monitored at several sites, although, again, more work is needed to refine and test the monitoring methods.

8.4.3 Existing and planned CO_2 capture and sequestration projects

Today, 4–5 Mt per year of CO_2 is captured and intentionally sequestered in deep geological formations at the Sleipner, Weyburn, In Salah, and Snøhvit projects [9] [10] [11]. The Sleipner and In Salah Projects were designed with CCS as their primary purpose. The Weyburn Project was designed initially as an enhanced-oil-recovery project, but has evolved in to a project that combines enhanced oil recovery with CO_2 sequestration. Today, over 32 years of cumulative experience has been gained from these projects.

Vast experience pumping CO_2 into oil reservoirs also comes from nearly 30 years of CO_2-enhanced oil recovery (CO_2-EOR), for which, today, nearly 40 Mt is injected every year. About 70 projects are under way worldwide, with the vast majority in the USA, in west Texas, Colorado, Wyoming, and most recently, Mississippi. When CO_2 is pumped into an oil reservoir, it mixes with the oil, lowering the viscosity and density of the oil. Under optimal conditions, oil and CO_2 are miscible, which results in efficient displacement of the oil from the pore spaces in the rock. An estimated increase in oil recovery of 10%–15% of the initial volume of oil-in-place is expected for successful CO_2-EOR projects. Not all of

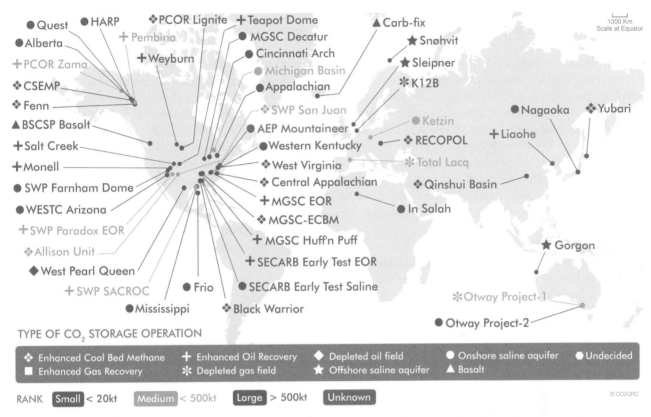

Figure 8.8. Locations of worldwide CCS projects in various stages of development (Courtesy of Peter Cook of CO2CRC).

the injected CO_2 stays underground, since 30%–60% is typically produced back with the oil. On the surface, the produced CO_2 is separated from the oil and re-injected back into the reservoir. If CO_2 is left in the reservoir after oil production stops, most of the CO_2 injected over the project lifetime remains stored underground.

The majority of CO_2-EOR projects today use CO_2 from naturally occurring CO_2 reservoirs. The high cost and limited availability of CO_2 has restricted deployment of CO_2-EOR to those areas with favorable geological conditions and a readily available source of CO_2. A few projects use CO_2 captured from industrial sources, notably the Weyburn Project discussed in Section 8.4.1 and the Salt Creek Project in Wyoming, which is injecting several million tonnes per year.

The extent to which CO_2-EOR will spur early investment and deployment in CCS remains uncertain. Oil-price volatility discourages investment in CO_2-EOR projects because of the comparatively high capital and operating costs. Prospects for obtaining tradable credits for storing CO_2 could make the economics of CO_2-EOR more favorable, but only if the cost of carbon credits is in excess of about $30\,\mathrm{t}^{-1}$ of CO_2.

Planned industrial and pilot-scale projects

A map showing existing and planned CCS projects is provided in Figure 8.8. Plans for new CCS projects are being announced at a rate of several each year, although many of the announced projects have been delayed significantly. The following are a few examples. In Australia, the Gorgon Project has proposed to produce liquefied natural gas (LNG) and store nearly 5 Mt of CO_2 per year in a deep saline formation [20]. In early 2006, a joint venture to produce 500 MW of electricity from petroleum-coke was announced by BP. Initially, the project was expected to be located in Long Beach, CA, but it was eventually moved to the Bakersfield, CA, area where the prospects for CO_2-EOR are more favorable. The GreenGen Project in China will produce electricity using IGCC technology. The second phase of the project, which is expected to be initiated by 2016, will include capture of CO_2 and sequestration in an oil field.

In addition to these industrial-scale projects, dozens of small-scale geological sequestration pilot projects are under way worldwide, and more are expected. For example, in the USA, the Department of Energy has sponsored seven Regional Sequestration Partnerships. Through 2014, these partnerships will conduct 25 pilot tests of sequestration in geological formations. These pilot and demonstration projects will help to assess the geographical extent and capacity of geological formations. Similar pilot tests are being carried out in Australia, Canada, Germany, Japan, the Netherlands, and Poland, and many more countries are expected to announce plans for pilot tests soon.

8.4.4 Health, safety, and environmental risks of CO_2 capture and sequestration

Carbon dioxide is used in a wide variety of industries, from chemical production to beverage carbonation and brewing, from enhanced oil recovery to refrigeration, and from fire suppression to inert-atmosphere food preservation. Because of its extensive use and production, the hazards of CO_2 are well known and routinely managed. Engineering and procedural controls for dealing with the hazards of compressed and cryogenic CO_2 are well established. Carbon dioxide capture and transportation pose no unique risks that are not managed routinely in comparable operations.

Although CO_2 is generally regarded as a safe and nontoxic, inert gas, exposure to elevated concentrations of CO_2 can lead to adverse consequences. In particular, because CO_2 is denser than air, hazardous situations arise when large amounts of CO_2 accumulate in low-lying, confined, or poorly ventilated spaces. Although the chances of this occurring are very low, if a large amount of injected CO_2 were to escape from a sequestration site, it could present risks to health and the local environment. Such releases could be associated with surface facilities, injection wells, or leakage from the sequestration formation itself. They could be small-scale diffuse leaks or leaks concentrated near the injection facilities. Leakage, if unchecked, could harm groundwater and ecosystems. Persistent leaks could suppress respiration in the root zone or result in soil acidification and eventually lead to tree kills such as those associated with soil gas concentrations in the range of 20%–30% CO_2 observed at Mammoth Mountain, CA, where volcanic outgassing of CO_2 has been occurring for several decades [21].

Analogous experience with gas and liquid injection derived from seasonal sequestration of natural gas [22], disposal of liquid wastes [23], acid-gas injection [24], and oil-field operations shows that underground injection activities can be carried out safely with appropriate precautions. In the unusual circumstances where leakage or surface releases occur, they are mostly caused by leakage from the injection well or leakage from wells that were drilled long ago and not properly sealed (so-called abandoned wells) [25]. Pumping cement into such leaking wells can reseal them. One of the greatest challenges to CCS in the USA and Canada, where many millions of wells have been drilled, is to locate, evaluate, and seal them before underground injection operations begin [26].

Extensive industrial experience with injection of CO_2 and gases in general indicates that risks from geological sequestration facilities are manageable using standard engineering controls and procedures. Regulatory oversight and institutional controls further enhance the safety of these operations and ensure that the site-selection and -monitoring strategies are robust. For CCS employed on a scale comparable to existing industrial analogs, the associated risks are comparable to those of today's oil and gas operations. Eventually, if CCS were to be deployed on the grand scale needed to significantly reduce CO_2 emissions (billions of tonnes annually), the scale of operations would increase to become as large as or larger than existing oil and gas operations [27]. In this eventuality, experience gained in the early years of CCS would be critical for assessing and managing the risks of very large-scale geological sequestration projects.

8.4.5 Monitoring and verification of CO_2 sequestration in deep geological formations

Although retention rates for well-selected and -managed sites are expected to be high, verification that the CO_2 remains in the sequestration reservoir is important for assuring effective containment. Various monitoring approaches are available, many of which were developed for application in the oil and gas industry. Monitoring methods such as seismic imaging, the observation of sound waves propagating through the Earth, have been used successfully to locate CO_2 injected at Sleipner, at Weyburn, and in several pilot projects [28] [29] [30]. Methods are also available to monitor the injection wells to ensure that injection rates and pressures stay within defined operating parameters [1]. Periodic inspection of an injection well using "well logs," information collected by lowering electronic instruments into a well, can be used to check that the construction and condition of the well is satisfactory [1].

It is also possible to directly monitor the ground surface to detect whether CO_2 is seeping back into the atmosphere. Eddy-covariance and flux-accumulation-chamber methods developed to study cycling of CO_2 between the atmosphere and the biosphere are expected to have the sensitivity needed to detect even small amounts of seepage [31]. Recently, a new method to detect CO_2 leakage using the stable isotopes ^{12}C and ^{13}C was developed [32]. Introduced tracers such as ^{14}C, fluorocarbons and cholorofluorocarbons may also be useful for tracking migration of CO_2.

Certainly, more work is needed to test, enhance, and validate the performance of monitoring technologies available today. In addition, research and development work is likely to discover more efficient and cost-effective approaches for monitoring and verifying the performance of geological sequestration projects. Nevertheless, even with today's technologies, prospects for reliable monitoring are good. Over time, as experience with CO_2 sequestration projects grows, standard protocols for monitoring and verification should be developed.

8.4.6 Cost of CO_2 capture and sequestration

With today's technology, estimated additional costs for generating electricity from a coal-fired power plant with CCS range from $50 to $100 per tonne of CO_2 avoided [1] [33] [34]. These costs are mainly dependent on the capture technology and concentration of CO_2 in the stream from which it is captured [1] [12]. This metric is useful for comparing the cost of CCS with other methods of reducing CO_2 emissions. Another metric is the increase in costs of electricity generation. Costs would increase by $0.01–0.05 per kW·h [8] depending on the design of the power plant and a number of site-specific factors, or the equivalent of about a 50%–100% increase in the costs of base-load power generation from a newly constructed power plant. Because consumers pay for generation plus other costs associated with the delivery and management of the electrical system, percentage increases to consumers associated with CCS would be somewhat less than this and would differ depending on the cost of delivered electricity. Capture and compression typically account for over 75% of the costs of CCS,

with the remaining costs attributed to transportation and underground sequestration. Pipeline transportation costs are highly site-specific, depending strongly on economy of scale and pipeline length.

In addition to the cost of CCS, the "energy penalty" for capture and compression must be considered. Post-combustion capture technologies use up to 30% of the total energy produced, thus significantly decreasing the overall efficiency of the power plant. Tight integration between the power-generation plant and the separation facility could lower these energy requirements, as could improvements in the performance of solvents [15]. Oxy-combustion has a similarly high energy penalty, although new materials might eventually lower the energy penalty by allowing higher-temperature and, consequently, more efficient combustion [1]. Pre-combustion technologies have the potential to lower energy penalties to 10%–15%, leading to higher overall efficiency and lower capture costs. Public and privately sponsored research and development programs are aggressively trying to lower the costs of CO_2 capture.

Table 8.3. The status of CCS technology

	Technology		
	Current	**In 10 years**	**In 25 years**
Key materials	Sorbents and solvents Gas turbines Contactor towers Compressors Combustion chambers Boilers	Sorbents and solvents Gas turbines High-temperature combustion chambers	Sorbents and solvents H_2 storage Fuel cells for more efficient electricity production from H_2
Cost of technology	$100–200 t^{-1} of CO_2 avoided	$50–100 t^{-1} of CO_2 avoided	$20–40 t^{-1} of CO_2 avoided
Cost of power produced	$0.10–0.15 per kW·h	$0.08–0.12 per kW·h	$0.06–0.08 per kW·h
Energy used	10%–30% of primary energy	10%–20% of primary energy	8%–15% of primary energy
Limitations of current technology	Cost of capture Energy use of capture Confidence in geological sequestration Lack of fully developed legal and regulatory framework for CCS	Cost of capture Energy use for capture Confidence in geological sequestration in a wide range of geological environments	Hydrogen storage for dealing with imbalance of supply and demand
Environmental roadblocks	Concerns about groundwater pollution and leakage of CO_2 back into the atmosphere	Concerns about groundwater pollution and leakage of CO_2 back into the atmosphere	
Key roadblocks to technology	Cost and energy requirements for capture	Cost and energy requirements for capture	

The fact that electricity generation is more costly with CCS than without it suggests that sustained policy initiatives will be needed to stimulate deployment. Governments around the world have now committed billions of dollars to stimulate early deployment, which is expected to reduce costs for future deployments through technological learning.

8.5 Summary

At first impression, CO_2 capture and sequestration in geological formations might appear to be a radical idea that would be difficult and perhaps risky to employ. Closer analysis, however, reveals that many of the component technologies are mature. Extensive experience with gasification, CO_2 capture, and underground injection of gases and liquids provides the foundation for future CCS operations. No doubt, challenges lie ahead for CCS. The cost of capture, the potential large scale of deployment of geological sequestration, and the process of adapting the existing energy infrastructure to accommodate CCS are significant hurdles to be overcome. For now, however, none of these difficulties seems to be insurmountable, and progress marches on through continued deployment of industrial-scale projects, research and development, and growing public awareness of this important option for lowering CO_2 emissions. Table 8.3 summarizes the current status of this technology and projections for future developments.

8.6 Questions for discussion

1. What sequestration options are available near where you live?
2. Do you think that CCS is likely to be viewed favorably in your community?
3. Are oil and gas reservoirs good analogs for CO_2 sequestration, and, if so, why?
4. Is CCS likely to be important in the coming transformation of our energy supply system, and, if so, why?
5. If renewable-energy technology advances and is deployed more quickly than expected, will CCS have a role to play in emission reductions?
6. What are the trade-offs and synergies between nuclear energy and CCS? To what extent do they compete with each other?
7. Will it be possible to reduce global CO_2 emissions quickly enough if CCS is not available to reduce emissions from fossil-fuel use?
8. What new approaches for CO_2 capture may significantly reduce the cost and energy penalty? What new materials could contribute to better capture technologies?

8.7 Further reading

- **B. Metz O. Davidson**, **H. C. de Coninck**, and **M. Loos**, 2005, *IPCC Special Report on Carbon Dioxide Capture and Storage*, Cambridge, Cambridge University Press.
- **IEA**, 2009, *Assessment of Carbon Capture and Storage Technology.*
- **US Department of Energy**, 2008, *North American Carbon Sequestration Atlas*, Washington, DC, US Department of Energy.
- **D. Thomas** and **S. M. Benson** (eds.), 2005, *Carbon Dioxide Capture for Storage in Deep Geologic Formations – Results from the CO_2 Capture Project, Vol. 1 and 2: Geologic Storage of Carbon Dioxide with Monitoring and Verification*, Oxford, Elsevier Publishing.
- **S. M. Benson** and **D. R. Cole**, 2008, "CO_2 Sequestration in deep sedimentary formations," in *Elements*, Vol. 4, pp. 325–331, doi: 10.2113/gselements.4.5.325.
- **World Resources Institute**, 2009, *CCS Guidelines.*

8.8 References

[1] **IPCC**, 2006, *Special Report on Carbon Dioxide Capture and Storage*, Cambridge, Cambridge University Press.
[2] **S. M. Benson**, 2005, *Carbon Dioxide Capture for Storage in Deep Geological Formations – Results from the CO_2 Capture Project. Volume 2: Geologic Storage of Carbon Dioxide with Monitoring and Verification*, Oxford, Elsevier.
[3] **K. O. Buesseler**, **J. E. Andrews**, **S. M. Pike**, and **M. A. Charette**, 2004, "The effects of iron fertilization on carbon sequestration in the Southern Ocean," *Science*, **304**, 414–417.
[4] **P. Kauppi**, and **R. Sedjo**, 2001, "Technological and economic potential of options to enhance, maintain, and manage biological carbon reserves and geo-engineering," in *Climate Change 2001: Mitigation: Contribution of Working Group III to the Third Assessment Report of the Intergovernmental Panel on Climate Change*, eds. **B. Metz**, **O. Davidson**, **R. Swart**, and **J. Pan**, Cambridge, Cambridge University Press pp. 301–343.
[5] **IEA**, 2010, *Energy Forecast.*
[6] **V. Smil**, 2010, *Energy Transitions: History, Requirements, Prospects*, Santa Barbara, CA, Praeger Press.
[7] **K. Riahi**, **F. Dentener**, **D. Gielen** *et al.*, 2011, "Energy transition pathways for sustainable development," in *Global Energy Assessment*, Cambridge, Cambridge University Press.
[8] **IEA**, 2004, *A Review of Global Capacity Estimates for the Geological Storage of Carbon Dioxide*, IEA Greenhouse Gas R&D Programme Technical Review (TR4).
[9] **T. A. Torp** and **J. Gale**, 2003, "Demonstrating storage of CO_2 in geological reservoirs: the Sleipner and SACS projects," in *Proceedings of the 6th International Conference on Greenhouse Gas Control Technologies (GHGT-6)*, eds. **J. Gale** and **Y. Kaya**, Amsterdam, Pergamon, vol. I, pp. 311–316.

[10] **F. A. Riddiford**, **A. Tourqui**, **C. D. Bishop**, **B. Taylor**, and **M. Smith**, 2003, "A cleaner development: the In Salah Gas Project, Algeria," *Proceedings of the 6th International Conference on Greenhouse Gas Control Technologies (GHGT-6)*, eds. **J. Gale** and **Y. Kaya**, Amsterdam, Pergamon, vol. I, pp. 601–606.

[11] **R. Moberg**, **D. B. Stewart**, and **D. Stachniak**, 2003, "The IEA Weyburn CO_2 Monitoring and Storage Project," in *Proceedings of the 6th International Conference on Greenhouse Gas Control Technologies (GHGT-6)*, eds. **J. Gale** and **Y. Kaya**, Amsterdam, Pergamon, vol. I, pp. 219–224.

[12] **D. C. Thomas** and **H. R. Kerr**, 2005, "The CO_2 Capture Project Introduction," in *Carbon Dioxide Capture for Storage in Deep Geologic Formations – Results from the CO_2 Capture Project, Volume 1: Capture and Separation of Carbon Dioxide from Combustion Sources*, ed. **D. C. Thomas**, London, Elsevier Science, pp. 1–15.

[13] **P. Chiesa**, **S. Consonni**, **T. Kreutz**, and **R. Williams**, 2005, "Co-production of hydrogen, electricity and CO_2 from coal with commercially ready technology. Part A: performance and emissions," *Int. J. Hydrogen Energy*, **30**, 747–767.

[14] **G. Wotzak**, **N. Z. Shilling**, **G. Simons**, and **K. Yackly**, 2005, "An evaluation of conversion of gas turbines to hydrogen fuel," in *Carbon Dioxide Capture for Storage in Deep Geologic Formations – Results from the CO_2 Capture Project, Volume 1: Capture and Separation of Carbon Dioxide from Combustion Sources*, ed. **D. C. Thomas**, London, Elsevier Science, pp. 427–440.

[15] **G. Choi**, **R. Chu**, **B. Degen** *et al.*, 2005, "CO_2 Removal from power plant flue gas – cost efficient design and integration study," in *Carbon Dioxide Capture for Storage in Deep Geologic Formations – Results from the CO_2 Capture Project, Volume 1: Capture and Separation of Carbon Dioxide from Combustion Sources*, ed. **D. C. Thomas**, London, Elsevier Science, pp. 99–116.

[16] **I. Miracca**, **K. I. Aasen**, **T. Brownsecombe**, **K. Gerdes**, and **M. Simmonds**, 2005, "Oxygen combustion for CO_2 capture technology," in *Carbon Dioxide Capture for Storage in Deep Geologic Formations – Results from the CO_2 Capture Project, Volume 1: Capture and Separation of Carbon Dioxide from Combustion Sources*, ed. **D. C. Thomas**, London, Elsevier Science, pp. 99–116.

[17] **S. Holloway** (ed.), 1996, *The Underground Disposal of Carbon Dioxide*, Final report of Joule 2 Project No. CT92–0031, British Geological Survey, Keyworth, Nottingham.

[18] **W. D. Gunter**, **S. Bachu**, and **S. Benson**, 2004, "The role of hydrogeological and geochemical trapping in sedimentary basins for secure geological storage for carbon dioxide," in *Geological Storage of Carbon Dioxide: Technology*, eds. **S. Baines** and **R. H. Worden**, London, Geological Society, pp. 129–145.

[19] **R. Hepple** and **S. M. Benson**, 2005, "Geologic storage of carbon dioxide as a climate change mitigation strategy: performance requirements for surface seepage," *Environmental Geol.*, **47**, 576–585.

[20] **P. M. Oen**, 2003, "The development of the Greater Gorgon Gas Fields," *APPEA J.*, **43**(2), 167–177.

[21] **B. Martini** and **E. Silver**, 2002, "The evolution and present state of tree-kills on Mammoth Mountain, California: tracking volcanogenic CO_2 and its lethal effects," in *Proceedings of the 2002 AVIRIS Airborne Geoscience Workshop*, Jet Propulsion Laboratory, California Institute of Technology, Pasadena, CA.

[22] **K. F. Perry**, 2005, "Natural gas storage industry experience and technology: potential application to CO_2 geological storage," in *Carbon Dioxide Capture for Storage in Deep Geologic Formations – Results from the CO_2 Capture Project, Volume 2: Geologic Storage of Carbon Dioxide with Monitoring and Verification*, ed. **S. M. Benson**, London, Elsevier Science, pp. 815–826.

[23] **J. Apps**, 2005, "The regulatory climate governing the disposal of liquid wastes in deep geologic formations: a paradigm for regulations for the subsurface disposal of CO_2," in *Carbon Dioxide Capture for Storage in Deep Geologic Formations – Results from the CO_2 Capture Project, Volume 2: Geologic Storage of Carbon Dioxide with Monitoring and Verification*, ed. **S. M. Benson**, London, Elsevier Science, pp. 1163–1188.

[24] **Bachu S.** and **K. Haug**, 2005, "In-situ characteristics of acid-gas injection operations in the Alberta basin, western Canada: demonstration of CO_2 geological storage," in *Carbon Dioxide Capture for Storage in Deep Geologic Formations – Results from the CO_2 Capture Project, Volume 2: Geologic Storage of Carbon Dioxide with Monitoring and Verification*, ed. **S. M. Benson**, London, Elsevier Science, pp. 867–876.

[25] **Benson, S. M.**, 2005, "Lessons learned from industrial and natural analogs for health, safety and environmental risk assessment for geologic storage of carbon dioxide," in *Carbon Dioxide Capture for Storage in Deep Geologic Formations – Results from the CO_2 Capture Project, Volume 2: Geologic Storage of Carbon Dioxide with Monitoring and Verification*, ed. **S. M. Benson**, London, Elsevier Science, pp. 1133–1141.

[26] **S. E. Gasda**, **S. Bachu**, and **M. A. Celia**, 2004, "The potential for CO_2 leakage from storage sites in geological media: analysis of well distribution in mature sedimentary basins," *Environmental Geol.*, **46**(6–7), 707–720.

[27] **R. Burruss** (ed.), 2004, *Geologic Storage of Carbon Dioxide in the Next 50 Years: An Energy Resource Perspective*, Washington, DC, Pew Center.

[28] **R. A. Chadwick**, **R. Arts**, and **O. Eiken**, 2005, "4D Seismic quantification of a growing CO_2 plume at Sleipner, North Sea," in *Petroleum Geology: North West Europe and Global Perspectives – Proceedings of the 6th Petroleum Geology Conference*, **A. G. Dore** and **B. Vining** London, Geological Society, 15 pp.

[29] **S. D. Hovorka**, **S. M. Benson**, **C. Doughty** *et al.*, 2005, *Measuring Permanence of CO_2 Storage in Saline Formations: the Frio Experiment, Environmental Geosciences* Special Issue.

[30] **D. White** (ed.), 2005, "Theme 2: prediction, monitoring and verification of CO_2 movements," in *Proceedings of the 7th International Conference on Greenhouse Gas Control Technologies (GHGT-7)*, ed. **M. Wilson** and **M. Monea**, Amsterdam, Pergamon, vol. III, pp. 73–148.

[31] **N. Miles**, **K. Davis**, and **J. Wyngaard**, 2005, "Detecting leaks from CO_2 reservoirs using micrometeorological methods," *Carbon Dioxide Capture for Storage in Deep Geologic Formations – Results from the CO_2 Capture Project, Volume 2: Geologic Storage of Carbon Dioxide with Monitoring and Verification*, ed. **S. M. Benson**, London, Elsevier Science, pp. 1031–1044.

[32] **S. C. Krevor**, **J.-C. Perrin**, **A. Esposito**, **C. Rella**, and **S. Benson**, 2010, "Rapid detection and characterization of surface CO_2 leakage through the realtime measurement of ^{13}C signatures in CO_2 flux from the ground," *Int. J. Greenhouse Gas Control*, doi:10.1016/j.ijggc.2010.05.002.

[33] **E. Rubin** and **A. Rao**, 2004, "Uncertainties in CO_2 capture and sequestration costs," in *Proceedings of Greenhouse Gas Control Technologies 6th International Conference (GHGT-6)*, ed. **J. Gale** and **Y. Kaya**, London, Elsevier Science, vol. 1, pp. 1119–1124.

[34] **H. Herzog**, 1999, "The economics of CO_2 capture," in *Greenhouse Gas Control Technologies, Proceedings of the 4th International Conference on Greenhouse Gas Control technologies*, eds. **B. Eliason**, **P. Reimer**, and **A. Wokaun**, Oxford, Pergamon.

PART 2

Nonrenewable energy sources

9 Petroleum and natural gas

Russell R. Chianelli,[1] Xiomara C. Kretschmer,[1] and Stephen A. Holditch[2]

[1]University of Texas at El Paso, El Paso, TX, USA. [2]Texas A&M University, College Station, TX, USA

9.1 Focus

Petroleum and natural gas have been the mainstay of energy production in developed countries. Global energy demand will continue to increase with "globalization." Oil and natural gas will continue to supply a majority of our energy in the near future and production will be from natural sources of petroleum, coal, and natural gas. For example, Energy Independence has reported that the USA has an estimated 260 billion tons of recoverable coal, equivalent to three or four times as much energy in coal as Saudi Arabia has in oil [1]. The needed increase requires the exploitation of conventional and unconventional reservoirs of oil and gas in an "environmentally friendly" manner. This necessitates advances in materials in the form of better catalysts to produce clean fuels and advanced materials for high-pressure, high-temperature, and high-stress processes.

9.2 Synopsis

The National Petroleum Council (NPC) in the USA recently published a report entitled *Facing the Hard Truths about Energy* that evaluates oil and gas supply and demand in the early part of the twenty-first century [2]. The report concluded that the total global demand for energy will grow by 50%–60% by 2030 due to the increase in world population and higher average standards of living in some developing countries. Clearly, for the next few decades, oil, gas, and coal will continue to be the primary energy sources. The energy industry will have to continue increasing the supply of hydrocarbon fuels to meet the global energy demand. There are ample hydrocarbon resources to meet the demand well into the twenty-first century. The volumes of oil and natural gas located in unconventional reservoirs are much larger than the conventional reservoirs currently used for what has been produced thus far.

Unconventional natural-gas deposits differ from conventional deposits because the gas is more difficult and expensive to extract than for conventional gas deposits. Examples of unconventional gas deposits are deep gas, tight gas, gas contained within shales, coalbed methane, geopressurized zones, and arctic and sub-sea hydrates.

The same is true for unconventional oils: they are more difficult and expensive to extract and may have a more negative environmental impact than conventional reserves. Examples of unconventional oil sources are extra heavy oil and oil sands, tight sands, and oil shale, to name a few.

The key to the future required growth is the development of new technology that allows the industry to produce oil and gas from unconventional reservoirs in an environmentally acceptable manner. Thus CO_2 sequestration and environmentally friendly operations will be a large part of developing new resources. Throughout the process, the development of materials, especially those that can withstand high-pressure, high-temperature, and high-stress conditions will be important to the entire industry. The needed research in energy materials has recently been reviewed in the *Materials Research Bulletin* [3].

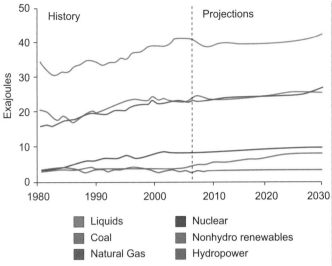

Figure 9.1. US energy consumption by fuel (1980–2030), [4]. Source: US Energy Information Administration (2010).

Table 9.1. Longevity of supply for selected countries [2]	
Country	Years remaining for current oil reserves producing at current oil flow rates
Saudi Arabia	75
Iran	87
Iraq	168
Kuwait	105
United Arab Emirates	70
Russia	20
Venezuela	52
USA	16

9.3 Historical perspective

Looking at US energy consumption by fuel type illustrates that, for the next few decades, oil, gas, and coal will continue to be the primary energy sources. Figure 9.1 will quickly show that the amount of all sustainable renewable energy in the USA is very small, with forecast increases providing less than 10% of our total use. The energy industry will have to continue increasing the supply of hydrocarbon fuels to meet global energy demand. This is particularly true for liquid transportation fuels, with hydrogen- and electric-powered cars well off in the future due to necessary improvements in hydrogen storage and battery materials that must be accomplished in order for these to be commercially viable.

9.4 Global energy oil and gas supply and demand

The global demand for oil in 2000 was 76 million barrels per day (bbls per day). Oil production is currently about 86 million bbls per day or 40,000 gallons per second or 31.4 billion barrels per year. The NPC estimates that the demand for oil will be 103–138 million bbls per day, or 37.6–50.4 billion bbls per year by 2030 [1].

"Global" conventional oil reserves are concentrated in the Middle East. A detailed history of the discovery and development of the petroleum reserves was written by Daniel Yergin [5]. The seven countries with the largest conventional oil reserves account for more than 70% of the world total. Saudi Arabia holds approximately 20% of the conventional reserves. Table 9.1 presents the ratio of current booked oil reserves divided by current oil-producing rate for selected countries. Notice that the

countries in the Middle East have sufficient reserves to produce at current flow rates for at least the first half of the twenty-first century. In addition, these countries are still exploring and working to increase both reserves and deliverability in existing fields. Owing to the use of new technology, world oil reserves have been increasing during the past 12 years.

As with oil, most of the reserves of natural gas are in the Middle East, approximately 44%. The remaining gas distribution is 18% in Europe and Eurasia (mostly Russia), 11% in Asia, 10% in Africa, and the rest in North America and South America. The largest natural-gas-reserve holders are Russia, Iran, Qatar, Saudi Arabia, the UAE, and the USA. In the coming decades, much of the natural gas in the Middle East will be converted to liquid natural gas (LNG) or liquid fuels and exported to Europe and North America.

9.4.1 Conventional oil

Colin Campbell studied the Petroconsultants worldwide database and concluded that, since inception, the industry had produced around 784 billion barrels of oil through the end of 1996 [3]. Campbell acknowledged that the records were not perfect, but he believed his estimate was very reasonable. Taking Campbell's value of 784 billion barrels through 1996, and the global oil-production values from the BP website http://www.bp.com/sectionbodycopy.do?categoryid=7500&contentid=7068481 for the past ten years, the oil industry has produced around 1063 billion or 1.063 trillion barrels of oil since the industry began in the late 1800s.

The industry believes it can produce another 1.25 trillion barrels from known fields. Thus, the industry has produced around 1 trillion barrels to date, and believes it can produce the second trillion in the twenty-first century. The Society of Petroleum Engineers recently held a research conference focusing on the technology needed to produce the third trillion barrels. Considering the information discussed above, it is clear that we will not be running out of oil any time soon, but environmental issues associated with producing and burning this fuel must be taken into consideration.

9.4.3 Conventional natural gas

The situation for natural-gas supply is even more optimistic than for oil. The world production of natural gas in 2000 was $6.9\times10^9\,m^3$ per day or $3\times10^{12}\,m^3$ per year. In 2030, the NPC projects that the demand will increase to $3.7–6\times10^{12}\,m^3$ per year. Most of the natural gas is currently used in North America, Europe, and Russia to generate electricity, to generate heat, and for a variety of residential and commercial uses.

The NPC suggests that the global "gas-in-place" number for natural gas is about $1.4\times10^5\,m^3$ largely unconventional gas. The NPC also suggests that the industry has already produced $8.5\times10^{13}\,m^3$ and has remaining reserves of $2\times10^{14}\,m^3$ which implies that the industry knows where the remaining reserves are located and that it can produce the gas economically.

9.5 Unconventional oil and gas

Nevertheless, the supply of hydrocarbon fuels must be increased to keep pace with increasing global energy demand. The volumes of oil and natural gas located in unconventional reservoirs are much larger than volumes in conventional reservoirs and energy independence will require significant effort to increase the supply of petroleum-based fuels from heavy petroleum supplies that are abundant in North America. See Box 9.1.

Box 9.1

Unconventional resources is a term commonly used to refer to reservoirs that produce oil or gas at very low flow rates due to low permeability, geological complexity or high fluid viscosity. Many of the low-permeability reservoirs that have been developed in the past are sandstone, but significant quantities of unconventional oil and gas are also produced from low-permeability carbonates, shales, and coal seams

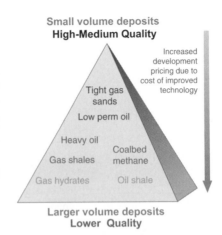

Figure 9.2. The resource triangle [2].

One way to define unconventional reservoirs is as follows: "natural gas or oil that cannot be produced at economic flow rates or in economic volumes unless the well is stimulated by a large hydraulic fracture treatment, a horizontal wellbore, multilateral wellbores, or an exotic technique such as steam injection."

The DOE recently reported that the available amounts of heavy hydrocarbons (tars, coal, and shale oils) in North America (the USA, Canada, and Mexico) are sufficient for 500 years at current rates of usage [6]. Technology to convert coal to liquid hydrocarbons is also a major potential source for petroleum-based fuels. Although large amounts of heavy petroleum products are available in North America, increasingly they occur in heavier and heavier forms. In Figure 9.2 we see the various types of heavy hydrocarbons and unconventional hydrocarbons found in large quantities in North America.

The high- and medium-quality petroleum sources are refined using conventional refinery techniques. However, as the petroleum sources become heavier, they contain more sulfur and better hydrodesulfurization (HDS) catalysts are needed. These catalysts based on MoS_2 and WS_2 have been the subject of basic and applied research for many years, and there is a deep well of basic and applied understanding that has recently been reviewed [7].

Unconventional oil and gas reservoirs in North America may have reservoir properties that are significantly different from those in South America or the Middle East. Thus the optimum drilling, completion, and stimulation methods for each well are a function of the reservoir characteristics and the economic situation. The costs of drilling, completing, and stimulating these wells, as well as the product prices and the market, affect how unconventional reservoirs are developed.

The volumes of oil and natural gas located in these larger unconventional reservoirs, such as heavy oil, oil shales, tight gas reservoirs, gas shales, and coal seams, require the development of new technologies that will allow the industry to produce oil and gas in an environmentally acceptable manner. These areas of developing technology will be discussed below.

9.5.1 Unconventional gas

Outside the USA, with a few exceptions, unconventional gas resources have largely been overlooked and understudied. They represent a potential long-term global resource of natural gas and have not been appraised in any systematic way. Unconventional gas resources – including tight sands, coalbed methane, and gas shales – constitute some of the largest components of remaining natural gas resources in the USA. Research and development into the geological controls and production technologies for these resources during the past several decades has enabled operators in the USA to begin to unlock the vast potential of these challenging resources. These resources are particularly attractive to natural-gas producers due to their long-lived reserves and stabilizing influence on reserve portfolios.

The magnitude and distribution of worldwide gas resources in gas shales, tight sands, and coalbed methane formations have yet to be understood. Worldwide estimates, however, are enormous, with some estimates higher than $9 \times 10^{14} \, \mathrm{m}^3$ [5]. The probability of this gas resource being in place is supported by information and experience with similar resources in North America. The $9 \times 10^{14} \, \mathrm{m}^3$ is likely to be a conservative estimate of the volume of gas in unconventional reservoirs worldwide, because there are fewer data to evaluate outside of North America. For example, Reuters reported that the amount of methane in the Gulf of Mexico is "astonishingly high" [8]. As more worldwide development occurs, more data will become available, and the estimates of worldwide unconventional gas volumes will undoubtedly increase.

Tight gas sand

From a global perspective, tight gas sand resources are vast, but undefined. This gas is found in tight formations underground that are unusually impermeable and nonporous. Because of this, more effort is required to extract this gas. Fracturing and acidifying can be used for gas extraction but are associated with higher costs. In a 2009 report, the EIA estimated that there were $9 \times 10^{12} \, \mathrm{m}^3$ of technically recoverable tight natural gas, representing over 17% of the total recoverable natural gas in the nation [9].

Coalbed methane

Coalbed methane is one of the best examples of how technology can have an impact on the understanding and eventual development of a natural-gas resource.

While gas has been known to exist in coal seams since the beginning of the coal-mining industry, only since 1989 has significant gas production been realized. Coalbed methane (CBM) was drilled through and observed for many years, yet never produced and sold as a resource. New technology and focused CBM research ultimately solved the resource-complexity riddle and unlocked its production potential. Coalbed methane now provides over $4.5 \times 10^{10} \, \mathrm{m}^3$ of gas production per year in the USA and is under development worldwide, including in Canada, Australia, India, China, and other countries [10].

Deposits of coal reserves are available in almost every country worldwide. Over 70 countries have coal reserves that can be mined and have potential CBM recovery. In 2005, over 5 billion tons of coal were produced worldwide. The top ten countries (China, the USA, India, Australia, South Africa, Russia, Indonesia, Poland, Kazakhstan, and Columbia) produced nearly 90% of the total. Estimates of gas in place around the world in coal seams range from $7.24 \times 10^{13} \, \mathrm{m}^3$ [8]. Using the USA as an analog, it is reasonable to expect that coal seams around the world hold potential for coalbed methane production. Worldwide coal resources are found in over 100 geological basins.

Shale formations

Shale formations act both as a source of gas and as its reservoir. Natural gas is stored in shale in three forms: free gas in rock pores, free gas in natural fractures, and adsorbed gas on organic matter and mineral surfaces. These different storage mechanisms affect the speed and efficiency of gas production. Shale gas reservoirs represent some of the most important development plays in North America. Undoubtedly, gas from shales around the world will be produced, and these reservoirs will become important assets in the coming decades.

Unconventional resources have been an important component of the US domestic natural-gas supply base for many years. From almost nonexistent production levels in the early 1970s, today unconventional resources, particularly tight sands, provide almost 30% of domestic gas supply in the USA. The volumes of gas produced from unconventional resources in the USA are projected to increase in importance over the next 25 years.

9.5.2 Unconventional oil

Unconventional oil is found in low-permeability formations, heavy oil deposits, tar sands, and oil shales [11]. Since they are found near the base of the resource triangle (Figure 9.2), both increased oil prices and better technology are required in order to make it viable to produce oil from these low-productivity reservoirs, as illustrated

Figure 9.3. Quantities and economic comparison of unconventional resources. Costs related to CO_2 emissions are not included. (MENA – Middle East and North Africa). Source: *World Energy Outlook 2008* © OECD/IEA, figure 9.10, page 218.

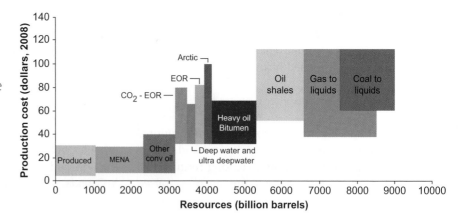

in Figure 9.3. The NPC study suggests that the world endowment of liquid hydrocarbons is on the order of 13–15 trillion barrels [1]. Much of this endowment is in unconventional, heavy oil reservoirs and will be difficult to produce.

When looking at Figure 9.3 it should be kept in mind that in 1979 there was an energy crisis that caused petroleum producers to commercialize technologies that developed shale oil and coal technologies to produce products assuming that petroleum would reach $35 per bbl or greater. The price then proceeded to collapse to the region of $10 per bbl and the technologies which were commercially available were then "shelved" but are still available today [12].

Conversion of heavy petroleum feedstocks

Heavy petroleum feedstocks such as tars and bitumen are heavy because they contain large amounts of asphaltenes. Asphaltenes are remnants of the original biological organisms that were buried and changed to petroleum through the petroleum-forming processes of digenesis and catagenesis. They are interesting materials in their own right. They have many potential applications, from producing carbon fibers to electrodes for fuel cells. They are called asphaltenes because, after extraction from the petroleum using solvent extraction, they are used to make asphalt by heating the asphaltene in the presence of steam, thus producing asphalt to pave roads.

Upgrading of heavy hydrocarbons therefore requires an understanding of the asphaltenes. Asphaltenes are micellar molecules containing aromatic cores with long saturated side chains that solubilize the asphaltene in the crude. They also contain most of the metals (vanadium (V) and nickel (Ni)) and much of the sulfur and nitrogen present in heavy hydrocarbons. These properties inhibit upgrading of heavy hydrocarbons through conventional processes because conventional catalysts are destroyed by depositing V and Ni metals. All petroleum throughout the world contains asphaltenes, except

for the very light crude oils. The more asphaltenes the crude oil contains, the heavier it is and the more difficult to convert. Thus, understanding and processing asphaltenes is the key to upgrading heavy petroleum products. The origin of these materials is unclear and remains an interesting fundamental problem in the origins of petroleum.

Upgrading heavy crude oils containing large amounts of asphaltenes requires either their destruction through thermal processing (coking) or hydroprocessing through hydrogen addition at elevated hydrogen pressure. In the industry coking is called "carbon rejection" and hydroprocessing is called "hydrogen addition." In thermal processing, which occurs at temperatures of 400–500 °C, valuable transportation liquids and gases are taken from heavy crude oil and the aromatic cores of the asphaltene are rejected as coke, which is mostly carbon. In the second case hydrogen at high pressure (>1000 psi of hydrogen) and high temperature (350–450 °C) is used in the presence of a catalyst to add hydrogen to the asphaltene, creating liquids and gases and reducing the asphaltene cores to useful liquids such as gasoline, diesel, and heating oil. Metals must be removed in this case and, as stated above, conventional catalytic processes are poisoned by the V and Ni deposited from the asphaltene.

There is always an economic and location-dependent choice between coking and hydroprocessing. Hydroprocessing at high pressure requires expensive reactors, which are located in some but not all refineries. If the petroleum producer is far from a hydroprocessing facility and the crude is too heavy (too viscous) to transport via pipeline, the choice to coke the crude oil at the production site may be made. In this case part of the crude is rejected as coke and used as low-value process heat.

Hydroprocessing catalysts

In hydroprocessing, Co- or Ni-promoted MoS_2 or WS_2 catalysts are used. These catalysts are traditionally supported on Al_2O_3, with approximately 20% active material. Recently, pure MoS_2 and pure WS_2 catalysts have been entering the refinery for upgrading petroleum. These catalysts with surface areas equal to or greater than those of conventional catalysts put more active material into the reaction zone, significantly improving performance. As sulfur regulations become stricter, a refiner has the choice of buying the more expensive "unsupported" catalyst of this type or building a new reactor. The least expensive choice is obvious and a triumph of modern catalytic materials science.

The hydrogenation function is dependent on the morphology of the layered MoS_2 nanoparticles. It has been determined that the *direct desulfurization* sites are located on the nanoparticle *edges* and that the hydrogenation sites are located on the *rims* [13]. This fundamental finding has allowed the control of selectivity of the desired hydrogenation/hydrodesulfurization function in practice. This aspect of the catalytic materials science is called *structure/function optimization*. The concept described above is illustrated in Figure 9.4 for MoS_2.

The sulfide catalysts have a wide variety of uses beyond those described above. For example, if sulfides are used in CO/H_2 reactions, long-chain alcohols are produced instead of alkanes/olefins. In a second example, if an olefin such as propene is oxidized in the presence of steam using a ruthenium (Ru) metal catalyst, only CO_2 and H_2O are produced. However, if RuS_2 is used, acetone is the product. The sulfides add selectivity to many reactions.

If we refer back to Figure 9.2 we see that many other hydrocarbon sources need upgrading and refining. This is the realm of catalytic materials science that has been discussed in detail by Gates *et al.* [14].

As stated in the introduction, hydrocarbon-based feedstocks will be used for many years to come. This is a major challenge to the catalytic materials science community. The result must be both economically viable and environmentally acceptable. Many different reactions will be studied and catalyst improvements made. In the case of many of the unconventional sources indicated in Figure 9.2, there will be a push to develop catalysts that can perform *down hole*. The production of clean liquid products performed at the source is one of the longer-term research goals. This means that the upgrading will be accomplished in the ground, removing the necessity for producing the feedstock except as clean liquid products. This represents a major effort that is only now beginning.

As explained earlier in this chapter, the industry has produced a little more than 1 trillion barrels of oil from conventional sources so far, and has another 1.25 trillion barrels booked as reserves. However, assuming that the

■ Edges optimization = HDS/HDN Optimization
□ Basal optimization = Hydroconversion/coking optimization

Figure 9.4. The figure represents schematically a nanoparticle and darker edge represents the MoS_2 sites.

industry will eventually unlock the secrets to producing heavy oil reservoirs, it is estimated that one-third of the world endowment of liquid hydrocarbons can be produced, which suggests that the ultimate recovery of liquid fuels could be as high as 4.5 trillion barrels, lasting well into this or the next century.

Much of the unconventional oil currently being produced is found in North America, South America, and Indonesia, but significant volumes can be found in other basins around the world. The industry has developed drilling and stimulation technologies to allow production of heavy oil in all three regions. Additional technology developments will be required, but, as demand increases, the incentives to develop such technologies will be there.

9.6 Technology requirements for producing unconventional resources

Regardless of the type of unconventional resource one is trying to develop, technology advancements are needed in virtually every technical category, including drilling, formation evaluation, reservoir engineering, and well-stimulation and completion methods. The technology must focus both on getting more oil and gas out of the reservoir and on reducing the costs of finding and producing the oil and gas. As discussed in the NPC report [1], most technologies in industry take 20 years to commercialize and become standard. Even in the oil and gas industry, commercialization of new technology takes an average of 16 years to progress from concept to widespread use in the industry.

9.6.1 Technology for reservoir evaluation

In reservoir evaluation, the ability to better locate productive sands and shales, and then delineate the "sweet spots" (those areas of the productive interval where permeability is highest and gas flow is least restricted) is one of the keys to economic development. Simply put, the sweet spot is where most of the production comes from and where operators make most of their profit.

Geophysical and petrophysical technologies are needed to help define the best part of the reservoirs. Some technologies are currently in use, such as amplitude-variation-with-offset (AVO) and other seismic techniques that can help find the gas zones [15]. In addition, a new class of nuclear magnetic resonance (NMR) logging tools has the potential to deliver the detailed permeability information needed to pinpoint sweet spots in the vertical section and optimize completions [16].

Finding unconventional oil sources is a challenge in itself. A new detection technique for this purpose is electroseismic hydrocarbon detection (EHD) [17]. The technique is fully field-tested and holds the promise of not only hydrocarbon detection, but also online monitoring and control of tar-sands production in the field. The technique is based on the electroseismic or electro-acoustic effect. An electric field, applied to a porous material that contains an ionic solution, causes a displacement between charges on the porous material and the ions in solution. This relative displacement induces pressure in the pore space and the porous matrix. In an AC applied electric field, the pressure response of the matrix takes the form of an acoustic, or seismic, wave. The nature of the detected signal is determined by the physical properties (ionic or conducting properties) of liquids in the pores. A rock filled with water gives a different response from that of a rock filled with oil. Thus, one of the "Holy Grails" of petroleum exploration has been discovered. It goes without saying that this technique has many other potential applications in materials science.

9.6.2 Technology for producing unconventional gas

Hydraulic fracture fluids are one of the more pressing technology needs confronting operators in unconventional gas reservoirs. The objective of pumping a fracture treatment is to fracture (crack) the reservoir rock around the borehole, and place propping agents into the crack to prop the fracture open, thus forming a permeable conduit that allows gas to flow to the well bore [18]. Polymer gel formulations are typically used to create the crack and carry the propping agents. However, gelled fluids can damage the fracture itself, particularly in formations where the temperature is less than 250 °F. The compositions of fracturing fluids, which are largely considered trade secrets by companies, vary widely, as do the additives used [19].

Propping agents

What the industry needs is a viscous fluid system that can transport propping agents deeply into a fracture, then cleanly break back to a low-viscosity fluid so it can be produced back from a low-temperature reservoir.

Although the industry is a long way from developing the ideal fracture fluid, new systems such as water-based viscoelastic surfactant fluids are offering solutions for some unconventional reservoirs [20]. Viscoelastic fluids are polymer-free. The surfactant forms worm-like micelles in the presence of brine to increase fluid viscosity. Continued advancement in propping-agent technology, such as the creation of strong, lightweight, inexpensive propping agents, is also an important need for the industry.

Unconventional reservoirs tend to be very thick, and any technology to enhance perforating, fracturing, and completing ultra-long intervals would have tremendous value, especially in multiple-stage treatments. In certain types of reservoirs, oriented perforating tools are proving highly effective, allowing operators to initiate hydraulic fractures in the preferred plane and facilitating orientation to natural fractures in the reservoir rock to improve proppant deliverability and enhance production.

9.6.3 Technology for unconventional oil

As with natural gas, technologies used to develop and produce unconventional oil must advance to allow economic production. The major production methods are open-pit mining (tar sands), cold production using horizontal wells, cyclic steam injection, steam flooding, and steam-assisted gravity drainage (SAGD) [21] [22].

Many hydrocarbons are located remotely in tar sands (Canada) or bitumen lakes (Trinidad) [23] [24]. These resources contain huge amounts of energy if they can be mined and converted into lighter oils. Extracting these hydrocarbons basically involves heating them with steam, and the recovered hydrocarbons are then further treated. These processes create huge environmental problems, including large amounts of contaminated water that must be re-injected into the original formation or otherwise treated. The hydrocarbons that are recovered are of low value, containing large amounts of sulfur and other pollutants that must be removed. Researchers are working on new catalytic processes to upgrade the hydrocarbons at the site of production. These upgraders will be modular and mobile, requiring novel materials to achieve the location flexibility required. Ultimately, there is a desire to upgrade the hydrocarbons *in situ*. Doing this will require new catalysts that can be injected into the formation and allowed to work for longer periods of time producing "pipeline-able" hydrocarbon fuel. Research is proceeding in this area. In steam-assisted gravity-drainage (SAGD) wells, production can be improved using solvents along with the steam. In addition, some *in situ* combustion methods have been used in heavy oil reservoirs. Another possible production method is downhole electric heating.

In steam wells and in deep gas plays around the world, the development of materials that can withstand high pressures (>15,000 psi), high temperatures (>400 °F) (HPHT), and corrosive environments (CO_2 and H_2S) will be an important technology area. In horizontal drilling, there is a need for downhole electronic equipment that can withstand these HPHT conditions. Many times, insulating these electronic tools is the key to providing long-lived service. In other cases, the metallurgical design of the downhole tubular goods that can withstand HPHT corrosive environments is the key to success.

9.7 Constraints to meeting energy demand

It appears evident that energy demand will continue to increase well into the twenty-first century. It is also apparent that there are ample hydrocarbon resources to meet the demand. It is particularly important to realize that, as shown in Figure 9.5, the USA and the North American continent have sufficient hydrocarbons to last well into the future [25]. However, there are significant constraints that will affect progress. The most important constraints are meeting CO_2 emission standards, limitations on research expenditure, capital requirements, manpower limitations, and access to drilling locations.

9.7.1 Environmental constraints

To continue to use hydrocarbon fuels, the world must develop the advanced technologies, plus the legal and regulatory framework, to enable carbon capture and sequestration. There needs to be a global framework for carbon management, including the establishment of a transparent and predictable cost for CO_2 emissions. It will be very important to develop clean-burning coal technology to allow the use of coal for the generation of electricity, freeing natural gas for use in more important applications.

In addition to issues relating to CO_2 emission, the industry must be cognizant of and solve problems involving land use, waste disposal, habitat, and noise [26]. Research on environmental friendly drilling and production methods, and of course advanced oil-spill avoidance and cleanup technologies, must continue. The industry should continue to develop green chemicals that can be used in wellbores. The cost of doing business involves being stewards of the environment wherever the industry is operating.

Figure 9.5. The possible locations for hydrocarbon supply in North America are shown as the colored arc and include heavy petroleum, shale oil, coal and methane.

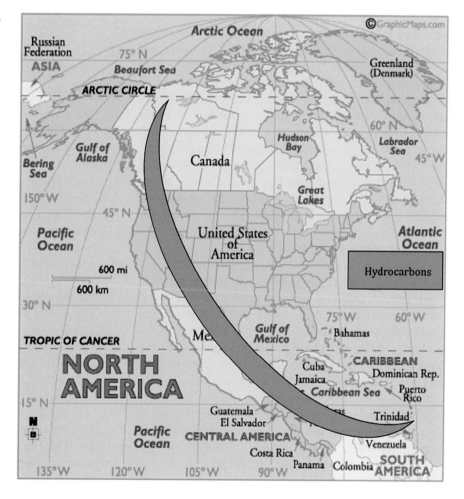

An interesting example of an open environment challenge is seen in NASA satellite images that show thousands of *flares*, particularly in Siberia and Nigeria. These are natural gas flares that are predominantly methane and CO_2. A recent article stated that the daily CO_2 release by the Siberian and Nigerian flares was greater than the daily release by all the transportation vehicles in the north-eastern USA [27]. This flaring occurs because the producing wells are too far from pipelines and therefore it is too expensive to transport the gas. Thus, besides adding a major amount of CO_2 to the atmosphere, a large amount of potentially useful energy is lost. The reforming reaction $CO_2 + CH_4 \Rightarrow$ liquid hydrocarbon is known, and developing an economic remote process for converting the gas stream into a liquid stream would be a major contribution. The need for catalytic research in this area is quite clear.

9.7.2 Technology constraints

The research intensity, measured as the research expenditure as a function of the percentage of sales, in the oil and gas industry is quite low compared with those in other industries, such as pharmaceuticals, transportation, and computers. It can be argued that some research is conducted on every well drilled and completed; however, not much learned on individual wells is recorded and put to use elsewhere. Most studies, such as the one recently performed by the NPC, suggest that the industry and governments in oil-producing countries should be putting more money into research and development. After all, developing and using better technology is the key to unlocking the oil and gas located near the bottom of the resource triangle. The research would involve many aspects of materials science: stronger and cheaper metals that resist corrosion, more effect catalysts for converting fuels to clean transportation liquids, and many others. A dream would be to convert the petroleum in the ground and have clean fuels "emerge" without further processing.

9.7.3 Manpower and capital constraints

As the industry continues to find more oil and gas from conventional reservoirs, in the arctic, in deep-water and in unconventional reservoirs, it means that many more wells will have to be drilled. That, of course, means more rigs, more equipment, more personnel, and more capital, as well as improved access to prospective lands and new pipeline infrastructure to market the oil and gas that are found. The NPC estimates that the worldwide investment in energy will be $20 trillion in the next 25 years. Someone will have to determine how to spend that money wisely.

In the next ten years, more than half of those currently in the industry will be eligible to retire. The NPC recommended that governments need to support young men and women seeking engineering and other technical degrees, both graduate and undergraduate, by increasing funding for scholarships and research at universities. The NPC also recommended that the tax laws be changed to allow those who have retired to work part-time without losing a portion of their retirement benefits.

9.7.4 Access to resources

In many areas, especially in North America, access to known or promising deposits of oil and gas is restricted due to environmental concerns or, in many cases, because the residents simply do not want to have to look at a drilling rig or even a wind mill. The NPC recommended that governments conduct national and regional basin-oriented resource and market assessments to identify opportunities to increase oil, gas, and coal supply. The industry should continue to create technology for environmentally friendly development of high-potential areas both onshore and offshore. The public must be educated about energy, its use, its benefits, and the need to have access to areas that hold large volumes of oil, natural gas, and coal.

9.8 Summary

On the basis of the information included in this article, much of which is contained in the recent NPC report [1], the following conclusions are offered.

1. The demand for energy globally will increase by 50%–60% in the next 25 years due to the increase in world population and the desire for all to increase their standard of living.
2. The cost of extraction and processing can greatly affect the total energy cost for heavy petroleum shale oil, coal and methane in terms of dollars per equivalent watt. This can ultimately determine the competitiveness of a particular resource.
3. Now and in 2030, the major source of energy will be hydrocarbon fuels – oil, gas, and coal. These fuels will supply around 80% of the energy in 2030.
4. The global endowments of oil, gas, and coal are ample to supply the world with fuel in most of the twenty-first century.
5. To meet demand, the energy industry will need technology improvements and the industry will need to improve how it burns or uses fossil fuels to reduce CO_2 emissions.
6. Constraints that could affect the supply of energy in the future consist of environmental, manpower, and technical issues, access to the deposits, and the capital needed to fund the projects.

7. Technology improvements in materials, such as polymers, chemicals, propping agents, metals, and composites, and electronics must occur in areas where steam injection is required, and in deep high-pressure, high-temperature reservoirs.

9.9 Questions for discussion

1. What are the primary unconventional oil and gas resources in the world today and where are they located?
2. What are the key incentives for unconventional oil and gas resources?
3. Describe constraints in producing *unconventional oil and gas*.
4. Which unconventional resources are set to grow strongly and where are these located?
5. What is the status of development of unconventional oil and gas across the continents of North America, South America, Europe, and Africa and in the Asia–Pacific region?

9.10 Further reading

- **Materialsforenergy** (http://materialsforenergy.typepad.com). A materials for energy blog sponsored by the Materials Research Society (MRS). This blog has multiple articles posted regarding research, commercialization, and environmental issues regarding materials for energy.
- **Materials UK (Materials for Energy)** (http://www.matuk.co.uk/energy.htm). A materials blog discussing the high priority of energy and the importance in sustaining research, development, and modeling of materials for energy applications.
- **S. A. Holditch** and **R. R. Chianelli**, 2008, "Factors That Will Influence Oil and Gas Supply and Demand in the 21st Century," *Mater. Res. Bull.*, **33**, 317–325.
- **J. M. Hunt**, 1979, *Petroleum Geochemistry and Geology*, New York, W. H. Freeman and Company. A basic textbook on geological factors affecting petroleum oil and gas deposits.
- **J. H. Gary** and **G. E. Handwerk**, 1994, *Petroleum Refining: Technology and Economics*, third edition, New York, Marcel Dekker. A basic textbook on refining technology for conventional oil.
- **J. R. Moroney**, 2008, *Power Struggle: World Energy in the Twenty-First Century* New York, Praeger Publishers. A textbook on history of oil and economic impacts of oil in the twenty-first century.
- **D. Yergin**, 2008, *The Prize: The Epic Quest for Oil, Money & Power*, New York, Free Press. A textbook covering the history of oil throughout the twentieth century and into the twenty-first.
- http://en.wikipedia.org/wiki/Petroleum.
- http://en.wikipedia.org/wiki/Natural_gas.

9.11 References

[1] http://www.americanenergyindependence.com/hydrocarbons.aspx.
[2] **NPC**, 2007, *Facing the Hard Truths about Energy – A Comprehensive View to 2030 of Global Oil and Natural Gas*, National Petroleum Council, Washington, DC, www.npc.org.
[3] **Campbell, Colin**: The coming oil crisis, multi-science Publishing Co. Ltd, Brentwood, Essex, U.K. (1997).
[4] **EIA**, 2010, *Annual Energy Outlook 2010 Early Release Overview*, US Energy Information Administration.
[5] **D. Yergin**, 1991, *The Prize: The Epic Quest for Oil, Money & Power*, Free Press.
[6] **University of Utah**, 2007, *A Technical, Economic, and Legal Assessment of North American Heavy Oil, Oil Sands, and Oil Shale Resources*, in response to energy policy act of 2005 section 369 (p), September 2007, available at http://fossil.energy.gov/programs/oilgas/publications/oilshale/HeavyOilLowRes.pdf.
[7] **R. R. Chianelli**, **G. Berhault**, and **B. Torres**, 2009, "Unsupported transition metal sulfide catalysts: 100 years of science and application," *Catalysis Today*, **147**, 275.
[8] http://www.reuters.com/article/2010/06/22/us-oil-spill-methane-idUSTRE65L6IA20100622.
[9] **EIA**, 2009, *Annual Energy Outlook 2009*, US Energy Information Administration.
[10] **I. Palmer**, 2008, "Coalbed methane wells are cheap, but permeability can be expensive!," *Energy Tribune*, March.
[11] **J. R. Dyni**, 2006, *Geology and Resources of Some World Oil-Shale Deposits*, USGS Scientific Investigations, available at http://pubs.usgs.gov/sir/2005/5294/pdf/sir5294_508.pdf.
[12] http://en.wikipedia.org/wiki/1979_energy_crisis.
[13] **M. Daage** and **R. R. Chianelli**, 1994, "Structure/function relations in molybdenum sulfide catalysts: the 'rim-edge' model," *J. Catal.* **149**, 414–427.
[14] **B. C. Gates**, **G. W. Huber**, **C. L. Marshall** et al., 2008, "Catalyst for emerging energy applications," *MRS Bull.*, **33**, 429–435.
[15] **A. Tura** and **D. Lumley**, 2000, "Estimating pressure and saturation changes from time-lapse AVO data," in *Fourth Wave Imaging, Offshore Technology Conference*, Houston, TX.
[16] **J. Bryan**, **A. Kantzas**, and **C. Bellehumeur**, "Oil-viscosity predictions from low-field NMR measurements," *SPE Reservoir Evaluation & Eng*, **8**, 44–52.
[17] **A. H. Thompson**, **S. Hornbostel**, **J. Burns** et al., 2007, "Field tests of electroseismic hydrocarbon detection," *Geophysics*, **72**, N1–N9.
[18] **R. W. Veatch Jr.**, **Z. A. Moschovidis**, and **C. R. Fast** "An overview of hydraulic fracturing," in *Recent Advances in Hydraulic Fracturing*, eds. **J. L. Gidley**, **S. A. Holditch**, **D. E. Nierode**, and **R. W. Veatch Jr.**, London, Society of Petroleum Engineers.

[19] **EPA**, 2004, "Hydraulic fracturing fluids," in *Evaluation of Impacts to Underground Sources of Drinking Water by Hydraulic Fracturing of Coalbed Methane Reservoirs*, EPA.

[20] **T. N. Castro**, **V. C. Santanna**, **A. A. Neto**, and **M. C. P. Moura**, 2005, "Hydraulic gel fracturing," *J. Dispersion Sci. Technol.*, **26**, 1–4.

[21] **SAGD**, 2010, "R&D for unlocking unconventional heavy-oil resources," *The Way Ahead*, **6**.

[22] **M. C. Herweyer** and **A. Gupta**, 2010, *Appendix D Tar Sands/Oil Sands*, available at http://www.theoildrum.com/node/3839.

[23] **M. B. Dusseault**, **A. Zambrano**, **J. R. Barrios**, and **C. Guerra**, 2008, "Estimating technically recoverable reserves in the Faja Petrolifera del Orinoco-FPO,"

presented at the World Heavy Oil Congress 2008, Edmonton, Alberta, Canada.

[24] **M. B. Dusseault**, 2001, *CHOPS: Cold Heavy Oil Production with Sand in the Canadian Heavy Oil Industry*, Alberta Department of Energy.

[25] **J. D. Hughes**, *Hydrocarbons in North America*, http://www.postcarbon.org/Reader/PCReader-Hughes-Energy.pdf.

[26] **J. A. Veil** and **J. J. Quinn**, 2010, *Water Issues Associated with Heavy Oil Production*, US Department of Energy, National Energy Technology Laboratory, available at http://www.ipd.anl.gov/anlpubs/2008/11/62916.pdf.

[27] 2007, "Russia Top Offender in Gas-Flare," *The Boston Globe*, June 21.

10 Advancing coal conversion technologies: materials challenges

Bryan D. Morreale, Cynthia A. Powell, and David R. Luebke

National Energy Technology Laboratory (NETL), Pittsburgh, PA, USA

10.1 Focus

The historic and current global energy portfolio is dominated by fossil fuels, an affordable and plentiful means of energizing the development of human civilization. The focus of this chapter is to give an overview introduction to coal and coal conversion processes, with a focus on the development of materials and strategies that promote the efficient use of coal in an environmentally friendly manner.

10.2 Synopsis

Coal is the altered remains of biomass, and can be considered a means of storing solar energy on very long time scales. Coal has been used for centuries as a means of an affordable and plentiful energy source, and in recent years has been exploited for these characteristics through two processes: the direct combustion of coal to produce heat and electricity though steam cycles; and the gasification process that produces a highly combustible gas or liquid fuels and chemicals through liquefaction processes.

However, as coal supplies and qualities diminish, and concern over environmental implications of fossil energy usage increases, the scientific community is focusing on developing technologies that promote the efficient and "green" use of coal. At the forefront of future coal conversion processes is the development of technologies, materials and strategies that allow operation at elevated temperatures and pressures and in harsher chemical environments to address the efficiency, economics, and environmental concerns that come with fossil energy usage.

Table 10.1. Example of coal classifications and corresponding characteristics

Class	Volatile matter (wt%)	Fixed carbon (wt%)	Heating value (MJ kg^{-1})
Anthracite	<8	>92	36–37
Bituminous	8–22	78–92	32–36
Sub-bituminous	22–27	73–78	28–32
Lignite	27.35	65–73	26–28

10.3 Historical perspective

Coal has fueled the development of human civilization, with its use being documented back 2000 years. Through the vast majority of that period, coal was predominantly used as a means to heat dwellings [1]. Since the eighteenth century, coal has been exploited as an affordable and plentiful energy supply that has contributed to global economic growth and improved quality of life [2].

The first records of coal in the USA date back to the late 1600s, while the documentation of coal mining dates back to the early-to-mid 1700s. In the early 1800s, coal was used as an industrial source of heat to produce salt and manufacture glass, and as fuel for locomotives and industrial equipment, as well as being converted to a combustible gas used for street lighting. Throughout the 1800s and into the 1900s coal continued to be utilized as a source of low-cost, indigenous, dense energy.

In the late 1800s, the concept of utilizing coal as a means of electric power generation was realized by Thomas Edison in the New York area, and continued to evolve over the next 100 years, with a major emphasis on mining and the development of uses. In the mid 1970s, the use of coal increased, and it became the national fuel source for power generation. Ever since the majority of the emphasis on coal has been associated with increasing efficiency while reducing environmental impact. The energy crisis associated with the OPEC oil embargo (1974), the Clean Coal Technology Act (1986), the Clean Air Act (1997), and the Energy Policy Act (2005) pushed the coal industry to "clean" technologies, specifically focused on reducing particulate, SO_x, NO_x, and heavy-metal emissions.

Environmental concerns continue to be at the forefront of coal utilization in the USA. Carbon dioxide concentrations in the atmosphere have steadily increased by 30% over the past 50 years; to a level of ~390 ppm by volume in 2010. The increase in the Earth's surface temperature has been attributed to the increased level of atmospheric carbon dioxide and other greenhouse gases. Currently, government, academia, and industry are aggressively conducting research and development focused on eliminating the impact associated with greenhouse gas emissions linked to fossil fuels. Specifically, technologies focused on increasing process efficiency and the capture of carbon dioxide from coal effluent streams, its permanent storage in geological reservoirs, and the conversion of carbon dioxide into valuable commodities are being investigated.

10.3.1 Coal origins and makeup

Fossil fuels, including coal, are the altered remains of prehistoric biomass. Coal can be considered a form of stored solar energy, similar to biomass, but on a very different time scale. Its conditions of formation, including biomass type, pressure, temperature, and age, contribute to the energy content and complex characteristics of coal [1] [2] [3]. A typical example of coal rank or classification is detailed in Table 10.1, which is obtained from either *proximate* analysis (moisture, volatile matter, fixed carbon, and ash content) or *ultimate* analysis (carbon, hydrogen, oxygen, sulfur, and nitrogen content). A comprehensive understanding of the importance of coal properties, specific test protocols, and the variance in chemical characteristics of coal are summarized in Higman and van der Burgt [3].

10.3.2 Coal resources

The World Coal Institute estimates that there are over 984 billion tons of proven coal reserves globally, which could last approximately 200 years at current consumption levels [2]. These coal reserves are predominantly located in the USA, Russia, China, and India, although all the continents and over 70 countries have identified coal deposits. Currently, coal supplies over 25% of global energy, over 40% of the world's electricity, and in excess of 90% of the electricity for some nations [2] [4]. Also, coal is one of the least expensive fossil fuels, costing approximately one-sixth as much as oil and natural gas on an energy basis. It is apparent, then, that these factors, together with dwindling supplies of other fossil fuels (petroleum, 41 years and natural gas, 65 years) and the relative immaturity and high costs of renewable energy will lead to coal continuing to play an important

Figure 10.1. Pulverized coal combustion system with post-combustion carbon dioxide capture and sequestration.

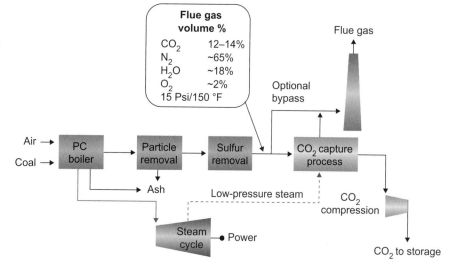

role in the economies of both developing and industrialized nations for the foreseeable future.

10.4 Coal conversion basics

10.4.1 Coal combustion processes

Currently, pulverized coal combustion (PCC) processes dominate power production, accounting for over 45% [5] of the USA's national production and 42% [6] globally. The combustion process is the highly exothermic chemical reaction of carbon and hydrogen in coal in the presence of excess oxygen, Equations (10.1) and (10.2), respectively:

$$C + O_2 = CO_2 \qquad \Delta H = -283 \text{ kJ mol}^{-1} \qquad (10.1)$$

$$H_2 + \frac{1}{2}O_2 = H_2O \qquad \Delta H = -242 \text{ kJ mol}^{-1} \qquad (10.2)$$

The heat generated from the combustion process is used to produce steam in a boiler. Steam expansion powers turbines coupled to generators, which produce electricity. A simplified schematic representation of the PCC process is detailed in Figure 10.1.

Subcritical power plants
Modern PCC technology is well developed, and today's subcritical coal-fired power plant, operating at a steam temperature of about 540 °C and a steam pressure of ~16.5 MPa, has a thermal efficiency of approximately 35%, on the basis of the higher heating value of the coal. More energy extracted per unit amount of coal means an increase in power-plant output and a relative decrease in emissions, and thus advanced combustion technologies seek to maximize the thermal efficiency of the power plant through increases in steam temperatures and/or pressures above the critical point.

Supercritical power plants
Generally, supercritical (SC) power plants are those operating with a steam pressure of ~22 MPa or higher, while ultra-supercritical (USC) power plants operate with steam conditions greater than 24 MPa and 593 °C. For commercial SC combustion plants, operating at 540–566 °C and 25 MPa, and commercial USC combustion plants, operating at 580–620 °C and 27–28.5 MPa, efficiencies of up to 41% and 46%, respectively, are reported. Additional increases in thermal efficiencies are being sought through even greater steam temperatures and pressures, but these will require further advances in materials technology in order for them to be realized on a commercial scale. In the USA, the Department of Energy is sponsoring research aimed at maximizing thermal efficiency from coal generation, which for coal combustion will require USC steam temperatures and pressures as high as 760 °C and 37.9 MPa. In Europe, the Thermie program is focused on USC steam conditions of 720 °C and 35 MPa.

Oxy-combustion
Combusting coal in oxygen, as opposed to air, can lead to a significant reduction in NO_x emissions, as well as simplifying CO_2 capture for sequestration, and, when combined with USC technology, can significantly reduce the environmental impact of PCC plants. To date, there are no commercial-scale oxy-fired PCC plants in operation, though several such new plant constructions have been announced recently in the USA. There is also the opportunity to retrofit existing PCC plants with oxy-fuel burners, thereby providing a carbon capture option for the existing power fleet. However, with no oxy-fired retrofit boilers currently in operation, there is little understanding of how the resulting change in operating environment will impact the boiler materials of

Figure 10.2. Pulverized coal oxy-combustion system with carbon dioxide capture and compression.

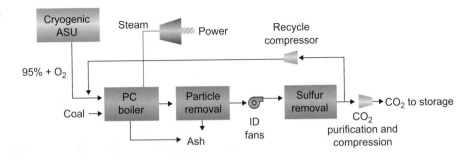

construction or operation of the plant as a whole. A simplified illustration of a coal-based oxy-combustion process is illustrated in Figure 10.2.

Fluidized bed combustion

Fluidized bed combustion (FBC) is another of the advanced combustion technologies that has evolved from efforts to develop an environmentally friendly combustion process that does not require external emission controls [5]. FBC systems burn coal and other carbon-containing feedstocks in a bed of heated particles suspended in flowing air. The fluidizing action promotes more complete coal combustion at lower flame temperatures, between 760 and 925 °C, which greatly reduces NO_x emissions. In addition, the particle bed includes a sulfur-absorbing sorbent such as limestone or dolomite, which can reduce SO_2 emissions by more than 95%. As a result, FBC systems can meet most environmental standards for SO_2 and NO_x, without the need for post-combustion pollution controls. Atmospheric-pressure FBCs are available in either a bubbling or a circulating FBC configuration, with power-generation efficiency similar to that of a standard PCC plant. There are now more than 400 FBC units operating worldwide. The technology is popular primarily because of its fuel flexibility (fluidized-bed combustors can burn any carbon-containing material, from low-rank coals to municipal waste) and environmental performance.

Pressurized fluidized-bed combustion (PFBC) operates at elevated pressures to produce a gas stream that can drive a turbine. This technology is less developed than standard PBC and has yet to find widespread market acceptance.

10.4.2 Coal gasification processes

The gasification process reacts a carbon-containing material, such as coal, with steam and controlled amounts of air or oxygen, at high pressures and temperatures to form a synthesis gas (or syngas) (Chapter 39) composed primarily of carbon monoxide and hydrogen [6]. Together with the combustion reactions shown

above (Equations 10.1 and 10.2), the chemical reactions listed below (Equations 10.3–10.8) dominate the gasification process [3].

$$CO + \frac{1}{2}O_2 = CO_2 \quad \Delta H = -283 \text{ kJ mol}^{-1} \quad (10.3)$$

$$C + CO_2 = 2CO \quad \Delta H = +172 \text{ kJ mol}^{-1} \quad (10.4)$$

$$C + H_2O = CO + H_2 \quad \Delta H = +131 \text{ kJ mol}^{-1} \quad (10.5)$$

$$C + 2H_2 = CH_4 \quad \Delta H = -75 \text{ kJ mol}^{-1} \quad (10.6)$$

$$CO + H_2O = CO_2 + H_2 \quad \Delta H = -41 \text{ kJ mol}^{-1} \quad (10.7)$$

$$CH_4 + H_2O = CO + 3H_2 \quad \Delta H = +206 \text{ kJ mol}^{-1} \quad (10.8)$$

Syngas can be used as a fuel for power generation or as a basic chemical building block for the formation of a variety of liquid fuels or other chemical products. The gasification of coal is not a new technology. Town gas (another name for syngas) was widely used in North America and Europe until it was replaced in the 1950s by natural gas, and Germany used this process to produce substantial amounts of liquid fuels during World War II. The technology's fuel and product flexibility, combined with its lower emissions of critical pollutants and the relatively straightforward nature of extraction of CO_2 from syngas, have brought renewed interest in gasification in recent years.

Prior to processing, syngas resulting from coal gasification often requires the removal of particulate matter and gaseous contaminants that may arise due to the trace impurities found in the coal. A typical coal-gasification process is illustrated schematically in Figure 10.3.

Water–gas shift

The water–gas shift (WGS) reaction (Chapter 39), Equation 10.7, is used to enhance hydrogen production, convert all syngas carbon into CO_2 for capture, or produce the required H_2/CO ratio for synthetic liquid fuel and chemical production. The WGS reaction, which is slightly exothermic, is often carried out in multiple catalytic reactors in the presence of Fe-, Cu-, or Mo-based catalysts and excess steam to minimize unreacted CO.

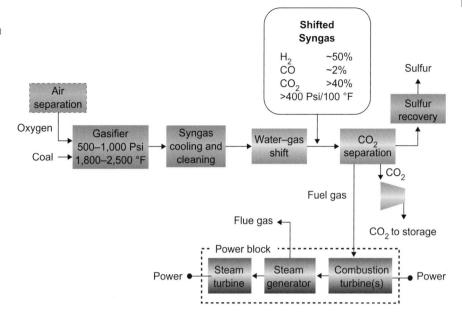

Figure 10.3. Coal gasification system with pre-combustion carbon dioxide capture and compression.

Hydrogen production

High-purity H_2 can be produced from syngas generated through the gasification process. The shifted syngas, predominantly composed of H_2 and CO_2, can be directed through several emerging separation technologies in an effort to produce pure H_2 and CO_2 streams. These technologies will be discussed in more detail later. Their integration with the aforementioned WGS reaction can increase overall process efficiency and circumvent the need for multiple reactor stages, added catalysts, and high steam-to-CO ratios, while allowing operation of the reaction at higher temperatures, which are thermodynamically unfavorable but kinetically desirable.

Power turbines

When the syngas produced by gasification is used to generate electricity, it is typically used as the fuel in an integrated gasification combined cycle (IGCC). In this case, cleaned synthesis gas is combusted in a combustionturbine to produce electricity, and the waste heat is used to in a steamturbine to generate additional electricity. The synergies associated with this type of processes lead to thermal efficiencies on the order of 40%. Through improvements in combustion-turbine technologies, including the development of materials and strategies that allow higher flame temperatures (hydrogen- and oxygen-rich feeds), thermal efficiencies over 50% can be realized.

Coal to liquid fuels

Liquid fuels can be generated from syngas through several processes including direct liquefaction, syngas-to-methanol-to-gasoline, and thermo-catalytic

Fischer–Tropsch (FT) synthesis, which generally revolve around the catalytic hydrogenation of carbon monoxide. Transportation fuels produced by FT are low in sulfur, particulates, and nitrogen oxides. Though South Africa has been producing coal-derived fuels since 1955, and currently produces more than 30% of that country's gasoline and diesel from indigenous coal, there is still room for improvement, particularly in the area of reaction chemistry. Catalyst systems that have historically been utilized in the FT process generally involve iron, cobalt, and ruthenium, while a multi-phase slurry bubble column reactor is often employed to mitigate the exothermicity of the reaction.

10.5 Current and future coal utilization technology needs

Technological advancement to address the world's growing demand for clean and affordable energy will require simultaneous advances in materials science and technology in order to meet the performance demands of new power systems, as well as the degrading quality of available fossil fuels. This is particularly true for coal-based technologies, where the drive for increased efficiencies and reduced environmental impact requires increased system operating temperatures and pressures, and thus materials that perform effectively in increasingly aggressive environments. Figure 10.4 illustrates the comparison of efficiencies of various coal conversion processes. The addition and utilization of alternative feedstocks, such as biomass, petcoke, and municipal wastes, can also place significant stress on the materials associated with fossil energy conversion processes. In an effort to accommodate such chemical variations

Figure 10.4. Efficiency comparison of various coal conversion processes with and without carbon capture and sequestration.

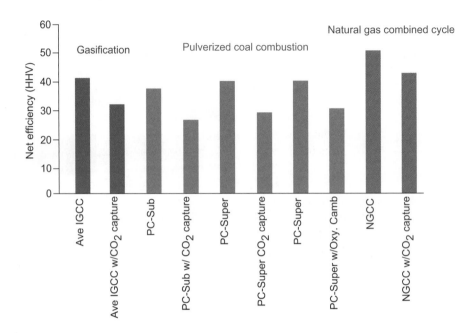

in feedstock, it is likely that materials with improved performance characteristics and the development of new materials-protection strategies are needed. Advances in all classifications of materials are required, including metals, ceramics, polymers, composites, semiconductors, and biomaterials, and will likely require successful implementation of emerging material concepts such as smart materials and nanotechnology. These advanced materials and concepts will be required in order to push materials beyond the mechanical and chemical limits that are currently achievable and pave the way for a revolution in energy conversion technology.

10.5.1 High-temperature materials

The elevated temperatures and harsh chemical environments associated with future energy conversion processes, including advanced gasification, USC steam cycles, hydrogen-fueled turbines, oxy-combustion, and co-firing, will require significant advances in materials technologies.

Ultra-supercritical steam cycles
Pushing steam cycles above 700 °C will require the further development of ferritic and austenitic alloys with increased mechanical strength and chemical robustness for boiler, heat-exchanger, and steam-turbine application. The alloys must be resistant to attack by sulfide, chloride, and other contaminants in carbonaceous feeds on the fireside (inside the boiler where combustion takes place), and require oxidation resistance on the steam side (inside the heat exchanger tubes where steam is generated) and in the steamturbine. For many alloys, a chemistry and microstructure optimized for creep resistance is not optimum for the required environmental resistance, and so effective coating and/or cladding strategies will also have to be considered. Beyond performance, processing and weldability must also be designed into the materials being developed. In all cases, because of the very large volume of materials required to construct a power plant, reliability must be obtained at an affordable cost.

Oxy-combustion
Perhaps the single biggest materials challenge associated with oxy-fuel combustion is in the production of oxygen, which will be covered in detail in following sections. Beyond oxygen production, the materials challenges associated with oxy-fuel combustion are similar to those for USC cycles, except that burning coal with oxygen will increase the radiant component of the flame, and as a result will increase metal temperatures inside the boiler. However, because an oxy-fired boiler has not yet been designed and built, the actual metal temperatures have still to be determined, but could be as high as the adiabatic flame temperatures of the feeds; >2,000 and 2,200 °C, for wood and coal, respectively. Retrofit oxy-combustion systems will likely be designed to match the heat-transfer characteristics of the original boiler by utilizing recycle streams, and so increased metal temperatures should not be an issue in such cases. However, recycling contaminants and increased levels of oxygen in the feed will significantly change the

fireside environment, and the environmental resistance of the materials of construction will have to be confirmed, or protection strategies developed, to ensure system reliability.

Advanced gasification

The materials challenges associated with gasification are concerned with the reliability of the gasifier itself, and with technologies associated with oxygen production and synthesis-gas processing [7]. The gasifier acts as a reaction vessel for the gasification reactions. Gasifiers can be classified into three main categories on the basis of the operating temperature: moving bed, 425–650 °C; fluidized bed, 900–1,050 °C, and entrained, 1,250–1,600 °C. Entrained gasification processes are well suited to deal with the decreasing quality of coals, since the principal advantage of this process is that it can accommodate practically any carbon-based feed to produce a tar-free syngas that can be utilized for a variety of products, making this an attractive option for a future coal conversion process. However, to exploit the entrained gasification process, advances in gasifier construction materials must be made in order to mitigate the extreme environments that consist of both flowing slag and corrosive gases.

Gasification-system reliability and optimization also requires an effective feedback system that can provide real-time information on process parameters, such as temperature, pressure, and gas composition. The challenge is the development of sensor materials that are sufficiently responsive and selective at temperatures greater than 500 °C, and sufficiently robust to withstand the harsh operating environments that may include exposure to flowing slags. Current research focuses primarily on optical silica and alumina-based materials for temperature and gas measurement, and metal oxide-based materials for gas sensing; however, most concepts are at the development stage and have yet to be proven in the field. In some cases, development of a robust packaging system that can ensure long-term stability in the service environment will be as important as development of the sensor material itself. Finally, significant challenges exist in implementation of these devices in the field, particularly with respect to communication of information from the sensors to the plant operator.

Turbines

As with other components in the advanced coal power plant, research continues into the development of next-generation stationary land-based combustion turbines with both increased efficiency and reduced impact on the environment. Turbines that are fuel flexible and capable of operating at temperatures in excess of 1,400 °C have the potential to increase the efficiency of the IGCC process by several percent. In the nearer term, these will be hydrogen turbines, capable of reliably burning fuel that contains a majority of hydrogen at high temperatures to maximize efficiency. In the longer term, it is expected that these will be oxy-fuel turbines, capable of combusting fuels consisting of nearly 100% hydrogen.

Realization of the next generation of land-based turbines will require significant advances in materials technology, likely coupled with innovations in cooling and thermal management strategies. Within the hot section of the turbine, construction materials will need to be resistant to oxidation, hot corrosion, creep, fatigue, and wear at temperatures in excess of 1,400 °C for long periods of operation. Current-generation nickel- and cobalt-based superalloys cannot withstand sustained metal temperatures greater than approximately 1,100 °C, requiring internal cooling as well as thermal barriers and oxidation-resistant coatings to meet today's turbine performance requirements. As a result, it appears likely that the next generation of land-based turbines will require the development of new substrate materials, combined with new coating strategies, to meet the substantially higher temperature requirements. With sufficient internal cooling, silicides, nitrides, and refractory metal-based alloys all have potential to meet the temperature requirements presented by the next-generation turbine; however, each has significant issues with regard to environmental stability, especially in the presence of moisture, which will likely require mitigation via some form of protective coating strategy. In addition, the production and processing of these next-generation materials will be non-trivial, especially at the scale required by land-based turbine systems. In some cases, totally new processing strategies may need to be developed. Regardless, because of the need for long-term reliability in these components, defects introduced during processing will have to be kept to an absolute minimum. Computational methods that link various length and time scales to define a complete materials chemistry, microstructure, and processing strategy could be key to speeding the development of these next-generation materials.

In addition to structural materials development and improved coatings, sensors that can provide real-time turbine process information (temperature, pressure, gas composition) to the operator are needed in order to fully optimize system performance. These requirements are very similar to those described previously for gasifier systems. In addition, *in situ* sensors that can monitor component reliability, and non-destructive methods for assessing component health during maintenance shutdowns, are needed in order to maximize materials performance in this application.

10.5.2 Gas separations

Oxygen production

Oxygen separation is central to a variety of coal-based processes and has been studied extensively for more than a century. Similarities in the size, shape, and properties of O_2 and N_2 make them extremely difficult to separate under ambient conditions. Industry currently depends on cryogenic distillation to produce the vast quantities of pure O_2 and N_2 which it consumes. Research is currently focused on developing ion-transport membranes, operating at 800–900 °C, to produce oxygen from compressed air. Ceramic membranes, consisting of proton-conducting perovskites, have the potential to produce highly pure oxygen, but brittleness, sealing difficulties, and relatively low permeability have prevented their widespread application. Mixed-matrix membranes, utilizing a polymer base coupled with a material that can increase the solubility or diffusivity properties of the composite, such as carbon nano-tubes or metal–organic frameworks, are also being investigated. Mixed-matrix membranes are generally classified as being highly permeable, but suffer from poor selectivity. Regardless of the material selected, the goal is to produce reliable membrane systems that are capable of producing up to 5,000 tons per day of oxygen, though the technology isn't expected to be viable before 2015.

Post-combustion CO_2 capture

In post-combustion systems, fully commercial gas-separation devices exist for removal of all major pollutants except CO_2. Post-combustion CO_2 capture for subsequent sequestration is a challenge for existing coal combustion plants (Chapter 8) without oxy-fuel firing, because it requires the separation of dilute concentrations of CO_2 (13–15 volume percent in a typical coal plant) from large volumes of low-pressure flue gas. Amine absorbents have been proven commercially for smaller-scale applications. These amines are capable of removing more than 90% of the CO_2 from the flue gas; however, regeneration of the solvents is an energy-intensive process that can consume as much as 25% of the energy produced in the plant. The energy intensity is partially due to the composition of the solvent mixture, which must be as much as 70% water by weight to avoid serious corrosion issues. In addition, amine solvents degrade in the presence of other flue-gas components, including O_2 and trace impurities, such as SO_x and NO_x. New separation materials and processes that can more efficiently remove CO_2 from the flue gas of conventional coal combustion plants will be needed if CO_2 sequestration is to become a reality for the existing coal combustion fleet.

Sorbent materials currently under development for post-combustion applications generally fall into either the liquid-absorbent or solid-adsorbent categories. Liquid-absorbent research often seeks to reduce or eliminate the water diluent and decrease the energy cost of solvent regeneration while improving contaminant resistance, diminishing volatility, and increasing hydrothermal stability. Liquid absorbents currently under development include advanced amines such as Mitsubishi Heavy Industry's KS-1 hindered amine and more novel materials such as the task-specific ionic liquids being developed at the University of Notre Dame and elsewhere [7].

Solid adsorbents avoid the problem of corrosion by placing CO_2-philic functionalities on high-surface-area solids instead of using an aqueous diluent. In this way liquid volatility is also eliminated, but loss of sorbent can still occur through contaminant poisoning, thermal degradation, or attrition of the support material. In addition, CO_2 capture tends to be a highly exothermic process, and it is more difficult to remove heat from solid sorbent systems than from liquid scrubbers. Considerable research has focused on the development of reactor designs that readily dissipate heat while minimizing sorbent attrition. Developmental sorbents can be based on either inorganic materials, including RTI's dry carbonate sorbent, or supported amines like those under development by NETL [8].

Membranes are a technology that avoids many of the pitfalls of both solid and liquid sorbents. Since membranes do not require regeneration, they avoid both expensive thermal cycling and complex process control; however, because they use a pressure differential as a driving force, they can be difficult to integrate into existing plants, which generate a partial pressure of only 0.15 atm in their flue gas. To make membranes practical for these separations, both advanced process-integration schemes like the one proposed by MTR and high-permeability materials are required [9].

Synthesis-gas separations

Separation of CO_2 and H_2 in synthesis gas can be accomplished either by the selective removal of CO_2 or by selective removal of H_2. Selective removal of H_2 has been the more common approach, primarily because traditionally the goal was to produce pure H_2 for use as a chemical feedstock or a transportation fuel, rather than H_2 for power and pure CO_2 for geological sequestration. In addition to the production of pure H_2, selective removal of H_2 has the added advantage of leaving the remaining synthesis-gas components at relatively high pressure, which is valuable since CO_2 must eventually be compressed to 2,200 psi for transportation to a storage site. Commercially, selective H_2 removal is accomplished by pressure-swing adsorption (PSA), usually on molecular sieves, silica gels, or activated carbon, but low selectivity leads to limited H_2 recovery. A great deal of research has been conducted on Pd-based membranes as an alternative to PSA. Metallic membranes have the advantages of near-perfect selectivity and acceptable fluxes, but significant problems exist with respect to their interactions with sulfur compounds. The membranes tend to form surface sulfide layers, which greatly reduce their flux. High-temperature polymer membranes, such as those based on polybenzimidazole (PBI) developed at Los Alamos National Laboratory (LANL), have also been studied, but materials-fabrication issues still exist with that technology [10].

Selective removal of CO_2 has the advantages of producing relatively pure CO_2 for sequestration and keeping the synthesis-gas H_2O with the H_2 for turbine expansion, which can result in a significant increase in IGCC efficiency. The commercial technologies for CO_2 removal from gasification-based power generation are the low-temperature physical solvents Selexol (40 °C) and Rectisol (−60 °C). Both solvents are highly efficient at separating CO_2 but require operating conditions with temperatures much below those of the low-temperature water–gas shift (260 °C), which is the most desirable location for capture in an IGCC process. Other solvent processes include Uhde's Morphysorb, which is currently under development for operation under similar conditions. Various technologies designed to operate at higher temperatures are also being developed. The technologies include inorganic sorbents, usually based on reversible metal carbonate formation, and facilitated-transport membranes.

10.5.3 Catalysis

The conversion of coal and other carbon-based feeds has historically utilized heterogeneous catalysis to make numerous processes techno-economically feasible, including reforming, the water–gas shift, and liquefaction. Similarly to the majority of the above-mentioned processes, future-generation catalysts will need to have higher activity and selectivity, as well as exhibiting increased thermal and chemical stability in an effort to accommodate process demands for increased efficiency and decreased environmental impacts. Current and future approaches to catalyst development will likely require an "atoms-up" approach using both fundamental experimental and computational techniques. For example, detailed physical and chemical characterization of active sites, and the synthesis and control of nanostructured catalysts, are keys in the development of multi-functional catalysts that promote desired products that will likely be utilized for future applications.

Additionally, conventional catalyst approaches utilize *ex situ* catalyst and product analysis and *in situ* experimentation under conditions far from industrial relevance in attempt to gain an insight to catalyst fundamentals. To efficiently develop catalyst systems for the future, it is imperative that computational and experimental approaches that allow characterization and examination for industrially relevant conditions and time scales are developed.

10.5.4 Environmental controls

Environmental concerns are at the forefront of fossil energy conversion use, specifically for coal. Coal conversion processes produce numerous materials that can impact the environment, with the most notable current concern being carbon dioxide. Several strategies are being developed to mitigate the emissions associated with carbon dioxide. For example, technologies for the capture of carbon dioxide from point sources currently exist, while geological sequestration yields a means of storing the carbon dioxide in geological formations. Additionally, the use of carbon dioxide in geological processes is currently employed to enhance the production of oil using "enhanced oil recovery."

Additionally, carbon dioxide generated from fossil conversion processes can also be utilized in numerous processes, including as a feedstock in the food industry, as a means of enhancing the geological recovery of oil, and as a necessary building block for photosynthesis. Plants, including algae, grown from fossil-fuel-linked carbon dioxide can have positive techno-economic impacts for the pharmaceutical, food, and energy industries.

10.6 Summary

Coal will continue to be an important part of the world energy mix at least through the remainder of the twenty-first century. Clean coal technologies are being developed that can make the conversion of coal an efficient and environmentally benign process; however, their successful realization will require simultaneous advancements in materials science and technology in a cost-effective manner. New materials that can reliably meet the demands of increased temperatures and pressures and increasingly severe service environments while maintaining required performance characteristics are needed, as are advanced processing technologies that can economically produce components at the required scale. Computational methodologies that can accurately link the various scales of materials development, materials processing, and materials performance will help speed the time from concept development to component reality.

10.7 Questions for discussion

1. As discussed within the chapter, the development of high-temperature materials is a critical path toward future fossil energy conversion processes. What are

the degradation mechanisms associated with metal, polymer, ceramic, and hybrid systems, and what are some of the historical means of overcoming these limitations?

2. Corrosion is a process that can degrade material performance. What are the characteristics of the gaseous environment and the metal systems that dictate corrosion, and how does it differ for pure metals and metal alloys? Predict the thermodynamic stability of pure Ni at various locations within (a) air-blown combustion, (b) oxygen-blown combustion, and (c) oxy-blown gasification processes.

3. What are the similar properties of N_2 and O_2 which make them difficult to separate in order to produce pure O_2 for oxy-combustion and gasification? What properties of the gases are substantially different and might be used as a basis for separating them?

10.8 Further reading

- **W. Callister**, 2003, *Materials Science and Engineering: An Introduction*, New York, John Wiley & Sons.
- **C. Higman** and **M. van der Burgt**, 2003, *Gasification*, Amsterdam, Elsevier.
- **C. H. Bartholomew** and **R. J. Farrauto**, 2005, *Fundamentals of Industrial Catalytic Processes*, New York, Wiley-Interscience.
- **R. F. Probstein** and **R. E. Hicks**, 1982, *Synthetic Fuels*, New York, McGraw-Hill.
- **K. L. Smith, L. D. Smoot, T. H. Fletcher**, and **R. J. Pugmire**, 1994, *The Structure and Reaction Processes of Coal*, New York, Plenum Press.

- **DOE/NETL**, 2010, *Carbon Dioxide Capture and Storage RD&D Roadmap*, http://www.netl.doe.gov/technologies/carbon_seq/refshelf/CCSRoadmap.pdf.

10.9 References

[1] **DOE**, 2008, *Fossil Energy: A Brief History of Coal Use in the United States*, available from http://fossil.energy.gov/education/energylessons/coal/coal_history.html.

[2] **World Coal Institute**, 2005, *The Coal Resource – A Comprehensive Overview of Coal*, Available from http://www.worldcoal.org.

[3] **C. Higman** and **M. van der Burgt**, 2003, *Gasification*, Amsterdam, Elsevier.

[4] **IEA**, 2009, *Key World Energy Statstics*, available from www.iea.org.

[5] **DOE**, 2010, *Net Generation by Energy Source: Total (All Sectors)*, available from http://www.eia.doe.gov/cneaf/electricity/epm/table1_1.html.

[6] **DOE**, 2010, *International Energy Outlook 2010 – Highlights*, available from http://www.eia.doe.gov/oiaf/ieo/highlights.html.

[7] **C. Cadena, J. L. Anthony, J. K. Shah** *et al.*, 2004, "Why is CO_2 so soluble in imidazolium-based ionic liquids?," *J. Am. Chem. Soc.*, **126**(16), 5300–5308.

[8] **M. L. Gray, J. S. Hoffman, D. C. Hreha** *et al.*, 2009, "Parametric study of solid amine sorbents for the capture of carbon dioxide," *Energy & Fuels*, **23**, 4840–4844.

[9] **T. C. Merkel, H. Lin, X. Wei**, and **R. Baker**, 2010, "Power plant post-combustion carbon dioxide capture: an opportunity for membranes," *J. Membrane Sci.*, **359**, 126–139.

[10] **J. R. Klaehn, C. J. Orme, T. A. Luther** *et al.*, 2006, "Polyimide and polybenzimidazole derivatives for gas separation applications," *ACS-PMSE Preprints*, **95**, 333–334.

11 Oil shale and tar sands

James W. Bunger

JWBA, Inc., Energy Technology and Engineering, Salt Lake City, UT, USA

11.1 Focus

Tar sands and oil shale are "unconventional" oil resources. Unconventional oil resources are characterized by their solid, or near-solid, state under reservoir conditions, which requires new, and sometimes unproven, technology for their recovery. For tar sands the hydrocarbon is a highly viscous bitumen; for oil shale, it is a solid hydrocarbon called "kerogen." Unconventional oil resources are found in greater quantities than conventional petroleum, and will play an increasingly important role in liquid fuel supply as conventional petroleum becomes harder to produce. With the commercial success of Canadian tar-sand production, and the proving of technology, these unconventional resources are increasingly becoming "conventional." This chapter focuses on the trends that drive increased production from tar sands and oil shale, and discusses the geological, technical, environmental, and fiscal issues governing their development.

11.2 Synopsis

Oil shale and tar sands occur in dozens of countries around the world. With in-place resources totaling at least 4 trillion barrels (bbl), they exceed the world's remaining petroleum reserves, which are probably less than 2 trillion bbl. As petroleum becomes harder to produce, oil shale and tar sands are finding economic and thermodynamic parity with petroleum. Thermodynamic parity, e.g., similarity in the energy cost of producing energy, is a key indicator of economic competitiveness.

Oil is being produced on a large commercial scale by Canada from tar sands, and to a lesser extent by Venezuela. The USA now imports well over 2 million barrels of oil per day from Canada, the majority of which is produced from tar sands. Production of oil from oil shale is occurring in Estonia, China, and Brazil albeit on smaller scales. Importantly, the USA is the largest holder of oil-shale resources. For that reason alone, and because of the growing need for imports in the USA, oil shale will receive greater development attention as petroleum supplies dwindle. Growth of unconventional fuel industries will be driven by continuing demand for liquid fuels, for which there are no non-fossil-fuel substitutes on a very large scale.

Economically, oil shale and tar sands are now competitive with petroleum. Current and future technology development seeks reliable, efficient recovery methods that keep emissions, discharges, and solids management within regulatory bounds. Once established, these unconventional oil industries will provide production assurances for decades, since there will be no decline in production, as with oil or gas. These assurances provide long-term social benefit, and support the sustainability of both the economy and the energy supply.

11.3 Historical perspective

Tar-sand and oil-shale deposits are located near the surface, and because of this humans very probably knew about both resources even before historical records were kept. In early times bitumen from tar sands was used to caulk boats and canoes. Oil shale has been used as a heating fuel by burning it directly.

The Green River Formation oil shale of Colorado, Utah, and Wyoming was well known in the pioneer days of the USA. Mining claims were made under the Mining Act of 1872. In 1912, President Taft designated the Office of Naval Petroleum and Oil Shale Reserves within the Department of Defense to manage and develop these resources, most of which occur on federal land. The Mineral Leasing Act of 1920 prohibited further private mining claims for oil shale, and in 1930 President Hoover issued Executive Order 5327, placing a moratorium on leasing of oil-shale lands. Although President Truman lifted the moratorium in 1952 with Executive Order 10355, and the US Department of the Interior took a tentative step in the mid 1970s to lease oil shale, as of the time of writing of this chapter there is no oil-shale leasing program for federal land in the USA. Some development is proceeding on state or privately owned land.

11.3.1 A commercial history of tar sands

The first successful venture in tar sands was the startup of the Great Canadian Oil Sands venture in 1967 at a location north of Ft. McMurray, Alberta, Canada. The Suncor operation of today is a continuation and expansion of this plant. The original design capacity was 50,000 barrels per day of synthetic crude oil. The GCOS venture pioneered the way for Syncrude, Canada, Ltd. to go onstream in 1978.

By the mid 1990s, the Province of Alberta and the government of Canada had realized that certain tax and royalty conditions were unnecessarily adding to investment risk, so these governments adopted fiscal policies that reduced this risk and promoted the growth of the industry. Today, well over 1 million barrels per day of syncrude and bitumen are being produced by a combination of surface and *in situ* technologies. Products are sold throughout Canada and the USA. The syncrude produced from the upgraders commands a premium price in relationship to conventional crude oil, largely because it contains no distillation residue. About 600,000 bbl per day of bitumen are produced from the Orinoco tar belt in Venezuela.

11.3.2 A commercial history of oil shale

The only continuing, commercially successful oil-shale operations existing today are found in Estonia, China,

and Brazil. Estonia has exhibited continuous operations for about 90 years. Each of these ventures is relatively small, with total worldwide production less than 100,000 bbl per day. Several attempts were made to commercialize oil shale in the USA, beginning in the mid 1970s but ending in 1991 with the closure of the Unocal plant in Colorado.

For the most part, failures to achieve sustained commercialization in the USA have derived from two causes. The most obvious factor, which stopped several major projects (Colony, in Colorado, and White River, in Utah, for example, and nearly caused the failure of Syncrude, Canada) was the unexpected, severe drop in oil prices of the early 1980s. The second factor has derived from poor technological reliability. Poor reliability at the startup almost finished the GCOS plant in Alberta, and, had senior management at Sun Oil (the owner) pulled the plug on this operation in 1967–8 when the magnitude of the problems became obvious, it is entirely possible that there would be no tar-sand industry in Alberta today.

The lessons for the future developments of these industries are, in some ways, simple. The investment requires fiscal certainty relative to prices and policies, and they need an efficient, robust technology. Such technological characteristics can be proven only through field experience, which itself requires a significant investment. Until the investor can see the potential for long-term growth, which also requires access to adequate resources in order to support growth, commercialization of oil shale will be constrained.

11.4 Origin and location of tar sands and oil shale

Tar sands contain bitumen, a viscous form of petroleum that does not flow at reservoir temperatures. In the USA *tar sand* is a sandstone containing a hydrocarbonaceous material with a gas-free viscosity of greater than 10,000 cP (centipoise). *Oil shale* is a fine-grained sedimentary rock that contains kerogen. For the most part kerogen is a solid that, upon heating, chemically converts to oil that is fluid at room temperatures.

11.4.1 Tar sands

The geochemical origins of tar sands are similar to those of petroleum; that is, remnants of prehistoric life, mostly algae, are co-deposited with sediment in fresh, brackish, or salt water. Certain chemical processes destroy proteins and carbohydrates, which are subjected to hydrolysis and biological activity. Fats resist these processes and persist as oil. Over time these sediments are buried ever deeper and are heated by heat from the Earth's core. Under high lithostatic pressure, these remaining "oils" are squeezed from the sediment and migrate to nearby sandstones that have greater porosity (larger void spaces between the grains), where the oil is found today. The process of deposition, chemical changes, squeezing of oil from the shale-like sediments, also known as primary migration, and secondary migration (flowing through porous media until a trap is encountered) is known as petroleum *maturation*. Tar-sand bitumen is the heavy ends of petroleum, left from near-surface deposition that has allowed evaporation of the light ends, and which has not been exposed to the higher temperatures of deeper deposition over longer periods of time. Hence, in one view, tar-sand bitumen can be considered an "immature" form of petroleum.

Over the years there has been some objection to the use of the word "tar" because the native bitumen differs significantly from pyrolysis products made from coal or petroleum residue, which historically have been called tar (as in "coal tar"). In Canada, the word "tar" was dropped several decades ago because of the negative impression this made when seeking investment. Even in the USA the petroleum geologists made a serious attempt in the 1960s to rename these deposits as "surface and shallow oil-impregnated rocks," but it is obvious why this arrhythmic definition failed to catch on.

11.4.2 Oil shale

In the simplest of terms, oil shale is an example of the original sedimentary deposits that made natural gas, petroleum, and tar sands. If oil-shale deposits were allowed the time to become buried and heated, the kerogen would be converted into petroleum and gas. As the oil migrates, the shale oil so produced is refined along its migration path, since the most polar, largest molecules are left absorbed on the sand/shale. Trapped in a reservoir, the resulting oil would be petroleum. Thus, kerogen-containing oil shale is geologically a very immature stage of petroleum, and is invariably found at shallow depths, from surface outcrops to about 3,000 feet in depth. All commercially viable oil-shale processes require the application of heat to speed the process of converting kerogen into hydrocarbon liquids, a process that would otherwise require the slow heat from the Earth. Deeper reservoirs of oil shale do, in fact, contain a liquid form of oil. The use of the word "oil shale" as applied to oil produced from deeper reservoirs such as the Bakken (North Dakota), or Eagle Ford (Texas) is a relatively recent event. For the purposes of this chapter all uses of the term "oil shale" refer to shallow occurrences where the kerogen is solid.

11.4.3 The significance of geological history

Geological history is significant for the following reasons.

- In order to recover oil from tar sands or oil shale, an extra process step is required that is not required for liquids and gases such as petroleum and natural gas.

- Because these oils have not migrated very far, most of the original organic material is still there, leading to quantities of bitumen and kerogen that equal or exceed the quantities of petroleum, at 2 trillion barrels or more. In other words, the sheer magnitude of the resource is significant.

- Because these resources occur in shallow deposits they are easily found and are readily measured, leading to low exploration risk. (In conventional oil production, finding the oil deposits presents a large investment risk, and the number of prospective places on Earth yet to explore is dwindling very rapidly.)

- Once the technology for recovery has been proven, there is a high degree of certainty regarding the quantity of oil that can be recovered. Even though the initial investment can be very large, the fact that the oil can be produced for 40 years or more, without decline in production rate, strongly factors into the investment decision.

11.5 Resources

When discussing unconventional resources, careful attention must be paid to the definition and use of selected terms. "In-place resource" is the amount of oil (bitumen or kerogen) that is actually found in the ground. "Proven reserves" or sometimes simply "reserves" are the amount of oil that can be recovered with today's technology and under current economic conditions. Other terms such as "measured" mean that there is assay data from coreholes or outcrops to support the quantity cited; "inferred" means that there is evidence that the geological province extends beyond the measured area, and that there is reason to believe the hydrocarbon content continues; "speculative" means that there is a rationale for the presence of hydrocarbon but that great uncertainty exists about the extent of the deposit and the concentration of hydrocarbon. Unfortunately, not all the literature rigorously follows this terminology and care must be taken when interpreting statistics on resources.

11.5.1 Tar-sand resources

According to the US Geological Survey (USGS), the resources of tar sand of Canada and Venezuela alone are 3.7 trillion barrels, in place. Nearly 70 countries have tar-sand resources, totaling about 4.5 trillion barrels. At present, it is not know how much of the in-place resource will eventually become "reserves." Canadian reserves are about 174 billion bbl. The USA contains measured resources of 21.6 billion barrels, with another 31.1 billion inferred resources. None of the US resources are classified as "reserves."

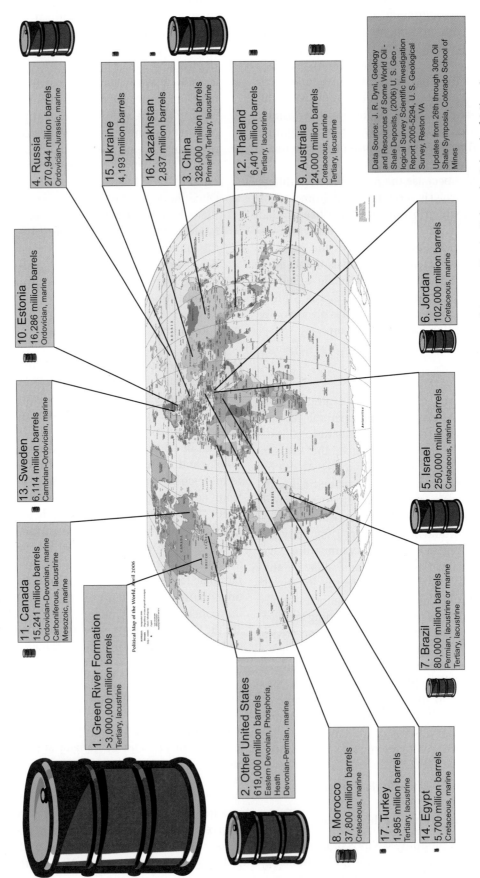

Figure 11.1. Estimated world oil-shale resources (courtesy of Jeremy Boak, Colorado School of Mines).

11.5.2 Oil-shale resources

According to J. R. Dyni, of the USGS, total world resources of oil shale are 4 trillion bbls, of which 3 trillion are found in the USA. Figure 11.1 shows the latest estimates of worldwide resources. The thickest resources are located in Colorado. There are locations in the depicenter of the Piceance (pronounced pē'-ăns) Creek Basin, where the thickness exceeds 1,000 ft at an average of 25 gallons per ton (gpt), potentially yielding more than 1.5 million barrels per acre. It is believed that this deposit is the greatest concentration of hydrocarbons on Earth.

Rich zones in Utah outcrop along the southern and eastern margins of the Uinta Basin. The depicenter of the Uinta Basin contains an intercept of 80 ft that averages 25 gpt. Oil shale in Wyoming is widespread, but generally leaner than that in Utah and Colorado. There are some near-surface deposits that could average 20 gpt that are of commercial interest. Because of their long-running industries, only Estonia, Brazil, and China can claim "proven reserves" of oil shale, albeit there are resources in the USA that possess characteristics of richness and accessibility similar to those of these "proven" deposits.

11.5.3 Classification of "resources" as "reserves"

In December of 2004, the *Oil and Gas Journal* made a determination that 174 billion barrels of Canadian oil sands would be reclassified from "resources" to "proven reserves." At the time it was a somewhat controversial determination, but is now generally accepted by the securities and exchange commissions when valuing company holdings. The significance of this reclassification cannot be overstated; it vaulted Canada from obscurity as a reserve-holder to second in the world, behind only Saudi Arabia. Canada is the single largest supplier of imported oil to the USA, greater than Saudi Arabia and Venezuela combined.

In August of 2004, the *Oil and Gas Journal* published an article that included the relationship of grade to resource quantity. This relationship is recast in Figure 11.2, and shows that there is a sizable amount of oil shale that is rich enough and thick enough to exceed the richness of Alberta oil sands, which are already considered conventional. If the USA could prove the commercial viability of a portion of these resources (400 billion barrels is plausible), it could become the world's largest holder of proven oil reserves.

11.6 Chemistry

11.6.1 Tar sands

Tar-sand bitumen (pronounced bitch'-u-men, if you are from the petroleum school, and bĭt-oo'-men, if you are from the coal school) resembles the heavy ends of petroleum. There is a remarkable similarity between southern California crude oils (Wilmington, for example) and Uinta Basin, Utah bitumen. Both of these substances are high in alicyclic saturates (naphthenic) hydrocarbons and both are high in nitrogen-containing compounds, primarily pyridinic and pyrrolic heterocyclics. Most bitumens of the world, including that of Alberta, Canada, are of marine (salt-water) origin and are higher in sulfur and lower in nitrogen, and contain more aromatic hydrocarbons, than those from lacustrine (fresh-water) origins.

11.6.2 Oil shale

When subjected to extraction by organic solvents, the kerogen contained in oil shale does not dissolve. The reason for the insolubility of kerogen is still subject to debate. One school holds that kerogen is actually a polymer of such high relative molecular mass (or, less formally, "molecular weight") that small-molecule organic solvents cannot dissolve it. Another school holds that the kerogen is so integrally mixed with the mineral particles that the solvents cannot access the kerogen to dissolve it. Yet another school holds that the kerogen is bonded to the minerals and solvents cannot break those bonds.

The facts show that kerogen is not converted to oils without subjecting it to cracking temperatures (generally above at least 575 °F, 300 °C). This observation would imply that there are some strong chemical bonds involved, probably in large organic molecules. Like bitumen from lacustrine deposits, oil shale is high in nitrogen content, also found in pyridinic and pyrrolic bonding structures, and low in sulfur content.

11.7 Technology

The key characteristics of unconventional oil-recovery technology are "effectiveness," "reliability," and regulatory "acceptability."

11.7.1 Recovery from tar sands

Mining and surface processing

The first successful commercial venture, Suncor, formerly Great Canadian Oil Sands, mixed mined ore with hot, caustic water to disengage the bitumen from the sand. The mixture was then subjected to a series of separation steps, including a final cleanup step that involved dissolving the bitumen in a solvent to reduce viscosity and separating the mineral fines by centrifuges, filtration, and the like. Even today, all surface mining processes use some variation on this theme.

Figure 11.2. Grade of Green River Formation USA oil shale (1 gal US/ton $= 4.17 \times 10^{-6}\,\mathrm{m^3/kg}$).

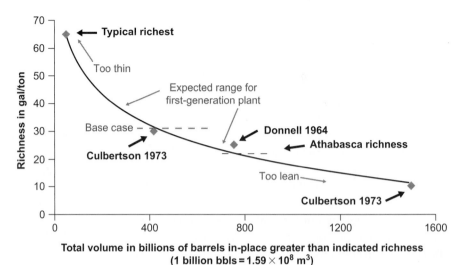

Total volume in billions of barrels in-place greater than indicated richness
(1 billion bbls $= 1.59 \times 10^8\,\mathrm{m^3}$)

In situ technologies

In later developments variations of *in situ* technologies have been used to recover bitumen. In general, *in situ* processes add heat to the reservoir to reduce the viscosity of the bitumen, allowing it to more easily flow to the production well.

Steam-assisted gravity drainage (SAGD, pronounced SAG-DEE) is the best-known commercially practiced *in situ* technology. Two slotted or perforated pipes are positioned horizontally in the bed, with one pipe overlying the other. Steam is injected into the upper pipe, which heats the bed, which lowers the viscosity of bitumen, allowing it to drain to the lower slotted pipe. Bitumen is collected and pumped to the surface for further processing.

Cyclic steam stimulation (CCS) is a variation of SAGD but intercepts the beds with vertical, slotted pipes. Steam is injected for a period of time and then pressure is released, allowing gases and low-viscosity fluids to migrate back to the well. This process is repeated many times, each time the area of influence of the heat grows. This technique is sometimes referred to as "huff and puff." Eventually the area of influence becomes large enough that it joins with heat from a neighboring well. At this point the system may revert to a "line-drive" in which steam is injected into a pattern of wells, and bitumen is recovered in a different set of wells.

Vapor extraction (VAPEX) uses light petroleum hydrocarbons (C_1–C_4) in place of steam. The light hydrocarbons dissolve in the bitumen, reducing its viscosity. Gases injected under pressure provide a drive for liquid once the pressure is released, as in "huff and puff."

Toe-to-heel air injection (THAI) deploys two perforated pipes laterally along the dip of the bedding plane, one near the top of the intercept and the other near the bottom. Air is injected into the upper pipe and a combustion front is ignited near the source of the air. A portion of the resource is burned to provide the heat that softens the bitumen, allowing it to drain to the lower pipe. The process works best when a near-vertical flame front is maintained as combustion progresses downdip through the bed.

In the Orinoco tar belt, bitumen is either diluted with condensate or subjected to heating with steam to reduce its viscosity and allow it to be pumped from wells.

11.7.2 Recovery from oil shale

Oil shale differs from tar sand primarily in the fact that the recovery process, commonly known as "retorting," performs the cracking simultaneously with recovery. With some mild stabilization to prevent fouling of pipelines and feed heaters, shale oil can be sold directly to petroleum markets. A price penalty may be paid if the oil contains large quantities of heteroatoms (N, S), but when Unocal fully upgraded its oil in the early 1990s, and removed these heteroatoms, they received a premium price for the product.

There are three basic approaches to retorting of oil shale. These are known as surface retorting, *in situ* (or sometimes true *in situ*) and modified *in situ*. Surface retorts may be configured as vertical cylinders, as with Paraho, Petrosix, Union, and Kiviter technologies. They also may be configured as rotating kilns, such as ATP and Enefit, formerly Galoter, practice. The advantages of surface retorts are the high degree of control and the high yields. The disadvantages are the high capital costs and long lead times from investment to revenue.

Until Shell proved otherwise, true *in situ* was not viewed as commercially viable. The conventional wisdom held that oil shale needed to be broken into rubble with access for hot gases to penetrate the bed and for oil products to escape. Shell heats the entire bed with resistance heaters, and the formation of oil and gas from the heating creates permeability in the bed that allows the products to flow to a well, where they are recovered. Shell's ICP process is now the best-known true *in situ* process.

The various technologies that employ a rubblized bed are generically referred to as modified *in situ* (MIS). The bed may be directly heated with combustion gas as Oxy and Geokinetics practiced, or indirectly heated with heat pipes as more recently proposed by Red Leaf Resources in their EcoShale process.

11.7.3 Conversion technology

For tar sand bitumen, coking or hydrocracking is needed to reduce the average relative molecular mass to ranges useful for matching the properties of petroleum products. Retorting fulfills this function with oil shale. Subsequent catalytic hydroprocessing may also be needed to render the product saleable for conventional petroleum refining.

Because of the rapid growth of SAGD and other *in situ* processes for recovery of bitumen, there has developed an economic incentive to defer this upgrading step to refineries downstream. This has led to the use of diluents to mix with the bitumen, in order to reduce the viscosity so that they can be pipelined. If natural-gas liquids (NGL) or naphtha are used as the diluent, the mixture is call "dilbit." If synthetic crude oil is used as a diluent, the mixture is called "synbit." There are economic pros and cons to each, but at the moment there is insufficient diluent to meet all of the demand to transport bitumen. Hence, in some cases a diluent return pipeline is used to "recycle" NGL or naphtha diluent. Once these products have reached a conventional petroleum refinery they can be co-mixed with petroleum or separately refined to gasoline, diesel, and aviation-turbine fuels.

11.8 Thermodynamics

In the end, all energy production and use (whether renewable or nonrenewable) is about thermodynamics, and thermodynamics will increasingly dictate the economic competitiveness of energy alternatives in the future. If the energy required to produce energy is high, the net energy available for end use must command a higher price, or subsidies, in order to compete. A prime example is the high energy cost of producing biofuels, which, even after 30 years of subsidies, could not exist without that financial help. Thermodynamic efficiency governs the minimum cost of producing energy.

11.8.1 Thermodynamic efficiency in the production of oil

M. King Hubbert famously said in 1982 "So long as oil is used as a source of energy, when the energy cost of recovering a barrel of oil becomes greater than the energy content of the oil, production will cease no matter what the monetary price may be."

In days past, the amount of energy needed to produce energy was small. Oil and coal, historically the major fossil energy resources, were near the surface and were easily recovered. Today, virtually all of the easily recovered fossil energy has been recovered. Tomorrow's oil and coal are deeper, higher in sulfur, more remote from the end use, heavier (in the case of petroleum), and lower in heating value per ton (in the case of coal). All of these trends cost more energy to provide specification fuel.

Ultimately, all nonrenewable resources will reach a point where unfavorable thermodynamics forces a cessation of production, and no amount of economic subsidy can overcome this fundamental condition. At that point, unless thermodynamic efficiency can be improved, there will be large residual quantities of hydrocarbon left in the Earth, never to be recovered.

11.8.2 Tar sands

For every 1 million Btu of heating value that enters the process, about 950,000 is in the form of bitumen, and about 50,000 is in the form of natural gas. For this 1 million Btu, 820,000 Btu is produced in the form of synthetic crude oil, and 180,000 Btu is lost as heat for power generation and thermal losses from equipment, or stockpiled as coke. Thus, the first-law efficiency of surface-mined Canadian oil sands is about 82% (820,000/1,000,000). The energy return on energy invested (EROI) defined by Hall is 4.6 (820,000/180,000). The thermodynamic efficiency of Alberta oil sands is currently increasing with time as mining and recovery processes wring out their inefficiencies. Eventually, however, when the mine is forced to recover leaner ore, or encounters greater overburden, the first-law efficiency will begin to decline.

11.8.3 Oil shale

The first-law efficiency for production of 25 gpt oil shale is calculated at 81%. Thermodynamic efficiencies will hold steady or improve with time, as long as the process is able to utilize a dependable grade of ore. Only when the ore grade decreases, as with tar sands, will the energy efficiency be forced to decline. The first-law efficiency for recovery of the full range of grades (barren-to-pay ratios no worse than 1 to 1) is provided in Figure 11.3.

11.8.4 Concepts of self-sufficient production

Professor Charles Hall coined the term energy return on energy invested (EROI), to quantify the demand a given process places on its surroundings. Hall's definition is meaningful for processes that compete for the same end-use energy demanded by alternative economic processes. Hall's definition is not as meaningful

Figure 11.3. First-law efficiency for surface production of shale oil. (1 gal US/ton $= 4.17 \times 10^{-6}$ m^3/kg).

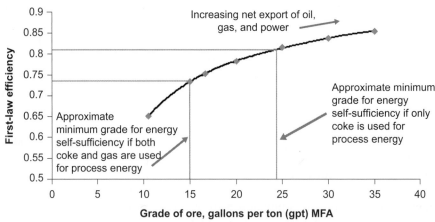

Figure 11.4. Production efficiency of world conventional and unconventional oil resources.

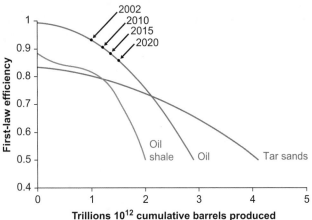

for processes that are self-sufficient in energy. With self-sufficient processes (operating in energy balance), little, or no, energy is purchased/imported from outside the process boundaries. For the most part oil shale and tar sands are energy self-sufficient. In cases of energy self-sufficiency, it is conceivable that a process could remain economical at an EROI of less than 1, but only if the cost per unit of energy consumed is much less than the sale value of the product, per unit of energy. It is generally believed that practical processes will require an EROI of at least 2, or first-law efficiencies of at least 70%.

11.8.5 Future trends in thermodynamic efficiency

If we take Hubbert's expression as a roadmap for the future, we can conclude that the ultimate limit of fossil energy recovery will occur at the point when the energy requirements equal the energy output. Taking what we know about tar sands, oil shale, and petroleum, we can construct the trendlines shown in Figure 11.4. Here we see that petroleum is about to enter a steep decline in recovery efficiency. More energy is needed for enhanced recovery methods, for exploration and production in remote or deep-water regions, and for upgrading of

heavier, higher-sulfur crudes. What is also sobering about this curve is the current rate at which we are depleting our petroleum reserves: roughly 1% of our total endowment, every year. Oil shale initially exhibits a higher thermodynamic efficiency than tar sands because of the many opportunities to recover resources in excess of 30 gpt.

If these projections hold true, there are about 2 trillion petroleum, 2 trillion oil-shale, and 4 trillion tar-sand barrels available for recovery. Economic and technological limitations may cause these ultimate numbers to be much smaller.

11.9 Products and markets

Marketable products from tar sand and oil shale include synthetic crude oil, natural gas, asphalts, chemicals, and electricity. Syncrudes produced from tar sand or oil shale are interchangeable with petroleum, and, indeed, refiners will pay a premium for these syncrudes, over comparable-quality petroleum feedstocks, because of their consistent quality and low residue content. Whereas conventional petroleum is becoming more variable in quality, due to the increasing numbers of different producing wells blended into a single pipeline, unconventional oils maintain their consistency over years of time. This factor alone, together with the ever improving thermodynamic efficiency, virtually guarantees the market value of unconventional oils. Political treatment and regulations could skew this trend, but such manipulation can only come at the expense of consumers and taxpayers.

11.10 Economics and fiscal regime

Getting an unconventional-fuels industry started is economically very difficult. There are technical, regulatory, and fiscal risks that must be overcome for any first-generation facility. The economics are driven by resource characteristics (grade and accessibility), process efficiency and reliability, timing of capital investment, product values, and tax and royalty treatment. Most observers believe that oil from oil shale and tar sands is economically viable at oil prices above about $60 per bbl. A significant deterrent to investment, especially for first-generation plants, is the prospect that the oil price could fall below this threshold. Nonetheless, for mature industries, capital investment is being made as rapidly as limitations on human and material resources allow.

Government tax and regulatory conditions will affect investment. If public policy determines that development is important to the economy and its citizens, then there are certain steps that can remove investment uncertainty. It has been shown in Canada that allowing investments to be expensed (as contrasted with depreciated), back-loading royalties until after investment payout, and encouraging research and development through tax regulations all accelerate investment, with little or no adverse long-term impact on public revenue. Investment responds to certainty, and policy, regulation, and statutes that are aimed at creating greater certainty help any investment decision.

11.11 Environmental and regulatory

For oil shale the most obvious environmental impact relates to surface disturbance. By mining, whether *in situ* or modified *in situ*, with the exception of some special cases of horizontal access in eroded canyons, the total surface is largely disturbed. In the case of oil shale, the disturbed acreage is small because of the great thickness of the beds. Whereas conventional petroleum extraction may recover 10,000 bbl per acre disturbed (taking Alaska North Slope as an example), oil shale will exhibit recoveries of 50,000 to 150,000 bbl per acre in Utah, and may exceed 1 million bbl per acre in Colorado.

Other environmental impacts relate to air emissions, where a permit is needed when emitting above certain regulatory thresholds. It is unlikely that technologies practiced today will discharge large amounts of water. By the time water has been cleaned for discharge, it will be useful for other process purposes such as irrigation during reclamation and dust control for haulage roads, mining, or ore preparation.

There is some question as to whether development of unconventional fuels will be inhibited by climate-change legislation. There is always a possibility that new costs will be added through regulation. However, with respect to carbon, as was shown in the section on thermodynamics above, oil shale and tar sands will be at parity with petroleum in just a few years, and thereafter may actually be superior to petroleum, relative to emission of CO_2, on a global basis.

The water-demand issues of tar-sand development are well documented by the Canadian experience. Over the years the net demand for water has gone down from about 5 bbls of water per bbl of oil to about 3. Oil shale, for which water is not used in the extraction process will likely require less water than oil sands. It is true that the western USA has no unused water. However, left to the marketplace, sufficient water rights can be purchased from current rights holders without an adverse impact on project economics and with net beneficial impact to local economies, albeit small ranching operations, for example, may be impacted.

Permitting and regulations for unconventional resources are still evolving. Unlike the situation in the 1970s, however, we now have well-established guidelines for impact on air, water, land use and reclamation, and other environmental values. The political discussion relative to CO_2 emissions is sure to be prominent, at least for the next few years, until society decides whether the problem is as serious as some pose, and, if so, what should be done about it.

It is unlikely that the CO_2 issue will remain in limbo for very long. Soon, shortages of fossil energy will obviate the projections of exponential growth in CO_2 emissions. Further, the world will come to recognize the magnitude of natural biosequestration occurring through accelerated plant growth. The combined effect of these two trends is to attenuate ultimate atmospheric CO_2 concentrations below levels of critical concern.

11.12 Socioeconomic factors

11.12.1 Permission to practice

Unconventional resources will be developed only when the local and state communities have given their "permission to practice." Because of historical, spectacular busts, communities are cautious and the hurdles are high. The most notable bust was the pullout of Exxon from its Colony project on May 2, 1982, known regionally as "Black Sunday," when the sponsors abandoned investment of more than $1 billion, and left thousands suddenly unemployed.

The situation would have been even worse if Colorado Governor Lamb and other officials had not insisted on prepayment for community infrastructure. Imagine the economic carnage, had local communities been left with bonding debt, in addition to the loss of economy. The lesson from this experience is that front-end money is needed to mitigate financial risk to the communities. Whether that money comes from project financing (making the return on investment more difficult), or from revenue sharing by the federal government (from mineral royalties collected on leases), local communities need to avoid bonding for infrastructure developments or operations.

Local communities have a strong incentive to see these unconventional-resource projects succeed, however. These projects have the same economic characteristics as a manufacturing business, or a mining business, where steady employment and revenue can be counted on for 50 years and more. Whereas traditional oil and gas suffer from production decline, unconventional oil will give enough long-term assurances that communities can grow and diversify their economies.

The Province of Alberta, Canada and the local community of Ft. McMurray are models for engaging communities in decisions relating to socioeconomic impact. There are several mechanisms in place to build consensus of interested parties, and in complex, high-impact developments, consensus can be difficult to achieve. The USA, which has no unconventional-resource industry today, would benefit by observing how this feature has been managed in Alberta's real-world situation. Alberta tar sands are pouring billions of dollars into provincial and federal coffers. In fact, tar sands directly and indirectly account for 17% of the total employment of Alberta.

11.12.2 National security and economy

Beneficial impacts of domestic production on economy and security are discussed throughout this book. One significant point should be made, however. If the USA could achieve the goal of reclassifying 400 billion barrels (about 25%) of its vast oil shale "resource" as "reserves," this would make the USA the holder of the largest proven reserve of hydrocarbons in the world. Additional benefits of unconventional-fuels production, whatever the host country, are a strengthening economy, greater energy self-sufficiency, and improved balance of payments.

11.12.3 World societal benefits

The fact that unconventional resources are so widely dispersed around the world means that these resources are broadly relevant to future human economy. It is easy to think simplistically about energy being the driver of wealth and living standards, and that, given available and affordable energy, the world economy can continue to grow. However, in the face of supply shortages, other factors such as the flow of energy, who holds the control, and impacts on civility move to the fore. It is readily understood why countries are becoming increasingly concerned about future supplies of energy and are looking to their unconventional resources for long-term supply assurances. Two that come to mind are Jordan and Israel, but there are many others.

11.13 Summary

Unconventional resources will become increasingly important additions to our fuel supplies as petroleum supplies become limited. The manufacturing nature of production lends itself to increasing efficiency and lower environmental impact as technological experience matures. The lack of a decline curve provides economic assurances to communities. The sheer magnitude of these resources provides incentive to pursue production.

11.14 Questions for discussion

1. What distinguishes unconventional oil from conventional oil?
2. What are the drivers that dictate the economic competitiveness of one form of energy over another (this can be the subject of an essay)?
3. How do products from tar sand and oil shale compare with petroleum?
4. Geologically, where would you find tar sands, and where would you find oil shale?
5. What is the importance of grade in the production of energy from tar sands and oil shale?
6. How does production of energy help the local, regional, and national economy?

11.15 Further reading

- An interesting summary of the history of Alberta tar sands can be found at http://www.syncrude.ca/users/folder.asp?FolderID=5657.
- A recent history of attempts to establish a US oil-shale leasing program, mandated by Section 369 of the Energy Policy Act of 2005, can be found at www.unconventionalfuels.org and http://www.ostseis.anl.gov/.
- World resources of heavy oil and natural bitumen can be found in US Geological Survey Fact Sheet 70–03, August 2003.
- http://en.wikipedia.org/wiki/Oil_sands#Reserves.
- http://pubs.usgs.gov/of/2007/1084/OF2007–1084v1.pdf.
- http://pubs.usgs.gov/sir/2005/5294/pdf/sir5294_508.pdf.
- More on thermodynamics can be found in J. W. Bunger and C. P. Russell, 2010, "Thermodynamics of oil shale production," in *Oil Shale: A Solution to the Liquid Fuel Dilemma*, eds. O. I. Ogunsola, A. Hartstein, and O. Ogunsola, Washington, DC, American Chemical Society, pp. 89–102.
- More on EROI can be found out from Charles Hall at http://www.esf.edu/efb/hall/#Top and http://www.esf.edu/efb/hall/documents/Energy_Intro10b.pdf.
- For a list of current oil-shale technologies and activities, see http://www.fossil.energy.gov/programs/reserves/npr/Secure_Fuels_from_Domestic_Resources_-_P.pdf.

12 Unconventional energy sources: gas hydrates

Carolyn A. Koh, E. Dendy Sloan, Amadeu K. Sum, and David T. Wu

Center for Hydrate Research, Department of Chemical Engineering,
Colorado School of Mines, Golden, CO, USA

12.1 Focus

Gas hydrates are typically formed when water and gas (e.g., light hydrocarbons) come into contact at high pressure and low temperature. Current estimates of the amount of energy trapped in naturally occurring gas hydrate deposits, which are found in ocean sediments along the continental margins and in sediments under the permafrost, range from twice to orders of magnitude larger than conventional gas reserves. This has led to gas hydrates being considered as a potential future unconventional energy source.

12.2 Synopsis

Gas hydrates (or clathrate hydrates) are icelike crystalline solids imprisoning gas molecules (e.g., methane, carbon dioxide, hydrogen) within icy cages. These fascinating solids present an attractive medium for storing energy: *naturally* in the deep oceans and permafrost regions, which hold vast quantities of energy waiting to be unlocked and used as an alternative energy supply; and *artificially* by manipulating synthetic clathrate materials to store clean fuel (natural gas or hydrogen). Conversely, the formation of these solids in oil and gas flowlines (the pipes through which oil and gas are transported, for example, from a well to a processing facility) can lead to blockage of the flowlines and disastrous consequences if not carefully controlled. This chapter on gas hydrates begins with an overview of the discovery and evolving scientific interest in gas hydrates, followed by a basic description of the structural and physical properties of gas hydrates and the different energy applications of gas hydrates. The main focus of this chapter is on surveying the potential prospect of producing energy in the form of clean gas from naturally occurring gas hydrates, which present a potential alternative energy resource and could be a significant component of the alternative energy portfolio. The paradigm shift from exploration to production of energy from gas hydrates is clearly illustrated by the production tests that have either been performed or are planned in the Mackenzie Delta in Canada, on the North Slope of Alaska, and off the coast of Japan.

12.3 Historical perspective

Gas hydrates were first reported by Sir Humphrey Davy [1] during his Bakerian lecture to the Royal Society in 1810, in which he briefly commented on chlorine hydrate (at that time, Cl_2 gas was called oxymuriatic gas), "The solution of oxymuriatic gas in water freezes more readily than pure water." However, it is possible that Joseph Priestley in Birmingham, UK, discovered gas hydrates as early as 1778. Priestley was performing cold experiments in his laboratory by leaving the window open on winter evenings before leaving at the end of the day. The next morning he observed that vitriolic air (SO_2) would impregnate water and cause it to freeze and refreeze, whereas marine acid air (HCl) and fluor acid (SiF_4) would not. Despite these early observations, Priestley's experiments were all performed below the ice point (the temperature at which ice will form at atmospheric pressure), so the system might not have actually been in the hydrate phase, and there is no record that he obtained any subsequent validation data. Therefore, Davy is cited as the first to discover gas hydrates [2].

After their discovery, researchers initially studied gas hydrates (clathrate hydrates) as a scientific curiosity, with the primary goals of identifying all of the compounds that would form clathrate hydrates and determining the structures and physical properties of these compounds, as well as the thermodynamic conditions under which they form. Then, in the mid 1930s, Elmer Hammerschmidt, [3] chief chemist of the Texoma Natural Gas Company, discovered that natural gas hydrates, rather than ice, were responsible for blocking gas transmission lines, which were frequently at temperatures above the ice point. This discovery led to a more pragmatic interest in gas hydrates and shortly thereafter to the regulation of the water content in natural gas pipelines to prevent formation of gas hydrates. Subsequently, research was focused on determining and predicting the conditions (temperature, pressure, gas composition) for hydrate formation and also searching for chemicals (salts, methanol, monoethylene glycol) to prevent hydrates from forming [2].

In the mid 1960s, Yuri F. Makogon [4][5] led a large research effort to investigate natural-gas hydrates as a potential substantial energy resource in the USSR. Makogon's work marks the beginning of gas hydrates being linked to nature, with subsequent discoveries of further hydrate deposits in deep oceans, lakes, and permafrost regions. Now, over four decades later, research groups from academia, industry, and energy agencies around the world are pursuing the prospect of gas production from natural hydrate deposits.

12.4 Gas hydrates as a potential energy source

12.4.1 Basics of gas hydrates

Clathrate hydrates (also known as **gas hydrates**, or sometimes just **hydrates**) are icelike crystalline solids that are typically formed when water and gas come into contact at high pressures and low temperatures [2]. Gas hydrates are composed of hydrogen-bonded water molecules forming water cages (**hosts**) that trap small gas molecules (**guests**). There is no chemical bonding between guest and host molecules, only weak physical interactions (van der Waals forces and repulsive interactions), which prevent the host water cage from collapsing on itself.

Structure

The most common gas-hydrate structures are known as structure I (sI), structure II (sII), and structure H (sH). Table 12.1 and Figure 12.1 illustrate the differences among sI, sII, and sH hydrates. All three structures have a common small cage, denoted 5^{12}, comprising 12 pentagonal faces in the cage. sI also has a large $5^{12}6^2$ cage, comprising 12 pentagonal and 2 hexagonal faces in the cage. sII has a large $5^{12}6^4$ cage, comprising 12 pentagonal and 4 hexagonal faces in the cage. sH has a mid-sized $4^35^66^3$ cage, comprising 3 square, 6 pentagonal, and 3 hexagonal faces in the cage, and also a large eicosahedral $5^{12}6^8$ cage, comprising 12 pentagonal and 8 hexagonal faces in the cage. All three hydrate structures are crystalline, with the lattice parameters given in Table 12.1. Note that the smallest repeating unit of a crystalline structure is called a unit cell.

The type of structure that is formed depends largely on the size of the guest(s): small molecules such as methane and ethane will form sI hydrates, larger molecules such as propane and isobutane will form sII hydrates, and even larger molecules such as methylcyclohexane and 2-methylbutane will form sH hydrates. (Note that sH formation requires a second "help gas," such as methane or xenon, that occupies and stabilizes the small and mid-sized cages.) For a more extensive list of gas molecules that form hydrate structures (also called gas-hydrate formers), see [2]. As shown in Table 12.1, an sII hydrate contains a larger number of small cages per unit cell than does an sI hydrate. This difference can explain why, for a mixture of sI-hydrate-forming gases, such as methane plus ethane, sII will tend to form when the ratio of methane to ethane increases above around 74 mol% (but remains less than around 99 mol%), because ethane prefers to occupy the large cages and the greater number of small methane molecules can be accommodated in the small cages [6]. For small molecules, such as nitrogen, oxygen, and hydrogen, there is a preference to occupy the small

Table 12.1. Structural properties of clathrate hydrates [2]

Property	Hydrate crystal structures						
	sI		sII		sH		
Cavity	5^{12} (S)	$5^{12}6^2$ (L)	5^{12} (S)	$5^{12}6^4$ (L)	5^{12} (S)	$4^35^66^3$ (M)	$5^{12}6^4$ (L)
Cavities per unit cell	2	6	16	8	3	2	1
Average cavity radius (nm)	0.395	0.433	0.391	0.473	0.394	0.404	0.579
Water molecules per cavity	20	24	20	28	20	20	36
Crystal system	cubic		cubic		hexagonal		
Space group	$Pm3n$		$Fd3m$		$P6/mmm$		
Lattice parameters	$a = 1.2$ nm		$a = 1.73$ nm		$a = 1.21$ nm, $c = 1.01$ nm		
	$\alpha = \beta = \gamma = 90°$		$\alpha = \beta = \gamma = 90°$		$\alpha = \beta = 90°$, $\gamma = 120°$		

Average cavity radius varies with temperature, pressure, and composition; determined from atomic coordinates measured using single-crystal X-ray diffraction. Lattice parameters are a function of temperature, pressure, and composition, and the values given are typical average values. S, small cage; L, large cage; M, mid-sized cage.

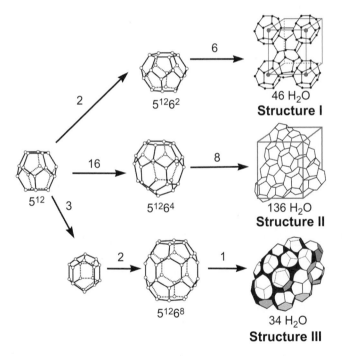

Figure 12.1. Cages of sI, sII, and sH hydrates (adapted from [2]).

cages, and, because sII hydrate contains more small cages than large cages, these molecules form sII hydrates. When pressures are very high (i.e., in the gigapascal range), multiple guest molecules [7][8], such as nitrogen and hydrogen, can occupy the large cavities of sII. Gas hydrates are also known as non-stoichiometric compounds, because some water cages can be vacant, while other cages will be occupied by a gas molecule. Typically (at low to moderate pressures), a maximum of one gas molecule will occupy a water cage.

Physical properties

When all the cavities of s1 or sII are occupied by a single guest species (e.g., methane or xenon), the stoichiometric number of water molecules per guest molecule is 5.75 or 5.67, respectively. Because both of these structures result in around 85 mol% water, a useful first approximation is to consider some properties of hydrates as variations from those of ice, as detailed in Table 12.2. Another reason for comparison of hydrates and ice is that the hydrogen bonds in hydrates average only 1% longer than those in ice, and the O−O−O angles of hydrates differ from the ice tetrahedral angles by 3.7° and 3.0° in structures I and II, respectively. As shown in Table 12.2, many of the physical properties of ice and hydrates are similar, with the exception of, for example, the thermal conductivity, which is significantly lower for hydrates [2].

12.4.2 Energy applications of gas hydrates

Figure 12.2 summarizes the key energy applications of gas hydrates. These areas include hydrates in flow assurance to prevent blockages in oil and gas pipelines, energy recovery from natural hydrated deposits containing vast amounts of untapped clean energy, and storage of energy (natural gas, hydrogen) in clathrate materials.

Flow assurance

In traditional oil and gas production, gas hydrates are considered a nuisance since they block oil and gas flowlines. Fluids (oil, water, and gas) within flowlines are typically at high pressures (e.g., >10 MPa) and can encounter low-temperature regions (e.g., sub-sea pipelines at

Table 12.2. Comparison of the physical properties of ice, sI hydrate, and sII hydrate [2]

Property	Ice	Structure I	Structure II
Isothermal Young's modulus at 268 K (10^9 Pa)	9.5	8.4[a]	8.2[a]
Poisson's ratio	0.3301	0.31403	0.31119
Bulk modulus (GPa)	9.097	8.762	8.482
Shear modulus (GPa)	3.488	3.574	3.6663
Compressional velocity, V_P (m s^{-1})	3870.1	3778	3821.8
Shear velocity, V_S (m s^{-1})	1949	1963.6	2001.14
Linear thermal expansion at 200 K (K^{-1})	56×10^{-6}	77×10^{-6}	52×10^{-6}
Thermal conductivity (W m^{-1} K^{-1})	2.23	0.49 ± 0.02	0.51 ± 0.02
Adiabatic bulk compression at 273 K (GPa)	12	14[a]	14[a]
Heat capacity (J kg^{-1} K^{-1})	1700 ± 200	2080	2130 ± 40
Density (g cm^{-1})	0.917	0.91	0.94

[a] Estimated.

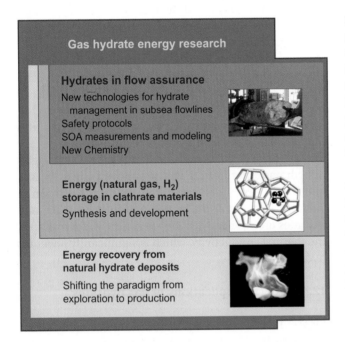

Figure 12.2. Energy applications of gas hydrates (hydrate plug in top figure, courtesy of Petrobras).

around 4 °C), yielding favorable conditions for hydrates to be thermodynamically stable. Flowline blockages due to gas hydrates prevent the flow of valuable commodities (oil and gas), hence halting energy (oil/gas) production and resulting in severe economic losses and safety risks. Gas hydrates are one of the solid phases commonly encountered in oil and gas production lines that disrupt flow. In the field of flow assurance, defined as

management of fluid transportation in multi-phase flow, gas-hydrate formation is the primary problem for the oil and gas industry, since hydrates tend to form much more quickly and often in larger volumes than other solid deposits (e.g., wax, asphaltenes) in pipelines [2]. Because most natural-gas flowlines contain gas compositions with propane and perhaps even heavier hydrocarbons, sII is the most common structure in industrial flow assurance.

In flow assurance, the conventional method for managing gas hydrates is to completely avoid hydrate formation by using thermodynamic inhibition methods, such as insulating the flowline to prevent the temperature of the fluids inside it from being low enough to enter into the hydrate stability zone, or thermodynamic inhibitors (methanol or monoethylene glycol), which act like antifreeze by shifting the thermodynamic (T, P) conditions required to form hydrates within the flowline to lower temperatures and/or higher pressures. However, in many cases, particularly new field developments where the flowlines are at water depths of several thousands of feet of seawater, thermodynamic inhibition becomes too costly and ecologically unsound (e.g., the required amounts of methanol or monoethylene glycol injection can be 40 vol% or greater of the water content). These concerns have paved the way toward the development of new technologies for hydrates in flow assurance that are based on risk management (rather than avoidance) to control gas-hydrate plug formation in flowlines. One of these technologies is the application of low-dosage hydrate inhibitors (LDHIs), which, at around 1 wt% in solution, have been shown to delay hydrate formation in the hydrate-prone section(s) of the flowline; the chemicals used in this approach are commonly known as kinetic hydrate inhibitors (KHIs). Another technology is to prevent hydrate particles from agglomerating so that hydrate particles can remain as a transportable slurry in the flowline; the chemicals used in this approach are commonly known as anti-agglomerants (AAs). These new technologies require detailed knowledge of the time-dependent properties of gas-hydrate formation, which is far more challenging than thermodynamic inhibition, which requires only knowledge of the temperature and pressure stability conditions for hydrates [2][9].

Energy storage

In the area of energy storage, one of the applications of gas hydrates is in the development of a transportation medium for natural gas from stranded or small gas fields. The storage of gases in hydrates is attractive because one volume of hydrates contains as much as 164 volumes of gas at STP (standard temperature and pressure conditions of 1 atm and 298 K). The pioneering company leading this effort, Mitsui Engineering & Shipbuilding Co., Ltd. (MES), has demonstrated a pilot plant for the production of natural gas hydrate (NGH) pellets with a capacity of 5 tons per day and near-term goals of 100 tons per day. These NGH pellets can then be transported in storage tanks by land or sea to end users. [10][11]. One of the attractive aspects for the transportation of natural gas in the form of gas hydrates is that NGH pellets can be made and stored under milder conditions than are needed for liquefied natural gas (LNG). The process developed by MES has shown that NGH pellets can be stored at $-20\,°C$ and atmospheric pressure, even though these conditions are well outside the hydrate stability conditions [11]. This unexpected stability of the NGH pellets is attributed to the self-preservation phenomenon [12], which, although not fully understood, is based on the formation of an ice layer around the NGH that prevents further dissociation of the hydrate for periods of up to several months.

In another area of energy storage, fundamental research is being performed to explore the utilization of clathrate hydrates for hydrogen (H_2) storage. Although H_2 was originally thought to be too small to be contained within clathrate water cages, molecular hydrogen was shown to form a sII hydrate by 1999 [13][14]. However, the formation of pure hydrogen hydrate requires pressures in excess of 200 MPa in order for the clathrate to be stabilized at near-ambient temperature, which is clearly impractical for storage applications. More recent studies have found that the addition of promoter molecules, such as tetrahydrofuran (THF), which act as coguest molecules with hydrogen in the clathrate hydrate framework, can significantly reduce the pressure conditions required for storage [15]. The demonstration that promoter molecules can reduce the pressures required for hydrogen storage in clathrate hydrates has stimulated a wealth of research around the world, with the overall goal of developing clathrate materials that will provide high storage capacity, moderate stability conditions in terms of temperature and pressure, and favorable kinetics of hydrogen release and recharge. The ongoing research in this area is advancing with the application of a diverse set of approaches, including high-pressure synthesis methods, spectroscopic and microscopic characterization, and computational modeling tools [16][17][18].

Unconventional energy resource

Gas hydrates occur naturally in marine sediments and sediments under the permafrost. The vast amounts of energy stored within these natural deposits present a potential alternative energy resource. Since most of the hydrates in nature comprise biogenic methane (formed from bacterial methanogenesis), sI is the predominant structure for these natural occurrences. As such, gas-hydrate deposits occurring in nature are often known and referred to as methane hydrates.

12.4.5 Energy resource potential of gas hydrates

Occurrence and estimated energy amounts

Natural-gas hydrate deposits occur under the permafrost and in ocean sediments, with some instances of seafloor hydrates (Figure 12.3). Figure 12.4 shows the different locations around the globe in which naturally occurring gas-hydrates have been identified by ocean drilling expeditions of cores or inferred from seismic signatures. As shown on the map in Figure 12.4, gas-hydrate occurrences are located all around the world along the continental margins and in permafrost regions. Over the past 30 years, researchers have compiled global inventories of estimates of the amount of energy trapped within gas-hydrate deposits. These vary by several orders of magnitude [19][20]. However, even the most conservative estimates place the amount of gas contained within hydrate deposits at 2–10 times larger than the global estimates of conventional natural gas of 4.4×10^{14} m^3 (~16,000 TCF, trillion cubic feet; all estimates reported at STP). Considering that the US annual gas consumption for 2008 was 6.5×10^{11} m^3 (23 TCF), the amount of energy contained within natural hydrate deposits is significant [21]. Global estimates typically project the amount of gas in gas hydrates in ocean sediments to be approximately 100 times greater than that in permafrost regions [2]. Recent assessments of the US gas resource from gas hydrates reported potential mean gas resources (technically recoverable resources) of approximately 2.4×10^{12} m^3 (85.4 TCF [22]) on the

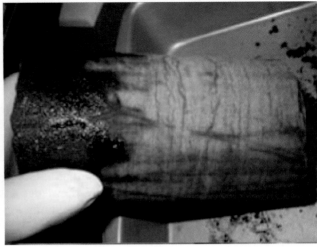

Figure 12.3. Recovered core samples of gas hydrates from the KG Basin, India (marine, left; courtesy of M. R. Walsh) and Mt. Elbert Well, North Slope, Alaska (permafrost, right; reprinted with permission from R. Boswell and T. Collett, 2006).

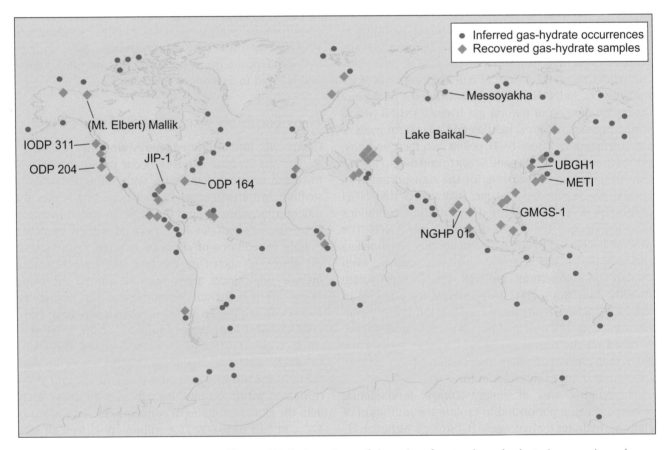

Figure 12.4. Locations of deposits of natural-gas hydrate in oceanic and permafrost regions (reprinted with permission from K. Kvenvolden).

North Slope of Alaska (permafrost) and approximately 607×10^{12} m³ (21,000 TCF [23]) in the Gulf of Mexico (oceanic; in-place resources). In comparison, the estimated resource potential of gas hydrates occurring in ocean sediments of the Eastern Nankai Trough [24] off Japan is 1.14×10^{12} m³ (40 TCF).

Typical depths of gas-hydrate deposits found in sediments under the permafrost are around $\leq 1,200$ m below the surface, whereas for marine environments methane hydrates are found at relatively shallow depths of ≤ 500 m below the seafloor (but in water depths of around 1,200 m). It should also be noted that gas-hydrate deposits typically

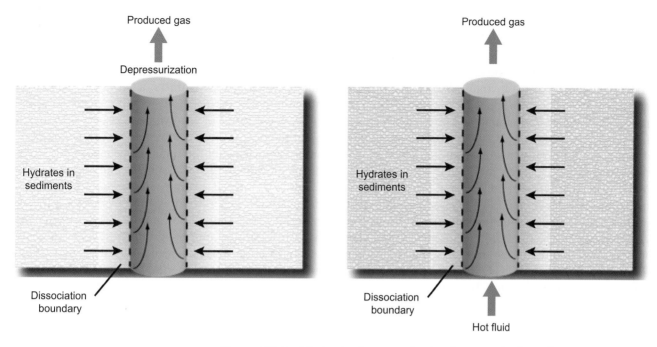

Figure 12.5. Methods of gas production from gas hydrate deposits [9]. Depressurization (left) or injecting hot fluid (right) both lead to hydrate dissociation, and subsequent production of gas from the hydrate-bearing sediments. Reprinted with permission from C.A. Koh, A.K. Sum, E.D., and Sloan, *J. Applied Physics*, **106** (2009) 061101. © 2009, American Institute of Physics.

are more concentrated in sediments found under the permafrost (containing higher hydrate saturations) than in marine sediments, which are typically more disperse and contain lower hydrate saturations.

Detecting naturally occurring gas hydrates

A number of techniques can be employed to detect the presence of naturally occurring gas-hydrate deposits, including seismic reflection surveys and electromagnetics. Correspondingly, the two most important physical properties for detecting and quantifying hydrate deposits are the primary-wave (P-wave) velocity and the electrical resistivity [21]. The **P-wave velocity** is the sonic compressional velocity (cf. V_P in Table 12.2) or the speed of sound passing through materials of different compositions.

When gas hydrate, a solid, is present in sediment, it replaces the pore fluid, thereby reducing the sediment porosity. Because porosity and acoustic impedance are inversely related, the impedance is typically much higher for hydrate-bearing than for non-hydrate-bearing sediments. This physical change can lead to differences in the sediment's P-wave velocity, which can be detected by remote-sensing methods, such as **seismic reflection surveys**.

Bottom simulating reflectors (BSRs) based on this technique have been used since the 1970s to detect the presence of hydrate, specifically the base of the hydrate stability zone (BHSZ). This is where the P-wave seismic velocities decrease from high values (attributed to methane hydrate being present) to low values (attributed

to free gas being present below the BSR). It should be noted that, although BSRs can provide a first start to exploration studies, they are not sufficient to determine the amount and location of the BHSZ.

Analogously, **electromagnetics** can be used to detect hydrates through the difference in resistivity of rocks/sediment containing hydrates. Specifically, when the pore fluid of sediment is saline (and hence conductive), formation of methane hydrate in the pores results in a decrease in conductivity (because the hydrate structure incorporates only pure water and excludes salt ions). Therefore, the bulk resistivity of the hydrate-bearing sediment is increased because of the behavior of methane hydrate as an electrical insulator. Consequently, from Archie's law for pore-filling hydrates, sediments detected to have higher resistivities are those that are more likely to contain methane hydrate [21]. It should be noted that the above seismic and resistivity methods of detecting gas hydrates can be used independently in survey mode or during logging of drilling holes.

Methods to produce gas from methane-hydrate deposits

Methods to produce gas from methane hydrate deposits include depressurization, thermal stimulation, carbon dioxide/methane exchange, and inhibitor injection (which is generally not considered viable because of concerns about environmental contamination). Both depressurization and thermal stimulation methods (see Figure 12.5) have been tested in reservoir-simulation calculations for both permafrost and oceanic hydrate deposits. These two methods, which are based on decreasing the pressure or increasing the temperature, displace the conditions in the reservoir from the hydrate stability zone so that the hydrates dissociate into water and gas, with gas being produced from the reservoir; the produced water is undesirable and is a major concern for production (see the previous discussion of flow assurance). According to reservoir simulations and production tests, such as the Mallik well test in the Mackenzie Delta, Canada, in 2007 and 2008, depressurization appears to be the most economical and viable approach for the production of gas from hydrate deposits [2][25]. The third production method is a novel but unproven approach [26] that is also potentially very attractive. This method involves injecting liquid carbon dioxide (a pollutant) into the reservoir containing hydrate-bearing sediment, resulting in carbon dioxide being sequestered in the hydrate cage framework while simultaneously releasing methane (fuel). These production methods will be tested over a long-term period on the North Slope of Alaska in 2011–2012.

Issues and future directions for producing energy from gas-hydrate deposits

The key issues that must be addressed to better assess the potential for gas hydrates to be a viable unconventional energy source are as follows.

A better understanding is required of the physical and geomechanical properties of gas-hydrate-bearing sediments. The structural integrity of the reservoir and the wellbore mechanical stability are important factors to be considered in gas production from gas-hydrate deposits. Better understanding of the reservoir's response to production requires both laboratory and field experiments to be performed over extended periods. In order to perform meaningful laboratory-scale tests, samples synthesized in the laboratory need to be analogous to those formed in nature and/or the issues associated with scaling up need to be clearly understood.

The environmental impacts of producing gas from gas-hydrate-bearing sediments need to be determined through long-term production field tests.

Because marine hydrates are the prize in terms of the large amounts of energy locked within these deposits, the technology for gas production from these deposits needs to be developed, particularly insofar as production of marine deposits will be within more hostile environments than that of permafrost hydrates.

12.5 Summary

As discussed in this chapter, gas hydrates play an important role in several energy applications (flow assurance, energy storage, and unconventional resources). It is recognized that the thermodynamics of gas hydrates are generally well established. On the other hand, gaining understanding of time-dependent phenomena for gas hydrates, such as the kinetics of formation and dissociation, remains challenging and is the major thrust of gas-hydrate research. Specifically, all of the energy applications of gas hydrates require a sound knowledge of the transient behavior of hydrates during formation in flow assurance and energy storage, energy recovery if hydrate reformation occurs, and dissociation in energy storage and recovery during gas release. The potential for producing gas from gas-hydrate deposits appears promising, with no major technical hurdles having been revealed by the research and production tests performed to date. However, issues of mechanical stability of hydrate-bearing sediments and environmental impacts during production need to be further investigated and understood (see Table 12.3).

12.6 Questions for discussion

1. What are the key factors determining gas-hydrate stability? Evaluate the significance of knowing whether gas hydrates are stable.
2. How do gas hydrates differ from ice? What are the best tools to distinguish gas hydrates and ice? Can you suggest other tools that could be applied to characterize gas hydrates that have not commonly been used to date? Comment on the benefits of applying these new tools to gas-hydrate characterization.
3. How are gas hydrate deposits detected? What are the advantages and disadvantages of these current technologies? Can you suggest other methods that could improve the accuracy and efficiency of detection of gas hydrate deposits?
4. Discuss the different methods that could be used to recover gas from gas-hydrate deposits. What are the advantages and disadvantages of these methods? Suggest alternative methods and explain why these alternative methods might be effective.
5. Discuss whether gas hydrates present a viable alternative energy resource, and include in your discussion the potential environmental, societal, and economic impacts of this activity.
6. Discuss the different technological applications of gas hydrates and include in your answer the potential environmental impact(s) of these applications. Consider other novel applications of gas hydrates in which environmental cleanup could be enhanced.

Table 12.3. Gas-hydrate technologies

Technology assessment table (estimates)	Controlling gas hydrates in flow assurance	Gas recovery from natural-gas hydrate deposits	Problems	Limitations
Current	Traditional avoidance methods moving to risk management	Exploration moving to pilot-scale production in permafrost regions	Flow assurance: avoidance methods are costly with environmental consequences	Flow assurance: risk management requires firm understanding of time-dependent hydrate properties in multiphase flow
			Gas recovery: lacking longer-term production data	Gas recovery: limitations on amount of funding for field trials
In 10 years	Deepwater developments with harsher T and P conditions will require new strategies, e.g., cold/stabilized flow, electrical heating	Pilot-scale production of oceanic deposits; larger-scale production in permafrost regions (if earlier tests successful) Environmental assessments of hydrate-recovery activities		
In 25 years	Advanced hydrate warning and development design incorporating in-built mitigation strategies for hydrate "risk" zones	If above successful, potential commercial recovery of gas from hydrates Detailed economic, environmental assessments		

12.7 Further reading

- **E. D. Sloan** and **C. A. Koh**, 2008, *Clathrate Hydrates of Natural Gases*, 3rd edn., Boca Raton, CRC Press/Taylor & Francis. Provides a comprehensive description of gas hydrates. http://www.crcnetbase.com/isbn/9781420008494.
- **E. D. Sloan**, **C. Koh**, and **A. K. Sum**, 2010, *Natural Gas Hydrates in Flow Assurance*, Amsterdam, Elsevier. Describes the state-of-the-art in flow assurance.
- **T. S. Collett**, **C. Knapp**, and **R. Boswell**, 2010, *Natural Gas Hydrates: Energy Resource Potential and Associated Hazards* (USGS Contribution to AAPG Memoir 89). Presents the resource potential and hazards of producing gas from gas hydrates. http://energy.usgs.gov/other/gashydrates/.
- **J. Grace**, **T. Collett**, **F. Colwell** *et al.*, 2008, *Energy from Gas Hydrates: Assessing the Opportunities and Challenges for Canada*, Report of the Expert Panel on Gas Hydrates, Council of Canadian Academies. Canadian report assessing hydrates as a resource. http://www.science-advice.ca/en/assessments/completed/gas-hydrates.aspx.
- **C. Paull**, **W. S. Reeburgh**, **S. R. Dallimore**, *et al.*, 2010, *Realizing the Energy Potential of Methane Hydrate for the United States*, National Academies NRC Report. US report assessing hydrates as a resource. http://www.nap.edu/catalog.php?record_id=12831.

12.8 References

[1] **H. Davy**, 1811, "On a combination of oxymuriatic gas and oxygene gas," *Phil. Trans. Roy. Soc. Lond.*, **101**, 155–162.
[2] **E. D. Sloan** and **C. A. Koh**, 2008, *Clathrate Hydrates of Natural Gases*, 3rd edn., Boca Raton, FL, CRC Press/Taylor & Francis.
[3] **E. G. Hammerschmidt**, 1934, "Formation of gas hydrates in natural gas transmission lines," *Ind. Eng. Chem.*, **26**, 851–855.

[4] **Y. F. Makogon**, 1965, "A gas hydrate formation in the gas saturated layers under low temperature," *Gazov. Promst.*, **5**, 14–15.

[5] **Y. F. Makogon**, 1974, *Hydrates of Natural Gas*, Moscow, Nedra, (in Russian); English translation 1981 by **W. J. Cieslesicz**, Tulsa, OK, PennWell Books.

[6] **S. Subramanian**, 2000, *Measurements of Clathrate Hydrates Containing Methane and Ethane Using Raman Spectroscopy*, Ph.D. Thesis, Colorado School of Mines, Golden, CO.

[7] **B. Chazallon** and **W. F. Kuhs**, 2002, "*In situ* structural properties of N_2^-, O_2^-, and air-clathrates by neutron diffraction," *J. Chem. Phys.*, 117, 308–320.

[8] **H. Hirai**, **T. Tanaka**, **K. Kawamura**, **Y. Yamamoto**, and **Y. Yagi**, 2004, "Structural changes in gas hydrates and existence of a filled ice structure of methane hydrate above 40 GPa," *J. Phys. Chem. Solids*, **65**, 1555–1559.

[9] **C. A. Koh**, **A. K. Sum**, and **E. D. Sloan**, 2009, "Gas hydrates: unlocking the energy from icy cages," *J. Appl. Phys.*, **106**, 061101.

[10] **J. S. Gudmundsson**, and **A. Borrehaug**, 1996, "Frozen hydrate for transport of natural gas," in *Proceedings of the Second International Conference on Gas Hydrates*, Toulouse, pp. 415–422.

[11] **T. Nakata**, **K. Hirai**, and **T. Takaoki**, 2008, "Study of natural gas hydrate (NGH) carriers," *Proceedings of the Sixth International Conference on Gas Hydrates*, Vancouver.

[12] **L. S. Stern**, **S. Circone**, **S. H. Kirby**, and **W. B. Durham**, 2001, "Preservation of methane hydrate at 1 atm," *Energy Fuels* 15, 499–501.

[13] **Y. A. Dyadin**, **E. G. Larionov**, **A. Yu. Manakov**, *et al.*, 1999, "Clathrate hydrates of hydrogen and neon," *Mendeleev Commun.*, **5**, 209–210.

[14] **W. L. Mao**, **H.-K. Mao**, and **A. F. Goncharov** *et al.*, 2002, "Hydrogen clusters in clathrate hydrate," *Science*, **297**, 2247–2249.

[15] **L. J. Florusse**, **C. J. Peters**, **J. Schoonman**, *et al.*, 2004, "Stable low-pressure hydrogen clusters stored in a binary clathrate hydrate," *Science*, **306**, 469–471.

[16] **T. S. Strobel**, **K. C. Hester**, **C. A. Koh**, **A. K. Sum**, and **E. D. Sloan**, 2009, "Properties of the clathrates of hydrogen and developments in their applicability for hydrogen storage," *Chem. Phys. Lett.*, **478**, 97–109.

[17] **T. Sugahara**, **J. C. Haag**, **P. S. R. Prasad**, *et al.*, 2009, "Increasing hydrogen storage capacity using tetrahydrofuran," *J. Am. Chem. Soc.*, **131**, 14616–14617.

[18] **M. R. Walsh**, **C. A. Koh**, **E. D. Sloan**, **A. K. Sum**, and **D. T. Wu**, 2009, "Microsecond simulation of spontaneous methane hydrate nucleation and growth," *Science*, **326**, 1095–1098.

[19] **K. A. Kvenvolden**, 1988, "Methane hydrates and global climate," *Global Biogeochem. Cycles*, **2**, 221–229.

[20] **A. V. Milkov**, 2004, "Global estimates of hydrate-bound gas in marine sediments. How much is really out there," *Earth Sci. Rev.*, **66**, 183–197.

[21] **C. Paull**, **W. S. Reeburgh**, **S. R. Dallimore**, *et al.*, 2010, *Realizing the Energy Potential of Methane Hydrate for the United States*, National Academies NRC Report.

[22] **T. S. Collett**, **W. F. Agena**, **M. W. Lee**, *et al.*, 2008, *Assessment of Gas Hydrate Resources on the North Slope, Alaska*, Reston, VA, US Geological Survey, 4 pp.

[23] **M. Frye** 2009, *Gas Hydrate Resource Evaluation: U.S. Outer Continental Shelf. Presentation to the Committee for the Assessment of the Department of Energy's Methane Hydrate Research and Development Program: Evaluating Methane Hydrates as a Future Energy Resource*, Golden, CO, US Department of Energy.

[24] **T. Fujii**, **T. Saeki**, **T. Kobayashi**, *et al.*, 2008, "Assessment of gas hydrate resources on the North Slope, Alaska," in *Proceedings of Offshore Technology Conference*, Houston, TX, pp. 1–15, doi:10.4043/19310-MS.

[25] **G. J. Moridis**, **T. S. Collett**, **R. Boswell**, *et al.*, 2009, "Toward production from gas hydrates: current status, assessment of resources, and simulation-based evaluation of technology and potential," *SPE Reservoir Evaluation Eng.*, **12**, 745–771, doi:10.2118/114163-PA.

[26] **J. C. Stevens**, **J. J. Howard**, **B. A. Baldwin**, *et al.*, 2008, "Experimental hydrate formation and gas production scenarios based on CO_2 sequestration," in *Proceedings of the Sixth International Conference on Gas Hydrates*, Vancouver, ed. **P. Englezos**, pp. 6–10.

13 Nuclear energy: current and future schemes

Christopher R. Stanek,[1] **Robin W. Grimes,**[2]
Cetin Unal,[1] **Stuart A. Maloy,**[1] **and Sara C. Scott**[1]

[1]Los Alamos National Laboratory, Los Alamos, NM, USA
[2]Imperial College London, London, UK

13.1 Focus

Nuclear power has been a reliable source of electricity in many countries for decades, and it will be an essential component of the mix of energy sources required to meet environmental goals by reducing greenhouse-gas emissions, reducing the dependence on fossil fuels, and enabling global access to energy. Materials science will play a key role in developing options in nuclear power, including new reactors with improved safety (especially in the light of the Fukushima Daiichi nuclear accident), reliability, and efficiency; technology to help minimize proliferation (discussed in Chapter 14); and viable, safe, long-term options for waste management (discussed in Chapter 15). Such efforts will provide opportunities to address broader challenges associated with nuclear energy, including public opinion and the investment risks associated with building new nuclear power plants.

13.2 Synopsis

The energy density (i.e., the quantity of useful energy stored per unit volume) of uranium fuel used today in light-water reactors (the most prevalent type of nuclear reactor) is already orders of magnitude larger than that of other energy sources. For example, one reactor fuel pellet produces approximately as much heat energy as 150 gal of fuel oil or 1 ton of high-grade coal. Moreover, the utilization of the energy density is being increased further through improvements in fuel technology and by developments in reactor design. This chapter focuses on the materials science challenges that exist in a fission reactor, that is, those related to the nuclear fuel, the cladding, and the structural materials, which are exposed to extremely high temperatures, moderate pressures, and an intense radiation field. Technical issues that extend beyond the workings of reactors, namely nuclear non-proliferation and nuclear waste, are addressed in Chapters 14 and 15, respectively.

From a materials science perspective, nuclear energy is especially interesting, challenging, and unique because of radiation effects. For materials that are used in nuclear reactors, it is not sufficient to understand only pre-irradiation properties and behavior. Rather, to optimize the performance of nuclear reactors, it is critical to understand how the properties of materials evolve during irradiation. The relatively recent advent of computational materials science provides an additional tool with which to explore the influence of radiation on materials properties and behavior. Indeed, the combination of computational methods and unit mechanism experiments will, in the future, yield a deeper understanding of the processes that govern materials' properties and behavior under irradiation. Ultimately, close coupling of computational, experimental, and manufacturing advances with real-world experience strengthened by scientific, industrial, and international partnering will further enhance the impact of materials science on nuclear energy.

13.3 Historical perspective

Nuclear fission, the process by which a heavy nucleus splits into two or more lighter nuclei plus some byproducts (e.g., free neutrons and photons), was discovered in 1938 [1][2]. Because fission results in the release of large amounts of energy, efforts were soon made to harness its power through controlled nuclear reactions. The first reactor to achieve a sustainable fission chain reaction was Chicago Pile 1, which was built under an abandoned football stadium at the University of Chicago, and "went critical" on December 2, 1942 [3]. The first nuclear reactor to generate electricity was the Experimental Breeder Reactor I (EBR-I) in Idaho in 1951. In 1954, the USSR's Obninsk Nuclear Power Station became the first nuclear reactor to generate electricity for a power grid. Two years later, the Calder Hall nuclear power station in Cumbria, UK, became the first reactor to produce commercial quantities of electricity. The Shippingport reactor in Pennsylvania, which operated from 1957 to 1982, was the first US reactor to generate civilian nuclear power for electricity generation.

However, the acceptance of nuclear energy has been tempered by several factors, including a small number of significant accidents. Among the first of these was the Windscale accident at the Sellafield facility in the UK in 1957. The cause of this accident was the uncontrolled release of so-called Wigner energy (increased potential energy due to the displacement of atoms in a solid caused by neutron irradiation) that occurred during a procedure that had been intended to release this energy in a controlled manner [4]. In 1979, at the Three Mile Island facility in Pennsylvania, a reactor suffered a partial core meltdown due to a small-break loss-of-coolant accident. No significant radiation release occurred in either of these accidents. However, in the 1986 Chernobyl accident in the USSR (Chernobyl is now in Ukraine), an ill-advised system test at low power ultimately resulted in a ruptured pressure vessel and numerous chemical explosions. More than 30 people died immediately either from the Chernobyl explosion or from radiation exposure during rescue or firefighting efforts, and contamination from the radiation released during the explosions extended over much of Europe. The ultimate number of deaths attributable to the Chernobyl disaster (e.g., due to chronic illness from radiation exposure) is a matter of controversy (see, e.g., [5]). Nevertheless, this incident has had a significant and lasting impact on public opinion. Most recently, reactors at the Fukushima Daiichi Nuclear Power Plant were damaged as a result of an earthquake and subsequent tsunami. It is worth noting that both the Three Mile Island accident and the Chernobyl accident were due to the fuel being subjected to off-normal conditions as a result of human error and in the case of Fukushima due to inadequate protection against natural disaster; they did not involve materials degradation during normal operation. In addition to affecting public opinion, the Three Mile Island and Chernobyl incidents also led to a revision of safety philosophy, best exemplified by the defense-in-depth approach, in which nuclear facilities are designed to prevent accidents that release radiation or hazardous materials. Opportunities for further improvements in safety, in the context of the Fukushima Daiichi nuclear reactor accident, are currently being explored.

Beginning in the late 1970s, concerns about safety and several other issues, including economics, regulatory policy, waste disposal, and non-proliferation, led to over two decades of stagnation in nuclear power. However, various factors are now leading to renewed interest in nuclear energy: recognition of nuclear energy's benefits, significant success in addressing identified challenges relating to nuclear energy, an improved understanding of the risks and limitations associated with other energy sources, a growing societal desire to reduce CO_2 emissions, and the lack of other viable options for providing needed levels of base-load electricity. As of October 2010, there were 441 civilian nuclear reactors operating in 30 countries, with an installed net capacity of about 375 GW_e [6]. Multiple studies predict global nuclear energy generating capacity to expand to as much as 750 GW_e by 2030. Consistently with this rate of expansion, the International Atomic Energy Agency (IAEA) reports that 44 nuclear reactors, with a total capacity of almost 40 GW_e, are currently under construction worldwide. In addition, at least 70 new units are being planned for the next 15 years, and another 250 have been proposed [7]. The stated interest in nuclear energy in the context of the recent events in Japan reinforces the importance of an international focus on nuclear safety. "Although the radiological releases in Japan will have no direct impacts of significance on the United States, the events at Fukushima are certain to affect attitudes toward nuclear technology here and abroad. Even if the health consequences of the Fukushima accident prove to be small compared to the direct impacts of the earthquake and tsunami, economic ramifications – including the permanent loss of contaminated land and six costly reactors – and the potential danger of a nuclear disaster remain abiding public concerns. These concerns must be directly and forthrightly addressed. . . . The capacity to pursue nuclear technology in the United States will depend to a large extent on other countries' success in achieving a high level of safety performance." [Blue Ribbon Commission on America's Nuclear, Future Draft Report to the Secretary of Energy, July 29, 2011.]

13.4 Materials challenges for nuclear energy

13.4.1 The nuclear fission process

The basic principles behind fission-based and fossil-fuel-based electricity generation are similar: both heat water into pressurized steam, which runs through turbines, which, in turn, power electrical generators. The key difference, of course, is in the method of heating the water.

In a nuclear reactor, the heat is generated through fission reactions, in which a heavy nucleus, typically uranium-235, absorbs a neutron and splits into two or more lighter nuclei, releasing kinetic energy, gamma radiation (high-energy electromagnetic radiation), and additional neutrons. These additional neutrons can be absorbed by additional heavy nuclei, inducing further fission, leading to a self-sustaining chain reaction. For example, the following reaction describes a typical example of the fission of ^{235}U in a nuclear reactor:

$$^{1}_{0}\text{n} + ^{235}_{92}\text{U} \rightarrow ^{236}_{92}\text{U}^* \rightarrow ^{141}_{56}\text{Ba} + ^{92}_{36}\text{Kr} + 3^{1}_{0}\text{n}. \quad (13.1)$$

Here n stands for neutron and the (left) super and subscript give atomic mass (protons + neutrons) and atomic number (protons) for all terms. There are, in fact, many possible fission reactions, leading to a plethora of elements being deposited within spent nuclear fuel (as discussed in Section 13.4.4).

The reason for the high energy density of nuclear fuels, as described in Section 13.2, is the high total energy released by a fission event, which can be expressed as [8]

$$\Delta E = \left(M_{\text{o}} - \sum_i M_i \right) c^2, \quad (13.2)$$

where M_{o} is the mass of the original nucleus (^{236}U* in the example given by Equation 13.3, so $M_{\text{o}} = 236.046$ amu (atomic mass units)) and M_i represents the masses of the resulting nuclei (from Equation 13.1, $\sum M_i = 3M_{\text{neutron}}$ (3×1.008664) + M_{Ba} (140.914) + M_{Kr} (91.926)). The example fission reaction given by Equation 13.1 results in an energy of approximately 170 MeV from the mass change of 0.180 amu. The deposition of this energy in nuclear materials results in radiation damage (atomic defects) that leads to the degradation of materials' properties (as discussed in Section 13.4.4). Both the fission products and radiation influence the properties of materials used in nuclear reactors. Understanding the behavior of these materials is what makes nuclear energy technology challenging.

13.4.2 Fuel-cycle options

Before discussing nuclear energy-related materials issues, it is worthwhile to summarize the context of these issues in terms of current and potential future **nuclear fuel cycles**. These include the **open fuel cycle, closed fuel cycle**, and **modified-open fuel cycle** [9]. Although this chapter focuses on materials science challenges that exist in a reactor, solutions for these challenges must be considered in the broader context of the nuclear energy system consisting of the reactor, any processing of the initially irradiated fuel that occurs, waste forms produced, interim fuel- and/or waste-storage strategies, the ultimate disposal pathway, safeguards and security technologies, and other non-proliferation frameworks and institutions applied.

Open fuel cycle

This fuel cycle (see Figure 13.1) is referred to as open, or once-through, because the fuel is used only once in a reactor and is disposed of without chemical processing. That is, after removal from the reactor, the used fuel is initially cooled in water prior to application of the method determined for long-term disposal, for example, in a geological disposal facility (see Chapter 15). Currently, the commercial nuclear industries of most countries rely exclusively on a once-through fuel cycle.

Closed fuel cycles

France, Japan, and Russia employ a closed fuel cycle in some or all of their nuclear facilities [10]. In a closed fuel cycle, the used nuclear fuel is recycled to achieve two primary goals: improved fuel utilization (approximately 95% of the fissile content of current fuels remains after burning once in common reactors) and long-term waste management, as shown in Figure 13.2. In large part because of ultimately limited fuel resources, it has been claimed that fuel reuse will become an important issue if, as is currently anticipated, more countries choose to implement nuclear energy to address their energy security and environmental needs [11]. Conversely, it has also been noted that there is ample time to significantly improve (or create revolutionary new) technology options before a decision on whether and how to close the fuel cycle will become necessary [12].

The fuel-recycling process not only separates uranium from the used fuel but also separates fission products and **transuranics** (such as plutonium, americium,

Figure 13.1. A schematic representation of an open fuel cycle, as currently used by industry.

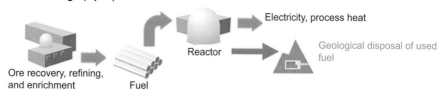

Once-Through (Open)

Ore recovery, refining, and enrichment

Fuel

Reactor

Electricity, process heat

Geological disposal of used fuel

Figure 13.2. A schematic representation of a full-recycling closed fuel cycle, as envisioned by the US Department of Energy.

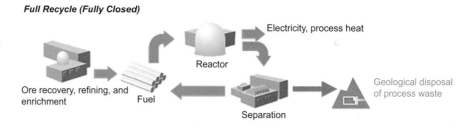

Full Recycle (Fully Closed)

Ore recovery, refining, and enrichment

Fuel

Reactor

Electricity, process heat

Separation

Geological disposal of process waste

neptunium, and curium). These transuranics contribute to the complexity of the disposal of used fuel, as discussed in Chapter 15. In addition, because these elements (specifically, their odd-numbered isotopes) are fissionable, removal of the unused uranium and transuranics from the fuel can not only reduce the radiotoxicity of waste and provide the opportunity for customized disposal strategies for specific components of the waste but also help extend the fissile material resource (through incorporation into new reactor fuels that are burned in either thermal or fast reactors). It has been proposed that the uranium resource could be further extended by intentionally maximizing the production of plutonium in a reactor (often referred to as a breeder reactor), extracting the plutonium, and incorporating it into new fuel that is then burned for energy [13].

A process to separate the actinides from the fission products is required as a part of a closed fuel cycle. Historically, "reprocessing" technology was developed to isolate plutonium for nuclear weapons; the plutonium was produced by irradiating uranium with neutrons in reactors (plutonium occurs naturally at only extremely low levels in uranium ores). The PUREX (plutonium–uranium reduction extraction) process, an organic liquid–liquid extraction method, was developed to recover the uranium and plutonium from the fission products produced when uranium is irradiated. This process was later employed to reprocess "used" or "spent" fuel from commercial nuclear power plants. PUREX plants reprocessing used fuel from commercial reactors are currently operating in France, Japan, Russia, and the UK [14]. No process other than PUREX has been used on a significant scale to separate the components of used commercial fuel. Diverse additional separation processes have been investigated and in some cases demonstrated with small amounts of used fuel, but none are currently deployed industrially. These separation processes include a variety of liquid–liquid extraction schemes, pyrochemical methods in molten salts, and volatilization of fluoride or chloride compounds. Future decisions regarding use of reprocessing technologies will require continued evaluation of a number of factors including separation efficiency, additional potential radiological hazard, effluent accumulation, and cost. Non-proliferation concerns have been and will be important considerations in this decision-making process as well – see Chapter 14 for an in-depth discussion.

In addition to requiring reprocessing, a closed fuel cycle could also involve reactor types unlike those at present employed to generate electricity. For example, if the intent is to reduce the radiotoxic inventory, then actinides could be separated, incorporated into new fuel, and burned in **fast-neutron spectrum reactors** (see Section 13.4.3). Other options for reducing the inventory of transuranics include accelerator-driven systems that operate in subcritical mode. In these systems, high-energy proton-beam irradiation of a heavy-metal spallation target generates fast neutrons that "transmute" long-lived transuranics into shorter-lived fission products, thereby reducing the actinide burden on a geological repository [15]. For example, recent studies indicate that the use of accelerator-driven systems in combination with fast reactors might be an efficient way of reducing americium (Am) inventories (which is desirable because burning large amounts of Am in fast-neutron spectrum reactors significantly complicates the reactor design).

Modified open fuel cycle

In addition to the open and closed fuel cycles, technical options are being explored for a "modified open" fuel cycle, a schematic representation of which is shown in Figure 13.3. The modified open cycle shares similar goals to the closed cycle, namely increased utilization of the fuel resource and a reduced quantity of actinides to be disposed of in used fuel. The modified open cycles would involve limited separation steps using technologies that aim to substantially lower proliferation risks. Innovative reactors, fuel forms, waste forms, and waste-management approaches will be important aspects of further developing these concepts.

13.4.3 Nuclear reactor options

In general, the challenges faced in designing and selecting materials for nuclear reactor applications are those associated with materials under the extreme conditions of high radiation flux, combined with high temperatures and chemical attack from coolant or from the accumulated fission products. However, specific problems may vary between reactor types; a brief review of current and proposed future reactor designs is provided here to enable an informed materials discussion in Section 13.4.4. For a detailed description of reactor physics (importantly

Figure 13.3. A schematic representation of a modified open fuel cycle, as envisioned by the US Department of Energy.

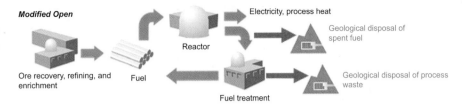

including thermal hydraulics and neutron transport, which are not discussed here) and the impact of reactor operating conditions on materials performance, the reader is directed to classic books in the field, e.g., [16][17][18][19]. Also, this section focuses on reactors primarily dedicated to (or proposed for) the generation of electricity. Research-scale reactors also exist, and, although not discussed in detail in this section, they are important for the materials research discussed in the next section.

It is convenient to categorize nuclear reactors by the type of neutron **moderator**, the type of **coolant**, and the type of fuel used. For example, the early reactors were moderated by either graphite or water, but cooled by gas (in the case of graphite) or light water (where "light water" contains the ^{1}H isotope of hydrogen, as opposed to heavy water, which contains deuterium, the ^{2}H isotope of hydrogen). Some reactor designs, particularly early designs, used uranium metal as the fuel. However, most current open-cycle reactors use light water as both the coolant and the neutron moderator in the reactor core, with fuel in the form of uranium dioxide, UO_2.

Light-water reactor (LWR) technology primarily uses two types of reactors: pressurized-water reactors (PWRs) and boiling-water reactors (BWRs). All civilian nuclear reactors operating in the USA to generate electricity (typically 1 GW_e per plant) are LWRs: 69 are PWRs and 35 are BWRs. However, because light water both slows (or "moderates") neutrons, thus enabling criticality, and also captures a small proportion of those neutrons, the uranium in the fuel needs to be slightly enriched. More specifically, the proportion of U-235, the isotope that undergoes fission, relative to U-238 must be increased from the natural abundance ratio of 0.7% to typically between 3% and 5%.

Outside the USA, other types of reactors are also operated for electricity production. For example, there are ~40 **heavy-water reactors (HWRs)** [most of which are CANDU (Canada deuterium uranium) reactors] generating electricity. As is the case for LWRs, water serves as both the moderator and the coolant in these reactors. However, an advantage of HWR-type reactors is that heavy water does not capture neutrons and, therefore, it is not necessary to use enriched uranium in the fuel.

In general, the form of the nuclear reactor fuels used by LWRs and HWRs is cylindrical pellets with diameters ranging from approximately 5 mm to 12 mm. These are inserted into zirconium-alloy tubes (typically 1–4 m long

and referred to as **clad**) in such a way that there is a small gap between fuel pellets and the zirconium-alloy clad; these fuel-filled tubes are typically referred to as **fuel rods** or **pins**. The most commonly used zirconium alloys contain a majority of zirconium, to which small amounts of alloying elements such as tin, niobium, iron, and/or chromium are added depending on the particular alloy. Multiple fuel rods comprise an assembly (or bundle), which is inserted into the reactor; the rods are held apart using grid spacers [typically Inconel, an austenitic nonmagnetic and corrosion-resistant NiCrFe-based superalloy] to induce mixing and turbulence in the coolant and to provide separation between the rods for coolant flow. In addition to holding the fuel pellets so that they can be inserted into the reactor core, the clad ultimately provides containment for the gaseous fission products (such as I, Kr and Xe) within an empty space above the fuel pellets called the plenum. Prior to irradiation, the pellet–clad gap and the plenum are filled with pressurized helium. As the fuel is burned (by fission), the clad provides the first safety barrier preventing release of radioactive fission products into the coolant and corrosion of the fuel by the coolant.

As discussed previously, **fast-neutron spectrum reactors** could be employed as an element of closed fuel cycles with the goal of improving fuel utilization and long-term waste management. For example, a fast-neutron reactor can be used to transmute the transuranics (e.g., plutonium, americium, neptunium, and curium) that are produced from the uranium-bearing fuel in an LWR. Fast-neutron spectrum reactors are used because the fast neutrons more often lead to fission of the transuranics (the thermal neutrons that predominate in LWRs are more likely to be captured by the uranium and produce more actinides). The intent of fissioning the transuranics is to produce energy while simultaneously transmuting these elements into ones with shorter half-lives and also enable options for more effective waste management. Fast-reactor designs typically have no moderator and use a liquid-metal coolant (such as sodium or lead–bismuth). Experimental fast reactors have been built and operated, but the technology has not been extensively deployed. Challenges for improving the viability of these reactors include the design and selection of the fuel and clad materials (i.e., materials that are more stable to radiation damage from the fast-neutron spectra and corrosion from the coolants used for these systems). Another potential long-term application of a fast-spectrum reactor is as a "breeder reactor," a plutonium reactor that could "produce more fuel than it consumes," significantly extending the uranium resource, as noted in Section 13.4.2. However, breeder reactors have not yet been commercialized; challenges and concerns include economics and the potential for producing plutonium that could be used in weapons [20].

Other advanced thermal reactor designs are also being pursued. For example, there is interest in **high-temperature gas-cooled reactors (HTGRs)**. These reactors, which are intended to help provide nuclear-energy options for supplying high-temperature process heat (for example, for use in coal gasification and hydrogen generation), are graphite-moderated and helium-cooled. Examples of these are the pebble-bed modular reactor (PBMR) and the modular high-temperature gas-cooled reactor (MHTGR). Although sustainable reactor outlet temperatures of up to 750 °C have been demonstrated, economic considerations (e.g., the ability to address a more diverse industrial customer base) are driving development of outlet temperatures up to 950 °C and possibly beyond [21]. From a materials science perspective, the challenges are significant: mechanical and environmental degradation are primary concerns, and fabrication, creep, and the effect of helium on properties such as tensile strength are important areas of investigation [22]. In addition, the coated particle fuel for HTGRs is particularly interesting. These particles are known as TRISO (tristructural-isotropic) particles and consist of a fissile kernel (UO_2 or UCO) that is coated with two carbon layers (porous and pyrolytic) and then with SiC to contain the fission products (effectively a clad layer) and a final outer pyrolytic carbon layer. Whereas SiC provides a suitable containment for temperatures that would be experienced in an HTGR with reactor outlet temperatures up to 750 °C, it would not be suitable in an HGTR with higher outlet temperatures. For these applications, alternative materials such as ZrC are being investigated [23].

A different reactor paradigm that is currently gaining attention is deployment of **small- and medium-sized reactors (SMRs)**, with <700 MW$_e$ output (SMR is also used to refer to small modular reactors with an output of <300 MW$_e$). Although some smaller reactors have been designed and built, there is renewed interest in these reactors because of their potential for options in fabrication, construction, and siting [24]. Additional advantages of these reactors include lower capital costs and modularity that could enable a "scalable" approach to building nuclear power plants (plants scaled to a particular need for electricity, with the ability to add units as energy demand increases and additional capital becomes available). Multiple SMR designs are being developed and promoted; however, none are at present being built to provide commercial electricity.

13.4.4 Materials science issues relating to nuclear energy

A nuclear power plant is constructed using a wide variety of different materials. The focus here is on materials within the **reactor pressure vessel (RPV)**, specifically the clad and fuel (rather than the RPV and the reactor internals), where the mechanical, chemical, radiation, and temperature conditions are the most extreme and the materials' performance is most critical to the operation of the reactor. There are, however, significant materials issues outside the RPV, including the degradation of concrete [25] and corrosion of heat exchangers and steam generators, cabling, and so on, as well as issues associated with fuel degradation once it has been removed from the reactor. In existing LWRs, it is critical for the establishment of safety margins to understand how the properties of materials evolve when they are exposed to high temperature and high neutron fluence (which is directly related to the fuel **burnup**, or fuel utilization, a measure of how much energy is extracted from a primary nuclear fuel source). Understanding this evolution presents a significant materials science challenge. The external surface of the clad is also exposed to a corrosive, flowing coolant at elevated temperatures (~300 °C), while the internal surface is subjected to the rapidly changing fuel composition as the fissile material is converted into fission products of significantly different chemistry. As future fuel cycles are pursued and efforts are made to increase the efficiency of power plants through higher operating temperatures, higher burnup, power uprates (increases in the maximum power level at which a commercial nuclear power plant is allowed to operate), and lifetime extension, the demands on the reactor materials will increase. Although the design of future reactors may be significantly different from that of existing LWRs, the materials' performance is still dictated at a fundamental level by mass and thermal transport under irradiation. Therefore, improved fundamental understanding of how materials' properties evolve under irradiation will enable the necessary materials advances required by future nuclear energy cycles [26]. It is also important to appreciate that, even between similar reactor designs, differences in operating profiles (that lead to different irradiation levels and temperatures) can result in quite different changes to materials' behavior.

Clad and structural materials

As previously mentioned, the clad serves as a barrier between the radioactive fission products and the reactor coolant. There are several mechanisms by which a fuel pin can fail. For example, under normal LWR operating conditions, fission-product gases (such as I, Kr and Xe) are either retained in the fuel grains or released to the rod free volume. If retained, the fuel can swell (which is also due to incorporation of solid fission products and thermal expansion) and mechanically interact with the clad [**fuel–clad mechanical interaction (FCMI)**]. If the gases are released, they increase the internal pressure on the clad walls, as well as the resistance to heat transfer. Furthermore, under accident conditions, the gas temperature can increase, even further increasing the internal pressure on the clad walls. The Three Mile Island loss-of-coolant accident (LOCA), the Chernobyl reactivity-insertion accident (RIA) and Fukushima Daiichi station-blackout accident were examples of fuel failure and fission-gas/product release were examples of fuel failure and fission-gas/product release. (However, it should be noted that, in these examples, the fuel was subjected to conditions beyond the design envelope, and industry has managed to reduce fuel failure under normal conditions from about 0.1% at the beginning of the nuclear-energy era to less than 0.001% today.) Other fission products can chemically attack, or corrode, the clad, decreasing its mechanical performance [known as **fuel–clad chemical interaction (FCCI)**]. Corrosion from fission products such as iodine can lead to stress corrosion cracking, which will further contribute to fuel failure [27]. Corrosive agents also exist within the cooling water, leading to corrosion on the clad walls and hydriding within the clad. Corrosion can lead to **stress corrosion cracking**, which will further contribute to fuel failure [27]. Furthermore, additional impurities in the coolant (for example, from the corrosion of steam generators) lead to the formation of a mixed oxide layer known as **CRUD (Chalk River unidentified deposit)**, which can impact the neutron reactivity and cause power shifts (due to nonuniform boron absorption and buildup), hindering the coolant from reaching the clad and accelerating corrosion [28]. Another fuel failure risk is **grid-to-rod fretting (GTRF)**, where the spacer grids that keep the fuel pins separated rub against the clad as the fuel pins vibrate during operation.

All of the phenomena mentioned in the preceding paragraph are further complicated by the exposure of the clad and structural materials to radiation. In a radiation-damage event, atomic-scale damage is imparted to the crystal lattice by neutrons from the fission process that have sufficient energy to displace atoms from their original lattice positions, subsequently resulting in a cascade of vacancies and interstitials, as can be seen in Figure 13.4. (Here we show an early example of a displacement cascade [29]. For a more modern representation see Figure 15.7.) The initial atom struck by the neutron is referred to as the primary knock-on atom (PKA). Radiation damage is typically measured according to the number of displacements per atom (dpa), which is the average number of times every atom in the system has been moved from its original lattice position. There are a variety of fates for these initially created defects,

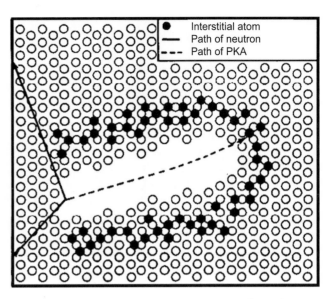

Figure 13.4. A schematic representation of a radiation collision cascade. Reprinted with permission from Brinkman, *Amer. J. Phys.*, **24**, 251 (1956), © 1956, American Association of Physics Teachers, [29]. Note: PKA denotes the primary knock-on atom. A more modern representation of a displacement cascade is shown in Figure 15.7.

depending on their mobilities and locations. For example, vacancies and interstitials can recombine with one another; alternatively, they can annihilate or accumulate at defect sinks, such as dislocations and grain boundaries [30][31]. The cascade itself is less important than the defects that remain after the cascade settles, since it is the remaining defects that lead to degradation of mechanical performance due to radiation-induced hardening (through dislocation loop and defect cluster formation) and creep, swelling from void formation, and embrittlement [32]. Furthermore, many issues contribute to these phenomena, including radiation-enhanced diffusion, radiation-induced segregation, and precipitation. Finally, the evolution of the microstructure as a result of these effects is complex and is a function of radiation dose and temperature. A more comprehensive description can be found elsewhere [33][34].

The Zircaloy alloys (Zr plus small additions of Fe, Cr, Sn, and/or other elements) are the most common zirconium alloys used in current LWRs. The original Zircaloy composition (Zircaloy-1 contains 2.5% Sn) has been subsequently alloyed with Cr and Nb, for example, to improve corrosion resistance. Its corrosion resistance and small neutron-capture cross section (meaning that the clad does not significantly absorb neutrons otherwise intended to participate in the fission process) are the main reasons why Zircaloy is used in LWRs. However, the incremental improvements to Zircaloy are reaching a performance limit. The Fukushima Daiichi

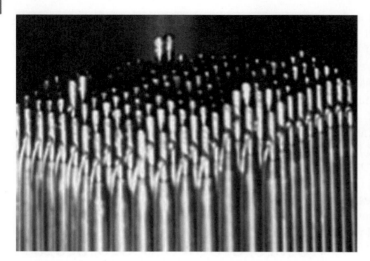
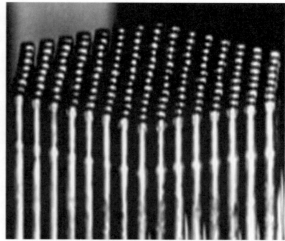

Figure 13.5. A comparison of swelling in D-9 (left) and HT-9 (right), after irradiation of ~70 displacements per atom (dpa) at the Fast Flux Test Facility (FFTF) in Hanford, WA, from [36]. Clearly D-9 is less stable under irradiation than is HT-9.

Nuclear Power Plant accident has focused renewed attention on the performance of Zr in LWRs under accident scenarios. Specifically, if no fresh water is introduce, the temperature of the rods will increase. Above 1200°Celsius, Zr will oxidize and liberate hydrogen from the water. If the hydrogen gas vented from the reactor core and containment vessel accumulates in sufficient quantities, it can explode. When proceeding to higher burnup in LWRs, the fracture toughness and ductility of Zircaloy decrease because of radiation damage, corrosion, and hydride formation. There is also higher likelihood of FCMI. If fast reactors are considered for a closed fuel cycle, the radiation environments are more severe still, in some cases involving damage up to hundreds of displacements per atom. Next-generation clad and structural materials must maintain their mechanical performance (i.e., strength, ductility, and fracture toughness) and dimensional stability by resisting the effects of creep and void swelling under such extreme radiation environments [35]. For fast reactors, the low thermal neutron-capture cross section for Zr is not needed, thus allowing the use of a wider range of materials for the clad. Consequently, various alloys have been considered to meet fast-reactor requirements. For example, Figure 13.5 [36] compares the dimensional stability of D-9 (an austenitic 15Cr–15Ni stainless steel stabilized with Ti) to that of HT-9 (a ferritic–martensitic steel) both irradiated to ~70 dpa in the Fast Flux Test Facility (FFTF, a sodium-cooled fast-neutron test reactor in Hanford, WA, that is no longer operational). Clearly, HT-9 exhibits superior dimensional stability than that of D-9 under these radiation conditions. However, HT-9 has degraded fracture toughness (after irradiation at low temperatures) and high-temperature strength, so other alloys are being considered [35]. In addition to composition variation, microstructure is another avenue being pursued during the search for radiation-tolerant structural materials and clad. A promising family of materials is **oxide-dispersion-strengthened (ODS) steels**, which have oxide nanoparticles embedded within ferritic alloys, such as MA957, a high-Cr ferritic alloy. The oxide nanoparticles in this material are precipitates of Y–Ti–O. These materials exploit the oxide nanoparticles as sinks for defects and He that would otherwise lead to void swelling. Additionally, the nanoparticles (through their small sizes and dense distribution) provide creep resistance by pinning dislocations. Such materials offer a potentially fruitful research area, and a detailed review can be found elsewhere [33].

Fuel

Uranium dioxide, UO_2, is the most widely used fuel in existing LWRs. Oxide fuels are chemically stable in the presence of the clad and have relatively high melting temperatures. UO_2 crystallizes in the fluorite structure, which is thermodynamically stable under typical reactor conditions. Furthermore, the fluorite structure (see Figure 15.8 in Chapter 15) is generally tolerant with respect to radiation [37] and readily incorporates fission products in unoccupied interstices within the crystal lattice. Fission products vary chemically and physically, and their solution behavior in UO_2 influences the physical properties of the fuel, e.g., thermal conductivity and melting point [38]. Figure 13.6 shows how the fission products can be classified according to chemistry and solution mechanisms. Particularly important fission products for fuel failures are the gases I, Kr and Xe. Many studies aimed at understanding fission-gas transport have been undertaken, e.g., [39]. However, questions still

Figure 13.6. The chemical states of fission products, where orange denotes volatile fission products, gray denotes metallic precipitates, blue denotes oxide precipitates, and green denotes products in solid solution. Elements labeled with more than one color indicate the possibility of an alternative chemical state, with the preferential chemical state denoted by the top color.

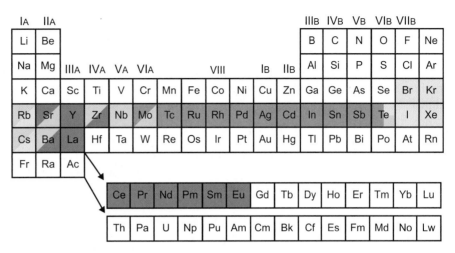

remain regarding their retention at intragranular sinks (such as dislocations), bubble evolution, and release along grain boundaries. In addition, and as previously mentioned, fission products, such as iodine, contribute to chemical attack of the clad through stress corrosion cracking corrosive.

The properties of oxide nuclear fuels are also strongly coupled to the non-stoichiometry of the fuel. Because of the variable oxidation states of uranium and plutonium and the corresponding charge-compensating oxygen defects, oxide fuels can exhibit remarkable deviations from stoichiometry. An example of the effect of non-stoichiometry on relevant materials' properties is that the melting temperature of UO_{2+x} has been shown to vary by more than 500 °C when x varies from 0.0 to 0.2 [40]. In addition to the 4+ oxidation state, uranium can exhibit 5+ and 6+ oxidation states, which correspond to hyperstoichiometric UO_{2+x}. Conversely, plutonium (and americium) exhibit only 3+ and 4+ oxidation states and can attain only hypostoichiometry, PuO_{2-x}. Understanding the atomistic mechanisms by which oxide fuels compensate for non-stoichiometry and the accommodation of fission products allows an improved understanding of the corresponding variations in materials' properties.

During burnup, the microstructure of the oxide fuel evolves. Fission-gas bubbles are formed, pores migrate, cracks emerge, and the grain structure develops spatially. Further, large radial temperature gradients develop across the fuel pellet. The pellet starts its life as a fairly uniform material and becomes more complex during burnup, forming regions that are microstructurally distinct. Thus, it is important to understand fuel behavior not only as a function of fuel composition (including non-stoichiometry and fission-product chemistry) but also as a function of microstructure.

Because the function of the fuel is to generate heat from the fission process and to transfer this heat to the coolant, a significant disadvantage of oxide fuels is their low thermal conductivity. Furthermore, the fissile density of oxide fuels is relatively low. Consequently, although

UO_2 has been used as a driver fuel for existing fast reactors, other fuel forms are being considered, including mixed-oxide (MOX) fuels, which can contain Pu and other actinides; metals (because of their high thermal conductivities and fissile densities); nitrides/carbides (because of their radiation tolerance, high thermal conductivities, and high melting points); and inert-matrix fuels (IMFs), which do not breed additional Pu. Research continues in the area of advanced fuels, particularly transmutation fuels for which actinides can be efficiently burned in fast reactors. However, because each fuel type has clear disadvantages, UO_2 remains the dominant fuel material in current reactor designs.

Finally, note that the alternative fuel cycle based on thorium, which is three times more abundant in nature than uranium, has not been pursued in earnest. Nevertheless, there is a renewed interest in Th-based fuels because of their potential for superior performance, reduced transuranic waste, and perceived proliferation resistance (because of the formation of ^{232}U, which has high-energy gamma decay that requires specialized remote-handling facilities). However, the performance of ThO_2-based fuels, especially at the high burnups expected of future fuels, is not well understood, and thus considerable effort will be required before regulators would be able to accept such materials in civil power reactors. This cycle has been particularly pursued by the Indian atomic energy program because India has significant reserves of thorium but little uranium [11].

13.4.5 Fundamental materials science research opportunities

Identification and control of multidimensional defects and their interaction

The challenging requirements for next-generation nuclear materials indicate a need for targeted research studies to develop optimized materials with improved performance. However, there are also opportunities for significant contributions from more general fundamental materials science research. A recurring theme encountered when endeavoring to optimize nuclear materials' performance is the ability to identify and control multidimensional defects and their interactions. Defects – deviations from perfect structures – can be categorized according to dimensionality. Zero-dimensional defects, known as point defects, are departures from perfect crystal structures involving only a single lattice site. Examples are vacancies, interstitials, and substitutions. Point defects can be either intrinsic (thermal) or extrinsic (non-stoichiometry; doping, fission products; and radiation damage). Examples of point-defect formation and associated implications for radiation-tolerant nuclear-waste forms are provided in Chapter 15.

In principle, all higher-order defects can be considered as being made up of point defects. One-dimensional defects are dislocations, edge, screw, or mixed; two-dimensional defects are surfaces, grain boundaries, and crystallographically specific boundaries such as twins or stacking faults; and three-dimensional defects include voids and bubbles. As early as 1952, Seitz noted that the interaction between defects was an area of rich physics [30]. Recent simulations of radiation damage have begun to reveal the importance of defect interactions [41]. For example, Figure 13.7 describes a recent atomic-scale simulation study in which Bai *et al.* predicted that grain boundaries serve not only as sinks for radiation defects, but also as sources, by emitting interstitials that can recombine with vacancies [42]. These results predict "self-healing" of the radiation-induced damage.

An experimental example of multidimensional defect interaction in nuclear fuel is provided in Figure 13.8 [43], where more fission gas (Kr) is released at high temperature in single-crystalline than in poly-crystalline UO_2. The implication of this result is that grain boundaries can trap fission-gas atoms, rather than only provide high-diffusivity pathways, under certain conditions. This result illustrates the general importance of microstructural features, such as grain boundaries, for materials' behaviors that govern nuclear fuel performance (rather than providing a systematic explanation of fission-gas release). That is, it is fundamentally difficult to apply bulk diffusivities of fission gases to accurately predict fission-gas release rates. The reason for this discrepancy is that the gas migration occurs through mechanisms that are more complex than lattice diffusion. If the diffusion models do not consider interactions of fission gas with multidimensional defects, then the release rates will not compare favorably with experimental release data. For example, gas atoms can become trapped at a number of different defect sites (such as dislocations, grain boundaries, and bubbles). For nuclear fuel, defects can alternatively be categorized according to whether they are present in as-fabricated fuel (i.e., so-called "natural defects," such as grain

boundaries, dislocations, pores, and impurities) or whether they are due to irradiation (such as vacancy clusters, dislocation loops, fission-gas bubbles, and solid fission-product precipitates) [44]. Improved understanding of the interaction between natural defects and defects from irradiation and the impact those defects have on fuel and clad performance is an important research area for advanced nuclear energy.

Multiscale modeling and simulation

Currently, the fuel performance simulation tools used by industry and regulators are heavily reliant on empirical fits to extensive experimental data. Despite this reliance on empiricism, the large number of nuclear reactors currently operating safely across the globe confirms that these codes are adequate within the parameter space for which they were developed. However, there are several compelling reasons to pursue modern multiscale modeling for nuclear materials, including (but not limited to) (1) improved understanding of fuel performance phenomena that can lead to improved safety margins, (2) the capability to predict fuel and clad performance for compositions other than those for which irradiation performance data already exist (e.g., UO_2 and Zircaloy), and (3) extension to reactor conditions or designs for which little experimental data exists and for experiments that are particularly demanding (e.g., transient conditions). A general requirement of fuel performance codes is the ability to define an "acceptability envelope" within which fuel and clad performance is both acceptable and predictable. Clearly, such models require detailed information on the evolution of relevant materials' properties during irradiation. Although there is significant promise in the development of multiscale, multi-physics fuel performance modeling (as outlined in [45][46]), significant challenges remain, especially when attempting to bridge extreme time and length scales. However, fundamental insight has been gained from explicit consideration of a wide range of phenomena at the atomistic, mesoscopic, and engineering scales. For example, the use of atomic-scale pair potentials led to seminal work on fission-product incorporation and transport in UO_2 [47][48] and has recently been extended to consider extended defects, such as grain boundaries [49]. Also at the atomic scale, density-functional theory (DFT) has emerged as a more quantitatively rigorous method and has recently been employed, for example, to predict defect behavior and mechanisms of non-stoichiometry in UO_2 [50][51]. Similar atomistic approaches have been widely used to predict the effects of radiation damage on the mechanical properties of clad materials [52]. Mesoscale activities have been attracting more attention recently for the consideration of phenomena such as diffusion (e.g., mass transport using kinetic Monte Carlo [53] and phase-field techniques for

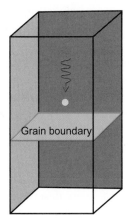

An energetic particle, such as a neutron, hits an atom ⬤ in the material, giving it a large amount of kinetic energy.

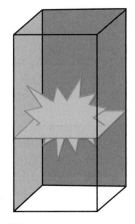

This atom displaces many other atoms in its path, creating a collision cascade, which overlaps with the grain boundary (GB).

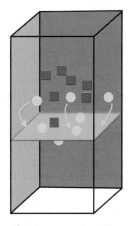

After the cascade settles, point defects – interstitials ⬤ and vacancies ◼ – remain. The interstitials quickly diffuse to the GB.

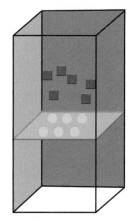

At this point, vacancies remain in the bulk and interstitials are trapped at the GB.

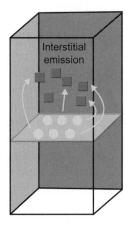

Surprisingly, these trapped interstitials can re-emit from the GB into the bulk, annihilating the vacancies on time scales much faster than vacancy diffusion.

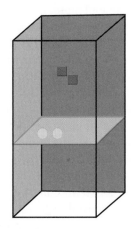

After the interstitial emission events have occurred, some vacancies that were out of reach persist. The system is now in a relatively static situation.

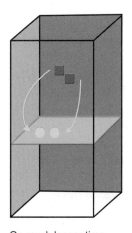

On much longer time scales, the remaining vacancies can diffuse to the GB, completing the healing of the material. At low temperatures this diffusion is exceedingly slow.

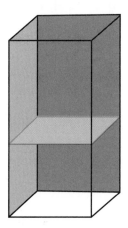

In the ideal case, the system returns to a pristine GB. At low temperatures, the only hope for reaching such a state is via the newly discovered interstitial emission mechanism.

Figure 13.7. Production and evolution of radiation damage near a grain boundary, from [42].

bubble/void formation and growth [54][55]). Finally, coupled, three-dimensional engineering-scale tools are emerging [56] and can serve as a destination for the improved materials insight gained from the types of lower-length-scale modeling described.

Despite these exciting results, coupling of the simulations at multiple time and length scales largely remains to be done. Because of the evolution of materials' properties upon exposure to radiation, a possible effective multiscale approach might be to focus on the interactions between multidimensional defects. Figure 13.9 describes a multiscale approach to address the role of microstructure in fission-gas behavior for

nuclear fuel that corresponds to the experimental result in Figure 13.8. First (the left image of Figure 13.9), DFT is employed to determine the diffusivity and diffusion mechanism of Xe in UO_2 [57]. Then, atomistic pair potentials are used to determine the interactions between fission-gas atoms and grain boundaries [58][59], dislocations [60], and other extended defects (middle image of Figure 13.9). Then, mesoscale models can be developed from the atomic results, and can subsequently be used to determine the role of more realistic microstructures in fission-gas behavior (right image of Figure 13.9). The ultimate goal, of course, is to develop a compositionally and microstructurally aware model of fission-gas retention and release to be used by a fuel performance code. In addition to fission-gas behavior, there are numerous similar examples of compositional and microstructural effects governing phenomena for fuel performance and synthesis, where lower-length-scale simulations not only provide improved fundamental understanding of the interaction between defects but also deliver components of constitutive relationships to engineering-scale codes used

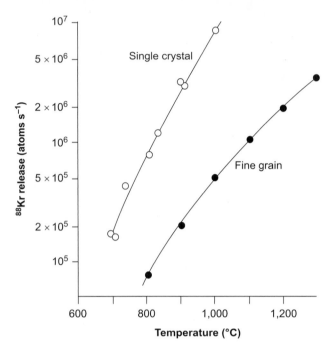

Figure 13.8. Comparison of Kr release rates from single-crystal and polycrystalline UO$_2$ at high temperature, from [43].

for fuel and clad performance. However, each computational approach has specific limitations that require knowledgeable application of the method. Therefore, it is of paramount importance to integrate these modeling activities with highly coupled experiments, both to validate calculations and also to guide research activities.

Unit mechanism experiments

Typically, nuclear materials are analyzed after a relatively long duration of exposure in a reactor. However, the phenomena that are studied post-irradiation are coupled and interdependent, as has been described. For this reason, it is not trivial to establish a path dependence of the materials' properties as a function of radiation exposure, i.e., access is limited to a $t = 0$ data point and a $t =$ high burnup data point with nothing in between. Further, burnup is a vague concept that doesn't describe the details of how the material arrived at the end state, in terms of thermal and stress history, damage spectrum, and so on. To fill in the gaps between these data points, it is possible to combine the modeling and simulation activities described in the previous section with directed experiments. By combining these activities, it should be possible to acquire a deeper understanding than if either route were chosen in isolation. We can think about the necessary experiments intended to decouple the integrated effects according to a "unit mechanism spectrum," which is nominally shown in Figure 13.10. This spectrum starts with experiments that are fully "separate" (see the left-hand side of the spectrum in Figure 13.10), meaning that there is no radiation. Moving toward the right-hand side of the spectrum, radiation is introduced in a controlled fashion through, for example, ion-beam irradiation, which

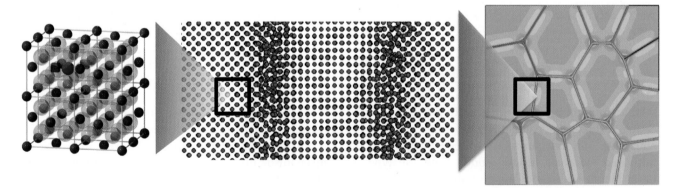

Figure 13.9. A schematic example of multiscale (atomistic to mesoscale) modeling for fission gas in UO$_2$, where the left image corresponds to bulk diffusion calculated by DFT [57], the middle image corresponds to gas segregation to grain boundaries (and dislocations) calculated with pair potentials [58] [59] [60], and the right image is a mesoscale result of fission-gas distribution based upon the results of the previous two images (D. A. Anderson, L. Casillas, B. P. Uberuaga, and C. R. Stanek, unpublished results).

Figure 13.10. A schematic representation of a unit mechanism spectrum. Note: rabbit tests involve a hydraulic system allowing for material to be inserted and removed from the reactor during operation.

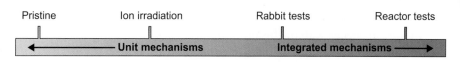

allows either the consideration of radiation damage only or the implantation of specific fission products. This is followed by very short irradiations in a reactor, for example, through the use of "rabbits" (a hydraulic system allowing material to be inserted and removed from the reactor during operation) in test reactors, to irradiate material in a stepwise manner in order to establish a path dependence of radiation effects. It is also important to note that this spectrum provides commentary on the nuclear-engineering relevance of the experiments. That is, just as models will not replace experiments, these unit mechanism experiments will not replace actual reactor tests. However, the knowledge gained from this approach will allow far more directed and informed (and possibly fewer, which will mean lower costs) reactor tests, and, importantly, will aid in the design of advanced materials.

Additionally, when designing unit mechanism experiments, it is crucial to initially prioritize on the basis of considerations of (1) which phenomena are most important and (2) which phenomena can be modeled and measured directly. The second criterion is additionally complex in that the "what" that is measured should ideally directly correlate to the "what" that is modeled, which is not always possible. Furthermore, to provide insight into unanswered experimental questions, it is important to recognize the difference between modeling validation needs and opportunities for modeling. If the relationship between modeling and experiment is only that experiments are intended to validate models, then an opportunity to improve our understanding by integrating modeling with experiments is lost. In that regard, there are advantages to considering modern simulation techniques as tools in an overall toolbox that are used to address innovative nuclear materials development.

Generally, these combined studies should be concerned with the role of crystal structure, composition, microstructure, and radiation effects (keeping in mind that radiation will affect composition and microstructure, and that there are limits to what can be measured) on defect behavior, mass transport, and thermal transport. A US Department of Energy report discussing separate-effects tests has been written and provides detailed information [61]. A wide range of specialized materials-characterization tools are necessary for such studies, and innovative characterization techniques not yet used for nuclear materials should be pursued. Ultimately, by shifting toward close coupling between experiments and modeling and simulations, the community will ensure the success of a science-based approach that provides materials science answers to enable next-generation nuclear energy.

13.5 Summary

This chapter has provided a snapshot of the past, present, and future of nuclear fission energy technology. The materials research opportunities in this area are numerous, and advances must contribute to the realization of a nuclear renaissance, since advanced materials are a crucial component of advanced nuclear energy development. Particularly, in nuclear reactors, materials are exposed to extreme radiation, temperature, and chemical environments. This becomes an even greater challenge under accident situations, which remain the ultimate test of a materials behavior. Challenges remain in understanding the effects of irradiation and an evolving microstructure on a range of materials' properties, especially the role of interfaces and boundaries in determining mechanical properties and mass and thermal transport. By understanding how to design materials to withstand these conditions, it will become possible to increase the efficiency of nuclear power.

13.6 Questions for discussion

1. To what extent will computer simulation be used to replace the need for expensive or hazardous experiments? Can, or should, efforts be made now to define suitable boundaries in order to avoid difficulties later in which it is not possible to satisfy regulatory requirements?

2. There has been significant emphasis on developing materials with enhanced radiation tolerance for reactors to greatly extended design lives. Is it realistic to expect that materials will fulfill such expectations when a detailed understanding of the radiation-enhanced degradation processes is lacking? Would it be better to simply design reactor systems that are entirely predicated on scheduled replacement and avoid the problem?

3. Multiscale simulations, in which information is passed between scales, represent a major modeling research thrust. To what extent is there a risk of missing the rate-determining but uncommon step when an understanding of basic radiation-damage processes that dominate on longer time scales is lacking?

4. Should there be a greater attempt for the fission community to engage with the fusion community despite the differences in neutron energy and fluences? What do the two communities have to offer to each other?

5. Heat is generated within fuel and passed across a series of interfaces, each of which acts as a heat-transfer barrier before the heat is exchanged with the coolant. To what extent could these interfaces be engineered to minimize their detrimental effects?

6. Accidents situations can result in materials being subjected to conditions in excess of those originally envisaged. How can materials science provide solutions to these challenges?

13.7 Further reading

- **J. A. Wheeler**, 1957, "Fission," in *Handbook of Physics*, eds. **E. U. Condon** and **H. Odishaw**, New York, McGraw Hill, Chapter 11. Provides a detailed account of the mechanism of the fission process.

- **D. R. Olander**, 1976, *Fundamental Aspects of Nuclear Reactor Fuel Elements*, TID-26711-P1, Springfield, VA, Technical Information Service, US Department of Commerce. Provides a systematic analysis of oxide fuel and structural materials' performance, motivated by an application of materials research to ultimately predict the performance and longevity.

- **J. T. A. Roberts**, 1981, *Structural Materials in Nuclear Power Systems*, New York, Plenum Press. A classic text dedicated to elucidating the requirements of structural nuclear materials.

- **G. S. Was**, 2007, *Fundamentals of Radiation Materials Science: Metals and Alloys*, Berlin, Springer. A treatise on the fundamental effects of radiation on metals and alloys.

- **W. J. Nuttall**, 2005, *Nuclear Renaissance: Technologies and Policies for the Future of Nuclear Power*, London, Institute of Physics Publishing. A relatively recent examination of the issues related to an expansion of nuclear energy.

13.8 References

[1] **O. Hahn** and **F. Strassmann**, 1938, "Concerning the existence of alkaline earth metals resulting from neutron irradiation of uranium," *Naturwissenschaften*, **26**, 755–756.

[2] **L. Meitner** and **O. R. Frisch**, 1939, "Disintegration of uranium by neutrons: a new type of nuclear reaction," *Nature*, **143**, 239.

[3] **E. Fermi**, 1946, "The development of the first chain reaction pile," *Proc. Am. Phil. Soc.*, **90**, 20–24.

[4] **L. Arnold**, 1992, *Windscale 1957. Anatomy of a Nuclear Accident*, New York, St. Martin's Press.

[5] **P. Finn**, 2005, Chernobyl's Harm Was Far Less Than Predicted, U.N. Report Says, *Washington Post*, September 25.

[6] **European Nuclear Society**, Nuclear Power Plants, World-wide, http://www.euronuclear.org/info/encyclopedia/n/nuclear-power-plant-world-wide.htm.

[7] **R. K. Lester** and **R. Rosner**, 2009, "The growth of nuclear power: drivers & constraints," *Daedalus*, **138**, 19.

[8] **N. Bohr** and **J. A. Wheeler**, 1939, "The mechanism of nuclear fission," *Phys. Rev.*, **56**, 426–450.

[9] **US DOE**, 2010, *Nuclear Energy Research and Development Roadmap, A Report to Congress*, US Department of Energy, Nuclear Energy.

[10] **IAEA**, 2005, *Country Nuclear Fuel Cycle Profiles*, 2nd edn., Vienna, IAEA.

[11] **R. W. Grimes** and **W. J. Nuttal**, 2010, "Generating the option of a two-stage nuclear renaissance," *Science*, **329**, 799.

[12] **MIT**, 2010, *The Future of the Nuclear Fuel Cycle, An Interdisciplinary MIT Study*, Cambridge, MA, Massachusetts Institute of Technology.

[13] **R. L. Murray**, 2001, *Nuclear Energy. An Introduction to the Concepts, Systems, and Applications of Nuclear Processes*, Boston, MA, Butterworth Heinemann.

[14] **I. S. Denniss** and **A. P. Jeapes**, 1996, "Reprocessing irradiated fuel," in *The Nuclear Fuel Cycle, From Ore to Waste*," ed. **P. D. Wilson**, Oxford, Oxford University Press, pp. 116–137.

[15] **R. L. Sheffield** and **E. J. Pitcher**, 2009, *Application of Accelerators in Nuclear Waste Management*, International Committee for Future Accelerators Newsletter, Los Alamos National Laboratory report LA-UR-09–05332.

[16] **J. J. Duderstadt** and **L. J. Hamilton**, 1976, *Nuclear Reactor Analysis*, New York, Wiley.

[17] **N. E. Todreas** and **M. Kazimi**, 1989, *Nuclear Systems Volume I: Thermal Hydraulic Fundamentals*, New York, Hemisphere.

[18] **N. E. Todreas** and **M. Kazimi**, 1989, *Nuclear Systems Volume II: Elements of Thermal Hydraulic Design*, New York, Hemisphere.

[19] **J. R. Lamarsh** and **A. J. Baratta**, 2001, *Introduction to Nuclear Engineering*, Englewood Cliffs, NJ, Prentice Hall.

[20] **T. B. Cochran, H. A. Feiveson, W. Patterson**, et al., 2010, *Fast Breeder Reactor Programs; History and Status*, International Panel on Fissile Materials.

[21] **H. D. Gougar, D. A. Petti, R. N. Wright** et al., 2010, *Current Status of VHTR Technology Development*, INL/CON-10–18035.

[22] **D. Buckthorpe**, 2009, "Materials for the Very High Temperature Reactor – results and progress within the Fifth and Sixth Framework Programmes," *Adv. Mater. Res.*, **59**, 243.

[23] **D. Petti, T. Abram, R. Hobbins**, and **J. Kendall**, 2010, *NGNP Fuel Acquisition Strategy*, PLN-3636, INL.

[24] **D. T. Ingersoll**, 2009, "Deliberately small reactors and the second nuclear era," *Prog. Nucl. Energy*, **51**, 589.

[25] **D. J. Naus**, 2009, "The management of aging in nuclear power plant concrete structures," *JOM*, **61**, 35.

[26] **R. W. Grimes, R. J. M. Konings**, and **L. Edwards**, 2008, "Greater tolerance for nuclear materials," *Nature Mater.*, **7**, 683.

[27] **G. S. Was**, 2007, "Role of irradiation in stress corrosion cracking," in *Radiation Effects in Solids*, eds. **K. Sickafus**, **E. A. Kotomin**, and **B. P. Uberuaga**, New York, Springer, p. 421.

[28] **R. Yang**, **B. Cheng**, **J. Deshon**, **K. Edsinger**, **O. Ozer**, 2006, "Fuel R & D to improve fuel reliability," *J. Nucl. Sci. Technol.*, **43**, 951.

[29] **J. A. Brinkman**, 1956, "Production of atomic displacements by high-energy particles," *Am. J. Phys.*, **24**, 251.

[30] **F. Seitz**, 1952, "Imperfections in nearly perfect crystals: a synthesis," in *Imperfections in Nearly Perfect Crystals*, ed. **W. Shockley**, New York, Wiley, pp. 3–77.

[31] **M. J. Demkowicz**, **R. G. Hoagland**, and **J. P. Hirth**, 2008, "Interface structure and radiation damage resistance in Cu–Nb multilayer nanocomposites," *Phys. Rev. Lett.*, **100**, 136102.

[32] **S. J. Zinkle** and **J. T. Busby**, 2009, "Structural materials for fission and fusion energy," *Mater. Today*, **12**, 12.

[33] **G. R. Odette**, **M. J. Alinger**, and **B. D. Wirth**, 2008, "Recent developments in irradiation-resistant steels," *Ann. Rev. Mater. Sci.*, **38**, 471.

[34] **T. R. Allen**, **H. Burlet**, **R. K. Nanstad**, **M. Samaras**, and **S. Ukai**, 2009, "Advanced structural materials and cladding," *MRS Bull.*, **34**, 20.

[35] **T. R. Allen**, **J. T. Busby**, **R. L. Klueh**, **S. A. Maloy**, and **M. B. Toloczko**, 2008, "Cladding and duct materials for advanced nuclear recycle reactors," *JOM*, **60**, 15.

[36] **F. A. Garner**, 1996, "Irradiation performance of cladding and structural steels in liquid metal reactors," in *Nuclear Materials*, ed. **B. R. T. Frost**, Weinheim, VCH Verlagsgesellschaft mbH, p. 420.

[37] **K. E. Sickafus**, *et al.*, 2007, "Radiation-induced amorphization resistance and radiation tolerance in structurally related oxides," *Nature Mater.*, **6**, 217.

[38] **H. Kleykamp**, 1985, "The chemical state of the fission products in oxide fuels," *J. Nucl. Mater.*, **131**, 221.

[39] **H. Matzke**, 1980, "Gas relase mechanisms in UO_2 – a critical review," *Radiat. Effects Defects Solids*, **53**, 219.

[40] **D. Manara**, **C. Ronchi**, **M. Sheindlin**, **M. Lewis**, and **M. Brykin**, 2005, "Melting of stoichiometric and hyperstoichiometric uranium dioxide," *J. Nucl. Mater*, **342**, 148–163.

[41] **B. D. Wirth**, 2007, "How does radiation damage materials?," *Science*, **318**, 923.

[42] **X. M. Bai**, **A. F. Voter**, **R. G. Hoagland**, **M. Nastasi**, and **B. P. Uberuaga**, 2010, "Efficient annealing of radiation damage near grain boundaries via interstitial emission," *Science*, **327**, 1631.

[43] **R. M. Carroll** and **O. Sisman**, 1966, "Fission-gas release during fissioning in UO_2," *Nucl. Appl.*, **2**, 142–150.

[44] **D. R. Olander**, 1976, *Fundamental Aspects of Nuclear Reactor Fuel Elements*, TID-26711-P1, Springfield, VA, *Technical Information Service*, US Department of Commerce.

[45] **R. Devanathan**, **L. Van Brutzel**, **A. Chartier** *et al.*, 2010, "Modeling and simulation of nuclear fuel materials," *Energy Environ. Sci.*, **3**, 1406.

[46] **M. Stan**, 2009, "Discovery and design of nuclear fuels," *Mater. Today*, **12**, 20.

[47] **C. R. A. Catlow**, 1978, "Fission gas diffusion in uranium dioxide," *Proc. Roy. Soc. Lond. A*, **364**, 473.

[48] **R. W. Grimes** and **C. R. A. Catlow**, 1991, "The stability of fission products in uranium dioxide," *Phil. Trans. Roy. Soc. Lond. A*, **335**, 601.

[49] **L. van Brutzel** and **E. Vincent-Aublant**, 2008, "Grain boundary influence on displacement cascades in UO_2: a molecular dynamics study," *J. Nucl. Mater.*, **377**, 522.

[50] **M. Freyss**, **T. Petit** and **J. P. Crocombette**, 2005, "Point defects in uranium dioxide: *ab initio* pseudopotential approach in the generalized gradient approximation," *J. Nucl. Mater.*, **247**, 44.

[51] **D. A. Andersson**, **J. Lezama**, **B. P. Uberuaga**, **C. Deo**, and **S. D. Conradson**, 2009, "Cooperativity among defect sites in AO_{2+x} and A_4O_9 (A = U, Np, Pu): density functional calculations," *Phys. Rev. B*, **79**, 0241100.

[52] **R. E. Stoller**, **G. R. Odette**, and **B. D. Wirth**, 1997, "Primary damage formation of BCC iron," *J. Nucl. Mater.*, **251**, 49.

[53] **D. A. Andersson**, **T. Wantanabe**, **C. Deo** and **B. P. Uberuaga**, 2009, "Role of di-interstitial clusters in oxygen transport in UO_{2+x} from first principles," *Phys. Rev. B*, **80**, 060101(R).

[54] **M. Stan**, **J. C. Ramirez**, **P. Cristea** *et al.*, 2007, "Models and simulations of nuclear fuel materials properties," *J. Alloys Comp.*, **444**, 415.

[55] **P. C. Millet**, **S. Rokkam**, **A. El-Azab**, **M. Tonks**, and **D. Wolf**, 2009, "Void nucleation and growth in irradiated polycrystalline metals: a phase-field model," *Modelling Simul. Mater. Sci. Eng.*, **17**, 064003.

[56] **C. Newman**, **G. Hansen**, and **D. Gaston**, 2009, "Three dimensional coupled simulation of thermomechanics, heat, and oxygen diffusion in UO_2 nuclear fuel rods," *J. Nucl. Mater.*, **392**, 6.

[57] **D. A. Andersson**, **B. P. Uberuaga**, **P. V. Nerikar**, **C. Unal**, and **C. R. Stanek**, 2011, "Xe and U transport in $UO_{2\pm x}$: density functional theory calculations," *Phys. Rev. B.* **84**, OS410.

[58] **P. V. Nerikar**, **K. Rudman**, **T. G. Desai** *et al.*, 2011, "Grain boundaries in uranium dioxide: scanning electron microscopy experiments and atomistic simulations," *J. Am. Ceram. Soc.*, **94**(6), 1893.

[59] **P. V. Nerikar**, **D. C. Parfitt**, **D. A. Andersson** *et al.*, 2011, "Xenon segregation to dislocations and grain boundaries in uranium dioxide," Submitted for publication.

[60] **D. Parfitt**, **C. L. Bishop**, **M. R. Wenman** and **R. W. Grimes**, "Strain fields and line energies of dislocations in uranium dioxide," *J. Phys. – Condens. Matter*, **22** (2010) 175004.

[61] **J. Carmack**, 2010, *Advanced Fuels Separate Effects Tests R&D Plan*, US Department of Energy, Advanced Fuels Campaign.

14 Nuclear non-proliferation

Siegfried S. Hecker,[1] Matthias Englert,[1] and Michael C. Miller[2]

[1]Center for International Security and Cooperation, Stanford University, Stanford, CA, USA
[2]Los Alamos National Laboratory, Los Alamos, NM, USA

14.1 Focus

Nuclear power holds the promise of a sustainable, affordable, carbon-friendly source of energy for the twenty-first century on a scale that can help meet the world's growing need for energy and slow the pace of global climate change. However, a global expansion of nuclear power also poses significant challenges. Nuclear power must be economically competitive, safe, and secure; its waste must be safely disposed of; and, most importantly, the expansion of nuclear power should not lead to further proliferation[1] of nuclear weapons. This chapter provides an overview of the proliferation risks of nuclear power and how they could be managed through a combination of technical, political, and institutional measures.

14.2 Synopsis

The million-fold increase in energy density in nuclear power compared with other traditional energy sources, such as chemical combustion, makes nuclear energy very attractive for the generation of electricity; however, it is exactly this high energy density that can be used to create weapons of unprecedented power and lethality. The development of commercial nuclear power has, since its inception, had to cope with the prospect of potentially aiding the spread of nuclear weapons. Although commercial nuclear power plants have not directly led to weapon proliferation, the technologies of the nuclear fuel cycle, namely fabricating and enriching fuel, operating the reactors, and dealing with the spent fuel, provides a means for countries to come perilously close to obtaining the fissile materials, ^{235}U and ^{239}Pu, which are required for nuclear weapons. Several countries have developed most of the technical essentials for nuclear weapons under the guise of pursuing nuclear power or research.[2]

The nuclear non-proliferation regime – a fabric of treaties, bilateral and multilateral agreements, organizations, and inspections designed to halt the spread of nuclear weapons while providing access to peaceful uses of atomic energy – has helped to limit the number of states with nuclear weapons. However, this regime is generally agreed to be under severe stress today; some say the world is approaching a "nuclear tipping point" that may usher in unchecked proliferation of weapons and increase the risk of a nuclear catastrophe [1]. Increased interest in expanding nuclear power around the globe, driven by humankind's insatiable demand for energy and concern about global climate change, compounds nuclear proliferation concerns. There is considerable disagreement about whether the risks of nuclear power are worth its benefits.

Whether the risks of a global expansion of nuclear energy can be managed depends not only on technical factors, but also on political, economic, and societal factors [2][3]. The technical challenges overlap with those for nuclear energy (see Chapter 13) and nuclear-waste management (see Chapter 15), except that the focus for nuclear non-proliferation is to examine fuel-cycle technologies and reactor designs and operations that reduce proliferation risk, for example, through greater use of advanced simulation and modeling. Proliferation depends on both capability and intent. Technical measures address only the former; they can mitigate but not eliminate proliferation risks. Hence, a comprehensive set of political, institutional, and technical measures is required in order to manage the incremental risk posed by a global expansion of nuclear power.

---

[1] The term *proliferation* in the nuclear context is used to describe the spread (horizontal proliferation) or further development (vertical proliferation) of nuclear-weapons between or by nation states. The term can also include the spread of nuclear-weapon-usable materials or sensitive nuclear technologies to produce those materials or the spread of sensitive information about nuclear weapons.

[2] The complex relationship between the civil and military use of nuclear technology will be discussed in more detail in supplementary online material available at www.cambridge.org/9781007000230.

14.3 Historical perspective

The awesome destructive power of nuclear weapons demonstrated at Hiroshima and Nagasaki, Japan,[3] convinced President Truman to seek international control of atomic energy shortly after the end of World War II. Toward that end, J. Robert Oppenheimer and colleagues authored the Acheson–Lilienthal Report [4], which warned that "the development of atomic energy for peaceful purposes and the development of atomic energy for bombs are in much of their course interchangeable and interdependent." On June 14, 1946, US diplomat Bernard Baruch presented a plan, based largely on the Acheson–Lilienthal Report, to the United Nations (UN) for the elimination of atomic weapons, together with effective safeguards and inspections [5]. The USSR rejected the Baruch Plan and joined the nuclear age with its first atomic explosion on August 28, 1949.

The Cold War, the British explosion of an atomic bomb, and US demonstration of the hydrogen bomb, a thousand times more powerful than the atomic bomb, led President Eisenhower to try again to rein in what he called the "fearful atomic dilemma" in his December 8, 1953, address to the UN in which he proposed the Atoms for Peace initiative. After initial skepticism, the USSR joined the initiative, leading to the opening of the secret world of nuclear energy to nations around the world that agreed to develop peaceful applications and forego military applications. It also led to the establishment in 1957 of the International Atomic Energy Agency (IAEA), which has the dual mission of promoting global civilian applications of atomic energy around the world and monitoring compliance to ensure their peaceful use.

The first 20 years of the nuclear era saw only France (1960) and China (1964) join the nuclear club. Nevertheless, many industrialized nations, including Sweden and Switzerland, explored the military potential of atomic energy. Initially, the greatest concerns were focused on the potential nuclear aspirations of Germany and Japan. President Kennedy feared that, by the end of the 1960s, the world might have 15–20 nations with nuclear weapons. In 1968, many of the non-nuclear-weapon states (NNWSs), alongside the USA, USSR, and UK, pushed for the adoption of the Treaty on the Non-Proliferation of Nuclear Weapons (NPT). One of the key features of the NPT (see Box 14.1) was the establishment of a comprehensive, full-scope safeguards system requiring inspections under the responsibility of the IAEA. It went into effect in 1970, although a number of states,

Box 14.1. The Treaty on the Non-Proliferation of Nuclear Weapons

The three pillars of the NPT (http://www.iaea.org/Publications/Documents/Infcircs/Others/infcirc140.pdf) reflect a bargain among the non-nuclear-weapon states (NNWSs) and between the NNWSs and the nuclear-weapon states (NWSs).

Non-proliferation. All NWSs agree not to transfer, or assist any state in acquiring, nuclear weapons, and all NNWSs agree not to manufacture nuclear weapons themselves and also not to receive such transfer or assistance (Articles I and II). To verify compliance with the treaty, NNWSs accept safeguards by the IAEA to be applied to all source or special fissionable materials in all peaceful nuclear activities within their territory (Article III).

Peaceful use. Article IV gives all NPT member states the right to develop and use nuclear energy for peaceful purposes in conformity with Articles I and II. Moreover, all members in a position to do so agree to cooperate in furthering the development of nuclear energy for peaceful purposes.

Disarmament. Each member agrees to pursue negotiations relating to the end of the nuclear arms race and to nuclear disarmament (Article VI).

Under Article IX.3 of this treaty, an NWS is one that manufactured and exploded a nuclear weapon or other nuclear explosive device prior to January 1, 1967. Hence, the USA, the USSR (nowadays Russia), the UK, France, and China are declared to be NWSs, and all others are NNWSs. The five NWSs are referred to as the P-5 states because, historically, they are also the five permanent members of the UN Security Council and, as such, hold veto power over any Security Council action.

notably China, France, India, Pakistan, Israel, Argentina, Brazil, and South Africa, did not sign at that time.

To date, the non-proliferation regime has had some laudable successes and some major failures. On the success side is the fact that many states use nuclear technology to supply electricity with little carbon release (constituting roughly 14% of the world's electricity production), to provide medical treatments that improve the lives of millions of people, and to conduct scientific research. The NPT has more signatories, 189 in all, than any other treaty, indicating that the non-proliferation norm has become accepted by the overwhelming majority of states. Argentina and Brazil gave up their nuclear weapon programs. Ukraine, Kazakhstan, and Belarus gave up the weapons

---

[3] These bombs had explosive yields equivalent to approximately 13 and 21 kilotons of TNT, and devastated the two cities, resulting in roughly 200,000 prompt and short-term fatalities.

Figure 14.1. Pathways to produce highly enriched uranium (HEU) and plutonium and typical material compositions. Fission products are neglected in the illustration. High burnup for commercial operation is 33 and 7.5 billion watt-days per ton of heavy metal (GW-d per tHM) for light- and heavy-water reactors (LWR and HWR), respectively; low burnup is 1 GW-d per tHM to produce weapons-grade plutonium. The HEU path here refers to weapons-grade HEU GCR (Gas Cooled Reactor).

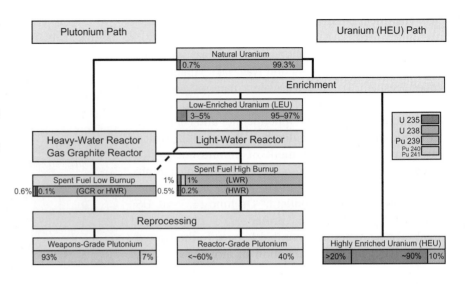

that they had inherited from the USSR, and South Africa gave up its indigenously developed weapons. In the 1960s, it was believed that 23 countries had nuclear weapons programs; by the 1980s, that number had dropped to 19; and today, it is believed to be 10 [6].

On the failure side, some nuclear weapon programs, such as that of Iraq, have evaded detection from international inspectors within the NPT framework and have procured equipment, materials, technology, and foreign assistance. Moreover, several states have not been integrated into the non-proliferation regime: Israel is believed to have the bomb, India and Pakistan have declared themselves nuclear-armed states, and North Korea withdrew from the treaty and subsequently also declared itself a nuclear-armed state. Concerns about the civil versus military ambiguity of sensitive nuclear technologies continue today, as claims to the right for peaceful use of nuclear energy, based on Article IV of the NPT, could be exploited for military purposes. Finally, progress toward nuclear disarmament, as called for in Article VI, has been insufficient.

14.4 Nuclear non-proliferation

14.4.1 From fissile materials to nuclear weapons

The difficulty of producing fissile materials constitutes the greatest barrier to building nuclear weapons because the two typical pathways (Figure 14.1) – enriching uranium for a highly enriched uranium (HEU) bomb and building reactors and reprocessing spent fuel for a plutonium bomb – are technologically demanding and time-consuming. The USA accomplished both during its crash program for the Manhattan Project in less than three years, but, historically, covert activities beyond the P-5 states, such as those in India, Pakistan, and North Korea,[4] have taken a decade or much longer.

The amount of fissile material needed for a bomb is very small compared with the amounts handled in commercial nuclear programs (see Box 14.2). The HEU for nuclear weapons and military naval reactors has been produced in dedicated enrichment facilities. None of the P-5 states is known to be producing more HEU for weapons today, but other states do (India, Pakistan). In addition, some small, compact reactors, such as those for research or naval vessels, are fueled with uranium enriched at levels above 20%. Dedicated weapons-grade plutonium production reactors use natural uranium fuel in heavy-water-moderated or graphite-moderated reactors (see Figure 14.1). These are operated for short burn cycles in order to yield the most attractive weapons-grade plutonium (higher ^{239}Pu content). All P-5 states produced their weapons-grade plutonium in dedicated production reactors before they built large commercial power reactors. In other countries, such as India, however, the reactors are multipurpose, that is, for research, medical-isotope generation, and plutonium production.[5] Commercial reactors could also in principle be operated to produce weapons-grade plutonium by changing the operation mode from long standing times of the fuel to a very uneconomic short irradiation of the fuel [10].

Information about basic designs of nuclear weapons is now readily available, yet weaponization is more challenging because it requires metallurgical experience with uranium or plutonium, precision machining, expertise

---

[4] The nuclear stockpile numbers for all nations have historically been classified secret. Only within the past year have the USA and UK declared the numbers openly. The Federation of American Scientists has historically estimated nuclear weapon stockpiles around the world (http://www.carnegieendowment.org/npp/index.cfm?fa=map&id=19238&prog=zgp&proj=znpp).

[5] More details on the dual-use characteristics of reactors to produce plutonium and electricity in the case of the first gas-cooled graphite-moderated reactors in France and the UK, reactors in Russia, and other cases of civil–military ambiguity can be found online at www.cambridge.org/9781107000230.

Box 14.2. Fissile and other nuclear-weapon-usable materials

A technical objective of IAEA safeguards is the timely detection of the diversion of nuclear material from peaceful nuclear activities to the manufacture of nuclear weapons. The IAEA defines a significant quantity (SQ) as the approximate amount of nuclear material for which the possibility of manufacturing a nuclear explosive device cannot be excluded. For plutonium and HEU, these amounts are 8 kg (with <80% ^{238}Pu, because of the high heat generation of this isotope) and 25 kg, respectively, per year [7]. However, declassified documents from the US Department of Energy state that 4 kg of plutonium or ^{233}U is hypothetically sufficient for one nuclear explosive device [8].

Plutonium. The isotope ^{239}Pu can be produced in reactors by neutron capture in fertile ^{238}U. The excited nuclide ^{239}U decays rapidly by double beta decay into ^{239}Pu. ^{239}Pu can capture neutrons to become ^{240}Pu. Subsequent absorption of neutrons leads to ^{241}Pu and ^{242}Pu:

$$^{238}\text{U}(n,\gamma) \longrightarrow {}^{239}\text{U} \xrightarrow{-\beta 23.5min} {}^{239}\text{Np} \xrightarrow{-\beta 2.3565d} {}^{239}\text{Pu},$$

$$^{239}\text{Pu}(n,\gamma) \longrightarrow {}^{240}\text{Pu}(n,\gamma) \longrightarrow {}^{241}\text{Pu}(n,\gamma) \longrightarrow {}^{242}\text{Pu}.$$

Terminology is Element and Left Superscript is atomic mass and Parenthesis is the particles released in the nuclear reaction (n = neutron, γ = gamma proton, β = electron, time = half life). The longer ^{238}U is exposed to a neutron flux the more plutonium is produced in total. However, the mixture of plutonium isotopes changes, with an increasing share of higher plutonium isotopes (^{240}Pu, ^{241}Pu, ^{242}Pu) and the buildup of ^{238}Pu by several neutron-capture and decay reactions from ^{235}U and ^{238}U:

$$^{235}\text{U}(n,\gamma) \longrightarrow {}^{236}\text{U}(n,\gamma) \longrightarrow$$

$$^{237}\text{U} \xrightarrow{-\beta 6.7d} {}^{237}\text{Np}(n,\gamma) \longrightarrow {}^{238}\text{Np} \xrightarrow{-\beta 2.1d} {}^{238}\text{Pu},$$

$$^{238}\text{U}(n,2n) \longrightarrow {}^{237}U \xrightarrow{-\beta 6.7d} {}^{237}\text{Np}(n,\gamma) \longrightarrow$$

$$^{238}\text{Np} \xrightarrow{-\beta 2.1d} {}^{238}\text{Pu}.$$

Weapons-grade plutonium is generally defined as >93% ^{239}Pu in the isotopic mixture, typically extracted from dedicated plutonium-production reactors in which the ^{239}Pu content is much higher than that of typical spent commercial nuclear fuel because of the short exposure to neutron irradiation. Longer exposure, such as in a commercial fuel cycle, produces more of the higher plutonium isotopes, neptunium, and americium.

There is still disagreement in the literature about the weapons utility of *reactor-grade plutonium* in which the percentage of the fissile isotopes ^{239}Pu and ^{241}Pu can be as low as 50% or so. The mix of plutonium isotopes makes reactor-grade plutonium somewhat less attractive for weapons from a nuclear physics standpoint and considerably more difficult for engineering and manufacturing because of the increased heat mainly from ^{238}Pu and radioactivity of ^{238}Pu and ^{240}Pu. Bathke and co-workers recently made the case that any plutonium produced in current fuel cycles and reprocessing schemes is sufficiently attractive for nuclear weapons, even if it is mixed with other transuranium elements (Np, Am, Cm), that it warrants safeguards and protection [9].

^{235}Uranium. *Highly enriched uranium (HEU)* is defined as ≥20% ^{235}U. Weapons-grade uranium is generally defined as roughly 90% ^{235}U, but any level above 20% can theoretically be used to make a bomb. **Low-enriched uranium (LEU)** is defined as <20% ^{235}U and natural uranium ore is 0.71% ^{235}U, with the remainder being ^{238}U.

^{233}Uranium. In addition to ^{235}U and ^{239}Pu, the IAEA includes the fissile isotope ^{233}U in the category of "special fissionable material" or "direct-use material" [7] requiring safeguards and protection. It can be produced in reactors in an analogous manner to ^{239}Pu, except starting with ^{232}Th instead of ^{238}U as the fertile reactor fuel:

$$^{232}\text{Th}(n,\gamma) \longrightarrow {}^{233}\text{Th} \xrightarrow{-\beta 22min} {}^{233}\text{Pa} \xrightarrow{-\beta 27d} {}^{233}U.$$

The IAEA defines 8 kg of ^{233}U as 1 SQ. However, there is no indication of any nation currently using ^{233}U in a nuclear arsenal.

Alternative nuclear materials. ^{237}Np, ^{241}Am, and ^{243}Am, which are present in typical spent fuel, are sometimes referred to as alternative nuclear materials, since they have nuclear properties that make them potentially suitable for nuclear weapons. [7]

Tritium can be used in small quantities to enhance the yield of a nuclear weapon. Tritium is not subject to IAEA safeguards, although it is subject to export controls in accordance with multilateral agreements and national-level controls.

Box 14.3. Explosive mechanisms of pure fission weapons

The first nuclear weapons built, and the only type ever used, were pure fission bombs. They used two different mechanisms, as shown in Figure 14.2, to achieve critical mass for a self-sustaining chain reaction.

Gun-type assembly. This weapon design is technologically straightforward: two-subcritical masses of HEU are impacted at high speed in a gun barrel. The bomb detonated over Hiroshima was an HEU-fueled gun-assembly device.

Implosion device. The gun-type assembly method turns out to be too slow for plutonium because enough neutrons are generated by spontaneous fission of the minor isotope ^{240}Pu in weapons-grade plutonium to pre-initiate a nuclear chain reaction, resulting in merely a "fizzle" rather than a full-scale explosion. Instead, a subcritical, spherical mass of plutonium is imploded with high explosives to reach much higher velocities and a full nuclear explosion. The bomb detonated at Nagasaki was a plutonium-fueled implosion device.

More advanced explosive mechanisms have since been developed and tested, including a fusion-boosted fission assembly method and the two-stage thermonuclear assembly method.

Figure 14.2. Mechanisms used to achieve critical mass for a self-sustaining chain reaction in a pure fission weapon: (left) gun-type and (right) implosion assembly methods. From [11].

Gun-type assembly method

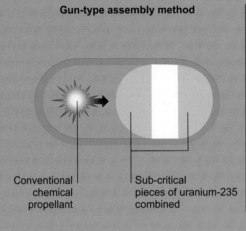

Conventional chemical propellant

Sub-critical pieces of uranium-235 combined

Implosion assembly method

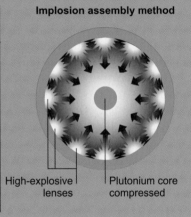

High-explosive lenses

Plutonium core compressed

with propellants or explosives, detonators, initiators to inject neutrons, and computational capabilities. Primitive bombs of the types exploded in Japan (see Box 14.3) are generally viewed as being within reach of technologically proficient nations once they have mastered the ability to make fissile materials. For example, South Africa and Pakistan developed HEU-fueled bombs, and North Korea opted for the plutonium route.

Historically, all nations developing a nuclear arsenal tested their early weapon designs, although there is considerable ambiguity for Israel and South Africa [12][13]. Testing is important for advanced designs. Modern thermonuclear warheads are among the most complex technological devices in the world. The miniaturization of a nuclear warhead so that it can be carried on an intercontinental ballistic missile (ICBM) constitutes a formidable technical challenge that only the P-5 states have mastered to date. Some of the other states with nuclear weapons might have mastered warheads small enough to mount on short- or medium-range missiles, but they do not possess ICBMs and are believed not to have produced ICBM-compatible miniaturized designs.

Sub-national groups or terrorists could instead settle for a van, boat, or airplane if they are able to acquire a nuclear device or make an improvised one.

14.4.2 Nuclear fuel cycle and proliferation concerns

Proliferation concerns depend on the specifics of the fuel cycle, particularly the front end (uranium enrichment) and back end (reprocessing), which provide opportunities for access to weapon-usable materials (Figure 14.1).

The method of choice today for uranium enrichment uses gas-centrifuge technology in which the lighter ^{235}U gas in the form of uranium hexafluoride is separated from the heavier ^{238}U in cascades of rapidly spinning centrifuges. Once a state acquires the capability to enrich uranium to the typical level needed for light-water reactors (LWRs), namely 3%–5% ^{235}U, the technical capability to continue enrichment to weapons-grade levels of roughly 90% inherently exists. Also, by producing LEU of 3%–5% enrichment, much of the separative work (approximately 70%–88%) necessary for getting to weapons grade is already done. Only a fraction of the cascades used for a full-scale commercial enrichment facility would need to be diverted or constructed to produce sufficient weapons-grade HEU for a few bombs per year. At the back end, for some fuel cycles, plutonium is extracted from spent fuel by reprocessing, typically using the PUREX (plutonium–uranium recovery by extraction) process (see Chapter 13) [14].

As explained in Chapter 13, the open, or once-through, fuel cycle with LWRs forms the backbone of the commercial nuclear power industry. The primary proliferation concerns are associated with uranium enrichment. The back end of the once-through fuel cycle is generally considered to be more proliferation-resistant than the back ends of other fuel cycles because the spent fuel is highly radioactive and is stored for eventual disposal without the separation of reactor-grade plutonium. This spent fuel is considered to remain self-protecting for 100 years or more.

Variations of the open fuel cycle include heavy-water reactors (HWRs) and gas-cooled reactors (GCRs). Heavy-water reactors can operate on natural uranium (0.7% ^{235}U), thus avoiding the need for enrichment (although some advanced HWRs might use 1%–2% ^{235}U). However, more plutonium, with higher ^{239}Pu content, is produced than for similar LWR burnups,[6] and spent fuel is removed from the reactor continuously, which makes safeguards more challenging than for LWRs.

Several countries have modified the open fuel cycle by reprocessing the spent fuel from LWRs and using the resultant plutonium in a uranium–plutonium mixed oxide (MOX) fuel that can be burned directly. Although the reactor-grade plutonium produced in such fuel cycles is not very attractive for weapons, its use in weapons is not impossible, and it must be safeguarded from potential diversion and protected during transportation.

Fast reactors operate at high neutron energy and are able to nearly fully consume the energy content of the fuel with several passes through the reactor. They can also operate in a "breeding" mode, in which more fissile material is produced than is consumed by the reactor. The proliferation concerns for such reactors stem from the fact that a great deal of plutonium is produced with high ^{239}Pu content and, in current schemes, it is separated, hence presenting significant security and safeguards challenges.

Thorium-fueled reactors are of interest as an alternative to the use of uranium. Proliferation concerns for

this technology stem from the capture of a neutron by fertile ^{232}Th, creating the fissile isotope ^{233}U. As ^{233}U is bred, small amounts (part-per-million quantities) of ^{232}U are also produced. With a half-life of only 73.6 years and strong gamma-ray emission from its daughters, most notably the 2.6-MeV gamma-ray from ^{208}Tl, ^{232}U could provide some proliferation resistance relative to the uranium fuel cycle as handling becomes more difficult [15].

14.4.3 Potential pathways to weapons from the commercial fuel cycle

The commercial fuel cycle and the associated research infrastructure can potentially lead to weapons proliferation through several pathways.

(1) A state's declared commercial or research facilities could be diverted to produce weapons-grade materials for bombs, either covertly or by leaving the NPT regime. The expertise and equipment from commercial operations or associated research programs could be used to develop clandestine facilities dedicated to the production of weapons material.

(2) Fissile material from the commercial fuel cycle or research facilities could be diverted or stolen to make nuclear weapons.

(3) A commercial nuclear infrastructure could constitute a latent nuclear-weapon capability; that is, a nation could produce fissile materials under the umbrella of a purely civilian program, or it could develop the necessary facilities and know-how to do so at a later time. Japan and Germany are examples of the first, and Iran is an example of the second.

Although none of the pathways can be ruled out, commercial nuclear power programs have not been used directly to develop nuclear weapons to date. Section 14.4.5 discusses how this situation could change as global nuclear power expands.

14.4.4 Current proliferation threats

Threats from emerging and aspiring nuclear-weapon states

To build a nuclear arsenal, countries need both capability and intent. Nuclear information has become broadly accessible, and the technological threshold for sensitive nuclear technologies (i.e., enrichment and reprocessing) is within reach of more countries. Rather than ushering in global peace, the end of the Cold War might have provided greater motivation for countries to develop their own nuclear weapons. Some countries saw their security guarantees evaporate with the dissolution of the Soviet bloc or the reordering of US foreign policy priorities. Others faced

---

[6] Burnup is a nuclear engineering term for the power released by fission events in a given mass of the fuel. Since most fission events in uranium fuel stem from fission of ^{235}U, the burnup is proportional to the amount of ^{235}U consumed to produce energy.

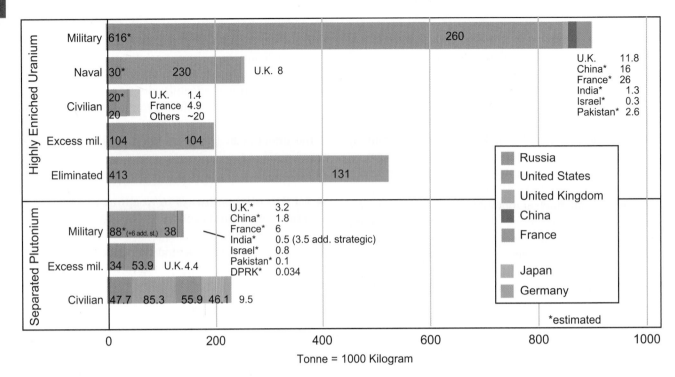

Figure 14.3. Estimated national stockpiles of highly enriched uranium and separated plutonium. Uncertainties in the military stockpiles for China, France, India, Israel, Pakistan, and Russia are on the order of 20%. DPRK, Democratic People's Republic of Korea. (See the report of the IPFM [11] for data and a detailed discussion.) *means the number estimated and grey is others.

new regional hostilities, and some felt threatened by a world apparently moving toward American hegemony.

During the past 12 years, India, Pakistan, and North Korea have declared themselves to be nuclear powers by testing nuclear devices. Following Pakistan's acquisition of the bomb, its leading scientist, A. Q. Khan, built a proliferation black-market network [16] that sold nuclear technologies and know-how to North Korea, Iran, and Libya. North Korea provided nuclear assistance to Syria and Libya, and possibly to Iran and Myanmar. For more details on some countries see Appendix 14.8.

These developments are not directly associated with commercial nuclear power, but they have rightfully caused alarm and concern that the international non-proliferation regime might be collapsing, despite the fact that there are fewer states with nuclear weapons than had been projected 50 years ago and fewer active nuclear-weapons programs than existed 40 years ago.

Threats from fissile-material stockpiles

Currently, enormous quantities of fissile material are available and pose a proliferation threat should they fall into the hands of nuclear terrorists or be used in weapon programs by nation states. Estimates of these

materials around the world (see Figure 14.3) have been provided by the International Panel on Fissile Materials (IPFM) [11].

Highly enriched uranium (enriched in ^{235}u) is not used in commercial nuclear fuel cycles. Most of the 1,850 t [1 tonne (t) = 1,000 kg] of HEU in the world is housed in military programs (in weapons programs, in naval reactors, or declared in excess of military needs). About 64 t of HEU, at levels of 36%–93%, is currently used in the civilian research-reactor fuel cycle, of which roughly 9 t is under IAEA safeguards in non-nuclear-weapon states. The recognition of the need to minimize the civilian use of HEU has led to several activities to improve safeguards, return fresh and spent HEU fuel to the original supplier country, and to convert reactors to use LEU (<20% enrichment).

Half of the 500 t of separated plutonium is weapons-grade plutonium and originated from military programs. This plutonium is very attractive for weapon purposes and is mostly in the possession of Russia and the USA, which have declared parts of it excess to their needs for military use. The other 250 t is civil separated reactor-grade plutonium produced in commercial power plants and subsequently reprocessed (see Figure 14.3). As indicated in Box 14.2, the isotopic composition of plutonium from commercial reactors is not ideal, and hence this plutonium presents significant, but not insurmountable, challenges for bomb manufacture.

More than 50 countries have spent fuel from commercial or research reactors. To the best of our knowledge, no commercial plutonium has been diverted for weapon use. Between 1993 and 2007, the IAEA reported 18 illicit trafficking incidents involving HEU and separated plutonium [17].

Nuclear terrorism threats

Concerns about separated plutonium and HEU from research reactors around the world highlight the increased threat of nuclear terrorism, particularly because HEU can be used in a gun-assembly nuclear device, which is technologically much simpler than an implosion device. Concerns about the potential nexus of international terrorism and sub-national groups and nuclear weapons have grown since the terrorist attacks on New York City and the Pentagon on September 11, 2001 (known as 9/11). The most serious threat is the acquisition of fissile materials, by either theft or diversion, by such groups and the subsequent building of an improvised nuclear device. Even an imperfect nuclear device of a few kilotons or lower detonated in one of the world's megacities could kill 10,000 or more people and cause global disruption on an unprecedented scale. The primary concern is the protection and safeguarding of these fissile materials to keep them out of the hands of terrorists. The global nuclear security summit held in Washington, DC, in 2010 demonstrated that the importance of doing so is now generally appreciated, but the technical difficulty of doing so is not [18].

Commercial nuclear power contributes rather little to the nuclear-terrorism threat today, because fissile materials from commercial operations are contained in spent fuel, well protected, and not very attractive to terrorists. However, there is considerable concern about the security of commercially separated plutonium. As noted above, transportation of separated plutonium or MOX is a security concern. For example, Japan regularly ships its spent fuel to France for reprocessing and receives separated plutonium and waste in return. Shipping the separated plutonium in the form of MOX fuel does not improve the proliferation resistance significantly. Such issues could be addressed by co-locating reactor facilities with fuel fabrication.

Nuclear power plants and fuel-cycle facilities can also be targets of nuclear sabotage or provide materials for radiological dispersal devices, so-called dirty bombs [19][20]. These events do not yield a nuclear explosion or damage on the scale of such an explosion. Although both sabotage and radiological dispersal could cause enormous disruption, they do not constitute nuclear proliferation and, therefore, are not considered further here.

14.4.5 Challenges posed by a global expansion of nuclear power

The resurgence of interest in commercial nuclear power around the world presents a challenge and an opportunity to this generation of scientists and engineers. In the USA, electric utilities have expressed interest in the construction of 26 new nuclear power plants – a dramatic change from the past 30 years. China, India, and Russia have ambitious plans for expansion of nuclear power in the coming decades. In addition, more than 60 IAEA member states have expressed interest in starting new nuclear power programs, although only a dozen or so are seriously pursuing new reactors at this time.

It remains to be seen how the Fukushima-Daiichi nuclear accident in March 2011, will affect global expansion of nuclear power. In China, for example, the Japanese reactor accident is predicted to delay China's ambitious nuclear reactor build-up, but not dramatically curtail it.

In assessing how the spread of nuclear power might impact global proliferation, it is important to differentiate the growth of nuclear power in countries that already use it from the spread of nuclear power to additional countries [21]. Aspirant nuclear states might have significantly poorer governance, resulting in less safe and secure nuclear power, and a much lower aggregate democracy score, increasing their likeliness to violate their NPT obligations. These states also rank high on the list of states having suffered serious terrorist attacks during the previous five years, thereby representing a serious proliferation challenge if they pursue nuclear power with fuel-cycle facilities.

A significant expansion of nuclear power could also lead to concerns about uranium shortages or tensions caused by the distribution of uranium resources, which, in turn, could lead more countries to pursue national enrichment and closed fuel cycles with breeder reactors, fueled with either uranium or thorium. India is a good case in point, insofar as its concern about nuclear fuel supply has led it to a strategy that is heavily reliant on plutonium breeder reactors, followed by thorium-fueled reactors, although other factors likely influenced this strategy.

In addition, an expansion of nuclear power could lead to a significant shift in nuclear-reactor and fuel-cycle suppliers around the world. As more countries enter the nuclear technology supply chain for commercial purposes, it might be more difficult to prevent clandestine sales of such equipment. Finally, a substantial increase in the number of states aspiring to nuclear power and fuel-cycle technologies would greatly overburden the monitoring and inspection capabilities of the IAEA. In addition to these non-proliferation issues, nuclear-reactor safety and protection of facilities from nuclear sabotage will become more challenging as the number of reactors and fuel-cycle facilities increases.

Proliferation risks must be mitigated by a combination of technical, institutional, and political measures. The residual risk must then be compared with the benefits achieved by expanded global nuclear power. The private sector will look primarily at economics, whereas governments must make the call for safety, security, and global climate change. The next section discusses potential technical measures with a focus on materials technologies.

14.4.6 Proliferation countermeasures

In general, one can categorize measures to improve fuel-cycle resistance to proliferation as intrinsic or extrinsic. Intrinsic measures rely on technical means, such as incorporation of material attributes (e.g., increased radiation dose from fission-product addition to nuclear fuel or removal of fissile-material precursors) that provide some additional barrier to illegitimate use or general technical design features of facilities. In addition, there are technical components to improving institutional barriers to proliferation – for example, better technologies for nuclear detection, monitoring, and nuclear forensics. Extrinsic measures rely on institutional and political measures to safeguard and secure material, control exports, and enforce international norms. Neither extrinsic nor intrinsic measures alone are sufficient to increase the confidence in or provide independent verification of a nation's peaceful application of nuclear energy.

The technical community has pursued proliferation-resistant reactors and fuel cycles for decades, but it is now generally acknowledged that none are proliferation-proof [22][23][24]. Yet, some are more proliferation-resistant than others, and hence methodologies to compare proliferation risks of specific technologies continue to be developed [25][26].

Examples of technical countermeasures and materials research opportunities

On the technical side, operating within institutional and political frameworks, are verification and accounting activities engaged in by both the host state and the international community, primarily the IAEA. Direct accounting of nuclear material requires a formal materials protection, control, and accountancy (MPC&A) program, which, in turn, relies on measurement techniques. Doyle provides a comprehensive treatment of the technical issues related to non-proliferation [27]. Radiation detection provides the basis for many of the techniques used in MPC&A and is covered in detail by Knoll [28].

Advanced measurement techniques. An important aspect of verification relies on non-destructive assay (NDA) measurements involving radiation detection

[29]. Uranium and plutonium (as well as thorium) are radioactive and present a variety of usable signatures. In particular, X- and gamma-rays and neutrons are used for detection and quantification because of their greater penetrating power as compared with alpha- and beta-radiation. ^{235}U has gamma-rays in the energy range of 100 keV to 200 keV, while isotopes of plutonium emit strongly in the 100–600-keV region. ^{238}U has a gamma-ray at 1 MeV. Plutonium, particularly the even isotopes, undergoes spontaneous fission. Thermal and fast neutrons can induce fission in ^{235}U, whereas ^{238}U undergoes neutron-induced fission only above a threshold of about 0.6 MeV. Fission creates multiple neutrons per event and is the basis of the most accurate NDA methods exploiting the neutron signature. Radiation detectors operate on the basis of charge creation and collection. X- and gamma-rays ionize matter directly through the Compton and photoelectric effects, whereas neutrons create charge by undergoing nuclear reactions (thermal) or elastic scattering (fast).

Most NDA methods based on neutron detection rely on the ^{3}He neutron-capture reaction, which has a higher cross section than that of either ^{10}B or ^{6}Li capture and little sensitivity to gamma-rays. Detector manufacturers have long relied on the supply of ^{3}He derived from the decay of tritium used in nuclear weapons. One outcome of nuclear arms reduction has been a reduced need for tritium for the remaining nuclear arsenal and, thereby, a reduced supply of ^{3}He. This, combined with a great increase in demand in response to the 9/11 attacks, has created a global crisis in ^{3}He supply. This shortage has implications for nuclear non-proliferation, particularly for safeguards, where neutron counting has been a core measurement technique for decades. In particular, time-correlation analysis based on thermal-neutron detection is the most accurate measurement technique for bulk samples [30].

Alternative technologies being investigated include gas proportional counters, and semiconductor and scintillation technologies utilizing primarily ^{10}B or ^{6}Li. Boron-lined ^{3}He proportional counters were developed for neutron detection in very high (100 roentgen per h) gamma-ray fields. An extension of this technology, where ^{3}He is replaced with a low-atomic-number (low-Z) charge-carrier gas is currently under development. The key issue is one of increasing the surface area (and therefore detection efficiency) and still maintaining both the ability to detect the alpha particle from the ^{10}B $(n,\alpha)^7$Li (α particle is the He nucleus) reaction and the long-term stability of the thin coating. Advanced materials for neutron detection also include semiconductors. The most mature of these devices uses a Si pin diode with etched channels containing ^{10}B or ^{6}Li as the active element. Advances in material science to control parameters such as purity, structure, and bonding will be needed in order to take these technologies out of the laboratory and into practical field application.

Improvements in gamma-ray detectors are also needed, primarily driven by the desire to achieve good energy resolution (for accurate nuclide determination) while avoiding the need for cryogenic operation (to simplify field use and cost). High-purity germanium (HPGe) is the standard in gamma-ray spectroscopy. It can be manufactured in large crystals with very high purity and has excellent energy resolution, but it requires liquid nitrogen (77 K) or equivalent mechanical cooling and is expensive. Sodium iodide is a room-temperature detector material with moderate energy resolution. Research on new materials, both scintillators and semiconductors, is needed and ongoing. One very promising new scintillator material is CLYL (Cs_2LiCl_6:Ce) [31], which not only has demonstrated better resolution than NaI but also provides neutron detection via ^{6}Li capture. Good pulse-shape discrimination for neutrons and gamma-rays has also been observed. In addition, the application of advanced modeling and simulation techniques is beginning to provide a more systematic approach to materials discovery [32][33]. This is a rich area to which advances in materials science have increasingly contributed.

When the highest possible accuracy or sensitivity is required, analytical chemistry-based measurement techniques are used. These methods are generically called destructive analysis because the sample being analyzed is consumed in the analysis process. Such techniques are best suited to situations where small portions of material can be removed and then analyzed, such as in bulk processing facilities (mining, fuel fabrication, reprocessing). The IAEA routinely extracts samples for chemical analysis as part of its verification activities. Under the Additional Protocol to the NPT (discussed further below), wide-area environmental sampling is allowed as part of the verification of the completeness of a country's declared nuclear program. Advances in materials science and associated chemical separation methods and instrumentation, such as fission-track thermal-ionization mass spectrometry and inductively coupled plasma mass spectrometry, have enabled analysis of individual particles. These methods offer the ultimate accuracy and sensitivity, but are extremely time-consuming and require extensive sample preparation. Continued development is needed in order to allow these and related techniques to be applied more extensively by moving them out of the laboratory and into the field.

Other technical issues. A host of other technical issues arise under the expansion and evolution of the nuclear fuel cycle. First-principles modeling and simulation will play an increasingly important role both for basic materials development and for optimization of the non-proliferation benefits of new fuel-cycle technologies. In general, advanced concepts involve materials that must withstand environments with higher radiation doses, higher temperatures and pressures, and higher corrosion [34]. Many of the materials issues associated with these advanced concepts and approaches are discussed in Chapters 13 and 15. Here, the non-proliferation aspects of such systems are addressed, specifically, small and medium-sized reactors, increasing fuel burnup, fast reactors, and advanced fuels and reprocessing.

In many cases, countries or locations that would benefit the most from access to nuclear power do not possess the infrastructure to support a traditional 1,000-MWe power reactor. Smaller, possibly sealed, reactors could be appealing in addressing this issue and offer an opportunity for strengthening the non-proliferation regime through integrating advanced materials, engineered systems, and advanced monitoring concepts. Interest in small modular reactors (SMRs) is also being revived in the USA especially because such reactors may help to reduce the enormous up-front financial barriers for new power-plant construction and might also provide an effective way for the USA to regain a global role in nuclear power export. In turn, these concepts place a major burden on materials. For example, a reactor that is delivered as a complete module to a site, operated for a number of years, and simply returned for replacement offers some obvious non-proliferation benefits (no access to fresh or spent fuel during normal operation, no requirement for enrichment or reprocessing capability in the receiving nation, reduction of safeguards verification to maintaining continuity of knowledge of an item and its sealed status) and requires advanced fuels and other materials to make the lifetime of such a system practical. Toward this end, advances in integrated monitoring, advanced sensors tied to global positioning, and central control are needed to enhance remote monitoring.

Increasing the utilization of uranium fuel also provides benefits to larger power reactors in terms of longer cycle times and reduced waste (see Chapters 13 and 15 for details), and much work has been done to examine these approaches. Reactor concepts that employ a pebble-bed approach and have fuel cycling through the reactor present a challenge to monitoring, because the volume is high and fuel is mixed with non-fuel materials, with both traveling in and out of the reactor core [35]. Inert-matrix fuels are being investigated as a means to burn plutonium in the absence of ^{238}U (which creates plutonium through neutron capture) and would be compatible with existing reactor designs, including those that utilize MOX fuel [36]. Advanced monitoring equipment that is capable of operating in real time and in a harsh environment is needed.

Fast reactors (Chapter 13) present a particular challenge for maintaining the continuity of knowledge in the case of liquid-metal cooling, where standard optical monitoring of the core cannot be implemented [37]. One technology that has been investigated is based on acoustic imaging of fuel bundles in liquid sodium. Sensors like this must survive and operate in a very harsh thermal–mechanical and radiation environment. Fundamental understanding of radiation damage leading to self-healing materials, for example, could play an important role in providing additional robustness for new fuel-cycle concepts.

Some countries are also exploring pyroprocessing instead of the typical aqueous PUREX process in an attempt to provide better proliferation resistance of the reprocessed actinide mixture. The US Advanced Fuel Cycle Initiative (AFCI) examined several schemes (aqueous and pyrochemical Chapter 13) that involve co-extraction of some of the minor actinides and fission products with plutonium. Some of these schemes represent better proliferation resistance in terms of material attractiveness, but none is proliferation-proof. Significant advances in enhancing proliferation resistance during reprocessing will most likely have to await the development of new reactors and new fuels for which better proliferation characteristics are designed in at the beginning, rather than trying to retrofit existing reactors and fuels.

The evolution of the nuclear fuel cycle includes a variety of advanced fuels and reprocessing techniques, which may present opportunities to apply technical advances in materials protection, control, and accounting. In particular, bulk processing facilities will need not only material-specific instruments and techniques, but also real-time, continuous operation in a variety of operating environments. Gains can also be realized in an integrated-systems sense by relating signature information across facilities. How to quantitatively measure the actinide content of spent fuel is being researched [38]. Many advanced concepts will require remote handling facilities for fuel fabrication.

Institutional and political countermeasures

Technical measures address mostly the supply side of nuclear proliferation. Effectively preventing proliferation requires control of supply, reduction of demand, and readiness to respond. Institutional and political measures must be combined with technical measures to address all three.

Controlling supply. Controlling the spread of enrichment and reprocessing facilities reduces proliferation risks. Such efforts must consider the full range of the fuel cycle, IAEA controls, and exports. Currently, the free-market system is working adequately to provide enrichment capacity, but attempts to restrict such facilities to countries that currently have them while expanding nuclear power globally are problematic.

Serious consideration of limiting the number of countries that should be allowed to run enrichment facilities has prompted Brazil, South Africa, Australia, and others to either declare their intent to establish enrichment facilities or expand what they have.

A variety of multinational arrangements have been proposed in recent years to ensure reliable access at competitive market prices without exacerbating proliferation concerns [39]. Three categories can be identified: (1) multinational ownership and operation of uranium enrichment facilities, (2) multinational or regional fuel procurement partnerships, and (3) multinational or international fuel reserves or fuel banks. Some multinational schemes for both front- and back-end facilities were proposed in the 1970s and 1980s but never came to fruition.

There has been no case in which an NPT State party has had to shut down a commercial reactor because it could not get fuel. Still, the insistence of some states on developing their own national enrichment facilities shows the shortcoming of the market approach. The reasons for this range from economic interests to domestic politics and state pride and from protection of technology to retention of a latent nuclear-weapon option. To find a lasting solution to reducing the incentives for countries to build their own enrichment plants, it is important to pay attention to what the user states consider to be adequate guarantees rather than simply having the supplier states dictate the solution.

Some of the options in Chapter 15 to deal with spent fuel are amenable to international solutions, which could increase proliferation resistance, for example, monitored interim storage of spent fuel or regional geological repositories. However, such schemes will most likely require having some defined solution to long-term disposition. In fact, the most attractive option for many countries might be a fuel-leasing service that provides both enriched fuel and take-back of the spent fuel and waste. Only Russia is capable both technically and legally of providing such services today. Cooperation on international or regional geological disposition would also be desirable but has not gained much support to date.

Reprocessing is the most proliferation-sensitive back-end technology. The total world capacity for reprocessing is expected to exceed demand unless plutonium recycling becomes more economically attractive [40]. All current reprocessing facilities are national facilities run by governments or state-controlled corporations, and there is no experience with multinational facilities.

Many in the non-proliferation community continue to stress that the best protection on the back end is to forego reprocessing – that is, to practice only an open fuel cycle [11][41]. However, much as was the case in the late 1970s, when the Ford and Carter administrations chose to end reprocessing in the USA, some of the key nuclear states see it very differently. Of particular note are France, Russia, Japan, India, and most likely China. Consequently, this problem must be considered from a greater global perspective to better understand what the principal driving forces are for other nuclear-power countries to pursue reprocessing.

Strengthening international safeguards continues to be a critical step to providing better proliferation countermeasures. In 2009, the IAEA applied safeguards in 170 countries with safeguards agreements in place. Eighty-nine of these states had both comprehensive safeguards agreements and additional protocols in place [42]. The objective is to detect diversion of significant quantities of nuclear material at declared facilities and clandestine nuclear-weapons activities under the Additional Protocol as early as possible, or at least to introduce a measure of uncertainty into a violating country's calculus about the probability of being detected. Such nations typically aim to develop a capability as quickly as possible while keeping the probability of detection as small as possible.

Iraq's nuclear-weapons program before the Gulf War demonstrated how comprehensive safeguards agreements were inadequate because they were directed at verifying only the correctness of a country's declaration of nuclear facilities and materials. The IAEA developed the Additional Protocol, which provides greater rights of access to suspect locations, allowing inspections on short notice, environmental sampling, and remote monitoring, to help verify the completeness of the declaration. However, the Additional Protocol remains voluntary and is in force in only 104 states as of December 2010, with many key countries not having ratified it.

Environmental monitoring is a powerful tool that can help to provide effective nuclear safeguards. However, there are rich opportunities for contributions from research in analytical and nuclear chemistry and materials science to address the associated technical challenges, including developing effective remote monitoring systems, better detectors, and improved nuclear forensics capabilities.

Strengthening existing mechanisms that control supply is imperative. This must be done in conjunction with the strengthening of export control laws in potential technology-supplier countries and the development of better domestic legislative bases for export control. The success of the A. Q. Khan network in circumventing existing supply controls demonstrates that one must constantly adapt control and legal mechanisms as these networks evolve [16]. Detection and attribution are important and require enhanced technical capabilities both in detector technology and nuclear forensics.

Reducing demand. Efforts to control the supply side of proliferation are necessary, but not sufficient. The motivations for countries to seek nuclear weapons must be better understood to limit the demand for these weapons. Three models have been suggested to analyze the nuclear-weapons aspirations of a state [43][44]: the security model, the domestic-politics model, and the norms model. The security model calls for states to build nuclear weapons to increase their security against foreign threats, especially nuclear threats. The domestic-politics model posits that nuclear weapons might serve the bureaucratic or political interests of individual actors or coalitions of actors, such as the military or the nuclear establishment, who can influence the state's decision making. The norms model views nuclear decisions as also serving important symbolic functions externally – both shaping and reflecting a state's identity in relation to the international community.

Security concerns have been the primary driving factors for the acquisition of nuclear weapons, although in cases such as the UK, France, and India international aspirations and domestic factors played substantial roles [1][45][46]. The principal approaches most studies highlight for reducing demand based on insecurity are security assurances, regional conflict-resolution initiatives, nuclear-weapons-free zones (five of which exist today, with four of them spanning the entire Southern Hemisphere; see http://www.armscontrol.org/factsheets/nwfz), and efforts by the P-5 states to reduce the salience of nuclear weapons. Domestic issues are more complex to ascertain and difficult to control, so a focus on domestic politics is unlikely to prove effective or efficient in reducing demand. In contrast, international norms do play an important role. Most of the non-proliferation community is concerned that the rules-based order of the nuclear non-proliferation regime is being steadily eroded, however, so efforts must be made to strengthen such norms.

Nuclear disarmament toward a nuclear-weapons-free world as stipulated in Article VI of the NPT is important and might help to reduce demand, but many consider this a distant goal. Even working toward this goal will pose formidable technical challenges in verifying nuclear arsenals and fissile-material stockpiles, as well as providing adequate security.

Strengthening enforcement and response. Enforcement and appropriate response continue to be the weakest link in the non-proliferation fabric. It is the IAEA Board of Governors' job to report non-compliance to all members and to the Security Council and General Assembly of the United Nations for action. Sanctions are one of the main tools, but often, they have not demonstrably changed nuclear ambitions, in part because the P-5 countries have different views of the effectiveness of sanctions and different strategic interests. However, most agree that a more effective consensus must be developed to implement serious penalties for countries violating their NPT obligations. Likewise, there is general consensus that withdrawal from the NPT should involve some additional legal hurdles for those who intentionally violate their obligations while accepting nuclear assistance and then simply withdraw to build nuclear weapons, as North Korea apparently did.

One interesting aspect that supports the non-proliferation regime is that countries that have well-developed and globally connected economies might simply have too much to lose to violate their obligations or break their agreements.

14.5 Summary

Nuclear non-proliferation issues, like those for nuclear energy and nuclear-waste disposal, depend on institutional and political considerations in addition to the technical issues that are the focus of this book. The technical non-proliferation challenges in general and the materials challenges in particular depend directly on the technical nature of the nuclear fuel cycle, reactor design and operations, and recycling and waste disposal, as discussed in Chapters 13 and 15. The institutional and political issues are similarly intertwined. There are technical measures that will increase proliferation resistance, but none will make reactors and fuel cycles proliferation-proof. Hence, nuclear non-proliferation requires detailed analysis of the interplay of technical, institutional, and political countermeasures to proliferation.

Despite the many serious challenges that the non-proliferation regime faces today, the non-proliferation treaty and regime appear to have a "hidden robustness" [1] that prevents the world from sliding toward a "tipping point" of unchecked proliferation. However, a global expansion of nuclear power can be managed only by the proper deployment of technical, institutional, and political countermeasures to make the benefits of an expansion of nuclear power outweigh the risks.

14.6 Questions for discussion

1. Which nuclear materials used or produced in a nuclear energy program might be used for nuclear weapons? Which materials properties make these materials attractive or render them unattractive for nuclear-weapon use. Explain the production mechanisms and the civil or possible military use of these materials.

2. Discuss the extent to which benefits of nuclear power with regard to proliferation might be outweighed by proliferation risks. How does the analysis change if nuclear power expands in the future?

3. Explain the threats that the use or production of weapon-usable nuclear materials can pose for international security.

4. Which choice of nuclear fuel-cycle facilities was most important for the capability to produce fissile materials in the cases of North Korea and Iran?

5. Which countries possess large civilian stockpiles of fissile materials? How and why was this material produced?

6. What technical or political and institutional mechanisms are proposed to reduce the risk of proliferation? How can the proliferation risks be managed?

7. In what regard can materials science research contribute to address proliferation risks?

8. Try to explain, analyze, and discuss the complex interaction of science, technology, and society that is typical for our modern world (climate, biotechnology, cyberspace), taking nuclear energy and proliferation as examples.

14.7 Further reading

- A comprehensive treatment of all aspects of the use of nuclear energy, especially with chapters on nuclear proliferation, is given in **D. Bodanski**, 2004, *Nuclear Energy: Principles, Practices, and Prospects*, 2nd edn., New York, Springer-Verlag, 2004.

- The standard textbook for processes developed to concentrate, purify, separate, and safely store fissile materials is **M. Benedict**, **T. H. Pigford**, and **H. W. Levi**, 1981, *Nuclear Chemical Engineering*, 2nd edn., New York, McGraw Hill.

- A comprehensive assessment on proliferation dangers and policies to control them, with data on countries that have or have given up nuclear weapons, is given in **J. Cirincione**, **J. B. Wolfsthal**, and **M. Rajkumar**, 2005, *Deadly Arsenals, Nuclear, Biological, and Chemical Threats*, 2nd edn., Washington, DC, Carnegie Endowment for International Peace.

- A comprehensive treatment of cutting-edge technologies used to trace, track, and safeguard nuclear material is available in **J. E. Doyle** (ed.), 2008, *Nuclear Safeguards, Security, and Non-proliferation: Achieving Security with Technology and Policy*, Burlington, MA, Butterworth-Heinemann.

- This book, which is fairly technical but understandable for nonscientists, gives an overview of how nuclear weapons explode and how nuclear reactors operate, and presents options on treating nuclear waste: **R. L. Garwin** and **G. Charpak**, 2001, *Megawatts and Megatons: A Turning Point in the Nuclear Age*, New York, Knopf.

- A discussion between two of the most well-known scholars on international relations concerning whether more or fewer nuclear weapons would be better: **S. Sagan** and **K. Waltz**, 2002, *The Spread of Nuclear Weapons, A Debate Renewed*, New York, W. W. Norton & Company.

- This comprehensive overview provides information about both the basic technologies and the international efforts to prevent the spread of nuclear weapons and sensitive technologies: **R. F. Mozley**, 1998, *The Politics and Technology of Nuclear Proliferation*, Washington, DC, University of Washington Press.

- This comprehensive study discusses the interrelated technical, economic, environmental, and political challenges facing a significant increase in global utilization of nuclear power: **MIT**, 2003, *The Future of Nuclear Power* Washington, DC, MIT (updated 2009), http://web.mit.edu/nuclearpower/.

- This comprehensive overview on uranium enrichment technology and its relevance for proliferation is still a must read in the field: **A. Krass**, **P. Boskma**, **B. Elzen**, and **W. A. Smit**, 1982, *Uranium Enrichment and Nuclear Weapon Proliferation*, Basingstoke, Taylor & Francis.

- Although the world inventories have changed, this book provides timeless insights and methods that are extremely useful for the assessment of inventories of fissile materials: **D. Albright**, **F. Berkhout**, and **W. Walker**, 1997, *Plutonium and Highly Enriched Uranium 1996. World Inventories, Capabilities and Policies*, Oxford, Oxford University Press.

- A collection of articles by world-leading scholars addressing the challenges of the use of nuclear power with several articles on various aspects of the interlinkage of nuclear power and proliferation is available in **S. E. Miller** and **S. D. Sagan** (eds.), 2009, *The Global Nuclear Future, Daedalus*, **138**(1:4), 1–167 S. E. Miller and S. D. Sagan, (eds.), 2010, *The Global Nuclear Future. Daedalus*, **139** (2:1), 1–140.

- The most comprehensive source of background and reference material on the Nuclear Non-proliferation Treaty and its associated regime is **J. Simpson**, **J. Nielsen**, and **M. Swinerd**, 2010, *The NPT Briefing Book*, Southampton, The Mountbatten Centre for International Studies.

- A bibliography of arms-control literature is available in **A. Glaser** and **Z. Mian**, 2008, "Resource letter PSNAC-1. Physics and society. Nuclear arms control," *Am. J. Phys.*, **76**(1), 5–14.

14.8 Appendix: Emerging or aspiring nuclear-weapon states

Twelve countries have developed or acquired nuclear weapons since the beginning of the atomic age [46]. All states that have built nuclear weapons appeared to have placed the need for nuclear weapons before the desire for commercial nuclear power, although India appears to have developed both, concurrently. Several other countries had active nuclear-weapon programs, but abandoned them before building such weapons.

Although commercial nuclear power played no direct role in any of these cases, most of them have links to the potential pathways described in Section 14.3. The list is not comprehensive and readers should refer to Cirincione *et al.* [6] for more details.

14.8.1 India

India used research facilities, which had been built ostensibly for civilian use with foreign assistance (e.g., the CIRUS reactor), to produce weapons-grade plutonium for its "peaceful" nuclear test in 1974. After a long hiatus, India conducted five nuclear tests in two days in 1998 and declared itself a nuclear power. India used its reprocessing plants in Kalpakkam and Tarapur for military and civilian purposes.

14.8.2 Pakistan

Pakistan used centrifuge blueprints, stolen from URENCO in Belgium by A. Q. Khan in 1976, and developed a wide web of nuclear suppliers, building a clandestine HEU weapon capability, which it demonstrated with six nuclear tests in response to India's tests in 1998 [46].

14.8.3 North Korea

North Korea used the Soviet Atoms for Peace assistance to train its nuclear scientists and establish nuclear research facilities, which was followed by building indigenous facilities for the full plutonium fuel cycle under the guise of a civilian program, after having signed the NPT. It eventually withdrew from the NPT and tested nuclear weapons in 2006 and 2009 [46]. In 2010, it also unveiled a modern, small industrial-scale centrifuge enrichment facility.

14.8.4 Iran

Iran used the A. Q. Khan network and its own worldwide procurement network to develop clandestine uranium centrifuge facilities. Once these facilities were discovered in 2002, Iran put in place all the capabilities to produce weapons-grade material by building enrichment facilities ostensibly for commercial use, claiming the right to peaceful nuclear energy under Article IV of the NPT. Combined with prior work on weaponization [46], Iran is believed to possess the latent-nuclear-weapon option, which it could exercise quickly if it chose to do so, albeit requiring abrogation of IAEA safeguards and its withdrawal from the NPT. Iran will also be capable of producing weapons-grade plutonium at the level of nearly two bombs per year with its heavy-water research reactor under construction in Arak.

14.8.5 Syria

It is suspected that North Korea provided Syria with a plutonium production reactor, which was destroyed by an Israeli air strike in 2007 [46]. Although it was built on Syrian soil, it remains unclear who the customer was for the plutonium that was to be produced by this reactor.

14.9 References

[1] **K. M. Campbell**, **R. J. Einhorn**, and **M. B. Reiss** (eds.), 2004, *The Nuclear Tipping Point: Why States Reconsider Their Nuclear Choices*, Washington, DC, Brookings Institution Press.

[2] **S. E. Miller** and **S. D. Sagan** (eds.), 2009, *The Global Nuclear Future, Daedalus*, **138** (1:4), 1–167.

[3] **S. E. Miller** and **S. D. Sagan** (eds.), 2010, *The Global Nuclear Future, Daedalus*, **139**(2:1), 1–140.

[4] **D. E. Lilienthal**, **C. I. Barnard**, **C. A. Thomas**, **J. R. Oppenheimer**, and **H. A. Winne**, 1946, *Acheson–Lilienthal Report: Report on the International Control of Atomic Energy*, Washington, DC, US Government Printing Office.

[5] **D. Kearn**, 2010, "The Baruch plan and the quest for atomic disarmament," *Diplomacy Statecraft*, **21**, 41–67.

[6] **J. Cirincione**, **J. B. Wolfsthal**, and **M. Rajkumar**, 2005, *Deadly Arsenals, Nuclear, Biological, and Chemical Threats*, 2nd edn., Washington, DC, Carnegie Endowment for International Peace.

[7] **IAEA**, 2002, *Safeguards Glossary 2001 Edition*, International Nuclear Verification Series No. 3, IAEA/NVS/3 Vienna, IAEA, Table II.

[8] **DOE**, 2002, *Restricted Data Declassification Decisions 1946 to the Present* (RDD-8), Washington, DC, US Department of Energy, Section II Materials, I Plutonium, §33, p. 23.

[9] **C. Bathke**, **B. Ebbinghaus**, **B. Sleaford** *et al.*, 2009, "Attractiveness of materials in advanced nuclear fuel cycles for various proliferation and theft scenarios," in *Proceedings of Global 2009*, paper 9544.

[10] **V. Gilinsky**, **M. Miller**, and **H. Hubbard**, 2006, "A fresh examination of the proliferation dangers of light water reactors," in *Taming the Next Set of Strategic Weapons Threats*, ed. **H. Sokolski**, Carlisle, PA, Strategic Studies Institute.

[11] **International Panel on Fissile Materials**, 2009, *Global Fissile Material Report 2009: A Path to Nuclear Disarmament*.

[12] **Los Alamos press release**, 1997, *Blast from the Past: Los Alamos Scientists Receive Vindication*, Los Alamos, NM.

[13] **D. Albright** and **C. Gay**, 1997, "Proliferation: a flash from the past," *Bull. Atomic Scientist*, **53**(6), 15–17.

[14] **D. Bodanski**, 2004, *Nuclear Energy: Principles, Practices, and Prospects*, 2nd edn., New York, NY, Springer-Verlag, p. 214.

[15] **International Atomic Energy Agency**, 2005, *Thorium Fuel Cycle – Potential Benefits and Challenges*, IAEA-TECDOC-1450, Vienna, IAEA.

[16] **C. Braun** and **C. Chyba**, 2004, "Proliferation rings: new challenges to the nuclear non-proliferation regime," *Int. Security*, **29**(2), 5–49.

[17] **Illicit Trafficking Database (ITBD)**, 2007, *Fact Sheet for 2007*, http://www-ns.iaea.org/downloads/security/ITDB_Fact_Sheet_2007.pdf.

[18] **S. S. Hecker**, 2006, "Toward a comprehensive safeguards system: keeping fissile materials out of terrorists' hands," *Ann. Am. Acad. Political Social Sci.*, **607**, 121–132.

[19] **C. D. Ferguson** and **W. C. Potter**, 2010, *The Four Faces of Nuclear Terrorism*, London, Routledge.

[20] **G. Bugliarello** (ed.), 2010, *Nuclear Dangers*, issue of *The Bridge*, **40**(2).

[21] **S. E. Miller** and **S. D. Sagan**, 2009, "Alternative nuclear futures," in S. E. Miller and S. D. Sagan (eds.) *The Global Nuclear Future. Daedalus*, **138**(1:4), 126–137.

[22] **NASAP**, 1980, Nuclear Proliferation and Civilian Nuclear Power. *Report of the Non-proliferation Alternative Systems Assessment Program (NASAP)*, Washington, DC, US Department of Energy, vols. 1–9.

[23] **IAEA**, 1980, *International Nuclear Fuel Cycle Evaluation. INFCE summary volume/International Nuclear Fuel Cycle Evaluation*, Vienna, IAEA.

[24] **Center for Global Security Research**, Lawrence Livermore National Laboratory, 2000, *Proliferation-Resistant Nuclear Power Systems: A Workshop on New Ideas*, UCRL-JC-137954, CGSR-2000–001.

[25] **Generation IV International Forum (GIF)**, 2006, *Evaluation Methodology for Proliferation Resistance and Physical Protection of Generation IV Nuclear Energy Systems*, GIF/PRPPWG-2006/005.

[26] **IAEA**, 2008, *Guidance for the Application of an Assessment Methodology for Innovative Nuclear Energy Systems*, vol. 1, IAEA-TECDOC-1575 Rev. 1, Vienna, IAEA.

[27] **J. E. Doyle** (ed.), 2008, *Nuclear Safeguards, Security, and Non-proliferation: Achieving Security with Technology and Policy*, Burlington, MA, Butterworth-Heinemann.

[28] **G. F. Knoll**, 2000, *Radiation Detection and Measurement*, 3rd edn., New York, Wiley.

[29] **D. Reilly**, **N. Ensslin**, and **H. A. Smith** (eds.), 1991, *Passive Nondestructive Assay of Nuclear Materials*, Los Alamos National Laboratory report LA-UR-90–732 NUREG/CR-5550, and 2007 Addendum, http://www.lanl.gov/orgs/n/n1/panda/index.shtml.

[30] **H. O. Menlove**, **J. B. Marlow**, and **M. T. Swinhoe**, 2010, ^{3}*He Neutron Detector Technical Requirements*

for Safeguards and Nuclear Non-proliferation, LA-UR-10–02648.

[31] **J. Glodo**, **W. A. Higgins**, **E. V. D. van Loef**, and **K. S. Shah**, 2008, "Scintillation properties of 1 inch Cs$_2$LiYCl$_6$:Ce crystals," *Trans. Nucl. Sci.*, **55**(3), 1206–1209.

[32] **A. Canning**, **R. Boutchko**, **A. Chaudhry**, and **S. E. Derenzo**, 2009, "First-principles studies and predictions of scintillation in Ce-doped materials," *Trans. Nucl. Sci.*, **56**(3), 944–948.

[33] **C. R. Stanek**, **M. R. Levy**, **B. P. Uberuaga**, **K. J. McClellan**, and **R. W. Grimes**, 2008, "Defect identification and compensation in rare earth scintillators," *Nucl. Instrum. Methods B*, **266**, 2657–2664.

[34] **R. W. Grimes**, **R. J. M. Konings**, and **L. Edwards**, 2008, "Greater tolerance for nuclear materials," *Nature Mater.* **7**(9), 683.

[35] **P. C. Durst**, **D. Beddingfield**, **B. Boyer** et al., 2009, *Nuclear Safeguards Considerations for the Pebble Bed Modular Reactor (PBMR)*, INL/EXT-09–16782, Idaho National Laboratory.

[36] **IAEA**, 2006, *Viability of Inert Matrix Fuels in Reducing Plutonium Amounts in Reactors*, IAEA-TECDOC-1516, Vienna, IAEA.

[37] **D. H. Beddingfield** and **M. Hori**, 2007, in *Proceedings of the JAEA–IAEA Workshop on Advanced Safeguards Technology for the Future Nuclear Fuel Cycle*, LA-UR-07–6878.

[38] **S. J. Tobin**, **M. L. Fensin**, **B. A. Ludeuig** et al., 2009, "Determining plutonium mass in spent fuel with non-destructive assay techniques – preliminary modeling results emphasizing integration among techniques," in *Proceedings of Global 2009*, Paper 9303.

[39] **T. Rauf**, 2010, "New approaches to the nuclear fuel cycle," in *Multinational Approaches to the Nuclear Fuel Cycle*, in eds. **C. McCombie** and **T. Isaacs**, Cambridge, MA, American Academy of Arts and Sciences, p. 25.

[40] **M. Bunn**, **S. Fetter**, **J. Holdren**, and **B. van der Zwaan**, 2003, *The Economics of Reprocessing vs. Direct Disposal of Spent Nuclear Fuel. Report for Project on Managing the Atom*, Cambridge, MA, Belfer Center for Science and International Affairs, Harvard Kennedy School.

[41] **E. Lyman** and **F. N. von Hippel**, 2008, "Reprocessing revisited: the international dimensions of the global nuclear energy partnership," *Arms Control Today*, April.

[42] **IAEA**, 2010, *Annual Report – 2009*, GC(54)/4.

[43] **S. D. Sagan**, 1996/7, "Why do states build nuclear weapons? Three models in search of a bomb," *Int. Security* **21**(3), 54–86.

[44] **S. D. Sagan**, 2000, "Rethinking the causes of nuclear proliferation: three models in search of a bomb," in *The Coming Crisis: Nuclear Proliferation, U.S. Interests, and World Order*, ed. **V. A. Utgoff**, Cambridge, MA, MIT Press, pp. 17–50.

[45] **S. Singh** and **C. R. Way**, 2004, "The correlates of nuclear proliferation: a quantitative test," *J. Conflict Resolution*, **48**, 859–884.

[46] **M. Fuhrmann**, 2009, "Spreading temptation," *Int. Security*, **34**(1), 7–41.

15

Nuclear-waste management and disposal

Rodney C. Ewing[1] and William J. Weber[2]

[1]University of Michigan, Ann Arbor, MI, USA
[2]University of Tennessee, Knoxville, TN, USA

15.1 Focus

The resurgence of nuclear power as an economical source of energy, plus as a strategy for reducing greenhouse-gas (GHG) emissions, has revived interest in the environmental impact of the nuclear fuel cycle. Just as GHG emissions are the main environmental impact of fossil-fuel combustion, the fate of nuclear waste determines whether nuclear power is viewed as an environmentally friendly source of energy. Not only must nuclear materials be designed that can operate under the extreme conditions within a reactor, but also new materials must be developed that can contain radionuclides for hundreds of thousands of years and prevent their release to the biosphere.

15.2 Synopsis

A variety of nuclear fuel cycles exists, each with its own types and volumes of nuclear waste [1]. An open nuclear fuel cycle envisions the direct geological disposal of the used nuclear fuel, mainly UO_2. A closed fuel cycle uses chemical processing of the used fuel to reclaim fissile material, namely ^{235}U and ^{239}Pu, which is then fabricated into a new, mixed-oxide (MOX) fuel that provides additional power in a second cycle of irradiation and fission in a nuclear reactor. Only a limited number of cycles of reprocessing can be applied, because of changes in the nuclide composition of the used fuel; thus, MOX fuel eventually requires geological disposal as well. In addition, the chemical processing itself creates waste streams that consist mainly of fission-product elements that are highly radioactive (e.g., ^{137}Cs) or have very long half-lives (e.g., ^{135}Cs), as well as transuranium elements (e.g., ^{237}Np) that form in the fuel as a result of neutron-capture and decay reactions. For all of these radionuclides, robust and extended isolation from the environment is required. More advanced "symbiotic" reprocessing technologies envision the separation of those radionuclides that can be transmuted in specially designed fast reactors (e.g., ones that utilize higher-energy neutrons to fission minor actinides, such as Np, Am, and Cm). Regardless of the type of fuel cycle envisioned, materials must be designed that can incorporate and isolate all residual radionuclides from the environment until they have decayed to safe levels of radioactivity. Nuclear-waste forms must be able to incorporate complex mixtures or individual radionuclides, survive the effects of radiation and self-heating caused by radioactive decay, and resist alteration and corrosion in a variety of geological environments over very long periods. The challenge for materials scientists is to design materials compatible with remote fabrication in high radiation fields. The response of these materials to radiation fields and to the geochemical and hydrological conditions in a geological repository must be well enough understood to project their performance over hundreds of thousands of years [2].

15.3 Historical perspective

15.3.1 Nuclear waste from defense programs

The first radioactive wastes were produced during World War II in the USA by the extensive array of nuclear reactors and reprocessing facilities built as part of the Manhattan Project to produce fissile nuclides, namely ^{239}Pu and enriched ^{235}U, for nuclear weapons, including those exploded over Hiroshima and Nagasaki, Japan, in August 1945. After World War II, the Cold War (1947–1991) saw a rapid expansion of the nuclear weapons complex in the USA. At its peak, the complex consisted of 16 major facilities, including vast tracts of land at the Nevada Test Site, the Hanford site in Washington, the Savannah River Site in South Carolina, and the Idaho National Laboratory near Idaho Falls [3]. At the Hanford site alone, there were nine plutonium-production reactors and five reprocessing plants. The USSR established a similar array of nuclear facilities. By the late 1980s, five countries had tested nuclear weapons, and the global inventory of nuclear weapons had grown to ~70,000. Since then, a series of Strategic Arms Reduction Treaties has significantly reduced the inventory of nuclear weapons (<5,000 in the USA), and the inventory of plutonium and highly enriched uranium from dismantled nuclear weapons has led to the need to develop materials for the immobilization and geological isolation of plutonium.

This environmental legacy from the Cold War is a wide variety of nuclear wastes [4]:

- highly radioactive fission products and transuranium elements from reprocessing, low-activity waste from laboratory tests and production, and hazardous chemical wastes;
- vast volumes of contaminated groundwater, surface water, and soils;
- nuclear materials, such as nuclear fuel from submarines and nuclear target materials; and
- facilities that are contaminated by radioactive elements.

Most of the environmental risk and cost are associated with the **high-level waste (HLW)** from reprocessing that is stored in single- and double-shell tanks and silos at Hanford, WA, Savannah River, SC, and Idaho National Laboratory. The single-shell tanks, originally built in the 1940s and 1950s, had an expected life of 10–20 years, and some of them are leaking, and some waste has been transferred to double-shell tanks. Present plans call for the **vitrification**, or transformation into a glass, of this HLW. At the Savannah River Site, approximately 3,000 metal canisters have been filled with vitrified HLW. At the Hanford site, a vitrification plant is under construction. The vitrified HLW will require geological disposal.

15.3.2 Nuclear waste from nuclear power plants

Although the legacy waste from defense programs accounts for a considerable volume and amount of radioactivity (Table 15.1), the largest source of radioactive material is the used nuclear fuel generated by commercial nuclear power plants. Typically, a nuclear reactor generates ~20 tonnes (t) of used nuclear fuel per year. Worldwide, the approximately 430 nuclear reactors have generated a global inventory of about 270,000 tonnes of heavy metal (tHM or mTHM), which increases by ~10,000 tHM per year. For typical burnups [40 megawatt-days (MW-d) per kg of U], approximately 1% of the uranium is converted into transuranium elements, mainly Pu. The global inventory of Pu is now ~2,000, growing by 70–90 t per year, mostly in used nuclear fuel [5]. The Pu and U can be reclaimed from the used fuel by chemical processing, to be used either to generate energy from a MOX fuel or to produce nuclear weapons. Removing the transuranium elements from the used fuel can also reduce the time frame considered to be important for geological disposal and change the thermal history of a repository. Other strategies, such as an open fuel cycle, reduce the proliferation risk by direct disposal of the used nuclear fuel without processing [6], but this requires long-term containment and an understanding of the behavior of actinides in the environment [7].

15.4 Nuclear-waste forms

15.4.1 Importance of nuclear-waste forms

Because of the long periods required for the isolation of nuclear waste from the environment, one of the tenets of nuclear-waste management is that this containment be accomplished by a series of barriers, engineered and natural. This "Russian-doll" concept of concentric barriers provides confidence in the overall performance of the disposal system over time scales of hundreds of thousands of years. **Engineered barriers** consist of the material incorporating the nuclear waste, the waste form, the metal waste package or canister, and an overpack or backfill of material designed to slow or prevent access of water to the metal waste package, and also to limit release of radionuclides from the repository. **Natural barriers** are the inherent geochemical and hydrological properties of the geology of a repository. Generally, this means selecting a site with redox conditions that reduce the solubility of radionuclides in groundwater, a high sorptive capacity of the mineral surfaces of the surrounding rock, and very low rates of groundwater flow.

Immobilization of radionuclides into chemically and physically durable materials increases the safety of

Table 15.1. Summary of estimated nuclear-waste inventories in the USA in 2010

Spent nuclear fuel (commercial)	61,800 tHM
	39,800 MCi
Spent nuclear fuel (weapons programs)	2,500 tHM
High-level waste (reprocessing)	380,000 m^3
	2,400 MCi
Buried waste (LLW)	6.2 million m^3
	50 MCi
Excess nuclear materials	
Highly enriched uranium	174 t
Plutonium	
Weapons-capable	38.2 t
Not weapons-usable without processing	14.3 t
Depleted uranium as UF$_6$	700,000 t
^{137}Cs and ^{90}Sr separated from HLW in capsules as CsCl and SrF$_2$	90 GCi
Uranium mine and mill tailings	438 million m^3
	3,000 MCi
Contaminated soil	30–80 million m^3
Contaminated water	1,800–4,700 million m^3

After [1]. HLW, high-level waste; LLW, low-level waste; ci, curie $= 3.7 \times 10^{10}$ Bq tons; tHM, tonnes of heavy metal.

geological disposal systems, because the radioactivity is initially entirely contained within the nuclear-**waste form**. Waste forms are important at every stage and facilitate the handling, transportation, storage, and disposal of radioactive wastes. Immobilization of HLW and actinides is achieved by atomic-scale incorporation into the structure of a suitable matrix (typically glass or a crystalline ceramic) so that the radionuclides are "captured" and thus unable to reach the biosphere. The key to assessing the long-term performance of nuclear-waste forms is to develop a fundamental understanding of waste form materials in dynamic environments in which the thermal and radiation fields are changing and are coupled to evolving geochemical and hydrological conditions.

15.4.2 Designing nuclear-waste forms

Waste forms serve a variety of purposes, from solidifying liquid and powdered waste in order to facilitate transportation and storage to providing the initial barrier to

radionuclide release in a geological repository. Designing materials for the latter purpose is the greatest challenge for materials scientists. The general requirements include the following.

Atomic-scale incorporation of radionuclides

Nuclear-waste streams can vary from being chemically complex (e.g., the HLW in the tanks at Hanford and Savannah River) to very pure single radionuclides (e.g., the ^{239}Pu from dismantled nuclear weapons). The waste form must be able to accommodate radionuclides that have a wide variety of oxidation states and atomic sizes. Additionally, radioactive decay will cause the composition of the waste to change over time; for example, ^{137}Cs decays to ^{137}Ba; ^{90}Sr decays to ^{90}Y and then to ^{90}Zr [8][9]. As a result of α-decay events, the He content of the material increases over time [8], in some cases forming bubbles. In other cases, atomic-scale incorporation is not possible; thus, coated or composite materials are used to encapsulate the radioactive phase.

High waste loading

The material must be able to accommodate a significant amount of waste (typically 20%–35 wt%) in order to minimize volume, thus reducing the space used in a geological repository. This is an important economic consideration, since a single waste package costs in excess of $1 million.

Radiation stability

Depending on the composition of the waste, the material will be subjected to a wide variety of radiation sources over a range of temperatures [10][11]. Principal sources of radiation in HLW are the β-decay of fission products (e.g., ^{137}Cs) and the α-decay of the actinide elements (e.g., U, Np, and Am). Both types of radiation can lead to physical and chemical changes in the waste form, but by very different processes, namely ionization and electronic excitation versus elastic, ballistic interactions: β-decay produces energetic β-particles, very-low-energy recoil nuclei, and γ-rays, whereas α-decay produces energetic α-particles, energetic recoil nuclei, and some γ-rays. The waste form should be highly tolerant with respect to radiation effects under expected temperature conditions [8].

Durability

Waste-form durability generally refers to physical and chemical integrity in the repository environment [8]. Physically, a waste form should have thermal phase stability, mechanical integrity, and a sufficiently high thermal conductivity to prevent unwanted high temperatures. Chemical durability is the waste forms' resistance to aqueous alteration and dissolution under repository-relevant conditions. The leach rate ($g\,m^{-2}$ per day) is a standard measure of radionuclide release into solution, but there are different release mechanisms. Ion exchange between the solid and liquid or incongruent dissolution (in which the composition of the solute in solution does not match that in the solid) can lead to the selective release of radionuclides. The low solubility of some radionuclides, particularly the actinides, can help control the release rate. Also, the corrosion and alteration mechanisms might change over time as the thermal and radiation fields decrease in intensity.

Natural and anthropomorphic analogs

Because direct laboratory testing of waste forms over relevant time scales is not possible, the availability of mineral or natural glass analogs of the waste forms can provide an important benchmark indication of the long-term performance of the material in the natural environment [12]. One of the unique features of this field of materials science is that it combines laboratory studies that extend for months to years with studies of natural materials that could be hundreds of millions of years old. In some cases, studies of ancient, man-made glasses also provide good analogs for waste-form glasses [13].

Avoidance of criticality

There must be careful consideration of the possibility of attaining a critical mass of **fissile material**, such as plutonium. This can be prevented by reducing actinide concentrations or by incorporating neutron absorbers, such as Gd and Hf, into the structure of the waste form.

Remote processing

For highly radioactive waste streams, such as HLW from reprocessing, much of the processing must be done remotely in a well-shielded environment. Fabrication must be accomplished under conditions that are easily maintained, using well-established methods to minimize both exposure of workers to radiation and the capital cost of the plant.

In general, the emphasis has been on processing technologies that can be operated remotely and handle large volumes of chemically complex liquid waste; hence, vitrification has been the standard processing technology in the USA, UK, France, Belgium, and Japan [14]. However, with the development of advanced fuel cycles and new reprocessing technologies, there is increased emphasis on designing new materials for particular waste streams or even individual radionuclides, such as ^{239}Pu, ^{99}Tc, and ^{129}I [15]. In this case, the durability of the waste form can be designed to match the specific characteristics of the radionuclide, including crystal chemistry, half-life, radiotoxicity, and mobility in the geological environment. Because the long-term performance of the waste form is sensitive to the geological conditions, there is also the opportunity to design waste forms whose properties match the geological conditions (groundwater composition, flow rate, and redox conditions), hence reducing radionuclide release and mobility.

15.4.3 Waste-form types

During the past 30 years, a wide variety of materials has been investigated as potential nuclear-waste forms [8][16].

Glass (silicate and phosphate)

Many different glass compositions have been developed, representing various compromises among a high solubility for radionuclides, a reasonably low melting temperature ($>1,200\,°C$) for production, and a low leach rate in repository environments. Most waste glasses are borosilicate glasses because the addition of boron lowers the melting temperature. Compositions are typically in the sodium borosilicate system, with minor additions of network-modifying oxides, such as those of Al, Li, Ca, and Zn. There has also been some work, mainly in Russia, on the use of phosphate glasses that have lower melting temperatures and viscosities than those of the borosilicate glass. A Pb–Fe phosphate glass was also developed [17][18]. Special glass compositions, such as the lanthanum borosilicate glass, have been developed for actinides. Because of their higher chemical durability, high-silica glasses have also been studied; however, because of their higher melting temperature, these glasses are usually produced by a sintering process [19].

Glass ceramics

Because some crystalline phases can incorporate specific radionuclides and are known to have a high durability, a polycrystalline glass ceramic can be made by adjusting the thermal treatment of the glass so as to cause the controlled bulk crystallization of specific phases. As an example, a **sphene** ($CaTiSiO_5$) glass ceramic has been proposed for the immobilization of HLW. Sphene is a durable phase that can incorporate a wide variety of radionuclides, such as U, Th, ^{90}Sr, and rare-earth elements [20].

Crystalline ceramics (single-phase and polyphase)

Since the earliest days of concern about the disposal and fate of nuclear waste, crystalline materials have been considered for the immobilization of a variety of nuclear-waste streams. The earliest proposal for the use of a crystalline ceramic, i.e., clay, as a nuclear-waste form was made in 1953 for the HLW at Hanford [21].

Single-phase waste forms can be used for the incorporation of separated radionuclides (e.g., ^{239}Pu from dismantled nuclear weapons) or for more chemically complex waste streams (e.g., HLW). As the waste stream composition becomes more complicated, it is generally necessary for the atomic structure of the waste-form phase to have multiple cation and anion sites that can accommodate a wider variety of radionuclides. **Zircon**, $ZrSiO_4$, has been proposed as a nuclear-waste form for the nearly pure Pu from dismantled nuclear weapons [22]. It is an extremely durable mineral that is commonly used for U/Pb age-dating, since high uranium concentrations (up to 20,000 ppm) can be present [2]. **Monazite**, $CePO_4$ or $LaPO_4$, on the other hand, can incorporate a wide variety of radionuclides and toxic metals into its structure. As a result, monazite has been proposed as a single-phase ceramic to incorporate chemically complex combinations of radionuclides [23].

Polycrystalline waste forms consist of an assemblage of crystalline phases. Individual phases are selected for the incorporation of specific radionuclides, and the proportions of these phases vary depending on the composition of the waste stream. **Synroc** is a dense, multi-phase titanate-based waste form designed for the immobilization of HLW, with phases based on mineral analogs [24]. The interest in this multi-phase crystalline

Table 15.2. Actinide waste-form phases that have been investigated [22][25]

Simple oxides	
zirconia	ZrO_2
uraninite	UO_2
thorianite	ThO_2
Complex oxides	
pyrochlore	$(Na,Ca,U)_2(Nb,Ti,Ta)_2O_6$
murataite	$(Na,Y)_4(Zn,Fe)_3(Ti,Nb)_6O_{18}(F,OH)_4$
zirconolite	$CaZrTi_2O_7$
Silicates	
zircon[a]	$ZrSiO_4$
thorite[a]	$ThSiO_4$
garnet[a]	$(Ca,Mg,Fe^{2+})_3(Al,Fe^{3+},Cr^{3+})_2SiO_4$
britholite[a]	$(Ca,Ce)_5(SiO_4)_3(OH,F)$
titanite	$CaTiSiO_5$
Phosphates	
monazite[a]	$LnPO_4$
apatite[a]	$Ca_{4-x}Ln_{6+x}(PO_4)_y(O,F)_2$
xenotime[a]	YPO_4

[a] Long-term durability can be confirmed by studies of minerals of great age, as indicated in the table.

ceramic stems from the robust nature of the individual phases. The original Synroc phase assembly included zirconolite ($CaZrTi_2O_7$), which incorporates rare earths and actinides; perovskite ($CaTiO_3$), which incorporates rare earths, actinides, and Sr; and hollandite ($BaAl_2Ti_6O_{16}$), which incorporates Cs, Rb, Ba, and rutile (TiO_2). Although the original Synroc was designed for the immobilization of HLW [24], new Synroc variants have been developed for actinides as well [25].

Recently, there was a resurgence of interest in crystalline nuclear-waste forms because of the need to develop durable materials for the stabilization and disposal of "excess" plutonium that results from dismantling nuclear weapons or that accumulates during the reprocessing of commercially generated nuclear fuels [8][16][26][27][28]. This resurgence in interest resulted in work on minerals such as apatite, monazite, zirconolite, zircon, and pyrochlore as potential nuclear-waste forms for actinides (Table 15.2). Another recent interest has been associated with the desire to create new nuclear materials as part of new or advanced nuclear fuel cycles that chemically process the used fuel and separate the radionuclides [15].

Further details on developments in waste-form research are available from a number of recent reviews [8][25][27][28][29][30][31][32][33][34].

15.4.4 Corrosion and alteration of used nuclear fuel

Spent fuel in a reactor

All nuclear fuel cycles are based on achieving criticality, a sustained nuclear chain reaction, with a **fissionable nuclide** in a reactor. Nature has provided fissile ^{235}U; hence, all nuclear fuel cycles are initially based on uranium. Uranium consists mainly of two isotopes: ^{235}U (0.72 at%), which is fissile, and ^{238}U (99.27 at%), which is **fertile**, meaning that it can be transmuted into fissile nuclides, such as ^{239}Pu. For most light-water reactors (LWRs), the uranium fuel is enriched to 3%–5% ^{235}U as a **low-enriched uranium** (LEU) fuel. Some reactors, such as research reactors, use **highly enriched uranium** (HEU, >20% ^{235}U). Weapons-grade material (>85% ^{235}U) from dismantled nuclear weapons can be blended down for use in LWRs. Some reactors, such as the Canadian CANDU reactors that use deuterium, ^{2}H, as a **moderator** (a material that slows the neutrons to energies at which fission of ^{235}U can occur), can use natural, non-enriched uranium as the fuel. Nature has provided two fertile nuclides, ^{232}Th and ^{238}U, that can breed fissionable ^{233}U and ^{239}Pu, respectively, by neutron-capture reactions and subsequent β-decay.

In a typical LWR, the fuel is exposed to a thermal (~0.03 eV) neutron flux that causes two principal types of nuclear reactions: fission,

$$_0^1 n + {}_{92}^{235}U \rightarrow \text{fission fragments} \atop + (2-3)_0^1 n (1-2\,\text{MeV}) + \text{energy}, \quad (15.1)$$

and neutron capture and β-decay,

$$_0^1 n + {}_{92}^{238}U \rightarrow {}_{92}^{239}U \rightarrow {}_{93}^{239}Np \rightarrow {}_{94}^{239}Pu. \quad (15.2)$$

The fission fragments form a bimodal distribution of elements (the fission yield) whose atomic masses are approximately half that of the fissioned, or split, uranium atom. Although many hundreds of fission-product isotopes are formed in the reactor, most have very short half-lives and decay within days to weeks of their formation. Neutron-capture reactions, followed by β-decay, lead to the formation of transuranium elements ($Z > 92$), of which Pu is the most abundant. Hence, the Pu concentration in the fuel increases with time, and its isotopes, such as ^{239}Pu, can also undergo fission, providing up to one-third of the energy generated in a nuclear

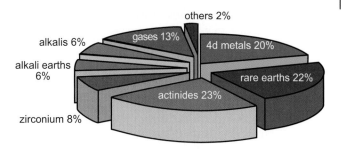

Figure 15.1. A pie diagram showing the relative proportions of the major types of fission products and transuranium elements found in spent fuel of moderate burnup (derived from [35]).

power plant. The energy of the neutron spectrum can be adjusted so that higher-energy, "fast" neutrons (>1 MeV) can be used to fission ^{238}U and the minor actinides, such as Np, Cm, and Am. The final composition of the fuel depends on the initial fuel type, the chemical composition, the level of enrichment of ^{235}U, the neutron energy spectrum, and the burnup or amount of fission. As a general guideline, a burnup of 40 MW-d per kg of U results in the conversion of 4% of the uranium to approximately 3% fission products and 1% transuranium elements (Figure 15.1). Typical burnups are in the range of 35–45 MW-d per kg of U, but there is an increasing tendency toward higher burnups.

After typical burnup, the radioactivity of the spent fuel has increased by a factor of 1 million [10^{17} becquerel per tonne of fuel (Bq t^{-1})]. One year after discharge from a reactor, the dose rate measured 1 m from the fuel assembly is one million millisieverts (mSv) per hour (the natural background dose is on the order of 3 mSv per year). A person exposed to this level of radioactivity at a distance of 1 m would receive a lethal dose in less than 1 min; hence, spent fuel must be handled remotely. This dramatic increase in radioactivity is caused by the presence of the 3–4 at% of fission products (e.g., ^{137}Cs and ^{90}Sr), transuranium elements (e.g., ^{239}Pu, ^{237}Np, and ^{241}Am), and activation products (e.g., ^{14}C, ^{60}Co, ^{63}Ni, and ^{94}Nb) in the metal of the spent fuel assemblies. The very penetrating ionizing radiation (β and γ) comes mainly from the short-lived fission products (^{137}Cs and ^{90}Sr, with half-lives of about 30 years). These fission products are mainly responsible for the thermal heat from the fuel (1,300 W t^{-1} after 40 years). The less penetrating radiation from α-decay events comes mainly from the very-long-lived actinides (e.g., ^{239}Pu and ^{237}Np, with half-lives of 24,100 years and 2.1 million years, respectively). The composition of the spent fuel, when it is initially removed from the reactor, is very complex because it contains hundreds of short-lived

Figure 15.2. Relative radioactivity of spent nuclear fuel with a burnup of 38 MW-d per kg of U as compared with the total activity of the uranium ore required to manufacture the nuclear fuel. The activity is dominated by fission products during the first 100 years and by actinides thereafter (derived from [36]).

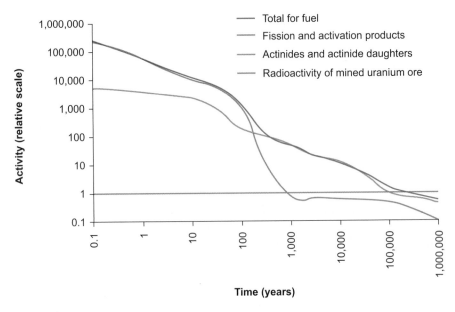

radionuclides. With time, the total radioactivity drops quickly, so that after 10,000 years the total activity is 0.01% of the activity one month after removal from the reactor [36]. The total radioactivity of the spent nuclear fuel equals the radioactivity of the uranium ore mined to create the nuclear fuel after several hundred thousand years (Figure 15.2) Over the periods relevant for geological disposal, namely hundreds of thousands of years, the radionuclides of interest are limited in number, including the isotopes of uranium and plutonium, ^{237}Np, and some of the long-lived fission products, such as ^{99}Tc and ^{129}I. The radionuclides of environmental interest will depend not only on their half-lives and toxicity, but also on their mobility under the geochemical and hydrological conditions at a specific repository site [7][36].

Structure and composition of spent fuel

Prior to irradiation, the fuel consists mainly of uranium as UO_2 or U metal. The fuel is less radioactive than the original uranium ore because all of the decay daughter products have been removed during chemical enrichment to create **yellow cake**, U_3O_8. Most commercial spent-fuel assemblies are rods composed of UO_2 compacted into cylindrical pellets 8–10 mm in diameter and 9–15 mm thick that are stacked in tubes of corrosion-resistant cladding materials, typically the zirconium alloy Zircaloy or stainless steel (Figure 15.3). The empty spaces, including the narrow annular gap between the fuel pellets and the surrounding cladding and the spaces at the ends of the fuel rods, are filled with helium gas. During reactor operation, the gap closes as the fuel expands slightly during irradiation. The fuel rods are grouped into different types of assemblies that depend on reactor design and can weigh as much as 500 kg per assembly.

The structure and composition of the various types of spent fuel have been investigated from the perspectives of both reactor operation and geological disposal [38][39]. At the end of the fuel's useful life in the reactor, about 95%–99% of the spent nuclear fuel still consists of UO_2. The balance consists of fission products, transuranium elements, and activation products (Figure 15.1), but these elements occur in many different forms [40]: (1) fission-product gases, such as Xe and Kr, that occur as finely dispersed bubbles in the fuel grains; (2) metallic fission products, such as Mo, Tc, Ru, Rh, and Pd, that occur as immiscible metallic precipitates, called ε-particles, that are micrometers to nanometers in size; (3) fission products that occur as oxide precipitates of Rb, Cs, Ba, and Zr; (4) fission products that form solid solutions with the UO_2 fuel, such as Sr, Zr, Nb, and the rare-earth elements; and (5) transuranium elements that substitute for U in the UO_2. The distribution of elements is not homogeneous within a single pellet (Figure 15.4) because of the step thermal gradient within the pellet (as high as 1,700 °C at the center of the pellet and decreasing to 400 °C at its rim). Thermal excursions during reactor operation can cause a coarsening of the grain size, as well as extensive restructuring and microfracturing. Volatile elements, such as Cs and I, migrate to grain boundaries, fractures, and the gap between the edge of the fuel pellet and the surrounding metal cladding. The burnup is also not uniform across the fuel pellet. The higher burnups at the edge of the pellet lead to higher concentrations of ^{239}Pu at the fuel edge, an increase in porosity, and polygonization of the UO_2 grains, resulting in a reduction in the size of individual grains (to ~0.15–0.3 μm), the so-called "rim effect" (Figure 15.4). Thus, the spent fuel has a complex chemistry that is the result of its initial composition, neutronic reactions, and thermal history.

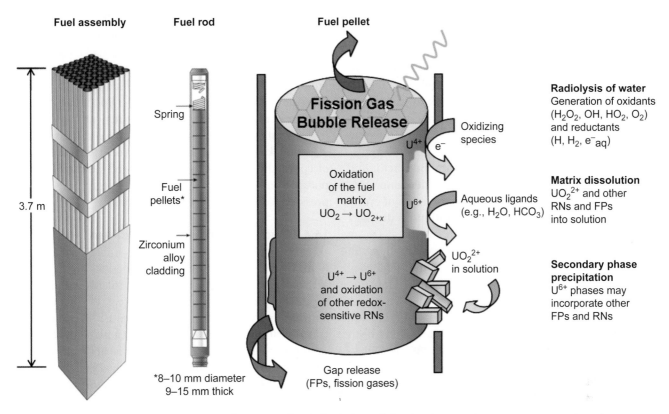

Figure 15.3. A depiction of the fundamental processes governing the alteration of spent nuclear fuel upon breach of the fuel rod (modified from [37]). FPs, fission products; RNs, radionuclides.

Figure 15.4. A schematic illustration of the microstructure of spent fuel and the distribution of actinides and fission products following burnup in a reactor [37]. Red lettering indicates elements that are released nearly instantaneously upon contact with water; blue lettering indicates elements released at lower rates. An, actinides; Ln, lanthanides. Adapted in [37] from [35] and [41].

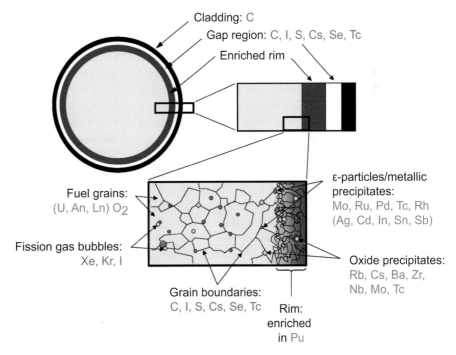

Corrosion and alteration of used nuclear fuel in a geological repository

From the perspective of geological disposal, spent nuclear fuel, UO_2, is a complex, redox-sensitive, semi-conducting, polycrystalline ceramic. Although the composition of the system is dominated by uranium, one is interested in the geochemical mobility of a limited, but chemically diverse, array of radionuclides (Figure 15.1). These radionuclides are heterogeneously distributed throughout the fuel assembly and occur in a variety of different types of phases, from inert gases to relatively stable oxides. The composition of the fuel changes with time as a result of radioactive decay (Figure 15.2), as do the thermal and radiation fields. Most assessments of the "success" of a geological repository depend on a calculation of the risk to humans at some distance from the repository at some time in the far future as stipulated by regulation. The science that supports such an analysis requires (1) a detailed knowledge of the fuel after it is removed from the reactor and as it evolves over time; (2) a thorough understanding of the potential release mechanisms as a function of evolving geochemical and hydrological conditions as the thermal and radiation fields decrease in intensity; (3) an understanding of the mobility of the radionuclides in the near- and far-field environment of the geological repository; and (4) the radiotoxicity of each radionuclide as a function of the expected pathways for human exposure. The mechanisms that cause the dissolution and alteration of spent nuclear fuel and their rates have been the focus of extensive research efforts [35][38][41][42].

The general approach in developing a "source-term" model for radionuclide release from spent nuclear fuel in a saturated medium involves the combination of many different processes that can be grouped into two stages [37].

The first stage is *instantaneous release at the time of waste-package failure*. This release is generally referred to as the instant-release fraction (IRF), which is the fraction of the inventory that is rapidly released when the metal canister is breached. The radionuclides of most interest during this rapid release are mainly the fission gases, such as Xe and Kr, and volatile elements, such as I, Cs, and Cl, that have migrated to grain boundaries and the gap between the fuel pellet and the metal cladding during reactor operation. The inventory and segregation of fission-product gases and volatile elements depends on the burnup of the fuel. The release fraction can vary significantly depending on the type of fuel. As an example, MOX fuel (a mixture of U and Pu) can experience much higher burnups along the rim of the pellet, leading to the formation of restructured Pu-rich agglomerates with much higher inventories of fission gases.

The second stage is *the much slower, long-term release that results from the alteration and dissolution of the fuel matrix, usually UO_2*. The important processes (Figure 15.3) on this time scale include (1) oxidation of the U($+4$) to U($+6$) and formation of higher-oxide structures on the fuel surface and at grain boundaries; (2) bulk dissolution of UO_2 and release of radionuclides (e.g., Pu and Np) that substitute for U; (3) dissolution of segregated oxides and immiscible metallic alloys in the fuel grains; and (4) formation of secondary alteration products, such as coffinite ($USiO_4$) under reducing conditions or U($6+$) phases under oxidizing conditions. All of these processes occur in a changing thermal and

radiation field. One of the important effects is radiolysis from the radiation, which breaks water into reactive species, such as H_2O_2, H_2, and H_3O^+. Finally, the chemical reactions depend on the groundwater compositions and flow rates. Thus, the release of radionuclides from the spent nuclear fuel can be understood only in the context of the coupled processes in the near field, which include interactions with the corrosion products of the waste package and the surrounding rock. As an example of coupled processes, the corrosion of the metal waste package will generate hydrogen that, in turn, might suppress the oxidizing effects of radiolysis.

15.4.5 Nuclear-waste glass for high-level waste

Reprocessing of used nuclear fuel generates highly radioactive waste streams that consist mainly of fission-product elements, some short-lived (e.g., ^{137}Cs and ^{90}Sr) and others long-lived (^{99}Tc, [135]Cs, and [129]I), and small concentrations, less than a fraction of a percent, of transuranic elements (e.g., [237]Np), as well as activation products (^{14}C) from the metal fuel assembly. During the past 50 years, most of the research on waste forms for HLW has focused on borosilicate glass. The ability of glass to accommodate wide variations in waste-stream composition and the ease of industrial-scale processing in heavily shielded, remotely controlled environments have been positive attributes of glass as a waste form. Over 40 years of industrial-scale experience in vitrifying radioactive waste has accrued in France, Belgium, the UK, Japan, and the USA. At the Savannah River site in South Carolina, by the end of 2010 over 3,000 canisters had been filled with vitrified high-level waste.

From a materials science perspective, the main issues have been the stability of glass in thermal and radiation environments and its long-term chemical durability. The response to thermal and radiation effects has been well summarized in two reviews [10][14]. In this section, we focus on the long-term chemical durability of glass.

Chemical durability of glass

The challenge for materials scientists is to be able to predict the behavior of waste forms over many hundreds of thousands of years in disposal environments that will evolve over time. The results of short-term experiments and computer simulations of corrosion mechanisms must be extrapolated to relevant disposal times that will stretch to hundreds of thousands of years before the level of radioactivity drops to inconsequential levels. The approach to arriving at the necessary understanding involves detailed experiments over a wide range of conditions that can last for several years [43], computational simulations of corrosion mechanisms at the atomic scale [44], geochemical modeling of coupled reactions between the waste form and the near-field geological environment [45][46], field studies of synthetic glasses that are buried and recovered after some extended period [47], and studies of the corrosion of natural glasses that are considered to be analogs for borosilicate glasses [12]. Estimates of the long-term behavior of glass in a geological repository are based on a synthesis of observations and simulations by all of these different observations and approaches.

From hundreds of experiments and studies of natural glasses, a general picture of the glass-corrosion process has been developed [48][49]. The initial glass–water reaction begins with the selective removal of alkali ions from the surface and near-surface region of the glass, by the coupled exchange of diffusing H^+ ions into the glass and the loss of alkali ions. The result is a silica-rich hydrated surface layer that is just a few nanometers thick, consisting of a mixture of silanol groups (—Si—OH) and molecular water. This hydrated glass network dissolves at a constant rate while the diffusion of water and ion exchange are occurring. Within just a few hours, these processes reach the same rate, resulting in congruent dissolution of the glass. Important species, including H_3BO_3, as well as alkali and alkaline-earth ions (e.g., ^{135}Cs, ^{137}Cs, and ^{90}Sr), are released from the outer surface of the glass. For these elements, the dissolution process is essentially congruent, but actinides and lanthanides, which have very low solubilities, precipitate onto the surface of the glass, either as oxyhydroxides or incorporated into the silicate structures of feldspathoids (cage structures) and clays (sheet structures). With increasing progress of the reaction, the dissolved silica in solution is resorbed onto the surface, forming a surface gel layer that grows in thickness with passing time. This gel layer also contains nanoscale crystallites of clays and zeolites. Once the gel layer has been formed, the reactive sites at the surface of the bulk glass can become blocked, and the compacted gel layer can become a diffusion barrier to the transport of dissolved silica from the glass–gel interface to the solution. The corroded glass surface consists of a series of distinct layers, each representing different mechanisms and stages in the glass-dissolution process (Figure 15.5).

Over the longer term, the dissolved glass components accumulate in the bulk solution, and this leads to a reduction in the reaction rates, to the so-called long-term rate. This lower, long-term rate has been attributed to two processes: (1) the silanol groups condensing onto the glass surface as a "saturation" effect; and (2) the gel layer acting as a diffusion barrier (Figure 15.6). Depending on the glass and groundwater compositions, both processes can act to lower the dissolution rate of the glass, and the long-term rate is typically four orders of magnitude less than the initial dissolution rate. Using this long-term rate, safety analyses predict that the vitrified waste will have a lifetime of at least 200,000 years [49]. Such long-term

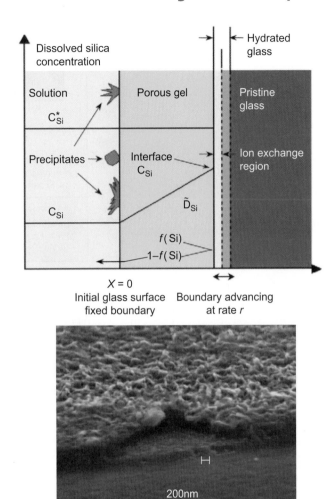

Figure 15.5. Evolution of the surface structure of a nuclear-waste borosilicate glass in contact with water: C_{Si} = Si concentration; C_{Si}^* = Si concentration at apparent saturation; C_{Si} interface = Si concentration at glass/gel interface; $\tilde{D}_{Si}$ = apparent Si diffusion coefficient; $f(Si)$ = fraction of Si remaining in gel layer. The inset is a scanning-electron-microscopy (SEM) image that shows the structure as it develops on the surface of a corroded borosilicate glass (courtesy of G. Calas after [59][60]).

extrapolations can be validated by studies of natural glasses. Basaltic glasses, such as those that erupt at mid-ocean ridges on Iceland, are closest to the borosilicate glasses in terms of their silica content. Reaction rates for basaltic glasses can be inferred from studies of alteration rind thicknesses and the age of dredged ocean samples. The general structure and mineralogy of these leached layers [12] and the reaction rates [52] are entirely consistent with experimental observations and measurements.

15.4.6 Radiation effects in crystalline ceramics

As described in Section 15.4.2, the radiation stability of nuclear-waste forms is of critical importance for long-term performance, particularly for crystalline phases

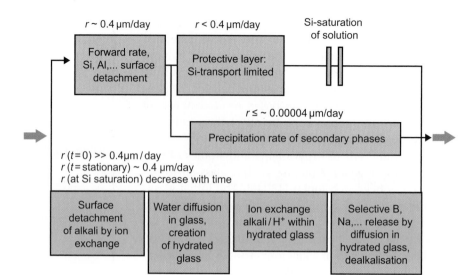

Figure 15.6. A schematic representation of the reaction mechanism of nuclear-waste borosilicate glass with water, coupling different sequential and parallel rate-controlling processes. Blue blocks indicate processes governing selective release of ions from the glass, whereas orange blocks indicate processes involved in the corrosion of the glass matrix itself. r values indicate average rates of release for a typical French nuclear-waste glass for different processes. Some processes are time-dependent (e.g., diffusion), so a range of values is given. The diagram can be interpreted as an electrical circuit of resistors and capacitors representing rates and solution saturation. In the case of parallel reaction rates, the highest rate (smallest resistance) governs the overall rate (overall resistance), whereas for sequential rates the slowest step is rate-controlling. The solution saturation capacitor initially (in the absence of saturation) causes no rate reduction, but, with time, saturation of radionuclides is reached (the capacitor is loaded), and the rate through this capacitor is zero. Adapted from [49].

that incorporate actinides. The α-decay of actinides produces energetic α-particles (4.5–5.5 MeV), energetic recoil nuclei (70–100 keV), and some γ-rays [11]. The α-particles lead to the buildup of He, and each α-particle creates several hundred isolated defects at the end of its trajectory. The recoil nuclei interact with surrounding atoms through a cascade of elastic collisions that cause the displacement of several thousand atoms. In crystalline materials, where the atomic-scale structure creates the "cage" for incorporated radionuclides, these damage cascades disrupt the local cages, and the damage volume accumulates, leading to (1) expansion of the unit cell due to the formation of isolated defects, (2) radiation-induced amorphization or decomposition, (3) a volume increase caused by the formation of amorphous material or helium bubbles that can lead to strain and microfracturing, and (4) a decrease in chemical durability (i.e., an increase in the leach rate of radionuclides). The extent of the change in microstructure that results from the radiation depends on the material properties, type of radiation, total radiation dose over the disposal time, and thermal history of the waste form.

Methods of simulating radiation damage

Because the effects of radiation extend over extremely long periods (Figure 15.2), various accelerated techniques are used in combination to understand and model the radiation response of waste forms over time [10][11]. For actinide α-decay events, three methods are used: (1) incorporation of highly radioactive, short-lived actinides, such as ^{238}Pu (half-life 87.7 years) or ^{244}Cm (half-life 18.1 years), that, in sufficient concentrations (0.2–3 wt%), produce relevant doses of 10^{18} α-decay events per gram in several years; (2) study of natural minerals that contain U and Th (up to 30 wt%) that have become "metamict" (amorphous) over periods of hundreds of millions of years; and (3) charged-particle irradiations (0.5–1.5 MeV) to simulate radiation effects from α-particles and α-decay recoil nuclei. In practice, all three of these techniques are required in order to establish the radiation response.

The effects of γ-ray interactions, which create energetic electrons in solids, and β-particles, which cause very few atomic displacements, can be studied by using ^{60}Co or ^{137}Cs γ-sources or electron beams from

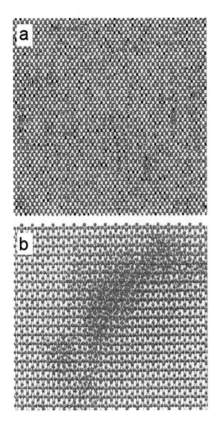

Figure 15.7. Projection along [100] of the primary damage state following 30-keV Zr recoil along [001] in (a) YSZ and (b) zircon. The initial velocity of the recoil is along the horizontal direction. Zr, Y, Si, and O atoms are represented as green/light gray, black, cyan/white, and red/dark gray spheres, respectively. Note that the damage recovery in zirconia is nearly complete, whereas in zircon, much of the displacement cascade is retained (from [53]).

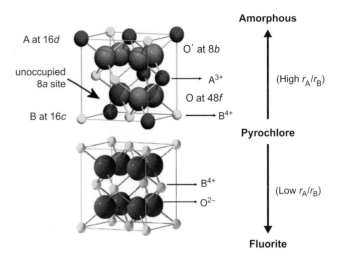

Figure 15.8. The pyrochlore structure (top) is a derivative of the simple fluorite structure (bottom). Disordering of the cations over the A and B sites and disordering of the anion vacancies create a disordered, defect fluorite structure [57]. Under irradiation, the pyrochlore structure can become amorphous, or it can become a disordered, defect fluorite structure. The ratio of the iconic radii of the A- and B-Site cations, rA/rB, is an important control on whether the pyrochlore structure becomes amorphous or a disordered, defect-fluorite structure on irradiation.

Figure 15.7 [53]. In YSZ, only a few point defects are created, whereas in zircon a highly disordered, amorphous-like state is produced in the core of the cascade.

Radiation effects in pyrochlore: case study

There is an extensive literature of radiation effects in glass and crystalline ceramic waste forms [10][11][25] [31][54][55]. Recently, pyrochlore has been the focus of attention because Zr- and Hf-rich compositions are "resistant" to radiation damage and do not become amorphous. As an example of how radiation effects are studied, this section focuses on the work on pyrochlore [28].

Structure of pyrochlore. There are over 500 synthetic compositions, including actinides, with the pyrochlore structure [56], as well as natural pyrochlores. Pyrochlore is isometric ($Fd3m$, $Z = 8$, $a = 0.9$–1.2 nm), and the structural formula is ideally $^{VIII}A_2{}^{VI}B_2{}^{IV}X_6{}^{IV}Y$ (Roman numerals indicate the coordination number), where the A and B sites contain metal cations and X ($= O^{2-}$) and Y ($= O^{2-}$, OH^-, F^-) are anions. The structure can be described in a variety of ways, most commonly in terms of the shapes and topology of the coordination polyhedra. Pyrochlore is closely related to the fluorite structure (AX_2), e.g., cubic UO_2, except that there are two cation sites and one-eighth of the anions are absent (Figure 15.8, top structure). The cations and oxygen vacancies are ordered. The loss of one-eighth of the anions reduces

accelerators or transmission electron microscopes. For γ-irradiations, dose rates up to 2.5×10^4 Gy h^{-1} can be obtained, and these dose rates are only 1–2 orders of magnitude greater than those experienced by commercial HLW glasses and 2–3 orders of magnitude larger than for defense HLW. Ballistic effects can also be studied by fast neutron irradiations or nuclear reactions, such as $^{10}B(n,\alpha)^7Li$ in borosilicate glasses, but these are less accurate simulations of the ballistic damage from the α-decay recoil nucleus.

While experimental methods provide data on the laboratory-scale radiation response of materials, they are limited in their ability to probe radiation-damage processes, such as single cascade events, occurring on nanometer length scales and picosecond time scales. Atomistic computer simulations can provide critical data and insights on radiation-damage processes on these length and time scales. For example, yttria-stabilized zirconia (YSZ) and zircon show markedly different responses to a 30-keV Zr recoil cascade, as shown in the coordination number of the B-site cation from eight to six. The X anion occupies the 48f position, and the Y anion occupies the 8b position. The A- and B-site coordination polyhedra are joined along edges, and the shapes of these polyhedra change as the positional parameter, x, of O_{48f} shifts to accommodate cations of different sizes. Thus, the positional parameter, x, of O_{48f} defines the polyhedral distortion and structural deviation from the ideal fluorite structure. The change in the structure depends on the ratio of the ionic radii of the A-site and B-site cations (r_A/r_B). As the A- and B-site cations become similar in size, the fluorite structure (Figure 15.8, bottom structure) is preferred; as the cations in the A site become larger or the B-site cations become smaller, the pyrochlore structure (Figure 15.8, upper structure) is more stable. With radiation damage, compositions that adopt the pyrochlore structure may become amorphous, while those with the ideal fluorite structure tend to remain crystalline.

Pyrochlore is an unusual oxide in that the order–disorder transformation occurs simultaneously on both the cation lattice and the anion lattice among three anion sites: 48f, 8a, and 8b. However, the cation and anion disordering can occur to different degrees and at different temperatures. Although the size difference between A and B sites is the driving force for cation ordering in the pyrochlore structure, no significant order–disorder structural transition occurs in the solid solution for which there is strong covalent bonding. This suggests that other factors (e.g., chemical bonding) have an important effect on the order–disorder transformation and the degree of disorder in the pyrochlore structure.

Radiation damage in pyrochlore. Pyrochlores that contain actinides will experience radiation damage from α-decay events. The α-particles dissipate most of their energy by ionization processes over a range of 16–22 μm, but undergo enough elastic collisions along their path to produce several hundred isolated atomic displacements that form Frenkel pairs (point defects formed when an atom or cation leaves its place in the lattice, creating a vacancy, and becomes an interstitial by lodging in a nearby location not usually occupied by an atom). The largest number of displaced atoms occurs near the end of the α-particle range. The more massive, but lower-energy α-recoil nucleus accounts for most of the total number of displacements produced by ballistic processes. The α-recoil loses 80% or more of its energy by elastic collisions over a very short range (20–30 nm) that produce energetic recoils, which, in turn, lose 60% or more of their energy in elastic collisions. As a result, about 50% of the energy of the recoil is deposited as "damage energy" in a highly localized collision cascade, where 500–2,000 atoms are energetically displaced some distance from their original atomic sites by elastic collisions and many more atoms can be collectively moved about their atomic positions.

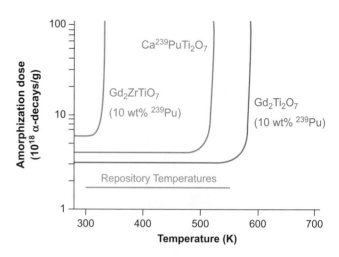

Figure 15.9. Predicted temperature dependence of amorphization in pyrochlore-related phases containing ^{239}Pu. The curves bend upward at elevated temperatures due to thermal annealing over geological time scales. Considering the curves from right to left, the temperature at which complete annealing occurs decreases, and the waste form remains crystalline despite a high α-decay dose. The range of potential repository temperatures is indicated by the horizontal line (derived from [28]).

Many pyrochlore compounds are susceptible to a radiation-induced crystalline-to-amorphous transformation as a result of α-decay events. The amorphous state is generally characterized by the loss of long-range order; however, short-range order is often retained. In general, the cumulative amount of amorphous material increases nonlinearly with dose, and the rate of amorphization decreases nonlinearly with temperature due to the kinetics of damage-recovery processes. The atomic-scale processes contributing to the production of the amorphous state control the functional dependence of the amorphous volume fraction on increasing dose. The competing processes of damage production and recovery control the temperature dependence of amorphization, and the kinetics of amorphization can be determined from the dependence of the amorphous fraction on the dose at different temperatures. In general, the dose for amorphization for specific compositions and irradiation conditions has most often been determined from electron-diffraction analysis of transmission-electron-microscopy (TEM) specimens for actinide-containing pyrochlores and related materials, minerals in the pyrochlore group, or for ion-irradiated pyrochlores. Complete amorphization will not occur if the amorphization rate is less than or equal to the damage-recovery rate. Thus, the temperature at which the rate of damage recovery equals the rate of amorphization defines the critical temperature, T_c, for amorphization, above which amorphization does not occur for a given set of irradiation conditions (Figure 15.9).

One of the more exciting outcomes from fundamental studies of irradiation effects using ion beams is the discovery of radiation-resistant $Gd_2Zr_2O_7$ and $Er_2Zr_2O_7$ pyrochlores [58][59]. These materials can readily accommodate Pu on the Gd (or Er) and Zr sites, and the zirconate pyrochlores are chemically durable. In the case of the $Gd_2(Zr_xTi_{1-x})_2O_7$ system, it has been shown that there is a systematic increase in the radiation resistance (decrease in T_c and increase in the baseline dose for complete amorphization at low temperatures) with increasing Zr content under 1.0-MeV Kr irradiation, which suggests a potential benefit for the immobilization of actinides in Zr-rich rather than Ti-rich pyrochlore compositions. These results are consistent with recent molecular-dynamics results that indicate direct formation of defect clusters that facilitate amorphization within displacement cascades in $Gd_2Ti_2O_7$, whereas in $Gd_2Zr_2O_7$, displacement cascades tend to produce only point defects [60].

Model predictions for pyrochlore. The two quantities most useful for predicting amorphization behavior in pyrochlores for the immobilization of actinides are the baseline amorphization dose, D_0, under ambient temperature conditions, preferably determined in studies employing short-lived actinides, and the critical temperature, T_c, due to thermal recovery processes. One of the predictions that can be provided for performance assessment is the temperature dependence of the amorphization dose for specific actinide host phases. The predicted temperature dependence of amorphization is shown in Figure 15.9 for amorphization in $Gd_2Ti_2O_7$ and Gd_2ZrTiO_7 containing 10 wt% ^{239}Pu and for $Ca^{239}Pu$-Ti_2O_7. The predictions clearly show that the critical temperature for ^{239}Pu-containing $Gd_2Ti_2O_7$ is above the expected repository temperatures; thus, amorphization for the titanate pyrochlore will occur without significant recovery. In the case of $Ca^{239}PuTi_2O_7$, amorphization will not occur for several decades until the temperature falls below 520 K. Amorphization of ^{239}Pu-containing Gd_2ZrTiO_7 will be delayed for a thousand years or more, until the temperature falls below about 330 K, at which time amorphization will progress at a much lower rate because of the higher dose required for amorphization.

15.5 Summary

One of the main challenges for nuclear power is the development of materials that can safely contain radioactive waste for up to hundreds of thousands of years. Material scientists must design waste forms that can be synthesized and produced in a high-radiation environment, incorporate specific radionuclides, have a high chemical durability, and resist the effects of radiation damage. The thermal and radiation fields of the waste form will decrease over time, but they will remain important for at least the first 1,000 years after disposal. The longer-term performance of the waste form, over hundreds of thousands of years, will depend critically on the geological conditions, mainly the geochemistry and hydrology, as they too evolve over time. The design, selection, and evaluation of nuclear-waste forms must consider the evolving and dynamic coupling of the properties of the material to the conditions of geological disposal. As illustrated by the discussion of pyrochlore as a waste form for actinides, it is now possible to match the material properties, e.g., thermal annealing of radiation damage, to the geological setting, e.g., use of the geothermal gradient or the repository ambient temperature to induce thermal annealing and retain the integrity of the waste form. The design of materials for these extreme conditions and very long periods of time remains a challenge for the next generation of materials scientists.

15.6 Questions for discussion

1. What are four main requirements for waste forms?
2. What are the three main types of waste forms?
3. Compare and contrast the main features of the corrosion process for spent fuel and waste glass.
4. How does radiation damage affect retention of radionuclides in nuclear-waste forms? What is the relation between radiation-damage accumulation and temperature for a nuclear-waste form?
5. On the basis of the crystal chemistry and geochemistry of Cs and Np, design a crystalline nuclear-waste form for each. Describe the structure of each waste form. How would you validate the long-term behavior of the waste form?

15.7 Further reading

- **D. Caurant**, **P. Loiseau**, **O. Majerus** *et al.*, 2009, *Glasses, Glass-Ceramics and Ceramics for Immobilization of Highly Radioactive Nuclear Wastes*, New York, Nova Science Publishers, Inc. This book provides the most recent summary of research on nuclear-waste forms.
- **I. W. Donald**, **B. L. Metcalfe**, and **R. N. J. Taylor**, 1997, "The immobilization of high level radioactive wastes using ceramics and glasses," *J. Mater. Sci.*, **32**, 5851–5887. This paper gives a good summary of the different types of waste forms and tables of their properties.
- **R. C. Ewing** (ed), 2006, *The Nuclear Fuel Cycle – Environmental Aspects. Elements*, **2**, 321–373. This special issue of *Elements* focuses on various aspects of nuclear-waste-form science.
- **W. Lutze** and **R. C. Ewing**, 1988, *Radioactive Waste Forms for the Future*, Amsterdam, North-Holland. This is the standard reference on waste forms. It has extensive chapters on glass waste forms, spent nuclear fuel, and Synroc.

15.8 References

[1] **R. C. Ewing**, 2004, "Environmental impact of the nuclear fuel cycle," in *Energy, Waste and the Environment: A Geochemical Perspective*, eds. **R. Gieré** and **P. Stille**, London, The Geological Society, pp. 7–23.

[2] **R. C. Ewing**, 2001, "The design and evaluation of nuclear-waste forms: clues from mineralogy," *Canadian Mineralogist*, **39**, 697–715.

[3] US Department of Energy, 1996, *Closing the Circle on the Splitting of the Atom*, Washington, DC, DOE, Office of Environmental Management.

[4] US Department of Energy, 1997, *Linking Legacies – Connecting the Cold War Nuclear Weapons Production Processes to Their Environmental Consequences*, Washington, DC, DOE, Office of Environmental Management.

[5] **D. Albright** and **K. Kramer**, 2004, "Fissile material – stockpiles still growing," *Bull. Atomic Scientists*, **14**, November/December, 14–16.

[6] **F. N. von Hippel**, 2001, "Energy – plutonium and reprocessing of spent nuclear fuel," *Science*, **293**, 2397–2398.

[7] **R. C. Ewing**, **W. Runde**, and **T. E. Albrecht-Schmitt**, 2010, "Environmental impact of the nuclear fuel cycle: fate of actinides," *MRS Bull.*, **36**(11), 859–866.

[8] **W. J. Weber**, **A. Navrotsky**, **S. Stefanovsky**, **E. R. Vance**, and **E. Vernaz**, 2009, "Materials science of high-level nuclear waste immobilization," *MRS Bull.*, **34** (1), 46–53.

[9] **C. Jiang**, **B. P. Uberuaga**, **K. E. Sickafus** *et al.*, 2010, "Using 'radioparagenesis' to design robust nuclear waste forms," *Energy Environ. Sci.*, **3**, 130–135.

[10] **W. J. Weber**, **R. C. Ewing**, **C. A. Angell** *et al.*, 1997, "Radiation effects in glasses used for immobilization of high-level waste and plutonium disposition," *J. Mater. Res.*, **12**(8), 1946–1975.

[11] **W. J. Weber**, **R. C. Ewing**, **C. R. A. Catlow** *et al.*, 1998, "Radiation effects in crystalline ceramics for the immobilization of high-level nuclear waste and plutonium," *J. Mater. Res.*, **13**(6), 1434–1484.

[12] **W. Lutze**, **G. Malow**, **R. C. Ewing**, **M. J. Jercinovic**, and **K. Keil**, 1985, "Alteration of basalt glasses: implications for modelling the long-term stability of nuclear waste glasses," *Nature*, **314**, 252–255.

[13] **A. Verney-Carron**, **S. Gin**, and **G. Libourel**, 2010, "Archeological analogs and the future of nuclear waste glass," *J. Nucl. Mater.*, **406**(3), 365–370.

[14] **W. Lutze**, 1988, "Silicate glasses," in *Radioactive Waste Forms for the Future*, eds. **W. Lutze** and **R. C. Ewing**, Amsterdam, North Holland, pp. 1–159.

[15] **M. T. Peters** and **R. C. Ewing**, 2007, "A science-based approach to understanding waste form durability in open and closed nuclear fuel cycles," *J. Nucl. Mater.*, **362**, 395–401.

[16] **I. Mueller** and **W. J. Weber**, 2001, "Plutonium in crystalline ceramics and glasses," *MRS Bull.*, **26**(9), 698–706.

[17] **B. C. Sales** and **L. A. Boater**, 1984, "Lead–iron phosphate glass: a stable storage medium for high-level nuclear waste," *Science*, **226**, 45–48.

[18] **B. C. Sales** and **L. A. Boatner**, 1988, "Lead–iron phosphate glass," in *Radioactive Waste Forms for the Future*, eds. **W. Lutze** and **R. C. Ewing**, Amsterdam, North-Holland, pp. 193–231.

[19] **S. Gahlert** and **G. Ondracek**, 1984, "Sintered glass," in *Radioactive Waste Forms for the Future*, eds. **W. Lutze** and **R. C. Ewing**, Amsterdam, North Holland, pp. 161–192.

[20] **P. J. Hayward**, 1988, "Glass-ceramics," in *Radioactive Waste Forms for the Future*, eds. **W. Lutze** and **R. C. Ewing**, Amsterdam, North-Holland, pp. 427–493.

[21] **L. P. Hatch**, 1953, "Ultimate disposal of radioactive wastes," *Am. Scientist*, **41**, 410–421.

[22] **R. C. Ewing**, **W. Lutze**, and **W. J. Weber**, 1995, "Zircon: a host-phase for the disposal of weapons plutonium," *J. Mater. Res.*, **10**(2), 243–246.

[23] **L. A. Boatner** and **B. C. Sales**, 1988, "Monazite," in *Radioactive Waste Forms for the Future*, eds. **W. Lutze** and **R. C. Ewing**, Amsterdam, North-Holland, pp. 495–564.

[24] **A. E. Ringwood**, **S. E. Kesson**, **N. G. Ware**, **W. Hibberson**, and **A. Major**, 1979, "Immobilization of high-level nuclear-reactor wastes in Synroc," *Nature*, **278**, 219–223.

[25] **R. C. Ewing** and **W. J. Weber**, 2010, "Actinide waste forms and radiation effects," in *The Chemistry of the Actinides and Transactinide Elements*, vol. 6, eds. **N. M. Edelstein**, **J. Fuger**, and **L. R. Morss**, New York, Springer, pp. 3813–3888.

[26] **V. M. Oversby**, **C. C. McPheeters**, **C. Degueldre**, and **J. M. Paratte**, 1997, "Control of civilian plutonium inventories using burning in non-fertile fuel," *J. Nucl. Mater.*, **245**, 17–26.

[27] **R. C. Ewing**, 1999, "Nuclear waste forms for actinides," *Proc. Natl. Acad. Sci.*, **96**, 3432–3439.

[28] **R. C. Ewing**, **W. J. Weber**, and **J. Lian**, 2004, "Pyrochlore ($A_2B_2O_7$): a nuclear waste form for the immobilization of plutonium and 'minor' actinides," *J. Appl. Phys.*, **95**, 5949–5971.

[29] **I. W. Donald**, **B. L. Metcalfe**, and **R. N. J. Taylor**, 1997, "The immobilization of high level radioactive wastes using ceramics and glasses," *J. Mater. Sci.*, **32**, 5851–5887.

[30] **G. R. Lumpkin**, 2001, "Alpha-decay damage and aqueous durability of actinide host phases in natural systems," *J. Nucl. Mater.*, **289**, 136–166.

[31] **G. R. Lumpkin**, 2006, "Ceramic waste forms for actinides," *Elements*, **2**, 365–372.

[32] **M. I. Ojovan** and **W. E. Lee**, 2005, *An Introduction to Nuclear Waste Immobilisation*, Amsterdam, Elsevier.

[33] **S. V. Yudintsev**, **S. V. Stefanovsky**, and **R. C. Ewing**, 2007, "Actinide host phases as radioactive waste forms," in *Structural Chemistry of Inorganic Actinide Compounds*, eds. **S. V. Krivovichev**, **P. C. Burns**, and **I. Tananaev**, Amsterdam, Elsevier, pp. 457–490.

[34] D. Caurant, P. Loiseau, O. Majerus, *et al.*, 2009, *Glasses, Glass-Ceramics and Ceramics for Immobilization of Highly Radioactive Nuclear Wastes*, New York, Nova Science Publishers, Inc.

[35] E. C. Buck, B. D. Hanson, and B. K. McNamara, 2004, "The geochemical behaviour of Tc, Np and Pu in spent nuclear fuel in an oxidizing environment," in *Energy, Waste, and the Environment: A Geochemical Perspective*, eds. R. Gieré and P. Stille, London, The Geological Society, pp. 65–88.

[36] A. Hedin, 1997, *Spent Nuclear Fuel – How Dangerous Is It?*, SKB Technical Report 97–13.

[37] J. Bruno and R. C. Ewing, 2006, "Spent nuclear fuel," *Elements*, 2 343–349.

[38] L. H. Johnson and E. W. Shoesmith, 1988, "Spent fuel," in *Radioactive Waste Forms for the Future*, eds. W. Lutze and R. C. Ewing, Amsterdam, North-Holland, pp. 635–698.

[39] V. M. Oversby, 1994, "Nuclear waste materials," in *Materials Science and Technology*, vol. 10B, eds. R. W. Cahn, P. Haasen, and E. J. Kramer, Weinheim, VCH Verlagsgesellschaft mbH, pp. 391–442.

[40] H. Kleykamp, 1985, "The chemical state of the fission products in oxide fuels," *J. Nucl. Mater.*, 131, 221–246.

[41] D. W. Shoesmith, 2000, "Fuel corrosion processes under waste disposal conditions," *J. Nucl. Mater.*, 282, 1–31.

[42] L. Johnson, C. Ferry, C. Poinssot, and P. Lovera, 2005, "Spent fuel radionuclide source-term model for assessing spent fuel performance in geological disposal. Part I: assessment of the instant release fraction," *J. Nucl. Mater.*, 346, 56–65.

[43] J. K. Bates, W. L. Ebert, X. Feng, and W. L. Bourcier, 1992, "Issues affecting the prediction of glass reactivity in an unsaturated environment," *J. Nucl. Mater.*, 190, 198–227.

[44] M. Aertsens and D. Ghaleb, 2001, "New techniques for modeling glass dissolution," *J. Nucl. Mater.*, 298, 37–46.

[45] B. Grambow and R. Müller, 2001, "First-order dissolution rate law and the role of surface layers in glass performance assessment," *J. Nucl. Mater.*, 298, 112–124.

[46] C. M. Jantzen, K. G. Brown, and J. B. Pickett, 2010, "Durable glass for thousands of years," *Int. J. Appl. Glass Sci.*, 1(1), 38–62.

[47] G. G. Wicks, 2001, "US field testing programs and results," *J. Nucl. Mater.*, 298, 78–85.

[48] P. Van Iseghem, M. Aertsens, S. Gin *et al.*, 2007, *GLAMOR – A Critical Evaluation of the Dissolution Mechanism of High Level Waste Glasses in Conditions of Relevance for Geological Disposal*, European Commission Report EUR 23097.

[49] B. Grambow, 2006, "Nuclear waste glasses – how durable?," *Elements*, 2, 357–364.

[50] E. Y. Vernaz, 2002, "Estimating the lifetime of R7T7 glass in various media," *Phys. Appl./Appl. Phys.*, 3, 813–825.

[51] E. Pélegrin, G. Calas, P. Ildefonse, P. Jollivet, and L. Galoisy, 2010, "Structural evolution of glass surface during alteration: application to nuclear waste glasses," *J. Non-Cryst. Solids*, 356, 2497–2508.

[52] X. Le Gal, J.-L. Crovisier, F. Gauthier-Lafaye, J. Honnorez, and B. Gramow, 1999, "Altération météorique des verres volcaniques d'Islande: changement du mécanisme a long terme," *C. R. Acad. Sci., Sci. Terre Planets*, 329, 175–181.

[53] R. Devanathan and W. J. Weber, 2008, "Dynamic annealing of defects in irradiated zirconia-based ceramics," *J. Mater. Res.*, 23(3), 593–597.

[54] R. C. Ewing, W. J. Weber, and F. W. Clinard, Jr., 1995, "Radiation effects in nuclear waste forms," *Prog. Nucl. Energy*, 29(2), 63–127.

[55] W. J. Weber and F. P. Roberts, 1983, "A review of radiation effects in solid nuclear waste forms," *Nucl. Technol.*, 60(2), 178–198.

[56] B. C. Chakoumakos, "Systematics of the pyrochlore struture types, ideal $A_2B_2X_6Y$." *J. Solid State Chemistry* 53 (1984), 120–129.

[57] K. E. Sickafus, R. W. Grimes, J. Valdez *et al.*, 2007, "Radiation-induced amorphization resistance and radiation tolerance in structurally related oxides," *Nature Mater.* 6, 217.

[58] S. X. Wang, B. D. Begg, L. M. Wang *et al.*, 1999, "Radiation stability of gadolinium zirconate: a waste form for plutonium disposition," *J. Mater. Res.*, 14, 4470–4473.

[59] K. E. Sickafus, L. Minervini, R. W. Grimes *et al.*, 2000, "Radiation tolerance of complex oxides," *Science*, 289, 748–751.

[60] R. Devanathan and W. J. Weber, 2005, "Insights into the radiation response of pyrochlores from calculations of threshold displacement events," *J. Appl. Phys.*, 98, 086110.

16 Material requirements for controlled nuclear fusion

Nathaniel J. Fisch, J. Luc Peterson, and Adam Cohen

Princeton Plasma Physics Laboratory, Princeton University, Princeton, NJ, USA

16.1 Focus

Controlled nuclear fusion has the potential to provide a clean, safe energy source with an essentially limitless supply of fuel, relatively few proliferation concerns (compared with those mentioned in Chapter 14), and substantially fewer of the waste-management concerns discussed in Chapter 15. Large experimental devices currently under construction are intended to demonstrate net fusion energy production, a key technological milestone on the way toward the commercial production of electricity. The economic practicality of energy from fusion processes, however, will still require other significant advances, including in the development of materials that can survive the harsh fusion environment.

16.2 Synopsis

The nuclear fusion of light elements is the energy source of the stars. A fusion-based power plant holds the prospect of a nearly limitless fuel source, without the concerns of greenhouse-gas emissions, nuclear proliferation, or serious waste management. While the release of enormous amounts of energy from this process has long been demonstrated in weapons, controlling and harnessing this energy for electricity production constitutes a technologically much more difficult problem. At present, the fusion community is exploring two major approaches to controlled nuclear fusion: magnetic confinement and inertial confinement. In the magnetic fusion energy (MFE) approach, powerful magnetic fields confine low-density hydrogen plasma as it is heated to very high temperatures. In the inertial fusion energy (IFE) approach, tiny pellets of solid hydrogen are compressed to very high densities and temperatures.

In each method, ions of deuterium and tritium, isotopes of hydrogen, reach high enough temperatures to overcome their mutually repulsive force, forming helium, and releasing in each fusion event mega-electron volts (MeV) of energy. Compared with the electron volts (eV) of energy released in a chemical event, this release of energy per unit mass of fuel is enormous. However, the assembly, the heating, and the confinement of the fusing particles require considerable energy. The aim in the next phase of experiments for both approaches is to demonstrate the possibility of net energy gain. Fusion reactors may then begin to accept a significant part of the world energy burden by the second half of the century.

Key to the commercialization of fusion power will be advances made in materials science. Neutrons produced in deuterium–tritium fusion reactions carry away most of the energy released, which must then be recovered as heat in a material blanket surrounding the reacting particles. In addition, since very little tritium is available in nature, these neutrons must be used to breed tritium through nuclear reactions in the blanket. Additionally, magnetic- and inertial-confinement fusion each present their own unique scientific and technological challenges, which, together with some material needs for a fusion power plant, will be discussed in this chapter.

16.3 Historical perspective

"For his contributions to the theory of nuclear reactions, especially his discoveries concerning the energy production in stars," Hans Bethe received the Nobel Prize in Physics in 1967 [1]. Bethe had proposed in 1938 that stars generated energy through the fusion of lighter elements into heavier ones. The energy released in stars is through the nuclear fusion of protons, or ions of hydrogen. The cross section for nuclear fusion of isotopes of hydrogen, namely deuterium and tritium, is much greater, so research has focused on this D–T reaction.

The first, and easier, application of nuclear fusion was military. The H-bomb, or hydrogen bomb, which was developed shortly after World War II, explodes through the uncontrolled release of nuclear fusion energy. On November 1, 1952, on the Eniwetok Atoll in the Marshall Islands, the USA detonated a device code-named Mike with a yield of about 10 megatons (MT), demonstrating conclusively that one could release a massive amount of energy from nuclear fusion [2].

Harnessing that enormous energy is necessary for peaceful energy production, yet controlling the fusion energy release is more difficult than releasing it all at once. The managed release of fusion energy is the goal of the international controlled-nuclear-fusion program. Such a program was anticipated early on by Edward Teller, also known as the father of the H-bomb, who called for academic discussions at Los Alamos on peacetime uses of fusion energy as early as 1942 [2].

The advantages of nuclear fusion power for peacetime purposes were apparent. A fusion power plant, like an operational fission plant (see Chapters 13–15) would release no greenhouse gases or air pollutants, and plants could be integrated into existing grid structures. However, nuclear fusion poses a relatively small proliferation and safety risk; since a plant would hold only a small amount of fuel at any time, runaway reactions are not possible; and the main fuels (deuterium and lithium) are abundantly found in seawater. Finally, the low-level radioactive waste generated by a fusion power plant could be safe within decades, instead of millennia, obviating most of the issues identified in Chapter 15. While early researchers were optimistic and expected a quick solution to fusion power, unforeseen challenges limited progress.

Recognizing that controlled nuclear fusion was proving to be very difficult to achieve, in September 1958, all nations working on the subject agreed to declassify their work at the Second International Atoms for Peace Conference in Geneva, beginning an era of international collaboration on nuclear fusion [3]. The earliest approaches sought to contain the hot plasma in a bottle of magnetic fields, so that the plasma could be heated up to hundreds of millions of degrees celsius, without contacting a material surface.

Research into many different magnetic-field configurations continues today. A few examples include the mirror, the stellarator, the field-reversed configuration, and the spheromak [4]. However, the most actively pursued method for the magnetic confinement of plasma is at present the **tokamak**, a toroidal plasma device with an axially induced current. The word *tokamak* is a Russian acronym for *to*roidalnaya *ka*mera i *ma*gnitnaya *k*atushka, or toroidal chamber and magnetic coil. The Russian Nobel Prize winners Igor Tamm and Andrei Sakharov had put forth this approach in 1950, with publication following the declassification [5]. In 1968, Russian physicists claimed spectacular plasma electron temperatures in excess of 11 million Kelvin in the T-3 Tokamak, stimulating research worldwide on the tokamak concept. Since 1968, tokamak plasmas have reached temperatures of over 500 million degrees and fusion power of over 10 megawatts [6]. The ITER tokamak (see Figure 16.1), currently under construction in southern France, is a multibillion-dollar magnetic-confinement fusion experiment involving the European Union, India, Japan, China, Russia, South Korea, and the USA. ITER is expected to demonstrate the feasibility of fusion power by producing 500 megawatts of fusion power using only 50 megawatts of input power. ITER is expected to begin operations by 2019 [7].

The compression of plasma to high density by lasers is the currently most advanced method of inertial-confinement fusion, which is most vigorously being pursued today at the Lawrence Livermore National Laboratory (LLNL). The invention of the laser in the 1960s enabled the compression of fusion fuel by lasers to thermonuclear temperatures, somewhat like in the original Teller–Ulam schemes for a fusion bomb [2], but designed for much smaller explosions. In 1972, the Long Path Laser at the LLNL delivered about 50 joules of energy to a fusion target; by 1989 the NOVA LLNL 10 beam laser was able to produce 20 kilojoules of infrared laser light during a nanosecond pulse [8]. Now research is beginning at the LLNL at the National Ignition Facility (NIF) (see Figure 16.2), which uses 192 laser beams to deliver over 1.8 megajoules of laser energy onto millimeter-size pellets, with the goals of heating the pellets to over 100 million Kelvin and pressures over 100 billion atmospheres and releasing more energy than the pellet absorbs [9][10].

In summary, controlled nuclear fusion has progressed steadily over the past 60 years. Now ITER is poised to be the first tokamak featuring self-sustaining fusion reactions, showing that controlled fusion is possible with magnetically confined plasmas.

Figure 16.1. The ITER tokamak, currently under construction in Cadarache, France, is the largest magnetic-fusion experiment in history. © ITER Organization.

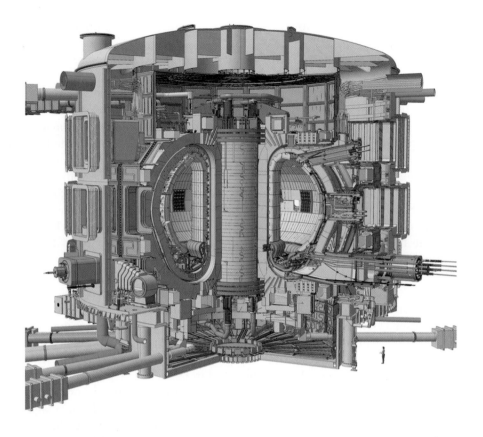

Figure 16.2. The National Ignition Facility (NIF) target chamber. The NIF at Lawrence Livermore National Laboratory in Livermore, California, aims to reach ignition by producing an energy gain from laser inertial fusion. In the center of the 10-m-diameter target chamber, a large boom, shown at the right, holds a 9-mm by 5-mm metal cylindrical target, or hohlraum, that houses a millimeter-sized D–T pellet. Lenses on the chamber wall focus 192 laser beams on the target. © Lawrence Livermore National Laboratory.

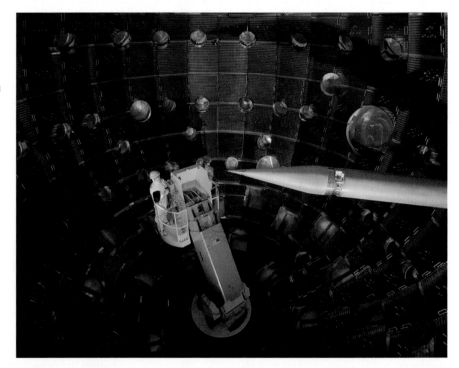

At the same time, the NIF is positioned to create miniature explosions of net energy gain. In two different approaches, each of these experiments represents a large step toward a demonstration fusion power plant.

16.4 The physics of nuclear fusion

16.4.1 The D–T fusion reaction

While many fusion reactions are possible under the right conditions, the first electricity generated by fusion

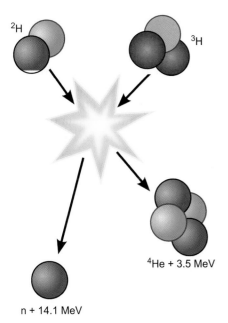

^{2}H

^{3}H

^{4}He + 3.5 MeV

n + 14.1 MeV

Figure 16.3. The D–T fusion reaction. The fusion of deuterium and tritium releases an alpha particle and a neutron. The neutron carries roughly 80% of the released energy in the reaction, 14.1 MeV. Helium, the other byproduct of the reaction, contains the additional 3.5 MeV.

reactors will almost certainly be based upon the D–T reaction, (See Chapter 14 Box 14.2 for notation).

$$^2_1\mathrm{D} + {}^3_1\mathrm{T} \rightarrow {}^4_2\mathrm{He}(3.5\,\mathrm{MeV}) + {}^1_0\mathrm{n}(14.1\,\mathrm{MeV}).$$

The helium nucleus (also called an alpha particle) is four times as massive as the neutron, so, by conservation of momentum, it carries away a quarter as much energy. This fusion reaction (see Figure 16.3) has the largest cross section, which peaks at lower energies of the colliding particles than for other fusion reactions. These two properties (a greater reaction rate at a lower energy) make the D–T fusion reaction attractive for energy production. This reaction peaks at temperatures of around 170 million Kelvin or 15 keV (one electron volt is equivalent to 11,604 K).

Since the reactions occur at such high energies compared with the ionization potential energy, the interacting particles must be fully ionized. Moreover, in order to have substantial power production, many ions must interact. A neutralizing charge of electrons prevents the ions from flying apart due to electrostatic Coulomb repulsion. Thus, the fusion reactions occur in an environment of many ions and electrons, or the state of matter called **plasma**. Moreover, since the cross sections for the fusing ions are so much smaller than the scattering cross sections for electrostatic Coulomb repulsion, the ions and electrons will have a nearly thermal distribution of energies, hence the term **thermonuclear fusion**.

Like other reaction rates, fusion reaction rates are proportional to the square of the density. The fusion power from this plasma can then be written as $P_\mathrm{f} = n^2\,g(T) = (nT)^2\,[g(T)/T^2]$, where n is the plasma density and $g(T)$ is a function that depends only on the plasma temperature T through the reaction cross section. The term nT is the plasma pressure, which is generally limited by the method of plasma confinement. The quantity $g(T)/T^2$ is a function of temperature only, and for the D–T reaction has a maximum at about 15 keV. Thus, D–T fusion power production is maximized, for constant pressure, at plasma temperatures of about 15 keV [6].

With 80% of the fusion energy and mean free paths much longer than those for the charged alpha particles, the energetic neutrons escape from the region in which fusion takes place and slow down through interactions in the surrounding blanket and structural materials. This energy, recovered as heat, is available to generate electricity and additional fuel. The energy of the alpha particles heats the background plasma and keeps the reaction self-sustaining.

The fuel for the D–T reaction is relatively abundant. Deuterium is found in seawater, in an essentially limitless supply [6]. Tritium, however, has a short half-life (12.3 years), so it must be bred by neutronic lithium bombardment, usually by one of the following two reactions:

$$^6_3\mathrm{Li} + {}^1_0\mathrm{n} \rightarrow {}^4_2\mathrm{He}(2.05\,\mathrm{MeV}) + {}^3_1\mathrm{T}(2.75\,\mathrm{MeV}),$$

$$^7_3\mathrm{Li} + {}^1_0\mathrm{n} \rightarrow {}^4_2\mathrm{He} + {}^3_1\mathrm{T} + {}^1_0\mathrm{n}.$$

While the lithium-6 reaction is exothermic, the lithium-7 process is endothermic, consuming 2.466 MeV of energy, but not the neutron. Both isotopes of lithium are abundant naturally in chemical compounds found in seawater and the Earth's crust. Estimated lithium reserves could provide enough tritium to meet the current global energy demand for 10,000–30,000 years [6][11].

16.4.2 Magnetic fusion energy

Plasma reaching thermonuclear temperatures cannot be contained directly in a material container. Since the wall temperature has to be much less than the plasma temperature, any container would itself become ionized or would cool the plasma. Thus, a *magnetic* container is used instead of a material container. Charged particles cannot move in a straight line perpendicular to a magnetic field and instead spiral along the magnetic lines of force. As such, a uniform magnetic field reduces the containment problem from three dimensions to one. There exist several magnetic confinement schemes that then solve the one-dimensional containment problem

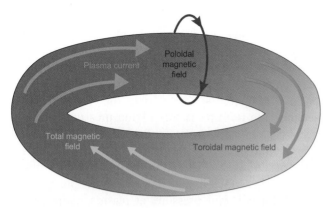

Figure 16.4. The tokamak magnetic fusion confinement concept. In a tokamak, external coils create a toroidal magnetic field. A toroidal current produced in the toroidally shaped plasma creates a poloidal magnetic field, which, added to the toroidal field, results in a twisting magnetic field.

along the field lines (see Chapter 13 of [4]). We focus here on the most actively pursued method: the tokamak.

Toroidal magnetic-confinement schemes exploit the fact that, if field lines wrap around on themselves in a large circle, charged particles circling the field lines would be forced into a large circle as well. These donut-shaped fields then solve the three-dimensional confinement problem. However, centrifugal forces in a purely toroidal magnetic field cause particles to drift out of the plasma. In the tokamak, this drift is canceled out by a poloidal magnetic field (see Figure 16.4). The twisted field that results circles both the donut hole and the center of the minor cross section (what is exposed when you lay a donut flat on a table and slice it vertically).

Generating these confining magnetic fields requires both power and materials. At several tesla, the toroidal field is the larger field, and is expected to be most economically provided by superconducting coils wrapped around the poloidal cross section. Protecting these superconductors from ionization radiation, neutron damage, and heat, while keeping the plant economically attractive, is a major materials issue in the tokamak concept. A toroidal current in the plasma provides the poloidal field. While some of this current can arise self-consistently through pressure gradients in the plasma, all the current cannot. The remaining toroidal current can be generated through coils that produce a time-varying magnetic flux that threads the donut hole. The time-varying magnetic field induces an electric field with curl. This toroidal electric field in the plasma then drives a toroidal current. This method is inherently pulsed, since the electric field persists only so long as the magnetic flux increases in time. Alternatively, for steady-state operation, the toroidal current might be

induced by waves or particle beams that are absorbed within the plasma [6].

Additionally, external sources of heating are necessary both to produce the high plasma temperatures required to start the fusion reaction and to control the reaction rate. These external sources inject radiofrequency waves that couple well to the plasma or inject beams of energetic ions that slow down by collisions in the plasma.

While the tokamak has shown excellent plasma confinement at thermonuclear temperatures and is the leading contender for next-generation magnetic fusion energy machines, many scientific and engineering challenges must be overcome to make a tokamak-based power plant viable. These include efficiently generating the necessary steady-state plasma current, controlling the confined plasma, mitigating the diffusion of heat through plasma turbulence, and providing for fuel (D–T) injection and ash (alpha particle) ejection. Many of these issues will be explored in ITER.

16.4.3 Inertial fusion energy

An alternative approach to avoiding a material container is to rely upon the inertia of the fuel to make it stay put while it is being heated to thermonuclear temperatures. In this approach, the plasma must be heated very quickly, leading to a very different regime of plasma parameters. Whereas, for the magnetic-confinement approach to controlled nuclear fusion, low-density plasma is maintained at thermonuclear temperatures for long periods of time, for the inertial fusion approach, plasma is compressed to very high densities for very short periods of time.

One way in which these conditions may be achieved would be to arrange a target with a fuel core of D–T gas, surrounded by a shell of solid or liquid D–T, surrounded by an outer shell of ablating material. When the ablating material is heated quickly to very high temperatures, it blows off. Then, like a rocket effect, the ablated material forces the inner shell of D–T to implode. As the shell of D–T implodes, it also heats the D–T gas to temperatures hot enough for fusion to occur. The energetic alpha particles produced in the implosion then slow down by collisions with the solid shell, the density of which is at that point 1,000 times the solid's density [8]. By making the fuel so dense, it is possible to make very small explosions of high efficiency.

The driver of the implosion might utilize the energy from magnetic fields, particle beams, lasers or X-rays. One actively pursued approach is **direct drive**, whereby the lasers compress the fuel pellet directly [12]. An alternative approach, and the one being actively pursued at the NIF is known as **indirect drive**, the fuel pellet is placed inside a **hohlraum**, a target cylinder whose walls are in equilibrium with an intense, essentially blackbody, radiation field [8]. Intense, nanosecond-duration laser pulses irradiate the hohlraum walls. As the walls reach a high temperature, they produce a uniform bath of X-ray radiation, which then ablates the target. Within nanoseconds, the target, through this sudden ablative inertia, dramatically compresses to fusion conditions.

The hohlraum walls are usually made of a heavy metal such as gold, because heavier walls more easily reach higher temperatures. Uniform compression is a necessity and a challenge. As the capsule is compressed, slight imperfections either in the manufacture of the pellet or in uniformity of the deposited laser power can cause asymmetric implosion. Hydrodynamic instabilities, such as the Rayleigh–Taylor instability, can grow quickly and cause the fusion fuel to burn inefficiently, for example by causing one location to fuse before the rest of the pellet. This can blow the pellet apart before the fusion fuel burns up [8].

Thus, the basic indirect drive configuration involves several subsystems: the high-power lasers, the millimeter-sized capsule of frozen D–T inside a centimeter-sized hohlraum, and a chamber in which a tiny explosion takes place, with walls capable of extracting the explosive release of neutronic energy. For a 3-GW thermal power plant, the energy released in each explosion might be about 300 MJ, requiring then a repetition rate of 10 Hz.

16.5 Materials for fusion energy production

A fusion-power-plant core, be it an MFE or an IFE system, is a source of high-energy neutrons. These neutrons must be converted into usable energy and additional fuel (tritium). The economical generation of energy by nuclear fusion will depend upon advances in materials, particularly those that can withstand the intense high-energy neutron and heat fluxes.

16.5.1 Common power-plant needs

Common to both the MFE and IFE approaches is the extraction of energy from a fusion core (Figure 16.5). The high-energy neutrons produced in the core pass through the **first wall**, into a **blanket module**, which absorbs the energy of the neutrons and breeds tritium. A one-meter-thick **neutron shield** prevents damage to sensitive components of the reactor. Heat exchangers extract the energy deposited by the neutrons in the first wall and blanket modules, as in a fission plant. Many of the materials issues for fission plants discussed in Chapter 13, such as void swelling, radiation embrittlement, and decay heat, also carry over to the fusion system. However, materials in a fusion power plant must

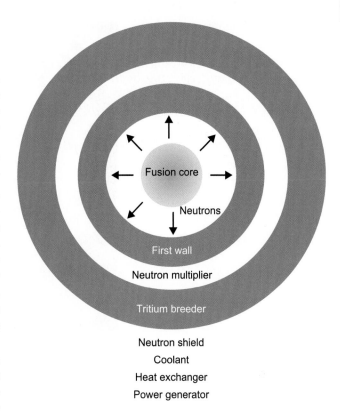

Figure 16.5. A conceptual fusion power plant. To generate electricity from an IFE or MFE fusion core requires the harnessing of energetic neutrons. They must pass through the first wall into the blanket, where they multiply and breed additional tritium to fuel the reactor. In the process the first wall and blanket heat up to several hundred degrees Celsius. This heat must be converted into usable electricity.

be able to withstand damage from higher-energy neutrons and heat fluxes that can approach those near the surface of the Sun.

Potential candidates for first-wall and blanket materials include low-activation steels[1] (such as ODS, oxide dispersion strengthened, ferritic or martensitic steels), refractory structural alloys (made with V, Mo, and W) and silicon carbide composites [13][14][15]. The successful material will provide high thermal efficiency and tensile strength, yet have low activation with resistance to damage from neutrons and heat [16]. Activation by neutrons can result in radioactive isotopes, such as Co-60. Undesirable alloying elements include niobium, molybdenum, Ni, Cu, Ti, Si, and Co. The blanket will also require a neutron multiplier (with a ratio of at least 1.1–1.4:1) to create neutrons to breed enough

---

[1] A principal approach to the minimization of radioactive waste for fusion reactors involves the selection of low-activation (LA) materials containing only elements for which the post-service activity decays at a sufficiently fast rate so that there is little if any residual radiation. Ideally you want a small cross section as well to avoid metallurgical problems.

Figure 16.6. The interior of the tokamak fusion test reactor (TFTR). The plasma from the TFTR produced 10 MW of power, spread out over the first wall tiles and experimental diagnostics. A power plant could produce 100 times as much power, exposing first-wall components to heat fluxes of several megawatts per square meter. © Dietmar Krause, Princeton Plasma Physics Laboratory.

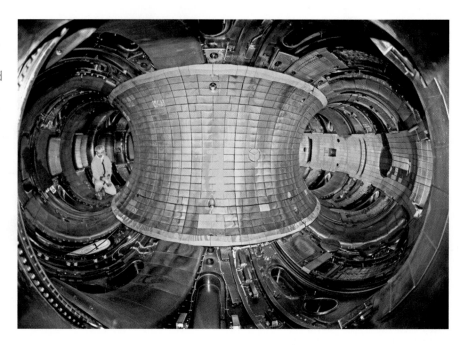

tritium both to sustain the fuel needs of the reactor and to provide the startup fuel for new reactors [17]. Likely materials include Be for the multiplication and Li for the breeding, perhaps in a fluoride-based molten salt (FLiBe, FLiNaK, FLiNaBe). In some designs [17][18], molten metals (such as PbLi) circulate through the entire first-wall–blanket module as coolant for the heat-exchange and power-generation systems; the current focus is on ceramic pellets and lead–lithium (PbLi) systems.

Materials for the first wall and blanket will be at very high temperatures (400–1,000 °C) and must withstand heat fluxes of 1–10 $MW m^{-2}$ and tens to hundreds of **displacements per atom** (dpa) over the wall lifetime [16][18]. While these requirements are similar to those of generation IV fission reactors (see Chapter 13), the energy of the neutrons makes the materials requirements more stringent. Since fusion neutrons have much higher energies than fission neutrons, their mean free path in a material is longer and the number of neutron–material interactions is greater. Thus, a fusion neutron produces about 100 times more helium through transmutations than do fission neutrons [16]. This extra helium-bubble creation can lead to structural defects, should the bubbles accumulate at grain boundaries. Further, the neutrons captured by materials other than the Li (to breed tritium, as discussed above) will activate the material. Radioactive materials must, of course, be handled carefully both during operations and after the end of the reactor's life. Therefore, it is preferable to use low-activation materials to minimize the generation of radioactive waste.

The development of the needed materials will be facilitated by increased understanding of the microstructural stability and deformation behavior of materials under the extreme conditions of high temperatures and heat fluxes, energetic neutrons and ion bombardment, helium impacts, surface erosion, and redeposition on surfaces. In addition, for the whole first wall and blanket, new manufacturing and joining techniques will be required to ensure the desired properties and structure (like in nanostructure- or oxide-dispersion-strengthened ferritic alloys).

16.5.2 Magnetic confinement

In addition to the issues concerning extracting energy from 14.1-MeV neutrons (see 16.4.1), an MFE system has its own unique materials requirements. Some of the more difficult materials issues to address include the interaction of the plasma with the first wall, the absorption and retention of tritium, and the protection from neutron damage of the equipment to heat and contain the plasma.

The first wall of an MFE system, such as that visible in Figure 16.6, actively interacts with the plasma. Plasma ions constantly collide with first-wall material. During such collisions, large quantities of material can ablate off the wall into the plasma. These ejections can cool and quench the fusion reaction. High-Z (Z is atomic number) materials can be especially harmful because they radiate away energy at greater rates than do low-Z materials. For these reasons, the first wall of an MFE reactor requires materials that can withstand large heat fluxes without ablating, such as tungsten, or low-Z materials (such as liquid lithium).

To generate the toroidal magnetic field with acceptable power losses, tokamak reactors will most likely employ superconducting coils. These magnets must be actively cooled with liquid helium, so the magnets must be insulated from the hot blanket and protected from neutron damage. ITER's superconducting coils are made of niobium, tin, or titanium (Nb_3Sn and NbTi) and are chilled with liquid helium to below 4 K [7]. Should they become commercially available in the quantity and quality needed for a reactor, high-temperature superconductors, operating at liquid-nitrogen (above 77 K) temperatures, might offer cheaper and more easily maintained magnets.

Additional materials requirements exist for MFE systems. Some arise from the large forces exerted by the magnetic fields, since sudden changes in the plasma current in strong magnetic fields can lead to large impulses on the support structures and blanket systems. Furthermore, the systems to heat, diagnose, and control magnetically confined plasma have their own materials challenges. For example, the efficient heating by radio-frequency (RF) waves often requires the placement of the RF antenna close to the hot plasma. Thus there exists a need for conductors that retain electrical properties under high heat and neutron loads. Beyond these few examples, many specific technological and material challenges exist for magnetic-confinement fusion power [19].

16.5.3 Inertial confinement

In IFE, unlike in MFE, the first wall does not affect the fusion interaction, but the reactor chamber will need to withstand high-frequency (1–20-Hz) pulsed operation. This complicates both the fueling and heating of the target, and the structural and maintenance requirements of the reactor.

To maintain the repetition rate, the IFE target is launched into the reactor chamber at hundreds of meters per second, actively tracked by the optics systems and compressed to fusion conditions. In each pulse, both debris and neutrons strike the reactor's first wall. Any ablated first-wall material does not impact the reaction, since the target has already exploded, but any debris in the chamber must be evacuated before the next pulse [20]. The first-wall and blanket systems will experience a short pulse of neutrons, so they must be constructed to maintain their strength under repetitive thermal loads.

Producing fusion fuel targets for an IFE plant is another great challenge. Each fuel pellet is a millimeter-scale sphere of D–T fuel. The fuel is held below its triple point in a hollow solid shell filled with a low-

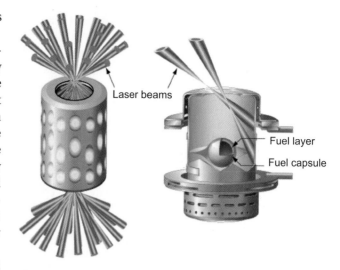

Figure 16.7. An NIF target capsule. In the NIF indirect-drive inertial fusion method, 192 laser beams enter and heat the half-centimeter-diameter cylindrical hohlraum, which emits X-rays that compress the pellet. © Lawrence Livermore National Laboratory.

vapor-pressure gas, at roughly 18 K. This D–T shell is typically contained within a low-Z shell (plastic, Be, or diamond-like high-density carbon) for ease of handling and efficient use of fuel. The D–T fuel itself is currently produced as a smooth frozen ice layer, via a process that uses the decay of tritium known as beta-layering. However, in a commercial plant the high-speed injection of pellets subjects them to large thermal and mechanical stresses. A D–T foam may be an alternative fuel, should these stresses be too great for an ice-layered pellet [20]. For indirect-drive IFE schemes (such as in the NIF and as shown in Figure 16.7), the fuel pellet rests within a hohlraum of high-Z (Au or Pb alloy) metal, designed to maximize laser-to-X-ray conversion efficiencies. Dimensional tolerances of the components range from $10 \, \mu m$ for the hohlraum size to $0.1 \, \mu m$ for its surface roughness. The D–T layer's inner roughness must be smooth to $1 \, \mu m$ and the pellet's outer surface to 0.01–$0.1 \, \mu m$ [20]. These fuel pellets must be made in vast quantities at low cost.

Pellets must also be tracked and targeted in real time. This requires a close interplay between targeting systems and optics, which must operate reliably in a neutron-rich environment. While the optics in the NIF are protected with a disposable first-wall lens that is replaced every 1–10 shots [9], replacement in a power plant must be less frequent.

The active medium of a laser enables the conversion of low-quality light (from a flashlamp or light-emitting diode) into high-quality, coherent radiation. This requires the doping of an active element (typically Nd or Yb) into a host material such as glass, which must be manufactured at high purity and uniformity at a large scale (20–40-cm-wide slabs). While traditional material choices had been limited to crystals, the recent development of transparent ceramics, such as Yb-doped yttrium aluminum garnet (YAG), $Y_3Al_5O_{12}$, could provide the performance necessary for IFE laser systems at power-plant scales [21].

The efficiency of the laser has a direct impact not only on the cost of IFE electricity, but also on the sustainability of the lasing material, as extraneous heat losses can alter the physical properties of the laser system. A promising driver option is to use arrays of laser diodes, which can operate at about 10% efficiency. Key materials problems in such systems (also faced by the photodiode lighting community; see Chapter 35) are the diode's thermal performance, operationally induced fatigue, and material creep.

16.6 The future of fusion energy

The hope of the research community is that by mid century an early generation of demonstration fusion power reactors will be built. But how can fusion become an integral part of the global energy economy more quickly? Certainly, its integration into the global economy would be hastened by solving the material problems outlined in this chapter in parallel with making the scientific advances pursued by contemporary fusion experiments.

One tempting but speculative possibility for a nearer-term role for fusion technology has been a fusion–fission hybrid reactor [10]. A hybrid reactor might use the fusion core as a source of high-energy neutrons, which would bombard low-grade fissile material, triggering secondary fission reactions in the blanket. This process could serve as a means of amplifying the energy output of the fusion core, or possibly as a method of burning up radioactive waste.

While the scientific and technological difficulty has delayed the commercial utilization of fusion energy, the benefits of a fusion-based energy source (no operational greenhouse-gas emissions or long-lived radioactive waste, few proliferation or nuclear runaway risks, and a nearly inexhaustible fuel supply) are still attractive. However, many technical and scientific obstacles must be overcome for fusion energy to become practical.

Considering those technical obstacles, such as the materials concerns introduced herein, and other externalities (e.g., the environmental costs of greenhouse-gas emissions), the timing for fusion as a practical energy source is uncertain. What is clear, though, is that, with next-generation experiments aiming to produce energy gain from controlled nuclear fusion, the time is ripe for filling research gaps that block the path to practical fusion energy. With additional investments in research and development, the commercialization of nuclear fusion might arrive sooner than expected.

16.7 Summary

Economical fusion energy will require significant advances in materials. The D–T fusion environment is harsh, with high fluxes of high-energy neutrons. These neutrons are used to breed reactor fuel and to generate electricity, but are extremely damaging to materials. With both the magnetic and the inertial fusion efforts moving toward reactor-relevant power regimes, developing materials suitable for the harsh reactor environment is an important step in the quest for controlled nuclear fusion.

16.8 Questions for discussion

1. Estimate the rate of tritium burnup (in grams per day) in a D–T power plant that produces 1 gigawatt of fusion power (when engineering efficiencies are taken into account, this plant may produce only hundreds of megawatts of electrical power). Assume that each tritium atom yields one 14.1-MeV neutron. Assume that you can ignore D–D reactions, if the D–D cross section for fusion is much less than the D–T cross section. How many kilograms of coal would be needed per day? Assume that the energy density of coal is approximately $20\ MJ\,kg^{-1}$.

2. Using your answer from question 1, how many pellets of fuel would an IFE plant consume in a year, assuming that each pellet contains $150\,\mu g$ of D–T gas? How many joules of energy are released per pellet? If one pellet is consumed per shot, at how many shots per second does the plant operate? If the fusion-to-electricity power efficiency of the plant is 50%, and electricity is sold at 5 cents per kilowatt-hour, estimate the value in dollars per year of the electricity sold. If the plant breaks even, with the cost of the shots exactly equal to the value of the electricity sold, what is the cost per shot? If only 1% of each shot's total cost is allowed to be from the fuel pellet, what is the cost per pellet? What does this mean for IFE target manufacturing? What challenges to IFE target production exist?

3. When a wire of length L carrying current I passes through a magnetic field B, it experiences a force, known as the **Laplace force**: $\bar{F} = L\bar{I} \times \bar{B}$. Assume that an MFE power plant carries a plasma current of around $10\,MA$ in a magnetic field of approximately $10\,T$. If the plasma can be approximated as a wire of length 30 m, estimate the magnitude of the Laplace force acting on the plasma. If the mass of the plasma is 1 g, how quickly would it accelerate under the influence of this force, using $\bar{F} = m\bar{a}$?

Falling objects near the Earth's surface experience an acceleration of approximately 9.8 m/s^2. How many times greater than this is the plasma's acceleration due to the Laplace force?

4. Under normal operation, the forces on magnetically confined plasma are balanced, but during a loss of confinement the balance can be disrupted and the Laplace force can accelerate the plasma. Why would this be undesirable? What are the implications of this for the materials of the reactor's first wall?

5. How does neutron damage and helium creation in a fusion environment differ from that in a fission environment (consider **atomic parts per million**, appm, He per dpa) (see Chapter 13)? If plant designs specify lifetime neutron doses of 100 dpa, what lifetime levels of He are expected in a fusion plant? What are some materials considered for the first wall? What considerations for the structure and composition of the materials facing the higher-energy fusion neutrons should be taken into account when designing or choosing the materials? How do the materials requirements for an IFE first wall differ from those for an MFE first wall?

6. What cryogenic materials exist in (a) an MFE plant and (b) an IFE plant? What are some challenges to having these cryogenically cooled systems?

7. Many of the challenges in fusion materials research stem from the energetic neutrons in a D–T fusion reaction. This has prompted interest in other fusion reactions, such as the deuterium–helium-3 and proton–boron-11 processes. Neither of these chains produces neutrons (although side D–D reactions in a D–He-3 reactor can produce lower-energy neutrons). These **advanced fuels** require much higher temperatures than for D–T in order to fuse (the D–He-3 cross section peaks at a four times greater temperature than does that for D–T, and with a smaller cross-section value, while p–B-11 would require ten times the temperature of D–T). This not only makes controlling the reaction more challenging, but also limits the energy obtained from the process: as temperatures increase, X-ray-radiation losses from the plasma become more pronounced. How could fusing advanced fuels change the materials requirements for a fusion plant? How would the necessary properties for the first wall, breeding blanket and power converter change if the energetic byproducts of the fusion reaction were all charged particles instead of neutrons? What advances in materials science and engineering, such as the invention of an X-ray mirror, could make advanced fuels more feasible for a reactor?

8. A hybrid fusion–fission power plant could use a fusion core as a source of high-energy neutrons to drive fission reactions in a blanket of low-grade fissile material or nuclear waste. How are the materials requirements for a hybrid plant different from those for (a) a pure fusion plant and (b) a pure fission plant? How do the proliferation and safety risks of fission, fusion, and hybrid plants differ? What would be the relative advantages and disadvantages of the three types of reactors?

9. How might a self-annealing conductor be useful in a tokamak reactor? (See [22] for how one might be constructed with nanoparticles.)

16.9 Further reading

- **T. K. Fowler**, 1997, *The Fusion Quest*, Baltimore, MD, Johns Hopkins University Press. In this history of fusion research, Fowler tracks not only historical progress, but also the technological and physical challenges facing fusion research. *The Fusion Quest* is more technical than other histories of fusion, yet written for a non-specialist.

- **S. Zinkle**, 2005, "Fusion materials science: overview of challenges and recent progress," *Phys. Plasmas*, **12**, 058101–9. This technical article represents a good introductory review to material science concepts important for fusion energy systems. The references contained therein serve as a good body of detailed technical information about material science issues in fusion reactors, beyond the scope of this introductory text.

- **Farrokh Najmabadi** and the ARIES Team 2006, "The ARIES-AT advanced tokamak, advanced technology fusion power plant," *Fusion Eng. Design*, **80**(1–4), 3–23. The ARIES studies are detailed analyses of fusion power plants, which not only document the scientific and engineering challenges to fusion, but also aim to design a working demonstration power plant on the basis of current knowledge. Readers will find technical details and economic estimates of a proposed fusion power plant.

- **E. I. Moses**, 2009, "Ignition on the National Ignition Facility: a path towards inertial fusion energy," *Nucl. Fusion*, **49**, 104022–30. This review article documents the start-up of the NIF, illustrates the scientific basis of inertial fusion energy, and proposes a next-step hybrid fusion–fission power plant, called *LIFE*, which uses fissile material to increase the energy gain from laser inertial fusion reactions.

- **J. Lindl**, 1998, *Inertial Confinement Fusion: The Quest for Ignition and Energy Gain Using Indirect Drive*, New York, AIP Press–Springer. Lindl details the scientific and historical progress of indirect-drive laser inertial-confinement fusion. With both technical detail and sufficient background material, it serves as both an introduction to the field and a reference for seasoned researchers.

- **R. P. Drake**, 2006, *High-Energy-Density Physics: Fundamentals, Inertial Fusion, and Experimental Astrophysics*, Berlin, Springer. An introduction to the broader science of high-energy-density physics, Drake's book examines inertial fusion not only as an energy source, but also as a laboratory for the study of such astrophysical phenomenon as supernovae.

- **W. Stacey**, 2010, *Fusion: An Introduction to the Physics and Technology of Magnetic Confinement Fusion*, Weinheim, Wiley–VCH. Stacey gives an introduction to the plasma physics and technology of magnetic-confinement fusion. This advanced undergraduate/graduate-level textbook develops a broad range of topics from the ground up, and serves as a comprehensive introduction to the field.

- **D. Rose** and **M. Clark, Jr.**, 1963, *Plasmas and Controlled Fusion*, Cambridge, MA, MIT Press and John Wiley and Sons. Despite being nearly 50 years old, Rose and Clark's overview of the naissance of fusion research still rings true today. While the introduction to plasma physics may be heavily detailed, the discussion of materials issues is concise and still relevant.

- **H. P. Furth**, 1995, "Fusion," *Scientific Am.*, **273**(3), 174–176. Furth, a pioneer in fusion research, briefly discusses the basics of nuclear fusion before examining the next 50 years of research and the future of fusion as an energy source.

- **US Department of Energy, Office of Fusion Energy Sciences**, 2009, *Research Needs for Magnetic Fusion Energy Sciences (ReNeW) Final Report*, http://burningplasma.org/renew.html. The ReNeW report details the scientific and technological challenges for magnetic-confinement fusion energy and suggests specific research paths to move MFE research from the current state of the art to commercial power production.

16.10 References

[1] *The Nobel Prize in Physics 1967*, http://nobelprize.org/nobel_prizes/physics/laureates/1967/index.html.

[2] **R. Rhodes**, 1995, *Dark Sun: The Making of the Hydrogen Bomb*, New York, Simon and Schuster.

[3] 1958 *Progress in Atomic Energy: Proceedings of the Second United Nations International Conference on the Peaceful Uses of Atomic Energy*.

[4] **J. P. Freidberg**, 2007, *Plasma Physics and Fusion Energy*, Cambridge, Cambridge University Press.

[5] **I. E. Tamm** and **A. D. Sakharov**, 1961, "Plasma physics and the problem of controlled thermonuclear reactions," in *Proceedings of the Second International Conference on the Peaceful Uses of Nuclear Energy*, ed. **M. A. Leontovich**, Oxford, Pergamon, vol. 1, pp. 1–47.

[6] **J. Wesson**, 2004, *Tokamaks*, 3rd edn., Oxford, Oxford University Press.

[7] *ITER – The Way to New Energy*, http://www.iter.org.

[8] **J. Lindl**, 1998, *Inertial Confinement Fusion: The Quest for Ignition and Energy Gain Using Indirect Drive*, New York, AIP Press–Springer.

[9] *National Ignition Facility and Photon Science*, http://lasers.llnl.gov.

[10] **E. I. Moses**, 2009, "Ignition on the National Ignition Facility: a path towards inertial fusion energy," *Nucl. Fusion*, **49**, 104022–30.

[11] **K. Schwochau**, 1984 *Extraction of Metals from Sea Water*, Berlin, Springer.

[12] **S. E. Bodner**, **D. G. Colombant**, **J. H. Gardner** *et al.*, 1998, "Direct-drive laser fusion: status and prospects," *Phys. Plasma*, **5**(5), 1901–1918.

[13] **R. L. Klueh**, **D. S. Gelles**, **S. Jitsukawa** *et al.*, 2002, "Ferritic/martensitic steels – overview of recent results," *J. Nucl. Mater.*, **307–311**, 455–465.

[14] **R. Kurtz**, **K. Abe**, **V. M. Chernov** *et al.*, 2000, "Critical issues and current status of vanadium alloys for fusion energy applications," *J. Nucl. Mater.*, **283–287**, 70–78.

[15] **B. Riccardi**, **L. Giancarli**, **A. Hasegawa** *et al.*, 2004, "Issues and advances in SiC_f/SiC composites development for fusion reactors," *J. Nucl. Mater.*, **329–333**, 56–65.

[16] **S. Zinkle**, 2005, "Fusion materials science: overview of challenges and recent progress," *Phys. Plasmas.* **12**, 058101–9.

[17] **A. Li Puma**, **J. L. Berton**, **B. Brañas**, *et al.*, 2006, "Breeding blanket design and systems integration for a helium-cooled lithium–lead fusion power plant," *Fusion Eng. Design*, **81**, 469–476.

[18] **A. R. Raffray**, **L. El-Guebaly**, **S. Gordeev** *et al.*, 2001, "High performance blanket for ARIES-AT power plant," *Fusion Eng. Design*, **58–59**, 549–553.

[19] **US Department of Energy, Office of Fusion Energy Sciences**, 2009, *Research Needs for Magnetic Fusion Energy Sciences (ReNeW) Final Report*, http://burningplasma.org/renew.html.

[20] **D. T. Goodin**, **N. B. Alexander**, **L. C. Brown** *et al.*, 2005, "Demonstrating a target supply for inertial fusion energy," *Fusion Sci. Technol.*, **47**(4), 1131–1138.

[21] **J. Kawanaka**, **N. Miyanaga**, **T. Kawashima** *et al.*, 2008, "New concept for laser fusion energy driver by using cryogenically-cooled Yb:YAG ceramic," *J. Phys.: Conf. Series*, **112**, 032058–61.

[22] **X.-M. Bai**, **A. F. Voter**, **R. G. Hoagland**, **M. Nastasi**, and **B. P. Uberuaga**, 2010, "Efficient annealing of radiation damage near grain boundaries via interstitial emission," *Science*, **327**, 1631–1634.

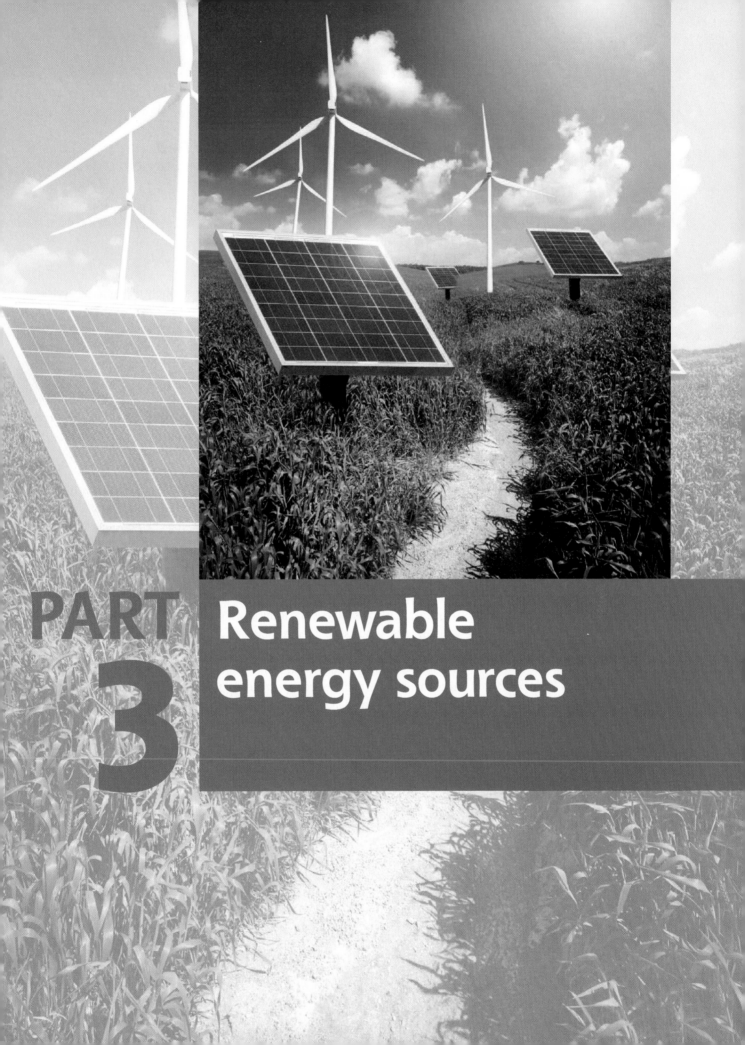

PART 3

Renewable energy sources

17 Solar energy overview

Miguel A. Contreras and Satyen Deb

Center for Basic Sciences, National Renewable Energy Laboratory, Golden, CO, USA

17.1 Focus

Utilization of solar energy on a terrawatt scale is a viable, environmentally conscious solution to the growing global demand for energy. Key solar technologies that can provide significant reductions in carbon emissions and environmental pollution, including conventional photovoltaics, concentrating photovoltaics, and solar thermal technologies, as well as passive solar technologies such as biofuels, biomass, and wind power, are highlighted here and discussed in detail in the following chapters.

17.2 Synopsis

As described in Chapter 3, energy, water, and food supply will all pose key challenges in the coming decades. Because of their ramifications in terms of socio-economic development, quality of life, and political relations, these mounting concerns could result in more conflict and global instability if not addressed promptly and effectively. This chapter summarizes the contributions that active and passive solar energy technologies could make toward addressing this crisis. Solar energy is a viable solution to both energy demand and environmental issues; however, the mass deployment of some solar technologies faces some real challenges that are not necessarily only technological in nature but in many cases are also economic and sociopolitical. Today, coal, natural gas, oil derivatives, and nuclear energy are the most cost-effective routes to large-scale electricity generation. However, these traditional technologies have environmental impacts that, in some instances, have been catastrophic not only for the environment but also for the people and the economy of the areas affected (e.g., the Chernobyl nuclear disaster of 1986, the Gulf of Mexico oil spill of 2010).

The subsequent chapters of this book (Chapters 18–28 and 30) discuss solar-energy-related technologies such as photovoltaics (PV), solar thermal [now more commonly referred to as concentrating solar power (CSP)], fuel production driven by solar energy (photosynthesis and artificial photosynthesis), thermoelectric materials, biomass, and wind power. The technical information that can be found in these chapters is intended to provide a view into the state of the art of these technologies, together with some insight into the advantages and current shortcomings for implementation. This introductory chapter on solar energy offers a general overview of global energy demand, the current costs of the different methods/technologies for electricity generation, some fundamental understanding of the solar resource, and a general account of the technologies that are covered in much more detail in subsequent chapters.

17.3 Historical perspective

Energy from the Sun is the basis of life on Earth, and, from time immemorial, humans have harnessed it with ever-evolving technologies to generate useful power. Prior to the discovery of oil and gas, ancient cultures used biomass (e.g., wood, peat, grasses, dried animal dung) to provide heat and light. The ancient Greeks, Chinese, and Romans designed dwellings and other buildings to effectively utilize sunlight for lighting and heating. For example, the Romans developed the first greenhouses around 100 CE, using transparent minerals and rocks such as mica to collect solar energy in order to provide a warm environment for the growth of fruits and vegetables. Likewise, the ancestral Puebloan peoples of what is now the southwestern USA (often called Anasazi), who lived from about 1200 BCE to 1300 CE, built their dwellings to use passive solar effects. Specifically, almost all of their homes were built facing southward to take advantage of the warmth of the winter sun. In addition, their later famous cliff dwellings were typically carved out of southward-facing cliffs that had large rock overhangs, so that, during the summer, when the Sun was high in the sky, their homes stayed cool in the shade of the overhang.

Similarly, the use of passive solar energy in the form of wind is well documented in history. The earliest example is the use of sails for water transportation, but windmills also date from at least as early as the seventh century CE. The earliest windmills, in Persia and China, used vertical sails rotating about a central vertical axis. In contrast, the modern, horizontal-axis windmill originated several centuries later in western Europe. In both cases, the initial uses of windmills were to automate the labor-intensive tasks of pumping water and grinding grain.

Thus, both active and passive solar energy practices have always been close to human development, and their utilization through new technological developments in materials science, chemistry, and physics has exploded into a variety of devices and processes for the conversion of solar energy into electricity and alternative forms of fuel.

17.4 Outlook on global energy demand and supply

Currently, the bulk of the energy supply (80%–90%) is based on the use of fossil fuels (coal, oil, and natural gas), but the threat of climate change and environmental pollution (Chapter 1–3) offers great opportunities but grand challenges for deploying renewables on a meaningful scale [1].

Mass deployment of renewable energy technologies on a global scale is a solution to balancing energy needs with environmental protection, and some of these technologies are beginning to be cost-competitive with conventional power generation. However, technological, scientific, and economic challenges remain and have prevented such technologies from reaching large-scale deployment (terawatts). To get an idea of how expensive renewables are in the global energy landscape today, one can use a measure such as the **levelized cost of electricity (LCOE)**, which is a number in cents per kilowatt hour at net present value and takes into account the cost complexities associated with the entire lifetime of a power plant from financing to end of life [2]. Table 17.1 compares LCOEs among different types of power sources. According to this table, technologies such as biomass are currently cost-competitive with coal and natural gas. In contrast, the fact that the current levelized costs of PV and solar thermal are four to five times higher than those of coal and natural gas prevents these technologies from penetrating the utility market on a mass scale. In addition, many solar technologies are intermittent or generate only during the day. Thus, at a very large scale, the cost of energy storage (discussed in detail in Chapter 44) will need to be integrated into the LCOE. Nevertheless, technological advances in efficiency and techniques to lower manufacturing costs continue to drive down the costs of renewable technologies toward competitiveness with conventional energy sources in large-scale electrical power generation.

For small- (1–100 kW) and medium-sized (100 kW–10 MW) applications, LCOE analyses indicate that solar technologies can already be cost-competitive. This is particularly true in cases in which the renewable resource (solar, wind, etc.) is plentiful, the grid is not available and would be cost-prohibitive to extend, and governmental credits or subsidies are available.

It is expected that renewable energy technologies will eventually have favorable LCOEs even in large-scale power generation. Moreover, two considerations that are excluded from LCOE arguments are human health and the environment. If factors such as human disease caused by pollution, loss of arable land, and contamination of drinkable water were considered, then there is no question that renewable energy technologies would be cost-competitive.

When properly designed, solar electric and hybrid systems (solar with diesel) can provide 100% of the electrical needs for a small village, a hospital, clinic, school, farm, ranch, or small commercial operation. Additional progress in energy storage to tackle the problem of intermittency in most solar power plants, combined with improved efficiencies, will help further solidify the advantages of solar technologies.

Table 17.1. Levelized costs of electricity for power plants using different sources [2]

US average levelized costs (2008 $ per MW·h) for plants entering service in 2016

Plant type	Capacity factor[a] (%)	Levelized capital cost	Levelized fixed O&M	Levelized variable O&M (including fuel)	Levelized transmission investment	Total system levelized cost
Conventional coal	85	69.2	3.8	23.9	3.6	100.4
Next generation coal	85	81.2	5.3	20.4	3.6	110.5
Next generation coal with CCS	85	92.6	6.3	26.4	3.9	129.3
Conventional combined cycle	87	22.9	1.7	54.9	3.6	83.1
Advanced combined cycle	87	22.4	1.6	51.7	3.6	79.3
Advanced CC with CCS	87	43.8	2.7	63	3.8	113.3
Conventional combustion turbine	30	41.1	4.7	82.9	10.8	139.5
Advanced combustion turbine (relevant for IGCC)	30	38.5	4.1	70.0	10.8	123.5
Advanced nuclear	90	94.9	11.7	9.4	3.0	119.0
Wind	34.4	130.5	10.4	0	8.4	149.3
Wind, offshore	39.3	159.9	23.8	0	7.4	191.1
Solar PV	21.7	376.8	6.4	0	13.0	396.1
Solar thermal	31.2	224.4	21.8	0	10.4	256.6
Geothermal	90	88.0	22.9	0	4.8	115.7
Biomass	83	73.3	9.1	24.9	3.8	111.0
Hydroelectric	51.4	103.7	3.5	7.1	5.7	119.9

[a] The capacity factor is defined as the ratio of the actual output of a power plant over a period of time to its output if it had operated at full maximum output the entire time.
CC, combined cycle; CCS, carbon capture and sequestration; O&M, operations and maintenance costs.

17.4.1 Solar energy

The solar resource is abundant, but also finite; therefore, before the introduction of specific solar technologies, a quantitative estimate of the solar resource is in order. The energy output emanating from the Sun (or any star) and the energy reaching Earth can be estimated by applying some fundamental laws of physics. First, assuming the Sun to be a blackbody emitter, the temperature of the Sun's surface can be estimated using Wien's displacement law for the wavelength distribution of thermal radiation from a blackbody:

$$\lambda_{max} = \frac{0.0029}{T}, \qquad (17.1)$$

where 0.0029 is Wien constant in units of Kelvin nanometers (K nm), T is the temperature of the Sun (or any other emitting body) in units of kelvin, and λ_{max} is the

Figure 17.1. The ASTM G173–03 spectral irradiance for extraterrestrial, global, and direct solar irradiation and for an ideal blackbody at a temperature of 5,800 K.

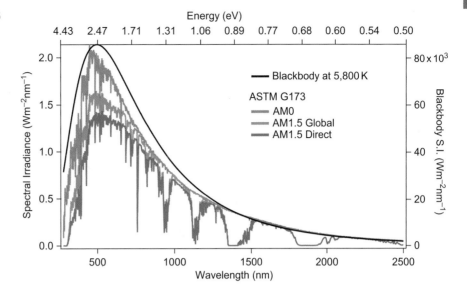

wavelength of maximum-intensity emission in units of nanometers. Measurements of the solar spectrum indicate that the highest intensity occurs at a wavelength of ~500 nm, which leads to a surface temperature for the sun of 5,780 K. Next, the total spectral irradiance, $P(\lambda)$ of an ideal (spherical) blackbody radiator held at a temperature of about 5,780 K is given by Planck's formula

$$P(\lambda) = 2\pi \frac{hc^2}{\lambda^5 (e^{hc/(k_B T \lambda)} - 1)} \ [\text{W m}^{-2} \text{ nm}], \qquad (17.2)$$

where h is Planck's constant (6.623×10^{-34} J s), c is the speed of light (2.998×10^8 m s^{-1}), λ is the wavelength, k_B is the Boltzmann constant (1.380×10^{-23} J K^{-1}), and T is the blackbody temperature. Integrating $P(\lambda)$ over all wavelengths gives a power density of 63.343 MW m^{-2} given off by the surface of the Sun. Furthermore, because the intensity of electromagnetic radiation (light) decreases monotonically with the inverse square of the distance from the source, it is possible to estimate the irradiance just outside the Earth's atmosphere as 1,367 W m^{-2}. Considering this value and the illuminated area of the Earth, one can estimate an average of 174,703 TW for the total irradiation arriving just above the Earth's atmosphere. This is an enormous amount of power; in fact, on the basis of these values, one could say that, each hour, the Earth receives more energy from the Sun than was consumed globally for the entire 2007 calendar year!

These calculations for an ideal blackbody apply to the region just outside Earth's atmosphere. However, because of scattering and optical absorption (by water vapor, nitrogen, oxygen, carbon monoxide, and other greenhouse gases) in the atmosphere, the solar spectrum on Earth's surface can be expected to be further attenuated and to have multiple strong and weak absorption features. Furthermore, the true irradiance at any point on Earth also depends on the optical path for sunlight (a function of latitude, time of year, and chemical composition of the atmosphere). Because of these variables, there is a need for a reference standard of the solar spectrum in order to allow quantitative and fair comparisons among PV (and other solar) products in terms of parameters such as energy-conversion efficiency, performance, and reliability and to facilitate studies in modeling and accelerated testing of various materials and components for solar and other products. The PV industry, in collaboration with the American Society for Testing and Materials (ASTM), developed such a standard, which is currently defined in the document ASTM G173–03. This standard defines two terrestrial spectral distributions: direct normal and global (see Figure 17.1). It also includes the latest refinement of the solar irradiance just above Earth's atmosphere, denoted as Air Mass 0 (AM0, by definition, refers to extraterrestrial irradiance), from which the other two spectra are derived. The direct normal spectrum is the direct component contributing to the total global (hemispherical) spectrum. The ASTM G173–03 spectra represent the terrestrial solar spectral irradiance on a surface of specified orientation (37° tilt angle toward the equator, also known as Air Mass 1.5) under one specified set of atmospheric conditions. The conditions selected were considered to be a reasonable average for the 48 contiguous states of the USA over a period of one year. The tilt angle selected is approximately the average latitude for the contiguous USA. In practice, the global spectrum is useful for 1-Sun-type studies and measurements (flat-plate thermal collectors, PV modules and cells), whereas the direct spectrum is better suited for concentrated solar power technologies.

Solar technologies such as PV and solar thermal use different portions of the solar spectrum and different photon energies. Solar thermal technologies typically aim to capture photons from the whole spectrum in an attempt to generate maximum thermal energy. PV technologies, on the other hand, can generate electric power only from photons with energies higher than the bandgap energy of their absorbing materials. Photons with energies lower than the bandgap are not absorbed by the PV cell and contribute to heating the solar cells when absorbed by the metallic back contact.

Solar technologies can be broadly classified as active or passive depending on the nature of energy conversion and utilization. Active solar technologies, for example, include PV and solar thermal systems that directly convert solar radiation into useful electrical and thermal energies, respectively. Passive solar, on the other hand, includes biomass (Chapters 24–28, especially Chapters 25 and 26), wind (Chapter 30), and hydropower (Chapter 29), which use the secondary effects of solar energy. Other passive applications include low-emissivity windows and Trombe walls (see Chapter 36), as well as passive solar water heaters.

17.4.2 Photovoltaics

The early stage of technology development in photovoltaics was directed toward space applications, for which cost was not a major concern. With the advent of the oil embargo in 1973 and increased awareness of global warming, the primary emphasis of PV technology is now on improving this technology so that it is cost-competitive with conventional means of power generation.

The basic principle of a PV device is relatively straightforward. The simplest device consists of two materials that differ in electronic properties (one electron-rich and the other electron-deficient). These materials are typically placed between two electrodes forming the cell front and rear contacts. When light strikes the two dissimilar materials, electron–hole pairs are generated. These photogenerated charge carriers are separated at the interface between the two materials by an internal electric field, resulting in the generation of electrical power. Individual cells are connected in series and parallel combinations to form modules and arrays to deliver desired levels of DC power to an external load. If DC-to-AC power conversion is needed, such as in the case of grid-tie systems, then the use of inverters is required. One of the major advantages of PV devices is their modularity, enabling the fabrication of power sources ranging from a few watts to megawatts depending on the application.

Since no energy conversion process is 100% efficient, a discussion of some of the limits of PV energy conversion is warranted. It is clear that maximizing the current and voltage outputs from PV cells is a main concern when attempting to increase conversion efficiency. The amount of current (and voltage) one can expect from a given solar cell material generally depends on (a) the material's energy bandgap, (b) the carrier-generation rate, and (c) the carrier-recombination rate. The energy bandgap value is material-dependent, since it is an intrinsic semiconductor property related to crystal structure, chemical composition, and, to a smaller degree, material structural properties (e.g., stress, particle size).

In practice, generation and recombination are the main physical phenomena that control solar-cell efficiency. A standard function that provides information on current generation and recombination of a given solar cell is the **quantum efficiency (QE)** or **spectral response** of the cell:

$$QE = QE(\lambda). \tag{17.3}$$

$QE(\lambda)$ is the ratio between the number of electrons (i.e., charge carriers) the cell yields and the number of photons illuminating the cell for a given unit of time. From this definition, a QE value of unity implies that every photon yields one electron–hole pair. In practice, $QE < 1$ for even the most efficient PV solar cells today, indicating that improvements to solar-cell efficiency are still possible. Figure 17.2 shows typical QE curves for selected PV materials. Classical theories assume that one photon can generate at most one electron–hole pair, but advocates of third-generation solar cells (particularly concepts such as intermediate-bandgap solar cells and impact ionization; see Chapter 19) challenge this viewpoint and suggest that values of $QE > 1$ are achievable.

If $QE(\lambda)$ for a given solar cell is known, then the maximum theoretical current density of the cell can be calculated for any given incident spectrum according to

$$J_{max} = q \int_{\lambda_0}^{\lambda_1} QE(\lambda)\Phi(\lambda)d\lambda, \tag{17.4}$$

where q is the electron charge, λ is the wavelength, λ_0 and λ_1 are the limits of the wavelength range within which the cell shows photoactivity, and $\Phi(\lambda)$ is the fluence of the spectrum illuminating the cell. $\Phi(\lambda)$ can be obtained for any given spectral irradiance, $P(\lambda)$, from

$$\Phi(\lambda) = \frac{P(\lambda)}{E(\lambda)} \text{ [number of photons m}^{-2}\text{ s}^{-1}\text{ nm].} \tag{17.5}$$

In Equation (17.5), $E(\lambda)$ is the photon energy, given by $E = hc/\lambda$, where h is Planck's constant, c is the speed of light, and λ is the photon wavelength. These equations allow the estimation of some upper limits on current generation in PV devices (cells and modules) and can also be used to understand some physical limitations and shortcomings of any PV device.

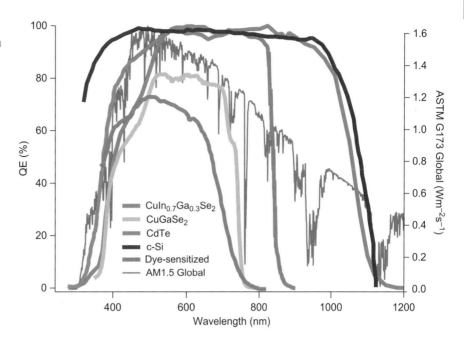

Figure 17.2. Quantum efficiency (carrier-generation rate per photon absorbed) of selected solar cells and a portion of the ASTM G173 global spectrum for reference.

Technology development in PV today is focused on increasing cell efficiency, enhancing durability, and reducing cost in order to achieve grid parity. The past three decades of intense technology developments have led to major advances in each of these areas. Efficiencies of research-type devices in the range of 20%–24.7% have been achieved in single-crystal and multicrystalline silicon-based solar cells; efficiencies of thin-film solar cells have reached 20%; ultrahigh efficiencies of over 40% have been achieved in multijunction devices based on III–V semiconductors; and potentially low-cost organic solar cells have exhibited efficiencies of over 6%. The efficiencies of commercial modules lag behind these values by a wide margin, and future technology development is directed toward narrowing this gap. The durability of the current generation of commercial devices based on single-crystal and polycrystalline silicon is projected to be 25–30 years. The durability of some of the commercial thin-film technologies is projected to be around 20 years and is being actively investigated. However, the durability of some of the future-generation technologies, which hold promise for lower cost, has not been clearly established. The most important parameter in the technology commercialization of a PV system is its cost, in terms of price per peak watt (US$ per W_p). The cost figure of merit ($ per W_p) is the ratio of manufacturing cost per unit area ($\$\,m^{-2}$) to the power output of the same area ($W\,m^{-2}$). Thus far, the cost of PV modules has followed the usual 80% learning curve, meaning that the price drops by 20% for every doubling of the cumulative production of PV modules. During the past three decades, the cost of PV electricity has decreased from about $3.65 per kW·h in 1976 to the current cost of about $0.25 per kW·h. If this trend continues, PV can achieve grid parity in the not-too-distant future. In view of numerous off-grid applications (see Chapter 23), which are particularly relevant to the economy of the developing world, PV is cost-competitive even now. The PV system costs include both modules and balance-of-system (BOS) costs, and typically the module costs amount to ~50% of the system costs. The BOS includes items such as inverters, chargers, batteries, and mounting hardware (see Chapter 18). At the present time, the cost of installing a 1-kW PV system ranges from $4,000 to $8,000, not including the life-cycle costs (LCC) and operations and maintenance costs.

A PV installation will generate clean power and produce no carbon emissions. However, the manufacturing of a PV module requires energy, as does that of each component in the BOS. Hence, a valid question is that of how long a PV module (or PV installation) must operate to generate as much energy as was used to make it. This is the concept of **energy payback time** [3] that can be used to compare cost issues of different PV technologies (silicon, thin films, etc.). Assuming (1) an energy requirement of 420 kW·h m^{-2} to manufacture multicrystalline silicon modules, (2) a similar amount of energy for the BOS, (3) a 12% module conversion efficiency, and (4) an availability of 1,700 kW·h m^{-2} per year of solar energy (the US average is 1,800 kW·h/m^2 per year), a payback of about 4 years for multicrystalline silicon modules can be estimated. Because PV systems come with a typical warranty of 25–30 years, it is safe to say that the energy investment in manufacturing the PV modules will provide 20+ years of clean electrical power.

Chapter 18 addresses PV materials and the physics of the solar cells introduced here, and Chapter 19 covers possible routes to enhancements in energy-conversion efficiency through novel concepts and materials. Solar-cell materials and efficiency are key topics in any attempt to further reduce the cost of PV technologies.

17.4.3 Concentrating photovoltaics

An alternative strategy for reducing the cost of solar energy is to use high-efficiency solar cells with relatively inexpensive concentrator optics and tracking systems. Concentrating photovoltaics (CPV) uses concentrating optics in the form of mirrors or lenses to focus solar radiation into small-area devices. thereby reducing the amount of solar-cell material needed and enabling the use of relatively expensive but highly efficient solar-cell materials. The optics can be designed for both low and high concentration, and usually, the low-concentration approaches ($<30\times$) apply more cost-effective PV cells such as those based on silicon or other thin-film devices. Low-concentration designs use linear concentrating optical elements that require only one-axis tracking (parabolic troughs or linear Fresnel lenses) and typically do not require active cooling of the cells. On the other hand, high-concentration optics ($>100\times$) are used in conjunction with more expensive, higher-efficiency multijunction cells and might require active cooling, depending on the concentration level. High-efficiency CPV uses parabolic dish reflectors or point Fresnel lenses that need a two-axis Sun-tracking system to optimize the power-generation efficiency.

Recently, remarkable progress has been made in achieving efficiencies around 41% under concentration in triple-junction solar cells consisting of GaInP/GaInAs/Ge grown monolithically by metal–organic chemical vapor deposition (MOCVD) on a Ge substrate. Several variations of this device structure, all involving different combinations of III–V semiconductor alloys, have shown similarly high efficiencies. These devices are fully space-qualified and are currently manufactured on a large scale for space applications. The key materials in wafer form for III–V alloys are too expensive to be cost-competitive for flat-plate terrestrial applications. However, in view of the major advances during the past decade in cell efficiencies that promise to reach 45%–50%, together with significant cost reductions in system technology, CPV based on III–V alloy semiconductors could become a serious contender for large-scale centralized power generation. Recent comparative cost analyses done on several PV technologies show that high-efficiency CPV can have the lowest cost. Currently, more than 30 companies worldwide are involved in the commercialization of CPV technologies, and several systems with power ratings of 1–100 kW have been deployed during the past few years. With the PV market experiencing growth at the gigawatt level, CPV is poised to make a strong market entry with a production capacity of hundreds of megawatts. Chapter 20 discusses at length the fundamentals of concentration optics and tracking systems, as well as the basic operating characteristics of multijunction PV devices under high concentration (500–1,000 Suns).

17.4.4 Solar thermal technologies

Solar thermal technologies utilize the longer-wavelength component of solar radiation and can be broadly categorized into low-, medium-, and high-temperature systems. Low- and medium-temperature systems ($<220\,°C$) generally use flat-plate collectors for a variety of residential and commercial applications, such as hot water for space heating, process heat, solar drying, and cooking that are outside the scope of this review. High-temperature solar thermal technologies, which require solar concentration, are generally used either to produce electricity through heat engines driving a generator or to drive chemical reactions to produce alternative fuels. A typical solar thermal energy-conversion system consists of three components: (1) an optical system of mirrors and lenses that captures and concentrates solar radiation; (2) a receiver onto which the concentrated radiation is directed and to which the heat is transferred; and (3) a working fluid to do mechanical work, generate electricity before or after storage, or produce fuels through thermochemical processes. One of the advantages of solar thermal power-conversion systems relative to PV is the option for thermal storage, thereby minimizing problems associated with the intermittent nature of solar energy. This time-shifting ability is key for this technology be employed at a large scale or in residential systems.

The technologies of concentrating solar power (CSP) used today employ linear- or point-focus concentrators such as parabolic troughs or dishes to heat a receiver. A large scale realization of the point of focus concept is that of the solar power tower which has an array of heliostats focused on a small area. (Figure 21.6(b)). Examples of CSP optics are shown in Chapter 21 in Figures 21.5(a) and 21.6(a). The engine/generator system takes the heat from the thermal receiver and uses it to produce electricity. Several methods for solar thermal-to-electrical energy conversion have been developed. The most common power-conversion cycles in power plants are the steam-based Rankine cycle and the air-based Brayton cycle, also known as the gas-turbine cycle. These two cycles can be combined to achieve higher efficiency by better utilizing the process heat. Another approach combines a parabolic concentrating dish with a Sterling heat engine driving an electric generator; it is currently the most efficient approach among CSP technologies, which have generally attained efficiencies in the range 20%–31%. Commercial

CSP power plants can generate several hundred megawatts of electricity. At the present time, the cost of solar thermal power is around $0.12–0.20 per kW·h and can be even lower than that of PV. However, CSP works well only in certain regions of the world with large amounts of direct sun, such as Arizona in the USA. The current worldwide production capacity is around 600 MW, which is expected to grow to several gigawatts in the near future. With continued growth in volume production and innovations in materials and system technologies, grid parity could be achieved within a decade or so by additional cost reductions affecting the overall levelized cost of electricity for CSP technologies.

As with PV technologies, CSP faces the issue of the intermittency of the solar resource. Indeed, this might be one of the greatest challenges facing solar technologies, especially for large power plants connected to the grid. Specifically, large power fluctuations make it difficult to dispatch energy and can be detrimental to the grid and the loads (and other generators) connected to it. Hybrid and/or energy-storage systems are a solution for this problem: whenever the solar resource is not available, the use of fossil fuels or stored heat could complement or supplement the energy needed for continuous and reliable operation of the plant.

Chapter 21 presents a detailed discussion of the evolution and current status of the main CSP technologies in use and the opportunities and challenges for future technology development. It also assesses the economics of CSP, with an eye toward predicting grid parity.

17.4.5 Solar fuel production

In addition to being used directly to generate electricity or heat, solar energy can be used to drive thermochemical and/or photochemical processes to generate hydrogen and other fuels (hydrocarbons and alcohols). The fuel could be stored for backup applications, transportation, and other industrial applications. Additionally, fossil fuels, such as coal and natural gas (CH_4), could be converted into cleaner fuels by endothermic processes such as gasification, reforming, and/or cracking. Hydrogen, for instance, can be obtained by using solar heat for the gasification of coal,

$$C + H_2O \xrightarrow{\text{heat}} CO + H_2, \tag{17.6}$$

the oxidation of CH_4,

$$2CH_4 + O_2 \xrightarrow{\text{heat}} 2CO + 4H_2, \tag{17.7}$$

or the oxidative steam reforming of methane,

$$3CH_4 + H_2O + O_2 \xrightarrow{\text{heat}} 3CO + 7H_2. \tag{17.8}$$

These thermochemical approaches typically yield higher energy- conversion efficiencies when performed at increased temperatures to drive the endothermic reactions. As shown in Equations (17.6)–(17.8), the two primary products derived from such approaches to fuel production are hydrogen and CO.

Another approach, perhaps more attractive for H_2 generation (because there are no carbon emissions), is the direct splitting of H_2O into H_2 and O_2. For this purpose, a few methods have been implemented. In principle, simple metal reactions such as

$$M + H_2O \xrightarrow{\text{heat}} MO + H_2 \tag{17.9}$$

produce hydrogen and metal oxide materials. In practice, H_2 generation has already been achieved using advanced mixed metal oxide ceramics and a two-step thermal process in which the first step is an endothermic process such as that in Equation (17.9) and the second (higher-temperature) step reduces and regenerates the metal for reuse in the first step,

$$2MO \xrightarrow{\text{heat}} 2M + O_2. \tag{17.10}$$

By analogy with naturally occurring energy-transfer reactions, such as photosynthesis, bioelectricity, and bioluminescence, photochemical processes might also be promising for hydrogen generation. Indeed, the photochemical process of photosynthesis not only is responsible for the biomass resource on Earth but also provides insight into the conversion of solar energy into chemical energy. Note that a key aspect of these biological processes is the development of specific, stable catalytic or photocatalytic systems. The field of artificial photosynthesis is focused on emulating and mimicking natural chemical reactions driven by sunlight and evolving them into efficient fuel-producing processes to provide additional routes to clean hydrogen and/or the production of other fuels. Interesting insights into both natural photosynthesis and artificial photosynthesis for the production of hydrogen and other solar fuels can be found in Chapters 24–28.

17.4.6 Biomass, geothermal, and wind power

Biomass is any organic material, produced by living organisms, that can be used for fuel. This wide definition opens a long list of possibilities for the biomass resource. Biomass materials of interest include agricultural residues, animal manure, sewage sludge, wood wastes from forest management and lumber mills, food and paper wastes, grasses (switchgrass), algae, sugarcane, beets, corn, wheat, soy, sunflower, vegetable oils from grains and trees (corn oil, palm oil, etc.), and others. Re-planting (and using wastes) means that biomass combustion does not introduce new carbon into the atmosphere. It is a carbon-neutral process in that it uses carbon that is already present in the natural cycle of carbon and found readily in the Earth's crust. In contrast, emissions from fossil fuels are not part of the natural cycle of carbon and, consequently, add net carbon into the atmosphere. Converting renewable resources such as cellulosic biomass, grains, sugarcane, and even grass into alcohols and biodiesel are notable examples of the potential of biofuels from biomass. The technology is fast evolving with new discoveries and improvements to processes and materials. Such developments will bring additional cost reductions and help develop biomass resources, preferably beyond those that are part of the human food cycle but rather alternatives that are of low value (or no value at all) as food crops. Biomass technology is a broad field, and its resources are vast. Energy output from biomass can be achieved in different ways, as mentioned earlier, including simply burning it to generate heat and consequently electricity. However, there are other outputs such as the production of gaseous and liquid fuels that might be more attractive solutions to clean energy needs and the needs of other solar technologies. Energy storage in the form of renewable fuels, for example, is a solution as backup power or for nighttime operations of CSP and other technologies. Chapters 25 and 26 provide detailed information on the technical aspects of biomass technology and the challenges it faces.

Geothermal energy is potentially a large resource globally and is similar in its application to solar thermal energy. Many of the same concerns about working fluid temperatures and storage are applicable. While some resources can be of high temperature ($>150\,^{\circ}$C) and can be used directly for power production, many are of moderate to low temperatures (90–$150\,^{\circ}$C) and can be used for space heating and heating of domestic hot water in buildings and dwellings using heat-pump technologies. The challenges, opportunities, and the materials used in geothermal energy are discussed in Chapter 29.

According to [2], wind power has been the fastest-growing source of new electricity generation for several years. Much like PV, wind power can be highly versatile because turbines can be made as small as a few hundred watts to a few megawatts in size. Many wind-power installations are decentralized, meaning that they contribute power to the grid but from rather small-scale installations; many small-scale wind turbines are also used in off-grid installations.

The power output of a wind turbine is proportional to the wind speed and the sweep area of the rotor. Doubling the rotor diameter increases the rotor sweep area by a factor of 4, which translates into four times as much power output. A typical utility wind generator rated at 600 kW has a rotor 44 m in size, and a 1.5-MW generator can be as large as 64 m. The trend today in utility-scale wind turbines is to build taller and larger turbines to reach wind resources found at greater heights. Such large structures present challenges to the materials used in the construction of the wind turbines and the control mechanisms necessary to prevent damage to turbines. State-of-the-art control mechanisms (assisted by computer control and code) help wind turbines shed some loads in extreme or very turbulent winds. Chapter 30 provides many more details on the economics, challenges, opportunities, and future direction of wind-power technology.

To date, biomass and wind power are among the most cost-effective renewable energy technologies. Their LCOEs for utility power generation are already competitive with those of fossil fuels and nuclear power. Thus, wherever the energy resources associated with these technologies are available, they present unique opportunities for energy production and energy independence. Each country around the globe knows best its own natural resources and opportunities. In this context, there are already successful examples of the application of renewable technologies to establish a sustainable future and meet domestic energy demands. For instance, Brazil is developing perhaps the first sustainable biofuels economy and becoming one of the largest exporters of ethanol in the world. Other similar examples demonstrate that biomass and wind power are technologies that can be embraced by many countries to help alleviate the political and environmental issues associated with traditional technologies.

17.5 Summary

The mass implementation of solar (and other renewable) energy technologies presents a viable path to clean energy and reductions in CO_2 emissions. Solar energy can also provide for energy independence and long-term sustainability. The most significant shortfall of most solar technologies today is their economics: technologies such as PV, solar thermal, and solar fuels are two to five times more expensive in terms of levelized cost of electricity than fossil-fuel (or nuclear) options.

The generation of electrical power from solar energy can take many forms. Whereas PV technologies provide for the direct conversion of solar energy into electricity, solar thermal and solar fuels technologies use indirect paths for power generation. Typically, the thermal energy from the Sun is converted into mechanical energy (as in solar thermal electricity) and then into electricity. That thermal energy from the solar resource can also be used to drive chemical reactions in the generation of environmentally benign fuels (hydrogen and others). Two of the most cost-effective renewable technologies today are biomass and wind power: they are already competitive with traditional technologies such as nuclear and natural gas in terms of LCOE. The technical issues and the physics behind these renewable energy technologies are described in detail in the subsequent chapters of this book.

17.6 Questions for discussion

1. Why is the demand for energy increasing globally? Discuss factors that affect energy demand.
2. Will it be possible to supply all of the energy needed by 2050 and beyond? What options are available to do so?
3. Explain the levelized cost of electricity and how it can be used to compare energy technologies.
4. Is the levelized cost of PV electricity competitive with that obtained from natural gas? What are the roadblocks which will have to be overcome in order for PV power to be competitive with electricity from fossil fuels?
5. Can solar and other renewable technologies ever be cost-competitive with fossil fuels? Explain.
6. Name some passive solar technologies.
7. How can environmentally friendly fuels (such as hydrogen) be produced using solar energy?
8. Define biomass and name some examples.
9. What determines the power output of a wind turbine?
10. A distant star emits visible light for which its maximum intensity is for photons with a wavelength of 600 nm. What is the temperature of the surface of that star? Plot Planck's formula for the spectral irradiance of this blackbody.
11. Mars is at an average distance of 227 million km from the Sun. What is the average solar irradiance (in $W\,m^{-2}$) arriving just outside the Martian atmosphere? Given that the radius of Mars is 3,396.2 km, what is the average insolation (in TW) that Mars receives just above its atmosphere? Hint: the calculated incoming irradiance just outside the Martian atmosphere is what Mars would see if it were flat, like a disk (although Mars is obviously not flat).
12. Assuming the quantum efficiency of a given solar cell to be unity, calculate the maximum current that could be obtained from a PV cell with a 1-eV bandgap under AM1.5 global irradiance.

17.7 Further reading

- Energy Information Agency at the US Department of Energy (http://www.eia.doe.gov/).
- Renewable energy technologies (http://www.nrel.gov/).
- Information about the Sun (http://www.nasa.gov/worldbook/sun_worldbook.html).
- Solar Energy Society in India: http://www.sesi.in/.
- Solar energy in India: http://www.solarindiaonline.com/index.php.
- Solar and renewable energy information for Europe: http://ec.europa.eu/dgs/jrc/index.cfm.

17.8 References

[1] 2005, *Basic Research Needs for Solar Energy Utilization*, Report of the Basic Energy Sciences Workshop on Solar Energy Utilization (OSTI ID: 899136x).
[2] 2009, *International Energy Outlook 2010. World Energy Demand and Economic Outlook*, DOE/EIA-0484; Source: Energy Information Administration, Annual Energy Outlook 2010, December 2009, DOE/EIA-0383.
[3] **Lazard**, 2009, *Levelized Cost of Energy Analysis*, version 3.0.

18 Direct solar energy conversion with photovoltaic devices

David S. Ginley,[1] Reuben Collins,[2] and David Cahen[3]

[1]Process Technology and Advanced Concepts, National Renewable Energy Laboratory, Golden, CO, USA

[2]Physics Department and Renewable Energy Materials Research Science and Engineering Center, Colorado School of Mines, Garden, CO, USA

[3]Department of Materials and Interfaces, Weizmann Institute of Science, Rehovot Israel

18.1 Focus

During the last decade the direct conversion of solar energy to electricity by photovoltaic cells has emerged from a pilot technology to one that produced 11 GW$_p$ of electricity generating capacity in 2009. With production growing at 50%–70% a year (at least until 2009) photovoltaics (PV) is becoming an important contributor to the next generation of renewable green power production. The question is that of *how we can move to the terawatt (TW) scale* [1].

18.2 Synopsis

The rapid evolution of PV as an alternative means of energy generation is bringing it closer to the point where it can make a significant contribution to challenges posed by the rapid growth of worldwide energy demand and the associated environmental issues. Together with the main existing technology, which is based on silicon (Si), the growth of the field is intertwined with the development of new materials and fabrication approaches. The PV industry, which was, until recently, based primarily on crystalline, polycrystalline, and amorphous Si, grew at an average annual rate of 50% during 2000–2010. This rate was increasing, at least until the 2008 economic crisis, with production of ~ 11 gigawatts (GW$_p$) per year in 2009 [2]. While this may seem a very large number, PV installations in total are still supplying only <0.03% of all the world's power needs (~ 14–15 TW) [2]. As production increases, increasing individual cell efficiency and translating that to modules, as well as reducing manufacturing expenses and increasing system lifetimes, are all critical to achieving grid parity, the point at which the cost of PV power is equal to the price of grid electricity.

Thin-film solar-cell approaches are now penetrating the PV market. At present polycrystalline CdTe thin-film solar cells have the lowest reported manufacturing cost, at <$0.80 per W$_p$ of any PV technology. Cells using alternative polycrystalline materials, or nanocrystalline and amorphous films, together with organic-polymer- and small-organic-molecule-based solar cells may be the best candidates for significant cost advantages in flexible cell and roll-to-roll processing. Module (cell) efficiencies of these alternatives to CdTe are as high as ~ 15.5% (~ 20%) for Cu(In,Ga)Se$_2$ (CIGS)-based cells, ~ 4% (~ 8%) for polymers and (~ 8%) for small-molecule cells [3][4][5][6]. Beyond this are the promises of significant increases in efficiency and/or reductions in cost through innovation in device architectures, such as multijunction-cell approaches and development of hybrid devices, organized at the nanoscale. Controlling materials at the nanoscale and engineering new thin films from nanostructures is receiving considerable attention for both present and future generations of solar cells. If price is not an obstacle, some of these approaches may reach conversion efficiencies >50%. There is also the potential for a significant sustainability advantage over CdTe. As technologies mature and our basic knowledge base grows, further improvements in efficiencies and/or in cost can be expected. Key is that the device is only part of the story and the balance of systems (BOS) must undergo a similar evolution. This chapter will look at the history of the field, the device basics, and the current state of the industry and what may be emerging just over the horizon.

18.3 Historical perspective

18.3.1 The origin of PV science

Photovoltaics as a large-scale industry worldwide is actually relatively new, but its history is quite old. Edmund Becquerel, a French experimental natural philosopher, discovered the photovoltaic effect in 1839 while experimenting with two metal electrodes in an aqueous solution [7]. That type of device structure would, today, be called a photoelectrochemical PV cell. Subsequently, in 1873 Willoughby Smith discovered photoconductivity in Se [8], and in 1876 Adams and Day demonstrated a photovoltaic effect in this material [9]. There were some important demonstrations of photoelectric processes and advances in understanding during the first decades of the twentieth century, including Einstein's explanation of the photoelectric effect [10] and development of photoelectric cells based on copper/cuprous oxide (Cu/Cu_2O) junctions [11] [12]. In 1918 the Polish scientist Czochralski demonstrated the method that bears his name for the growth of single-crystalline Si, which really enabled the emergence of the later electronics and solar-cell industries. In 1941, the photoelectric effect was first demonstrated in Si at naturally formed "defect" junctions [13][14]. Then 1954 was a breakthrough year, with the report by Reynolds *et al.* on a photovoltaic effect in CdS and work by Chapin, Fuller, and Pearson, who demonstrated the first modern Si cell (Figure 18.1), with an efficiency of 4.5% [15][16][17]. In 1956 Jenny, Loferski, and Rappaport showed 4%-efficient GaAs cells [18]. The emerging space industry needed power for spacecraft (Figure 18.2), and by 1955 cells were being sold by Western Electric and Hoffman Electronics. The latter announced a commercial product operating at 2% efficiency and the cost of the energy was $1,500 per W.

By 1959 Hoffman Electronics was reporting Si-cell efficiencies of up to 10% and the Vanguard I satellite was launched with these cells, which worked for 8 years. Array sizes grew both terrestrially and for space applications, but sizes were not more than 1 kW$_p$[1] up to about 1966 [19]. The first use of CdS cells was on satellites in 1968. By the early 1980s array sizes had grown to around 100 kW$_p$. In 1982 worldwide PV production exceeded 9.3 MW$_p$. This led to the modern era of PV and the development of other systems such as amorphous Si, concentrator III–V cells, $Cu(InGa)Se_2$ and CdTe cells (both incorporating CdS), leading to an overall production of >11 GW$_p$ in 2009. More recently, now that it is becoming possible to control materials at the nanoscale,

Figure 18.1. Chapin, Fuller, and Pearson demonstrating a Si photodiode at Bell Laboratories (John Perlin, 2002, *From Space to Earth: The Story of Solar Electricity*, Harvard University Press).

Figure 18.2. The NASA Mars Rovers use high-efficiency multijunction III–V GaAs PV technology for power (marsrovers.jpl.nasa.gov/gallery/artworkhires/rover2.jpg).

many new PV concepts that had lain dormant for some time, have started coming into their own, such as the so-called dye cell (first reported in 1990) [20] and organic cells (first reported in 1986) [21].

18.3.2 Applications of PV systems

As already noted, the first conventional photovoltaic cells, which were produced in the late 1950s and in the 1960s, served principally to provide electrical power for

---

[1] The subscript "p" stands for "peak" and means that the quantity refers to the power that is produced under full AM (air mass) 1.5 (1 kW m^{-2}) insolation. The equivalent continuous power is obtained, depending on the location, by dividing by 5–6.

Earth-orbiting satellites.[2] In the 1970s, improvements in manufacturing, performance and quality of PV modules helped to reduce costs and created opportunities for remote terrestrial applications, including battery charging for navigational aids, signals, telecommunications equipment and other critical, low-power needs. In the 1980s, PV became a popular power source for consumer electronic devices, including calculators, watches, radios, lanterns, and other small battery-charging applications. Following the energy crises of the 1970s, significant effort was directed at developing PV power systems for residential and commercial uses, both for stand-alone, remote power as well as for utility-connected applications. During the same period, applications of PV systems to power rural health clinics, for refrigeration, water pumping, and telecommunications, and for off-grid households in the developing world increased dramatically, and remain a major part of the present world market for PV products. Today, the industry's production of PV modules is growing at approximately 50% annually, and major programs worldwide are rapidly accelerating the implementation of PV systems on buildings and their connection to utility networks [19].

18.4 Photovoltaics basics

18.4.1 The usable part of the solar spectrum – why PV makes sense

As noted in Chapter 17, we can think of solar emission from the Sun in much same way as we understand emission of thermal radiation from the filament in an incandescent lamp, i.e., an idealized, perfectly absorbing and emitting ~5,780-K-hot object that emits a blackbody spectrum.

As shown in Figure 18.3 most "solar radiation" that reaches the Earth is in a small, but crucial part of the electromagnetic spectrum (cf. also Chapter 17, Figure 17.1).

Optical absorption within a solar material involves an electronic transition from a filled to an empty state and generally this implies there will be an energy threshold below which no photons are absorbed. For a molecular system this threshold corresponds to the energy difference between the highest occupied molecular orbital (HOMO) and the lowest unoccupied molecular orbital (LUMO). For a semiconductor the threshold is the bandgap and corresponds to the minimal photon energy that suffices to promote an electron from the semiconductor valence band to the conduction band.[3] It is clear that the absorber must have an optical absorption threshold that is optimized relative to the solar spectrum for

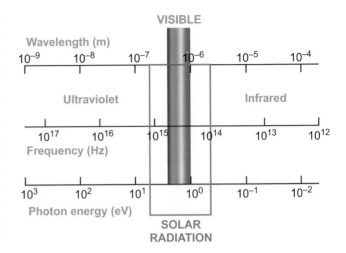

Figure 18.3. The solar spectrum in the electromagnetic spectrum, as a function of wavelength, frequency, and energy.

maximum energy conversion by the photovoltaic effect. It is also true that in molecular and polymer systems photons with suitable energy for the HOMO–LUMO excitation can have a much higher probability of being absorbed than in most semiconductor systems. A lower energy threshold increases the number of photons that are absorbed, while a higher energy threshold increases the useful energy that can be obtained from each photon. This leads to a maximum that, for a single-threshold system, is calculated to be between 1.1 and 1.5 eV, corresponding to an absorption onset wavelength that ranges from 825–1,125 nm, with a maximum solar-to-electrical power-conversion efficiency of ~30% for non-concentrated terrestrial solar illumination.

One way to circumvent this limit is to use a number of separate materials with different thresholds (E_{g1} to E_{g3} in Figure 18.4; see also Figures 19.6 and 20.11) to absorb different spectral regions of the solar irradiation. By splitting the spectrum in this way, each absorber can be better matched to its associated spectral region. If there is no concomitant increase in the effective collection area compared with the single-material case, then this can significantly increase the total possible conversion efficiency. Theoretically, the more absorbers used, the higher the potential efficiency. Practically, inclusion of more than three or four different absorbing materials with different absorption characteristics is difficult to achieve without introducing other loss mechanisms [22]. Multi-junction solar cells to achieve spectral splitting are discussed in more detail in Chapters 19 and 20.

18.4.2 The junction formation

A PV device in its most general form is shown in Figure 18.5. We can describe the device as a system that has two metastable states, a high- and a low-energy one, a

---

[2] Several Soviet satellites used nuclear power. In fact, a few US satellites also used nuclear power generation, as did non-"Earth-orbit" missions like Viking and Voyager.

[3] See, e.g., B. Van Zeghbroeck's 2007 introduction to semiconductor physics at http://ecee.colorado.edu/~bart/book/book/contents.htm.

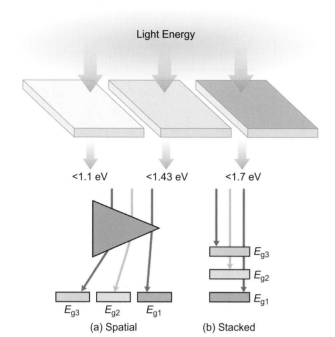

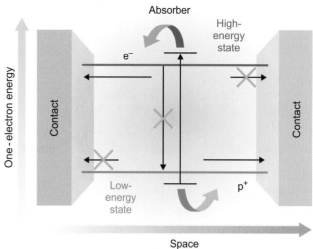

Figure 18.5. Basic principles for a PV device following the development of [23]. "Space" indicates a spatial coordinate across the device.

Figure 18.4. Conversion of solar energy with multiple absorbers instead of one. The cells can have a spectrum splitter (a) or be mechanically or electrically stacked (b) [22].

mechanism for photon absorption to transfer opposite electronic charges into these states, a way to separate these charges, and contacts that are selective for each of these charges, allowing them to be collected.

In nearly all cells, especially those developed commercially, the solar absorber is a semiconductor. In some devices based on organic absorbers, an excited electron–hole pair forms first as an intermediate, before the charge-separation and collection steps. If just one kind of semiconductor is used, the most common approach is to make a p–n junction with it (Figure 18.6) [22][24]. For that, the material needs to be "ambipolar," meaning that it can be doped to make it either p- or n-type, i.e., with either excess free holes or electrons. For Si such doping can be achieved by adding impurity atoms such as B or P to make p- or n-Si, respectively. The junction between the p- and n-forms of the same semiconductor is termed a homojunction. If two different semiconductors are used, one should be p-, the other n-type, and these are termed heterojunctions. Either p- or n-type material can be used in metal/ electrolyte (liquid or solid)/semiconductor cells, including so-called MIS (metal–insulator–semiconductor) and dye cells. Electrolytic and dye cells are discussed in more detail in Chapter 49. Returning to the most common cell type, the p–n junction, if p- and n-type materials are brought into contact so that they can reach electronic equilibrium, they form a junction (a diode or PV device), which can, because of the difference in electrochemical potential of the electrons

between the p- and n-type regions, separate photogenerated carriers (Figure 18.6).

Suitable external contacts then collect these photogenerated carriers and produce useful work (electric power) by the flow of current from high to low potential, through an external circuit.

18.4.3 Device basics

Since the PV device is basically just a diode, a solar cell can ideally be modeled as a simple electrical circuit with a current source, due to the light-generated current, in parallel with a diode. When there is no light present to generate any current, the cell behaves like a normal diode, with current growing exponentially in one voltage direction, and very little current for voltage applied in the other direction, as shown in Figure 18.7. As the intensity of incident light increases, increasing current is generated by the PV cell, as illustrated in Figure 18.7.

In Figure 18.7 the current is plotted with a sign convention that is the opposite of what is typically used for a diode; hence the forward-bias current is negative. This is done to follow the convention for solar cells, which shows photocurrent generation in the first quadrant. Here current is flowing in the opposite sense to the applied voltage, hence the device is generating rather than dissipating power. In an ideal cell, the total current I equals the photogenerated current I_L minus the diode current I_D that would be observed in the absence of illumination and is given by

$$I = I_e - I_D = I_C - I_0 \left(\frac{qV}{e^{kT} - 1} \right). \tag{18.1}$$

Here I_0 is the saturation current of the diode, q is the elementary charge 1.6×10^{-19} C, k_B is the Boltzmann

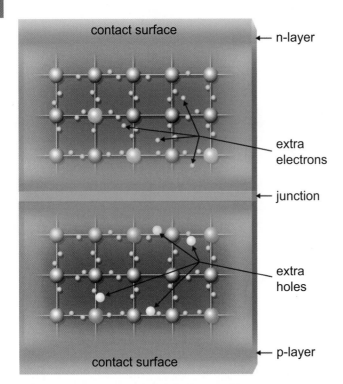

contact surface — n-layer

extra electrons

junction

extra holes

p-layer

contact surface

Figure 18.6. A schematic diagram of a p–n junction, formed from p- and n-type regions in intimate electronic contact. In the n-layer, substitution of atoms with more valence electrons than the host lattice atoms leads to an excess of electrons. In the p-layer the substitutional atoms have fewer valence electrons than the host lattice atoms. The missing electrons are referred to as holes and act as positive charge carriers. This difference in occupation of electronic energy levels creates an electrochemical potential difference between the n- and p-regions. Mostly, this is an electrical potential difference, which then leads to an electric field in the junction region that provides the driving force for separation of photogenerated carriers.

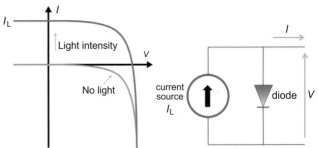

Figure 18.7. The current (I)–voltage (V) curve of a PV cell and the associated equivalent circuit.

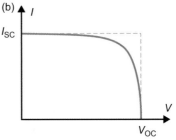

(a)

(b)

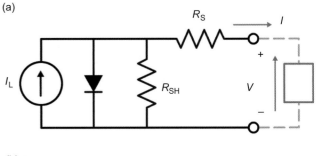

Figure 18.8. (a) The equivalent-circuit model for a PV cell including series and shunt resistances and an external load (blue box). (b) The I–V characteristic of an illuminated PV cell showing the open-circuit voltage (V_{OC}) and short-circuit current (I_{SC}).

constant (1.38×10^{-23} J K^{-1}), T is the cell temperature in Kelvins, and V is the voltage across the cell. Depending on the measurement configuration, this voltage is either generated by the cell (in quadrant one, called the power quadrant) or applied to the cell externally (voltage bias).

Real solar cells exhibit non-idealities not accounted for in this simple model, including series resistance at the contacts, generation and recombination of carriers at defects, and current leakage through the junction. A more accurate model would include an additional diode term, with the second diode often arising from one of the contacts. Here we concentrate on a single-diode model. Extending the analysis gives the simplified circuit model shown in Figure 18.8(a) and the associated equation

$$I = I_L - I_0 \left(\exp\left(\frac{q(V + I R_S)}{n k_B T} \right) - 1 \right) - \frac{(V + I)R_S}{R_{SH}}. \quad (18.2)$$

Here n is the diode ideality factor, which is typically between 1 and 2. For solar cells the ideality factor is probably best viewed as a fitting parameter that helps take into account such effects as recombination of photoexcited carriers in the junction region. R_S and R_{SH} represent the series and shunt resistances and are described further, later in the chapter.

Figure 18.8(b) is a schematic diagram of a typical current–voltage (I–V) characteristic for an illuminated solar cell in the power-generating quadrant. For a solar cell under illumination, changing the size of the resistive load connected to the cell will move the operating point of the cell along its I–V curve. A very large resistive load will cause nearly an open-circuit condition. A small load

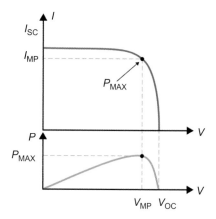

Figure 18.9. Determination of the maximum-power operating point from the *I–V* curve of an illuminated solar cell.

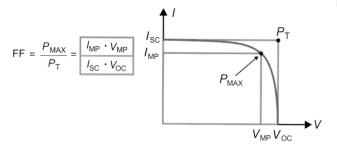

Figure 18.10. Determining the fill factor from the *I–V* curve.

will result in operation near zero voltage. In practice, the *I–V* characteristic of a solar cell is often determined by sweeping an external voltage or current source. In power-generating operation, electronics between the solar panel and the load (such as a bank of batteries or the grid) adjusts the operating point to maximize the power delivered to the load. Several important metrics describing the performance of a solar cell can be determined from the *I–V* curve, as will be described next.

The short-circuit current (I_{SC})
The short-circuit current I_{SC} corresponds to the (short-circuit) condition when the impedance of the load that the solar cell is driving is negligible. It is the current that flows when the voltage equals zero, i.e., I (at $V = 0$) = I_{SC}. I_{SC} occurs at the start of a forward-bias sweep and is the maximum current value in the power quadrant. For an ideal cell ($R_S = 0$ and $R_{SH} = \infty$), this maximum current value is the total current produced in the solar cell by photon excitation. $I_{SC} = I_L = I_{MAX}$ in the forward-bias power quadrant.

The open-circuit voltage (V_{OC})
The open-circuit voltage (V_{OC}) is the voltage across the device when no current passes through the cell, i.e., V (at $I = 0$) = V_{OC}. V_{OC} is also the maximum voltage difference across the cell for a forward-bias sweep in the power quadrant. $V_{OC} = V_{MAX}$ in the forward-bias power quadrant.

The maximum power (P_{MAX}), current at P_{MAX} (I_{MP}), and voltage at P_{MAX} (V_{MP})
The power (P) produced by the cell and delivered to the load, in watts, can be easily calculated along the *I–V* characteristic by applying the equation $P = IV$. At the I_{SC} and V_{OC} points, the power will be zero and the maximum value for the power will occur between the two (Figure 18.9). The voltage and current at this maximum-power point are denoted V_{MP} and I_{MP}, respectively.

The fill factor (FF)
The fill factor (FF) is essentially a measure of quality of the solar cell. It is calculated by comparing the maximum power with the theoretical power (P_T) that would be generated if the cell could simultaneously generate the open-circuit voltage and short-circuit current. The FF can also be interpreted graphically as the ratio of the rectangular areas outlined in blue and green in Figure 18.10.

Clearly, the higher the fill factor the better the solar cell, in terms of efficiency for solar-to-electrical power conversion. Graphically it will correspond to an *I–V* sweep that is more rectangular and less curved. Typical fill factors range from 0.5 to 0.82. The fill factor is also often represented as a percentage.

The conversion efficiency (η)
The solar-to-electrical power-conversion efficiency, η, is the ratio of the electrical power output P_{in}, to the solar power input, P_{in}, into the PV cell. P_{in} is the product of the irradiance of the incident light, measured in W m^{-2} and the surface area of the solar cell in m^2. The maximum efficiency occurs when the cell is operated at V_{mp} and I_{mp} with $P_{out} = P_{MAX}$,

$$\eta = \frac{P_{out}}{P_{in}} \Rightarrow \eta_{MAX} = \frac{P_{MAX}}{P_{in}}.$$

The maximum measured conversion efficiency (η_{MAX}) is the appropriate indicator of the performance of the device under test and is usually reported as the device efficiency without noting that the measurement is made at the maximum-efficiency operating point.

The shunt resistance (R_{SH}) and series resistance (R_S)
During operation, the efficiency of solar cells is reduced by the dissipation of power in internal resistances. These parasitic resistances can be modeled as a parallel shunt resistance (R_{SH}) and series resistance (R_S), as depicted in Figure 18.8(a). For an ideal cell, R_{SH} would be infinite and would not provide an alternative path along which current could flow, while R_S would be zero, resulting in no further voltage drop before the load.

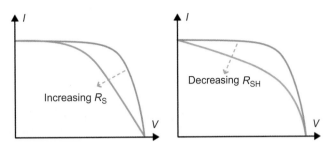

Figure 18.11. Effects of non-ideal R_S and R_{SH} values on the PV current–voltage characteristic.

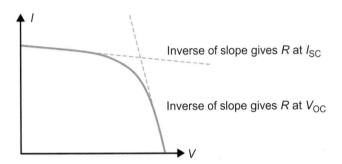

Figure 18.12. Illustration of how to extract series and shunt resistances from a PV cell's I–V curve. The actual slopes will be negative numbers with the absolute values of the inverse of the slopes giving the (approximate or relative) parasitic resistances.

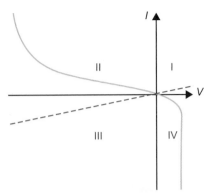

Figure 18.13. The I–V curve of a solar cell without light excitation.

It is important to note, however, that for real cells these resistances are often a function of the light level, and their light and dark values can differ.

Temperature-measurement considerations
All the parameters extracted from the I–V characteristics can be affected by ambient conditions such as temperature and the intensity and spectrum of the incident light. For this reason, PV cells are compared using similar lighting and temperature conditions. Standard test conditions are 25 °C, using an AM1.5 spectrum with 1,000 W m^{-2}. These illumination conditions correspond roughly to solar noon during the spring and fall equinoxes in the southern USA. In fact, it's also useful to know how cell performance changes with parameters such as temperature and illumination since this can help select optimal technologies for specific environments where, for example, the sky is often overcast or where the temperature can change dramatically throughout the day. It may also necessitate cooling or more sophisticated temperature control to maintain optimum efficiency. The properties of the materials used to make PV cells, are, like those of virtually all materials, temperature-dependent. In the case of semiconductors this dependence is well known and the effects on solar-cell performance can often be understood within this context. Thus, in a typical Si solar cell, V_{OC} decreases with increasing temperature while I_{SC} increases slightly. A key factor is that in Si the bandgap narrows with increasing temperature. Each PV material has its own distinct temperature dependence. So in Si the overall effect is a drop in efficiency of some 0.5% per °C. However, organic solar cells increase in efficiency with temperature even if V_{OC} decreases. This is usually associated with an improvement in the fill factor and hence the quality of the cell.

Decreasing R_{SH} and/or increasing R_S will decrease the fill factor (FF) and P_{MAX} as shown in Figure 18.11. For the R_S and R_{SH} values used to draw the figures, V_{OC} and I_{SC} do not appear to have changed. In fact, as R_{SH} decreases, V_{OC} drops while I_{SC} is not affected. Increasing R_S excessively can cause I_{SC} to drop instead, with V_{OC} remaining constant. The larger the deviations from ideality, the more apparent changes in V_{OC} and I_{SC} become.

It is possible to approximate R_S and R_{SH} from the absolute value of the V/I slopes of the I–V curve at V_{OC} and I_{SC}, respectively, as shown in Figure 18.12. The slope at V_{OC}, however, is at best proportional to the series resistance, but it is larger than the series resistance. The V/I slope at I_{SC} is a better approximation for R_{SH}.

If light is prevented from illuminating and exciting the solar cell, the dark I–V curve shown in Figure 18.13 can be obtained. This is essentially the dark I–V curve from Figure 18.7, but including the effects of non-idealities like finite R_{SH}. The absolute value of the inverse slope of the linear region of the curve in the second quadrant (reverse bias) is a continuation of the linear region in the fourth quadrant, which is the linear region which was used to calculate R_{SH} in Figure 18.12. It follows that R_{SH} can be derived from the I–V plot obtained with or without providing light excitation.

18.4.4 I–V curves for modules

For a module or array of PV cells, the shape of the I–V curve does not change from that of a cell. However, it is scaled, on the basis of the number of cells connected in series and/or in parallel. With n the number of cells connected in series and m the number of cells connected in parallel, the module's I–V characteristic is ideally that of a cell, but with a short-circuit current $I_{SC-M} = mI_{SC}$ and an open-circuit voltage $V_{OC-M} = nV_{OC}$, with I_{SC} and V_{OC} the values for individual cells.

18.4.5 Device types

Photovoltaic devices can be divided into the following types.

Homojunction

Crystalline Si (c-Si) is the primary example of this kind of cell. A single material – crystalline Si – is altered so that one side is p-type, with positive carriers, holes, as the dominant electronic charge species; and the other side is n-type, with the negatively charged electrons as dominant charge carriers. This is the type of device depicted in Figure 18.6. As discussed in Section 18.4.2, the work-function difference (the work function is the energy needed to move an electron from the Fermi level infinitely far away to the "vacuum" level) between the n- and p-type regions creates an internal electric field. When the cell is illuminated, excitations from the lower energy state (valence band) to the higher energy state (conduction band) in Figure 18.5 generate additional free electrons and holes. The p–n junction is located near the surface so that the maximum amount of light is absorbed near it. Carriers created in the junction region are separated by the internal electric field. Those generated by light absorbed deeper in the cell, outside the junction region, diffuse to the p–n junction where they are also separated. The photogenerated carriers then diffuse to the contacts to produce current if the material is of sufficiently high quality to prevent the carriers from recombining with each other (de-excitation) before they can be collected.

In this homojunction design, we may vary several aspects of the cell to increase the conversion efficiency:

- the depth of the p–n junction below the cell's surface,
- the density and distribution of dopant atoms on either side of the p–n junction, and
- the crystallinity and purity of the semiconductor.

Some homojunction cells have also been designed with the positive and negative electrical contacts both on the back of the cell. This geometry eliminates the shadowing caused by the electrical grid on top of the cell. A disadvantage is that the charge carriers, which are mostly generated near the top surface of the cell, must travel further – all the way to the back of the cell – to reach an electrical contact. For this to be possible, the semiconductor must be of very high quality, with a minimal density of defects, since defects can cause recombination of electrons and holes. The best-known examples of homojunction solar cells are the Si and GaAs cells [22][23][24].

Heterojunction

An example of this type of device structure is the CIGS cell, where the junction is formed by contacting two different semiconductors, CdS and CIGS. This structure is often chosen for producing cells from materials that absorb light much better than Si and can, thus, be used as thin films. Because only thin films (of thickness a few micrometers) are needed to absorb all of the incident solar radiation, the stringent materials purity requirements that are valid for Si can be relaxed. In practice, some of the most successful of these cells use thin films of *polycrystalline* material.

The two materials in a heterojunction device have different roles. The top layer, or "window" layer, is made of a material with a large bandgap, selected for its transparency to light. It is usually also made thin to further minimize absorption. CdS, for example, has an energy gap of ~2.4 eV, which is in the blue–green part of the visible spectrum. It is transparent to light in the yellow, red, and infrared spectral regions where a large fraction of solar energy lies. The window allows almost all incident light to reach the bottom layer, which is a lower-bandgap material that readily absorbs the light with greater than bandgap energy. This light then generates electrons and holes very near the junction, which again separates them before they can recombine.

Heterojunction devices have an inherent advantage over homojunction devices. Many PV materials can, practically, be doped only p-type or only n-type, but not both. Because heterojunctions allow us to forego the requirement of ambipolar materials, many promising PV materials can be investigated to produce optimal cells. Also, a high-bandgap window layer reduces the cell's series resistance, so a window material can be made highly conductive. In principle its thickness can be also be increased without reducing light transmittance. As a result, light-generated carriers can easily flow laterally in the window layer to reach an electrical contact. In practice, window layers tend to have some residual absorption in the part of the solar spectrum where the lower-bandgap material absorbs, creating a trade-off between thickness and performance that must be optimized for each material.

p–i–n and n–i–p devices

Typically, amorphous Si (actually hydrogenated amorphous Si, a-Si:H) thin-film cells have a p–i–n structure. The basic structure is as follows: a three-layer sandwich is created, with a middle intrinsic (i-type or undoped) layer between an n-type layer and a p-type layer. Light is incident on the p-layer. In this geometry the p- and n-type regions set up an electric field within the middle intrinsic resistive region. Light generates free electrons and holes in the intrinsic region, which are then separated by the electric field.

In the p–i–n a-Si:H cell, the top layer is p-type, the middle layer is intrinsic, and the bottom layer is n-type a-Si:H. Because amorphous Si has many atomic-level electrical defects when it is electrically conductive, very little current would flow if such a cell had to depend on diffusion. However, in a p–i–n cell, current flows because the free electrons and holes are generated in a region with an electric field, rather than having to move toward the field.

In contrast to amorphous Si, a CdTe cell can be considered an n–i–p heterojunction structure. The device structure is similar to the a-Si cell, except the order of layers is flipped upside down. Specifically, in a typical CdTe cell the top layer is n-CdS, the middle layer is intrinsic CdTe, and the bottom layer is a p-type contact layer such as zinc telluride (ZnTe) [25].

Excitonic solar cells

More recently a new class of solar cells has emerged in which the absorbing layer is made of organic (molecular or polymeric) materials. The materials have the advantages of high absorption coefficients, simplified processing, since they can be prepared and deposited from solutions of organic solvents, and being structurally flexible, which can simplify manufacturing of cells and panels. Since the way in which they convert light into current is different from that in conventional inorganic solar cells, the design of an efficient cell of this type is also quite different.

Organic materials have dielectric constants, ε, four to five times smaller than those of inorganic semiconductors like Si. As a result, absorption of a solar photon does not easily create a free electron and hole as it does in Si. Instead, the excited state can be thought of as consisting of a bound electron–hole pair known as an exciton. While such a bound pair also exists in inorganic materials, the binding energy of the pair is 10–20 times smaller (the reduction is proportional to $k_B T$, the thermal energy, at room temperature). Hence the bound carriers easily escape to become free carriers that can be collected. In organic materials, separation of excitons into free carriers generally relies on the presence of an interface between the organic absorber (often referred to as a donor material) and a second material (often referred to as an acceptor material). For properly aligned HOMO and LUMO energy levels of the organic donor and the acceptor, an exciton present at the interface can decompose into a hole in the HOMO level of the donor and an electron in the LUMO level of the acceptor. Device operation then involves optical excitation of the donor material (a polymer or dye), exciton diffusion to an interface between the donor and acceptor material (a fullerene or oxide for example), injection of an electron into the acceptor, with a hole remaining in the

donor, and then transport of the carriers to the contacts to produce current. Thus, in this case, we must be concerned about the diffusion of both the excitons and the carriers. The process is complicated by the short exciton diffusion length of ~10 nm. As a result, an interface must be located within approximately 10 nm of where the exciton is formed in order to avoid recombination before the exciton can decompose into free carriers. Fortuitously, it has been shown that a mixture of an appropriate polymer donor and a fullerene acceptor can phase-separate on this length scale so that most excitons form charge carriers and charge carriers can diffuse to the external contacts [26]. This nanostructured, phase-separated blend of two materials is referred to as a bulk heterojunction (BHJ) in analogy to the planar heterojunction designs discussed above.

Dye solar cells are another solar-cell design with strong similarities to organic (e.g., polymer) solar cells. Here solar radiation is absorbed by an organic dye that has been deposited onto a high-surface-area nanostructured oxide (typically TiO_2) contact layer. The dye-coated oxide is immersed in an electrolyte. The excited dye injects an electron into the oxide and the remaining hole is captured by an I^-/I_3^- redox couple within the electrolyte, which transports it to a contact electrode [27]. Dye cells are covered in more detail in Chapter 49.

Multijunction devices

This structure, also called a cascade or tandem cell, provides a way of splitting the solar spectrum up between multiple absorbers with different bandgaps as discussed in Section 18.4.1 and shown in Figure 18.4(b). Multijunction devices can achieve higher overall conversion efficiency than single-junction devices, by using several thresholds and thereby not only capturing a larger portion of the solar spectrum, but also overcoming the single-threshold Shockley–Queisser limit [28]. Multijunction approaches are discussed in more detail in Chapter 20, and we note in passing that nearly all the above systems can be incorporated into multijunction devices with more or less difficulty, depending on interfacial properties.

18.4.6 Device concerns for scaling to terrawatts (TW)

As PV solar energy conversion tries to move to the TW_p production scale there are a number of key concerns that must be addressed with respect to the ultimate viability of any technological approach. These include the following.

1. *Scalability.* This has to do with elemental abundance as well as the ability to process the materials quickly at low temperatures and with high yield.

2. *Sustainability.* Materials and processes must be sustainable at the TW_p scale.

3. *Cost.* Prices must approach the cost of base-load-generated power (4–12 cents per kWh) up to about 15% of net power production. At some point beyond this level of production the intermittency of solar electricity will become an issue for grid use, and storage will be required in order to bridge the time periods during which the Sun is not shining. It will then be necessary to meet base-load-generated power costs including the cost of storage.

4. *Lifetime.* Since most of the cost of a PV power system is paid up front, the average cost of PV-produced electricity depends on how long the system continues to produce power at its rated level. On the basis of the costs of materials and manufacturing equipment, energy used in the processing of the devices, and balance-of-systems expenses (e.g., inverters, installation costs, etc.) the lifetime required to reach a cost equivalent to conventional electric power could be in the range 5–30 years. Clearly longer life is advantageous.

5. *Reliability.* The reliability of the panels and of the balance of the system influences both lifetime and cost. Minimizing maintenance and replacement of panels and components further reduces electricity costs.

Some of the key challenges in addressing these concerns are as follows.

- Materials that can easily be fabricated and deployed on the massive scale required to reach TW_p power generation level are rare. Each PV technology being developed for large-scale electricity production must demonstrate sufficient abundance of its elemental constituents to achieve TW_p production levels.

- Device and system designs that have a small energy and environmental footprint are even rarer. While we tend to think of PV as a green technology, the manufacturing of cells and systems has its own energy usage and environmental impact. This means that the energy input and pollution and carbon dioxide output associated with the processing and development of the materials, the cell, and its support structures must be considered when comparing various PV technologies and when comparing PV with conventional power generation.

- Cell manufacturing must involve easy processing of materials, so as to enable the high throughput that will be required for any technology that is to reach TW_p power generation; therefore, multi-step, highly complex processes should be avoided as much as possible.

18.5 Current photovoltaic technologies

18.5.1 Introduction

In 2007, according to US Energy Information Agency numbers, the net global electrical power provided by electrical power-production plants for general consumption was 2.15 TW-yr (the energy equivalent of a 2.15 TW power plant running continuously for one year) [29]. This electrical power could, in principle, be produced if a 270-km $\times$ 270-km area distributed across high-solar-insolation desert regions of the Earth were covered with solar panels with a power-conversion efficiency of 15%. In reality a good goal is to achieve about 25% total power from solar electricity by 2050, which would require 180 km $\times$ 180 km (since the electrical power capacity is projected to increase by some 85% by then). For reference the US Department of Energy (DOE) estimates that the land needed to produce 1 GW for a reasonable efficiency range is 10–50 km^2 for PV, 100 km^2 for wind, 1,000 km^2 for biomass, and 20–50 km^2 for solar thermal or PV concentrators (http://www1. eere.energy.gov/solar/pdfs/35097.pdf). The rapid expansion of manufacturing capability in PV and deployment of concentrated solar power (CSP) systems offers the potential for supplying a significant fraction (15% or more without need of storage) of this energy with minimal environmental impact using solar power. The key to reaching this potential is reducing the cost of solar-produced electricity to achieve grid parity.

Solar modules are generally rated in units of a watt peak (W_p), where a 1 W_p panel will produce 1 W under AM1.5 standardized conditions, which roughly correspond to the Sun directly overhead at a mid latitude on a 25-°C day (cf. footnote [1] in Section 18.3.1). To meet the US DOE cost goal of \$0.33 per W_p leading to 0.05–0.06\$ per kWh for utility-scale production, these modules would need to be manufactured at a cost of \$50 per m^2 or less. To reach these objectives will require significant improvements in cell performance as well as in the additional components making up the balance of systems. Chapter 17 discussed the continuous reduction in the price of PV-produced electricity that has occurred over the last several decades. As noted below, the incremental improvements that are responsible for this trend might not, however, be sufficient to reach grid parity. Even if grid parity can be achieved with today's technology, for reasons that range from environmental and national security concerns to economic competitiveness the industry is driven to reach this point much faster. To truly accelerate the pace at which grid parity is achieved will require materials-science-driven, not just incremental, economies of scale-related, cost reductions.

Worldwide production levels for terrestrial solar-cell modules have been growing rapidly over the last several

Figure 18.14. Global PV cell production by year, from *PV News* May 2010 and L. R. Brown, *World on the Edge: How to Prevent Environmental and Economic Collapse* (New York, W.W. Norton & Company, 2011).

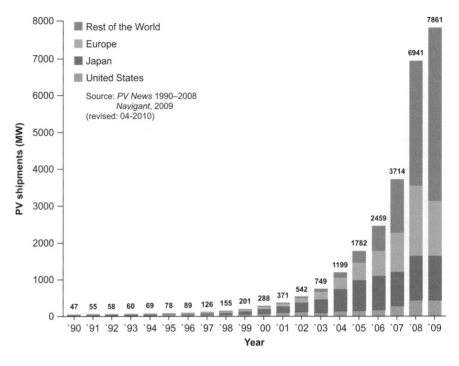

Figure 18.15. Cost efficiency analysis for first-, second-, and third-generation PV technologies (from [31]). The present limit is the Shockley–Queisser limit for a single-junction device. The generations can be thought of as follows: the first is current technologies, the second is near-commercial or initially commercialized technology, and the third is emerging technologies.

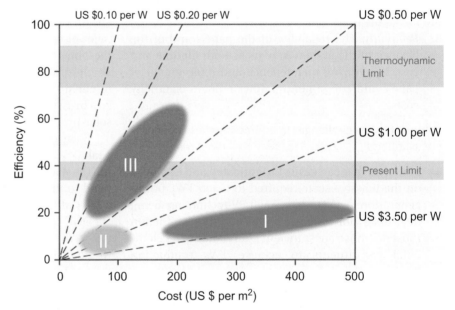

years, with China recently taking the lead in total production volume (Figure 18.14).

This production has historically been dominated by c-Si modules, which include both large-grain polycrystalline material, commonly referred to as multicrystalline Si, and single-crystalline materials, such that c-Si represented ~85% of the 2009 market. CdTe-based modules have quickly moved into second place and a broad range of potential replacement solar technologies is under active research and early-stage development. Progress in efficiencies of research-scale PV devices for each of these technologies over the last several decades is shown in Figure 4.1 of Chapter 4. In nearly every technology, better understanding of materials and

device properties has resulted in a continuous increase in efficiency.

These PV approaches are often named first-, second-, and third-generation technologies (a terminology coined by Martin Green, UNSW), as illustrated in Figure 18.15 [30].

Single-crystalline or multicrystalline Si-based modules represent the first generation of PVs. The second generation is based on thin-film cells composed, for example, of a-Si:H, CIGS, CdTe, and even thin polycrystalline Si absorbers. Thin-film cells are fabricated by high-throughput techniques at low temperature on low-cost substrates such as glass and metal foils rather than being cut as wafers from bulk Si. They seek to significantly reduce cost without sacrificing too much in

Figure 18.16. Historical costs for wafer crystalline and multicrystalline Si PV modules versus their cumulative production (in megawatts). Extrapolations for future technologies are also shown (from [32][33]).

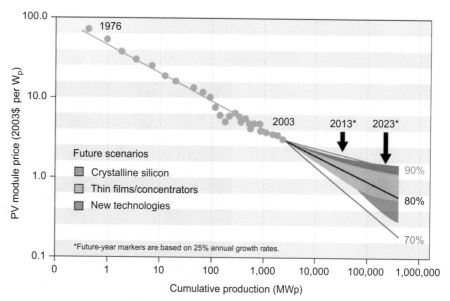

efficiency, leading to an overall reduction in module cost per W_p. Third-generation approaches exploit novel materials (e.g., quantum dots), device designs, and physical phenomena to significantly boost device efficiency and further lower solar-electricity costs. Third-generation concepts have just begun to be investigated and have a long materials and development path ahead. Many of them are covered in Chapter 19. The boundary lines between these definitions are often somewhat blurred. Organic solar cells, for example, are thin films that can potentially be produced at very low cost, which is analogous to a successful second-generation approach. However, at the same time, their excitonic nature as discussed above is relatively unexplored, but might lead to new modes of energy harvesting or third-generation approaches (Chapter 19). Even with this blurring, the organization into various generations provides a useful way to categorize PV technologies.

Given the historical advantage of crystalline and multicrystalline Si over other cell types, for second- and ultimately third-generation technologies to displace c-Si technology they must demonstrate significant cost reductions. One key to the development of any PV technology is the cost reduction associated with economies of scale. This has been very evident in the case of c-Si PV. Figure 18.16 shows the decrease in the cost of c-Si PV modules as the production rate has increased. The steady decline in module cost is due to the incremental improvements and economies of scale mentioned earlier in this chapter. Figure 18.16 also shows predicted future costs for both wafer-based c-Si and emerging technologies [32][33].

The student is cautioned, though, that predictions of future prices and production depend on assumptions about materials costs and market growth and hence can be unreliable. A shortage of Si feedstock that occurred around 2005 slowed the decline in module

prices. Retail prices were then above \$4.00 per W_p (http://www.solarbuzz.com/). The economic downturn of 2008 coupled with an increase in feedstock created in response to the earlier shortage then drove module prices down quickly, with wafer-based Si module producers reporting manufacturing costs of ~\$1.50 per W_p or less in 2010 [34][35] with retail prices of \$2.00–3.00 per W_p (http://www.solarbuzz.com/). In roughly the same time-frame, CdTe solar cells manufactured by First Solar penetrated the market with a dramatically reduced manufacturing cost, which in 2010 was reported to be \$0.76 per W_p [36]. The learning curve for CdTe would probably best be represented as a new line in Figure 18.16 below the historical Si learning curve. A key question concerns the slope of the CdTe learning curve and whether CdTe can maintain its cost advantage relative to c-Si and multicrystalline Si. Although it is difficult to determine the minimum price either wafer Si or thin-film PV modules can reach, it has been argued that the contribution of Si feedstock to the cost of wafer-based technologies cannot be reduced indefinitely, and at some point in the next 10 years prices for wafer-based Si panels will level off above the \$0.33 per W_p target. This has driven investments in alternative technologies such as thin-film CdTe and CIGS. We note that reaching grid parity will require substantial reductions in cell/module cost but will also necessitate that the balance-of-systems (BOS) (transport, installation, supports, maintenance, etc.) cost be substantially reduced, as currently, for Si, module and BOS are about equal in cost.

The rest of this chapter will focus on the strengths and weaknesses of the several first- and second-generation approaches that have been proposed for solar electric power generation as well as organic solar cells, together with the fundamental materials science understanding required for each of these technologies to succeed.

Figure 18.17. The photo shows the 2-MW PV system at Denver International Airport; the airport has now added an additional 1.6-MW array.

Chapter 19 addresses the development of third-generation technologies, Chapter 20 treats concentrator approaches, while Chapter 49 considers electrochemical systems, including dye cells. Together these represent the major PV technologies being researched, developed, and manufactured. As we look to the future it is clear that materials science will play a critical role in each of these areas. While this chapter focuses on the cell- and device-level materials science, it is important to keep in mind the broad range of materials considerations required for cost-effective solar conversion systems. For example, solar systems must have installation lifetimes and warranties of at least 25 years. To accomplish this, of the system components must be long-lived, low-maintenance, and stable. This requires solar-cell packaging, contacting (bus structures), and support structures to be stable in a wide variety of climates with extremes in temperature, humidity, wind, etc. Si and stabilized amorphous Si have been able to meet these challenges. To date the other thin-film technologies have not been commercially available for long enough to evaluate their lifetimes, making the ability to do accelerated aging on modules to predict stability a critical emerging area of solar materials science. Coupled closely to the development of improved cost-effective PV is the eventual development of low-cost energy-storage solutions, which will be discussed in other chapters. Also, as with all other technologies, PV modules contain valuable materials that will need to be recycled. This is generating a "cradle to grave" approach by some solar companies whereby they install and will eventually remove and recycle their products. The natural resource needs, environmental issues of the various technologies, and impacts of recycling on the efficacy of technologies that use scarce or toxic materials are also important considerations as the solar energy sector moves forward.

18.5.2 First-generation technologies

Figure 18.17 shows a 2-MW$_p$ first-generation system installed at Denver International Airport in Colorado, USA. First-generation technologies are primarily c-Si, including both large-grain multicrystalline and single-crystalline approaches.

These are single-junction devices, which are limited by thermodynamic considerations (discussed in more detail in Chapter 19) to a maximum theoretical power-conversion efficiency of ~30% under direct AM1.5 sunlight [37] and an ideal Si solar cell operating under direct sunlight would convert ~29% of the illuminating solar radiation into electric power. Real cells suffer from parasitic losses and exhibit lower efficiencies. With Si, however, efficiencies are moving close to the theoretical limit, with a record cell efficiency without solar concentration of 25% [5].

Current Si solar-cell design represents a considerable evolution beyond that of a simple single-junction device, incorporating passivation of the surfaces, light trapping, contacts that block less light, and sophisticated antireflection coatings to help absorb most of the light in the wavelength range accessible to Si and ensure that each absorbed photon leads to a carrier in the external circuit. Manufactured cell designs can yield efficiencies very close to record efficiencies. As an example, Figure 18.18 illustrates a Sanyo high-efficiency Si HIT (heterojunction with intrinsic thin layer) solar cell, which has an efficiency of up to 22.3% [38]. SunPower has reported a large-area cell with 24.2% efficiency [39]. Given how close modern cell efficiencies

Figure 18.18. A schematic diagram of the Sanyo HIT cell that is c-Si-based and has demonstrated an efficiency of 22.3%.

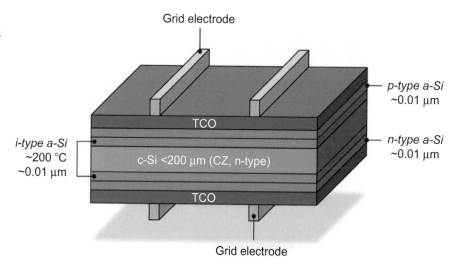

are to their theoretical limit, it is, in fact, methods to get significant cost reductions in manufacturing or alternative sophisticated structures that are needed in order to reduce the costs of modules made with them.

The current module designs are rapidly evolving toward manufacturing costs of $1 per W_p (see Figure 18.18 and the associated discussion). Key issues are Si feedstock supply, losses involved in preparing Si wafers, and developing lower-cost, high-throughput processing. The cost of the Si itself is a significant component of the cost of a cell, an issue that is compounded by Si's indirect bandgap, which causes it to have a lower absorption coefficient, and means that thicker absorber layers are required than with other PV materials. To this end, some companies are exploring the use of ribbon-based technologies whereby c-Si is grown as a thin sheet directly from molten Si. This avoids the materials loss associated with cutting blocks of polycrystalline Si or boules of single-crystalline Si into wafers. This is known as kerf loss. To date the materials produced by the ribbon approach do not yield the efficiency of wafer-based single-crystalline materials, but efficiencies are reasonable for commercial viability. Both ribbon and boule technologies aim at achieving thinner cells. A cell of thickness ~30 μm (current cells are of thickness 100–200 μm) would significantly reduce materials costs and could be designed to avoid compromising efficiency significantly. This would require new ways to process, contact, and handle such thin materials. For example, contacting may be done by inkjet non-contact printing instead of the screen-printing approach employed currently.

18.5.3 Second-generation technologies

Over the last decade, there has been considerable effort directed at thin-film, "second-generation" technologies that do not require the use of Si-wafer substrates and may allow significantly reduced manufacturing cost

per W_p [40]. Steady progress has been made in laboratory efficiencies, as can be seen in Figure 4.1 for devices based on, for example, CdTe, CIGS, and amorphous Si. These devices are fabricated using large-area techniques such as sputtering, physical vapor deposition, and chemical vapor deposition, which are well-understood manufacturing techniques in the coatings, display, and microelectronics industries. Unlike Si solar panels, panels composed of these materials are well below their potential maximum efficiencies and the reported record research efficiencies have not easily translated into production efficiencies. Reductions in manufacturing cost and improvements in panel efficiencies are both routes to better cost-per-W_p performance.

Amorphous-Si:H, amorphous-SiGe:H, and nanocrystalline and polycrystalline Si

Until the recent growth of CdTe technology, a-Si:H solar cells were the most commercially successful thin-film PV technology to date, with ~5% of the 2009 PV market. Their ability to be fabricated at relatively low cost and be integrated into electronics and building roofing material was central to their success.

Devices are often a single- or two-junction design put down in multiple layers by vacuum deposition processes such as sputtering and plasma-enhanced chemical vapor deposition. Junctions typically have a p–i–n structure. More complex triple-junction designs are being developed, as shown in Figure 18.19. They exhibit stabilized laboratory efficiencies of 12% and represent a materials science *tour de force* [41]. The continued development of these multijunction cells is a materials challenge pushing the limits of the materials synthesis and processing, which in and of itself is still not fully understood.

An amorphous material has no long-range crystalline order, with atoms being bonded to each other in a random way. Window glass is an example of an optically useful amorphous material. In Si, this absence of

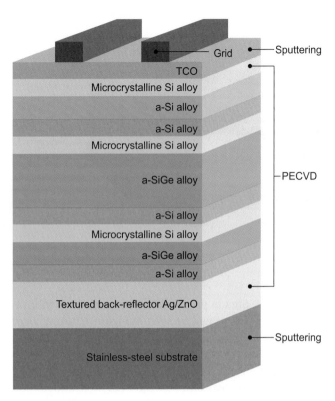

Figure labels (top to bottom): Grid — Sputtering; TCO; Microcrystalline Si alloy; a-Si alloy; a-Si alloy; Microcrystalline Si alloy; a-SiGe alloy; a-Si alloy; Microcrystalline Si alloy; a-SiGe alloy; a-Si alloy; Textured back-reflector Ag/ZnO; Stainless-steel substrate; PECVD; Sputtering

Figure 18.19. A triple-junction amorphous and microcrystalline Si solar cell as produced on a flexible steel substrate. PECVD is plasma-enhanced chemical vapor deposition.

order is attractive because it creates a bandgap that is "quasi-direct," which leads to a larger optical absorption coefficient and, hence, thinner absorbing layers and less materials cost than for c-Si. Alloying with Ge allows the optical absorption to be tuned across a useful range of the solar spectrum. Amorphous Si, however, suffers from a light-induced instability known as the "Staebler–Wronski" [42] effect, which causes the cell efficiency to degrade with time. While the effect cannot, so far, be eliminated, the size of the degradation can typically be reduced to 10%–20% of the as-manufactured and un-aged efficiency. This degradation has had a major effect on the a-Si solar-cell architecture. The effects of the degradation are less in thin layers; hence p–i–n structures with a thin intrinsic layer are preferred. Incomplete absorption in the thin absorber is then mitigated through formation of the multijunction stacks described above. Maintaining the initial efficiency of a-Si:H would have a major impact on the technology and is one of the great materials science challenges for a-Si:H.

The growth techniques used to deposit amorphous Si can also be used to grow films consisting of nanocrystalline or microcrystalline Si, either with or without an accompanying amorphous matrix. These materials show crystalline regularly ordered regions on a small length scale and usually form as the process temperature is increased. As the grain size grows, the film begins

to have more of the characteristics of polycrystalline Si, with the energy gap becoming more indirect, i.e., lower optical absorption and, thus, thicker layers required for complete absorption, and passivation becoming more important. Large-grain-size polycrystalline Si thin films produced, for example, by solid-phase crystallization of amorphous Si are also being investigated, and are now in production. Combinations of poly-Si, amorphous Si, and microcrystalline Si are being explored to develop low-cost, thin-film Si solar cells on glass, which could substantially reduce costs for large-area production.

CdTe and CIGS: chalcogenide-based thin-film solar cells

Figure 18.20 illustrates typical cross sections for polycrystalline CIGS (or for $CuInSe_2$, CIS) and CdTe solar cells. Both are nominally heterojunction devices with a junction formed between a thin n-type CdS layer and a p-type CdTe or CIGS absorber layer. For both devices in Figure 18.20, light enters from the top of the figure. With CdTe this means light enters through the glass on which the films making up the device are deposited. For CIGS cells the light enters from the face opposite the substrate. These are referred to as superstrate and substrate configurations, respectively. The record efficiency for a CdTe cell is a bit above 16%, while laboratory efficiencies of CIGS cells have now reached 20% (Figure 4.1) [43][44].

Many companies around the world are developing a variety of manufacturing approaches aimed at low-cost, high-yield, large-area-device processes that maintain laboratory-level efficiencies in these systems. As noted above, CdTe panels from First Solar with a compellingly low manufacturing cost have recently emerged as one of the most successful second-generation PV approaches to date. In 2009 CdTe modules represented roughly 12% of the market.

Materials challenges exist in each of the layers shown in Figure 18.20 and in the interactions between layers. Thus, there is research into improved transparent conducting oxide (TCO) contacts. These need to be made from low-cost, plentiful elements, must have high conductivity and high transparency in the visible, and must allow easy electrical isolation of devices. Another example is provided by the thin CdS contact layer in both devices, which also functions as a window for solar radiation. Since CdS absorbs in the blue, making this layer thin (or developing an alternative larger-bandgap material) is important. When it is too thin, however, pinholes between the TCO contact and the absorber layer create shorts. This is especially problematic for CdTe cells, where diffusion of S into the CdTe during post-growth anneals further decreases the CdS layer thickness. The inclusion of thin buffer layers between the TCO and CdS, such as a highly resistive transparent oxide, improves efficiency [44][45]. The exact role of the

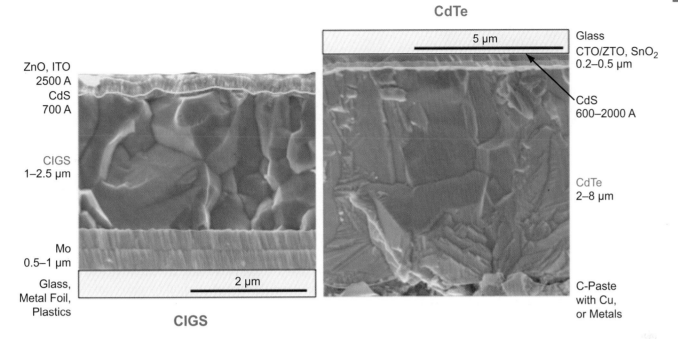

CdTe

ZnO, ITO
2500 A
CdS
700 A

CIGS
1–2.5 µm

Mo
0.5–1 µm

Glass,
Metal Foil,
Plastics

2 µm

CIGS

Glass
CTO/ZTO, SnO₂
0.2–0.5 µm

CdS
600–2000 A

CdTe
2–8 µm

C-Paste
with Cu,
or Metals

5 µm

Figure 18.20. Cross-sectional SEM images for a CIGS solar cell on the left and a CdTe cell on the right. The absorber layer thicknesses are 1–2.5 µm and 2–8 µm thick, respectively. CTO/ZTO stands for $Cd_2SnO_4/ZnSnO_x$.

buffer layer, whether it simply introduces resistance into shorts or changes the interfacial energetics, is not well understood, and optimizing this interface is a critical need. As can be seen in Figure 18.21, although the absorbing CIGS and CdTe films are quite thin (ideally 1 and 5 µm, respectively) the grain size for efficient devices is a large fraction of the thickness. In the case of CdTe this is achieved by a post-deposition anneal in the presence of $CdCl_2$ and O_2, which promotes low-temperature grain growth. Finding controllable and manufacturable methods for doing this anneal, or, better still, incorporating it directly into the growth process, is an active area of research and development. The highest-efficiency CIGS films are produced by physical vapor deposition and the grain microstructure is defined by a complex evolution from Cu-rich to In- and Ga-rich phases during growth [46]. Reproducing this growth path with other, more manufacturable deposition options is a major challenge that several of the startup companies producing CIGS cells have undertaken. Incorporation of Na^+ ions, which occurs naturally during growth on soda lime glass substrates, is also important for optimizing the CIGS device efficiency. Although there may be no one explanation for the beneficial effects of Na, it is thought to catalyze surface oxidation, which reduces the resistivity of the film; possibly, it also increases the grain size [47]. The defect structure of grain boundaries in both CIGS and CdTe is just beginning to be understood and is believed to play a dominant role both in determining the minority-

carrier lifetime and in carrier collection. Other absorber-layer materials and processing issues that affect cell costs include the need to lower growth and processing temperatures, which at present exceed 500 °C for the highest-efficiency films, and the desire to use thinner films without loss of absorption. The latter requires methods for increasing the optical path length through back reflection, texturing, or light trapping. Moving attention from the absorber layer to the back contact, significant challenges exist in understanding and improving the electronic and structural characteristics of these layers in CIGS and CdTe devices. Delamination issues, for example, in CIGS cells are typically correlated with problems in the molybdenum back contact layer.

While the basic CIGS and CdTe devices have demonstrated good operational lifetimes when glass/glass encapsulated, the devices are quite sensitive to moisture. The current encapsulation approach adds cost, increases system weight, and does not entirely eliminate failures. We note that required certification for commercialization of thin-film PV takes testing at 85 °C and 85% relative humidity. This is a very demanding test, and nearly all photovoltaic devices and their associated contacts, antireflection coatings and encapsulation will ultimately need to be improved to achieve lifetimes >25 years. A thin-film encapsulation approach, which provides a moisture barrier without compromising efficiency, could be a significant advance. Beyond the simple thin-film cells illustrated in Figure 18.20 is the potential to combine thin-film materials like Si, a-Si,

Figure 18.21. A schematic representation of a bulk heterojunction excitonic solar cell. The orange and black active layer is a two-phase mixture of a polymeric donor and a fullerene (black phase) acceptor.

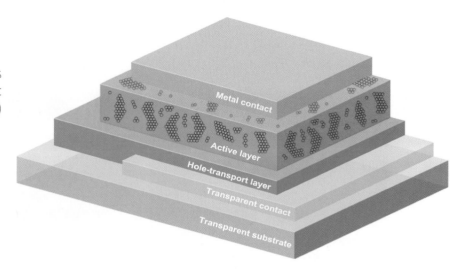

CIGS, and CdTe to form tandem cells that capture more of the solar spectrum through multijunction structures. This places even more stringent requirements on understanding and control of the interfaces between dissimilar materials.

A broader materials issue with both systems concerns their elemental composition. There are residual concerns about the environmental effects of Cd, and questions exist about the cost and availability of such elements as In and Te, which are relatively rare in the Earth's crust. Scarcity can be particularly important as the industry attempts to scale PV power generation from present levels to the TW levels needed to meet a significant fraction of the world's electricity needs. There is considerable debate about how much we should be concerned about toxicity or whether a technology based, for example, on Si, which is plentiful, has a long-term advantage over a CdTe technology as production levels grow. At the same time, methods for dealing with these issues are being implemented or investigated. First Solar, for example, has a "cradle to grave" approach for the panels they manufacture. Their price structure includes end-of-life recycling to capture the Cd and Te for reuse in new panels, an approach that can mitigate both toxicity and scarcity concerns. Efforts to reduce the width of the active layer for a given efficiency also reduce the amount of material needed. Alternative materials, such as ZnS to replace CdS in CIGS cells, are being developed to address some of these concerns.

Organic photovoltaic cells

Organic PV (OPV) is a rapidly emerging device technology with the potential for low-cost, non-vacuum processed devices [48]. OPV has rapidly moved from very low efficiency to a certified efficiency >8% (2010) for a small-area device from Solarmer Energy Inc. and for at least 1-cm^2 devices from several other companies [49]. In addition, module efficiencies for large-area roll-to-roll production are approaching 4%. Key is that these devices combine polymers, small molecules, and/or inorganic nanometer-sized (nano)structures to build an excitonic solar cell. At the heart of the cell is an absorber layer composed of an organic donor that absorbs most of the light (small organic molecules, polymers, etc.) and an electron acceptor (fullerene, inorganic quantum dot, nanowire, etc.). High-efficiency devices use a bulk heterojunction (BHJ) blend of the donor and acceptor and the most successful is a polymer–fullerene blend. A typical cell configuration for a BHJ device is shown in Figure 18.21. Light generally enters through the transparent substrate hence this is a "superstrate" configuration device.

Figure 18.22 illustrates a typical energy-level diagram like that in Figure 18.5, but specific to an organic solar cell. In this case the cell is composed of a poly(3-hexylthiophene) (P3HT) polymer donor and a phenyl-C_{61}-butyric acid methyl ester (PCBM) acceptor. The acceptor is essentially a C_{60} fullerene with an attached organic group to increase its solubility in solvents. The P3HT/PCBM bulk heterojunction is a commonly used test system for OPV development. We can think of this diagram as describing one the many P3HT/PCBM interfaces in a bulk heterojunction blend. In operation, light is absorbed by the P3HT and, as described in the discussion above on excitonic solar cells, the exciton that is created diffuses to the P3HT/PCBM interface, where it decomposes into a hole in the HOMO level (5.2 eV) of the polymer and electron in LUMO level (4.2 eV) of the fullerene. The carriers then move in an energetically "downhill" fashion with holes collected at the ITO (indium tin oxide) contact and electrons at the Ca-doped Al contact. The PEDOT:PSS layer is an organic film that transports holes much easier than electrons and minimizes recombination of electrons and holes, before they reach the electrodes. The maximum V_{OC} is the difference between the HOMO for the polymer and the LUMO for the fullerene. While the excitonic mechanism differs from that in conventional PV devices, the theoretical efficiency is the same.

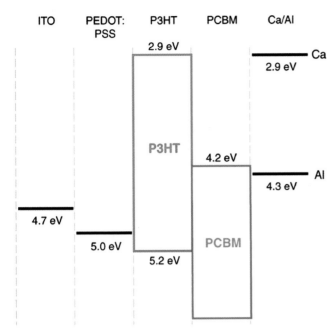

Figure 18.22. The energy-level band diagram for a typical bulk heterojunction solar cell. The numerical values refer to the energy difference between the specific level and the vacuum level (essentially the work function) (not shown). Such devices achieved efficiencies of ~5% and, with alternate materials, ~8% (2010 data).

What makes OPV attractive is the possibility of efficiencies comparable to those of existing second-generation devices, but with significant reductions in cost per W_p, because of plastic processing technologies. The development of materials for OPV leverages the emergence of an organic electronics industry based on displays (organic light-emitting devices, OLEDs) and transistors [50][51] [52]. A key aspect of OPV is that the organic small-molecule and polymer materials that are being investigated are inherently inexpensive, typically have very high optical absorption coefficients, permitting very thin films, are compatible with flexible plastic substrates, and can be fabricated using high-throughput, low-temperature approaches using low-capital-cost roll-to-roll processes. A manufacturing line for organic solar cells might be more akin to a newspaper printing process than to a c-Si solar-cell production line (See Figure 18.23). The potentially low production cost of organic solar cells can even relax constraints on the efficiency and module lifetime required to allow them to be cost competitive with other first- and second-generation approaches.

Another important aspect of organic materials is flexibility to synthesize molecules and, especially, to synthesize variants of basic molecules to alter a wide range of properties, including relative molecular mass, energetics, wetting and structural properties, and doping. This ability to design and synthesize molecules and then integrate them into organic–organic and

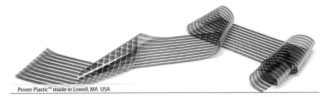

Figure 18.23. An OPV bulk heterojunction solar-cell ribbon produced by Konarka Technologies Inc.

inorganic–organic composites provides a unique path to new device materials. This has in fact been a key factor in the rapid advancement in device efficiency in the field, where polythiophene poly(3-hexylthiophene) (P3HT) has been replaced by poly[[9-(1-octylnonyl)-9H-carbazole-2,7-diyl]-2,5-thiophenediyl-2,1,3-benzothiadiazole-4,7-diyl-2,5-thiophenediyl] (PCDTBT) (Figure 18.24), which gives efficiencies approaching 7%. Key is that combining molecular modeling with synthesis allows the development of new acceptors and donors and enhanced morphologies that produce much higher effiencies. Additionally, with the ability to alter the device color, fabricate devices on flexible substrates, and potentially print them in any pattern, OPV may be integrated into existing structures and into new commercial products more easily than other cell types, e.g., into window units for building integrated PV systems.

A number of key issues must be addressed, though, for OPV to succeed. An active research area involves development of molecules that absorb further into the red (or even the infrared) than current materials in order to use more of the solar spectrum, but without losing V_{OC}. An example of the directions of the evolution is a system in which PCDTBT absorbs nearly 100 nm further into the red than does P3HT. It is also necessary to develop optimized interfaces at the nanoscale, which allow exciton splitting, charge transfer, and optimization of the morphology of the organic constituents. This is closely connected with controlling the nanometer-scale organization of BHJ blends that strongly affects device performance and must often be re-optimized for new organic absorber molecules or acceptors. This is leading to a whole new field of looking into the details of the substituent chemistry for an electronic polymer and its subsequent effects on the morphology, carrier and excitonic lifetimes, V_{OC}, and stability. Directly related to this are studies of the organic/inorganic interface and its effects on these same parameters.

A key aspect of the studies is that they are leading to an understanding of the factors that determine the stability of OPV devices. For example, if the devices are of the conventional bulk heterojunction structure then they are stacked as transparent contact (ITO)/PEDOT: PSS/bulk heterojunction/Ca/Al, where holes go to the ITO and electrons to the Al. This means that you have a very reactive metal near the exposed surface of

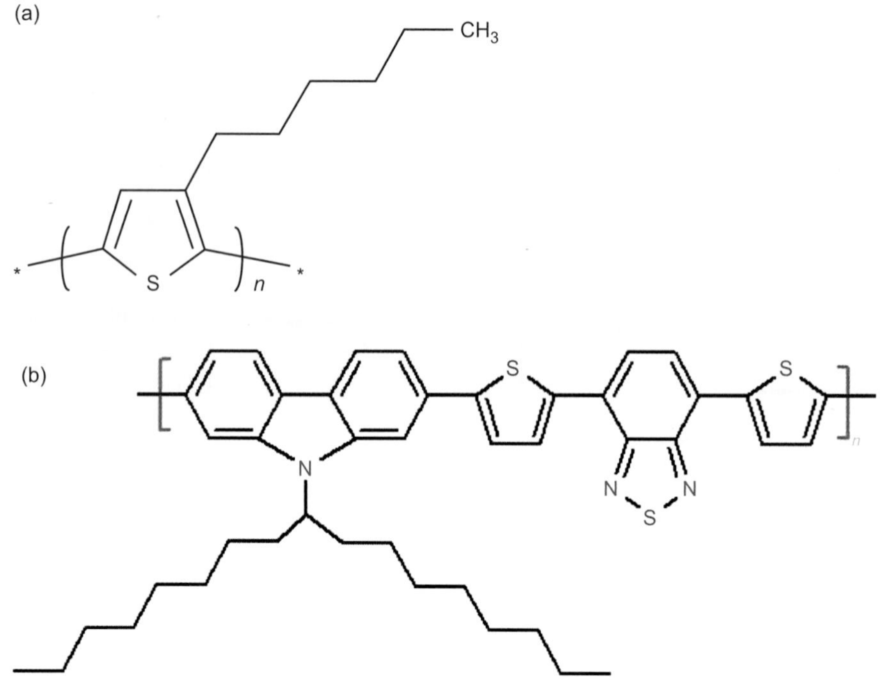

Figure 18.24. Illustrations of the development of new polymers based on altering molecular structure and substituents: (a) P3HT, maximum efficiency about 5% and (b) PCDTBT, maximum efficiency >7%.

the device that is exposed to the atmosphere. This has produced lifetime issues and requires very good encapsulation. However, inverting the device by stacking it as ITO/ZnO/bulk heterojunction/PEDOT:PSS/Ag, the carriers can be made to flow in the opposite direction. Here the exposed contact is the less reactive metal, Ag, and the devices can be made in air. These "inverted" devices are also more amenable to roll-to-roll processing on flexible substrates. They also demonstrate much better lifetimes, even for unencapsulated devices.

Studies on OPV lifetime are giving encouraging results [53] with encapsulated devices showing nearly 10,000 hours for bulk heterojunctions and >30,000 hours for small-molecule devices. Inverted structures have shown up to 5,000 hours for devices with no encapsulation. While the mechanisms of device degradation are still open questions, they may stem largely from changes in morphology, loss of interfacial adhesion, and interdiffusion of components, as opposed to strictly chemical decomposition of the molecular constituents. Like many of the newer technologies, OPV is more sensitive to air and moisture than conventional Si technology. Thus, careful design, materials engineering, and improved encapsulation are being explored to improve device lifetimes, and the encapsulation may need to have two or three orders of magnitude less in-diffusion of oxygen and water. Encapsulation is becoming a inceasingly active area of research because it is its critical for a wide variety of emerging opto-electronic technologies. Stability of encapsulants for 10–25 years with no yellowing and no diffusion of O_2 or H_2O, yet having low cost and easy processibility is a real challenge. Increasingly, as in the OPV devices themselves, there

is a focus on nanomaterial–polymer composite encapsulants where composition and structure are designed to scavenge impurities and slow down diffusion. There are also efforts to develop new polymers with very low diffusion coefficients and photosensitivity. One question that has been raised by the OPV community but not resolved to date is that of whether, if the technology is cheap enough, you could get by with a 5–10-year lifetime and replace modules rather then have a conventional 25-year lifetime.

Manufacturing and balance of systems

Our discussion thus far has primarily focused on the current and emerging device technologies and some discussion of integrating the devices monolithically or discretely into modules.

A key factor to the ultimate production of PV at the terawatt scale is that costs will need to reach grid parity. This includes not only those for the cell and module but also all of the balance-of-systems (BOS) costs. The left-hand side of Figure 18.25 illustrates the current breakdown in costs between module and BOS for installed ground-based (large) and rooftop systems. The BOS contributes 40%–50% of the total installed cost. The right-hand side of Figure 18.25 illustrates the breakdown of the BOS costs. Clearly there are some efficiencies of scale in large ground-based systems. Nearly all portions of the BOS detail can be improved. Also not included in this is essentially a reliability factor considering the ability of other system components to match the 25-year warranty of the cell/module. We also note that in distributed systems where storage is important, this significantly increases BOS and

Figure 18.25. Balance-of-systems cost for current PV systems from a recent study by the Rocky Mountain Institute [54].

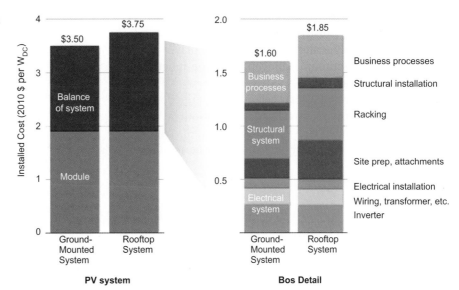

reliability issues. To be able to achieve grid parity nearly all parts of the PV supply chain need to be improved, from the production of raw materials to the ultimate labor for device installation and maintenance. There is thus the opportunity to significantly reduce the BOS cost by improving performance and reducing the cost of inverters, tracking systems, structural installation, and smart wiring to the grid. Ultimately, as PV installations become a significant power source, energy storage and its integration must also be brought into the BOS equation. To optimize the BOS cost will require an assessment of the whole system to understand optimum module size and even device size/performance. A potential way to achieve a significant reduction in BOS cost is to focus on low-cost manufacturing techniques such as roll to roll, solution processing, development of smart processing with *in situ* diagnostics for improving the yield of large-area panels, looking at lower-cost substrates and encapsulation, and improving the integration of modules into support structures and their installation. As cells become more efficient and cheaper, it is in fact the BOS cost that begins to dominate. There is a great deal of materials research that must go into all of the items above. This will need to be technology-dependent since Si wafers and OPV require very different manufacturing approaches and they may have very different substrates etc., which changes the BOS requirements.

18.6 Summary

This chapter has discussed the history and growth of solar PV. Though the industry is currently based on c-Si and multicrystalline Si wafers, ultimately these might not be able to meet cost goals required to achieve grid parity. This has opened the door to second-generation

thin-film devices. While significant materials challenges still exist for these devices, they are nearing large-scale manufacturability, as has been shown by the recent growth of CdTe production. New high-efficiency or low-cost technologies such as multijunction and organic-based devices are advancing rapidly and may have second- and third-generation embodiments. Finally, new very-high-efficiency third-generation approaches, discussed in more detail in the following chapter, offer potential in a long-term time-frame to produce devices that can convert much larger portions of the solar spectrum. Given the predicted market growth, many of these approaches should be investigated in parallel to meet the demand.

Within the limits of this chapter the discussions of various PV approaches were of necessity brief, and it was also impossible to give adequate credit to the many scientists who are contributing and have contributed to this field. For more information, interested readers are referred to the suggested texts for further reading, which give a broad understanding of the area, and to the specific references.

18.7 Questions for discussion

1. What is the difference between power and energy insofar as it pertains to a PV system?
2. What is the difference between direct and diffuse solar radiation?
3. What environmental factors are important in PV production and deployment?
4. If a 1-W_p solar panel has an efficiency of 15%, costs $1.50, requires an additional $1.50 to install and connect to the grid, and will last 25 years, how much does the electricity it produces cost per kilowatt hour?

5. What is the difference between PV applied in a central and in a distributed manner?

6. What is a rectifier and what role does it play?

7. What is the optimum bandgap for a one-junction solar convertor? Why?

8. Name the parts of a solar cell and their functions.

9. The balance-of-systems (BOS) cost is about 50% of the total installed cost of the solar system. Which part of the BOS cost do you think can be reduced most easily?

10. How does temperature affect the performance of a PV system?

11. If the same solar panel were installed for a year in Saudi Arabia or in Vermont, USA, in which location would it produce more energy and about how much more?

12. If you consider a 10%-efficient solar cell, a reduction factor of 5 for the variation of solar light intensity during a day–night cycle, and an electric power-to-fuel conversion efficiency of 60% for electrolysis, how large must a PV panel be to provide for the energy consumption of 6 kW of an average European?

18.8 Further reading

- **A. L. Fahrenbruch** and **R. H. Bube**, 1983, *Fundamentals of Solar Cells*, New York, Academic Press. Although somewhat dated, this is probably still the best all-round textbook.
- **S. J. Fonash**, 1981, *Solar Cell Device Physics*, New York, Academic Press.
- **H. J. Moeller**, 1993, *Semiconductors for Solar Cells*, New York, Artech House.
- **J. Nelson**, 2003, *The Physics of Solar Cells*, London, Imperial College Press.
- **M. A. Green**, 2003, *Third Generation Photovoltaics: Advanced Solar Energy Conversion*, Berlin, Springer.
- **M. D. Archer** and **R. Hill** (eds.), 2001, *Clean Electricity from Photovoltaics*, London, Imperial College Press.
- **R. H. Bube**, 1998, *Photovoltaic Materials*, London, Imperial College Press.
- **P. Würfel**, 2009, *Physics of Solar Cells: From Basic Principles to Advanced Concepts*, Weinheim, Wiley-VCH.

18.9 References

[1] **W. A. Hermann**, 2006, "Quantifying global exergy resources," *Energy*, **31**(12), 1685–1702.

[2] **Shyam Mehta**, "26th Annual Data Collection Results: Another bumper year for manufacturing Masks turmoil." *PVNews*, **29**(5) (May 2010), 11–14.

[3] **B. von Roedern**, **K. Zweibel**, and **H. S. Ullal**, 2005, "The role of polycrystalline thin-film PV technologies for achieving mid-term market-competitive PV modules," in *Proceedings of the 31st IEEE Photovoltaics Specialists Conference and Exhibition*, Lake Buena Vista, FL.

[4] **D. Ginley**, **M. Green**, and **R. Collins**, 2008, "Solar energy conversion toward 1 terawatt," *MRS Bull.*, **33**, 355–364.

[5] **M. A. Green**, **K. Emery**, **Y. Hishikawa**, and **W. Warta**, 2011, "Solar cell efficiency tables (version 37)," *Prog. Photovoltaics: Res. Appl.*, **19**, 84–92.

[6] **L. Kazmerski**, *Solar Efficiency Tables*, NREL, http://en.wikipedia.org/wiki/File:PVeff(rev100921).jpg.

[7] **E. Becquerel**, 1839, "Mémoires sur les effets électriques produits sous l'influence des rayons," *Comptes Rendues*, **9**, 561–567.

[8] **W. Smith**, 1873, "Effect of light on selenium during the passage of an electric current," *Nature*, February 20, 1873.

[9] **W. G. Adams** and **R. E. Day**, 1877, "The action of light on selenium," *Proc. Roy. Soc. A*, **25**, 113.

[10] **A. Einstein**, 1905, "Über einen die Erzeugung und Verwandlung des Lichtes betreffenden heuristischen Gesichtspunkt" ["On a heuristic viewpoint concerning the production and transformation of light"], *Ann. Phys.* **17**, 132–148.

[11] **E. H. Kennard** and **E. O. Dietrich**, 1917, "An effect of light upon the contact potential of selenium and cuprous oxide," *Phys. Rev.*, **9**, 58–63.

[12] **L. O. Grondahl**, 1933, "The copper–cuprous oxide rectifier and photoelectric cell," *Rev. Mod. Phys.*, **5**, 141–168.

[13] **R. S. Ohl**, 1946, "Light-sensitive Electric Device," US Patent No. 2,402,622.

[14] **R. S. Ohl**, 1948, "Light-sensitive Device Including Silicon," US Patent No. 2,443,542.

[15] **P. Rappaport**, 1954, "The electron voltaic effect in p–n junctions induced by beta-particle bombardment," *Phys. Rev.*, **93**, 246.

[16] **D. M. Chapin**, **C. S. Fuller**, and **G. L. Pearson**, 1954, "A new silicon photocell for converting solar radiation into electrical power," *J. Appl. Phys.*, **25**, 676.

[17] **D. C. Reynolds**, **G. Leies**, **L. L. Antes**, and **R. E. Marburger**, 1954, "Photovoltaic effect in cadmium sulfide," *Phys. Rev.*, **96**, 533.

[18] **D. A. Jenny**, **J. J. Loferski**, and **P. Rappaport**, 1956, "Photovoltaic effect in GaAs p–n junctions and solar energy conversion," *Phys. Rev.*, **101**, 1208.

[19] **J. Loferski**, 1993, "The first forty years: a brief history of the modern photovoltaic age," *Prog. Photovoltaics: Res. Appl.*, **1**, 67–78.

[20] **B. O'Regan** and **M. Grätzel**, 1991, "A low-cost, high-efficiency solar cell based on dye-sensitized colloidal TiO_2 films," *Nature*, **353**(6346), 737–740.

[21] **C. W. Tang**, 1986, "Two-layer organic photovoltaic cell," *Appl. Phys. Lett.*, **48**, 183–185.

[22] **P. Würfel**, 2009, *Physics of Solar Cells: From Basic Principles to Advanced Concepts*, Weinheim, Wiley-VCH.

[23] **M. A. Green**, 2002, "Photovoltaic principles," *Physica E*, **14**, 11–17.

[24] **J. Nelson**, 2003, *The Physics of Solar Cells*, London, Imperial College Press.

[25] **M. A. Green**, 2007, "Thin film solar cells: review of materials, technologies and commercial status," *J. Mater. Sci. Mater. Electron.*, **18**, 515.

[26] **S. Günes, H. Neugebauer**, and **N. S. Sariciftci**, 2007, "Conjugated polymer-based organic solar cells," *Chem. Rev.*, **107**, 1324–1338.

[27] **M. Grätzel**, 2003, "Dye-sensitized solar cells," *J. Photochem. Photobiol. C: Photochem. Rev.*, **4**, 145–153.

[28] **M. Yamaguchi, T. Takamoto, K. Araki**, and **N. Ekins-Daukes**, 2005, "Multi-junction III–V solar cells," *Solar Energy*, **79**(1), 78–85.

[29] **DOE**, 2009, *Annual Energy Review 2009*, http://www.eia.doe.gov/oiaf/ieo/electricity.html.

[30] **M. A. Green**, 2001, "Third generation photovoltaics: ultra-high conversion efficiency at low cost," *Prog. Photovoltaics: Res. Appl.*, **9**, 123–135.

[31] **M. A. Green**, 2001, "Photovoltaics for the 21st century II," *Electrochem. Soc. Proc.*, **2001**(10), 30–45.

[32] **T. Surek**, 2005, "Crystal growth and materials research in photovoltaics: progress and challenges," *J. Cryst. Growth*, **275**(1–2), 292–304.

[33] **R. M. Margolis**, 2003, "Photovoltaic technology experience curves and markets," in *NCPV Solar Program Review Meeting*.

[34] **Trina Solar**, Third quarter 2010 supplemental earnings call presentation, http://media.corporate-ir.net/media_files/IROL/20/206405/Supplemental_Presentation_Q3_2010_nov_30_1730hr_v5_cy_ir_ty.pdf.

[35] **SunPower Corp.**, 2010, Analyst day presentation, November 18, 2010, http://files.shareholder.com/downloads/SPWR/1159074174x0x420687/fefb453b-9d20-42f8-a27f-a22d31b6c0df/2010%20Analyst%20Day%20Final%20v4.pdf.

[36] **First Solar, Inc.**, Second quarter 2010 financial results, http://phx.corporate-ir.net/phoenix.zhtml?c=201491&p=irol-newsArticle&ID=1454084&highlight=.

[37] **W. Shockley** and **H. J. Queisser**, 1961, "Detailed balance limit of efficiency of p–n junction solar cells," *J. Appl. Phys.*, **32**(3), 510–519.

[38] **S. Taira, Y. Yoshimine, T. Baba** *et al.*, 2007, "Our approaches for achieving HIT solar cells with more than 23% efficiency," in *Proceedings of the 22nd EU-PVSEC*, pp. 932–936.

[39] **SunPower Corp.**, 2010, Press release, June 23, 2010, http://investors.sunpowercorp.com/releasedetail.cfm?ReleaseID=482133.

[40] **A. Slaoui** and **R. T. Collins**, 2007, "Advanced inorganic materials for photovoltaics," *MRS Bull.*, **32**(3), 211–218.

[41] **B. Yan, G. Yue, J. M. Owens, J. Yang**, and **S. Guha**, 2006, "Over 15% efficient hydrogenated amorphous silicon based triple-junction solar cells incorporating nanocrystalline silicon," in *Conference Record of the 2006 IEEE 4th World Conference on Photovoltaic Energy Conversion*, pp. 1477–1480.

[42] **D. L. Staebler** and **C. R. Wronski**, 1980, "Optically induced conductivity changes in discharge-produced hydrogenated amorphous silicon," *J. Appl. Phys.*, **51**(6), 3262–3268.

[43] **M. A. Contreras, M. J. Romero**, and **R. Noufi**, 2006, "Characterization of $Cu(In,Ga)Se_2$ materials used in record performance solar cells," *Thin Solid Films*, **511–512**, 51–54.

[44] **X. Wu**, 2004, "High-efficiency polycrystalline CdTe thin-film solar cells," *Solar Energy*, **77**(6), 803–814.

[45] **M. A. Contreras, B. Egaas, K. Ramanathan** *et al.*, 1999, *Prog. Photovoltaics: Res. Appl.*, **7**(4), 311–316.

[46] **J. AbuShama, R. Noufi, Y. Yan** *et al.*, 2001, "$Cu(In,Ga)Se_2$ thin-film evolution during growth from $(In,Ga)_2Se_3$ precursors," *Mater. Res. Soc. Symp. Proc.*, **664**

[47] **L. Kronik, D. Cahen**, and **H. W. Schock**, 1998, "Effect of Na on $CuInSe_2$ and its solar cell performance," *Adv. Mater.*, **10**, 31–36.

[48] **A. Romeo, M. Terheggen, D. Abou-Ras** *et al.*, 2004, "Development of thin-film $Cu(In,Ga)Se_2$ and CdTe solar cells," *Prog. Photovoltaics: Res. Appl.*, **12**(2–3), 93–111.

[49] **S. E. Shaheen, D. S. Ginley**, and **G. E. Jabbour**, 2005, "Organic-based photovoltaics: toward low-cost power generation," *MRS Bull.*, **30**(1), 10–19.

[50] **Solarmer Energy, Inc.**, 2010, Press release, July 2010, http://www.solarmer.com/newsevents.php.

[51] **G. Collins**, 2004, "Next stretch for plastics electronics," *Sci. Am.*, **291**, 74–81.

[52] **D. M. de Leeuw**, 1999, "Plastic electronics," *Phys. World*, **12**, 31–34.

[53] **C. J. Brabec, S. Gowrisanker, J. J. M. Halls** *et al.*, 2010, "Polymer–fullerene bulk-heterojunction solar cells," *Adv. Mater.*, **22**(34), 3839–3856.

[54] **L. Bony, S. Doig, C. Hart, E. Maurer**, and **S. Newman**, 2010, *Achieving Low-Cost Solar PV: Industry Workshop Recommendations for Near-Term Balance of System Cost Reductions*, Rocky Mountain Institute, p. 5.

Future concepts for photovoltaic energy conversion

Jean-François Guillemoles

Institut de Recherche et Développement sur l'Energie Photovoltaïque (IRDEP), Chatou, France

19.1 Focus

Photovoltaic (PV) conversion of solar energy could be much more effective than it is currently, using basic p–n junctions. The approaches required to reach theoretical conversion limits (~ 90%) are very challenging. Some have already been demonstrated, such as multijunction devices. Others, with considerable improvements in the description of excited states in condensed matter and in nanoscience, might be on the verge of a breakthrough. Given current knowledge about solar energy conversion and materials science, is it possible to achieve the ultimate solar cell?

19.2 Synopsis

Photovoltaic solar cells are now commonplace, and their development has taken advantage of the progress that has been made in electronics. Yet, they are still expensive and far from the performances that could, in principle, be achieved. This chapter first describes the limitations of PV devices as currently designed, before considering the various options being investigated to overcome these limitations. This description is put in perspective with achievable efficiencies according to thermodynamics. Then, we describe various options to experimentally approach the limits set on conversion efficiency of solar energy. While multijunctions are essentially presented in Chapter 20, here the emphasis is on other physical mechanisms than the ones currently operating in p–n diodes. These, first, include purely photonic conversion processes such as up and down conversion combined with a regular PV device, which provides additional power with relatively benign technology changes: an additional functional layer disconnected from the electrical circuit. The focus here is mainly on the optical properties of the additional materials. A second approach relies on the possibility of the absorber being able to harvest more energy from the solar spectrum than what semiconductors and molecules provide today: process of multi-generation of electron–hole pairs with a single photon or, conversely, processes by which several small-energy photons can contribute to the formation of an electron–hole pair. This not only requires new functional materials but also will change the device technology. Finally, we also look at approaches trying to tap into the heat produced upon absorption of photons to generate additional power. These devices are not isothermal, and, on top of optical and electronic properties, one needs to consider heat transfer between sub-parts of the system, and therefore take into consideration their thermal properties.

All these approaches are described in terms of principles of operation and potential. Also included are discussions of the state of the art in terms of experimental work and challenges involved in efforts to develop very-high-efficiency solar devices. Those challenges are setting stringent constraints on the materials and on their targeted properties that would be needed in order to make these new devices. Looking at the materials available today gives us a good idea of the distance still to be covered.

19.3 Historical perspective

Efficiency has always been a concern with energy conversion in general and with PV more specifically. Indeed, efforts to improve efficiency have driven developments in solar energy technology ever since the inception of the field. The link between efficiency and cost in solar technology is illustrated in Figure 19.1.

First-generation technologies consist primarily of cells based on crystalline Si (c-Si). Such cells are reliable, perform well (efficiencies of up to 20% commercially), and comprise the bulk (>80%) of PV production today (cf. Chapter 18).

Second-generation technologies are those that have demonstrated practical conversion efficiencies at potentially lower costs per watt than c-Si, but have a market penetration still below 25% at present. Commercialization of these approaches is typically in the early manufacturing phase. This represents a fairly broad base of PV technologies based on thin-film processes, and includes Si-based, inorganic compounds as well as the emerging field of organic PV. Thin-film solar cells have already enabled an essential step to be taken in the medium term by lowering the costs of PV considerably, albeit with the penalty of somewhat inferior performances than those attainable with c-Si, thus making solar technology economically viable for a considerable fraction of users. However, it might be extremely difficult to attempt to reduce production costs far beyond what is currently being obtained with inorganic thin-film systems. Indeed, there is not much space to the left (i.e., toward lower cost) of the oval representing second-generation technology in Figure 19.1, especially if one considers the need to use materials of sufficient quality and strength to last for decades. Finally, the gain in the surface cost of PV modules has been somewhat at the expense of their efficiency. A lower efficiency of modules will mean a higher cost of the rest of the system (balance of system or BOS) and a longer time for the return on investment. The efficiency factor is therefore particularly important when the cost of the solar kWh is considered. From another perspective, there is a great deal of room at the top of Figure 19.1, between the present limit of conventional converters of the diode type and the solar energy conversion limit of around 87% ("the multicolor solar cell limit," see below).

Thus, in the longer term, PV development needs to be based on major technological breakthroughs regarding the use of processes and materials at very low cost or/and on the engineering of devices offering far higher performance, capable of best exploiting the available solar energy (Figure 19.1). This leads to the concept of the "third-generation solar cell" [1] [2], which was originally coined to describe an "ultimate" thin-film solar-cell technology. Desired features included high

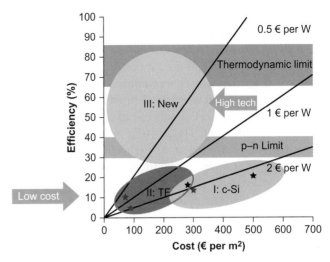

Figure 19.1. Relation of the cost per W_p of solar energy to the cost per area of manufacturing solar devices (modules) and the device efficiency. The regions are I, first generation, II, second generation and III, third generation solar technologies. The red zone marks the anticipated range for very-high-efficiency devices. For comparison, limiting efficiencies derived from thermodynamic constraints are also indicated (low range, no concentration; high range, maximal concentration). Stars indicate industrial production costs estimated in 2010 from available data: black stars are for c-Si modules and gray ones for thin- film technologies (adapted from [1] and updated).

efficiency (derived from operating principles that avoid the constraints on the performance of single-junction cells) and the use of abundant, nontoxic, and durable materials. The term has subsequently been used to describe any advanced PV technology, ranging from organic cells to multijunction cells used in concentrator systems, but will be used here in its original meaning.

Returning to efficiency, the correct framework to study energy conversion is thermodynamics (see Appendix A). Yet the link between PV devices and steam engines is far from obvious and kept many a scientist busy understanding where the limits really are in terms of efficiency. These questions were essentially settled at the end of the twentieth century. To get a better description of PV action it has been necessary to add relevant variables, such as current and electric bias, to standard thermodynamic variables such as temperature, pressure, and volume, and to describe the thermodynamic cycle producing work from the radiative energy flux. The next step was to put all these elements together. This started in the early 1960s with the work of Shockley and Queisser in the framework of the detailed-balance theory [3]. Detailed balance is a consequence of the analysis of equilibrium, as proposed in the 1920s, whereby each process is exactly compensated at equilibrium by a reverse process. Many refinements that came after that enabled one to understand better what the real limit to solar energy conversion is and what can be done, beyond simple p–n junctions, to approach this limit. This led to the concept of a so-called "third-generation solar cell" [1][2] with which one should search for a combination of high efficiency, low-cost processes (presumably of the thin-film type), and abundant compounds to arrive at affordable and sustainable solar energy conversion (Figure 19.1).

While the first generation (c-Si solar cells; Chapter 18) was able to demonstrate reliability and reasonable efficiency, the second generation (thin films; Chapter 18) is demonstrating low cost of manufacturing, thereby enabling development on a still larger scale. If it can be achieved, the implementation of third-generation solar cells is likely to be a requirement to enable truly massive deployment of PV systems that will be both economically and environmentally sustainable. Put in this form, it is essentially a materials science problem.

19.4 New approaches to high-efficiency PV conversion: science and technology

19.4.1 Limits of PV conversion

The thermodynamic limit to solar energy conversion should be rather high as the following argument heuristically shows. Like any system of energy conversion, a PV generator has its performance limited by the laws of thermodynamics. In a first approach, one can regard the sun as a hot source with a temperature of $T_H = 5{,}780$ K (according to various authors, the Sun's surface temperature used in evaluating the thermodynamic limit is between 5,500 and 6,000 K), the Sun's surface temperature, with the cold source being Earth ($T_C \sim 300$ K). Using the equation for the **Carnot efficiency** of an ideal heat engine (Appendix A),

$$\eta_{Carnot} = 1 - \frac{T_C}{T_H}. \tag{19.1}$$

This analogy suggests that the efficiency of solar energy conversion should, in theory, approach 95%. This does not take into account that heat is transported radiatively from the Sun and absorbed by the converter. A more reasonable limit (the endo-reversible or "thermal limit") is given by

$$\eta_{therm} = \left(1 - \frac{C}{C_m}\frac{T_a^4}{T_s^4}\right)\left(1 - \frac{T}{T_a}\right), \tag{19.2}$$

where the solar flux is concentrated on the absorber by a factor C, the concentration ratio ($C = C_m = 46{,}200$ at

maximum concentration), which heats up the absorber (at temperature T_a, which can also re-emit blackbody radiation), and a Carnot engine operates between T_a and the cold sink, at $T \sim 300$ K (see also Figure 19.12 later). The maximum efficiency is ~85%, which is obtained for $T_a \sim 2{,}500$ K [1]. A slightly better efficiency, ~87% (the "multicolor limit"), can be obtained by separating the light into its different components and having each wavelength drive a solar cell whose gap E_g matches perfectly the energy of the incident photons [1]. A still higher limiting efficiency of 93.3% can in principle be achieved, using non-time-reciprocal optics to convert light re-emitted by the absorbers [1].

The difficulty in achieving such highly efficient conversion of solar energy in practice can be summarized as the difficulty of optimizing the conversion of a whole range of photons whose energy goes from the infrared (IR) to the ultraviolet (UV) region, with only one active material whose optical properties can be optimally adapted only to one given photon energy (the absorption threshold, or bandgap energy, E_g). To understand why this is a problem in more detail, we must return to the principles of PV conversion [3].

Principles

Quantum mechanics imposes that the energy levels of bound electron states (e.g., when the electron is confined to a bounded domain) are discrete. Not all energies are available to these electrons. The electrons can jump from a lower level to a higher level under a perturbation such as electromagnetic radiation, if the perturbing energy, which is quantized, at least matches the energy separation of the levels. This will enable transfer of energy from electromagnetic radiation (light) to electrons, and lies at the basis of PV action. Two levels play a special role, the highest occupied level [the highest occupied molecular orbital (HOMO) in molecules and the top of the valence band (VB) in semiconductors] and the lowest unoccupied level [the lowest unoccupied molecular orbital (LUMO) in molecules and the bottom of the conduction band (CB) in semiconductors].

Photovoltaic conversion can be divided into three main stages (Figure 19.2). (a) Absorption of photons of energy $h\nu > E_g$ creates populations of electrons and holes that are out of equilibrium. (b) Each type of carrier very quickly reaches a steady state, defined by a **quasi-Fermi level** of E_{Fn} for the electrons and E_{Fp} for the holes, whose difference, $E_{Fn} - E_{Fp} = qV$, is the recoverable free energy from photon absorption (where q is the elementary charge and V is the photovoltage). (c) The carriers, collected by the contacts before they can recombine, participate in the photocurrent.

In such a device, the photon energy must be greater than the bandgap energy (i.e., $h\nu > E_g$) because of the

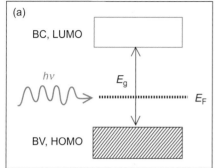 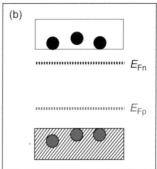 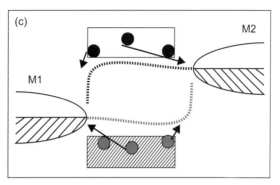

Figure 19.2. Principles of photovoltaic conversion, divided into three principal stages. (a) Absorption of a photon and promotion of an electron to a higher, empty level. (b) Establishment of two populations of electrons and holes in internal quasi-equilibrium (through fast, picosecond, relaxation processes), but unbalanced with one another. (c) Preferential collection of one type of charge carrier at each contact, where M1 and M2 are the two contacts, depicted here as metallic, for the holes and the electrons, respectively (adapted from [4]).

properties of absorption. In addition, the bandgap energy must be greater than the difference in the quasi-Fermi levels of the charge carriers (i.e., $E_g > E_{Fn} - E_{Fp}$). Indeed, when $E_{Fn} - E_{Fp}$ becomes close to E_g, population inversion between electrons and holes occurs, and stimulated emission of radiation eventually becomes predominant and opposes any further increase in qV. Although population inversion of defect levels is possible, due to their low total density of states (DOS), one should note that inversion of CB and VB occupations is incompatible with the large DOS required for efficient absorption of solar radiation (and therefore would not occur with photon densities available with non-concentrated insolation).

A solar cell presents therefore analogies in some sense to an electrochemical battery or, rather, a capacitor charged with light within picoseconds and discharged in nanoseconds to milliseconds, where the very same material is used as both anode and cathode (indeed, the anode and cathode consist of different energy levels of the same material).

Impact of solar concentration

Because the quasi-Fermi levels depend on the concentration of charge carriers, which, in turn, increases with the absorbed light flux, it follows that the recoverable free energy per incident photon (i.e., qV) increases with their flux: that is, it is advantageous to concentrate the solar flux before conversion. This increase is in general logarithmic,

$$E_F = E_{F0} + k_B T \ln(n), \qquad (19.3)$$

and the potential efficiency gain is ~3% per decade of solar flux concentration. This is not negligible, but it currently brings a significant added complexity.

Note that the maximum concentration of solar flux is limited to a factor of 42,600. Indeed, the maximum concentration factor is 2π divided by the solid angle of the Sun, i.e., when the exposed surface of the cell can see only the Sun. At higher concentration factors, the focal point would have a higher radiation temperature than the source, which is thermodynamically impossible: the image of an object cannot be hotter than the object itself. Otherwise, one could reversibly transport heat from a cold body toward a hot body, something forbidden by the second law of thermodynamics.

It should be stressed also that, because qV is related to the concentration of photogenerated carriers, if a given number of photons can be absorbed in a smaller volume of material (say, by a clever photon-management scheme, all else being equal), more free energy is available and can, in principle, be recovered. It follows that using thinner photoactive materials not only saves material, but can also, in principle, lead to higher conversion efficiencies. This is of special importance for the approaches being discussed, in which the conceptual devices would operate under conditions further away from equilibrium (as compared with normal p–n junctions), and photogenerated carriers might have to be removed from the absorber faster than is acceptable in today's devices.

How thin can a device be? This is set by the Yablonovitch limit [5] when geometrical optics is valid, but seems to be similar even in more complex cases [6]. Today, semiconductors as thin as 50 nm could become effective in capturing most of the Sun's power [7][8], using either photonic crystals or plasmonics. There are different ways by which plasmons could enhance radiative processes, and they can be grouped into three main effects [9]: light diffusion, local field increase, and resonant coupling to the absorber.

Limits

The fraction of energy that can effectively be converted by an absorbed photon, $qV/(h\nu)$, is therefore much smaller than its maximum value, because the free energy obtained from each photon is independent of its energy (the remainder goes to the lattice heat within a few picoseconds, during carrier thermalization). Including the fraction of the photons that is not absorbed, the efficiency is limited to ~30% for the optimal gap E_g under standard conditions of illumination [3]. In short, this limit can be derived, provided that the following idealizing assumptions are made so as to focus on intrinsic losses and neglect extrinsic losses (those that can, in principle, be made vanishingly small).

- Charge transport can occur with negligible ohmic losses in devices; i.e., a high carrier mobility is assumed.
- Optical losses such as reflection or parasitic absorption can be made very small in practice. Moreover, if a fraction of the photons of a given energy is not used to create an electron–hole pair, this is clearly sub-optimal; therefore, the absorptivity of the active material is designated as either 0 (under the absorption threshold) or 1 (above).
- Losses from non-radiative processes, whereby an electron and a hole recombine without emitting a photon, can be made very small in some systems. Such recombinations contribute to heat dissipation, and are therefore also non-optimal; note that radiative recombination is mandatory for consistency (the detailed-balance rule) and has to be accounted for.
- In a given material, several absorption processes might be possible for a given photon (e.g., intraband and interband). However, for each photon, one assumes that only one process is allowed, namely the most efficient process for energy conversion.
- Because of the dissipation of power (thermalization), PV devices heat up under operation, and this is, for most device types, detrimental to the conversion efficiency (see Chapters 20–22). Yet, in principle, the device could be efficiently coupled to a thermostat and its temperature maintained uniform and fixed at a given value; therefore variations in temperature are ignored.

The system studied entails some intrinsic losses.

- Some photons are not absorbed.
- Some photons are absorbed, leading to the reverse process of photon emission and therefore radiative losses. Losses are incurred by establishing an equilibrium between the generated photocarriers (whose average kinetic energy is generally significantly larger than $\frac{3}{2}kT$) and the lattice. These are generally called "thermalization losses."

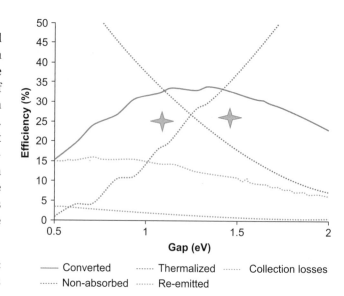

Figure 19.3. Limiting efficiency of p–n junction solar cells as a function of the bandgap of the absorbing material, taking into account only unavoidable losses, under AM1.5 illumination at 300 K. The crosses mark the efficiencies achieved for Si and GaAs cells. The blue continuous line is the Shockley–Queisser limit, i.e., the maximum fraction of the incident flux (here, AM1.5) that can be converted. The green dashed line gives the fraction of the incident power that is not absorbed by the photoactive material. The red dashed line gives the fraction of the incident power that is thermalized as a result of the excess initial energy of the electron–hole pair formed after photon absorption versus its equilibrium value (that is, $h\nu - E_g - 3k_BT$). The blue dashed line gives the fraction of the incident power that is lost through radiative recombination. The purple dashed line gives the fraction of the incident power that is lost during the process of carrier collection in the photoactive material or in the contacts (that is, $E_g + 3k_BT - qV$, where $E_g + 3k_BT$ is the average energy, potential and kinetic, of electron–hole pairs after thermalization and qV is the work effectively extracted). These collection losses are sometimes divided into "Carnot" and "Boltzmann" losses [10], and they depend on the illumination level of the solar cell.

- The free energy collected per electron–hole pair (the cell voltage) is lower than the bandgap of the semiconductor (i.e., the energy of the pair). This is called "contact losses."

When these simplifying (but physically consistent) assumptions are taken into account, we arrive at the so-called Shockley–Queisser limit [3], which is the maximum value of the electrical power that a p–n junction can generate under solar illumination, Figure 19.3.

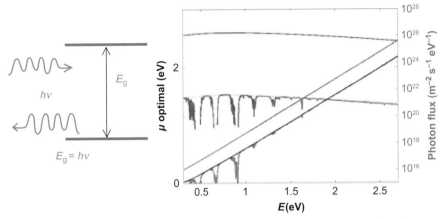

Figure 19.4. (a) The scheme of the two-level systems considered. (b) The recoverable optimal free energy μ_{opt} for an absorbed photon, i.e., at the maximum electric power extraction (ordinate, left axis), according to both the photon energy (abscissa axis) and the incident photon flux (different curves). The blue curve corresponds to a 6,000-K blackbody spectrum, the red curve to the terrestrial solar intensity, and the black curve to that of the solar flux above the atmosphere. Green curves show the photon fluxes on a log scale (right axis) for the same spectra. From [4].

Detailed balance

The principle of detailed balance has been stated in many different forms [11]. In essence, it stems from the fact that no work can be extracted from a system at equilibrium. It follows that, at equilibrium, each microscopic process has to be compensated. This compensation could be global (say, optical generation of an exciton compensated by non-radiative recombination). The detailed-balance principle states that each process is compensated at equilibrium by its reverse process.

A consequence is that, if an elementary process leading to energy flow between states is possible, then its exact reverse is also possible and occurs with the same probability: the rates of a process and its time-reversed equivalent are therefore interdependent (for instance, the transition-matrix elements are equal). If then only radiative recombination processes are considered, the balance on photons gives the balance on charge carriers as

$$I = q(\Phi_{\mathrm{abs}} - \Phi_{\mathrm{em}}), \tag{19.4}$$

where

$$\Phi_{\mathrm{abs}} = \int A(E) N_{\mathrm{inc}}(E) dE,$$

$$\Phi_{\mathrm{em}} = \int \varepsilon(E) N_{\mathrm{em}}(E) dE,$$

N_{inc} and N_{em} are, respectively, the incident and emitted photon flux densities, and $A(E)$ is the photoactive material's absorptivity. It actually follows from detailed balance that $A(E) = \varepsilon(E)$, the emissivity. This is also called Kirchoff's law.

Equation (19.4) shows that the electron flux I is simply equal to the difference between the fluxes of photons absorbed, Φ_{abs}, and photons emitted, Φ_{em}. When the electron population of the absorber (which is also an emitter by virtue of the detailed-balance principle) is in thermal equilibrium, Φ_{em} expresses the **law of blackbody radiation**. When it is not, but the electron and hole populations can each be ascribed to a quasi-Fermi level, $N_{\mathrm{em}}(E)$ is given by [12]:

$$N_{\mathrm{em}}(E) = 2\pi h^3 c^2 \frac{E^2}{e^{(E-\mu)k_{\mathrm{B}}T} - 1}, \tag{19.5}$$

where $\mu = E_{\mathrm{Fn}} - E_{\mathrm{Fp}} = qV$.

The electrical power extracted from a PV device is

$$P = IV = \mu[\Phi_{\mathrm{abs}} - \Phi_{\mathrm{em}}(\mu)]. \tag{19.6}$$

From Equations (19.4) and (19.6), using the AM1.5 spectrum for the incident photon spectrum, we can recover the Shockley–Queisser limiting curve as in Figure 19.3.

Another situation of interest is that of a monochromatic photon flux, incident on a two-level system tuned to absorb it (see scheme Figure 19.4(a)). This case is of general and fundamental significance because it enables the calculation of the maximum value of free energy that can be extracted for each absorbed photon, μ_{opt} (Figure 19.4), independently of the conversion system [1]. This free energy depends solely on the energy of the photon and the incident flux and it can be computed using Equations (19.4) and (19.6) applied to a two-level system [13]. The conversion efficiency of a two-level system under monochromatic light can be read from Figure 19.3; it is simply $\mu_{\mathrm{opt}}(E)/E$.

With the help of Figure 19.4, it can be calculated that the maximum efficiency ("multicolor limit") of a PV device is 67% under standard solar illumination and ~87% under

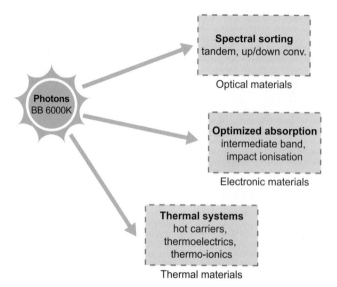

Figure 19.5. The relationships between main conversion concepts and material properties and a possible classification.

maximum concentration as the weighted (by the photon flux) average of μ_{opt} divided by the incident power. The fact that no such efficiencies have been approached yet is because the materials and technologies were not ready to meet the severe specifications necessary (as described in the assumption list above). Clearly, following the way suggested by the method used in deriving the result to realize a device would be very difficult in practice.

19.4.2 Very-high-efficiency pathways

There are many possible strategies for achieving very-high-efficiency solar devices, many of which have ultimate conversion efficiencies close to 85%. All concepts essentially fall into three main groups (Figure 19.5) according to the requirements on materials' properties needed to put them into action.

(1) **Photonic devices**. If all solar energy were concentrated in a narrow spectral band, the current devices would already be able to convert somewhat over 50%. One can thus try to adapt the spectrum incident on one or more photodiodes. This essentially translates into requirements on the optical properties of the materials that make up these solar cells.

(2) **Optimized absorption materials**. These include materials having intermediate electronic levels or materials allowing the generation of several electron–hole pairs by sufficiently energetic photons. Optimization of these processes leads to stringent requirements on the electronic structures of the materials that make up these solar cells.

(3) **Heat engines**. In these devices, the absorption of light leads to the production of heat that can still be converted into electric output. In this case, one needs to look closely at the thermal and "phononic" properties of the solids considered.

A summary of the prospective capabilities of various concepts for improved efficiency is presented in Table 19.1 and discussed in detail in the next section. The table shows that there are three groups of technologies presenting increasing opportunities, which we will discuss next.

(1) An initial improvement compared with conventional systems (by up to 10% ultimately) can be obtained with relatively simple means, not necessarily using concentration. The physical phenomena implemented in the conversion process are relatively well known and have experimental bases.

(2) A second group of technologies should make it possible to improve efficiency levels substantially (by up to 20% ultimately) without any major addition in terms of complexity and can be expected to approach or even surpass in practice the 40% efficiency level. Pertinent experimental data on these approaches are scarce. Concentration might be necessary for both technical (up-conversion) and economic (triple-junction) reasons.

(3) Finally, there is a third group near the thermodynamic efficiency limit that allows us to consider, with a fair margin of maneuver, efficiencies that in practice would be >50% (an improvement by up to 45% ultimately). A distinction can be made between approaches resulting in a considerable increase in system complexity, with an increasing number of parameters to be optimized to approach the limit [e.g., multiple-exciton generation (MEG), see below], and those that remain relatively simple and of the thermal type (hot carriers, thermionics, thermoelectrics, thermophotovoltaics; see below), but for which the possibility of finding appropriate materials is questionable. A very high concentration might be necessary for technical reasons, especially in the case of "thermal" approaches.

Table 19.1 lists the absolute limits required by thermodynamics. From the practical point of view, one can expect to reach 80% of these values in real devices with an R&D effort. For conventional PV cells, 85% (in the laboratory) and 75% (for c-Si modules) of the thermodynamic efficiency limits have already been reached, much like for many energy-production techniques that have reached maturity.

Photonic devices

Here, the difficult task is either to sort photons to be sent to junctions adapted to the corresponding part of the spectrum (multijunction devices) or to change their energy so as to lead to a "bunched" distribution, before

Table 19.1. Thermodynamic limits of the various approaches: possible efficiency gains for a concentration range of 1× to 46,000× (the maximum possible concentration)

Type of cell	Efficiency range AM1.5 at 46,000×	Notes
Group 1		
Thermo-ionic, thermoelectric	54%–85%	In practice, 20% achieved [14]
Impact ionization (optimal QE)	44%–86%	Extremely constrained electronic band structure
Hot carriers	67%–86%	Not demonstrated in practice
Tandem, $N > 100$	68%–87%	Extremely complex, 43% achieved for six junctions [15]
Group 2		
Rectification of light waves	<48%	Complex and not demonstrated
Impact ionization (QE = 2)	38%–52%	Small gain demonstrated in practice
Down conversion	39%–52%	Effect demonstrated in practice
Tandem, $N = 2$	43%–56%	In practice, 32% achieved [16]
Intermediate level	48%–63%	Not demonstrated in practice
Up conversion	48%–63%	Small gain demonstrated in practice
Tandem, $N = 3$	49%–64%	In practice, 42% achieved [17]
Group 3		
Simple p–n junction	31%–41%	In practice, 29% achieved [16]
QE, quantum efficiency.		

the photons are harvested by a standard diode adapted to the narrowed spectrum.

Multijunction devices

In a traditional device, the photons whose energy conversion is most effective are those whose energy is just above the threshold of absorption (the bandgap energy E_g). For those photons, efficiencies of about 60% are reached in experiments, which is rather close to expectations (Figure 19.6). The use of several different cell materials, each one with a bandgap optimized for a different part of the solar spectrum, makes it possible to increase the output power using several configurations.

For a given number of cells, N, and an incident spectrum, there is an optimal choice for the gaps that give the highest output [1]. For example, for three cells under maximum concentration, the maximum theoretical yield is 63% (49% without concentration). As a general rule, the stack efficiency goes as (see also [18])

$$\eta(N) = \frac{\eta^\infty}{1 + a/N}, \tag{19.7}$$

where $a \approx 1$ and depends slightly on solar concentration and η^∞ is the efficiency at large N, for a given incident flux (i.e., $\eta^\infty = 68\%$ for AM1.5 or 86.8% at 46,200× concentration). This formula expresses that, for a multijunction device composed of N junctions to operate optimally, the incident spectrum should be partitioned into $N + 1$ portions containing about the same fraction of photons, with the upper limits of the first N regions giving the optimal gaps of the absorber materials to be chosen for the junctions.

Yet, it must be realized from Equation (19.7) that the incremental gain in power from the addition of a cell in a stack including N junctions varies as $1/N^2$. If one takes into account the electrical and optical losses that are incurred, the expected net gain from another cell is close to zero after the fourth cell (i.e., the practical limit on N is 4).

Optical converters

Alternatively, one could change the wavelength of the photons before they reach the photodiode, to obtain a narrower spectrum. Such an approach has been explored for Si-based solar cells [19].

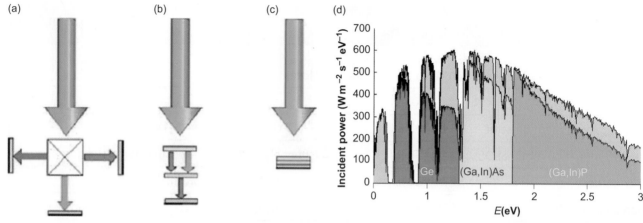

Figure 19.6. The principle of multijunction (tandem) conversion (specifically for a triple junction) and possible practical configurations thereof. (a) An optical device separates sunlight by means of filters into three beams, denoted blue, green, and red, which are converted by three cells whose bandgaps were adapted for these three spectral bands. One can simplify the device as indicated in (b), noticing that the blue cell does not absorb wavelengths higher than its threshold of absorption. It can thus be used as a filter for the green and red cells. In the same way, the green cell lets the red beam pass. It is necessary, of course, that the substrates of the first two cells be transparent. Finally, in (c) a transparent electrical contact between the cells is realized. One can achieve this stack by deposition of the various cell materials onto the same substrate. To avoid the creation of a reverse diode at the interface between two cells (a p–n diode if n–p diodes are piled up), one connects them by a tunnel junction. In case (c), the cells are connected in series, whereas in cases (a) and (b), one remains free to use the electricity produced in the most effective way. A typical structure is a stack of about 20 layers, epitaxially grown on a crystalline substrate (e.g., Ge, from which the bottom junction is made) forming p–n junctions of adjusted bandgaps for photon conversion (generally GaAs and GaInP$_2$) and tunnel junctions to connect them. These structures are described in Chapter 20. (d) The solar spectrum and its partition between three junctions as in present-day highly efficient devices. The gray area is the AM1.5 global power density; the colored area corresponds to the maximal power fraction extracted from each of the three junctions in a Ge/GaInAs/GaInP triple junction (i.e., each photon yields a power less than the cell's bandgap). More details on multijunction (or tandem) devices can be found in Chapter 20.

In the **"photon-addition"** approach, the photons whose energy is too low to be used directly by a photodiode could be converted by nonlinear optics into a smaller number of photons of higher energy. An illustration of such a device is given in Figure 19.7, where, most simply, the converter is placed behind the solar cell that then has to be bifacial (as in multijunction devices) and only lets through the low-energy light to be up-converted. If all low energy could be optimally up-converted (considering only two-photon processes), significant gains could be obtained under concentration (up to 63% under maximal concentration, Figure 19.7) using PV devices having bandgaps E_g between 1.5 and 2 eV [20], depending on the solar concentration ratio.

Notably, this concept results in much faster increases in efficiency with increasing incident power (solar concentration factor) than for a standard cell. This can be understood as a two-photon process, so its kinetics should be of second order. Thus the converted power (in the visible range) is proportional to the square of the incident power (in the low-energy range). Although the projected efficiency can be large under maximal concentration, it is normally negligible without concentration or a light-management system.

Most data on this approach have been obtained with materials doped with rare-earth elements, since these have excellent radiative properties thanks to their shielded f-shell electrons.

(a)

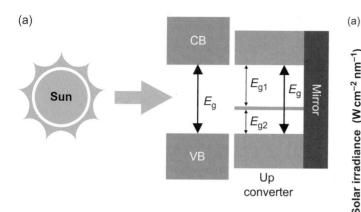

(b)

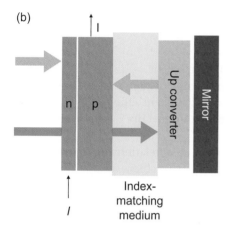

(a)

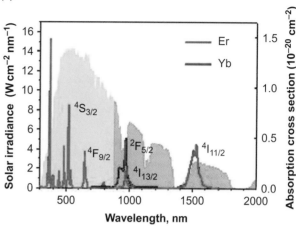

(b)

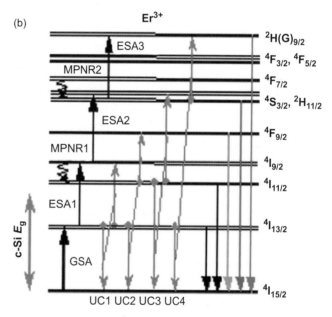

Figure 19.7. (a) The band energy diagram of an up-conversion system. All solar rays are incident on the solar cells, but only the high-energy part of the spectrum ($h\nu > E_g$) can promote electrons from valence band (VB) to the conduction band (CB). Sub-bandgap photons are transmitted to the up converter, where two sub-bandgap photons of energies $>E_{g1}$ and $>E_{g2}$, respectively, create an excitation in the up converter that can radiatively decay as a photon of energy $h\nu > E_g$. A back mirror ensures that all emitted light goes to the cell. (b) A possible scheme of construction of the system (adapted from [20]).

In these systems, infrared radiation is absorbed by a rare-earth ion, and its energy is transferred either to another lanthanide that is able to emit effectively at close to twice the frequency [21] by a sequential energy-transfer process or to an ion of the same species by a cross-relaxation mechanism. Excited-state absorption is also a possible mechanism in these systems. All of these mechanisms can result in the absorption of more than two photons (three- and four-photon absorptions have a sizable probability of occurrence). All of these mechanisms are illustrated in Figure 19.8 for the elements erbium (Er, atomic number 68) and ytterbium (Yb, atomic number 70).

The drawbacks of this system are (1) the relatively low optical cross sections of the f–f transitions, on the

Figure 19.8. The mechanism of up conversion (courtesy of F. Pellé). (a) A representation of the solar spectrum together with the optical capture cross sections of Er and Yb ions. (b) The various intra-atomic energy transfers involved in the operation of an up-conversion device. Fast non-radiative processes are represented by wiggly arrows (NR 1 and 2). Ground-state and excited-state absorption processes (GSA and ESA 1–3) and radiative recombinations (labeled with the emission wavelength) are indicated with straight arrows. Interatomic cross-relaxation processes of up conversion (UC 1–4) are also included.

order of 10^{-20} cm^2, and (2) the relatively narrow range of absorption (band width below 100 nm). Both result in difficulties in covering the IR range efficiently (Figure 19.8(a)); nevertheless, relatively efficient systems have been found using either Er doping (for conversion of the band around 1.5 μm) or Yb sensitizer ions coupled to Er (for conversion of the band around 1 μm).

To date, a few attempts have been made [22][23][24] to prove the feasibility of the concept, but with limited

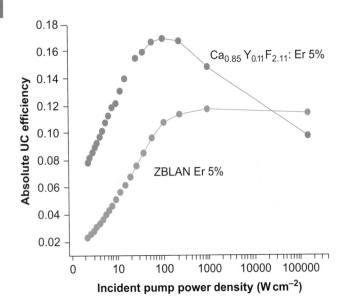

Figure 19.9. Absolute power-conversion efficiencies of two up-converter materials based on rare-earth elements [25] in the 1-μm wavelength band as a function of the excitation power density of the 1.5-μm laser used for the two materials doped with Er (ZBLAN and $Ca_{1-x}Y_xF_{2+x}$). The efficiency peaks at 17% for the fluoride, when heating up of the material causes a decrease of the radiative efficiency. (Courtesy of F. Pellé.)

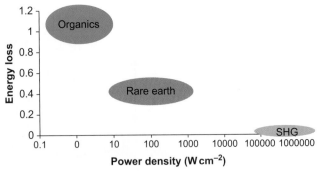

Figure 19.10. The relation between excitation intensity and energy relaxation in up-conversion process efficiency for various up-conversion systems. SHG, second-harmonic generation.

success in terms of increases in absolute effectiveness. Approximately 3% of the IR photons in the narrow band of absorption of the rare-earth elements used, under an illumination equivalent to 250 Suns, were actually converted into photons of higher energy. In the best case, approximately 17% of the absorbed photon energy in the IR region was converted into photons of twice the energy [25][26], being today the highest value for the absolute power-conversion efficiency by up conversion (Figure 19.9). This high conversion efficiency is obtained at power densities equivalent to 20,000 Suns (i.e., this is the concentration of the solar energy needed to have 100 $W\,cm^{-2}$ in the band around 1.5 μm useful for the up-conversion process).

Optical materials (used for second-harmonic generation), that have polarizabilities of the second or third order, can convert over broader frequency bands unfortunately because they are based on non-linear effects they are relatively inefficient, so what appear to be unreasonable illumination intensities would be necessary.

A last category of processes is represented by the phenomenon of excited-state absorption (ESA), which can be observed in solids [27], as well as in many molecular species [28] [29] [30]. In particular, systems presenting a small relaxation of the first excited state, which sometimes makes it possible to increase the lifetime considerably and, thus, the probability of a second absorption, such as the passage to a triplet state, are particularly interesting.

Finally, the most efficient systems are probably those relying on molecular triplet–triplet annihilation [31], since they show significant efficiencies already at quite low concentration levels (10×), even though they do not exhibit stability.

Interestingly, there is a relationship between the energy density required to achieve a significant efficiency and the energy lost in the process (i.e., entropy generation), as can actually be expected from fundamental principles (Figure 19.10).

Another principle, "**photon cutting**," consists, conversely, of absorbing photons of high energy in a luminescent converter to emit several photons of lower energy, to increase the current output of the photodiode. The fluorescent converter can be placed in front of the device (Figure 19.11). With good optical confinement (for example, using a down converter with a high optical index), this will ensure that most of the emitted light is absorbed by the underlying solar cell. The converter can also be placed at the rear of the solar cell if that cell can be made transparent for the high-energy photons, as is possible for organic devices.

Down conversion is less interesting in principle than up conversion in terms of the maximum efficiency achievable. The maximum theoretical yield is <40% in the absence of concentration for the ideal converter (within the radiative limit and with an absorptivity of 1). The influence of solar concentration is opposite to that for up conversion. For instance, under maximum concentration, a down converter on the front side does not provide any advantage over a cell without a converter. Conversely, this means that this concept could be well suited to solar cells working without concentration. Interestingly, the ideal gap for the best possible performance is 1.1 eV, which is near the gap of Si.

Fluorescent converters have already enabled an improvement in performance of solar cells, while shifting toward the red the photons poorly used by these cells [32]. However, these materials, with quantum

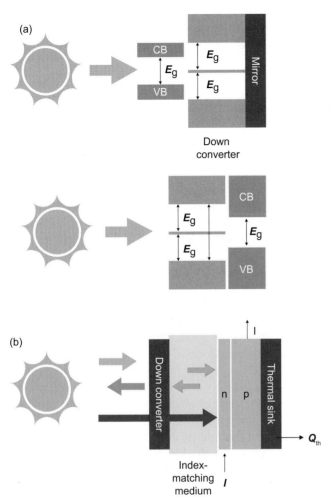

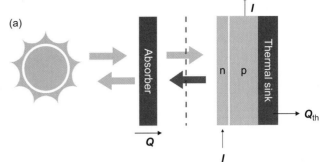

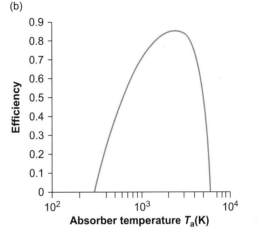

Figure 19.12. (a) Schemes for thermophotovoltaic (TPV) solar energy conversion, showing energy fluxes (thermal fluxes Q and Q_{th}; photon fluxes as arrows) and electron current I in the p–n solar cell. The filter is wavelength-selective and lets through photons with energies just above the bandgap of the solar cell while reflecting all others. Adapted from [36][37]. (b) The ideal efficiency of a TPV system computed as a function of the absorber temperature for maximally concentrated ($C = 46,200$) sunlight.

Figure 19.11. (a) The energy scheme of a down-conversion system composed of a down converter and a photovoltaic device. The down converter can be placed at the rear of the solar cell if the latter has a suitably narrow range of absorption (top) or in front, in which case there are fewer requirements for the absorption range of the solar cell, but then the down converter will interfere with cell operation (bottom). (b) The principle of construction of such a device with the down converter in front. In all cases, the threshold of absorption of the converter must be twice that of the solar cell for optimal operation. VB, valence band; CB, conduction band.

efficiency (QE) < 1, are more an incremental improvement of a non-optimal device rather than a new concept that is able to address the fundamental limits of PV conversion. For the latter, QE > 1 is needed.

As it turns out, fluorescent materials with quantum yields greater than 1 were discovered in recent years. They are based on rare-earth elements (as for up conversion) or on nanostructured materials, namely YF$_3$:Pr with QE = 1.4 [33], LiGdF$_4$ with QE = 2 [34][35], and nanocrystals of PbSe with QE = 2 for $hv > 3\,E_g$ [36]. The fluorescent converters based on rare-earth elements that have been proposed, such as YF$_3$:Pr and LiGdF$_4$:Eu,

are efficient only with UV light, where little energy flux is available. These systems using rare-earth elements are essentially relying on the same type of energy transfer as in Figure 19.8. However, although high quantum yields (close to 2) have been reached, it should be noted that the energetic efficiency is not very good, at less than 70%, as can be inferred from the values of the energies of incoming and outgoing photons [33][34][35].

Thermophotovoltaics/thermophotonics
Another method, which is simple in principle, can be used to modify the spectrum incident on a cell: **thermophotovoltaic (TPV)** solar energy conversion. In this case, the photovoltaic cell is not directly illuminated by solar radiation, but rather, an intermediate absorber is heated by absorbing solar radiation, and its emitted radiation is converted into electrical energy by a solar cell. These cells can directly convert any near IR radiation into

Figure 19.13. Recombination and generation paths in several materials, illustrated with schematic band structures: (a) standard semiconductor, (b) semiconductor having intermediate levels or an intermediate band, and (c) ferromagnetic semiconductor in which the band and level have significant polarization. VB, valence band; CB, conduction band.

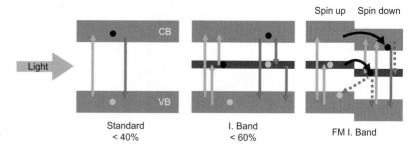

electricity such as waste heat in power stations. Figure 19.12(a) illustrates the principle involved. If the cell can receive only radiation of energy that is slightly above its bandgap, this arrangement can allow the reduction of losses occurring in conventional solar energy conversion in solar cells by both non-absorption of photons with sub-bandgap energy and thermalization of electron–hole pairs generated with energies greater than the bandgap.

Thermophotovoltaics can be achieved in several ways, such as by making the emitter selective, by placing a filter between emitter and solar cell, and, to a lesser degree, by using a mirror behind the solar cell that reflects all photons with energy below the bandgap for which the solar cell is transparent [38][39]. Photons that are not used by the solar cell are not lost, but are reflected back onto the emitter and help to keep the intermediate absorber/emitter at a high temperature. Even the loss due to the photons emitted by the solar cell through radiative recombination is avoided, because the emitter also reabsorbs these photons and recycles their energy. This is quite idealized though, since the emitter has to emit in a narrow band the power that it receives from the Sun at all wavelengths. This can be achieved only if the emitter surface facing the solar cell and the solar cell itself have a larger area than that of the absorber facing the Sun [38], and may actually be a severe limitation for achieving high efficiencies with this concept. The only loss occurring in an ideal TPV process is caused by the emission of the intermediate absorber toward the outside. The limiting efficiencies for solar TPV energy conversion for concentrated and non-concentrated sunlight have been determined to be 85.4% (i.e., the "thermal limit" for solar energy conversion) and 54%, respectively [13].

Thermophotonics is a variant of this scheme [37] [40]. In the thermophotonic configuration, the absorber/emitter is a diode having the same gap as the solar cell. The emissivity of the semiconductor leads naturally to emission with a narrow energy spectrum, which might make it possible to obtain better efficiencies in practice [37].

With respect to selective emitters, which are an essential part of the system, near-field radiators have recently been proposed and demonstrated [41][42]

[43]. These have radiative rates of transfer that are several orders of magnitude greater than the blackbody rates (which constitutes the limit in the far-field region).

Optimal absorbers and intermediate-level devices

In addition to the transitions from the valence band to the conduction band that take place in semiconductors under illumination, semiconductors can absorb photons of energy lower than the bandgap through intermediate levels located in the forbidden band, which play the role of a "ladder for electrons" (Figure 19.13(b)) [44]. Such a device would be similar to a multijunction device with respect to efficiency, but with the simplicity of a simple junction. The main issue is achieving a strong absorption of the intermediate levels without increasing the rate of non-radiative recombination excessively. There are many ways of obtaining intermediate levels or bands, for example, by introducing extended defects or impurities or a superlattice of **quantum dots** (a quantum dot is a semiconductor whose excitons are confined in all three spatial dimensions, so that its electronic properties are intermediate between those of bulk semiconductors and those of discrete molecules).

Closely related schemes were first proposed by Wolf [45], and were revived in the early 1990s after an experimental measurement artifact in a c-Si solar cell led scientists [46] to attribute cell performance to absorption of sub-bandgap light and electron excitation from the VB to the CB in two steps via localized defects, impurities, or crystal defects. This spurred renewed experimental and theoretical work on the subject [47][48]. We note that the bandgap of Si is actually too small to lead to any enhancement, even in the most idealized case.

In principle, efficiencies close to 50% can be achieved under AM1.5 illumination (Figure 19.14), and gaps of 1.5–3 eV are compatible with efficiencies above 40% (the optimal position of the intermediate band with respect to either of the band edges is then approximately given by $0.54E_g - 0.4$ eV). Under full concentration (~46,200×), the limiting efficiency is 63%, the same as can actually be expected from a series-connected triple junction. Indeed, in both cases, the solar spectrum is split into three parts that are used for three different transitions.

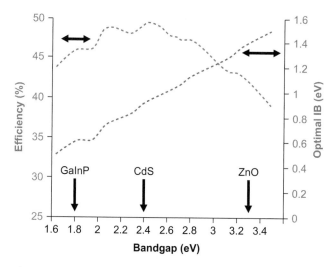

Figure 19.14 The limiting efficiency of intermediate-band solar cells as a function of the bandgap of the material (left axis and blue curve) (the largest transition in Figure 19.13(c)) and the optimal position of the intermediate band (right axis) under AM1.5 illumination.

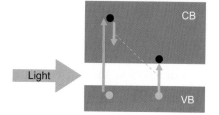

Figure 19.15. Energy band diagram showing potentially improved carrier excitation compared to the standard case through the use of impact ionization (i.e., the cross relaxation mechanism, illustrated by the green arrows). If sufficient energy of the charge carrier is used, impact ionization enables transfer of excess energy from the carrier to an electron, which becomes promoted from the valence band (VB) to the conduction band (CB).

One of the main issues is to achieve efficient absorption through the intermediate level or band. This requires the intermediate level to be located at the Fermi level and to have large numbers of both empty states (required as the final state for the transition from the valence band) and filled states (required as the initial state for the transition to the conduction band): therefore, these compounds should be metallic. Finally, optical cross sections should be large and, at any rate, larger than the non-radiative cross sections. For this to occur, and because optical cross sections are usually small, one needs specific selection rules for the transitions to favor optical rates. Because impurities, and localized levels in general, do not have selection rules on the momentum of electrons, these are thought to be less favorable systems than bands.

There is now some experimental evidence that none of these systems can produce any extra current from the utilization of IR radiation [48][49], while very large recombination losses, strongly affecting values of the open-circuit voltage, have always been observed concomitantly.

Other intermediate-level systems have appeared more recently. Ferromagnetic semiconductor compounds, similar to GaAs:Mn, could present reduced non-radiative decay, because the selection rules on spin could impede some recombination processes [50], as in Figure 19.13(c), where some recombinations through the intermediate level are spin-forbidden in the absence of a strong spin–orbit coupling. Other selection rules acting on phonon emission have also been postulated [49] to hinder non-radiative recombination rates.

"Multiple-exciton-generation" materials

The absorption of photons with energy exceeding twice that of the gap makes it possible to consider other mechanisms, whereby the excess energy can be used to create a second electron–hole pair. This phenomenon is called **impact ionization** (Figure 19.15). The output of devices with impact ionization can be of practical interest, provided that the process is effective in the vicinity of the physical threshold ($\sim 2E_g$). Work at the Max Planck Institute of Stuttgart, Germany, predicted that Si–Ge alloys could show a measurable effect on cell performance [51]. However, this was found to be lower than 1% of the output power of cells made of this material. High values of impact ionization have so far been observed only for photon energies higher than $3E_g$ in bulk semiconductors (such as Si–Ge alloys) [52].

Impact ionization becomes quite efficient for photons with energies well above the bandgap, and this is used in detectors and scintillators (especially for X- and γ-ray detection). What is needed for PV applications is materials in which these processes are operative and efficient already for visible light and have a low threshold energy. In its simplest form, photons with energy above threshold (at least $2E_g$) would give two electron–hole pairs, i.e., a quantum efficiency of two for a photon of energy $h\nu$ above the threshold E_T (QE = 2 for $h\nu > E_T > E_g$). In principle (and this is the case for most materials), if the photon energy is high enough, more electron–hole pairs can be produced. For improved PV conversion, one should have QE = N, when $NE_g < h\nu < (N+1)E_g$. The limiting value for the conversion efficiency using such a process is obtained when E_g becomes vanishingly small (see Table 19.1).

It has also been proposed that these processes might be more efficient in quantum dots made from materials such as Si, InAs, CdSe, and PbSe [36][53], although this feature of quantum dots has not been confirmed by most recent experiments [54] and is still being debated. Where it does occur, multiexcitons tend to recombine fast (and the more excitons in a quantum dot, the faster their annihilation), and extraction of the carriers from the dot is not a simple thing to do.

Heat engines

As described earlier, loss of incident photon energy to waste-heat production is a significant fraction of the losses during PV conversion. In c-Si cells, this is the main loss by far. Ideally, it would be preferable to make use of this heat somehow. Such a scheme could be realized, for instance, by using thermoelectric (TE) materials to convert the heat flow into electricity. (Thermoelectric conversion is discussed in Chapter 22.) This would mean that the solar cells would have to operate at a somewhat higher temperature than ambient for the heat flow to be useful, thereby decreasing the PV efficiency of conversion, but generating additional power, possibly offsetting the losses.

Another option would be to include both PV and TE effects in the same device (see Chapter 22). In principle, this is possible [55], especially given that PV and TE devices are both based on p–n junctions of semiconductors. However, heat conductivity by the lattice and conflicting requirements on device geometry and material doping make this scheme difficult to implement. An elegant alternative to this approach would be to keep the lattice cold while the electrons become much hotter. This is a common phenomenon in semiconductors under high injection levels and leads to the concept of **hot carriers** (see below).

Yet another possibility is to ensure thermal insulation by having a gap between the electrodes. This leads to the concept of **thermionics** [56] (i.e., the heat-induced flow of charge carriers), which is similar to PV (i.e., the light-induced flow of charge carriers) but carried out with a vacuum photodiode instead of a semiconductor p–n diode. Such systems were studied in the 1960s in a completely different context (nuclear energy and space applications) and led to efficiencies in the 20% range [14]. Nevertheless, operation of these systems was cumbersome, and, because the electrode emitting electrons in vacuum (the cathode) had to operate at very high temperatures (1,500–2,200 K), they would not last long, even using refractory metals such as tungsten, irridium, nobium, and rhenium. Moreover, at the large current densities needed to achieve high efficiencies, development of space charges became a severe issue.

Interest in this type of approach revived recently after utilization of GaN at 400 °C was found to yield a substantial quantum efficiency [57]. This concept is based on photo-enhanced thermo-ionic emission

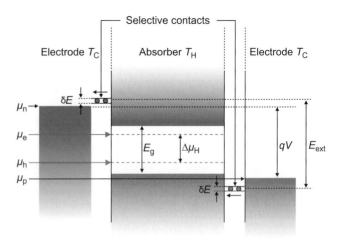

Figure 19.16. Energy band diagram of a hot-carrier cell. The solar photons are absorbed in the central part of the cell, where a gas of hot electrons and holes is formed. Because of their excess kinetic energy, the charge carriers can be collected at an energy higher than that of the edge of the conduction band (for the electrons) or the valence band (for the holes). The energy selectivity of the contact (given by the energy width, δE) allows minimal heat transfer of hot carriers to the contacts. From [58].

(PETE), and benefits from the high stability of cesiated (Cs-covered) GaN surfaces. The cathode is cesiated p-type GaN, where photogenerated electrons can be produced by solar illumination, which would make up for a large fraction of the energy needed to emit electrons in vacuum while the remainder, to overcome the material's electron affinity, could be provided as heat originating from the unused part of the spectrum (that which is not absorbed by GaN). This concept (no efficiency has been measured as yet) may provide an elegant solution to the previous conundrum of thermo-ionic conversion, with efficient conversion at low temperature.

Hot-carrier solar cells

In a hot-carrier solar cell, the carriers generated in the absorber are not thermalized instantaneously with the system at temperature T_C; rather, in a transitory way, they form a hot gas of electrons and holes with a distribution corresponding to a temperature $T_H > T_C$. If these carriers can be collected quickly through selective energy levels (Figure 19.16), the heat flow through the contacts is minimal, and the conversion of the kinetic energy of the hot gas into electrical potential energy is optimal [59]. Calculations of the limiting conversion efficiency give values of 67% under AM1.5 illumination and 86% under full concentration (Figure 19.17). These are very close to the values calculated for a multijunction device containing an infinite

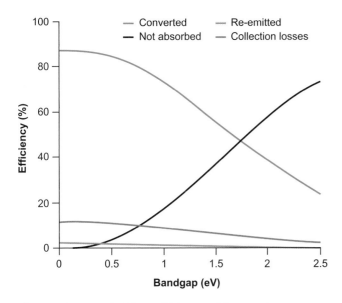

Figure 19.17. Limiting efficiency of hot-carrier solar cells under fully concentrated solar radiation, as a function of the absorber bandgap. Losses due to non-absorption, emission, and collection are also represented.

number of cells, with each one adapted to a fraction of the spectrum (68% and 87%, respectively), but with a much simpler system. It thus could be considered, to some extent, as the ultimate device of solar energy transformation.

No cell with hot carriers has yet been built; however, experiments have shown that hot-carrier thermalization rates can be reduced under strong excitation in nanostructures [60], to the point that conversion efficiencies above 50% are possible under concentration [61]. The main mechanism operative in cooling the electron gas is LO phonon scattering of the hot carriers (Auger scattering is also important in other respects but conserves the energy of the electron–hole plasma). Under high light flux, a hot-phonon bottleneck forms [60][61] and the cooling rate of the electrons is considerably slowed down.

Such outputs are sensitive to the energy width of the contacts beyond a few meV, because the contacts introduce a thermal loss by transfer of heat. Nevertheless, more than decreasing the thermalization rate, the practical realization of selective contacts is likely to be the delicate point in the fabrication of these devices. Directions of research here have focused on quantum dots embedded in a dielectric to provide a resonant tunneling barrier [61], such as Si quantum dots in a silica barrier.

Fortunately, it was recently shown that very selective contacts might not be necessary [58], since semi-selective contacts (i.e., barriers that prevent low-energy electrons from entering the contact but with no filtering

of the high-energy ones) already enable significant gains, even in the presence of thermalization in the absorber. Note that semi-selective or nonselective contacts actually provide another thermalization path for hot electrons, because some hot electrons can flow to the cold contacts and cold electrons can enter the absorber. The achievable efficiencies in hot-carrier solar cells will be dependent on the rates of thermalization of the excess carrier kinetic energy and on the height and selectivity of the barrier to carrier collection. For instance, an absorber with a bandgap of 1 eV under full solar concentration, for which the extraction energy is set to its optimal value, will undergo an efficiency drop as a function of the selectivity of the contact, δE (see Figure 19.16), of the order of 10% this drop would be more if the thermalization rate in the absorber is very low and less if this rate is close to or larger than the best values reported [58], which would still keep the system more efficient than the Schockley–Queisser limit in favorable cases.

Rectennas

Given that light is an alternating electromagnetic field, it produces an alternating current of the same frequency in an antenna. In principle, this alternating current could be rectified to produce a direct current with an excellent efficiency, and several proposals for such rectifying antennas, or **rectennas**, have been made [63][64]. Of course, the frequency of light is much higher than the operating frequency of any diode made to date, but it seems that, in principle, it could be done, probably using very small devices able to cope with such high frequencies.

The feasibility of such a scheme is, in fact, highly questionable. Actually, sunlight comes from an incoherent and broadband source, whereas the idea of a rectenna is suited for coherent and monochromatic radiation, which presents quite a different problem actually.

The influence of such (blackbody-type) radiation on an antenna would be to create random currents that could be seen as "noise." Noise in a conductor has an associated temperature. The efficiency of conversion of this thermal noise by rectification of the current in a conductor was analyzed in depth by Corkish and co-workers [65][66] and was found to be limited to 48% [67]. Moreover, such a conversion would require the design of antennas that were able to collect a broad band of radiation and of rectifiers that were operative at optical frequencies, which would be extremely challenging.

19.5 Summary

This chapter has discussed the nature of the limits on the efficiency of PV conversion, as an introduction to the presentation of the various ways proposed to approach these limits. In essence, the factors that make today's solar cells perform far below achievable efficiencies are well understood. They stem essentially from the broadband nature of the solar radiation. Many a creative way has been proposed to turn a larger fraction of the incoming solar energy into electrical work. The proposals include smart ways, which are possibly realizable in relatively simple devices that could convert a large fraction of the incoming solar energy, over 50% for many concepts. Most of these, however, still lack thorough experimental investigation in spite of a recent interest in this line of research, and this should be a focus and a challenge for the future.

Will systems that are radically different from diodes be developed for high-performance PV conversion? There are very many possibilities through which such an end could be achieved and it is impossible to predict which, if any, will succeed.

In the short run, multijunction devices should continue to demonstrate significant progress. In the somewhat longer term, they could even have a significant impact on the production of PV electricity, through the use of solar tracking and concentration.

Finally, in the more distant future, cells with photon interconversion, intermediate levels, or hot carriers, which present the greater scientific and technological challenges, could approach the ultimate performance with relatively simple and robust devices.

19.6 Questions for discussion

1. What are the unavoidable losses in a solar cell? Which, if any, are reduced under light concentration? Turning to non-radiative recombination, which mechanisms will be of reduced importance and which of increased importance, when the illumination is increased?

2. Consider the optimal μ curve in Figure 19.4(b) (left). Sketch the curves that would correspond to a single p–n junction and then those which would correspond to a tandem cell.

3. Explain the efficiency dependence of up conversion on the incident intensity. What would you expect for a three-photon process? Why does it saturate in practice under strong illumination? Hint: think of the lifetimes of the states involved and of possible temperature effects.

4. Draw the equivalent circuits for a series-connected triple junction, an intermediate-band solar cell, and an up-conversion device. Those turn out to have rather similar limiting efficiencies. Discuss why. The intermediate-band and up-conversion systems have strong conceptual similarities, but rely on quite different properties of materials. In view of

material constraints, which of the two seems the easiest to realize?

5. Discuss the spectral sensitivity of multijunction devices. Discuss the temperature sensitivity (take into account dependences of the bandgap energy and open-circuit voltage on junction temperatures, but always keep the simplifying assumption that the absorptivity is unity for $hv > E_g$ and 0 otherwise). How does it affect monolithically integrated multijunctions? Compare this case with hot-carrier solar cells and intermediate-band concepts.

6. The endo-reversible limit turns out to be the limit for hot carriers and multiple exciton generation (both in the limit of very-small-gap absorbers). Discuss why this should be so. Compare the material properties required for the two approaches.

19.7 Further reading

- **M. A. Green**, 2005, *Third Generation Photovoltaics: Advanced Solar Electricity Generation*, Berlin, Springer. This is, as of today, probably the most comprehensive book on advanced concepts for PV conversion.
- **P. Würfel**, 2005, *Physics of Solar Cells: From Basic Principles to Advanced Concepts*, New York, Wiley. This book describes in very general terms the physics of solar cells and the fundamental limits of their operation. It has a short chapter introducing the new concepts for PV conversion.
- **J. Nelson**, 2003, *The Physics of Solar Cells*, London, Imperial College Press. This book is an alternative to the above.
- **A. Luque** and **S. Hegedus** (eds.), 2003, *Handbook of Photovoltaic Science and Engineering*, New York, Wiley. This volume contains a chapter on the theoretical limits of photovoltaic conversion, with a discussion of entropy generation in solar cells, and several chapters on high-efficiency conversion, including concentration cells and space cells.
- **A. De Vos**, 1992, *Endoreversible Thermodynamics of Solar Energy Conversion*, Oxford, Oxford University Press. This book has recently been re-edited. It gives a purely thermodynamic approach to the limits of solar energy conversion. It is written in a very lively style, and is quite accessible.
- **M. D. Archer** and **R. Hill** (eds.), 2001, *Clean Electricity from Photovoltaics*, vol. 1, London, Imperial College Press. This is another handbook on PV, with a chapter on advanced concepts for high-efficiency conversion.

Acknowledgments

The author would like to thank A. Le Bris and F. Pellé for some of the figures used.

19.8 References

[1] **M. A. Green**, 2003, *Third Generation Photovoltaics: Advanced Solar Electricity Generation*, Berlin, Springer.

[2] **M. A. Green**, 2000, "Third generation photovoltaics: advanced structures capable of high efficiencies at low cost," *Proceedings of the 16th EC Photovoltaic Solar Energy Conference*, London James & James, p. 51.

[3] **W. Shockley** and **H. J. Queisser**, 1961, "Detailed balance limit of efficiency of p–n junction solar cells," *J. Appl. Phys.*, **32**, 510–519.

[4] **J. F. Guillemoles**, 2010, "The quest for very high efficiency in photovoltaic energy conversion," *Europhys. News*, **41**(2), 19–22.

[5] **E. Yablonovitch**, 1982, "Statistical ray optics," *J. Opt. Soc. Am.*, **72**, 899–907.

[6] **Z. Yu, A. Raman**, and **S. Fan**, 2011, "Fundamental limit of nanophotonic light trapping in solar cells," *Proc. Natl. Acad. Sci.* (early edition), www.pnas.org/cgi/doi/10.1073/pnas.1008296107.

[7] **E. Ruben** *et al.*, 2010, *Appl. Phys. Lett.*

[8] **S. Collin, C. Sauvan, C. Colin** *et al.*, 2010, "High-efficient ultra-thin solar cells," in *25th EU PVSEC*, Valencia, pp. 265–268.

[9] **H. A. Atwater** and **A. Polman**, 2010, "Plasmonics for improved photovoltaic devices," *Nature Mater.*, **9**, 205–213.

[10] **L. C. Hirst**, and **N. J. Ekins-Daukes**, 2010, "Fundamental losses in solar cells," *Prog. Photovoltaics: Res. Appl.*, doi:10.1002/pip.1024.

[11] **L. Onsager**, 1931, "Reciprocal relations in irreversible process. I," *Phys. Rev.*, **37**, 405–426.

[12] **P. Würfel**, 1982, "The chemical potential of radiation," *J. Phys. C*, **15**, 3967.

[13] **A. De Vos**, 1992, *Endoreversible Thermodynamics of Solar Energy Conversion*, Oxford, Oxford University Press.

[14] **C. Greaves**, 1968, "The direct conversion of heat into electricity. Thermoelectric conversion and thermionic conversion," *Phys. Education*, **3**, 330–337.

[15] www.renewableenergyaccess.com/rea/news/story?id=49483&src=rss.

[16] **M. A. Green**, 2010, "Solar cell efficiency tables," *Prog. Photovoltaics: Res. Appl.* **16**, 346–352.

[17] www.ise.fraunhofer.de/press-and-media/press-releases/press-releases-2009/World-record-41.1.

[18] **A. Brown**, 2003, PhD thesis, University of New South Wales.

[19] **C. Strümpel, M. McCann, G. Beaucarne** *et al.*, 2007, "Modifying the solar spectrum to enhance silicon solar cell efficiency – an overview of available materials," *Solar Energy Mater. Solar Cells*, **91**(4), 238–249.

[20] **T. Trupke, M. A. Green**, and **P. Würfel**, 2002, "Improving solar cell efficiencies by up-conversion of sub-band-gap light," *J. Appl. Phys.*, **92**(7), 4117–4122.

[21] **P. Gibart, F. Auzel, J. C. Guillaume**, and **K. Zahraman**, 1996, "Below band-gap IR response of substrate-free GaAs solar cells using two-photon up-conversion," *Jap. J. Appl. Phys.*, **35**, 4401.

[22] **P. Gibart** *et al.*, 1995, "IR response of substrate-free GaAs solar cells using two-photon up-conversion," in *Proceedings of the 13th Photovoltaic Solar Energy Conference*, Nice.

[23] **A. Shalav, B. S. Richards, T. Trupke** *et al.*, 2003, "The application of up-converting phosphors for increased solar cell conversion efficiencies," in *3rd World Conference, on Photovoltaic Energy Conversion*, pp. 248–250.

[24] **J. C. Goldschmidt, S. Fischer, P. Löper** *et al.*, 2010, "Upconversion to enhance silicon solar cell efficiency – detailed experimental analysis with both coherent monochromatic irradiation and white light illumination," in *Proceedings of the 25th EU PVSEC*, Valencia.

[25] **S. Ivanova** and **F. Pellé**, 2009, "Strong 1.53 μm to NIR–VIS–UV upconversion in Er-doped fluoride glass for high efficiency solar cells," *J. Opt. Soc. Am. B*, **26**, 1930–1938.

[26] **F. Pellé, S. Ivanova**, and **J.-F. Guillemoles**, 2010, "Improved c-Si solar cell efficiency by upconversion in Er^{3+} doped fluoride-based materials,"

[27] **B. Ullrich** and **R. Schroeder**, 2002, "Two-photon-excited green emission and its dichroic shift of oriented thin-film CdS on glass formed by laser deposition," *Appl. Phys. Lett.*, **80**, 356.

[28] **D. Dini, M. Hanack**, and **M. Meneghetti**, 2005, "Non-linear optical properties of tetrapyrazinoporphyrazinato indium chloride complexes due to excited-state absorption processes," *J. Phys. Chem. B*.

[29] **W. Wenseleers, F. Stellaci, T. M. Friedrichsen** *et al.*, 2002, "Five orders-of-magnitude enhancement of two-photon absorption for dyes on silver nanoparticle fractal clusters," *J. Phys. Chem. B*, **106**, 6853–6863.

[30] **R. R. Islangulov, D. V. Kozlov**, and **F. N. Castellano**, 2005, "Low power upconversion using MLCT sensitizers," *Chem. Commun.*, 3776–3778.

[31] **S. Baluschev, T. Miteva, V. Yakutkin** *et al.*, 2006, "Up-conversion fluorescence: noncoherent excitation by sunlight," *Phys. Rev. Lett.*, **97**, 143903 (2006).

[32] **T. Maruyama** and **R. Kitamura**, 2001, "Transformations of the wavelength of the light incident upon solar cells," *Solar Energy Mater. Solar Cells*, **69**, 207–216.

[33] **J. L. Sommerdijk, A. Bril**, and **A. W. de Jager**, 1974, "Two-photon luminescence with ultraviolet excitation of trivalent praseodymium," *J. Lumin.*, **8**, 341–343.

[34] **R. T. Wegh, H. Donker, K. D. Donker**, and **A. Meijerink**, 1999, "Visible quantum cutting in Eu^{3+}-doped gadolinium fluorides via downconversion," *J. Lumin.*, **82**, 93–104.

[35] **R. T. Wegh, H. Donker, K. D. Oskam**, and **A. Meijerink**, 1999, "Visible quantum cutting in $LiGdF_4$:Eu^{3+} through downconversion," *Science*, **283**, 663.

[36] **R. D. Schaller** and **V. I. Klimov**, 2004, "High efficiency carrier multiplication in PbSe nanocrystals: implications for solar energy conversion," *Phys. Rev. Lett.*, **92**, 186601.

[37] **N.-P. Harder** and **M.A. Green**, 2003, "Thermophotonics," *Semicond. Sci. Technol.*, **18**, S270–S278.

[38] **N. Harder**, 2003, "Theoretical limits of thermophotovoltaic solar energy conversion," *Semicond. Sci. Technol.*, **18**, S151–S157.

[39] **P. Würfel**, 2003, "Theoretical limits of thermophotovoltaic solar energy conversion," *Semicond. Sci. Technol.*, **18**, S151–S157.

[40] **K.R. Catchpole**, **K.L. Lin**, **M.A. Green** *et al.*, 2002, "Thin semiconducting layers and nanostructures as active and passive emitters for thermophotonics and thermophotovoltaics," *Physica E*, **14**, 91–95.

[41] **G. Chen**, **A. Narayanaswamy**, and **C. Dames**, 2004, "Engineering nanoscale phonon and photon transport for direct energy conversion," *Superlattices Microstructures*, **35**, 161–172.

[42] **Laroche**, 2005, Thèse Ecole Centrale de Paris.

[43] **R. Carminati** and **J.J. Greffet**, 1999, "Near-field effects in spatial coherence of thermal sources," *Phys. Rev. Lett.*, **82**, 1660–1663.

[44] **A. Luque** and **A. Martí**, 1997, "Increasing the efficiency of ideal solar cells by photon induced transitions at intermediate levels," *Phys. Rev. Lett.*, **78**, 5014–5017.

[45] **M. Wolf**, 1960, "Limitations and possibilities for improvement of photo-voltaic solar energy converters," *Proc. IRE*, **48**, 1246–1263.

[46] **J. Li**, **M. Chong**, **J. Zhu** *et al.*, 1992, "35% Efficient nonconcentrating novel silicon solar cell," *Appl. Phys. Lett.*, **60**, 2240–2242.

[47] **M.J. Keevers** and **M.A. Green**, 1994, "Efficiency improvements of silicon solar cells by the impurity photovoltaic effect," *J. Appl. Phys.*, **75**, 4022–4031.

[48] **M.J. Keevers** and **M.A. Green**, 1996, "Extended infrared response of silicon solar cells and the impurity photovoltaic effect," *Solar Energy Mater. Solar Cells*, **41–42**, 195–204.

[49] **A. Luque**, **A. Martí**, **A. Bett** *et al.*, 2005, "FULLSPECTRUM; a new PV wave making more efficient use of the solar spectrum," *Solar Energy Mater. Solar Cells*, **87**, 467–479.

[50] **P. Olsson**, **C. Domain**, and **J.F. Guillemoles**, 2009, "Ferromagnetic compounds for high efficiency photovoltaic conversion: the case of AlP: Cr," *Phys. Rev. Lett.*, **102**, 227204.

[51] **J.H. Werner**, **S. Kolodinski**, and **H.J. Queisser**, 1994, "Novel optimization principles and efficiency limits for semiconductor solar cells," *Phys. Rev. Lett.*, **72**, 3851–3854.

[52] **M. Wolf**, **R. Brendel**, **J.H. Werner**, and **H.J. Queisser**, 1998, "Solar cell efficiency and carrier multiplication in $Si_{12x}Ge_x$ alloys," *J. Appl. Phys.*, **83**, 4213–4221.

[53] **A. Shabaev**, **A.L. Efros**, and **A.J. Nozik**, 2006, "Multiexciton generation by a single photon in nanocrystals," *Nano Lett.*, **6**(12), 2856–2863.

[54] **J.J.H. Pijpers**, **R. Ulbricht**, **K.J. Tielrooij** *et al.*, 2009, "Assessment of carrier-multiplication efficiency in bulk PbSe and PbS," *Nature. Phys.* **5**, 811–814.

[55] **S. Kettemann** and **J.F. Guillemoles**, 2002, "Thermoelectric field effects in low dimensional structure solar cells, *Physica E*, **14**, 101–106.

[56] **G.P. Smestad**, 2004, "Conversion of heat and light simultaneously using a vacuum photodiode and the thermionic and photoelectric effects," *Solar Energy Mater. Solar Cells*, **82**, 227–240.

[57] **Schwede** *et al.*, 2010, "Photon-enhanced thermionic emission for solar concentrator systems," *Nature Mater.*, **9**(9), 762 and supplementary materials.

[58] **A. Le Bris** *et al.*, 2010, *Appl. Phys. Lett.*

[59] **P. Würfel**, 1997, "Solar energy conversion with hot electrons from impact ionisation," *Solar Energy Mater. Solar Cells*, **46**, 43–52.

[60] **Y. Rosenwaks**, **M.C. Hanna**, **D.H. Levi** *et al.*, 1993, "Hot-carrier cooling in GaAs: quantum wells versus bulk," *Phys. Rev. B*, **48**, 14675.

[61] **G.J. Conibeer**, **J.F. Guillemoles**, *et al.*, 2005, *20th EPVSEC*, Barcelona.

[62] **G. Conibeer**, **C.W. Jiang**, **M. Green**, **N. Harder**, and **A. Straub**, 2003, in *Proceedings of the 3rd World PV Conference*, Osaka, vol. **3**, p. 2730.

[63] **B. Berland**, 2001, "Optical rectenna for direct conversion of sunlight to electricity," in *Proceedings of the National Center for Photovoltaics Program Review Meeting*, NREL, p. 323

[64] **B. Berland**, *Photovoltaic Technologies Beyond the Horizon: Optical Rectenna Solar Cell*. B. Berland ITN Energy Systems, Littleton, CO, Inc.

[65] **R. Corkish**, **M.A. Green**, and **T. Puzzer**, 2002, "Solar energy collection by antennas," *Sol. Energy*, **73**, 395.

[66] **R. Corkish**, **M.A. Green**, **T. Humphrey** and **T. Puzzer**, 2003, "Efficiency of antenna solar collection," *Proceedings of the 3rd World PV Conference, Osaka*

[67] **I.M. Sokolov**, 1998, "On the energetics of a nonlinear system rectifying thermal fluctuations," *Europhys. Lett.*, **44**, 278.

20 Concentrating and multijunction photovoltaics

Daniel J. Friedman

National Renewable Energy Laboratory, Golden, CO, USA

20.1 Focus

Sunlight has two fundamental characteristics that make its use as a cost-effective large-scale source of electricity challenging: its low power density and its broad spectrum. The first characteristic means that sunlight has to be collected over large areas to gather significant amounts of power, and the second characteristic means that conventional solar cells are inherently limited in converting this power to electricity. The "concentrator photovoltaics" (CPV) approach using multijunction solar cells addresses these two challenges head-on.

20.2 Synopsis

Sunlight shines with a power of about 1 kW m^{-2}, on average, onto the Earth's surface. Although this might feel considerable to a beachgoer on a hot sunny day, for electrical power generation, this power density is actually inconveniently small. For perspective, consider that a typical electrical power plant generates 1 GW, enough to supply the needs of a rather small city and about 0.1% of the total electricity generation capacity in the USA. The very best conventional solar cells are of efficiency about 20%, so to make a 1-GW solar photovoltaic power plant would require at least 5 km^2 of solar cells. To gather sunlight over such large areas and convert it to electricity economically is a fundamental challenge of photovoltaics. One approach, discussed in other chapters, is to reduce the cost of the solar cells. In contrast, this chapter describes an alternative approach that reduces the *amount* of cells needed: CPV.

In CPV, sunlight is gathered using optical elements and then focused or "concentrated" onto much smaller solar cells than are used in conventional "flat-plate" photovoltaics. Thus, CPV addresses the low power density of sunlight by increasing its power density. Concentration factors of 500–1,000 "suns" are currently attracting the most attention. At these high concentrations, the cell cost as a fraction of the system cost is reduced to such an extent that one can afford to use more expensive cells than are viable for flat-plate photovoltaics (although the cost of the newly added elements must be sufficiently low to generate a net lowering of the system cost per unit of energy produced). This change in economics makes viable the use of multijunction solar cells, which are much more expensive but much more efficient than conventional cells.

This chapter first describes how light is concentrated, starting with the optics and proceeding to the theoretical and practical limitations on concentration. (Similar issues apply to solar thermal energy production, which is discussed in Chapter 21.) Then, it explores how the concentration of the sunlight affects the photovoltaic conversion of this light to electricity, and how the efficiency of this conversion can be improved using multijunction solar cells.

20.3 Historical perspective

The development of CPV started not long after the 1954 invention of the silicon solar cell (see Chapter 18). The basic idea behind CPV is to use low cost optics to concentrate light on efficient but expensive cells. The total area from which energy is collected is the same but the active semiconductor area is much less. The factor by which sunlight is concentrated is often referred to in units of "suns," which is the ratio of the intensity of the light striking the solar cell to the standard intensity of unconcentrated sunlight, 1 kWm^{-2}. By 1966, Ralph [1] had documented the concept of concentrated sunlight to lower the costs of photovoltaic power generation, and had demonstrated the concept experimentally. Ralph's system used conical reflectors to deliver a relatively low concentration of about three suns to silicon solar cells 4.5 cm in diameter. Under concentration, the cells operated at ~50–65 °C, with an efficiency of 11% at the lower end of this temperature range. A major research program conducted by Sandia National Laboratories led to the development in the mid 1970s of more sophisticated systems, in which the issues of tracking and thermal management were carefully addressed. As a participant in the Sandia program, Martin Marietta Corporation developed point-focus Fresnel-lens CPV systems of significant size, leading to the installation in 1981 of a 350-kW system in Saudi Arabia and a 225-kW system in Arizona [2]. Some CPV systems were developed and deployed using a variety of different technical approaches, including both point- and line-focus configurations and using both lens-based and mirror-based concentrating optics. A whole new field of "non-imaging" optics was developed to address the special needs of concentrator systems [3].

To address the need for solar cells compatible with concentrator operation, in the mid to late 1980s, Swanson and colleagues at Stanford University developed advanced silicon solar cells that could operate at concentrations above 100 suns with unprecedented high efficiencies of 28% [4]. In parallel, multijunction solar cells were being developed. The concept of the multijunction solar cell appears to have been first described as far back as 1958, only a few years after the 1954 invention at Bell Laboratories of the first modern solar cell. The early multijunction-cell research focused on two-junction cells, including AlGaAs/Si and GaAs/GaSb device structures. As early as 1989, a remarkable 32.6% efficiency at 100 suns was demonstrated by Fraas at Boeing for a GaAs junction mechanically stacked on a GaSb junction [5]. In 1984, Olson at the National Renewable Energy Laboratory invented the GaInP/GaAs two-junction cell. With further development by Spectrolab and others, this structure became the basis for the modern GaInP/GaAs/Ge three-junction cell, which has demonstrated efficiency in excess of 40% at concentrations approaching 1,000 suns. Because this cell can be grown monolithically (giving it a cost and engineering advantage over mechanical stacks) and with reproducibly high quality, it has become the industry standard.

After this promising start, CPV as an industry stagnated from about 1980 to 2005, but, more recently, it has grown. There are now dozens of CPV companies throughout the world [6]. Figure 20.1(a) shows a photograph of a typical modern lens-based system by the CPV company Amonix. The system, which produces 53 kW from each pedestal-mounted array of lenses, uses an array of Fresnel lenses to concentrate light to 500 suns onto multijunction solar cells. Figure 20.1(b) shows a mirror-based system by the CPV company REhnu. This system uses much larger optical elements (the mirror, in this case) than the optical elements in the Amonix system. This comparison illustrates the wide range of configurations possible for concentrator systems. It is not yet clear whether one configuration will ultimately prove more commercially viable.

20.4 Concentration of sunlight

20.4.1 Direct and diffuse components of sunlight

Not all of the light that comes from the Sun goes directly through the atmosphere to the Earth's surface. Some of the light is absorbed or reflected by the atmosphere and, therefore, never reaches the surface of the Earth. Some of the light that does get through the atmosphere to the Earth's surface traverses an uninterrupted path directly from the Sun; this component of the sunlight is known as the "direct" component. However, not all sunlight is direct: there is a "diffuse" component of the sunlight that is scattered by the atmosphere on its way toward the Earth's surface. The direct light comes from the very small area in the sky where the Sun is, whereas the diffuse light comes from the entire sky. Because of these very different angular distributions of the incoming light, high concentration of the direct light can be achieved, whereas very little concentration of the diffuse light is possible. Therefore, concentrator systems generally collect only the direct component. (The direct and diffuse components differ not only in their angular distribution but also in their spectral content; the diffuse light is more weighted toward shorter wavelengths, one consequence of which is the blue color of the sky.) The amount of diffuse compared with direct light depends on the atmospheric conditions: the clearer and less cloudy the atmosphere, the greater the amount of direct sunlight. Because concentrator systems generally collect only the direct light, they are best situated in areas of high direct sunlight such as the desert southwest of the USA.

(a)

Figure 20.1. (a) A photograph of three Amonix 7700 CPV systems. These systems use an array of Fresnel lenses to concentrate light to 500 suns onto multijunction solar cells. Each system is 23.5 m × 15 m in size and produces a nominal 53 kW of AC power. The pictured systems are installed at the Southern Nevada Water Authority (SNWA) River Mountains Water Treatment Facility in Henderson, NV. Photograph courtesy of Carla Pihowich, Amonix. (b) A photograph of a REhnu CPV system that uses mirrors to focus light onto multijunction solar cells. Photograph courtesy of Roger Angel, REhnu.

(b)

20.4.2 Types of concentrator optics configurations

There are many approaches to concentrating sunlight, each with advantages and disadvantages for obtaining a reliable, high-performance, inexpensive system.

Reflective versus refractive optics

To focus the direct light from the Sun, it is necessary to change the path of the light rays. Two common types of optical elements are available for this purpose, as shown in Figure 20.2: lenses, which refract light as it passes through the lens material, and mirrors, which reflect

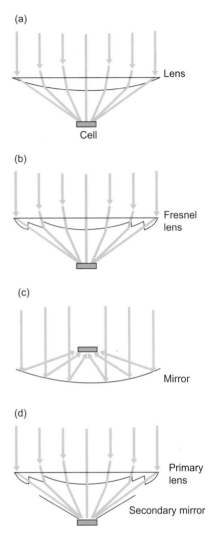

Figure 20.2. Schematics of commonly used concentrator optical elements: (a) lens, (b) Fresnel lens, (c) mirror, and (d) primary lens with secondary mirror.

the light from the mirror surface. Concentrator lenses are typically of the Fresnel configuration, essentially the plano-convex lens of Figure 20.2(a) collapsed down to a thin structure as illustrated in Figure 20.2(b).

Both refractive and reflective optics have been used successfully in real-world systems. The system shown in Figure 20.1(a) uses Fresnel lenses, whereas a system using mirrors is shown in Figure 20.1(b). Concentrator systems frequently incorporate a second optical stage after the primary concentrating optics. The primary and secondary stages are often of different types; for example, one could have an optical system with a refractive primary stage and a reflective secondary stage, as illustrated in Figure 20.2(d).

Materials requirements for optical elements

For the optical elements to perform well in a real-world concentrator system, the materials used to make them must meet several challenging requirements without being impractically expensive. In a concentrator system, the optics will be exposed to a variety of environmental conditions, including moisture, large variations in temperature, and exposure to sunlight. It is critical that the optical elements do not degrade significantly over the desired ~30-year lifetime of the concentrator system. Mirror optics are usually made by starting with a hard, smooth surface such as glass and depositing layers of highly reflective materials such as silver, which are, in turn, coated to protect the reflective layer from the environment. The coating process and materials must provide resistance to delamination of the reflective film, pitting (which is especially important in space and desert applications), oxidation, hydrolysis, and any other degradation of the coating that would reduce its reflectance. Lens optics, in contrast, are made from materials with high optical transmission such as acrylics, e.g., poly(methyl methacrylate), or glass. The material must be rugged and resistant not just to pitting, but also to discoloration, which reduces the optical transmission. The resistance to discoloration is especially challenging for acrylics, which tend to be subject to "yellowing" after long-term exposure to the ultraviolet photons of the solar spectrum. (Cell encapsulants can suffer a similar yellowing, as discussed in Chapter 18.) To act as a lens, the lens material must also be readily formed into the desired shape and must have an index of refraction greater than 1. Typical indexes of refraction for glass and acrylic are ~1.5, which is adequate, although higher values would confer significant advantages (see Section 20.4.3). An additional constraint for the lens material is that the index of refraction should be as nearly wavelength-independent as possible, so that the optical properties of the lens are the same for the different photon energies in the solar spectrum. In practice, this wavelength-independence is not generally achieved for real-world materials. Lenses made from materials whose index is wavelength-dependent are subject to a type of lens defect known as chromatic aberration; in a concentrator system, this aberration would direct photons of different energies to slightly different locations on the solar cell, thus potentially reducing the power output of the cell. (Chromatic aberration is especially problematic for multijunction solar cells. Careful optics design can minimize this effect, and cells can also be designed in such a way as to mitigate its impact on the cell performance.) Clearly, the materials challenges for optical elements are diverse and challenging, and the perfect optical material has yet to be developed. See Box 20.1 for a case study.

Optical geometries

A variety of configurations can be used for the focusing of light. Light can be focused in one dimension, with the

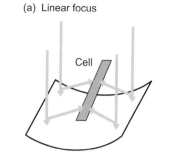

(a) Linear focus

Cell

(b) Point focus

(c) Heliostat

Box 20.1. Case study: silicone-on-glass Fresnel lenses

To make durable Fresnel lenses with high performance and reasonable cost, it would be desirable to find a material that combines the durability of glass with the lens-shaping advantages of acrylic. One common and effective approach is to use glass coated with silicone as illustrated in Figure 20.3, using a mold to form the shape of the lens elements in the silicone. The glass, which faces outward toward the environment, provides the ruggedness and durability, whereas the readily molded silicone provides an inexpensive way of forming lens shapes. Suitably chosen formulations of silicone have high optical transmission and are resistant to discoloration.

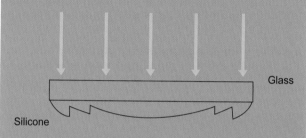

Glass

Silicone

Figure 20.3. The cross section of a Fresnel lens fabricated from silicone-coated glass. The lens shapes are molded into the silicone. The glass faces outward toward the environment.

Figure 20.4. Examples of several common concentrator configurations: (a) linear focus, (b) point focus, and (c) heliostat.

focused light illuminating a line, or in two dimensions, with the light focused down to a point. Higher concentration is possible for point focusing than for linear focusing. Examples of linear- and point-focus configurations are shown in Figure 20.4. (Photographs of line-focus systems are shown in Figure 21.5 of Chapter 21.) Most high-concentration CPV systems are of point-focus type. Another configuration, also illustrated in Figure 20.4, is the heliostat, in which an array of ground-mounted mirrors focuses light onto a central receiving tower. (See Figure 21.6(b) of Chapter 21 for a photograph of a heliostat system.)

20.4.3 Tracking and the limits of concentration

Pointing and tracking

In conventional, non-concentrating photovoltaics, the panels of solar cells can sit in a fixed orientation and still collect most of the sunlight over the course of a day. For CPV, the situation is more complex, since the lens must be oriented carefully with respect to the Sun to get the spot of focused light to fall at the desired location. More generally, the higher the level of concentration desired, the more precisely the concentrating optics must be pointed toward the Sun. Because the position of the Sun in the sky changes during the course of the day, the concentrating optics must be made to continually point at, or "track," the position of the Sun. Modern computer controls, combined with the use of global positioning systems to calculate precisely the location of the tracker, have made the computation of the Sun's position in the sky straightforward. Even with these aids, however, tracking still adds design and mechanical complexity, moving parts, and the necessity to guard against misalignment of the optics. The more precise the tracking and alignment must be, the more complex and expensive the resulting system will be. It is therefore vital to understand when tracking is necessary, how much tracking error can be tolerated, and how to relax the tracking requirements.

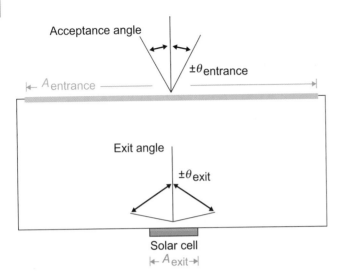

Figure 20.5. Geometry of an idealized linear concentrator with an entrance aperture of area $A_{entrance}$, an exit aperture of area A_{exit}, a light acceptance angle of $\theta_{entrance}$, and a light exit angle of θ_{exit}. The concentration is equal to $A_{entrance}/A_{exit}$.

Fundamental limits of concentration

A full discussion of concentrator optics is beyond the scope of this chapter. However, a tremendous amount of insight can be obtained from one very important conservation theorem of non-imaging optics [3]. (Note that this discussion applies only to geometric concentrators; fluorescent concentrators operate under different principles and are not included here.) Consider ideal, generalized concentrator optics that can use any imaginable combination of lenses and mirrors to change the path of the light. First consider the case of a linear focus optics, which focuses the light in one dimension. The optics has an entrance aperture of area $A_{entrance}$ and focuses the light down to an exit aperture of area A_{exit}, where the solar cell will sit. The concentration ratio C is thus $C = A_{entrance}/A_{exit}$. The light coming into the optic has an angular spread of $\pm\theta_{entrance}$ (or, equivalently, the acceptance angle of the optics is $\pm\theta_{entrance}$). The optics bends the light toward the focus, so the light exiting the optics at the focus has the new angular spread $\pm\theta_{exit}$. This geometry is illustrated schematically in Figure 20.5.

The receiver cell is embedded in a material having an index of refraction n. The theorem [3] states that, for cells that can accept light from one side, the highest possible concentration is

$$C_{1D} = n\frac{\sin(\theta_{exit})}{\sin(\theta_{entrance})}.$$

For point-focus optics in which the light is concentrated in two dimensions, the highest possible concentration is

$$C_{2D} = \left[n\frac{\sin(\theta_{exit})}{\sin(\theta_{entrance})}\right]^2.$$

(The case of cells that can accept light from both sides is slightly different and is not considered here.) The concentration is highest for $\theta_{exit} = 90°$, i.e., the incoming rays are bent so that they are at grazing incidence on the solar cell. The maximum concentrations are then

$$C_{1D} = \frac{n}{\sin(\theta_{entrance})}, \quad C_{2D} = \left[\frac{n}{\sin(\theta_{entrance})}\right]^2. \quad (20.1)$$

These equations indicate that, for a given collection angle $\theta_{entrance}$, there is a fundamental limit to how much the light can be concentrated, no matter how ingenious and complex the optical system. Note that these equations are equally relevant for systems that gather light for solar thermal applications as discussed in Chapter 21. Equations (20.1) allow the calculation of the highest degree of concentration possible for concentrator optics that collect all of the direct sunlight. Because the Sun's disk covers a half-angle in the sky of roughly 0.27° for a clear sky, the optics must collect light over the same angle in order to collect precisely all of the light from the Sun's disk. Substitution of this value for $\theta_{entrance}$ into Equations (20.1) gives maximum possible concentrations of $C_{2D} = 46{,}000n^2$ for a point-focus concentrator and $C_{1D} = 215n$ for a linear-focus concentrator. For a point-focus concentrator using only air ($n = 1$), the maximum concentration is $C_{2D} = 46{,}000$. These results, showing that sunlight cannot be concentrated to an arbitrarily high degree, are a dramatic illustration of both the power and the perhaps unintuitive implications of Equations (20.1).

Pointing requirements and tracking tolerance

The preceding discussion implicitly assumed perfectly aligned optics that perfectly track the Sun. However, such perfection cannot be achieved in real-world systems; in practice, it is necessary to increase the acceptance angle to allow for imperfections and misalignment of the optics, as well as some margin of error in their pointing and tracking. This margin of error is usually expressed as the difference between the system's acceptance angle, including the tracking error, and the angle required for perfect pointing (e.g., $\theta_{entrance} \approx 0.27°$ for ideal collection of rays from the solar disk); for example, a system with an acceptance angle $\theta_{entrance} = 0.37°$ has a pointing tolerance of $0.37° - 0.27° \approx 0.1°$. Systems with small margins for pointing error are expensive and prone to malfunction or misalignment, so a large pointing tolerance provided by a large acceptance angle is desirable. However, as indicated by Equations (20.1), this increased acceptance angle necessarily lowers the concentration that can be achieved. This trade-off between concentration and tracking tolerance is typical of the choices facing the concentrator-system designer. A 0.1° pointing tolerance, while achievable, is challenging. A much more desirable pointing tolerance is 1°. However, to accommodate a 1° pointing tolerance, the concentrator would need to have an acceptance angle $\theta_{\mathrm{entrance}} = 0.27° + 1° = 1.27°$. This increased acceptance angle lowers the maximum possible point-focus concentration by a factor of $[\sin(1.27°)/\sin(0.27°)]^2 \approx 22$, from $C_{2D} = 46{,}000n^2$ to $C_{2D} \approx 2{,}000n^2$. Real-world high-concentration systems typically use concentrations in the range 500–1,000, so 1° tracking tolerances are achievable for such systems. Practical limitations to concentration are discussed further in Chapter 21.

20.4.4 Solar cells for concentrators

In the preceding sections of this chapter, we described the use of optics to concentrate sunlight onto a solar cell, and we noted that concentrations in the range of 500–1,000 suns are typical for high-concentration systems. The power density of sunlight at 1,000-suns concentration is roughly 100 W cm^{-2}. Under such high concentrations, the performance of a solar cell might be expected to differ from its performance at one sun. In this section, we discuss how concentration affects the operating characteristics of solar cells, and how cells can be designed to perform optimally under concentration.

Current and voltage of a solar cell under concentration
We can gain considerable insight into the behavior of an ideal solar cell as a function of concentration from the equation for an ideal solar cell (Equation (18.1)) in Chapter 18:

$$I(V) = I_0[\exp(eV/(k_B T)) - 1] - I_{SC}, \qquad (20.2)$$

which describes how the current I of the cell depends on the voltage V across the cell as a function of the dark current I_0, the temperature T, and the short-circuit current I_{SC} generated by the sunlight illuminating the cell. The power output of the cell is $P = IV$, and the maximum value of the power output determines the efficiency of the cell. The open-circuit voltage of the cell (i.e., the voltage when no current is flowing) is determined from Equation (20.2) by setting $I = 0$, to obtain

$$V_{oc} \approx (k_B T/e)\ln(I_{SC}/I_0), \qquad (20.3)$$

where we have made the very good approximation $I_{SC}/I_0 - 1 \approx I_{SC}/I_0$ because $I_{SC}/I_0 \gg 1$ for practical cells.

To understand how the ideal cell behaves under concentration, we need to understand how the parameters depend on concentration. First consider the short-circuit current I_{SC} for a cell of area A. Recall that I_{SC} is determined by the cell's quantum efficiency, QE(λ), and by the spectral flux, $\Phi(\lambda)$, of light incident on the cell:

$$I_{SC} = eA \int_0^\infty QE(\lambda)\Phi(\lambda)d\lambda, \qquad (20.4)$$

which simply says that we need to multiply the number of photons at wavelength λ by the probability, QE(λ),

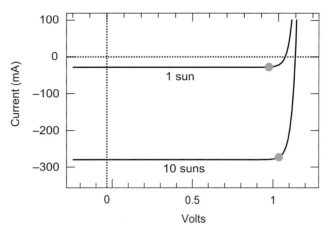

Figure 20.6. The I–V characteristic from Equation (20.2) for a cell at $T = 300$ K with $I_0 = 10^{-16}$ mA and $I_{SC} = 28$ mA. (These values are typical for a high-quality GaAs cell of area 1 cm^2.) The figure also shows the I–V characteristic for the same cell at a concentration of $C = 10$, i.e., $I_{SC} = 280$ mA. The maximum-power points, which determine the fill factors, are indicated with circular markers on the curves.

of such photons being collected as current, and then integrate the result over all wavelengths. The quantum efficiency of the cell, QE(λ), can vary with concentration; however, for the types of high-performance cells used in concentrator systems, the quantum efficiency is to a good approximation independent of concentration over concentration ranges from 1 to over 1,000 suns. Furthermore, when an ideal concentrator system concentrates the sunlight by a factor C, the light intensity at each wavelength increases by the same amount C, so the spectrum $\Phi_C(\lambda)$ at a concentration level C is proportional to the spectrum $\Phi_{C=1}(\lambda)$ at 1 sun: $\Phi_C(\lambda) = C\Phi_{C=1}(\lambda)$. Therefore, Equation (20.4) indicates that I_{SC} is simply proportional to the concentration: $I_{SC}(C) = CI_{SC}(1)$, where $I_{SC}(C)$ is the short-circuit current at concentration ratio C, and $I_{SC}(1)$ is the value of I_{SC} at 1 sun (i.e., no concentration, $C = 1$).

To get a sense of how concentration affects the cell's performance, Figure 20.6 shows a plot of the I–V characteristic from Equation (20.2) for a cell at $T = 300$ K with $I_0 = 10^{-16}$ mA and $I_{SC}(1) = 28$ mA. (These values are typical for a high-quality 1-cm^2 GaAs cell at room temperature.) Figure 20.6 also shows the I–V characteristic for the same cell at a concentration of $C = 10$. Recall from Chapter 18 that the efficiency η of a solar cell is determined from the I–V curve as $\eta = V_{OC} \times I_{SC} \times$ FF/P_{in}, where FF is the fill factor and P_{in} is the solar power led into the cell. Figure 20.6 shows that V_{OC} and the fill factor, and hence the efficiency, increase with concentration. Figure 20.7 plots these parameters as a function of concentration, showing that the increase in V_{OC} with concentration is logarithmic.

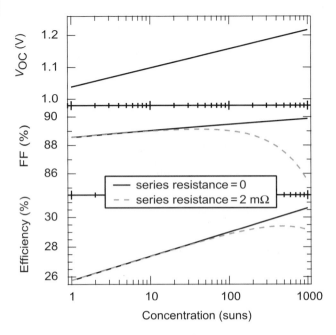

Figure 20.7. V_{OC}, the fill factor (FF), and the efficiency as functions of concentration for the idealized GaAs cell of Figure 20.6 (solid black lines) and for the same cell in series with a resistance of 2 mΩ.

We can get a quantitative and general result for the increase of V_{OC} with concentration from Equation (20.3). Denoting the value of V_{OC} at concentration ratio C as $V_{OC}(C)$, we can make the concentration dependence of Equation (20.3) explicit by writing it as

$$V_{OC}(C) = (k_B T/e)\ln(I_{SC}(C)/I_0). \tag{20.5}$$

On substituting $I_{SC}(C) = C\ I_{SC}(1)$, and using $V_{OC}(1) = (k_B T/e)\ln(I_{SC}(1)/I_0)$, we find that

$$V_{OC}(C) = V_{OC}(1) + (k_B T/e)\ln(C). \tag{20.6}$$

Thus, concentration increases V_{OC} by the amount $(k_B T/e)\ln(C)$. At room temperature (25 °C), this corresponds to an increase of about 0.06 V for each factor of 10 increase in concentration, or about 0.18 V for 1,000-suns concentration – an increase that has a noticeable effect on the corresponding cell efficiency. In the example of Figure 20.7, the cell efficiency increases from 26% to more than 30% as the concentration is raised from 1 to 1,000 suns.

Resistive power losses

The preceding analysis of cell performance vs. concentration was based on Equation (20.2), which describes an ideal cell. The behavior of real cells can diverge from ideality in various ways, but, in particular, all real cells inevitably have series resistances not described by Equation (20.2). These series resistances play a critical role in the performance of concentrator solar cells, because the current densities in cells used under concentrated light are much higher than the current

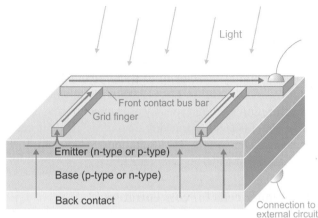

Figure 20.8. A schematic cross section of a solar cell, showing the metal contacts on both sides of the p–n junction. The top layer of the p–n junction is conventionally referred to as the emitter, and the bottom layer as the base. The current flows in the emitter, and grid fingers are indicated by arrows. The figure is not drawn to scale.

densities at 1 sun. Recall from basic circuit analysis that the power dissipated in a resistor of resistance R carrying a current I is given by I^2R: the power loss is proportional to the square of the current. For example, a resistive power loss that might be negligible at 1 sun will be 10^6 times greater at 1,000-suns concentration. Figure 20.7 compares solar-cell parameters with and without series resistance, calculated from equation 18.2 illustrating how series resistance becomes increasingly significant at increasing concentration. As the concentration increases, the efficiency eventually rolls off and starts to decrease with further increasing concentration.

Mitigation of series resistance is thus far more critical for concentrator cells than for 1-sun cells. For the high-efficiency cells used in modern high-concentration systems, one of the major sources of series resistance is related to the extraction of current from the cell. To extract the current from the cell, electrical contacts made of a highly conducting metal such as silver are fabricated on the top and bottom of the cell, as indicated schematically in Figure 20.8. The top layer of the p–n junction is conventionally referred to as the emitter, and the bottom layer as the base. The electrical contact on the top of the cell cannot cover the entire front surface, because that would prevent any light from entering. Instead, the front contact is made of an array of metal fingers that cover only a fraction (typically less than 10%) of the top surface. The electrical current must therefore flow laterally through the emitter to get to the grid fingers, as indicated by the arrows in Figure 20.8. The current then flows along the grid fingers to a bus bar, where the external electrical contact to the cell is made. These current flows result in I^2R losses because of the nonzero resistivities of the emitter and grid fingers. Minimizing these resistivities lowers the associated I^2R losses.

To minimize the grid-finger resistivity, grid fingers are usually made from a highly conductive material such as silver. Furthermore, the resistive losses in the array of grid fingers can be reduced simply by increasing the number of grid fingers, at the cost of letting less light into the cell. The design of the grids must therefore find the optimal trade-off between the resistive and light-blocking losses. An analogous trade-off applies in the design of the emitter: the thicker and more heavily doped the emitter, the lower its resistance, but also the lower its optical conductivity. In the design of a concentrator cell, great care is taken to optimize the grids and emitter to minimize the power losses in the cell. A typical concentrator cell based on the GaInAsP semiconductor materials system (see sec. 20.4.5) might have an emitter of thickness 0.1 μm with an n-type doping density of 10^{18} cm^{-2} and 5-μm-wide grid fingers that are spaced 100 μm apart. These values are intended to give a sense for orders of magnitude; the actual values can vary depending on the details of the cell design.

Temperature
The efficiency of a solar cell depends on its operating temperature (see Chapter 18). For multijunction cells under high concentration, the efficiency decreases with increasing temperature. A typical magnitude for this efficiency decrease is ~0.06% (absolute) per °C. To put this in perspective, for a cell with this temperature coefficient and an efficiency of 40% at 25 °C, the efficiency at 125 °C would be reduced to 34%, amounting to an unacceptable loss. To keep the efficiency losses due to elevated temperature within reasonable bounds, cell cooling is critical; a typical goal for the cell operating temperature in a CPV system would be less than 80 °C. In CPV, cell cooling requires special attention because heating of the cell by the sunlight is increased when the light is concentrated. Two common approaches to heat management are passive cooling using heat sinks to conduct heat away from the cell and active cooling using a heat-transfer fluid. Active cooling is more effective but more costly; for small cells with well-designed heat sinks, passive cooling is usually adequate. In principle, it is possible to capture and make use of the waste heat. Such approaches are known as co-generation because both electricity and usable heat are generated. In practice, however, the value of the heat energy is not always high enough to justify its capture.

20.4.5 Multijunction cells for concentrator systems

Because the cost of the solar cell used in a high-concentration photovoltaic system is a small fraction (typically 10%–30%) of the system cost, in CPV systems one can afford to use more expensive cells than are viable for conventional flat-panel systems. Investment in a costly concentrator solar cell can be worthwhile if the cell's efficiency is significantly greater than the efficiency of a conventional solar cell. This section describes multijunction solar cells, which are more efficient than conventional cells and are capable of operating under very high concentrations. As a result, multijunction cells have become the standard solar cells for high-concentration systems.

The multijunction concept
The energy in sunlight is distributed over a broad spectrum of photon energies from $hv \approx 0$ eV (infrared light) to about $hv \approx 3.5$ eV (ultraviolet light). This poses a fundamental limitation for conventional solar cells, which are characterized by a single bandgap energy, E_g. As described in Chapter 18, photons with energy $hv < E_g$ are not absorbed by the solar cell, which therefore collects none of their energy. In contrast, photons with energy $hv > E_g$ are absorbed by the solar cell, but the excess energy, $hv - E_g$, is lost as heat. (The endeavor to create solar cells that can harvest the excess energy is described in Chapter 19.) The detailed-balance efficiency limit [8] for a single-junction cell operating at 298 K under the unconcentrated global terrestrial spectrum is about 34% [9]. (This value of 34% differs from the often-quoted value of 30% in Shockley and Queisser's classic paper [8] because that paper considers a spectrum that differs from the terrestrial spectrum.) This is a fundamental limitation – and a serious one because it restricts conventional photovoltaics to collecting a rather small fraction of the energy in sunlight. This limitation can be overcome, both conceptually and in practice, by dividing the solar spectrum into spectral bands and sending each band to its own p–n junction with an appropriate bandgap. Such a cell is known as a multijunction solar cell.

Spectrum splitting and monolithic series connection
Implementing the multijunction concept requires a means of directing the various spectral bands of sunlight's spectrum to their corresponding junctions in the solar cell. An obvious way to do so would be to pass the sunlight through a prism, as shown in Figure 18.4 of Chapter 18. However, a much more elegant and practical approach is to use the junctions themselves as optical filters. Recall that a photon of energy hv is absorbed by a junction of bandgap E_g only if $hv \geq E_g$; if $hv < E_g$, the photon is not absorbed and simply passes through the junction. Thus, if the junctions of a multijunction cell are arranged in a stack with the bandgaps decreasing downward into the stack, then a photon entering the cell from the top of the stack will pass

Figure 20.9. The cross section of a monolithic series-connected three-junction solar cell, indicating the distribution of the different regions of the solar spectrum to the corresponding junctions of the cell. The locations of the tunnel-junction (TJ) interconnects are indicated.

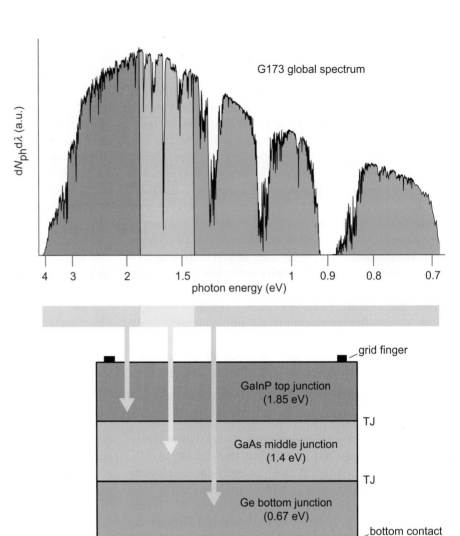

through the stack until it reaches the proper junction. This concept is illustrated in Figure 20.9, which shows the solar spectrum directed to a three-junction solar cell. The stacked configuration has highly desirable features. First, the stacking of the junctions can be accomplished in practice by growing all of the junctions, one after another, on a single substrate. This "monolithic" approach eliminates the need for mechanically stacking the separate junctions, as well as the expense of performing several separate growth steps, each on its own substrate. The junctions are connected together in series, with contacts only at the top and bottom of the stack. The electrical interconnections between adjacent photovoltaic junctions are accomplished by growing a tunnel junction between the junctions (indicated as "TJ" in Figure 20.9). A tunnel junction is a thin p–n junction that is doped so heavily that electrons move across the junction by quantum-mechanical tunneling, providing what is effectively a low-resistance connection [10]. In the standard multijunction solar cell, the tunnel junction provides a non-rectifying, low-resistance contact between the base of one junction and the emitter of the junction underneath. The resulting monolithic,

series-connected, two-terminal multijunction cell configuration simplifies both the fabrication of the cell and the incorporation of the cell into a photovoltaic system. For these reasons, all commercially available multijunction cells are monolithic series-connected two-terminal devices. Because the junctions are series connected, the V_{OC} of the resulting multijunction cell is the sum of the V_{OC}s of the individual junctions. For the GaInP/GaAs/Ge three-junction cell illustrated in Figure 20.9, a typical observed V_{OC} is 2.7 V, which is much higher than for any single-junction cell. Another consequence of the series connection of the junctions is that the current in any series-connected multijunction device is constrained to be the lowest of the currents of the individual junctions, and therefore efficiencies of series-connected junctions don't add as if they were independent. To maximize the efficiency, multijunction cells are therefore designed so that the junctions all generate the same currents. Because the spectrum of sunlight changes throughout the day, the junction currents will not always meet this criterion of being equal to each other. Fortunately, the effect on yearly energy production of the multijunction cell is minimal.

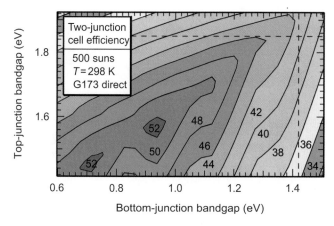

Figure 20.10. The calculated detailed-balance efficiency of a series-connected two-junction cell at a temperature of 298 K under the standard G173 terrestrial direct spectrum concentrated to 500 suns, as a function of the bandgaps of the junctions. The case of a (1.85, 1.42) eV bandgap combination corresponding to the lattice-matched GaInP/GaAs materials combination is indicated by the dashed lines.

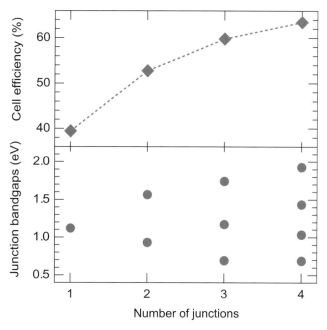

Figure 20.11. Optimal bandgaps and corresponding efficiencies as a function of the number of junctions in series-connected multijunction cells, for the same conditions as in Figure 20.10.

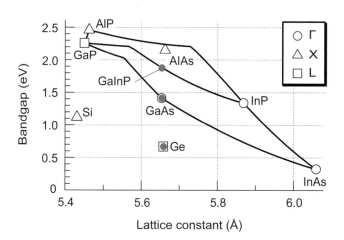

Figure 20.12. Bandgap versus lattice constant for compounds and alloys in the AlGaInAsP materials system. Values for Si and Ge are indicated as well. The legend indicates the location of the bandgap in the Brillouin zone. The Ge, GaAs, and $Ga_{0.5}In_{0.5}P$ compositions used in the $Ga_{0.5}In_{0.5}P$/GaAs/Ge three-junction cell are indicated by red dots.

Multijunction cell efficiency

The efficiency of a multijunction cell depends on the bandgaps of the individual junctions. Figure 20.10 shows the detailed-balance efficiency [8] of a series-connected two-junction cell at a temperature of 298 K (i.e., room temperature) as a function of the bandgaps of the junctions at 500-suns concentration under the standard terrestrial direct spectrum. The maximum efficiency is 53%, which is attained for top and bottom junction

bandgaps of 1.57 and 0.93 eV, respectively. [There is also a comparable local maximum at a (top, bottom) bandgap combination of (1.46, 0.71) eV.] In contrast, for a single junction under the same conditions, the maximum efficiency is only 40%. Even higher efficiencies are possible with more junctions. Figure 20.11 shows the optimal bandgaps and corresponding efficiencies as a function of the number of junctions in a series-connected multijunction cell, for the same conditions as in Figure 20.10. Figure 20.11 shows that the efficiency gains diminish with increasing number of junctions but are still significant for four junctions. We can see that bandgaps in the range of about 0.7–2 eV are sufficient to satisfy the requirements of up to four junctions. The challenge in multijunction cells is to obtain these efficiency advantages in an actual, practical device. We next turn our attention to this challenge.

Creating a practical multijunction solar cell

The creation of an actual monolithic multijunction cell requires semiconductor materials with the desired bandgaps. It is also very desirable to be able to fabricate all of the various layers of the junctions of the solar cell in one growth reactor, as already mentioned, using readily available substrates on which to grow the layers. The materials also must be capable of being doped n- and p-type to form the junctions. Fortunately, these desired characteristics are all available in the materials system composed of the semiconductor alloy $Al_zGa_xIn_{1-x-z}As_yP_{1-y}$. Figure 20.12 shows the bandgap as a function of the lattice constant for this materials system, as well as those for Si and Ge. It might then appear that we could simply select the alloy compositions that provide the ideal bandgap combinations from Figure 20.11 to make a high-efficiency multijunction cell. However, the bandgaps are not the only parameter critical to the performance of a solar cell: it is also necessary that the materials have minority-carrier transport lengths greater than the cell thickness, in order to collect the photogenerated carriers. When a semiconductor layer is epitaxially grown on another having a different lattice constant, the resulting strain causes dislocation defects in the material that can have a very detrimental effect on minority-carrier transport and, hence, on solar-cell performance.

For example, for a two-junction cell, Figure 20.10 indicates an optimal bandgap pair of (1.57, 0.93) eV for the two junctions. Figure 20.12 shows that there is no composition of GaInAsP that has a bandgap of 0.93 eV and has the same lattice constant as GaAs or Ge substrates, which are readily available. In contrast, the two-junction bandgap combination of (1.8, 1.4) eV, although not optimal according to Figure 20.10, can be attained with the materials $Ga_{0.5}In_{0.5}P$ and GaAs, whose lattice constants are matched to each other and to that of Ge. Similarly, for a three-junction cell, the $Ga_{0.5}In_{0.5}P$/GaAs/ Ge materials combination with bandgaps of (1.8, 1.4, 0.7) eV can be grown lattice-matched to GaAs or Ge substrates, whereas the ideal bandgap combination of (1.86, 1.34, 0.93) eV from Figure 20.11 cannot. This GaInP/GaAs/Ge three-junction cell structure can be grown by conventional metal–organic vapor-phase epitaxy with very high performance (see Box 20.2). Its first commercial application, starting in the mid 1990s,

Box 20.2. Materials growth challenges for the GaInP/ GaAs/Ge three-junction cell

Although the multijunction cell concept as illustrated in Figure 20.9 is straightforward, actual multijunction structures are quite complex. Figure 20.13 shows a detailed schematic cross section of a typical GaInP/ GaAs/Ge three-junction cell structure. The compositions and other basic materials properties of the layers are indicated in the figure. Figure 20.13 shows that the cell structure includes many layers and interfaces, whose properties must be precisely controlled during the growth of the structure in order for the resulting device to work as designed. The first step of the growth, and the first challenge, is to prepare the surface of the Ge substrate so that it is free of oxide and other contaminants and has the atomic surface structure needed for growth of high-quality III–V materials. One aspect of this

Box 20.2. (cont.)

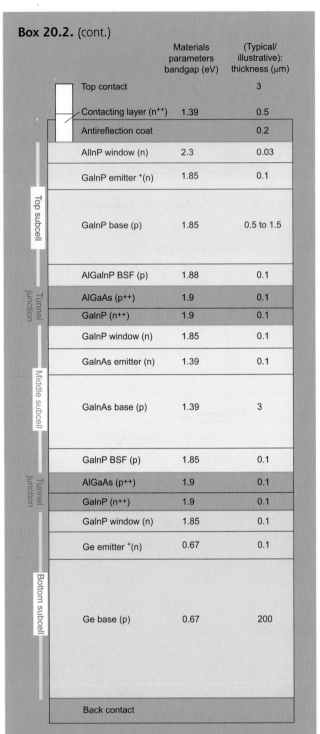

Layer	Materials parameters bandgap (eV)	(Typical/ illustrative): thickness (μm)
Top contact		3
Contacting layer (n++)	1.39	0.5
Antireflection coat		0.2
AlInP window (n)	2.3	0.03
GaInP emitter +(n)	1.85	0.1
GaInP base (p)	1.85	0.5 to 1.5
AlGaInP BSF (p)	1.88	0.1
AlGaAs (p++)	1.9	0.1
GaInP (n++)	1.9	0.1
GaInP window (n)	1.85	0.1
GaInAs emitter (n)	1.39	0.1
GaInAs base (p)	1.39	3
GaInP BSF (p)	1.85	0.1
AlGaAs (p++)	1.9	0.1
GaInP (n++)	1.9	0.1
GaInP window (n)	1.85	0.1
Ge emitter +(n)	0.67	0.1
Ge base (p)	0.67	200
Back contact		

Subcell labels: Top subcell, Tunnel junction, Middle subcell, Tunnel junction, Bottom subcell.

Figure 20.13. A detailed cross-sectional schematic diagram of a GaInP/GaAs/Ge three-junction solar cell. For each of the layers, basic materials properties are indicated, including composition, bandgap, and doping type (n or p). Layer thicknesses are indicated as well; note that the drawing is not to scale. Certain details are omitted for simplicity; for example, the front contact is typically a stack of several different metals, and the antireflective (AR) coat is a stack of at least two dielectric layers of different refractive indices.

Box 20.2. (cont.)

challenge is that III–V materials are polar whereas Ge is nonpolar, potentially leading to antiphase domains (APDs, a type of defect in which the group-III and group-V atoms do not alternate positions as they would in the bulk crystal) when the III–V layer growth is initiated on the Ge. Minimizing APDs requires the careful optimization of growth parameters such as the relative fluxes of the group-III and group-V source materials, as well as the use of Ge substrate orientations tilted a few degrees away from the (100) direction to provide a suitable array of atomic steps as templates for the III–V epitaxy.

Furthermore, as the initial group-III and group-V atoms are deposited on the substrate, some of these atoms diffuse into the Ge, where they act as dopants. If done correctly, this growth initiation forms a diffused p–n junction in the Ge that provides the bottom junction of the three-junction solar cell. The location and diffusion of dopants must be precisely controlled not just at the III–V/Ge interface but at every location in the cell structure. This control is especially challenging in the tunnel junctions, which require the doping to switch from extremely heavily n-type to p-type within a few atomic layers in order for the proper electron tunneling behavior to be obtained.

Once the growth of the III–V material has been initiated, great care must be taken to ensure that the resulting epilayers are free of structural defects such as dislocations, as well as contaminants such as oxygen that would form point defects within the semiconductor. To minimize the formation of dislocations during the growth of alloys such as $Ga_xIn_{1-x}P$, the composition x must be precisely maintained to keep the lattice constant of the $Ga_xIn_{1-x}P$ the same as those of the Ge and GaAs layers in the structure. The growth of GaInP has the additional, intriguing materials-science complication of spontaneously forming a GaP–InP superlattice under certain growth conditions. This ordering phenomenon lowers the bandgap of the material by more than 0.1 eV compared with that of the random alloy, a large enough difference to affect (and to be used to optimize) the performance of a solar cell.

Box 20.3. Materials availability and energy payback time

Gallium, indium, and germanium are all relatively rare materials, so it is worth asking how their rarity may constrain the use of multijunction concentrator cells that are based on these materials (see Chapter 18). Fortunately, the ~ 1,000 × reduction in required cell area in a high-concentration system reduces the cells' materials use commensurately. It appears that materials availability will not limit the growth of CPV for the foreseeable future [6].

Another key criterion for the scalability of a photovoltaic system is the energy payback time – the length of time the system would need to operate to generate the amount of energy needed for its fabrication. The developers of the Amonix system shown in Figure 20.1 estimate (http://www.amonix.com/content/sustainability) an energy payback time of 0.7 years for their system operating in the high-direct-sunlight areas of the southwest USA, which is comparable to the best energy payback times for flat-plate phovoltaics.

on the order of 100 suns in both cases [11].) These cells are available commercially from several manufacturers.

See Box 2.3 on issues concerning the availability of materials.

Next-generation multijunction cells

Because the bandgaps of GaInP/GaAs/Ge cell junctions are not optimally matched to the solar spectrum, even higher efficiencies are, in principle, attainable using the optimal bandgaps as summarized in Figure 20.11. In practice, developing junctions that have both the desired bandgaps and sufficiently long minority-carrier transport lengths has been challenging: as noted above, an AlGaInAsP semiconductor alloy lattice-matched to GaAs or Ge cannot have a bandgap less than 1.4 eV, as would be needed for the optimal-bandgap combinations. One approach is to manipulate the bandgaps of the materials. For example, by introducing "quantum dots" into GaAs, it is possible to lower the bandgap of the resulting material without changing the lattice constant. A second approach is to employ junction materials having lattice constants different from that of the substrate. In this case, it is necessary to mitigate the effect of the dislocation defects that result from the resulting strain. The development of advanced multijunction solar cells, including designs with four or more junctions, is a very active area of device and materials research, and future multijunction cell efficiencies well in excess of 40% appear attainable [12].

was for use in spacecraft applications such as communications satellites. The high cell efficiency translates into reduced weight and size for the solar panels, which are critical advantages for spacecraft applications. More recently, concentrator versions of this cell have been developed. (In comparison, the very highest single-junction efficiencies are about 28% for very sophisticated silicon cells and 29% for GaAs cells, at a concentration

Table 20.1. An assessment of progress and challenges in concentrating photovoltaics technology

	Energy cost ($ per kWh)	Production and installation system cost ($ per W)	Installed capacity (GW)	Limitations	Roadblocks	Environmental impact
Present	0.25[a]	6[a]	0.02[b]	Absence of proven large-scale operation	Lack of demonstrated long-term large-scale operation makes financing difficult	Negligible
In 10 years	0.08[a]	2[a]	6[a]	Viable system size is > 1 MW	Absence of CPV manufacturing component standards	Large-area footprint in desert areas
In 25 years	?	?	?	Suitable only in regions of high direct sunlight	Challenge of integrating CPV energy input into the grid	Possible need for new power transmission lines

[a] Interpolated from estimates in [13].
[b] From [6].

20.5 Summary

Concentrating photovoltaics uses optical elements to increase the rather low power density of sunlight, thereby reducing the area of solar cells required by as much as a factor of 1,000 or more. We demonstrated that the performance of solar cells is a function of the level of concentration and showed that the efficiency of photovoltaic power conversion can be increased using multijunction solar cells. The resulting CPV systems, installed in regions of high direct sunlight, hold promise for lowering the cost of solar electricity. See Table 20.1 for a summary assessment.

20.6 Questions for discussion

1. Is there an optimal level of concentration for a concentrator system? What factors would determine this?
2. The power production of non-concentrating flat-panel solar modules also benefits from tracking. Why?
3. In the mirror configuration in Figure 20.2(c), the cell blocks some of the light from reaching the mirror. Is there a way to modify the mirror shape and light path to avoid this problem? How?
4. Can Equations (20.1) be rearranged to tell us the maximum acceptance angle that can be achieved for a specific concentration? If so, what are the rearranged forms of these equations?

5. During the course of a year, does the Sun occupy every location in the sky? If not, what are the implications for being able to concentrate the direct component of sunlight without tracking?
6. Would the efficiency advantage of multijunction cells over single-junction cells still hold when converting a monochromatic light source, such as the light from a laser? Why or why not?
7. GaInP/GaAs/Ge three-junction cells cost roughly $5 per cm^2. Estimate the cost of the cell per watt of electrical power produced, for a system using these cells without concentration, and at 1,000-suns concentration. What assumptions and approximations enter into this estimate?

20.7 Further reading

- **R. M. Swanson**, 2003, "Photovoltaic concentrators," in *Handbook of Photovoltaic Science and Engineering*, eds. **A. Luque** and **S. Hegedus**, New York, John Wiley and Sons. A remarkably readable, complete, authoritative survey of concentrator photovoltaics.
- **A. Luque**, 1989, *Solar Cells and Optics for Photovoltaics Concentration*, Bristol, Adam Hilger. Covers concentrator solar cells and – especially – optics in great technical depth.
- **V. M. Andreev**, **V. A. Grilikhes**, and **V. D. Rumyantsev**, 1997, *Photovoltaic Conversion of Concentrated Sunlight*,

New York, John Wiley and Sons. A detailed technical discussion of concentrator solar cells and systems.

- R. Winston, J. C. Miñano, and P. Benítez, 2005, *Non-imaging Optics*, Burlington, MA, Elsevier. An updated version of a pioneering exposition of optics for concentrators.
- "X marks the spot," *Photon International*, April 2007, p. 122. A survey of concentrator systems and companies. Slightly dated but gives a good sense of the range of CPV technologies.
- J. M. Olson, D. J. Friedman, and S. R. Kurtz, 2003, *High-efficiency III–V Multijunction Solar Cells*, in *Handbook of Photovoltaic Science and Engineering*, eds. A. Luque and S. Hegedus, New York, John Wiley and Sons. A detailed technical exposition of the principles and practicalities of modern multijunction solar cells.

20.8 References

[1] E. L. Ralph, 1966, "Use of concentrated sunlight with solar cells for terrestrial applications," *Solar Energy*, 10, 67–71.

[2] A. A., Salim, F. S. Huraib, N. N. Eugenio, and T. C. Lepley, 1987, "Performance comparison of two similar concentrating PV systems operating in the US and Saudi Arabia, in *Proceedings of the 19th IEEE Photovoltaic Specialists Conference*. p. 1351–1357

[3] R., Winston, J. C. Miñano, and P. Benítez, 2005, *Nonimaging Optics*, Burlington, MA, Elsevier.

[4] R. A. Sinton and R. M. Swanson, 1987, "An optimization study of Si point-contact concentrator solar cells," in *Proceedings of the 19th IEEE Photovoltaic Specialists Conference*. p. 1201–1208

[5] L. M. Fraas, J. E. Avery, V. S. Sundaram et al., 1990, "Over 35% efficiency GaAs/GaSb stacked concentrator cell assemblies for terrestrial applications," *21st PVSC*, pp.190–195.

[6] S. R. Kurtz, 2009, *Opportunities and Challenges for Development of a Mature Concentrating Photovoltaic Industry*, Golden, CO, National Renewable Energy Laboratory.

[7] G. Smestad, H. Ries, R. Winston, and E. Yablonovitch, 1990, "The thermodyamic limits of light concentrators," *Solar Energy Mater.*, 21, 99–111.

[8] W. Shockley and H. J. Queisser, 1961, "Detailed balance limit of efficiency of p–n junction solar cells," *J. Appl. Phys.*, 32, 510.

[9] S. P., Bremner, M. Y. Levy, and C. B. Honsberg, 2008, "Analysis of tandem solar cell efficiencies under AM1.5G spectrum using a rapid flux calculation method," *Prog. Photovoltaics*, 16, 225–233.

[10] S. M., Sze, 1969, *Physics of Semiconductor Devices*, New York, Wiley.

[11] M. A. Green, K. Emery, Y. Hishikawa, and W. Warta, 2010, "Solar cell efficiency tables (version 35)," *Prog. Photovoltaics*, 18, 144–150.

[12] D. J. Friedman, 2011, "Progress and challenges for next-generation high-efficiency multijunction solar cells," *Curr. Opin. Solid State Mater. Sci.*, 14, 131–138.

[13] S. Grama, E. Wayman, and T. Bradford, 2008, *Concentrating Solar Power – Technology, Cost, and Markets*, Chicago, IL, Prometheus Institute.

21

Concentrating solar thermal power

Abraham Kribus

Tel Aviv University, Tel Aviv, Israel

21.1 Focus

Power generation from solar energy by thermomechanical conversion is a major path for creating clean renewable power, while building on the mature technology base of conventional power plants. This solar technology was the first for which it was possible to demonstrate full-scale power plants (using Luz parabolic troughs built in California during the 1980s). With plants generating several thousands of megawatts currently in operation and under construction around the world, concentrating solar thermal power is fast becoming a mainstream solar power technology.

21.2 Synopsis

Solar thermal power generation includes three conversion steps: from solar radiation to heat, from heat to mechanical work, and from work to electricity. The last two steps are well known from conventional power plants, with the leading technologies being heat engines based on the steam cycle and the gas turbine cycle. A solar thermal plant can use these mature heat-engine technologies with the replacement of fuel-fired heat by "solar-fired" heat. Providing heat from solar energy at the appropriate temperatures requires concentration of the sunlight, because, otherwise, heat losses to the environment are too high. The leading concentration methods are linear concentrators (parabolic trough and linear Fresnel), towers with heliostat fields, and parabolic dishes, each of which is suitable for a range of operating temperatures and types of heat engine. Considering the inherent energy losses in the process of concentration, the overall solar plant efficiency (from collected solar radiation to electricity) is typically in the range 15%–25%, with the best systems reaching around 30%.

Today, concentrating solar power (CSP) plants are mostly constructed with the linear-concentrator approach and use steam-cycle conversion at moderate temperature, since these choices offer lower risk. The competing technologies of solar towers and parabolic dishes are also making initial steps. At the end of 2010, about 1,000 MW of solar thermal power was in operation in the USA and Spain, with about 10,000 MW more under construction or development. However, CSP suffers from the same affliction as most renewable energy technologies: it is still too expensive to compete directly against conventional power generation. Therefore, plants are currently built only in locations where an adequate form of government support is available. Several paths are being pursued toward the goals of lower cost and higher performance of CSP technologies. These include increased efficiency through higher operating temperature, innovative designs and materials, and cost reduction in plant manufacturing and installation.

This chapter presents the evolution of solar thermal power generation, the state of the art of the main technologies now in use, and the promise and challenges of emerging technology directions.

21.3 Historical perspective

The first attempts to use sunlight as the energy source for powering machinery took place during the nineteenth century. Since then, several scientists have tinkered with the gadgets of that time, such as "burning-mirror" solar concentrators and "hot-box" solar collectors, and found that they can generate steam just as well as burning coal can. A particularly successful pioneer was Augustin Mouchot, a French mathematics professor turned inventor and engineer. After developing several small models, he demonstrated at international exhibitions in Paris in 1878 and 1882 two large solar collectors that generated steam and operated steam engines to drive an ice-making machine and a printing press (Figure 21.1). Mouchot and his assistant, Abel Pifre, developed both parabolic-trough and parabolic-dish concentrators that operated on the same principles as the advanced collectors in use today. Unfortunately, because of the availability of cheap coal at the time, the French government deemed solar energy uneconomical and discontinued funding for their research. However, their inventions did inspire other scientists and inventors to pursue the promise of solar concentrators. Another well-known pioneer was the prolific Swedish-American inventor John Ericsson (1803–1889), who designed and built several variants of line- and point-focus concentrators, showing that they could operate hot-air engines. Ericsson's collectors, likewise, never materialized into an industrial success.

In 1912, American engineer and inventor Frank Shuman (1862–1918) built the first industrial-scale parabolic-trough plant in Egypt, which successfully and reliably powered a 50-kW steam engine driving an irrigation pump (Figure 21.2). Again, the circumstances were not favorable: the outbreak of World War I and

Figure 21.1. Mouchot's solar steam generation collector at the Paris exhibition in 1878. (Source: Conservatoire Numérique des Arts et Métiers.)

Figure 21.2. Shuman's solar thermal parabolic trough plant in Egypt, c. 1913.

the discovery of abundant cheap oil in North Africa left this solar technology behind, only to be rediscovered and exercised with much greater success in the 1980s.

Many decades passed without much interest in solar power generation, despite continuing advances in academic research and, in particular, the efforts, inventions, and achievements of pioneers such as Giovanni Francia (Italy), Harry Tabor (Israel), and Roland Winston (USA). After this long hiatus, the oil crises during the 1970s renewed interest and investments in renewable energy in general, including solar thermal power. Several large-scale CSP research and demonstration facilities were constructed around the world. In 1984, Luz International built in California its 14-MW SEGS (solar electricity generation system) plant, which employed parabolic-trough concentrators generating steam at $307\,^\circ$C and operating a steam turbine. SEGS-I was the first of nine plants that have demonstrated and validated solar thermal generation as a mature and reliable utility-scale technology. Subsequent plants built by Luz have been larger (up to 80 MW) and operated at higher temperatures (up to $390\,^\circ$C), leading to higher efficiency and lower cost per unit energy. These plants are still in operation today, supplying up to 354 MW to the California grid.

Legislation enacted in Spain in 2004 that established preferred feed-in tariffs for solar power ushered in the current wave of large-scale CSP implementation. Numerous 50-MW parabolic-trough plants, as well as several tower plants, are operating or under construction in Spain. In the USA, government actions such as tax credits and renewable portfolio standards have spurred the development of many trough, tower, and dish plants in California and other southwestern states. Additional projects are under construction and development in North Africa, India, and a few other locations around the world. Together with a resurgence in academic research in related areas, these efforts are pushing CSP across the threshold from a research concept to an industrial reality and an accepted technology in the renewable energy arena.

21.4 Solar thermal power generation

21.4.1 The conversion process

Solar thermal power generation includes conversion from solar radiation to heat, from heat to mechanical work, and from work to electricity. Although this might seem more complex than the one-step photovoltaic conversion (from solar radiation to electricity), this approach has the advantage of relying on mature areas of science and engineering with centuries of accumulated technology base: optics, heat transfer, and thermodynamics. Another major distinction between thermal and photovoltaic generation is that the intermediate form of energy, namely thermal energy, offers opportunities for energy storage and for hybridization (providing heat from an alternative source when solar heat is insufficient). Intermittency is a major obstacle to the implementation of solar energy, and the potential for smoothing this intermittency through storage and hybridization is a distinct advantage of solar thermal conversion.

The efficiency of conversion from solar radiation to electricity can be expressed as the product of the efficiencies of all of the steps in the process: the optical collection efficiency η_{opt}, the radiation-to-heat (receiver) conversion efficiency η_{rec}, and the heat-to-electricity ("power-block") conversion efficiency η_{PB}:

$$\eta = \eta_{\mathrm{opt}}(C) \cdot \eta_{\mathrm{rec}}(C, T_{\mathrm{rec}}) \cdot \eta_{\mathrm{PB}}(T_{\mathrm{rec}}). \tag{21.1}$$

The main parameters that affect the efficiency are the concentration level at the receiver, C, and the temperature of the working fluid exiting the receiver, T_{rec}. Increasing the concentration typically increases the receiver efficiency (reducing the receiver area responsible for losses), but reduces the optical efficiency. Conversely, increasing the operating temperature improves the power-block efficiency, but reduces the receiver efficiency. Therefore, careful optimization is needed to determine the best values of C and T_{rec} for a solar plant. Obviously, efficiency is not the only criterion for solar plant optimization, and the cost of producing and installing different solutions is a major consideration in the selection of an optimal solution.

24.1.2 Solar concentrators

Basics of concentration

Sunlight arrives at the Earth's surface at very low **flux** (power per unit area): the direct component of sunlight can reach about 800–1000 $\mathrm{W\,m^{-2}}$ on a clear day at good locations. Fortunately, it is possible to increase the radiation flux by optical concentration, making use of the directional nature of light. Direct sunlight is available within a narrow cone of directions, defined by the half-angle subtended by the Sun from Earth's point of view: $\theta_{\mathrm{s}} \approx 5$ mrad (0.27°). A concentrator manipulates the directional distribution of the light by reflection (using a mirror) or refraction (using a lens), causing a larger amount of radiation to reach the target (receiver) from a larger range of directions (Figure 21.3). Concentration in CSP systems usually employs mirrors and can be done in one or two dimensions, leading to a line focus or a point focus, respectively (Figure 21.4). Theoretically, the maximum concentration ratio (ratio of the flux incident on the receiver to the flux of normal sunlight) can

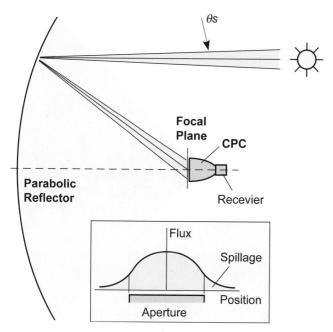

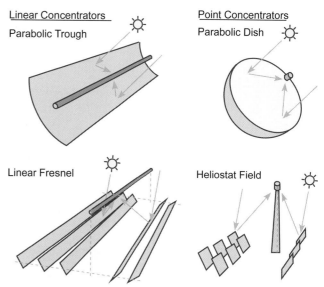

Figure 21.3. Concentration using a parabolic reflector and a CPC secondary non-imaging concentrator; insert: flux distribution at the focal plane. CPC is compound parabolic concentrator.

Figure 21.4. Types of solar concentrator systems: parabolic trough and linear Fresnel representing linear (1-D) concentrators, parabolic dish and heliostat field representing point (2-D) concentrators.

reach the **thermodynamic limit** for one- and two-dimensional concentrators:

$$C_{1D,max} = \frac{1}{\sin(\theta_s)}, \qquad C_{2D,max} = \frac{1}{\sin^2(\theta_s)}. \qquad (21.2)$$

Practical concentrators reach much lower values, because of many factors such as geometric inaccuracy in the manufacturing and assembly of the optical surfaces, imperfect material properties (reflectivity less than 100%), and mutual interference among neighboring elements of a concentrator (**shading** and **blocking losses**). The distribution of incident flux on the receiver aperture plane is usually broad, with long tails of gradually diminishing flux, as a result of inherent optical aberrations and spreading due to geometric imperfections. Trying to collect all of the radiation would require a large receiver aperture, leading to a very low average concentration. Therefore, the receiver collects only a part of the radiation, and the rest is lost as **spillage** (Figure 21.3). Typical concentration values in practice are 30–80 for linear concentrators and 300–1,000 for point-focus devices.

Effective high concentration of direct sunlight can occur only when the optical axis of the concentrator points in an accurate relation to the direction of the Sun, requiring motion of the concentrator (tracking) to follow the apparent motion of the Sun in the sky. Linear concentrators can operate with tracking in one axis, whereas point-focus systems for high concentration require tracking in two axes in order for them to point in the required direction with sufficient accuracy.

Tracking is done using well-known astronomical formulas that define the direction of the Sun at any time and at any location on Earth [1]. The need for a tracking mechanism introduces additional complexity and cost, as well as possible loss of energy when the tracking accuracy is not sufficient. When tracking error is present, some of the collected solar radiation can fall outside of the aperture of the receiver, leading to an increase in spillage losses.

Achieving high concentration with a single optical element is often difficult, so it can be beneficial to add a secondary optical element in front of the receiver. There are many possible shapes for the secondary [2], among which the **compound parabolic concentrator (CPC)** is perhaps the most well known. The secondary concentrator is designed to allow the entrance of radiation at its inlet aperture from a specific range of angles (the **acceptance angle**) and to redirect this radiation to a smaller exit aperture. The acceptance angle should match the directional distribution of the radiation incident on the secondary concentrator as determined by the size and distance away of the primary concentrator.

The secondary concentrator increases the incident flux on the receiver and increases the receiver's efficiency. However, it also incurs losses due to imperfect reflectivity (absorption of some of the reflected radiation) and possibly rejection of some radiation that does not reach the exit aperture. The increase in receiver efficiency must therefore be high enough to justify the additional cost of the secondary optics.

Additional discussion of solar concentration optics can be found in Chapter 20.

Figure 21.5. Line-focus concentrators: (a) a Luz parabolic trough collector at Kramer Junction, CA; (b) Ausra Linear Fresnel collector in Bakersfield, CA.

(a)

(b)

Line-focus concentrators

The **parabolic-trough collector** [3] uses the basic optical property of a parabola, as shown in Figure 21.3. A pipe carrying a heat-transfer fluid is installed at the focus of the parabola, to absorb the radiation and transfer the energy to the fluid, which carries the energy away. The reflector is usually made from glass, bent in one dimension into a parabolic shape and silver-coated on the back surface similarly to a standard mirror. As discussed later, several companies are also investigating polymer-film-based reflector designs. Typically, a trough aperture width of 6–7 m is divided into four curved reflector segments, supported by a stiff metal structure, as shown in Figure 21.5(a). The entire structure carrying the reflector and the receiver tube rotates about an axis to follow the apparent daily motion of the Sun from east to west.

The **linear Fresnel concentrator** forms a line focus similar to the parabolic trough, but has two notable differences. First, the reflector is segmented into parallel strips installed on a planar structure close to ground level, as seen in Figure 21.5(b), instead of forming a continuous parabolic surface. Each mirror strip rotates about its own axis and redirects solar radiation toward the receiver tube. The mirror strips can be planar or slightly curved, and are subject to lower wind forces because they are always close to the ground; therefore, linear Fresnel collectors are simpler and cheaper than parabolic troughs. The second difference is that the receiver is stationary, rather than moving with the tracking concentrator. The amount of energy collected by a linear Fresnel collector, however, is 20%–30% lower than that collected by a parabolic trough. When the Sun is low in the sky, the linear Fresnel concentrator offers a smaller area of reflectors projected toward the Sun than does the trough, which fully rotates to present its full area toward the incident radiation.

Point-focus concentrators

The **parabolic dish collector** is a complete plant with concentrator, receiver, and heat engine integrated and installed on a single tracker, as shown in Figure 21.6(a).

Figure 21.6. Point-focus concentrators: (a) a Solar Energy Systems dish-Stirling collector; (b) the eSolar towers and heliostat field in Lancaster, CA.

(a)

(b)

The concentrator is made from mirror segments curved in two dimensions, forming a paraboloid of revolution with a common focal point. A receiver and an engine are suspended at the focal point. A two-axis tracker with suitably high-accuracy tracking motors moves the dish such that its optical axis always points toward the Sun. The parabolic dish can, in principle, create high concentrations of several thousand times, but, in most cases, the average concentration on the receiver is kept below 1,000, which is sufficient for reasonable receiver efficiency and avoids excessive difficulties with heat transfer inside the receiver. Because of its high concentration, a parabolic dish can easily reach high receiver temperatures and, given a suitable heat engine, will provide a high overall conversion efficiency. A dish system currently holds the efficiency record in solar thermal electricity generation of 31.25%. Dish collectors are limited in size, however, because they need to track the Sun as a single unit. Typical sizes involve aperture areas of around 100 m^2, and the largest dish to date is about 400 m^2. This size limit narrows the selection of the heat engine, which can be up to about 100 kW. Currently, the only choice for a dish system is a Stirling engine, which is not a mainstream power-plant technology (as discussed in Section 21.4.4 below). The use of very small gas turbines (microturbines) in a parabolic dish has also been proposed, but not implemented; common steam and gas turbines are inefficient and not widely available at this small size.

An alternative point-focus concentrator system is the **solar tower** (also called the "power tower" or central receiver system). A field of mirrors (called **heliostats**), each of which independently tracks the Sun, collects and concentrates sunlight into the receiver on top of the tower, as seen in Figure 21.6(b). The heliostats do not point toward the Sun; rather, the tracking system of each heliostat points the normal to the heliostat surface at a direction that bisects the angle between the direction toward the Sun and the direction toward the target. Each heliostat can be flat or slightly curved, with the focal distance of the curvature matching the distance from the heliostat to the receiver. The heliostat size varies widely: many are in the range 100–200 m^2 and are made from many curved mirrors attached to a common support frame, whereas the smallest are made from a single mirror of about 1 m^2 size, as can be seen in Figure 21.6(b). The field of heliostats around a tower can create a concentration of a few hundred times at the receiver aperture, and, therefore, a solar tower plant can reach high temperature and high conversion efficiency similarly to a parabolic dish. The added advantage is that the heliostat field is not restricted to small size and large high-efficiency industrial heat engines such as steam and gas turbines can be used. On the other hand, the optical efficiency of a heliostat field is significantly lower than that of a parabolic dish, because of several effects: there is mutual interference of the heliostats (shading, blocking); the effective mirror area projected toward the Sun is significantly smaller than the actual mirror area (this is often called the "cosine effect"); and significant optical aberrations increase the loss by spillage around the receiver aperture.

21.4.3 Solar high-temperature receivers

Receiver energy balance

The solar receiver in a CSP system intercepts and absorbs the concentrated solar radiation, and transfers the absorbed energy to a **heat-transfer fluid (HTF)** passing through the absorber; this should be done while minimizing energy losses to the environment. The receiver has an aperture to admit the incident radiation, and usually most of the energy losses occur through this aperture as well. The net thermal power provided by the receiver to the HTF is

$$\dot{Q}_{net} = CI_{DNI}A_{ap} - A_{ap}\left[(1-\alpha)CI_{DNI} - \varepsilon\sigma T_r^4 - h(T_r - T_a)\right] \\ - UA_{cav}(T_r - T_a) = \dot{m}(i_{ex} - i_{in}). \qquad (21.3)$$

The first term is the power incident on the receiver aperture, where C is the average flux concentration, I_{DNI} is the direct normal insolation flux, and A_{ap} is the aperture area. The next three terms are losses through the aperture: reflection of the incident radiation, where α is the effective absorptivity; emission of radiation from the hot absorber surfaces, where ϵ is the emissivity, σ is the Stefan–Boltzmann constant, and T_r is the average temperature of the receiver surface; and convection to the outside air, where h is the convection coefficient and T_a is the ambient temperature. The last term is loss through external surfaces of the receiver that are not exposed to incident sunlight; these surfaces are insulated, and the energy loss is by conduction to the environment through these surfaces, where U is the heat-transfer coefficient and A_{cav} is the surface area. The HTF at mass flow rate $\dot{m}$ absorbs the net thermal power and increases its specific enthalpy from i_{in} at the inlet to i_{ex} at the exit. The receiver efficiency is then defined as

$$\eta_{rec} = \frac{\dot{Q}_{net}}{CI_{DNI}A_{ap}}. \qquad (21.4)$$

Tubular receivers for linear-focus concentrators

The receiver of a linear concentrator, sometimes called the **heat-collection element (HCE)**, is a pipe installed along the line focus. The typical parabolic-trough HCE structure shown in Figure 21.7(a) includes an internal stainless steel pipe, a radiation-absorbing coating on the exterior of the pipe, an external glass pipe that maintains a vacuum around the steel pipe to reduce convection losses (Equation (21.3)), a bellows at the end of the pipe to compensate for differences in thermal expansion between the glass and the metal, and a getter to maintain the vacuum by absorbing gases that permeate into the evacuated space.

The absorber coating is spectrally selective to reduce emission losses (Equation (21.3)): it has high spectral emissivity for radiation in the solar spectrum (wavelengths of up to 2 μm), and low emissivity for wavelengths that characterize the emission from the hot receiver (2–10 μm), as shown in Figure 21.7(b). A typical coating provides a high overall absorptivity of 0.93–0.96 for solar radiation, together with a low overall emissivity of about 0.05 at room temperature and 0.1–0.15 at operating temperature of 400–500 °C. A wide range of materials and microstructures can produce such spectrally selective behaviour [4]. The most

(a)

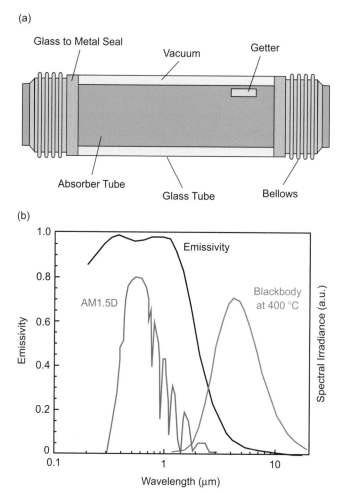

(b)

Figure 21.7. Heat Collection Element (HCE) receiver for parabolic troughs: (a) HCE structure, (b) typical spectral emissivity of absorber coating, with representative spectral irradiance of terrestrial direct radiation (AM1.5D, D is for direct radiation only) and of a blackbody at 400°C.

(a)

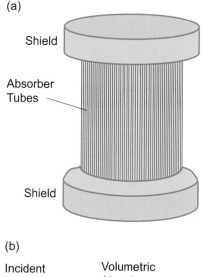

(b)

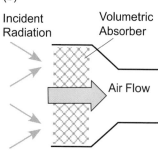

Figure 21.8. Schematic structure of typical central receivers: (a) external tubular receiver, (b) volumetric receiver.

popular solutions are cermets (composites of ceramic matrix with metal particles) with two layers, or one layer with a graded metal concentration, over a back metallic infrared (IR) reflector. The main limitation of the selective coatings is stability at high temperature, particularly if the coating is exposed to air, and they tend to fail if the vacuum in the HCE is breached.

The receiver in a linear Fresnel system can be a tube or tubes inside a downward-facing cavity, where the back wall of the cavity can be insulated or fitted with a second reflector. Because convection losses are reduced by the cavity structure, the receiver can operate at reasonable efficiency without the need for a vacuum in lower-temperature systems (Equation (21.3)). However, the receiver performance might suffer without a vacuum in higher-temperature systems.

The HTF in a parabolic-trough receiver is typically a synthetic fluid such as Therminol VP-1 that can operate at up to 400 °C. The hot HTF is used to generate steam

for the power-generation cycle in a separate boiler. **Direct steam generation (DSG)** in the receiver pipe is also an option, allowing higher steam temperature, but this requires the pipe to sustain higher pressure and needs a selective coating that can be stable at higher temperatures; DSG receiver tubes have been demonstrated in linear Fresnel and trough collectors [5], but have not yet been adopted for commercial use. Molten nitrate salts have also been considered for parabolic-trough plants [6], offering higher temperature than the synthetic oil and lower pressure than DSG.

Tubular receivers for point-focus concentrators
Tubular receivers in point-focus systems have an arrangement of pipes carrying the HTF in a compact structure at the focal region of the concentrator. Typical structures include external receivers, where arrays of pipes are placed around a cylindrical volume on top of the tower, see Figure 21.8(a); and cavity receivers, where arrays of tubes are located along the inner walls of an insulated cavity. The tubes can be arranged very close to each other, forming a nearly continuous curtain that intercepts all incident radiation; in that case, the backs of the tubes are not illuminated and receive heat only by conduction through the tube wall. Alternatively, the tubes can be spaced and the wall behind the tubes fitted

with a diffusely reflecting surface that redirects the radiation toward the backs of the tubes. The tubes are usually made from a high-temperature-resistant steel or a nickel alloy such as Inconel. The metal is coated with a black coating or paint to increase its absorptivity. The coating can be spectrally selective, although, at receiver temperatures of 700 °C and higher, spectral selectivity is less effective, given that the solar spectrum and the emission spectrum have a large overlap. In some cases, the desired receiver temperature is too high for metals, and the tubes can then be made from ceramics such as SiC.

The HTF in the tubular central receiver can be water/steam, a liquid such as molten salt or liquid sodium, or a gas such as helium or air. Each type of HTF has certain limitations regarding the range of temperatures and pressures over which it can be used. In a direct steam receiver, the receiver is usually separated into two sections: boiler tubes that carry a two-phase flow and can be subjected to higher incident radiation flux and superheater tubes, with a single-phase vapor flow that has lower convective heat transfer and must experience a lower incident flux. In the case of molten-salt HTFs, the flow is always in the liquid phase, but the maximum temperature is limited by the stability of the salt: the common nitrate salt allows up to about 550 °C. The tube diameter is kept relatively small to increase the flow velocity and the convective heat-transfer coefficient, and dividing the HTF flow among many parallel tubes increases the surface area available for convection. Typically, the fluid must make several passes through the receiver before it reaches the required exit temperature.

Volumetric receivers

Achieving very high temperature in the heating of a gas such as air is difficult in a tubular receiver, given that typical tube materials do not have a high thermal conductivity and the high heat flux through the wall of the tube requires a high temperature difference. This increases thermal losses and might exceed the tube material's temperature limitation. A volumetric absorber is a porous structure that is exposed to the concentrated radiation and absorbs radiation through its volume, rather than at an external surface. The absorber transfers heat by convection to the HTF that flows through the pores, as shown in Figure 21.8(b). The volumetric structure provides a large surface area for absorption and for convection, while eliminating the need for conduction heat transfer through the absorber material. The volumetric receiver can be open (in which case the HTF is ambient air drawn into the receiver through an open aperture) or closed (in which case the HTF is mechanically forced into the receiver and the receiver's aperture is sealed by a transparent window).

Materials suitable for a volumetric absorber include high-temperature metals and ceramics. Porous structures for use as volumetric absorbers include metallic

and ceramic foams [7], metallic wire mesh and wire pack, ceramic grids and honeycombs, and ceramic pin-fin arrays [8].

21.4.4 Heat-to-power conversion cycles

The two most common power-conversion cycles in conventional power plants are the steam-based **Rankine cycle** and the air-based open **Brayton cycle** (also called the **gas turbine cycle**) [9]. Schematic representations of these two cycles are shown in Figure 21.9. In both cycles, mechanical power is produced by manipulation of pressure and temperature of the working fluid, namely water or air. In both cycles, the fluid is compressed (from point 1 to point 2 in Figure 21.9) and then heated (from point 2 to point 3) directly or indirectly by a solar receiver. Figure 21.9 also shows the option of hybridizing the

(a)

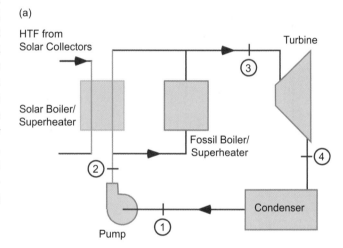

(b)

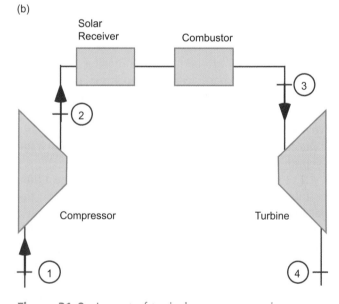

Figure 21.9. Layout of typical power conversion cycles in a solar hybrid plant: (a) Rankine (steam) cycle with an auxiliary fossil fuel boiler, (b) Brayton (gas turbine) cycle with an inline combustor.

cycle, by adding heat from combustion of a fossil fuel in parallel to the solar heat. The hot fluid then expands through a turbine (from point 3 to point 4) and produces mechanical work, which is converted into electricity by a generator. The part of the heat that is not converted to work is discarded to the environment by a condenser (Rankine cycle, point 4 to point 1) or an exhaust stream (open-loop Brayton cycle, point 4). A more detailed description of the operation of these standard cycles can be found in many thermodynamics textbooks. A combination of these two cycles, the **combined cycle** (CC), uses the hot exhaust stream from the gas turbine, point 4 in Figure 21.9(b), to generate the steam for the Rankine cycle, thus exploiting a larger fraction of the heat input and reaching a higher efficiency.

Steam turbines are available in sizes ranging from a few megawatts to hundreds of megawatts, and gas turbines are available in sizes from 60 kW to hundreds of megawatts. Typically, the larger turbines in each category tend to provide higher conversion efficiencies and lower specific costs per megawatt. The combined cycle is available only at the higher end of the size range.

The efficiency of these heat engines in conversion from heat to mechanical work is constrained by the Carnot efficiency:

$$\eta_{\text{Carnot}} = \frac{\dot{W}_{\text{net}}}{\dot{Q}_{\text{in}}} \leq 1 - \frac{T_C}{T_H} = \eta_{\text{Carnot}} \qquad (21.5)$$

where $\dot{W}_{\text{net}}$ is the net mechanical power delivered by the engine, $\dot{Q}_{\text{in}}$ is the thermal power input needed to heat the working fluid, and T_H and T_C are the temperatures at which the heat is supplied ("hot") and removed to the environment ("cold"), respectively. Typically, real heat engines are capable of achieving about 50%–70% of the ideal Carnot efficiency corresponding to their operating temperatures.

Another well-known power-conversion cycle is the **Stirling cycle**, which is currently available only in small engines of up to 25 kW. This cycle theoretically can reach the Carnot efficiency, Equation (21.5), although practical engines fall short of this ideal value. Stirling engines currently offer the highest efficiency available at small scale (>40%), in contrast to Rankine and Brayton turbines, which reach good efficiency only at much larger scales. However, Stirling engines have a reputation for being unreliable, and are currently much more expensive (per unit power) than steam and gas turbines.

Today, all large CSP plants use the Rankine cycle, whereas small dish engine units use the Stirling cycle. Typical operating conditions for a Rankine-cycle CSP plant are a steam pressure of 60–100 bar, a superheated-steam temperature of 390 °C (in a standard trough) to 550 °C (in a tower), and a turbine efficiency of 38%–42%. The overall plant efficiency from solar radiation to electricity, including optical, receiver, and heat-engine losses, can be in the range 15%–20%; CSP plants based on the Brayton cycle may reach higher plant efficiencies of around 25%. The main reason for the absence of gas turbines in current industrial CSP installations is that they require air at elevated pressure (e.g., 3–20 bar) to be heated to very high temperatures (800–1,400 °C) that are difficult to attain with solar power. Thus, CSP plant developers are opting for the less risky steam cycle. However, the use of gas turbines at both small and large scales, as well as the coveted high-efficiency combined cycle, has been gaining interest and is currently under development.

21.4.5 Thermal storage

Solar radiation is an intermittent resource, providing energy at the mercy of local weather, whereas the electric grid is driven by consumer demand and cannot adjust to the availability of sunlight. This mismatch of supply and demand can be bridged in a solar thermal power plant by heat storage: collecting and storing an excess of thermal energy while sunlight is available and using the stored heat to generate electricity upon demand. Heat can be stored as an increase in the internal energy of a storage material, using one of the following three processes: increasing the temperature (sensible heat), causing a phase change from solid to liquid (latent heat), and performing an endothermal chemical reaction (chemical storage, which is discussed in detail in Chapter 48). Extracting the heat for power generation is performed by reversing each of the three processes.

Sensible heat storage is currently the most common method and is usually done using a molten nitrate salt. In plants where the molten salt is also the HTF, storage is simply accomplished by drawing low-temperature salt from a "cold" tank, heating it in the receiver, and storing the hot salt in a hot tank. Heat extraction is performed by drawing hot salt though the steam generator and back to the cold tank. Alternatively, a single tank with hot and cold zones can be used, with the region separating the two zones (the **thermocline**) moving back and forth upon charging and discharging. Storage of molten salt is also used in plants with a different HTF. This requires an additional heat exchanger to transfer the heat from the HTF to the salt. Another method of sensible heat storage is to use a porous matrix of solid storage material, such as crushed rock or ceramic pellets, and to pass the heated HTF through the matrix to transfer heat to the solid and charge the storage. Heat is extracted by passing cold HTF in the reverse direction through the hot solid matrix. This approach has been considered with air [10] and with a molten salt [11] as the HTF. Another variation of solid-medium storage is an ordered structure of HTF pipes passing through a tank filled with a solid medium such as concrete (chosen for low cost) or graphite (chosen for high thermal conductivity).

The latent-heat approach offers the possibility of storing large amounts of heat without a change in temperature. This is especially attractive when storing heat for evaporation of water into steam, which is an isothermal process as well. Several suitable **phase-change materials (PCMs)** with phase-change temperatures relevant for steam power plants (around 300 °C) are available, such as nitrate salts. The main obstacle is the low thermal conductivity of these materials, which severely limits the rate of heat charge and discharge. Current methods to overcome this difficulty include various designs of fins to enhance the heat transfer from the steam pipes into the PCM [12].

Chemical storage is based on a closed loop with an endothermic reaction at one end for heat charging and an exothermic reaction at the other end for heat discharging. Proposed chemical systems include methane reforming with CO_2 and dissociation of ammonia. Further discussion of chemical storage can be found in Chapter 48.

21.4.6 Hybridization

Another approach to address the intermittency of solar radiation is to provide a second heat source that can provide the needed heat when solar heat is insufficient or unavailable. Typically the additional heat is needed during morning startup and during cloud transients, and can be used also to extend operation into evening peak hours. This has been implemented in solar steam plants by introducing a standard fossil-fuel boiler alongside the solar boiler [13], providing a variable amount of steam as needed to complement the solar boiler, as shown in Figure 21.9(a). In solar gas-turbine cycles, hybridization typically involves an inline combustor located between the solar receiver and the turbine, as shown in Figure 21.9(b) [14]. The combustor needs to be adapted to accept high-temperature inlet air without damage and to provide highly variable power down to a minimal level (a pilot flame) or down to zero (which requires hot re-ignition ability). In solar Stirling systems the addition of an auxiliary heat source is more difficult due to the compact integration of the receiver and the engine, but some experimental designs have been proposed and tested for this system as well.

Use of fossil fuels, and in particular natural gas, to hybridize CSP plants has been criticized due to the relatively low conversion efficiency of the solar plants. It is argued that natural gas can be used in conventional CC plants at an efficiency of 55%, and should not be consumed in solar plants at a conversion efficiency of around 40%, both from an economic point of view and from an environmental point of view (the solar plant may end up creating higher emissions due to the lower efficiency). This stresses the obvious need to improve solar power-plant efficiency beyond the current state of the art. The possibility of alternative hybridization with renewable heat sources such as biomass or biomass-derived fuels has been investigated, and this is clearly preferable from the emissions perspective. However, the availability of biomass in regions relevant to CSP is limited, and these alternative sources entail additional costs and additional complexity that need to be addressed.

21.4.7 Deployment of CSP

Geographical distribution

For CSP one requires direct solar radiation, rather than diffuse radiation; therefore, it is suitable for geographical locations where the overall amount of direct radiation is high. Figure 21.10 shows the global distribution of average direct radiation, which is also called beam radiation or **direct normal insolation (DNI)**. Regions with high

Figure 21.10. Map of candidate regions for CSP, based on annual direct insolation, and neighboring regions within a reasonable distance for electricity transmission (adapted from [16]).

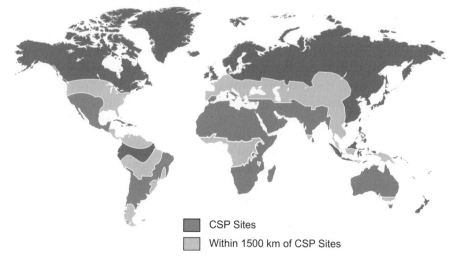

CSP Sites

Within 1500 km of CSP Sites

average DNI values are distributed roughly in two strips around latitudes 20–40° both in the Northern and in the Southern Hemisphere; these strips are often called "sunbelts." The most attractive regions include North Africa, the Mediterranean, and the Middle East through India; the southwestern USA and northern Mexico; and Australia, southern Africa, and some locations in South America. As a general rule, locations where the daily average DNI is higher than 5 kW·h m^{-2} per day (equivalent to 1,800 kW h m^{-2} per year), can be considered as candidates for CSP.

The suitability of a specific location depends not only on the DNI, but also on many other aspects, including topography (a CSP plant requires a large contiguous area of flat or nearly flat land), access (convenient access to transportation and to the electrical grid is needed), availability of water (water is needed for plants that use water cooling), and fuel availability (if plants are designed to operate in hybrid mode, a backup fuel is consumed when sunlight is not available). Each of these site-specific aspects can have a major impact on the feasibility and cost-effectiveness of the CSP plant and must be considered carefully when selecting a site for the plant [15].

CSP economics

The most common economic criterion for evaluating solar plants is the **levelized energy cost (LEC)**. This is the cost to produce one unit of energy (kW·h), when taking into account both one-time expenditures (initial investment costs) and continuing expenses (operation and maintenance). The LEC can be compared across different technologies, including conventional generation with fossil fuels. On the basis of the LEC comparison, policies can be set, for example, regarding which technologies should be supported and how much support is needed for a technology to reach implementation.

The main principle in LEC analysis is discounting all income and expenses to the common basis of present value, using a relevant **discount rate**, which represents the changes in the value of money. The LEC can be computed using the expression

$$\text{LEC} = \frac{\sum_{i=1}^{N} (\text{IC}_i + \text{OM}_i + \text{FC}_i)(1+r)^{-i}}{\sum_{i=1}^{N} E_i (1+r)^{-i}}, \tag{21.6}$$

where r is the discount rate, N is the number of years of plant operation, IC_i is the investment cost incurred in year i (including costs of land, equipment, construction labor, engineering and other services, permits, financing, etc.), OM_i and FC_i are the routine operating and maintenance costs and fuel cost in year i, respectively, and E_i is the net energy produced and sold in year i. This expression assumes that the discount rate is fixed

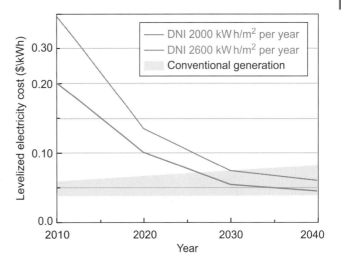

Figure 21.11. Estimate of typical levelized electricity cost for CSP plants, for moderate and high levels of DNI, compared to a cost forecast of conventional electricity generation Adapted from [16].

throughout the lifetime of the plant and that expenses and electricity generation can be predicted with a reasonable accuracy.

Estimated LECs for CSP plants can vary widely, depending on location, technology, and the assumed value of the discount rate. Figure 21.11 shows a forecast of typical values for CSP, based on data corresponding to a large parabolic trough, which is currently the leading CSP technology [16]. The forecast shows a gradual decrease from an LEC of $0.20 per kW·h or higher today to around $0.08–0.11 per kW·h in 2025, and to $0.04–0.05 per kW·h by around 2050. This trend assumes continued technological improvements and increased deployment of large-scale plants, leading to significant progress along the "learning curve" for CSP technology. For comparison, the LECs of power generation with conventional fossil fuels and with nuclear power are around $0.04–0.08 per kW·h, meaning that CSP can become competitive at some high-DNI sites around 2025 and at other sites a few years later. This goal is called **grid parity**, where the cost of solar generation equals that of conventional generation. Other studies provide even lower LEC values, predicting grid parity earlier than 2025 [17].

The LEC of solar power generation is not low enough today for solar energy to be competitive as a commercial product. To bring about the cost reductions presented in Figure 21.11, many governments are employing incentives that bridge the gap between the true generation costs and the electricity-market reality. These incentives can be in the form of investment grants, tax breaks, or a direct subsidy to the price of a solar kilowatt-hour fed into the grid. In some cases, the government might set a target or a mandate for the percentage of electricity to be produced from solar energy, with a penalty to the local utility if it fails to meet this mandate. These incentives are understood to be temporary measures to push solar technology toward grid parity.

Deployment barriers

A CSP plant requires a large area of land compared with a fossil-fuel plant – although this comparison ignores the land area needed for the extraction of fossil fuels. Typical land requirements lead to CSP generation in the range of 30–50 MW km^{-2}, meaning that, in most cases, an insignificant fraction of a country's total land area would be required in order to build CSP plants sufficient to satisfy its energy needs. However, in many countries, most of the land is already allocated to other uses, even when it is still open and uninhabited land. Re-allocation of such areas for solar power plants is a major issue and often runs into public opposition.

The availability of water is another major obstacle, since CSP plants consume water for keeping the concentrators clean and for removal of waste heat (steam condensation), which is frequently done with evaporative cooling towers. Replacing wet cooling by dry cooling (in which turbine exhaust steam is condensed by air-cooled condensers) produces a penalty of a 2%–4% decrease in efficiency and a 5%–8% increase in LEC [18]. Unfortunately, most sites that have high direct radiation are also short of water resources. Finding approaches that provide effective dry cooling is therefore a key issue for CSP.

Another barrier is the need for dispatchable power, that is, the ability to contribute to the grid whenever electricity is needed, regardless of the availability of solar radiation. If the contribution of CSP increased to a significant fraction of the overall generation capacity, then a lack of dispatchability would become a major threat to the stability of the grid. This can be resolved by two approaches: hybridization (using an alternative heat source to operate the solar plant when sunlight is insufficient) and storage (collecting and storing extra heat during daytime, for use during cloud outages or nighttime). Hybridization is currently problematic because turbines in solar plants tend to be less efficient than those in conventional plants, leading to inefficient use of fossil fuel. Cost-effective heat storage is still an emerging technology and has yet to be proven on an industrial scale. Nevertheless, the prospects of developing effective methods for dispatchable power from CSP plants are perceived as a major potential advantage of CSP technologies compared with photovoltaics (in which the solar power is directly converted to electricity), which cannot be hybridized and does not offer reasonable storage options.

21.4.8 Materials technology challenges in CSP

Reflector materials

A CSP concentrator mirror needs to provide high specular reflectivity (>90%) and durability over the lifetime of a plant (>20 years). To date, most of these mirrors have been produced from curved thick glass with a back coating of silver and protective layers behind the silver. This solution requires curving of the glass at high accuracy, which can be done by thermal sagging (for short focal lengths) or elastic deformation (for long focal lengths) and can be expensive for both options. Thin glass (<1 mm) with the same back silver coating is easier to curve and offers a lower cost, but such glasses are fragile and difficult to handle. Several alternative front-surface reflector systems have been proposed as alternatives to glass-based reflectors. One uses an enhanced aluminum or aluminum/silver coating on a polished aluminum substrate, with a top protective layer of anodization or polymer. Another is based on a polymer substrate coated with silver and protected by poly (methyl methacrylate), a transparent weather-resistant acrylic, and with a back adhesive for attaching to a metal substrate. These systems show promise and might reach a reasonable durability approaching 10 years in the field, but this is not yet sufficient to replace the reliable old solution of a glass back-surface reflector [19]. The challenge is then still that of finding a low-cost reflector material system to replace back-silvered glass, providing easy curving and structural rigidity once curved, high specular reflectivity, and long-term durability under tough outdoor conditions.

Heat-transfer fluids

The main difficulty in increasing the operating temperature of a steam CSP system is the HTF. The synthetic oil currently used in trough plants limits the steam temperature to about 390 °C, with a significant impact on cycle efficiency. Direct steam generation allows higher temperatures and is used in tower and linear Fresnel systems, but the application of this complex two-phase flow system in parabolic-trough plants is difficult [5] and has not yet been accepted by industry. The use of a molten nitrate salt (typically a mixture of $NaNO_3$ and KNO_3) has been demonstrated in tower systems, allowing steam temperatures of up to 560 °C. Molten salt has also been proposed for use in troughs [6], and this approach is being tested on a large scale. Using a salt as the HTF also enables relatively easy heat storage without the need to transfer heat to a separate medium. However, the nitrate salt freezes at about 240 °C, leading to major difficulties in salt management and protection against its freezing during system outages. Formulating alternative salts, or other liquids, with a lower freezing temperature is then a major challenge for materials research, considering that the list of requirements also includes stability at high temperature (at least $>500\,°C$), reasonably high thermal conductivity and specific heat, reasonably low viscosity across the temperature range, chemical compatibility with standard tube materials, and obviously very low cost. Other proposals for alternative HTFs include pressurized CO_2 and some ionic liquids, but these are still in the early research phase.

Absorber/receiver materials

The desire to increase the operating temperature of the solar plant, and thereby increase the conversion efficiency, places high demands on the materials used to construct the high-temperature components and, in particular, the receiver. High-efficiency Rankine steam plants are expected to be near-critical or supercritical, requiring temperatures of $>600\,°C$ and pressures in the range 200–300 bar. The materials available for tubes under these conditions include Inconel and similar high-quality stainless steel or nickel alloys, which tend to be very expensive. Tube materials for high-temperature receivers are therefore a major challenge in the development of high-efficiency solar steam plants.

Another challenge is the development of a spectrally selective absorber coating that can operate at high temperature. Today's coatings operate well in the range of 400–500 °C, but they require protection by a vacuum around the receiver tube. Future direct steam systems that operate at $>600\,°C$ in tower-top receivers will not have the luxury of a vacuum envelope, and the coating will have to be stable under an oxidizing and humid atmosphere at high operating temperature. Many materials and structures have been proposed for high-temperature selective coatings, including multilayer and cermet composites [4]. However, a successful solution has yet to be found.

A different set of challenges exists in receivers intended for air (gas turbine) plants, which operate at considerably higher temperatures of 800–1,400 °C. A receiver heating air to this temperature range will also have to maintain a pressure of up to 30 bar. Tubular receivers made from high-temperature metals (e.g., nickel-based) are high in cost and will be able to operate only at the lower end of this range, so ceramic tubes will be needed for higher temperatures. Because another crucial requirement is high thermal conductivity of the tube wall, materials such as SiC have been considered. The list of requirements also includes mechanical strength and robustness under thermal shock and various mechanical loads (wind forces, thermal expansion, etc.). A method of joining and sealing the ceramic elements under these demanding conditions is also needed.

The volumetric receiver approach for air heating to high temperature poses another challenge in materials and mechanics: the need for a transparent window that can sustain the high internal pressure and temperature. Currently, the only practical material known to survive such an environment is fused quartz, and it does so only when shaped in an appropriate convex form that can maintain the internal stresses in compression only [20]. The window is heated mostly by absorption of IR radiation emitted from the volumetric absorber, and if it reaches a high enough temperature, then the material undergoes devitrification (loss of transparency), especially if there are traces of alkali metals or other pollutants in contact with the glass. The challenges therefore include delaying devitrification to a higher temperature; reducing heating of the window (e.g., by use of a selective coating that rejects IR radiation); and possibly employing alternative glassy materials, if suitable materials could be found at reasonable cost.

Storage materials

Currently, the most common thermal storage method utilizes the sensible heat of a molten nitrate salt in the range of 280–560 °C. This range is too restricted: the lower limit should be reduced to prevent inadvertent freezing of the salt, and the upper limit should be increased to allow cycles operating at higher temperature and higher conversion efficiency. Alternative materials and mixtures (eutectics) with a broader range of liquid phase are sought, subject to the additional requirements of a reasonably high specific heat and thermal conductivity, a reasonably low viscosity, stability under repeated temperature cycling, and chemical compatibility with pipe and tank materials such as stainless steel.

Thermal storage in a solid medium requires materials that offer a combination of high thermal conductivity and specific heat, stability under thermal cycling, and low cost. Materials in use today address these requirements only partially: concrete and crushed rock offer low cost but low thermal conductivity; graphite [12] offers excellent thermal conductivity but at higher cost. Finding better materials, as well as developing better heat-transfer methods for charging and discharging the storage medium, are current topics of active research.

Thermal storage in a PCM for steam generation requires a solution to the main challenge of increasing the thermal conductivity of the storage material. Some gains have been made using the approach of a composite PCM: mixing the PCM with particles of high-conductivity metals or graphite, leading to materials similar to composite PCMs based on paraffin with graphite flakes for low-temperature thermal storage in applications such as solar water heating and increasing the thermal mass of building walls. Significant challenges remain in ensuring the stability and homogeneity of the composite material under thermal cycling, handling volume changes, further increasing the thermal conductivity, and, of course, reducing the costs to a range acceptable by the power industry.

21.5 Solar water heating

Domestic water heating (DWH) systems using solar energy are very common in many countries. By the end of 2008 about 150 GW of solar water-heating systems had been installed worldwide, and annual growth of this market was around 20% [21] (Chapter 36, Section 36.5.3). Solar water heaters are usually simple, inexpensive and reliable, and offer excellent economic return when installed in regions with abundant solar radiation. Three types of solar water heaters are common: vacuum tubes, glazed flat-plate collectors, and unglazed flat-plate collectors.

Solar water-heating collectors include pipes carrying the water to be heated, with either forced flow (using a pump) or natural convection flow (thermosyphoning – water that is heated rises due to lower density, making way for new colder water to enter the pipe). The external surface of the pipes is coated to produce high absorption of solar radiation, and is often attached to absorber plates that increase the area of intercepted and absorbed solar radiation. The water that is heated in the pipes is usually stored in an insulated tank and can be used on demand. A solar water-heating system consists then of an array of collectors, a storage tank, an optional pump, and additional optional elements such as an auxiliary heater (typically electric or natural-gas-fired) to provide hot water when solar heating is not sufficient. A typical layout of a solar domestic water-heating system is shown in Figure 21.12. In some climates it is necessary to protect the collector system against freezing of water in the pipes during winter by adding an antifreeze component, and therefore the circuit between the collector and the storage tank is separated from the user circuit and operated as a closed circuit, and the heat is transferred to the water in the user's tank via a coiled-pipe heat exchanger.

The efficiency of a water-heating collector is determined by several loss mechanisms: reflection of incident radiation, convection from the exposed hot surface to the ambient air, thermal emission of radiation from the hot surface, and, if some of the surface area is insulated, conduction through the insulation. The collector efficiency varies according to the following model:

$$\eta = F_{\mathrm{R}}\left(\eta_0 - U_{\mathrm{L}} \cdot \frac{T_{\mathrm{Ci}} - T_{\mathrm{a}}}{I}\right), \tag{21.7}$$

where F_{R} is the collector's heat-removal factor, η_0 is the collector's optical efficiency (representing reflection losses), U_{L} is an overall heat-loss coefficient (representing convection, conduction and thermal-emission losses), T_{Ci} and T_{a} are the collector-inlet and ambient temperatures, respectively, and I is the radiation flux incident on the collector. The ratio of temperature difference to incident radiation is often denoted by X and then the collector efficiency is expressed as a linear function of X.

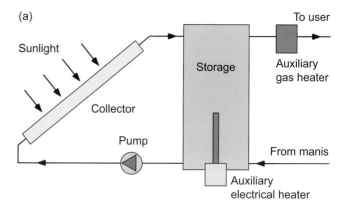

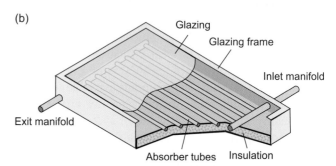

Figure 21.12. (a) Layout of a typical solar domestic water heating system with forced circulation and auxiliary heaters. (b) Structure of a flat plate glazed collector. (c) Solar water heating systems with flat plate collectors on a rooftop.

Collectors based on vacuum tubes employ two principles to reduce the main losses and increase the efficiency: spectrally selective surface coatings to reduce reflection and emission losses, and a vacuum envelope to reduce convection losses. This is similar to the design of evacuated-tube receivers in parabolic-trough solar power plants. Flat-plate glazed collectors employ a similar spectrally selective coating, but instead of vacuum they use an air gap between the collector plate and a parallel glass plate to suppress convection losses (Figure 21.12). Unglazed collectors have an unprotected absorber exposed to the environment and incurring higher losses, and are therefore used only at lower temperatures for applications such as heating of water in swimming pools. Water-heating collectors are usually stationary and do not track the apparent movement of the Sun, in order to keep down their cost and complexity; therefore, the amount of solar radiation that they collect is lower than for tracking collectors. Stationary collectors are usually installed with a slope toward the equator of approximately the latitude angle, in order to optimize the amount of radiation collected over a year.

21.6 Summary

The term CSP describes a group of solar thermal power-plant technologies that are based on the concentration of sunlight, conversion of the radiation to thermal energy at high temperature, and conversion of the heat to electricity in a heat engine. Some CSP plants are already operating on a large scale in Spain and the USA, and they are under development in other countries. Currently, CSP electricity offers moderate conversion efficiencies and uncompetitive costs and therefore requires government incentives. Many opportunities are available for improving CSP: increasing the operating temperature, which, in turn, would increase the conversion efficiency and reduce specific costs (per unit energy); reducing component and plant costs by developing improved materials and manufacturing practices; and adding thermal storage to provide true dispatchability. These improvements require major advances in materials, thermodynamics, heat transfer, and other disciplines, as well as further development and demonstration. It is expected that improved CSP technologies will reach grid parity around 2025, leading to implementation on a much larger scale.

21.7 Questions for discussion

1. What are the possible advantages and disadvantages of the different concentration methods: parabolic trough, linear Fresnel, parabolic dish, and tower?
2. Is it preferable to develop the less expensive low-temperature, low-efficiency CSP technologies or the more expensive high-temperature, high-efficiency alternatives? Why?
3. What are the upper limits to the operating temperature of a solar thermal receiver due to material limitations? Discuss working fluids (water, salt, air) and structural materials (metal, ceramic, glass) and suggest alternative materials that might increase the temperature limits.
4. What are the advantages and disadvantages of thermal storage within a CSP plant, compared with storage of electricity outside the plant (e.g., by batteries)?
5. Hybrid solar power plants, where a significant part of the energy is provided by fossil fuels, are currently discouraged by solar incentive schemes. Is this a good policy? What are the advantages of the hybrid approach compared with purely solar plants?
6. If a photovoltaic (PV) system can convert sunlight to electricity at an efficiency of 25% (which is possible with concentrating PV), are CSP plants that operate at 20% efficiency still needed? Explain.

21.8 Further reading

- **D. R. Mills**, 2001, "Solar thermal electricity," in *Solar Energy. The State of the Art*, ed. **J. Gordon**, London, James & James, pp. 577–651. A chapter reviewing the state of CSP as it was about a decade ago.
- **R. Winston**, 2001, "Solar concentrators," in *Solar Energy. The State of the Art*, ed. **J. Gordon**, London, James & James, pp. 357–436. A chapter on solar concentrator optics including concentrators that are used in CSP.
- **A. Rabl**, 1985, *Active Solar Collectors and their Applications*, Oxford, Oxford University Press, 1985. A basic text on solar energy and solar collector technologies.
- **D. R. Mills**, 2004, "Advances in solar thermal electricity technology," *Solar Energy*, **76**, 19–31. A review article on CSP technologies and R&D directions.
- **A. Gil**, **M. Medrano**, **I. Martorell** *et al.*, 2010, "State of the art on high temperature thermal energy storage for power generation. Part 1 – concepts, materials and modellization," *Renewable Sustainable Energy Rev.*, **14**, 31–55. A review article on thermal storage technologies and R&D directions.
- http://www1.eere.energy.gov/solar/csp_program. html. Information on CSP technologies and on CSP research and development at the US Department of Energy.
- http://www.solarpaces.org/. Information on the CSP industry, projects, and R&D worldwide, provided by the SolarPACES group of the International Energy Agency.

21.9 References

[1] **I. Reda** and **A. Andreas**, 2004, "Solar position algorithm for solar radiation applications," *Solar Energy*, **76**, 577–589.
[2] **R. Winston**, **J. C. Minano**, and **P. G. Benitez**, 2005, *Nonimaging Optics*, New York, Academic Press.
[3] **A. Fernandez-Garcia**, **E. Zarza**, **L. Valenzuela**, and **M. Perez**, 2010, "Parabolic-trough solar collectors and their applications," *Renewable Sustainable Energy Rev.*, **14**, 1695–1721.
[4] **C. E. Kennedy**, 2002, *Review of Mid- to High-Temperature Solar Selective Absorber Materials*, NREL, http://www.nrel. gov/docs/fy02osti/31267.pdf

[5] **E. Zarza** *et al.*, 2004, "Direct steam generation in parabolic troughs final results and conslusions of the DISS project," *Energy*, **29**, 635–644.

[6] **D. Kearney** *et al.*, 2004, "Engineering aspects of a molten salt heat transfer fluid in a trough solar field," *Energy*, **29**, 861–870.

[7] **T. Fend**, **R. Pitz-Paal**, **O. Reutter**, **J. Bauer**, and **B. Hoffschmidt**, 2004, "Two novel high-porosity materials as volumetric receivers for concentrated solar radiation," *Solar Energy Mater. Solar Cells*, **84**, 291–304.

[8] **J. Karni**, **A. Kribus**, **R. Rubin**, and **Doron P.**, 1998, "The 'Porcupine': a novel high-flux absorber for volumetric solar receivers," *J. Solar Energy Eng.*, **120**, 85–95.

[9] **W. B. Stine** and **M. Geyer**, 2001, *Power From the Sun*, http://www.powerfromthesun.net/book.htm.

[10] **H. W. Fricker**, 2004, "Regenerative thermal storage in atmospheric air system solar power plants," *Energy*, **29**, 871–881.

[11] **J. E. Pacheco**, **S. K. Showalter**, and **W. J. Kolb**, 2002, "Development of a molten-salt thermocline thermal storage system for parabolic trough plants," *J. Solar Energy Eng.*, **124**, 1–7.

[12] **W. D. Steinmann** and **R. Tamme**, 2008, "Latent heat storage for solar steam systems," *J. Solar Energy Eng.*, **130**, 011004–1.

[13] **L. Schnatbaum**, 2009, "Solar thermal power plants," *Eur. Phys. J. Special Topics*, **176**, 127–140, doi:10.1140/epjst/e2009–01153–0.

[14] **P. Schwarzboezl**, **R. Buck**, **C. Sugarmen**, **A. Ring**, **M. J. Crespo**, **P. Altwegg**, and **J. Enrile**, 2006, "Solar gas turbine systems: design, cost and perspectives," *Solar Energy*, **80**, 1231–1240.

[15] **C. Breyer** and **G. Knies**, 2009, "Global energy supply potential of concentrating solar power," in *SolarPACES 2009*, Berlin.

[16] **C. Philibert**, 2010, http://www.iea.org/papers/2010/csp_roadmap.pdf.

[18] **M. J. Wagner** and **C. F. Kutscher**, 2010, "Assessing the impact of heat rejection technology on CSP plant revenue," in *SolarPACES 2010*, Perpignan, France, available at http://www.nrel.gov/docs/fy11osti/49369.pdf.

[17] **L. Stoddard**, **J. Abiecunas**, and **R. O'Connell**, 2006, *Economic, Energy, and Environmental Benefits of Concentrating Solar Power in California*, NREL.

[19] **C. E. Kennedy** and **K. Terwilliger**, 2005, "Optical durability of candidate solar reflectors," *J. Solar Energy Eng.*, **127**, 262–269.

[20] **J. Karni**, **A. Kribus**, **B. Ostraich**, and **E. Kochavi**, 1998, "A high-pressure window for volumetric solar receivers," *J. Solar Energy Eng.*, **120**, 101–107.

[21] **W. Weiss** and **F. Mauthner**, 2010, *Solar Heat Worldwide: Markets and Contribution to the Energy Supply 2008*, IEA Solar Heating & Cooling Programme.

22

Solar thermoelectrics: direct solar thermal energy conversion

Terry M. Tritt,[1] Xinfeng Tang,[2] Qingjie Zhang,[2] and Wenjie Xie[2]

[1]Department of Physics and Astronomy, Clemson University, Clemson, SC, USA
[2]Wuhan University of Technology, Wuhan, China

22.1 Focus

The Sun's radiation can be modeled as a 6,000-K blackbody radiator. Whereas photovoltaics (PV) can convert the part of the Sun's spectrum to electrical energy, over 40% of that spectrum, namely, the infrared (IR) range, is lost as heat. In solar thermoelectrics (TE), the thermal energy from the IR range is converted directly into electricity. Therefore, a solar PV–TE hybrid system would have access to the entire spectrum of the Sun.

22.2 Synopsis

With respect to solar energy conversion, PV devices utilize the UV region, whereas TE devices utilize the IR region (which is waste heat with respect to the PV devices) to generate electricity. In a solar PV–TE hybrid system, a high-efficiency solar collector would turn the sunlight (from the IR spectrum) into heat that would then be transformed by TE devices into usable electricity. In addition, the solar thermal energy could be stored in a thermal bath, or TE devices could be used to charge batteries that could then provide electricity when the Sun was not shining. Such a TE system would need to operate at around 1,000 K (~ 700 °C), and the materials would need to exhibit high ZT values around this temperature.

22.3 Historical perspective

The field of thermoelectricity began in the early 1800s with the discovery of the **Seebeck effect**, one of three reversible phenomena known collectively as **thermoelectric effects**. Specifically, German physicist Thomas Seebeck [1] found that, when the junctions of two dissimilar materials are held at different temperatures, a magnetic needle was deflected. It was later learned that an electrical current is generated in this configuration, which increases with increasing difference in temperature (ΔT). This effect, which is called the "Seebeck effect," can be measured as a voltage (ΔV) that is proportional to ΔT, i.e., $\Delta V = -\alpha \, \Delta T$, where the proportionality constant α is called the Seebeck coefficient or thermopower. In fact, the thermocouple (see, e.g., [2] for a list of thermocouple tables and materials), a simple temperature sensor consisting of two dissimilar metals joined together at one end, is based on the Seebeck effect in an open-loop configuration. When the loop is closed, this couple allows the direct conversion of thermal energy (heat) into useful electrical energy. The conversion efficiency, η_{TE}, is related to a dimensionless quantity called the thermoelectric figure of merit, ZT, which is determined from three main material parameters: the thermopower (α), the electrical conductivity (σ), and the thermal conductivity (κ). Heat is carried by both electrons (electronic thermal conductivity κ_{e}) and **phonons** (phonon thermal conductivity κ_{ph}), with $\kappa = \kappa_{\mathrm{e}} + \kappa_{\mathrm{ph}}$. ZT itself is defined as

$$\mathrm{ZT} = \frac{\alpha^2 \sigma T}{\kappa_{\mathrm{e}} + \kappa_{\mathrm{ph}}} \qquad (22.1)$$

and the thermoelectric efficiency, η_{TE}, is given by

$$\eta_{\mathrm{TE}} = \eta_{\mathrm{C}} \left(\frac{\sqrt{1 + \mathrm{ZT}} - 1}{\sqrt{1 + \mathrm{ZT}} + T_{\mathrm{C}}/T_{\mathrm{H}}} \right) \qquad (22.2)$$

where η_{C} is the Carnot efficiency, $\eta_{\mathrm{C}} = 1 - T_{\mathrm{C}}/T_{\mathrm{H}}$, with T_{H} and T_{C} being the hot and cold temperatures, respectively, with T in Equation (22.1) being the average of the two. However, Equation (22.2) does not take into account the effect the various parasitic losses such as those due to contact resistance, thermal heat sinking, and radiation of the materials under a large ΔT. Instead, it considers only the ZT of the materials, not the effective ZT of the device in question. Furthermore, in addition to a high ZT value, a significant temperature difference is also needed in order to generate sufficient electrical energy; and fortunately, as can be seen in Figures 22.1 and 22.2, the IR region of the solar spectrum can supply the needed high temperature, T_{H}.

It was not until the mid 1900s, with the advent of semiconductor materials research, that TE materials and devices became important [3][4] Semiconducting materials permit band tuning and control of the carrier

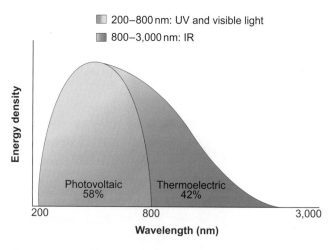

Figure 22.1. The Sun radiates energy as a 6,000-K blackbody radiator, where part of the energy is in the UV-visible spectrum and part in the IR spectrum.

concentration, thus allowing optimization of a given set of materials. A TE couple is made up of n-type and p-type materials, and, in typical TE devices, numerous couples are then connected electrically in series and thermally in parallel. Thermoelectric devices convert thermal gradients directly into electricity and operate by a solid-state conversion process that is quiet, has no mechanical parts, and provides long-term stability. The devices can be used either for cooling (the Peltier effect) or for power generation (the Seebeck effect). Thus, heat (typically waste heat) can be converted directly into useful electrical energy. In a pure PV device the IR spectrum is therefore waste heat that is not being utilized by the device. The solar–TE device would take advantage of that heat and convert it into electrical energy.

Thermoelectric applications are quite broad. Thermoelectric materials had their first decisive long-term test with the start of intensive space research. For example, TE materials were responsible for the power supply for the Voyager missions. Furthermore, radioisotope TE generators (RTGs) are the power supplies (~350 W) for all deep-space missions beyond Mars. Recently, the Cassini satellite was launched and had three RTGs using ^{238}Pu as a thermal energy source and SiGe as the TE conversion material, [5]. Earth-orbit satellites can use solar panels to generate electricity because of their much closer proximity to the Sun and are able to take advantage of the Sun's radiation, whereas the deep-space missions cannot. In addition, smaller self-powered systems, such as TE-powered radios using a kerosene lamp as the heat source were manufactured in Russia around the mid 1900s. Currently, millions of TE climate-controlled seats are being installed in luxury cars to serve either as a seat coolers and as a seat warmers. Furthermore, millions of TE coolers are able to provide us with cold beverages. Even wrist watches

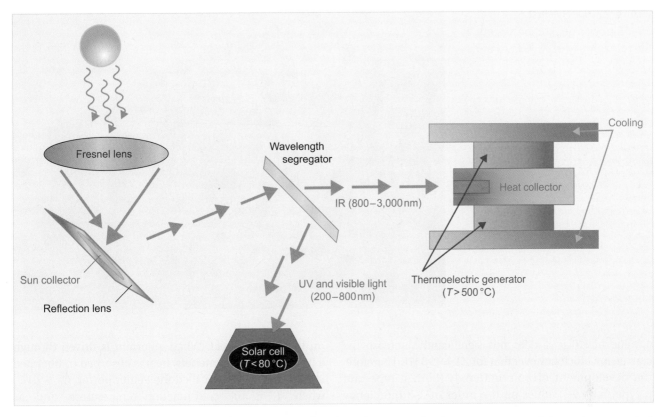

Figure 22.2. After the solar energy has been concentrated, it can be converted into electricity both by photovoltaics (the UV spectrum) and by thermoelectrics (the IR spectrum).

marketed by Seiko and Citizen and bio-TE-powered pacemakers are powered by the very small temperature differences within the body or between body heat and the ambient temperature are also being considered to serve as power supplies for remote sensors in many areas where temperature gradients exist and they can operate without the need of excess electrical wiring. Therefore, TE devices and technologies are receiving a much broader acceptance as potential power sources in areas that had been previously under utilized.

Current state-of-the-art TE materials have a value of $ZT \approx 1$, and the corresponding TE devices exhibit low efficiency (~7%–8%) [6] and have been used primarily in niche markets. However, much broader applications are opening up. For example, TE devices are being investigated for such applications as **waste-heat-recovery** systems that derive benefit from an automobile's exhaust to provide additional electrical energy to the automobile and increase the overall efficiency. Even though the conversion efficiency is low, this technology would allow low-quality waste heat to be turned into useful electrical energy. New materials with ZT values of 2–3 are needed in order to provide the desired conversion efficiencies. For example, a TE power-conversion device with $ZT = 3$ operating between 500 and 30 °C (room temperature) would yield about 50% of the Carnot efficiency. As shown in Figure 22.3,

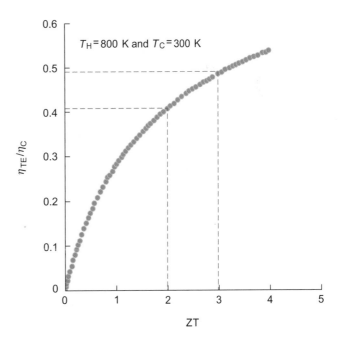

Figure 22.3. The ratio of thermoelectric efficiency to Carnot efficiency (η_{TE}/η_C) as a function of the figure of merit, ZT. The maximum efficiency, or Carnot efficiency, for a TE system operating between 500 and 30 °C (approximately 800 and 300 K, respectively) is given by $\eta_C = (T_H - T_C)/T_H = (800\ \text{K} - 300\ \text{K})/800\ \text{K} = 63\%$.

Figure 22.4. The first hybrid PV–TE power-generation system in China (courtesy of X. F. Tang and Q. Zhang).

a value of ZT > 4 does not significantly increase the conversion efficiency over that for ZT ≈ 2–3 [7]. Therefore, the development of bulk materials (both n-type and p-type) with ZT values on the order of 2–3 (efficiency ≈15%–20%) that exhibit low parasitic losses (i.e., contact resistance, radiation effects, interdiffusion, etc.), have low manufacturing costs, and consist of elements that are readily abundant, would mark a significant achievement for TE technologies. Furthermore, if TE generators could achieve efficiencies of 15%–20% (ZT ≈ 2–3), then they would probably be manufactured and marketed primarily as self-contained units and likely not in conjunction with a higher-cost solar–TE hybrid unit.

Current research into nanostructured materials provides hope that significant increases in the figure of merit and, thus, the overall TE efficiency might be possible within the next decade. Several groups have observed that more than a 50% improvement in p-type BiSbTe can be achieved by forming the materials, by either melt spinning or ball milling, such that they have an inherent nanostructure that exists within the bulk material. The nanostructures lead to an overall reduction in the thermal conductivity with little or no degradation of the electronic properties, thus leading to an increase in the ZT value [7][8].

22.4 Solar thermoelectric hybrid systems: challenges and opportunities

Discussions about using solar–TE hybrid systems have been going on for over 50 years [9], soon after the discovery of Bi_2Te_3 TE materials [4] and the identification of other semiconductor materials [5]. In addition, TE materials can be used in devices for solid-state refrigeration based on the **Peltier effect**. When a current is driven through a junction of two materials, heat is absorbed or liberated at that junction. It is called the Peltier power, $\dot{Q}_P = \alpha I T_c$, where T_c is the cold-junction temperature, and, as one can see, it is proportional to the current. There have also been discussions about using solar energy to power these TE coolers for refrigeration applications, for example they could also be used to cool the solar panels themselves to make them more efficient. [10][11]

Researchers in China hope to eventually produce solar PV–TE hybrid systems that would be capable of generating several kilowatts. The first hybrid PV–TE power-generation system has been jointly developed by China and Japan. This system has been serviced for more that three years in Weihai city, Mongolia province. A photo of such a system is shown in Figure 22.4. Many researchers in China are extensively researching higher-efficiency solar cells as well as higher-efficiency TE materials and devices.

If a large number of TE generator or refrigerator systems were needed in the marketplace, then one would have to use materials other than Pb- or Te-based TE materials. This is due to the harmful effects of Pb and the scarcity of Te materials. Thus, many researchers are working on materials based on Mg, Si, Ge, and Sn alloys [12]. Typically, the SiGe materials used in deep-space probes have a composition somewhere between $Si_{70}Ge_{30}$ and $Si_{80}Ge_{20}$. Recently, researchers at Massachusetts Institute of Technology and Boston College were able to develop much-higher-efficiency p-type thermoelectric silicon-based materials by forming precursor nanomaterials using high-energy ball-milling procedures, and these materials required a much lower concentration of the more expensive Ge, specifically $Si_{95}Ge_5$ [13]. These SiGe nanocomposites exhibited a

Figure 22.5. Development of thermoelectric materials over the past 60 years shown in a timeline plot of ZT_{Max} versus year. The development of early materials is illustrated, and one can see that the recent successes in the past 15 years have yielded both thin-film and bulk materials that are able to break the $ZT \approx 1$ ceiling that had been in place for over 40 years.

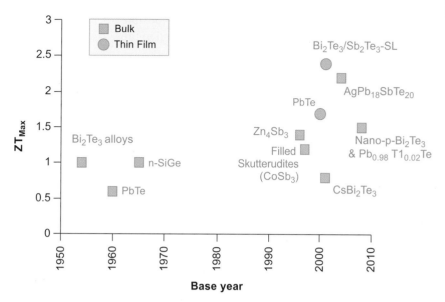

ZT value approximately twice that of the current state-of-the-art bulk p-type $Si_{1-x}Ge_x$ materials.

22.4.1 The future of thermoelectric materials

The future expansion of TE energy-conversion technologies is tied primarily to higher performance of TE materials, together with better thermal management of TE modules. Efficient thermal management and heat-sinking technologies and challenges are especially critical in many TE applications, and this is especially so in the solar–TE hybrid devices. Over the past 15 or so years, significant progress has been made in the field of TE materials. This is illustrated in Figure 22.5, where the timeline of ZT_{Max} for various TE materials is shown. One can see that the recent successes in the past 15 years have yielded both thin-film and bulk materials that have been able to break the $ZT \approx 1$ ceiling that had been in place for over 40 years. This progress is based on two main themes. First, in the early 1990s, American physicist Glen Slack contended that the best TE material would behave as a so-called "phonon-glass–electron crystal," i.e., electrons would be minimally scattered, as in a crystalline material, whereas phonons would be highly scattered, as in an amorphous material [14][15]. Materials researchers are now investigating several systems of materials, including typical narrow-gap semiconductors, oxides, and cage-structure materials (skutterudites and clathrates); see, e.g., [16]. Second, in the mid 1990s, Hicks and Dresselhaus contended that more exotic structures that exhibit reduced dimensionality could yield higher ZT values than those of their bulk counterparts as a result of quantum-confinement enhancement of the Seebeck coefficient and reduced thermal conductivity due to interface scattering [17]. Therefore, nanostructured TE materials, including

superlattices, quantum dots, and nanodot bulk materials, have been the focus of many recent research efforts [18][19][20][21]. Certainly, one of the more recent paradigms has been the pursuit of high-efficiency TE nanocomposites, a mixture of nanomaterials or nanostructures that can exist within the framework of a bulk matrix, thus being something of a combination of both the phonon-glass–electron-crystal concept and the low-dimensional-materials concept. The emerging field of these TE nanocomposites appears to be one of the most promising recent research directions. A very important aspect will be the stability of the nanostructures in these bulk TE nanocomposites if they are to operate effectively at elevated temperatures.

22.5 Summary

The most stable, long-term, and readily available worldwide energy source for our future energy demands is solar energy. Indeed, the field of solar TE is well positioned to contribute to future energy supplies, in both stand-alone and hybrid systems. Over the past decade, TE materials have seen increases in ZT by a factor of 2 over previous materials. Another 50% increase in the value of ZT (to $ZT \approx 3$) with the appropriate material characteristics and costs will position TE materials and devices as a significant contributor to future energy needs, especially in solar energy conversion.

22.6 Questions for discussion

1. Why must one be concerned about the stability of nanostructured materials, if they are to be operated at high temperature, say, $T > 800$ K?
2. Why would one expect to observe an enhanced Seebeck effect in low-dimensional materials?

3. What material parameter was primarily responsible for the enhanced ZT values that were reported in the low-dimensional TE materials and nanocomposites discussed herein?

4. What is the limit of the TE efficiency if ZT $\gg$ 1? Why?

5. Why is a value of ZT $\approx$ 2–3 regarded as the desired goal in bulk TE materials? Why not ZT $\approx$ 10?

6. What are the relative availabilities of some of the elements that go into the TE materials discussed herein? Would a large market be able to be sustained for, say, Te-based TE materials?

22.7 Further reading

- **G. S. Nolas**, **J. Sharp**, and **H. J. Goldsmid**, 2001, *Thermoelectrics: Basic Principles and New Materials Development*, New York, Springer. This book reviews the basic TE principles and electrical and thermal transport phenomena in TE materials, as well as the efficiency and operation of TE devices and a selection of optimization strategies for new materials developments.

- **D. M. Rowe**, (ed.) 2006, *Thermoelectrics Handbook: Macro to Nano*, Basingstoke, Taylor & Francis. This book, with over 40 chapters written by prominent TE researchers from around the world, reviews the basic TE principles and electrical and thermal transport phenomena in TE materials, including many of the most recent developments and new materials.

- **T. M. Tritt** and **M. A. Subramanian** (eds.), 2006, "Harvesting energy through thermoelectrics: power generation and cooling," *MRS Bull.* **31**(3), 188–229 (six articles and all references therein), available at the Materials Research Society website (www.mrs.org). This focused *MRS Bulletin* contains a summary of TE phenomena, plus reviews of bulk, low-dimensional, and nanoscale TE materials, as well as an article on space and automotive applications.

22.8 References

[1] **T. J. Seebeck**, 1823, "Magnetische Polarisation der Metalle und Erze durch Temperatur-Differenz," *Abh. K. Akad. Wiss. Berlin*, 265.

[2] **Omega Engineering**, *Thermocouples – An Introduction*, http://www.omega.com/prodinfo/thermocouples.html.

[3] **H. J. Goldsmid** and **R. W. Douglas**, 1954, "The use of semiconductors in thermoelectric refrigeration," *Brit. J. Appl. Phys.*, **5**, 386–390.

[4] **A. F. Ioffe**, 1957, *Semiconductor Thermoelements and Thermoelectric Cooling*, London, Infosearch.

[5] **J. Yang** and **T. Caillat**, 2006, "Thermoelectric materials for space and automotive power generation," *MRS Bull.*, **31**(3), 224–229.

[6] **T. M. Tritt** and **M. A. Subramanian**, 2006, "Thermoelectric materials, phenomena and applications: a bird's eye view," *MRS Bull.*, **31**(3), 188–194.

[7] **B. Poudel**, **Q. Hao**, **Y. Ma**, *et al.*, 2008, "High-thermoelectric performance of nanostructured bismuth antimony telluride bulk alloys," *Science*, **320**, 634–638.

[8] **W. Xie**, **X. F. Tang**, **Y. Yan**, **Q. Zhang**, and **T. M. Tritt**, 2009, "Unique low-dimensional structure and enhanced thermoelectric performance for p-type $Bi_{0.52}Sb_{1.48}Te_3$ bulk material," *Appl. Phys. Lett.*, **94**, 102111–102113.

[9] **M. Telkes**, 1954, "Solar thermoelectric generators," *J. Appl. Phys.*, **25**, 765–777.

[10] **Y. J. Dai**, **R. Z. Wang**, and **L. Ni**, 2003, "Experimental investigation on a thermoelectric refrigerator driven by solar cells," *Renewable Energy*, **28**, 949–959.

[11] **D. S. Kim** and **C. A. Infante Ferreira**, 2008, "Solar refrigeration options – a state-of-the-art review," *Int. J. Refrigeration*, **31**(1), 3–15.

[12] **D. M. Rowe** (ed.), 2006, *Thermoelectrics Handbook: Macro to Nano*, Basingstoke, Taylor & Francis.

[13] **G. Joshi**, **H. Lee**, **L. Hohyun** *et al.*, 2008, "Enhanced thermoelectric figure-of-merit in nanostructured p-type silicon germanium bulk alloys," *Nanoletters*, **8**, 4670–4674.

[14] **G. A. Slack**, 1995, Chapter 34, in *CRC Handbook of Thermoelectrics*, ed. **D. M. Rowe**, Boca Raton, FL, CRC Press, pp. 407–440.

[15] **G. A. Slack** and **V. Tsoukala**, 1994, "Some properties of semiconducting $IrSb_3$," *J. Appl. Phys.*, **76**, 1665–1671.

[16] **G. S. Nolas**, **S. J. Poon**, and **M. Kanatzidis**, 2006, "Recent developments in bulk thermoelectric materials," *MRS Bull.*, **31**(3), 199–205.

[17] **L. D. Hicks** and **M. S. Dresselhaus**, 1993, "Effect of quantum-well structures on the thermoelectric figure of merit," *Phys. Rev. B*, **47**(19), 12727–12731.

[18] **R. Venkatasubramanian**, **E. Siivola**, **T. Colpitts**, and **B. O'Quinn**, 2001, "Thin-film thermoelectric devices with high room-temperature figures of merit," *Nature*, **413**(6856), 597–602.

[19] **T. C. Harman**, **P. J. Taylor**, **M. P. Walsh**, *et al.*, 2002, "Quantum dot superlattice thermoelectric materials and devices," *Science*, **297**(5590), 2229–2232.

[20] **M. S. Dresselhaus**, **G. Chen**, **M. Y. Tang** *et al.*, 2007, "New directions for low-dimensional thermoelectric materials," *Adv. Mater.*, **19**, 1043–1053.

[21] **M. Kanatzidis**, 2010, "Nanostructured thermoelectrics: the new paradigm?," *Chem. Mater. Rev.*, 12 pages.

23 Off-grid solar in the developing world

Tiffany Tong,[1] Wali Akande,[1] and Winston O. Soboyejo[2]

[1] Princeton Institute of Science and Technology of Materials (PRISM) and Department of Electrical Engineering, Princeton University, Princeton, NJ, USA

[2] Princeton Institute of Science and Technology of Materials (PRISM) and Department of Mechanical and Aerospace Engineering, Princeton University, Princeton, NJ, USA; and Department of Materials Science and Engineering, The African University of Science and Technology, Abuja (AUST-Abuja), Federal Capital Territory, Abuja, Nigeria

23.1 Focus

This chapter introduces a wide array of sustainable approaches that use solar energy technology to address challenges faced by communities in developing regions of the world. Special emphasis is placed on how rural electrification efforts can provide communities with an off-grid power supply that can stimulate technological development through the improvement of a wide range of resources including infrastructure, health care, and education.

23.2 Synopsis

In many areas of the developing world, even basic energy access is still a privilege, not a right. Excessive demand from rapid urbanization often leads to unreliable electricity supplies in the urban areas. However, the problem is most acute in rural and nomadic communities, where lower population densities and income levels make it less practical to establish the necessary infrastructure to connect these communities to the electrical grids that power the cities and megacities. In this context, there is a growing need for innovative energy solutions tailored to the particular needs and demands of these communities, if wide-scale **rural electrification** that extends energy access to rural regions is to be possible.

Consequently, many communities in developing regions are turning increasingly to solar energy as a means of providing off-grid energy access to their rural populations. Improved solar-cell efficiencies and falling prices have combined with increased innovation to provide a wide array of commercially available products that can be used on a personal, household, or community-wide scale.

In most parts of the developing world, the initial efforts to develop applications of solar energy focused largely on the use of conventional solar cells that were connected to batteries, controllers, and inverters. However, such systems were often too expensive for local people, who often earn between $1 and $2 per day. Furthermore, many aid agencies and non-governmental organizations often gave people the systems, without requiring payment for use or ownership of the systems. This resulted in poor attitudes to the maintenance and upkeep of such systems, leading ultimately to the failure of many well-intentioned programs.

Consequently, the focus has shifted recently to the development of affordable solar-powered devices that address aspects of peoples' needs. These include the scaling of solar power to match the income and needs of people regarding lighting, mobile-phone chargers, refrigeration, and street lighting. Popular products include solar-powered flashlights and lanterns, mobile-phone chargers, and street lights. Such products can be sold to people and local governments in rural areas, in ways that develop a sense of appreciation of the solar-energy-based solutions that address the basic needs of people. They can also be linked to social-entrepreneurship-based strategies that are sustainable.

Although solar technology might not yet be mature enough to replace traditional fossil fuels on a worldwide scale, it has already proven to be an important stepping stone that can meet the modest energy demands of those who are most in need of basic energy access. This is a promising development, since the performance of solar technology will continue to improve and can be integrated with other energy sources to eventually provide a comprehensive and holistic approach to rural electrification. This chapter will provide a holistic perspective of a potential framework for the application of solar-energy-based products in rural parts of the developing world.

23.3 Historical perspective

The Industrial Revolution in the late eighteenth and early nineteenth centuries had a massive impact on the degree of technological efficiency with which societies operate and function. The incorporation of heavy machinery and automated processes, replacing tasks previously done by hand, stimulated the growth of economies while increasing workers' leisure time and improving their quality of life. However, these improvements were not distributed evenly among societies. Specifically, countries in Asia initially saw very little growth in their economies, until the region surged in the 1980s. In contrast, African nations have experienced very little of the positive impact of industrialization. Consequently, the standards of living of industrialized, newly industrialized, and emerging countries are strongly stratified and largely separated by disparities in industrial, technological, and digital capabilities. For those at the lowest end of the spectrum, a disproportionate amount of time is spent on inefficient tasks, such as the collection of water or firewood, to meet basic human needs.

In industrialized nations, some people are skeptical of the feasibility of solar energy as a cost-effective alternative to fossil fuels without additional price incentives, given the current efficiency rates. While autonomous solar panels might not present a significant energy offset compared with the energy grid, in the **off-grid communities** of developing regions the energy needs are far lower. Hence, a far greater reward can be achieved at much lower costs. In these environments, being able to power a small refrigerator to preserve food, drinks, or medicine is a significant achievement. Similarly, the shift from kerosene-based lamps, which have adverse health and economic effects, to electric lights, which are safer and have longer lifetimes,

would constitute a major improvement. Socioeconomic groups have often been described in terms of a "pyramid" scheme, with a small upper class supported by a sizeable middle class and an even larger lower class. At the **base of the pyramid**, where over 2.5 billion people live on less than $2.50 (World Bank Economic Indicators 2008) a day, the wide consumer base creates a promising market environment, despite the low purchasing power of individuals.

In this environment, solar-energy-based products aimed at providing consumers in developing regions with small-scale power-generation capabilities have emerged as a particularly promising application of technology. Although the degree of market penetration varies according to location, over a million solar panels peek out over rooftops in rural areas of developing regions around the world ([1], p. 315). Similarly, small and inexpensive solar panels have been fitted onto personal electronics such as mobile phones to enable mobile charging in communities that would otherwise be unable to charge the batteries for their personal electronics. The success of solar technologies in these regions has been helped by a somewhat convenient strategic advantage, insofar as much of the developing world lies in low-latitude areas that receive higher-than-average amounts of solar radiation throughout the year (Figure 23.1).

Nevertheless, the successful market penetration of solar energy technology will be possible only through enhanced marketing and distribution channels that can help potential consumers find suppliers and vice versa.

23.4 Off-grid solar for developing regions

Solar energy clearly has the potential to make a substantial impact in the developing world. In this section, we explore the potential market for solar energy in developing

(a)　　　　　　(b)　　　　　　(c)

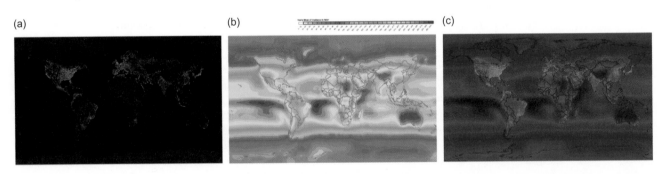

Figure 23.1. The convergence of developmental needs and promising environmental conditions. (a) A map of the world at night highlights the uneven levels of development in different regions of the world (assuming that light pollution can be taken as a signature of development). (Source: C. Mayhew and R. Simmon, (NASA/GSFC), NOAA/ NGDC, DMSP Digital Archive.) (b) A color-coded map shows the mean irradiance that reached the ground in the world from 1990–2004. (Source: Mines ParisTech/Armines 2006.) (c) Overlaying the two maps demonstrates the potential impact that solar technology can have in meeting developmental goals.

regions, present a handful of examples of existing solar products and applications, and discuss the implications of these examples for future growth and development in off-grid rural areas.

23.4.1 The market for solar energy in developing regions

As the market for renewable energy in developing regions expands, rural electrification efforts will continue to evolve beyond the traditional domain of donors to include a diverse array of stakeholders such as manufacturers, community groups, and entrepreneurs ([1], p. 310). The adoption of technologies in developing regions is driven primarily by functionality and affordability, combined with proper distribution and marketing channels that deliver the products to the consumers. One-quarter of the world lives on less than $1.25 per day ([2], p. 2), yet a family of four in a rural village in Kenya can be forced to travel for hours while spending $4–6 every month on kerosene to meet their lighting needs alone. Even for those who are able to afford more advanced technologies, such as car batteries or generators, the long distances between villages and cities can make it prohibitively difficult to fix or replace lost, broken, or stolen parts. Spending an overwhelming proportion of their time and money to meet fundamental needs means that these people are unable to pursue higher-return income-generating activities and are trapped in a cycle that features only marginal economic growth prospects.

One example of a successful company that has developed effective marketing strategies for getting solar energy to rural areas is SELCO, which is a solar energy company that is based in India. SELCO specializes in providing solar energy for rural populations through a microfinancing scheme. This financing scheme is a crucial component of SELCO's business plan because the hand-to-mouth economic habits of its clientele make it difficult, if not impossible, to amass the long-term savings necessary to pay the entire cost of the units up-front. Instead, the recurring payments break down the economic barriers and make each payment manageable. In just over 20 years, SELCO has provided over 75,000 solar units (~$450 for a 40-W system) to 300,000 people, generating over $3 million per year in revenue despite working with a client base in rural India with a limited income. SELCO's success demonstrates the global demand and potential for such initiatives in other parts of the developing world. It also shows that the proper combination of production costs, financial incentives, and distribution methods can have a strong impact on the diffusion of solar energy in rural areas.

23.4.2 Case studies

The issues described in this chapter have stimulated a number of Grand Challenge projects designed to provide new insights into how sustainable solutions can be developed to address problems involved in implementing solar energy in the developing world. These include projects designed to develop sustainable ways of providing scalable solar energy solutions for lighting, mobile-phone charging, and the refrigeration of vaccines for people in rural parts of the developing world. Selected examples of these projects are presented here, together with insights from ongoing USA–Africa collaborations in materials research. These include off-grid solar projects on energy for rural housing and vaccine refrigeration in Mpala, a village in the Laikipia District of Kenya. The third project explores the potential for using solar energy to address the basic lighting needs in a six-bedroom home in Abeokuta, a town in Ogun State in Nigeria. All three projects provide new insights into developing sustainable solutions for the energy needs of people living in off-grid rural communities.

The Mpala lighting project
Through the Mpala rural lighting project, a newly developed solar lantern has been introduced into an off-grid community that lives within the Mpala Research Centre. The solar lantern, shown in Figure 23.2, was manufactured by Roy Solar of Shanghai, China, and is

Figure 23.2. (a) The Roy Solar lantern and accompanying solar panel. (b) A villager in Mpala, Kenya, compares the solar lantern with his traditional kerosene lantern. (Source: Tiffany Tong, Princeton University.)

(a)

(b)

an example of the rapidly growing off-grid solar lighting industry.

These portable lanterns are a healthier and more cost-effective alternative to the kerosene lamps that have become ubiquitous worldwide. Whereas kerosene lanterns give off hazardous fumes, these solar lanterns are safe and environmentally friendly and can be transported or used anywhere in the world that receives ample amounts of sunlight. The lanterns consist of light-emitting-diode (LED) bulbs that are powered by a 6-V battery that is connected to a 3-W crystalline-silicon solar panel. The system supplies enough energy to provide 10–12 h of energy to power lights, cell phones, or a radio, thus providing health, education, and socioeconomic benefits. With an average of 4–5 h of usage in such a manner, each lantern charge can therefore last for up to 2–3 nights of use.

The lanterns were sold to members of the community for $50, which is about one-tenth of the cost of a typical lighting system with conventional 40–80-W panels connected to 12-V batteries and a controller–inverter system. To make the lanterns even more attractive and affordable, a financing system that made the lanterns available at a cost as low as $4 per month – equal to the typical cost of kerosene for these families – was used. Moreover, the solar lanterns present the additional advantages of eliminating the need to travel long distances to obtain kerosene and avoiding the fumes from burning kerosene that cause pulmonary problems in most people in the rural parts of the developing world.

Following their introduction in February 2010, the entire initial order of 100 lanterns was purchased by the villagers within the first two months. The decision to sell, rather than donate, the lanterns was a crucial cornerstone of the project, and it was observed that, because the villagers had paid for the lanterns, they were very careful about maintaining them and protecting them from damage. In fact, many indicated that they would not be willing to invest in the lanterns if no spare parts were available for proper maintenance. This level of attention is typically not seen when solar energy products are given to people for free. Hence, one can conclude that the payments led to a strong sense of ownership of the products. Furthermore, with the rapid diffusion of the initial batch of solar lanterns into Mpala, demand developed for another 100 solar lanterns, which are being financed by the payments being made by the Mpala villagers for the original 100 lanterns. Above all, the project is sustainable in the long term because the funds derived from the sales and financing are being used to provide additional solar energy solutions for members of the Mpala community. It therefore has the potential for scale-up and a larger-scale impact in many rural villages across the world.

The Mpala solar refrigeration project

Another example of a successful off-grid solar project is the Mpala solar camel refrigeration system that was developed by a team from Princeton University in Princeton, NJ; the Art College Center for Design (ACCD) in Pasadena, CA; and the Nomadic Peoples Trust (NPT) in the Laikipia District of Kenya. The NPT provides comprehensive medical care (e.g., child health services, prenatal care, family planning, HIV screening/counseling, and general health services) to a community of over 300,000 people living in the Laikipia District. Because many of the people are nomadic, the clinic relies on the camels to reach communities that cannot be reached by wheeled vehicles. However, because of a lack of refrigeration, the clinic simply stored the vaccines under ice packs, meaning that the vaccines could not be preserved once the ice packs had melted.

Hence, in an effort to address these needs, a team of scientists, engineers, designers, and medical practitioners from Princeton, the ACCD, and the NPT worked together to develop a solar-powered vaccine-refrigeration system that can provide enough energy to power a vaccine refrigerator for 2–3 days, even under overcast conditions. The final design, which is shown in Figure 23.3, consists of a flexible amorphous-silicon solar panel that is draped over a bamboo saddle design that was inspired by the traditional saddles normally used by the nomadic people of Laikipia.

The 12-V battery and controller are also integrated into the saddle design, which has been tested successfully in the Laikipia District of Kenya. The successful implementation of the project illustrates the need for teams that include local and international experts who can address all aspects of the design, integration, and application of off-grid solar energy solutions. The project has the potential for scale-up to other regions where camels are widely used by the local people. The solar refrigeration systems can also be integrated with other animals, such as donkeys and horses, depending on the local culture of the people and their experience with animals. In this way, the technology can be readily adapted to different cultures in off-grid rural communities.

The Abeokuta solar power project

Although the town of Abeokuta in Ogun State in Nigeria is connected to the grid, the supply of electricity to homes in Abeokuta is often interrupted by power failures. Hence, one approach to developing a reliable power supply would be to assume that the homes are not connected to the unreliable grid. This approach to "off-grid" lighting was explored by conducting an experiment on a six-bedroom project home that represents the limit one might consider in a rural/urban environment in the developing world. The experiment was conducted over a 12-month period.

Figure 23.3. A solar camel refrigeration system consisting of flexible amorphous-silicon solar panels draped on a bamboo saddle attached to a refrigeration unit, batteries, and charge controller during tests conducted in the Laikipia district of Kenya. (Source: Edward Weng, Princeton University.)

Figure 23.4. The team of Princeton University students, technicians, and engineers, and local Nigerian technicians who worked on the installation and testing of the 600-W/3-kWh system in Abeokuta, Ogun State, Nigeria. (Source: Meghan Leftwich, Niyi Olubiyi, and Michael Vocaturo, Princeton University.)

The system that was developed included three 200-W panels (Figure 23.4), an inverter and a charge controller, and a bank of 4–8 24-V/200-Ah batteries. The system was installed by a team of local technicians, who worked closely with students and technicians from Princeton during the design and installation process. To facilitate the efficient use of energy, the 60-W incandescent bulbs were replaced by 20-W efficient light-emitting devices in all of the rooms. These provided the same amount of light as the incandescent bulbs at a third of the energy cost. An efficient 100-W refrigeration system, with an integrated control system, was also used to power the fridge, with power being used only when the fridge temperature rose above the desired range between 2 and 8 °C. Finally, the system was used to power a stereo system and up to six cell-phone chargers.

With approximately 5–6 h of effective Sun, the system provided stable power for 3–4 h of light at night. The power was also sufficient to power about 12 h of refrigeration, 3–4 h of television, 1–2 h use of a stereo system, and up to six cell-phone chargers, provided that the fans in the house were not turned on. However, the 100-W standing fans destabilize the system, since they consumed all of the stored energy within a few hours of use. Since most homes in the tropics require about 12 h of fan use, the energy budget for such systems remains a challenge for future work.

Hence, this experiment showed that larger-scale solar-energy-based systems can provide basic lighting, some entertainment (television and radio), and refrigeration, provided that careful attention is paid to isolating other components that can destabilize the systems. This can be engineered by separating the wiring of the solar-powered systems from the rest of the wiring of the homes. When this is done, such mid-scale power systems can be used to provide all of the basic energy needs for people in rural/urban areas, such as Abeokuta.

23.4.3 The need for research and development

Before closing, it is important to highlight the need for basic and applied research and development to address off-grid solar energy needs in the short, medium, and long terms. In the short term (1–2 years), the solutions could involve local innovations that help add value and provide a sense of ownership. For example, local materials such as bamboo could be used in the local design and assembly of bamboo solar lanterns. However, in the medium term (2–5 years), solar energy factories must be built in the developing world, where the capacity to manufacture deep-cycle batteries, controllers, inverters, and wires already exists. The financing of such factories is clearly a challenge for local banks, development agencies, and investment banks, but could yield higher returns by helping to stimulate local economies and enhancing industrial capabilities.

Finally, in the long term, there is a need for research and development on low-cost solar energy approaches that could make clean energy solutions even more affordable than those based on silicon technologies. These could include the development of organic/flexible electronic structures that have the potential to cost as little as one-tenth the cost of silicon-based solar cells. Such goals have inspired the efforts of a USA–Africa collaboration on materials research and education based at US Universities such as Princeton University and Rutgers University (US–Africa Materials Institute, the National Science Foundation Clean Energy IGERT) that have already resulted in the development of organic solar cells and LED devices [3][4]. Such collaborations might also provide the long-term basis for the development of off-grid solar energy solutions for the energy needs of people in the developing world.

23.5 Summary

This chapter presents an introduction to the potential for off-grid solar energy approaches to addressing the energy needs of people in the developing world. The scope of the energy gap between the developed and developing worlds is identified, together with the need for sustainable solutions that can be scaled up to address the diverse energy needs of people in off-grid rural communities. A few examples of emerging off-grid solar projects are presented to illustrate the key components that are needed for the sustainability of such projects. These include the need for local ownership through payments and financing schemes that are consistent with existing income levels and energy budgets in rural areas of the developing world. The chapter concludes with examples of short-, medium-, and long-term research and development needs that can be tackled by local/global teams.

23.6 Questions for discussion

1. How much energy could be generated globally and in off-grid areas from solar energy?
2. What can be learned from the SELCO model about how to provide solar energy to off-grid areas?
3. Why shouldn't solar energy products be given away for free to people in rural areas?
4. In what ways are the needs of people in rural areas and rural/urban areas different?
5. What are the barriers to the scale-up of the three examples presented in the case studies?
6. What could be the advantages and disadvantages of organic solar cells over conventional silicon solar cells in rural off-grid applications?

23.7 Further reading

- **M. Hankins**, 1995, *Solar for Africa*, London, Commonwealth Science Council and Harare, AGROTEC. This book provides practical examples of how to design and install solar energy systems in off-grid areas in Africa.
- **R. Taylor** and **W. Soboyejo**, 2008, "Off-grid solar energy," *MRS Bull.*, **33**, 368–371. This article presents selected examples of off-grid solar energy applications.
- **Nigerian Academy of Science,** 2007, *Mobilizing Science-Based Enterprises for Energy, Water, and Medicines in Nigeria*, Washington, DC, National Academies Press. This book provides case studies of how to develop science-based enterprises that can provide sustainable solutions to problems faced by people in the developing world.

23.8 References

[1] **E. Martinot**, **A. Chaurey**, **D. Lew**, **J. R. Moreira**, and **N. Wamukonya**, 2002, "Renewable energy markets in developing countries," *Ann. Rev. Energy Environment*, **27**, 309–348, http://pysolar.wordpress.com/world-solar-heat-map/.

[2] **S. Chen** and **M. Ravallion**, 2008, *The Developing World is Poorer Than We Thought, But No Less Successful in the Fight Against Poverty*, World Bank Policy Research Working Paper No. 4703, World Bank Development Research Group.

[3] **T. Tong**, **B. Babatope**, **S. Admassie** *et al.*, 2009, "Adhesion in organic electronic structures," *J. Appl. Phys.*, **106**, 083708–8.

[4] **W. Akande**, **O. Akogwu**, **T. Tong**, and **W. O. Soboyejo**, 2010, "Thermally induced surface instabilities in polymer light emitting diodes," *J. Appl. Phys.*, **108**, 023510–6.

24 Principles of photosynthesis

Johannes Messinger[1] and Dmitriy Shevela[2]

[1]Department of Chemistry, Chemical Biological Centre (KBC) Umeå University, Umeå, Sweden
[2]University of Stavanger, Norway

24.1 Focus

Photosynthesis is the biological process that converts sunlight into chemical energy. It provides the basis for life on Earth and is the ultimate source of all fossil fuels and of the oxygen we breathe. The primary light reactions occur with high quantum yield and drive free-energy-demanding chemical reactions with unsurpassed efficiency. Coupling of photosynthesis to hydrogenases allows some organisms to evolve H_2. Research into understanding and applying the molecular details and reaction mechanisms of the involved catalysts is well under way.

24.2 Synopsis

Life needs free energy. On our planet this free energy is mostly provided by the Sun. The sunlight is captured and converted into chemical energy by a process known as **photosynthesis** (from Greek, *photo*, "light," and *synthesis*, "putting together"). This process occurs in plants and many bacteria. The "big bang" of evolution was the development of oxygenic photosynthesis. In this process sunlight is employed to split the abundant water into the molecular oxygen we breathe. The protons and electrons gained are employed by the organism within complex reaction sequences to reduce CO_2 to carbohydrates. The widespread availability of the electron source water allowed oxygenic organisms to spread and diversify rapidly. The O_2 produced was initially toxic for most species, but those which learned to cope with the emerging oxygen-rich atmosphere were able to gain additional energy by "burning" organic matter.

Photosynthesis is a complex process [1]. From a biophysical perspective photosynthesis involves a sequence of several steps. First sunlight is captured by pigments in the photosynthetic antenna system. This converts the energy of photons into excitation energy by promoting electrons of the absorbing pigments from the ground state into a higher energy state. This excitation energy is subsequently transferred to a special pigment arrangement that allows the conversion of the excitation energy into a charge pair. These initial charge separations occur within the photosynthetic reaction centers. In order to be able to utilize the charge-separated state for driving comparably slow chemical reactions it needs to be stabilized by a series of electron-transfer steps that reduce the energy gap between the positive and negative charges, and simultaneously increase their distance.

Splitting water with visible light requires a special trick, because for the first one-electron oxidation of water to the hydroxide radical almost twice as much free energy is needed as that contained in a visible photon. To circumvent this problem a catalyst is used. The catalyst first stores four oxidizing equivalents and only then oxidizes two water molecules to molecular oxygen and four protons. The oxidation potential of the water-splitting catalyst is kept almost constant during the stepwise catalytic cycle by coupling its stepwise one-electron oxidations with deprotonation reactions that prevent charge building up on this cofactor.

The electrons gained by charge separation are sent on a journey toward a second photosystem, where they gain in a second photoreaction enough free energy to eventually reduce CO_2 to carbohydrates (for an overview of biological building materials see Box 24.1), or protons to hydrogen. It is via these chemical reactions that the energy of photons is finally stored within energy-rich chemical bonds, i.e., as chemical fuel. Additional free energy is gained by performing the above electron-transfer reactions with complexes that are integrated into a membrane in a directional fashion. This allows coupling of the electron-transfer reactions with proton transfers across this membrane. The thus-created "proton-motive force" is also used for the synthesis of high-energy bonds.

Photosynthesis employs Earth-abundant elements and a modular design for the synthesis of the light-harvesting complexes and the reaction centers. The redox potentials of chemically identical or highly similar cofactors are tuned to the required levels via changes in the surrounding protein matrix (Box 24.1). Short- and long-term regulatory processes are employed in photosynthesis to minimize damage to the reaction centers caused by too much light, and repair mechanisms are in place to fix those centers that become damaged nonetheless. It is an important message of this short introductory chapter that the basic principles of photosynthetic conversion of light into chemical fuels apply similarly to man-made systems and thereby provide useful blueprints for their design. In the following sections we describe many of these aspects in more detail.

Box 24.1. Biological building materials

Proteins. Biopolymers formed from a set of 20 different amino acids, which can be grouped into nonpolar, uncharged polar, and negatively and positively charged amino acids. Amino acids consist of the elements H, C, O, N, and S. The amino acids are linked together to form a linear string at enzymes known as ribosomes, according to a sequence encoded in the DNA (see below). After synthesis at the ribosomes, this linear string (primary structure) folds into domains known as α-helices, β-sheets, and random coils (secondary structure), and further into defined three-dimensional structures (tertiary structure). In this process, metal ions or molecules (quinones, chlorophylls, etc.) are often bound. Only in its folded state can the protein fulfill its function, and misfolding can lead to illnesses such as mad cow disease. Frequently, several protein subunits come together to form protein complexes. Proteins and protein complexes can be either water-soluble or lipophilic, i.e., able to integrate into a biological membrane formed by lipids (see below). Most enzymes in nature are proteins or protein complexes.

Lipids. Lipids are molecules that have a water-soluble head group connected to nonpolar (hydrophobic) side chains (fatty acids). Lipids are formed from the elements H, C, O, N, and P. In water, these molecules cluster together so that the hydrophobic side chains come together and only the head groups are exposed to water. In this way, micelles and biological membranes are formed. Depending on the head groups, which can be polar (sugars) or positively or negatively charged, and side chains, which can vary in length and number of double bonds, the membranes have different properties with respect to protein binding, proton transport, and fluidity at a given temperature.

Deoxyribonucleic acid (DNA). DNA is a linear biopolymer that carries genetic information, i.e., the building plan for an organism. It is made up of four bases: adenine, guanine, cytosine, and thymine. Each base consists of the head group that gives it its name, a sugar molecule (ribose), and a phosphate group (elements H, C, O, N, and P). Through the phosphate and sugar groups, the bases are connected into a linear string that pairs up with a complementary string. A sequence of three bases on this string encodes for one amino acid. Similarly to proteins, DNA does not stay linear but folds into a double helix and chromosomes. For the genetic information to be read, for example, so that proteins can be synthesized, the DNA has to unwind, and then a selected region is transcribed into ribonucleic acid (RNA). RNA is very similar to DNA, but has a modified sugar group and one other base. The RNA is then translated at the ribosomes into proteins.

Carbohydrates (Carbs). Carbs amount to most of organic substance on Earth because of their extensive functions in all forms of life. They serve as *energy storage* in the form of starch (in plants) or glycogen (in animals) (for instant energy generation, sugars and starch are perfect fuel), *structural components* (as polysaccharides) in the cell walls of bacteria and plants, *carbon supply* for synthesis of other important compounds, and the *structural framework* (as ribose and deoxyribose) *of DNA and RNA*. Carbs are built from monosaccharides, small molecules that typically contain from three to nine carbon atoms and vary in size and in the stereochemical configuration at one or more carbon centers (elements H, C, and O). These monosaccharides may be linked together to form a large variety of oligosaccharide structures.

Cofactors. Cofactors are organic molecules or ions (often metal or halide) or clusters thereof (e.g., iron–sulfur clusters) that are bound to proteins and are required for the protein's biological activity.

24.3 Historical perspective

24.3.1 Development of photosynthesis

Earth formed approximately 4.5 billion years ago. The first forms of life developed through abiotic processes from molecules that were present in water pools. Geochemical data suggest that simple, anoxygenic (not oxygen-producing) photosynthetic organisms first performed carbon fixation about 3.8–3.4 billion years ago. Survival of these first photobacteria was dependent on electron donors such as H_2, H_2S, NH_3, CH_4, organic acids, and ferrous iron – substances that were in limited supply on Earth's surface compared with the huge water pool. Then, 3.4–2.3 billion years ago, some cyanobacteria-like organisms managed to utilize H_2O as an electron and proton source for reduction of CO_2.

By about 2.3 billion years ago, photosynthetic organisms capable of splitting water had begun to dominate, and traces of free O_2 started to appear in Earth's atmosphere (Figure 24.1). However, O_2 was toxic for most organisms, and only those that either found O_2-free niches or developed protective mechanisms were able to survive. Most successful were those organisms that managed to employ O_2 for efficiently burning organic matter within mitochondria (see Box 24.2), since this liberates at least 15 times more free energy from organic matter than do oxygen-free processes. The further rapid development of oxygenic photosynthesis

Figure 24.1. The changing concentration of atmospheric oxygen as a function of geological time in billions of years. Relationships between estimated oxygen concentrations and biological complexity are based on data given in [2] [3]. Only selected events of evolutionary diversification are included.

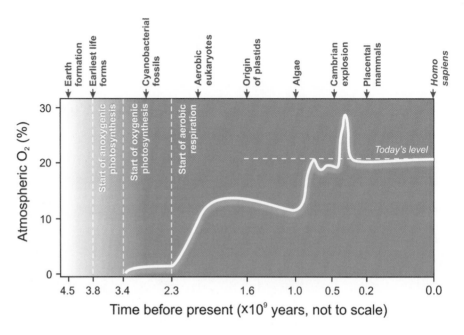

> **Box 24.2.** Biological structures
>
> *Cells.* The cell is the basic unit of living organisms. It is surrounded by a cell wall and/or membrane and contains the complete genetic information and all required organelles and enzymes. One can distinguish between prokaryotic and eukaryotic cells. There are unicellular and multicellular organisms. In the latter, the functions of the individual cells usually vary among cells.
>
> *Organelles.* Organelles are large substructures of cells with specific functions. Examples are (1) the nucleus that contains the DNA; (2) mitochondria, which are often referred to as the powerhouse of the cell because they "burn" (by reaction with oxygen) compounds such as sugars to gain ATP; and (3) chloroplasts, in which photosynthesis occurs. Unicellular precursors of chloroplasts and mitochondria were taken up by early eukaryotic cells. As organelles, they have lost much of their autonomy, but still retain DNA that encodes proteins that are important for their function.
>
> *Membranes.* Membranes allow Nature to form specific reaction environments. There are outer membranes that enclose the whole cell, but also membranes that enclose organelles or that form other substructures such as the thylakoids in chloroplasts. Membranes are mostly made of lipids, but often have a high protein content. These proteins can, for example, be channels for certain molecules or enzymes such as the photosynthetic reaction centers.
>
> *Enzymes.* An enzyme is (usually) a protein or protein complex that accelerates biological reactions by lowering the transition-state energy of a reaction. Enzymes often form special binding pockets with active sites in which the substrates can be brought into close contact and the electronic structure of the reactants can be tuned favorably by charged amino acids or by binding to transition-metal ions. Frequently, it is also important for the reaction that these groups accept or donate protons or electrons.

created today's aerobic atmosphere with about 20% O_2. The increased O_2 levels also led to the formation of the protective ozone layer that absorbs a large part of the ultraviolet radiation from the Sun. The present global net amount of carbon being fixed has been estimated to be about 104 Gt per year [4]. This corresponds to the yearly fixation of about 380 Gt of CO_2 and the release of about 280 Gt of O_2 by photosynthetic organisms (the effect of respiration by these organisms is taken into account). Assuming that the CO_2 was fixed as glucose, this corresponds to 4×10^{18} kJ or 1×10^{15} kWh per year.

24.3.2 Photosynthesis as an energy source

Humans thrive on the products of photosynthesis: we eat plants and animals that live on plants, and we require O_2 for breathing. From the very beginning, we collected biomass (e.g., fruits, nuts, wood) for food, fire, tools, and buildings. By 8000 BCE, efforts to optimize photosynthesis for human advantage had already begun, through the cultivation and improvement of certain plants. Moreover, because fossil fuels originate from the transformation of residual organic matter within sedimentary rocks over millions of years, use of fossil fuels is another means by which humans exploit the products of photosynthesis. The use of coal as a fuel can be traced back to 3000–2000 BCE, but it became particularly important during the Industrial Revolution (in the late eighteenth century). Similarly, fossil oil has been known since ancient times, but its use as a fuel became widespread only after the invention of its distillation to kerosene in 1850. Natural gas was possibly first used as a fuel in China in about 500 BCE; in Europe and the USA, pipelines for its broad distribution were built in around 1900.

It is therefore almost inevitable that efforts to further exploit the power of photosynthesis are now under way. For example, scientists are using modern genetic tools to improve the quantity of biomass production per unit land area, increase the production of valuable products by plants, and improve the properties of biomass for certain industrial processes (see Chapters 25, 26, and 28). Further, researchers are interested in employing solar energy to generate hydrogen as a fuel using natural photosynthetic organisms such as green algae (see Chapters 25 and 28) and artificial photosynthetic systems that mimic the principles or even the details of the corresponding biological process [5] (see Chapter 27). Key to all such applications of solar energy is an understanding of the underlying process of photosynthesis.

24.4 Basics of photosynthetic energy conversion

24.4.1 Overall organization of the photosynthetic processes

The overall chemical reaction of oxygenic photosynthesis is

$$H_2O + CO_2 \xrightarrow[\text{enzymes}]{\text{light energy}} -(CH_2O)- + O_2 \quad (24.1)$$

where $-(CH_2O)-$ represents a one-carbon unit of carbohydrate (note that carbohydrate molecules have more than one carbon).

In algae and higher plants, the reactions of photosynthesis occur within a special cell organelle, the chloroplast. The chloroplast has two outer membranes, which enclose the stroma. Inside the stroma is a closed membrane vesicle, the thylakoid, which harbours the lumen. Whereas the stroma is the site where the **CO$_2$ fixation** occurs,

$$CO_2 + 2NADPH + 3ATP^{4-} + 3H_2O \xrightarrow{\text{dark}}$$
$$-(CH_2O)- + H_2O + 2NADP^+ + 3ADP^{3-} + 3HPO_4^{2-} + H^+ \quad (24.2)$$

the thylakoid membrane is the site of the **light reactions** in which solar energy is converted into chemical bond energy:

$$2H_2O + 2NADP^+ + 3ADP^{3-} + 3HPO_4^{2-} + H^+ \xrightarrow[8h\nu]{\text{light}}$$
$$O_2 + 2NADPH + 3ATP^{4-} + 3H_2O \quad (24.3)$$

The steps involved in photosynthetic energy conversion are catalyzed by four protein complexes that, together with lipids, form the thylakoid membrane. These complexes are **photosystem II (PSII)**, **cytochrome b_6f (Cyt b_6f)**, **photosystem I (PSI)**, and **adenosine-5′-triphosphate synthase (ATP synthase)** (Figure 24.2). In addition, there exist **light-harvesting complexes** that differ significantly between species. PS II, Cyt b_6f, and PSI contain redox-active cofactors that allow photo-induced electron transfer from water to NADP$^+$ (the oxidized form of nicotinamide adenine dinucleotide phosphate) through the thylakoid membrane. Figure 24.3 displays this linear electron transport arranged in the form of the **Z-scheme** proposed by Hill and Bendall in 1960 [6]. The mobile electron carriers **plastoquinone (PQ)** and **plastocyanine (PC)** transfer the electrons between PSII and Cyt b_6f, and between Cyt b_6f and PSI, respectively.

Water-splitting and unidirectional proton transport across the thylakoid membrane by PQ lead to an accumulation of protons in the lumen (Figure 24.2). The thus-formed electrochemical potential difference across the thylakoid membrane (proton-motive force) is utilized by the ATP synthase for phosphorylation of ADP to ATP (Figure 24.2) [7]. The free energy stored in the chemical bonds of NADPH and ATP is subsequently used for the reduction of CO$_2$ to sugars within a cyclic metabolic pathway (the Calvin–Benson cycle, see [8] and references therein) that occurs in the stroma and does not directly require light (photosynthetic "dark" reactions), Equation (24.2) and Figure 24.2.

24.4.2 Capturing the energy of sunlight

The initial event in photosynthesis is the absorption of light (*photons*) by **light-harvesting complexes** (LHCs) located around PSII and PSI.

These protein complexes bind chromophores (chlorophylls, phycobillins, carotenoids) in high concentration and special geometric arrangements. As a result of light absorption, electrons within these chromophores are excited into a higher energy level. This excitation energy is distributed over the whole antenna and the reaction center either by hopping from pigment to pigment or in a coherent way [1]. Once the absorbed light energy has been delivered to the **chlorophylls (Chl)** in the reaction centers, denoted P, the excited singlet state ($^1P^*$) is formed. From this excited state charge separation takes place between $^1P^*$ and the primary electron acceptor, denoted A, to form the radical pair $P^{\bullet+}A^{\bullet-}$. Subsequently, $P^{\bullet+}$ is reduced by the electron donor D. These processes are schematically depicted in Figure 24.4. From $A^{\bullet-}$ the electron is further

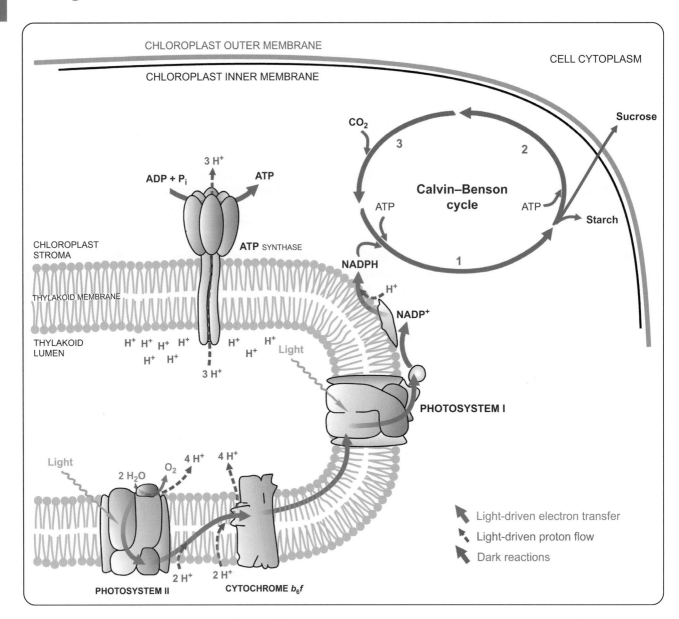

Figure 24.2. A schematic representation of the cofactor–protein complexes involved in the light-induced electron- and proton-transfer reactions in the thylakoid membrane (oxygenic photosynthesis). The products of the light reactions, NADPH and ATP, drive the CO_2-fixing dark reactions of the Calvin–Benson cycle in the stroma that lead to the production of carbohydrates. The Calvin–Benson cycle can be conveniently divided into three phases: reduction (1), regeneration (2), and carboxylation (3).

transferred along the electron-transfer chain that is designed to stabilize the initial charge separation by both increasing the distance between the charges and reducing their potential-energy difference (Figure 24.3).

24.4.3 Photosynthetic reaction centers

Classification of reaction centers and their origin
Every photosynthetic organism contains a **reaction center (RC)** that performs the trans-membrane charge separation in response to light excitation. All photosynthetic RCs are composed of a membrane-integral protein complex (Figure 24.2) of homodimeric or heterodimeric nature to which pigments (carotenoids and chlorophylls) and redox-active cofactors (such as chlorophylls, quinones, and transition-metal complexes) are bound. The strong similarities in overall construction of all known RCs suggest a common evolutionary origin [1]. Because of differences in the nature of the initial electron acceptors, the RCs can be classified into two types. The RCs

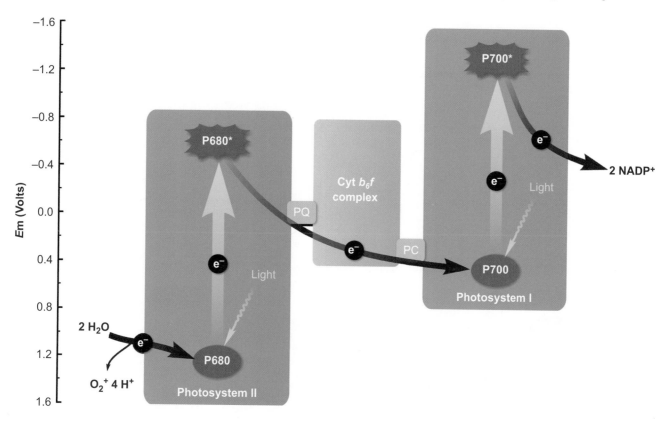

Figure 24.3. The Z-scheme of oxygenic photosynthesis for linear electron transfer from water to NADP$^+$ plotted on a redox potential scale. Black arrows, electron transfer; straight yellow arrow, excitation of an electron into a higher energy state within the reaction-center chlorophyll *a* (Chl a) molecules (P680 and P700) induced by excitation energy transfer from the antenna or by direct absorption of a photon (wiggly yellow arrow). The numbers 680 and 700 are the wavelengths of the red absorption maxima for the special reaction center Chl *a* molecules. Em is midpoint potential vs SHE (Appendix B)

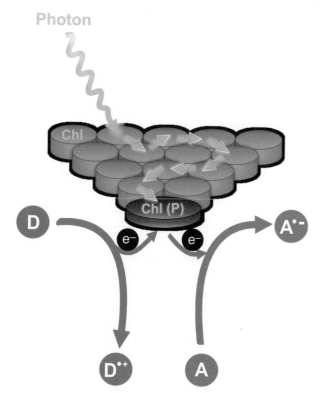

of green sulfur bacteria, heliobacteria, and PSI belong to the **iron–sulfur type (RC1 type)**, whereas the RCs of purple bacteria, green non-sulfur (green filamentous) bacteria, and PSII belong to the **pheophytin–quinone type (RC2 type)** [1]. This classification is shown schematically in Figure 24.5.

Whereas all anoxygenic photosynthetic bacteria have either RC1 or RC2, all oxygen-evolving photosynthetic organisms contain both types of RCs (PSI and PSII). Such an arrangement is a prerequisite for efficiently bridging the large energy gap between H_2O and NADPH with the energy of visible light. However, to oxidize water to molecular oxygen, two further changes

Figure 24.4. Photosynthetic light harvesting and charge separation: the energy of absorbed photons is passed through chlorophyll molecules until it reaches the reaction-center chlorophyll (P). The excited reaction-center chlorophyll (^{1}P*) donates its energized electron to an electron acceptor (A). The electron vacancy of the chlorophyll (P$^{\bullet+}$) is filled by the electron donor (D).

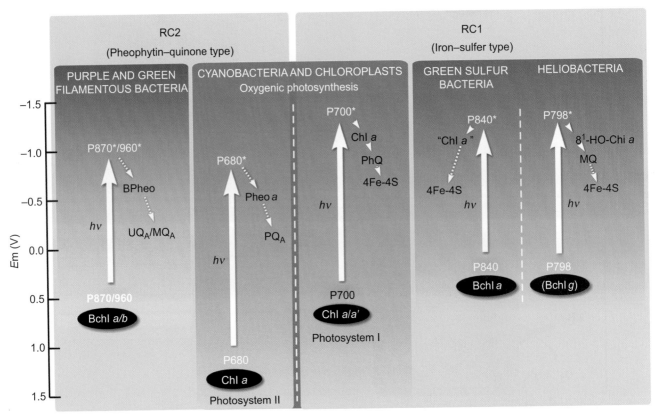

Figure 24.5. Photosynthetic reaction centers (RCs). For each RC the initial electron donor (P) and the primary and secondary electron acceptors are plotted on a redox scale (adapted from [1]). The star symbolizes the electronically excited state that the P attains after light absorption or energy transfer from the antenna. The numbers behind P relate to the long-wavelength absorption maxima of the reaction-center chlorophylls, and are inversely related to the absorbed energy (length of upward arrow). BPheo, bacteriopheophytin; Pheo, Pheophytin; Bchl, bacteriochlorophyll; Fe-S, iron–sulfur center; MQ, menaquinone; PhQ, phylloquinone; UQ, ubiquinone; PQ, plastoquinone.

compared with anoxygenic RCs were required: a water-splitting catalyst and an RC pigment with a very high oxidizing potential [9]. It is remarkable that the RC Chl molecules in PSII and PSI are chemically identical and that the required much stronger oxidizing power in PSII was achieved simply by changes in the surroundings of the RC Chl molecules. An important design feature of the RC is also that the special RC Chl molecules are absorbing as far as possible down to the low-energy end of the visible light range. While this is wasting part of the free energy of higher-energy photons as heat, it allows the utilization of most visible photons for driving photosynthesis. This is an important advantage, especially under light-limiting conditions.

Photosystem II, the water:plastoquinone oxidoreductase

Photosystem II is a large multimeric pigment–protein complex that is present in the thylakoid membranes

of all oxygenic photosynthetic organisms. It has a total relative molecular mass of 350 kDa and consists of about 30 protein subunits and close to 80 cofactors, among which are 35 Chl a molecules, 25 integral lipids, 12 β-carotene molecules, 4 Mn atoms, 2 Cl$^-$ ions, and 1 Ca^{2+} ion. Figure 24.6(a) is a schematic depiction of the PSII core proteins and central cofactors, and Figure 24.6(b) shows the spatial arrangement and chemical structures of these cofactors derived from X-ray crystal structures.

Driven by light-induced charge separations in the RC, PSII catalyzes the oxidation of water to molecular oxygen and protons on its donor side and the reduction of **plastoquinone** (PQ) to **plastohydroquinone** (PQH$_2$) on its acceptor side. It thereby couples the fast (picosecond) one-electron photochemistry with both the two-electron–two-proton reaction on the acceptor side (the Q$_B$ side; hundreds of microseconds) and the slow (1–2 ms) four-electron–four-proton water-splitting chemistry on the

Figure 24.6. (a) A schematic view of PSII in higher plants and algae (only core proteins are included). Black arrows show the direction of electron transfer, while red arrows indicate the migration of the excitation energy toward P680. D1 and D2 are the central RC proteins, which bind all electron-transfer cofactors. However, the electron transfer occurs mainly on the D1 side of the PSII RC (active branch). The symmetrically related cofactors located on the D2 side do not participate in linear electron flow through PSII (inactive branch), but some of them play protective roles against photo-induced damage of PSII. (b) Central cofactors from the PSII crystal structure of the cyanobacterium *Thermosynechococcus elongatus* at a resolution of 2.9 Å [10]. For cyanobacteria the surrounding of the thylakoid membrane is called cytoplasm instead of stroma. CP43 and CP47, internal antenna subunits; Mn_4CaO_5, inorganic core of the water-oxidizing complex; 17, 23, and 33, extrinsic proteins that stabilize the Mn_4CaO_5 cluster (numbers are their relative molecular masses in kDa); Y_D, redox-active tyrosine D; Cyt b_{559}, heme of cytochrome b_{559}; Cyt c_{550}, small extrinsic protein in cyanobacteria (PsbV, 17 kDa); Car, carotenoid.

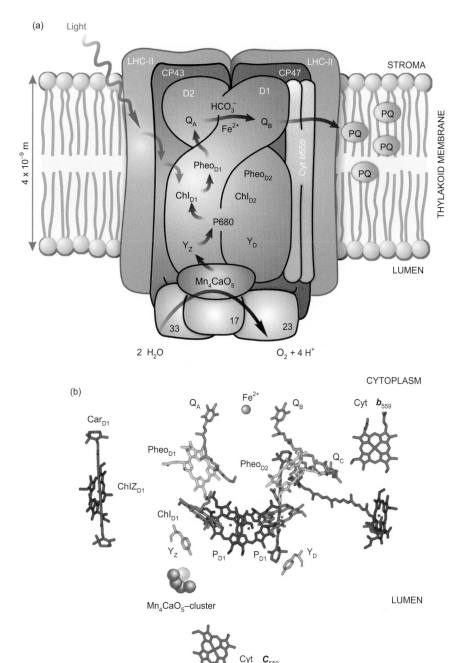

donor side (the Mn_4CaO_5 cluster). To accomplish this with high quantum yield (up to 90%), all reaction steps must be finely tuned (the maximum energy efficiency, however, is 16% (see Box 24.3) [11]). The primary charge separation occurs between chlorophyll molecules in the D1 protein (Chl_{D1} and P680) and the pheophytin molecule ($Pheo_{D1}$) of the same subunit, and results in the formation of $P680^{\bullet+}Pheo^{\bullet-}$. Then $Pheo^{\bullet-}$ transfers the electron on a 300-ps time scale to Q_A, which is a plastoquinone molecule that acts as one-electron acceptor owing to its protein environment. Then $Q_A^{\bullet-}$ donates the electron to Q_B, which, after a second charge separation, accepts a second electron. It then takes up two protons from the stroma, and the PQH_2 formed leaves the binding pocket, which is filled by another PQ molecule from the PQ pool in the thylakoid membrane [10]. On the other side, $P680^{\bullet+}$ is first reduced with multiphasic kinetics (nanosecond to microsecond) by the redox-active Tyr Z (Y_Z), which, in turn, abstracts electrons from the Mn_4CaO_5 cluster. The rate constant of the latter reaction varies from 30 μs to 1.3 ms depending on the redox state of the Mn_4CaO_5 cluster (see below). This description shows that only the left branch in Figure 24.6 is active in the normal photochemistry and

Box 24.3. Energy efficiency

The maximum energy efficiency of photosynthetic complexes (PSI and PSII) is a product of three factors: (1) the light-harvesting efficiency, which includes the formation of the excited state in the primary donor; (2) the percentage of the free energy of the excited state that is converted into useful chemical energy (the energy-conversion efficiency); and (3) the probability (quantum yield) of formation of the energy products. Below, these general terms are discussed using the example of PSII.

Light-harvesting efficiency. PSII is connected to light-harvesting complexes that contain several pigments (chlorophylls, carotenoids) that absorb well across the visible spectrum (300–700 nm), but somewhat less efficiently in the green. Because a minimum energy is required to form ^{1}P680*, PSII has a relatively sharp cutoff for the light-harvesting efficiency at frequencies below 680 nm. Under low-light conditions, the excitation energy transfer from the antenna to P680 occurs with very high efficiency. On this basis, it has been estimated that the upper limit for the light-harvesting efficiency of PSII is about 34% of the total solar radiation spectrum [11].

Energy-conversion efficiency. As described in the main text, the formation of ^{1}P680* is the prerequisite for the primary charge separation, which is subsequently stabilized by a cascade of electron-transfer reactions that separate the charges in space and reduce their free-energy difference. This is required for coupling of the fast photochemistry with the slow water-splitting chemistry. The free energy that is stored by PSII due to the formation of O_2 and plastohydroquinone is about 50% of that of ^{1}P680*.

Quantum yield. This quantum yield is related to the miss parameter of S_i state turnover, which can be calculated from the flash-induced oxygen-evolution pattern. Under optimal conditions of temperature, pH, and light, the overall quantum yield can be as good as 90%.

From these estimates, a maximum light-to-chemical energy-conversion efficiency of 16% can be calculated for PSII. This value neglects the contribution of the PSII reactions to establishing the proton-motive force across the thylakoid membrane that leads to ATP synthesis by the ATP synthase. When comparing the PSII efficiency with the efficiency of solar cells, one should keep in mind that the efficiencies of the latter refer merely to the conversion of light into electricity, whereas the above efficiency of 16% includes the water-oxidation chemistry.

In most organisms, PSII and its antenna are optimized for limiting light conditions. Under full light, the organism actively reduces the efficiency of light harvesting by several regulatory mechanisms to minimize damage to PSII. This can significantly reduce the above value. Similarly, for biomass or H_2 production, the efficiencies are significantly lower (a few percent at best), because energy losses also occur in the Calvin–Benson cycle (dark reactions) and other metabolic steps. In addition, the organism also requires energy for growth, repair, and reproduction.

electron-transfer reactions. However, also the other side has a function: if the normal reaction path is blocked, the normally inactive cofactors donate electrons to P680$^{\bullet+}$ and thereby protect PSII from damage. This is important since P680$^{\bullet+}$ is such a strong oxidant that it otherwise oxidizes surrounding protein chains. Despite this protective mechanism PSII is damaged over time, for example by side reactions that lead to the formation of reactive oxygen species at the Mn_4CaO_5 cluster. Under full sunlight the D1 protein is therefore replaced every 20–30 min through a specific repair cycle.

How does PSII split water?

Although P680$^{\bullet+}$ has the highest known oxidizing potential in nature (about 1.25–1.3 V) [12], it is not strong enough to directly abstract electrons from water (the formation of HO$^{\bullet-}$ requires more than 2 V in solution). Therefore, a catalyst that can first store four oxidizing equivalents before water-splitting occurs in a concerted four-electron reaction is required [13]. This role is fulfilled in PSII by a complex of four manganese, one calcium, and five bridging oxygen atoms (the Mn_4CaO_5 cluster). This property leads to a periodicity of four in O_2 evolution, if well dark-adapted PSII samples are illuminated with a series of short, saturating flashes. This was first discovered by Pierre Joliot and co-workers in 1969 [14] on the basis of polarographic oxygen measurements and described in a kinetic scheme, the Kok cycle, in 1970 by Kok and co-workers [15] (Figure 24.7). In terms of this scheme, the redox intermediates of the Mn_4CaO_5 cluster are referred to as S_i states, where the index gives the number of stored oxidizing equivalents. The first maximum of oxygen evolution occurs after the third flash, since S_1 is the dark-stable state. The damping of the oscillation in Figure 24.7(a) is caused mainly by redox equilibria between the cofactors shown in Figure 24.6.

On the basis of extended X-ray absorption spectroscopy, X-ray crystallography, and DFT calculations [10][16][17], very detailed models of the Mn_4CaO_5 cluster and its protein ligands in the different S_i states are available. The best model at present is shown in Figure 24.8 [17]. Except for one histidine, all other ligands are

(a)

(b)

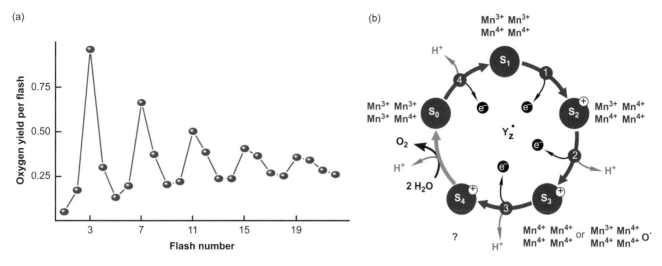

Figure 24.7. (a) Oxygen evolution per flash of dark-adapted spinach thylakoids induced by a train of saturating xenon flashes (a very similar pattern was obtained by Joliot in 1969). (b) The Kok cycle that describes the O_2 pattern shown in (a). Currently discussed oxidation states of the four Mn ions in the S_i states are indicated.

carboxylates or water-derived ligands. The oxidation states of the Mn ions in the S_0 state are $Mn^{3+}Mn^{3+}Mn^{3+}Mn^{4+}$, and Mn^{3+} to Mn^{4+} oxidations occur on the following oxidation steps. A possible exception is the $S_2 \rightarrow S_3$ transition, where a ligand- (oxygen-bridge-)centered oxidation is also considered possible.

Time-resolved mass-spectrometry experiments have shown that at least one substrate water molecule is already bound in the S_0 state, and both substrate water molecules are bound in the S_2 and S_3 states. During the S_i-state cycle, the substrate molecules (or other groups at the active site) are deprotonated and thereby prepared for O—O bond formation in the S_4 state. These deprotonation reactions (see Figure 24.7 (b)) are crucial also for the energetics of the S_i-state turnovers [18]. They prevent the accumulation of positive charges on the Mn_4CaO_5 cluster, thereby avoiding an increase of its oxidation potential during the S_i-state cycle. This is important, because every charge separation in the RC generates the same oxidizing power.

In Figure 24.8, the two substrate "water" molecules, labeled W_s and W_f, are shown in orange in their proposed positions just before O—O bond formation in the S_4 state. Both are fully deprotonated, and W_f is a terminally bound oxygen radical. It can also be seen that Ca is involved in binding W_s.

24.5 Coupling photosynthesis with H_2 production

Jules Verne was among the first to write about the potential of water for the production of H_2 as fuel. In 1874 he noted in his novel *Mysterious Island* "Yes, my friends, I believe that water will one day be employed as fuel, that hydrogen and oxygen which constitute it, used singly or together, will furnish an inexhaustible source of heat and light, ..."

More than 60 years ago, Gaffron and Rubin [20] reported that a green alga, *Scenedesmus obliquus*, produced molecular hydrogen under light conditions after being dark-adapted in the absence of carbon dioxide (made anaerobically). This was the first indication of the existence of an enzyme that converts protons and electrons into molecular hydrogen in photosynthetic organisms. This enzyme is referred to as **hydrogenase**. Almost 40 years after Gaffron and Rubin's discovery, the hydrogenase from *Chlamydomonas reinhardtii* was purified, and it was determined that *ferredoxin* (Fd) is its direct electron donor [21]. Today, many cyanobacteria, in addition to green algae, are known to be able to generate molecular hydrogen using light as the driving force to extract electrons from water to produce strong reductants (Fd in the case of green algae and NADPH in the case of cyanobacteria). The pathways of photobiological molecular-hydrogen production are depicted in Figure 24.9. It is to be noted that photobiological H_2 production competes with the Calvin–Benson cycle for reducing equivalents and that most hydrogenases are irreversibly inhibited by O_2. In wild-type cells, photobiological H_2 production therefore occurs only under special conditions and with low yield. Metabolic engineering is being employed with the aim of redirecting as much electron flow to the hydrogenases as possible.

Figure 24.8. A model of the Mn_4CaO_5 cluster (only bridging oxygens are counted) in the S_4 state with some surrounding amino acids [19]. W_s and W_f signify the slowly and the rapidly exchanging substrate "water" molecules (they are generally referred to as substrate waters independently of their protonation state). In Siegbahn's mechanism W_s is bound in all S_i states, while W_f binds to the indicated Mn in the S_3 state. After O_2 release, W_s is replaced by an OH^- ion from the medium. Purple, manganese; yellow, calcium; red, oxygen; light blue, nitrogen; gray, carbon.

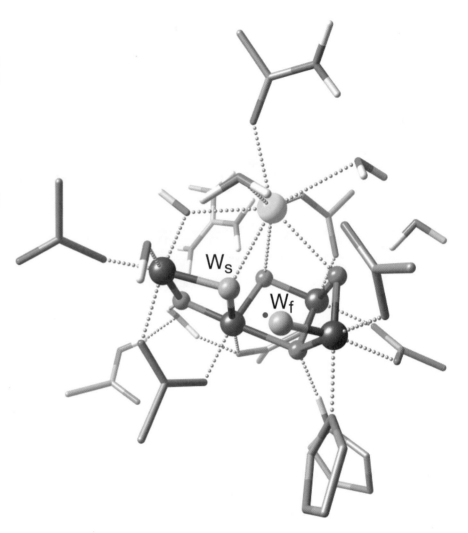

In addition to hydrogenases, **nitrogenases** can also produce H_2 and are found in oxygenic cyanobacteria and in anoxygenic purple bacteria. However, they require ATP for this process, and, therefore, the theoretically achievable photon-to-H_2 ratio is significantly worse (for a review, see [22]).

24.6 Summary

Photosynthesis is Nature's way of converting sunlight into chemical energy. This process had a dramatic effect on evolution on Earth. It also created the organic material that was converted, through geological processes, into coal, oil, and natural gas. Researchers have developed a thorough understanding of the various levels of the photosynthetic processes, including molecular details of the light and dark reactions, regulatory processes at the cell/chloroplast level, factors that affect plant growth and the production of certain chemicals, and complex interactions of plant ecosystems with the environment and their effects on the climate. This chapter has focused primarily on only a very small section of that knowledge: the photosynthetic light reactions.

Photosynthesis will likely be able to contribute in two ways to solving the current and upcoming energy crises: first, by finding ways to increase biomass and bio-H_2 production (see Chapters 25, 26, and 28); and, second, by providing blueprints for an effective water-splitting catalyst that is made of Earth-abundant materials and that works effectively under ambient conditions (see Chapter 27). While the biological process is still under intensive study by many research groups, the past several years have brought about a tremendous improvement of our understanding. This has already inspired many scientists to apply the principles of photosynthesis in the search for new ways to efficiently explore solar energy for the production of biomass or fuels. This is described in the next four chapters.

24.7 Questions for discussion

1. In which ways did photosynthesis change life on Earth?
2. Describe the dependence of humans on photosynthesis.
3. How is sunlight converted into chemical energy in photosystem II?

Figure. 24.9. H_2- photoproduction pathways in green algae (top) and cyanobacteria (bottom). In these organisms, the production of molecular H_2 is catalyzed by hydrogenases. WOC, Water-oxidizing complex.

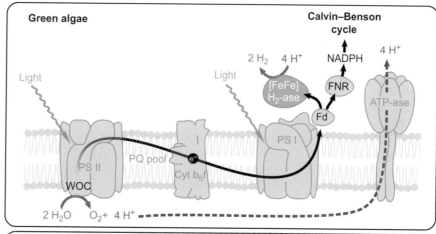

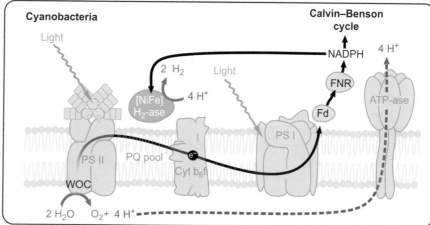

4. Discuss the energy efficiency of photosystem II in comparison with that of whole plants and solar cells.

5. In which ways can photosynthesis – directly or indirectly – contribute to solving the climate and energy problems on Earth?

24.8 Further reading

- **L. Taiz** and **E. Zeiger**, 2010, *Plant Physiology*, 5th edn., Sunderland, MA, Sinauer Associates. A well-received, established textbook that provides a good introduction to photosynthesis and the physiological processes in plants in general.
- **N. Lane**, 2004, *Oxygen: The Molecule That Made the World*, New York, Oxford University Press Inc. An interesting introductory book devoted to evolution on Earth and oxygen biochemistry.
- **O. Morton**, 2008, *Eating the Sun: How Plants Power the Planet*, New York, HarperCollins Publishers. A popular science book that focuses on photosynthesis and its relation to plant life, animal life, and the history of life and the climate.
- **R. E. Blankenship**, 2002, *Molecular Mechanisms of Photosynthesis*, Oxford, Blackwell Publishing. A comprehensive, well-written book that includes all aspects of photosynthetic molecular mechanisms.

- **W. Lubitz, E. J. Reijerse**, and **J. Messinger**, 2008, "Solar water-splitting into H_2 and O_2: design principles of photosystem II and hydrogenases," *Energy Environmental Sci.*, **1**, 15–31. This review outlines the principles of the oxidation of water in photosystem II and of hydrogen production by hydrogenases in order to facilitate the design of artificial catalysts for solar fuel production.

24.9 References

[1] **R. E. Blankenship**, 2002, *Molecular Mechanisms of Photosynthesis*, Oxford, Blackwell Publishing.
[2] **P. G. Falkowski**, 2006, "Tracing oxygen's imprint on Earth's metabolic evolution," *Science*, **311**, 1724–1725.
[3] **L. P. Kump**, 2008, "The rise of atmospheric oxygen," *Nature*, **451**, 277–278.
[4] **C. B. Field, M. J. Behrenfeld, J. T. Randerson,** and **P. Falkowski**, 1998, "Primary production of the biosphere: integrating terrestrial and oceanic components," *Science*, **281**, 237–240.
[5] **W. Lubitz, E. J. Reijerse**, and **J. Messinger**, 2008, "Solar water-splitting into H_2 and O_2: design principles of photosystem II and hydrogenases," *Energy Environ. Sci.*, **1**, 15–31.

[6] **R. Hill** and **F. Bendall**, 1960, "Function of the 2 cytochrome components in chloroplasts – working hypothesis," *Nature*, **186**, 136–137.

[7] **W. Junge**, **H. Sielaff**, and **S. Engelbrecht**, 2009, "Torque generation and elastic power transmission in the rotary F_OF_1-ATPase," *Nature*, **459**, 364–370.

[8] **W. Martin**, **R. Scheibe**, and **C. Schnarrenberger**, 2000, "The Calvin cycle and its regulation," in *Photosynthesis: Physiology and Metabolism*, eds. **R. C. Leegood**, **T. D. Shakey**, and **S. von Caemmerer**, Dordrecht, Kluwer Academic Publishers, pp. 9–51.

[9] **J. M. Olson** and **R. E. Blankenship**, 2004, "Thinking about the evolution of photosynthesis," *Photosynth. Res.*, **80**, 373–386.

[10] **A. Guskov**, **J. Kern**, **A. Gabdulkhakov** *et al.*, 2009, "Cyanobacterial photosystem II at 2.9-angstrom resolution and the role of quinones, lipids, channels and chloride," *Nat. Struct. Molec. Biol.*, **16**, 334–342.

[11] **H. Dau** and **I. Zaharieva**, 2009, "Principles, efficiency, and blueprint character of solar-energy conversion in photosynthetic water oxidation," *Acc. Chem. Res.*, **42**, 1861–1870.

[12] **F. Rappaport**, **M. Guergova-Kuras**, **P. J. Nixon**, **B. A. Diner**, and **J. Lavergne**, 2002, "Kinetics and pathways of charge recombination in photosystem II," *Biochemistry*, **41**, 8518–8527.

[13] **J. Messinger** and **G. Renger**, 2008, "Photosynthetic water splitting," in *Primary Processes of Photosynthesis, Part 2 Principles and Apparatus*, ed. **G. Renger**, Cambridge, RSC Publishing, pp. 291–351.

[14] **P. Joliot**, **G. Barbieri**, and **R. Chabaud**, 1969, "Un nouveau modèle des centres photochimiques du système II," *Photochem. Photobiol.*, **10**, 309–329.

[15] **B. Kok**, **B. Forbush**, and **M. McGloin**, 1970, "Cooperation of charges in photosynthetic O_2 evolution," *Photochem. Photobiol.*, **11**, 457–476.

[16] **J. Yano**, **J. Kern**, **K. Sauer** *et al.*, 2006, "Where water is oxidized to dioxygen: structure of the photosynthetic Mn_4Ca cluster," *Science*, **314**, 821–825.

[17] **P. E. M. Siegbahn**, 2009, "Structures and energetics for O_2 formation in photosystem II," *Acc. Chem. Res.*, **42**, 1871–1880.

[18] **H. Dau** and **M. Haumann**, 2008, "The manganese complex of photosystem II in its reaction cycle – basic framework and possible realization at the atomic level," *Coord. Chem. Rev.*, **252**, 273–295.

[19] **P. E. M. Siegbahn** and **M. R. A. Blomberg**, 2009, "A combined picture from theory and experiments on water oxidation, oxygen reduction and proton pumping," *Dalton Trans.*, 5832–5840.

[20] **H. Gaffron** and **J. Rubin**, 1942, "Fermentative and photochemical production of hydrogen in algae," *J. Gen. Physiol.*, **26**, 219–240.

[21] **P. G. Roessler** and **S. Lien**, 1984, "Activation and *de novo* synthesis of hydrogenase in *Chlamydomonas*," *Plant Physiol.*, **76**, 1086–1089.

[22] **M. L. Ghirardi**, **A. Dubini**, **J. Yu**, and **P.-C. Maness**, 2009, "Photobiological hydrogen-producing systems," *Chem. Soc. Rev.*, **38**, 52–61.

25 Biofuels and biomaterials from microbes

Trent R. Northen

Joint Bioenergy Institute (JBEI), Emeryville, CA, USA and Department of GTL Bioenergy and Structural Biology, Life Sciences Division, Lawrence Berkeley National Laboratory, Berkeley, CA, USA

25.1 Focus

The development of "carbon-neutral" biofuels and biomaterials is critical for stabilizing atmospheric carbon dioxide levels and reducing the current dependence on petroleum. Microbes are self-replicating "biocatalytic" systems that can convert solar energy and plant biomass into a wide range of molecules that can be used for biofuels and biomaterials. However, major technical challenges need to be addressed before the approaches become economically viable. Among these challenges are the recalcitrance of lignocellulosic feedstocks, the low cell density of algal production systems, and the scaling needed to minimize impacts on freshwater/arable land.

25.2 Synopsis

Biofuels and biomaterials are among the diverse portfolio of technologies considered essential to address concerns over an excessive dependence on fossil fuels and their impact on the environment, such as through CO_2 emissions. Whereas the burning of fossil fuels increases the amount of CO_2 in the atmosphere, biofuels and biomaterials have the potential to be carbon neutral or negative. Because the carbon sequestered in biofuels is eventually returned to the atmosphere upon burning, these are primarily carbon-neutral technologies, although some crops accumulate carbon in the soil and have the potential to be carbon-negative. The use of biomaterials, on the other hand, can decrease the atmospheric concentration of carbon dioxide by fixing it in useful materials.

Microorganisms provide complex adaptive systems that can be used to address the need to help balance the Earth's carbon cycle by replacing fossil fuels and chemical feedstocks with biofuels and biomaterials derived from renewable resources. Over billions of years, micro-organisms have evolved to inhabit every ecological niche on Earth, and, in the process, they have developed a tremendous repertoire of metabolic and functional diversity. Metabolic processes found in these organisms could provide new routes to the production of biofuels and biomaterials by harnessing catalytic solutions that nature has evolved. However, these organisms have evolved to be able to exist in their specific natural environments. Thus, although organisms that produce molecules suitable for biofuel and biomaterial production have been discovered, they will likely need to be optimized to use resources (land, water, energy) efficiently and to increase yields.

Fortunately, genomic technologies provide a "toolbox" to meet these needs: genome sequencing provides a basis for understanding the inner functions of micro-organisms. Genome-scale computational models are used to determine how best to alter existing processes to increase sustainable biofuel production; synthetic biology utilizes this information to adapt/design new biological systems to maximize the production of the desired products and the overall robustness of the system. Directed evolution approaches are used to discover other solutions for biofuel production and for fine-tuning of these systems. The integration of these tools is essential to creating sustainable and economically viable biofuels [1].

25.3 Historical perspective

Plant and animal biomass have been used as fuels (wood and oil) and materials (leather, fiber, etc.) for millennia [2]. The recent trend of building vehicles to run on bio-diesel and ethanol is actually nothing new (please see Further reading, Songstad *et al.* for citations and additional detail). In fact, diesel and automotive engines have a long history of running on biofuels: Rudolf Diesel (Figure 25.1) designed his diesel engine for plant oils and demonstrated it at the 1900 Paris Exposition running on pure peanut oil.

Likewise, Henry Ford's (Figure 25.1) famed Model T was designed to run on ethanol – which he called the "fuel of the future" – and the car itself included plant-derived materials. For example, the Model T had coil cases made of wheat gluten resin mixed with asbestos fibers. In the 1920s, the Ford Company used soybean oil in automotive paints, enamels, and rubber substitutes, and cross-linked soy meal combined with plant fibers resulted in a material that was lighter than steel and capable of withstanding sledge-hammer blows without denting.

Despite these precedents, biomass proved to be a more expensive feedstock than petroleum, and society is only now recognizing the external costs associated with combustion of petroleum and returning to the concept of using renewable fuels and materials. Biofuel-powered vehicles (e.g., flex-fuel vehicles) are becoming more common, and biomaterials are finding new applications, as in the Toyota Prius carpeting constructed from polylactic acid derived from corn. Despite these recent developments, weaning society off oil is a Herculean task, as today's high standard of living depends on petroleum and petroleum products. How can these needs be met with biofuels, especially considering that developing countries (which represent most of the world's population) are seeking to expand their economies and quality of life?

Periods of war provide examples of societies forced to find alternatives to petroleum: during World Wars I and II, the supply of petroleum was greatly reduced in many countries, necessitating the development of alternative fuels and materials. Thermochemical processes were developed during World War II and used extensively to convert solid fuels into gasoline – in fact a larger fraction of the third Reich's automative fuel was produced by this process. In the first step, the biomass (or coal) is gasified through controlled treatment with oxygen and steam to produce hydrogen and carbon monoxide (syngas), which is then converted into alkanes using cobalt or iron catalysis (the Fischer–Tropsch process).

In addition, the second-largest industrial fermentation process technologies (after ethanol production) were developed during this period for the production of butanol and acetone. Butanol is a promising biofuel and was first industrially produced by bacterial fermentation in 1916. This process was pioneered by Chaim Weizmann (1874–1952), whose work was based on that of Louis Pasteur, who isolated the relevant microbe, *Clostridium acetobutylicum*. Acetone was actually the desired product of the fermentation, but, for every kg of acetone formed, twice the mass of butanol was produced. This process was continued in the USSR (Figure 25.2), resulting in a process that would convert 100 t of molasses/grain into 12 t of acetone and 24 t of butanol [3]. However, increased molasses prices, frequent bacteriophage outbreaks, and competition from petroleum resulted in the steady decline from a peak in the 1950s to obsolescence by the 1980s. Considering this example, the development of economically viable and sustainable fuels against the backdrop of an ever-increasing demand for energy is a truly urgent and difficult challenge.

Figure 25.1. Biofuel pioneers: (left) Rudolf Diesel (1858–1913) and (right) Henry Ford (1863–1947).

Figure 25.2. Two fermenters used to produce acetone, butanol, and ethanol in the USSR. From [3].

25.4 Biofuels

25.4.1 Biofuel basics

The Sun is the ultimate source of energy for production of all liquid fuels: solar energy is used by plants and microbes to fix atmospheric CO_2 as sugars, polymers of sugars (cellulose and glycogen), lipids, and hydrocarbons. In fact, fossil fuels are a result of photosynthetic activity, whereby ancient biomass has been converted into hydrocarbons under conditions of high temperature and pressure deep below the Earth's surface. Generally speaking, herbaceous and woody plant materials resulted in coal (e.g., lignite), whereas planktonic (i.e., algal) biomass was converted into petroleum [4]. Consequently, it is not surprising that two types of biomass are being developed for biofuels. However, developing approaches that supplant what nature has accomplished in millenia is going to be a tremendous challenge.

One recent report [5] comparing clean energy technologies suggests that biofuels are one of the least expensive technologies to implement at the gigaton scale (the scale required to reduce CO_2 equivalent emissions by 1 Gt per year, roughly 2% of global CO_2 emissions): to reach the gigaton scale by 2020, biofuels would require an investment of $383 billion, compared with $61 billion for building efficiency, $445 billion for construction materials, $919 billion for geothermal, $1.27 trillion for nuclear, $1.38 trillion for wind, $1.71 trillion for solar photovoltaic, and $2.24 trillion for concentrated solar.

Although this analysis is very useful in scaling the impact of clean energies/materials, clearly many of these technologies are not interchangeable. Only biofuels have sufficient energy density for point-to-point transportation (cars and trucks) of goods over long distances. For example, biodiesel has 400 times the energy density of a lead–acid battery. Although higher-energy batteries exist in the market, none are comparable to biofuels in energy density, not to mention cost. Therefore, it is critical to develop biofuel alternatives to petroleum fuels. This will largely be driven by favorable government policies: increasing the allowable biofuel content in mixed fuels, creating a carbon tax, and providing government grants and subsidies are all critical to stimulating growth in the biofuels market [1].

Physical and chemical properties of fuels

Broadly speaking, biofuels are being targeted for three types of engines: diesel, Otto (gasoline), and jet (turbine). Given the multi-trillion-dollar refining/transportation infrastructures, it is desirable to produce biofuels with physical and chemical properties similar to those of existing (petroleum-based) fuels. Therefore, the physical and chemical properties of petroleum products must be considered in the development of biofuel alternatives. As shown in Table 25.1, ethanol has a much lower energy content and requires a lower air-to-fuel ratio than does gasoline. Most automobile engines can handle 10% ethanol (the current US standard) (Chapter 31). However, higher percentages (e.g., E85) require increased jet diameters (carbureted engines) or larger injectors and higher fuel-rail pressures (electronic fuel-injection systems), among other changes noted below. Moving to higher biofuel percentages in transportation fuels (for example, from the current 10% corn ethanol) is an

Table 25.1. A comparison of the properties of fuels

	Gasoline	Ethanol	No. 2 diesel	Biodiesel
Energy content (kW-h/l)[a]	8.99	5.91	9.94	9.25
Autoignition temperature (°C)	257	423	~300	177
Air/fuel ratio	14.7	9.0	14.7	13.8
Vapor pressure (Pascal at 38 °C)	0.55 to 10.3×10^6	1.6×10^5	$< 1.4 \times 10^3$	$< 1.4 \times 10^3$
Water solubility in fuel at 21 °C	Negligible	Miscible	Negligible	Negligible
Freezing point (°C)	−40	−114	−40 to −34	−3 to 19

Source: US DOE EERE Alternative Fuel and Advanced Vehicle Data Center (http://www.afdc.energy.gov/afdc/) and tsocorp technical document (http://www.tsocorp.com/stellent/groups/corpcomm/documents/tsocorp_documents/msdsbiodiesel.pdf)
[a] Lower heating value.
[b] Cloud point.

important goal and will drive growth in biofuels by increasing the size of the domestic US market.

Converting a diesel vehicle to run on plant oils typically requires some sort of heating system to maintain liquid biodiesel in the tank, as well as fuel lines to compensate for the higher freezing temperature compared with that of No. 2 diesel. Chemical conversion (transesterification) of vegetable oils produces a fuel (biodiesel) that has comparable properties to conventional diesel. It is miscible with No. 2 diesel and can therefore be blended over a wide range of compositions, such as B2, 2% biodiesel; B20, 20% biodiesel (the current US limit); and B100, 100% biodiesel. However, high percentages typically require replacing seals because many elastomers are incompatible with high-percentage biodiesel.

Another critical property is the solubility of water in the fuel. Because water has virtually no solubility in gasoline and diesel (Table 25.1), the existing petroleum infrastructure has not been designed to be resistant to water. This property is more challenging than energy density because ethanol often contains some percentage of water and can damage existing infrastructure (e.g., oil pipelines). Even if all of the water is removed, ethanol will absorb moisture (hygroscopically); therefore, it is impractical to maintain water-free ethanol. This means that pipes might need to be replaced with corrosion-resistant materials before being switched to ethanol, which is possible for vehicles but very expensive for oil-distribution systems.

Jet engines, unlike diesel and automobile engines (which can handle a range of fuel compositions), require the highest standards for fuel performance. This is due to the low temperatures to which engines are exposed, the ultrahigh performance (and cost) of the engines, and the consequences if the engines fail. For these reasons, a fuel must pass extremely strict standards to be certified as a jet fuel, making this the most difficult fuel to replace with a biofuel.

Economic, land, and water-use considerations

Currently, all biofuels, with the possible exception of sugarcane ethanol in Brazil, are more expensive than gasoline [6]. Brazilian ethanol is a good illustration of what it takes for biofuels to be economically viable: the climate allows production of a feedstock (sugarcane) at a very high density (tonnes per hectare); the sugar is easily liberated in a form suitable for conversion to ethanol through fermentation by yeast; and the sugarcane bagasse remaining is burned to provide inexpensive energy for ethanol purification (distillation), with excess energy being sold as electricity to the grid. Additionally, there is an existing infrastructure for transporting the fuel, and the industry has received extensive government subsidies and favorable tax policies.

Because all biofuels originate from the photosynthetic conversion of CO_2 into reduced carbon compounds, biofuel production will ultimately be intimately linked to agriculture. Consistently with this line of thinking, biofuel production should be designed around the different types of climate, soil types, and feedstocks that are available at a local level. Figure 25.3 shows the highly varied land use across the USA. Whereas the farming of corn is highly efficient and the sugar and starch is efficiently used for biofuel production, there is an overall negative impact because this approach puts biofuels in direct competition with food production for limited arable land, fresh water resources, and so on. Therefore, an intensive effort is being made to utilize non-food feedstocks that don't complete with resources used for food production. For example, in the USA, the coasts could be used for tree biomass, whereas the midwest could be used for producing grass (e.g., *Miscanthus* and switchgrass) and agricultural waste (corn stover). In contrast, the desert southwest has limited forest and arable land, but vast areas with saline water reservoirs and intense sunlight where saltwater algae could be developed for biofuel production.

Land-use shares

■ Cropland
■ Grassland pasture and range
■ Forest-use land
■ Special uses/urban/other land

Figure 25.3. The US distribution of land uses by region in the 48 contiguous states in 2002. The size of the pie charts is proportional to the land area in each state. Shares for Alaska are 25% in forest-use land, 75% in special uses/urban/other land, and less than 0.5% in all other uses. Shares for Hawaii are 5% in cropland, 24% in grassland pasture and range, 38% in forest use, and 33% in special uses/urban/other land. From http://www.ers.usda.gov/data/majorlanduses/map.htm.

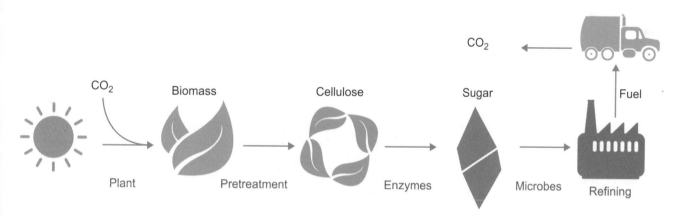

Figure 25.4. Conversion of plant biomass into biofuels.

25.4.2 Feedstocks for producing biofuels

Plant feedstocks

Limited arable land, fresh water, and negative effects on food supply have led to major efforts to develop energy crops from non-food plants and trees that can grow on marginal, degraded, and other lands unsuitable for agriculture. The resulting biomass has the potential for large-scale conversion into biofuels and biomaterials [7], for example, using a bioreactor in which microbes convert the cellulose into sugars that can be metabolized (by microbes) to produce biofuels, as shown in Figure 25.4.

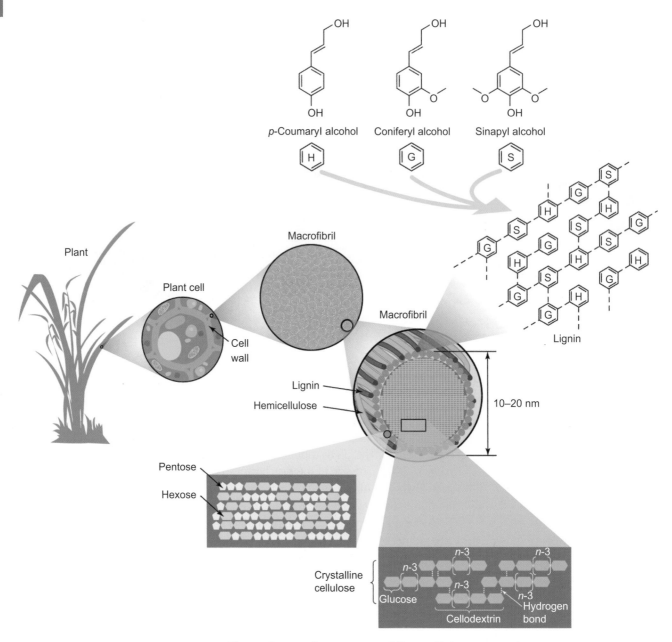

Figure 25.5. The structure of lignocellulose. From [8].

Plant biomass is primarily composed of lignocellulose [8] (Figure 25.5) (cf. Chapter 26), a structural material containing cellulose, hemicellulose, and lignin in a ratio of approximately 5:3:2. In nature, cellulosic biomass resists decomposition, with the result that degradation is slow. This recalcitrance is a result of several factors in combination, including the crystallinity of cellulose, the hydrophobicity of lignin, and the encapsulation of cellulose by the lignin/hemicellulose matrix.

As discussed in detail in Chapter 26, there are many technologies for increasing the accessibility of cellulose and hemicellulose, including treatment with acids at high temperature and conversion to amorphous cellulose using ionic liquids [9]. These pretreatments serve to open up the plant cell wall, through hydrolysis of the hemicellulose and redistribution of the lignin [10], and, in the case of ionic liquids, to convert the cellulose from its crystalline form to an amorphous state [11][12]. This dramatically increases the rate of subsequent enzymatic digestion through increased accessibility to the cellulose. However, because of the capital and/or material costs of these methods [11], as well as the formation of compounds that are inhibitory to downstream fermentation, these techniques are currently not economically feasible.

Once the bonds have been exposed, hydrolytic enzymes (derived from microbes) can efficiently convert the cellulose into sugars (i.e., glucose). The sugars can then be made into alcohols using fermentative metabolism (see Section 25.4.3). Soluble alcohols (e.g., ethanol) are isolated using distillation. Because this is an energy-intensive

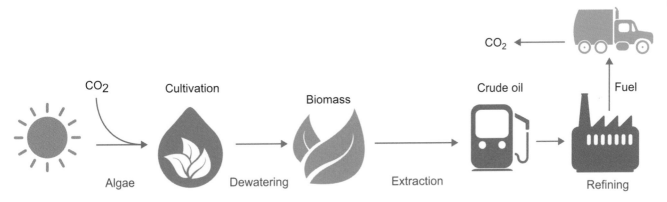

Figure 25.6. An overview of the algal biofuel process.

process, it is desirable to produce insoluble fuels. For example, higher alcohols (e.g., butanol) exhibit limited solubility in water and will phase separate from the fermentation broth, reducing the energy required for their recovery. However, there are many challenges to this approach: obligate anaerobes such as *Clostridia* are difficult to grow, making it desirable to produce these fuels in other microbes, ideally *E. coli* (*Escherichia coli*) given its short generation time, model organism status, and tractable genetics. However, butanol is more toxic than ethanol, making it difficult to produce the high concentrations required for efficient production. (See Chapter 26.)

Algal feedstocks

Microalgae are photosynthetic organisms that can convert sunlight and CO_2 into a range of fuels (Figure 25.6). The most famous example is *Botryococcus braunii*, which is known to produce some 50% of its dry mass as long-chain hydrocarbons [13]. Not surprisingly, this is a very slowly growing organism (with a doubling time of many days) – no doubt because it is using so much energy making hydrocarbons. However, because of limited fresh water supplies, biofuel production in this and other fresh water algae is a nonstarter. Instead, a solution that is more likely to be viable is biofuel production in halophilic (salt-loving) species that can be grown using saline water (potentially ocean water). There are many rapidly growing marine/halophilic algae known to produce large quantities of lipids [13], fatty acids, and triacylglycerols, often in response to nutrient stress [14].

The two major approaches for the cultivation of algae are open ponds and closed bioreactors. Each has advantages: open ponds are less expensive, whereas closed bioreactors are more productive and have fewer issues with contamination [13]. In either case, the efficient removal of water is a major technical challenge. To illustrate this point, consider the difference in obtaining oils from plants compared with algae. Both have high oil contents, yet it is vastly easier to isolate the plant seeds from the plants and the oil from the seeds (typically by pressing) than it is to isolate the few-micrometer-diameter algae from the pond

or surrounding medium (i.e., through centrifugation) and the oil from the algae (typically using hexane extraction). To avoid the need for dewatering, an alternative approach might involve metabolically engineering organisms that can produce and excrete insoluble compounds, similarly to *B. braunii* but in faster-growing marine strains. A related approach, taking advantage of the low solubility in water, is biohydrogen production accomplished through alteration of the photosynthetic process [15][16]. Clearly, this process requires closed bioreactors to capture the hydrogen, and this fuel is not yet compatible with today's infrastructure for fuel transportation and storage.

Life-cycle assessment

Given the many options for producing biofuels (feedstocks, products, etc.), the enormous scale required to meet world transportation demands, and the limited pilot-scale production experiments to date, it is difficult to predict economic viability. Life-cycle assessment (see Chapter 41) is one attempt to take all of the aspects of fuel production into consideration [17]. The types of factors that must be considered include capital and operating costs, land, water, nutrients, and energy versus the value of the products produced.

On the basis of this type of assessment, it is unlikely that the Brazilian model of biofuels can be applied generally in other regions of the world – there are many other factors that complicate efficient bioethanol production. This consideration is reflected by the higher estimated price of ethanol in the USA (largely due to a higher glucose cost) than in Brazil. Even with this economic conclusion, US agriculture is heavily dependent on petroleum to drive tractors, make fertilizers, and so on. It is therefore critical to consider not just the value of the products produced but all of the energy inputs, material costs, capital costs, waste products, and so on when evaluating the viability of a biofuel proposition [17].

The scale at which biofuels must be implemented to replace all petroleum fuels is enormous. Figure 25.7 shows land-requirement estimates for replacing US gasoline use for the major approaches discussed. Although

Figure 25.7. A comparison of land requirements for several leading biofuel approaches. Note: AMOPs are aquatic microbial oxygenic photoautotrophs, more commonly known as algae, cyanobacteria, and diatoms. From [18].

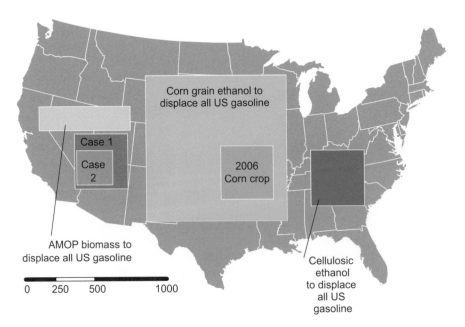

cellulosic and algal feedstocks have smaller land-use requirements than that for corn ethanol, the scale is still unfeasible given current technologies. Major innovations are required in order to decrease land use by increasing the productivity of biofuel crops. This will likely be achieved by engineering microbes [7] and plants [2] that are optimized for biofuel production.

25.4.3 Metabolic pathways for making biofuels and biomaterials

Consider the conventional approaches used to make fuels and materials: raw feedstocks (i.e., metal ores, petroleum, and fuels, e.g., coal and natural gas) are isolated from the environment. The fuels are combusted to generate electricity and provide energy to drive the production of central precursors, such as ethylene. These precursors can then be converted into a vast array of products, again using power/electricity. Similarly, through many decades of biochemical research (see further reading on biochemical pathways for details and citations), we know that microbes take raw materials and energy from the environment and use them to produce reducing equivalents (think electricity) such as the reduced form of nicotinamide adenine dinucleotide (NADH); readily available energy [adenosine-5′-triphosphate (ATP)]; and central metabolic precursors, namely pyruvate and acetyl coenzyme A (acetyl-CoA). Microbes use these critical metabolites to produce reduced carbon compounds that can be used as biofuels or can be converted into biofuels. In many ways, these reactions are similar to the conversion of ethylene into a vast array of products through polymerization, cyclization, oxidation, and reduction. Indeed, engineers and scientists can learn a great deal by studying the "molecular machines" (enzymes) that orchestrate the exquisite specificity and selectivity of these reactions. Enzymes are already extensively used in biotechnology to make drugs, fuels, and materials, and are inspiring the development of new nanotechnologies. The following subsections introduce major metabolic pathways and key enzymes for biofuel production, and then present the toolbox for altering these pathways for biofuel and biomaterial production.

Fermentation

Catabolic processes, defined as the breakdown of molecules to produce cellular energy and metabolites, are driven by electron acceptors. Respiration uses external electron acceptors such as oxygen, which is a very favorable acceptor, allowing complete oxidation of glucose to carbon dioxide and thus providing ~38 ATP molecules per glucose molecule. Fermentative metabolism often occurs in the absence of an external electron acceptor, under which conditions the cell utilizes internal electron acceptors and incomplete oxidation of glucose. Hence, this process is much less favorable than respiration and produces only two ATP molecules per glucose molecule. The simplest fermentation is lactic acid fermentation (Figure 25.8, also known as homo-lactic fermentation). In a multi-step reaction (glycolysis), glucose ($C_6H_{12}O_6$, six carbons) is converted into two molecules of pyruvate ($C_3H_4O_3$, three carbons), producing two ATP molecules and reducing two NAD^+ ions to two NADH molecules with the overall stoichiometry given in the following equation:

$$C_6H_{12}O_6 + 2ADP + 2Pi + 2NAD^+ \rightarrow 2C_3H_4O_3 + 2ATP + 2NADH + 2H^+ \tag{25.1}$$

(Pi stands for inorganic phosphate). In this case, the glucose goes through the intermediate fructose 1,6-bisphosphate, which is cleaved into dihydroxyacetone

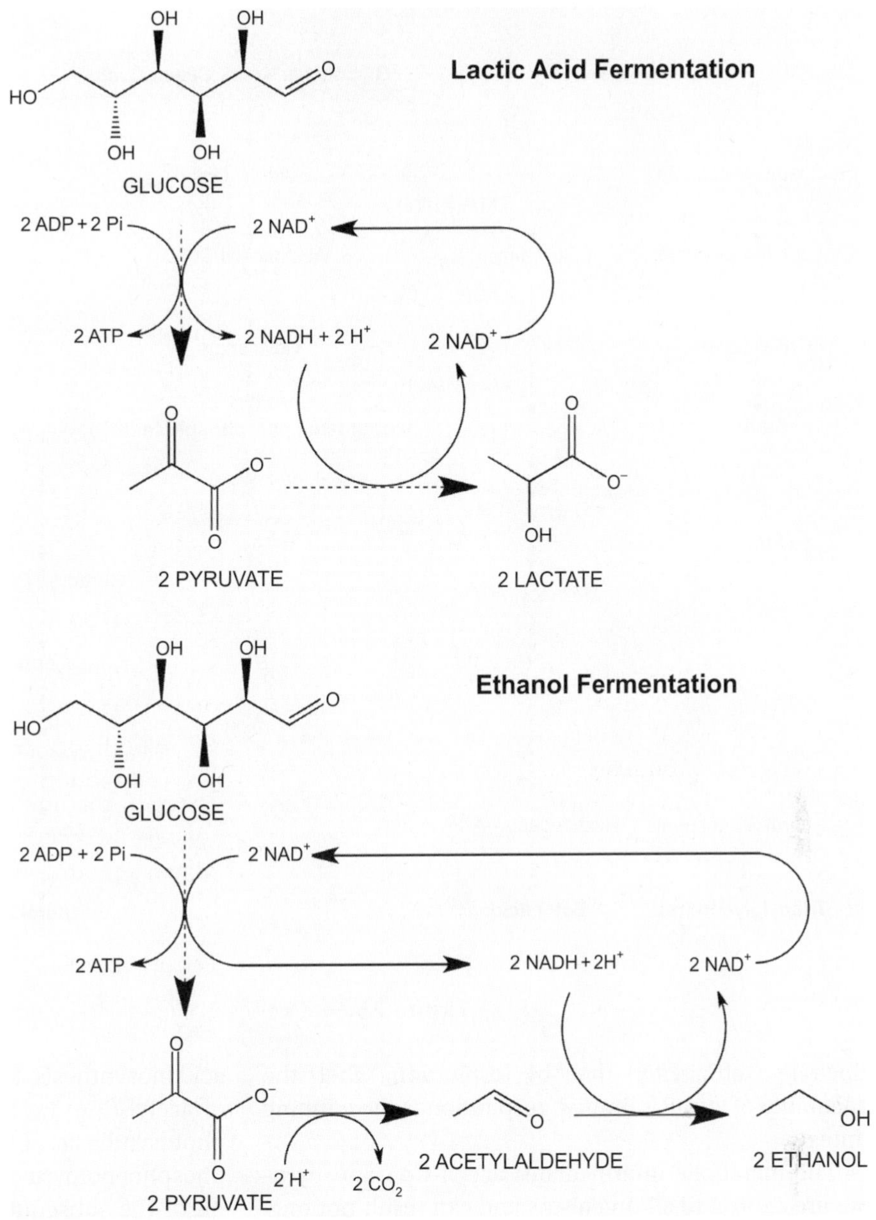

Figure 25.8. Fermentation of glucose to ethanol and lactate.

phosphate and glyceraldehyde 3-phosphate – both of which are subsequently converted to pyruvate. Because this reaction depends on NAD^+, the cell must have a mechanism of re-oxidizing NADH to NAD^+. In lactic acid fermentation, this is accomplished by transferring electrons from NADH to pyruvate, resulting in the formation of lactic acid. In fact, sore muscles following vigorous exercise result from the production of lactic acid by this process in oxygen-deprived muscles.

Ethanol is produced by a similar reaction (Figure 25.8), although the details of the reaction differ between eukaryotes (e.g., yeast) and prokaryotes (i.e., *E. coli* and *Clostridia*; also see Chapter 26). In many eukaryotes, the key enzyme in this pathway is pyruvate decarboxylase, which, as the name suggests, catalyzes the release of carbon dioxide from pyruvate, producing

acetylaldehyde, which is subsequently reduced by NADH to produce ethanol. Prokaryotes typically first convert pyruvate into acetyl-CoA through pyruvate formate lyase (also producing formate). This acetyl-CoA is then converted to acetylaldehyde, which is then reduced to ethanol. However, *Clostridia* can also use acetyl-CoA to produce higher alcohols (e.g., butanol) by essentially acetylating one acetyl-CoA molecule with another through acetyl-CoA acetyltransferase to produce acetoacetyl-CoA. This four-carbon acetoacetyl group undergoes multiple rounds of reduction, releasing the CoA and producing butanol. Interestingly, yeast (*Saccharomyces cerevisiae*) ferments even under aerobic conditions, presumably because it gives a competitive advantage: when there is an abundance of sugar, fermentation allows yeast cells to produce ATP faster

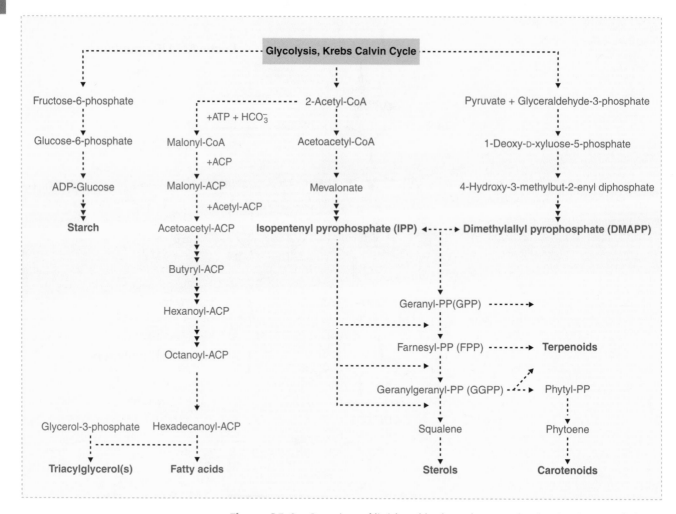

Figure 25.9. Overview of lipid and hydrocarbon production in plants and algae.

(albeit less efficiently) than by respiration; when the availability of sugar is limited, respiration is the winning strategy.

The metabolic intermediates acetyl-CoA and pyruvate are central to all organisms and can result not only from glycolysis but also from photosynthesis in plants and algae. As shown in Figure 25.9, a wide range of biofuel-related molecules are produced from these intermediates, for example, fatty acids and triacylglycerols that are chemically converted into biodiesel (transesterification). Hydrocarbons are also produced directly through the polymerization of isopentenyl pyrophospate (IPP) and dimethylallyl pyrophosphate, resulting in compounds including sterols (e.g., cholesterol), terpenoids (e.g., limonene), and carotenoids (e.g., vitamin A), many of which have the potential to be valuable co-products (also see Chapter 26).

Fatty-acid biosynthesis

In general, fatty-acid biosynthesis begins with the action of acetyl-CoA carboxylase (ACC), which converts some of the acetyl-CoA into malonyl-CoA (Figure 25.9). Generally speaking, this is the first committed step in fatty-

acid biosynthesis. In plants and algae, the major sources of acetyl-CoA are a result of the Calvin cycle (used in photosynthesis; cf. Chapters 24 and 28) or conversion of phosphoenolpyruvate coming from respiration (dark). All of the subsequent reactions comprise a repetitious but tightly controlled polymerization process in which the carbon chain is attached to acyl carrier proteins (ACPs) and extended two carbons at a time, after which it is subjected to four reduction steps. This process is repeated to produce C_{16} and C_{18}-ACP saturated chains (C18:0 and C16:0). Unsaturated bonds are a result of stearoyl-ACP desaturase [19]. Fatty acids are produced from these acyl-ACPs through hydrolysis of the thioester-ACP linkage by acyl-ACP thioesterase. Omega-3 fatty acids such as α-linolenic acid have commercial value as nutritional supplements and can be valuable co-products to increase the overall viability of the process.

Triacylglycerol biosynthesis

In addition to hydrolysis of the acyl-ACPs to form fatty acids, acyl-ACPs can be linked to glycerol (Figure 25.9). Because glycerol has three hydroxyls, it is possible to add between one and three acyl chains. Typically, plant oils

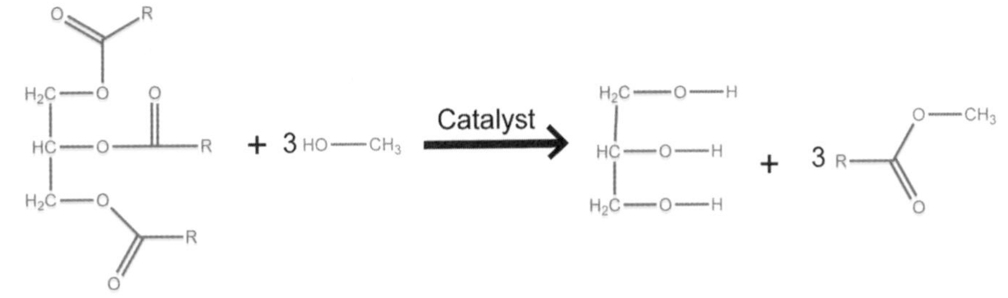

Figure 25.10. Transesterification of triacylglyerols with methanol to produce fatty-acid methyl esters (biodiesel).

(i.e., vegetable oil) have three fatty acids (triglycerides). The process of linking acyl chains to glycerol proceeds by reacting the acyl-ACP to glycerol-3-phosphate or mono-acylglycerol-3-phosphate [20]. This process occurs in the chloroplast of plants and algae (cytosol of cyanobacteria). Diacylglycerols (DAGs) are formed by dephosphorylation of diacylglycerol-3-phosphate. Triacylglycerols (TAGs) are produced by adding yet a third acyl group to the alcohol of the DAGs through the action of diacylglycerol acyltransferase. Although this is the predominant pathway for TAG production in plants and algae, it is also possible to produce TAGs from the condensation of DAGs and phospholipids [21]. In plants, a large portion of the fatty acids produced inside the plastid (chloroplast) is exported to the cytosol, especially the endoplasmic reticulum for conversion into glycerol lipids [20].

The high freezing point of acylglycerols is a challenge for their direct use in biodiesel vehicles, often requiring two fuel tanks, one for petroleum diesel and a second for vegetable oil. In other cases, vehicles have heated gas tanks and fuel lines to "melt" the fuel. Conversion of acylglycerols into methylesters (biodiesel) through transesterification (Figure 25.10) results in fuels with performance comparable to that of No. 2 diesel. Here, glycerol esters are converted to methyl esters using three mole equivalents of methanol for a TAG and a strong base catalyst (e.g., potassium hydroxide) [13][19]. Note that this reaction produces glycerol as a waste product that can be metabolized by many microbes and, therefore, has potential as a low-cost feedstock for additional biofuel production (although it is unlikely to be the only one).

Terpenoid biosynthesis

Isoprenoids are the oldest and most diverse class of metabolites, having been recovered from sediments up to 2.5 billion years old and comprising over 30,000 known compounds [22]. These high-energy hydrocarbons have excellent potential as biofuels. Although all isoprenoids are formed through condensations of isopentenyl pyrophosphate (IPP, a C_5 hydrocarbon), there are two routes by which to form IPP in plants and algae, the 1-deoxy-D-xylulose-5-phosphate (DXP or non-mevalonate) pathway and the mevalonate pathway (MVA) [23], as shown in Figure 25.9. In the DXP pathway,

IPP is made from the condensation of pyruvate and glyceraldehyde-3-phosphate. In the mevalonate pathway, it is formed from acetyl-CoA. Typically, IPP generated through the mevalonate pathway is cytosolic, whereas IPP generated through the DXP pathway is localized in the plastid [24]. As shown in Figure 25.9, IPP is interconverted into dimethylallyl pyrophosphate (DMAPP) through isopentenyl pyrophosphate isomerase. The condensation of IPP and DMAPP by dimethylallyltransferase results in the formation of geranyl pyrophosphate (GPP, a C_{10} branched hydrocarbon). GPP is further transformed into monoterpenoids (i.e., limonene, "orange oil") or can be extended with the addition of another IPP unit through geranyltranstransferase to form farnesyl pyrophosphate (FPP, a C_{15} branched hydrocarbon). This, in turn, is transformed into sesquiterpenes and sterols (through squalene), which are used to produce the carotenoid staphyloxanthin, or extended with an additional IPP to form geranylgeranyl pyrophosphate (GGPP, a C_{20} branched hydrocarbon). A tremendous diversity of molecules can be made from the resulting GGPP, including diterpenoids, higher terpenoids, and virtually all of the carotenoids.

Carotenoid biosynthesis

Carotenoids are another high-energy hydrocarbon produced in microbes and plants. These materials are largely linear molecules again derived from IPP and DMAPP (Figure 25.9), similar to natural rubber (but much shorter). They are often highly unsaturated and conjugated compounds involved in photoprotection and photosynthesis. In algae and higher plants, carotenoids are formed from GGPP through conversion to phytoene through phytoene synthase, which catalyzes the head-to-tail condensation of two GGPP molecules. Phytoene (C_{40}) is a colorless carotenoid and is thought to be a key regulatory point for determining flux into carotenogenesis [25].

Subsequent reactions in the carotenoid pathways largely involve the desaturation (increasing the conjugate bond length, thereby changing the color), cyclizing the ends, and further oxidization steps. Phytoene is converted to phytofluene and ζ-carotene through two sequential dehydrogenation reactions by phytoene desaturase (PDS). Both in cyanobacteria and in plants, the subsequent dehydrogenation of ζ-carotene into neutosporene occurs through ζ-carotene desaturase (ZDS). The majority of the carotenoids are formed by further oxidation into lycopene by ZDS. Lycopene is the last linear (branched but not cyclized) carotenoid and is then transformed into a number of the major well-known carotenoids such as lutein, α- and β-carotene, zeaxanthin, and astaxanthin. Many of the carotenoids are valuable compounds. For example, β-carotene (vitamin A) and astaxanthin (salmon feed) are extremely high-value products, are commercially produced using algae, and could be side products providing valuable co-products for biofuel ventures – although these relatively niche markets could quickly become saturated, especially at the scale required to replace petroleum fuels.

25.4.4 Biomaterials and industrial chemicals

Because plastics and industrial chemicals are largely obtained from petroleum feedstocks, it will be critical to find renewable alternatives as these feedstocks become more expensive. They can either be produced directly in the microbe or subsequently chemically converted. Biomaterials have the potential to sequester carbon as useful materials and, therefore, to be carbon-negative. For example, CO_2 fixed by plants and microbes converted into stable materials such as recyclable bioplastics results in a net decrease in atmospheric CO_2. Therefore, widespread sequestration of carbon in biomaterials will also be important for stabilizing atmospheric CO_2 levels.

Bioplastics can be made in microbes in three ways: (1) some species of bacteria produce plastics directly as a means of storing carbon and energy, (2) some microbial metabolites can be further processed into polymers, and (3) microbes can now be engineered to make new feedstocks for polymer production. An example of the first case is *polyhydroxyalkanoates* (PHAs), which represent a diverse class of linear polymers produced fermentatively by bacteria as storage compounds. *Ralstonia eutropha* has been the most widely used wild-type strain for the industrial production of PHB and could grow to very high cell densities: in one case, cell densities of 200 $g\,l^{-1}$ were achieved with an 80% PHB content [26]. They are thermoplastics and can be processed using injection molding, similarly to polypropylene; and they are UV stable compared with other bioplastics. The second case typically utilizes sugar-derived chemicals synthesized by micro-organisms (e.g., succinic acid, glycerol, hydroxypropionic acid, lactic acid, and levulinic acid) as "platform chemicals" (feedstocks) for industrial processes. For example, polylactic acid (PLA) is a thermoplastic with properties similar to those of poly(ethylene terephthalate) (PETE). It is biodegradable and typically made by fermenting cornstarch into lactic acid, which is then converted to lactide and polymerized to PLA. Because PLA is biodegradable, it is widely used for making compostable plastics (e.g., cups) and biomedical applications (e.g., sutures, stents, tissue scaffolds).

Another famous example of this approach is the "biocatalytic" production of nylon using monomers produced in microbes. In this case, metabolically engineered *E. coli* was used to convert glucose to muconic acid with 22% yield. The muconic acid is then hydrogenated to adipic acid [27], a monomer that is conventionally produced industrial from petroleum. This process has the potential to significantly reduce the reliance on petroleum to supply the $\sim 2.2 \times 10^9$ kg of adipic acid used each year to produce nylon 6,6.

Finally, as an example of the third case, it is possible to engineer microbes to make novel precursors using biotechnology (as described in the next section). For example, a cyanobacterium (*Synechococcus elongatus* PCC7942) has been engineered to produce a feedstock (isobutyraldehyde) that can be used to make isobutylene (a polymer precursor) [28]. In addition to these organic materials, microbes can also fix CO_2 as inorganic materials that might be useful for carbon capture and storage [29], such as bioconcrete. Considering the complex structures constructed by calcifying algae (e.g., Figure 25.11), it would seem that nanomaterials that are much more advanced than concrete are possible, although engineering systems to produce nanomaterials will require new insights into the process of microbial calcification.

25.4.5 A toolbox for optimizing microbial biofuel production

Improving both the practical and the economic viability of biofuels will depend on increasing the efficiency of growth, the yield/concentration of biofuel production, and the efficiency of biofuel harvesting and refinement. Because microbes have not evolved to be turned into biofuels, optimizing production is a major challenge. Fortunately, scientific developments over the past several decades have greatly improved human understanding of microbes to the point that it is now possible to construct microbes largely *de novo* [36]. These developments are essential in order to optimize microbes for the production of affordable biofuels at the scale needed to address climate change and petroleum replacement.

Genomics
Genomes are the complete set of hereditary information and include thousands of genes each of which encodes one or more functions, such as the 4.6-mega-base-pair (Mbp) *E. coli* genome and the 3,200-Mbp haploid

Figure 25.11. Complex calcium carbonate-based structures produced by the calcifying alga *Reticulofenstra sessilis*. From [30].

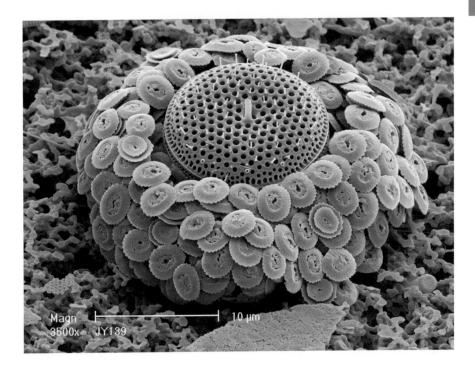

Figure 25.12. *E. coli* K-12 strain MG1655. From [31].

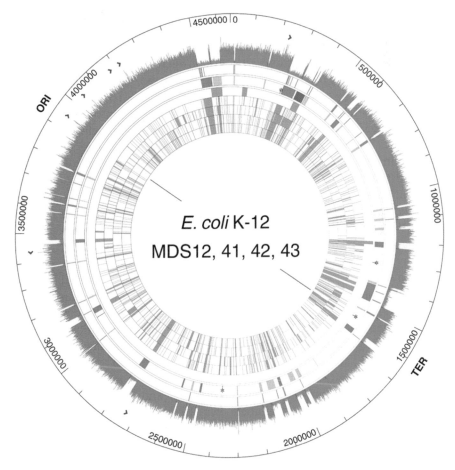

human genome. Figure 25.12 shows an example of an *E. coli* genome that has been refined using synthetic biology [31] (discussed below). This represents the precise determination of the sequence of millions of nucleotides and the association of their function on the basis of sequence similarity to known genes. Amazingly, it is now routine to sequence an entire microbial genome in a matter of weeks or months (depending on the complexity).

Figure 25.13. Operon structure. Five genes of the *E. coli* trp operon (trpA, -B, -C,-D, and -E) are transcribed as one messenger RNA (mRNA) molecule under the control of a promoter. This simple solution for the coordinated control of related genes is common in bacteria.

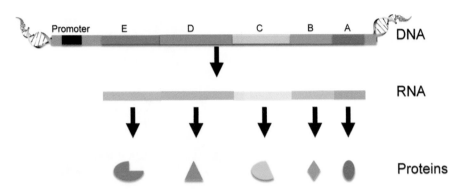

However, typically, some 30–50% of an overall sequence does not have any known function, making it difficult for the sequences to be used for biofuel production. The *E. coli* strain K-12, on the other hand, has one of the best-characterized genomes, with 87% of genes having functional assignments [31], (either from homology based predictions or biochemical characterization) which makes it one of the most popular platforms, or "chassis," for genetic engineering.

Bacterial genomes are typically organized in circular chromosomes (Figure 25.12). In addition, many genes, such as those encoding antibiotic resistance, are found on one or several extrachromosomal DNA molecules (plasmids). Genes are often grouped together as operons (Figure 25.13) on the basis of their function under the control of specific regulatory elements (promoters). This allows genes to be expressed in concert. For example, the tryptophan biosynthetic pathway (Figure 25.13) has five enzymes in the pathway, all of which are turned on through the binding of a specific protein (transcription factor) to the promoter. These genes are transcribed into an RNA molecule and subsequently translated into individual proteins. The transfer of information from DNA to RNA to protein is known as the "central dogma" of biology.

Many microbes have evolved to exchange plasmid DNA as a means of "sharing" genes between microbes. This fact has been exploited to engineer microbes: provided that one knows the genome sequence and how the gene is turned on (i.e., the identity of the promoter), it is possible to insert foreign plasmids into the microbe (transformation) and have the microbe express the gene (make the desired protein). Some microbes (e.g., many cyanobacteria) are naturally transformable and will perform uptake of DNA on their own. However, most require disruption of the cellular membranes to provide an efficient means for DNA uptake. This is typically accomplished using either electrical discharge (electroporation) or small-particle bombardment (a gene gun) [32].

It is also common to add and delete genes from native DNA molecules (as opposed to foreign plasmids), for example, increasing biofuel production by adding foreign biosynthetic genes and/or deleting genes from competing pathways [32]. This is accomplished using the inherent sequence-specific affinity of DNA. Sequences to be inserted are flanked by specific sequences that target the region where one or more genes are to be inserted (or deleted). This "vector" is then integrated through homologous recombination into the host DNA by taking advantage of the microbe's own DNA repair mechanisms.

In all cases, genes of interest are conferred a competitive advantage by being co-localized with a selectable marker, the integration of the foreign DNA elements is a relatively rare occurrence. How does one locate "clones" that have successfully been transformed? This is accomplished using either selection or screening [33]. Selection is preferable because it is highly specific (and easy). For example, insertion of a gene that encodes a protein that confers antibiotic resistance allows effective selection of transformed cells because such cells are the only ones that survive in the presence of the antibiotic. Screening requires more effort. In this case, the genetic element inserted includes a gene that allows the identification of transformed cells. Green fluorescent protein (GFP) is commonly used for this purpose, and allows screening of transformed cells. It is also useful for determining the conditions for gene expression, for example, to confirm that the genes are expressed under the desired conditions.

As illustrated in Figure 25.13, genomics is not the whole story. The existence of a gene does not necessitate its expression, i.e., that it folds into an active protein and produces the desired metabolic effect. The development of "systems" approaches that utilize global RNA (transcriptomics), protein (proteomics), and metabolites (metabolomics) is critical to re-wiring metabolic networks for biofuel production [34].

Flux-balance analysis

Metabolic networks are highly complex and interconnected, with many parts that must fit and work together, making it extremely difficult to determine *a priori* which genes to add and delete, much less how the whole system should be regulated. How, then, can one determine which genes to add and delete to increase the

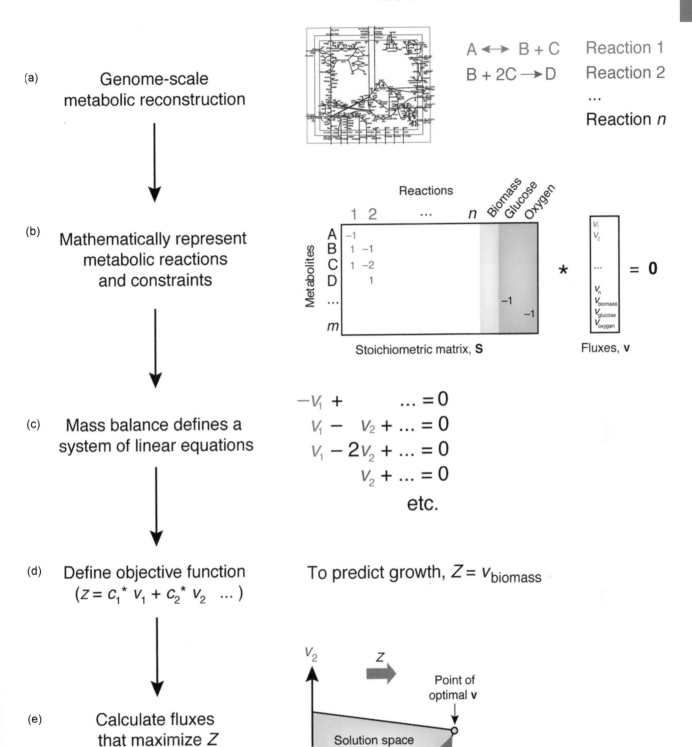

(a) Genome-scale metabolic reconstruction

$A \longleftrightarrow B + C$ Reaction 1
$B + 2C \longrightarrow D$ Reaction 2
...
Reaction n

(b) Mathematically represent metabolic reactions and constraints

Reactions

$$
\begin{array}{c}
\text{Biomass} \\
\text{Glucose} \\
\text{Oxygen}
\end{array}
$$

Metabolites

$$
\begin{array}{c|cccccc}
 & 1 & 2 & \cdots & n & & \\
A & -1 & & & & & \\
B & 1 & -1 & & & & \\
C & 1 & -2 & & & & \\
D & & 1 & & & & \\
\cdots & & & & & -1 & \\
m & & & & & & -1
\end{array}
$$

Stoichiometric matrix, **S**

$* \begin{bmatrix} V_1 \\ V_2 \\ \cdots \\ V_n \\ V_{\text{biomass}} \\ V_{\text{glucose}} \\ V_{\text{oxygen}} \end{bmatrix} = 0$

Fluxes, **v**

(c) Mass balance defines a system of linear equations

$$-V_1 + \quad\quad \cdots = 0$$
$$V_1 - V_2 + \cdots = 0$$
$$V_1 - 2V_2 + \cdots = 0$$
$$V_2 + \cdots = 0$$

etc.

(d) Define objective function $(z = c_1{}^* v_1 + c_2{}^* v_2 \ \cdots)$

To predict growth, $Z = v_{\text{biomass}}$

(e) Calculate fluxes that maximize Z

V_2 z

Point of optimal **v**

Solution space defined by constraints

V_1

Figure 25.14. Flux-balance analysis. From [35].

production rate (flux) of a desired biofuel molecule? Genome-scale models can be constructed on the basis of the similarity of genes to those of known function, coupled with detailed refinement and updating of bioinformatics databases. Using the stoichiometric constraints of biochemical reactions and the assumption that every reaction in the microbe is designed to optimize an objective, for example biomass production (growth), one can develop models of metabolic flux [35]. Flux-balance analysis (Figure 25.14) represents the metabolic network as a system of linear equations that can be solved to predict the flux through individual reactions that is required in order to maximize growth. This approach is being used to determine the modifications (gene additions and deletions) that are required in order to maximize biofuel production. For example, it might be necessary to reduce flux to competing pathways that act as "sinks" for flux through the desired pathway. It is often necessary to add multiple copies of a biosynthetic gene to compensate for less efficient enzymes (slow reactions).

Synthetic biology

It is now routine to design and construct cells with functions that are not found in nature [33]. Recently, researchers have begun compiling useful "parts" into databases of genes (Figure 25.15). These parts are grouped in terms of their function, for example, biosynthetic, regulatory, reporter genes. Each part in the parts registry has a datasheet describing sequence information, function, and other related information. Parts are constructed following a BioBrick standardization that requires common sites for "sticking" (ligation sites) and "cutting" (restriction sites) parts to allow efficient assembly. Thinking of these parts analogously to car parts (engine, turn signal, etc.), one of the most fundamental parts is the "chassis" for the car around which the parts are assembled. As shown in the lower left of Figure 25.15, only a few model microbes are commonly used for synthetic biology (*E. coli*, yeast, etc.), and parts that are compatible with a selected chassis can be selected [36]. The small number of chassis is a major limitation for synthetic biology, and the development of new chassis is extremely critical to the field.

Directed evolution

Rational design and engineering of microbes still have many limitations. Even the best-characterized microbes do not always behave as expected, largely because some of the essential functions of genes are still not known. How does one engineer a system that one does not understand for a specific function? One answer is to take advantage of evolution/natural selection. If the selection process is designed such that the fittest organisms are the ones that best do whatever is desired, then the organisms can be allowed to optimize themselves. This process of directed evolution [37] (also called adaptive evolution, Figure 25.16) is a very successful and widely used approach. Take, for example, biofuel tolerance. Ideally, a microbe should be able to make high concentrations of biofuel, but, in most cases, biofuels are toxic at high concentrations. Therefore, transporters must be added to quickly remove the biofuel from the microbe. Yet, even this is not enough, because, at some concentration, the fuel begins to act as a solvent, lyzing the cell

membranes. In contrast, growth of the microbes under increasing concentrations of the biofuel could select for those that can (somehow) survive. Often, evolution works too slowly for human needs, so it is common to increase the number of mutations and even shuffle whole sections of DNA between generations.

As more is learned about how to engineer metabolic pathways and enzymatic catalytic steps, it should become possible to design micro-organisms that produce specified blends of fatty-acid alkyl esters and hydrocarbons as drop-in fuels for direct use as diesel, gasoline, or jet fuel. Ideally, it will also become possible to develop secretion systems that allow the cells to release the biofuel molecules to the surrounding medium for efficient collection.

Process integration

Given the challenges of biofuels replacing any significant portion of petroleum fuels, it is important to consider integrated processes that can further improve their economic viability. This will likely include the production of co-products and innovative approaches to improve production efficiency, for example, using anaerobic digestion of waste streams from cellulosic biofuel plants to generate energy for ethanol distillation. One proposal for an integrated process is shown in Figure 25.17. This design takes advantage of the saline water that is displaced when CO_2 from a power plant is injected into the ground for geological sequestration. It is critical that this water not be allowed to pollute the aquifer (fresh water). However, this water can be used to grow halophilic (salt-loving) algae to produce biofuels. In fact, CO_2 from the power plant increases the growth of algae, allowing CO_2 from fossil fuels to be sequestered by algae as biofuels, as well as co-products. Provided that there is an economic cost of producing CO_2 (and therefore of sequestering it), this integrated process would have a number of revenue streams (geologically sequestered carbon, carbon captured by algae, algal products, etc.), thus improving its economic viability.

25.5 Case study

Cost is a critical barrier to the viable production of biofuels. One of the best case studies of how synthetic biology can reduce production costs is the production of the drug artemisinin. Artemisinin derivatives are the key ingredients in artemisinin combination therapies (ACTs), the most effective treatments for *Falciparum malaria* disease, which results in some million deaths per year. In 2004, this highly effective treatment was too expensive for many patients at $2.40 per dose. This cost was largely due to the fact that the drug was produced in *Artemisia annua*, a slow-growing, low-production plant. To make this drug affordable, the Bill and Melinda Gates

DATA SHEET FOR BIOBRICKS
Part BBa_K274100
Part BBa_K274110
(Test in *E.coli* strain MG1655)

Key words

lycopene, red, coloured pigment, carotenoid, colour output, *E. coli* MG1655, CrtE (geranylgeranyl pyrophosphate synthase), CrtB (phytoene synthase), CrtI (phytoene dehydrogenase).

Basic Information

Registry Entry	Sequence information	Length	Remarks
BBa_K274100	CrtE CrtB CrtI	3385 bp	Convert FPP to lycopene.
BBa_K274110	R0011 CrtE CrtB CrtI	3448 bp	Put BBa_K274100 under constitutive promoter.

Pigment Biochemistry
Lycopene

Molecular formula: $C_{40}H_{56}$

Chemical structure:

Main absorbance wavelength: ~470 nm

Natural sources: tomatoes, red carrots, watermelons, papayas, etc.

Related compounds: carotenoid, e.g. β-carotene (see Part Bba_K274200).

Synthesis in *E. coli*

E. coli naturally synthesises IPP and DMAPP via Non-mevalonate Pathway, which are readily converted to FPP (colourless). FPP is sequentially converted by enzymes CrtE, CrtB and CrtI to lycopene.

*IPP: Isopentyl pyrophosphate. DMAPP: dimethylallyl pyrophosphate. FPP: farnesyl pyrophosphate.

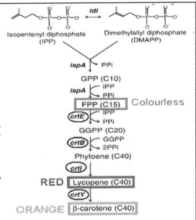

Isopentenyl diphosphate (IPP)   Dimethylallyl diphosphate (DMAPP)

ispA — PPi
GPP (C10)
ispA — IPP
— PPi
FPP (C15) Colourless
crtE — IPP
— PPi
GGPP (C20)
crtB — GGPP
— 2PPi
Phytoene (C40)
crtI
RED Lycopene (C40)
crtY
ORANGE β-carotene (C40)

Expression in *E.coli* strain MG1655

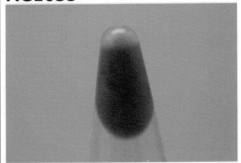

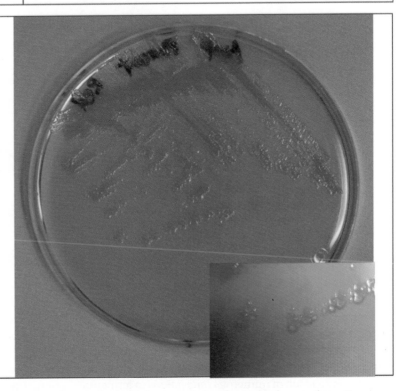

Top: Cell pellet of *E. coli* strain MG1655 transformed with BBa_K274110 (from 200mL LB culture at 37°C for 24 hours).

Right: Growth of *E. coli* strain MG1655 transformed with BBa_K274110 on agar plate. Results after overnight incubation at 37°C. Small insert shows single colonies (red, indicate production of lycopene).

Figure 25.15. The synthetic biology registry of standard parts (http://partsregistry.org/Catalog).

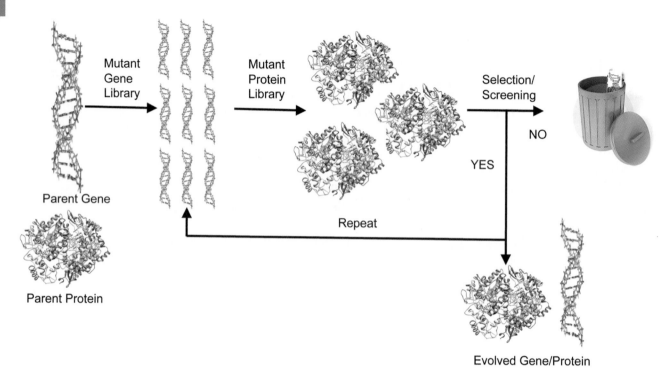

Parent Gene

Parent Protein

Mutant Gene Library

Mutant Protein Library

Selection/Screening

NO

YES

Repeat

Evolved Gene/Protein

Figure 25.16. Directed evolution workflow. (Adapted from [37].)

Figure 25.17. Integrated carbon sequestration and biofuel production. (Adapted from [38].)

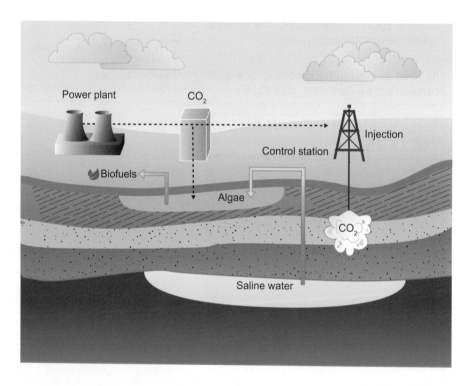

Power plant

CO_2

Injection

Control station

Biofuels

Algae

CO_2

Saline water

Foundation funded a joint project involving Amyris Biotechnologies, the University of California at Berkeley, and the Institute for OneWorld Health to synthesize the drug using microbes. This effort ultimately reduced the cost to $0.25 per dose. This dramatic reduction in cost was largely a result of numerous innovations in fermentation processing and synthetic biology [39]: increasing substrate and eliminating product loss through volatilization using a two-phase partitioning bioreactor increased yields by a factor of 20 to $0.5\,\mathrm{g\,l^{-1}}$ for the key artemisinin precursor amorpha-4,11-diene [40], (a terpenoid, see Section 25.4.3). Replacing key enzymes with more active enzymes from *Staphylococcus aureus* more than doubled production. Carbon and ammonia restriction together, with many other innovations, resulted in production of over $25\,\mathrm{g\,l^{-1}}$ in *E. coli*.

25.6 Summary

This chapter has presented the major feedstocks and metabolic pathways for production of biofuels and bio-materials. Although a wide variety of chemicals with the physical and chemical properties required for use as transportation fuels can be derived from microbes, it is a grand challenge to do this at the scale and cost required to replace petroleum, given economic and land-use considerations.

Lignocellulosic biofuels must overcome the inherent recalcitrance of lignocellulosic biomass (Chapter 10), whereas algal biofuels require the development of efficient harvesting and dewatering technologies. Both require high-yield production of fuel molecules that can be isolated with minimum energy inputs. Microbes already produce a wide range of fuels, including alcohols through fermentation, terpenes and isoprenoids that can be used for making new fuels for automobile engines, and fatty acids and acylglycerols that can be converted into biodiesel. In any case, biofuel processes will generate extensive co-products that improve the overall viability of the proposition. However, no existing microbes have the ideal growth and production properties, making it necessary to engineer them. Fortunately, genomic technologies that provide the toolbox for adapting and engineering microbes to meet these needs have been developed over the past decades. The microbe's genomics provides a roadmap of cellular functions, which, combined with flux-balance analysis, can be used to define the modifications required in order to increase flux to the desired product. The necessary parts can be selected from parts databases in a standardized format to optimize biofuel production. However, in many cases, it is necessary to harness evolutionary processes using directed evolution. This "synthetic-biology"-driven approach can be applied to a wide range of microbes to efficiently convert a wide range of feedstocks into sustainable biofuels.

25.7 Questions for discussion

1. Give three examples of materials that can be made from byproducts of biofuel production.
2. What types of useful nanomaterials might be produced by engineering calcifying algae?
3. What are the social implications of large-scale biofuel production using genetically engineered microbes?
4. List five environmental impacts that must be considered when farming a biofuel crop, such as switchgrass or algae.
5. Estimate the transportation costs associated with shipping ethanol from the primary ethanol-producing regions to coastal areas where it is consumed.
6. There is often a delay for biotechnology companies between needing money to do the research and making profit. Amyris used funding from the Gates Foundation to pioneer the technology for using microbes to produce the anti-malarial drug artemisinin, a technology that is now being employed to produce biofuels. Other companies such as Solazyme are trying to use green algae to produce expensive neutraceuticals as an initial product and then transition to large-scale production of biofuels. Are these viable business models? What should the government's role in such efforts be?

25.8 Further reading

- National algal biofuels technology roadmap (US DOE).
- The Gigaton Throwdown Initiative (http://www.gigatonthrowdown.org).
- **H. Alper** and **G. Stephanopoulos**, 2009, "Engineering for biofuels: exploiting innate microbial capacity or importing biosynthetic potential?," *Nature Rev. Microbiol.*, **7**(10), 715–723.
- **J.L. Fortman S. Chhabra, A. Mukhopadhyay** *et al.* 2008, "Biofuel alternatives to ethanol: pumping the microbial well," *Trends Biotechnol.*, **26**(7), 375–381.
- **J. Sheehan, T. Dunahay, J. Benemann**, and **P. Roessler**, 1998, *A Look Back at the US Department of Energy's Aquatic Species Program – Biodiesel from Algae*, report no. NREL/TP-580–24190, Golden, CO, National Renewable Energy Laboratory.
- 2009, *Expanding Biofuel Production: Sustainability and the Transition to Advanced Biofuels: Summary of a Workshop*, Washington, DC, US National Academy of Sciences.
- **Wang, H.H. Isaacs, F.J. Carr**, *et al.* Programming cell by multiplex genome engineering and accelerated evolution. Nature, **460**, 1–6.
- For more detail and citations on the history of biofuels please see D.D. Songstad & P. Lakshmanan & J. Chen *et al.* 2009, Historical perspective of biofuels: Learning from the past to rediscover the future. *In Vitro Cell. Dev. Biol. – Plant*, **45**, 189–192 DOI 10.1007/S11627-009-9218-6
- For more details on biochemical pathways including citations to numerous seminal works: Biochemical pathways: *An Atlas of Biochemistry and Molecular Biology*, Editor Gerhard Michal. 1st edition 1998, ISBN-10: 0471331309

25.9 References

[1] **P. Koshel** and **K. McAllister**, 2010 *Expanding Biofuel Production: Sustainability and the Transition to*

Advanced Biofuels: Summary of a Workshop, Washington, DC, National Academies Press, p. 2.

[2] **J. Gressel**, 2008, "Transgenics are imperative for biofuel crops," *Plant Sci.*, **174**(3), 246–263.

[3] **V. V. Zverlov, O. Berezina, G. A. Velikodvorskaya,** and **W. H. Schwarz**, 2006, "Bacterial acetone and butanol production by industrial fermentation in the Soviet Union: use of hydrolyzed agricultural waste for biorefinery," *Appl. Microbiol. Biotechnol.*, **71**(5), 587–597.

[4] **I. Suárez-Ruiz** and **J. C. Crelling** (eds.), 2008, *Applied Coal Petrology. The Role of Petrology in Coal Utilization*, New York, Academic Press, p. 318.

[5] **The Gigaton Throwdown Initiative**, http://www.gigatonthrowdown.org.

[6] **S. Spatari** and **D. Kammen**, 2009, "Redefining what's possible for clean energy by 2020," in *Job Growth, Energy Security, Climate Change Solutions*, ed. **N. Wishner**, San Francisco, CA, The Gigaton Throwdown Initiative, pp. 27–44.

[7] **E. J. Steen, Y. Kang, G. Bokinsky** *et al.*, 2010, "Microbial production of fatty-acid-derived fuels and chemicals from plant biomass," *Nature*, **463**(7280), 559–562.

[8] **E. M. Rubin**, 2008, "Genomics of cellulosic biofuels," *Nature*, **454**(7206), 841–845.

[9] **L. da Costa Sousa, S. Chundawat,** and **V. Balan**. 2009, "'Cradle-to-grave' assessment of existing lignocellulose pretreatment technologies," *Current Opin. Biotechnol.*, **20**(1), 1–9.

[10] **C. E. Wyman, B. E. Dale, R. T. Elander** *et al.*, 2005, "Coordinated development of leading biomass pretreatment technologies," *Bioresource Technol.*, **96**(18), 1959–1966.

[11] **A. Dadi, S. Varanasi,** and **C. Schall**, 2006, "Enhancement of cellulose saccharification kinetics using an ionic liquid pretreatment step," *Biotechnol. Bioeng.*, **95**, 904–910.

[12] **T. Eggeman,** and **R. T. Elander**, 2005, "Process and economic analysis of pretreatment technologies," *Bioresource Technol.*, **96**, 2019–2025.

[13] **Y. Chisti**, 2007, "Biodiesel from microalgae," *Biotechnol. Adv.*, **25**(3), 294–306.

[14] **J. Sheehan, T. Dunahay, J. Benemann,** and **P. Roessler** 1998, *A Look Back at the US Department of Energy's Aquatic Species Program – Biodiesel from Algae*, report no. NREL/TP-580-24190, Golden, CO, National Renewable Energy Laboratory.

[15] **D. B. Levin, L. Pitt,** and **M. Love**, 2004, "Biohydrogen production: prospects and limitations to practical application," *Int. J. Hydrogen Energy*, **29**(2), 173–185.

[16] **A. Melis, M. Seibert,** and **M. L. Ghirardi**, 2007, "Hydrogen fuel production by transgenic microalgae," in *Transgenic Microalgae as Green Cell Factories*, Berlin, Springer, pp. 108–121.

[17] **H. Kim, S. Kim,** and **B. Dale**, 2009, "Biofuels, land use change, and greenhouse gas emissions: some unexplored variables," *Environmental Sci. Technol.*, **43**(3), 961–967.

[18] **G. J. Dismukes, D. Carrieri, N. Bennette, G. M. Ananyev,** and **M. C. Posewitz**, 2008, "Aquatic phototrophs: efficient alternatives to land-based crops for biofuels," *Current Opin. Biotechnol.*, **19**, 235–240.

[19] **D. Antoni, V. V. Zverlov,** and **W. H. Schwarz**, 2007, "Biofuels from microbes," *Appl. Microbiol. Biotechnol.*, **77**(1), 23–35.

[20] **J. Ohlrogge** and **J. Browse**, 1995, "Lipid biosynthesis," *Plant Cell*, **7**(7), 957–970.

[21] **A. Dahlqvist, U. Ståhl, M. Lenman,** *et al.*, 2000, "Phospholipid:diacylglycerol acyltransferase: an enzyme that catalyzes the acyl-CoA-independent formation of triacylglycerol in yeast and plants," *Proc. Nat. Acad. Sci.*, **97**(12), 6487–6492.

[22] **B. M. Lange, T. Rujan, W. Martin,** and **R. Croteau**, 2000, "Isoprenoid biosynthesis: the evolution of two ancient and distinct pathways across genomes," *Proc. Nat. Acad. Sci.*, **97**(24), 13172–13177.

[23] **T. Kuzuyama**, 2002, "Mevalonate and nonmevalonate pathways for the biosynthesis of isoprene units," *Biosci. Biotechnol. Biochem.*, **66**(8), 1619–1627.

[24] **J. D. Newman** and **J. Chappell**, 1999, "Isoprenoid biosynthesis in plants: carbon partitioning within the cytoplasmic pathway," *Crit. Rev. Biochem. Mol. Biol.*, **34**(2), 95–106.

[25] **C. K. Shewmaker, J. A. Sheehy, M. Daley, S. Colburn,** and **D. Y. Ke**, 1999, "Seed-specific overexpression of phytoene synthase: increase in carotenoids and other metabolic effects," *Plant J.*, **20**(4), 401–412.

[26] **G.-Q. Chen**, 2009, "A microbial polyhydroxyalkanoates (PHA) based bio- and materials industry," *Chem. Soc. Rev.*, **38**(8), 2434–2446.

[27] **W. Niu, K. M. Draths,** and **J. W. Frost**, 2002, "Benzene-free synthesis of adipic acid," *Biotechnol. Prog.*, **18**(2), 201–211.

[28] **S. Atsumi, W. Higashide,** and **J. C. Liao**, 2009, "Direct photosynthetic recycling of carbon dioxide to isobutyraldehyde," *Nature Biotechnol.*, **27**(12), 1177–1180.

[29] **C. Jansson** and **T. Northen**, 2010, "Calcifying cyanobacteria – the potential of biomineralization for carbon capture and storage," *Current Opin. Biotechnol.*, **21**(3), 365–371.

[30] **J. Young, M. Geisen, L. Cros** *et al.* (eds.), 2003, *A Guide to Extant Coccolithophore Taxonomy, J. Nanoplankton Res.* special issue 1, 1–123.

[31] **G. Pósfai, G. Plunkett III, T. Fehér** *et al.*, 2006, "Emergent properties of reduced-genome *Escherichia coli*," *Science*, **312**(5776), 1044–1046.

[32] **H. Alper** and **G. Stephanopoulos**, 2009, "Engineering for biofuels: exploiting innate microbial capacity or importing biosynthetic potential?," *Nature Rev. Microbiol.*, **7**(10), 715–723.

[33] **E. Andrianantoandro**, **S. Basu**, **D. K. Karig**, and **R. Weiss**, 2006, "Synthetic biology: new engineering rules for an emerging discipline," *Mol. Syst. Biol.*, **2**, 2006.0028.

[34] **R. Baran**, **W. Reindl**, and **T. Northen**, 2009, "Mass spectrometry based metabolomics and enzymatic assays for functional genomics," *Current Opin. Microbiol.*, **12**(5), 547–552.

[35] **J. D. Orth**, **I. Thiele**, and **B. Ø. Palsson**, 2010, "What is flux balance analysis?," *Nature Biotechnol.*, **28**(3), 245–248.

[36] **D. G. Gibson**, **J. I. Glass**, **C. Lartigue** *et al.*, 2010, "Creation of a bacterial cell controlled by a chemically synthesized genome," *Science* doi:20.1126/science.1190719.

[37] **J. D. Bloom**, and **F. H. Arnold**, 2009, "In the light of directed evolution: pathways of adaptive protein evolution," *Proc. Natl. Acad. Sci.*, **16** (106 Suppl. 1), 9995–10000.

[38] **C. Jansson**, 2011, "Cyanobacteria for biofuels and carbon sequestration," in *Progress in Botany*, vol. 73, ed. **D. Francis**, New York, Springer, at press.

[39] **H. Tsuruta**, **C. J. Paddon**, **D. Eng** *et al.*, 2009, "High-level production of amorpha-4,11-diene, a precursor of the antimalarial agent artemisinin, in Escherichia coli," *PLoS One*, **4**(2), e4489.

[40] **J. D. Newman**, **J. Marshall**, **M. Chang** *et al.*, 2006, "High-level production of amorpha-4,11-diene in a two-phase partitioning bioreactor of metabolically engineered *Escherichia coli*," *Biotechnol. Bioeng.*, **95**(4), 684–691.

26 Biofuels from cellulosic biomass via aqueous processing

Jian Shi,[1] Qing Qing,[1,2] Taiying Zhang,[1] Charles E. Wyman,[1,2] and Todd A. Lloyd[3]

[1]Center for Environmental Research and Technology, University of California, Riverside, CA, USA
[2]Department of Chemical and Environmental Engineering, University of California, Riverside, CA, USA
[3]Mascoma Corporation, Lebanon, NH, USA

26.1 Focus

Thermochemical aqueous processing of cellulosic biomass requires depolymerization of long chains of carbohydrate molecules into fragments that can be metabolized by micro-organisms or catalytically converted to fuels and chemicals. This chapter focuses on such processes for carbohydrate depolymerization and their integration with subsequent product-formation steps in an effort to produce ethanol and other biofuels.

26.2 Synopsis

Cellulosic biofuels, which once were widely used but whose usage dropped sharply upon the introduction of refined petroleum products to the energy supply, can be a cost-effective fuel with applications in vital areas. Current strategies focus on maximizing the efficiency of conversion of cellulosic biomass waste into energy-rich products, especially liquid fuels, such as alcohols and other hydrocarbons. Recent research on the chemical and biological pretreatment of cellulosic feedstock materials shows promise for surpassing thermal processes in catalyzing the breakdown of cellulose and lignin, which is a crucial first step in the production of useful fuels. Chemical pretreatments include autohydrolysis, application of low and high pH (i.e., acids and bases), exposure to ammonia, and treatment with organic solvents and ionic liquids. Each of these methods is effective at breaking cellulose down so that it can be more easily digested enzymatically. These techniques generally offer good yields from a variety of feedstocks and therefore should be broadly applicable. In particular, it is expected that feedstocks will include waste materials such as food-crop residues (e.g., corn stover and sugarcane bagasse) and dedicated energy crops (e.g., switchgrass) that can be grown on otherwise agriculturally poor land. This aspect is particularly important in terms of minimizing the societal and environmental impacts of biofuels technology. For example, use of such feedstocks is intended to eliminate competition with food crops for arable land, which could lead to sharp increases in food prices. It should also help minimize the issue of indirect land-use change (see Chapter 2) that could actually result in increased CO_2 emissions.

After pretreatment, the initial products can be further broken down chemically or fermented into alcohols. Chemical breakdown is typically achieved using enzymes, whereas micro-organisms (such as common yeast) can be used to produce alcohols through fermentation. Further development of these techniques is expected to result in processes that can be scaled up to a level that can meet the growing demand for liquid fuels. In addition, a biorefinery model, in which valuable co-products are obtained along with the desired biofuels, is likely to make biofuel production even more cost-effective.

26.3 Historical perspective

Since prehistoric times, cellulosic biomass has been burned to provide heat for warmth and for the preparation of food. However, with the dawning of the industrial age came the need for fuel sources with higher energy density, and with the advent of the internal combustion engine for transportation came the need for energy-rich liquid fuels.

Ethanol, which had been used for lamp oil and cooking since ancient times, was considered a promising option. Fermentation of carbohydrates to ethanol had also been known for millennia, so efforts were made to commercialize processes to convert cellulosic biomass to fuels and chemicals. France and Germany were especially proactive in pursuing ethanol as a fuel option. For example, the first commercial ethanol production process was developed in Germany in 1898. It applied dilute sulfuric acid to hydrolyze the cellulose in wood, generating 7.6 l of ethanol per 100 kg of wood waste (18 gal per short ton). On an industrial scale, the optimized process achieved yields of around 21 l per 100 kg of biomass (50 gal per short ton) [1]. The French ministry of agriculture encouraged research into ethanol by offering prizes for the best alcohol-fueled engines and household appliances and holding a fair in Paris in 1902 featuring hundreds of such devices [2]. In 1914, the outbreak of World War I led to shortages in oil supplies, causing many more countries to take an interest in ethanol fuel. In fact, between World Wars I and II at least 40 industrial nations adopted the French and German model of blending ethanol with engine fuel [2].

At that time, fuels and chemicals derived from biomass were competitive with those from the new petroleum industry, and it was unclear which feedstock would come to dominate. However, over the ensuing century, petroleum recovery and processing improved at a rapid pace, continually driving down the cost and making biomass-derived fuels and chemicals uncompetitive.

With depletion of world petroleum reserves accelerating and production rates stalled over the past several years, many believe that the maximum production rate has been reached and that this finite resource will only become scarcer. This maximum, known as "**peak oil**," was reached in 1970 for the USA, as predicted by geophysicist M. King Hubbert in 1956 [3]. Figure 26.1 shows actual oil production and imports from 1920 to 2005. According to Hubbert's model, the production rate of a limited resource will follow a roughly symmetric logistic distribution curve (sometimes incorrectly compared to a bell-shaped curve) based on the limits of exploitability and market pressures. Various modified versions of his original logistic model are used, with more complex functions inserted to account for real-world forces. On the basis of Hubbert's model and others, it is believed

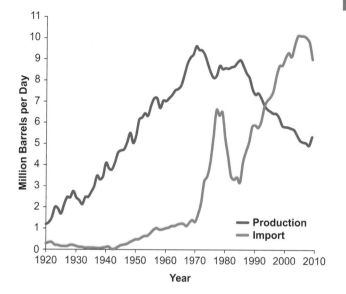

Figure 26.1. Daily average US oil production and oil imports, 1920–2009 (from data published by the Energy Information Administration of the US Department of Energy).

that world oil production will peak by the middle of this century and then begin to decline, although predictions for the peak range from 2021 to 2112 [4].

As oil production declines and energy demand continues to grow, alternative sources of fuels will be needed in order to meet the demand, and cellulosic biomass is a unique resource for sustainable production of liquid organic fuels with advantages such as being carbon-neutral and renewable, which are unmatched by other options [5]. In addition, with significant advances in enzyme technologies in the past three decades, biological biomass-conversion processes are cost-competitive with acid processes, and research and development efforts using biological conversion are accelerating.

26.4 Fuels derived from cellulosic biomass

With world oil prices in constant flux and subject to geopolitical circumstances as much as to the market influences of supply and demand, the need for economically stable feedstock alternatives will only continue to increase in the future. The energy density of **cellulosic biomass** is about 20 MJ kg^{-1}, a little less than half the roughly 45 MJ kg^{-1} for petroleum (crude oil). Of greater relevance, on a cost per unit energy basis, is that cellulosic biomass selling for $66 per Mt is equivalent to crude oil (139 kg per bbl) selling for $23 per bbl. Thus, as an energy source, cellulosic biomass is clearly cost-competitive with oil at today's market prices (over $90 per bbl in early 2011), and conversion to fuels on a large scale could help to stabilize the price of oil.

26.4.1 Cellulosic raw materials

Three constituents typically make up the largest fractions of cellulosic biomass, which is also called lignocellulosic biomass: **cellulose, hemicellulose**, and **lignin**. According to a joint study by the US Department of Energy and the US Department of Agriculture, over 1.3 billion tons of cellulosic biomass could be available in the USA annually, which, if converted to biofuels such as ethanol, could displace more than two-thirds of current transportation fuel needs [6]. Included as potential feedstocks are managed timber; forest industry byproducts; food-crop residues such as corn stover, sugarcane bagasse, and oil-palm waste; and dedicated energy crops such as miscanthus, switchgrass, and hybrid poplar.

The recalcitrance of cellulosic biomass

Cellulose, hemicellulose, and lignin occur in plant biomass in a ratio of approximately 5:3:2. Cellulose is a linear polymer consisting of linear chains of several hundred to over 10,000 $\beta(1\rightarrow4)$-linked D-glucose units arranged in a crystalline matrix, whereas hemicellulose is an amorphous, branching class of heteropolymers consisting of sugars such as xylose, arabinose, mannose, and galactose. The third major component of plant biomass, lignin, is a complex cross-linked phenylpropanoid network that strengthens the plant cell wall and provides resistance to microbial and fungal degradation. Since plants have evolved to be recalcitrant to breakdown, their conversion into biofuels is no easy task and requires a significant amount of energy.

In nature, the primary mechanism of cellulosic biomass decomposition occurs through **hydrolysis** of the carbohydrate and lignin polymers that make up cell walls by enzymes secreted by micro-organisms, and the structure of the cell wall (specifically, the highly compact structure of the crystalline cellulose combined with hemicellulose and the highly cross-linked network of lignin) naturally limits the access of enzymes to hydrolysis sites. The success of a cellulosic-biomass-conversion industry rests on the ability to **depolymerize** carbohydrates inexpensively, and some promising options to accomplish this task are described in the next few sections.

26.4.2 Overview of conversion pathways

There are two primary conversion pathways from cellulosic biomass to fuels and chemicals: thermal and biological.

Thermal conversion processes

A thermal route that can effectively convert solid biomass into a gaseous intermediate that is subsequently converted to liquid fuels is gasification [7]. The resulting product gas, "**synthesis gas**" or simply "syngas," consists mainly of CO, H_2, CO_2, H_2O, N_2, and hydrocarbons and has a low to medium energy content, depending on the gasifying agent. Biomass gasification (see also Chapter 48) is well understood, but minor syngas components, including tars, sulfur and nitrogen oxides, alkali metals, and particulates must be removed or their presence mitigated for successful downstream conversion.

Pyrolysis is another thermal process, in which biomass is heated in the absence of air, which converts it to medium-energy-density liquids called **bio-oils**. Bio-oils can be burned in boilers to generate steam for power production but are not suitable as transportation fuels without further processing because of the presence of high-viscosity tars and their tendency to harden when allowed to sit. Another thermal conversion process under development is liquefaction, in which heat is applied to biomass under pressure to generate intermediates that can be upgraded to fuel products.

Biological conversion processes

Biological conversion utilizes micro-organisms to convert cellulosic biomass into fuels and chemicals. Two processes in series are typically applied: hydrolysis of polymeric cellulose and hemicellulose to short-chain fermentable sugars and **metabolic conversion** of those sugars into products. In nature, enzyme proteins catalyze this hydrolysis reaction, and, because the β-1,4 glycosidic bonds between sugar units that make up cellulose and hemicellulose are more stable than the α-1,4(6) glycosidic bonds in starch, the hydrolysis rates for cellulose and hemicellulose are lower by a factor of 4 than that of starch. Furthermore, treatment of biomass prior to enzymatic hydrolysis is required in order to achieve high yields in a reasonable amount of time.

26.4.3 Pretreatment

As mentioned above, breakdown of cellulosic biomass to sugars is slow as a result of the natural resistance or recalcitrance of plants to biological degradation. These very low rates are a barrier to commercialization of biological conversion processes, and sugar release rates must be increased by an order of magnitude or more over those found in nature. Research has shown that disrupting the lignin/hemicellulose structure in the cell wall by thermochemical treatments increases subsequent enzyme hydrolysis rates [8], and this disruption step is called "**pretreatment**."

The role of pretreatment

Pretreatment is costly, and it impacts virtually all other process operations in an integrated biorefinery [9]. For example, a pretreatment process must be compatible with the choice of feedstocks, since the effectiveness of a pretreatment process can vary significantly depending on the feedstock. Pretreatment also influences downstream processing. The effectiveness of enzymatic hydrolysis of the pretreated feedstock depends on pretreatment conditions and the extent to which the biomass structure has been disrupted. Thus, enzyme formulations and loadings must be tailored to match the type and severity of pretreatment applied. Furthermore, pretreatment impacts fermentation configurations, e.g., the choice of micro-organisms, requirements for hydrolyzate conditioning, and the mode of reactor operation. Pretreatment must open the biomass structure to increase access to enzymes while minimizing sugar losses and creation of enzyme inhibitors.

Pretreatment options

Pretreatments that lead to high downstream yields have been developed at both laboratory and pilot-plant scales over the past few decades, and can be grouped into three categories: physical, biological, and thermochemical. Physical pretreatments primarily rely on size-reduction to increase surface area and access to hydrolysis sites by enzymes. To be cost-effective, size reduction equipment, and the cost of power to run it, must be offset by an increase in conversion yields, but yields are typically increased by only about 10%–20%, with little or no gain below a certain size reduction.

In biological pretreatments, enzymes secreted by micro-organisms solubilize lignin and increase access to cellulose. Biological pretreatment is an inexpensive, environmentally benign, and low-energy alternative to other pretreatments. However, biological pretreatments are usually slow and not easily controlled, and further development is needed before biological pretreatment can be used commercially.

Because conversion yields can be high, thermochemical pretreatments are favored by many. Thermochemical pretreatments can be divided into four subcategories: water-only, acid, alkaline, and solvent.

Autohydrolysis pretreatment

Water-only pretreatment, also called "**autohydrolysis**" because acetic acid liberated from the hemicellulose at elevated temperature is thought to catalyze hydrolysis, requires no addition of reagents other than water and yet can be very effective for some, but not all, feedstocks. Typical water-only pretreatment temperatures range from 180 to 240 °C, with residence times of 1–30 min. Steam pressures range from 0.96 to 2.1 mPa and the treated feedstock is usually explosively discharged to atmospheric pressure, further disrupting cell-wall structure. The result is solubilization, or destruction of a high fraction of the hemicellulose, with much of it preserved as short chains of sugars, known as oligomers, and, to a lesser extent, monomeric xylose and other sugars that comprise hemicellulose. However, overall hemicellulose recovery tends to be low at about 65% of the maximum possible, and high enzyme doses are needed in hydrolysis to convert a high fraction of the cellulose and hemicellulose left in the pretreated solids.

Low-pH pretreatment

Hemicellulose hydrolysis is accelerated and sugar yields are improved by impregnating biomass with acid prior to pretreatment to lower the pH. **Dilute sulfuric acid** is most widely used because of its low cost, although many other mineral or organic acidic agents, including gaseous sulfur dioxide, hydrochloric acid, phosphoric acid, hydrofluoric acid, and oxalic acid, have also been evaluated for improving sugar yields. Reaction temperatures tend to be lower and yields higher for dilute acid pretreatment than for autohydrolysis. Prior to enzyme addition, residual acid must be neutralized, and inhibitors formed or released in pretreatment must be removed or reduced in concentration. Lime (calcium oxide) is often used to achieve both ends. However, lime precipitates some sugar along with inhibitors, and the precipitated neutralization products require removal and disposal. Although it is more costly than lime, ammonium hydroxide can be used as the neutralizing base, minimizing sugar losses and reducing waste handling.

High-pH pretreatment

Pretreatment can also be effective at basic pH. Effective bases include **lime**, calcium carbonate, potassium hydroxide, and sodium hydroxide, all of which tend to remove a high fraction of lignin, but remove much less hemicellulose than do acids. For instance, processing with lime in the presence of oxygen at high temperatures (100–200 °C) and high pressure for 10 min to 6 h, depending on the specific biomass type, can create a very digestible solid residue with downstream yields above 85%. Another version of lime pretreatment uses low temperatures (25–55 °C) and air instead of oxygen. Although this process is relatively inexpensive, conversion rates are low, and residence times of several weeks are required.

Percolating aqueous ammonia through a bed of biomass in a reactor or simply subjecting the biomass to **soaking in aqueous ammonia (SAA)** at temperatures of about 160 °C is an effective pretreatment. A typical percolation process solubilizes about half of the total xylan at optimal conditions and retains more than 90% of the cellulose content, while removing lignin and swelling the residual cellulose. In another process, ammonia and moist biomass are combined and loaded into a reactor at temperatures of 60–200 °C with residence times of 5–45 min. At the end of the reaction period, pressure is suddenly reduced to atmospheric, causing a rapid expansion of gas or "explosion." This process, known as **ammonia fiber explosion (AFEX)**, is believed to decrystallize cellulose and disrupt lignin, while partially hydrolyzing hemicellulose. However, unlike other pretreatments, little hemicellulose, cellulose, or lignin is solubilized, and the resulting solids are readily hydrolyzed by enzymes to monomer sugars for downstream processing.

Organosolv pretreatment

All of the above pretreatment options alter the structure of the biomass and make the resulting solids more amenable to enzymatic digestion, although the reaction products might contain soluble hemicellulose sugars and/or soluble lignin. In contrast, another pretreatment option uses the addition of nonreactive solvents to fractionate whole cellulosic biomass into lignin-, cellulose-, and hemicellulose-enriched streams. Processes using solvents are known collectively as **organosolv pretreatment**.

The organosolv process applies organic solvents, usually alcohols, to solubilize lignin in biomass in the presence of an acidic catalyst that breaks down hemicellulose–lignin bonds. Operating temperatures range between 90 and 120 °C for grasses and between 155 and 220 °C for woods, with residence times ranging from 25 to 100 min. In another version of organosolv pretreatment, fractionation is accomplished using sequential extractions with phosphoric acid, acetone, and water at temperatures of around 50 °C. This process solubilizes hemicellulose, lignin, and acetic acid, and decrystallizes cellulose to make it very digestible by enzymes. However, as with all organosolv processes, cost-effective recovery and recycling of the solvents is challenging.

Ionic liquid pretreatment

Ionic liquids (ILs) are organic salts that usually melt below 100 °C. Since their discovery as pretreatment reagents, ILs have become attractive as "green solvent" alternatives to volatile and unstable organic solvents [10]. Ionic liquids have some very interesting properties such as solubilizing whole cellulosic biomass or selectively dissolving one component of cellulosic biomass, e.g., lignin. In addition, ILs can dissolve cellulose, as well as entire lignocellulosic materials such as corn stalks, rice straw, bagasse, pine wood, and spruce wood. The versatile solvent ability of ILs is conceptually attractive for segregating major components in cellulosic biomass and decreasing the crystallinity of cellulose to make it more easily broken down to sugars by enzymes.

After IL treatment, the solubilized carbohydrate can be precipitated with water, methanol, or ethanol. Studies have shown that IL-dissolved and reconstituted cellulose is less crystalline, has greater accessibility to cellulases, and is more digestible by enzymes than native cellulose. Extracted lignin has potential uses as a binder, dispersant, emulsifier, and blending agent in commodity plastics. Despite their interesting features for pretreating lignocellulosic biomass, ionic liquids are not at all economically viable yet, and major improvements in the recycling of ILs are required.

Effect of pretreatment

The main compositional changes incurred by lower-pH pretreatments (e.g., autohydrolysis, dilute acid, and SO_2) are removal of a major portion of the hemicellulose and some lignin and cellulose. High-pH pretreatments (e.g., lime and SAA, but not AFEX), however, remove mostly lignin, although the extents of these effects vary with substrate and pretreatment system. AFEX pretreatment usually results in virtually no compositional change from that of the original biomass except for limited xylan dissolution and minor acetyl removal. Although the compositions of the solids resulting from pretreatment vary significantly with the technologies applied, AFEX, dilute sulfuric acid, lime, SAA, and sulfur dioxide all achieve good yields from many feedstocks, with significantly increased accessible surface area and altered lignin structures. Lime and sulfur dioxide have given high yields for all feedstocks tested in controlled experiments: poplar wood, corn stover, and switchgrass [11].

26.4.4 Aqueous processing to produce reactive intermediates

The pretreatment concept can be extended to more general technologies for aqueous thermochemical processing of cellulosic biomass to produce a wider range of reactive intermediates in addition to sugars for conversion to fuels and biomaterials, as shown in Figure 26.2. Starting from cellulosic biomass, the reactive intermediates generated from pretreatment are cellulose, hemicellulose, lignin, and carboxylic acids. Once liberated, sugars can be further reacted to furfural, 5-hydroxymethyl furfural (5-HMF), formic acid, and levulinic acid. Although only the sugars can be fermented to ethanol and other products, all of these compounds can be catalytically reacted to provide hydrocarbon fuels that are more compatible with the existing petroleum-based infrastructure [12] (see Section 26.4.7).

26.4.5 Hydrolysis and fermentation to ethanol

Process overview

After pretreatment of the biomass, cellulase produced by fungus or other organisms can be added to the residual solids to hydrolyze insoluble cellulose and hemicellulose to fermentable sugars. Thereafter, microorganisms, namely yeast, bacteria, or fungi, can be added to ferment the sugars to ethanol and carbon dioxide. Many such micro-organisms can readily ferment glucose and other six-carbon sugars (hexoses);

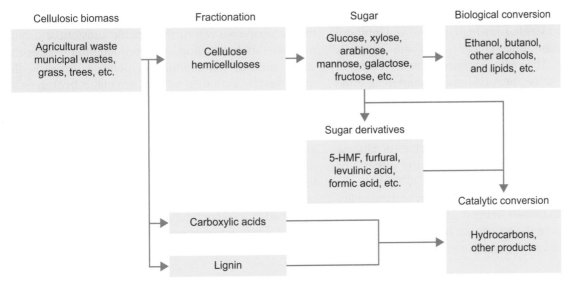

Figure 26.2. Conversion of major components in cellulosic biomass to reactive intermediates.

however, only in the past few decades have researchers been able to ferment the five-carbon sugars xylose and arabinose to ethanol, through use of genetically modified micro-organisms [13]. The ethanol-containing broth generated in fermentation is then distilled to recover pure ethanol. The solid residue after distillation, mainly lignin and unconverted carbohydrate, can be burned to generate all the heat and electricity needed to power the process; in fact, some processes will be net power exporters. In addition to ethanol, sugars produced in this process can be biologically or chemically converted to other valuable products and chemicals.

Enzymatic hydrolysis
Efficient saccharification of polymeric carbohydrates in cellulosic biomass is the key to commercializing a process, and the understanding of enzyme hydrolysis is still far from complete. Enzymatic hydrolysis of cellulose involves a heterogeneous reaction at the liquid–solid interface, and various substrate- and enzyme-related features directly or indirectly affect this reaction, including the cellulose crystallinity, degree of polymerization, effective surface area, and enzyme denaturation or inhibition. Also, hemicellulose and lignin residues can hinder cellulose hydrolysis by unproductively binding and physically blocking cellulase enzymes.

Hydrolyzing enzymes can be produced by several types of micro-organisms, including aerobic and anaerobic fungi, aerobic *actinomycetes*, and anaerobic bacteria. In the aerobic fungal enzyme system, depolymerization of pretreated cellulose requires at least three classes of cellulase for high yields [14]:

endoglucanases (EGs), cellobiohydrolases (CBHs), and β-glucosidase. Endoglucanase reduces the degree of polymerization of the substrate by randomly hydrolyzing intrachain bonds, creating chain ends. Cellobiohydrolases react with chain ends to cleave off cellobiose (glucose dimer). Finally, β-glucosidase cleaves cellobiose into two glucose monomers. Any residual hemicellulose requires hemicellulases for hydrolysis, and their action is analogous to that of cellulases.

The aerobic filamentous fungus *Trichoderma reesei* is used commercially for the manufacture of cellulase enzymes. It was discovered in the hot, humid jungles of the South Pacific during World War II, and *T. reesei* cellulases were found to be responsible for rapid degradation of cotton fabrics, reducing canvas and clothing to rags. Today, *T. reesei* enzymes are used in the textile industry for making "faded" apparel such as "stonewashed" jeans, and are being developed for commercial cellulosic ethanol processes. *T. reesei* produces two types of cellobiohydrolases and at least five endoglucanases that work together to hydrolyze cellulose cooperatively by binding to the cellulose surface and hydrolyzing cellulose into smaller fragments. *T. reesei* enzymes have a distinct structure [15] that contains a cellulose-binding module (CBM) and a catalytic domain (CD).

Hexose fermentation
The glucose released from cellulose hydrolysis can be fermented to ethanol by various bacteria, yeasts, and fungi according to the equation

$$C_6H_{12}O_6 \Rightarrow 2CO_2 + 2C_2H_5O. \qquad (26.1)$$

Most industrial ethanologens can convert glucose to ethanol with high yields, have high ethanol and inhibitor tolerance, function anaerobically, and are generally regarded as safe. The yeast *Saccharomyces cerevisiae*, or "baker's yeast," is the most common commercial ethanologen, and its use dates to before recorded history. *S. cerevisiae* has high ethanol productivity, high inhibitor tolerance, and natural robustness in industrial processes [16]. Although *S. cerevisiae* readily ferments glucose and other strains can also ferment the other six-carbon sugars, the wild type is not able to ferment the pentoses arabinose and xylose prevalent in hemicellulose. Other ethanologens being developed for biological conversion include the anaerobic bacteria *Zymomonas mobilis*, *Clostridium thermocellum*, and *Thermocellum saccharolyticum*.

Pentose fermentation

Lignocellulosic hydrolysate produced by either chemical or enzymatic treatments contains hexoses, typically mostly D-glucose with lesser amounts of D-galactose and D-mannose. However, many types of plants also have large amounts of pentoses including D-xylose and L-arabinose. Because the pentose fractions can account for about 8%–28% of the raw feedstock, their conversion to ethanol is important for commercial economic viability. However, traditional micro-organisms, such as wild-type *Saccharomyces cerevisiae* and *Zymomonas mobilis*, can metabolize only hexose. Fortunately, genetic engineering techniques (see Chapter 25) have allowed the development of recombinant hexose- and pentose-fermenting bacteria and yeasts, and high fermentation yields are now likely.

Processing configurations

Typically, four steps are involved in the biological processing of cellulosic materials: cellulase and hemicellulase enzyme production, enzymatic hydrolysis of cellulose and hemicellulose, hexose fermentation, and pentose fermentation. These four steps can be arranged in various process configurations, including separate hydrolysis and fermentation (SHF), simultaneous saccharification and fermentation (SSF), simultaneous saccharification and co-fermentation (SSCF), and consolidated bioprocessing (CBP) [17].

Cellulase is strongly inhibited by its hydrolysis products glucose and cellobiose, so that, as the concentrations of these products increase, the rate of hydrolysis decreases. The SSF process was developed to mitigate cellulase inhibition by rapidly converting these sugars to ethanol, which is less inhibitory, thereby allowing hydrolysis rates to remain high. Compared with SHF, SSF supports higher yields, rates, and ethanol concentrations with less enzyme, shorter fermentation times, reduced contamination risk, and less equipment.

The utilization of gene-modified or recombinant micro-organisms can further enable substantial capital and operational savings by consolidating processing steps, such as fermentation of pentose sugars in the same vessel as that in which enzymatic hydrolysis of cellulose and glucose fermentation to ethanol is brought about through the SSCF of pentose. Similarly to SSF, SSCF offers advantages of reducing process complexity and end-product inhibition. Finally, CBP uses micro-organisms that both make hydrolyzing enzymes and ferment the resulting sugars, eliminating separate cellulase production. While research supports the feasibility of this advanced configuration, the key challenge lies in the development of organisms that can convert sugars to desired products in high yields.

26.4.6 Biological conversion to other fuels and chemicals

Making ethanol from cellulosic biomass has historically received the most attention and, consequently, is closest to commercialization. However, research has recently focused on other fermentation products for third-generation biofuels such as C_3–C_5 alcohols, isoprenoid-derived fuels, and fatty-acid esters [18]. The C_3–C_5 alcohols found as minor co-products in ethanol fermentations include isopropanol, 1-propanol, 1-butanol, isobutanol, 3-methyl-1-butanol, 2-methyl-1-butanol, and isopentenol [19]. These options are very promising and open avenues for biofuels production beyond ethanol. In this section, we discuss recent developments in some of these interesting pathways, including those for making butanol, fatty acids, and mixed alcohols.

Butanol

Butanol has properties that give it advantages over ethanol for integration into the existing fuel infrastructure. First, its energy density is 30% higher than that of ethanol and almost the same as that of gasoline, and it has a low vapor pressure. It is also less hygroscopic, less soluble in water, less hazardous to handle, and less flammable than ethanol, and it can be blended with gasoline over a wide range. Butanol can be dehydrated to alkenes, hydrogenated, or polymerized to yield a variety of compounds that can be used as fuels, fuel additives, or commodity chemicals. However, yields and concentrations are low because butanol is far more toxic to fermentative organisms than is ethanol. Thus, new efforts are focused on improving process economics through novel reactor designs and improved microbial strains.

Butanol fermentation is one of the oldest commercial fermentation processes (see Chapter 25). However, in the 1950s, a process to derive butanol inexpensively from petroleum was developed, and fermentative butanol production dropped rapidly. Interest in fermentative butanol has increased recently, however, because of high oil prices and the development of new butanol fermentation technologies. The most studied natural microbial strains for butanol fermentation are *Clostridium acetobutylicum* and *Clostridium beijierinckii*, both of which were used for commercial solvent production in South Africa (Germiston) until the early 1980s. Metabolic regulation of butanol production in *Clostridia* is based on controlling pathways and shifting between acidogenesis and solventogenesis in a reducing environment.

Because of butanol's toxicity, batch fermentations result in low yields and poor productivity and thus are not likely to be commercially viable. Typical batch production rates vary from 0.07 to 0.35 g l^{-1} h^{-1}, but alternative process strategies such as free-cell continuous fermentation, cell recycling, and immobilized cell reactors have all shown great improvements in butanol yields and productivity [20]. For instance, production rates of 6.5–15.8 g l^{-1} h^{-1} were achieved in an immobilized cell reactor, 40–50 times higher than those achieved in batch reactors. Also, reducing butanol toxicity by keeping butanol concentrations low can improve production rates, with simultaneous product-recovery techniques, including gas stripping, adsorption, liquid–liquid extraction, perstraction (permeation followed by extraction), pervaporation (separation by partial vaporization through a membrane), and reverse osmosis.

Further development of butanol fermentation technology will most likely rely on improved butanologens.

Lipids

Biodiesel based on fatty-acid methyl ester (FAME) is commonly produced through the transesterification of vegetable oils or animal fats (see Section 25.4.3). However, compared with carbohydrate in cellulosic biomass, vegetable oils and animal fats are expensive and in limited supply. In contrast, fat made by microbes using abundant cellulosic biomass could have greater impact, and many yeasts, fungi, molds, and prokaryotic bacteria can convert sugars to triacylglycerols (TAGs), which can be converted to liquid fuels. TAGs are nonpolar, water-insoluble triesters of glycerol with fatty acids and are important precursors for the synthesis of biodiesel and jet fuels.

TAGs occur in many eukaryotic organisms, such as yeast, fungi, plants, and animals, whereas only a few prokaryotic bacteria produce them, including *Mycobacterium*, *Streptomyces*, *Rhodococcus*, and *Nocarda* [21]. It is generally believed that periods of metabolic stress, e.g., starvation, stimulate TAG accumulation in yeast or fungi. However, in most bacteria, accumulation of TAGs and other neutral lipids is usually stimulated when the carbon source is in excess and nitrogen is limiting. Recently, a yield of approximately 38% TAGs by batch

culture of *R. opacus* with glucose was reported, along in addition to the finding that pH control and a suitable carbon-to-nitrogen ratio in the defined medium were critical for high fatty-acid production. Also a new fungus, *Gliocladium roseum* (NRRL 20072), that can grow on woody biomass and produces a range of long-chain hydrocarbon molecules virtually identical to the fuel-grade compounds in petroleum was discovered [22]. This fungus has a hydrocarbon profile that contains a number of compounds normally associated with diesel fuels; thus, the term "myco-diesel" was coined for its product. Although work with this organism is just beginning, it serves as another example of the diversity of products that can be made by biological conversion of the sugars in cellulosic biomass. However, as interesting as these new technologies are, yields are low, and significant development efforts are required before commercialization can be realized.

The MixAlco process

Instead of fractionating cellulosic biomass into sugars and fermenting them to ethanol or other fuels, cellulosic biomass sugars can be converted microbially to carboxylic acids, which are further converted to fuels. In one such process, developed by Lefranc in 1927, cellulosic biomass waste was fermented to butyric acid and then neutralized with calcium carbonate [23]. The recovered calcium butyrate was then thermally converted to ketones for use as motor fuel. A recent development for mixed alcohol products (MixAlco) starts with lime pretreatment of biomass, followed by inoculation with a consortium of acid-producing micro-organisms from rumen or terrestrial sources [24]. Ammonium, iodoform, or other inhibitors are introduced into the anaerobic digestion to stop methanogenic reactions, which would normally prevail, and allow accumulation of mixed carboxylate salts. The carboxylic acids can then be regenerated from their salts using acid springing technology and used to make a series of fuels and chemicals and reactive intermediates. For example, carboxylic acids can be esterified and then hydrogenated to alcohols. Alternatively, catalysts such as zirconium oxide can convert them into ketones that can be converted into alcohols.

Two types of lime pretreatment have been developed for this process. The more rapid approach is carried out at a temperature of ~100 °C and a residence time of 1–3 h, with a lime loading of ~0.1 g of Ca(OH)$_2$ per g of dry biomass added to conventional reactors. In the slower process, temperatures are only ~50 °C, and residence times can be 4–6 weeks, making pretreatment in piles instead of reactors more likely. Lime pretreatment enhances anaerobic digestion of carbohydrate polymers in biomass primarily through solubilizing lignin and neutralizing acetyl groups in hemicellulose. Potential benefits of the MixAlco process include the

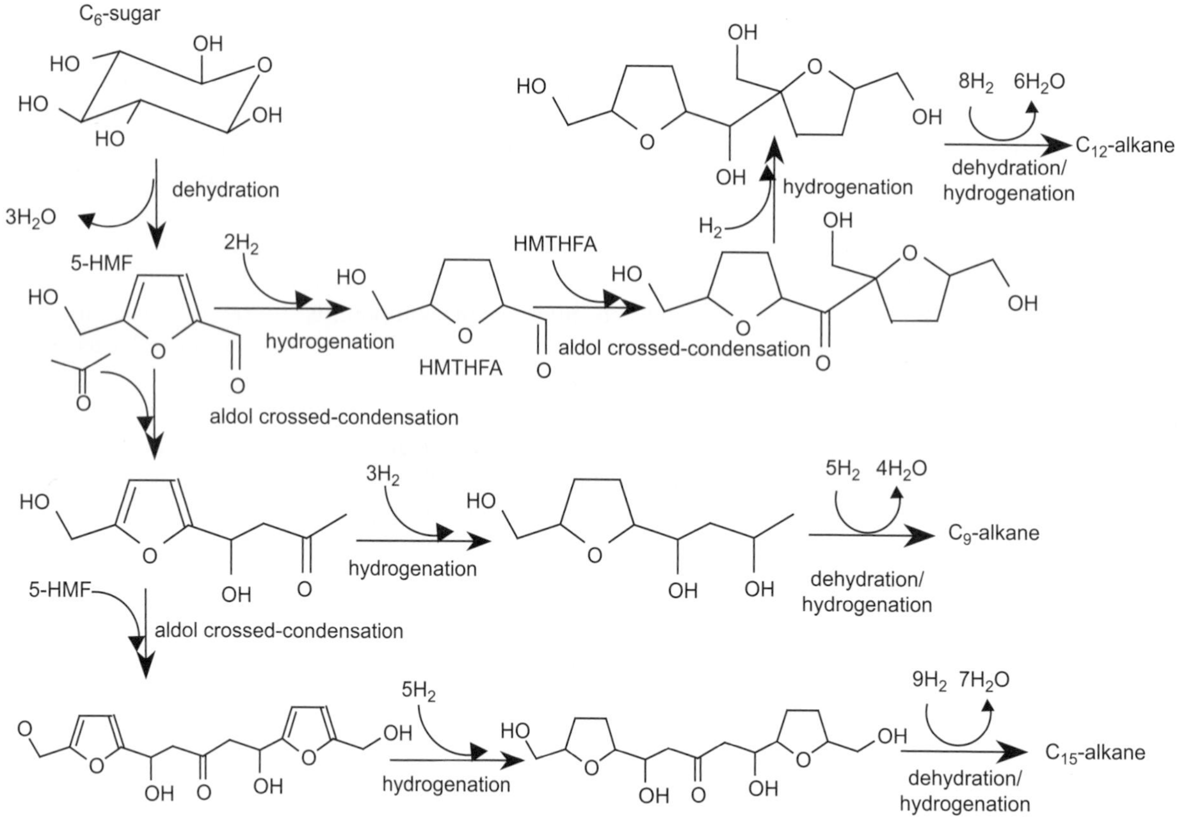

Figure 26.3. Reaction pathways for making alkanes from glucose and other six-carbon (C_6) sugars. 5-HMF, 5-hydroxymethyl furfural; and HMTHFA, 5-hydroxymethyl-2-furaldehyde.

fact that there is no need for sterility, its adaptability to a variety of feedstocks, the low capital expenditure needed, the fact that there is no endogenous enzyme, the fact that no extra nutrients are needed, and higher energy recovery.

26.4.7 Reactive intermediates and catalytic processing to hydrocarbons

In addition to fermenting sugars to alcohols and organic acids, liberated carbohydrates can be catalytically converted to C_5 and C_6 alkanes directly or by condensation and hydrogenation reactions to higher-relative-molecular-mass alkanes. Also, insoluble lignin residue can be converted to higher-relative-molecular-mass alkanes by hydrogenation.

Cellulose platform
Several routes have been used to make alkanes from the glucose in cellulosic biomass. The first of these relies on glucose dehydration to 5-hydroxymethyl-2-furaldehyde which, through condensation and hydrogenation reactions, can be transformed into C_9, C_{12}, and C_{15} alkanes.

Dehydration is catalyzed with 4 wt% $Pt/SiO_2–Al_2O_3$ at 523–538 K. The condensation reaction is catalyzed with $Mg–Al_2O_3$ and NaOH at room temperature. Hydrogenation can occur either in water or methanol, or in a mixture of the two in the presence of a Pd/Al_2O_3 catalyst and at a hydrogen pressure of 52–60 bar. The reaction sequence is shown in Figure 26.3.

An alternative is to convert 5-hydroxymethyl furfural into levulinic acid, with the first few steps being as shown in Figure 26.4. First, as before, glucose dehydration produces 5-hydroxymethyl-2-furaldehyde with further scission to levulinic and formic acids. Humins are undesirable degradation products that can be formed from the reactive intermediates and must be avoided as much as possible. Levulinic acid can then be dehydrated and hydrogenated to γ-valerolactone (GVL), which can be hydrolyzed to pentenoic acid for production of 5-nonanone [25]. In this process, a solid acid catalyst, SiO_2/Al_2O_3, is used for both transformations in the presence of water in a single fixed-bed reactor at pressures ranging from ambient to 36 bar.

A third pathway is based on conversion of glucose to sorbitol by hydrogenation, followed by deoxygenation to

Figure 26.4. The reaction pathway from cellulose to levulinic and formic acids.

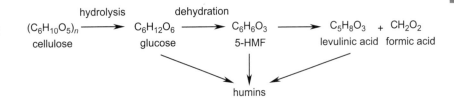

$$(C_6H_{10}O_5)_n \xrightarrow{\text{hydrolysis}} C_6H_{12}O_6 \xrightarrow{\text{dehydration}} C_6H_6O_3 \longrightarrow C_5H_8O_3 + CH_2O_2$$

cellulose   glucose   5-HMF   levulinic acid formic acid

humins

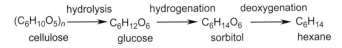

$$(C_6H_{10}O_5)_n \xrightarrow{\text{hydrolysis}} C_6H_{12}O_6 \xrightarrow{\text{hydrogenation}} C_6H_{14}O_6 \xrightarrow{\text{deoxygenation}} C_6H_{14}$$

cellulose   glucose   sorbitol   hexane

Figure 26.5. Conversion of glucose to hexane.

hexane or glycerol, depending on the catalyst. Using Pd or Pt on SiO_2 or Al_2O_3 as the catalyst at 225–265 °C with 26–58 bar of hydrogen, D-sorbitol is converted to n-hexane [26] in high yield (Figure 26.5), and, using Ru/SiO_2 as the catalyst at 180–240 °C and 80–125 bar of hydrogen pressure, glycerol is produced.

Glycerol can then be converted to liquid alkanes by an integrated process involving catalytic conversion to H_2–CO gas mixtures (synthesis gas) combined with Fischer–Tropsch synthesis [27], as shown in Figure 26.6. It was reported that concentrated glycerol feed solutions at 275 °C and 1–17 bar over a 10 wt% Pt–Re/C catalyst (Pt/Re ratio of 1:1) could produce alkanes at high rates and selectivities (H_2:CO ratio between 1:1.0 and 1:1.6).

In addition to the above routes, other reaction pathways are possible. Clearly, glucose is a versatile feedstock with a wide variety of fuels and chemicals potentially derivable from it.

The hemicellulose platform

Analogously to the glucose-to-fuels and chemicals processes just described, hemicellulose sugars (primarily xylose) have many conversion pathways. Xylose from hemicellulose can be converted to levulinic acid, which can then be converted to alkanes through γ-valerolactone as shown in Figure 26.6. In the process, xylose is dehydrated to furfural, which is then hydrogenated to furfural alcohol and converted to levulinic acid by heating to 60–100 °C under atmospheric pressure in the presence of water and a strong non-oxidizing acid [28].

In an alternative approach, furfural is converted to methylfuran in high yields using copper-based catalysts at 200–300 °C through dehydration of xylose followed by hydrogenation [29]. Subsequently, methylfuran can be converted to pentanol and pentanone by additional hydrogenation. The pentanol yield depends on the catalyst and reaction conditions used, and conversion of 2-methylfuran can exceed 80% over a commercial catalyst (Cu:Zn:Al:Ca:Na = 59:33:6:1:1) at 300 °C. Similarly to

the conversion of glucose to intermediates and ultimately alkanes, xylose can be reacted to give intermediates and alkanes through dehydration, hydrogenation, and aldol condensations, with end products and yields largely dependent on the catalyst(s) used.

26.4.8 Co-products

It is envisioned that a mature biomass-to-fuels and -chemicals industry will employ "biorefineries" that are analogous to today's petroleum refineries, in which a suite of intermediates and final products will be derived from feedstocks in a central, albeit large, processing facility [30]. In the biorefinery, feedstock is processed to fuels, chemical intermediates, and finished chemicals. A biorefinery could have both aqueous and thermal processes. Upstream, carbohydrate in the feed might be converted to sugar and then to fermentation products, or to alkanes and other chemicals as described in the previous section. Downstream, the solid residue from aqueous processing, lignin, might then be thermally converted to syngas, alkanes (through hydrogenation), or bio-oil. Many configurations are possible, and process selection will be driven by the market.

Commercialization efforts will likely evolve into the biorefinery model for a number of reasons.

- Revenues from high-value co-products reduce the selling price of the primary product.
- Economies of scale for a full-sized biorefinery lower the processing costs of low-volume, high-value co-products.
- Less fractional market displacement is required for cost-effective production of high-value co-products as a result of the economies of scale made possible by the primary product.
- Biorefineries maximize value generated from heterogeneous feedstocks, making use of all component fractions.
- Common process elements are involved in producing fermentable carbohydrate, regardless of how many products are produced.
- Co-production can provide process-integration benefits by meeting process energy requirements with electricity and steam co-generated from process residues.

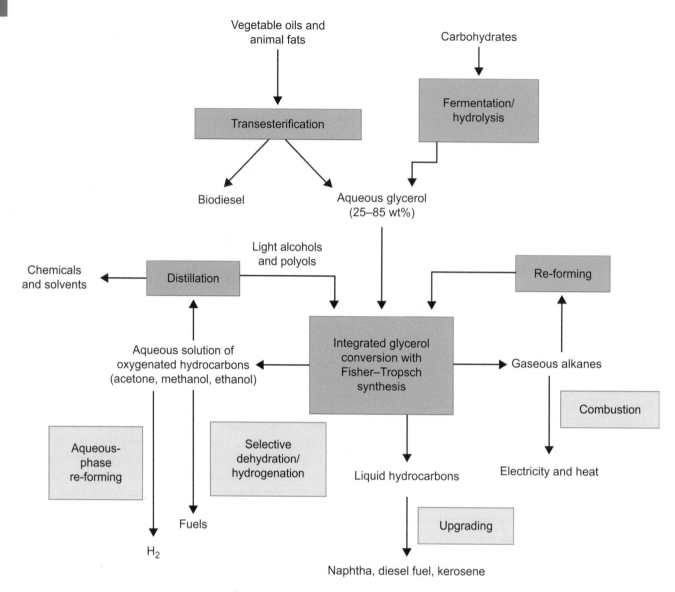

Figure 26.6. Process pathways for production of liquid fuels from biomass by integrated glycerol conversion to synthesis gas and Fischer–Tropsch synthesis.

26.5 Summary

This chapter describes water-based processing strategies to convert the substantial carbohydrate fraction in lignocellulosic biomass to liquid transportation fuels and commodity chemicals. Unlike the anhydrous thermal processes of pyrolysis and gasification, aqueous platforms first hydrolyze the carbohydrate polymers in biomass to sugars, which in turn can be converted into other intermediates or used directly for conversion to final products in downstream aqueous chemical and/or biological processes. This chain of reactions means that high yields of intermediates are essential to realizing high final-product yields.

Although there are challenges to low-cost processing, alternatives to petroleum feedstocks must play a major role in our transition to a sustainable fuels and chemicals future. Furthermore, aqueous carbohydrate conversion processes are poised to provide many of the products currently derived from petroleum or substitutes for those products, reducing the consequences that could result from petroleum shortages.

26.6 Questions for discussion

1. Briefly discuss the necessity to find alternatives to petroleum and what the major advantages of cellulosic ethanol are.
2. What are the differences between gasification and pyrolysis?
3. What are the key effects of pretreatment? Why does cellulosic biomass need to be pretreated prior to enzymatic hydrolysis?

4. List the four processing configurations of biological conversion of cellulosic materials and compare their pros and cons.

5. What are the main catalytic reactions for making alkanes from glucose or other six-carbon sugars?

26.7 Further reading

- C. E. Wyman (ed.), 1996, *Handbook on Bioethanol: Production and Utilization*, Washington, DC, Taylor and Francis.
- O. Kitani and C. W. Hall (eds.), 1989, *Biomass Handbook*, New York, Gordon and Breach Science Publishers.
- D. Yergin, 1991, *The Prize: The Epic Quest for Oil, Money, and Power*, New York, Free Press.

26.8 References

[1] R. Katzen and D. J. Schell, 2006, "Lignocellulosic feedstock biorefinery: history and plant development for biomass hydrolysis," in *Biorefineries – Industrial Processes and Products*, eds. B. Kamm, P. R. Gruber, and M. Kamm, Weinheim, Wiley-VCH, vol. 1, pp. 129–138.

[2] W. Kovarik, 2006, "Ethanol's first century:Fuel blending and substitution programs in Europe, Asia, Africa and Latin America," presented at *XVI International Symposium on Alcohol Fuels*, Rio de Janeiro, available at http://www.radford.edu/wkovarik/papers/International.History.Ethanol.Fuel.html.

[3] S. Udall, C. Conconi, and D. Osterhout, 1974, *The Energy Balloon*, New York, McGraw-Hill Book Co.

[4] J. H. Wood, G. R. Long, and D. F. Morehouse, 2003, "World conventional oil supply expected to peak in 21st century," *Offshore*, 63, 4.

[5] Bringezu, S. H. Schütz, M. O'Brien, L. Kauppi, and R. W. Howarth, 2009, *Assessing Biofuels: Towards Sustainable Production and Use of Resources*, Nairobi, United Nations Environment Programme.

[6] R. Perlack, L. Wright, A. Turhollow et al., 2005, *Biomass as Feedstock for a Bionergy and Bioproducts Industry: The Technical Feasibility of a Billion-Ton Annual Supply*, DOE/GO-102005-2135 and ORNL/TM-2005/66, Washington, DC, US Departments of Energy and Agriculture and Oak Ridge, TN, Oak Ridge National Laboratory.

[7] G. Huber, S. Iborra, and A. Corma, 2006, "Synthesis of transportation fuels from biomass: chemistry, catalysts, and engineering," *Chem. Rev.*, 4044–4098.

[8] N. Mosier, C. Wyman, B. Dale et al., 2005, "Features of promising technologies for pretreatment of lignocellulosic biomass," *Bioresource Technol.*, 96, 673–686.

[9] B. Yang and C. E. Wyman, 2008, "Pretreatment: the key to unlocking low-cost cellulosic ethanol," *Biofuel Bioprod. Biorefining*, 2, 26–40.

[10] R. Sheldon, R. Lau, M. Sorgedrager, F. van Rantwijk, and K. Seddon, 2002, "Biocatalysis in ionic liquids," *Green Chem.*, 4, 147–151.

[11] C. E. Wyman, B. E. Dale, R. T. Elander et al., 2009, "Comparative sugar recovery and fermentation data following pretreatment of poplar wood by leading technologies," *Biotechnol. Prog.*, 25, 333–339.

[12] C. E. Wyman, 2008, "Cellulosic ethanol: a unique sustainable liquid transportation fuel. *MRS Bull.*, 33, 381–383.

[13] A. Asghari, R. J. Bothast, J. B. Doran, and L. O. Ingram, 1996, "Ethanol production from hemicellulose hydrolysates of agricultural residues using genetically engineered *Escherichia coli* strain KO11," *J. Indust. Microbiol.*, 16, 42–47.

[14] L. R. Lynd, P. J. Weimer, W. H. van Zyl, and I. S. Pretorius, 2002, "Microbial cellulose utilization: fundamentals and biotechnology," *Microbiol. Molec. Biol. Rev.*, 66, 506–577.

[15] M. E. Himmel, S.-Y. Ding, D. K. Johnson et al., 2007, "Biomass recalcitrance: engineering plants and enzymes for biofuels production," *Science*, 315, 804–807.

[16] R. den Haan, S. H. Rose, L. R. Lynd, and W. H. van Zyl, 2007, "Hydrolysis and fermentation of amorphous cellulose by recombinant *Saccharomyces cerevisiae*," *Metabolic Eng.*, 9, 87–94.

[17] L. R. Lynd, C. E. Wyman, and T. U. Gerngross, 1999, "Biocommodity engineering," *Biotechnol. Prog.*, 15, 777–793.

[18] M. R. Connor, and J. C. Liao, 2009, "Microbial production of advanced transportation fuels in non-natural hosts," *Curr. Opin. Biotechnol.* 20, 307–315.

[19] M. Rude and A. Schirmer, 2009, "New microbial fuels: a biotech perspective," *Curr. Opin. Microbiol.*, 12, 274–281.

[20] N. Qureshi and H. P. Blaschek, 2001, "Recent advances in ABE fermentation: hyper-butanol producing *Clostridium beijerinckii* BA101," *J. Indust. Microbiol. Biotechnol.*, 27, 287–291.

[21] H. Alvarez and A. Steinbuchel, 2002, "Triacylglycerols in prokaryotic microorganisms," *Appl. Microbiol. Biotechnol.*, 60, 367–376.

[22] G. A. Strobel, B. Knighto, K. Kluck et al., 2008, "The production of myco-diesel hydrocarbons and their derivatives by the endophytic fungus *Gliocladium roseum* (NRRL 50072)," *Microbiology*, 154, 3319–3328.

[23] L. Le Franc, 1927, Manufacture of butyric acids and other aliphatic acids, US patent 1625732.

[24] M. T. Holtzapple, M. K. Ross, N. S. Chang, V. S. Chang, S. K. Adelson, and C. Brazel, 1997, "Biomass conversion to mixed alcohol fuels using the MixAlco process," Fuels and Chemicals from Biomass, *ACS Symposium Series 666*, (B. C. Saha and J. Woodward, Eds.), *American Chemical Society*, Washington, D.C., pp. 130–142.

[25] J. Q. Bond, D. M. Alonso, D. Wang, R. M. West, and J. A. Dumesic, 2010, "Integrated catalytic conversion of

gamma-valerolactone to liquid alkenes for transportation fuels," *Science* **327**, 1110–1114.

[26] **G. W. Huber**, **R. D. Cortright**, and **J. A. Dumesic**, 2004, "Renewable alkanes by aqueous-phase reforming of biomass-derived oxygenates," *Angew. Chem. Int. Ed.*, **43**, 1549–1551.

[27] **D. A. Simonetti**, **J. Rass-Hansen**, **E. L. Kunkes**, **R. R. Soares**, and **J. A. Dumesic**, 2007, "Coupling of glycerol processing with Fischer–Tropsch synthesis for production of liquid fuels," *Green Chem.* **9**, 1073–1083.

[28] **B. Capai** and **G. Lartigau**, 1992, Preparation of laevulinic acid, US Patent 5,175,358.

[29] **H. Y. Zheng**, **Y.-L. Zhu**, **L. Huang** *et al.*, 2008, "Study on Cu–Mn–Si catalysts for synthesis of cyclohexanone and 2-methylfuran through the coupling process," *Catal. Commun.*, **9**, 342–348.

[30] **T. Werpy**, and **G. Petersen** (eds.), 2004, *Top Value Added Chemicals from Biomass – Volume 1 – Results of Screening from Sugars and Synthesis Gas*, Golden, CO, National Renewable Energy Laboratory.

27 Artificial photosynthesis for solar energy conversion

Boris Rybtchinski[1] and Michael R. Wasielewski[2]

[1]Department of Organic Chemistry, Weizmann Institute of Science, Rehovot, Israel
[2]Department of Chemistry, Northwestern University, Evanston, IL, USA

27.1 Focus

In natural photosynthesis, organisms optimize solar energy conversion through organized assemblies of photofunctional chromophores and catalysts within proteins that provide specifically tailored environments for chemical reactions. As with their natural counterparts, artificial photosynthetic systems for practical production of solar fuels must collect light energy, separate charge, and transport charge to catalytic sites where multielectron redox processes occur. Although encouraging progress has been made on each aspect of this complex problem, researchers have not yet developed self-ordering components and the tailored environments necessary to realize a fully functional artificial photosynthetic system.

27.2 Synopsis

Previously, researchers used complex, covalent molecular systems comprising chromophores, electron donors, and electron acceptors to mimic both the light-harvesting (antenna) and charge-separation functions of natural photosynthetic arrays. These systems allow one to derive fundamental insights into the dependences of electron-transfer rate constants on donor–acceptor distance and orientation, electronic interaction, and the free energy of the reaction. However, significantly more complex systems are required in order to achieve functions comparable to natural photosynthesis. Self-assembly provides a facile means for organizing large numbers of molecules into supramolecular structures that can bridge length scales from nanometers to macroscopic dimensions. To achieve an artificial photosynthetic system, the resulting structures must provide pathways for the migration of light excitation energy among antenna chromophores, and from antennas to reaction centers. They also must incorporate charge conduits, that is, molecular "wires" that can efficiently move electrons and holes between reaction centers and catalytic sites. The central challenge is to develop small, functional building blocks that have the appropriate molecular-recognition properties to facilitate self-assembly of complete, functional artificial photosynthetic systems.

Current strategies to create photofunctional assemblies use covalent building blocks based on chemically robust dyes, biomimetic porphyrins (or chlorophylls), redox relays, and metal complexes. These approaches take advantage of the shapes, sizes, and intermolecular interactions – such as π–π and/or metal–ligand interactions – to direct the formation of supramolecular structures having enhanced energy-capture and charge-transport properties. Modern analytical methods are critical for developing the deeper understanding of structure–function relationships in artificial photosynthetic assemblies required for their rational design. Although significant progress has been made in the creation of fairly complex artificial photosynthetic modules, their integration into a fully functional system capable of sustained solar-light-driven fuel production remains a key challenge, especially to get stable conversion efficiencies competitive with other renewable energy approaches .

27.3 Historical perspective

Natural photosynthesis that uses solar energy to power the biological world and store energy in the form of fossil fuels poses a great challenge to science: is it possible to create artificial systems that use solar light to make fuels in a sustainable way? The essence of artificial photosynthesis was formulated at the beginning of the twentieth century [1][2] by the eminent Italian chemist Giacomo Ciamician (Figure 27.1(a)) in an attempt to investigate whether "fossil solar energy" was the only type available or whether other methods of production could be found that rivaled the photochemical processes of the plants. Ciamician noted that

> ... the solar energy that reaches a small tropical country ... is equal annually to the energy produced by the entire amount of coal mined in the world!" He suggested that "For our purposes the fundamental problem from the technical point of view is how to fix the solar energy through suitable photochemical reactions ... By using suitable catalyzers, it should be possible to transform the mixture of water and carbon dioxide into oxygen and methane, or to cause other endo-energetic processes ... On the arid lands there will spring up industrial colonies without smoke and without smokestacks; forests of glass tubes will extend over the plants and glass buildings will rise everywhere; inside of these will take place the photochemical processes that hitherto have been the guarded secret of the plants, but that will have been mastered by human industry which will know how to make them bear even more abundant fruit than nature, for nature is not in a hurry and mankind is.

Ciamician's vision has inspired generations of researchers, but it has yet to be realized.

One of the first scientists to use the term "artificial photosynthesis" and make seminal contributions to the field was Melvin Calvin (Figure 27.1(b)), who received a Nobel Prize for his research on photosynthetic carbon fixation. Calvin, who was educated as a physical organic chemist, explored the idea of using synthetic "artificial" chromophores and catalysts as well as self-assembly to rationally design a system that uses sunlight to initiate an electron-transfer sequence that eventually converts light energy and abundant, inexpensive chemicals into energy-rich compounds that can be utilized as fuels [3].

Joseph J. Katz (Figure 27.1(c)) was a contemporary of Calvin, whose work on the physical chemistry of chlorophylls and other cofactors found in photosynthetic proteins paved the way for future work in artificial photosynthesis. Katz's fundamental studies on the self-assembly of chlorophylls and the properties of their radical ions provided critical information for the future design of complex systems for photochemical solar energy conversion. Perhaps most importantly, more than 40 years ago, Katz had a vision that has now become common in this field. Katz predicted the development of an artificial photosynthetic system, which he termed the "synthetic leaf," in which the key light-absorption and charge-separation components of photosynthesis could be assembled into a robust, protein-free biomimetic system to carry out the production of solar fuels and/or electricity in cells that are similar to Si-based inorganic solar cells [4].

27.4 The pursuit of artificial photosynthesis

In this chapter, **artificial photosynthesis** is defined as a sunlight-driven process leading to the formation of energy-rich compounds that can be used as fuels.

(a)　　　　　　　　(b)　　　　　　　　(c)

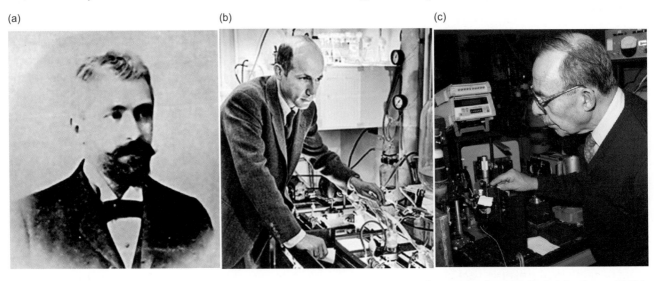

Figure 27.1. (a) Giacomo Ciamician (from [1]). (b) Melvin Calvin (from C&EN, 2008, 86, 60–61). (c) Joseph J. Katz (Argonne National Laboratory, 1982).

Natural photosynthesis is carried out by organized assemblies of photofunctional chromophores and catalysts within proteins that provide specifically tailored environments to optimize solar energy conversion [5]. The function of photosynthetic systems depends critically on their spatial organization, which provides efficient channels for energy, electron, and proton transfers. As a consequence, the photosynthetic proteins that carry pre-organized assemblies of natural chromophores (e.g., chlorophylls) and redox cofactors (e.g., chlorophylls, pheophytins and quinones) are in turn embedded in membranes that are responsible for further spatial organization of the components. Photosynthesis begins with absorption of light by chlorophyll chromophores within light-harvesting or "antenna" proteins, which act to increase the light-collection efficiency of the organism. Absorption of light by the chlorophyll produces its lowest-energy, electronically excited singlet state. This excited state then migrates between chlorophylls within single light-harvesting proteins and then between such proteins until it is eventually trapped by a pair of neighboring chlorophylls known as a "special pair" within a protein called the reaction center. The primary charge-separation chemistry occurs within the reaction center when the excited special pair of chlorophylls donates an electron to an adjacent third chlorophyll, initiating a cascade of electron transfers between cofactors with the protein. This sequence of electron transfers ultimately leaves a single positive charge (hole) on one chlorophyll of the special pair, which in some organisms is shared by both chlorophylls of the pair, and a negative charge (electron) at the opposite side of the protein trapped on a quinone molecule. The reaction-center protein is positioned in the membrane such that the hole and electron end up on opposite sides of the membrane, about 4 nm apart. In anoxygenic (bacterial) photosynthesis, a single type of reaction center is employed because water is not oxidized, whereas in oxygenic photosynthesis (plants, algae, cyanobacteria), two photosystems are used in series to provide the oxidizing (hole) and reducing (electron) equivalents, which are used to oxidize water to oxygen and protons and to reduce carbon dioxide to sugars, respectively. The resultant sugars store energy that can be used when needed. Overall, the conversion of light into chemical energy is a multi-step, multielectron process requiring several subsystems: *light-harvesting systems* that efficiently absorb and transfer excitation energy to the *sites where charge transfer occurs*, and *charge- and proton-transferring moieties* that supply electrons, holes, and protons to *catalytic sites* responsible for the production of energy-rich compounds from appropriate substrates.

In nature, photosynthesis supplies energy in a form suitable for biological systems, providing the energy source for all living organisms. The analogous processes in artificial photosynthesis need to store solar energy in the form of fuels that can be used when needed, providing an optimal solution for energy storage. The production and use of solar fuels can furnish a sustainable energy source that uses only sunlight, water, and CO_2, and does not produce undesirable side-products. However, unlike its natural counterpart, artificial photosynthesis is not restricted to a specific set of chromophores, redox cofactors, and catalysts. A broader selection of molecules and materials is available for the design and optimization of an artificial photosynthetic system, including robust organic dyes, metal and semiconductor nanoparticles, polymers, metal and nanoparticle catalysts, and so on. However, the organization and adaptivity of natural systems – the results of biological evolution – generate complexity and functionality yet to be achieved in artificial systems. Thus, given an excellent choice of building blocks, their (adaptive) assembly is one of the main challenges in artificial photosynthesis research.

27.4.1 Energy-storing reactions

Since most energy-releasing reactions involve oxidation, artificial photosynthesis targets the production of energy-rich compounds that are highly reduced and can release energy via reaction with atmospheric oxygen. Given the importance of reduction processes, artificial photosynthetic systems necessarily need to provide a source of electrons, so oxidation of a readily available substrate, providing the electrons required for reductive synthesis of fuels, is equally important. By its very nature, photo-induced single-electron transfer produces both positive and negative charge carriers (holes and electrons) that can drive oxidation and reduction reactions, respectively. Both species must be consumed in order to restore the system to its initial state, ready to execute a new cycle; the rate at which each cycle is completed is called the **turnover rate**. Most fuel-forming reactions require more than one hole or electron to carry out the requisite oxidation or reduction chemistry, so efficient, sustainable systems must deliver holes or electrons over many cycles to produce the maximal number of energy-rich chemical bonds per photon of light to attain a high quantum yield (defined as the number of redox equivalents or chemical bonds produced per photon). Two key energy-storage reactions considered in artificial photosynthesis are splitting of water (to give H_2 and O_2) and reduction of carbon dioxide (accompanied by oxidation of water) that utilize the most abundant and inexpensive substrates, which are also used in natural photosynthesis. We note that reduction of protons to produce hydrogen and reduction of carbon dioxide using sacrificial sources of electrons

Figure 27.2. A schematic representation of a photodriven molecular assembly for artificial photosynthesis using water-splitting and proton-reduction catalysts [12]. D, electron donor; C, chromophore; A, electron acceptor.

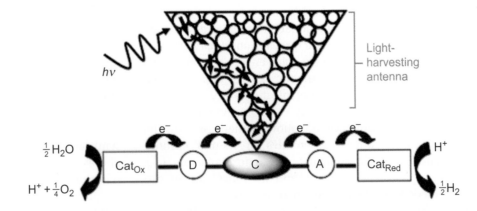

other than water can also be considered; however, such processes are costly and create undesirable side-products.

Currently there are no known molecular or hetero-geneous catalysts for the reduction of CO_2 to fuels that have the combination of rates, overpotentials, selectivities, cost, and lifetimes that would be needed for solar fuel production. Methane is a product of exhaustive CO_2 reduction,

$$2H_2O + CO_2 \xrightarrow{8h\nu} CH_4 + 2O_2, \qquad (27.1)$$

which can be stopped at an earlier stage, e.g., two-electron reduction to CO that can be used in well-established syngas-based processes to produce either the alkanes that comprise diesel fuel [6][7][8] or methanol [9][10][11]. The availability of syngas through partial reduction of CO_2 to CO may preclude the need for reducing CO_2 all the way to CH_4, but ultimately both its scientific feasibility and the economics will dictate whether formation of CH_4 is a desirable endpoint. Although none of the known catalysts for these CO_2 reduction processes is optimal, some are sufficiently promising to make it worthwhile to begin studies on integrating them with solar energy capture and conversion systems to produce solar fuels.

An energy input of 1.23 eV is needed to drive splitting of water,

$$2H_2O \xrightarrow{4h\nu} 2H_2 + O_2, \qquad (27.2)$$

which can be supplied by the energy of sunlight that, for example, has an energy of 2.25 eV in the middle of the visible spectrum at 550 nm. The hydrogen fuel produced by this overall process can react with oxygen to release the energy again; thus, hydrogen is a "solar fuel" analogous to conventional hydrocarbon fuels that release energy when oxidized (burned) with oxygen. Reactions with oxygen lead to oxidation of the fuels and recycling of the substrates (carbon dioxide and water), creating a sustainable energy supply cycle that needs only sunlight as an energy source.

Reactions (27.1) and (27.2) can be split into two half-reactions (oxidation and reduction). For example, in the case of water splitting:

$$2H_2O \rightarrow O_2 + 4H^+ + 4e^-, \qquad (27.3)$$

$$4H^+ + 4e^- \rightarrow 2H_2. \qquad (27.4)$$

Importantly, water splitting and carbon dioxide reduction are multiphoton, multielectron processes coupled to proton transfer. A fundamental understanding of such reactions is only starting to emerge, while control over multielectron processes remains a major challenge. Yet, as was stated by Alstrum-Acevedo *et al.* [12],

> Resolution into half-reactions provides the basis for a "modular" approach to artificial photosynthesis. As in natural photosynthesis, separate half-reactions can be addressed separately and combined at a later stage into a single device. Complexity is unavoidable because of multifunctional requirements (light absorption, energy transfer, electron transfer, redox catalysis). There is a need to arrange and integrate functional groups and to provide an overall structural hierarchy.

See Figure 27.2.

27.4.2 Artificial photosynthesis strategies

Conversion of solar into chemical energy can be based on several approaches. This section provides a brief description of selected strategies involving both inorganic and organic materials.

Solar-powered electrolysis

Solar photons can be converted into fuels using a two-stage approach in which (1) a solar cell powers electrolysis (2) to split water into hydrogen and oxygen or to perform other energy-storing reactions. The advantage of such a scheme lies in the fact that both solar cells and electrolysis technologies are readily available, and significant effort is currently being devoted to making more efficient solar cells and electrolytic systems. The latter are improved by developing new electrocatalysts that have redox potentials matched to those of the reaction of interest and significantly speed up the chemical reaction. Progress has been made in the development of CO_2 reduction electrocatalysts [13], and stable water oxidation electrocatalysts operating at low overpotentials have recently been described [14][15]. It should be noted that redox catalysts are involved in all artificial photosynthesis schemes (see below), and therefore, their development is of paramount importance to the field.

Photoelectrochemical cells based on inorganic materials (Chapter 49)

Fujishima and Honda [16] showed in 1972 that photoelectrochemical cells based on titanium dioxide (TiO_2) split water upon irradiation with light, but only with ultraviolet light that matches the TiO_2 bandgap. This system triggered research on direct water splitting without use of electrical connections. Thus, following this discovery, inorganic materials that absorb visible light and are capable of splitting water have been targeted. For example, Zou et al. [17] reported that indium–tantalum oxide photocatalysts doped with nickel, $In_{1-x}Ni_xTaO_4$ ($x = 0$–0.2), directly split water into stoichiometric amounts of oxygen and hydrogen under visible-light irradiation with a quantum yield of about 0.66%. It was also reported that a solid solution of gallium nitride and zinc oxide, $(Ga_{1-x}Zn_x)(N_{1-x}O_x)$, modified with rhodium and chromium mixed-oxide nanoparticles promotes catalytic splitting of water with a modest number of turnovers [18]. Although they are not very efficient, these systems represent an entry into light-driven water splitting using stable inorganic materials.

In photoelectrochemical cells, half-reactions at the electrodes are directly powered with light, requiring integration of chromophores into the cell. Two systems that exemplify the developments in the field are shown schematically in Figures 27.3 and 27.4. Khaselev and Turner [19] have shown that water can be split in a cell with a semiconductor cathode and metal anode (Figure 27.3). The semiconductor absorbs light, separates charge, and provides a catalytic surface as well. In a design by Youngblood et al. [20] (Figure 27.4), which is based on a dye-sensitized solar cell (DSSC), a ruthenium tris(bipyridyl) molecular chromophore is bound to nanoporous TiO_2 to produce a photocathode. Iridium oxide coupled to the ruthenium complex serves as a water-oxidation catalyst. Although both systems have rather low efficiencies, they nevertheless constitute a proof of concept regarding the feasibility of artificial photosynthesis based on photoelectrochemical cells.

Self-assembly of organic materials for artificial photosynthesis

Complex, covalent molecular systems comprised of chromophores, electron donors, and electron acceptors

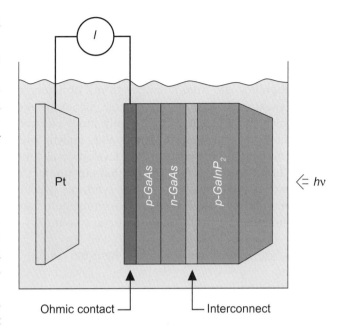

Figure 27.3. A photoelectrochemical cell [19].

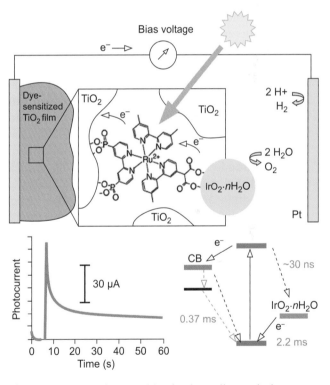

Figure 27.4. A dye-sensitized solar cell coupled to a water-splitting catalyst [20].

that mimic both the light-harvesting and the charge-separation functions of photosynthetic assemblies have been used to study the energy- and electron-transfer processes relevant to artificial photosynthesis. Despite significant progress in the development of these covalent arrays, no complete artificial photosynthetic systems are known. Self-assembly provides a facile means for organizing large numbers of molecules into supramolecular structures that can bridge length scales from nanometers to macroscopic dimensions. The resulting structures must provide pathways for the migration of light excitation energy among antenna chromophores and from antennae to reaction centers. They must also incorporate charge conduits (i.e., molecular "wires") that can efficiently move electrons and holes between reaction centers and catalytic sites. The central scientific challenge is to develop small, functional building blocks, having a minimum number of covalent linkages (because covalent synthesis is costly and time-consuming), that can self-assemble into complete, functional artificial photosynthetic systems.

Energy transport in self-assembled light-harvesting systems

Following photoexcitation of an antenna chromophore array, one or more energy-transfer steps must funnel the excitation energy to a site at which charge separation occurs. Covalent light-harvesting chromophoric arrays designed to funnel energy to a central site often involve significant synthetic complexity. Ever since the crystal structure of the antenna protein from purple photosynthetic bacteria revealed that its bacteriochlorophylls are arrayed in a ring structure (Figure 27.5(a)) [21], considerable effort has been invested in making biomimetic chromophore rings, many of which offer insights into the photophysics of energy collection and transfer in these proteins. [22][23][24][25]. However, a more effective strategy for multichromophore array construction is desirable for developing artificial photosynthetic systems for solar energy conversion. The 20-position of chlorophyll a (Chl a) offers a new site for rigid attachment of ligands that can be used to build supramolecular Chl structures based on metal–ligand binding [26]. Figures 27.5(b) and (c) show covalent Chl_3 and Chl_4 arrays, respectively, in which the energy-transfer pathways between the Chls have been found to be very fast. For example, the Chl–Chl energy-transfer rate in Chl_3 was found to be $(6 \text{ ps})^{-1}$ and is comparable to that seen in antenna proteins [26]. The asymmetric spatial relationship between Chl pairs within Chl_4 allows an analysis of the differential energy-transfer rates involving the various Chl–Chl distances [27]. Such well-defined covalent building blocks allow detailed information to be acquired regarding the fundamental dependence of energy-transfer rates on distance and molecular orientation. Given this information, they can be used to develop more complex self-assembling systems.

To explore this strategy, a simple 20-(4-pyridinyl)Chl derivative (Chl-py) that self-assembles into a cyclic tetramer, as shown SAXS and WAXS studies in solution, was prepared (Figure 27.5(d)) [28]. This cyclic tetramer was found to exhibit an ultrafast intramolecular energy-

transfer rate of $(1.3 \text{ ps})^{-1}$. Figure 27.5(e) shows the structure of such a self-assembled trimer of Chl_3 molecules as determined by SAXS and WAXS techniques [29]. Once again, analysis of energy-transfer rates showed that the assembly has two dominant energy-transfer processes, a faster one involving adjacent Chls $(7 \text{ ps})^{-1}$ and a slower one involving energy transfer across the large ring structure $(152 \text{ ps})^{-1}$. Antenna structures need not be restricted to **biomimetic** rings, but can have significantly different geometries. Figure 27.5(f) shows a prismatic structure in which three Chl dimers (Chl_2) assemble into a trigonal prismatic structure through the agency of two molecules of a three-fold-symmetric ligand having three pyridines [30]. Once again, energy transfer between the Chl_2 faces of the prism was found to be very fast $(6 \text{ ps})^{-1}$ and to depend strongly on the distance between the Chl_2 faces of the prism.

Self-assembled photodriven charge-separation and -transport systems

Chlorophylls and other biomimetic pigments, such as the related porphyrins, constitute only a small fraction of the vast number of organic dyes that could have possible applications as photoactive redox agents in artificial photosynthetic systems. For example, perylene-3,4:9,10-bis(dicarboximide) (PDI) dyes and their many derivatives provide a robust platform on which to develop **bio-inspired** charge-separation and -transport systems for solar energy conversion. They absorb in the central range of the visible spectrum, are powerful photo-oxidants, and are easy to reversibly reduce. Consequently, PDI derivatives have attracted significant interest as active materials for light harvesting [31][32], photovoltaics [33][34][35], and studies of basic photoinduced charge- and energy-transfer processes [36][37]. In one such case, it was found that attaching pyrrolidine rings to the 1,7-positions of PDI gives a molecule (5PDI) that absorbs light at wavelengths similar to those at which Chl absorbs and has a brilliant green color [38]. When two 5PDI molecules are stacked in a face-to-face geometry, absorption of a photon by one 5PDI molecule results in ultrafast formation of the charge-separated radical ion pair $5PDI^{+\bullet}$–$5PDI^{-\bullet}$ [39]. Taking advantage of this unusual result, the self-assembly properties of PDI were used to bring two 5PDI molecules close to one another to provide a light-harvesting antenna array. The similar molecule $5PDI\text{-}PDI_4$ self-assembles into π-stacked dimers $(5PDI\text{-}PDI_4)_2$ in solution (Figure 27.6) [40]. This dimeric array demonstrates that self-assembly of a robust PDI-based artificial light-harvesting antenna structure induces self-assembly of a functional special pair of 5PDI molecules that undergoes ultrafast, quantitative charge separation. Electron transfer occurs only in the dimeric system, not in the disassembled monomer, thus mimicking both the antenna and the special pair

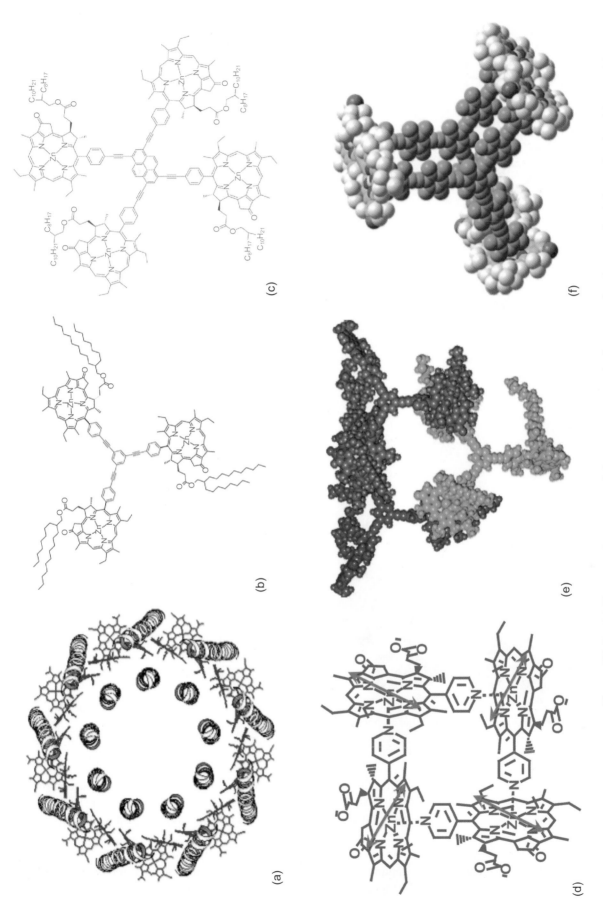

Figure 27.5. (a) The crystal structure of the antenna protein from purple photosynthetic bacteria. (b) Chl₃. (c) Chl₄. (d) Self-assembled cyclic (Chl-py)₄. (e) Self-assembled trimer of Chl₃. (f) Self-assembled prismatic structure with three butadiyne-linked Chl₂ faces.

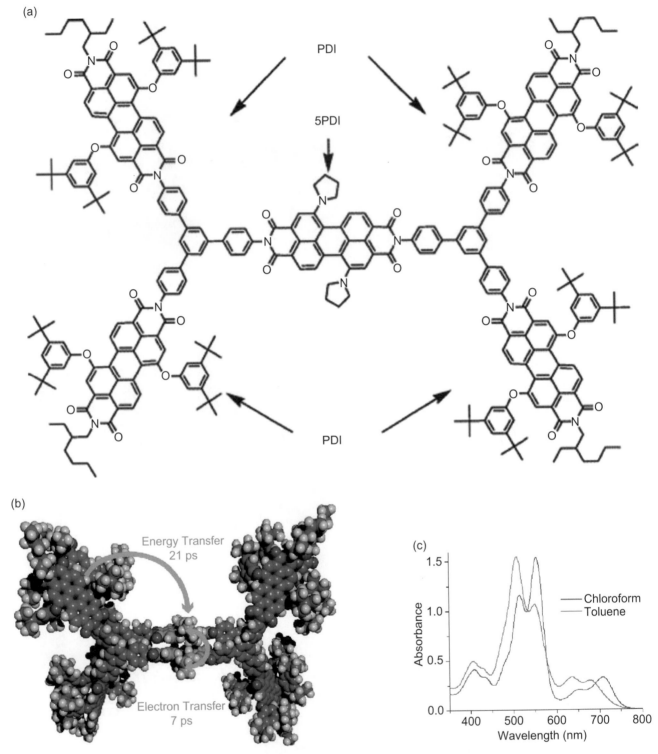

Figure 27.6. (a) The chemical structure of 5PDI-PDI$_4$. (b) The self-assembled structure of (5PDI-PDI$_4$)$_2$ from modeling the SAXS/WAXS data. (c) Ultraviolet–visible spectra showing dimerization in toluene (monomers in CHCl$_3$) and excellent solar spectral coverage [40].

function of photosynthesis. The charge separation observed in (5PDI-PDI$_4$)$_2$ is an example of **emergent behavior** that does not occur when monomeric 5PDI-PDI$_4$ is photoexcited.

The strong tendency of PDI molecules to self-assemble can be used to develop extended assemblies based on a particular covalent building block designed to yield segregated stacks of electron donors and

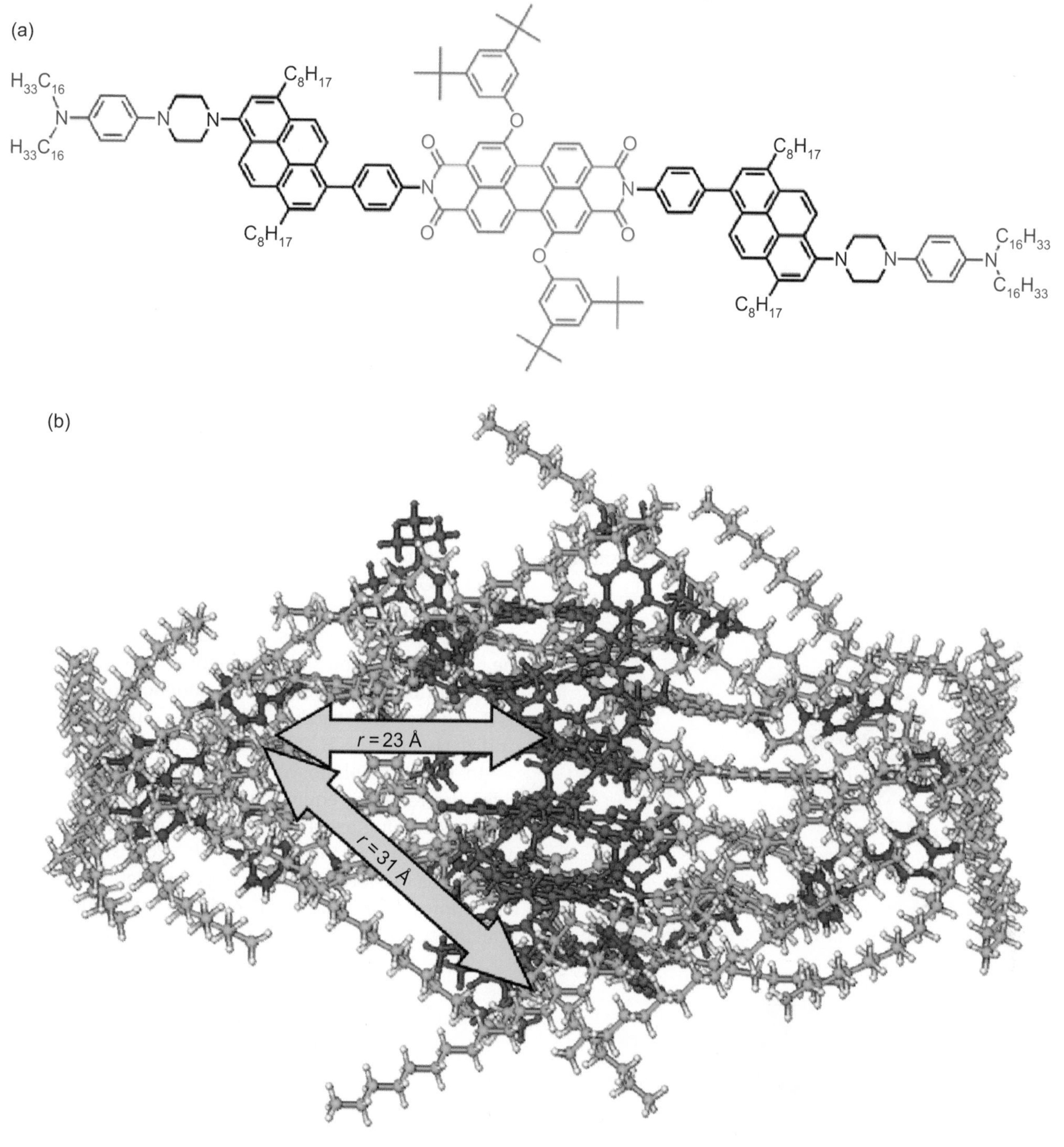

(a)

(b)

Figure 27.7. (a) DAB-Py-PDI-Py-DAB monomer. (b) The structure (side view) of its self-assembled hexamer in methylcyclohexane as determined by SAXS/WAXS [41].

acceptors to serve as charge conduits. Recently, a donor–acceptor triad building block consisting of a PDI chromophoric acceptor, an aminopyrene primary donor (APy), and a diaminobenzene (DAB) secondary donor was synthesized (Figure 27.7) [41]. This building block self-assembles into a helical hexamer in methylcyclohexane (Figure 27.7). Photoexcitation of the PDI chromophore in this protein-sized assembly produces

$DAB^{+\bullet}$–$PDI^{-\bullet}$, which lives for about 40 ns, during which the electron freely migrates through the π-stacked PDI acceptors (having good electronic communication due to the overlap of their π-systems). This is the first example of non-covalent secondary charge transport in a π-stacked system being competitive with charge recombination that depletes the concentration of energy-storing charge-separated species. Such behavior is important for developing both artificial photosynthetic systems and organic photovoltaics for solar energy conversion.

An important set of experiments has shown that charge separation within an artificial photosynthetic system can be coupled to a biochemical energy-transduction process. Gust *et al.* showed that the enzyme adenosine-5′-triphosphate (ATP) synthase, which is a key part of the photosynthetic machinery, can be powered by light using an artificial system that is capable of long-lived charge separation [42]. Thus, the latter mimics a photosynthetic reaction center. Importantly, all components of the system are embedded in a membrane, akin to natural photosynthesis (Figure 27.8). The porphyrin (P) in the C–P–Q triad absorbs light, and two-step charge separation occurs to generate a $C^{+\bullet}$–P–$Q^{-\bullet}$ charge-separated state having a lifetime of 110 ns. The quinone radical anion of the triad that is near the outer membrane surface reduces a molecule of quinone (Q_S) to the semiquinone anion, which accepts a proton from water. The resulting neutral semiquinone radical is oxidized back to the quinone by $C^{+\bullet}$ near the inner membrane surface. The protonated quinone is a strong acid that releases the proton into the inner volume of the liposome, creating a proton gradient that powers ATP synthase (Figure 27.9).

27.4.3 Light-driven catalysis for water splitting

Water-oxidation catalysts

Once photo-induced charge separation occurs, multiple charges must be transported to catalytic centers where fuels are produced. Water oxidation in photosystem II (PSII) of green plants uses a multimetallic Mn complex to produce O_2 and H^+ (Figure 27.8; see also Chapters 28 and 47) [43]. While the cooperation of several metal centers situated in close proximity as been shown to be necessary in a number of model H_2O-oxidation catalysts, there is an ever growing number of monometallic catalysts that show great promise [44]. Almost all photo-induced-H_2O-splitting systems studied thus far employ reaction schemes in which only one of the oxidation ($2H_2O - O_2 + 4H^+ + 4e^-$) and reduction ($2H^+ + 2e^- \ H_2$) half-reactions occurs. These systems all require sacrificial electron donors or acceptors, which negate the economic and environmental advantages of H_2O splitting. Thus, practical production of H_2 from H_2O requires two catalysts acting in tandem, one to oxidize H_2O to $O_2 + H^+$ and another to reduce H^+ to H_2.

In oxygenic photosynthetic organisms, H_2O provides the reducing equivalents for CO_2 fixation, while itself undergoing oxidation catalyzed by an oxomanganese cluster in PSII (Figure 27.8) [45]. It is attractive to use H_2O as the source of reducing equivalents for artificial photosynthesis as well, owing to its widespread availability and the possibility of creating a completely

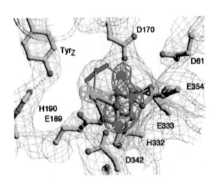

Figure 27.8. A model of the Mn_4 water-oxidation catalyst of PSII obtained from X-ray diffraction data [43].

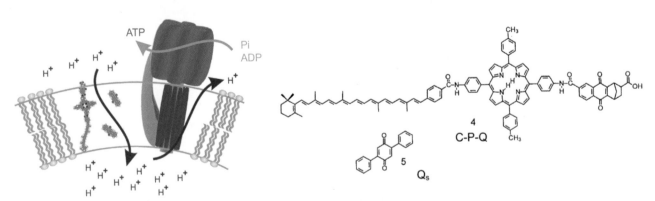

Figure 27.9. A schematic representation of an artificial photosynthetic membrane. The lipid bilayer of a liposome vesicle contains the components of a light-driven proton pump: a vectorially inserted C–P–Q triad and "shuttle" quinone Q_S. Illumination of the triad leads to transport of hydrogen ions into the liposome interior, establishing a proton-motive force. The membrane also contains a vectorially inserted ATP synthase enzyme. The flow of protons out of the liposome through this enzyme drives the production of ATP [42].

renewable energy conversion cycle. One of the first examples of an active homogeneous O_2-evolving catalyst based on Mn is the terpyridine complex [Mn(III, IV)$_2$O$_2$(terpy)$_2$(H$_2$O)$_2$]$^{3+}$, (R = H or Ph, Figure 27.10) [46]. This dimanganese catalyst evolves O_2 with hundreds of turnovers using oxone as the primary oxidant and has been studied in great detail mechanistically. Although the origin of the oxygen atoms in the O_2 produced was verified using H$_2$ ^{18}O, competing mechanisms involving oxone (potassium peroxymonosulfate) result in partial incorporation of oxygen from oxone into the observed O_2. Substituting Ce(IV) for oxone results in H$_2$O providing all the observed O_2, with the key shortfall being the use of a one-electron sacrificial oxidant. A key challenge is to provide this catalyst with a photodriven source of oxidizing equivalents to replace Ce(IV).

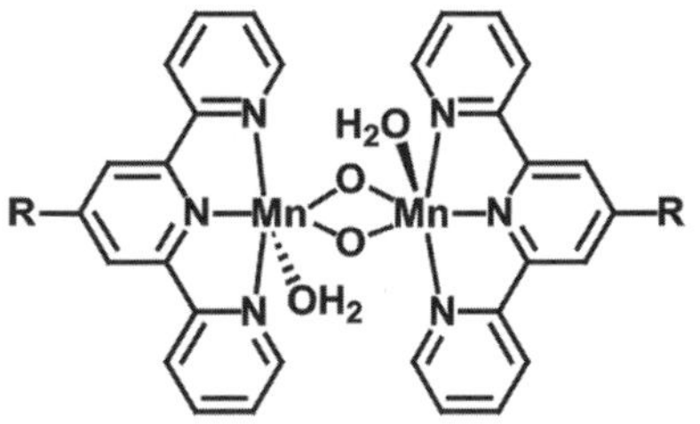

Figure 27.10. A dimanganese complex with water-splitting catalytic activity.

Figure 27.11. Manganese–ruthenium trisbipyridinyl-dinaphthalenediimide triad PSII synthetic mimic [47].

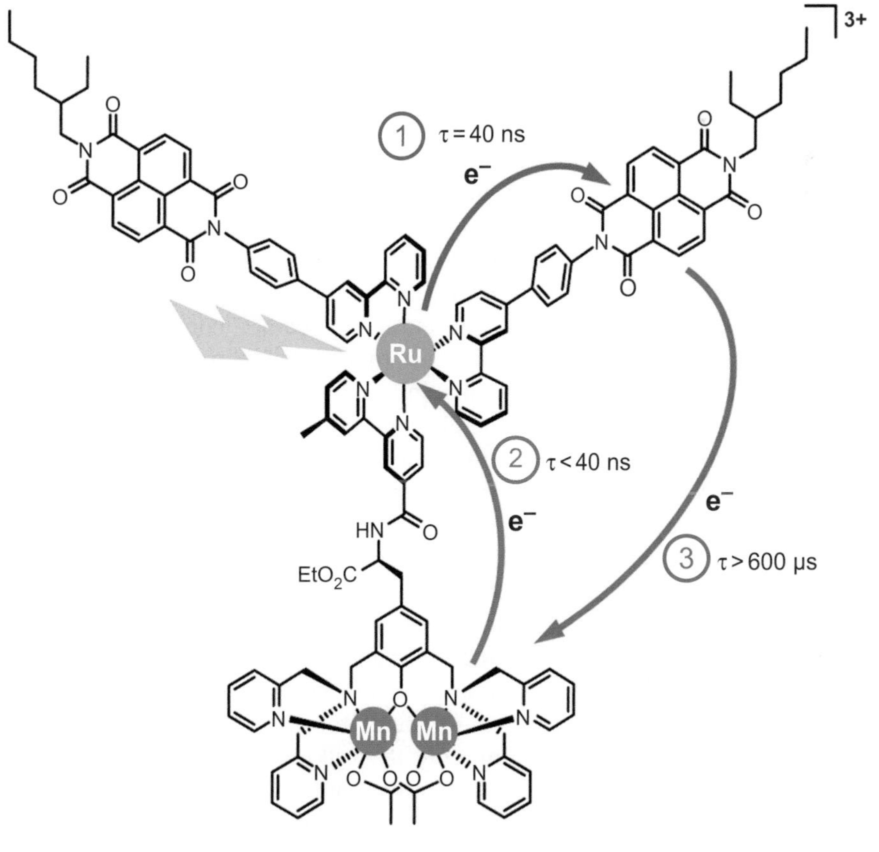

In the mechanism of catalyst function, oxygen–oxygen bond formation is proposed to occur through attack of H$_2$O on a terminal oxo species bound to a high-valent Mn ion, formally Mn(V)=O, a reaction analogous to current mechanistic proposals in PSII. A great advantage of this system over other homogeneous H$_2$O oxidation catalysts (reviewed in [44]) is the use of an abundant metal, Mn, making any successful catalyst viable on a large scale. We note that the exact mechanism of water oxidation in PSII has yet to be fully understood, being the subject of much current research.

Photodriven oxygen evolution from water requires the accumulation of multiple oxidative equivalents at a catalyst, which introduces additional mechanistic complications. Mimics of the water-splitting manganese–calcium complex in PSII have been synthesized and linked to photosensitizers (light-absorbing molecules) to systematically determine how molecular structure and charge-transfer thermodynamics and kinetics influence accumulative electron-transfer processes (Figure 27.11) [47]. Although the generation of a fully functional, efficient, molecular-level mimic has not been synthesized, this research illustrates the importance of creating a molecular-scale mimic to gain basic mechanistic understanding before increasing the system's complexity.

One of the major hurdles for useful and efficient oxidation of water is stability. One way in which natural

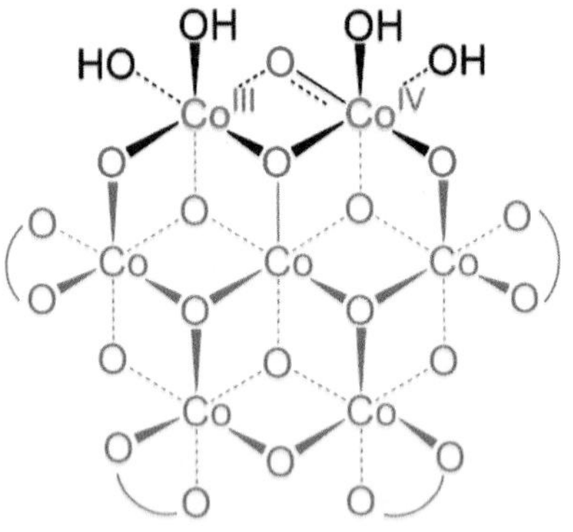

Figure 27.12. A self-healing cobalt water-oxidation catalyst [48].

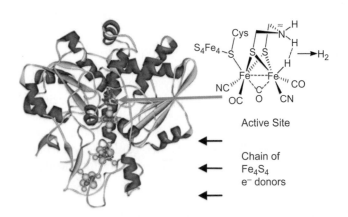

Figure 27.13. Hydrogenase and its active site and electron-transfer chain.

photosynthesis overcomes this problem is by regenerating a key protein in the oxygen-evolving complex every ~30 minutes. Therefore, achieving stable artificial photosynthesis requires materials and/or mechanisms in place that account for long-term chemical stability. Consequently, the discovery of a "self-healing," cobalt–phosphate water-oxidation catalyst that operates under neutral conditions (and a wide pH range) introduces a new catalyst with increased stability, while operating under benign conditions and with Earth-abundant materials (Figure 27.12) [14][48]. This catalyst also self-assembles *in situ*, which minimizes synthetic complexity and reduces cost.

Proton-reduction catalysts for H_2 production

Once photodriven water oxidation has been achieved, the resulting protons must be reduced to H_2 in the second half-reaction using a very different catalyst. Nature again provides an important example: hydrogenase enzymes that readily carry out the two-electron reduction of protons to H_2 at high turnover rates under physiological conditions (Figure 27.13) [49]. Although H_2 production is only weakly coupled to photosynthesis in Nature, the hydrogenase enzymes provide models based on abundant first-row transition metals with catalytic activities rivaling those of the best known platinum catalysts [50][51]. Crystallographic information on diiron hydrogenase (Figure 27.13) [52][53]. and synthetic strategies for a variety of diiron active-site motifs [54] make the diiron hydrogenase active site an attractive target for integration into light-harvesting supramolecular assemblies. In general, such systems consist of a proton-reduction catalyst, a chromophore that serves to harvest light and transfer reducing equivalents to the catalyst, and a sacrificial electron donor and proton source. Diiron complexes modeled on the active site of the diiron hydrogenases having the general formula $[Fe_2(\mu-R)(CO)_{6-n}(L)_n]$, where commonly R = alkyl or aryl dithiolate and L = CO, CN⁻, or PR_3, are a promising class of catalysts for use in photodriven H_2 production,

so there have been numerous studies of assemblies consisting of diiron complexes in combination with photosensitizers such as $Ru(bpy)_3^{2+}$ and zinc porphyrins [55][56][57]. To date, however, photo-induced electron transfer from a chromophore to a diiron hydrogenase model complex has been observed in only a handful of cases [58][59][60], and the dynamics of this process remain elusive. In many other instances, electron transfer is inferred from the quenching of the chromophore fluorescence in the presence of the diiron complex, but this is an unreliable indicator of electron transfer because a variety of other photophysical processes can be responsible for decreases in fluorescence intensity. Indeed, one of the main limitations in this field is that photo-induced electron transfer is often thermodynamically unfavorable because of the very negative reduction potentials (less than –1.2 V vs. SCE) of most diiron complexes.

A recent example that addresses these issues is the diiron complex NMI–$Fe_2S_2(CO)_6$ (NMI = naphthalene monoimide) and its covalently linked, fixed-distance zinc porphyrin dyad (Figure 27.14) [61]. The electron-withdrawing nature of the NMI group makes the diiron complex only about 80 mV harder to reduce than the natural diiron cofactor in hydrogenases, and this complex is among the most easily reduced hydrogenase mimics reported to date. Selective photoexcitation of the zinc porphyrin yields transient absorption spectra that show a distinct peak at 616 nm, which has been assigned to $(NMI–Fe_2S_2(CO)_6)^{-\bullet}$ on the basis of spectroelectrochemical measurements on the diiron complex alone. Photoexcitation of the zinc porphyrin–NMI–$Fe_2S_2(CO)_6$ dyad in the presence of acid was shown to generate H_2. Further optimization of this system and related biomimetic systems will be required if we are to achieve the goal of efficient photodriven proton reduction to H_2.

Recently, Kohl *et al.* showed that a single metal center acting cooperatively with a ligand capable of

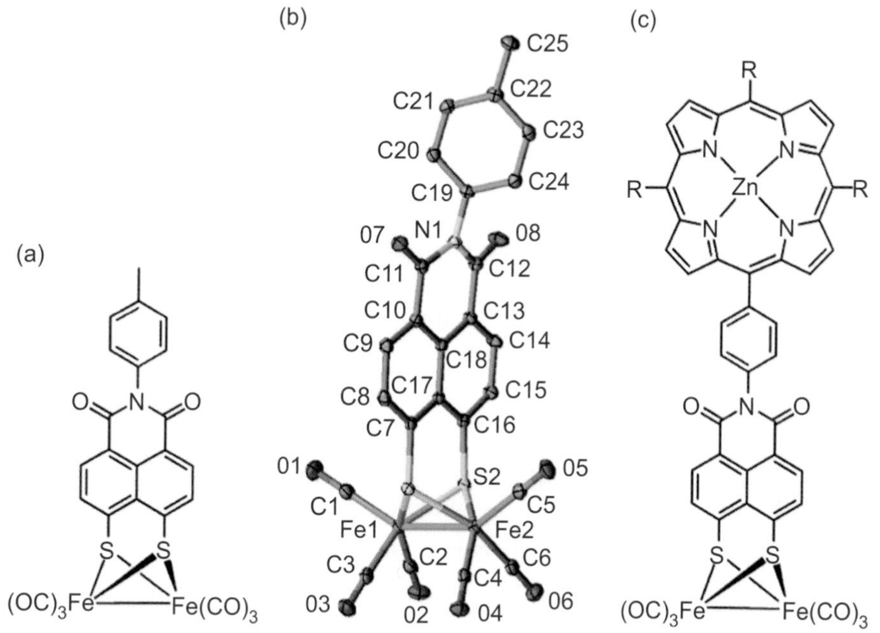

Figure 27.14. (a) The chemical structure of NMI-Fe$_2$S$_2$(CO)$_6$. (b) The X-ray structure of NMI-Fe$_2$S$_2$(CO)$_6$ with thermal ellipsoids set to 50% probability. (c) The chemical structure of Zn porphyrin–NMI-Fe$_2$S$_2$(CO)$_6$ (R = n-C$_5$H$_{11}$) [61].

(a)

(b)

(c)

proton transfer can promote a reaction that leads to splitting of water in consecutive heat- and light-driven steps (Figure 27.15) [62]. Initial reaction of water at 25 °C with a dearomatized Ru(II) pincer complex, **1**, yields a monomeric aromatic Ru(II) hydrido-hydroxo complex, **2**, which, upon further reaction with water at 100 °C, releases H$_2$ and forms a *cis*-dihydroxo complex, **3**. Irradiation of this complex with 320–420-nm light liberates O$_2$ and regenerates the starting hydrido-hydroxo Ru(II) complex, **2**, probably by elimination of hydrogen peroxide, which rapidly disproportionates. Labeling experiments with H$_2{}^{17}$O and H$_2{}^{18}$O showed unequivocally that the process of O—O bond formation is intramolecular. Although the reaction is stoichiometric rather than catalytic, it reveals a previously elusive fundamental step toward O$_2$-generating homogeneous catalysis and shows the feasibility of both reduction and oxidation of water by the same metal complex.

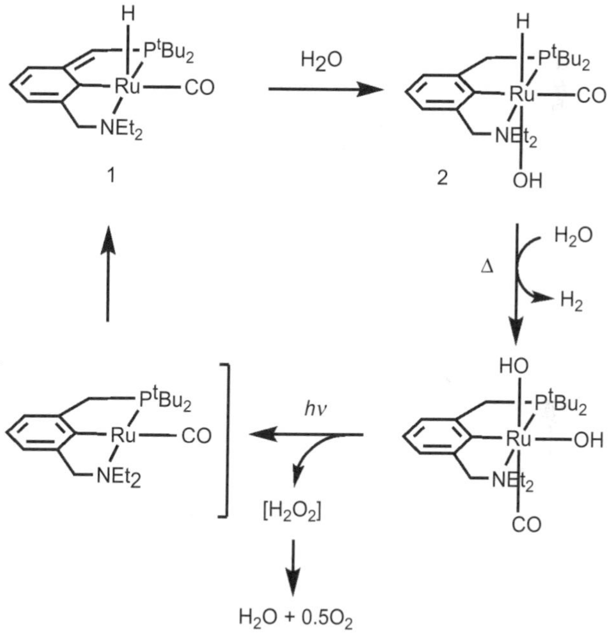

27.5 Summary

The primary processes of natural photosynthesis, such as photo-induced energy and charge transfer and redox catalysis, have been achieved in artificial systems. These constitute individual functional modules that need to be integrated into a complete photosynthetic array. Recent insights into natural photosynthesis provide important guidelines for the design of artificial systems, emphasizing the importance of the precise organization of the functional components. First steps toward mimicking the complexity of biological systems have been made by employing self-assembly to form photofunctional

Figure 27.15. Splitting of water by a well-defined mononuclear ruthenium complex. Adapted from [62].

nanoscale supramolecular systems. Importantly, a plethora of robust synthetic photoactive and redox-active building blocks is available, including a broad variety of organic dyes, polymers, nanoparticles, and metal complexes. The challenge lies in their integration into fully functional systems capable of photo-induced fuel production. Self-assembly appears to be the most promising methodology for the creation of such systems.

27.6 Questions for discussion

1. What methods can be used to optimize light capture, charge generation, and charge transport in artificial photosynthetic systems?
2. How can the properties of interfaces at which charge generation, separation, transport, and selective chemical reactions occur be controlled to optimize performance?
3. How can the problem of accumulating multiple charges at a catalyst be solved, given that only one electron–hole pair can be generated for every photon absorbed?
4. How can the use of sacrificial electron donors and acceptors in artificial photosynthetic systems be avoided?
5. Can organic and organic/inorganic materials be made sufficiently robust to carry out artificial photosynthesis efficiently for long times?

27.7 Further reading

- N. J. Turro, V. Ramamurthy, and J. C. Scaiano, 2009, *Principles of Molecular Photochemistry: An Introduction*, Herndon, VA, University Science Books. This text has an excellent introduction to the basic principles of energy and electron transfer.
- DOE BES Workshop, 2005, *Basic Research Needs for Solar Energy Utilization*, http://www.er.doe.gov/bes/reports/list.html. This is a broad and insightful description of the basic science issues facing the research community as it seeks to develop artificial photosynthetic systems.
- DOE BES, 2007, *Directing Matter and Energy: Five Challenges for Science and Imagination*, http://www.er.doe.gov/bes/reports/list.html. This report lays out major research directions that it is necessary to address if significant progress in developing fundamentally new materials for energy production is to be realized.
- M. R. Wasielewski, 2009, "Self-assembly strategies for integrating light harvesting and charge separation in artificial photosynthetic systems," *Acc. Chem. Res.*, 42, 1910–1921. This is a recent short review that discusses the challenges and strategies for integrating functional components into complete artificial photosynthetic systems.

Acknowledgment

Development of this manuscript at Northwestern (MRW) was supported as part of the ANSER Center, an Energy Frontier Research Center funded by the U.S. Department of Energy, Office of Science, Office of Basic Energy Sciences, under Award Number DE-SC0001059. Work at Weizmann (BR) was supported by grants from the Weizmann Institute Alternative Sustainable Energy Research Initiative and Yossie and Dana Hollander Foundation.

27.8 References

[1] M. Venturi, V. Balzani, and M. T. Gandolfi, 2005, "Fuels from solar energy. A dream of Giacomo Ciamician, the father of photochemistry," in *Proceedings of the 2005 Solar World Congress*.

[2] G. Ciamician, 1912, "The photochemistry of the future," *Science*, 36, 385–394.

[3] M. Calvin, 1987, "Artificial photosynthesis," *J. Membrane Sci.*, 33, 137–149.

[4] J. J. Katz, and M. R. Wasielewski, 1978, "Biomimetic approaches to artificial photosynthesis," *Biotechnol. Bioeng. Symp.*, 8, 433–452.

[5] R. E. Blankenship, 2002, *Molecular Mechanisms of Photosynthesis*, Oxford, Blackwell Science.

[6] M. E. Dry, 2001, "High quality diesel via the Fischer–Tropsch process – a review," *J. Chem. Tech. Biotech.*, 77, 43–50.

[7] M. E. Dry, 2002, "The Fischer–Tropsch process: 1950–2000," *Catal. Today*, 72, 227–241.

[8] M. E. Dry, 1996, "Practical and theoretical aspects of the catalytic Fischer–Tropsch process," *Appl. Catal. A*, 138, 319–344.

[9] L. J. Shadle, D. A. Berry, and M. Syamlal, 2004, "Coal gasification," in *Kirk–Othmer Encyclopedia of Chemical Technology*, 5th edn., vol. 6, pp. 771–832.

[10] M. V. Twigg, and M. S. Spencer, 2001, "Deactivation of supported copper metal catalysts for hydrogenation reactions," *Appl. Catal. A* 212, 161–174.

[11] A. Baiker, 2000, "Utilization of carbon dioxide in heterogeneous catalytic synthesis," *Appl. Organometall. Chem.*, 14, 751–762.

[12] J. H. Alstrum-Acevedo, M. K. Brennaman, and T. J. Meyer 2005, "Chemical approaches to artificial photosynthesis," 2, *Inorg. Chem.*, 44, 6802–6827.

[13] E. E. Benson, C. P. Kubiak, A. J. Sathrum, and J. M. Smieja 2009, "Electrocatalytic and homogeneous approaches to conversion of CO_2 to liquid fuels," *Chem. Soc. Rev.*, 38, 89–99.

[14] M. W. Kanan and D. G. Nocera, 2008, "*In situ* formation of an oxygen-evolving catalyst in neutral water containing phosphate and Co^{2+}," *Science*, 321, 1072–1075.

[15] M. Dinca, Y. Surendranath, and D. G. Nocera, 2010, "Nickel-borate oxygen-evolving catalyst that functions under benign conditions," *Proc. Natl. Acad. Sci.*, 107, 10337–10341.

[16] A. Fujishima and K. Honda, 1972, "Electrochemical photolysis of water at a semiconductor electrode," *Nature*, 238, 37–38.

[17] Z. G. Zou, J. H. Ye, K. Sayama, and H. Arakawa, 2001, "Direct splitting of water under visible light irradiation with an oxide semiconductor photocatalyst," *Nature*, 414, 625–627.

[18] K. Maeda, K. Teramura, D. L. Lu et al., 2006, "Photocatalyst releasing hydrogen from water – enhancing catalytic performance holds promise for hydrogen production by water splitting in sunlight," *Nature*, **440**, 295–295.

[19] O. Khaselev and J. A. Turner, 1998, "A monolithic photovoltaic–photoelectrochemical device for hydrogen production via water splitting," *Science*, **280**, 425–427.

[20] W. J. Youngblood, S.-H. A. Lee, K. Maeda, and T. E. Mallouk, 2009, "Visible light water splitting using dye-sensitized oxide semiconductors," *Acc. Chem. Res.*, **42**, 1966–1973.

[21] G. McDermott, S. M. Prince, A. A. Freer, et al., 1995, "Crystal structure of an integral membrane light-harvesting complex from photosynthetic bacteria," *Nature*, **374**, 517–521.

[22] S. Rucareanu, A. Schuwey, and A. Gossauer, 2006, "One-step template-directed synthesis of a macrocyclic tetraarylporphyrin hexamer based on supramolecular interactions with a C_3-symmetric tetraarylporphyrin trimer," *J. Am. Chem. Soc.*, **128**, 3396–3413.

[23] O. Shoji, H. Tanaka, T. Kawai, and Y. Kobuke, 2005, "Single molecule visualization of coordination-assembled porphyrin macrocycles reinforced with covalent linkings," *J. Am. Chem. Soc.*, **127**, 8598–8599.

[24] Y. Nakamura, I.-W. Hwang, N. Aratani et al., 2005, "Directly meso–meso linked porphyrin rings: synthesis, characterization, and efficient excitation energy hopping," *J. Am. Chem. Soc.*, **127**, 236–246.

[25] J. Li, A. Ambroise, S. I. Yang et al., 1999, "Template-directed synthesis, excited-state photodynamics, and electronic communication in a hexameric wheel of porphyrins," *J. Am. Chem. Soc.*, **121**, 8927–8940.

[26] R. F. Kelley, M. J. Tauber, and M. R. Wasielewski, 2006, "Intramolecular electron transfer through the 20-position of a chlorophyll a derivative: an unexpectedly efficient conduit for charge transport," *J. Am. Chem. Soc.*, **128**, 4779–4791.

[27] V. L. Gunderson, T. M. Wilson, and M. R. Wasielewski, 2009, "Excitation energy transfer pathways in asymmetric covalent chlorophyll a tetramers," *J. Phys. Chem. C*, **113**, 11936–11942.

[28] R. F. Kelley, R. H. Goldsmith, and M. R. Wasielewski, 2007, "Ultrafast energy transfer within cyclic self-assembled chlorophyll tetramers," *J. Am. Chem. Soc.*, **129**, 6384–6385.

[29] V. L. Gunderson, S. M. Mickley Conron, and M. R. Wasielewski, 2010, "Self-assembly of a hexagonal supramolecular light-harvesting array from chlorophyll a trefoil building blocks," *Chem. Commun.*, **46**, 401–403.

[30] R. F. Kelley, S. J. Lee, T. M. Wilson et al., 2008, "Intramolecular energy transfer within butadiyne-linked chlorophyll and porphyrin dimer-faced, self-assembled prisms," *J. Am. Chem. Soc.*, **130**, 4277–4284.

[31] K. Muthukumaran, R. S. Loewe, C. Kirmaier et al., 2003, "Synthesis and excited-state photodynamics of a perylene-monoimide-oxochlorin dyad. A light-harvesting array," *J. Phys. Chem. B*, **107**, 3431–3442.

[32] S. E. Miller, Y. Zhao, R. Schaller et al., 2002, "Ultrafast electron transfer reactions initiated by excited CT states of push–pull perylenes," *Chem. Phys.*, **275**, 167–183.

[33] B. A. Gregg, and R. A. Cormier, 2001, "Doping molecular semiconductors: n-type doping of a liquid crystal perylene diimide," *J. Am. Chem. Soc.*, **123**, 7959–7960.

[34] E. E. Neuteboom, E. H. A. Beckers, S. C. J. Meskers, E. W. Meijer, and R. A. J. Janssen, 2003, "Singlet-energy transfer in quadruple hydrogen-bonded oligo(p-phenylenevinylene)perylene-diimide dyads," *Org. Biomol. Chem.*, **1**, 198–203.

[35] C. W. Tang, 1986, "Two-layer organic photovoltaic cell," *App. Phys. Lett.*, **48**, 183–185.

[36] F. Würthner, C. Thalacker, and A. Sautter, 1999, "Hierarchical organization of functional perylene chromophores to mesoscopic superstructures by hydrogen bonding and ι–ι interactions," *Adv. Mater.*, **11**, 754–758.

[37] A. P. H. J. Schenning, J. van Herrikhuyzen, P. Jonkheijm et al., 2002, "Photoinduced electron transfer in hydrogen-bonded oligo(p-phenylene vinylene)-perylene bisimide chiral assemblies," *J. Am. Chem. Soc.*, **124**, 10252–10253.

[38] Y. Zhao and M. R. Wasielewski, 1999, "3,4:9,10-Perylenebis(dicarboximide) chromophores that function as both electron donors and acceptors," *Tetrahedron Lett.*, **40**, 7047–7050.

[39] J. M. Giaimo, A. V. Gusev, and M. R. Wasielewski, 2002, "Excited-state symmetry breaking in cofacial and linear dimers of a green perylenediimide chlorophyll analogue leading to ultrafast charge separation," *J. Am. Chem. Soc.*, **124**, 8530–8531.

[40] B. Rybtchinski, L. E. Sinks, and M. R. Wasielewski, 2004, "Combining light-harvesting and charge separation in a self-assembled artificial photosynthetic system based on perylenediimide chromophores," *J. Am. Chem. Soc.*, **126**, 12268–12269.

[41] J. E. Bullock, R. Carmieli, S. M. Mickley, J. Vura-Weis, and M. R. Wasielewski, 2009, "Photoinitiated charge transport through pi-Stacked electron conduits in supramolecular ordered assemblies of donor–acceptor triads," *J. Am. Chem. Soc.*, **131**, 11919–11929.

[42] D. Gust, T. A. Moore, and A. L. Moore, 2001, "Mimicking photosynthetic solar energy transduction," *Acc. Chem. Res.*, **34**, 40–48.

[43] K. N. Ferreira, T. M. Iverson, K. Maghlaoui, J. Barber, and S. Iwata, 2004, "Architecture of the photosynthetic oxygen-evolving center," *Science*, **303**, 1831–1838.

[44] C. W. Cady, R. H. Crabtree, and G. W. Brudvig, 2008, "Functional models for the oxygen-evolving complex of photosystem II," *Coord. Chem. Rev.*, **252**, 444–455.

[45] J. P. McEvoy and G. W. Brudvig, 2006, "Water-splitting chemistry of photosystem II," *Chem. Rev.*, **106**, 4455–4483.

[46] **J. Limburg**, **J. S. Vrettos**, **L. M. Liable-Sands** *et al.*, 1999, "A functional model for O—O bond formation by the O$_2$-evolving complex in photosystem II," *Science*, **283**, 1524–1527.

[47] **A. Magnuson**, **M. Anderlund**, **O. Johansson** *et al.*, 2009, "Biomimetic and microbial approaches to solar fuel generation," *Acc. Chem. Res.*, **42**, 1899–1909.

[48] **M. W. Kanan**, **Y. Surendranath**, and **D. G. Nocera**, 2009, "Cobalt–phosphate oxygen-evolving compound," *Chem. Soc. Rev.*, **38**, 109–114.

[49] **W. Lubitz** and **W. Tumas**, 2007, "Hydrogen: an overview," *Chem. Rev.*, **107**, 3900–3903.

[50] **R. Cammack**, 1999, "Hydrogenase sophistication," *Nature*, **397**, 214–215.

[51] **J. W. Tye**, **M. B. Hall**, and **M. Y. Darensbourg**, 2005, "Better than platinum? Fuels cells energized by enzymes," *Proc. Natl. Acad. Sci.*, **102**, 16911–16912.

[52] **Y. Nicolet**, **C. Piras**, **P. Legrand** *et al.*, 1999, "*Desulfovibrio desulfuricans* iron hydrogenase: the structure shows unusual coordination to an active site Fe binuclear center," *Structure*, **7**, 13–23.

[53] **J. W. Peters**, **W. N. Lanzilotta**, **B. J. Lemon**, and **L. C. Seefeldt**, 1998, "X-ray crystal structure of the Fe-only hydrogenase (CpI) from *Clostridium pasteurianum* to 1.8 angstrom resolution," *Science*, **282**, 1853–1858.

[54] **F. Gloaguen** and **T. B. Rauchfuss**, 2009, "Small molecule mimics of hydrogenases: hydrides and redox," *Chem. Soc. Rev.*, **38**, 100–108.

[55] **C. Tard** and **C. J. Pickett**, 2009, "Structural and functional analogues of the active sites of the [Fe]-, [NiFe]-, and [FeFe]-hydrogenases," *Chem. Rev.*, **109**, 2245–2274.

[56] **M. Wang**, **Y. Na**, **M. Gorlov**, and **L. Sun**, 2009, "Light-driven hydrogen production catalysed by transition metal complexes in homogeneous systems," *Dalton Trans.*, 6458–6467.

[57] **L. Sun**, **B. Åkermark**, and **S. Ott**, 2005, "Iron hydrogenase active site mimics in supramolecular systems aiming for light-driven hydrogen production," *Coord. Chem. Rev.*, **249**, 1653–1663.

[58] **Y. Na**, **J. Pan**, **M. Wang**, and **L. Sun**, 2007, "Intermolecular electron transfer from photogenerated Ru(bpy)$^{3+}$ to [2Fe2S] model complexes of the iron-only hydrogenase active site," *Inorg. Chem.*, **46**, 3813–3815.

[59] **Y. Na**, **M. Wang**, **J. Pan** *et al.*, 2008, "Visible light-driven electron transfer and hydrogen generation catalyzed by bioinspired [2Fe2S] complexes," *Inorg. Chem.*, **47**, 2805–2810.

[60] **X. Li**, **M. Wang**, **S. Zhang** *et al.*, 2008, "Noncovalent assembly of a metalloporphyrin and an iron hydrogenase active-site model: photo-induced electron transfer and hydrogen generation," *J. Phys. Chem. B*, **112**, 8198–8202.

[61] **A. P. S. Samuel**, **D. T. Co**, **C. L. Stern**, and **M. R. Wasielewski**, 2010, "Ultrafast photodriven intramolecular electron transfer from a zinc porphyrin to a readily reduced diiron hydrogenase model complex," *J. Am. Chem. Soc.*, **132**, 8813–8815.

[62] **S. W. Kohl**, **L. Weiner**, **L. Schwartsburd** *et al.*, 2009, "Consecutive thermal H$_2$ and light-induced O$_2$ evolution from water promoted by a metal complex," *Science*, **324**, 74–77.

28 Engineering natural photosynthesis

Huub J. M. de Groot

Leiden Institute of Chemistry, Leiden University, Leiden, the Netherlands

28.1 Focus

What is called in this chapter the "engineering" of natural photosynthesis has been performed by evolution over several billions of years. Its optimization, against thermodynamic and other selection criteria from the biological environment, has led to a remarkably limited set of molecular and supramolecular motifs. Photosynthesis starts with absorption of light by mutually interacting chlorophyll (Chl) and related molecules. These are embedded in protein matrices that promote rapid transport of energy by excitons and lower the energy of transition states for charge separation and multielectron catalysis. An understanding of engineered natural photosynthesis is the underpinning of the design of artificial solar-to-fuel devices.

28.2 Synopsis[1]

Light energy is abundant and evenly spread over the surface of the Earth, and virtually all organisms depend on the conversion and chemical storage of light energy by natural photosynthesis. The advent of natural photosynthesis opened up new energy conversion pathways for progression toward thermodynamic equilibrium between the very hot Sun and the much colder Earth. The first-principles thermodynamic and mechanistic analysis of how natural photosynthesis is engineered by biological evolution in this chapter shows that combining different functionalities of light harvesting, charge separation, and catalysis in small molecules or at a single narrow interface is incompatible with high solar-to-fuel conversion efficiency, and that possible solutions to the problem of accumulation of solar energy in chemicals require complex device topologies. This led biology to a modular design approach that is the basis of photosynthesis. It is the result of an intensive evolutionary effort of shaping complex structures, driven by the abundant solar energy spread over the planet.

The molecular basis of photosynthesis is in protein complexes that contain chlorophylls, molecules with good absorbance and a long excited-state lifetime. Photons are captured by light-harvesting antenna complexes, and the energy is rapidly transferred to reaction centers to produce electrical charges of opposite polarity and then an electrochemical membrane potential. Photosynthesis can be divided into the oxygenic (O_2-producing) photosynthesis processes of plants, green algae, and cyanobacteria, and non-oxygenic photosynthesis that occurs in purple bacteria and green sulfur bacteria, for example. Photosynthetic processes separate H^+ and e^- for synthesis of ATP, NAD(P)H and ultimately CO_2 fixation, to produce carbohydrates such as starch and glycogen to store H^+ and e^- in CO_2-neutral fuels. In plants, the photosynthesis machinery is embedded in the thylakoid membrane inside organelles called chloroplasts, whereas in bacteria it is part of the plasma membrane (see Chapter 24).

A modular structure with compartments for light harvesting and charge separation provides flexibility for systems design and adaptation against the natural environmental constraints. Biology uses a small set of active components, most notably interacting chlorophylls to initiate the photochemistry and metal ions for multielectron catalysis, while maintaining the underlying design principles.

---

[1] Most of the concepts, acronyms, and abbreviations employed are defined within the chapter, but the reader is advised to consult also Chapter 24, in case of doubt.

28.3 Historical perspective

Evolution proceeds by the interaction between diversification at the basis of the biological hierarchy, namely the molecular structure and supramolecular organization, and selection from the top levels, namely the physics, chemistry, and biology of the environment. Driven by the energy they themselves collected from the Sun, the molecular machines of natural photosynthesis have been engineered by evolution to adapt to the biological, hierarchical, order and the environmental settings [1]. However, although photosynthesis occurs in different ways, its basic architectures and operational principles have been largely preserved during the diversification into higher organisms and its basic engineering principles are shared across taxonomic boundaries.

Early photosynthetic species were not able to generate the high redox potential to oxidize H_2O and used H_2, H_2S, and other anaerobic electron donors as sources of electrons. The first photosynthesizers that extract electrons from water have been found, by micropaleontologists, in stromatolites, bacterial fossils, which show that oxygen-evolving cyanobacteria appeared around 2.6 billion years ago, and these started to oxygenate the atmosphere around 2.4 billion years ago (Figure 28.1). About a billion years ago, eukaryotic organisms started symbiotic relationships with these cyanobacteria. The chloroplasts in modern plants are the descendants of these ancient endosymbiotic events [2]. They excrete simple sugars such as glucose and produce the raw construction materials from which plants manufacture many other substances such as complex carbohydrates like starch, lipids, proteins, and lignocellulose.

In recent years structural biology, biophysics, and molecular phylogenetics have provided insight into the modular structure, the operational mechanisms, and the evolution of the molecular machinery of the photosynthetic apparatus. Photosynthesis uses antennae compartments for light harvesting and reaction-center complexes for charge separation. The oxygenic photosynthesis in higher organisms requires two light reactions and two photosystems, photosystem I (PSI) and photosystem II (PSII) [3]. The PSII water-oxidation catalysis can proceed on the millisecond time scale in high light for up to half an hour before the complex is irreversibly damaged, which means that the natural complex can oxidize water at high efficiency for hundreds of thousands of cycles. In contrast, the anoxygenic bacteria at the early stages of the evolution of photosynthesis have only one photosystem, either the type 1 found in green bacteria that is similar to PSI, or the type 2 found in purple bacteria that is similar to PSII (Figure 28.2) [4]. Evidence is currently converging to indicate that oxygenic photosynthesis evolved from lateral gene transfer across species that produced the two important symbiotic fusion events in the evolutionary history of engineered natural photosynthesis: the formation of a cyanobacterium and the development of the chloroplast [1].

28.4 Discussion

In plants and green algae, the photochemical conversion network is in the thylakoid membrane and its biochemical environment. The photosynthetic conversion begins with the harvesting of light, and proceeds via

Figure 28.1. Stromatolites are the oldest fossils, of cyanobacteria that started to emerge several billion years ago and oxygenated the atmosphere. From Wikipedia commons, http://commons. wikimedia.org/wiki/File: Stromatolites_in_Shark_Bay.jp.

(a)

L-polypeptide   M-polypeptide

(b)

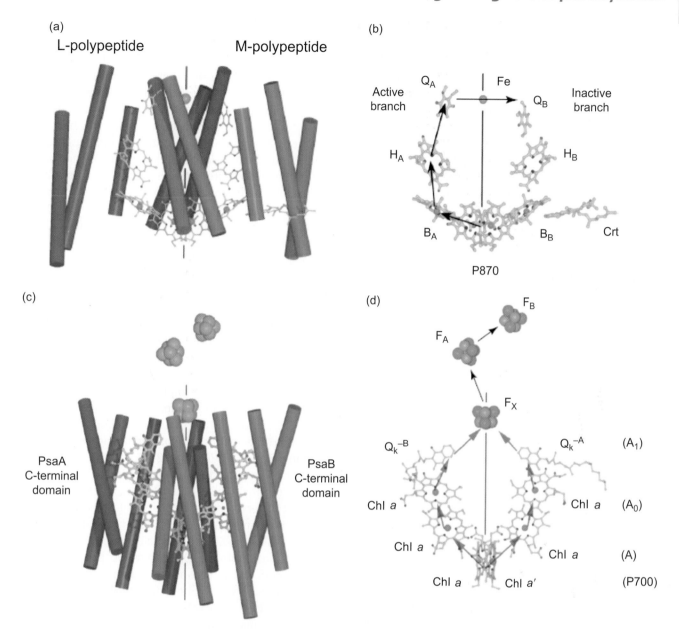

Figure 28.2. The transmembrane helix homology and scaffolding of the cofactors in type-2, (a) and (b), and type-1, (c) and (d), reaction centers. From [1].

trans-membrane charge separation, photochemical oxidation of water, and electron and proton transport, to the Calvin–Benson cycle whereby carbohydrates are synthesized from dilute atmospheric CO_2. The two photosystems, PSII and PSI, operate in tandem, and incoming photons generate the excited states PSII* and PSI*. Both act as strong reducing agents, and leave the reaction-center chlorophyll in an oxidized form. The PSII chlorophyll oxidizes the tetramanganese O_2-evolving complex that accumulates positive charge by extracting electrons from water,

$$2H_2O \longrightarrow O_2 + 4H^+ + 4e^-,$$

and produces O_2. The Z-scheme of photosynthesis in Figure 28.3 indicates how the electrons pass through a redox gradient over the pheophytin (Pheo) and quinones (Q_A and Q_B) to the cytochrome b_6f complex (the box in the diagram) and on to the photo-oxidized PSI [5]. The electron from PSI passes through quinones (A_0 and A_1) and iron–sulfur clusters ($FeS_{x,A\&B}$) to reach ferredoxin (Fd) that reduces $NADP^+$. In an alternative pathway, the electrons from ferredoxin are transferred back to the plastocyanin (Pc) electron-transfer protein via the cytochrome b_6f complex. This cyclic electron transport, which does not require the input of free energy by PSII, results in a trans-membrane electrochemical gradient

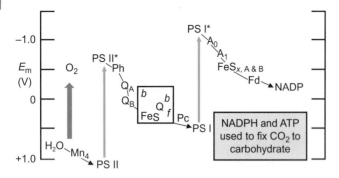

Figure 28.3. The Z-scheme and midpoint electrochemical potentials E_m of electron transfer in oxygenic photosynthesis in plants. Adapted From [5].

that can be used to produce ATP. NADPH and ATP are used to reduce CO_2 to carbohydrates in the subsequent dark reactions of the Calvin–Benson cycle. Thus, with a second photochemical step coupled to the dark reactions

$$CO_2 + 4H^+ + 4e^- \longrightarrow H_2CO + H_2O,$$

the electrons and protons are fed into the metabolic network of the organism, and are utilized for the synthesis of ATP and for the Calvin–Benson cycle that chemically balances the production of energy with the energy demand from the organism with a carbohydrate "fuel" reserve.

Each of the two half-reactions requires four photons, and together they lead to the net production of carbohydrate fuel via transported protons and electrons, using ATP and NADPH as intermediate energy carriers. To reduce one molecule of CO_2 requires a minimum of eight quanta of light. However, if the additional energy needs for cell maintenance and for concentrating CO_2 are also considered, 10–12 quanta are required.

The principal molecular design of the helix and cofactor arrangement of photosynthetic reaction centers is highly conserved in evolution and is used both by micro-organisms and by the chloroplasts of plants and algae [6]. The trans-membrane helix homology and the similarities in the positioning of the cofactors between type-2 and type-1 reaction centers reveal a common origin in the biological evolution (Figure 28.2). The left panels of Figure 28.2 show the trans-membrane helix homology between a type-2 bacterial reaction center, RC, in (a), and PSI, in (c); and the right panels illustrate the homology in the cofactor arrangement: both type-1 and type-2 reaction centers contain six chlorophyll-type rings and two quinone-type molecules. The arrows indicate the directions of the electron transfers. Photosystem I contains three Fe–S clusters. In contrast to the bacterial type-2 RC, PSII has in addition the Mn_4Ca O_2-evolving complex that catalyzes the water-oxidation reaction.

The PSI core is a type-1 reaction center consisting of 11–13 protein subunits. An X-ray structure of PSI has been obtained at 2.5-Å resolution [7]. The two largest subunits, PsaA and PsaB, comprise a heterodimer that binds the majority of the reaction-center cofactors and core antenna pigments. Photosystem I contains an integral antenna system consisting of about 90 Chl a molecules and 22 carotenoids. The antenna pigments can be divided into three regions. Some antenna pigments surround the inner core, and there are two peripheral regions where chlorophylls form layers on the stromal and luminal sides of the membrane. Excitation of the antenna pigments results in a rapid equilibrium distribution of the energy among the antenna chlorophylls with a lifetime of 4–8 ps [7]. The rate of energy transfer from the antenna system to P700 RC varies between 20 and 35 ps^{-1} and depends on the organism and the antenna size.

The PSII core complex consists of 19 proteins (Figure 28.4) [4]. There is a type-2 RC complex inside, which contains four Chl a molecules that are coupled to form the P680 oligomer. In addition, there are two pheophytin a molecules and two plastoquinones in a heterodimeric protein scaffold, formed by two subunits, denoted D1 and D2, that match the L and M helix and cofactor arrangement of the bacterial reaction center in the upper panels of Figure 28.2 [8]. Excitons, formed in the light-harvesting stage, become delocalized over the P680 chlorophylls within 100–500 fs. An intermediate charge-transfer state $Chl^{+\bullet}Pheo^{-\bullet}$ is formed between a chlorophyll donor and a pheophytin acceptor on a time scale of 1.5 ps [9]. Electron transfer proceeds further to the first plastoquinone Q_A within 200 ps; $Q_A^{-\bullet}$, then doubly reduces the secondary quinone acceptor, Q_B, with the possible involvement of a non-heme Fe, located on the pseudo-C_2 axis, with time constants of 0.2–0.4 ms and 0.6–0.8 ms for the first and second reductions, respectively. After receiving two protons, Q_B then leaves its binding pocket as a plastoquinol molecule. The plastoquinol then diffuses out of the protein to be oxidized by cytochrome b_6f (Figure 28.3).

Purple non-sulfur bacteria contain a very-well-studied photosynthetic apparatus, which consists of two light-harvesting (LH) pigment–protein complexes (LH1 and LH2) and a type-2 reaction center [10]. The pigment–protein complexes are membrane-bound and utilize bacteriochlorins and carotenoids for light harvesting [11]. Green sulfur bacteria use large aggregates of up to 100,000 bacterochlorophyll c, d, or e molecules that are contained in vesicles called chlorosomes [12]. The energy harvested by these chlorosomes is transferred via a Fenna–Matthews–Olson antenna protein complex to the membrane-bound reaction center [13]. Cyanobacteria use large peripheral phycobilisomes as their major light-harvesting system. The phycobilisomes

Figure 28.4. The architecture of the photosystem II water-splitting enzyme complex from *Thermosynechococcus elongatus* determined by X-ray crystallography. From [8].

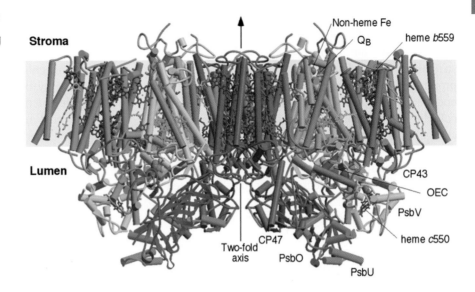

funnel absorbed energy down into the membrane and supply excitation energy to PSI and PSII [14].

The solar energy that is harvested and converted sustains the biological steady state of a photosynthetic organism, and the stability that arises from this homeostasis allows a photosynthetic species to probe and alter its genome and transcriptome for evolution. In this way mutation and development work to engineer natural photosynthesis for a particular environment. Photosynthetic organisms partition their resources between electron transport and metabolism, so that these component processes co-limit photosynthesis. While the photosynthetic reaction centers are highly conserved across species, other parts of the photosynthetic apparatus, such as the antenna (LH) system, are not. The composition of the photosynthetic apparatus is affected by adaptation to the light environment or the chemical environment, such as Fe availability for marine organisms. Changes can occur at the level of the concentration and interaction of participating molecules, the dynamic functional structure of the thylakoid membrane, the number of cells, the leaf architecture, and the way the plants and other photosynthetic organisms are designed and store the harvested energy in a sink or reservoir.

28.4.1 The thermodynamics of photosynthesis is a matter of buying time

To understand how thermodynamic principles constrain engineered natural photosynthesis, it is instructive to look briefly at photovoltaic (PV) conversion by solar cells (cf. Chapters 18 and 19). In a silicon solar cell, photons are harvested by transfer of electrons from the valence band into the conduction band. In a short time the electrons are thermalized to the 1.1-eV bandgap energy,

the energy that will determine what can be measured as the open voltage of a solar cell when it is not producing current [15]. The excited electrons in the conduction band of the silicon can decay to the ground state by emission of a photon, unless there is a piece of equipment connected to the solar cell that takes the electrons out of the conduction band and uses the energy (cf. Chapter 18).

Four loss mechanisms limit the maximum power efficiency of a solar cell under non-concentrated blackbody solar radiation to ~31%, the Shockley–Queisser or detailed-balance limit [15]. First, photons with energy below the bandgap energy are not absorbed. Second, thermalization of electrons excited by photons with energy greater than the band gap leads to energy losses by vibrational cooling. Third, the maximum free-energy storage resulting from light absorption is always less than the bandgap energy. Fourth, radiative-recombination losses occur at the absorber due to partial thermodynamic reversibility [16].

Natural photosynthesis uses molecular absorbers and thus is subject to the Shockley–Queisser limit. They convert solar energy in the wavelength region between 400 and 700 nm for storage into chemicals. Photosynthetic energy conversion is engineered to favor the forward reaction for chemical storage of energy against the back reaction and recombination losses in the chlorophyll by fluorescence [17]. For power conversion into chemical storage, photoexcited electrons have to be extracted rapidly from the chlorophyll absorbers. To achieve this, natural photosystems optimize photosynthetic performance with specific arrangements of cofactors and protein modules. The understanding of how this works leads to design principles, fundamental ideas that can serve to guide the design of artificial photosynthesis mimics. One principal element is modularity, the concept that photosystems and conversion pathways are composed of modules, protein–cofactor matrices that self-assemble into three-dimensional biological structures. They change their physical and chemical properties in a dynamic response mechanism that is encoded by the self-assembly of the polypeptides and cofactors to overcome the barriers to energy transfer, charge separation, and multielectron catalysis.

In photosynthesis, nearly always chlorophyll molecules are the most abundant photon absorbers. Upon excitation, a localized electron–hole pair is created by moving an electron from the highest occupied molecular orbital (HOMO) to the lowest unoccupied molecular orbital (LUMO) in a molecule or in an oligomer of strongly interacting molecules P, to form the excited state P^*. The HOMO–LUMO energy difference $h\nu_0$ plays the role of a semiconductor bandgap. The chlorophyll a in plants operates with a HOMO–LUMO gap of $h\nu_0 = 1.8$ eV, while the bacteriochlorophylls in photosynthetic micro-organisms have $h\nu_0 = 1.4$ eV, and this will be the internal energy of the electrons after thermalization by vibrational cooling.

In its ground state P the molecules are at a chemical potential

$$\mu = \mu_0 + k_B T \ln p, \tag{28.1}$$

with the chemical potential of P^* given by

$$\mu^* = \mu_0^* + k_B T \ln p^*. \tag{28.2}$$

These chemical potentials are defined on a molecular scale in electron volts (eV). Here μ_0 represents the energy of P and μ_0^* the energy of P^*, while p and p^* are the molecular fractions of P and P^*, respectively, which are proportional to the concentration of each species in the bulk [18]. The difference in chemical potential between P* and P is $\Delta\mu_{abs} = \mu^* - \mu$. In the absorber $\mu_0^* - \mu_0 = h\nu_0$, the HOMO–LUMO gap. This leads to

$$\Delta\mu_{abs} = h\nu_0 + k_B T \ln\left(\frac{p^*}{p}\right). \tag{28.3}$$

Here the second term represents the entropy of mixing, which is due to the mixing of the ground state and the excited state in the same molecular volume [17]. In the dark the excited and ground state are in thermodynamic equilibrium and $\Delta\mu_{abs} = 0$, leading to $p^*/p = \exp[-h\nu_0/(k_B T)]$, the Boltzmann distribution [18].

When exposed to sunlight, the absorber is excited at a rate g by solar photons, while the excited state is quenched via fluorescence at a decay rate τ^{-1} according to

$$P \underset{1/\tau}{\overset{g}{\rightleftarrows}} P^*.$$

These processes will shift the population distribution, according to $p^*/p = g\tau$; another activation/deactivation

equilibrium is then established, with $\Delta\mu_{abs} \neq 0$, and thus free energy can be extracted from the chlorophyll donor absorber [18].

Since the chlorophyll excited state decays to the ground state by the fluorescence in τ in a short time, on the order of nanoseconds for chlorophylls in proteins, the type-1 and type-2 RCs need to transfer excited electrons in a shorter time from P^*, into an excited-state acceptor sink Q to produce an electron–hole pair,

$$P \underset{1/\tau}{\overset{g}{\rightleftarrows}} P^* \underset{I_b}{\overset{I_f}{\rightleftarrows}} P^+ Q^-,$$

with the electron on Q and the hole on P. In the dark, the molecular fraction of electron–hole pairs is $p_h q_e$, with

$$\mu_{pair} = \mu_{0,pair} + k_B T \ln(p_h q_e) \tag{28.4}$$

for the chemical potential for the donor–acceptor pair. For the excited pair

$$\mu_{pair}^* = \mu_{0,pair}^* + k_B T \ln(p_h^* q_e^*), \tag{28.5}$$

with $p_h^* q_e^*$ the concentration of electron–hole pairs after absorption of a photon [18]. The physical limit for energy storage by the pair is

$$\Delta\mu_{st} = h\nu_0 + k_B T \ln\left(\frac{p_h^* q_e^*}{p_h q_e}\right). \tag{28.6}$$

Here the entropic term measures the concentration of electron–hole pairs in the charge-separated state, i.e., the probabilities of electron and hole occupation of the acceptor and donor states, in the light relative to the concentration in the dark.

Upon excitation with light, a charge-separated state is produced at a forward transfer rate I_f, in steady-state equilibrium with the backward transfer rate I_b, and

$$\frac{p_h^* q_e^*}{p_h q_e} = \frac{I_f}{I_b}. \tag{28.7}$$

The backward rate is the inverse of the lifetime of the storage reservoir $I_b = t^{-1}$, while the forward rate is constrained by the fluorescence lifetime of the absorber $I_f = \tau^{-1}$. This leads to a time-dependent limit for photochemical storage of energy

$$\Delta\mu_{st} = h\nu_0 - k_B T \ln\left(\frac{t}{\tau}\right). \tag{28.8}$$

The longer the energy needs to be stored, the more entropy of mixing needs to be generated in the chemical network for storage in order to prevent loss of energy due to fluorescence at the start of the conversion chain in competition with the storage [19]. Equation (28.8) is the consequence of time-reversal symmetry, and is rooted deeply in the first principles of the physical world. (Hamiltonians are Hermitian and time development operators in quantum theory are therefore unitary.

This leads to time-reversal symmetry and the second law of thermodynamics.)

In oxygenic photosynthesis, the fluorescence lifetime is also very short, compared with the required storage times for water-oxidation catalysis, which is the most difficult reaction in natural photosynthesis. Water oxidation is a four-electron reaction that is rate-limited at a time scale of milliseconds, and, according to Equation (28.8) the storage of electrons on this time scale requires a mixing entropy of at least ~0.6 eV [20]. This is a significant part of the incoming $h\nu_0 = 1.8$ eV for the chlorophyll a absorber. The four positive charges have to be accumulated in a small catalytic site to split the water and form the O—O bond on the Ångström scale, which is very small compared with the wavelength of optical absorption of 700 nm. To achieve this, natural photosynthetic systems are engineered to harvest the photons in antenna systems, separate charges in reaction centers by efficient quantum delocalization and electron tunneling in a redox gradient, and concentrate charges into a catalytic site by a proper match of time scales and length scales in dedicated photosynthetic membrane topologies [21]. From Equation (28.8) it follows that, for the desired storage times to balance production and demand in human energy use, in the range of days to years, around 50% of the energy after thermalization cannot be extracted [20]. This is a principal difference between PV and photochemistry, insofar as PV-produced electricity can be transported rapidly to locations where energy is required, which eliminates the need for storage and the accompanying losses. On the other hand, natural photosynthesis at its best is remarkably efficiently engineered. Calculations project conversion efficiency as high as 9% for prokaryotic micro-organisms which includes thermalization and long-term storage (cf. Chapter 24) [22].

28.4.2 Engineered light-harvesting modules

Natural light-harvesting antennae permit an organism to increase greatly the absorption cross section for light and transport of energy on a very short time scale. The intensity of sunlight is dilute, and every single pigment molecule absorbs at most a few photons per second. To sustain the catalysis for photosynthesis at the limiting rate of 10^{-3}–10^{-4} s, several hundred pigments are incorporated into supramolecular assemblies of antenna units. They cover large photosynthetic membrane surfaces to ensure that photons striking any spot on the surface will be absorbed and concentrated for feeding energy quanta into the reaction center where charge separation and electron transport take place. An example of a photosynthetic membrane topology is shown in Figure 28.5(a), for bacterial photosynthesis, which is based on three membrane protein complexes,

two light-harvesting antennae, denoted LHI and LHII, and a reaction-center complex RC [11]. The antenna units transfer the energy from light by exciton migration over long distances to the RCs. They surround the RCs, optimizing the energy-transfer efficiency by use of multiple antenna–RC connections [23].

The LHII antenna, which is the most abundant one, contains two chlorophyll species, B800 with maximum absorption at $\lambda_{\max} = 800$ nm and B850 with $\lambda_{\max} = 850$ nm. In the ground state the LHII antennae are highly ordered, almost-crystalline assemblies of trans-membrane polypeptides with many pigment-binding sites. They increase the absorption cross section by containing densely packed dye molecules with inhomogeneously broadened optical absorption profiles. Chlorophylls are moderately sized molecules that are sterically crowded in the side chains to the central chlorin macro-aromatic cycle. This makes for an almost flat energy landscape of the macro-aromatic cycle, and the distortion of the chlorin ring is primarily along only the six lowest-frequency normal coordinates [24]. The optical broadening in an otherwise extremely homogeneous structure can be enhanced by induced misfits, namely local spots of structural frustration that are established in the folding and self-assembly process of the complex [25][26]. The structural frustration leads to an energy landscape that favors thermally activated modulation of the energy levels among the chlorophylls within a single antenna or reaction center complex on time scales that are long compared with those of the energy-absorption and charge-transfer processes and produces inhomogeneous broadening of the electronic transitions due to ensemble averaging [27]. For B850 there is considerable physical frustration in the histidine ligands to the central Mg^{2+} ion of the B850 chlorophylls. These histidines are hydrogen bonded to the protein, and stabilize the ring V keto functionality of a neighboring chlorophyll in an anti-parallel orientation. They exhibit an anomalous electronic hybridization and stabilize positive charge due to stress and strain exerted by the protein complex [28].

The chlorophylls in the B800 and B850 rings strongly overlap, and, when they absorb light, collective electronic excitations that are promoted by quantum delocalization over the electronically coupled pigments are formed [29]. However, the exciton–phonon coupling leads to classical confinement by structural reorganization of the pigments and to symmetry breaking by a polaronic contribution to the excitons that is of chlorophyll–chlorophyll charge-transfer character, while the modulation of the electronic transition energies by slow conformational changes produces electronic disorder. Both lead to more localized exciton wavefunctions that extend over four to six chlorophylls in the rings [27]. The excitation dynamics in the antennae is a superposition of coherent motion of delocalized

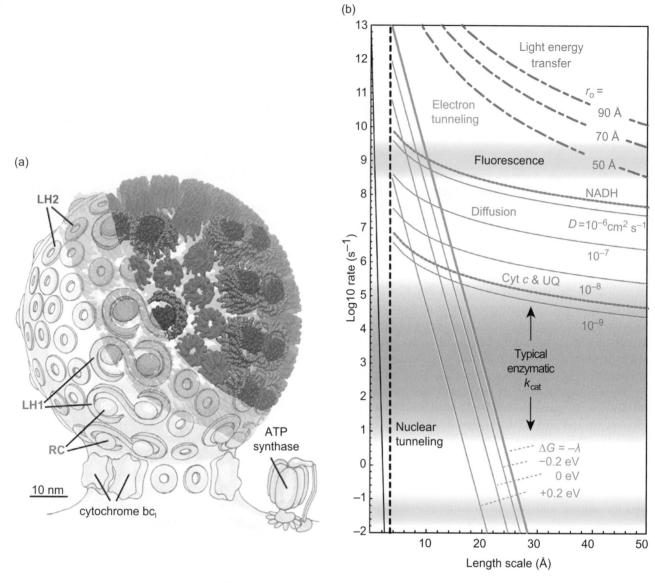

Figure 28.5. (a) A model of the photosynthetic unit of purple bacteria showing the arrangement of light harvesting units 1 and 2 (LHI and LHII) and the reaction center (RC). The structures of LHII and the RC have been determined by X-ray crystallography, and LHI is simulated by analogy to LHII. From [11] (b) Time scales and length scales of energy and electron transfer in photosynthesis. From [21].

excitons on a time scale of 100 fs, hopping on a time scale of 350 ps, and pinning of localized excitations. In single LHII complexes it is possible to observe the switching between these regimes. Apparently the antennae in bacterial photosynthesis are engineered to accommodate mutually interacting chlorophylls that produce a heterogeneous manifold of excited states, including excitons with a high degree of delocalization in combination with more localized excitations due to the presence of weakly coupled pigments. The basic mechanism of photosynthetic light harvesting for the delivery of excitation energy to the reaction centers

includes fast relaxation between exciton states on a femtosecond time scale within strongly coupled chlorophyll clusters and slower energy migration on a picosecond time scale between clusters or monomeric sites [27].

The major type of peripheral antennae in green plants, LHCII, is a trimer, with different chlorophyll species that mutually interact and carotenoids that have a strong excitonic coupling with the chlorophyll pigments to achieve efficient energy transfer as accessory light-harvesting pigments by absorbing light energy in the visible spectrum that is unavailable to chlorophylls (Figure 28.6) [30]. As for the bacterial antennae, clusters

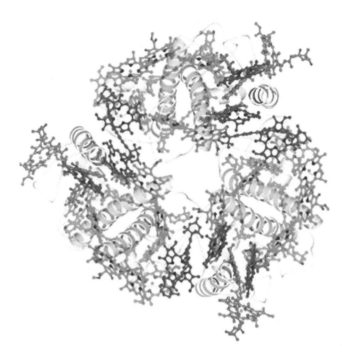

Figure 28.6. The crystal structure of the LHCII antenna trimer from spinach. From [30].

NPQ on a time scale of a few hundred picoseconds [32]. Alternatively, a channel for energy dissipation by transfer to a bound carotenoid can open up [33][34].

To support long-range excitation transfer, groups of tightly coupled pigments are coupled to other groups and RCs by the Förster resonant-transfer (FRET) mechanism with first-order energy-transfer rate

$$k_e = (r_0/r)^6/\tau, \tag{28.9}$$

due to Coulomb dipole–dipole interactions [35]. The migration rate is inversely proportional to the fluorescence lifetime and r_0 is the distance at which the transfer efficiency is 50%. Figure 28.5(b) shows the transfer ranges for $\tau = 10^{-9}$ s and $r_0 = 50$, 70, or 90 Å [21]. For conservation of energy, strong optical overlap between the donor and acceptor complexes is beneficial.

The length scale of exciton migration in the antenna is considerable, and the exciton lifetimes should be long enough to allow photons striking any part of the antenna to reach the RC. Once excitons have been trapped in the RC, the process of charge separation should take place faster than back-transfer to the antenna units. In photosynthetic assemblies this is achieved by having a relatively large antenna–RC distance versus a short distance for the redox pigments in the RC, in a topology such that RCs are surrounded by assemblies of antenna pigments giving a spatial arrangement in which energy transfer is optimized by multiple contact points for the transfer of excitons to the RC (Figure 28.5(a)).

There is an exception, however, in chlorosome antennae that are engineered by biological evolution to operate under environmental conditions where light is a limiting factor. Chlorosomes are anomalous in the sense that they are engineered to sustain extended polarons, where quantum delocalization is combined with symmetry breaking of the lowest exciton state to gain polaronic charge-transfer character while at the same time extending the system across many chlorophylls [36]. Chlorosomes are the largest and fastest light-harvesting antennae found in nature and are constructed from hundreds of thousands of self-assembled bacteriochlorophylls c, d, or e. The pigments are closely packed in stacks, with the stacks aligned to form sheets that self-assemble into coaxial tubes [12]. Following efficient excitation in the direction of the stacks, the biological light-harvesting requirement is fulfilled by providing ultrafast helical exciton-delocalization pathways with polaronic character along helical arrangements of BChl molecules with aligned electric dipoles that are interconnected by polarizable hydrogen bonds (H-bonds) (Figure 28.7) [12][36]. These H-bonds are strained and do not contribute to the stabilization of the aggregate, but transform the helical bacteriochlorophyll array into a material with a high dielectric constant. Consequently,

of strongly coupled pigments depopulate on a short time scale by a combination of coherent transfer and hopping into a sink formed by chlorophyll a on the outside of the trimer to feed the energy into plant PSI or PSII. Since the performance and survival of plants in natural environments rely on their ability to actively adapt to variable light intensity, an important additional design feature of the LHCII photosynthetic complexes is their built-in mechanism for the dissipation of excess energy [31]. In the shade, light is efficiently harvested in photosynthesis. However, in full sunlight, much of the energy absorbed cannot be processed and photoprotection regulation mechanisms have evolved to protect the conversion chain against photo-oxidative damage. The redox state of the cytochrome b_6f complex induces state transitions, in which the LHCII shuttles between PSI and PSII, balancing the electron transport. In addition, when overproduction of protons by the photosystems leads to acidification of the lumen, the LHCII antenna can rapidly switch to a conformational state that safely dissipates the excess energy as heat by non-photochemical quenching (NPQ) mechanisms involving excited states of chlorophylls or carotenoids [32][33][34]. The deprotonation of a pH-sensing protein, PsbS, controls reversible oligomerization of the LHCII via a chemical conversion of violaxanthin carotenoids to stabilize the LHCII assemblies. This can give rise to symmetry breaking whereby a polaronic contribution to the excitons becomes dominant and leads to the formation of localized chlorophyll–chlorophyll charge-transfer states with an enhanced coupling to the ground state to produce

Figure 28.7. The bacteriochlorophylls in chlorosome antenna form stacks and sheets that self-assemble into concentric nanotubes. Upon excitation with light, extended excitons are formed in the direction of the aligned electric dipoles, facilitated by activated hydrogen bonds in the C=O···H—O···Mg motifs in the structure.

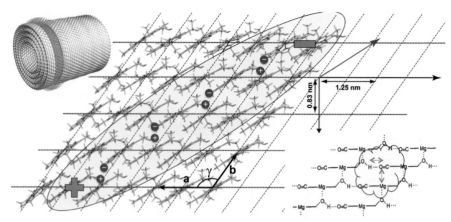

electric-field screening tends to reduce the Coulomb interaction between electrons and holes. This favors the symmetry breaking of excitons into extended polarons by the release of mechanical strain in the molecular array of activated hydrogen bonds. This leads to a collinear proton-coupled electron-transfer (PCET)-type mechanism, whereby the symmetry breaking of the exciton is synergistic with the stabilization of the H-bond helix by reorganization of the activated hydrogen bonds. This dielectric response produces a transient shift of the molecular energy levels. The chlorosome is an organelle found in early photosynthesizers and it nicely illustrates how mutually interacting chlorophylls can provide a versatile engineering material, with both energy-transfer and charge-transfer characteristics.

28.4.3 Smart matrices for electron transfer and electrochemistry

Natural photosynthesis achieves a quantum efficiency of 99% under low-light conditions, which is possible only if the forward current rate I_f is faster than the fluorescence decay rate of 10^9 s^{-1} [17]. The quantum efficiency is the ratio of the number of separated e$^-$ and holes to the number of photons in the photosynthetically active region of the spectrum incident on the photosynthetic assembly. In transition-state theory, rates of chemical reactions are described along a potential-energy surface. As reactants gain energy from thermal collisions, they overcome an activation-energy barrier E_a to the transition state, after which they spontaneously decay to the product. The net energy transfer with the surroundings is zero, and the reaction is adiabatic [37]. In solar cells, charge separation is facilitated by an electric field over the depletion region of a p–n junction that separates the light-generated minority and majority carriers. In photosynthetic reaction centers, excitons that enter from the antenna produce excited states P* that are unstable with respect to symmetry breaking to form first chlorophyll–chlorophyll charge-transfer states and then full charge separation, P$^+$Q$^-$, over ~40 Å, the

thickness of the biological membrane. The charge separation is mediated by weak electronic overlap between the donor and acceptor and driven by a downhill redox gradient as shown in Figure 28.3 [38]. This is accomplished by placing multiple redox cofactors so that they are separated by an edge-to-edge distance less than 0.6 nm to ensure rapid forward transfer, on a time scale of ~10 ps or less, by downhill dissipative electron tunneling, while increasing the storage time for every step [39]. The electron tunneling rates in weakly coupled donor–acceptor pairs are described well by

$$I_f = \frac{2\pi}{\hbar} V_{PQ} FC, \qquad (28.10)$$

with V_{PQ} the electronic matrix coupling element between the donor and the acceptor, and FC a Franck–Condon factor that describes the structural changes that are required in order to overcome the activation barrier in the donor–acceptor pair in the electron-transfer reaction. The structural rearrangements are generally described classically by a harmonic potential to depict how the energy of the precursor complex depends on its nuclear configuration prior to the reaction, while another single-potential surface is used to describe the product complex. This leads to the expression

$$FC = (4\pi\lambda k_B T)^{-1/2} \exp[-E_a/(k_B T)], \qquad (28.11)$$

where E_a has a quadratic dependence on the reorganization energy λ and on the standard Gibbs free-energy difference between the products and reactants $\Delta G°$ according to

$$E_a = -(\Delta G° - \lambda)^2/(4\lambda). \qquad (28.12)$$

For the electron-transfer reaction in proteins the $\Delta G° < 0$ and $\lambda = 0.7$ eV [39]. In the "normal" region the current I_f increases when $-\Delta G°$ increases, until $-\Delta G° = \lambda$, for which there is no barrier to electron-transfer and the electron current reaches a maximum. One of the key features of electron-transfer theory is that it predicts an "inverted region" where the rate of an electron-transfer reaction will slow down when the free energy of the reaction becomes very large, $-\Delta G° > \lambda$ [40].

An analysis of electron-transfer processes in proteins in the classical harmonic approximation reveals a common $\hbar\omega = 70$ meV, which is well above the Boltzmann energy $k_{\mathrm{B}}T = 25$ meV available at room temperature [39]. This is important, since the Marcus theory is derived in the high-temperature limit. Thus, the description given above is self-consistent and subject to refinement. For instance, the electron transfer may be accompanied by proton transfer in a synchronous or asynchronous PCET mechanism [41].

The electron-transfer mechanism is quite insensitive to the thermodynamic driving force, in particular when there is a global match $-\Delta G° \sim \lambda$ [39]. With charge separation over a downhill redox gradient the $\Delta G < 0$, and the electron tunneling current is to a good approximation described by

$$\log I_{\mathrm{f}} = 15 - 0.6r - \frac{3.1(\Delta G + \lambda)^2}{\lambda}. \qquad (28.13)$$

For a specific time $\tau_{\mathrm{f}} = I_{\mathrm{f}}^{-1}$ there is a maximum distance r over which tunneling can proceed, depending on the redox gradient [39]. The fastest transfer rate is obtained for $-\Delta G° = \lambda$, and the red lines in Figure 28.5(b) depict the tunneling behaviour [21]. In the reaction center the tunneling processes serve to bridge the six orders of magnitude between the time scale of fluorescence and the time scale of catalysis (Figure 28.5(b)). At the time scale of catalysis, 10^{-3} s, the tunneling distance is ~20 Å. Thus, a biological membrane of thickness ~40 Å is a good tunneling barrier for the accumulation of charge for catalysis. However, the time scale of fluorescence, 10^{-9} s, requires that efficient charge separation proceeds on a time scale of a few picoseconds. This translates into a tunneling barrier of ~5 Å. Thus, for a high quantum efficiency the distance between P^+ and Q^- needs to be 5 Å, while for the accumulation of charge for catalysis the distance between P^+ and Q^- needs to be in excess of 20 Å. Natural photosynthesis is engineered to solve this dilemma with multiple intermediates and a sequence of transfer steps with increasing edge-to-edge distance between cofactors and increasing storage times to avoid fluorescence losses by quantum delocalization and tunneling from the product states backwards into the reactant.

In oxygenic photosynthesis, oxidation of water proceeds following the accumulation of four positive charges in the oxygen-evolving complex (OEC), an Mn_4Ca cluster attached to the PSII charge-separation chain (Figures 28.4 and 28.8). It provides a structural framework for describing the water-splitting chemistry of PSII and therefore is of major importance for designing artificial catalytic systems for reproducing this chemistry. While the multielectron catalysis is rate-limited at 10^{-3}–10^{-4} s at high light, a large dynamic range with respect to the incoming photon flux implies

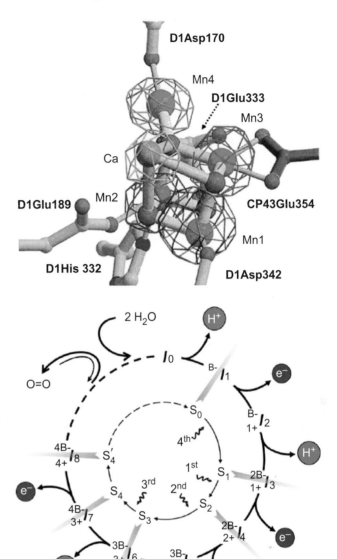

Figure 28.8. The oxygen-evolving complex in *Thermosynechococcus elongatus* [8] and the four-stage proton-coupled electron-transfer catalytic cycle [42].

potentially very long accumulation and storage times of catalytic intermediates in the four-electron water-oxidation cycle when the incoming photon flux is low. Very long accumulation times are possible, provided that the intermediate storage of holes is coupled to the rapid irreversible release of protons from the catalytic site. Here the thermodynamic irreversibility of proton release generates mixing entropy and prevents wasteful back reactions.

The protein environment adjusts the proton chemical potential to the OEC catalytic sites and the resting state of the PSII is the S1 state, with one electron extracted from the OEC, stabilized by proton release (Figure 28.8(b)). To maintain charge neutrality throughout the S-cycle for leveling of the redox potential, release of electrons into the P680 chlorophylls alternates with the release of H^+ into the protein environment [42][43].

Oxidation of water proceeds with a redox potential difference of at least 1.23 V (vs. NHE). If the catalysis is rate-limited on a time scale of 10^{-3} s, and the oxidation potential needs to be generated with an absorber with a lifetime on the nanosecond scale, the mixing-entropy losses will be equivalent to at least 0.6 eV, as is explained above from the coupling of time scales and energy scales. This leads to a minimal "bandgap" for photon absorption of 1.8 eV. Natural oxygenic photosynthesis employs two reaction centers operating in tandem to compensate for the nonlinear response of the thylakoid membrane assembly and redox gradients associated with downhill electron transport in the catalysts, the cytochrome, and the energy carriers like NADPH [43]. This leads to the Z-scheme for photosynthesis depicted in Figure 28.3.

According to Equation (28.8) the energy difference between the absorber and the OEC is constrained by $hv_0 - \Delta\mu_{st} = k_B T \ln(t/\tau)$, and the mixing entropy represents an unavoidable overpotential in the operation of the multielectron catalyst in natural photosynthesis [19]. While the stabilization of intermediates in the charge accumulation can be very long due to irreversible proton release, the overpotential from the mixing entropy can be put to use for the confinement of holes on the time scale of the rate-limiting step in the oxygen-evolution process.

28.5 Summary

The large temperature difference between the Sun and the Earth provides a powerful engine for work on the evolution of the biological steady state and is the only source of sufficient energy for engineering the biodiversity on the planet. The principal molecular design of the helix and cofactor arrangement of photosynthetic reaction centers is highly conserved in evolution and is used both by micro-organisms and by the chloroplasts of plants and algae.

Chemical storage of solar energy requires "buying time" by taking away electrons rapidly at the expense of mixing entropy in the chemical conversion, to prevent the loss of harvested energy by radiative recombination of electron–hole pairs.

Since electrons and holes are very light particles with high tunneling probabilities, natural photosynthetic reaction centers are engineered for charge confinement by phonon dressing in the antenna for energy transfer by neutral excitons, while in the RCs large tunneling barriers prevent leakage from storage sinks back into the absorber and fluorescence losses on the time scales of catalysis.

Light harvesting by chlorosomes exploits the spatial anisotropy in the materials properties of extended chlorophyll aggregates. They combine photon absorption into symmetric excitons along the stacking direction with symmetry breaking into polaronic excitons along the H-bond helix. The latter can extend over tens of monomers for barrierless ultrafast energy transport in less than 200 fs to the RC complex.

Photosynthetic donor–acceptor–protein complexes are "smart matrices" that have been engineered by evolution to stabilize the transition state. Upon excitation with light, strain is released in the form of classically coherent (nuclear) motion with $\hbar\omega = 70$ meV to reorganize the donor–acceptor pair along a trajectory that leads to the matching of the high $\lambda = 0.7$ eV.

In the modular design of engineered natural photosynthesis the tunneling bridge provided by the RC and tuning of the proton chemical potentials for proton-coupled electron transfer at the OEC allow the efficient integration of storage with catalysis for oxidation of water. In particular, the incorporation of an insulating tunneling bridge between the OEC and the electron donor P680 is a critical element in the modular design, since it ensures that the mixing entropy for photoelectrochemical catalysis is utilized to produce an overpotential for the confinement of charge in the OEC catalytic site on the time scale of catalysis.

The information on how photosynthetic organisms are able to use solar energy to extract electrons from water to produce hydrogen or reduced carbon compounds presented in this chapter can be used to design modular photocatalytic systems capable of using solar energy to produce fuels directly from sunlight using water as a raw material (see also Chapter 47). Many scientists are convinced that, by using state-of-the-art knowledge from photosynthesis, the value of current technology for conversion of solar energy into electricity can be raised by adding a storage function, as in natural photosynthetic assemblies, leading to the development of novel robust photoelectrochemical solar fuel producers (see Chapter 49), or "artificial leaves (Chapter 27)."

28.6 Questions for discussion

1. Use Equation (28.8) to estimate the maximum storage time of solar energy by photochemical conversion. Is this longer or shorter than the lifetime of the solar system?

2. The transfer efficiency of the Förster hopping process is $E = k_e/(\tau^{-1} + k_e)$. What is the maximum distance r in units of r_0 for 95%, 90%, and 50% efficiency?

3. Using Equation (28.13) for the long-distance electron transfer in proteins, calculate how many cofactors are needed to span a 40-Å tunneling barrier in 100 μs.

4. If you had a fast water-oxidation catalyst that operates on the microsecond time scale, what would be the thickness of the tunneling barrier for charge confinement in the catalyst? How many cofactors would be needed to bring the electrons to the catalyst when they come from a molecular absorber with a lifetime of 10^{-8}s?

5. Write an essay on how engineered natural photosynthesis inspires the design of artifical solar-to-fuel conversion.

28.7 Further reading

- http://en.wikipedia.org/wiki/Photosynthesis. A short introduction to the biological design and diversity of photosynthesis for the novice reader that is complementary to the materials-engineering perspective of this chapter.

- **R. E. Blankenship**, 2002, *Molecular Mechanisms of Photosynthesis*, Oxford, Blackwell Science. For a comprehensive and in-depth overview of the principles of photosynthesis on an advanced level, this book is an excellent starting point.

- **P. Fromme**, 2008, *Photosynthetic Protein Complexes: A Structural Approach*, Weinheim, Wiley-VCH. This book provides a heuristic overview of how the structures relate to mechanisms of function on the level of the molecular machines involved in the processes of photosynthesis.

- **A. Pandit**, **H. J. M. de Groot**, and **A. Holzwarth**, 2006, *Harnessing Solar Energy for the Production of Clean Fuels*, Strasbourg, (European Science Foundation, http://www.esf.org/publications/lesc.html). This text, on which the present chapter is partly based, provides additional information on how engineered natural photosynthesis inspires the production of clean fuels.

28.8 References

[1] **G. Giacometti** and **G. M. Giacometti**, 2010, "Evolution of photosynthesis and respiration: which came first?," *Appl. Magn. Reson.*, **37**(1), 13–25.

[2] **P. G. Falkowski** and **A. H. Knoll** (eds.), 2007, *Evolution of Primary Producers in the Sea*, San Diego, CA, Academic Press.

[3] **L. N. M. Duysens**, **J. Amesz**, and **B. M. Kamp**, 1961, "Two photochemical systems in photosynthesis," *Nature*, **190**, 510–151.

[4] **I. Grotjohann**, **C. Jolley**, and **P. Fromme**, 2004, "Evolution of photosynthesis and oxygen evolution: implications from the structural comparison of Photosystems I and II,". *Phys. Chem. Chem. Phys.*, **6**(20), 4743–4753.

[5] **R. C. Prince** and **H. S. Kheshgi**, 2005, "The photobiological production of hydrogen: potential efficiency and effectiveness as a renewable fuel," *Crit. Rev. Microbiol.*, **31**, 19–31.

[6] **R. E. Blankenship**, 2002, *Molecular Mechanisms of Photosynthesis*, Oxford, Blackwell Science.

[7] **P. Fromme**, **P. Jordan**, and **N. Krauss**, 2001, "Structure of photosystem I," *Biochim. Biophys. Acta – Bioenergetics*, **1507**(1–3), 5–31.

[8] **K. N. Ferreira**, **T. M. Iverson**, **K. Maghlaoui**, **J. Barber**, and **S. Iwata**, 2004, "Architecture of the photosynthetic oxygen-evolving center," *Science*, **303**(5665), 1831–1838.

[9] **L. M. Yoder**, **A. G. Cole**, and **R. J. Sension**, 2002, "Structure and function in the isolated reaction center complex of photosystem II: energy and charge transfer dynamics and mechanism," *Photosynthesis Res.*, **72**(2), 147–158.

[10] **C. N. Hunter**, 2009, *The Purple Phototrophic Bacteria*, Dordrecht, Springer.

[11] **M. K. Sener**, **J. D. Olsen**, **C. N. Hunter**, and **K. Schulten**, 2007, "Atomic-level structural and functional model of a bacterial photosynthetic membrane vesicle," *Proc. Natl. Acad. Sci.*, **104**(40), 15723–15728.

[12] **S. Ganapathy**, **G. T. Oostergetel**, **P. K. Wawrzyniak**, *et al.*, 2009, "Alternating syn–anti bacteriochlorophylls form concentric helical nanotubes in chlorosomes," *Proc. Natl. Acad. Sci.*, **106**(21), 8525–8530.

[13] **Y. F. Li**, **W. L. Zhou**, **R. E. Blankenship**, and **J. P. Allen**, 1997, "Crystal structure of the bacteriochlorophyll *a* protein from *Chlorobium tepidum*," *J. Mol. Biol.*, **271**(3), 456–471.

[14] **R. MacColl**, 1998, "Cyanobacterial phycobilisomes," *J. Struct. Biol.*, **124**(2–3), 311–334.

[15] **W. Shockley** and **H. J. Queisser**, 1961, "Detailed balance limit of efficiency of p–n junction solar cells," *J. Appl. Phys.*, **32** 510–519.

[16] **M. C. Hanna** and **A. J. Nozik**, 2006, "Solar conversion efficiency of photovoltaic and photoelectrolysis cells with carrier multiplication absorbers," *J. Appl. Phys.*, **100**(7), 074510.

[17] **R. T. Ross** and **M. Calvin**, 1967, "Thermodynamics of light emission and free-energy storage in photosynthesis," *Biophys. J.*, **7**(5), 595–614.

[18] **T. Markvart** and **P. Landsberg**, 2002, "Thermodynamics and reciprocity of solar energy conversion," *Physica E*, **14**(1–2), 71–7.

[19] **H. J. M. de Groot**, 2010, "Integration of catalysis with storage for the design of multi-electron photochemistry devices for solar fuel," *Appl. Magn. Reson.*, **37**, 497–503.

[20] **H. J. van Gorkom**, 1986, "Photochemistry of photosynthetic reaction centres," *Bioelectrochem. Bioenergetics*, **16**, 77–87.

[21] **D. Noy**, **C. C. Moser**, and **P. L. Dutton**, 2006, "Design and engineering of photosynthetic light-harvesting and electron transfer using length, time, and energy scales," *Biochim. Biophys. Acta*, **1757**(2), 90–105.

[22] **M. Janssen**, **J. Tramper**, **L. R. Mur**, and **R. H. Wijffels**, 2003, "Enclosed outdoor photobioreactors: light regime, photosynthetic efficiency, scale-up, and future prospects," *Biotechnol. Bioeng.*, **81**(2), 193–210.

[23] S. Bahatyrova, R. N. Frese, C. A. Siebert et al., 2004, "The native architecture of a photosynthetic membrane," Nature, 430(7003), 1058–1062.

[24] J. A. Shelnutt, X.-Z. Song, J.-G. Ma et al., 2010, "Nonplanar porphyrins and their significance in proteins," Chem. Soc. Rev., doi:10.1002/chin.199818321.

[25] A. Pandit, F. Buda, A. J. van Gammeren, S. Ganapathy, and H. J. M. De Groot, 2010, "Selective chemical shift assignment of bacteriochlorophyll a in uniformly [^{13}C–^{15}N]-labeled light-harvesting 1 complexes by solid-state NMR in ultrahigh magnetic field," J. Phys. Chem. B, 114(18), 6207–6215.

[26] A. Pandit, P. K. Wawrzyniak, A. J. van Gammeren, et al., 2010, "Nuclear magnetic resonance secondary shifts of a light-harvesting 2 complex reveal local backbone perturbations induced by its higher-order interactions," Biochemistry, 49(3), 478–486.

[27] R. van Grondelle and V. I. Novoderezhkin, 2006, "Energy transfer in photosynthesis: experimental insights and quantitative models," Phys. Chem. Chem. Phys., 8, 793–807.

[28] P. K. Wawrzyniak, A. Alia, R. G. Schaap et al., 2008, "Protein-induced geometric constraints and charge transfer in bacteriochlorophyll–histidine complexes in LH2," Phys. Chem. Chem. Phys., 10(46), 6971–6978.

[29] R. van Grondelle, J. P. Dekker, T. Gillbro, and V. Sundstrom, 1994, "Energy-transfer and trapping in photosynthesis," Biochim. Biophys. Acta – Bioenergetics, 1187(1), 1–65.

[30] Z. Liu, H. Yan, K. Wang et al., 2004, "Crystal structure of spinach major light-harvesting complex at 2.72 angstrom resolution," Nature, 428(6980), 287–292.

[31] B. Robert, P. Horton, A. A. Pascal and A. V. Ruban, 2004, "Insights into the molecular dynamics of plant light-harvesting proteins in vivo," Trends Plant Sci., 9(8), 385–390.

[32] M. G. Müller, P. Lambrev, M. Reus et al., 2010, "Singlet energy dissipation in the photosystem II light-harvesting complex does not involve energy transfer to carotenoids," ChemPhySchem, 11(6), 1289–1296.

[33] A. V. Ruban, R. Berera, C. Ilioaia et al., 2007, "Identification of a mechanism of photoprotective energy dissipation in higher plants," Nature, 450(7169), 575–U22.

[34] T. K. Ahn, T. J. Avenson, M. Ballottari, et al., 2008, "Architecture of a charge-transfer state regulating light harvesting in a plant antenna protein," Science, 320(5877), 794–779.

[35] T. Förster, 1965, "Delocalized excitation and excitation transfer," in Modern Quantum Chemistry, ed. O. Sinanoğlu, New York, Academic Press, pp. 93–137.

[36] V. I. Prokhorenko, D. B. Steensgaard, and A. R. Holzwarth, 2003, "Exciton theory for supramolecular chlorosomal aggregates: 1. Aggregate size dependence of the linear spectra," Biophys. J., 85(5), 3173–3186.

[37] J. H. Golbeck, 2003, "Photosynthetic reaction centers: so little time, so much to do," Biophysics Textbook Online, p. 31.

[38] R. A. Marcus, 1956, "On the theory of oxidation–reduction reactions involving electron transfer. 1," J. Chem. Phys., 24(5), 966–978.

[39] C. C. Moser, J. M. Keske, K. Warncke, R. S. Farid, and P. L. Dutton, 1992, "Nature of biological electron transfer," Nature, 355(6363), 796–802.

[40] G. L. Closs and J. R. Miller, 1988, "Intramolecular long-distance electron transfer in organic molecules," Science, 240(4851), 440–447.

[41] S. Y. Reece and D. G. Nocera, 2009, "Proton-coupled electron transfer in biology: results from synergistic studies in natural and model systems," Ann. Rev. Biochem., 78, 673–699.

[42] M. Haumann, P. Liebisch, C. Müller et al., 2005, "Photosynthetic O_2 formation tracked by time-resolved X-ray experiments," Science, 310(5750), 1019–1021.

[43] H. Dau and I. Zaharieva, 2009, "Principles, efficiency, and blueprint character of solar-energy conversion in photosynthetic water oxidation," Acc. Chem. Res., 42(12), 1861–1870.

29 Geothermal and ocean energy

Jiabin Han,[1] J. William Carey,[1] and Bruce A. Robinson[2]

[1]*Earth and Environmental Sciences Division, Los Alamos National Laboratory, Los Alamos, NM, USA*
[2]*Civilian Nuclear Programs, Los Alamos National Laboratory, Los Alamos, NM, USA*

29.1 Focus

Geothermal energy is a renewable energy source that taps the heat contained within the Earth. In principle, it is environmentally friendly, with a relatively small footprint and few emissions to the environment. Life-cycle costs of geothermal energy make it competitive with conventional fuels. The energy potentially available from geothermal resources is enormous (e.g., in the USA it is ~2,000 times current annual energy use [1]). This chapter presents an overview of the current and next-generation technology available for utilizing geothermal energy. Current proven geothermal technologies include geothermal heat pumps (utilizing constant near-surface ground temperature for heating and cooling), direct use of hydrothermal water for building, heating, and industrial applications, and the generation of electricity from the highest-temperature geothermal resources. We review engineering challenges including scaling, corrosion of metallic materials, and degradation of cement. Enhanced geothermal systems (EGSs) represent the next generation of geothermal technology and have the potential to make a very significant impact on the energy sector. These resources are deep, and we discuss the required technology development, focusing on drilling, well completion, and heat-transmission media. This chapter also looks at the various ways in which ocean energy can be converted to electricity. This is a vast resource. While current technology lags behind that for geothermal, it is being rapidly developed.

29.2 Synopsis

Geothermal resources in the US have the potential to provide 100 GW of low-CO_2 electrical energy by 2050 (10% of current electrical use) for the US and 1 TW worldwide total accessible resource and could play a major role in developing low-carbon-intensity energy supplies. The geothermal resources with the greatest technical potential, known as EGSs, will require technological development if they are to become significant energy sources. The key materials and technological requirements necessary to realize this next-generation technology include the following [1].

1. Drilling technology
   a. More durable and faster-cutting drill bits and development of more rapid drilling methods.
   b. More durable cements for containing geothermal fluids.
   c. Improved design and durability of well casing and development of longer casing strings.
   d. Improved, thermally stable, down-hole sensors.
2. Power-plant design
   a. More efficient thermal-transfer materials and processes for recovering thermal energy from geothermal fluids.
3. Reservoir and well-system management
   a. Better methods and materials for creating and maintaining consistent fractures and permeability.
   b. Improved management of scaling (mineral precipitation) in fractures and in well systems.
   c. Reduced rates and consequences of corrosion of steel and degradation of cement.
   d. Development of enhanced geothermal fluids (e.g., supercritical CO_2) for improved reservoir penetration and heat transfer.

The development of ocean energy technology is advancing rapidly and is being explored via a number of approaches which include: wave energy, tidal energy, marine current energy, thermal energy conversion, and osmotic energy from salinity gradients. We discuss several experimental pilots that have been built to explore the utilization of ocean energy.

29.3 Historical perspectives

Archeological findings show that the first human use of geothermal resources dates from 10,000 years ago in North America and involved the direct use of hot water. Since that time, geothermal energy has evolved into three primary technologies. The direct use of hot geothermal water for heat occurs in a variety of applications, including agriculture, horticulture, aquaculture, industrial processes, bathing and spas, roads, bridges, and heating of buildings [2]. Geothermal heat pumps (GHPs) use conventional vapor compression (refrigerant-based) to extract low-grade heat from the shallow Earth to heat building spaces during the winter and, conversely, provide sinks for heat to the Earth during the summer. This simple technology is one of the most efficient heating–cooling systems [3]. The indirect use of geothermal energy to produce electricity began with the first 10-kW geothermal power plant built in 1904 in Larderello, Italy (Figure 29.1) [4]. The USA installed its first experimental geothermal power plant in 1920, and Pacific Gas and Electric Company has successfully operated the oldest geothermal electric power plant in the USA at The Geysers in California since 1960 [5]. As the world's largest dry-steam geothermal steam field, The Geysers served 1.8 million people at its peak production in 1987. The production has gradually declined due to diminishing underground water resources and the field currently serves 1.1 million people. New techniques employed by Calpine Corporation and the Northern California Power Agency have extended the production life of The Geysers field. Treated waste water from the neighboring community is injected to recharge the reservoirs and supplement underground water resources. This is expected to provide an additional 85 MW of power [6].

In 2008, the USA had 2,930 MW_e geothermal capacity installed and about 2,900 MW_e of new geothermal plants under development. Geothermal energy generated 14,885 gigawatt-hours (GWh) of electricity in 2007, accounting for 4% of renewable energy [7].

29.4 World scope

In 2004, 72 countries have deployed direct-use technologies utilizing geothermal energy (76 TWh per year). Twenty-two countries produce geothermal electricity, 57 TWh per year (Figure 29.2) [8]. Five of these countries obtained between 15% and 22% of their total electricity production from geothermal sources. The USA produced 27.6% of the world's total geothermal electricity. World geothermal electricity production has the potential to achieve 10–70 GW using current technology and potentially up to 140 GW applying advanced technology such as EGS [8].

Tables 29.1 and 29.2 show geothermal energy development in the world's leading geothermal-energy-producing countries in 2005–2010, indicating that there has been an overall 20% growth of installed capacity [9].

29.5 Resource abundance and sustainability

Geothermal energy, which is mainly generated by decay of radioactive isotopes, the original heat of the formation of the earth and volcanic activity, is abundant and generally sustainable. The temperature in the Earth's interior reaches temperatures greater than 4,000 °C, and usable temperatures at accessible depths are present virtually everywhere. This heat energy continuously flows to the Earth's surface and is estimated to be about 10^{13} EJ (exajoules). As a theoretical limit, this can provide 10^9 years of energy demand at the current world

Figure 29.1. The first geothermal power plant, 1904, Larderello, Italy [4].

Table 29.1. New geothermal energy capacity installed in the five leading geothermal producers between 2005 and 2010 [9] (electrical power only)

Country	Installed capacity (MW)
USA	530
Indonesia	400
Iceland	373
New Zealand	193
Turkey	0.62

Table 29.2. The top five countries in terms of percentage increase in installed geothermal capacity between 2005 and 2010 [9] (electrical power only)

Country	Increase (%)
Germany	2,774
Papua New Guinea	833
Australia	633
Turkey	308
Iceland	184

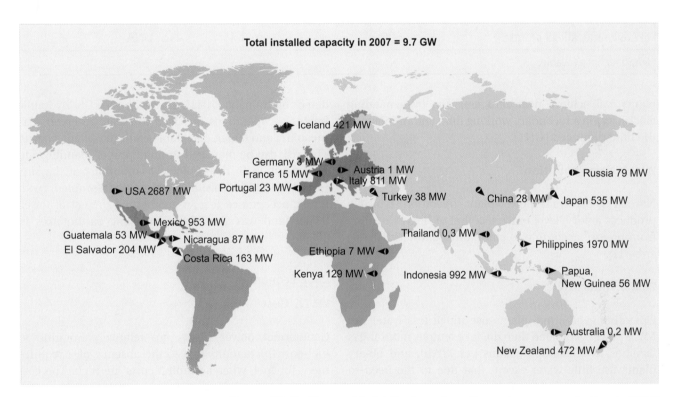

Figure 29.2. The world distribution of geothermal power production in 2007 (total installed capacity 9.7 GW) [8]. Colors are for illustrative purposes only.

consumption of 40 million MW of thermal energy [10]. The geothermal resources in the USA to a depth of **3 km** are estimated to be 3 million quads ($\sim 0.9 \times 10^{15}$ MW), and are **"equivalent to a 30,000-year energy supply at our current rate for the United States!"** [2].

29.6 Environmental impact

Compared with coal and nuclear power plants, geothermal energy production requires less physical mining, less distance processing and transportation, less radioactive disposal, and less mass storage space. Environmental issues that should be considered and managed include air emissions, groundwater usage,

and contamination during production, and safety, noise, seismicity, visual impact, and land use during drilling/production [1].

29.6.1 Emissions

Geothermal electricity generation does not directly produce emissions because no combustion is required as with fossil-fuel power plants. It does, however, indirectly generate a relatively small amount of gaseous emissions, including CO_2, NO_x, SO_2, and other gases and particulates. However, the emissions are minimal compared with those of other nonrenewable energies (Table 29.3) including coal, oil, and gas. Binary geothermal power

Table 29.3. A comparison of average emissions per unit of electricity generated among selected types of power-generating plant [1]

Plant type	Emissions (kg MWh^{-1})			
	CO$_2$	SO$_2$	NO$_x$	Particulates
Coal-fired	1,000	5	2	1
Oil-fired	800	5	2	NA
Gas-fired	600	0.1	1	0.06
Hydrothermal: flash-steam	30	0.2	0	0
Hydrothermal: dry steam field	40	0.0001	0.0005	Negligible
Hydrothermal: closed-loop binary	0	0	0	Negligible
EPA average, all US plants	600	3	1	NA
NA, data not available.				

plants can achieve near zero emission by generating electricity via a secondary working fluid and re-injecting the fluid produced [1][11].

29.6.2 Noise pollution

Normal geothermal power-plant operation typically produces a noise level much like that of "leaves rustling from breeze" according to common sound-level standards [11].

29.6.3 Water use

Operating geothermal plants use much less water per MWh power generated than do, for example, natural-gas facilities (5 versus 361 gallons per MWh), and binary plants use little water except that due to the need to replace evaporative losses. In many cases, waste water from surrounding communities is used for re-injection, both reducing surface water pollution and increasing the resilience of the geothermal reservoir via pressure maintenance (e.g., The Geysers, California) [1][11].

29.6.4 Land use

Over a 30-year life cycle, a geothermal facility's land use per GW is one-tenth that of a coal facility. This impact is further reduced by the relatively low visual impact of geothermal facilities compared with the presence of cooling towers and tall smoke stacks [1][11].

29.6.5 Induced seismicity

Geothermal production and injection can result in "micro-earthquakes," generally at a level difficult to

detect by humans [1][11]. However, for EGS, hydraulic fracturing operations are used to create reservoir permeability and larger, detectable earthquakes may occur. This risk can be managed through proper planning and consideration of the geological conditions of the site. However, induced seismicity can result in problems with public acceptance (e.g., a geothermal project in Basel, Switzerland was shut down following earthquakes in 2006).

29.7 Economics

29.7.1 Cost

Geothermal power plants are relatively immune to fuel-cost fluctuations due to the absence of a requirement for fuel, whereas capital costs are high. The cost of a geothermal power plant depends on a number of factors, including power-conversion technology, power-plant size, development costs, financing charges, operating and maintenance (O&M) costs, and resource credits. The cost of geothermal electricity is in the range of USD $0.06–0.10 per kWh [7]. The levelized power cost ranges from USD $0.055–0.075 per kWh. Taking an example for a 20-MW geothermal plant, the total development cost is about USD $80 million (Table 29.4) [7].

29.7.2 Contribution

Geothermal power plants make a significant contribution to local communities by providing employment as well as to city, state and federal governments in taxes [12]. A 50-MW geothermal power plant has an output with a value of 770 million dollars (Table 29.5).

Table 29.4. Cost summary (USD, 2008) for a typical 20-MW geothermal power plant [7]

Development stage	Cost (millions of $)
Exploration and resource assessment	8
Well-field drilling and development	20
Power plant, surface facilities, and transmission	40
Other costs (e.g., fees, operating reserves, and contingencies)	12
Total	80

Table 29.5. Contribution benefit of a typical 50-MW$_e$ geothermal power plant over 30 years in the USA [12]

Contribution	Amount
Employment	212 full-time jobs
Economic output	$750 million
To the Federal government	$5 million
To the state	$11 million
To the county	$5 million
Total capital	$770 million

Figure 29.3. A schematic diagram of geothermal systems driven by a magmatic heat source. Hydrothermal reservoirs tap permeable regions with circulating water/steam. Hot dry rock may exist in the region of impermeable bedrock above the magma chamber. The magma itself represents a geothermal resource but its exploitation is technically infeasible due to the very high temperatures involved (Credit: US Department of Energy [13]).

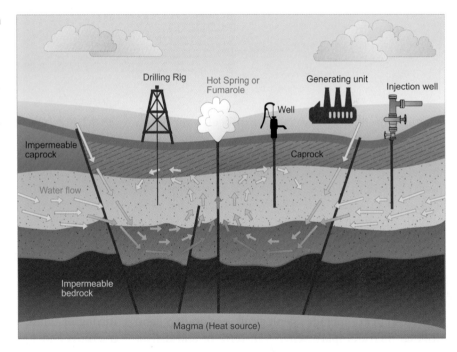

29.8 Geothermal industry processes

There are four different forms of geothermal reservoirs (not including ground-source heat pumps): hydrothermal reservoirs (steam or hot water), hot dry rock reservoirs (largely impermeable rocks), geopressurized reservoirs (deeply buried brines), and magma (molten rock) [13]. After a favorable geothermal site has been found, production and re-injection wells are drilled. After the wells have been completed and cemented, the production well extracts heat, usually by producing water or steam. The hot water or steam is pumped to heat-conversion facilities to generate power and distribute heat. The heat-exchanged water is often re-injected into the reservoir to complete the circle (Figure 29.3) [13].

Conventionally, geothermal electricity is generated using geothermal heat by four commercial types of plants: (a) single-flash, (b) binary, (c) flash–binary combined, and (d) dry-steam power plants (Figure 29.4). A single-flash power plant takes a mixed-phase liquid–vapor resource (>182 °C), and separates the steam to drive a turbine generator (Figure 29.4(a)). Binary (or binary cycle) geothermal power plants are generally used in liquid-dominant geothermal systems at moderate temperature (<200 °C), and pump the hot geothermal fluid through a heat exchanger to heat a secondary working fluid (e.g., a hydrocarbon) that is used to drive a turbine (Figure 29.4(b)). Flash–binary geothermal power plants combine single-flash and binary techniques. The steam phase of the geothermal fluids is directly used to drive a power generator,

Figure 29.4. Schematic drawings of the four main types of commercial geothermal power plants (Credit: Geothermal Energy Association [3]). These include single-flash plants that use superheated water that flashes to steam and drives a turbine; binary geothermal plants that use hot water to vaporize a secondary, working fluid; flash–binary plants that combine these two processes; and relatively rare dry-steam geothermal resources to drive a turbine.

(a) Single-flash geothermal power plant

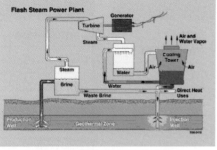

(b) Binary geothermal power plant

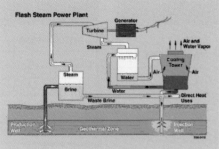

(c) Flash–binary geothermal power plant

(d) Dry-steam geothermal power plant

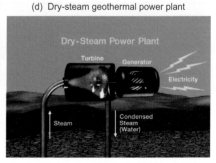

whereas the liquid phase is used to heat a secondary fluid to drive a second tubine (Figure 29.4(c)). Dry-steam geothermal power plants use steam/vapor produced from a geothermal reservoir to directly drive the power generator (Figure 29.4(d)). This kind of reservoir is rare. Larderello and The Geysers are the most famous dry-steam fields.

29.9 Current technology challenges

The technologies required to develop current geothermal resources include geological exploration and drilling, fluid-pumping systems to retrieve and deliver geothermal water/steam, heat-exchangers and binary cycle systems, and finally turbine generators for electricity. In this section, we cover technical challenges associated with maintaining the fluid-delivery system. Geothermal fluids are generally rich in dissolved minerals and can be corrosive to steel piping. Fluid extraction and energy recovery results in temperatures and pressures that cause mineral precipitation. When precipitation occurs within the reservoir, the fluid permeability is reduced, fractures close, and the efficiency of extraction of thermal energy is reduced. Precipitation can also occur within the production lines and is known as scaling. Examples of such problems include the failures of a 110-MW turbine in Mexico in 2005 [14], and a geothermal steam diffuser in The Geysers, California in 1980 [15]. Geothermal fluids can also be chemically aggressive and can create significant metallic corrosion.

29.9.1 Scaling

Scaling is one of the causes of facility failure [16]. Geothermal fluids flow through the Earth's crust, resulting in mineral dissolution at high geothermal temperatures. This saturated fluid is transported through wells to distribution pipelines where the drop in temperature and pressure can result in super-saturation of minerals and precipitation on the pipe walls. This can eventually clog the transportation paths and cause under-deposit corrosion of the steel pipe. Precipitation can also occur in the reservoir due to changes in temperature, pressure or mixing of fluids and can result in sealing of hydraulic fractures and lower fluid production. During flashing operations, pressure reduction and venting of noncondensible gases can also lead to thermodynamic conditions conducive to precipitation.

The prediction and control of scaling processes are critical to successful reservoir operation. These numerical calculations require heat- and mass-transfer fluid flow codes to represent fluid movement and temperature in the reservoir and in the pipelines; and they require geochemical codes to predict dissolution and precipitation processes. The fluids in geothermal reservoir systems are typically of high salinity and are therefore strongly non-ideal, requiring application of sophisticated fluid-composition modeling. For example, Pitzer's model for solution chemistry has been widely applied and has the capability to accurately predict thermodynamic properties of geo-fluids [17]. (See Appendix 1 for a detailed explanation of the Pitzer methodology.) Pitzer's model has been implemented in a variety of geochemical and transport computer codes, including FLOTRAN (Los Alamos National Laboratory), TEQUIL, EQ3/6NR, GMIN, REACT, FREEZCHEM, ESP, TUFFREACT, and SCAPE2 [18].

Scaling control

A conventional way to control scaling within the geothermal reservoir is to renew permeability by hydraulic fracturing by pumping high-pressure fluid to create conductive fractures. An alternative method is a chemical treatment or acidic (HCl or H_2SO_4) flush to dissolve scales. This can be used both in reservoir fractures and in the piping system, but can cause severe corrosion of the well casing, particularly at elevated temperatures. The use of scale inhibitors (polymeric dispersants, weak complexing acids, or organic inhibitors) is another approach employed to prevent scale formation. While scale can be removed mechanically, the freshly created surface suffers worse corrosion rates. An alternative approach is to extract the minerals from the fluids and recover potentially valuable compounds, including silica, zinc, lithium, gold, silver, cesium, rubidium, NaCl, Na_2SO_4, H_2O, and $CaCl_2$, by acid leaching, biochemical leaching, sorption, evaporation, and precipitation. These technologies diminish but do not prevent scaling and improve the efficiency of geothermal energy plants [19].

29.9.2 Corrosion

Corrosion problems have significant impacts on the cost of geothermal power plants, adversely impacting power-plant lifetime and electricity output (due to down-time and replacement/refurbishment). The commonly seen forms of corrosion in geothermal systems are uniform corrosion, localized corrosion, intergranular corrosion, erosion corrosion, crevice corrosion, and stress corrosion cracking (SCC). The materials used in geothermal systems are carbon steel (90%), copper and its alloys, corrosion-resistant alloys (e.g., stainless steels), other metallic materials, and nonmetallic materials (including polymers).

Corrosive species in geothermal systems

Corrosion in geothermal fluids occurs via oxygen, protons, carbonic acid (CO_2), bicarbonates, hydrogen sulfide species (hydrogen sulfide, bisulfide ions, and sulfide ions), chloride ions, ammonia species (ammonia and ammonium ions), sulfate ions, particulates [20], and microbes.

Oxygen is extremely corrosive both in alkaline and in acidic environments [21]:

$$O_2 + 2H_2O + 4e^- \rightarrow 4OH^- \text{ (alkaline)}, \quad (29.1)$$

$$O_2 + 4H^+ + 4e^- \rightarrow 2H_2O \text{ (acidic)}. \quad (29.2)$$

Corrosion induced by oxygen is generally limited by oxygen transfer and the corrosion rate (CR) in mm per year is expressed as [22]:

$$CR = 1.155\, i^d_{\lim(O_2)} = 1.155 \times 4k_{m,O_2}F \times c_{O_2} \quad (29.3)$$

where $i^d_{\lim(O_2)}$ is the limiting current density controlled by mass transfer, k_{m,O_2} is a mass-transfer coefficient, c_{O_2} is the concentration of O_2 (mol m^{-3}), and F is the Faraday constant. A rule-of-thumb concentration used in industrial corrosion-control calculations is a maximum of 20 ppm oxygen.

The **proton's** cathodic reaction is shown in Equation (29.17). At lower pH (high proton concentration), corrosion is severe for steel and will also result in severe degradation of cement. Above pH 8, the proton-induced corrosion rate decreases significantly [21].

Dissolved carbon dioxide forms carbonic acid and acidifies the solution generating corrosive protons. In addition, carbonic acid directly reacts with and corrodes steel [21], causing more severe corrosion compared with deaerated (O_2-free) solution at the same pH.

Chloride affects the steel anodic reaction and accelerates the kinetics in deaerated brine systems. If oxygen is present, chloride can reduce corrosion rates due to decreasing the solubility of oxygen by the "salting-out" effect. Salt also decreases corrosion rates in CO_2-saturated brines, although it is not clear whether this is caused by the "salting-out" effect or some other mechanism. The better known and more significant corrosive effect of chloride is through its promotion of localized corrosion by damaging passive films, which is especially problematic with stainless steels as well as other corrosion-resistant alloys [23].

Bicarbonates (HCO_3^-) are considered to react directly with steels, being in this regard similar in effect to carbonic acid but with much slower kinetics. A more significant effect of bicarbonates is through the formation of carbonate scale [20].

Hydrogen sulfide species (hydrogen sulfide, H_2S, bisulfide ions, HS^-, and sulfide ions, S^{2-}) react with steel to form a compact FeS film that resists further corrosion. However, mechanical erosion (spallation) or damage of this film can result in localized corrosion. Sulfide is highly corrosive to copper- and nickel-containing alloys. In addition, sulfide-induced cracking is another serious concern [20].

Microbes: sulfate-reducing bacteria (SRB) are the main micro-organisms responsible for microbially induced corrosion (MIC). By synergetic interactions between microbial activity and the electrochemical corrosion reaction, MIC accelerates corrosion, and, even worse, causes localized corrosion under biofilm colonies [22].

Sulfate ions primarily cause degradation/corrosion of cements and facilitate growth of SRB.

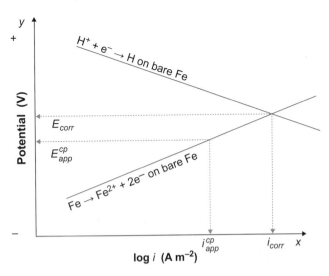

Figure 29.5. A schematic diagram illustrating the cathodic protection mechanism. The corrosion potential of the system is given by the intersection of reduction of hydrogen and oxidation of iron at a potential of E_{corr} generating a corrosion current of i_{corr}. Cathodic protection involves applying a more negative potential (E_{app}^{cp}) to the system, which results in a much lower corrosion current (i_{corr}^{cp}).

Ammonia species (ammonia, NH_3, and ammonium ions, NH_4^+) cause SCC of some copper-based alloys [20].

Corrosion control

Corrosion can be controlled by material selection, chemical treatment, cathodic protection, and isolation (coating) [22]. Material selection is a critical method to control corrosion that requires balancing of the costs of the material, maintenance, and degradation. Carbon steel is cost-effective and resistant to localized corrosion, but is not resistant to uniform corrosion and can have high maintenance or degradation costs. Corrosion-resistant alloys have excellent corrosion-resistance properties, but are expensive materials and susceptible to localized corrosion.

Injecting chemical inhibitors to retard corrosion is problematic, since they usually contain toxic compounds and their use may be restricted by environmental regulations for subsurface applications in geothermal reservoirs. Nontoxic organic inhibitors, which decompose without contamination of the environment, are under development.

Coatings on steel can provide a physical barrier to corrosive geothermal fluids. However, defects or damage of the coating will result in severe localized corrosion.

Cathodic protection is achieved by imposing on the system a more negative potential than the corrosion potential (Figure 29.5). The initial corrosion current density i_{corr} at a particular corrosion potential (E_{corr}) is

decreased to i_{corr}^{cp} by application of a negative cathodic potential (E_{app}^{cp}).

29.10 Next-generation technology: enhanced geothermal systems

Conventional hydrothermal geothermal energy resources are limited in geographical availability and require particular geological conditions that trap steam or pressurized hot water in permeable geological formations. Much more abundant resources can be found by looking deeper into the Earth's crust and taking advantage of the Earth's natural thermal gradient. Heat energy in the Earth is maintained by radioactive decay and results in a thermal energy flow to the surface of 44.2 TW [24]. Potential resources are present at depths between 3 and 10 km and are known as EGSs, sometimes called hot dry rock. Heat is actively "mined" from the formation by circulating fluid through the rock to extract the heat. There are significant obstacles to the development of EGSs because of the greater depth and because the reservoirs generally have low permeability and lack extractable, replenishable pore fluids. The key areas of technology development include reservoir fracturing, drilling technology, robust wellbore completions (including steel and cement materials), and the design and use of heat-transport fluids.

29.10.1 Reservoir fracturing

Three different concepts for recovering thermal energy from EGS reservoirs are illustrated by projects conducted at (a) Fenton Hill by Los Alamos National Lab, (b) Cornwall, UK by Camborne School of Mines, and (c) Soultz, France by the European Hot Dry Rock project (Figure 29.6). At Fenton Hill, the design involved injection of high-pressure water, either to create new fractures or to stimulate previously closed natural fractures (hydraulic fracturing). A second production well was drilled to intersect the fractures. The Camborne model utilizes a naturally occurring fracture network. Water is used to stimulate the existing fractures in order to facilitate heat extraction by fluids. At Soultz, a permeable fault system was exploited to connect the injection and production wells. Hydraulic fracturing was used to connect each well to the fault system [25].

29.10.2 Drilling

Well drilling and construction is one of the most expensive aspects of EGSs, and technology to reduce these costs is needed [9]. With EGSs one recovers heat from deeper reservoirs, typically at depths 3–10 km compared with less than 1 km in conventional geothermal energy production. The temperature is generally higher than

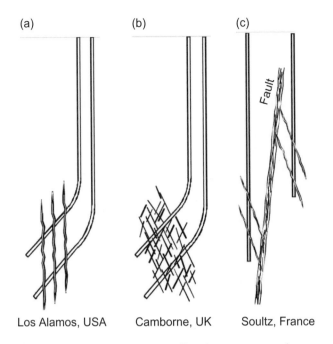

(a) Los Alamos, USA   (b) Camborne, UK   (c) Soultz, France

Figure 29.6. Concepts regarding how to extract heat from EGS systems are being tested in three different locations [25]: (a) hydraulic fracturing creates a network; (b) a natural fracture network; and (c) hydraulic fracturing connects wells to a natural fault system.

that of a conventional reservoir field. There is relatively little experience for geothermal wells at depths in excess of 3 km, particularly for the hard, crystalline rocks and high temperatures in a typical EGS reservoir [26][27]. Technology development is needed for drilling, the casing material, cementation, and electronic monitoring [27].

The most durable drilling bits currently used are made of polycrystalline diamond compact (PDC). These bits seldom have a diameter greater than 31 cm, while a deep geothermal well typically requires larger diameters. A drilling bit made of PDC with a diameter of 60 cm is under development [27].

To complete a well, cement is used to seal the annulus between the steel well casing and surrounding rock, or between concentric casing strings. A typical EGS environment consists of a high-temperature ($\sim$100–320 °C), high-salinity brine, and contains reactive species (e.g., CO_2 and H_2SO_4). Conventional cements are susceptible to reaction with hot CO_2 and H_2SO_4, causing their deterioration. This in turn causes a loss of protection of steel casing and a compromised seal between the formation and the well. A few new cements have been developed for geothermal applications, including calcium aluminate phosphate (CaP), magnesium–potassium phosphates (ceramicrete), and sodium silicate-activated slag (Ultra seal® SSAS) [28].

CaP cement is composed of CaO–Al_2O_3–P_2O_5–H_2O and Na_2O–CaO–Al_2O_3–SiO_2–P_2O_5–H_2O systems.

Crystalline hydroxyapatite [$Ca_5(PO4)_3(OH)$], boehmite (γ-$AlOOH$), hydrogarnet ($Ca_3Al_2(OH)_{12}$, and analcime ($NaAlSi_2O_6 \cdot H_2O$) phases control the strength and density of the cement. This cement is also resistant to CO_2 and mild acid, and exhibits improved pumpability, low density, toughness, and durability of bonding to the casing pipe's surface [28].

Ceramicrete is of very low porosity and low permeability, being composed of a mixture of hydrated magnesium–potassium phosphates. Extensive testing has been conducted to demonstrate the feasibility of using this material in wellbore systems (in terms of its setting time, ultimate strength, permeability, etc.) [29].

The SSAS cement combines crystalline calcium silicate hydrate ($CaO \cdot SiO_2 \cdot xH_2O$, CSH) and tobermorite ($5CaO \cdot 6SiO_2 \cdot xH_2O$) phases, which result in improved strength and lower permeability. The addition of class-F fly ash (a byproduct from coal combustion?) enhances resistance to acid [28].

Ultra Seal® cement provides exceptional annular sealing performance in high-temperature and high-pressure environments. Adding industrial epoxy resin gives the cement the capability of sealing when conventional cement fails [30].

29.10.3 Advanced heat-transfer fluids

The use of supercritical CO_2 as a geothermal working fluid was first proposed by Brown [31]. This concept was further developed by coupling geothermal energy recovery with geological sequestration of CO_2. Supercritical fluids are used to create fracture networks in hot dry rock as well as serving as the thermal-energy-transfer fluid to generate electricity. CO_2 has favorable thermodynamic and transport properties. One advantage of this technology is that the buoyancy of the heated gas reduces pumping costs. In addition, supercritical CO_2 is thought to be less reactive with the reservoir rock, potentially reducing scaling and corrosion [32]. Although it is recognized that the heat capacity of supercritical CO_2 is not as good as that of water, the overall heat-conversion efficiency using supercritical CO_2 was demonstrated to be 50% higher than when water is used as the heat-transmission mediaum [33]. However, if supercritical CO_2 is used, any contamination with water significantly accelerates corrosion of steel material [34]. This risk must be evaluated before this concept can be deployed. Additionally, there is a non-trivial phase-change issue near the surface as CO_2 condenses that has not yet been solved.

29.11 The history of ocean energy

People have captured ocean energy since the late eighteenth century. The first patent for wave-energy

Table 29.6. Estimated global resources of ocean energy [37]

Form of ocean energy	Estimated global resource (TW per year)
Tides	300+
Waves	8,000–80,000
Marine current	800+
Thermal resources	10,000
Osmotic energy	2,000

conversion was received by Monsieur Girard in 1799 [35]. The largest tidal-range power station in the world was built in St. Malo, France in 1966 and continues to generate electricity today [35]. The plant provides 240 MW of electricity using tides of height 2.4 m. Other tidal plants have been operated in the USA, China, Russia, and Canada [36].

29.12 Resources for ocean energy

Oceans cover 70% of Earth's surface and provide huge energy resources in five forms: wave, tidal range, marine current, ocean thermal energy conversion (OTEC), and osmotic energies [37] (Table 29.6). Among them, ocean current resources in the USA are primarily in the Gulf Stream. Ocean water has an energy density 800 times that of air (wind energy). If 1/1,000 of the Gulf Stream current energy were captured, it could provide 35% of the demand of the state of Florida [38]. The utilization of OTEC requires a minimum temperature difference of 20 °C, as in tropical and subtropical oceans within a depth of 1 km [39]. However, given the inherent small temperature differences in any practical system, OTEC systems will always suffer from very low Carnot efficiency compared with conventional thermal energy resources. Finally, tidal current and wave energy is estimated to potentially provide 10% of the USA's current national energy demand [40].

29.13 Environmental impact of ocean energy

Ocean energy technologies are new, unproven, and under development. Their cumulative environmental impacts are not known [41], but apparently, little negative impact on the environment is expected.

29.14 Ocean energy technologies

Five types of ocean energy conversion exist: wave energy, tidal energy, marine current energy, OTEC and osmotic energy.

Wave energy is the energy carried by ocean surface movements. More than 1,000 patents regarding how to tap this resource have been filed. The first such commercial power plant, with a capability of 2.25 MW, was installed in Portugal in 2008. It was shut down two months later due to technical and financial problems. The wave-energy map (Figure 29.7) shows that the highest energy is to be found above 40 degrees latitude.

The energy (in $kW\ m^{-3}\ s^{-1}$) carried by waves where the water depth is larger than half of the wavelength can be calculated as follows [43]:

$$P = \frac{\rho g^2}{64\pi} H_w t_w \approx 0.5 H_w t_w \qquad (29.4)$$

where P is the wave energy flux, H_w is the significant wave height, t_w is the wave period, ρ is the density, of seawater, and g is the gravity constant.

The energy flux passing through a perpendicular plane during wave propagation is given by [43]

$$P = \frac{1}{16} \rho g H_w^2 t_w c_g, \qquad (29.5)$$

where P is wave energy flux, ρ is the density of seawater, g is the gravity constant, H_w is the wave height, t_w is the wave period, and c_g is the group velocity [43].

Tidal energy is driven by the gravitational force between the Earth and the Moon (and, to a lesser extent, the Sun). The available potential energy is governed by the height difference of water between low tide and high tide, which must be at least 5 m for viability (16 feet; Figure 29.8). Tidal energy is periodic but predictable, in contrast to the variability of wind energy.

One method of power generation from tides works by storing high-tide water behind a "tidal barrage" and releasing it at low tide. The potential energy carried by a volume of water is [44]

$$E = \frac{1}{2} A_{\text{barrage}} \rho g h_w^2, \qquad (29.6)$$

where E is the potential energy, A_{barrage} is the horizontal surface area of the barrage basin, and h is the vertical tidal range.

The power output from a turbine engine can be expressed as [44]

$$p = \frac{\xi \rho A_{\text{turbine}} V_w^3}{2}, \qquad (29.7)$$

where P is power generation/flux, ξ is the turbine efficiency, A_{turbine} is the sweep area of the turbine, and V is the flow velocity.

Tidal energy technology is still under development. To capture tidal energy, a tidal barrage is used to allow tidal water flow in both directions.

Current energy is generated due to ocean water flow. Current energy is more stable than tidal range since the water current flow is steady throughout the

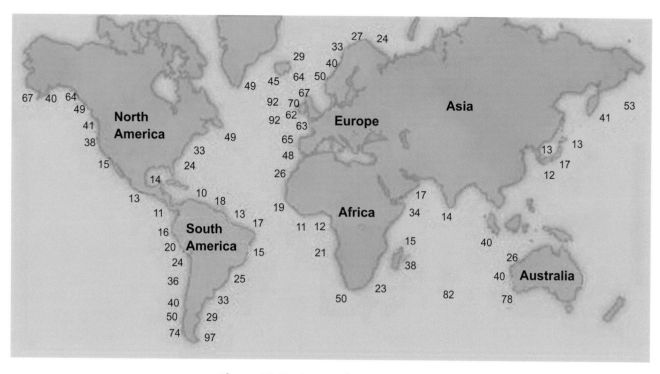

Figure 29.7. A map of wave energy around global coastlines. Note: the unit of the numbers is kW m^{-1} crest length. (Credit: European Ocean Energy Association [42].)

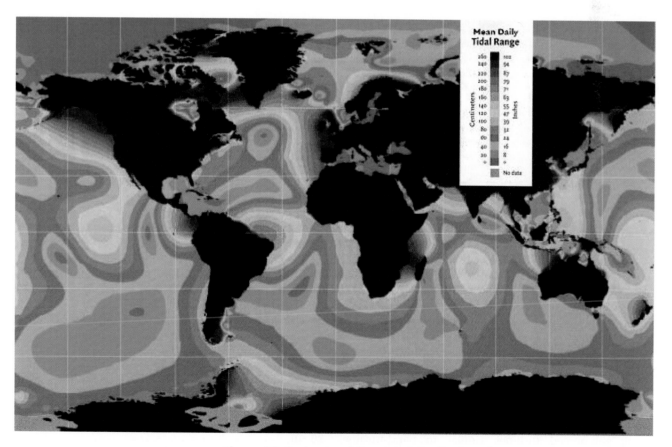

Figure 29.8. A map of the global distribution of mean tidal range. (Credit: the Electric Power Research Institute.) Tidal energy is feasible where the range exceeds 5 m [42].

Figure 29.9. A map of global major ocean currents. The red arrows represent stronger ocean currents and the light blue arrows represent weaker ones. (Credit: National Oceanic and Atmospheric Administration [42].)

Figure 29.10. A map of global ocean temperature differences between the ocean surfaces and 1,000 m depth. (Credit: National Renewable Energy Laboratory, US Department of Energy [42].)

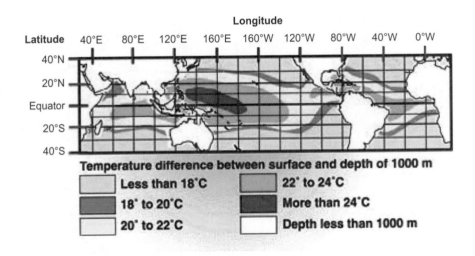

year. Figure 29.9 shows the global distribution of major ocean currents. Most coastal countries have access to strong ocean currents. Near the USA, the Gulf Stream flows out of the Caribbean and into the North Atlantic. The speed of the current reaches 7.4 km h^{-1} off the coast of Florida, and this flow has an extractable energy of 1 kW m^{-2} near the surface. Kinetic energy can be captured using submerged turbines similar to wind-turbine designs; however, these can be deployed only at shallow depths at present. The challenge of ocean current energy conversion is to properly design the turbine to be durable and efficient. Current approaches need significant improvements in order to be viable [35].

With **OTEC** one uses ocean temperature differences from the surface to the depths to extract energy through a heat engine. To be efficient, OTEC requires a temperature difference of at least 20 °C. A map shows that most of the usable energy is located near the equator (Figure 29.10).

Three types of heat-conversion technologies under development are closed-cycle, open-cycle, and hybrid-cycle systems.

Table 29.7. An assessment of geothermal technologies for current and future challenges

	Limitations	Energy-capture materials	Energy-conversion materials	Operational challenges	Reservoir engineering challenges
Current	Efficiency and operation	Water	Water steam	Scaling and corrosion	Shallow reservoir: no challenges
>10 years	Operation	CO_2 (other?)	Organic fluids (e.g., isobutane or pentafluoropropane)	Cement scaling corrosion	Deep reservoir: fracture opening

Closed-cycle systems use a working fluid with a low boiling point, such as ammonia, to drive an electricity-generation turbine. In the closed-cycle system, the working fluid is pumped through a heat exchanger and vaporized. This vaporized fluid drives the turbine and generates electricity.

In the open-cycle system, the warm surface water is pressurized in a vacuum chamber and converted to steam. This steam drives a turbine to generate electricity.

In a hybrid-cycle system, combining the features of both closed-cycle and open-cycle systems, warm seawater enters a vacuum chamber and is evaporated into steam in an open-cycle loop. The steam is used to vaporize a working fluid in a closed-cycle loop, and the vaporized fluid drives a turbine to produce electricity [45].

Osmotic energy depends on a salinity gradient due to a salt-concentration differential between seawater and fresh water, using osmosis with an ion-specific membrane to generate power. This technology has been confirmed in the world's first 4-kW-capacity power plant operated in Tofte, Norway [46].

29.15 Summary

Geothermal energy is a sustainable and green energy resource. Utilization of geothermal energy has been experiencing rapid growth internationally, especially in the past five years. Geothermal energy production creates fewer environmental problems, including emissions, land use, and noise, but does have the potential to induce seismicity. Geothermal energy is exploited in the forms of direct use, geothermal heat pumps, and electricity generation. The extraction of hot water from geothermal reservoirs presents problems in scaling and corrosion that must be addressed. The next-generation geothermal technology is EGSs, which are deep geothermal resources with low intrinsic permeability and fluid availability that are developed through hydraulic fracturing techniques and fluid injection. Frontier research

areas include fracture techniques, well drilling and completion in deep, harsh environments, and novel working fluids such as supercritical CO_2.

Ocean energy is a vast energy source available in the forms of wave energy, tidal energy, ocean currents, thermal gradients, and salinity gradients. Several commercial and demonstration plants have been developed to utilize this energy; however, the cost of production is higher than with traditional fossil and hydro energies. New technologies for tapping this large potential resource are currently under development in order to achieve more efficient and economic energy production (Table 29.7).

29.16 Questions for discussion

1. Why is geothermal energy considered to be sustainable?
2. What are the environmental impacts of geothermal energy?
3. What are the other technical challenges for developing conventional geothermal energy systems mentioned and not mentioned in the chapter?
4. What are the challenges for EGSs?
5. Given the widespread availability of geothermal resources, what practical and political limitations limit its greater use?
6. What are the different types of geothermal energy and what are their distinguishing characteristics?
7. Why is cement important in geothermal systems, and what modifications can make it more durable in geothermal environments?
8. How does a geothermal heat pump both heat and cool a building to be sufficiently warm in the winter and sufficiently cool in the summer?
9. How do EGSs differ from conventional geothermal energy?
10. What does fracturing achieve in EGSs and why is it necessary?
11. How does scaling impact upon geothermal heat and power production?

12. Why is Pitzer's theory required to understand scaling?
13. In what part(s) of the geothermal energy production cycle does corrosion usually occur?
14. What would you propose in order to reduce the present costs of ocean energy technologies?
15. What are the major forms of ocean energy, and which is most likely to have a significant impact within 10 years?

29.17 Further reading

- **Massachusetts Institute of Technology**, 2006, *The Future of Geothermal Energy: Impact of Enhanced Geothermal Systems (EGS) on the United States in the 21st Century*, Cambridge, MA, Massachusetts Institute of Technology.
- **M. Kaltschmitt**, **W. Streicher**, and **A. Wiese**, 2007, *Renewable Energy, Technology, and Environment*, Berlin, Springer.
- **L. Blodgett** and **K. Slack**, 2009, *Geothermal 101: Basics of Geothermal Energy Production and Use*, Washington, DC, Geothermal Energy Association.
- **P. C. Hewlett**, 1998, *Lea's Chemistry of Cement and Concrete*, 4th edn., London, Arnold, co-published New York, John Wiley & Sons Inc.
- **J. O'M. Bockris**, **A. K. N. Reddy**, and **M. E. Gamboa-Aldeco**, 2002, *Modern Electrochemistry 2A: Fundamentals of Electrodics*, 2nd edn., Dordrecht, Kluwer Academic Publishers.
- **K. S. Pitzer**, 1991, *Activity Coefficients in Electrolyte Solutions*, 2nd edn., Boca Raton, FL, CRC Press.

29.18 Appendix 1. Pitzer's theory

Pitzer developed an activity model based on ion interactions. The thermodynamic properties of the electrolytes can be predicted within experimental error for non-ideal electrolyte solutions. If the third virial terms are neglected, Pitzer's model can be expressed as [17]

$$\varphi - 1 = \frac{2}{\sum_i m_i}\left[\frac{-A_\phi I^{3/2}}{1+1.2I^{1/2}} + \sum_c\sum_a m_c m_a (B_{ca}^\varphi + ZC_{ca})\right.$$

$$+ \sum_{c<}\sum_{c'} m_c m_{c'}\left(\Phi_{cc'}^\phi + \sum_a m_a \psi_{cc'a}\right)$$

$$+ \sum_{a<}\sum_{a'} m_a m_{a'}\left(\Phi_{aa'}^\phi + \sum_c m_c \psi_{ca'a}\right)$$

$$+ \sum_n\sum_c m_n m_c \lambda_{nc} + \sum_n\sum_a m_n m_a \lambda_{na}$$

$$\left.+ \sum_{n<}\sum_{n'} m_n m_{n'} \lambda_{nn'} + \left(\frac{1}{2}\right)\sum_n m_n^2 \lambda_{nn} + \cdots\right], \tag{29.8}$$

$$\ln\gamma_M = z_m^2 F + \sum_a m_a (2B_{Ma} + ZC_{Ma})$$

$$+ \sum_c m_c\left(2\Phi_{Mc} + \sum_a m_a \psi_{Mca}\right)$$

$$+ \sum_{a<}\sum_{a'} m_a m_{a'} \psi_{Maa'} \tag{29.9}$$

$$+ z_M \sum_c\sum_a m_c m_a C_{ca} + 2\sum_n m_n \lambda_{nM} + \cdots,$$

$$\ln\gamma_X = z_X^2 F + \sum_c m_c (2B_{cX} + ZC_{cX})$$

$$+ \sum_a m_a\left(2\Phi_{Xa} + \sum_c m_c \psi_{cXa}\right)$$

$$+ \sum_{c<}\sum_{c'} m_c m_{c'} \psi_{cc'X} + |z_X|\sum_c\sum_a m_c m_a C_{ca} \tag{29.10}$$

$$+ 2\sum_n m_n \lambda_{nX} + \cdots,$$

$$\ln\gamma_N = 2\left(\sum_c m_c \lambda_{Nc} + \sum_a m_a \lambda_{Na} + \sum_n m_n \lambda_{Nn}\right), \tag{29.11}$$

where φ is the osmotic coefficient of water m is the concentration expressed as molality, z is charge, a and X are different anions, c and M are different cations, n and N are different neutral species, and i is species i.

I is the ionic strength and can be expressed as [17]

$$I = \frac{1}{2}\sum_i m_i z_i^2. \tag{29.12}$$

A_φ is the Debye–Hückel slope for the osmotic coefficient of water, which can be expressed as [17]

$$A_\varphi = \frac{1}{3}(2\pi N_0 d_W)^{1/2}\left(\frac{e^2}{4\pi\varepsilon\varepsilon_0 k_B T}\right)^{3/2}, \tag{29.13}$$

where N_0 is Avogadro's number, d_W is the density of the solvent (water), e is the electronic charge, k_B is Boltzmann's constant, ε is the dielectric constant or the relative permittivity of water, ε_0 is the permittivity of free space and T is the absolute temperature.

Then

$$Z = \sum_i m_i |z_i|, \tag{29A.7}$$

$$F = -A_\varphi\left[\frac{I^{1/2}}{1+1.2I^{1/2}} + \frac{2}{1.2\ln(1+1.2I^{1/2})}\right]$$

$$+ \sum_c\sum_a m_c m_a B_{ca}' + \sum_{c<}\sum_{c'} m_c m_{c'} \Phi_{cc'}^\phi \tag{29.14}$$

$$+ \sum_{a<}\sum_{a'} m_a m_{a'} \Phi_{aa'}^\phi$$

The activity of water can be calculated from [17]

$$a_{H_2O} = \exp\left(-\frac{M_{H_2O}\phi}{1,000}\sum_j m_j\right), \tag{29.15}$$

where M_{H_2O} is the molecular mass of water and j labels all the species in the electrolytes, including neutral species.

For further details of the meanings of the various symbols etc., see [17].

29.19 Appendix 2. Corrosion basics

Aqueous corrosion is a spontaneously occurring electrochemical process composed of anodic and cathodic reactions. For example, in the corrosion of iron by an acid such as HCl, the anodic reaction is

$$Fe(s.) \rightarrow Fe^{2+}(aq.) + 2e^- \qquad (29.16)$$

and the cathodic reaction is

$$2H^+(aq.) + 2e^- \rightarrow H_2(g.) \qquad (29.17)$$

with an overall reaction of

$$Fe(s.) + 2H^+(aq.) \rightarrow Fe^{2+}(aq.) + H_2(g.). \quad (29A.12)$$

The open-circuit potential (E_{rev}) of the above reaction follows the Nernst equation [22],

$$E_{rev} = E^\circ + \frac{RT}{nF} \ln\left(\frac{\prod_j a_j^{ox}}{\prod_k a_k^{red}}\right), \qquad (29.18)$$

where E° is the standard-state open-circuit potential and the remaining terms correct the potential for the actual solution composition.

The corrosion rate (CR) is linearly related to the corrosion current density as defined by [22]

$$CR = constant \times i_{corr}, \qquad (29.19)$$

where the corrosion current density (i_{corr}) is be controlled by mass transfer, electrochemical kinetics, and chemical reaction. A general formulation to calculate corrosion is [22]

$$\frac{1}{i_{corr}} = \frac{1}{i_r} + \frac{1}{i_L}, \qquad (29.20)$$

where i_r is the kinetically controlled electrode current density and i_L is the mass transfer or chemical-reaction-limited current density.

The kinetically controlled corrosion current density is obtained using the Butler–Volmer equation for the net current density (i_{net}) as [47]

$$i_r = i_a - i_c$$
$$= i_0 = \left[\exp\left(\frac{(1-\alpha)nF}{RT}\eta\right) - \exp\left(\frac{-\alpha nF}{RT}\eta\right)\right] \quad (29.30)$$
$$= i_0(10^{-\eta/b_a} - 10^{-\eta/b_c}), \qquad E > E_{rev} \quad or \quad \eta > 0$$

and

$$i_r = i_c - i_a = i_0(10^{-\eta/b_c} - 10^{-\eta/b_a}), \qquad (29.31)$$
$$E < E_{rev} \quad or \quad \eta < 0,$$

where the Tafel constants are defined for cathodic and anodic reactions, respectively, as [47]

$$b_c = 2.303\frac{RT}{\alpha nF} \quad and \quad b_a = 2.303\frac{RT}{(1-\alpha)nF}. \quad (29.32)$$

For mass-transfer-limited corrosion, the mass-transfer-limited corrosion current density, i_L, can be calculated from [48]

$$i_{\lim(i)} = nk_{m,i}F \times c_i, \qquad (29.33)$$

where n is the charge transfer during the electrochemical reaction, c_i is the concentration, and $k_{m,i}$ is the mass-transfer coefficient for species i, which can be obtained from [48]

$$Sh = \frac{k_{md}}{D} = constant \times Re^x Sc^y \qquad (29.34)$$

where Re is the Reynolds number and Sc is the Schmidt number [48],

$$Re = \frac{dv\rho}{\mu}, \qquad (29.35)$$

$$Sc = \frac{\mu}{\rho D}. \qquad (29.36)$$

The overall cathodic and anodic currents should be balanced when steady state is reached [47]:

$$\sum i_a = \sum i_c. \qquad (29.37)$$

See [47][48] for explanations of the meanings of the various symbols.

29.20 References

[1] **Massachusetts Institute of Technology**, 2006, *The Future of Geothermal Energy: Impact of Enhanced Geothermal Systems (EGS) on the United States in the 21st Century*, Cambridge, MA, Massachusetts Institute of Technology.

[2] **B. D. Green** and **R. G. Nix**, 2006, *Geothermal – The Energy under Our Feet: Geothermal Resource Estimates for the United States*, Report NREL/TP-840–40665, NREL.

[3] **L. Blodgett** and **K. Slack**, 2009, *Geothermal 101: Basics of Geothermal Energy Production and Use*, Washington, DC, Geothermal Energy Association.

[4] **J. W. Lund**, 2004, "100 years of geothermal power production," *GHC Bull.*, 11–19.

[5] **L. McLarty** and **M. J. Reed**, 1992, "The U.S. geothermal industry: three decades of growth," *Energy Sources*, **14**, 11–19.

[6] **DOE**, Office of Energy Efficiency and Renewable Energy, 2004, *Geothermal Technologies Program*, pp. 4–5.

[7] **National Renewable Energy Laboratory**, 2008, *Geothermal Tomorrow 2008*, Report DOE/GO-102008–2633.

[8] **R. Bertani**, 2009, "Geothermal energy: an overview on resources and potential, in *International Geothermal Days*, Conference & Summer School, Slovakia.

[9] **GEA news release**, 2010, *US Maintains Geothermal Power Lead; Germany Fastest Growing, New Industry Report Concludes.*

[10] **L. Rybach**, 2007, Geothermal sustainability, *GHC Bull.*, 1–7.

[11] **A. Kagel**, **D. Bates**, and **K. Gawell**, 2007, *A Guide to Geothermal Energy and the Environment*, Washington, DC, Geothermal Energy Association.

[12] **A. Kagel**, 2006, *A Handbook on the Externalities, Employment, and Economics of Geothermal Energy*, Washington, DC, Geothermal Energy Association.

[13] **Energy Information Administration**, Office of Coal, Nuclear, Electric and Alternate Fuels, US Department of Energy, 2001, *Background Information and 1990 Baseline Data Initially Published in the Renewable Energy Annual 1995 Revised, Updated and Extended with the Renewable Energy Annual 1996*, DOE/EIA-0603(95), http://www.eia.doe.gov/cneaf/solar.renewables/renewable.energy.annual/backgrnd/tablecon.htm.

[14] **J. Kubiak**, **J. G. González**, **F. Z. Sierra** *et al.*, 2005, "Failure of last-stage turbine blades in a geothermal power plant," *J. Failure Analysis Prevention*, **5**(5), 26–32.

[15] **R. McAlpin** and **P. F. Ellis II**, 1980, *Failure Analysis Report Geothermal Steam Muffler Diffusers The Geysers*, California, Report DCN# 80-212–003–11.

[16] **D. W. Shannon**, 1975, *Economic Impact of Corrosion and Scaling Problems in Geothermal Energy Systems*, Report BNWL-1866 Uc-4.

[17] **K. S. Pitzer**, 1991, "Ion interaction approach: theory and data correlation," in *Activity Coefficients in Electrolyte Solutions*, 2nd edn., ed. **K. S. Pitzer**, Boca Raton, FL, CRC Press pp. 75–154.

[18] **N. Møller** and **J. H. Weare**, 2008, *Enhanced Geothermal Systems Research and Development: Models of Subsurface Chemical Processes Affecting Fluid Flow*, Technical report.

[19] **W. L. Bourcier**, **M. Lin**, and **G. Nix**, 2005, *Recovery of Minerals and Metals from Geothermal Fluids*, Report UCRL-CONF-215135.

[20] **P. F. Ellis II**, 1985, *Companion Study Guide to Short Courses on Geothermal Corrosion and Mitigation in Low Temperature Geothermal Heating Systems*, DCN 85–212–040–01.

[21] **S. Nešić**, 2007, "Key issues related to modeling of internal corrosion of oil and gas pipelines – a review," *Corrosion Sci.*, **49**, 4308–4338.

[22] **E. Bardal**, 2004, *Corrosion and Protection*, London, Springer.

[23] **J. Han**, **J. W. Carey**, and **J. Zhang**, 2011, "A coupled electrochemical–geochemical model of corrosion for mild steel in high-pressure CO_2–saline environments," *Int. J. Greenhouse Gas Control*, doi:10.1016/j.ijggc.2011.02.005.

[24] **V. Vacquier**, 1998, "A theory of the origin of the Earth's internal heat," *Tectonophysics*, **291**(1–4), 1–7.

[25] **M. Kaltschmitt**, **W. Streicher**, and **A. Wiese**, 2007, *Renewable Energy, Technology, and Environment*, Berlin, Springer.

[26] **X. Xie**, **K. K. Bloomfield**, **G. L. Mines**, and **G. M. Shook**, 2005, *Design Considerations for Artificial Lifting of Enhanced Geothermal System Fluids*, Report INL/EXT-05–00533.

[27] **Y. Polsky**, **L. Capuano Jr.**, **J. Finger** *et al.*, 2008, *Enhanced Geothermal Systems (EGS) Well Construction Technology Evaluation Report*, Report SAND2008–7866.

[28] **T. Sugama**, 2006, *Advanced Cements for Geothermal Wells*, Report BNL-77901–2007-IR.

[29] **A. Wagh**, 2004, *Chemically Bonded Phosphate Ceramics: Twenty-First Century Materials with Diverse Applications*, Amsterdam, Elsevier.

[30] **CSI Technologies**, 2007, *LLC, Supercement for Annular Seal and Long-Term Integrity in Deep, Hot Wells "DeepTrek,"* DOE Award No. DE-FC26–06NT41836.

[31] **D. W. Brown**, 2000, "A hot dry rock geothermal energy concept utilizing supercritical CO_2 instead of water," in *Proceedings of the Twenty-Fifth Workshop on Geothermal Reservoir Engineering*, Stanford University, pp. 233–238.

[32] **A. D. Atrens**, **H. Gurgenci**, and **V. Rudolph**, 2009, "CO_2 Thermosiphon for competitive geothermal power generation," *Energy Fuels*, **23**, 553–557.

[33] **K. Pruess**, 2006, "Enhanced geothermal systems (EGS) using CO_2 as working fluid – a novel approach for generating renewable energy with simultaneous sequestration of carbon," *Geothermics*, **35**(4), 351–367.

[34] **J. Han**, **J. W. Carey**, and **J. Zhang**, 2010, "Assessing the effect of cement–steel interface on well casing corrosion in aqueous CO_2 environments," in *Ninth Annual Conference on Carbon Capture & Sequestration*, Pittsburg PA, paper no. 361.

[35] **T. J. Gary** and **O. K. Gashus**, 1972, *Tidal Power*, New York, Plenum Press (cited in **K. Burman** and **A. Walker**, 2009, *Ocean Energy Technology Overview*, DOE/GO-102009–2823).

[36] **R. Pelc** and **R. M. Fujita**, 2002, "Renewable energy from the ocean," *Marine Policy*, **26**, 471–479.

[37] **J. Bard**, 2008, *Ocean Energy Systems Implementing Agreement. An International Collaborative Programme*, IEA-OES, http://www.iea.org/work/2008/neet_russia/Jochen_Bard.pdf.

[38] **US Department of the Interior Minerals Management Services**, 2006, *Ocean Current Energy Potential on the U.S. Outer Continental Shelf.*

[39] **R. Murray**, 2006, *Review and Analysis of Ocean Energy Systems Development and Supporting Policies*, International Energy Agency.

[40] **R. Bedard**, **M. Previsic**, and **G. Hagerman**, 2007, "North American ocean energy status March 2007,"

in *Tidal Power*, vol. **8**, Electric Power Research Institute.

[41] **W. Musial**, 2008, *Status of Wave and Tidal Power Technologies for the United States*, Technical Report NREL/TP-500–43240.

[42] **Federal Energy Management Program (FEMP)**, 2009, *Ocean Energy Technology Overview*, DOE/GO-102009–2823.

[43] **O. M. Phillips**, 1977, *The Dynamics of the Upper Ocean*, 2nd edn., Cambridge, Cambridge University Press.

[44] **B. Kirke**, 2006, *Developments in Ducted Water Current Turbines*, http://www.cyberiad.net/library/pdf/bk_tidal_paper25apr06.pdf.

[45] **Ocean thermal energy conversion**, http://en.wikipedia.org/wiki/Ocean_thermal_energy_conversion.

[46] **W. Moskwa**, 2009, Norway Opens World's First Osmotic Power Plant, Reuters, http://www.reuters.com/article/idUSTRE5AN20Q20091124.

[47] **J. O'M. Bockris**, **A. K. N. Reddy** and **M. Gamboa-Aldeco**, 2000, *Modern Electrochemistry 2A: Fundamentals of Electrodics*, 2nd edn., Dordrecht, Kluwer Academic/Plenum Publishers.

[48] **D. Landolt**, 2002, "Introduction to surface reactions: electrochemical basis of corrosion," in *Corrosion Mechanisms in Theory and Practice*, 2nd edn., ed. **P. Marcus**, New York, Marcel Dekker, Inc.

30 Wind energy

Michael Robinson, Neil Kelley, Patrick Moriarty, Scott Schreck, David Simms, and Alan Wright

NREL's National Wind Technology Center, Golden, CO, USA

30.1 Focus

During the last 30 years, wind energy technology has emerged as the leading renewable alternative to electrical power production from fossil fuels. Commercial development and deployment, driven by lower capital costs, technical innovations, and international standards, continue to facilitate installed capacity growth at a rate of 30%–40% per year worldwide [1]. Utility-class machines exceed 2 MW, with robust designs providing 95%–98% availability. Future technology advances will focus on lowering the cost of land-based systems and evolving next-generation technology for ocean deployments in both shallow and deep water.

30.2 Synopsis

Wind energy technology is poised to play a major role in delivering carbon-free electrical power worldwide. Advanced technology and manufacturing innovations have helped the cost of wind energy drop from $0.45 per kW·h 30 years ago to $0.05–$0.06 per kW·h, thus positioning wind energy to be directly competitive with fossil-fuel power generation. In 2009, wind technology accounted for 39% of all new electrical generation in the USA [2]. Worldwide, wind deployment continues to penetrate new markets, with power-plant installations spanning months instead of years. In the European Union, cumulative wind power capacity increased by an average of 32% per year between 1995 and 2005, reaching 74,767 MW by the end of 2009 [3]. The USA leads the world in total installed capacity, while India and China are emerging as major potential markets. Wind energy can no longer be considered European-centric and has become an international alternative to fossil-fuel power generation.

The maturation of wind technology has been driven by numerous factors, including significant R&D investments internationally, system integration and design optimization, and the adoption of international design standards. Today's technology is dominated by three-bladed, upwind platforms capable of energy-extraction efficiencies approaching 90% of theoretical limits and availabilities exceeding 98%. Utility installations utilizing 2.5-MW turbines in land-based systems are routine. Turbines mounted on towers more than 80 m in height and configured with rotors spanning 90 m in diameter have proven ideal in deployments where agriculture and power production coexist in rural locations that support wind plants ranging in size up to 500 MW.

Investments in R&D are moving away from improving individual turbine performance to addressing the greater challenge of optimizing wind-farm performance in both land and offshore applications. Multi-megawatt wind farms arranged in multiple arrays, often located in complex terrains, offer a plethora of challenges. Quantifying wind-resource modifications, inter-array and downwind microclimatology environments, and turbine performance in highly turbulent flow fields requires advanced computational modeling capabilities validated with field measurement campaigns. The goals are to accurately predict wind-plant power generation prior to development, to achieve 25-year wind-turbine operational performance, and to attain improved reliability at lower capital costs.

30.3 Historical perspective

Extracting energy from the wind has changed very little in principle since the mid 1800s when rotors constructed from wood and canvas were geared mechanically to power millwheels and drive belts for an array of applications. The mid 1900s brought fundamental changes to wind technology through advances in materials, improved engineering methods, and a better understanding of the wind resource. Often overlooked are the substantive changes made in just the past 20 years. Wind-turbine technology actually downsized from multi-megawatt prototypes developed in the 1980s to smaller, less risky commercial machines that helped to establish today's modern industry. From humble roots, commercial turbines have once again evolved into the multi-megawatt turbines used by utilities in wind farms. An extensive historical review of wind-turbine technology, from ancient times to present day, is given by Spera [4].

Since the 1980s, average wind-turbine diameters have grown, almost linearly over time, with a substantial number of new designs rated at 2 to 2.5 MW and rotor diameters of 90 to 100 m (Figure 30.1). Each new and larger wind-turbine design was predicated on achieving the ultimate in cost and reliability. However, each larger design also demonstrated a reduction in life-cycle cost

of energy. Most of the energy-capture enhancements can be attributed to larger rotors requiring taller towers. The more energetic wind from wind shear increases both the wind-turbine capacity factor and energy production. Optimization is achieved from a balance between enhanced energy capture and the increased costs associated with larger rotor and tower structure.

A natural limitation on turbine size, from a cost perspective, is based on the simple assumption of a "square-cube law." Energy output increases as the rotor-swept area (length squared) increases, while the total turbine mass (material weight) increases as the length cubed. At some optimal size, the material costs grow faster than the resulting energy-capture potential, thereby making the economics unfavorable for further growth in size. Engineers have successfully avoided this limitation with innovations in design, removing material or using material more efficiently to trim additional weight and cost. These changes have often resulted in more dynamically active turbines due to reduced stiffness and a need for active rather than passive control schemes. The enhanced knowledge gained through an understanding of aeroelastic loads, use of advanced finite-element models, and calculations of structural-dynamics codes drove the evolution of commercial wind turbines to the robust, less expensive models we see today.

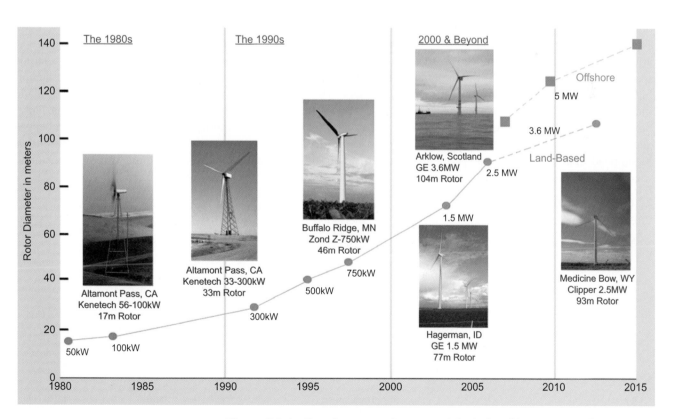

Figure 30.1. Development of commercial wind turbines. Images from NREL historical publications.

30.4 Wind energy technology

30.4.1 Wind-turbine technology

Integrated turbine systems

Modern utility-class wind turbines function as integrated systems comprised of highly optimized components that produce the lowest possible cost of energy (COE). Designs have become more or less standardized as three-bladed, upwind platform configurations based on market acceptance and optimized engineering. Important design criteria include performance, reliability, operation, maintenance, capital cost, and transportation and logistics when designing new machines or modifying existing machines. Figure 30.2 shows the major elements of standard upwind turbine architectures.

Energy is extracted aerodynamically through the rotor/inflow interaction. The ultimate efficiency of conversion from wind momentum to mechanical torque is determined by the aerodynamic design of the rotor blades. Blade wind loading is transferred as torque through the rotor hub to the main or low-speed shaft. The rotor allows individual blades to pitch, thereby modifying the rotor performance for turbine control. The integrated blade–rotor–pitch system determines the ultimate turbine performance and energy-capture potential.

The low-speed shaft transfers torque from the rotor to the drivetrain assembly. Rotors on conventional multi-megawatt designs rotate at speeds less than 20 rpm in order to maintain acceptable acoustic emission levels. The gearbox, usually a three-stage planetary configuration, increases the rotation speed to the 1,000–1,800-rpm range, and is coupled to the generator through a high-speed shaft. Variations in drivetrain designs include direct-drive machines that eliminate the gearbox as well as multiple generators closely coupled to a gearbox with fewer stages. A disk brake located on the high-speed shaft is designed to stop the rotor in an emergency and to lock the rotor during routine maintenance. Full-power braking of megawatt machines is done by blade pitching, since using the brake under such conditions would ruin the gearbox.

Generator and grid interconnection systems utilize various design approaches. Generators range from 400 to 690 V, operate at 50 and 60 Hz, and are designed with both synchronous and induction architectures. Turbine-to-wind-farm grid connections vary, and can include direct coupling of the generator to the grid or coupling through power electronics. Wind-farm distribution/collection grids vary from 12 kV for older systems to 34.5 kV for more modern installations. A step-up transformer, usually located at each turbine base, provides the appropriate interconnection voltage for the turbine power system and collection. Likewise, power within a farm is collected at one or more substations, wherein the wind-farm grid voltage is increased to accommodate utility transmission voltages between 110 and 765 kV.

Blades

Horizontal-axis wind-turbine (HAWT) blades interact with the local inflow (wind inflow combined with relative wind caused by blade rotation) to generate aerodynamic forces that are converted to shaft torque. It is important to note that blade aerodynamic force production relies on maximal lift force coupled with minimal drag force [5]. To optimally achieve this objective under operating conditions unique to wind turbines and to maximize energy capture, airfoil families have been specially designed for use in HAWT blades [6][7][8].

Although two-dimensional airfoil design and analysis has matured to a large degree, accurate prediction remains elusive for aerodynamic loads produced by blade flow fields under generalized operating conditions. Flow-field complexity and the associated modeling challenges arise due to the prevalence of rotational augmentation and dynamic stall, which entail three-dimensional,

Figure 30.2. A modern upwind turbine configuration.

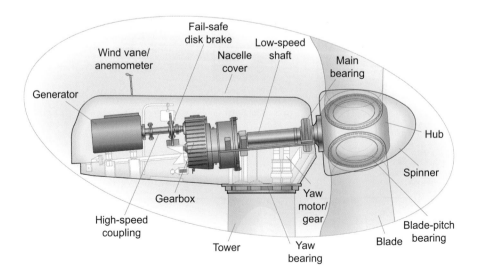

Figure 30.3. An unsteady, separated flow field similar to those produced by operating HAWT blades (from [10]).

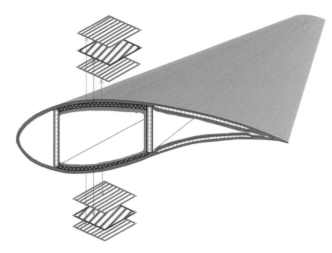

Figure 30.4. Typical blade structure, showing semi-monocoque internal structure, fiber layers, and blade twist and taper with radial location.

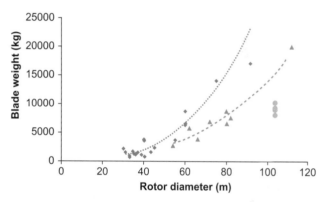

- ● Four preliminary designs from this work
- ◆ Earlier designs from WindStats
- ▲ Latest designs from WindStats
- ··· Power (earlier designs from WindStats)
- -- Power (latest designs from WindStats)

Figure 30.5. Advanced designs and stronger materials have slowed blade weight growth with increasing length (from [5]). Dashed lines are polynomial fits to data points.

unsteady, separated fluid dynamics in a rotating reference frame [5][9][10]. Flows similar to that visualized in Figure 30.3 occur routinely on wind-turbine blades, prompting short-lived though intense aerodynamic loads that can severely curtail blade structural life.

In addition to carrying the aerodynamic loads generated across the turbine operating envelope, blades also must comply with weight limits imposed by turbine design weight and cost budgets. Accordingly, blade geometry has evolved to semi-monocoque structural configurations, in which the skin and internal members share the structural load. From root to tip, blades twist and taper with increasing radius, as shown in Figure 30.4.

These structures generally are fabricated using glass and fiber composite layups, which allow the introduction of twist and taper, balancing strength-to-weight ratio, stiffness, and fatigue life with cost. Though superior from the standpoint of structural properties, carbon-fiber composite layups are more expensive, and so must be used judiciously to contain blade cost. As shown in Figure 30.5, blade growth for larger wind turbines has been facilitated by improved structural efficiency

and advanced materials, which have enabled longer blades with lower penalties in weight and cost [11].

Rotors

The rotor assembly is the most critical turbine component, being responsible for overall machine performance, including wind-energy capture and turbine control. As shown in Figure 30.6, wind-turbine operation is divided into three performance regions. Region 1 falls below the wind velocity required to overcome frictional and other parasitic losses to produce net positive power. In region 2, the variable-speed turbine rotors operate at the optimum aerodynamic efficiency and extract the maximum wind energy possible for a specific rotor

Figure 30.6. Schematic of turbine power vs. wind speed.

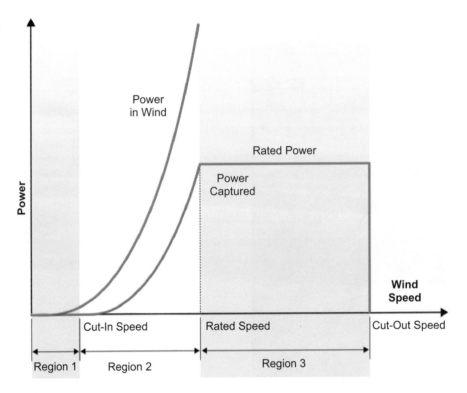

design. Turbine rated power is defined as the upper limit that the machine is capable of delivering. This limit is usually set by the electrical component ratings of the generator and/or power electronics. The rated speed point is the speed at which the turbine transitions from variable to rated, or constant, speed. In region 3, rated power is maintained by maintaining the rotor's rotational speed nearly constant and actively pitching the blades to shed excess energy. In high winds, the turbine shuts down at the rated cut-out speed, a limit set to avoid tower buckling failure from excessive rotor thrust loads or rotor–tower impact from excessive blade deflections.

The total power contained in the inflow and the turbine power delivered by the rotor can be calculated as follows:

$$P_{wind} = \frac{1}{2}\rho AV^3, \tag{30.1}$$

$$P_{rotor} = C_p\left(\frac{1}{2}\rho AV^3\right), \tag{30.2}$$

where ρ is the local air density, A is the area swept by the rotor, V is the wind velocity, and C_p is the rotor power coefficient. As shown in Figure 30.6, the actual wind power available increases as the third power of the wind speed and greatly exceeds the turbine's power-conversion capability. The difference between resource availability and turbine delivery is a function of the theoretical extraction or Lanchester–Betz limit and the actual aerodynamic performance of the turbine.

The Lanchester–Betz limit as explained by Bergey [12] can be derived from either the fluid-mechanics energy or momentum equations and has a value of 16/27 (0.593). This theoretical limit establishes the upper energy-extraction potential of an idealized rotor immersed in air flowing at constant speed. The first "reality check" of actual rotor performance is to verify that the rotor power coefficient (C_p) does not exceed the Lanchester–Betz limit.

Performance is quantified by the rotor power coefficient (C_p) as a function of the rotor blade tip speed ratio TSR (tip speed/inflow velocity; $\Omega R/V$). Figure 30.7 shows a typical rotor C_p vs. TSR performance curve that establishes the aerodynamic control envelope. All modern variable-speed rotors have an optimal C_{pmax} at one TSR_{opt} for any given pitch setting. During variable-speed operation in region 2, turbine rpm is adjusted with varying wind speed in order to maintain this optimum operating point for maximum energy capture. Operation in region 3 requires blade load (and energy) to be shed in order to operate at a constant power. The rotor accomplishes this in two ways: (1) operating at constant rpm with increasing winds reduces the rotor C_p; and (2) blade pitch control is used in region 3 to maintain constant turbine power or speed, thus limiting turbine loads.

The impact of rotor performance on turbine annual energy production (AEP) is shown in Figure 30.8. The analysis begins with a typical Weibull wind distribution (green), which reflects the probabilistic distribution of wind speed over a one-year period. Then, the total

Figure 30.7. The rotor power coefficient (C_p) vs. the tip speed ratio (TSR).

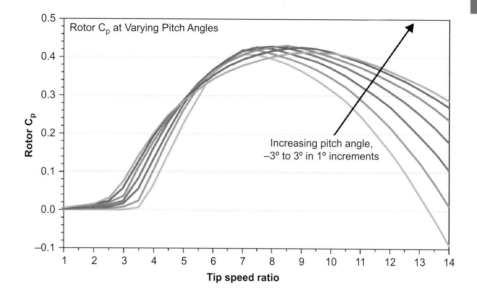

Figure 30.8 Rotor energy capture.

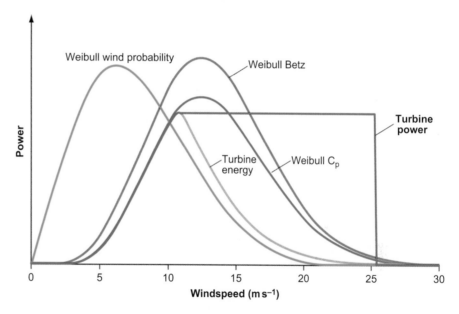

theoretical Weibull–Betz annual power (red) is obtained by multiplying P_{wind} at each wind speed by the Lanchester–Betz limit. Likewise the idealized Weibull-C_p power (purple), from a rotor operating at optimal C_p without operational limits is obtained by multiplying P_{wind} at each wind speed by C_{pmax}. The area between the two represents the energy loss from aerodynamic design. When the rated power (cut-in and cut-out operating constraints) is taken into account (blue), the actual energy produced as indicated by the area under the curve shows the additional energy reduction. Thus, the actual AEP is a function of the site-specific wind resource and turbine selection.

Drivetrain

Several generator types and interconnection configurations are used in modern wind turbines. These architectures fall into four major types, as indicated in

Figure 30.9. Type 1 utilizes squirrel-cage induction generators directly connected to the grid and operated at constant speed. Soft-starters are used to limit in-rush currents during startup. Type 2 introduces variable slip, allowing up to 10% compliance in rotor rpm during operation. Compliance helps mitigate power spiking from inflow turbulence by allowing the rpm to vary and thereby reducing rotor mechanical loads delivered to the drivetrain through the low-speed shaft.

The addition of power-conversion technology decoupled turbine rotation speed from the grid and facilitated variable-speed operation. Type 3 machines use doubly fed induction generator (DFIG) topologies that allow variable-speed operation over a limited (≈50%) rpm range. Type 4 configurations have full power conversion and variable-speed operation from zero to rated rpm. This architecture includes both induction generators and permanent-magnet synchronous generators (PMSG),

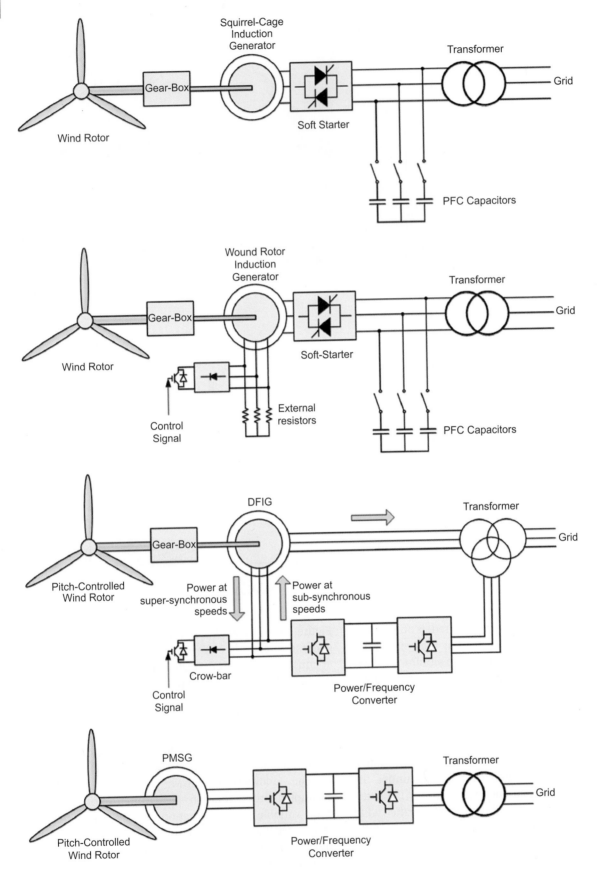

Figure 30.9. Turbine-to-grid interconnection.

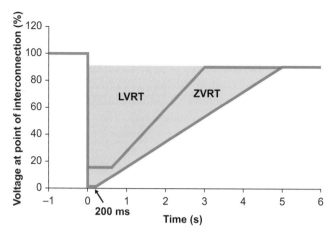

Figure 30.10. Definition of FERC low-voltage ride-through (LVRT) and zero-voltage ride-through (ZVRT) requirements.

with and without gearboxes, respectively. Type 3 and 4 architectures are the industry-preferred designs owing to both their enhanced energy-capture potential and the ability to adjust output reactive power.

Recent grid interconnection standards, including US Federal Energy Regulatory Commission (FERC) Order 661-A issued in December 2005, have had a direct influence on topology selection. The requirements for low-voltage ride through (LVRT) and the even more stringent zero-voltage ride through (ZVRT) shown in Figure 30.10 require turbines to remain online in the event of system faults. Unlike others, type 3 and 4 architectures can achieve this requirement by proper control of the power converter.

Towers

Taller wind-turbine towers position the rotor higher in the planetary boundary layer (PBL), where wind speeds are faster and the wind resource is more energetic. However, prospects for greater energy capture and associated economic advantages are hindered by challenges associated with transporting and erecting towers taller than approximately 80 m. Though these challenges are significant, lucrative prospects for enhanced energy capture and reduced cost of energy have prompted investigation of technologies to overcome these tall-tower transportation and installation challenges.

Transportation of tubular towers is critically constrained by diameter. Diameters larger than 4.4 m result in prohibitive total vehicle heights. Achieving tall tubular towers with diameters fixed at the transportation limitation is possible, but will result in less efficient material usage, increased tower costs, and increased weight. Realistically addressing this limitation is likely to prompt examination of tower-design approaches that rely on more intensive field assembly and onsite fabrication. Field assembly approaches include quartered

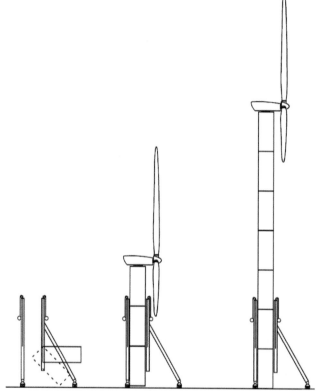

Figure 30.11. The jack-up method of self-erection for constant-diameter towers.

tubular towers, truss towers, guyed towers, and combined tripod–tubular towers [13]. Onsite fabrication methods could encompass all-concrete or hybrid concrete–steel towers [14], wound composite structures, or flexible manufacturing methods for metal forming [15].

For erecting tall wind-turbine towers, economically and technologically viable alternatives to high-reach, heavy-lift cranes include various self-erection approaches. The economics of self-erection are sensitive to the wind-shear exponent, the complexity of the terrain, crane cost and availability, and other factors. According to a preliminary analysis, these factors have the potential to reduce the COE for larger turbines, particularly in complex terrain where significant disassembly of the large conventional cranes would be required in order to change turbine locations. Most concepts for self-erection of wind turbines can be grouped into four categories: (1) telescoping towers, (2) tower-climbing devices, (3) jack-up devices, and (4) secondary lifting structures [16][17]. Figures 30.11 and 30.12 show concepts 1 and 3.

30.4.2 Integrated system control

Control strategy overview

A wind-turbine control system consists of sensors, actuators, and a system that ties these elements together

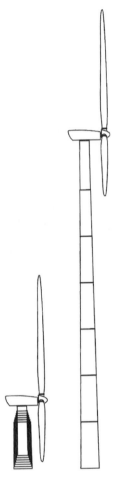

Figure 30.12. The telescoping method of self-erection for tapered towers.

(Figure 30.13). The main goal of the controller is to modify the operating states of the turbine: to maintain safe turbine operation, maximize power, mitigate damaging fatigue loads, and detect fault conditions. A supervisory control system performs starting and stopping of the machine, yawing the turbine when there is a significant yaw misalignment, for detecting fault conditions, for performing emergency shut-downs, and for dealing with similar conditions.

Wind turbines have employed different control actuation and strategies to maximize power in region 2 and limit power or turbine speed in region 3. Some turbines have achieved control through passive means, such as in fixed-pitch, stall-control machines. In these machines, the blades are designed so that power is limited in region 3 through blade stall. In region 2, the generator speed is fixed. Typically, control of these machines involves only starting and stopping the turbine.

To maximize power output in region 2, the rotational speed of the turbine must vary with wind speed in order to maintain a constant TSR. Currently, most large commercial wind turbines are variable-speed, pitch-regulated machines. This allows the turbine to operate at near optimum C_p and maximize power over a range of wind speeds in region 2. Blade pitch control is used in region 3 to limit power or keep the turbine speed constant.

Actuator technologies

There is a limited number of types of actuators available in most modern turbines for control purposes. These concern blade pitch, generator torque, and nacelle yaw.

Control actulators

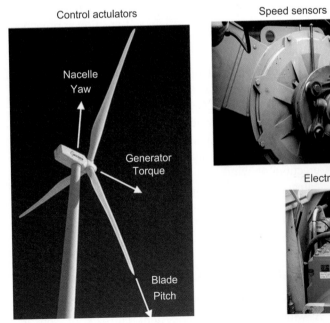

Speed sensors

Accelerometers

Anemometers and wind vanes

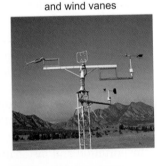

Electrical power sensors

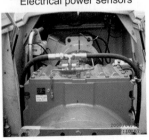

Strain gages

Figure 30.13. Wind-turbine control actuators and typical sensors.

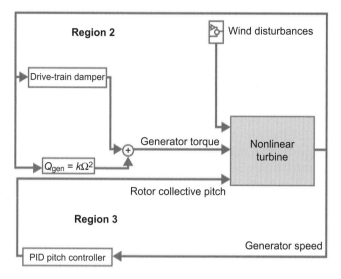

Figure 30.14 A typical industry controller.

Most large commercial wind turbines employ active yaw control to orient the machine into the wind. A yaw-error signal from a nacelle-mounted wind-direction sensor is used to calculate a control error. The control signal is usually just a command to yaw the turbine at a slow constant rate in one direction or the other. The yaw motor is switched on when the yaw error exceeds a certain amount, and is switched off when the yaw error is less than some prescribed amount.

Typical large commercial wind turbines are variable-speed machines, and control generator torque in region 2 to maximize power and pitch in region 3 to maintain constant turbine power. The controls for such machines are typically designed using simple classical control design techniques such as proportional-integral-derivative (PID) control for pitch regulation in region 3 [18]. A typical controller for such a machine is shown in Figure 30.14. Generator torque is controlled using $Q_{gen} = k\Omega^2$ in region 2 as shown in the upper control loop of the figure. The measured control input is usually the generator speed. In region 2, blade pitch is held constant.

The lower loop in Figure 30.14 shows the region 3 pitch control. Classical PID control design techniques are typically used to design the blade-pitch controller for region 3, as explained in [18]. Referring to the lower loop in Figure 30.14, the generator or rotor speed is measured and passed to the pitch controller. The goal is to use PID pitch control to regulate turbine speed in the presence of wind-speed disturbances. The expression for the blade-pitch command is

$$\Delta\theta(t) = K_P \, \Delta\Omega(t) + K_I \int \Delta\Omega(t)\mathrm{d}t + K_D \, \Delta\dot{\Omega}(t), \quad (30.3)$$

where K_P represents the proportional feedback gain, K_I the integral feedback gain, and K_D the derivative feedback gain. $\Delta\Omega$ is a generator or rotor rotational speed error (formed by subtracting the instantaneous speed

from the desired speed set-point), and $\Delta\theta$ is the commanded blade pitch. The values for the gains K_P, K_I, and K_D give the "desired" tracking of rotor speed to a desired "set-point," and closed-loop stability characteristics must be determined by classical control design methods and verified through turbine simulation and field tests.

In region 3, blade-pitch actuation can be performed either collectively, in which case each blade is pitched by the same amount, or independently, in which case each blade is pitched individually. Collective pitch is usually effective for regulating turbine speed in the presence of wind disturbances, which engulf the entire rotor (i.e., are uniform across the rotor disk). Independent blade pitch is useful for mitigating the effects of wind variations across the rotor disk, such as those due to steady wind shear or turbulent wind fluctuations [19]. For example, in the presence of a vertical wind shear, the pitch angle of a blade when it is straight up must be larger than when it is straight down in order to minimize cyclic blade loads due to shear.

The wind inflow displays a complex variation of speed and character across the rotor disk, depending on atmospheric conditions. As turbine rotor sizes increase, these localized inflow structures cause the loads to vary dramatically and rapidly along the rotor blade. Pitching the entire rotor blade might not optimize load mitigation. For rapid and localized wind-speed variations across the rotor disk, more localized fast-acting blade actuators located at different blade-span locations may be needed. New sensors are also needed, in order to measure the flow at different span-wise blade positions.

Some of the innovative actuators include such devices as trailing-edge flaps, micro-tabs, and adaptive trailing-edge devices. New sensors being investigated include localized flow-measuring devices and embedded fiber-optic sensors. The goal is to develop "smart" rotor blades with embedded sensors and actuators that control local aerodynamic effects. Recent advances in this controls technology are described in [20]. Future research will include further development of these light-weight actuators and sensors, and demonstration of their effectiveness through wind-tunnel and full-scale tests.

Simple classical controllers can be used to regulate turbine speed or power in region 3. Other control objectives include regulation in the presence of stochastic wind disturbances, accounting for turbine flexibility, mitigating loads and deflections. When additional control objectives must be met, additional control loops must be added to the standard controllers mentioned in the last section. These additional loops can destabilize the machine, if they are not carefully designed. Figure 30.15 shows some of the turbine flexible modes that can be destabilized.

(a)

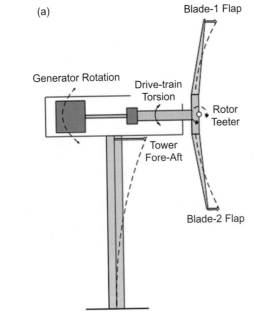

(b)

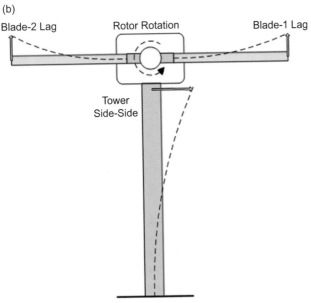

Figure 30.15. Typical flexible modes of a wind turbine.

Modern control architectures

Modern control designs using state-space methods more adequately address these issues, because the controller uses a model to determine system states. Controllers can be designed not only to maximize power or regulate speed but also to add damping to important flexible modes. Using all available turbine actuators in a single control loop to maximize the load-alleviating potential is advantageous. Such advanced control algorithms are designed on the basis of linear time-invariant wind-turbine models [21]. These linear models can be represented in the form

$$\Delta\dot{\mathbf{x}} = \mathbf{A}\,\Delta\mathbf{x} + \mathbf{B}\,\Delta\mathbf{u} + \mathbf{B}_d\,\Delta\mathbf{u}_d,$$
$$\Delta\mathbf{y} = \mathbf{C}\,\Delta\mathbf{x} + \mathbf{D}\,\Delta\mathbf{u} + \mathbf{D}_d\,\Delta\mathbf{u}_d, \qquad (30.4)$$

where $\Delta\mathbf{x}$ is the state vector, $\Delta\mathbf{u}$ is the control input vector (blade pitch or generator torque), $\Delta\mathbf{u}_d$ is the disturbance input vector (such as uniform or turbulent wind fluctuations), and $\Delta\mathbf{y}$ is the measured output (such as generator speed, tower-top acceleration, blade bending moments). $\mathbf{A}$ represents the state matrix, $\mathbf{B}$ represents the control input gain matrix, $\mathbf{B}_d$ represents the disturbance input gain matrix, $\mathbf{C}$ relates the measured output $\Delta\mathbf{y}$ to the turbine states, $\mathbf{D}$ relates the measured output to the control input, and $\mathbf{D}_d$ relates the measured output to the disturbance states. $\Delta\dot{\mathbf{x}}$ represents the time derivative of $\Delta\mathbf{x}$.

Full-state feedback is used in advanced wind-turbine control to actively damp various turbine modes [21]. The control law is formulated as $\Delta\mathbf{u}$, a linear combination of the turbine states:

$$\Delta\mathbf{u}(t) = \mathbf{G}\,\Delta\mathbf{x}(t), \qquad (30.5)$$

where $\mathbf{G}$ is the "gain" matrix. State estimation is used to estimate the states needed in these models from selected feedback measurements, eliminating the need to measure every state contained in (Equation 30.5).

Disturbance-accommodating control (DAC) is a way to reduce or counteract persistent disturbances [22] using full-state feedback. Its basic idea is to augment the usual state-estimator-based controller to re-create disturbance states through an assumed-waveform model; these disturbance states are used as part of the feedback control to reduce (accommodate) or counteract any persistent disturbance effects, such as turbulent wind fluctuations.

Another issue in wind-turbine control design is periodicity of the wind-turbine dynamic system. For a real wind turbine, the state matrices ($\mathbf{A}$, $\mathbf{B}$, $\mathbf{B}_d$, etc.) in Equation (30.4) are not really constant, but vary as the blade rotates. In [23], methods to develop controllers that have periodic gains using time-varying linear–quadratic regulator (LQR) techniques are shown for a two-bladed teetering hub turbine operating in region 3.

Field testing the advanced controllers is the real proof of control-system performance. In [24], multivariable controls were tested in the two-bladed CART. Through implementation and field tests, researchers have shown that generator torque and blade-independent pitch can be used together in a single multiple-input, multiple-output (MIMO) control loop to add active damping to the tower modes and mitigate the effects of shear across the rotor disk [24]. Figure 30.16 shows a comparison between measured field-test results from a state-space controller and a simple PID controller for region 3. Results are normalized with respect to baseline PID controller results. As can be seen in Figure 30.16, the advanced state-space controller reduces fatigue loads in each component significantly compared with the standard region 3 PID controller.

Figure 30.16. Advanced control field-test results.

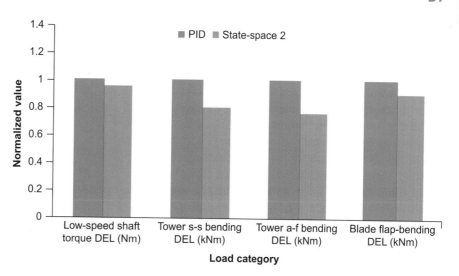

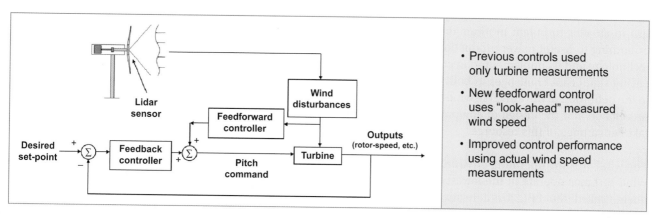

Figure 30.17. Future feedforward controls using "look ahead" wind-speed measuring sensors. Courtesy of Dr Lucy Pao.

The nonlinear behavior of a wind turbine can make control design difficult. For example, the aerodynamics are highly nonlinear. To address this issue, multiple controllers can be designed at different turbine operating points. As the turbine's operating point varies from one region to another, the controllers can be switched [25].

Another control technique to account for changing gains is adaptive control, in which the control gains "adapt" to changing conditions. Bossanyi [26] researched an adaptive scheme, consisting of a time-varying state estimator using optimal control, applied to take varying gains into account. Other more recent work in adaptive controls has been performed by Johnson [27], who reported using an adaptive control approach to improve energy capture in region 2. Additional work has been done by Frost et al. [28] in the use of nonlinear adaptive controls for speed regulation in region 3.

Future innovations

Currently, most control algorithms depend on measured turbine signals in the control feedback loop for load mitigation. Often, these turbine measurements are unreliable or too slow and the turbine controls must react to these complex atmospheric disturbances after their effects have been "felt" by the turbine. A big advantage in load-mitigating capability could be attained by measuring these phenomena upwind of the turbine before they are felt by the turbine rotor. The needed control actuation signals could then be prepared in advance and applied right as the inflow structure enters the turbine disk, providing a significant improvement in load mitigation.

Future controls research must explore the use of "look-ahead" wind-measuring sensors for improved turbine control. Such sensing techniques as LIDAR (light detection and ranging) and SODAR (sonic detection and ranging) [29] should be investigated for use in advanced controls. The measured wind characteristics can be used in a feedforward approach, as shown in Figure 30.17, resulting in improved load mitigation and better wind-turbine performance. Initial studies have documented some of the advantages of using LIDAR to sense the wind-shear characteristics upwind of the turbine for use in feedforward independent blade-pitch control algorithms [30][31].

30.4.3 The wind resource

Wind is the "fuel" that provides the reservoir of energy that is extracted and converted into electricity by modern wind-turbine technology. The source of the wind is the general circulation of air driven by a combination of the Sun's unequal heating of the Earth's surface and by the Earth's rotation. The distribution of this renewable resource for power production includes utilizing scales of motion ranging from the planetary dimensions to local distances of less than 1 km; however, the physical structure of wind is dynamically influenced by small, random velocity variations called turbulence. Thus, wind turbines must be designed to efficiently extract energy from scales of motion that are roughly equivalent to the overall turbine dimensions, while at the same time remaining dynamically insensitive to smaller turbulent motions that contribute little or nothing to useful power production. Modern turbine technology and its implementation currently are experiencing significant challenges in both categories, that is, in the definition of the available wind resource for power production and the design of equipment that optimally takes advantage of this resource.

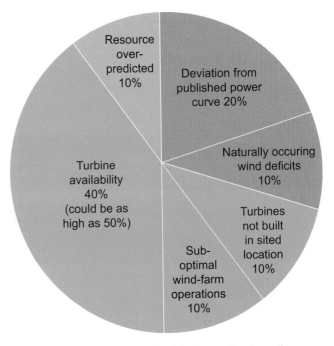

Figure 30.18. Estimated level of contribution of sources of wind-farm underproduction. Source: American Wind Energy Association.

The wind-turbine operating environment
Wind turbines operate in the lowest part of the atmosphere, called the PBL. Until relatively recently, most turbines operated in the lowest portion of the PBL, which is called the *surface layer* (SL). It is in this layer that the influence of the frictional drag induced by the ground is most significant. The SL is characterized by a strong diurnal variation in wind speed, with height and turbulence characteristics that are a consequence of heat being added to the atmosphere during the day and removed at night. The daytime or *convective* SL is generally deeper than its nocturnal counterpart, because of the intense turbulent mixing by large buoyant eddies, which results in the vertical transport (fluxes) of heat and momentum being nearly constant with height. The *nocturnal* SL is usually much shallower. At some point, the surface drag can become great enough to cause the winds above to decouple and accelerate relative to those near the surface. This creates significant velocity shears and, at times, a low-level maximum in the wind profile. The vertical distribution of temperature and buoyancy determines the behavior of the vertical air motions. The SL is considered statically *unstable* if the motions increase in intensity with height, and *stable* if they are suppressed.

As the size and height of wind turbine rotors have increased the turbines also have penetrated deeper into the PBL and what is known as the *mixed* or *residual* layer (ML). This layer extends from the height where the vertical fluxes can no longer be considered constant to the top of the PBL. This height can vary from a few

hundred meters to as much as 2 km. Most important is that the turbulence characteristics of the ML can be quite different from those in the SL. The scaling in this layer is not as well understood. The consequences of this transition are that the standard turbulence certification models in current use for turbine design are based on SL scaling and characteristics and therefore do not adequately reflect the wind and turbulence characteristics that occur in the ML.

Turbine design and operational challenges related to the atmosphere
Recent wind-farm performance studies conducted in Europe and the USA have revealed a systemic power underproduction in the neighborhood of 10% accompanied by higher than expected maintenance and repair costs. Several possible causes of power underproduction have been identified, together with their contributions [32][33][34][35], most importantly, unexpectedly low turbine availability (40%–50%), inaccurate estimation of the turbine's power curve (20%), overprediction of the local wind resource (10%), and naturally occurring variations in wind strength (10%), as shown in Figure 30.18. A commercial energy consulting firm has assembled an extensive database of the details of wind-farm design, construction, and operations based on more than 20 years of record. From this information, they have calculated annual turbine maintenance and repair costs. A typical trend for a turbine installed in 2008 is charted in Figure 30.19. For the first two years after installation, the

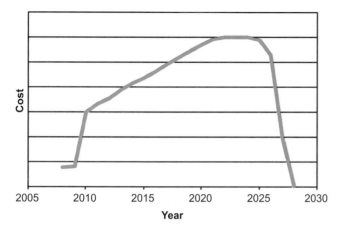

Figure 30.19. A model of the expected trends of annual maintenance and repair costs over a 20-year service lifetime for a turbine installed in 2008 based on existing wind farms and wind turbines. Source: Matthias Henke, Lahmeyer International, presented at Windpower 2008.

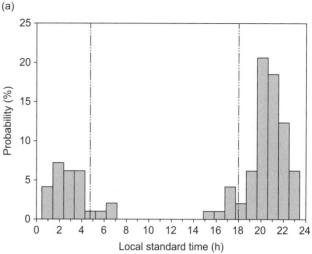

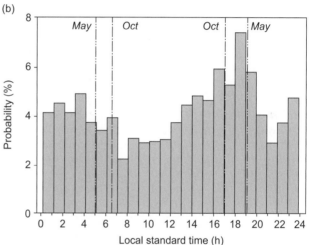

Figure 30.20. Diurnal probability variation of high blade fatigue loads for turbines: (a) in row 37 of a 41-row Southern California wind farm, and (b) at the National Wind Technology Center. In the lower panel, "May" boundaries show sunrise and sunset in May, and "Oct" boundaries show sunrise and sunset in October.

turbine is under warranty and costs are low and predictable. However, after this initial period, a cumulative cost trend that continues for the next 11 years is evident. After this time, only minor repairs are made prior to a turbine's planned replacement.

A case can be made that the majority, if not all, of the potential sources of the observed power underproduction in Figure 30.18 are related to the wind resource in which the turbines operate, and that these influence the shape of the cost curve in Figure 30.19. In fact, one interpretation of the cumulative portion of the curve in Figure 30.19 is that the current designs used for wind turbines are not fully compatible with their operating environments, which results in higher repair costs. Such an incompatibility also can be a major influence on turbine-availability issues that are believed to account for almost half of the observed wind-farm power underproduction as shown in Figure 30.18. Thus, not only is the characterization of the available wind resource important, but so is a better understanding of the role that PBL processes play in the operational aspects of wind energy technology.

Interaction of PBL turbulence with wind-turbine dynamics

Research has shown that the highest turbine-blade fatigue loads tend to occur during the day-to-night transition of the PBL and again during the early morning hours. Figure 30.20 demonstrates this diurnal variation for turbines installed in row 37 of a 41-row wind farm in Southern California and also at the National Wind Technology Center (NWTC) near Boulder, CO. The operating conditions between 20 and 23 h local standard time (LST) at the NWTC often exceeded operational

limits for the turbine, which accounts for the apparent inconsistency with the results for the multi-row wind farm. The highest levels of blade fatigue loads both in the California wind farm and at the NWTC were found to occur in just slightly stable conditions, as shown in Figure 30.21. Further investigation found that the largest loading cycles were associated with organized, or coherent, patches of turbulence that developed in a slightly stable boundary layer in the presence of significant vertical wind shear. These conditions occurred at the same time as the peak loads shown in Figure 30.20 [36][37]. How much cumulative fatigue damage occurs on a given turbine as a result of the interaction with coherent turbulence depends on how many hours the turbine operates under such conditions.

(a)

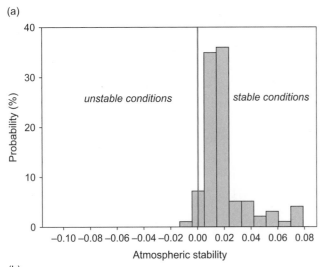

(b)

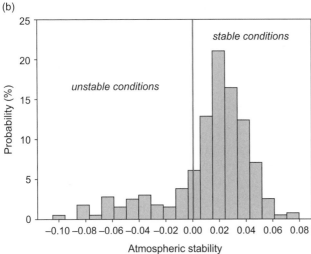

Figure 30.21. The atmospheric stability probability associated with high levels of blade fatigue loads with the red line indicating neutral conditions: (a) in row 37 of the 41-row Southern California wind farm, and (b) at the National Wind Technology Center. Atmospheric stability is quantified on the horizontal axis by the Richardson number.

Low-level jets: an atmospheric structure that produces coherent turbulent conditions

Nocturnal low-level jets (LLJs) are particularly important for wind energy applications in the Great Plains of the USA, as well as in the area surrounding the Baltic Sea and in other locations in northern Europe during the spring and summer months. Turbulence dynamics associated with a nocturnal LLJ are shown in Figure 30.22. This representation was generated from specially processed observations of a coherent Doppler LIDAR sensor on the high plains of southeast Colorado [38]. Intense coherent turbulent structures develop in the region with the most significant mean wind shear, which coincides with the height layer occupied by the 1.5-MW wind turbines installed at this location. Bursts of coherent turbulent activity were observed for periods of 10–30 minutes, two or three times between local sunset and sunrise. During these periods, the turbines were exposed to conditions that can induce significant dynamic structural responses. Figure 30.23 emphasizes the close relationship between the excellent wind resource at 80 m elevation in the Great Plains and the observed frequency of LLJs. The US National Renewable Energy Laboratory (NREL) developed a stochastic turbulence simulator, called TurbSim, that is based on extensive field measurements at this site [39][40]. It can recreate inflows similar to those observed in Figure 30.22 for use with the NREL winds-turbine design codes. This simulator also provides characteristic inflows associated with the high-turbulence environment at the NWTC and the California wind farm discussed above.

Summary

Current wind-resource-assessment and turbine-design practices do not account for the presence, frequency of occurrence, and other characteristics of atmospheric structures that can generate coherent turbulence. These structures have been shown to induce potentially damaging responses in wind turbines. Statistical studies of the productivity and operating costs of the current fleet of wind farms have shown that there is systemic power underproduction coupled with higher than expected maintenance and repair costs. Turbine availability is estimated to account for almost half of this loss, of which downtime for repairs is a major constituent. The cumulative nature of the turbine-lifetime trend in maintenance and repair costs strongly suggests that consideration of the operational environments in the design of wind turbines is deficient. One example of such a deficiency is the number of hours a particular turbine design is exposed to coherent turbulent structures generated by the frequent presence of LLJs.

30.4.4 Wind-plant technology challenges

While wind turbines are often thought of as stand-alone devices, utility-sized turbines are most often placed in large groups, known as wind farms or plants, with combined power ratings on the order of 100 MW or more. These wind plants can contain hundreds of individual turbines and may also cover hundreds of square kilometers. The design and operation of wind plants requires detailed analysis of many different variables. The challenge for developers is to optimally lay out the wind plant for a given land area such that turbines are spaced far enough away from each other to minimize power losses from turbine–wake interactions, while limiting turbine spacing to mitigate costs of roads and transmission lines. Wind-plant designers must rely on a variety of tools to estimate the various impacts of

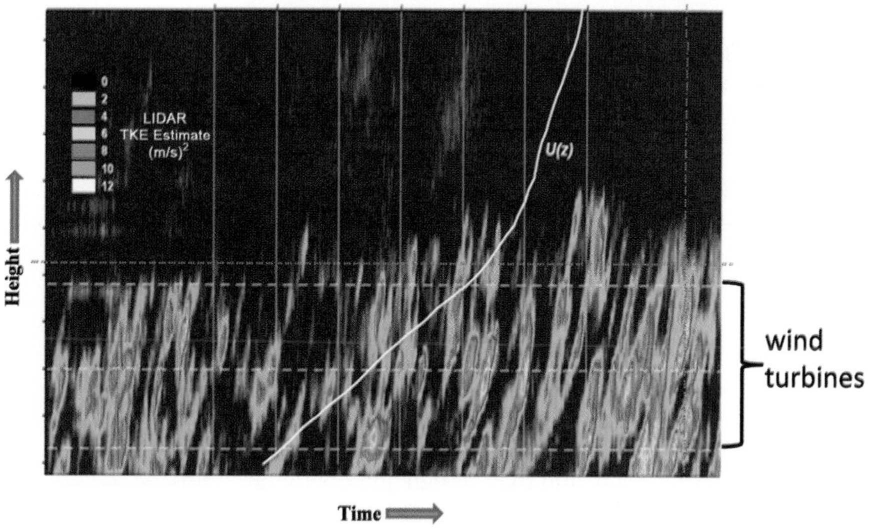

Figure 30.22. Coherent turbulent structures and the mean vertical wind-speed profile *U(z)* observed for a period of slightly less than 10 minutes with a coherent Doppler LIDAR sensor beneath a nocturnal LLJ in southeast Colorado. The dashed lines indicate the lower extent, hub, and upper extent of a General Electric 1.5-MW wind-turbine rotor installed at that location. Bright reds and yellows indicate the most intense turbulent structures. LIDAR provided by the National Oceanic and Atmospheric Administration Earth System Research Laboratory (NOAA/ESRL).

Figure 30.23. Probability of occurrence of low-level jets in the central and eastern USA overlaid on the 80-m wind-resource estimates. Criteria for jet detection are a wind-speed maximum of 16 m s^{-1} that decreases by at least 8 m s^{-1} at the next higher minimum or at an altitude of 3 km, whichever is lower. Source of LLJ data: W. D. Bonner, 1968, "Climatology of the Low-level Jet," *Monthly Weather Rev.* 96 (12) 833–850. Wind-resource estimates developed by AWS Truewind, LLC for windNavigator, http://navigator.awstruewind.com

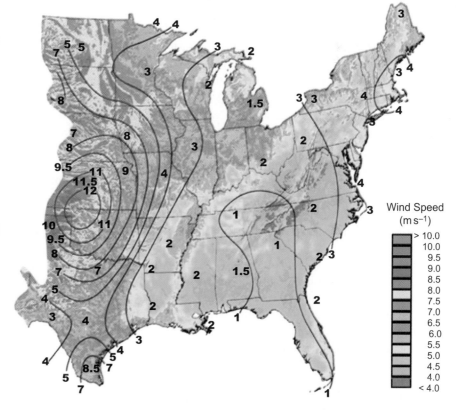

turbines, surface roughness, terrain, vegetation, and wind resource on the optimal energy production of the plant. They also must consider the potential environmental impacts of wind plants due to noise and shadow flicker, impacts on wildlife, and sociological acceptance within surrounding communities.

Technically, there are many challenges for the development of wind plants. The operation and maintenance of such plants is dependent on understanding the smaller-scale interactions between the turbines themselves and the local atmosphere. As more wind plants are constructed in close proximity to each other, they may have broader-scale impacts on the local wind resource and even the local and regional climate. These large-scale impacts and smaller-scale interactions are currently poorly understood and represent technological challenges for efficient operation of wind plants in the future. Finally, as wind energy becomes more prominent, new control and layout schemes will be needed in order to continue to lower the cost of energy from these wind plants.

Turbines as integrated components

Since the beginning of the modern wind industry in the 1980s, most wind turbines have been designed as single entities. Indeed, the international body that maintains the wind industry's certification standards for wind-turbine design only recently added specific requirements for wind-plant effects to its design standard [41]. These impacts include increased turbulence intensity inside wind plants that lead to higher design loads compared with those on turbines outside of wind plants. This shift in thinking reflects the realization that in the future the wind industry will be designing wind turbines more for operation within wind plants and less for isolated operation. In this new paradigm, turbines become components in an overall power plant, and their design and operation may change significantly to reflect that. This philosophy will present its own challenges and opportunities.

One of the more significant future research opportunities is in wind-plant control and operation. Under current methodologies, wind turbines within wind farms operate so that each individual turbine tries to maximize its own energy capture on the basis of the local wind speed. Under many conditions and depending on the plant layout, this may, or might not, result in the highest energy capture for the plant as a whole. Control systems also do not consider that turbines operating in the wake of one another will experience higher fatigue loads and, therefore, have higher maintenance costs. Improving the performance and reducing the operational costs of a wind plant, as a complete system, will require new control systems that reflect overall plant behavior. This may entail operating some turbines suboptimally to increase the overall plant performance. Recent research has shown [42] that this approach can lead to better overall performance, although testing in the field has shown less favorable outcomes.

There are many other opportunities in wind-farm design that may also optimize the wind power plant for a given site, in addition to control. The number of layout design variables may be expanded beyond turbine location to include rotor diameters, hub heights, numbers of blades, blade designs, upwind or downwind orientations, and rotational directions, all of which vary across the plant. These, of course, will likely be site-specific and dependent on quantities such as the local wind, terrain, vegetation, and even different atmospheric conditions across the plant. However, to make such optimizations possible, there is a need for greater understanding both of the wind-plant aerodynamics and of climate impacts, which are the subjects of the next sections.

Wind-plant aerodynamics

Owing to economies of scale, both wind turbine size and the number of turbines within wind plants continue to

Figure 30.24. Natural flow visualization of turbine wakes in the Horns Rev Wind Farm off the west coast of Denmark. Horns Rev 1 is owned by Vattenfall. Photograph by Christian Steiness.

grow. As the relevant scales have grown larger, the impact of wind plants on the local atmosphere and climate has increased to a point at which the aerodynamic behavior of the wind turbine/plant and the atmosphere are now fully coupled. Plant operation influences the local atmospheric physics, which, in turn, impacts the operation of individual turbines within the plant and also the interaction among turbines. Unfortunately, modeling tools and the basic understanding of these physical processes were developed when turbines were smaller and often stood alone [43] [44]. Hence many wind-turbine arrays are experiencing underperformance issues; wind farms are producing at least 5% less energy than originally expected [45]. Turbines within farms are also experiencing extremely high component failure rates [46], and observations have shown that wind turbines within wind farms sustain extreme loading levels that are 33% higher than those outside of farms [47]. These underperformance and component-failure issues result in millions of dollars less revenue per year than expected in the wind industry.

One major cause of lower energy capture and increased failure rates is turbine–wake interaction, the physics of which currently is poorly understood. Turbines in wind plants are in close proximity to one another, causing many of them to operate in the wakes of upstream turbines (see Figure 30.24). By definition, wakes are areas of lower wind velocity and also increased turbulence intensity. Thus, operation within a wake causes a decrease in power output, and an increase in mechanical loading, translating into less energy and greater maintenance costs.

Because wind-turbine plants are operating in the lower PBL, the behavior of the PBL within which the

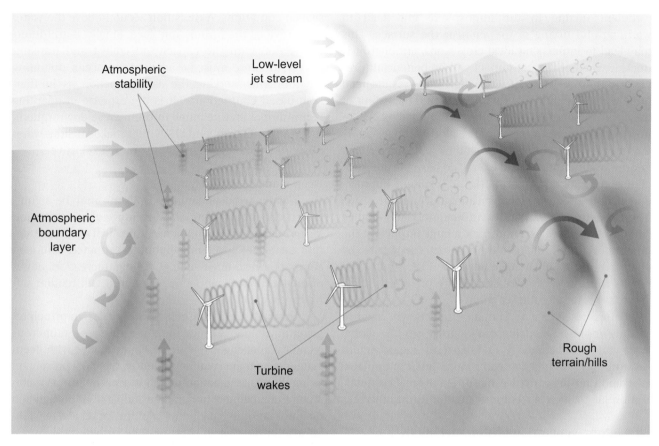

Figure 30.25. A schematic representation of wind-farm interactions with the atmospheric boundary layer, terrain, and vegetation.

array resides can have a large impact on array performance and turbine loads (see Figure 30.25) [48]. For example, an unstable, highly turbulent boundary layer can increase array performance by diffusing the turbine wakes more quickly than in stable conditions, but the atmospheric turbulence itself imposes loads on the turbines. Furthermore, stable boundary layers can produce LLJs with large wind shear across rotor planes, resulting in large asymmetric loading.

Another issue with wind-plant–PBL interaction is that large turbine plants extract enough energy from the PBL to slow the entire mean flow field, again reducing energy capture. This is similar to the impact of a forest on the PBL, where the boundary-layer thickness will grow as though the wind plant is a step change in roughness at ground level. This effect has been observed in operating offshore wind plants and some simple models have been developed to account for it [49], but a more thorough understanding of this effect is needed.

The impact of terrain is another area of uncertainty. It is well known that wind speeds increase at the tops of ridge lines, which is why wind turbines are often placed in these windiest locations, but the downwind impacts of terrain on potential wind sites are largely unknown, and could be significant [50]. These impacts

are also tightly coupled with local vegetation and diurnal changes in atmospheric stability.

The impact of vegetation could be of prime importance, but it is not well understood. Wind-farm developers have often sought to avoid sites with large amounts of tall vegetation due to increasing complexity of the flow field. However, often vegetation is unavoidable and wind plants will be built in such areas in the future. Local growth of forests or tree felling may have a long-term impact on wind-plant energy production. Therefore, further understanding of the impact of local vegetation on mean wind speeds and turbulence will continue to be important.

Macro- and micro-climate impacts

With large-scale penetrations of wind energy expected in the next 20 years [51], many questions remain unanswered as to the impact of large-scale deployment of wind energy on local, regional, and global climates. However, it is believed that the impact is significantly less than that from traditional fossil-based energy sources. This problem is particularly worrisome in the USA, where the greatest potential wind resource and the bulk of the agricultural lands coincide.

Initial modeling studies of local climate impacts [52] have shown that wind plants can enhance vertical turbulent mixing in the atmosphere and lead to warming and drying of the air near the surface. This may lead to impacts on local crop lands, although it is unclear whether the net impact will be beneficial (preventing frost) or detrimental (drying). These results need to be verified with field tests to investigate the extent of the local impact. On a regional and global scale, the impacts are more uncertain. One global impact study employing climate-model simulations [53] showed that enormous amounts of wind energy do have the ability to change turbulent transport in the atmosphere, but have a negligible effect on global-mean surface temperature. The impact of natural and man-made climate change from other sources also may have a large impact on the amount of wind energy available in the future by changing winds on a regional scale [54]. Again, the results from both of these studies are preliminary and these issues require further study.

Future research into climatic impacts will need to include improved modeling of PBL processes on a global scale, and establishment of long-term measurement networks, as was highlighted in more detail at the DOE Workshop on Research Needs for Resource Characterization [55], which is recommended for further reading.

30.4.5 Next-generation offshore technology

Offshore wind, especially deep-water deployment, is one of the next technology-development challenges. Although the cost of energy is significantly higher offshore, the close proximity to coastal load centers and excellent wind resources continues to foster development and drive deployment both domestically and internationally. The predominant market remains Europe, where population density and significant shallow water resources make offshore development attractive. In the USA, overcoming transmission constraints and higher wholesale energy rates continue to attract developers', state, and local government interest. Compared with land-based deployment, offshore development remains relatively small (≈ 4 GW) with a significant number of technical, economic, and social challenges remaining.

Shallow-water technology
Shallow-water installation (<30 m) is the only available commercial option, with Europe leading development, demonstration, and deployment. Today's offshore turbine reflects a "marinized" version of conventional architectures, including enhancement for servicing (e.g., helicopter and boat landing pads), consideration of extremely corrosive environments through design accommodations, and investigation to enhance operation and maintenance (O&M). The principal technology difference is the sea foundation, usually a monopile or gravity-based design dependent upon the local

seabed conditions. A transition section above the water line provides the turbine interface. Finally, a grid interconnection with the capacity to collect and transmit energy to shore must be adapted for ocean operating environments, serving the same function as in their land-based counterparts. It is the expense of adapting conventional architectures to the marine environment, the sea foundation, and O&M that make shallow-water offshore technology significantly more expensive.

Deep-water technology
Deep-water wind technology (>30 m), may be significant in the longer term. The advantage of locating wind farms beyond the shoreline viewshed will overcome significant issues hindering public acceptance, although competing resource use (e.g., fishing, recreation, environmental impact, and shipping lanes) will remain significantly challenging.

Although the fundamental turbine design principles governing energy extraction remain the same, numerous technical challenges (Figure 30.26) associated with deep-water deployment require resolution; the most prominent being the sea-foundation design approach. Several architectures are being considered, including fixed bottom and floating platforms, spar buoys, multiple jacket and tension-leg platforms – all impact the final turbine design in a different way. Various computational models and analysis tools are in development internationally, with code comparisons being conducted through international collaboration [56]. As these capabilities evolve and more data are collected from test prototypes [57], a more prominent technology-development pathway will emerge.

30.4.6 Test, certification, and standards

Utility-scale production of wind energy is relatively new compared with other established conventional sources of energy. In 2009, the global installed wind capacity of 158 GW accounted for 1.7% of world electricity production, compared with less than 13.6 GW 10 years earlier according to the European Wind Energy Association (www.ewea.org) and the Global Wind Energy Council (www.gwec.net). The 100-m-diameter rotors of today's 2.5-MW-rated wind turbines were half that size in the late 1990s, when typical machines were approaching 1 MW. Through the last 10 years of dramatic industry growth, international standards and test procedures for wind turbines have matured, but still have a long way to go. Independent wind-turbine-certification agencies require test data acquired by independent organizations in accordance with international standards to evaluate designs and satisfy the demands of manufacturers, developers, operators, lenders, insurance companies, safety organizations, permitting agencies, and other interested industry stakeholders.

Figure 30.26. Deep-water offshore design considerations.

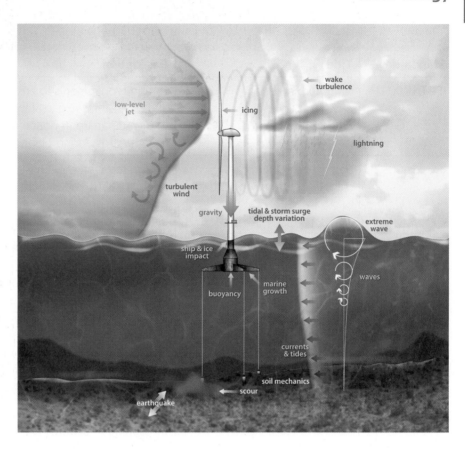

The development of testing methodologies and standards that are specific to wind turbines began about 20 years ago. They were initiated by nationally funded R&D organizations such as the US Department of Energy's National Renewable Energy Laboratory (www. nrel.gov) and similar European laboratories such as Risø (www.risoe.dtu.dk) in Denmark, the CENER (www.cener. com) in Spain, and the ECN (www.ecn.nl) in the Netherlands. Test standards have evolved in accordance with the needs of the industry and are currently at different stages of maturity. Existing standards satisfy some basic certification requirements and were crafted to avoid undue burden on the COE competitiveness of a fledgling industry; however, they are certainly in their infancy compared with those of other, more established, energy technologies. Standards' development has often been motivated by industry and investor needs for performance standardization, such as power production and noise emissions, and component-reliability issues such as mechanical loads and blade testing.

Given that the bulk of the existing fleet of wind turbines has been in operation for less than half of their intended 20-year design lifetime, it is very likely that unanticipated reliability issues have yet to be discovered. As the wind industry matures, there will undoubtedly be many opportunities to optimize and refine testing capabilities and standards accordingly.

IEC certification requirements

International standards for wind turbines are created by committees of researchers and industry members under the auspices of the ISO (the International Organization for Standardization, www.iso.org) and IEC (the International Electrotechnical Commission, www.iec.ch). National bodies that are members of the ISO or IEC participate through technical committees focusing on particular fields of wind-turbine technology. International standards are drafted in accordance with the ISO and IEC rules. Draft international standards adopted by the technical committees are circulated to participating countries for voting. Publication as an "international standard" requires approval by at least 75% of the member bodies casting a vote. Several approved international standards dictate design specifications for wind manufacturers seeking turbine certification. Many of the standards also specify testing procedures. Some IEC standards specific to wind turbines are listed in Table 30.1. In addition, many other general IEC and ISO standards apply, such as electrical, safety, and manufacturing standards.

Blade-test validation

IEC standard 61400–23 defines requirements for full-scale wind-turbine blade testing. Two types of blade testing are specified: static and fatigue. Blade static

Table 30.1. Approved international standards specific to wind turbines

Standard	Title	Description
IEC 61400–1	Wind-turbine design requirements	Design requirements to ensure engineering integrity of turbines and subsystems (e.g., safety, electrical, mechanical, support) to provide an appropriate level of protection against damage from hazards during planned lifetime
IEC 61400–2	Safety of small turbines	Design requirements and testing techniques associated with safety, quality, and integrity of small wind turbines (under 200 m^2) to provide an appropriate level of protection against damage from hazards during planned lifetime
IEC 61400–3	Design of offshore wind turbines	Specifies additional requirements for assessment of the external conditions at an offshore wind-turbine site and specifies essential design requirements to ensure the engineering integrity of offshore wind turbines
ISO 81400–4	Design and specification of gearboxes for wind turbines	Establishes the design and specification of gearboxes for wind turbines with power capacities ranging from 40 kW to 2 MW
IEC 61400–11	Acoustics	Uniform methodology to ensure consistent and accurate measurement of wind-turbine acoustic emissions
IEC 61400–12	Power performance	Uniform methodology to ensure consistent and accurate measurement of wind-turbine electrical power performance as a function of wind speed
IEC 61400–13	Measurement of mechanical loads	Guide describing methodology and techniques for the experimental determination of wind-turbine mechanical loading used for verification of codes and/or direct measurement of structural loading
IEC 61400–21	Power quality	Uniform methodology to ensure consistent and accurate power-quality measurement of grid connected wind turbines (e.g., voltage, harmonics)
IEC 61400–23	Full-scale structural testing of rotor blades	Guidelines for the full-scale structural testing of wind-turbine blades and for the interpretation or evaluation of results, including static strength tests, fatigue tests, and blade-property tests
IEC 61400–24	Lightning	Common framework for appropriate protection of wind turbines against lightning
IEC 61400–25	Communications/SCADA	Uniform communications for monitoring and control of wind power plants

proof-load testing can be accomplished by attaching the blade root to a test stand and loading the blade along its span with distributed suspended weights or winch–cable pull systems as shown in Figure 30.27. Blade dynamic-fatigue tests are designed to accelerate lifetime fatigue loading by oscillating the blade (using forced or resonant methods) over millions of cycles, each cycle at an exaggerated load deflection as shown in Figure 30.28. Static and dynamic tests can exercise the blade in the in-plane (edge) or out-of-plane (flap) direction, or in both directions simultaneously. A dual-axis forced hydraulic fatigue test is shown in Figure 30.29. Blades are extensively instrumented while undergoing testing to measure the input load and response, including deflection and strain.

Drivetrain dynamometer validation

At the time of publication, no international standard for wind-turbine drivetrain testing exists, nor are there standards for testing of individual wind-turbine drivetrain

Figure 30.27. Deflection of a cantilevered 45-m blade under distributed static load testing.

Figure 30.28. One-second time-phased photo (three composite images) of a 45-m blade undergoing laboratory resonant fatigue testing (the tip deflection is approximately 7 m).

components, such as gearboxes, generators, and power converters. Although most manufacturers require proof-testing of individual components, it is much more costly and difficult to test the entire integrated drivetrain system. Such testing necessitates the use of a specialized dynamometer test facility designed specifically for the purpose of testing wind-turbine drivetrains. Owing to the cost of producing high torque at very low rotation speeds, a challenge unique to wind turbines, there are only a few such facilities in the world, most of which are subsidized with government funding. Figure 30.30 shows an example of a wind-turbine drivetrain undergoing testing in the 2.5-MW dynamometer test facility (up to 1380 kN·m torque over the range of 16–33 rpm) at the DOE/NREL National Wind Technology Center (www.nrel.gov/wind). Sometimes non-torque loading is also applied at the low-speed shaft to simulate radial and axial rotor-induced loading.

Field performance validation

Many of the existing IEC standards define uniform testing methodologies aimed at providing consistent cost-effective measurements on full-scale turbines. For example, the requirements of a wind-turbine power performance test conducted in accordance with IEC Standard 61400–12–1 stipulate

- use of two IEC-qualified anemometers (primary and secondary) calibrated in a wind tunnel as specified in the standard
- mounting of the primary anemometer at hub-height on a meteorological tower between two and four rotor diameters from the turbine
- mounting of the anemometer above the top of the tower to minimize anemometer free-stream flow

Figure 30.29. A dual-axis forced-hydraulic fatigue test on a truncated 37-m wind-turbine blade. Note the vertically mounted hydraulic cylinders in the foreground. Direct flap loading is achieved through the hydraulic cylinder on the left; edgewise loading is achieved through the hydraulic cylinder and bell crank (yellow) on the right.

Figure 30.30. Testing of a wind-turbine drivetrain in the NWTC 2.5-MW dynamometer test facility. The test article on the left includes the bedplate (white, attached to the foundation), low-speed shaft (gray), gearbox (blue), generator (blue, far left), and power-conversion system (behind the drivetrain). The dynamometer low-speed drive is connected to the test article at the blue shaft and flexible coupling. The dynamometer table (suspended between concrete buttresses on the right) can be moved up/down and tilted to accommodate different test-article configurations.

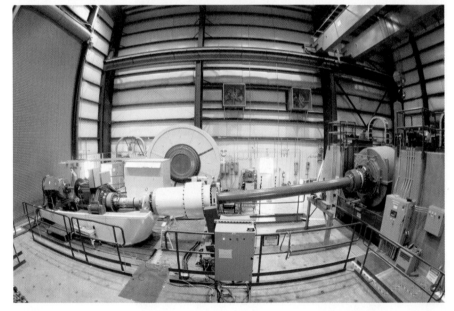

disruption induced by tower shadow as shown in Figure 30.31

- mounting of a secondary reference anemometer at least 1.5 m below the primary
- mounting of a qualified wind-direction sensor at approximately hub-height (at least 1.5 m below the primary)
- qualified barometric pressure and temperature sensors to calculate air density
- installation of qualified power transducers on the output of the power-conversion system

- use of a data-acquisition system calibrated by an accredited laboratory, with data paths calibrated via known signals injected at sensor locations
- acquisition of time-series data in wind-speed bins of width 0.5 ms^{-1} collected in 1- or 10-minute-average sets (depending on turbine size), with multiple data sets in each bin
- standardized processing of data to calculate power-curve statistics.

Specific test procedures, as defined in the standards, are the result of testing experience coupled with practical

Cost-of-energy calculation

The principal elements that govern wind COE are related in Equation (30.6) below, which expresses the COE in cents per kW·h or dollars per MW·h [4]. Common conventions include stating the COE as a "levelized" average that disregards annual variations, and adjusting dollars back to a chosen base year to exclude inflation. Also common is omitting the federal production tax credit, due to its potentially transitory nature. The equation is

$$\text{COE} = \frac{\text{FCR}\,(\text{TCC} + \text{BOS}) + \text{LRC} + \text{O\&M}}{\text{AEP}}, \quad (30.6)$$

where FCR is the fixed charge rate, TCC is the turbine capital cost, BOS is the balance of the station, LRC is the levelized replacement cost, O&M is the operations and maintenance cost, and AEP is the annual energy production. The terms are explained in the following sections.

Turbine capital costs

Turbine capital costs include those for the rotor, which comprises the blades, hub, and aerodynamic control system, as well as any labor costs for factory assembly of rotor components shipped as integrated subsystems. Also included in capital costs are the nacelle, which is made up of the main shaft, main bearings, couplings, gearbox, generator, brake subsystem, bedplate, yaw subsystem, nacelle cover, and any factory assembly costs. The TCC also encompasses the costs for the power electronics, controls, and electrical systems, again including costs for factory assembly. The tower also is categorized within the TCC, although onsite assembly costs are not captured here. Finally, the TCC also entails shipping costs (including permits and insurance), warranty costs (including insurance), and manufacturers' mark-ups (including profit, overhead, and royalties) [58].

Balance-of-station costs

The BOS consists of all costs associated with installing the wind plant and establishing the initial operating capability. Specifically, the costs involve resource assessment and feasibility studies, surveying, site preparation (grading, roads, fences), the electrical collection system and substation, wind-turbine foundations, operation and maintenance facilities and equipment, wind-plant control and monitoring equipment, and the initial spare-parts inventory. Also included are the receipt, installation, checkout, and startup of all equipment. Permits and licenses, legal fees, project engineering, project management, and construction insurance fall into the BOS, as well [58].

Figure 30.31. Atmospheric sensor configuration for an IEC Power Performance test. At the top, the IEC-qualified primary anemometer (at turbine hub-height elevation) is unobstructed by the tower structure. The lower secondary anemometer and wind-direction vane are obstructed depending on the wind direction (filtered out during data processing).

cost limitations within the capabilities of industry test providers. Although the IEC procedures help to improve the consistency of basic turbine performance or load characterization, they are not intended to be comprehensive. More detailed testing is often conducted during the development of new turbines and for research purposes.

30.4.7 The cost of energy

To be commercially viable in the competitive US market, wind energy must be cost effective for consumers and offer power producers realistic prospects for profit. Thus, reliable COE computations have been crucial to the commercial success of wind energy, and will remain so as the market continues to evolve and wind technology continues to advance. Methods for the computation of wind COE vary considerably, depending on the available data and the intended application. However, many methods have in common several of the principal elements summarized below.

Operation and maintenance costs

This category encompasses all labor, parts, and supplies for both scheduled and unscheduled turbine maintenance. Similarly, parts and supplies for maintenance of equipment and facilities are included in O&M costs, as are labor for administration and support of the wind plant. If not captured elsewhere, land-lease costs are carried in this cost category as an annual operating expense. However, property taxes and insurance generally are not allocated to O&M costs, because these are calculated separately and included in the FCR [58].

Levelized replacement costs

The LRC distributes the cost of major replacements and overhauls throughout the projected service life of wind-plant facilities. In most models, funds to cover replacements and overhauls are allocated to a reserve fund during years preceding maintenance/overhaul events. Following the event, the outlay is depreciated on a linear schedule. Thus, both the reserve-fund allocations and outlay depreciations are incorporated in the LRC [58].

Annual energy production

The AEP can be predicted prior to wind-plant construction or measured during wind-plant operation. Preconstruction estimates are prepared using wind-resource assessments, which are adjusted for turbine hub-height, and then decremented for losses due to factors like turbine operation in multiple-row arrays, blade soiling, controls and system losses, and energy dissipation in the wind-plant collection system. Energy-capture measurements are taken during operation at the collection system input for single turbines or at the substation for wind plants. In either case, the AEP is impacted by turbine availability, which is reduced by both planned and unanticipated outages [58].

The fixed charge rate

The FCR is a multiplier applied to the TCC and BOS cost to account for diverse ancillary costs like financing, financing fees, return on debt and equity, depreciation, income tax, property tax, and insurance. However, the federal production tax credit is not accounted for in the fixed charge rate, due to the potentially transitory nature of the tax credit, as explained above.

Use of the FCR in this manner greatly simplifies COE estimations, but it implies that some type of discounted cash-flow analysis has been completed in order to arrive at an FCR. Moreover, a highly accurate COE projection requires that a discounted cash-flow analysis be carried out in substantial detail. Four market-based non-utility approaches used by the wind community include project finance, balance-sheet (GenCo) finance, portfolio finance, and all-equity finance. These analysis approaches and other essentials regarding wind cost of energy analysis are documented in [59].

30.5 Summary

This chapter provided a brief overview of wind technology including COE and wind-resource considerations that played a significant role in the evolution of the technologies used in wind energy deployments worldwide. Wind-turbine maturation through innovation, advanced engineering, and manufacturing expertise has produced three-bladed, upwind architectures with 95%–98% availability and 90% energy-extraction efficiency, at costs rivaling those for fossil-fuel power generation. Future wind-technology challenges will focus on *in situ* performance from turbines operating in multi-megawatt wind farms, aligned in multiple arrays, and located in complex terrains and ocean environments. In time, new technology is expected to emerge wherein turbines, as elements of integrated systems, will optimize energy-plant performance, achieving accurate pre-installation production estimates, 25-year design life performance, and improved reliability, all at lower capital costs. See Table 30.2.

30.6 Questions for discussion

1. Since the 1980s, the cost of wind energy has decreased significantly with time, while turbine size and rated capacity have grown. What influences have wind energy technology development, economic and market factors, and simple economies of scale had on this trend?

2. What technologies from other industries (e.g., aerospace, automotive, construction, marine, ...) might be transferable to the wind energy industry? What obstacles might hinder successful technology transfer?

3. How will future wind plants be integrated with other traditional and renewable sources of electricity on the grid? What are the issues associated with co-location of other renewable sources (biomass, solar, etc.) with wind plants?

4. What is the maximum amount of energy extracted from the atmosphere by a single turbine? How much can be extracted by a large-scale wind plant? What is the estimated overall percentage of total energy in the atmospheric boundary layer? On the basis of this estimate, will wind plants have a major impact on climate? Would a similar impact occur from major reforestation efforts?

Table 30.2. Wind energy technologies (USA only, 2010 dollars)

	Electrical energy cost ($ per MW · h)[a]	Production and installation cost ($ per MW)[b]	Installed capacity (GW)[c]	Materials	Problems	Limitations
Current	71–82	1.78M	35	Glass/carbon composites, steel, copper, aluminum concrete	Cost of energy, energy capture, machine service, life	Transmission availability, commodity-price volatility
In 10 years	66–74	1.69M	150	Above, processed to enable onsite manufacturing and marine applications	Large-scale exploitation of Great Plains and offshore wind resources	Long-range transmission availability, characterization of wind resource
In 25 years	59–66	1.55M	360	Above, formulated for minimal environmental impact, maximal recyclability	Large-scale integration of variable wind resource	National-grid robustness and real-time control

[a] From [51], COE computed using inputs from Table B-10 for wind classes 4–6.
[b] From [51], Section B.3.4. [c] From [51], Section A.4.

30.7 Further reading

- **Anon.**, *AWEA Annual Wind Industry Report*, American Wind Energy Association, available at http://www.awea.org/publications.
- **R. Wiser** and **M. Bolinger**, 2009, *2008 Wind Technologies Market Report*, TP-6A2–46026, Golden, CO, National Renewable Energy Laboratory.
- **J. S. Rohatgi** and **V. Nelson**, 1994, *Wind Characteristics: An Analysis for the Generation of Wind Power*, Canyon, TX, Alternative Energy Institute, West Texas A&M University.
- **J. F. Manwell**, **J. G. McGowan**, and **A. L. Rodgers**, 2002, *Wind Energy Explained: Theory, Design, and Application*, ed. John Wiley & Sons, Ltd. West Sussex, Chapter 2.
- **R. B. Stull**, 2000, *Meteorology for Scientists and Engineers*, 2nd edn., Pacific Grove, CA, Brooks/Cole Publishing.
- **N. Kelley**, **M. Shirazi**, **D. Jager** *et al.*, 2004, *Lamar Low-Level Jet Project: Interim Report*, NREL/TP-500–34593, Golden, CO, National Renewable Energy Laboratory.
- **R. M. Banta**, **R. K. Newsom**, **J. K. Lundquist** *et al.*, 2002, "Nocturnal low-level jet characteristics over Kansas during CASES-99," *Boundary Layer Meteorol.*, **105**, 221–252.
- **T. Burton**, **D. Sharpe**, **N. Jenkins**, and **E. Bossanyi**, 2001, *Wind Energy Handbook*, Chichester, John Wiley & Sons.
- **D. Spera**, 2009, *Wind Turbine Technology – Fundamental Concepts of Wind Turbine Engineering*, 2nd edn., New York, ASME.
- **S. Schreck**, **J. Lundquist**, and **W. Shaw**, 2008, *U.S. Department Of Energy Workshop Report: Research Needs for Wind Resource Characterization*, NREL/TP500–43521, Golden, CO, National Renewable Energy Laboratory.
- **A. Crespo**, **J. Hernandez**, and **S. Frandsen**, 1999, "Survey and modeling methods for wind turbine wakes and wind farms," *Wind Energy*, **2**, 1–24.
- **S. Frandsen**, 2005, *Turbulence and Turbulence-Generated Fatigue Loading in Wind Turbine Clusters*, Roskilde, Risø National Laboratory.
- **European Wind Energy Association**, 2009, *Wind Energy – The Facts: A Guide to the Technology, Economics and Future of Wind Power*, London, Earthscan.

30.8 References

[1] **BTM Consult ApS**, 2010, *World Market Update 2009, Forecast 2010–2014*, Ringkøbing, BTM Consult ApS.
[2] **American Wind Energy Association**, 2009, *Annual Market Report, Year Ending 2009*, http://www.awea.org/learnabout/publications/reports/AWEA-US-Wind-Industry-Market-Reports.cfm.
[3] **European Renewable Energy Council (EREC)**, http://www.erec.org/renewable-energy/wind-energy.html.
[4] **D. Spera** (ed.), 2009, *Wind Turbine Technology – Fundamental Concepts of Wind Turbine Engineering*, 2nd edn, New York, ASME, chapters 1–3.

[5] **D. Spera** (ed.), 2009, *Wind Turbine Technology – Fundamental Concepts of Wind Turbine Engineering*, 2nd edn., New York, ASME, chapter 5.

[6] **J. Tangler** and **D. Somers**, 2000, *Evolution of Blade and Rotor Design*, NREL/CP-500-/28410, Golden, CO, National Renewable Energy Laboratory.

[7] **W. Timmer** and **R. van Rooij**, 2003, *Summary of the Delft University Wind Turbine Dedicated Airfoils*, AIAA-2003–0352.

[8] **P. Fuglsang**, **C. Bak**, and **I. Antoniou**, 2004, "Design and verification of the Risø-B1 airfoil family for wind turbines," *J. Solar Energy Eng.*, **126**(4), 1002–1010.

[9] **M. O. L. Hansen**, **J. N. Sørensen**, **S. Voutsinas**, **N. N. Sørensen**, and **H.Aa. Madsen**, 2006, "State of the art in wind turbine aerodynamics and aeroelasticity," *Prog. Aerospace Sci.*, **42**, 285–330.

[10] **S. Schreck** and **M. Robinson**, 2007, "Horizontal axis wind turbine blade aerodynamics in experiments and modeling," *IEEE Trans. Energy Conversion*, **22**(1), 61–70.

[11] **D. Griffin**, 2001, *WindPACT Turbine Design Scaling Studies Technical Area 1 – Composite Blades for 80- to 120-Meter Rotor*, NREL/SR-500–29492, Golden, CO, National Renewable Energy Laboratory.

[12] **K. Bergey**, 1979, "The Lanchester–Betz limit," *J. Energy*, **3**(6), 382–384.

[13] **K. Smith**, 2001, *WindPACT Turbine Design Scaling Studies Technical Area 2: Turbine, Rotor and Blade Logistics*, NREL/SR-500–29439, Golden, CO, National Renewable Energy Laboratory.

[14] **M. LaNier**, 2005, *LWST Phase I Project Conceptual Design Study: Evaluation of Design and Construction Approaches for Economical Hybrid Steel/Concrete Wind Turbine Towers*, NREL/SR-500–36777, Golden, CO, National Renewable Energy Laboratory.

[15] **J. Jones**, 2006, *Low Wind Speed Technology Phase II: Design and Demonstration of On-Site Fabrication of Fluted-Steel Towers Using LITS-Form(TM) Process*, NREL/FS-500–39549, Golden, CO, National Renewable Energy Laboratory.

[16] **Anon.**, 2001, *WindPACT Turbine Design Scaling Studies Technical Area 3 – Self-Erecting Tower and Nacelle Feasibility*, NREL/SR-500–29493, Golden, CO, National Renewable Energy Laboratory.

[17] **Anon.**, 2002, *Addendum to WindPACT Turbine Design Scaling Studies Technical Area 3 – Self-Erecting Tower and Nacelle Feasibility*, NREL/SR-500-29493-A, Golden, CO, National Renewable Energy Laboratory.

[18] **E. A. Bossanyi**, 2000, "The design of closed loop controllers for wind turbines," *Wind Energy*, **3**(3), 149–163.

[19] **E. A. Bossanyi**, 2004, "Developments in individual blade pitch control," in *EWEA Conference "The Science of Making Torque from Wind,"* DUWIND, Delft University of Technology, pp. 486–497.

[20] **International Energy Agency (IEA)**, 2008, *The Application of Smart Structures for Large Wind Turbine Rotor Blades, Preliminary Proceedings of the 56th IEA Topical Expert Meeting, Organized by Sandia National Laboratories*, Albuquerque, New Mexico.

[21] **H. Kwakernaak** and **R. Sivan**, 1972, *Linear Optimal Control Systems*, New York, Wiley Interscience.

[22] **M. J. Balas**, **Y. J. Lee**, and **L. Kendall**, 1998, "Disturbance tracking control theory with application to horizontal axis wind turbines," in *Proceedings of the 1998 ASME Wind Energy Symposium*, Reno, NV, pp. 95–99.

[23] **K. Stol** and **M. Balas**, 2003, "Periodic disturbance accommodating control for blade load mitigation in wind turbines," *ASME J. Solar Energy Eng.*, **125**(4), 379–385.

[24] **A. Wright**, **L. Fingersh**, and **K. Stol**, 2010, "Testing further controls to mitigate loads in the controls advanced research turbine," in *29th ASME Wind Energy Conference*, Orlando, FL.

[25] **I. Krann** and **P. M. Bongers**, 1993, "Control of a wind turbine using several linear robust controllers," in *Proceedings of the 32nd Conference on Decision and Control*, San Antonio, TX.

[26] **E. A. Bossanyi**, 1987, "Adaptive pitch control for a 250 kW wind turbine," in *Proceedings of the 9th BWEA Wind Energy Conference*, Edinburgh, pp. 85–92.

[27] **K. E. Johnson**, **L. Fingersh**, **M. Balas**, and **L. Y. Pao**, 2004, "Methods for increasing region 2 power capture on a variable-speed HAWT," *ASME J. Solar Energy Eng.*, **126**(4), 1092–1100.

[28] **S. A. Frost**, **M. J. Balas**, and **A. D. Wright**, 2009, "Direct adaptive control of a utility-scale wind turbine for speed regulation," *Int. J. Robust Nonlinear Control*, **19**(1), 59–71.

[29] **N. D. Kelley**, **B. J. Jonkman**, **G. N. Scott**, and **Y. L. Pichugina**, 2007, *Comparing Pulsed Doppler LIDAR with SODAR and Direct Measurements for Wind Assessment*, NREL/CP-500–41792, Golden, CO, National Renewable Energy Laboratory.

[30] **M. Harris**, **M. Hand**, and **A. Wright**, 2006, *Lidar for Turbine Control*, NREL/TP-500–39154, Golden, CO, National Renewable Energy Laboratory.

[31] **J. Laks**, **L. Pao**, **A. Wright**, **N. Kelley**, and **B. Jonkman**, 2010, "Blade pitch control with preview wind measurements," in *29th ASME Wind Energy Conference*, Orlando, FL.

[32] **S. Jones**, 2007, "How have projects performed? Comparing performance with pre-construction estimates," presented at the American Wind Energy Association Wind Resource Assessment Workshop September 2007, Portland, OR.

[33] **A. Tindal**, 2007, "Validation and uncertainty predictions by comparison to actual production," presented at the American Wind Energy Association Wind Resource Assessment Workshop September 2007, Portland, OR.

[34] **C. Johnson**, 2008, "Summary of actual vs. predicted wind farm performance – recap of windpower 2008," presented at the American Wind Energy Association Wind Resource Assessment Workshop September 2008, Portland, OR.

[35] **E. White**, 2009, "Operational performance: closing the loop on pre-construction estimates," presented at the American Wind Energy Association Wind Resource Assessment Workshop September 2009, Minneapolis, MN.

[36] **N. D. Kelley, R. M. Osgood, J. T. Bialasiewicz,** and **A. Jakubowski**, "Using wavelet analysis to assess turbulence/rotor interactions," *Wind Energy*, **3**, 121–134.

[37] **N. D. Kelley, B. J. Jonkman,** and **G. N. Scott**, 2005, *The Impact of Coherent Turbulence on Wind Turbine Aeroelastic Response and Its Simulation*, NREL/CP-500–38074, Golden, CO, National Renewable Energy Laboratory.

[38] **R. M. Banta, Y. L. Pichugina, N. D. Kelley, B.Jonkman,** and **W. A. Brewer**, 2008, "Doppler lidar measurements of the Great Plains low-level jet: applications to wind energy," *IOP Conf. Series: Earth Environmental Sci.*, **1** (2008), doi:10:1088/1755–1307/1/1/012020.

[39] **N. D. Kelley** and **B. J. Jonkman**, 2007, *Overview of the TurbSim Stochastic Inflow Turbulence Simulator: Version 1.21*, NREL/TP-500/41137, Golden, CO, National Renewable Energy Laboratory.

[40] **B. J. Jonkman**, 2009, *TurbSim User's Guide: Version 1.50*, NREL/TP-500–46198, Golden, CO, National Renewable Energy Laboratory.

[41] **Anon.**, 2005, *Wind Turbines – Part 1: Design Requirements*, International Electrotechnical Commission (IEC), IEC/TC88, 61400–1 ed. 3.

[42] **G. P. Corten** and **P. Schaak**, 2004, "More power and less loads in wind farms," in *"Heat and Flux," the European Wind Energy Conference*, London.

[43] **J. F. Ainslie**, 1988, "Calculating the flow field in the wake of wind turbines," *J. Wind Eng. Industrial Aerodynamics*, **27**, 213–224.

[44] **N. O. Jensen**, 1983, *A Note on Wind Generator Interaction*, Roskilde, Risø National Laboratories.

[45] **C. Johnson**, 2008, "Summary of actual versus predicted wind farm performance – a recap of WINDPOWER 2008," presented at the 2008 AWEA Wind Resource Assessment Workshop, Portland, OR.

[46] **W. Musial, S. Butterfield,** and **B. McNiff**, 2007, "Improving wind turbine gearbox reliability," presented at the 2007 EWEC Conference, Milan.

[47] **U. Hassan, G. J.Taylor,** and **A. D. Garrad**, 1988, "The dynamic response of wind turbines operating in a wake flow," *J. Wind Eng. Indust. Aerodynamics*, **27**, 113–126.

[48] **L. E. Jensen**, 2007, *"Analysis of array efficiency at Horns Rev and the effect of atmospheric stability,"* presented at the 2007 EWEC Conference, Milan.

[49] **W. Schlez** and **A. Neubert**, 2009, "New developments in large wind farm Modeling," presented at the 2009 EWEC, Marseilles.

[50] **K. Ayotte, R. Daby,** and **R. Coppin**, 2001, "A simple temporal and spatial analysis of flow in complex terrain in the context of wind energy modeling," *Boundary-Layer Meteorol.*, **98**(2), 277.

[51] **US Department of Energy**, 2008, *20% Wind Energy by 2030: Increasing Wind Energy's Contribution to U.S. Electricity Supply*, Washington, DC, US Department of Energy.

[52] **S. Baidya Roy, S. W. Pacala,** and **R. L. Walko**, 2004, "Can large wind farms affect local meteorology?," *J. Geophys. Res.*, **109**, D19101.

[53] **D. W. Keith, J. F. DeCarolis, D. C. Denkenberger** *et al.*, 2004, "The influence of large-scale wind-power on global climate," *Proc. Nat. Acad. Sci.*, **101**, 16115–16120.

[54] **S. C. Pryor, R. J. Barthelmie,** and **E. Kjellström**, 2005, "Analyses of the potential climate change impact on wind energy resources in northern Europe using output from a regional climate model," *Climate Dynamics*, **25**, 815–835.

[55] **S. Schreck, J. Lundquist,** and **W. Shaw**, 2008, *U.S. Department Of Energy Workshop Report: Research Needs for Wind Resource Characterization*, NREL/TP500–43521, Golden, CO, National Renewable Energy Laboratory.

[56] **Anon.**, 2009, *IEA Wind Energy Annual Report 2008*, Paris, International Energy Agency, pp. 46–52, available at http://www.ieawind.org/AnnualReports_PDF/2009.html.

[57] **Anon.**, 2005, *Offshore Wind Experiences*, Paris, IEA Publications, available at http://www.iea.org/textbase/papers/2005/offshore.pdf.

[58] **S. Schreck** and **A. Laxson**, 2005, *Low Wind Speed Technologies Annual Turbine Technology Update (ATTU) Process for Land-Based Utility-Class Technologies*, NREL/TP-500–37505, Golden, CO, National Renewable Energy Laboratory.

[59] **K. George** and **T. Schweizer**, 2008, *Primer: The DOE Wind Energy Program's Approach to Calculating Cost of Energy*, NREL/SR-500–37653, Golden, CO, National Renewable Energy Laboratory.

PART 4 Transportation

31 Transportation: motor vehicles

Jerry Gibbs,[1] Ahmad A. Pesaran,[2] Philip S. Sklad,[3] and Laura D. Marlino[3]

[1]U.S. Department of Energy, Vehicle Technologies Program, Washington, DC, USA
[2]National Renewable Energy Laboratory, Golden, CO, USA
[3]Oak Ridge National Laboratory, Oak Ridge, TN, USA

31.1 Focus

Motor vehicles consume about 19% of the world's total energy supplies, with 95% of this amount being petroleum, accounting for about 60% of the total world petroleum production [1]. In the USA about 80.5% of the motorized transportation energy is consumed by road vehicles [2]. The recent increase in petroleum prices, expanding world economic prosperity, the probable peaking of conventional petroleum production in the coming decades, regulations to increase fuel economy standards, concerns about global climate change, and the recent release of significant quantities of oil as a result of the failure of the deep-sea well in the Gulf of Mexico all suggest the need to focus efforts to increase the efficiency of the use of, and develop alternatives for, petroleum-based fuels used in road transportation. Efforts to increase the energy efficiency of a vehicle will require improvements in materials and processes for propulsion systems and structures, new advanced propulsion systems, batteries, and alternative fuels.

31.2 Synopsis

In many industrial countries, road transportation accounts for a significant portion of the country's energy consumption. In developing countries, the use of energy for transportation is on the rise. Most studies indicate that 70%–80% of the energy usage during the life cycle of a road transportation vehicle is in the use phase, including maintenance. The remainder is energy usage in the production of the vehicles, including the production of the materials, supply of the fuel, and disposing of the vehicles. Fuel economy and greenhouse-gas-emission regulations in North America, Japan, and Europe are forcing manufacturers to look into reducing fuel consumption in any cost-effective manner possible. Thus, advances in many materials and processes will be required in efforts to increase the energy efficiency of motorized vehicles for road transportation.

In today's internal-combustion-engine (ICE) vehicles, only about 15% of the fuel consumed is actually used to propel the vehicle and support the accessory loads (such as air conditioning, radio, and lights) [3]. The useful work done by the fuel's energy is used to overcome the vehicle's inertia, tire rolling resistance, and wind drag. The rest of the fuel's energy is lost as heat and friction due to inefficiencies of the engine and drivetrain and idling. The energy efficiency of a vehicle can be improved in several ways: lightweighting the vehicle's structure and powertrain using advanced materials and designs, improving the energy efficiency of the ICE, reducing tire rolling resistance, reducing aerodynamic drag, electrification, and hybridization (recapturing kinetic and frictional losses, reducing stop/idle losses, and reducing engine size while providing launch and acceleration assist with a more efficient electric drive system). Plug-in hybrid electric vehicles (PHEVs), battery electric vehicles, and fuel-cell electric vehicles use an electric drive system that is much more efficient than conventional ICEs; however, they are generally heavier and more costly. It is envisioned that fuel-cell powertrains might, at some time, replace the ICE powertrains. In the last couple of years, there has been significant interest in plug-in hybrid electric, extended-range electric, and all-electric vehicles. As a result, there has been a considerable increase in research, development, deployment, manufacturing activities, and funding for batteries, power electronics, electric machines, and electric-drive vehicles. For example, the American Recovery and Reinvestment Act of 2009 provided about $2 billion for battery and electric-drive manufacturing and deployment.

31.3 Historical perspective

From the time of Leonardo da Vinci to the present time, inventors and engineers have dreamed of the perfect self-propelled ground transportation vehicle that would provide a means of conveyance for goods and people. Much effort has been expended over five and one-half centuries and great progress has indeed been made, such that the vehicles of today represent a highly sophisticated blend of performance, safety, passenger comfort, and aesthetics. Trends toward increased efficiency will certainly result in a continuation of the evolution. That is not to say that this progress has been a matter of continuous evolution, for, as the acceptance and popularity of all classes of vehicles has increased, some original concepts have been discarded as not practical or too costly. Along the evolutionary path almost every known construction material has been tried, with varying degrees of success and longevity. Early vehicles were constructed primarily of wood with metal parts for propulsion components, but performance requirements soon dictated that stronger, more durable materials be used. The Ford Model T (in production 1908–1927) included a vanadium steel front axle, wooden artillery wheels, brass radiator and headlights, and a cast-iron engine block. The expansion of manufacturing capabilities around the world during and after World War II, and the increased prosperity which followed, resulted in a rapid expansion of the automotive industry to meet increased demands. From the early 1900s until the oil embargos of the 1970s, most ground transportation vehicles were manufactured primarily of steel and powered by gasoline and/or diesel. Nonetheless, at various times demonstration vehicles were produced to highlight particular materials, e.g., aluminum, magnesium, polymer composites, and even titanium. Most of these were not practical due to the costs involved in manufacturing such vehicles in large numbers. Similar diversity also has been shown in propulsion systems. Whereas early vehicles were propelled by steam engines, and some success was demonstrated with battery-powered vehicles, the range and reliability provided by the ICE soon made it the most prevalent choice. In what seems to have been a predictor of future trends, the Ford Model T was also capable of running on gasoline or ethanol.

The oil embargos of the 1970s led to the passage of the Energy Policy and Conservation Act of 1975 in the USA, which established standards for fuel economies of automobiles and light-duty trucks. In response, the Partnership for a New Generation of Vehicles (PNGV) was established in 1993 to conduct joint, pre-competitive R&D involving the three traditionally US-based automakers, DaimlerChrysler Corporation, Ford Motor Company, and General Motors (GM) Corporation, and eight US Federal agencies and to address the efficiency losses in the engine and drivetrain as shown in Figure 31.1. The goal of the partnership was to develop and demonstrate mid-size, family automobiles seating five or six with a fuel efficiency of 33 km l^{-1} (80 miles per gallon) without sacrificing affordability, performance, safety, or recyclability and meeting applicable emission standards. As a result of this partnership the three automakers in 2000 produced concept vehicles that they proposed to develop to meet the established goals. All three vehicles, with different hybrid electric powertrain technologies, met or came close to the fuel-economy, performance, safety, and emission requirements, but with predicted cost premiums (at high production volumes) of 10%–15%. The concept cars used a combination of advanced technologies for powerplants, variable transmissions, lightweight aluminum and composite body and chassis structures, aerodynamic designs, low-coefficient-of-friction tires, and other technologies. During the same period, Japanese manufacturers, with support from their government, investigated the development of hybrid electric vehicles (HEVs) and produced their own HEVs that eventually entered the world and US market in 1999 and 2000 with the Toyota Prius and Honda Insight.

In 2002 the PNGV was replaced by the FreedomCAR Partnership between the Department of Energy and the United States Council for Automotive Research (USCAR, the joint consortium of Chrysler, Ford, and GM). The long-term goal of the FreedomCAR and Fuel Partnership (FreedomCAR was expanded in 2003 to include five oil companies and became the FreedomCAR and Fuel Partnership) is to enable the transition to a transportation system "that uses sustainable energy resources and produces minimal criteria or net carbon emissions on a life cycle or well-to-wheel basis." The current plan envisions a pathway initially involving more fuel-efficient ICEs, followed by increasing use of HEVs and, ultimately, transition to an infrastructure for supplying hydrogen fuel to fuel-cell-powered vehicles [4]. Research and development for other advanced automotive technologies through the continuation of key enabling research on advanced ICEs and emission-control systems, lightweight materials, power electronics and motor development, high-power/energy-battery development, and alternative fuels is continuing. The FreedomCAR lightweighting goal is even more aggressive than the PNGV's, namely 50% reduction with respect to the average 2002 light-duty vehicle; thus, magnesium (Mg) and carbon-fiber-reinforced polymer composites are emphasized for the long term and aluminum (Al) and advanced high-strength steels (AHSSs) for the medium term. Goals have also been established to guide future developments in advanced propulsion systems, which will include fully electric systems, hybrids that combine

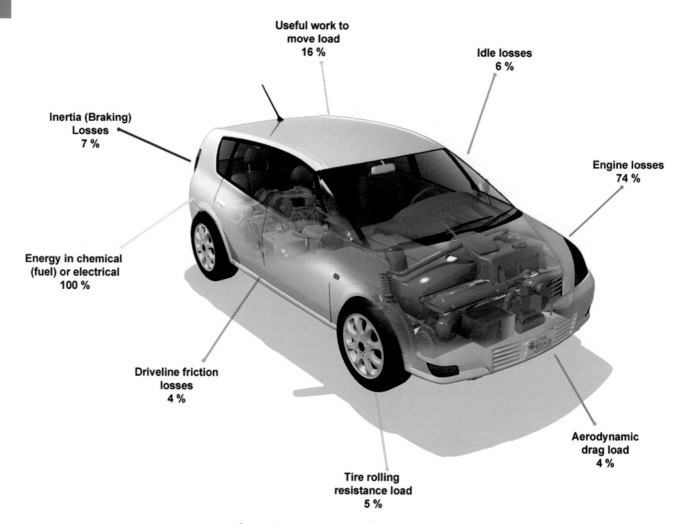

Useful work to move load 16 %

Idle losses 6 %

Inertia (Braking) Losses 7 %

Engine losses 74 %

Energy in chemical (fuel) or electrical 100 %

Driveline friction losses 4 %

Aerodynamic drag load 4 %

Tire rolling resistance load 5 %

Figure 31.1. In a typical conventional gasoline vehicle, approximatley 15% of the fuel's energy is transferred to the wheels; the rest is lost in the engine as heat, in idling, and in powering accessories. (Source: Oak Ridge National Laboratory; data from *On the Road in 2035*, a report by the Laboratory for Energy and the Environment, Massachusetts Institute of Technology.)

battery electric drives with ICEs, aftertreatment technologies, and fuel cells.

In 2000, the 21st Century Truck Partnership was announced as a partnership between 15 industrial companies and 4 government agencies. The details of the activities envisioned for this partnership can be found in the 21st Century Truck Partnership Roadmap/Technical White Papers document [5]. Since they are aimed at developing the technologies for trucks and buses that could safely and cost-effectively move larger volumes of freight and greater numbers of passengers while emitting little or no pollution, with a dramatic reduction in dependence on imported oil, many of the activities are similar to the activities of the FreedomCAR partnership. Through various R&D efforts the partnership aims to promote research into engines, combustion, exhaust aftertreatment, fuels, and advanced materials in order to achieve both significantly higher efficiency and lower emissions.

In 2010 the Department of Energy announced awards for R&D for "Systems Level Technology Development, Integration, and Demonstration for Efficient Class 8 Trucks (SuperTruck) and Advanced Technology Powertrains for Light-Duty Vehicles (ATP-LD)." The goal of the SuperTruck solicitation is to develop and demonstrate a 50% improvement in overall freight efficiency on a heavy-duty Class 8 tractor–trailer combination measured in ton-miles per gallon. Over the 3–5-year period of this activity, the selected participants will develop, test, and ultimately demonstrate these advanced technologies on a full-scale vehicle.

Since 2000, production HEVs have evolved, and more models have been introduced. The Toyota Prius, which has gone through three generations, has been the most popular and successful; more than a million have been sold. Today, such models as the Ford Escape Hybrid, Ford Fusion Hybrid, Honda Insight, Honda Civic

Hybrid, Toyota Highlander Hybrid, Toyota Prius Hybrid, and GM two-mode hybrids are being offered. Both GM's Volt and Nissan's Leaf were introduced to the market in 2011.

31.4 Lightweighting materials

The most obvious role for materials in modern vehicle design and manufacture is to enable the use of light-weight structures, thereby improving fuel economy. It has been estimated that, for every 10% decrease in vehicle weight, fuel economy increases by 6%–8% (with all other factors held constant) [6]. This estimate includes the effect of mass decompounding, which means that, in addition to the increase in fuel economy due to direct weight reduction of components or structures, further gains can be obtained by correctly sizing other vehicle components such as the powertrain, suspension, and braking system to provide equal performance in the lighter vehicle. A second role of materials is to compensate for the added weight and cost per unit of power of future vehicles that will contain alternative powertrains, such as fuel-cell vehicles, HEVs, or fully electric vehicles. However, it is imperative that the use of these materials be cost-effective if they are to be implemented in large volumes, and care must be taken to ensure that these advanced material technologies exhibit performance, reliability, and safety characteristics comparable to those of conventional vehicle materials. The entire system must be optimized to reach these competing goals, which will require that materials are chosen to take maximum advantage of their properties in each application. Utilizing the right materials for the right application will result in an optimized multi-material vehicle.

The term lightweighting refers to reduction in the overall weight of a vehicle and its components. This term is used to emphasize the fact that some weight-reduction strategies rely on design changes and the use of thinner-guage, higher-strength materials to meet targets as opposed to simply replacing incumbent materials with lower-density alternatives. For example, new AHSSs are of the same density as standard plain-carbon steels, but when used in thinner guage can meet performance requirements with a net weight reduction.

The use of lightweighting materials is governed by several factors. First and foremost are the inherent properties of the materials under consideration. Of primary concern for structural, or even semi-structural, applications are the stiffness or modulus of elasticity, the yield strength, and the ductility of the materials. Depending on the specific component, other properties may also come into consideration. Thus, for external components that may be exposed to water and ice-clearing chemicals, the corrosion behavior must also be considered, whereas for components that are near the engine, where temperatures will be higher, resistance to creep deformation is a necessary requirement. It is not within the scope of this chapter to provide a detailed description of the physics and/or chemistry which governs these properties, description of the techniques for measurement, or the interrelationship between them. That information can be found in other textbooks devoted to the subject. (See, for example, the 3rd, 4th & 5th entry in Section 31.12.)

When trying to choose a material for a specific application, it can be misleading to base the decision on direct comparison of values of a given property. Clearly the tensile strength of a high-strength automotive steel such as DP 980 (~980 MPa) is much greater than the tensile strength of aluminum alloy Al6061-T6 (~310 MPa). It is much more appropriate to compare the specific strengths of the materials, i.e., the strength divided by the density of the materials. When compared this way, the properties of DP 980 steel (125 MPa per $g\,cm^{-3}$) and the aluminum alloy (115 MPa per $g\,cm^{-3}$) are much more similar, and it is clear that aluminum alloys can perform as well as steel in certain applications. Table 31.1 summarizes the specific strengths and rigidity (modulus or stiffness) for some lightweighting materials.

Other factors that influence the use of lightweighting materials include the potential for mass reduction, cost, availability, and, which is becoming increasingly important in light of concern over sustainability, recyclability. Other non-technical factors such as design compromises resulting from lack of familiarity with newer materials, capital investment in existing materials or technologies, which can impose constraints simply due to the inertia which must be overcome in order for change to occur, and cost variability on world markets also affect the ability to use these materials. Table 31.2 compares some of the lightweighting materials being considered for automotive and heavy-truck applications in terms of mass reduction and relative costs. The AHSSs, aluminum, magnesium, and glass- and carbon-reinforced polymer matrix composites offer the greatest opportunities for substantial weight reductions.

In the decades since the oil embargo in 1975, auto companies have pursued various strategies, including increased use of some of the advanced materials listed in Table 31.2, in an attempt to reduce the mass of their vehicles and increase fuel efficiency (see Figure 31.2).

Despite those efforts, the average vehicle weight in 2004 was virtually the same as it was in 1977, due primarily to vehicle upsizing, addition of customer-focused features (audio, video, navigation systems), or performance factors mandated by regulation (safety, pollution abatement, etc.) (see Figure 31.3).

Recent mandated changes in Corporate Average Fuel Economy (CAFE) standards have placed increased emphasis on "lightweighting" as a necessary component

Table 31.1. Specific strengths and rigidity for some lightweighting materials

Material	Density (g cm^{-3})	Elastic modulus (GPa)	Specific rigidity (GPa per g cm^{-3})	Yield strength (MPa)	Tensile strength	Elongation (%)	Specific strength (MPa per g cm^{-3})	Cost ($ per lb)
Al6061-T6	2.7	70	25.93	275	310	12–17	114.8	1.0
Al6061-T4	2.7	70	25.93	145	240	20–25	88.9	1.0
Al5754-H26	2.7	70	25.93	245	290	10	107.4	1.0
Al5754-O	2.7	70	25.93	100	215	25	79.6	1.0
Mg AZ31-H24	1.74	45	25.86	214	290	15	166.4	1.5
Mg AZ31-O	1.74	45	25.86	145	179	20	103.0	1.5
Mild steel	7.8	211	27.05				0.0	0.5
DP 980	7.8	211	27.05		980	10	125.6	0.6?
3rd GEN-1	7.8	211	27.05		1,000	15	128.2	?
3rd GEN-2	7.8	211	27.05		1,400	25	179.5	?

Sources: Al, J. R. Davis (ed.), 1993, *Aluminum and Aluminum Alloys*, New York, ASM, http://asm.matweb.com/search/SpecificMaterial.asp?bassnum = MA6061T6;
Mg, http://www.magnesium-elektron.com/data/downloads/DS482.pdf; AHSSs, Roger Heimbuch, A/SP, private communication.
(1 dollar/lb is ~ 0.45 dollar/kg)

Table 31.2. Comparison of lightweighting materials being considered for cars and heavy trucks

Lightweight material	Material replaced	Mass reduction (%)	Relative cost (per part)
High-strength steel	Mild steel	10–25	1
Aluminum (Al)	Steel, cast iron	40–60	1.3–2
Magnesium	Steel or cast iron	60–75	1.5–2.5
Magnesium	Aluminum	25–35	1–1.5
Glass FRP composites	Steel	25–35	1–1.5
Graphite FRP composites	Steel	50–60	2–10+
Al-matrix composites	Steel or cast iron	50–65	1.5–3+
Titanium	Alloy steel	40–55	1.5–10+
Stainless steel	Carbon steel	20–45	1.2–1.7

Source: W. F. Powers, 2000, *Adv. Mater. Processes*, **157**(5), 38041. Reprinted with permission of ASM International. All rights reserved. www.asminternational.org

in vehicle design and development, both in North America and in the rest of the world. Complicating efforts to address this need are new regulations focused on improved passenger safety. If auto companies are to reach the stated goal of a 50% reduction in vehicle weight, efforts to utilize lightweighting materials must be significantly increased. Recent announcements by various vehicle manufacturers indicate that there has been some success in meeting the 50% weight-reduction target. However, the cost of these vehicles makes them

Figure 31.2. Materials in a typical family vehicle in different model years. Hi/Med Strength, high/medium-strength. (Source: American Metal Market.)

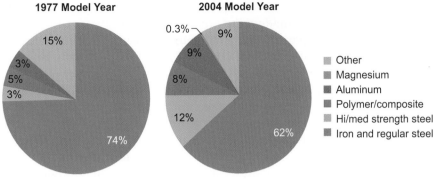

Figure 31.3. Light-duty vehicle trends – fuel economy, weight, and acceleration. (Source: US Environmental Protection Agency, 2004, *Light Duty Automotive Technology and Fuel Economy Trends: 1975 through 2004.*)

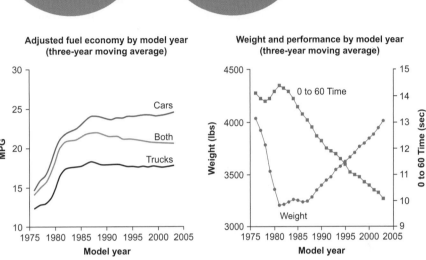

low-volume, expensive, luxury vehicles that are not affordable to the general public.

The various materials identified with potential to decrease vehicle weight are associated with corresponding technical challenges that can limit their implementation in high-volume manufacturing.

In general, the sets of materials that are considered for light-duty applications (passenger cars, utility vehicles, and light trucks) and for heavy-duty applications (vocational trucks and long-haul tractor–trailer combinations) are the same. However, the performance requirements, which are defined by the specific duty cycles, require different approaches to design, as well as manufacturing. Light-duty designs are stiffness-driven, whereas heavy-duty designs are driven by considerations of strength and durability, including corrosion and fatigue. In addition, production volumes for light-duty vehicles are in millions per year, whereas the annual production volume for heavy-duty vehicles is approximately 300,000. The lower volume of heavy-duty vehicles may allow consideration of manufacturing and assembly techniques that are slower and, perhaps, consideration of materials and processes that are more costly.

Regardless of whether the intended application is for light- or heavy-duty vehicles, a number of barriers to the commercial application of lightweighting materials exist.

- **Cost**. The greatest single barrier to the viability of advanced lightweighting materials for ground transportation is their relatively high cost. In general,

the contributions to the high costs come from the initial cost of the basic materials or alloys, as well as the costs associated with processing. Development of alternative, more energy-efficient technologies for the production of the materials of most interest, e.g., carbon fiber, magnesium, aluminum, and titanium, can reduce the costs of these materials substantially.

- **Manufacturing**. Methods for cost-competitive production of components and assemblies from advanced lightweight materials are not sufficiently well developed, except for some applications of cast aluminum and magnesium. In order for these materials to be competitive with the incumbent materials, technologies that yield the required component shapes in a cost-effective, rapid, repeatable, and environmentally conscious manner must be developed. For example, the ability to produce large, high-integrity magnesium castings would enable the use of such components in structural applications; the development of processing technologies such as injection molding, injection compression molding, pultrusion,[1] resin transfer molding, non-thermal curing methods, and automated materials-handling systems would enable high-volume manufacturing of reinforced

---

[1] Pultrusion is a continuous process of manufacturing of composite materials with constant cross section whereby reinforced fibers are pulled through a resin, possibly followed by a separate performing system, and into a heated die, where the resin undergoes polymerization.

carbon-fiber composites; and the development of alternative high-volume forming processes that are optimized for the properties of aluminum and magnesium wrought materials, as well as AHSSs, will lead to more rapid implementation.

- **Design methodologies**. Adequate design data such as databases of material properties, test methodology, analytical tools such as predictive models, and durability data are needed for widespread applications. Uncertainties about material properties in regimes corresponding to realistic applications often result in overdesign of components. To best take advantage of the properties of polymer composites and lightweight metals in structural components, a significant shift in component-design philosophy must be made. Additionally, the differences in properties of the materials under consideration require the development of analytical methodologies to predict the responses of the materials after long-term loading, under exposure to different environments, and in crash scenarios.

- **Joining**. High-volume, high-yield joining technologies for lightweighting materials are insufficient or not fully developed. The joining methods required for joining nonferrous materials, polymer composites, and AHSSs are significantly different from those used for incumbent steel materials. New technologies must be developed to provide rapid, affordable, reproducible, and reliable joints that meet at least the same level of safety as that which currently exists. In addition, as design strategies are developed for optimizing weight reduction, manufacturability, and performance, the ability to join dissimilar materials will take on increased significance. In many cases, joining techniques such as resistance spot welding are not possible when joining materials with widely different melting temperatures, e.g., steel and aluminum. Therefore, greater emphasis must be placed on solid-state joining techniques or fastening technologies, which, in general, are more expensive.

- **Recycling and repair**. Technologies for recycling and repair of advanced materials, such as carbon-fiber-reinforced composites, are not sufficiently well developed to meet the demands which will arise from large-scale use of these materials. Technologies for purification of in-plant prompt magnesium scrap as well as for post-consumer sorting of aluminum, magnesium, and steel scrap need to be developed to the point of commercial viability. In addition, robust methods for rapidly and reliably repairing aluminum, magnesium, and composite structures will be required. The cost of these repair technologies relative to the cost of replacement of components must be factored into design and joining strategies.

31.5 Powertrain materials

The conventional powertrain consists of the mechanical systems that provide power to move a vehicle. The systems include the ICE, the transmission, the driveline, the differential, and axle assemblies. These systems may be combined into a single subassembly; for example, the transaxle of a front-wheel-drive car consists of the transmission, differential, and axle assembly. The ICE and driveline are the result of over 100 years of continual innovation, but during the period from 1880 to 1900 there was much controversy as to which technologies, combustion cycles, and fuels would dominate vehicle powertrains. By 1900 the competition among steam, electric, and ICE powertrains was a dead heat, with each having 33% of the market. A poll taken at the 1900 National Automobile Show in New York City indicated that people's order of preference for powertrains was (1) electric, (2) steam, and (3) the ICE. On the basis of this poll, electric cars similar to the 1898 Baker (see Figure 31.4) should have dominated the early automobile industry.

Figure 31.5 shows the volumetric energy content of a number of potential fuels and batteries that could be used to power vehicles. Eventually, the availability of low-cost petroleum products and the range potential they provided allowed the ICE to become the dominant power train of the twentieth century.

Today's vehicle powertrains demonstrate high reliability, good durability, good power density, good range, reasonable fuel economy, low exhaust emissions, and extreme cost-competitiveness.

Figure 31.4. The 1898 Baker electric car.

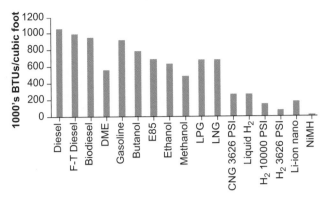

Figure 31.5. Work at the SNL (Sandia National Laboratories) on the volumetric energy content of various fuels, from *Advanced Propulsion Daily*, January 17, 2006. Li-ion batteries have four times the energy density of lead–acid batteries and two to three times the energy density of nickel–cadmium and NiMH batteries. Conversion: Btu per cubic foot $\times$ 37.259 $=$ kJ m^{-3}.

Table 31.3 A comparison of the Carnot efficiency and the Carnot efficiency with irreversible heat-transfer processes as it relates to the potential ideal efficiency of a heat engine

	Flame temperature (K)	Carnot efficiency (%)	Carnot efficiency with irreversibility (%)
Ethanol	2193	83.0	58.8
Diesel	2327	84.0	60.0
Gasoline	2302	83.8	59.7
Methanol	2143	82.6	58.3
Natural gas	2063	81.9	57.5
Propane	2263	83.5	59.4

Vehicle manufacturers are facing a changing landscape very similar to that seen in 1900. Many are questioning the future viability of the ICE, and every day a new fuel or technology is touted as the powertrain of the future. A survey of the changing landscape reveals several factors that lead to uncertainty on the future of conventional power trains. These include the problematic future availability of petroleum fuel, new fuel-economy regulations, emissions regulations, and greenhouse-gas emissions with their associated implications for global climate change. However, a 2010 report to the Transportation Research Board by Lin and Greene indicated that the ICE is expected to remain an integral part of the powertrain mix for vehicles at least until 2045.

Regardless of the fuel used, it is in the best interest of society that future vehicles and powertrains utilize fuel in the most efficient manner possible. Today's gasoline engines have a thermal efficiency near 30% and diesel engines are about 40% efficient; however, this does not represent the theoretical limits of the internal-combustion heat engine. Thermodynamics (Appendix A) shows that the Carnot thermal efficiency of an ideal engine is a function of the hot and cold temperatures of the system. For an ideal engine the Carnot efficiency n of the engine is

$$n = 1 - \frac{T_{cold}}{T_{hot}},$$

where T_{cold} is the temperature of the cold reservoir and T_{hot} is the temperature of the heat source, with the temperatures in Kelvin.

So the potential efficiency of an engine using a fuel as a heat source and the ambient air as a cold source is 1 minus the coolant temperature divided by the adiabatic flame temperature of the fuel. However, an ICE is not an ideal heat engine, there are major irreversibility issues, and to account for this a variation of the Carnot efficiency can be applied when the processes of heat transfer are not reversible; in this case the potential efficiency of the engine can be calculated as

$$n = 1 - \frac{\sqrt{T_{cold}}}{T_{hot}},$$

where T_{cold} is the coolant temperature and T_{hot} is the adiabatic flame temperature of the fuel [7].

According to the data in Table 31.3, the heat engine has a potential ideal efficiency of near 80% with reversible and near 59% with irreversible processes. Further, when one examines the adiabatic flame temperatures of the potential replacement fuels, the maximum efficiencies vary by only about 3% from one fuel to another.

Historically, increases in ICE efficiencies have been closely related to compression ratios and peak cylinder pressures. The theoretical Carnot-efficiency estimates discussed earlier match well with Figure 31.6 illustrating the relationship between the indicated thermal efficiency and the compression ratio for stoichiometric and lean-burn Otto cycles.

Historical engine-performance data also show this trend; as engine efficiency increases so does the peak cylinder pressure. Figure 31.7 shows historical performance trends for heavy-duty diesel engines and extrapolations to higher engine-efficiency levels. Similar data for the compression ratio and peak cylinder pressure (PCP) for European heavy- and light-duty diesels can be found in *MTZ Worldwide Magazine*, Volume 66

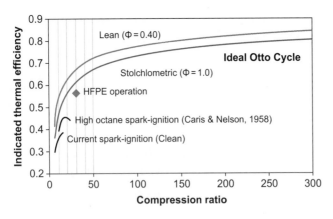

Figure 31.6. Data extracted from "Potential efficiency impacts of low-temperature combustion modes," presented by Paul Miles from Sandia National Laboratories at the Combustion Engine Efficiency Colloquium, March 3–4, 2010, Southfield, MI. HFPE is hydraulic free piston engine.

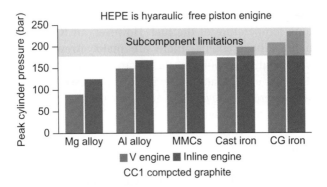

Figure 31.8. Peak cylinder pressures of different cast materials used in inline and V-engine geometries. CG is compacted graphite.

(German issue). The dip in actual efficiency starting in 2000 is a result of increased regulation of exhaust emissions; in future engines one will need to carefully balance efficiency against its impact on regulated exhaust emissions. From the trends illustrated in the graph, as engine efficiencies reach 50% one may expect peak cylinder pressures to approach 250 bar, and engines with 60% thermal efficiency could be expected to have peak cylinder pressures near 350 bar.

Materials are going to be a critical enabler if future vehicles are to achieve their maximum potential. Figure 31.8 illustrates the peak cylinder pressures of different cast materials used in either inline or V engine geometries. Note that the cast-iron and subcomponent limitations of 190 bar closely match where the historical diesel-engine peak cylinder pressures level off. This would lead one to infer that current heavy-duty diesel-engine efficiencies are materials-constrained.

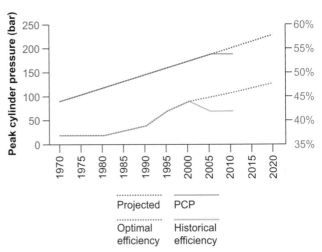

Figure 31.7. Historical data on heavy-duty diesel-engine trends to 2008 provided by Cummins Engine Company, with future trends extrapolated.

A number of new technologies will be necessary in order to maximize the efficiency of the ICE; each of these technologies will have specific materials requirements and will be introduced through either heavy-duty compression-ignition or light-duty spark-ignition engines. Heavy-duty engines are closer to the maximum theoretical efficiency and have much higher loading, so the technologies will begin pushing the limits for fuel-injection and cylinder pressures sooner.

In order to maximize the efficiency of the internal combustion engine technologists will need to optimize airflow, fuel-injection events, and ignition or combustion events, while minimizing friction and thermal losses.

To optimize airflow, engineers will need to perfect the interaction among the intake manifold mechanical boosting mechanisms, the valve operational events, and piston movement. Boosting mechanisms such as a superchargers or turbochargers will need to minimize friction and weight-induced dynamic response lag. The valve operational events can be optimized for the engine operating parameters through variable valve-actuation strategies such as valve cam phasing or independent valve actuation. The mass of the valves will have a negative impact on the functionality of variable valve-actuation systems, so there may be a drive to reduce the weight of the valve train. At the same time it is expected that the exhaust-gas temperature will increase, so the operating temperature of valves and valve seats will also need to increase.

To maximize efficiency, advanced engines will need to deliver fuel to the combustion chamber as quickly and efficiently as possible. There is a very narrow time window available to deliver the fuel for optimum combustion efficiency, so the fuel-injection event will need to be very rapid. Further, the fuel-injection system will need to inject the fuel into a cylinder under high compression. These demands will push manufacturers to

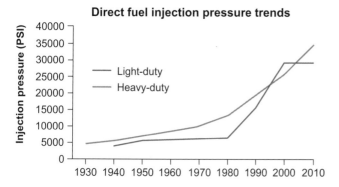

Figure 31.9. Peak injection pressure vs. model year (Source: http://www.dieselnet.com/tech/diesel_fi.html). PSI is 6.9 kilopascal.

develop fuel-injection systems that operate at very high pressures (see Figure 31.9). These new fuel-injection systems will push the limits of the current materials used for lines, housings, plungers, and fuel-injector tips.

Ignition or combustion events can be optimized through two mechanisms; the first is to limit when there is a combustible air–fuel mixture in the cylinder and the second is to dramatically reduce the time required for the ignition or spark event. The air–fuel mixture can be controlled by the use of direct injection, or directly injecting the fuel into the combustion chamber at the precise instant when it is needed for the combustion event; this will also eliminate compression knock in spark-ignition engines. The time required for the ignition event in spark-ignition engines can be dramatically reduced using one of several approaches, such as plasma ignition, which produces a broad ignition field and is an order of magnitude faster than spark plugs, or laser ignition, which produces a line of ignition points across the combustion chamber. Each of the potential new ignition systems will have materials issues, such as high-temperature erosion, which must be addressed before they can become commercially viable.

As the combustion process becomes more efficient the relative losses to friction will appear to increase. The ICE experiences friction losses whenever there is relative movement between components; there is rubbing friction at the piston-ring–cylinder-wall interface, there is sliding friction at each of the bearings on the crank shaft, and there are significant frictional losses in the valve train. These frictional losses may be addressed in a number of ways, such as the use of surface texturing, low-friction coatings, and modifications to the engine lubricant. Each of these approaches will have significant materials, materials–lubricant interaction, and materials-processing issues that will need to be overcome if the system is to reach commercial viability.

Thermal management will be critical to achieving the maximum efficiency from tomorrow's ICE. As combustion takes place the piston, cylinder wall, cylinder head, and valves are exposed to the full heat of

combustion as the energy in the expanding gases is converted into mechanical work. Without engine coolant these metal components would rapidly approach their melting point and the engine would fail. However, excessive heat transfer to the engine coolant represents a loss of energy in the exhaust that could be converted to work or used in the exhaust aftertreatment system. With the advent of devices such as turbochargers, thermoelectric devices, and Rankine-cycle energy-recovery systems that convert exhaust energy to work, any heat lost to the cooling system represents a potential loss of system efficiency. In addition to withstanding much higher peak cylinder pressures, the materials used in pistons, cylinder blocks, valves, and cylinder heads will need to withstand higher operating temperatures and minimize excessive heat loss to the engine coolant.

In addition to materials needs resulting from changing combustion regimes, these new engines are expected to expanding their use of exhaust aftertreatment devices such as catalysts and particulate filters. The long-life-cycle requirements of these devices will require detailed knowledge of fatigue life behavior of filters, ceramic substrates, and catalyst coatings. Further, life-cycle costs will push the industry to find replacements for precious metal-based catalysts.

31.5.1 Economics

The average mid-size car of 2010 costs $7.38 per pound for a curb weight of 3,560 pounds and costs about $26,300. This figure includes manufacturing, assembly, shipping, and profit; so any materials and processing techniques are going to have very stringent cost targets. Assuming that consumers are logical, any new technologies will need to break even on a life-cycle basis. Assuming a baseline fuel economy of 28 miles per gallon (0.425 kilometers/l) with the new technologies resulting in double the fuel economy, gasoline at $5.00 per gallon (~ 1.3 per liter), annual vehicle miles traveled (VMT) of 12,000 miles per year, and a 4-year ownership period, the maximum incremental cost of the new technologies would be $4,285 (including profit margins). So a car using the new technology would have a maximum cost of $30,585; that would be $8.58 per pound with a curb weight of 3,560 pounds or $11.45 per pound with a curb weight of pounds 2,670 (assuming a 50% body-weight reduction). Clearly, maintaining costs within an acceptable level will be one of the most difficult requirements for the high-performance materials required for achieving the maximum efficiency potential of the light-duty ICE.

A similar approach can be applied to heavy-duty trucks. Using a Class 8 over-the-road truck as an example we can establish the following baseline. The initial cost is about $150,000 and the truck weighs about 19,000 pounds; this makes the cost per pound about $7.90 when using the following criteria: fuel at $5.00 per gallon, an annual VMT of 100,000 miles, 1 mile = 1.61 km) a baseline engine efficiency of 40%, a baseline fuel economy of 6 mpg, and a 3-year payback period. The maximum incremental cost of a Class 8 over-the-road truck offering a 25% increase in fuel economy (50% thermally efficient engine) would be $50,000 or $10.50 per pound. Clearly, users of heavy-duty vehicles will find converting to more efficient vehicles attractive and will have a slightly higher cost tolerance than users of light-duty vehicles. This explains why the heavy-duty-vehicle industry is often an early adopter of advanced engine technologies and is a logical platform for advanced ICE research.

31.6 Hybridization and electric drives

Hybridization is achieved by adding an electric drive system to a conventional vehicle or even a fuel-cell vehicle [8]. The electric drive system (Figure 31.10) can include a motor, generator, power electronics (inverters and convertors), and a means for storing electrical energy (batteries, fuel cells and/or ultracapacitors). Hybrid electric vehicles achieve lower fuel consumption than that of conventional vehicles with internal combustion engines in several ways [9]. By recapturing the kinetic energy of a decelerating vehicle (while the engine speed is decreasing) and using regenerative braking instead of friction breaking, the fuel economy of a typical vehicle can be increased by 10% [10]. Regenerative braking works as follows. As a driver brakes in a hybrid

or electric car, the kinetic energy, which would normally be dissipated as heat, instead supplies. Electrical energy that charges the energy-storage system every time the brakes are applied. The stored energy is subsequently used to propel the vehicle, requiring less fuel use and thus leading to more efficient operation. The availability of an electric motor for launch assist and quick engine restarts allows the engine to turn off during stops, thus reducing fuel consumption at idle by 5%–10%. Because the combination of a motor and an engine provides power for acceleration, the engine could be downsized to meet only top-speed and grade requirements; this could provide another 10%–20% improvement in fuel economy [11].

Overall, hybridization could improve fuel economy by 20%–50%. For example, the combined EPA-rated city/highway fuel economy of a 2007 Toyota Camry Hybrid at 39 mpg is 30% higher than that of a conventional 2007 Camry at 27 mpg [12]. The EPA-rated city-driving fuel economy of a 2011 Ford Fusion Hybrid is 41 mpg, whereas the non-hybrid version's fuel economy is about 23 mpg. In 2011, GM introduced a range-extender electric vehicle (a series PHEV) called the Volt, which has a nominal 40-mile electric range with electricity charged from a 110-V plug. After exhausting the battery energy and reaching the end of the all-electric range, the Volt operates as a charge-sustaining HEV and can go another 400 miles. Several all-electric vehicles have entered the market recently: the Tesla Roaster with an impressive acceleration (0–6 mph in 4.2 s), the compact Mitsubishi iMiEV, the compact Mini Cooper EV,

Figure 31.10. Components in a plug-in hybrid vehicle. (Without the plug and the on-board charger, the components represent a hybrid electric vehicle. Removing the engine represents an all-electric vehicle.)

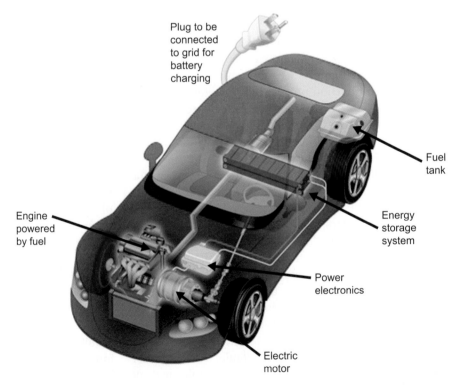

Plug to be connected to grid for battery charging

Fuel tank

Energy storage system

Power electronics

Engine powered by fuel

Electric motor

and the Nissan Leaf, a mid-size sedan with a nominal electric range of 100 miles.

Each of the major hybridization or electrified components entails corresponding materials challenges.

31.7 Energy storage

The electrical-energy-storage systems for HEVs must be smaller, lighter, longer-lasting, more powerful, more energy dense, and less expensive than today's batteries if these advanced vehicles are to significantly expand into markets around the world. The all-electric vehicles require energy storage that has high energy density and specific energy, which can last 10 years and undergo at least 1,000 deep discharge–charge cycles. Thus, the energy-storage systems offer the most challenging materials issues, particularly those for PHEVs and battery electric vehicles. Materials in PHEV energy-storage systems have the more challenging task of withstanding many shallow charge–discharge cycles (as in HEVs) and many deep charge–discharge cycles (as in electric vehicles) [13]. Chapter 44 covers the general topic of electrical-energy-storage materials issues, but some challenges are related specifically to transportation applications, namely safety, calendar life, and cold-start operation. Many automotive researchers believe that lithium-ion batteries are the choice for electric-drive applications for the next 10–15 years. The focus has been on lower manufacturing cost, increasing energy and power, and improving their low-temperature performance and safety [14].

The major components of batteries are cathodes, anodes, separators, and electrolytes, this is discussed in more detail in Chapter 44; below is a short summary relevant to vehicles. To improve the safety of lithium-ion batteries, researchers and developers are investigating new cathodes such as nanophase iron phosphate [15], manganese oxide, nickel–cobalt–manganese, or mixed oxides and new anodes such as new forms of carbon/graphite that are less active than today's carbon materials. Silicon as anode has received a lot of attention recently because it has a much higher capacity than those of graphite or carbon, but the challenge that researchers are studying is the large volume expansion of silicon. Other components to be improved include lithium titanate ($Li_4Ti_5O_{12}$) anodes [16]; separators with better melt integrity, lower shrinkage, and shutdown functionality, such as ceramic-coated Teflon; and nonflammable electrolytes [14]. Given that the use of alternative cathodes, anodes, separators, and electrolytes might reduce the performance and/or life of the battery, other researchers are investigating the possibilities of depositing atomic layers of protective materials onto existing high-performing anodes and cathodes and using additives to reduce the flammability of electrolytes [17]. Still, internal electrical shorts that develop over many months, although rare, pose the most

challenging materials and design issues for lithium-ion batteries in vehicles [18]. Researchers are developing models and test methods to understand this failure mode in order to propose solutions [19].

At temperatures below freezing, the power capability and, to a certain extent, the available energy of lithium-ion batteries decrease. For consumer acceptance, a lithium-battery-powered vehicle must be able to start at $-30\,°C$. Some of the limitation is attributed to the interface between electrolyte and active materials [20]. In general, changes in the active material and electrolyte have not, to date, significantly improved the poor low-temperature performance. Therefore, to improve performance, electrolyte formulations based on $LiBF_4$ in carbonate and ester mixtures are being investigated [21]. Moreover, methyl butyrate is being tested as an additive that can decrease the viscosity of the electrolytes to enhance their wettability properties in the separator (allowing better electrolyte coverage), thus leading to improved low-temperature performance [14].

One of the factors that influence the life of batteries is high temperature. In order to minimize the adverse impact of high temperatures on batteries in electric-drive vehicles, battery thermal management is needed. In recent years, the industry has started moving from simple air cooling to liquid cooling and the use of materials with high thermal conductivity, but low electrical conductivity. The structure of the battery container should be light, strong, and fire-resistant. Because of the safety concern with lithium-ion batteries, the use of fire-retardant components may be necessary.

31.8 Electric machines and power electronics

Vehicle electrification and hybridization requires the development of multiple interactive modules and components involving various engineering skill sets. Successful electrical, mechanical, thermal, and material design and integration are all necessary in achieving a high-quality product. The parts must be reliable, of low weight, and with reduced volume footprints, all the while realizing low manufacturing costs. Vehicle electrification can include, at a minimum, one traction motor and an inverter (including the associated low-voltage control circuitry). Additionally, the system may include a DC–DC boost converter, a generator and its associated drive components, and supplemental gearing, as well as any auxiliary cooling systems necessitated by the high heat fluxes generated. Figure 31.11 shows a block diagram highlighting the typical modules necessary to achieve vehicle electrification.

Different vehicle architectures will generally dictate the power level and specifics of the necessary components. An HEV needs to operate only for intermittent periods using electrical power whereas a PHEV or fully

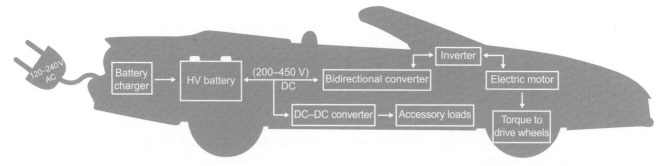

Figure 31.11. Electronics and motors for vehicle electrification.

Figure 31.12. The engine compartment's thermal profile (Delphi Delco Electronic Systems).

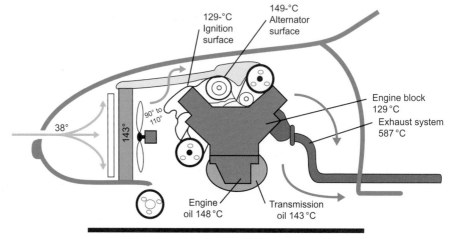

electric vehicle (EVs, without an ICE) must be sized for considerably more electrical power. Exacerbating the situation for EVs and PHEVs is the fact that the electronics and the electric motor are operating much more of the time during a drive cycle and hence are subject to more extreme and prolonged operational temperatures. There are multiple original-equipment-manufacture (OEM)-specific architectures, but in general they can be segregated into the following classifications; parallel, mildly parallel, power-split series-parallel, series-hybrid, plug-in hybrid, fuel-cell and purely electric vehicles. Each of these drivetrain configurations involves varying levels of electrification of the vehicle which will determine the rigors imposed upon the power electronics and machinery by dictating the duty cycle of its usage. Referring to Figure 31.11, the components designated perform the following functions.

Battery charger: required of PHEVs and EVs, and can reside off or onboard the vehicle. It is expected that for residential charging these chargers will utilize 240-V AC where available instead of 110 V AC for greater efficiency and higher charge rates to charge the energy-storage batteries in the vehicle.

Bidirectional converter: utilized to step up the battery voltage from what is supplied by the battery to the DC link voltage for running the motor;

also serves to provide an avenue for regeneration from the wheels to charge the battery during braking.

Inverter: converts the direct current from the battery into alternating current to provide phased power for vehicle traction motors.

DC–DC converter: provides power for auxiliary accessories, i.e., radio, lights, air conditioning, brake assist, power steering, etc.

Electric motor: converts electrical power into mechanical power to provide torque to the wheels to drive the vehicle.

All of these components must operate under harsh environmental conditions, including elevated-temperature operation, high vibration levels, and exposure to fluids. Figure 31.12 details some of the temperature extremes seen within a vehicle with an ICE, while Table 31.4 presents maximum temperatures and vibration levels.

Achieving high-efficiency performance with power electronics and motors under these conditions while simultaneously meeting volume, weight, and cost targets presents significant challenges. Hybrid vehicles still account for only about 2%–3% of all vehicles sold in the USA [22]. Currently, most consumers have concluded that the fuel savings are not sufficient to offset the higher cost of a hybrid vehicle. The level of market penetration is

Table 31.4 Maximum temperatures and vibration levels in a typical internal combustion engine

Location	Typical continuous maximum temperature (°C)	Vibration level, G_{rms}	Fluid exposure
On engine on transmission	140	Up to 10	Harsh
At the engine (intake manifold)	125	Up to 10	Harsh
Underhood near engine	120	3–5	Harsh
Underhood remote location	105	3–5	Harsh
Exterior	70	3–5	Harsh
Passenger compartment	70–80	3–5	Harsh

expected to remain low until hybrids become more cost-competitive with conventional vehicles. Achieving a market share that is large enough to justify large-scale manufacturing will require reducing costs to a level at which consumers can justify their purchase in economic terms. The materials utilized both in the electronics and in the electric machinery in these vehicles are dictated ultimately by reliability and cost considerations.

One path to reducing the overall cost of the traction drive system (i.e., the electric machine and power electronics) is to eliminate extra cooling loops within the vehicles. At present, most on-the-road HEVs utilize a 65–70-°C auxiliary coolant loop. Researchers are investigating approaches to use the existing ICE water–ethylene glycol coolant directly out of the engine radiator, at approximately 105 °C, to cool the power electronics. Higher-temperature coolants impose significant operational and reliability challenges to the electrical modules, components, and materials used in these systems, which are already taxed by high internal self-heating due to their elevated operational currents.

Figure 31.13 depicts the typical components of an automotive inverter, with approximate cost shares for each constituent. New materials that can result in cost savings with similar or better performance than what has traditionally been utilized are continually being sought in order to bring down the cost differential for these vehicles. Currently, it is estimated that the power electronics represents 20% of the material costs for a hybrid vehicle [23], and that percentage is expected to grow significantly for PHEVs and EVs with the addition of charger circuitry.

The power modules are comprised of silicon switches, commonly insulated gate bipolar transistors (IGBTs), and diodes, which are packaged in a configuration similar to that shown in Figure 31.14 [24].

The IGBTs are a slice of silicon (typically 10 mm × 10 mm × 0.1 mm for a switch rated at 300 A/600 V), which is a complicated construction made by microfabrication

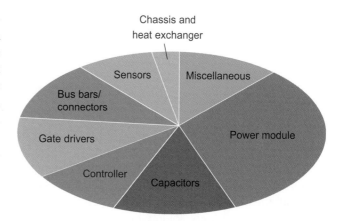

Figure 31.13. Proportional cost breakdown for an inverter (which converts the direct current out of the battery to alternating current).

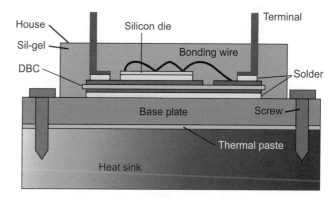

Figure 31.14. Traditional insulated-gate bipolar transistor (IGBT) packaging.

technologies. The packaging of the power module must provide electrical interconnection between the dies and the terminals to the outside world. Additionally, it must provide thermal management for the dies, due to the large amount of heat generated during their operation. It is notable that so many diverse materials are involved in the packaging structure, namely metals,

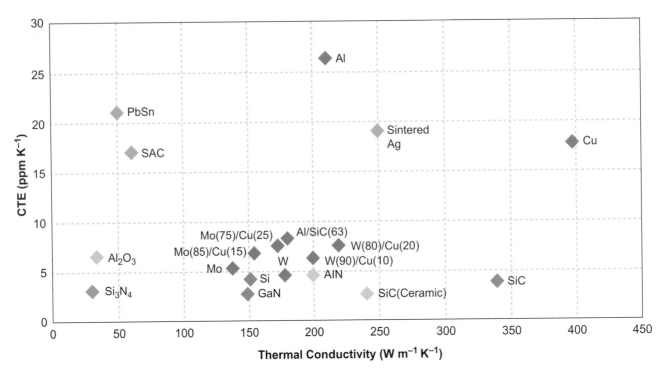

Figure 31.15. The thermal conductivity vs. coefficient of thermal expansion (CTE) of typical power-module packaging materials.

semiconductors, and dielectrics, organic and inorganic materials, in addition to the joining agents necessary to bond all the parts together.

The silicon dies are soldered down to an insulated substrate. Usually it is a directly bonded copper (DBC) substrate, which is comprised of a preformed piece of copper braised to an insulating ceramic on both sides. Al_2O_3 is generally utilized because of its cost advantages over alternatives. The wire bonds are comprised of aluminum wires, which are ultrasonically bonded to the dies. Solders have traditionally contained 36%–95% lead for eutectic (183 °C) and higher-temperature solders. Today there is a growing requirement to go to lead-free solders. The baseplates are generally copper, and are attached to the heat sink by screws via a thermal interface material, such as the paste shown in Figure 31.14. From the baseplate up the whole assembly is encapsulated by polymer materials, such as silicone gel, and is epoxy molded. This traditional stacking of the power module has evolved from low-voltage packaging concepts. As automotive high-power requirements continue to be fleshed out, innovative materials and packaging methods are surfacing.

One of the major issues with this method of packaging for the silicon switches is the disparity between the coefficients of thermal expansion (CTEs) of the materials which leads to field failures being brought about through power and temperature cycling. Figure 31.15 presents the typical thermal-conductivity and CTE data of power-module packaging materials. The difference between the CTEs of the die and the solder can lead to voids being created through extended periods of power cycling, ultimately resulting in increased resistance to the vertical current flow in the package and ultimately module failure. Additionally, the large solder area between the DBC and the baseplate and their corresponding CTE mismatch becomes problematic over time. Advances are being made in work on power-module packaging materials and process technology. AlN and Si_3N_4 ceramics offer closer CTE matches to silicon and are being selectively utilized, though they are more expensive. Directly bonded aluminum (DBA) is also being adopted in some packages, allowing increased tolerance of large thermal cycling [25].

Alternative baseplate materials to copper are AlSiC, Cu/Mo, Cu/W, and other MMCs (metal matrix composites) such as carbon-fiber-reinforced metal (Cu, Al) [26]. The specific design must consider the trade-offs among cost, thermal performance, reliability, and component performance.

There is a growing trend toward sintering technologies for replacing solders. Semikron is applying this technology extensively to their power modules, and claims that their sintering process [27][28] for die attachment leads to a 5% lower thermal resistance leading to a 2–3 times higher power-cycling capability. To date, the materials utilized in die-attach sintering operations are composed of silver pastes, which under temperature and pressure processes, form a eutectic bond with the die. Table 31.5 details a comparison of the properties of sintered Ag and Sn–Ag lead-free solder.

Table 31.5. A comparison between solder and silver sintering bonds

Properties	Sn–Ag solder layer	Ag-diffusion sinter layer
Melting temperature (°C)	221	962
Thermal conductivity (W m^{-1} K^{-1})	70	240
Electrical conductivity (MS m^{-1})	8	41
Layer thickness (µm)	~90	~20
Coefficient of thermal expansion (ppm K^{-1})	28	19
Tensile strength (MPa)	30	55

Traditionally the top interconnection of the dies to the insulated circuit substrate has been made by ultrasonic bonding of thick (0.25 to 0.5 mm in diameter) aluminum wires on the top Al pad on the dies as shown in Figure 31.15. To improve its power-cycling capability, aluminum ribbon, copper (Cu) wires [27][28] and sintered or soldered metal straps have been intensively studied and applied in advanced power modules.

Polymer materials are traditionally used in power-module packages as encapsulates for electrical insulation and for mechanical protection. Thermal interface materials (TIMs) are used to increase the thermal conductivity between the power module and heat sink. The TIM represents the largest thermal resistance in the package stack-up and there considerable effort is being made to integrate the baseplate with the heat sink by directly bonding them together [29], thereby eliminating TIMs in the module assembly.

The silicon switches shown in Figure 31.14 are used to perform switching functions in the inverter to shape the current and voltage waveforms to the motor. They can consume upto 40% of the power module's cost. One method to reduce the content of these costly components in the package is to improve the efficiency of their electric power conversion. Another direction is to push up their operational temperature capability. Automotive power devices typically can be operated at junction temperatures of up to 125 °C, and high-power silicon devices capable of continuous operation at 150 °C are coming onto the market. Utilizing higher coolant temperatures reduces the ability to cool the silicon devices, so either the current through them has to be reduced or more silicon devices are required. Alternatively, a wide-bandgap semiconductor such as silicon carbide or gallium nitride could be used at higher temperatures with higher efficiency than with silicon. In addition, these devices are capable of operating at higher frequencies. By going to higher switching frequencies passive components in the system can be reduced, so the volume, size, and weight as well as material costs can be decreased correspondingly. However, these wide-bandgap technologies are still in their infancy, and considerable development and cost reduction needs to occur before they will be ready for inclusion in the extremely cost-sensitive automotive market.

The low-voltage gate drive circuits necessary to control the inverter switches are still temperature-limited and may not survive future automotive environments that are either air cooled or cooled from the ICE radiator. At present the gate drivers are commercial or automotive-grade silicon devices on standard FR [4] printed-circuit boards. Silicon-on-insulator (SOI) technology incorporates an embedded oxide layer just below the silicon surface. This results in a reduction in leakage current and performance advantages over devices designed and built in bulk silicon wafers. Gate drive circuits with SOI technology that will enable operation at temperatures up to and beyond 200 °C are being developed. However, due to the limited-volume production they are still more expensive than traditional silicon alternatives. Growing markets in solar and wind inverter technologies may leverage market opportunities for these new devices, leading to ultimate cost reductions. Automotive environments at extreme temperatures will need to utilize new circuit-board materials. Low-cost materials capable of prolonged operation while enduring high vibration levels without cracking or delamination are needed.

Capacitors, utilized on the DC bus (common rail) in the inverter, are needed to protect the batteries from current ripple fed back from the motor, which can significantly limit the batteries' lifetimes. Capacitor technologies are currently too bulky, expensive, and commercially available only to 105 °C. Additionally, there are issues with reliability and performance at higher temperatures. As temperatures increase their current-ripple-handling capability is reduced, necessitating the addition of even more capacitors to achieve the same performance levels. This adds to cost and volume.

Currently, polypropylene capacitors are the mainstay of the automotive inverter industry for use as DC link capacitors. They have greater longevity than electrolytic capacitors, possess an established benign failure mode, and are less expensive than other film alternatives. Improved film materials and resins used in capacitor packaging are needed as well as low-resistance and -inductance terminal metals. Research on alternative capacitor technologies is being carried out. Ceramic capacitors have higher-temperature capabilities as well as a higher energy density, which results in a lower volume.

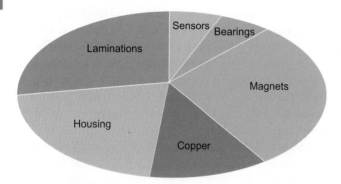

Figure 31.16. Proportional cost breakdown for a traction motor.

However, they are more expensive and have an unacceptable failure mode, which severely limits their usage in automotive applications. High-dielectric-constant glass ceramic material is another potential candidate for bus capacitor applications in inverters.

While the above components and materials are needed, it is recognized that flexible and scalable processes to produce them must be developed. Cooling systems for vehicles may be different in the near and the far term, and the ability of materials and components to survive a broad range of environmental extremes will ensure a greater range of applications as well as a growing supplier base.

In contrast to the power electronics the choice of materials utilized in the electric machinery for HEVs and EVs is considerably smaller. Currently, the motor of choice with OEMs for hybrid technologies is the brushless permanent-magnet machine. Typically, these machines are constructed with either internal or surface-mounted rare earth permanent magnets, silicon–steel laminations and copper windings enclosed by the motor housing. These machines are preferred to alternative motor technologies such as induction and switched-reluctance machines due to their higher efficiency and greater power density. However, they come with a higher cost, primarily due to the rare-earth magnet materials used for the magnets. Figure 31.16 depicts a generic permanent-magnet motor's cost breakdown.

The rare-earth magnets are composed of a variety of alloys and are the strongest permanent magnets manufactured. Neodymium–iron–boron (NeFeB) magnets yield the greatest magnetic field strength and energy product of the available rare-earth magnets. However, other components must be added to the mix to enable operational enhancements. The addition of dysprosium can enable higher-temperature operation and cobalt can increase the flux density. Commercially available NeFeB magnets are specified up to a continuous operational temperature of 150 °C. As operational temperatures rise above that the capability of the magnet decreases. If these magnets are exposed to excessively high temperatures

during operation they can exceed their Curie point and demagnetize. Permanent demagnetization can also occur below the Curie point when the load line is below the knee of the magnet's demagnetization curve. Samarium–cobalt rare-earth magnets are capable of operating at higher temperatures than the NeFeB magnets but have weaker field strength and come with a higher price tag. Additional alloys are being experimented with in order to enhance the magnets' operational capabilities and increase their resistivity in order to reduce losses.

These rare-earth magnetic materials are generally sintered in different combinations to achieve the desired energy densities and temperature characteristics, resulting in high-strength magnets that are utilized in the rotor to generate the motor's torque. Nearly all the rare-earth elements in the world are currently mined in China and many analysts believe that an overall increase in the Chinese electronics manufacturing will consume this entire supply by 2012 [30].

Bonded magnets are attractive from a manufacturing perspective because they can be injection molded to desired shapes with possibly lower manufacturing costs. However, due to the epoxy resins required to hold the flakes together, resulting in a reduced fill factor, bonded magnets have a lower energy product than that of the same volume of sintered magnets and might not be qualified for operation at as high a temperature.

The rotor and stator cores of the motors are composed of silicon–steel laminations, which are typically punched or laser cut to form the desired shapes. Silicon steel is an iron alloy designed for low losses and high permeability. Silicon is utilized in the alloy because it increases the electrical resistivity of the steel, resulting in lower losses. Laminations are available in various thicknesses with different annealing treatments, which can result in different B–H curves. Each lamination is coated with a surface insulation material to increase the electrical resistance between the laminations and prevent corrosion.

Soft-magnetic-core (SMC) materials are being considered for use in stators. These allow net shaping of the stator rather than using laminations, enabling lower manufacturing costs. These materials are bonded and their use allows flux to flow in three dimensions rather than being constrained to two dimensions by electrically isolated, thin laminations. They have lower permanence than silicon steel, however, and utilizing them in traditional motor designs results in machines of larger volume to obtain comparable performance. These SMC materials can enable new design geometries, which are currently just beginning to be researched for potential benefits.

Motor windings are traditionally made from copper; however, aluminum windings have been considered. The conductivity of aluminum wires is only two-thirds that of copper and their resistivity is higher, which leads to increased I^2R losses. However, this option comes with a considerably lower price. The windings' insulation is critical to the motor's reliability. The insulation material comes in different class ratings. Class H is probably the most used in industry and is rated to 180 °C. Above that temperature the cost of the insulation increases rapidly. Insulation rated to 240 °C can cost approximately three times that of Class H insulation. Resins contain the windings in the stator and are generally temperature-matched to the insulation rating.

Motor housings in HEVs and EVs are traditionally made of aluminum due to cost and weight, as opposed to the case of motors in industrial applications, which can utilize cast-iron or rolled-steel frames for the housings. Motor shafts are typically composed of high-strength steel, as are the motor bearings. In addition to the selection of appropriate materials, manufacturing improvements are needed. The cost of permanent-magnet motors could be reduced by design and manufacturing improvements, and by eliminating or reducing the amount of magnets and/or lamination material. As rare earth magnet costs rise and supplies are choked by China's reduction in exports, as their internal consumption grows, shifts are being made to different types of motors. However, improvements will be necessary in alternatives to achieve similar performance to these machines. Increasing the temperature of the coolant, although beneficial from a cost perspective, will create many challenges for electric propulsion systems.

31.9 Tires and rolling resistance

Road transportation depends heavily on tires. Key ingredients of tires are natural rubber (which comes from the rubber tree), synthetic rubber/carbon black (from petroleum), and steel [31]. Sixty percent of the expense of tire production is attributed to petroleum prices. Materials research on tires is directed toward improving the rolling resistance while maintaining the life and handling, reducing or reusing the natural-rubber content, and reducing the energy required in production. One of the key ways to improve automotive efficiency is to reduce the rolling resistance of vehicle tires. This is not a measure of a tire's traction or "grip" on the road surface, but rather simply indicates how easily a tire rolls down the road, minimizing the energy wasted as heat between the tire and the road, within the tire sidewall itself, and between the tire and the rim [32]. Detailed modeling has indicated that a 10% reduction in tire rolling resistance should yield fuel savings of about 1%–2%, depending on the driving conditions and vehicle type [33]. According to research for the California Energy Commission, about 1.5%–4.5% of total gasoline use could be saved if all replacement tires in use had low rolling resistance [34]. This translates roughly into average savings of up to 30 gallons of gasoline per vehicle per year, or $2.5–7.5 billion worth of national-average gasoline savings in the USA.

31.10 Summary

In this chapter, we discussed how climate change, petroleum issues (prices and peaking), energy security, and government regulation for increases in vehicle fuel economy or decreases in CO_2 emission have collectively resulted in the need for transformation of the power-train and fuel choices in the twenty-first century. Road vehicles have started to become more energy-efficient, are emitting lesser amounts of gases and particulate matters, will depend on various sources of fuel, and will utilize hybridization and electrification. Lightweighting and hybridization must play key roles in further improving fuel economy. In addition to new and improved designs, new materials or improvements in existing materials for vehicle structures, powertrains, energy storage systems, motors, power electronics, and tires will make reducing fuel consumption and pollution possible. Most proposed energy-efficiency and lightweighting technologies come with additional cost, since initially the vehicle manufacturers will depend on less expensive fuel-saving alternatives to meet the regulations, but in the future fuel-economy regulations can only be satisfied with the introduction of advanced vehicles that are lightweight, hybridized, or electrified.

31.11 Questions for discussion

1. What factors impact the fuel consumption of a vehicle?
2. Why there is a renewed interest in vehicles with the least possible dependence on gasoline or diesel?
3. What is the difference between lightweighting and "lightweight" materials?
4. What are the various material choices for lightweighting? Which one will be the winner?
5. What is the difference between a hybrid electric vehicle and an all-electric vehicle?
6. What components are similar in electric, hybrid, and plug-in hybrid vehicles?
7. Name various components referred to as power electronics in electric-drive vehicles?
8. Why is there interest in changing the current materials in tires?
9. What would be the major change in vehicles in use in 2020 as compared to with those in current use?

31.12 Further reading

- **H. Heisler**, 2002, *Advanced Vehicle Technology*, 2nd edn., Amsterdam, Elsevier.
- **I. Husain**, 2010, *Electric and Hybrid Vehicles: Design Fundamentals*, 2nd edn., Boca Raton, FL, CRC Press.
- **G. R. Dieter**, 1961, *Mechanical Metallurgy*, New York, McGraw-Hill Book Company, Inc.

- **W. F. Smith**, 1986, *Principles of Materials Science and Engineering*, New York, McGraw-Hill Book Company, Inc.
- **W. D. Callister**, 2001, *Fundamentals of Materials Science and Engineering*, New York, John Wiley & Sons.
- **S. Boschert**, 2006, *Plug-in Hybrids: The Cars That Will Recharge America*, Philadelphia, PA, New Society Publishers.
- **G. Pistoia**, 2010, *Electric and Hybrid Vehicles: Power Sources, Models, Sustainability, Infrastructure and the Market*, Amsterdam, Elsevier.
- **D. B. Sandalow**, 2009, *Plug-In Electric Vehicles: What Role for Washington?*, Washington, DC, The Brooking Institution.
- **A. E. Fuhs**, 2008, *Hybrid Vehicles and the Future of Personal Transportation*, Boca Raton, FL, CRC Press.
- **C. D. Anderson** and **J. Anderson**, 2010, *Electric and Hybrid Cars: A History*, 2nd edn., Jefferson, NC, McFarland & Co., Inc.

31.13 References

[1] **International Energy Agency**, 2007, "Stastics" and "Key stastics," in *Key World Energy Stastics 2007* (International Energy Agency, Paris), www.iea.org/Textbase/nppdf/free/2007/key_stats_2007.pdf.

[2] **S. C. Davis** and **S. W. Diegel**, 2007, *Transportation Energy Data Book*, 26th edn., Oak Ridge, TN, Oak Ridge National Laboratory, 3.1, and 3.2.

[3] **US Government Fuel Economy Website**, *Advanced Technologies & Energy Efficiency*, www.fueleconomy.gov/feg/atv.shtml.

[4] **US DOE**, *FreedomCAR and Vehicle Technologies Multi-Year Program Plan*, Washington, DC, US Department of Energy, Energy Efficiency and Renewable Energy, available at http://www.eere.energy.gov/vehiclesandfuels/resources/fcvt_partnership_plan.pdf.

[5] *21st Century Truck Partnership Roadmap/Technical White Papers*, available at http://www1.eere.energy.gov/vehiclesandfuels/pdfs/program/21ctp_roadmap_2007.pdf.

[6] 2002, *Effectiveness and Impact of Corporate Average Fuel Economy (CAFE) Standards 2002*.

[7] **H. B. Callen**, 1985, *Thermodynamics and an Introduction to Thermostatistics*, 2nd edn., New York, John Wiley & Sons, Inc.

[8] **M. Westbrook**, 2001, *The Electric and Hybrid Electric Car*, Warrendale, PA, SAE International.

[9] **K. Hellman** and **R. Heavenrich**, 2003, *Light-Duty Automotive Technology and Fuel Economy Trends*, Report EPA420-R-03–006, Washington, DC, US. Environmental Protection Agency.

[10] **F. An** and **D. Santini**, 2004, "Mass impacts on fuel economies of conventional vs. hybrid electric vehicles," *J. Fuels Lubricants*, SAE 2004-01-0572.

[11] **M. R. Cuddy** and **K. B. Wipke**, 1997, "Analysis of the fuel economy benefit of drivetrain hybridization," in *SAE International Congress & Exposition*, Detroit, MI.

[12] **Government Fuel Economy Website**, "Find a car" (listing of fuel economy of vehicles of various model years), www.fueleconomy.gov/feg/findacar.htm.

[13] **T. Markel** and **A. Simpson**, 2007, "Cost–benefit analysis of plug-in hybrid electric vehicle technology," *World Electric Vehicle Assoc. J.*, **1**, 1–8.

[14] **US DOE**, 2008, *Annual Progress Report for Energy Storage Research and Development*, Washington, DC, US Department of Energy, available at www1.eere.energy.gov/vehiclesandfuels/pdfs/program/2007_energy_storage.pdf.

[15] **A. Chu**, 2007, "Nanophosphate lithium-ion technology for transportation applications," in *23rd Electric Vehicle Symposium*, Anaheim, CA.

[16] **T. Tan**, **H. Yumoto**, **D. Buck**, **B. Fattig**, and **C. Hartzog**, 2007, "Development of safe and high power batteries for HEV," *23rd Electric Vehicle Symposium*, Anaheim, CA.

[17] **US DOE**, 2007, *Annual Progress Report for Energy Storage Research and Development*, Washington, DC, US Department of Energy, available at www1.eere.energy.gov/vehiclesandfuels/pdfs/program/2006_energy_storage.pdf.

[18] **D. Doughty**, **E. Roth**, **C. Crafts** *et al.*, 2005, "Effects of additives on thermal stability of Li ion cells," *J. Power Sources*, **146**, 116.

[19] **G.-H. Kim** and **A. Pesaran**, 2007, "Analysis of heat dissipation in Li-ion cells and modules for modeling of thermal runaway," in *Proceedings of the 3rd International Symposium on Large Lithium Ion Battery Technology and Application*, Long Beach, CA.

[20] **A. N. Jansen**, **D. W. Dees**, **D. P. Abraham**, **K. Amine**, and **G. L. Henriksen**, 2007, "Low-temperature study of lithium-ion cells using a LiSn micro-reference electrode," *J. Power Sources*, **174**, 373.

[21] **Z. Chen** and **K. Amine**, 2006, "Tris(pentafluorophenyl) borane as an additive to improve the power capabilities of lithium-ion batteries," *J. Electrochem. Soc.*, **153**, A1221.

[22] 2009, http://www.hybridcars.com/hybrid-sales-dashboard/december-2009-dashboard.html.

[23] **P. Roussel**, 2009, *Power Electronics in Electric & Hybrid Vehicles*, report, Lyon, Yole Développement, www.yole.fr.

[24] **U. Scheuermann** and **U. Hecht**, 2002, "Power cycling lifetime of advanced power modules for different temperature swings," in *Proceedings of PCIM*, Nuremberg, pp. 59–64.

[25] **T. G. Lei**, **J. N. Calata**, **K. D. T. Ngo**, and **G.-Q. Lu**, 2009, "Effects of large-temperature cycling range on direct bond aluminum substrate," *IEEE Trans Device Materials Reliability*, **9** (4), 563–568.

[26] **C. Zweben**, 2006, "Thermal materials solve power electronics challenges," *Power Electronics Technol.*, February, 40–47.

[27] **C. Göbl**, 2009, "Sinter technology for power modules," *Power Electronics Europe*, issue 4, 28–29.Ref.c. By **Carl Zweben**,

[28] **C. Zweben**, 2006, "Thermal materials solve power electronics challenges," *Power Electronics Technol.*, February, 40–47.

[29] **N. Nozawa**, **T. Maekawa**, **S. Nozawa**, and **K. Asakura**, 2009, "Development of power control unit for compact-class vehicle," in *SAE World Congress & Exhibition.*

[30] **Synthesis Partners**, 2010, *Rare Earth Materials and Rare Earth Magnets*, Reston, VA, Synthesis Partners' LLC, also available in US DOE, 2010, *Critical Materials Strategy*, www.energy.gov/news/documents/critical-materialsstrategy.pdf.

[31] **H. B. Pacejka**, 2005, *Tire and Vehicle Dynamics*, 2nd edn., Warrendale, PA, SAE International.

[32] ***Encyclopedia Britannica On-Line***, 2008, "Tire materials," www.britannica.com/eb/article-7283/tire.

[33] **T. Markel**, **A. Brooker**, **V. Johnson**, *et al.*, 2002, "ADVISOR: a systems analysis tool for advanced vehicle modeling," *J. Power Sources*, **110**(2), 255.

[34] **California Energy Commission**, "Fuel-efficient tires and *CEC Proceedings* documents page," www.energy.ca.gov/transportation/tire_efficiency/documents/ index.html.

32

Transportation: aviation

Robin G. Bennett, Linda A. Cadwell Stancin, William L. Carberry, Timothy F. Rahmes, Peter M. Thompson, and Jeanne C. Yu

The Boeing Company, Everett, WA, USA

32.1 Focus

The technology of flight provides immeasurable benefits for today's society: promoting global trade and commerce, providing humanitarian relief, and connecting people. In the next millennium, progressive environmental considerations will play a key role in our ability to continue to provide these benefits seamlessly. As with other transport, the consumption of petroleum-based fuels and materials draws from the Earth's finite natural resources. To move toward fully sustainable aviation, there must be a continued focus on reducing the environmental footprint over the product life cycle.

32.2 Synopsis

To ensure a balance between the social and economic benefits of aviation and the energy and environmental impacts, the aviation industry is working on improvements across the entire life cycle of its products and services. Opportunities for environmental improvement lie in advanced materials and manufacturing technologies, improved aerodynamics systems and engine efficiency, alternative fuels, increased fleet operational efficiency, and aircraft recycling.

The design of aircraft is highly dependent on materials and technologies that can meet the stringent performance requirements established by manufacturers and aviation authorities to ensure safe flight. Additionally, aircraft manufacturers focus on technologies that can improve fuel efficiency, which reduces the quantity of fuel required, thereby lessening the impact on the environment. Technologies can provide improvement for new aircraft, but also can provide improvement in the performance of existing aircraft. Some technologies modify the airplane itself and other technologies improve operational systems, helping aircraft to fly the most fuel-efficient routes. Both result in fuel conservation.

In addition to conservation of fuel, technologies are also focused on the fuel itself and sustainable alternatives. Sustainable fuels based on renewable resources provide long-term viability and can reduce environmental impact if the source of the fuel is plant-based and offsets aircraft carbon dioxide emissions through photosynthesis.

At the end-of-service, the aircraft-recycling industry has captured the high-value materials found in aircraft for processing reuse by other industries. As the material composition of aircraft changes over the years, to be fully sustainable the technologies will need to evolve to utilize those new materials and ultimately target reincorporating them into new aircraft. Ideally, the goal is a closed-loop materials cycle that optimizes material resource utilization, minimizes the energy required for processing and minimizes environmental impacts over the entire life cycle of an aircraft.

Please note that, throughout this chapter, numerous Boeing examples are provided to illustrate general industry trends.

Figure 32.1. Early aircraft construction, illustrating hand-applied fabric lay-up on wood (left). Aircraft construction in 2005 of the first full-scale one-piece fuselage section for the Boeing 787 started with computerized lay-down of composite tape on a mold (right). This section is 22 feet (7 m) long and nearly 19 feet (6 m) wide.

32.3 Historical perspective

Materials development has played a major role in the success of the aviation industry. Each generation of lighter, stronger materials enables the achievement of new levels of aircraft performance and efficiency. In the early 1900s airplanes were commonly built using wood-and-linen structures. The resulting airplanes were capable of carrying small payloads, usually one or two people, over short distances, at low speeds.

As aviation grew, there was a need to fly higher, further, faster, and with greater payload. The increased stresses of the new flight regimes required new materials. By the 1910s simple wood fuselages were being replaced by laminated wood, capable of carrying higher loads and increasing efficiency. Metal came into use in the 1920s, allowing airplanes to carry more passengers more efficiently than ever before.

By the 1950s material advances had enabled the first commercially viable jet-powered airplanes. Airplanes such as the de Havilland Comet, the Boeing 707, and the Douglas DC-8 were capable of carrying more than 100 passengers thousands of miles. On a passenger-mile basis this new generation of airplanes, built of stronger, lighter aluminum alloys, achieved new levels of fuel efficiency.

The newest commercial aircraft, such as the Boeing 787, are utilizing carbon-reinforced epoxy composite materials in their wing and fuselage skins (Figure 32.1).

32.4 Aviation's influence on energy and environmental sustainability

Since the beginning of the jet age nearly 40 years ago, advances in technology have allowed the industry to achieve incredible reductions in the environmental impact of airplanes. These advancements in technology have resulted in a 70% reduction in fuel consumption and carbon dioxide emissions (which are directly proportional to the amount of aviation fuel consumed) (Figure 32.2). New technologies include material, structural, aerodynamic, systems, and engine improvements.

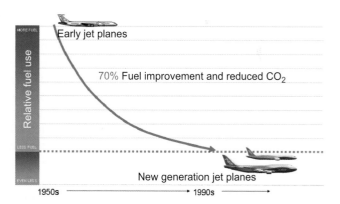

Figure 32.2. Significant changes to aircraft design have netted an approximately 70% improvement in overall aviation fuel efficiency.

32.4.1 Advanced materials and manufacturing processes

High-performance materials

The aviation industry is pursuing advanced materials and processes for environmental sustainability through a series of technology emphasis areas: (1) high-performance materials that improve structural efficiency (strength/weight), (2) high-performance materials that offer

Figure 32.3. The expanded use of composites in the new Boeing 787.

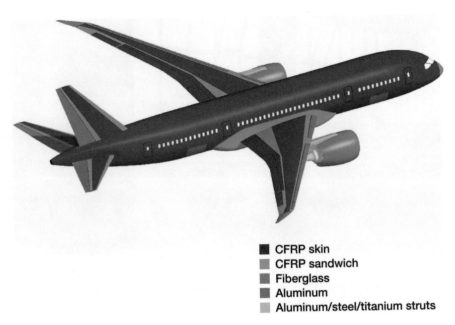

- CFRP skin
- CFRP sandwich
- Fiberglass
- Aluminum
- Aluminum/steel/titanium struts

aerodynamic benefits, (3) reduced material consumption to produce aircraft parts, (4) replacement of older processes and materials with more environmentally progressive options, and (5) reducing the energy required to produce both raw materials and the aircraft components.

An example of a major material change is from traditional aluminum alloys to toughened epoxy prepregs (pre-impregnated materials). Boeing led the commercial implementation of structural composites from the extensive use of fiberglass epoxy composites on aerodynamic control surfaces, fairings, and trailing edge panels on the 747 to increased applications on the 757 and 767 (main landing- and nose-gear doors, engine cowls, wing spoilers and ailerons, and empennage trailing-edge panels, rudder, elevators, and fin and stabilizer tips). The 777 continued the trend, utilizing toughened epoxy prepreg in a more primary structural application for the empennage stabilizer and fin torque box. With the 787 design, the shift from aluminum to composites reduces weight and thus reduces fuel burn and emissions (Figure 32.3). It also reduces the amount of chemical processing required to provide corrosion protection for aluminum airplanes, decreases the need for maintenance processes, and increases aircraft lifespan.

Since the application of composites on a large scale is relatively new, there is a promise of continued performance improvement for weight reduction of structural composites. Work is occuring worldwide to improve structural composites. Figure 32.4 shows the development trend in carbon-fiber strength and modulus.

Improvements in metallic materials also support weight-efficient structures for aviation. Usage of aluminum–lithium alloys is increasing to take advantage of the reduced density and improved stiffness and corrosion resistance. It may also enable a reduction in the use of chromium-containing finishes. Aluminum–lithium alloys are being used in commercial aircraft, including applications on the Bombardier C-Series, Airbus A380, and Boeing 787.

Material reduction and processing changes

To reduce material consumption, there are active technology efforts toward reducing "buy to fly." "Buy to fly" is a term used to describe how much material must be acquired versus how much material is in the finished part. As an example, Boeing is implementing near net shape thermoplastic fittings, see Figure 32.5, to replace machined metal components. These reinforced thermoplastic materials are recyclable, and the process used to manufacture the parts produces a net shape part and reduces energy consumption. In addition, the parts are lighter and thus help reduce overall weight and improve aircraft efficiency.

The development of more monolithic process methods also reduces the number of joins and fasteners. Having fewer parts results in less structural weight and more manufacturing efficiency, reducing energy consumption and waste.

The aviation industry has been working diligently to reduce the environmental impact of material processes used to produce aircraft components. In the titanium-production area, Figure 32.6 shows that greenhouse-gas emissions are reduced as manufacturing processes are optimized.

32.4.2 Engine-material improvements

One of the primary improvements for fuel efficiency of aircraft engines is lighter-weight high-temperature materials. As with airframe structures, lower-density

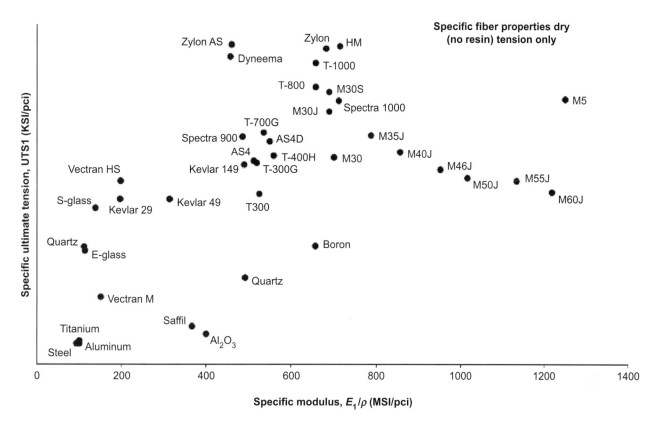

Figure 32.4. Composite fibers show improvements in structural capabilities [1]. KSI is thousand pounds per square inch, pci is pounds per cubic inch and MSI is mega pounds square inch, 1 psi/pci = 0.25 pascal/kgm^3.

Figure 32.5. Left: near net shape thermoplastic short fiber fitting replacing metal fittings in Boeing's 787. Right: a 787 test fuselage barrel is all one piece, reducing the number of parts and process complexity.

materials, such as titanium aluminide intermetallic alloys, can reduce engine core weight, and just as importantly, higher service temperatures can enable improved engine thermodynamic efficiencies. Nickel-based superalloys that take advantage of special solidification processes to orient grain structures to give maximum creep strength are being developed to this end.

Composite advancements also factor in, but the high-temperature environment of the engine has historically limited the average composite makeup of engines. However, recent advances in polymeric composites have increased use temperatures from 120 to 290 °C and are thus seeing expanded use in nacelle

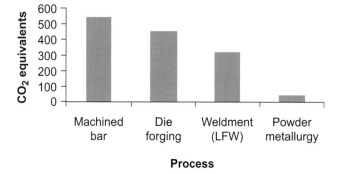

Figure 32.6. Titanium processes influence greenhouse-gas emissions [2]. LFW is linear friction welding.

design. These material technologies, in concert with improved engine design, yield substantial weight savings and enhanced performance. More improvements are still possible, and alternatives such as metal matrix composites and ceramic matrix composites are currently in development.

32.4.3 Reduce fuel consumption through aircraft operations

There are three primary ways to improve the fuel efficiency of worldwide airplane fleets. First, airplane modifications, such as retrofit winglets, advanced avionics, and system upgrades for lower power consumption. Second, maximized airline operations can be realized through optimized fleet management, scheduling maintenance for the most efficient aircraft performance, and specific on-ground practices, such as use of a single engine during aircraft taxiing. Third, airspace optimization ensures that aircraft fly as efficiently as possible. Examples of airspace optimization include adjusting flight paths for reduced fuel burn, optimized and real-time air-traffic management, and specific flight practices, such as continuous-descent arrivals.

32.4.4 Sustainable alternative fuels for aviation

Aviation fuels consist of refined hydrocarbons derived from conventional sources including crude oil, natural-gas liquid condensates, heavy oil, shale oil, and oil sands. Kerosene-type fuels typically are approved for use in gas-turbine engines and aircraft by a certifying authority subsequent to formal submission of evidence as part of a type-certification program for that aircraft and engine model. The two commonly used fuels allowable are Jet A and Jet A-1, which are burned in combustion chambers by vaporizing fuel into a rapidly flowing stream of hot air. The hot gases are diluted to lower temperatures for the turbine stages. The performance of these fuels must meet conditions on various properties, including thermal stability, combustion characteristics, density, net heat of combustion, fuel atomization, viscosity, freezing point, materials compatibility, storage stability, lubricity, flash point, electrical conductivity, and water and particulate content.

A new standard, ASTM D7566, has paved the way for future bio-derived SPK (synthetic paraffinic kerosene), from plants, animal tallow, etc. and acceptance by the aviation industry [3]. See Chapter 25 for more information on sustainable biofuel feedstocks.

There are several plant-derived oils that are sustainable fuel sources, if implemented properly. Owing to the high oil production of algae, and their ability to reproduce up to several times per day, oil from algae remains

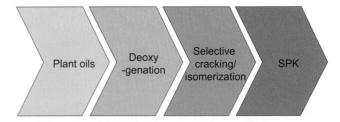

Figure 32.7. A possible bio-derived oil-to-SPK conversion process path.

Table 32.1. Approval of fuels specifications
Fuel properties vary within an expected range, but are never exactly the same. These property limits are the concern of standards committees
The first aviation biofuels to be approved will be SPK from the Fischer–Tropsch process as well as selective cracking and isomerization processes
Other conversion paths expected in the coming years include synthesis via alcohols, pyrolysis, and perhaps microbial production and other routes

the key promising source to supplement and perhaps eventually satisfy the world's need for high-energy liquid fuel products for aviation as well as other industries. As of 2010, four key sustainable feedstocks have been highlighted by biofuel flight and engine-test programs, namely camelina, algae, jatropha, and babassu [4].

In addition to the feedstock identification, significant progress has been made in the last several years toward identifying chemical processing methods that can produce these jet fuels. Sustainable bio-derived oils (triglycerides and free fatty acids) are being converted to the various fuel types that a variety of industries, including aviation, need. The use of the same types of molecules as those typically found in Jet A and Jet A-1 requires no changes to the aircraft systems or infrastructure. A simplified process is shown in Figure 32.7.

However, as of 2010, current processing does not produce the aromatic compounds which are typically present in conventional jet fuel and allowable up to 25% by volume (per specification). Thus, blending of the bio-SPK with conventional jet fuel is necessary in order to meet specifications for aviation turbine fuel, namely ASTM D 1655, ASTM D 7566, and DEFSTAN 91–91, and the military JP-8 specification for various fuel properties [3][4][5][6][7]. See Table 32.1.

Increasing the bio-derived fraction of jet-fuel blends could offer economic, environmental, and performance benefits. Therefore, continued research into aspects to enable higher blends (greater than 50%) such as materials compatibility, renewable aromatics, chemical composition limitations, and density, is needed. Figure 32.8 shows an example of acceptable 50% blending, but also clearly indicates that use of 100% SPK is beyond current specification limits.

32.4.5 Aircraft recycling

When an aircraft is no longer operable, it still retains value. It is usually parked in a remote location where parts can be removed and packaged for inspection, repair and re-sale, and the airframe cut up for scrap [9] (see the photo sequence in Figure 32.9).

Generally, the highest-value components of a retired aircraft are the engines, the landing gear, and the auxiliary power unit. Other parts have value depending on the vintage of the aircraft and the number of units from that aircraft family of similar vintage still in revenue service.

Once the aircraft's value has diminished, it is torn down for scrap. However, another set of challenges work against optimized recovery of the many expensive materials that went into its original manufacturing. Unlike automobiles, home appliances, and building

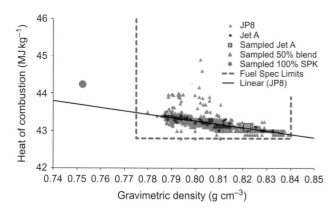

Figure 32.8. A comparison of the heat of combustion with density of one SPK blend is shown with historical data [8]. Data for several thousand JP-8 and Jet A fuels constitute the scatter plot.

scrap, which are generally collected and processed in "scrap yards" near the large population centers that consumed the items, aircraft are parked and scrapped in remote areas far away from any recycling facilities.

Aircraft are among the most complex of the commonly produced commodities from heavy manufacturing, and the variety of materials used in aircraft manufacturing combined with the extreme durability that is required of the finished product makes aircraft one of the most challenging commodities to recycle. For example, a commercial aircraft contains upwards of 10–15 alloys aluminum. Aircraft aluminum alloys have very precise proportions of zinc, copper, magnesium, etc. to meet very specific performance requirements characteristic of various parts of an aircraft's structure (strength, flexibility, weldability, etc.).

Just as plastics recyclers go to great lengths to separate plastics by category, if aluminum is not sorted by alloy series then all of the alloying ingredients end as a mix. The ingot from mixed aircraft aluminum scrap retains none of the original performance characteristics from any of its constituent alloys, which now effectively become contaminants that collectively reduce the subsequent suitable manufacturing applications. A primary research focus in aircraft recycling today is the search for an effective, efficient, and affordable technology that separates scrap aircraft aluminum alloys to maximize recyclate value. As use of structural composites increases in commercial aircraft, the effective reclamation of these materials also becomes important.

32.5 Summary

In concert with university and industry researchers, aerospace engineers and scientists are developing new technologies, such as advanced lightweight materials, efficient and low-impact manufacturing processes, improved engine technology, alternative fuels, and aircraft recycling. These technologies, together with improvements in operations, will be essential in order to progress toward fully sustainable aviation. Although

Figure 32.9. The sequence of end-of-life aircraft processing: parked and components scavenged; parting-out; cutting up for scrap.

new aircraft will have much less impact than their predecessors, the near doubling in their number will result in competition from other users of fuel and materials unless actions are taken. Hence, the industry is committed to carbon-neutral growth from 2020 [10].

32.6 Questions for discussion

1. What will be the characteristics of the materials which will enable the next leap in aviation efficiency and sustainability?
2. Are fully sustainable materials for aviation possible? What would be required? How would materials need to change?
3. What is the potential applicability of solar-, wind-, hydrogen-, or other renewable energy-based technologies for future aircraft?
4. How can recycled materials reduce overall energy consumption and life-cycle CO_2 emissions over virgin-based materials?
5. How does material selection affect the life-cycle footprint of aviation?

32.7 Further reading

- **D. Stinton**, 1998, *The Anatomy of the Airplane*, 2nd edn., Reston, VA, AIAA. A technical perspective of aircraft in terms of operational, design, and natural requirements.
- **A. A. Baker**, **S. Dutton**, and **D. Kelly**, 2004, *Composite Materials for Aircraft Structures*, 2nd edn., Reston, VA, AIAA. This book introduces all aspects of the composite-materials technology in aeronautical design and structure.
- **L. Nicolai** and **G. Carichner**, 2010, *Fundamentals of Aircraft and Airship Design, Volume I: Aircraft Design*, Reston, VA, AIAA. This book focuses on the design of aircraft from the conceptual phase to reiteration.
- http://www.greenaironline.com/. An independent website with information on aviation and the environment.
- http://www.enviro.aero/default.aspx. The Air Transport Action Group (ATAG) website with information on aviation and the environment.
- **M. Pinkham**, 2008, "Engine builders are running red hot," *American Metal Market*, **117**(10), 54–56.

32.8 References

[1] **L. Cadwell Stancin**, and **B. Roeseler** (The Boeing Company), 2009, *Composites Structures: Aerospace History and Our Future*, Composites Design Tutorial, Stanford University, Stanford, CA.

[2] **J. Cotton**, **R. Boyer**, **Briggs, Slattery, K.**, and **G. Weber** (The Boeing Company), 2010, "Titanium Development Need for Commercial Airframes: An Update," in *AeroMat 2010*, Bellevue, WA.

[3] **American Society for Testing and Materials International**, 2009, *Aviation Turbine Fuel Containing Synthesized Hydrocarbons*, ASTM D 7566.

[4] **T. F. Rahmes**, **J. D. Kinder**, **M. Henry** *et al.*, 2009, "Sustainable bio-derived synthetic paraffinic kerosene (bio-SPK) jet fuel flight tests and engine program results," in *9th AIAA Aviation Technology, Integration and Operations Conference (ATIO)*, Hilton Head, SC, American Institute for Aeronautics and Astronautics, pp. 1–19.

[5] **American Society for Testing and Materials International**, 2006, *Standard Specification for Aviation Turbine Fuels*, ASTM D 1655.

[6] **Ministry of Defence**, 2006, *Turbine Fuels, Aviation Kerosine Type, Jet A-1*. Defence Standard 91–91; NATO Code: F-35; Joint Service Designation: AVTUR.

[7] **Department of Defense**, *Turbine Fuels, Aviation Kerosene Types*, MIL-DTL-83133F.

[8] **Defense Energy Support Center (DESC)**, 2009, *Petroleum Quality Information System Fuels Data (2008)*.

[9] **W. Carberry**, (The Boeing Company), 2010, "Trash to Cash – Reclamation of Retired Aircraft," in *Defense Manufacturing Summit*, Miami, FL.

[10] **Air Transportation Action Group (ATAG)**, 2008, *Aviation Industry Commitment to Action on Climate Change*, http://www.enviro.aero/Aviationindustryenvironmentaldeclaration.aspx.

33 Transportation: shipping

David S. Ginley

Process Technology and Advanced Concepts, National Renewable Energy Laboratory, Golden, CO, USA

33.1 Focus

Shipping delivers huge numbers and amounts of goods to consumers worldwide, whether that be through container ships full of automobiles, tankers full of oil, or trawlers full of fish. Despite being such a central component of the global economy, shipping is not regulated by the Kyoto Protocol, and recent studies project that shipping will produce between 400 Mt and 1.12 Gt of CO_2 by 2020, which would be more than aviation and up to 4.5% of global CO_2 (http://www.guardian.co.uk/environment/2008/feb/13/climate-change.pollution). This chapter focuses on the need for shipping to change in a carbon-sensitive world and possible changes that would allow shipping to reduce its environmental impact while still delivering increasing amounts of goods efficiently worldwide.

33.2 Synopsis

Since oceans cover approximately 70% of the Earth's surface, the development of shipping was inevitable. In addition to allowing human communities on different land masses to engage in the crucial activity of trading commodities such as spices and gold, shipping nucleated the cross-fertilization of groups by transporting people as well. These activities have grown over time to the point that shipping now transports over 90% of the total goods worldwide. In fact, this amount is still growing as international trade continues to expand [1].

As shipping has evolved from dugout canoes to supertankers, it has also become both a major user of energy and a significant generator of pollution. This is especially problematic because shipping is nearly unregulated in terms of CO_2 emissions, and very poor statistics are kept worldwide. The industry has begun to realize that such a model is untenable and is beginning to look at ways to remedy the negative environmental impacts of shipping while still continuing to grow. Most of the approaches to reducing emissions are materials-centered, and thus the basic materials science of shipping is beginning to change dramatically, from its wooden and iron foundation to the current state of the art using composites and smart materials. This revolution in materials usage includes the overall ship structure, as well as the propulsion and ancillary systems. The shipping industry is just beginning to think creatively, and, with a current commercial fleet of nearly 70,000 commercial vessels, registered in over 150 nations, change will be slow.

Although shipping generally remains in the background of public awareness, being overshadowed by more conspicuous concerns (e.g., automobile transportation and power generation), it is critical to the distribution of goods, food, *and energy* in a complex international way, with far-reaching ramifications for the world economy. Further, although shipping has traditionally had a "green" aura, the realization is slowly emerging that this massive industry has substantial environmental impacts. Within this context, this chapter discusses some of the near- and longer-term changes that are beginning to occur in shipping to address these concerns.

33.3 Historical perspective

Shipping has a very long history, dating back to ancient times. In fact, the earliest commercial use of shipping was probably around 3000 BCE in Egypt. Archaeological evidence indicates that the early Egyptians developed several shipbuilding techniques, constructing vessels of wooden planks (or bundles of papyrus reeds) that were lashed together with woven straps or fastened with wooden pegs. Seams between the planks could be filled with reeds or grass or caulked with pitch to prevent leaks. Figure 33.1(a) shows an image of an Egyptian ship dating from approximately 2500 BCE and measuring 43.6 m (143 ft) long and 5.9 m (19.5 ft) wide. It was built largely of Lebanon cedar, with pegs and other small parts made from native tree species such as acacia and sycamore. Whereas this ship was equipped with oars, use of sails was also common for travel along the Nile, originally as an adjunct to oar power. Sails were also favored in China, where they were used in conjunction with the Chinese-developed stern-mounted rudder in the classic junk sailing-vessel design (Figure 33.1(b)). Even though this design was first developed during the Han Dynasty (220 BCE–200 CE), such vessels are in use virtually unchanged today. They were used extensively throughout Asia and opened up sea travel and trade.

In Europe, the most skilled early seafarers were the Norse explorers known as Vikings (c. 800–1100 CE). Although they are best known today for their warships, the Vikings were also merchants and developed cargo ships for carrying out trade by sea. Their ships were generally constructed using the clinker technique, in which the wooden planks (preferably oak or pine) making up the hull of a boat were attached to each other, using either wooden dowels or iron rivets, so that they overlapped along their edges. These ships were typically powered by sail, supplemented with a limited number of oars.

Later European merchant ships included the cog (tenth to twelfth centuries, built of oak, with a single mast and a square-rigged single sail) and the carrack (fifteenth century). The carrack was a three- or four-masted sailing ship developed in Western Europe for use in the Atlantic Ocean: so it had to be large enough to be stable in heavy seas and roomy enough to carry provisions for long voyages in addition to cargo. This basic design remained essentially unchanged until the mid nineteenth century, throughout the period dominated by sailing ships.

The construction designs of ships slowly began to change as new materials were introduced. Wooden hulls were first supplemented and then replaced by copper and especially iron. In turn, these materials were replaced by steel when it became available in the 1880s, providing great savings in terms of weight and,

(a)

(b)

Figure 33.1. (a) The famous Royal Ship of King Cheops (also known as Khufu), who ruled Ancient Egypt from 2589 to 2566 BCE. This ship was discovered in a pit at the foot of the Great Pyramid of Giza in 1954 and carefully reassembled from 1,224 pieces. (Image obtained from http://www.kingtutshop.com/freeinfo/egyptian-boats.htm, which, in turn, is from *The Boat beneath the Pyramid: King Cheops' Royal Ship* by Nancy Jenkins.) (b) A photograph of a modern Chinese junk against the Hong Kong skyline (http://www.billiewalter.com/, 2010).

therefore, cost. Initially, structures were formed by riveting the metal plates together to make them watertight, but this technique had been replaced by welding by the 1940s. Beginning in the 1950s, specialized steels

Figure 33.2. The rate of increase in ship size vs. time over the last century. Ships have grown to the point that cargo ships gross over half a million tonnes and even passenger ships are over 300,000 tonne.

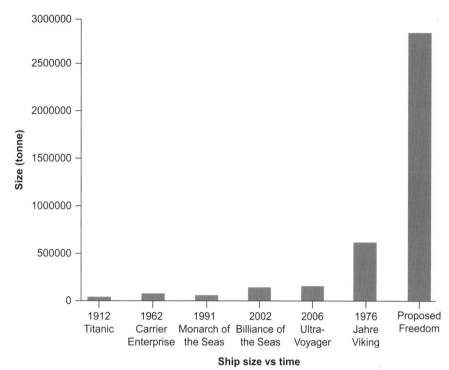

with properties tailored for use in shipbuilding (e.g., increased fracture toughness) became available.

Alongside changes in construction materials came changes in motive power. In the early 1800s, sails began to be replaced by coal-based steam power, typically driving propellers or paddlewheels. Initially, steam power was slow to catch on in commercial shipping because of the cost of the fuel compared with free wind. However, the advantages of steam power – speed, reliability, and shorter routes that did not depend on the trade winds – soon gave it the advantage. Although coal power did improve the reliability of ocean transport and lead to a great increase in shipping, it was the growth of the oil industry in the early 1900s that promoted the development of the current generation of fuel-oil-based shipping, including the construction of ships as large as supertankers. The rate of growth for ships is plotted vs. time from the Titanic on in Figure 33.2, where the scale has increased to the point that the future *Freedom*, a virtual ocean-going city, would be over 2 million tons.

33.4 Approaches to the future of shipping

33.4.1 The need for CO$_2$ reductions in shipping

Shipping has historically moved both people and goods, and was the key method of transport from about 1500 CE until the mid twentieth century, when it was supplanted by the advent of air travel for moving people [1]. For goods, however, the cost of air transport is much greater than that of shipping. Thus, shipping is responsible for transporting nearly 90% (by weight) of goods worldwide.

Another key consideration in comparing modes of transport is the relative amount of CO$_2$ generated. In terms of megatonnes of CO$_2$ emitted per kilometer traveled, air transport generates 70 times more CO$_2$ than modern shipping (Figure 33.3); however, the 70,000 ships in the commercial shipping fleet cover a much larger number of kilometers than commercial cargo planes (of which there were 1,755 in 2009 [2]). A report from the International Chamber of Shipping shows that, from 1989 to 2008, in terms of total mass carried and distance traveled, sea transport increased from 21,000 to 33,000 Gt·mile (i.e., from 34,000 to 53,000 Gt·km), and the rate of growth is increasing, leading potentially to more fuel use and higher emissions [2].

As a result, the shipping of goods at present generates approximately 4%–5% of global CO$_2$. More importantly, because it uses high-sulfur fuels in many cases, shipping is one of the worst overall polluters: ship emissions represent more than 16% of sulfur emissions from world petroleum use and more than 14% of nitrogen emissions from global fuel-combustion sources [3][4]. Currently, the 70,000 ships in the commercial merchant fleet burn about 200 Mt of fuel per year. Considering that about 20,000 new ships are currently on order [5], this amount is expected to grow to 350 Mt per year by 2020. Some studies have shown that container movement will have increased seven-fold worldwide in 2050 compared with 2005 [6]. As attention is slowly being directed toward the fuel consumption by and

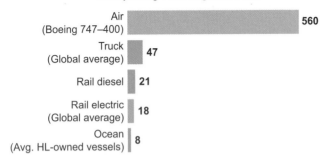

Grams of CO₂ emitted by transporting 1 ton of goods 1 Km

Air (Boeing 747–400)	560
Truck (Global average)	47
Rail diesel	21
Rail electric (Global average)	18
Ocean (Avg. HL-owned vessels)	8

Figure 33.3. Comparison of different modes of transportation in terms of amounts of CO₂ emitted (in grams) upon transporting 1 t of goods a distance of 1 km (based on data from the Network for Transport and Environment, Sweden).

emissions from the shipping industry, it is clear why the shipping industry will need both to increase efficiency and to reduce emissions. Presently, the shipping industry is virtually unregulated for greenhouse gas emissions. There are no international policies or incentives related to fostering new technologies and approaches. Nonetheless, the industry is not stagnant, and work that will help define the future of ocean transport is proceeding. The remainder of this chapter discusses the possible means for accomplishing this end. The other major motivation for the development of more efficient ships is that fuel costs have become the largest operating cost for ships, with an increase of 300% over 10 years, such that they amount to 41% of the total cost of shipping [7].

33.4.2 Requirements for modern ships

Whereas ships might usually be considered as a fairly simple technology, the reality is that, to optimize performance and lifetime, ships must meet a complex set of materials and performance requirements. These requirements can be divided into two main categories: materials issues related to construction and materials issues related to performance and efficiency.

Materials issues related to construction include factors such as overall strength (i.e., resistance to failure, given by the applied load per unit area), toughness (i.e., the energy required to crack a material), durability, weldability/processability, formability, availability, and affordability [8][9].

Materials issues related to performance and efficiency include weight; fracture toughness; and resistance to corrosion, biological attack, fire, wave loading (constant low-frequency, high-cycle fatigue stress), vibration, and hull damage. The shipping environment is quite stressful for materials, since there are

tremendous mechanical loads on a ship during transit. In addition, the chemically corrosive environment of salt water and sea spray impacts all ship components.

With an average vessel lifetime of 25–35 years, these constraints become even more important. In the modern world, steel is an obvious choice for the structural components of ships: it is fairly inexpensive, easily formable, and weldable; it has great strength; and it has reasonable corrosion resistance and lifetime. However, steel is heavy, it *does* corrode, and it is a good substrate for undesired organic growth (biofilm). Thus, alternatives are now of interest.

33.4.3 Ship structures

Historically, ships started out being made from easily accessible lightweight woods and reeds. Later, as demands on shipping increased and manufactured materials became available, shipbuilding transitioned into iron and then high-grade steels, especially for high-performance ships [10]. More recently, there has been increasing work on the possibility of new materials for all or parts of the ship, interestingly including a return to wood for some applications. There is now also a focus on the use of composites, such as glass-reinforced polymer (GRP) composites, also called fiber-reinforced plastic (FRP), in ship construction, because this can potentially increase strength and lifetime and improve corrosion resistance while reducing weight. However, to date, the cost of implementing an all-composites ship has been prohibitive.

Nevertheless, some hybrid composites such as sandwich structures or metal foams could be used broadly in ship construction. Recent studies at the Fraunhofer Institute in Chemnitz, Germany, have shown that use of Al or Ti metal foams could reduce the weight of an average-sized freight vessel by up to 30%, saving more than 1,000 tons of its deadweight. [11][12].

Efforts have also been made to replace specific parts of ships with lightweight high-performance materials. For example, interior watertight doors are quite heavy, and new lattice block materials (LBMs), comprising three-dimensional metal lattices (Figure 33.4), are being investigated to replace the conventional steel used for these doors [13]. For LBMs, the variability of the internal triangular matrices (Figure 33.4) controls the density, internal angles, overall dimensions, and characteristics of the final structure. Because LBMs are cast from molten metal using precisely engineered molds, they can be constructed in numerous shapes and sizes and with various radii of curvature. They can also be made with integral skins and can incorporate complementary material systems in the interior voids of their structure. Similarly to the case for the doors, the entire superstructure (the portion of the ship above the hull) could be

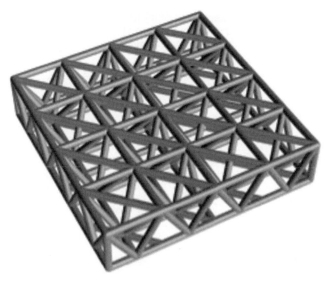

Figure 33.4. A schematic representation of lattice block material.

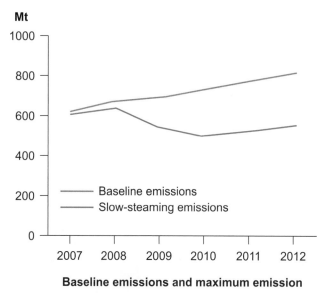

Baseline emissions and maximum emission

reductions with slow steaming

Figure 33.6. Baseline emissions and maximum emission reductions obtainable with slow steaming [19].

Figure 33.5. The superstructure portion of a ship (outlined in blue) that could be replaced by a module made of composite materials. From D. McGeorge, "Lightweight and load-bearing composite structures: a case study on superstructures," presented at Skibsteknisk Selskab, Copenhagen, November 19, 2008.

replaced with a composite structure, especially on large ships such as supertankers (see Figure 33.5). This could reduce the superstructure's weight by 60%–70% (less than 1% of the loaded weight) and lower the center of gravity, which, together, can reduce fuel consumption by 0.5%–3% [14][15][16,17]. However, because GRP materials can support combustion (unlike metals) and lose strength when subjected to high temperatures, additional precautions (such as use of insulation) must be applied to satisfy fire-safety regulations when GRP components are developed for ships [2].

Overall, it is clear that the materials challenge is significant, both in integrating new materials into next-generation ships and regarding retrofitting existing ships.

Indeed, shipping is a highly demanding application. As indicated above, many factors must be considered when integrating new materials into the complex environment of a ship. Combined with the key factors of cost and lifetime, this has been very difficult. Over the next 10 years, it is projected that there will be a significant increase in the use of composites and novel materials, mostly to maintain strength and safety while reducing weight. The challenge that remains is also one of scale, given the size of modern ships and the large number of active commercial vessels. Finally, there is the possibility that smart materials could have an impact on future ships. For example, it is possible to imagine hulls that reshape themselves as a function of speed to optimize the hydrodynamics, which would dramatically reduce fuel consumption. Likewise, making hulls that were self-healing and self-cleaning, such as by using environmentally responsive polymer coatings, would improve both efficiency and lifetime [18].

33.4.4 Motive power

The previous section discussed the possible evolution of just the construction of shipping vessels. This section considers current trends toward improving efficiency in terms of motive power. **Slow steaming (reducing sailing speed)** is a major new thrust in trying to save fuel and reduce emissions, although it is a somewhat controversial approach, because it increases the time required for the delivery of goods and, thus, increases inventory needs. However, as Figure 33.6 shows, the reduction in emissions can be as much as 30%, with net savings in

(a)

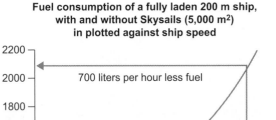

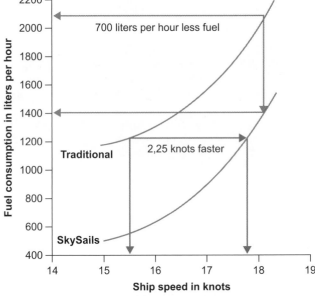

Figure 33.8. Fuel consumption of a fully laden 200-m ship, with and without SkySails (5,000 m²), as a function of ship speed. As shown in the graph, at a speed of 18 knots, the SkySail-assisted ship would use 700 l h⁻¹ less fuel, whereas at a constant fuel consumption of 1,200 l h⁻¹, the SkySail-assisted ship could travel 2.25 knots faster. Note: 1 knot = 1 nautical mile per hour = 1.825 km h⁻¹.

(b)

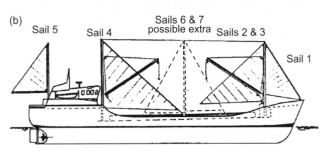

Figure 33.7. (a) Illustrates SkySails towing kite attached to a cargo ship. (b) Illustrates a similar concept using fixed sails in conjunction with normal propulsion.

fuel of millions of gallons for the shipping industry as a whole [19][20]. If shipping by slow steaming is priced accordingly, the lower costs might be sufficiently attractive to manufacturers to balance inventory concerns.

Another approach to increasing efficiency is to employ **wind assistance**. Figure 33.7 illustrates two approaches that could be used to take advantage of wind power to assist overall propulsion. For the SkySails concept in Figure 33.7, Figure 33.8 shows a plot of the potential savings for a 200-m ship as a function of speed in knots. Clearly, although a return to three-masted

schooners is not realistic, the idea of assisting propulsion with a renewable resource is of increasing interest. One approach is to combine fixed sails in a hybrid drive. The other is to use renewable resources to add to convential power. This raises the potential of an electrical assist with wind or photovoltaic power for large-surface-area vessels. Although very little research has been published on this possibility, some preliminary experiments performed by a Japanese company have confirmed that solar photovoltaics can assist in propulsion and provide electricity to power onboard systems for a freight ship [21]. Moreover, the solar power system proved capable of withstanding the harsh weather conditions encountered at sea. Such results should encourage further research into improving system output.

An alternative to wind–fuel hybrids is the potential of **nuclear-powered shipping**. Although some test cases have shown promise, the political ramifications of non-military, non-stationary nuclear reactors, as well as high operating costs, have severely restrained the field. In addition, given the huge number of ships in the merchant fleet, a great deal of nuclear material would be required in order to have a reasonable effect on total emissions, and this would generate appreciable quantities of nuclear waste. Nonetheless, nuclear-powered

Figure 33.9. *N.S. Savannah*, a nuclear-powered cargo ship operated from 1962 to 1970 as part of the US Atoms for Peace Program.

Figure 33.10. The Japan-based Nippon Yusen KK (NYK) has designed a ship using H_2 or methanol fuel cells and augmented by solar and wind energy.

shipping could potentially be a viable solution with no atmospheric emissions for very large ships [22][23][24] [25]. Figure 33.9 shows a picture of the *N.S. Savannah*, which was constructed by the USA as part of the Atoms for Peace program. Although it was a technical success, it was decommissioned after less than 10 years for economic reasons. At today's fuel prices, however, operation of this ship would have been cost-competitive with current shipping practices.

33.4.5 Beyond current shipping

Given the discussion in the previous section, one might ask what is coming in the next 10 years. Clearly, many of the concepts above will be implemented, and, to some extent, shipping will remain generically similar to its current state, with an increase in vessel size. There are, however, some new technologies and practices coming along that could be transformative.

Enhanced navigation
Cooperative shipping is, as described in Thomas Friedman's book *The World is Flat* [26], now a coupled worldwide industry with major players located throughout the world. As a consequence, improved navigation assisted by the global positioning system (GPS) and improved weather prediction will provide an opportunity for the industry to develop sophisticated, coupled route selection and cooperation to minimize travel and emissions while decreasing delivery times.

Alternative power sources
Fuel cells are being developed for use in automobiles (see Chapters 31 and 46), and, although they are more efficient than internal combustion engines, they still need much development to be technologically ready for real-world use. However, the environment on a ship

is very different from that in an automobile and is actually well suited to large solid-oxide-fuel-cell (SOFC, see Chapter 46) installations such as those that are currently used for building power. In fact, just recently, the first fuel-cell-powered ship, the Norwegian *Viking Lady*, which uses a molten carbonate fuel cell burning liquefied natural gas, began its initial runs [27]. If adequate fuel (such as H_2, liquefied natural gas, or methanol) could be stored on board, and if the fuel cell could be coupled to an electric drive, this configuration could be more efficient than the present power sources of propulsion, as well as being virtually non-emitting [28]. The energy density of the mentioned fuels, however, is low, and thus stores could take up a large volume on the ship, which would displace cargo. A preferable option would be to generate fuel or energy onboard using solar or wind technology, thus avoiding the need to store enough fuel for the entire journey. Finding sufficient area for solar cells to make a significant contribution to a ship's energy needs is difficult, however, even on the deck of a supertanker. Thus, significant effort will be required in order to implement a hybrid supership such as that shown in Figure 33.10.

Return to green
Finally, there is an emerging industry that could be termed "green" shipping. Three-masted sailing ships are being used to deliver ecofriendly goods. The shipping time is appreciable longer, but for some products, such as wine, this appears to be a sufficient selling point for the industry [28].

33.5 Summary

This chapter has provided some indication of the size, complexity, and evolving nature of the global shipping industry. It is an industry that will be increasingly scrutinized as society addresses concerns over energy and environment, because it produces significant quantities of CO_2 and uses large amounts of energy, but is also vital to world commerce. In addition, in its current form, it is not sustainable. For example, the use of fuel oil as the primary energy source will become problematic in the future. One way to solve this issue could be a return to coal, liquefied natural gas, or even wind power. Alternatively, nuclear-powered shipping could become an option, if it can be implemented rapidly enough and used safely, as could hydrogen-fuel-cell-powered ships. In the meantime, it is a key industry priority to minimize emissions and optimize fuel use even if delivery times must increase. Primary transport of goods by ships is a critical part of the global economy, and adapting the shipping industry to improve its efficiency and reduce its environmental impact will be a key global transportation challenge of the next 50–100 years.

33.6 Questions for discussion

1. Compare and contrast the various approaches to reducing emissions for large ocean transports. What solutions could be implemented in the near term, and what is the best long-term solution?
2. A supertanker of 550,000 t deadweight capacity must sail 3,000 miles. If a H_2 fuel cell is 50% efficient, how much H_2 is needed, and how will it fit in the ship?
3. Discuss how cooperative shipping for transport of goods might work. Why does it make economic sense?
4. Iron is abundant and cheap, but requires a great deal of energy to process. Can it be displaced by a more sustainable approach? What would that be?
5. Discuss the advantages and disadvantages of the nuclear ship.
6. Does wind power on a supertanker actually help? How can its effect be optimized?

33.7 Further reading

- **T. L. Friedman**, 2008, *Hot, Flat, and Crowded: Why We Need a Green Revolution – and How It Can Renew America*, New York, Farrar, Straus and Giroux.
- **P. Lorange**, 2009, *Shipping Strategy: Innovating for Success*, Cambridge, Cambridge University Press.
- **N. Wijnolst** and **T. Wergeland**, 2008, *Shipping Innovation*, Fairfax, VA, IOS Press.
- **E. J. Sheppard** and **D. Seidman**, 2001, "Ocean shipping alliances: the wave of the future," *Int. J. Maritime Economics*, **3**, 351.
- **International Chamber of Shipping**, http://www.marisec.org/shippingfacts/.
- **International Chamber of Shipping**, 2009, *Shipping, World Trade, and the Reduction of CO_2 Emissions*, London, International Chamber of Shipping, available at http://www.shippingandco2.org/CO2%20Flyer.pdf.
- **A. Miola**, **B. Ciuffo**, **E. Giovine**, and **A. Marra**, 2010, *Regulating Air Emissions from Ships – The State of the Art on Methodologies, Technologies and Policy Options*, European Commission.

33.8 References

[1] **C. E. Fayle**, 2006, *Business History, A Short History of the World's Shipping Industry*, London, Routledge.
[2] **Boeing**, 2009, *World Air Cargo Forecast 2010–2011*, Seattle, WA, Boeing, available at http://www.boeing.com/commercial/cargo/index.html.
[3] **International Chamber of Shipping**, 2009, *Shipping, World Trade, and the Reduction of CO_2 Emissions*, London, International Chamber of Shipping, available at http://www.shippingandco2.org/CO2%20Flyer.pdf.
[4] **K. P. Capaldo**, **J. J. Corbett**, **P. Kasibhatla**, **P. Fischbeck**, and **S. N. Pandis**, 1999, "Effects of ship emissions on sulphur cycling and radiative climate forcing over the ocean," *Nature*, **400**, 743–746.
[5] **J. J. Corbett** and **P. S. Fischbeck**, 1997, "Emissions from ships," *Science*, **278**(5339), 823–824.
[6] **J. Vidal**, 2007, CO_2 Output from Shipping Twice as Much as Airlines, March 3, http://www.guardian.co.uk/environment/2007/mar/03/travelsenvironmentalimpact.transportintheuk.
[7] **Ocean Policy Research Foundation Japan**, 2009, *Maritime Society in the Era of Global Warming – A Message from the Year 2050*, http://www.sof.or.jp/en/report/pdf/200910_report01.pdf.
[8] **D. Biello**, 2010, "World's first fuel cell ship docks in Copenhagen," *Scientific American*, December, News section, http://www.scientificamerican.com/article.cfm?id=worlds-first-fuel-cell-ship.
[9] **T. C. Gillmer**, 1977, *Modern Ship Design*, 2nd edn., Annapolis, MD, Naval Institute Press.
[10] National Shipbuilding Research Program, Advanced Shipbuilding Enterprise, www.NSRP.org.
[11] **D. J. Eyres**, 2006, *Ship Construction*, Amsterdam, Elsevier.
[12] **T. W. Clyne** and **F. Simancik** (eds.), 2000, *Metal Matrix Composites and Metallic Foams*, Weinheim, Wiley-VCH.
[13] **H. W. Seeliger**, 2004, "Aluminium foam sandwich (AFS) ready for market introduction," *Adv. Eng. Mater.*, **6**, 448.
[14] **A. G. Evans**, 2001, "Lightweight materials and structures," *MRS Bull.*, **26**, 790–797.
[15] **A. P. Mouritz**, **E. Gellert**, **P. Burchill**, and **K. Challis**, 2001, "Review of advanced composite structures for naval ships and submarines," *Composite Structures*, **53**, 21.
[16] **T. Calvert**, 2009, "Composite superstructures offer weight and cost benefits," *Reinforced Plastics*, **53**, 34.
[17] **B. Hayman**, **A. T. Echtermeyer**, and **D. McGeorge**, 2001, "Use of fibre composites in naval ships," in

Proceedings of the International Symposium, WAR-SHIP2001, London, RINA.

[18] **D. McGeorge**, **J. Lilleborge**, **B. Høyning**, and **G. Eliassen**, 2003, "Survivable composite sandwich superstructures for naval applications," in *Proceedings of the 6th International Conference on Sandwich Structures*, Fort Lauderdale, FL.

[19] **Anon.**, 2009, "Developments in fire protection of FRP composite vessels," presented at the RINA conference *Innovation in High Speed Marine Vessels*, Freemantle, Australia, available at http://www.composite-super-structure.com/RINA%20Conference%20Paper.pdf.

[20] **M. Schwartz**, (ed.), 2008, *Smart Materials*, Boca Raton, FL, CRC Press.

[21] **J. Corbett**, **H. Wang**, and **J. Winebrake**, 2009, "The effectiveness and costs of speed reductions on emissions from international shipping," *Transportation Res. D*, **14**, 539–598.

[22] **H. N. Psaraftis** and **C. A. Kontovas** 2010, "Balancing the economic and environmental performance of maritime transportation," *Transportation Res. D*, **15**, 458.

[23] *Freight Ship "Auriga Leader" Completes Seven Months of Voyages Assisted by Solar Power*, http://www.solarserver.com/solar-magazine/solar-news/current/freight-ship-auriga-leader-completes-seven-months-of-voyages-assisted-by-solar-power.html.

[24] **R. M. Adams**, 1995, "Nuclear power for commercial ships," in *Propulsion 95*, New Orleans, LA.

[25] **G. A. Sawyer** and **J. A. Stroud**, 2008, *Analysis of High-Speed Trans-Pacific Nuclear Containership Service*, London, Royal Institute of Naval Architects.

[26] **T. L. Friedman**, 2005, *The World is Flat: A Brief History of the Twentieth-First Century*, New York, Farrar, Straus and Giroux.

[27] **K. Adamson** and **L. C. Jerrram**, 2009, "Nicke transport survey," *Fuel Cell Today* (www.fuelcelltoday.com).

[28] http://www.environmentalgraffiti.com/businesspolitics/sailing-into-the-future-of-shipping/832.

34 Transportation: fully autonomous vehicles

Christopher E. Borroni-Bird and Mark. W. Verbrugge

General Motors Research & Development, Warren, MI, USA

34.1 Focus

A new automotive "DNA" based on electrification and connectivity will be required in order to address the associated energy, environment, safety, and congestion challenges. What are the potential materials and design implications if future vehicles can sense and communicate with each other and the surroundings, drive autonomously, and do not crash?

34.2 Synopsis

Transformational change is coming to the automobile. Amid growing concerns about energy security, the environment, traffic safety, and congestion, there is an increasing realization that the 120-year-old foundational "DNA" of the automobile is not sustainable. In response, auto manufacturers are introducing a wide range of propulsion, electronics, and communications technologies. These will create a new automotive DNA, based on electrification and connectivity, and will profoundly affect personal mobility. Future vehicles are likely to be tailored to a range of specific uses, from short urban commutes to long-distance cargo hauling, and they will likely be energized by electricity and hydrogen. Unlike gasoline or diesel fuels, electricity and hydrogen can be made from primary energy sources that are diverse and can be made renewably. Future vehicles will be propelled with electric motors, perhaps in the wheels, and the braking, steering, and driving functions will be controlled electronically.

Most automotive researchers would agree that future automobiles will be able to operate semi-autonomously (and eventually autonomously) to avoid collisions, park themselves, and optimize traffic management. An enabling technology for fully autonomous operation is the Mobility Internet, whereby future vehicles will function as nodes on a connected transportation network, communicating wirelessly with each other and the infrastructure, and knowing precisely where they are in relation to other vehicles and the roadway. With the development of vehicles that can sense and communicate with each other and their surroundings, one can rationally postulate that it should be possible for future vehicles to drive autonomously and to avoid collisions to a much better extent than is the case with the human-driven vehicles of today. The ramifications in terms of vehicle designs, materials selection, and manufacturing have largely been overlooked, so this chapter will focus on the possible materials and design implications for a potential future state when vehicles can be driven autonomously and do not crash.

34.3 Historical perspective

To set the stage, we first look at present-day automobiles and their associated technology. These continue to be largely energized by petroleum, powered by the internal combustion engine, controlled mechanically, and operated solely by the driver. They are designed to meet almost all conceivable needs for moving people and cargo over long distances. Because of this design for occasions that occur only a small fraction of the time, a typical automobile must be able to travel at least 300 miles without refueling and at speeds of at least 100 miles per hour. These requirements drive substantial cost, energy, mass, and space inefficiency into the vehicle. For example, a typical automobile weighs 20 times as much as its driver, occupies more than 150 square feet for parking, and is parked more than 90% of the time. To summarize this current state, a depiction of technology drivers in terms of existing challenges and corresponding goals for automobile manufactures is provided in Table 34.1.

As we electrify and connect automobiles, dramatically different vehicles will begin to appear, and might not even be viewed as automobiles [1]. Figure 34.1 reflects changes to the automotive DNA that are consistent with the goals of Table 34.1. All of this new capability can make our future vehicles cleaner, more energy-efficient, safer, smarter, more innovative, and more fun to drive and ride in than today's cars and trucks.

The reinvention of the automobile promises to be exciting and is critical as vehicle ownership grows rapidly around the world, fueled by population increase and a growing affluence in emerging markets. There are an estimated 950 million automobiles in the world today, which means that only about one in seven of the world's population enjoys the benefits and freedom enabled by personal transportation. Given the growing wealth in emerging markets and the universal aspiration for automobiles, we expect ownership rates to continue to increase so that by 2020 there should be over a billion vehicles on the planet. China alone is projected to have 100 million automobiles in use, representing about one-tenth of all the automobiles. The challenge for the auto industry, therefore, is to make our future vehicles more sustainable in terms of energy, the environment, safety, congestion and parking (particularly in urban centers). All this should be done while maintaining the affordability and the pleasure that customers obtain from owning and using an automobile today.

There is another trend that can stimulate the transition toward autonomous driving. In the USA, the average age of a new-automobile buyer has been increasing at a faster rate than the population. In 1990, the median age for the new-vehicle buyer was just over 40 years old, whereas the corresponding age is nearly 55 today. As the

Table 34.1. Drivers of automobile technology	
Challenges	**Stretch goals**
Energy	Low-cost renewable energy
Emissions	No tailpipe environmental impact
Safety	Vehicles that don't crash
Congestion	Congestion-free routing. Megacity parking
Affordability	Vehicle for every purse and purpose

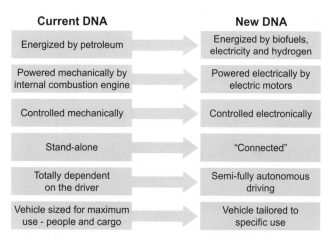

Figure 34.1. Convergence of new automotive technologies. The progression from current passenger vehicles to the future passenger vehicle that is the subject of this chapter is depicted.

population ages in developed and emerging markets, it is reasonable to expect that autonomous driving will be increasingly appreciated as a means to preserve independence and freedom of movement.

We end this introduction with a question: if vehicles could be driven autonomously and did not participate in crash events, how could they be designed differently and what new materials usages can we predict? The next section of this chapter provides a basis for the conjecture that future vehicles could some day be designed to drive themselves autonomously and in such a way that they would not crash. The following sections address the mass-reduction opportunities and vehicle-design implications resulting from this conjecture.

34.4 Vehicles that drive themselves and do not crash

Autonomous operation will be enabled by precise sensing and positioning capability and by precise actuation of braking, steering, and driving functions, such as electronic stability control, that may best be achieved by electronic controls and electro-mechanical actuation.

Sensor-based features are being introduced into production automobiles, such as blind-spot detection systems that can sense objects that are not normally visible to the driver and warn the driver of their presence. Another example is lane-keeping systems that will maintain the vehicle's lateral position so that it does not unintentionally drift into another lane.

What can be achieved when sensing technology is combined with a Global Positioning System (GPS) and enhanced digital maps was demonstrated in November 2007 when several self-driving, or autonomous, vehicles completed a 60-mile (96 km) race in a mock urban environment in the Defense Advanced Research Projects Agency (DARPA) Urban Challenge without incident, following all the rules of the road and even interacting safely with manually driven vehicles. The winner, which was developed by Carnegie Mellon University with

Figure 34.2. The Chevrolet Tahoe "Boss" Autonomous Vehicle, winner of the 2007 DARPA Urban Challenge.

Figure 34.3. The V2X approach includes vehicles communicating with each other (V2V), with the roadside infrastructure (V2I), and with pedestrians and cyclists (V2P).

support from General Motors (GM) and several other companies, is shown in Figure 34.2.

Alternatively, vehicles will know the location and speed of an approaching vehicle by wirelessly communicating and exchanging information with it. Such vehicle-to-vehicle (V2V) communications will give the vehicle a "sixth sense" as it broadcasts its position and velocity to its "neighborhood" and continuously monitors the status of surrounding vehicles within a quarter-mile radius. Prototype V2V communications systems have been demonstrated by GM and other automakers that support automated safety features like lane-change alert, blind-spot detection, sudden stopping, forward collision warning with automatic braking, and intersection collision warning. This approach offers very robust performance even in extreme weather conditions.

It is expected that future V2V and vehicle-to-infrastructure (V2I) communications will lead to automated, cooperative driving because vehicles may be accurately located (e.g., by GPS technology) and may be aware of all significant objects around themselves through some combination of sensing and wireless communications. They may also communicate with the roadside infrastructure and even be able to drive themselves. These autonomous vehicles will sense what is around them and will either avoid a crash or be able to decelerate to a low enough speed to reduce the potential harm to pedestrians, cyclists, or vehicle occupants. Moreover, since the wireless transponder will soon be small enough to fit inside a handheld device, it could be carried by pedestrians and cyclists to increase their safety when they are using the road. This capability could in time virtually eliminate vehicle collisions at speeds and forces that cause harm (figure 34.3).

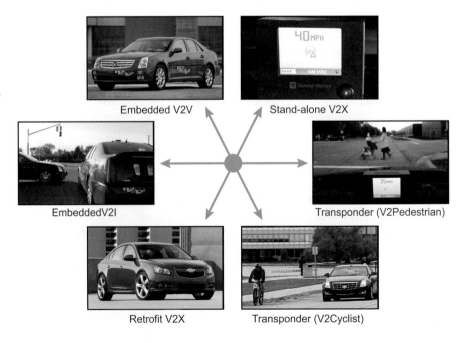

Embedded V2V   Stand-alone V2X

EmbeddedV2I   Transponder (V2Pedestrian)

Retrofit V2X   Transponder (V2Cyclist)

Figure 34.4. GM's EN-V (Electric Networked-Vehicle).

Figure 34.5. A roadmap to autonomous driving.

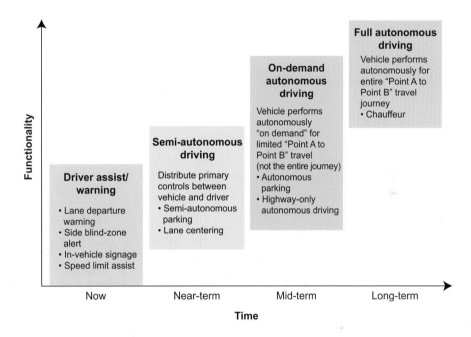

Looked at separately, sensing and communications each have their own advantages and disadvantages. Sensors add mass and cost to the vehicle, and might not be able to detect an object in conditions of poor visibility or a vehicle that is not in the line of sight. Conversely, relying solely on communications requires that the other nearby vehicles must be capable of responding back. The two approaches are complementary and could be combined to provide a more cost-effective and robust solution for a wide range of real-world conditions. This combination has recently been demonstrated in GM's Electric Networked-Vehicle (EN-V) concepts, shown in Figure 34.4. The future vision behind these concept vehicles is to address the energy, environment, safety, congestion, parking, and affordability challenges associated with personal urban mobility; hence, they are small, lightweight, battery electric-powered, and have the necessary sensing, GPS, and wireless communications technologies to enable connected and autonomous operation.

General Motors expects that future vehicles will be able to drive autonomously and has outlined a roadmap for how we will get there, which is shown in Figure 34.5. Building blocks will include semi-automated parking, adaptive cruise control, and lane keeping. Further into the future, the fusion of adaptive cruise control and lane centering could enable autonomous operation on a highway lane. The ultimate objective is for the vehicle to drive autonomously whenever and wherever the "driver" wants. The rest of this chapter will describe the implications for vehicle design, with particular emphasis on materials selection, if future vehicles did not crash, and, therefore, the safety standards and regulations relating to occupant protection in crashes were no longer needed.

34.5 Mass-reduction opportunities with autonomous vehicles

We begin with an energy analysis of today's gasoline-engine-powered vehicles. The relevant energy flow is depicted in Figure 34.6 [2]. About 85% of the energy in the fuel (gasoline) is lost by the powertrain and accessories, and 15% reaches the axle for vehicle traction

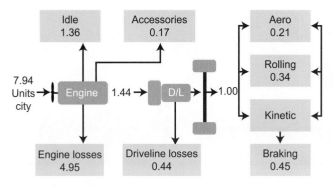

Figure 34.6. Representative energy flows for today's mid-sized vehicles.

(Chapter 31). The traction power losses P can be represented as [3]

$$P_{\text{traction}} = P_v + P_W + P_g + P_a.$$

The aerodynamic losses P_v attributable to the vehicle's velocity v relative the surrounding air can be expressed as

$$P_v = \frac{1}{2}\rho_{\text{air}}\overline{C_D A}v^3,$$

where $\overline{C_D A}$ reflects drag losses based on the frontal area A, with C_D representing a drag coefficient (the overbar denotes an average value), and the density of air corresponds to ρ_{air}. The wheel losses P_W attributable to friction and associated rolling resistance are expressed as

$$P_W = f_R mg(\cos\theta)v,$$

where θ is the angle of the vehicle relative to the road ($\theta = 0$ for level driving), m refers to the vehicle mass, g denotes the Earth's gravity, and f_R is the friction coefficient. Inertial losses P_a associated with acceleration a (and deceleration d) can be represented as

$$P_a = \begin{cases} \lambda_a mav & \text{for } a \geq 0, \\ \lambda_d mav & \text{for } a < 0, \end{cases}$$

where the coefficients λ_a and λ_d are near unity. Last, gravity losses P_g attributable to inclines impact the vehicle power consumption if $\theta > 0$:

$$P_g = mg(\sin\theta)v.$$

This very simple set of equations makes it clear that mass plays a central role in energy consumption, since all of the traction power losses are directly proportional to mass, with the exception of the aerodynamic losses P_v. About a 7% improvement in fuel economy is seen for a 10% reduction in mass [4] (Chapter 31). Put another way, about a 0.4-mpg improvement is observed for every 100 lb of mass removed for a 3,500-lb vehicle, or a 0.5-km l^{-1} improvement per 100-kg weight reduction for a 1,500-kg vehicle.

Table 34.2. Subsystem mass as a fraction of curb mass for a sedan passenger vehicle (e.g., curb mass about 3,500 lb (1,590 kg))

System	Mass fraction
Body and exterior	0.50
Body structure	0.227
Body, non-structural	0.204
Closures	0.046
Bumpers	0.022
Powertrain	0.23
Engine	0.118
Transmission	0.067
Fuel and exhaust	0.040
Chassis	0.20
Tires and wheels	0.065
Front suspension	0.049
Rear suspension	0.044
Braking	0.032
Steering	0.014
Other	0.07
Electrical	0.046
HVAC	0.027
Sum	1.00

HVAC, heating, ventilation, and air conditioning.

The masses of the various subsystems that make up a passenger vehicle of today are approximated in Table 34.2. A depiction of materials used in today's passenger vehicles is provided in Figure 34.7. Table 34.3 indicates how much mass could be removed from a vehicle structure if it were converted from conventional steel to various low-density materials, and Table 34.4 provides representative materials properties. These data can be extracted from [5][6][7].

The assumption that we have made up to this point is that vehicles would not need to be designed to accommodate crash events, and we have provided a baseline for conventional vehicle materials and subsystem masses. Given such information, how best can we answer the question posed at the end of the introduction to this

Figure 34.7. Materials in a typical passenger vehicle in 1977 (left) and today (right).

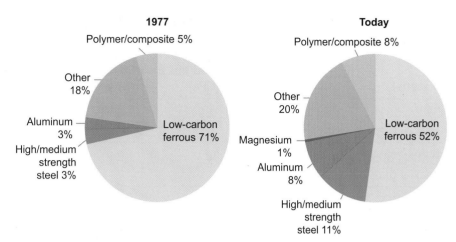

Table 34.3. Potential percentage mass reduction vs. conventional steel for various materials

	Body structure	Body closures	Chassis
High-strength steel	25	15	25
Aluminum	40	45	50
Magnesium	55	55	60
Polymer composite			
Carbon	>60	>60	60
Glass	25	25	35
Titanium	NA	NA	50
Metal-matrix composite	NA	NA	60
NA, not applicable.			

with time, engine blocks are already dominantly aluminum, a substantially lightweight material, and the rest of the powertrain constitutes minimal mass relative to the vehicle mass, which is also the case for the category "other."

With these assumptions, we arrive at the following calculations. The body structure and bumper masses in Table 34.2 are reduced by 60%, as are the chassis systems with the exception of the braking system, and the closure masses are reduced by 25%. Next, we conservatively estimate that, through redesign and mass decompounding, the powertrain and braking subsystems that were left unaltered to this point are now reduced in mass by 30%. (Note that, for conventional mass decompounding, for every unit of mass removed from a vehicle, one additional unit of mass can be removed through redesign of the vehicle [6][10][11][12].)

In addition to the opportunities for reducing the mass of the structure, if vehicles were never involved in crashes, mass could also be taken out of the vehicle's interior by removing passive safety components that are currently needed in order to meet safety standards. It is estimated that by eliminating airbags and the energy-absorbing foam in padded interior surfaces and by reducing the required structure that attaches the seats to the floor (to accommodate crash safety requirements) it may be possible to remove approximately 40 kg of mass from a typical vehicle interior.

The end result could be a 50% mass reduction from that depicted in Table 34.2 (i.e., 3,500 lb (1,590 kg) is reduced to 1,750 lb (795 kg)). Clearly, many ambitious assumptions have been made in order to arrive at this approximation, chief among these being the possibility that, if crashes were no longer to occur, vehicles will not need to satisfy crash requirements. These calculations are for a compact sedan. For a vehicle such as the EN-V, a much lighter concept would emerge. Again, we stress that the point of this exercise is to provide a futuristic vision and allude to the opportunity, not to render a specific projection.

chapter? We shall satisfy ourselves with a very crude approximation procedure here, since it is the intent of this chapter to stimulate work on this important topic rather than to provide a specific answer. First, we leap to the assumption that non-structural thermoplastic olefins can be used for closures. Nanoclay-filled poly(olefins) [8][9] approach glass-fiber mass-reduction capabilities for closures, which we approximate as a 25% mass reduction from today's closures (see Table 34.3). For the body structure, Mg can satisfy requirements that are less demanding in that crash energy management need not be considered, resulting in an approximation of a 60% mass reduction for the body and chassis systems. At this point, we make no further assumptions about the powertrain and "other" elements listed in Table 34.2. While conventional internal combustion engine powertrain masses will come down

Table 34.4. Representative structural and semi-structural material properties

Material	Density ρ (g cm^{-3})	Young's modulus E (GPa)	Strength σ_y (MPa)	Fracture toughness K_{IC} (MPa m$^{1/2}$)	E/ρ	Specific strength σ_y/ρ
CFRP*, 58% uniaxial C/epoxy	1.5	189	1,050	32–45	126	700
GFRP**, 50% uniaxial glass polyester	2.0	48	1,240	42–60	24	620
Steel (AHSS >550 MPa)	7.8	207	400–1,000	180–30	27	50–130
Al alloys	2.8	71	50–600	50–20	25	20–215
Mg alloys	1.8	45	70–300	30–10	25	40–170
Titanium	4.5	104	925	50	23.1	205
Epoxy resins	1.2–1.4	2.1–5.5	40–85	0.6–1.0 (20 °C)	1–5	28–70
Polyesters	1.1–1.4	1.3–4.5	45–85	0.5 (20 °C)	1–4	32–75
Polypropylene	0.91	1.2–1.7	50–70	3.5 (20 °C)	1–2	55–77

*CFRP carbon fiber reinforced polymer **GFRP glass fiber reinforced polymer.

34.6 Vehicle-design implications for autonomous vehicles

In addition to the many possible opportunities for reducing the mass of the vehicle's structure, as explained in the previous section, there may be indirect means to further reduce mass or energy consumption by fundamentally redesigning the vehicle if measures to accommodate crash events were no longer required because it did not crash. For example, it may be possible to redesign the interior with lighter seats that can be made thinner than those of today. This could increase the interior spaciousness for an automobile of given size or reduce the vehicle's exterior packaging without compromising interior space. The benefit of shrinking the exterior envelope is that it can enable cost and mass reduction and aerodynamic-drag improvements (by decreasing the frontal area) and lead to improved fuel economy. A third potential outcome is that the seats could retain their existing size but have more functionality embedded into them while maintaining the same mass (e.g., swivel and extreme recline motions or additional heating and cooling capability). These thoughts highlight the choices that autonomous vehicles can give the designer: one option being to take advantage of the mass, cost, and space savings by developing thinner, lighter, more energy-efficient designs, with the other option being to add content

and functionality while maintaining the same mass, cost, and size.

Seats and instrument panels in today's automobiles are significant heat sinks in the cabin, and the need to cool them down quickly after a hot soak means that the size of an automotive heating, ventilation, and air-conditioning (HVAC) system has to be similar to that of the system needed to cool a small house. If the mass of padded foam can be reduced and if the internal structures of these components can be downsized then the cabin will not only be lighter in mass but also will have lower thermal mass. For a given cool-down time, this will allow the HVAC system to be smaller and lighter (contributing to improved fuel economy) as well as being more energy efficient as a subsystem.

Another interesting opportunity for reinventing the automobile is in the areas of exterior lighting and glazing. New exterior lighting technologies could also be possible if lighting requirements are changed, and this could realize further savings in electrical energy. Glazing systems, in particular, can have a major impact on a vehicle's energy efficiency since they affect not only mass but thermal loading and aerodynamic drag. For example, it might be possible to reduce light transmissibility through windshield and other glazing surfaces if the vehicle had the capability to "see" other vehicles and to prevent collisions. If this were to occur, then one could use new glazing materials that are not only lighter and less expensive but also designed to reduce heat transfer into the vehicle so that less demand is placed on cabin heating and cooling systems. Alternatively, it may enable the use of an electro-chromic windshield that could change color when needed (for privacy and to reduce heat build-up), but which could be heavier than the existing glazing. Although some constraints on glazing may be relaxed, the new automotive DNA may introduce some new constraints. For example, with any new glazing material that is used one will need to ensure that there is no disruption of wireless signals such as those sent by cell phones, police and emergency communication devices, and GPS.

As propulsion systems become more efficient and increasingly electrified, the management of thermal loads will be increasingly important to improving real-world fuel economy, and these developments in glazing could be very beneficial, depending on the climate. New glazing materials could also be designed to reduce transmission of ultraviolet (UV) light, which might offer opportunities to develop new interior fabrics, without concern for UV degradation, that are lighter or less expensive. This combination of a reduced heat load (due to changes in light transmission through glazing) and a reduced heat sink (through changes in the thermal mass of the seats and interior trim) could transform how occupant thermal comfort is achieved in the vehicle. The existing paradigm of a centralized forced-air system could even tip toward a decentralized human-contact system (heated/cooled seating) that does not require additional forced-air support and is significantly more efficient.

As noted previously, there is also the potential for energy savings from reducing the aerodynamic drag through thinner, smaller-cross-section automobiles, since less-energy-absorbing materials would be needed for side-impact protection in the door, for example if crashes were nonexistent. Less obvious is the potential for improving the windshield's aerodynamic optimization if windshield angles are not so critically defined by transmissibility and visibility requirements in a "non-crash" world.

In addition to the potential changes to the rake angle, it may also be possible to change other vehicle-design proportions, such as by reducing the front and rear overhangs through elimination of crush space. Also changes to the roofline and tumblehome (the slope at which the body side and glazing meets the roof) may be possible. From a materials perspective, optimization of surfaces through adaptive solutions to fine-tune design and aerodynamics on the basis of real-time conditions, such as grille openings and ride height, may be easier to achieve if safety constraints can be relaxed. This could further stimulate the development of smart materials, for example. The need for side mirrors and pillars (e.g.,

wrap-around windshields) could also be rethought if the vehicle has the ability to "see" everything that the driver needs to see and the ability to avoid crashes; eliminating side mirrors and pillars could provide substantial aerodynamic-drag and mass savings. The driver could still see the outside world but with small cameras mounted all around the vehicle that relay images back into the cabin.

Although the changes to the exterior may be dramatic, it is perhaps in the interior that the most radical changes may occur if a future vehicle did not have to be designed to meet crash standards because crashes were nonexistent. The decentralized HVAC system mentioned earlier might be realized if thermal and structural loads can be significantly reduced. The by-wire enabled human–machine interface (steering wheel, pedals, displays, and controls) could also be connected directly to the driver's seat so that it can move together with the seat. Reconfigurable seats could be made to easily translate across the floor since there is greater flexibility with seating arrangements and seat track travel. More flexibility with the location of seat attachments can enable useful storage under the seat. Swivel seats may also be easier to implement, given the elimination of the B-pillar and the increased space available due to thinner headers and roof rails (no roof-rail airbags).

There is extra design freedom for the instrument panel (IP) since it could be pushed forward to create more leg room. Moreover, if airbags were no longer needed this would allow a stationary steering-wheel center hub and the ability to place gauges and an LCD screen in this location. It would also be possible to transform the IP functionality to enable new types of movements for the front seats and to provide ample storage both for the driver and for a front passenger. New types of cross-car structural materials and designs, perhaps even visible to the occupants, could also be possible (this would be similar to the way in which a motorcycle's structure is exposed and becomes part of the vehicle's styling and brand character). Thinner doors that admit more storage are possible if there is no need for doors to contain space for energy-absorbing foam (e.g., 75 mm for pelvic EA foam) or side-impact beams. With the relaxation of crash requirements it becomes easier to optimally accommodate (in terms of spaciousness, visibility, ergonomics, comfort, etc.) all types of passengers and to implement universal design principles. An example would be the capability to provide greater adjustments in the seat position, in accord with occupants' needs.

New trim materials such as wood, metals, and ceramics could be integrated more prominently into parts of the interior where they are not used today because of head-impact concerns, for example. (However, these materials might add mass to the interior versus the status quo.)

These are just some of the ideas for how the vehicle's design might be transformed if there were no vehicle crashes and vehicles did not have to be designed to meet today's stringent crash requirements. These design changes may, or may not, have an impact on materials selection or energy consumption but are mentioned as examples to illustrate the breadth of "reinvention" that might be possible.

34.7 Summary

A note of caution should be sounded on the expected vehicle-efficiency improvements that could occur if crash-safety constraints were relaxed for a future state in which crashes did not occur. Although it may be possible to take advantage of such changes to make smaller, lighter, and more efficient designs without sacrificing comfort, the same enablers could allow vehicles to maintain their existing size and efficiency while providing dramatically enhanced functionality (particularly in the interior). This is somewhat analogous to improvements in engine-efficiency technology allowing the vehicle's fuel economy to remain approximately constant because new content for safety, performance, comfort, and utility was demanded by the customer.

Extremes in energy efficiency, cost reduction, multifunctionality, and fashion statements may all be possible, but probably not at the same time. For example, while all vehicles could contain the same foundational DNA (electrification and connectivity), jewel-like materials and electro-chromic windshields for projecting a display could cater to those who desire luxury vehicles, while canvas or open tops and no-frills handheld displays could satisfy cost-conscious vehicle owners.

What is probably true is that changes in safety standards, when vehicles can drive autonomously without crashing, could provide huge opportunities to tailor the designs and materials selection for different applications. In summary, the shift toward autonomous vehicles that can drive themselves and do not crash will open the way for more compelling designs and rides than is possible today with vehicles that are manually driven. It can also pave the way for significantly lighter and more energy efficient vehicle designs with greater flexibility in choice of materials.

34.8 Questions for discussion

1. How will the property requirements for structural and non-structural materials change if vehicles do not crash?
2. How will the extra cost of hardware and software required to make vehicles drive autonomously compare with the cost of hardware or materials that can be removed?

3. How much energy efficiency can be gained through vehicle connectivity versus vehicle electrification versus vehicle lightweighting?
4. Will autonomous operation affect the distances people drive in a vehicle, further impacting energy usage?
5. Will the development of small, electric networked vehicles for urban use lead to more diversity in vehicle design or less diversity?

34.9 Further reading

- **W. J. Mitchell**, **C. E. Borroni-Bird**, and **L. D. Burns**, 2010, *Reinventing the Automobile. Personal Urban Mobility for the 21st Century*, Cambridge, MA, MIT Press. This book outlines the challenges facing personal mobility in cities, recommends a variety of solutions and explains their societal and consumer benefits.
- **A. I. Taub**, 2006, "Automotive materials: technology trends and challenges in the 21st century," *MRS Bull.*, **31**, 336. This review paper does a nice job of overviewing trends in materials used in automobiles since the beginning of the industry alongside with an enumeration of current challenges.
- **M. Verbrugge**, **T. Lee**, **P. Krajewski** *et al.*, 2009, "Mass decompounding and vehicle lightweighting," *Mater. Sci. Forum*, **618–619**, 411. This paper focuses on how one can remove additional mass from the vehicle once an initial mass reduction has been made.

34.10 Acknowledgment

The authors appreciate insightful discussions with and suggestions of their GM colleagues Alan Taub and Lerinda Frost.

34.11 References

[1] **W. J. Mitchell**, **C. E. Borroni-Bird**, and **L. D. Burns**, 2010, *Reinventing the Automobile. Personal Urban Mobility for the 21st Century*, Cambridge, MA, MIT Press.
[2] **A. I. Taub**, 2006, "Automotive materials: technology trends and challenges in the 21st century," *MRS Bull.*, **31**, 336.
[3] **G. Lechner** and **H. Naunheimer**, 1999, *Automotive Transmissions. Fundamentals, Selection, Design, and Application*, Berlin, Springer.
[4] **M. R. Cuddy** and **K. B. Wipke**, 1997, *Analysis of the Fuel Economy Benefit of Drivetrain Hybridization*, SAE 970289.
[5] **A. I. Taub**, **P. E. Krajewski**, **A. A. Luo**, and **J. N. Owens**, 2007, "The evolution of technology for materials processing over the last 50 years: the automotive example," *JOM*, **59**, 48.

[6] **M. Verbrugge**, **T. Lee**, **P. Krajewski** *et al.*, 2009, "Mass decompounding and vehicle lightweighting," *Mater. Sci. Forum*, **618–619**, 411.

[7] **M. W. Verbrugge**, **P. E. Krajewski**, **A. K. Sachdev** *et al.*, 2010, "Challenges and opportunities relative to increased usage of aluminum with the automotive industry," in *Proceedings of the Minerals, Metals & Materials Society (TMS) 2010* Annual Meeting, Seattle, WA.

[8] **W. R. Rodgers**, 2009, "Automotive industry applications of nanocomposites," in *Industry Guide to Nanocomposites*, 1st edn., Bristol, Plastics Information Direct, a division of Applied Market Information, Ltd., pp. 273–308.

[9] **S. K. Basu**, **A. Tewari**, **P. D. Fasulo**, and **W. R. Rodgers**, 2007, "Transmission electron microscopy based direct mathematical quantifiers for dispersion in nanocomposites," *Appl. Phys. Lett.*, **91**, 053105.

[10] **D. E. Malen** and **K. Reddy**, 2007, *Preliminary Vehicle Mass Estimation Using Empirical Subsystem Influence Coefficients*, http://www.asp.net/database/custom/Mass%20Compounding%20%20Final%20Report.pdf.

[11] **C. Bjelkengren**, 2008, *The Impact of Mass Decompounding on Assessing the Value of Vehicle Lightweighting*, M.Sc. Thesis MIT.

[12] **C. Bjelkengren**, **T. M. Lee**, **R. Roth**, and **R. Kirchain**, 2008, "Impact of mass decompounding on assessing the value of lightweighting," presented at the *Materials Science and Technology Conference and Exhibition, 2008 Annual TMS Meeting*, New Orleans, LA.

PART 5

Energy efficiency

35

Lighting

Dandan Zhu and Colin J. Humphreys

Department of Materials Science and Metallurgy, University of Cambridge, Cambridge, UK

35.1 Focus

Electricity generation is the main source of energy-related green-house-gas emissions, and lighting uses one-fifth of its output. Solid-state lighting (SSL) using light-emitting diodes (LEDs) is poised to reduce this value by at least 50%, so that lighting will then use less than one-tenth of all electricity generated. The use of LEDs for lighting will provide reductions of at least 10% in fuel consumption and carbon dioxide emissions from power stations within the next 5–10 years. Even greater reductions are likely on a 10–20-year time scale.

35.2 Synopsis

Artificial lighting is one of the factors contributing significantly to the quality of human life. Modern light sources, such as incandescent light bulbs (a heated tungsten wire in a bulb that is evacuated or filled with inert gas) and compact fluorescent lamps (a phosphor-coated gas discharge tube), use electricity to generate light. Worldwide, grid-based electric lighting consumed about 2650 TW·h of electricity in 2005, some 19% of total global electricity consumption [1]. Using an average cost of $2.8 per megalumen-hour (Mlm·h), the International Energy Agency estimated that the energy bill for electric lighting cost end-users $234 billion and accounted for two-thirds of the total cost of electric-lighting services ($356 billion), which includes lighting equipment and labor costs as well as energy. The annual cost of grid-based electric lighting is about 1% of global gross domestic product.

In addition to consuming a significant amount of energy, lighting is also extremely inefficient. Incandescent light bulbs convert about 5% of the electricity they use into visible light; the rest is lost as waste heat. Even energy-saving compact fluorescent lamps are only about 20% efficient. These low efficiencies contrast starkly with the efficiencies of most electric household appliances. For example, electric ovens, clothes dryers, and toasters convert electricity to useful heat with a typical efficiency of at least 70%. Electric motors for fans are typically 90% efficient. Indeed, lighting is so inefficient, and it consumes so much energy, that there is probably more potential for large energy savings in this field than in any other area.

Solid-state lighting using LEDs is an emerging technology that promises higher energy efficiency and better performance than with traditional artificial lighting. In the past decades, LED technology has advanced tremendously, with many applications and a fast-growing global market. The next generation of home and office lighting will almost certainly be LEDs.

This chapter provides a brief introduction to various lighting options with a focus on LED-based SSL, the current technological challenges, the key materials research needs, and the future of LED lighting.

35.3 Historical perspective

In the world of lighting, the past 130 years have been the age of the incandescent light bulb. Such bulbs make light by heating a metal filament wire to a high temperature until it glows. Although many individuals contributed to early investigations of incandescent lighting, the first commercial successes were realized by the British physicist and chemist Joseph Swan and the American inventor Thomas Edison, both in approximately 1880. The success of this light-bulb technology hinged on obtaining a sufficiently high vacuum (subsequently replaced by inert gas) in the glass bulb and developing effective and long-lived filaments.

However, a major change in the world of lighting is imminent: the next five years or so are expected to be the era of the **compact fluorescent lamp (CFL)**. This change is being precipitated by the realization by governments around the world that incandescent light bulbs are extremely inefficient, converting only about 5% of the input electrical energy to visible light, and that banning the sales of these bulbs is therefore one of the easiest and most effective ways of saving energy and reducing carbon dioxide emissions. For example, in September 2007, the UK government announced that shops would stop selling 150-W filament light bulbs by January 2008, 100-W bulbs by January 2009, 60-W bulbs by January 2010, and 40-W bulbs by December 2011. What form of lighting is available now to take over from incandescent light bulbs? Long fluorescent tubes have already largely become the standard in offices, commercial buildings, and factories, and any remaining incandescent bulbs in these buildings are likely to be replaced rapidly by such tubes. Regarding lighting in houses, CFLs appear to be the only realistic replacement for incandescent bulbs at the present time. However, CFLs have an environmental drawback: each CFL contains about 5 mg of mercury, which is a highly toxic cumulative heavy-metal poison. Therefore, CFLs are likely to be a stop-gap measure to replace incandescent lamps, lasting until a more efficient, nontoxic source of white light is available at a reasonable cost.

Light-emitting diodes (LEDs) use thin layers of semiconductors (either organic or inorganic) to convert electricity into light and can, in principle, be 100% efficient. Organic LEDs are known as OLEDs, whereas inorganic LEDs are simply called LEDs. With their recent dramatic improvements in efficiency, OLED and LED devices have become potential nontoxic replacements for current light sources. However, the significant penetration of these devices into the general illumination market (especially for home and office lighting) still requires further improvements in performance and significant reductions in cost. Both technologies must overcome some major technical barriers, and materials science will play an important role in achieving this goal. Despite some impressive research results, the development of OLEDs for lighting has lagged behind that of LEDs, and it has not been possible so far to achieve high brightness at high efficiency with a long lifetime. Whereas LEDs are currently starting to be used for general illumination purposes, OLEDs are not.

Nevertheless, SSL technology is advancing so fast that there is good reason to believe that these barriers will disappear in the next few years and that LEDs (inorganic, organic, or both) will dominate the future of lighting.

35.4 Solid-state lighting and materials research challenges

35.4.1 Lighting, energy, and carbon dioxide emissions

It is not widely realized that lighting is one of the largest causes of greenhouse-gas emissions. The energy consumed to supply lighting throughout the world entails greenhouse-gas emissions of 1,900 Mt of CO_2 per year, assuming an energy mix based on the 2005 world electricity-generation values of 40% coal, 20% natural gas, 16% hydropower, 15% nuclear, 7% oil, and 2% renewables other than hydropower [2]. This is equivalent to 70% of the emissions from the world's cars and over three times the emissions from aircraft. In addition, many developing countries do not have grid-based electricity and instead use oil lamps for lighting. Use of these lamps consumes 3% of the world's oil supply but is responsible for only 1% of global lighting, while accounting for 20% of CO_2 emissions due to lighting [1]. Worldwide demand for lighting is increasing at a rapid rate. A conservative estimate is that the global demand for artificial lighting will be 80% higher by 2030 than the 2005 level described above [1]. It is therefore very important to move toward more efficient lighting to contribute to a sustainable world.

35.4.2 Lighting and the human eye

The human eye is sensitive only to light in the visible spectrum, ranging from violet (with a wavelength of ~400 nm) to red (with a wavelength of ~700 nm). Figure 35.1 shows the modified **eye-sensitivity function**, $V(\lambda)$, used to describe the sensitivity of the human eye as a function of wavelength [3], and the luminous efficacy. As can be seen in Figure 35.1, the maximum sensitivity of the human eye is to green light with a wavelength of 555 nm. The **lumen (lm)** is the unit of light intensity perceived by the human eye, and the **luminous efficacy** of optical radiation is defined as the luminous flux (lm) per unit optical power (W). This definition takes into

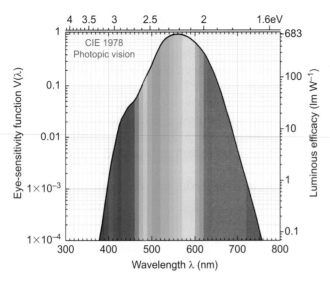

Figure 35.1. The eye-sensitivity function, $V(\lambda)$, and luminous efficacy (in lm W^{-1}) as functions of wavelength, λ. The maximum sensitivity of the human eye is to green light with a wavelength of 555 nm (Data after 1978 CIE).

account the sensitivity of human vision, so that green light contributes more strongly to efficacy than blue or red light, and ultraviolet (UV) and infrared (IR) wavelengths do not contribute at all.

For a light source, the **luminous efficiency** (lm W^{-1}) corresponds to the light power output (as perceived by the human eye and measured in lumens) relative to the electrical input power (measured in watts). The **power efficiency** of a light source is the light power output, *not* adjusted for the response of the human eye, divided by the electrical power input. Power efficiency is dimensionless and is usually given as a percentage. In lighting, instead of luminous efficiency, the term luminous efficacy of a light source (or, confusingly, just efficacy) is also often used. Strictly speaking, the luminous efficiency is equal to the luminous efficacy of radiation only when the light source has 100% power efficiency. Nevertheless, most references use the term efficacy to mean luminous efficiency and the term efficiency to mean power efficiency; for consistency with the literature, we also use this convention here.

The maximum possible theoretical efficacy of a light source is 683 lm W^{-1}, for the case of monochromatic 555-nm green light. However, this monochromatic light source would render well only green objects that reflect at its wavelength. An ideal light source is one that can reproduce colors perfectly and, thus, has a **color-rendering index (CRI)** of 100. The CRI is an internationally accepted measure of how well a light source renders colors. The theoretical maximum efficacy of an ideal white light source is 240 lm W^{-1} (because the human eye is less sensitive to wavelengths other than green).

Another important characteristic of a white light source is the **correlated color temperature (CCT)**, given in units of kelvins. The CCT of a white light source is defined as the temperature of a Planckian blackbody radiator whose color is closest to that of the white-light source. For conventional lighting technologies, the CCT spans a wide range, from 2,700 K to 6,500 K. "**Warm white,**" such as from incandescent lamps, has a lower CCT (~2,700–3,500 K), whereas "**cool white,**" which is a more blueish white, has a higher CCT (~3,500–5,500 K). Warm white is the most common lamp color used in residential lighting in the USA and Europe. The maximum possible efficacy of an LED warm-white-light source with a CCT of 3,000 K and a CRI of 90% is 408 lm W^{-1} [4]. It is useful to bear this value in mind when reading this chapter, since it represents a theoretical maximum efficacy for excellent-quality white lighting.

In addition to the luminous efficacy, CRI, and CCT, other important attributes of a white-light source include the light distribution, intensity, spectrum, controllability, lifetime, and cost. Understanding these performance attributes is also necessary before one can fully appreciate the great potential of LEDs as a revolutionary light source.

35.4.3 The range of lighting options

Incandescent light bulbs

An incandescent light bulb uses electricity to heat a coiled tungsten wire in a glass bulb that is either filled with inert gas or evacuated. The wire has a temperature of about 3,500 K and glows white-hot, radiating white light. Incandescent light bulbs convert only 5%–10% of the input power to light, and the rest is emitted as heat (IR radiation). Despite their low efficacy (typically 15 lm W^{-1}) and short lifetime (1,000 h), incandescent light bulbs still persist as a major light source, especially for home lighting. One important reason is that the spectrum of radiation emitted from an incandescent bulb fills the entire wavelength range (400–700 nm) perceived by the human eye and, therefore, renders colors extremely well (CRI $\approx$ 100).

Fluorescent tubes

The first fluorescent tube was made by General Electric (GE) in 1937. Fluorescent tubes consist of a glass tube filled with an inert gas, usually argon, and small amounts of mercury, typically 3–15 mg. Ultraviolet light is created by passing an electric current between electrodes at each end of the tube, which excites electrons in the mercury vapor. When the excited electrons relax, UV light is emitted. The UV light excites phosphors coating the inner surface of the glass tube, which then emit visible light.

Fluorescent tubes have much longer lifetimes than do incandescent light bulbs. White fluorescent tubes use line-emitting phosphors in the blue, green, and red spectral regions, but this produces light of lower CRI than that from incandescent bulbs. However, because of the advantages of much longer lifetimes (7,500–30,000 h) than incandescent light bulbs (1000 h), higher efficacy (60–100 lm W^{-1}), and higher efficiency (25%), combined with reasonable costs, fluorescent tubes rapidly replaced incandescent lighting and now dominate lighting in the workplace, especially in offices and public buildings.

Compact fluorescent lamps

Compact fluorescent lamps (CFLs) were first commercialized in the early 1980s. They usually consist of two, four, or six small fluorescent tubes, which can be straight or coiled. Their efficacy is typically 35–80 lm W^{-1}, and their efficiency is typically 20%. The color rendering of CFLs has improved considerably, and warmer white light is now available, with a CRI of up to 90. Usually CFLs have a lifetime of between 6,000 and 15,000 h. However, switching CFLs on and off for short periods of time will reduce their lifetime. In addition, the brightness of CFLs drops dramatically at low and high temperatures.

At the present time, CFLs are the only realistic replacement for incandescent bulbs in homes. The main advantage of CFLs over incandescent bulbs is efficiency. Over its expected 10,000-h life (given the right operating conditions), a CFL will save 470 kW·h of electricity compared with an equivalent 60-W incandescent bulb. This results in the emission of over 330 kg less carbon dioxide, as well as reductions in emissions of 0.73 kg of nitrogen oxides and 2.0 kg of sulfur dioxide. Each incandescent bulb replaced by a CFL will save the householder, over the 10,000-h assumed lifetime of the CFL, about $50 in the USA and about $100 in the UK (assuming typical 2007 household electricity costs of $0.1 per kW·h and $0.2 per kW·h, respectively). For these reasons, sales of CFLs are soaring. In the USA, for example, CFL shipments have grown tremendously, from 21 million lamps in 2000 to 397 million lamps in 2007 [5].

However, CFLs have an environmental drawback, insofar as each CFL contains about 5 mg of mercury. If the CFLs are not recycled or properly disposed of, the mercury will spill out and contaminate the refuse, which will then enter incinerators and landfill sites. From incinerators, the mercury will enter the atmosphere, and from landfill sites, it can leech out into water supplies and into vegetables subsequently grown on the land. A report by Yale University researchers in 2008 [6] concluded that the mercury leaked from the use of CFLs could actually surpass the amount saved by using less electricity from power plants, where mercury is released into the air from the use of different fossil-fuel sources.

There are 130 million houses in the USA, each containing 45 light bulbs on average. If these are replaced by CFLs, then there will be over 29 t of mercury in CFLs in US houses. The potential for serious mercury contamination is therefore significant. Mercury levels in many countries are already dangerously high. It is therefore important to minimize exposure to mercury. Hence, it is desirable to use low-energy light bulbs that do not contain mercury or other toxic materials. Fortunately, LEDs will provide a safe nontoxic alternative to CFLs and fluorescent tubes.

Inorganic LEDs

Inorganic materials (in particular, semiconductor compounds of elements from groups III–V) emitting red light were first demonstrated in 1962 [7]. Since then, researchers have continued to improve the technology, achieving higher efficiencies and expanding the range of emission wavelengths to cover the whole visible spectrum, as well as the IR and UV regions, by engineering III–V materials systems based on compounds such as GaAs, InP, and GaN. In particular, the first bright-blue LED was announced on November 12, 1993 [8]. White light can also be produced by combining a blue gallium nitride (GaN) LED with a phosphor. Such LEDs are called "white LEDs." White LEDs can last up to 100,000 h and are not affected by frequent switching on and off, unlike CFLs. The best white LEDs available commercially have an efficiency of 30% and an efficacy of over 100 lm W^{-1}. The performance of LEDs continues to improve with the maturing of this technology, and an efficacy of over 200 lm W^{-1} for a white LED was recently achieved in a laboratory [9]. Sales of high-brightness (HB) LEDs have had a compound annual growth rate of over 46% since 1995. In 2010, total sales of LEDs grew by an amazing 68%, reaching $9.1 billion, according to LED market reports. However, white inorganic LEDs are currently too expensive for general household lighting.

Organic LEDs

Light-emitting organic materials were first demonstrated in 1963 [10], within one year of the demonstration of light emission from inorganic materials. Subsequent development led to a range of colored OLEDs based on polymeric or molecular thin films.

A typical OLED comprises (1) a substrate (plastic, glass, or foil) for support; (2) a transparent anode and cathode to inject holes and electrons, respectively, when a current flows through the device; (3) a conducting layer made of organic plastic molecules (e.g., polyaniline) to transport the holes from the anode; and (4) an emissive layer, also made of organic plastic molecules (e.g., polyfluorene), to transport electrons from the

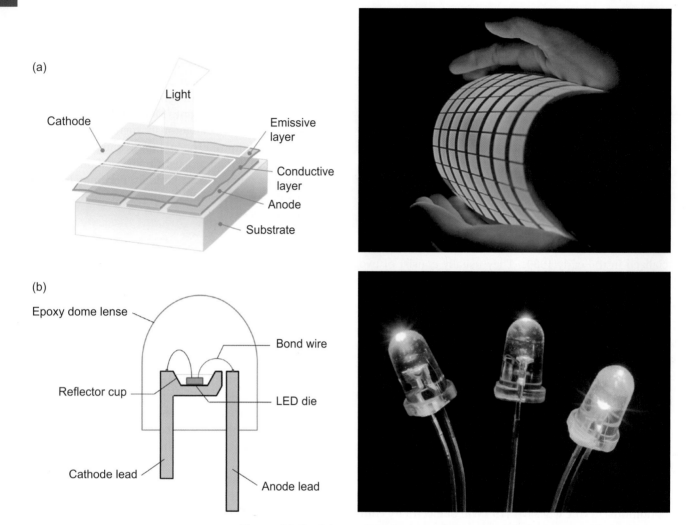

Figure 35.2. Schematic structures of (a) a top-emitting OLED, where the cathode is transparent, and (b) a traditional low-power LED, together with pictures of a blue-emitting OLED panel from GE and green-emitting low-power LEDs.

cathode and generate light. The color of the light emitted from an OLED can be tuned by the structure of the organic molecule in the emissive layer. White light can be generated from separate sources such as blue OLEDs plus yellow phosphors, or from a single source by the blending of different compounds.

Figure 35.2 shows the schematic structure of a top-emitting OLED and a picture of blue-emitting OLEDs from GE, together with the schematic structure of a low-power LED and a picture of such green-emitting devices for comparison.

Although both OLEDs and LEDs can be used for the same applications, such as displays and general illumination, they actually represent two very different light sources. One obvious difference is that an OLED is a less intense diffuse light source, whereas an LED is an intense point source. Other interesting characteristics of OLEDs include their light weight, larger area, and

foldability, whereas a single LED is rigid and small. Currently, OLEDs are mostly used in small-screen devices such as cell phones; OLED-based, ultraslim televisions can also be found on the market. In theory, OLEDs also have a number of attractive properties for SSL, including ease of processing and low cost, as well as the ability for device properties to be tuned by chemically modifying the molecular structure of the organic thin films [11]. However, the development of OLEDs for lighting has lagged behind that of LEDs. State-of-the-art white OLED lighting devices have been reported to have an efficacy of 56 lm W^{-1} [12], which is a breakthrough for OLEDs but still much lower than that for inorganic LED lighting devices. So far, OLEDs have not yet become a practical general illumination source. In contrast, inorganic LED light bulbs, light tubes, recessed lights (also called downlights), and street lights have already started to replace conventional incandescent and florescent light

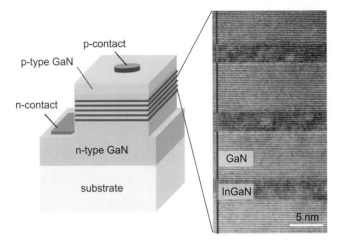

Figure 35.3. A schematic InGaN/GaN quantum-well LED structure together with a high-resolution transmission electron microscope (0002) lattice fringe image of three InGaN quantum wells separated by GaN barriers.

sources in real applications. Thus, the remainder of this chapter will focus on inorganic LED-based SSL, in particular GaN-based LEDs.

35.4.4 Light-emitting diodes for solid-state lighting

Light-emitting diodes are semiconductors, consisting of a p–n-junction diode with a thin active region at the junction. Most high-efficiency LEDs routinely use quantum wells (QWs) in the active region, which provide additional carrier confinement in one direction, improving the radiative efficiency, which is also called the **internal quantum efficiency (IQE)**. Quantum wells consist of a very thin (a few nanometers) layer of a lower-bandgap material, such as InGaN, between higher-bandgap barriers, such as GaN (see Figure 35.3). The QW active region is sandwiched between two thicker layers of doped GaN, one of which, called n-type GaN, is rich in electrons, whereas the other, called p-type GaN, is rich in (positively charged) holes. When a voltage is applied to the LED (called forward bias), the electrons injected from the n-type GaN and holes injected from the p-type GaN exist in the InGaN QWs at different energy levels separated by an energy bandgap. When the electrons and holes subsequently meet and recombine, the energy released is given out as light with an emission wavelength close to the bandgap energy. This results in the emission of light of a single color, such as red, green, or blue, as approximately monochromatic light. This color can be changed by varying the composition and/or changing the thickness of the quantum well, allowing the fabrication of LEDs that emit light of many desired colors. This tailor-made lighting has become possible only recently because of some

fundamental advances in materials science, and it is revolutionizing the field of lighting.

The importance of gallium nitride

Gallium nitride (GaN) is probably the most important new semiconductor material since silicon. The main materials that emitted light before the era of GaN are summarized in Figure 35.4(b). As shown, many different materials were required in order to go from the IR (InAs) to blue (ZnSe). However, despite major research efforts worldwide, ZnSe blue-light-emitting devices have never worked well, one problem being that defects in the material quench the light emission and reduce the lifetime. Green LEDs have also been a problem because, even though both GaP and AlAs emit green light, both have an indirect bandgap (which means that momentum must be supplied to the electrons and holes in order for them to recombine), so the light emitted is weak. Hence, before the advent of GaN, bright LEDs ranged from the IR to yellow, but bright green, blue, and white LEDs were not available.

Gallium nitride, and its sister materials indium nitride (InN) and aluminum nitride (AlN), changed the situation dramatically. As Figure 35.4(a) shows, InN has a bandgap of about 0.7 eV and emits in the IR range, while GaN has a bandgap of 3.4 eV and emits in the near-UV range, at the other end of the visible spectrum. By mixing InN and GaN, which have the same hexagonal crystal structure, any color in the visible spectrum can, in principle, be produced, in addition to IR and near-UV radiation. For example, blue and green light can be produced from $In_xGa_{1-x}N$ with $x \approx 0.10$ and 0.20, respectively. In practice, however, the intensity of light emission from InGaN is low at high indium contents, for reasons that are not fully understood. This means that, although InGaN emits blue light strongly, its green-light emission is less intense, and its red-light emission is weak (see Figure 35.5).

If GaN is mixed with AlN (bandgap 6.2 eV) to form $Al_xGa_{1-x}N$, then UV light ranging from the near-UV to the deep-UV can be produced. The focus in this chapter is on visible lighting, but it is worth noting that deep-UV radiation has many important applications, including water purification, air purification, and detection of biological agents. InN, GaN, and AlN are collectively known as III-nitrides. These materials do not exist in nature, and the creation of this semiconductor family that emits light over such a huge range of important wavelengths is a major breakthrough in materials science. Indeed, the fact that GaN-based LEDs emit light at all is due to extreme scientific good fortune (see below).

Most commercial LEDs are grown on *c*-plane Al_2O_3 (sapphire) substrates, and the lattice mismatch (difference in atom spacing) between GaN and sapphire is a massive 16%. This misfit results in a high density of

Figure 35.4. (a) Bandgap energies of InN, GaN, AlN, and the ternary alloys In$_x$Ga$_{1-x}$N and Al$_y$Ga$_{1-y}$N. The laser energies for reading and writing compact discs, digital video discs, and Blu-ray discs are also shown, together with a rainbow depiction of the visible spectrum. (b) Main materials that were known to emit light prior to the advent of GaN. (Solid circles denote direct-bandgap materials, and open circles denote indirect-bandgap materials; solid lines denote direct-bandgap alloys, and dashed lines denote indirect-bandgap alloys.)

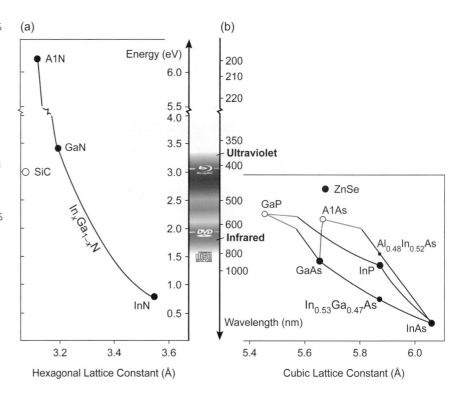

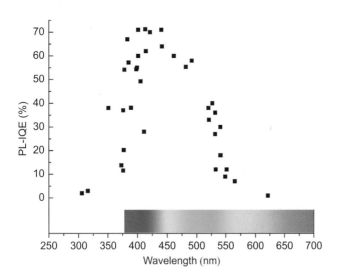

Figure 35.5. A plot of the internal quantum efficiency (IQE), measured using photoluminescence (PL), versus the emission wavelength of c-plane III-nitride quantum-well structures. The points plotted are the highest IQE values reported in the literature [13][14][15][16][17][18][19].

defects called threading dislocations: a typical density of threading dislocations passing through the active InGaN quantum wells is 5 billion per square centimeter (5×10^9 cm^{-2}), as shown in Figure 35.6. Dislocations in GaN are known to be non-radiative recombination centers that should quench the light emission. Indeed, if the dislocation density in other semiconductors, for

example, GaAs, exceeds 1,000 cm^{-2}, the light emission is quenched.

Why do highly defective GaN-based materials emit brilliant light?

The consensus now emerging [20] is that the high brightness of GaN-based LEDs is due to two factors. The first is a fortunate and unexpected interface effect evidenced by transmission electron microscopy [21] and three-dimensional atom-probe studies [22]: monolayer-height interface steps on the InGaN quantum wells. Because the QWs are strained, and because of the high piezoelectric effect in GaN, a monolayer interface step produces an additional carrier-confinement energy of about $3k_BT$ at room temperature, where k_B is the Boltzmann constant and T is the temperature. This is sufficient to localize the electrons [23]. Recent three-dimensional atom-probe studies also confirmed that InGaN is a random alloy [24]. Calculations show that random alloy fluctuations on a nanometer scale localize the holes. Thus, the above two mechanisms localize both the electrons and the holes, and prevent them from diffusing to dislocations.

Generation of white light with LEDs

Whereas LEDs emit light of a single color in a narrow band of wavelengths, white light is required for a huge range of applications, including LED backlighting for large liquid-crystal displays (LCDs) and general home and office lighting. There are various ways of obtaining white light from LEDs (see Figure 35.7).

Figure 35.6. Transmission electron micrographs showing the high density of threading dislocations resulting from the growth of GaN on sapphire. The lattice mismatch between GaN and (0001) sapphire is 16%, which gives rise to a dislocation density in the GaN of typically 5×10^9 cm^{-2}, unless dislocation-reduction methods are used.

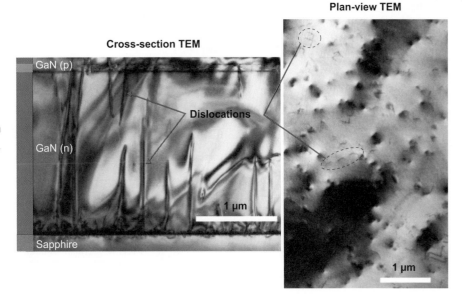

Figure 35.7. Various ways of generating white light from LEDs: (a) blue LED plus yellow phosphor; (b) near-UV LED plus red, green, and blue phosphors; (c) red plus green plus blue LEDs; and (d) red, green, and blue quantum dots in a single LED.

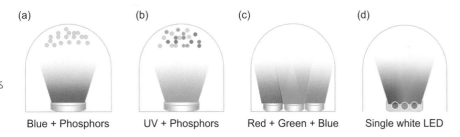

Blue LED and yellow phosphor. The first commercially available white LED was based on an InGaN chip emitting blue light at a wavelength of 460 nm that was coated with a cerium-doped yttrium aluminum garnet (YAG) phosphor layer that converted some of the blue light into yellow light [25]. Nearly all white LEDs sold today use this method. The phosphor layer is sufficiently thin that some blue light is transmitted through it, and the combination of blue and yellow produces a cool white light. This is fine for many applications (e.g., displays, key rings, and cell phones), but the quality of light is probably not good enough for home lighting, for which a warmer white light containing some red light is desirable. To generate warm white light, red phosphors are typically added [26].

Near-UV LED plus red, green, and blue phosphors. Because the efficiency with which existing red phosphors are excited is much less using blue light than using near-UV light, a better route to generate warm white light might be to use a near-UV LED plus red, green, and blue or more colored phosphors. There are no dangers in using a near-UV LED, since thick phosphor layers would be used so that no near-UV light would be transmitted, in much the same way as the phosphor coating on fluorescent tubes and CFLs prevents the transmission of UV light. The drawback of this method is the intrinsic energy loss from a near-UV photon to a lower-energy visible photon.

Red plus green plus blue LEDs. Mixing red, green, and blue (RGB) LEDs is the obvious way to produce white light. However, there are three basic problems with this method. The first is that the efficiency of green LEDs is much less than that of red and blue (this is known as the "green gap" problem), which limits the overall efficiency of this method. Second, the efficiencies of red, green, and blue LEDs change over time at different rates. Hence, if a high-quality white light is produced initially, the quality of the white light degrades noticeably over time. This process is slow, however, and can be corrected using automatic feedback. Third, because the emission peaks of LEDs are narrower than those of most phosphors, the combination of red, green, and blue LEDs will give a poorer color rendering than would the combination of red, green, and blue phosphors. This problem can be minimized by a careful choice of LED emission wavelengths and, of course, by using more than three different color LEDs for better coverage of the visible spectrum. In particular, using four LEDs [red, yellow, green, and blue (RYGB)] can give a good color rendering.

Red, green, and blue quantum dots in a single LED. It is possible to produce a single LED with InGaN quantum dots of different sizes and compositions so that white

Figure 35.8. (a) The schematic structure of a high-power LED package with good optical efficiency and thermal management as required for high-power LED chips. (b) Cross section of a high-power flip-chip LED, illustrating the complex structure of state-of-the-art white LEDs for illumination. (c) A picture of a high-power LED package that can deliver >5 W of power. (After Philips Lumileds.)

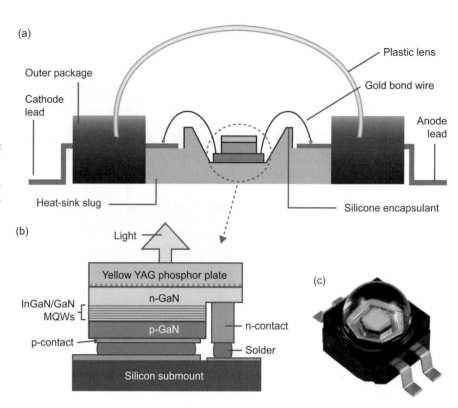

light is emitted. This is a recent development, and the efficiency, reproducibility, and lifetime of these LEDs are not yet known.

High-power LED packages

Recall that LEDs have a high internal quantum efficiency (see Figure 35.5). However, efficient generation of light alone is not enough for an efficient device. For most III–V semiconductors, the refractive index is quite high (e.g., ~2.4 for GaN). This means that most of the light generated in the QW active region is trapped within the crystalline structure and only a very small fraction (~3.4% for GaN) can escape directly from the top surface into the air. To increase the light extraction, several methods, such as chip shaping, photonic crystal structures, and surface texturing, have been developed. An example of high-power chip design for high extraction efficiency is called flip-chip geometry [27], as shown in Figure 35.8(b). In this LED chip design, the sapphire substrate is removed, and the n-type GaN is thinned and its surface roughened to increase the light extraction. The p-contact layer is chosen to be highly reflective at the wavelength of the emitted light, so that the light traveling toward it will be reflected back and escape from the n-type GaN side. The light extraction in today's highest-performing GaN-based LEDs is >80%.

Figure 35.8(a) shows a schematic structure of a high-power LED package, where a high-power flip-chip LED is employed to deliver high-flux white light (blue LED plus yellow phosphor) when driven at high currents

(>350 mA). The heat is removed by including a heat-sink slug in the package.

High-power LED packages are very different from traditional low-power LEDs (as shown in Figure 35.2). Low-power LEDs are commonly 3–8 mm in size. They are typically operated at low current (~20 mA), produce a small amount of light (no more than 30–60 mW, <10 lm), and generate very little heat. High-power LEDs are more complex, using larger LED chips to produce more light (>1 W, 100–1,000 lm). They are operated at high current (>350 mA) and require sophisticated packaging to remove the heat. The applications of small low-power LEDs include cell-phone keypad backlights, signs, and traffic lights. For applications such as television (TV) and computer screen backlights, large displays, car headlamps, and general illumination, high-power white LEDs are required.

35.4.5 Limitations and key materials-research challenges of white LEDs in general illumination applications

Although existing markets for LEDs are large, the real prize is home and office lighting. However, the widespread use of GaN-based white LEDs in these areas is being impeded by five main factors: efficiency, heat management, color rendering, lifetime, and cost.

Efficiency

Table 35.1 compares the efficiencies and efficacies of various forms of lighting, including white LEDs. For high

Table 35.1. Efficiencies and efficacies of various forms of commercially available lighting as of 2009

Type of light source	Efficiency (%)	Efficacy (lm W^{-1})
Incandescent light bulb	5	16
Long fluorescent tube	25	80
Compact fluorescent lamp (CFL)	20	60
Sodium lamp (high-pressure)	45	130
High-power white LEDs (350 mA drive current at 25 °C)	30	132 (cool white) 83 (warm white)
Low-power white LEDs (20 mA drive current at 25 °C)	55	170
High-power white LEDs (5-year target)	55	188 (cool white) 138 (warm white)

light output, LEDs need to be run at a high drive current (typically 350 mA for high-power white LEDs). The maximum efficiency of a commercial high-power white LED is currently about 30% (>100 lm W^{-1} efficacy for cool-white LEDs). This is six times greater than the efficiency of a filament light bulb and 50% better than that of a CFL. However, at a lower drive current of 20 mA, and therefore less total light output (~10 lm), LEDs with an efficiency of 50% are commercially available.

Although it is clear from Figure 35.5 and Table 35.1 that existing blue LEDs are already efficient light sources, the IQE of InGaN-based LEDs is still considerably below 100% and decreases with increasing current density ("**efficiency droop**"), as well as with increasing emission wavelength from the blue to green spectral region (the "**green gap**"). The factors currently limiting the IQE of GaN LEDs are complex and not well understood. Improving efficiencies further at all wavelengths and at higher drive currents requires some in-depth fundamental research.

Dislocation reduction. As previously described, blue InGaN-based LEDs emit brilliant light even though the dislocation density is high. However, the efficiency is even higher if the dislocation density is reduced. Free-standing low-dislocation-density GaN substrates are expensive, so a key research challenge is to reduce the dislocation density substantially *in situ* in the growth reactor by optimizing such factors as the growth conditions and the use of interlayers [28].

Nonpolar and semipolar GaN. Virtually all commercial GaN-based LEDs are grown in the [0001] direction. This is a polar direction, so there is an electric field across the InGaN quantum well that keeps the electrons and holes apart and, hence, reduces the rate of radiative recombination and light emission. Growth in a nonpolar (or semipolar) direction should eliminate this effect,

thus enhancing the efficiency of light emission. However, unless expensive free-standing nonpolar GaN substrates are used, the defect density increases, so the IQE decreases instead of improving. Further research into how to grow nonpolar GaN with low defect densities is required.

Improved p-GaN. The holes in p-type GaN have a low mobility and a low concentration. This reduces the LED efficiency. Further research is required in order to improve the quality of p-GaN.

Novel wide-bandgap semiconductors. Fundamental research on novel wide-bandgap semiconductors is required, since these materials might be even more efficient than GaN-based devices. ZnO is currently receiving much attention, but efficient p-doping of ZnO is proving to be a major problem. ScN is a potentially promising novel nitride [29], but further work is required on new wide-bandgap semiconductors.

Heat management
Currently, many commercial high-power white LEDs operate at an efficacy of about 80 lm W^{-1} at 350 mA, which is about 25% efficient. This means that, although LEDs stay cool relative to filament light bulbs, 75% of the input power is still dissipated as heat. If the LED active region becomes too hot, then the LED lifetime decreases. Heat management is therefore an important issue in many applications of high-power LEDs [30].

Fortunately, this problem will largely disappear when commercial white LEDs reach an efficacy of 150 lm W^{-1}, corresponding to 50% efficiency, because less heat will then be emitted; this is expected to occur in the next few years. Hence, higher LED efficiency is important not only for saving energy and carbon emissions, but also for greatly facilitating heat management.

Color rendering

Nearly all commercial white LEDs are blue LEDs coated with a yellow phosphor, which emit a cool-white light that might not be acceptable for home lighting. If the various routes to higher-quality white LED lighting mentioned earlier prove to be successful, such LEDs could then mimic the visible spectrum of sunlight, giving high-quality "natural" lighting in homes and offices. Apart from the public preference for this natural lighting, there might well be health benefits, for there is increasing evidence [31] that both light intensity and light color are important factors controlling our internal body clock (the circadian rhythm), which determines not only sleeping patterns, but also such functions as brainwave activity, hormone production, and cell regeneration. At current rates of progress, white LEDs that are highly efficient and have excellent color rendering should be available within the next 10 years. The progress is likely to be incremental, with light of better quality becoming available each year.

LED lifetime

Red LEDs are known to have a lifetime of 100,000 h (11 years). Many manufacturers claim a 100,000-h lifetime for their white LEDs (blue LED plus yellow phosphor) as well, but this claim is often not true. Whereas the basic blue-emitting chip does have a lifetime of at least 100,000 h, the problem usually lies in the packaging, in the control electronics, and in the use of cheap, poor-quality components.

One problem is that the epoxy resin used in the LED encapsulation process can become very inflexible when exposed to heat and to blue light. When the LED is switched on, the temperatures of all of the components increase, and thermal expansion can cause the contact wires to move relative to the LED chip and frame. If the cured epoxy resin is inflexible, the contact wires can become detached. The use of a high-grade silicone polymer instead of epoxy resin eliminates this problem and others. With good packaging, as is now being employed by the best LED manufacturers, the lifetime of white LEDs should be about 100,000 h (although the control electronics often have a shorter lifetime).

Cost

Cost is probably the major factor limiting the widespread use of white LEDs in homes and offices. GaN-based LEDs are significantly more expensive than filament light bulbs and CFLs. However, the cost is continuously decreasing. It should be noted that the total ownership cost of lighting takes into account energy savings and replacement cost, which makes LEDs much more competitive, compared with conventional lighting technologies. Nevertheless, to achieve significant market penetration, the initial cost (per kilolumen) of LEDs needs to be reduced by a factor of 10 to be comparable to the cost of CFLs. To achieve the required cost reduction, many aspects of the manufacturing process will need to be addressed in parallel. Considering the material growth (the first step in manufacturing LEDs), increasing the diameter of the currently used substrates, 2-in. (50-mm)-diameter sapphire (Al_2O_3) and SiC, will reduce costs by increasing the usable wafer area. A more dramatic cost reduction might be achieved by growing GaN-based LEDs on silicon wafers, which have the advantage of being readily available in large diameters [e.g., 6 in. (150 mm) or larger] and at low cost. In addition, such wafers are compatible with existing processing lines commonly used in the electronics industry. There are several significant technological challenges in growing GaN on Si substrates, but research is overcoming these. A further point to note is that the improvement in IQE and thus efficacy, especially at high current density, is also very important in LED cost reduction. With further research and development, substantial reductions in the future cost of white LEDs can be expected. Finally, the availability of the raw materials required for LED production, such as substrate materials and indium/gallium metal–organic sources, presents a further limitation of cost reduction. Although short-term demand might create shortages of these raw materials, studies show that the constituent elements are sufficiently abundant to meet the long-term requirements for SSL.

35.4.6 Solid-state lighting

Present applications

Light-emitting diodes are compact, efficient, long-lasting, and controllable, and they are already widely used (see Figure 35.9), for example, as backlighting in cell phones and LCD TVs and displays; as traffic signals; as large displays; as bulbs in flash lights; and as interior and exterior lighting in aircraft, cars, and buses. Because of their long lifetimes, LEDs are also being fitted on airport runways, where the operating cost can be significantly lowered: traditional lighting on runways lasts for only about six months, and the runway has to be closed to replace it, at considerable cost. Further, the performance of LEDs improves at lower temperatures, making them perfect for illuminating refrigerated displays in supermarkets, where CFLs give poor performance because of their very low efficiency in cold environments. Architectural lighting also favors LEDs, which combine art with energy savings and ecofriendliness.

Markets for white LED lighting: the next 5–10 years

With the maturing of ultrahigh-brightness white LEDs, the automotive lighting LED market is growing rapidly. An LED lighting system could save a medium-sized car

Figure 35.9. Examples of current and emerging applications of LED lighting. (a) One of the largest LED displays in the world, on Fremont Street in Las Vegas, NV. (Photo by Scott Roeben, reproduced with permission.) The picture changes continuously. The initial display contained 2.1 million filament light bulbs, which constantly needed to be replaced because of failure. The new display, opened in 2004, contains 12.5 million LEDs. (b) LED traffic lights. (c) The 2010 Audi A8 with LED headlights produced by Osram and (d) a closeup of the LED daytime running lights. (Photo from *LEDs Magazine*, http://www.ledsmagazine.com/news/7/1/2, reproduced with permission.)

>400 kg of CO_2 emissions over 10 years. The close-to-daylight quality of LED front headlamps and the low-power daytime running lights (Figure 35.9(c) and (d)) also have the potential for reducing accidents.

The use of white LEDs for home and office lighting is currently extremely small. Warm-white LEDs based on blue LEDs plus yellow and red phosphors are now available. However, these warm-white LEDs are less efficient than cool-white LEDs because of the low efficiency of the red phosphors excited by blue light. Higher-quality warm-white LEDs, with almost perfect color rendering, need further research but should be widely available in the next five years.

As previously mentioned, a key issue for the use of LED lighting in homes and offices is cost. The total cost of ownership of a light bulb, combining the purchase cost plus running costs, at 8 h per day, is listed in Table 35.2 for one and five years, assuming the current costs of electricity for households in the USA and UK. Although most people consider only the capital cost when buying light bulbs for homes, some will consider capital plus running costs over time. It is clear that warm-white LED

lighting is not yet economically attractive for home lighting. Therefore, to start with, LEDs will mainly be fitted in locations where the 100,000-h lifetime is important, such as those that are difficult to access.

In 2007, the US Department of Energy and the Opto-electronics Industry Development Association (OIDA) published a Technology Roadmap for solid-state lighting, which gave the 2006 cost of light bulbs in dollars per kilolumen as $0.3 for incandescent light bulbs, $3.5 for CFLs, and $90 for warm-white LEDs [32]. The 2009 update of this Roadmap [33] gave the cost of warm-white LEDs as $40 per klm, a 50% reduction in two years. In fact, the field is moving so fast that some of the values in this chapter were out of date by the time it was published. In addition, white LEDs contain no toxic materials such as mercury, and, if legislation bans the use of mercury in light bulbs, as was done for thermometers, then this would accelerate the use of LED lighting. One can therefore expect that white LEDs will enter the general lighting market strongly in the next five years, provided that an efficacy of about 130 lm W^{-1} and a cost of $4 per kilolumen can be achieved.

Table 35.2. Total costs of ownership of a light bulb for one and five Years

	USA		UK	
Type of light source	1 year	5 years	1 year	5 years
60-W incandescent	$18	$90	$36	$180
Compact fluorescent lamp (CFL)	$6	$28	$11	$50
60-lm W^{-1} warm-white LED bulb	$18	$36	$23	$58

Note: values calculated assuming light bulbs used for 8 h per day at electricity costs of $0.1 per kW·h (USA) and $0.2 per kW·h (UK), cost of incandescent bulb $0.5, cost of CFL $2, and cost of LED $14.

In the next 10 years, it is reasonable to expect that the efficiency of white LEDs will continue to increase, with dramatic energy savings. For example, at an efficacy of 150 lm W^{-1}, a white LED with light output equivalent to that of a 60-W incandescent light bulb can be left on all year, for 24 h per day, at an electricity cost of only $5 (at today's US electricity cost of $0.1 per kW·h). Also, in the next 10 years, one can expect today's warm-white LEDs to be replaced by higher-quality white LEDs giving out natural lighting similar to sunlight. Further research is needed in order to develop these, but they are likely to be popular with customers for the health reasons mentioned earlier.

White LED technology for lighting: the next 10–20 years

The 2007 Roadmap [32] predicted the efficacy of a commercial cool-white LED to be 113 lm W^{-1} in 2010, 135 lm W^{-1} in 2012, and 168 lm W^{-1} in 2015. In fact, the prediction of 113 lm W^{-1} in 2010, published in May 2007, was exceeded later in 2007, with the aforementioned commercial availability of a cool-white LED having an efficacy of 132 lm W^{-1} at 350 mA. Recently, a commercial white LED with an efficacy of 160 lm W^{-1} at 350 mA was released [34]. Again, this indicates the rapid progress being made with inorganic LEDs, and it raises expectations that the other future predictions for inorganic LEDs will also be met sooner than expected.

The updated 2009 Roadmap [33] predicts efficacy improvements to 164 lm W^{-1} in 2012 and 188 lm W^{-1} in 2015 for commercial cool-white LEDs and to 114 lm W^{-1} in 2012 and 138 lm W^{-1} in 2015 for warm-white LEDs. Efforts are, indeed, on course for achieving these milestones, and 208 lm W^{-1} for a cool-white LED was recently achieved in a laboratory. It is unlikely that 250 lm W^{-1} will be achieved using an LED-plus-phosphor combination because of efficiency losses in the phosphor as well as in converting high-energy photons from the LED into lower-energy photons (the Stokes shift). The solution to this problem is to avoid phosphors and

to produce white light by mixing red, yellow, green, and blue (RYGB) LEDs. This cannot be done efficiently at present because, as mentioned earlier, green LEDs are much less efficient than blue and red (the so-called "green gap").

If the green-gap problem can be solved, then it should be possible to produce warm-white light from RYGB LEDs with an efficacy of at least 200 lm W^{-1}. The savings in energy, and carbon dioxide emissions, will then be even greater than the values given at the start of this chapter, which assume an LED efficacy of 150 lm W^{-1}. In addition, this white-light source will be color-tunable, enabling people to wake up to a blue-white light, for example, and go to bed with a red-white light.

35.5 Case study: using an LED luminaire to illuminate a 4.5-m × 4.5-m kitchen in the UK[1]

General lighting requirements for kitchens are as follows.

1. Illumination >130 lm m^{-2} (lux) on the work surfaces for food preparation.
2. Good color rendering to ensure that food looks natural and to see when food is cooked or whether it has gone bad.
3. Fixed lights, since cables from freestanding lights are a potential hazard.

Typically, the lighting in kitchens is achieved using halogen downlighters that have a CRI of >90. However, halogen downlighters are not very efficient and have a relatively short lifetime. In this case study, LED luminaires are used to replace the halogen downlighters to illuminate a kitchen room of size 4.5 m (W) × 4.5 m (L) × 2.7 m (H) (14.8 ft × 14.8 ft × 8.9 ft).

The application employs nine units in a 3 × 3 array of the LED luminaires specified in Table 35.3 (see Figure

---

[1] Data from PhotonStar LED Ltd.

Table 35.3 Features of LED luminaires used in this case study

Fixed recessed downlight with glass diffuser to provide gentle but effective illumination.
Contains seven high-brightness LEDs
Illumination >55 luminaire lumens per circuit watt, more energy-efficient than CFL and halogen downlighters
A color-rendering index (CRI) of 86
Maintenance-free operating life of 50,000 h at high ambient temperatures up to 55 °C
Dimmable and instant on

Table 35.4. A simulation for the arrangement of the LEDs, with height of room 2.7 m (8.9 ft) and mounting height 2.7 m (8.9 ft)

Surface	Reflectance	Average illuminance (lux)	Minimum illuminance (lux)	Maximum illuminance (lux)
Workplace	–	130	70	162
Floor	20	110	64	139
Ceiling	80	29	22	32
Walls (×4)	50	64	32	93

Figure 35.10. Features of the LED luminaires used in the example.

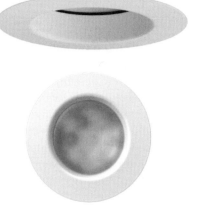

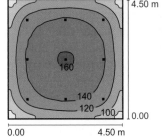

Figure 35.11. Simulations and, on the right, the finished kitchen illuminated by LED luminaires.

35.10), spaced roughly 1.5 m apart. Figure 35.11 and Table 35.4 show the simulation of the arrangement of the LEDs (black spots) on the kitchen ceiling (left top image) and a density plot indicating the uniformity (left bottom image) of the illuminance across a work plane 1 m (3.3 ft) above the floor. The average illuminance on the work plane is 130 lux. A picture of the finished kitchen illuminated by LED luminaires is also given in Figure 35.11.

A comparison of energy cost and CO_2 emissions between LED luminaires and other downlighters giving equivalent illumination performance in this case is given in Table 35.5.

35.6 Summary

Light-emitting-diode-based SSL promises to provide a high-quality and energy-efficient light source for general use. A summary of this technology is given in Table 35.6. If the materials challenges outlined in this chapter can be met, then in 10–20 years time, LED lighting will be the dominant form of lighting in homes, offices, cities,

Table 35.5. A comparison of energy cost and CO_2 emissions for various types of lamp including the LED luminaires used in the example

	65-W Incandescent lamp (reference)	14-W Compact fluorescent lamp	35-W Halogen downlighters	8-W LED luminaires used in this case
Number of lamps	9	9	9	9
Power used per lamp (W)	65	14	35	8
Total power of all lamps (W)	585	126	315	72
Energy cost over 50,000 h ($)	5,850	1,260	3,150	720
Relative energy savings over 50,000 h of operation ($)	–	4,590	2,700	5,130
Annual energy cost ($)	85.4	18.4	46.0	10.5
Carbon dioxide emitted over 50,000 h of operation (kg)	11,700	2,520	6,330	1,440
Relative amount of carbon dioxide saved over 50,000 h of operation (kg)	–	9,180	5,400	10,260

Values based on energy costs in the UK of $0.2 per $kW \cdot h$ and production of 0.4 kg of carbon dioxide per $kW \cdot h$. A modest domestic usage of kitchen lighting of 2 h per day was used to calculate the annual energy cost. In addition, over the 50,000 h of operation of the LED luminaires, the consumer would most likely buy, change, and dispose of a staggering 450 incandescent light bulbs or 72 low-energy light bulbs!

and transport throughout the world. Light-emitting diodes are expected to shine the way to a bright future, providing a light source that is

- ultra-energy-efficient (15 times more than incandescent light bulbs and 5 times more than CFLs),
- ultra-long-lived (100,000 h compared with 10,000 h for CFLs and 1,000 h for incandescent light bulbs),
- environmentally friendly, with no toxic mercury,
- inexpensive,
- similar to sunlight (supporting health and well-being), and
- tunable (with the ability to have mood lighting of any desired color or shade of white).

35.7 Questions for discussion

1. Discuss the advantages and disadvantages of various light sources for home and office lighting, such as incandescent light bulbs, fluorescent tubes, CFLs, and LEDs.
2. Why is the efficacy lower for CFLs and LEDs with higher CRI values? Why is a high CRI preferred in most lighting?

3. Discuss the limitations today of white LEDs for home and office lighting.
4. What improvements are required in order to achieve higher-efficiency white LEDs, and what benefits might such LEDs bring?
5. Light-emitting diodes can provide versatile lighting solutions for many applications. List five of these, and discuss the advantages of using LEDs for each one.

35.8 Further reading

- **E. F. Schubert**, 2006, *Light-Emitting Diodes*, 2nd edn., Cambridge, Cambridge University Press. A comprehensive introduction to LEDs, including many relevant topics such as color rendering.
- **B. Gil**, 1998, *Group III Nitride Semiconductor Compounds: Physics and Applications*, Oxford, Clarendon Press. A comprehensive description of both experimental and theoretical developments in the field of III-nitrides.
- **S. Nakamura** and **G. Fasol**, 1997, *The Blue Laser Diode: GaN Based Light Emitters and Lasers*, Berlin, Springer. Detailed descriptions of some of the early developments of GaN LEDs and laser diodes.

Table 35.6. A summary of the technology

Technology	Current	In 10 years	In 20 years
Cost of warm-white LED lamp ($ per klm)	>80	<9	<5
Efficacy (lm W^{-1})	<80	>150	>200
Electrical power cost in USA ($ per Mlm·h)	>1.2	<0.7	<0.5
Key materials	III–V Compounds	III–V Compounds	III–V Compounds
Limitations of current technology	Efficiency		
	Heat management		
	Color rendering		
	Lifetime		
	Cost		
Key materials-research challenges	Efficiency droop		
	Green gap		
	Dislocation reduction		
	Nonpolar and semipolar GaN		
	Improved p-GaN		
	Novel wide-bandgap semiconductors		
Environmental impact (USA alone) relative to no usage of solid-state lighting			
Reduction in electricity consumption for lighting(%)	<0.8	>6.8	>25
Reduction in carbon dioxide emissions (Mt)	<4.2	>36	>114

Note: Values calculated assuming warm-white LED lamp of high CRI (76–90) at an electricity cost of $0.1 per kW·h (USA) and carbon dioxide production of 0.6 kg per kW·h (USA). Values were calculated for the environmental impact based on the predictions given in [35].

35.9 References

[1] **International Energy Agency**, 2006, *Light's Labour's Lost: Polices for Energy-Efficient Lighting*, http://www.iea.org/textbase/npsum/lll.pdf.

[2] **International Energy Agency**, *Key World Energy Statistics 2007*, http://www.iea.org/Textbase/nppdf/free/2007/key_stats_2007.pdf.

[3] **J. J. Vos**, 1978, "Colorimetric and photometric properties of a 2-deg fundamental observer," *Color Res. Appl.*, **3**, 125.

[4] **J. M. Phillips**, **M. E. Coltrin**, **M. H. Crawford** *et al.*, 2007, "Research challenges to ultra-efficient inorganic solid-state lighting," *Laser Photon. Rev.*, **1**(4), 307–333.

[5] **US Department of Energy**, 2009, *CFL Market Profile 2009*, http://www.energystar.gov/ia/products/downloads/CFL_Market_Profile.pdf.

[6] **E. Engelhaupt**, 2008, "Do compact fluorescent bulbs reduce mercury pollution?," *Environ. Sci. Technol.*, **42** (22), 8176.

[7] **N. Holonyak** and **S. F. Bevacqua**, 1962, "Coherent (visible) light emission from Ga(As$_{1-x}$P$_x$) junctions," *Appl. Phys. Lett.*, **1**(4), 82–83.

[8] **S. Nakamura**, **M. Sonoh**, and **T. Mukai**, 1993, "High-power InGaN/GaN double-heterostructure violet light emitting diodes," *Appl. Phys. Lett.*, **62**(19), 2390–2392.

[9] **Cree**, 2010, *Cree Breaks 200 Lumen Per Watt Efficacy Barrier*, http://www.cree.com/press/press_detail.asp?i=1265232091259.

[10] **M. Pope**, **H. P. Kallmann**, and **P. Magnante**, 1963, "Electroluminescence in organic crystals," *J. Chem. Phys.*, **38**, 2042.

[11] **OIDA**, 2002, *Organic Light Emitting Diodes (OLEDs) for General Illumination Update 2002*; Washington, DC, OIDA, http://lighting.sandia.gov/lightingdocs/OIDA_SSL_OLED_Roadmap_Full.pdf.

[12] **GE**, 2010, *White OLED Outlook Brightens with Efficiency Breakthrough*, http://www.genewscenter.com/Press-Releases/White-OLED-Outlook-Brightens-with-Efficiency-Breakthrough-2969.aspx.

[13] **D. M. Graham**, **P. Dawson**, **G. R. Chabrol** *et al.*, 2007, "High photoluminescence quantum efficiency InGaN multiple quantum well structures emitting at 380 nm," *J. Appl. Phys.*, **101**, 033516.

[14] **D. Fuhrmann**, **T. Retzlaff**, **U. Rossow**, **H. Bremers**, and **A. Hangleiter**, 2006, "Large internal quantum efficiency of In-free UV-emitting GaN/AlGaN quantum-well structures," *Appl. Phys. Lett.*, **88**, 191108.

[15] **D. Fuhrmann**, **U. Rossow**, **C. Netzel** *et al.*, 2006, "Optimizing the internal quantum efficiency of GaInN SQW structures for green light emitters," *Phys. Status Solidi C*, **3**(6), 1966–1969.

[16] **T. Akasaka**, **H. Gotoh**, **T. Saito**, and **T. Makimoto**, 2004, "High luminescent efficiency of InGaN multiple quantum wells grown on InGaN underlying layers," *Appl. Phys. Lett.*, **85**, 3089–3091.

[17] **A. Hangleiter**, **D. Huhrmann**, **M. Grewe**, *et al.*, 2004, "Towards understanding the emission efficiency of nitride quantum wells," *Phys. Status Solidi A*, **201**(12), 2808–2813.

[18] **J. Harris**, **T. Someya**, **K. Hoshino**, **S. Kako**, and **Y. Arakawa**, 2000, "Photoluminescence of GaN quantum wells with AlGaN barriers of high aluminum content," *Phys. Status Solidi A*, **180**(1), 339–343.

[19] **Y. Sun**, **Y. Cho**, **H. Kim**, and **T. W. Kang**, 2005, "High efficiency and brightness of blue light emission from dislocation-free InGaN/GaN quantum well nanorod arrays," *Appl. Phys. Lett.*, **87**, 093115.

[20] **R. A. Oliver** and **B. Daudin** (eds.), 2007, *Intentional and Unintentional Localization in InGaN, Phil. Mag.*, **87**(13), 1967–2093 (special issue).

[21] **D. M. Graham**, **A. Soltani-Vala**, **P. Dawson** *et al.*, 2005, "Optical and microstructural studies of InGaN/GaN single-quantum-well structures," *J. Appl. Phys.*, **97**, 103508.

[22] **M. J. Galtrey**, **R. A. Oliver**, **M. J. Kappers**, *et al.*, 2008, "Compositional inhomogeneity of high-efficiency $In_{1-x}Ga_xN$ based multiple quantum well ultraviolet emitters studied by three dimensional atom probe," *Appl. Phys. Lett.*, **92**, 041904.

[23] **C. J. Humphreys**, 2007, "Does In form In-rich clusters in InGaN quantum wells?," *Phil. Mag.*, **87**, 1971–1982.

[24] **M. J. Galtrey**, **R. A. Oliver**, **M. J. Kappers**, *et al.*, 2007, "Three-dimensional atom probe studies of an $In_{1-x}Ga_xN$/GaN multiple quantum well structure: assessment of possible indium clustering," *Appl. Phys. Lett.*, **90**, 061903.

[25] **P. Schlotter**, **R. Schmidt**, and **J. Schneider**, 1997, "Luminescence conversion of blue light emitting diodes," *Appl. Phys. A*, **64**(4), 417–418.

[26] **R. Mueller-Mach**, **G. O. Mueller**, **M. R. Krames**, and **T. Trottier**, 2002, "High-power phosphor-converted light-emitting diodes based on III-nitrides," *IEEE J. Selected Topics Quant. Electron.*, **8**(2), 339–345.

[27] **M. R. Krames**, **D. A. Steigerwald**, **F. A. Kish** *et al.*, 2003, III-Nitride light-emitting device with increased light generating capability, US Patent No. 6,521,914.

[28] **M. J. Kappers**, **M. A. Moram Y. Zhang**, *et al.*, 2007, "Interlayer methods of reducing the dislocation density in gallium nitride," *Physica B: Condens. Matter*, **401**, 296–301.

[29] **M. A. Moram**, **Y. Zhang**, **M. J. Kappers**, **Z. H. Barber**, and **C. J. Humphreys**, 2007, "Dislocation reduction in gallium nitride films using scandium nitride interlayers," *Appl. Phys. Lett.*, **91**, 152101.

[30] **L. X. Zhao**, **E. J. Thrush**, **C. J. Humphreys**, and **W. A. Phillips**, 2008, "Degradation of GaN-based quantum well light-emitting diodes," *J. Appl. Phys.*, **103**, 024501.

[31] **L. A. Newman**, **M. T. Walker**, **R. V. Brown**, **T. W. Cronin**, and **P. R. Robinson**, 2003, "Melanopsin forms a functional short-wavelength photopigment," *Biochemistry*, **42**(44), 12734–12738.

[32] **Navigant Consulting Inc and Radcliffe Advisors for the US Department of Energy**, 2007, *Multi-Year Program Plan FY08-FY13, Solid-State-Lighting Research and Development*, http://apps1.eere.energy.gov/buildings/publications/pdfs/ssl/ssl_mypp2007_web.pdf.

[33] **US Department of Energy**, 2009, *Solid-State Lighting Research and Development: Manufacturing Roadmap*, http://apps1.eere.energy.gov/buildings/publications/pdfs/ssl/ssl-manufacturing-roadmap_09–09.pdf.

[34] **Cree**, 2010, *Cree's New Lighting-Class LEDs shatter Industry Performance Standards*, http://www.cree.com/press/press_detail.asp?i=1289396994146.

[35] **US Department of Energy**, 2010, *Energy Saving Potential of Solid-State Lighting in General Illumination Applications 2010 to 2030*, http://apps1.eere.energy.gov/buildings/publications/pdfs/ssl/ssl_energy-savings-report_ 10–30.pp.

36 Energy efficient buildings

Ron Judkoff

National Renewable Energy Laboratory, Golden, CO, USA

36.1 Focus

Materials advances could help to reduce the energy and environmental impacts of buildings. Globally, buildings consume 30%–40% of primary energy and account for 25%–33% of CO_2 emissions. Building energy consumption emanates from a variety of sources, some of which are related to the building envelope or fabric, some to the equipment in the building, and some to both. Opportunities for reducing energy use in buildings through the application of innovative materials are therefore numerous, but there is no one system, component, or material whose improvement can alone solve the building energy problem. Many of the loads in a building are interactive, and this complicates cost–benefit analysis for new materials, components, and systems. Moreover, components and materials for buildings must meet stringent durability and cost–performance criteria to last the long service lifetimes of buildings and compete successfully in the marketplace.

36.2 Synopsis

The world's buildings account for about one-third of greenhouse-gas emissions. The authors of a number of studies have concluded that one of the most cost-effective ways to reduce carbon emissions is to increase the energy efficiency of the existing building stock via energy retrofits, and to require a high degree of efficiency in new buildings. Several nations and states have even set goals requiring new buildings to have "zero net energy" consumption by established dates in the future. Some of the increases in efficiency can be achieved with currently available technology by using advanced building energy simulation and optimization techniques to establish technology packages that deliver maximum energy savings for minimal cost. The technology packages will vary by building type and climate type, and will also need to differ according to the level of development in various parts of the world. To realize further efficiency gains, new and better materials and technologies will be needed. This presents a unique opportunity and challenge for materials scientists. Success will depend not only on mastery of the usual disciplines associated with materials science, but also on a holistic understanding of the comfort, construction, and energy systems in buildings. This chapter explores some foundational concepts about how energy is used in buildings, and also highlights areas where materials-science advances would be most beneficial.

36.3 History

Protection from the elements is fundamental to human survival, and architecture originated from the need for shelter. Archeological findings from ancient Greece indicate that the houses were oriented to the south, and Socrates observed that, "In houses that look toward the south, the Sun penetrates the portico in winter." The Roman architect Vitruvius, who lived during the first century BC, documented many of the principles of bioclimatic building design and urban planning in his ten-volume treatise on architecture. At that time, an advance in materials science allowed wealthier Romans to use mica and glass to glaze openings on south façades for the purpose of better providing heat in winter. The name for such a specially designed room was *heliocaminus*, meaning "sun furnace." The rooms were made of high-heat-capacity materials with decorative carvings in the mass to increase both the surface area and the effective surface heat-transfer coefficient, to better move the heat in and out of the mass. Other ancient examples of environmentally adapted architecture are found in the Anasazi ruins of the American southwest and the great mud-walled cities of the Sahara.

Ancient mud-wall Saharan city near Marrakech, Morocco. (Picture credit Ron Judkoff)

36.4 Energy and environmental impact of buildings

The buildings sector accounts for about 40% of primary energy consumption, 70% of electricity use, and 40% of greenhouse-gas emissions in developed countries [1]. Globally, buildings account for about 30%–40% of primary energy use, 57% of electricity demand, and 25%–33% of greenhouse-gas emissions [2][3]. The differences can be explained because in less developed countries many buildings do not provide the levels of thermal and luminous comfort, and air quality, common in

Table 36.1. Current and projected annual rates of global building energy consumption and carbon dioxide emissions under a business-as-usual scenario (projected growth rate 1.3% per year) 1 quad is 1.055 EJ

Year	Energy consumption (quads per year)	CO$_2$ (Gt per year)
2005	116	8.8
2030	154	14
2050	204	20.5

developed nations. Also, many buildings in less developed countries have little or no access to electricity. The correlation between primary energy use and carbon emissions depends on the mix of end uses, energy sources, and climate in any given country. For example, the carbon emissions from coal-fired power plants are different from those from gas-fired plants, nuclear plants, and hydro-electric plants. In North America and in European nations, heating is the largest energy end use in buildings, whereas in India domestic cooking is the dominant use.

Currently, the total annual world primary energy demand from human activities is about 466 quadrillion Btu (quads) (487 EJ), which is expected to grow by about 1.6% per year to 675 quads (705 EJ) by 2030 under a business-as-usual scenario. Corresponding world anthropogenic carbon dioxide emissions amount to about 30 gigatons of carbon dioxide equivalent per year (Gt CO$_2$e per year) and 45 Gt per year by 2030, also expected to grow by 1.6% per year [3][4] (Chapters 1 and 8). Determining the energy and carbon emissions attributable to the world's building stock is more difficult, and some of the source documents differ substantially on the data. Table 36.1 shows approximate annual global building primary energy use and carbon dioxide emissions projected to the year 2050 (my calculations plus data from [2][3][5][6]). By 2030, world building energy consumption and emission rates are projected to grow by about 38 quads per year and 12 Gt CO$_2$e per year.

These energy and carbon quantities are important because climate studies indicate that global temperatures are correlated to greenhouse-gas concentrations in the atmosphere. Long-term CO$_2$e values above 450 ppm will result in unfavorable global temperature increases, according to the United Nations Intergovernmental Panel on Climate Change (IPCC), as shown in Table 36.2.

Studies by the International Energy Agency, the World Business Council for Sustainable Development, and McKinsey & Company among others, indicate that an improvement in energy efficiency of new and existing buildings is the most cost-effective way to achieve emissions reductions [2][7][8]. The McKinsey report, for

Table 36.2. Atmospheric CO₂e concentrations and temperature

CO₂e (ppm)	Temperature rise (°C)	Emission reduction needed by 2050 to stabilize at concentrations in column 1 (%)
445 to 490	2 to 2.4	−85 to −50
490 to 535	2.4 to 2.8	−60 to −30
535 to 590	2.8 to 3.2	−30 to 5.0
590 to 710	3.2 to 4.0	10 to 60

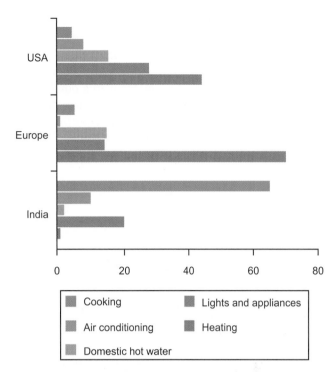

Figure 36.1. Residential site energy end uses (%).

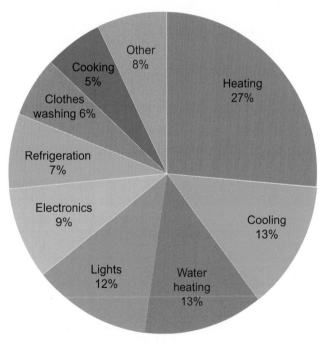

Figure 36.2. Residential building primary energy end uses.

example, concludes that primary energy consumption in the USA could be reduced by about 12 quads per year by 2020, yielding a reduction of about 0.7 Gt per year in emission of CO₂e, with an investment of $354 billion and a return on investment in avoided energy costs of $685 billion. Annual energy savings in 2020, compared with business as usual, would be $78 billion per year. All of the efficiency measures included in the evaluation to attain these savings are currently available and all were economically positive from a net-present-value (NPV) perspective. Nevertheless, this is not nearly enough to achieve the carbon abatements necessary for stabilizing greenhouse-gas concentrations below 450 ppm. All the reports concluded that an emphasis on developing new, cost-effective, highly efficient building materials and technologies is essential.

Figure 36.1 shows the relative average disaggregated end uses of energy in residential buildings for the USA, Europe, and India [4][8]. It is readily apparent that not all new materials or components will be equally applicable across regions. This is also true across building types as demonstrated by Figures 36.2 and 36.3 comparing primary energy end uses for residential and commercial buildings in the USA [4].

In winter and summer, energy leaks through walls and windows via conduction, radiation, and convection. The heating, ventilating, and air conditioning (HVAC) system then consumes energy to maintain comfortable temperatures and acceptable air quality for occupants. Energy is also provided for the purposes of heating water, cooking, and powering electrical devices, among other uses. Energy expended for heating and cooling can be reduced by improving the thermal fabric of the building, or by increasing the efficiency of the heating and cooling systems. The thermal fabric is defined as the building's envelope or shell, including windows, plus all

other materials of which the building itself is made that can affect the heating and cooling loads. For example, the heat capacitance and moisture capacitance of interior walls, partitions, floors, ceilings, and furnishings can affect the building's load profile in addition to the effect from the exterior skin. Table 36.3 shows aggregate component loads for all residential buildings in the USA [4].

From Figures 36.1, 36.2, 36.3 and Table 36.3, it is evident that there are no "silver-bullet" solutions to the building energy problem. To radically reduce energy consumption in the building sector, all uses and losses must be addressed. Many of the loads in a building are interactive. For example, improved insulation for walls reduces energy use for heating, as does a higher-efficiency furnace. The benefit-to-cost ratios of these two measures are both interdependent and dependent on the order in which they are applied to the building. In this case, whichever is applied first will appear relatively better than the one that is applied second. The aggregate energy savings will be less than the sum of the independent energy savings for each efficiency measure. These interactions can be either detrimental or advantageous. Smart holistic energy design attempts to take advantage of these interactions, as well as the interactions between the building and the local climate; however, the parameter space for building energy optimization is extremely large and difficult to model. Computers commonly available to architects and engineers have only recently become powerful enough to begin to address the notion of whole-building energy and cost optimization [9].

One can improve building energy performance by improving the individual technologies, or by optimizing the mix and interaction among all the parameters that affect building energy use. Figure 36.4 shows the savings from an energy-optimized house compared with the base case of a typical house built to code. The x axis shows the percentage of source energy savings from the base case, and the y axis shows the sum of the annual utility-bill costs plus any extra annual mortgage costs for energy-savings features. At the 0 savings point (y axis), all costs are utility-bills. As one moves to the right on the savings curve, the utility-bill savings outweigh the additional mortgage cost for energy-efficient features, causing the curve to slope downward. The neutral cost point is the point at which the utility-bill savings just balance the additional mortgage cost. In this proposed house in a continental climate at latitude 40° N, energy optimization provides nearly 60% energy savings using off-the-shelf technology at no additional cash-flow burden to

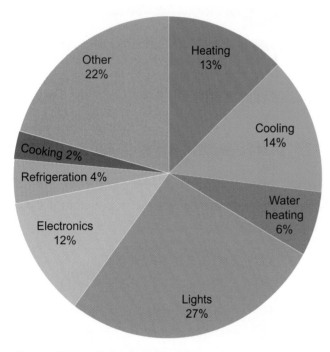

Figure 36.3. Commercial building primary energy end uses.

	Heating (quads)	**Heating (%)**	**Cooling (quads)**	**Cooling (%)**
Roofs	−0.65	12	0.16	14
Walls	−1.00	19	0.11	10
Foundation	−0.76	15	−0.07	Ground heat transfer reduces cooling load
Infiltration	−1.47	28	0.19	16
Windows (conduction)	−1.34	26	0.01	1
Windows (solar gain)	0.43	Reduces heating load	0.37	32
Internal gains	0.79	Reduces heating load	0.31	27
Total load	−4.00	100	1.08	100

Table 36.3. Residential building aggregate component loads and percentages

Note: infiltration is the leakage of outside air into the living space. Internal gains are heat gains to the living space from lights, equipment, and electrical devices. In conditioned spaces, these heat gains reduce the heat load on the furnace in winter, but increase the cooling load on the air conditioner in summer.

Figure 36.4. Savings from an energy-optimized house in Greensburg, Kansas, compared with a typical house built to code (2,000 ft², two-story house with 16% window-to-floor area ratio and unconditioned basement). IECC, International Energy Conservation Code; BEopt is the name of a building energy-optimization software package developed by the National Renewable Energy Laboratory.

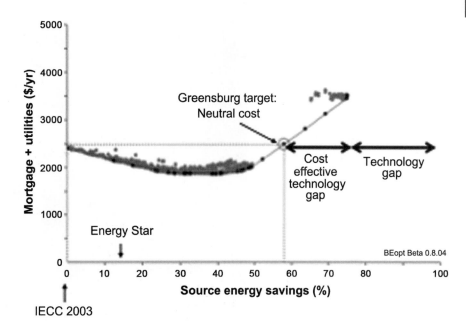

the building owner. Of course, from the owner's economic perspective, the minimum cost point at the nadir of the curve may be considered most favorable. However, from a societal perspective, the neutral cost point suggests a means to capture enormous energy and carbon savings in a manner that is favorable to the building's owner, the lending institution, and the societal goal of reducing greenhouse-gas emissions. The owner has a more comfortable building and is cushioned from energy price increases in the future, and the lending institution has made a larger, lower-risk loan with less overhead per dollar loaned. Local and national economies may also benefit from job growth that cannot be outsourced. The point at which the curve becomes linear is where current efficiency technology becomes more expensive than adding photovoltaic (PV) panels on the roof. With the addition of a PV system, savings of about 80% can be achieved, with some extra cost and cash-flow burden to the owner. The curve stops at 80% because no more roof area is available for further PV panels. Of course, if materials scientists succeed in increasing the efficiency of PV systems at acceptable cost, then 100% or more of the load could be met with the PV panels on the roof, yielding, in this case, a net zero consumption of energy or an energy-producing building. The neutral-cost technology package for this building includes (a) an R-22 wall assembly (2×6 studs, R-19 batts plus foam sheathing) (the R-value is a measure of thermal resistance and is discussed further in the next section), (b) R-50 attic insulation, (c) an R-10 basement, (d) very airtight construction (two air changes per hour at a pressure difference of 50 Pa between the inside and outside of the pressure envelope of the building), (e) low-emissivity and low-solar-heat-gain-coefficient double glazing with an argon-gas fill (0.28 U-value and

0.37 SHGC)[1], (f) 80% compact fluorescent lighting (CFL), (g) a SEER 18 air conditioner, (h) a 90+ AFUE condensing gas forced-air furnace, (i) a gas tankless water heater with an energy factor of 0.8, (j) tight ducts (mastic-sealed 5% leakage) completely inside the conditioned space, (k) Energy Star appliances, and (l) a 1.5-kW PV system. This package added about $26,000 to the initial cost of the house, but was cash-flow neutral assuming a 30-year 7% mortgage.

Figure 36.4 shows that, to go any further toward a zero-energy building (100% savings on the diagram), new technology is needed. In this case, more efficient PV technology or more cost-effective efficiency measures would provide further savings. The best technology advances would move the neutral cost point further to the right. To understand what is happening, think of designing a residential building in a cold sunny climate with greater window area on the south side (in the Northern Hemisphere) and correspondingly less window area on the other orientations. As a result, because of solar energy flowing through the larger area of south-facing windows, the heating load would be decreased at no additional cost and with no need for new technology. However, creating materials that allow windows to transmit heat when needed and reject heat when not needed would facilitate even more energy savings and allow greater design freedom.

Figure 36.5 shows a zero-energy "Habitat for Humanity" home in a high-plains climate at latitude 40° N designed by the Buildings R&D Program at the National Renewable Energy Laboratory (NREL). After

---

[1] SHGC is solar heat gain coefficient, SEER is seasonal energy efficiency ratio, AFUE is annual fuel utilization efficiency, R is thermal resistance and U is overall heat transfer coefficient, the inverse of R.

Figure 36.5. Zero-energy house.

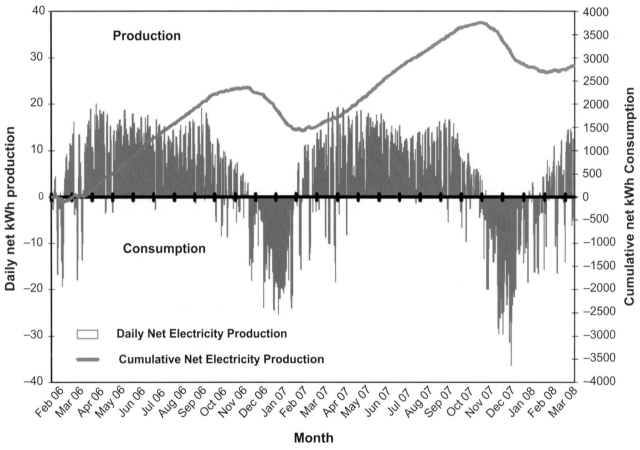

Figure 36.6. Data for the zero-energy house.

two years of monitoring, the building was actually a net energy producer, as shown in figure 36.6. Energy features of the building include *R*-40 walls, an *R*-60 roof, double-glazed low-emissivity windows (high solar-heat-gain coefficient on the southern face), southern roof overhang for summer shading, a ventilation heat-recovery system, a 4-kW PV system on the roof, and solar water-heating panels on the roof. In contrast to the other house above, there is sufficient roof area on the Habitat house to achieve zero net consumption of energy because the house is a single story and the shell is more efficient. Houses such as this one demonstrate

Figure 36.7. Zion National Park Visitor Center, (a) north façade and (b) south façade.

(a)

(b)

that it is technically possible to design and construct zero-energy buildings now. However, the cost of the Habitat house was not optimal, being well above the previously described neutral cost point. New, better-performing, and less expensive materials are needed in order to help solve the cost-effectiveness issue.

Figures 36.2 and 36.3 displayed some of the differences in loads between residential and commercial buildings. Despite these differences it is possible to take a similar optimization approach for commercial buildings, even though the parameter space for commercial buildings is substantially larger and more complex than that for residential buildings. Figures 36.7(a) and (b) show an ultra-energy-efficient commercial building designed by the NREL and the US National Park Service (the Zion National Park Visitor Center). The building uses 65% less energy than is used by a comparable code-compliant typical building and has an energy intensity of 27 kBtu per ft^2 per year (85 kWh m^{-2} per year). Energy-saving features include passive downdraft evaporative cooling towers, automated natural-ventilation cooling, clerestories for day-lighting, overhangs for summer shading, photosensors and occupancy sensors

to turn off electric lights when they are not needed, a thermal storage wall (non-circulating Trombe wall) to provide delayed passive heating in winter, internal mass for additional heat capacitance, 7 kW of PV electricity generation, and a computerized control system to optimally control HVAC functions.

Figure 36.8 shows an optimization curve for a typical mid-sized retail box store in the climate of Boulder, CO, USA. The curve is interpreted like the previous residential curve except that the criteria on the y axis are those generally used by that industry for economic decision making. The analysis indicates that, for a new store, 50% energy savings could be achieved from optimization alone under these economic criteria in this climate. Numerous optimization runs performed at the NREL using the OptEPlus and BEopt software for a variety of building types in a variety of climates have shown that, in general, 30%–60% savings would be possible for new buildings at neutral cost.

Buildings have long life cycles and the retirement rates (rates at which buildings are taken out of service, destroyed, or replaced) are much slower than in the transportation sector, for example. Table 36.4 shows

Table 36.4. Lifetimes of buildings and associated energy components

Component	Years
Incandescent light bulbs	0.5 to 1
Fluorescent light bulbs	5 to 10
Office equipment: computers, printers, copiers . . .	2 to 7
Consumer electronics	3 to 15
Consumer appliances	5 to 20
Residential water-heating equipment	5 to 20
Residential HVAC	10 to 30
Commercial HVAC	10 to 25
Building envelope components	20 to 50
Cars	5 to 15
Whole commercial and residential buildings	40 to 120
Patterns of transport links and urban development	40 to >200

Data from C. Philibert and J. Pershing, 2002, *Beyond Kyoto: Energy Dynamics and Climate Stabilisation*, Paris, IEA and OECD.

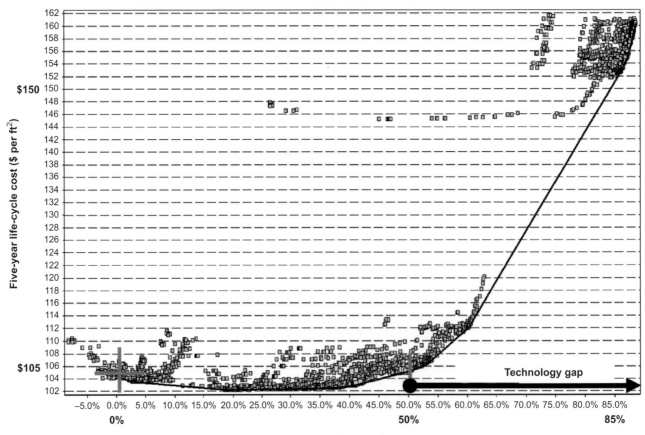

Figure 36.8. Curve generated with the NREL OptEPlus commercial building energy-optimization software.

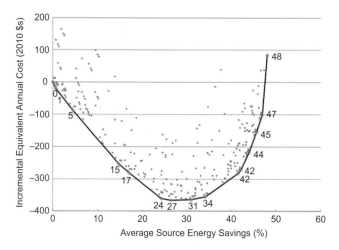

Figure 36.9. Optimization curve for a residential retrofit in a cold climate (Chicago, IL, USA) assuming a typical 1960s-vintage single-family detached house as the base case using the NREL BEopt software.

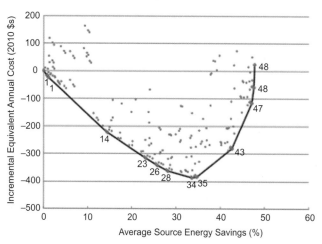

Figure 36.10. Optimization curve for a residential retrofit in a hot climate (Phoenix, AZ, USA) assuming a typical 1960s-vintage single-family detached house as the base case using the NREL BEopt software.

the average lifetimes of buildings and their associated major energy-consuming components. It is evident that, for major reductions in energy consumption and carbon emissions in the building sector, the energy retrofitting of existing buildings must be addressed because attrition will not suffice to replace the existing building stock with higher-efficiency new buildings rapidly enough to resolve the greenhouse-gas problem.

Figures 36.9 and 36.10 show optimizations performed with a new version of BEopt adapted for residential retrofit analysis. These two graphs are to be read in a similar manner to the previous graphs, except that the base case for each (the zero savings point on the x axis) is the pre-retrofit building instead of a conceptual "equivalent code-compliant building." Many experts have surmised that energy-efficiency retrofits would be relatively uneconomical compared with designing energy efficiency into new buildings. Figures 36.9 and 36.10 suggest that the economics of energy retrofits would be comparable to those for new buildings, and the magnitude of savings can be larger because these buildings start with a more elevated energy intensity than that of new, or newer, buildings that are, or were, subject to more stringent energy codes. The key point for cost-effective and relatively deep energy savings is that the energy retrofit must be addressed as an optimized package with financing provided to cover the initial costs. The longer the term of the loan, the more efficiency measures can be included in the package without negatively impacting the owner's cash flow. Public and private programs that harvest the "low-hanging fruit" such as, for example, by distributing CFL bulbs, can actually inhibit the ability to do cost-effective deep retrofits in the future. The low-cost measures offset the more expensive measures in the optimized package. Also, there is extra overhead cost in

retrofitting a building several times with individual measures, versus retrofitting it once with an optimized package.

It is important to note some of the special problems associated with the retrofit market sector. There are health and safety issues associated with the energy retrofit of existing buildings. Air sealing, for example, can make a bad situation worse with regard to air quality and combustion gases where non-sealed combustion equipment is present. Other issues may involve the presence of friable asbestos, lead paint, mold, moisture problems, and electrical hazards, among others. Another difficulty with energy retrofits is that the market is quite dispersed in hundreds of millions of buildings across the globe and among billions of devices and components. The overhead associated with trying to reach such a dispersed market is considerable, and the design of successful programs and business models for deployment to the retrofit sector is non-trivial.

As in the other building sectors discussed above, to proceed past the neutral cost point into deeper savings, a technology gap will have to be overcome. Better, more cost-effective materials are needed for energy retrofits, and some of the materials properties needed are specific to this market. In new buildings, insulation, gaskets, seals, and caulking can be applied before the relevant construction details are blocked or completely obscured by other construction assemblies. In retrofits, materials that can be inexpensively applied without direct access to the construction assembly of interest are needed. For example, a common method of insulating walls is called "drill and fill," wherein small holes are drilled in the wall to access the stud cavities with a "fill tube" through which chopped fiberglass or chopped cellulose is blown into the wall. The stud cavities have many obstructions such as wiring, electrical boxes, nail ends, and fire blocking. Current materials tend to hang up, creating insulation voids that are the equivalent of "energy leaks" in the building shell. Insulation materials with non-clogging properties suitable for retrofit applications would, of course, be highly desirable.

A challenge for materials scientists beyond creating superior materials is the difficulty in determining the potential impact a new material or component will have on overall energy use, costs, and environmental performance over the life of the building. Some of the previously mentioned simulation and optimization tools can be of help in this regard. Other challenges have to do with the durability, low maintenance, and safety requirements for materials that are in close proximity to humans. Despite these difficulties, the opportunities for energy savings through advanced materials are significant.

36.5 Potential for materials advances

36.5.1 Walls, roofs, and floors (building envelope or fabric)

In the building envelope, it would be beneficial to have materials with high thermal resistance, good moisture-management properties, and variable opaque surface characteristics. High thermal capacitance can also be beneficial when used correctly in the right climates.

Materials intended for use in the building envelope should have a high thermal resistance, or a high resistance to heat transmission in minimal thicknesses. In the building industry, thermal resistance is typically reported per unit thickness and per unit area of material in terms of R-values, which have units of hour-square foot-degrees Fahrenheit per Btu per inch [(ft^2 °F h)/(Btu in)] or meter-kelvin per watt (m K W^{-1}). Current thermal resistances of building insulation materials are on the order of R-3 to R-7 (ft^2 °F h)/(Btu in), or 18.3–44.1 m KW^{-1}. Materials with much higher values would be beneficial. Table 36.5 lists a few of the most common current building insulation materials [10].

Researchers have developed evacuated insulation materials for buildings because of the very high R-values theoretically possible with such materials. This is because, in a vacuum, the heat-conduction and -convection mechanisms are eliminated, leaving only radiation heat transfer. Moreover, the radiation is highly attenuated because the metallic surfaces on the inside of the component have low infrared emissivity (see Figure 36.11).

In practice, laboratory prototypes have been limited to R-values on the order of 15–50 (ft^2 °F h)/(Btu in), or 94.5–315 m KW^{-1}, because of thermal short circuits from structural spacers and edge losses associated with reasonably sized modules for building applications [11]. Other problems include the difficulty of maintaining a high vacuum for in excess of 50 years and the danger of penetrations compromising the vacuum during or after

Table 36.5 Insulation R-values for common building insulation materials

Insulation material	R-value BRITISH UNITS, (SI)
Glass-fiber batts and blankets	4 (26)
Chopped cellulose	3–4 (22–26)
Rigid expanded perlite	3 (19)
Rigid extruded polystyrene	4–5 (30–35)
Rigid cellular polyisocyanurate	6 (40)
Rigid cellular polyisocyanurate with gas-impermeable facers	7 (50)
Polyurethane foam, spray applied	6 (40)
Infrared reflective membrane ($\varepsilon^a < 0.5$) in the center of a 0.75-in. (1.9-cm) cavity^b	3 (0.57)

^{a} ε is the emissivity.
^{b} Effective R-value for the assembly.

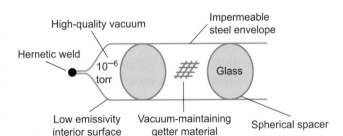

Figure 36.11. Compact vacuum insulation.

construction. Alternative approaches have included evacuated pouches containing low-thermal-conductivity powders [12]. At present, although some products are available, no evacuated insulation products have substantially penetrated the building construction market.

An insulation material is a layer in a wall or roof system that must also manage moisture at both wall surfaces and within the wall. Ideally, the envelope should present a perfect weather and water barrier. In reality, detailing is imperfect, and the interior layers of the envelope have a substantial probability of becoming wet from rain, vapor condensation, or plumbing failures at some time during their service lifetime. Materials that are impermeable to liquid water, but with passively variable vapor permeability, would help prevent wetting and also facilitate drying after a wetting event.

Other favorable properties in terms of moisture management include the retention of a high *R*-value by the insulation layer when wet and when subsequently dried, as well as resistance to mold growth, rot, and structural damage in the presence of moisture for all materials in the wall assembly. Some proprietary materials exist that address some, but not all, of these issues. One such material is a 2-mil polyamide film that becomes more permeable to water vapor at relative humidity levels above 60% [13]. This material is appropriate for cold and mixed climates, but it cannot be used in hot humid climates because, under conditions of high humidity outside with air conditioning inside, vapor would pass through the membrane and condense within the wall cavity wherever a surface temperature was below the dew point. Developing a material that could be used in all climates would therefore be beneficial.

Solar energy incident on the opaque exterior surfaces of a building can provide beneficial heat in winter or create additional need for cooling in summer. Exterior surface materials with variable surface optical properties could have a high solar-spectrum absorptivity and low infrared emissivity in winter to help heat the building, but a high solar-spectrum reflectivity and high infrared emissivity in summer to help cool the building. Such surfaces would have to retain these properties for many years of weather exposure and would have to be aesthetically appropriate for the architectural context of the building. There exist light-colored and low-emissivity exterior and interior surface products that incorporate such materials as ceramic microspheres, microscopic metallic particles or flakes, or porcelain enamel compositions with glass and cerium oxide, but none have variable properties [14]. Materials with variable properties would be beneficial in all seasons and all climates. For interior surfaces, low-emissivity paint already exists, helping to reflect infrared radiation back into the building in winter and resisting the emission of infrared radiation into the building in summer. The long-term durability of these exterior and interior surface properties is not well known.

Materials with high volumetric heat capacitances have the ability to store a great deal of energy in a small volume. Materials such as water and masonry can store many more units of energy per unit of volume than can lighter materials such as wood and plasterboard. Such heavy materials can reduce energy use and peak loads by smoothing out the diurnal load profile of the building and by retaining warmth and coolness in passive buildings designed to effectively utilize the mass. Constructing high-thermal-capacitance buildings is expensive because of the need for a great deal of heavy and thermally conductive material with extensive surface area exposed to the internal spaces of the building. Modern construction tends to be lightweight, especially for internal walls. It would be beneficial to have inexpensive phase-change materials that could be impregnated into common interior surface materials such as gypsum board.

Phase-change materials store and release a large quantity of heat when they transform from one phase to another such as from liquid to solid and back again (Chapter 21). These phase-change materials would need to have transition temperatures within, or near, the human comfort range and be easy to contain, nontoxic, fire-retardant, and aesthetically pleasing when integrated with interior surfaces. They would also need to be non-destructive to the matrix material in which they are integrated and contained. Various phase-change materials and containment systems have been studied for building applications by researchers such as Maria Telkes (formerly of the Massachusetts Institute of Technology) [15], David Benson from the NREL [16], Ivol Salyar and Kelly Kissock from the University of Dayton, and Jan Kozny from Oak Ridge National Laboratory. The work of these researchers and others has been compiled into a reference list by D. Buddhi [17]. Examples of these materials include eutectic salts and paraffins. Eutectic salts or salt hydrates such as sodium sulfate decahydrate and calcium chloride hexahydrate suffer from containment and/or separation problems after cycling. Paraffins such as n-hexadecane and n-octadecane are free of many of these problems and have good thermal properties, but they have difficulty meeting flame-spread safety codes when used at the interior wall surface. Some paraffin-based products have been developed for use as a layer buried in the wall or attic insulation, but this decouples the mass from the interior living space, thus reducing its effectiveness, especially for passive solar buildings.

Solid-state phase-change materials have the advantage of storing large amounts of energy per unit mass without the disadvantages of melting and the associated problems of leaking or separation [16]. These solid-state phase-change materials belong to the class of compounds called polyalcohols or polyols. Pentaerythritol, pentaglycerine, and neopentyl glycol were determined to be of primary interest for building applications. Several experiments demonstrated favorable thermal properties from mixtures of these materials; however, small dimensional changes in transition eventually damaged the matrix materials upon cycling [16].

36.5.2 Windows

For windows, advances in materials science related to low-emissivity coatings, electro-chromic materials, and high-thermal-resistance glazings would be desirable. Current low-emissivity windows use a variety of spectrally selective and infrared reflective coatings. Such coatings are microscopically thin layers of metallic oxide, usually bonded to the surfaces of glass that face the cavity in a double-glazed window. However, these windows do not change properties from summer to winter. The best window-property selections for winter do not perform optimally in summer and vice versa. It would be advantageous to have selective, switchable, and/or tunable glazing materials that transmit solar energy and reflect infrared energy emanating from within the building in winter, but reject solar energy and heat in summer. The glazing should also provide an adequate view or visible transmittance in all modes.

Electro-chromic windows that can be darkened or lightened to reduce or increase solar transmissivity as needed are available on the market [18]. Advantages of electro-chromics over blinds are that they involve no moving parts and are easily adapted to automated control. SAGE Electrochromics, Inc. (www.sage-ec.com), a manufacturer of electro-chromic glass in the USA, describes their process as sputter coating of multiple microscopically thin layers of metal oxides through which lithium ions pass when subjected to an electric charge. Migration of the ions in one direction causes the window to darken, and migration in the other direction causes it to lighten or bleach. Although, when darkened, these glazings reduce transmissivity and heat gain compared with a clear window, they still absorb substantial amounts of solar energy in the darkened state, resulting in unwanted summer heat gains. Improved materials would be controllable for both transmissivity and reflectivity at the outer window surface, thus providing precisely the right amounts of light and heat transmission and rejection needed at any point in time.

The highest thermal resistance for windows available under current technology is R-4 (ft^2 °F h per Btu) or 0.76 m^2 K W^{-1} (effective resistance for the entire assembly). Such windows are triple glazed and have low emissivity on two surfaces, with krypton gas filling the sealed spaces between the panes of glass. Such windows require very good thermal breaks and high-thermal-resistance frames to impede heat loss and gain at glazing edges and through frames. Several researchers have attempted to develop evacuated glazings to increase window thermal resistance to R-10 (1.9 m^2 KW^{-1}) and above [19][20]. Difficulties with such efforts included creating a good edge seal, maintaining a high vacuum for in excess of 30 years, and needing to support the panes of glass with spacers while providing satisfactory visual clarity through the window.

36.5.3 Solar equipment

The building industry would benefit from advances in solar technology, particularly in terms of polymer solar domestic hot-water systems and combined PV–thermal solar systems (Chapters 21 and 49). Current-generation solar water systems are generally fabricated from glass, copper, and aluminum, and have so far been too expensive for mass-market penetration in the USA. China is the world's largest producer of glass evacuated-tube collectors. Systems based on these tubes have gained traction in countries with high energy costs, limited energy availability, cheap labor, and/or favorable government policies including Germany, Greece, and Israel. The USA is the world's largest producer of unglazed polymer water heaters for swimming pools. Such water heaters can operate at relatively low temperatures of around 70–80 °F (21–27 °C), which presents no problem for ultraviolet (UV)-stabilized polymers. For use in heating domestic hot water, higher temperatures are required.

Polymer systems are favorable because they lend themselves to cost-cutting mass-production techniques (see Figures 36.12 and 36.13). A few polymer-based systems are currently being marketed internationally in non-freezing climates. In the USA, a freeze-protected system was recently introduced primarily for mild climates, and at least one other polymer-based system is under development [21]. No polymer materials to date have all the properties needed for wide application in all climates because the combination of pressure, temperature, and UV exposure for solar collectors is at the limit of "off-the-shelf" polymer material properties. Improvements in the properties of both glazing materials and absorbers for polymer-based solar domestic hot-water systems are desirable. Ideally, one would want a glazing material that is highly transmissive in the solar spectrum, resistant to UV and weathering for more than 30 years, and dimensionally stable and structurally sound through a wide temperature range from below freezing to a stagnation temperature of about 400 °F (200 °C). Stagnation is a fault mode that occurs when the collector is exposed to sunlight while empty of heat-transfer fluid. For an absorber, desirable material properties include a high absorptivity in the solar spectrum, low infrared emissivity, resistance to UV and weathering for more than 30 years, dimensional stability, and structural soundness through a wide temperature range from below freezing to a stagnation temperature of about 400 °F (200 °C) under pressures of about 200 psi, and not being subject to damage from a hard freeze.

Building technology would also benefit from the development of combined PV–thermal solar systems. In typical terrestrial PV arrays, only about 10%–20% of the incident solar energy is converted to electricity. The remainder of the energy is lost as heat. For building-integrated PV arrays, it would be beneficial to utilize the waste heat for purposes such as domestic hot water. This would increase the overall efficiency of the PV system, reduce building energy use, and maximize the usefulness of the limited PV-appropriate roof areas on buildings. All current commonly available PV materials

Figure 36.12. Prototype polymer collector.

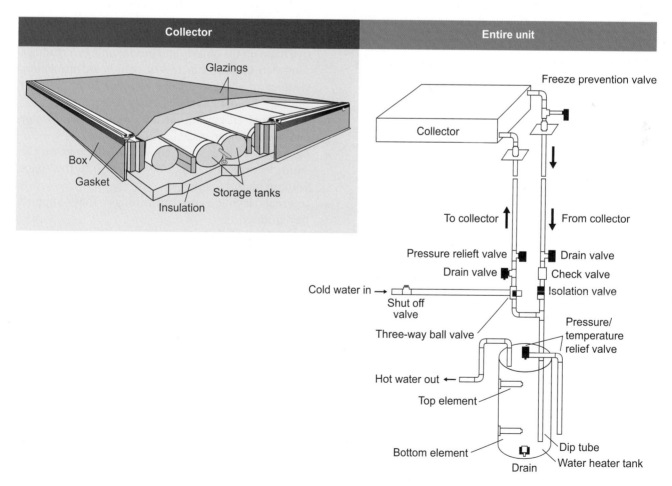

Figure 36.13. The integral collector–storage (ICS) system.

Figure 36.14. The NREL Research Support Facility (RSF), a 220,000-ft^2 ultra-efficient zero-energy office building.

have negative temperature coefficients, so their performance degrades as the temperature of the PV material rises. Thus, for combined PV–thermal systems, it would be beneficial to have a PV material exhibiting little or no degradation in performance up to about 200 °F (93 °C).

36.6 Case study: the NREL Research Support Facility, a 220,000-ft^2 zero-energy office building

Construction of a 220,000-ft^2 (2,040-m^2) zero-energy office building was completed on the campus of the NREL near Golden, Colorado in August of 2010. The building, called the Research Support Facility (RSF), is now in the "move-in" phase. Building scientists from the Buildings R&D Program at the NREL were involved in the "energy design" of the building from its inception, and will monitor the actual energy performance of the building for several years. Energy simulation and optimization software developed at the NREL was used, together with data collected from studying a number of energy-efficient commercial buildings across the USA and around the world, to set the overall energy concepts and energy specifications for the RSF. A bottom-line energy intensity target of 25 kBtu per ft^2 per year (79 kWh m^{-2} per year) was established for the building, assuming 750 occupants. An additional 6 kBtu per ft^2 per year (19 kWh m^{-2} per year) was allowed for a large data center that will serve the entire campus, and an additional 3 kBtu per ft^2 per year (9 kWh m^{-2} per year) was allotted for designing to a higher occupancy efficiency allowing an extra 150 occupants without increasing the floor area. This represents about a 50% energy saving versus a comparable code-compliant building. Primary energy efficiency features of the RSF include the following:

- a predominant cross section of no more than 60 feet (18 m) to facilitate daylighting and natural ventilation cooling;
- elongation of the building along the east and west axis to maximize the south and north-wall area;
- pre-cast wall panels with 2 inches of concrete for thermal mass on the outside and inside surfaces, creating a sandwich around 4 inches of rigid-board insulation, yielding an effective R-value of about 20 ft^2 °F h per Btu 1 ft^2 °F h per Btu = 0.176 m2k W-1;
- properly sized glazing for the Colorado front-range climate (not over-glazed);
- light louver clerestories above the windows on the south to project daylight 30–40 feet into the space without glare;
- internal layout and furnishings that support daylighting;
- carefully shaded glazing on the south, west, and east faces of the building;
- under-floor, low-pressure, evaporatively cooled, fresh-air supply;
- radiant heating and cooling in the ceiling slab;
- high-efficiency campus chilled-water plant;
- solar-wall fresh-air solar-heating system;
- foundation thermal-mass storage system (the "labyrinth");
- a 1,327-kW$_{peak}$ PV system on the roof;
- careful selection of best-in-class office equipment and electronics to minimize internal gains; and
- intelligent controls to turn things off when they don't need to be on.

See Figures 36.14–36.17 (images courtesy of RNL Design, Stantec, Architectural Energy Corporation, and Haseldon Construction).

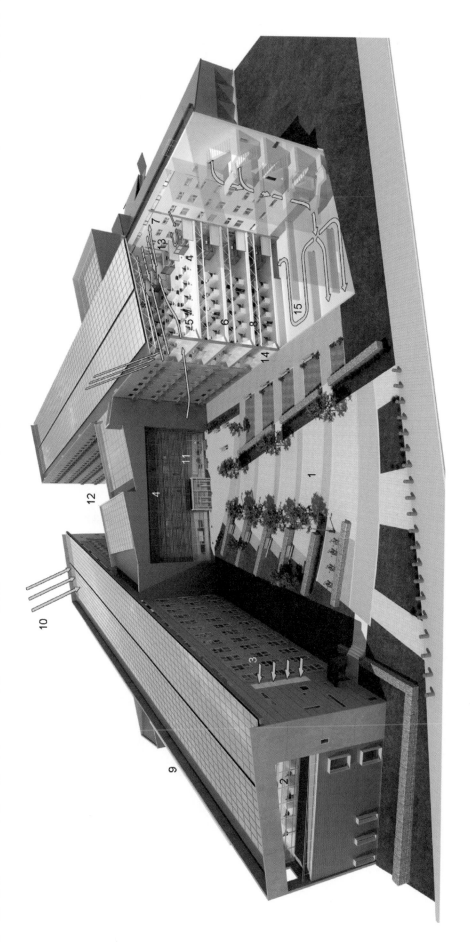

1. Permeable landscaping
2. Thermochromic east-facing glazing
3. Fresh air intake
4. Louvered sunshade
5. Low-profile open offices
6. Underfloor air, data, and power
7. Reflective interior paint, flooring, workstations
8. Radiant floor slab

9. Transpired solar collector on south facade (not shown)
10. 1.8 MW of photovoltaics
11. Sculptural beetle-kill wood wall
12. Electrochromatic west facing windows (not shown)
13. Open-ceiling offices
14. Repurposed natural gas pipe
15. Basement thermal mass labyrinth

Figure 36.15. A cutaway view of the RSF showing a number of energy-efficiency features.

Research Support Facility

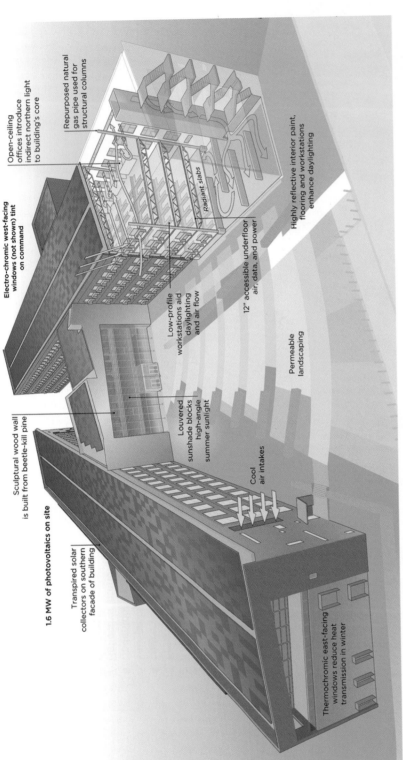

Sculptural wood wall is built from beetle-kill pine

Electro-chromic west-facing windows (not shown) tint on command

Open-ceiling offices introduce indirect northern light to building's core

Repurposed natural gas pipe used for structural columns

1.6 MW of photovoltaics on site

Transpired solar collectors on southern facade of building

Radiant slabs

Low-profile workstations aid daylighting and air flow

12" accessible underfloor air, data, and power

Highly reflective interior paint, flooring and workstations enhance daylighting

Louvered sunshade blocks high-angle summer sunlight

Permeable landscaping

Cool air intakes

Thermochromic east-facing windows reduce heat transmission in winter

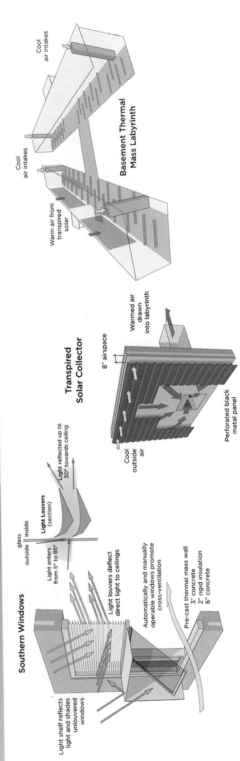

Cool air intakes

Cool air intakes

Warm air from transpired solar

Basement Thermal Mass Labyrinth

Transpired Solar Collector

8" airspace

Warmed air drawn into labyrinth

Cool outside air

Perforated black metal panel

Southern Windows

glass | inside
outside |

Light Louvers (section)

Light reflected up to 30° towards ceiling

Light enters from 5° to 85°

Light louvers deflect direct light to ceilings

Automatically and manually operable windows promote cross-ventilation

Pre-cast thermal mass wall
3" concrete
2" rigid insulation
6" concrete

Light shelf reflects light and shades unlouvered windows

Figure 36.16. Building geometry and orientation facilitates daylighting and natural ventilation.

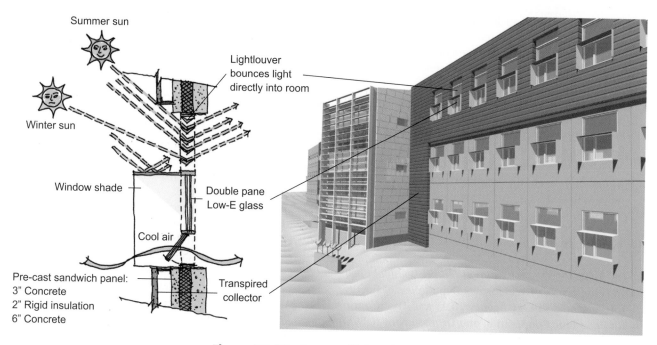

Figure 36.17. Energy-efficient façade features of the RSF.

36.7 Summary

There are many opportunities for materials advances to reduce the energy use and atmospheric emissions associated with the building sector. The energy and cost performance of walls, roofs, windows, mechanical systems, and onsite renewable electrical and thermal systems can all be improved through advances in materials. Specifically, materials that improve the performance of thermal insulation, thermal storage, vapor retarders, weather barriers, glazings, solar thermal collectors, and PV generators could all have a profound impact on the overall energy efficiency and sustainability of buildings. Buildings have relatively long life cycles compared with automobiles and most manufactured products, so materials for buildings must be highly durable, nontoxic, aesthetically pleasing, comfortable, and safe for human proximity. Materials that reduce energy use both in new construction and in retrofitting and refurbishment projects are needed. A challenge for building scientists and materials scientists is the difficulty of assigning quantitative energy savings to any given materials improvement. The elements of a building are highly interactive in their energy performance and also dependent on the surrounding climate, building type, and usage patterns in the building. Building scientists at the NREL in the USA have begun to develop sophisticated computer tools to address this issue, and those tools will improve as computer power increases. Because buildings are so numerous, even relatively small energy reductions on an individual-building basis can have a large impact globally.

36.8 Questions for discussion

1. Why is it difficult to ascertain the energy, carbon dioxide, and energy-cost savings associated with a given materials and technology advance in buildings?
2. What are the key issues to consider in developing more energy-efficient opaque building shell components and systems?
3. What are the key issues to consider in developing more energy-efficient window and glazing systems?
4. What are the key issues to consider in developing better, more cost-effective solar heating systems?
5. What are the key issues to consider in developing better "building-integrated" thermal storage systems?

36.9 Further reading

- **D. R. Roberts** and **R. Anderson**, 2009, *Technical Approach for the Development of DOE Building America Builders Challenge Technology Information Packages (Revised)*, NREL Report No. TP-550–44687. This report presents a brief summary of the BEopt optimization tool and the results of applying the tool to a prototype new residential building in a variety of climate zones. Appendix D of the report is especially interesting in that it presents the optimization curves, and specifies the package of measures for the minimum cost point and the neutral cost point.
- **J. Krigger** and **C. Dorsi**, 2004, *Residential Energy: Cost Savings and Comfort for Existing Buildings*, Helena, MT, Saturn Resource Management, Inc. This book

is a good, mostly qualitative, overview of building physics in a residential-retrofit context, with lots of good illustrations of construction details and diagrams of the inner workings of household mechanical equipment.

- **J. D. Burch**, 2006, "Polymer-based solar thermal systems: past, present and potential products," in *ANTEC 2006 Plastics: Proceedings of the Annual Technical Conference*, Brookfield, CT, Society of Plastics Engineers (SPE); also available as NREL Report No. CP-550–39461, 1877 (2006). A good overview of the issues and potential for low-cost polymer-based solar heating systems.

36.10 References

[1] **US DOE**, 2007, *International Energy Outlook 2007*, Washington, DC, United States Department of Energy.

[2] **International Energy Agency (IEA) and Organization for Economic Cooperation and Development (OECD)**, 2008, *Energy Technology Perspectives: Scenarios and Strategies to 2050*, Paris, IEA and OECD.

[3] **International Energy Agency (IEA)** and **Organization for Economic Cooperation and Development (OECD)**, 2007, *World Energy Outlook 2007*, Paris, OECD and IEA.

[4] **US DOE**, 2006, *2008 Building Energy Data Book*, Washington, DC, and United States Department of Energy D&R International, Ltd.

[5] **US DOE**, 2007, *Energy Information Administration Report DOE/EIA-0484*, Washington, DC, United States Department of Energy.

[6] **US Energy Information Administration (EIA)**, 2003, *Greenhouse Gases, Climate Change and Energy*, Washington, DC, EIA.

[7] **H. C. Granade, J. Creyts, A. Derkach** *et al.*, 2009, *Unlocking Energy Efficiency in the US Economy*, McKinsey Global Energy and Materials.

[8] **World Business Council for Sustainable Development**, 2009, *Energy Efficiency in Buildings: Transforming the Market*, Geneva, World Business Council for Sustainable Development.

[9] **C. Christensen, R. Anderson, S. Horowitz, A. Courtney**, and **J. Spencer**, 2006, *BEopt(TM) Software for Building Energy Optimization: Features and Capabilities*, NREL Report No. TP-550–39929, 21.

[10] **American Society of Heating, Refrigerating and Air-Conditioning Engineers (ASHRAE)**, 2005, *ASHRAE Handbook of Fundamentals*, Atlanta, GA, ASHRAE.

[11] **D. K. Benson** and **T. F. Potter**, 1992, *Compact Vacuum Insulation*, US Patent No. 5,157,893.

[12] **B. T. Griffith** and **D. Arasteh**, 1992, *Advanced Insulations for Refrigerator/Freezers: The Potential for New Shell Designs Incorporating Polymer Barrier Constructions*, report *LBNL* 33376.

[13] US Patent 6808772 B2.

[14] US Patent 7157112 and associated art.

[15] **M. Telkes**, US Patent 4585572.

[16] **D. K. Benson, J. D. Webb, R. W. Burrows, J. D. O. McFadden**, and **C. Christensen**, 1985, *Materials Research for Passive Solar Systems: Solid-State Phase-Change Materials*, SERI (now NREL) Report SERI/TR-255–1828: **Atul Sharma, V. V. Tyagi, C. R. Chen, D. Buddhi**, 2009. *Review on thermal energy storage with phase change materials and applications, Renewable and Sustainable Energy Reviews*, **13**, 318–345.

[17] **D. Buddhi**, *A Selected List of References in Twentieth Century (1900–1999) on Phase Change Materials and Latent Heat Energy Storage Systems*, Thermal Energy Storage Laboratory, School of Energy and Environmental Studies, Devi Ahilya University, Indore, India.

[18] **F. Pichot, S. Ferrere, R. J. Pitts**, and **B. A. Gregg**, 1999, "Flexible solid-state photo-electrochromic windows," *J. Electrochem. Soc.*, **146**(11), 4324–4326; also available as NREL Report No. JA-590–26316.

[19] **D. K. Benson, C. E. Tracy**, and **G. J. Jorgensen**, 1984, *Evacuated Window Glazing Research and Development: A Progress Report*, NREL Report No. PR-255–2578, 11.

[20] **J. Carmody, S. Selkowitz, E. Lee, D. Arasteh**, and **T. Willmert**, 2004, *Window Systems for High Performance Buildings*, New York, W. W. Norton and Company.

[21] **J. D. Burch**, 2006, "Polymer-based solar thermal systems: past, present and potential products," in *ANTEC 2006 Plastics: Proceedings of the Annual Technical Conference*, Brookfield, CT, Society of Plastics Engineers (SPE); also available as NREL Report No. CP-550–39461, 1877.

37

Insulation science

Leon R. Glicksman[1] and Ellann Cohen[2]

[1]Departments of Architecture and Mechanical Engineering, Massachusetts Institute of Technology, Cambridge, MA, USA

[2]Department of Mechanical Engineering, Massachusetts Institute of Technology, Cambridge, MA, USA

37.1 Focus

Energy efficiency has been recognized as the most effective near-term means to meet the energy and environmental crisis we face today. In the USA, buildings are the largest energy-consumption sector of the economy. Residential and commercial buildings combined consume over 40% of the primary energy and over two-thirds of the total electricity [1]. Heating and cooling are the largest portions of this. Demonstration homes have shown that the heating consumption can be reduced by as much as 90% by the proper application of very thick thermal insulation in the walls, roof, and windows [2]. One challenge is the development of very thin economical insulation materials that provide the same performance.

37.2 Synopsis

Thermal insulations comprise a wide variety of materials whose primary function is the reduction of heat and mass transfer. These insulations are made from foams, fibers, and other fine-structured solids that encapsulate a gas or are held in vacuum. In buildings, insulation improves energy efficiency by reducing heat loss in winter and heat gain in summer. Even modern windows have been designed to act as insulators to improve building performance. Appliances such as refrigerators and ovens use insulation to maintain temperature and to be more energy-efficient. Insulations are also used in industrial operations such as furnaces for metal and glass manufacture as well as as a means to control silicon-chip formation. In space and on Earth, insulations are used for protection in harsh environments. The development of the next generation of insulations requires an understanding of the physics of heat transfer and of the role advanced materials play in limiting heat transfer by the mechanisms of conduction, radiation, and convection.

37.3 Historical perspective

Early humans discovered the value of thermal insulation for survival and comfort. Animal skins were used as protective clothing and leaves and twigs were used in the construction of shelters. As technologies advanced, thatched roofs were used in Europe and in the Pacific Islands, as shown in Figures 37.1 and 37.2. Europeans used straw to create a thick roof with a large number of open spaces that trapped air. Pacific Islanders used dried reeds for a similar function. In the southern part of North America as well as in the Middle East and North Africa, thick adobe walls provided both resistance to heat transfer as well as thermal mass to smooth out cyclic climatic variations. Some colonial buildings in America were built with a cavity between exterior and interior walls. Although these buildings were most often left empty, the cavities were sometimes filled with brick noggin. Brick does slow the spread of fire through walls, but brick is not a very good insulator, although it does provide thermal mass. Pictures of brick noggin in buildings from Colonial Williamsburg are shown in Figure 37.3 and Figure 37.4. In pre-Tudor times in England and in colonial America, wall cavities were made with wattle and daub – twigs plastered with mud and straw shown schematically in Figure 37.5. Ancient masonry construction of castles and churches frequently did without the wall cavity, and such buildings were drafty and cold in the winter due to their poor insulative properties. Decorative tapestries hung on the walls and rugs on the floor provided a very modest level of insulation. Other ancient insulation in walls suffered from degradation due to moisture penetration and decay (a common fault in modern construction as well!) and allowed large air leaks from the outside.

More recent insulations such as those used in the early twentieth century include Cabot's Quilt developed by Samuel Cabot, Inc., which was simply dried seaweed sandwiched between two layers of paper and stitched together like a quilt, Temlock board, which was ground pulpwood held together with glue, fiberglass batting, first manufactured by Owens Corning, and asbestos, which is no longer used due to health hazards [3][4][5].

The performance of insulation is governed by the physics of heat transfer. The first notable contribution to this field came from Joseph Fourier, an official in France during Napoleon's time, who can be seen in Figure 37.6.

Figure 37.2. A close-up view of decorative scalloping above eaves on thatching.

Figure 37.1. A traditional rural byre-dwelling with thatch roofing.

Figure 37.3. The interior of Everard House in Colonial Williamsburg during renovation. Notice the nogging in the right wall. (Source: The Colonial Williamsburg Foundation.)

Figure 37.4. A close-up view of nogging between studs of Everard House in Colonial Williamsburg after the exterior siding was removed. (Source: The Colonial Williamsburg Foundation.)

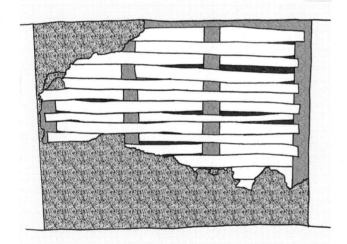

Figure 37.5. Wattle and daub.

Figure 37.6. Joseph Fourier. (Etching from *Portraits et histoire des hommes utiles, collection de cinquante portraits*, Société Montyon et Franklin 1839–1840.)

37.4 Basics

Heat is the exchange of energy across a body due to a temperature difference. Heat transfer across a building's exterior wall results in energy loss during the winter from the warm interior to the exterior. Similarly, heat transfer from a warm interior space, such as a kitchen, to the interior of a refrigerator causes an energy gain in the refrigerator and increases power consumption by the cooling system. The primary function of thermal insulation, then, is to minimize these energy exchanges. The insulation must reduce heat transfer due to three primary mechanisms: thermal conduction, thermal radiation, and convection heat transfer. We will present a brief explanation of each.

Conduction heat transfer across a body is due to molecular motions. It can be expressed using Fourier's equation,

$$\mathbf{q} = -kA \operatorname{grad} T, \qquad (37.1)$$

where $\mathbf{q}$ is the rate of heat transfer in watts, k is a material property (the thermal conductivity in W m^{-1} K^{-1}) that is a measure of a material's ability to transfer energy, A is the cross-sectional area normal to the direction of the transfer (which is parallel to the gradient of

He published his *Théorie analytique de la chaleur* (*Analytical Theory of Heat*), in which he formulated a general theory of heat conduction, in 1822. The late nineteenth and twentieth centuries witnessed a growing body of research in convection and radiation heat transfer, the latter including Planck's quantum theory to describe the observed spectrum of visible and infrared radiation. Modern insulations use a variety of means to limit all of these forms of heat transfer. Thus, reviewing the principles of heat transfer is a key to understanding insulation performance and providing a guide to its improvement.

Figure 37.7. Common material and insulation thermal properties. In the USA, the R-value in IP units per inch is the most common number used to describe the heat-transfer properties of a material. Other places use the thermal conductivity in SI units and not as a function of thickness. The relation between the two is given in Equation (37.2) and the conversion between the two is 1 ft^2 °F h per Btu = 0.176 m^2 K W^{-1} [6][7][8][9][10].

	Material	R-Value per Inch (ft² °F h / Btu)
Common Materials	Concrete	
	Wood	
	Air	
Loose-Fill	Cellulose	
	Expanded Polystyrene	
Blankets	Fiberglass	
	Rock Wool	
Foamed-in-Place	Closed-Cell Phenolic	
	Open-Cell Phenolic	
	Polyisocyanurate	
Rigid Board	Expanded Polystyrene (EPS)	
	Extruded Polystyrene (XPS)	
	Polyurethane Foam	
	Polyisocyanurate Board–Foil-faced	
	Polyisocyanurate Board–Unfaced	
	Phenolic Foam	
Aerogel	Aspen Aerogel Blanket	
	Cabot Nanogel Granules	
	Cabot Thermal Wrap Blanket	
	Cellulose Aerogel	
Vacuum Panels	Glacier Bay Ultra-R w/ Aerogel Core	
	ThermoCor	

R-Value per Inch (ft² °F h / Btu): 0 5 10 15 20 25 30 35 40 45 50

Thermal Conductivity (mW m^{-1} K^{-1}): 29 14 9.6 7.2 5.8 4.8 4.1 3.6 3.2 2.9

temperature), and T is the temperature. Where conduction can be considered one-dimensional (for example, when the thickness of a homogeneous material is much smaller than its other two dimensions) and the process is steady, Fourier's law can be expressed simply using k, or alternatively the R factor, as

$$q = -kA\frac{\Delta T}{\Delta x} = -A\frac{\Delta T}{R}, \qquad (37.2)$$

where Δx is the thickness of the material. Figure 37.7 shows the values of thermal conductivity and the R-value per inch for a range of insulations and common materials. Note that solid materials that are good electrical conductors have a high thermal conductivity since the free electrons aid in the energy transfer. Energy is also transferred in a solid by virtue of lattice vibrations referred to as phonon transfer. Materials such as wood and polymers that are good electrical insulators have a relatively low thermal conductivity. In gases, energy transfer takes place by the motion of the gas molecules. The thermal conductivity of a gas varies inversely with the relative molecular mass (less formally, molecular weight) and is generally far less than the values for solids. Common insulation materials have a high volume fraction of gas to make use of its good insulating properties. The gas is surrounded by a solid to limit convection heat transfer. In advanced insulations currently under development, the heat transfer in gas is suppressed when the mean free path of the gas molecules is of the same order

Figure 37.8. A schematic representation of air flow in a wall cavity with a thermal gradient.

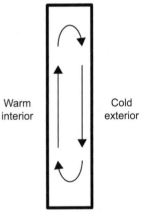

Warm interior   Cold exterior

as the cavity dimensions (more on this later). Removing the gas altogether as in a vacuum panel reduces the heat transfer significantly, but very low pressures (<100 Pa) must be maintained.

Convection heat transfer takes place in gases and liquids, and is the combination of conduction and fluid motion. In closed spaces such as a cavity in an exterior wall, between the interior and exterior building surfaces, the fluid motion is set up by buoyancy forces. Figure 37.8 shows an example of a wall cavity during the winter. The interior surface is warm, while the exterior surface can be much colder. Air in the cavity close to the cold wall will be much denser than the air close to the warm wall. The cold air will tend to move down under gravity, while the warm air rises. This sets up an energy exchange

Figure 37.9. The convective heat-transfer coefficient (*h*) vs. cavity width for different temperature differences [11].

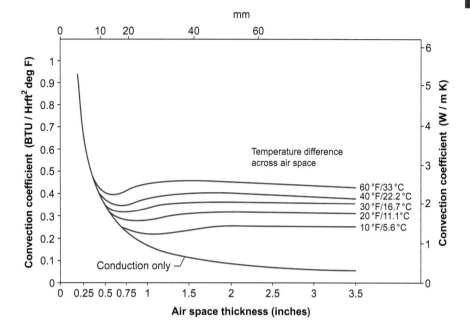

due to the fluid circulation, which substantially augments the exchange due solely to molecular motion.

The amount of fluid motion is determined by a balance between the buoyancy forces, which are proportional to the temperature difference, and the viscous forces resisting the fluid motion. When the width of the cavity is small, there is a large viscous force between the upward and downward flows that limits buoyancy-induced motion. The net convective heat transfer can be expressed in terms of *h*, the convective heat-transfer coefficient measured in $W\,m^{-2}\,K^{-1}$, which is defined by the equation

$$q = hA(T_{hotwall} - T_{coldwall}). \tag{37.3}$$

This equation includes both effects of conduction in the gas and convective motion that sets up heat transfer across the cavity. Figure 37.9 shows the experimental results for the heat-transfer coefficient due to convection (with radiation effects omitted) for different cavity widths [11]. The cavity walls are vertical in a manner similar to the orientation shown in Figure 37.8. At small cavity widths, convection motion is suppressed and *h* is due solely to molecular conduction across the cavity, $h = k_{gas}/w$, where *w* is the width. As the cavity becomes wider, convective motion becomes important. Note that, when the temperature difference is larger, there are larger air velocities and the heat-transfer coefficient is enhanced. These results are directly applicable to the design of windows with multiple glass layers. The ideal spacing between the glass layers is near the minimum point of *h* on Figure 37.9.

For foam or fibrous materials, there is the possibility of convection within the interior of the foam cells or within the open spaces between the fibers. This is governed by the same balance between buoyancy and viscous forces as for the wall cavity. The ratio of these

forces is expressed by a dimensionless number called the Rayleigh number, defined as $g\beta(\Delta T)x^3/(v\alpha)$, where *g* is the force of gravity, β is the thermal expansion coefficient, ΔT is the temperature difference across the cavity or cell, *x* is the width of the cavity, *v* is the kinematic viscosity, and α is the thermal diffusivity. When the Rayleigh number is less than 10^3, convection is suppressed [12]. For foams, convection effects can be ignored when the cell diameter is less than 1.0 mm. Also, fibrous insulations are manufactured with enough fine fibers per unit volume to suppress convection under most conditions.

Thermal radiation consists of electromagnetic waves emitted from a body by virtue of its temperature level. Radiation heat transfer is of importance in two wavelength regimes for building insulation and windows, namely the visible and near-infrared regime of 400–2,000 nm, and the long-wavelength infrared regime of 8,000–24,000 nm. The former is the spectrum of solar radiation, while the latter represents radiation emitted from bodies at room temperature.

Radiation emitted by plane glass windows in the long-wavelength infrared regime approaches backbody behavior. Blackbodies absorb all incoming radiation and they emit the maximum amount of radiation of any body at that temperature. When two planar blackbodies are parallel and the space between them is transparent, the net thermal radiation between them is given as

$$q = A\sigma(T_1^4 - T_2^4), \tag{37.4}$$

where σ is the Stefan–Boltzmann constant (5.67×10^{-8} $W\,m^{-2}\,K^{-4}$) and T_1 and T_2 are the temperatures of the two bodies in Kelvin. To limit infrared radiation between two glass layers in a double-glazed window, a low-emissivity, high-reflectivity coating, such as a thin vapor-deposited metallic layer, is used on the surfaces. The radiation heat transfer becomes

$$q = A\sigma \frac{T_1{}^4 - T_2{}^4}{1/\varepsilon_1 + 1/\varepsilon_2 - 1}, \quad (37.5)$$

where ε is the surface emissivity which is generally equal to one minus the surface reflectivity.

For moderate temperature differences, Equation (37.4) or (37.5) can be linearized as

$$q = A4T_{\mathrm{M}}^3\sigma \frac{T_1 - T_2}{1/\varepsilon_1 + 1/\varepsilon_2 - 1}, \quad (37.6)$$

where T_{M} is the mean temperature between the two surfaces.

37.4.1 Insulated windows

Insulating windows combine multiple layers of glass or other transparent films to form a series of separate gas layers, each spaced to limit convection on the basis of the results of Figure 37.9. In addition, for each pair of glass layers, one surface has a low-emissivity coating to minimize long-wavelength infrared radiation. Advanced windows may have up to three or four glass layers forming two or three gas layers, respectively. Designers of windows for northern climates strive to maximize direct solar gains to the building interior by using a glass that has a high transmissivity in the visible and near-infrared wavelengths; while in southern, cooling-dominated regions, the glass is designed to block the near-infrared radiation from the Sun while still allowing visible wavelengths through. In northern climates, we want to maximize transmission of solar radiation while minimizing the passage of long-wavelength-range infra-red radiation that contributes to heat loss. Advanced windows for northern climates have a net energy gain because the solar energy gains through the window exceed heat-transfer losses through the window from the warm interior to the cold ambient air. A challenge is to find coatings for northern climates that increase solar transmission while maintaining low emissivity in the long-wavelength infrared range.

Some advanced windows contain a vacuum between the inner and outer glass layers to eliminate conduction. Issues concerning the long-term life of seals between the glass layers and distortion of the view due to deflection of the glass under the pressure forces are being addressed in these advanced window designs.

37.4.2 Foam and fiber insulations

In porous fiber insulations such as fiberglass or foam insulations, radiation can still be an important mechanism of heat transfer. In this case, the radiation can be emitted from one boundary and travel a short distance

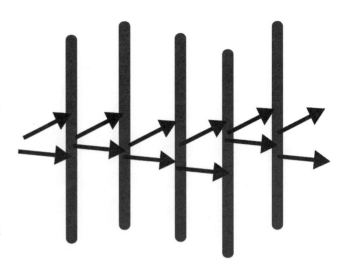

Figure 37.10. The idealized path of radiation in fibrous insulations.

(proportional to the mean free path for radiation) before it is absorbed by a fiber or other solid element. At that location, radiation is re-emitted and some of that moves to the next location, as shown in Figure 37.10.

When the radiation mean free path is much smaller than the insulation thickness, the radiation transfer is analogous to the diffusion process in conduction heat transfer. The mean free path is a function of the spacing between fibers and the fiber size. The radiant heat transfer is proportional to the mean free path of the radiation, l_{MF} and the gradient of the blackbody emission, σT^4. This relation is known as the Rosseland equation, which is written as

$$q = -\frac{4}{3}Al_{\mathrm{MF}}\frac{d\sigma T^4}{dx} = -\frac{4}{3K_R}A\frac{d\sigma T^4}{dx} = -\frac{16}{3}\frac{\sigma T^3}{K_R}\frac{dT}{dx}, \quad (37.7)$$

where K_{R} is the Rosseland mean absorption coefficient. Note that, in the final form, the equation has the same dependence on the temperature gradient as conduction, and the two mechanisms can be combined into a single effective conductivity.

As the density of the foam or fiber material is increased it becomes more opaque, reducing the mean free path of the radiation. However, as the proportion of solid material in the insulation is increased, conduction through the solid becomes more important. In addition, the cost of the insulation goes up. For developing countries, the cost of conventional insulations is prohibitively expensive for much of the population. Research to develop insulation panels out of waste material such as straw is under way [13]. The panels should be strong enough to be used in the interior of roofs or walls. A major hurdle is the development of low-cost, environmentally acceptable adhesives to bind the straw together.

Foam insulation is made up of small cells enclosed by a thin polymeric material such as polyurethane,

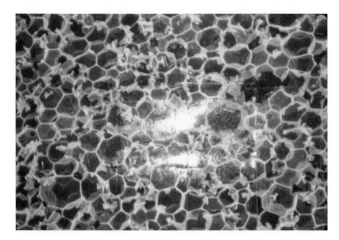

Figure 37.11. An image through a microscope of a foam cross section.

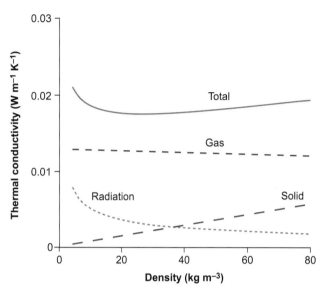

Figure 37.12. Predicted influence of foam density (kg m^{-3}) on contributions of radiation, solid conduction, and gas conduction to the total thermal conductivity (W m^{-1}K^{-1}) of closed-cell foam filled with a high-relative-molecular-mass blowing agent.

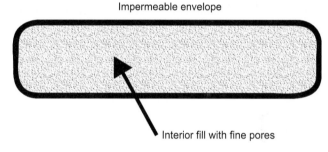

Figure 37.13. A schematic representation of a vacuum insulation panel.

polystyrene, or polyisocyanurate. Most insulating foams are closed cells and in some instances contain high-relative-molecular-mass gases that have a thermal conductivity less than that of air. In the past, these gases were chlorofluorocarbons (CFCs) or hydrochlorofluoro-carbons (HCFCs), but they were banned for environmental reasons. Today hydrocarbons such as pentane and hydrofluorocarbons (HFCs) as well as CO_2 are used [14]. Figure 37.11 shows an enlargement of a foam cross section. Typically, each cell has a diameter less than 1.0 mm. Figure 37.12 shows how thermal radiation, conduction through the gas, and conduction along the solid material in the cell walls contribute to the overall foam conductivity. As the foam density increases, radiation is reduced because the cell walls become more opaque. This is partially offset by an increase in the conduction within the solid material, leading to a minimum in the sum of radiation and conduction at an intermediate density as shown in Figure 37.12. Some researchers have investigated the use of opaque particles that can be embedded in the solid material to increase its opacity [15]. In the past, many closed-cell insulations contained CFC gas. These have been phased out due to their impact on the ozone layer. Replacement gases are safer for the environment, but also exhibit higher thermal conductivity.

37.4.3 Vacuum insulations

To further improve the performance of insulations and obtain higher thermal resistance for a given thickness, attention has been focused on vacuum insulation systems. A typical vacuum panel is shown in Figure 37.13. It consists of an impermeable envelope that maintains a very low gas pressure or vacuum condition within the panel for the life of the insulation. Low-emissivity interior surfaces and an interior fill material with small pores, like a fine silica-based powder or aerogel for

example, limits radiation heat transfer. The reduced pore size of the fill material also limits conduction through the gas remaining within the panel. To explain this, we must first consider a simple model for conduction heat transfer through a gas under normal conditions.

Heat transfer through a gas, in the absence of convective motion, takes place by molecular motion. In an enclosure with parallel hot and cold surfaces on either side, gas molecules near the hot surface gain energy by interaction with that surface. They are constantly in motion and some move toward the cold wall, where they interact with less energetic molecules. These, in turn interact with molecules still closer to the cold surface. The result is a net exchange of energy from hot to cold surface facilitated by the molecular motions. A simplified model from kinetic theory for the thermal conductivity of the gas gives

$$k_{gas} \sim n\lambda, \tag{37.8}$$

where n is the number of gas molecules per unit volume, which increases with the gas pressure, and λ is the mean free path for gas molecules meaning the average distance a molecule travels before encountering another molecule [16]. When the mean free path is much smaller than the space between neighboring surfaces, λ varies as $1/n$. Thus, the product $n\lambda$ and the resulting gas conductivity are independent of the gas pressure.

As the gas pressure is decreased, the mean free path increases. When λ approaches the interstitial spacing or pore size in a vacuum panel the average distance a gas molecule travels is limited by the pore size. With further reduction of the gas pressure the distance term in Equation (37.8) approaches a constant while the number of gas molecules continues to decrease. For rarified gases in which λ is much larger than the average spacing between surfaces, the gas conductivity becomes linearly proportional to the gas pressure.

In vacuum panels, as the gas pressure is reduced, initially the conductivity remains constant until the mean free path of the gas molecules approaches the spacing of internal pores. After this critical pressure has been reached, the contribution of gas conduction decreases with further reductions in pressure. The critical pressure at which this reduction is first seen is a function of the pore size. As the pore size between individual grains of powder or fibers is reduced, the critical pressure is raised. An important design issue for vacuum panels is the identification of an impermeable envelope material that will not allow significant diffusion of gas into the panel interior, raising the pressure over the life of the insulation. Metallic foils meet this requirement. However, thick metallic films will allow "thermal bridges" with substantial heat flow by conduction from the hot side to the cold side around the circumference of the panel. A perfect barrier would prevent all gas diffusion and would not allow these thermal bridges, but such a barrier does not yet exist.

Aerogels represent a limiting case of media with fine pore size. The pore dimensions of aerogels are in the nanometer range. At one atmosphere pressure, some of the pores are smaller than the mean free path of the air molecules, and the conductivity of the aerogel can be much lower than the conductivity of air. Further reduction of pressure improves the performance of the aerogel as an insulator. Excellent insulation performance can be achieved at pressures well above those required for earlier vacuum insulation systems. Figure 37.14 shows typical variation of the conductivity of aerogels samples with pressure. At a pressure of one-tenth of an atmosphere aerogels shows substantial improvement in insulating properties while conventional vacuum panels using powders of much larger size require a pressure several orders of magnitude lower to achieve high insulation levels. Sample aerogels are shown in Figure 37.15.

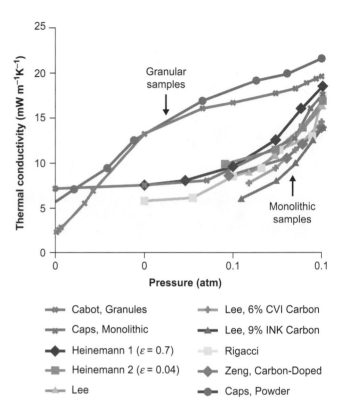

Figure 37.14. Variation of aerogel thermal conductivity with pressure. All aerogels are monolithic except for the Cabot, Granules and the Caps, Powder [8][17][18][19][20][21].

Figure 37.15. Aerogel samples.

37.5 Applications

There are many different applications for insulation, from the most basic example of a blanket on a bed to the space-age materials used on the space shuttle. A novel use of insulation is for advanced residences such as the Passivhaus (passive house) concept. This concept includes very thick insulation in the walls, together with triple-glazed windows and heat recovery in ventilation systems. A thick wall section is shown in Figure 37.16. Test homes have documented heating savings in northern Europe of 80%–90% compared with conventional house designs [2]. Use of advanced insulations such as

Figure 37.16. A thick wall section in Swedish home yielding high insulation levels

Figure 37.18. An adobe wall under construction in Taos, New Mexico. (Image courtesy of Ed Darrell.)

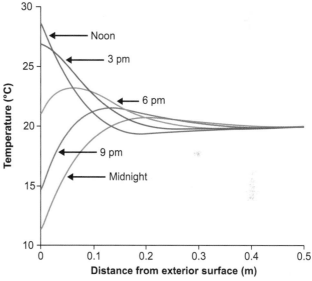

Figure 37.19. Predicted temperature behavior within an adobe wall when the exterior temperature varies as a cosine function over 24 h.

Figure 37.17. The space shuttle used insulating tiles against extreme conditions. (Image courtesy of NASA.)

aerogels should permit the same performance with much thinner walls. Currently, aerogel manufacture is not economically competitive with conventional insulations in such applications.

A more extreme application of insulation is on the space shuttle (Figure 37.17). It is covered in insulated tiles that protect the aluminum skin during the heat of reentry. Tiles on the surface contain fine silica fibers. The fibers limit both radiation and conduction from the hot outer surface to the structural skin. They must withstand high temperatures and rapid expansion and contraction during liftoff and reentry.

Thick adobe walls (Figure 37.18) combine moderate insulation properties with substantial thermal mass. In climates where there is substantial fluctuation between day and night temperatures, such as desert areas, the combined properties serve to damp out the fluctuating temperature so that the interior of the adobe wall remains close to the average of day and night extremes. Figure 37.19 shows the predicted progress of a thermal wave from the exterior wall to the interior, assuming that the daytime temperature fluctuated sinusoidally. Adobe walls (see Figure 37.18) are most advantageous in locations where the average diurnal temperature is within the comfortable range. In colder seasons, the moderate insulating properties of the adobe must be supplemented by additional forms of insulation; colder climates minimize the advantage of thermal mass. In this case, a much higher level of insulation is called for. A recent development is insulation impregnated with phase-change material that acts as a thermal mass.

In modern building construction, high levels of insulation are used in walls and roofs similar to the Passivhaus. The surface is also covered by a continuous flexible fabric-like material that blocks air leaks through the wall due to wind or buoyancy forces. Special care must be taken to prevent moisture accumulation within the assembly by use of a vapor barrier and a path for water drainage parallel to the building surface. The problem can be more severe in new tight assemblies since there is not a flow of air to help dry out any accumulated moisture. There have been numerous cases of mold and moisture damage in improperly designed or poorly constructed walls.

37.6 Summary

Thermal insulation acts to limit heat transfer by conduction, radiation, and convection. Conduction is limited by including a large volume of insulating gas in the insulation. It is further reduced under vacuum conditions. Convection is suppressed by reducing the size of foam cells or voids between powders or fibers and limiting the spacing between glass layers in windows. Low-emissivity coating and multiple opaque surfaces formed by fibers, powders, or foams reduce radiation. The next generation of insulation requires new materials and forming methods to produce products ranging from aerogels with nanosize pores to panels made from crop wastes for the developing world.

37.7 Questions for discussion

1. Why are thermal insulations important in modern buildings?
2. Explain how an insulated window improves the comfort for an occupant siting close to the window in winter.
3. Give examples of insulation used a century or more ago in different parts of the world.
4. Compare the insulating performance of solid wood with that of a structure made using a paper honeycomb.
5. The walls of modern energy-star-rated refrigerators are quite thick (check for yourself). How could you reduce the wall thickness without compromising the refrigerator's efficiency?

37.8 Further reading

- **American Society of Heating, Refrigerating and Air-Conditioning Engineers (ASHRAE)**, 2009, *ASHRAE Handbook: Fundamentals*, Atlanta, GA, ASHRAE. This is a comprehensive source of data on common building insulation systems.
- **R. T. Bynum**, 2001, *Insulation Handbook*, New York, McGraw-Hill. Provides a history of insulation and general advantages and drawbacks of current insulations.
- **L. Glicksman**, 1994, "Heat transfer in foam," in *Low Density Cellular Plastics: Physical Basis of Behaviour*, eds. **N. C. Hilyard** and **A. Cunningham**, London, Chapman and Hall, pp. 104–152. Provides a detailed analysis of the various heat transfer mechanisms in foam insulation.
- **J. H. Lienhard IV** and **J. H. Lienhard V**, 2004, *A Heat Transfer Textbook*, 3rd edn., Cambridge, MA, Phlogiston Press. A good overview of the different basic heat-transfer mechanisms.

37.9 References

[1] **US Department of Energy's** US Energy Information Adminstration, http://www.eia.doe.gov.
[2] **Passive House Institute**, http://www.passivehouse.com.
[3] **S. Wyllie-Echeverria** and **P. A. Cox**, 1999, "The seagrass (*zostera marina* [*sosteraceae*]) industry of Nova Scotia (1907–1960)," *Economic Botany*, **53**(4), 419–426.
[4] *The Billboard* October 5, 1946, **96**. – Temlock Board
[5] **Owens** *Corning History*, available at http://www.owenscorning.com/acquainted/about/history/index.asp.
[6] **R. T. Bynum**, 2001, *Insulation Handbook*, New York, McGraw-Hill.
[7] **Aspen Aerogel**, http://www.aerogel.com/.
[8] **Cabot Corp. Aerogel**, http://www.cabot-corp.com/Aerogel.
[9] **Glacier Bay Ultra Insulation**, http://www.glacierbay-technology.com/products/ultra-r-db-insulation/.
[10] **ThermoCor Vacuum Insulation Panels**, http://www.thermocorvip.com/.
[11] **H. E. Robinson, F. J Powlitch**, and **R. S. Dill**, 1954, *The Thermal Insulation Value of Airspaces. Housing Research Paper 32*, Washington, DC, Housing and Home Finance Agency.
[12] **W. M. Rohsenow, J. P. Harnett**, and **E. N. Ganic**, 1985, *Handbook of Heat Transfer Fundamentals*. 2nd edn., (New York, McGraw-Hill.
[13] **Z. Ali**, 2007, *Sustainable Shelters for Post Disaster Reconstruction, An Integrated Approach for Reconstruction after the South Asia Earthquake*, B.S. Thesis, Massachusetts Institute of Technology.
[14] **US Environmental Protection Agency**, *Foam Blowing Agents: Alternatives/SNAP*, available at www.epa.gov/ozone/snap/foams/index.html.
[15] **L. Glicksman**, 1994, "Heat transfer in foam," in *Low Density Cellular Plastics: Physical Basis of Behaviour*,

eds. **N. C. Hilyard** and **A. Cunningham**, London, Chapman and Hall, pp. 104–152.

[16] **J. H. Jeans**, 1960, *An Introduction to the Kinetic Theory of Gases*, Cambridge Cambridge University Press.

[17] **R. Caps** and **J. Fricke**, 2004, "Aerogels for thermal insualation," in *Sol–gel Technologies for Glass Producers and Users*, eds. **M. A. Aegerter** and **M. Mennig**, Boston, MA, Kluwer Academic Publishers, pp. 349–353.

[18] **U. Heinemann**, **R. Caps**, and **J. Fricke**, 1996, "Radiation–conduction interaction: an investigation on silica aerogels," *Int. J. Heat Mass Transfer*, **39**(10), 2115–2130.

[19] **D. Lee**, **P. C. Stevens**, **S. Q. Zeng**, and **A. Hunt**, 1995, "Thermal characterization of carbon-opacified silica aerogels," *J. Non-Crystalline Solids*, **186**, 285–290.

[20] **A. Rigacci** and **M. Tantot-Neirac**, 2005, "Aerogel-like materials for building super-insulation," in *2nd International Symposium on Nanotechnology in Construction*, pp. 383–393.

[21] **S. Q. Zeng**, **A. Hunt**, and **R. Greif**, 1995, "Geometric structure and thermal conductivity of porous medium silica aerogel," *J. Heat Transfer*, **117** (4), 1055–1058.

38

Industrial energy efficiency: a case study

Joe A. Almaguer

The Dow Chemical Company, Midland, MI, USA

38.1 Focus

Industry accounts for a large segment of the energy consumed globally and, as a result, advances made by industry toward increased energy efficiency have a significant influence on the global energy and environmental outlook. This chapter offers an overview of strategies, methodologies, and resources industry can use to address one of the greatest global challenges of our time: how to foster economic growth while also addressing energy-supply issues and the consequences of our dependence on fossil fuels.

38.2 Synopsis

Over the past decades, three factors have dramatically changed the way the world thinks about sustainable energy.

- Economic issues: affordable and less volatile energy pricing is critical to economic investment and growth.
- Security issues: supply may be impacted by geopolitical issues and aggravated by political instability in some of the world's largest oil- and gas-producing regions.
- Environmental issues: growing concerns about escalating greenhouse-gas (GHG) emissions and their impact on the planet.

How can industry meet this growing energy and environmental challenge? There are several options. Increased development of oil and natural-gas fields will increase energy supply. Alternative and renewable energy solutions, such as nuclear, solar, wind, and biofuels, offer a way to reduce dependence on fossil fuels. Both of these options need government and nongovernment support around the world.

But the simplest, most accessible, and least expensive option for industry is increasing energy efficiency and conservation. It is not only the cleanest option, it is also the easiest to implement and the quickest way to extend energy supplies while reducing carbon emissions.

In practical terms, in order for this option to work for industry in both the short and the long term, it must be a value-adding, sustainable proposition, representing a low-risk investment where savings flow directly to the bottom line. This scenario is not only possible, but is currently being exemplified by industries around the world.

This chapter provides an overview of resources, tools, strategies, and methodologies that have successfully been used to increase energy efficiency and conservation in an industrial setting. A case study is provided on recent strides in energy efficiency and conservation by The Dow Chemical Company, one of the largest industrial consumers of energy in the world.

Table 38.1. Manufacturing energy consumption in the USA, 2002

Manufacturing group	Btu (trillions)	Manufacturing group	Btu (trillions)
Food	1,123	Plastics and rubber products	351
Beverage and tobacco products	105	Nonmetallic mineral products	1,059
Textile mills	207	Primary metals	2,120
Textile product mills	60	Fabricated metal products	388
Apparel	30	Machinery	177
Leather and allied products	7	Computer and electronic products	201
Wood products	377	Electrical equipment, appliances, and components	172
Paper	2,363	Transportation equipment	429
Printing and related support	98	Furniture and related products	64
Petroleum and coal products	6,799	Miscellaneous	71
Chemicals	6,465	TOTAL	22,666

Source: [4] 1 Btu is 1.055 kJ.

38.3 Historical perspective

In 2005, the manufacturing sector accounted for 33% of global energy consumption, the highest of all end-use sectors [1], p. 16. That same year the USA was responsible for approximately 22% of total global energy consumption, down from approximately 25% in 1995, but still the highest among individual nations [2]. Breaking US energy consumption down among major sectors (residential, commercial, industrial, and transportation) shows industry as the highest energy consumer historically (Figure 38.1). Although there were several dips in industrial energy consumption in the early and late 1970s, the consumption curve maintained an upward trend until the mid 1990s, when a modest decline was seen. Various factors contributed to this decline in industrial energy consumption, including a slowing economy, increased fuel costs, and outsourcing of industrial production to other countries.

38.3.1 Chemical industry

A further breakdown of energy-consumption data [3] highlights the role played by the chemical industry. According to 2002 data, the chemical industry was one of the largest consumers of energy among US manufacturers that year, exceeded only by the petroleum/coal industry (Table 38.1). An article published in 2000 [5]

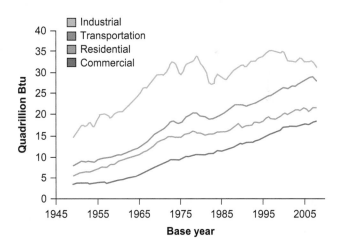

Figure 38.1. Energy consumption in the USA by sector, 1949–2008.

offers data on segments of the chemical industry, including industrial organics, industrial inorganics, plastic materials and resins, nitrogenous fertilizers, industrial gases, and alkalies and chlorine.

The International Energy Agency (IEA) defines the chemical industry as facilities that produce bulk or specialty compounds by chemical reactions between organic and/or inorganic materials. The petrochemical industry is defined as facilities that create synthetic organic products from hydrocarbon feedstock, oil, and natural gas [6], p. 61. The chemical and petrochemical industry consumed 30% of total global industry final energy use in 2004, increasing from 15.5% in 1971. North America held the largest share of the energy consumed by the petrochemical and chemical industry, at 41%.

38.4 Industrial energy management standards

Among industrial energy consumers who have been successful at institutionalizing the efficient and effective use of energy, a fundamental first step has been the adoption of a set of principles or standards by which to manage energy performance. These standards serve to establish and institutionalize the way in which an organization values, prioritizes, measures, and monitors its energy performance, and how it acts to systematically improve performance.

The Energy Star program recommends a seven-step process for energy management that has been successfully followed by many companies:

1. Make Commitment
2. Assess Performance
3. Set Goals
4. Create Action Plan
5. Implement Action Plan
6. Evaluate Progress
7. Recognize Achievements

The Dow Chemical Company modeled its Energy Efficiency and Conservation system using both American National Standards Institute (ANSI) standards and Energy Star guidelines. A case study is given later in this chapter. More details of management tools are given below.

Historically, individual nations have developed their own energy management standards, followed guidelines developed by international organizations, or followed guidelines prepared by major industrial nations, such as the USA. In 2008, the International Organization for Standardization (ISO) announced the development of an international standard on energy management (ISO 50001) to establish a framework for industry, commercial facilities, and other organizations to manage energy [7].

38.4.1 ISO 50001

The scope of ISO 50001 is stated as "Standardization in the field of energy management, including for example: energy efficiency, energy performance, energy supply, procurement practices for energy using equipment and systems, and energy use" [8]. The standard is believed to have the potential to impact 60% of the world's energy use, including industry, commercial, and institutional sectors [9].

The first working draft was available in September 2009. It is intended that development of the standard will move very quickly, with completion planned for the third quarter of 2011. The standard will provide a framework for integrating energy efficiency into management practices, enabling multinational organizations to implement energy efficiency across their organizations [7]. Other listed objectives include the following.

- Assist organizations in making better use of their existing energy-consuming assets.
- Offer guidance on benchmarking, measuring, documenting, and reporting energy-intensity improvements and their projected impact on reductions in GHG emissions.
- Create transparency and facilitate communication on the management of energy resources.
- Promote energy-management best practices and reinforce good energy-management behaviors.
- Assist facilities in evaluating and prioritizing the implementation of new energy-efficient technologies.
- Provide a framework for promoting energy efficiency throughout the supply chain.
- Facilitate energy-management improvements in the context of GHG-emission-reduction projects.

38.4.2 National standards

The American National Standards Institute
The American National Standards Institute (ANSI, http://www.ansi.org) serves as the administrator and coordinator of the US private-sector voluntary standardization system. In addition, the ANSI works with the international community in the adoption of standards. The ANSI and the Associação Brasileira de Normas Técnicas (Brazil) comprise the secretariat for the development of ISO 50001.

The ANSI provides a search engine for searching its system for standards [10]. The only listed standard for energy management is ANSI/MSE 2000:2008, A Management System for Energy, which covers the purchase, storage, use, and disposal of primary and secondary energy resources. The ANSI also provides information on other national standards programs [11].

38.5 Industrial energy-performance-improvement practices

There appears to be agreement that, despite advances made thus far, industry still has significant opportunities for increased energy performance [6], p. 3, [12], p. 163, [13]. Predictions for global improvement potential include 17.8% for chemicals and petrochemicals, 9%–40% for iron and steel, 11%–40% for cement, 15%–18% for pulp and paper, and 6%–8% for aluminum [12], p. 166 and references therein.

Although establishing a comprehensive energy-management system is a requirement in order to achieve sustained energy-performance success, it is not sufficient. To improve energy performance, changes that result in either reduction or elimination of wasted energy, or both, and increased productivity of the energy consumed must be made.

Energy is wasted when its consumption is not necessary to achieve the desired output. Stopping the waste of energy, often through raised awareness and behavior change, is many times the first step in improving energy performance. Some examples of wasted energy are lights left on in unoccupied rooms, buildings, or work areas; pumps left running in a total-recirculation loop for extended periods; idling engines, hot standby process heaters or boilers; and running air compressors venting the air to the atmosphere. Much of the waste in energy results from attempts to compensate for unreliable or poorly designed systems, lack of proper maintenance, or poor operating discipline.

Increasing the productivity of energy involves reducing the amount of energy required to achieve the desired output. Exactly how this is done depends on the energy-consuming processes, systems, and equipment that are used in the manufacturing process. Significant or large energy uses, having the greatest impact on overall energy intensity, should always take priority when seeking improvement opportunities. In general, it is best to build efficiency into a system while in the design stage rather than after it has been built and is operational. This can be done through selection of high-energy-efficiency equipment and arrangement of system components in ways that maximize the reuse of energy that might otherwise be discharged as waste heat or high-energy fluids. In large multi-system facilities, process heat integration is widely utilized to exchange thermal energy between streams as a way to increase energy productivity.

In existing manufacturing systems, the process of seeking and identifying opportunities for improving energy performance is often referred to as conducting an energy assessment of the facilities. Energy assessments are planned to make use of tools and techniques suitable for the type of systems and equipment in use, but all aim at reducing waste and improving efficiency and energy productivity in order to reduce the overall energy intensity of the facility. In the next section several useful, widely applicable, energy-assessment tools and guides are described.

38.5.1 Energy-assessment and -management tools

Several authors offer assessment guidance, especially for complex industrial systems [14][15]. However, in a more general sense, many tools and guides are user-developed or are available online. They are intended to help organizations identify efficiency gaps or waste and evaluate potential improvement opportunities. Improvement opportunities may exist in how an exiting system or piece of equipment is operated and maintained. These are usually addressed by behavior changes specified in operating/maintenance procedures. Other efficiency gaps or opportunities might reflect the current design of the system or the selection and installation of equipment, in which case modifications and/or equipment replacement are probably required in order to bring about improved energy performance.

A number of international and US-based sites offer tools and guidebooks that are publicly available, often at no cost. The Intelligent Energy eLibrary, sponsored by Intelligent Energy Europe (IEE), offers a long list of tools available from sources around the world [16]. The US-based Energy Star program provides a tool (the Energy Management Assessment Matrix) that enables users to compare their energy-management practices with those outlined in the Energy Star guidelines [17]. The US Department of Energy (DOE) offers software to help industrial facilities identify and analyze energy-system savings opportunities across an entire plant and also for specific applications [18]. Below are descriptions of some widely used tools.

Plant-wide

The **Industrial Facilities Systems Assessment Tool (IFSAT)** helps users analyze energy-use scenarios in industrial facilities where building energy is being consumed, such as enclosed and conditioned process buildings. It includes building and equipment types that are applicable to manufacturing. The current software is a beta version.

The **Quick Energy Profiler** helps plant managers improve energy management at industrial facilities. Enables users to establish a baseline for how energy is being used at their plant, identify best opportunities to save energy and money, and calculate emissions. Its application can be completed in an hour. It focuses on major energy-consuming systems.

The **Integrated Tool Suite** is similar to the Quick Energy Profiler but can be downloaded as a stand-alone tool and does not require an Internet connection. Also features system-specific scorecards for quickly estimating savings opportunities.

Motor-driven

AIRMaster+ helps users analyze energy use and savings opportunities in industrial compressed-air systems. It baselines existing and model future system operations improvements and evaluates energy and dollar savings from many energy-efficiency measures.

The **Fan System Assessment Tool (FSAT)** quantifies energy use and savings opportunities in industrial fan systems. It helps users to understand how well fan systems are operating, determine the economic benefit of system modifications, and establish which options are most economically viable when there exist multiple opportunities for system modification.

MotorMaster+ is a motor-selection and -management tool that supports motor and motor-systems planning by identifying the most efficient action for a given repair or motor-purchase decision.

MotorMaster+International includes many of the capabilities and features of MotorMaster+. It helps users evaluate repair/replacement options on a broader range of motors, including 60-Hz motors tested under the Institute of Electrical and Electronic Engineers (IEEE) standard and 50-Hz motors manufactured and tested in accordance with International Electrotechnical Commission (IEC) standards.

The **Chilled Water System Analysis Tool (CWSAT)** helps users evaluate changes to existing equipment, including chillers, pumps, and towers, and calculates the energy and cost savings gained by implementing those changes.

The **Pumping System Assessment Tool (PSAT)** helps users assess the efficiency of pumping-system operations. It calculates potential energy and associated cost savings. It also enables users to save and retrieve log files, default values, and system curves for sharing analyses with other users.

Steam

The **Steam System Scoping Tool** helps users perform initial self-assessments of steam systems. Users can profile and grade steam-system operations and management and evaluate steam-system operations against best practices.

The **Steam System Assessment Tool (SSAT)** helps users develop approximate models of real steam systems. It also helps quantify the magnitude of key potential steam improvement opportunities.

3E Plus calculates the most economical thickness of industrial insulation for user input operating conditions. It uses built-in thermal performance relationships of generic insulation materials or supplied conductivity data for other materials.

Process heating

The **Combined Heat and Power (CHP) Application** evaluates the feasibility of using gas turbines to generate power and turbine exhaust gases to supply heat to industrial systems. It analyzes three typical system types: fluid heating, exhaust-gas heat recovery, and duct burner systems.

The **NOx and Energy Assessment Tool (NxEAT)** helps plants in the petroleum refining and chemical industries analyze nitrogen oxide (NO_x) emissions and application of energy-efficiency improvements. It enables users to inventory emissions from equipment that generates NO_x and compare how various technology applications and efficiency measures affect overall costs and reduction of NO_x emissions.

The **Process Heating Assessment and Survey Tool (PHAST)** helps users survey process heating equipment that consumes fuel, steam, or electricity, and identifies the most energy-intensive equipment.

Buildings

The **Industrial Facilities Scorecard** provides a snapshot of the energy used by all buildings at a plant. It is used to identify potential energy-saving measures on the basis of annual energy use and costs that can be considered for further investigation.

The **Buildings Cooling, Heating, and Power Systems (BCHP) Screening Tool** assesses the economic potential of these systems in commercial buildings. Features databases for heating, ventilating, and air-conditioning (HVAC) equipment, electric generators, thermal storage systems, prototypical commercial buildings, and climate data. It calculates heating, cooling, and electrical loads.

Data centers

The **Data Center Profiler Tool Suite** helps users identify and evaluate energy efficiency opportunities in data centers. It includes a profiling tool and a set of system-assessment tools to perform energy evaluations on specific areas of a data center.

38.5.2 Combined heat and power

Combined heat and power (CHP), also known as cogeneration, is a system that produces electricity and heat from a single fuel source. It is generally recognized as one of the most effective methods for improving energy efficiency and reducing carbon emissions [19]. It is an integrated energy system that can be adapted as needed by the end user. The most common system configurations are gas turbine or engine with heat recovery unit, and steam boiler with steam turbine. The US Environmental Protection Agency (EPA) sponsors a partnership program that promotes the use of CHP [20].

38.5.3 Active energy management

Active energy management is the concept of generating power outside of peak usage times and storing it for later, more efficient use. Across a power grid, the energy economics would be improved and environmental effects would be minimized by avoiding the use of more highly polluting energy generators to boost power supply during peak times. Various versions of power "smart grids" are under development or consideration in the major industrialized areas of the world [21][22]. Smart grids typically involve the overall power system from utilities across a broad swath of consumers.

The development of active energy management is being driven by information technology (IT) tools that were not available a generation ago. Automated demand response (OpenADR) is defined as "an open-standards based communications system to send signals to customers to allow them to manage their electric demand in response to supply conditions, such as prices or reliability, through a set of standard, open communications" [23]. The Demand Response Research Center (DRRC) at the Lawrence Berkeley National Laboratory is conducting research on the use of OpenADR in industrial facilities with the following stated goals [23]:

- increase knowledge of what, where, for how long, and under what conditions industrial facilities will shed or shift load in response to an automated signal;
- develop a better understanding of the dynamics of maximizing load-reduction savings without affecting operations;
- facilitate deployment of industrial OpenADR that is economically attractive and technologically feasible;
- more effectively target efforts to recruit industrial OpenADR sites;
- evaluate the opportunities to combine advanced controls and continuous measurement for optimal energy efficiency and demand response.

38.6 Industry/government/nongovernment resources

There are many resources available from industry, government, and nongovernment organizations that help define policy and guidelines, provide energy measurement and conservation tools, and offer funding. Below are summaries of major sources of information.

38.6.1 International

European Commission Energy

The European Commission Energy site (http://ec.europa.eu/energy/index_en.htm) states the focus of European Union (EU) energy policy as "creating a competitive internal energy market offering quality service at low prices, developing renewable energy sources, reducing dependence on imported fuels, and doing more with a lower consumption of energy." The site describes policies on a wide variety of energy topics, as well as providing other resources.

European Commission Intelligent Energy Europe

The purpose of Intelligent Energy Europe (IEE) (http://ec.europa.eu/energy/intelligent/index_en.html) is to fund actions that will enable Europe to save energy and encourage the use of renewable energy sources. The organization provides the Intelligent Energy eLibrary, which brings together tools and guidebooks on energy efficiency, renewable-energy applications, and sustainable mobility. The Energy Efficiency in Industry segment has an exhaustive list of tools and guidebooks specifically designed for industry, from general tools to energy audits, energy management/benchmarking, and energy-saving measures. A related booklet has been published [24].

The European Bank for Reconstruction and Development

The European Bank for Reconstruction and Development (EBRD) is an international financial institution with projects in 29 countries from central Europe to central Asia. The EBRD promotes entrepreneurship and fosters transition toward open and democratic market economies. The EBRD's energy-efficiency and climate-change team develops energy-efficiency and renewable-energy credit lines, promotes energy efficiency in public buildings and industries, and works to build a carbon-credit market in the countries in which it operates. The EBRD's program on industrial energy efficiency [25] works with project teams and clients to identify energy-saving opportunities, which are developed for inclusion in the client's investment program.

The Federal Ministry of Economics and Technology

The Federal Ministry of Economics and Technology (Bundesministerium für Wirtschaft und Technologie) has the responsibility for the formulation and implementation of energy policy in Germany [26]. The site describes policies and activities regarding energy and climate and the renewable-energies export initiative, as well as related publications.

IHS, Inc.

IHS provides information and insight for global business customers on energy, product life cycle, security, and environment, all supported by macroeconomics. The section on energy provides solutions, news, and resources [27].

The International Energy Agency

The International Energy Agency (IEA) (http://www.iea.org) is an intergovernmental organization based in Paris, France, which acts as energy-policy advisor to 28 member countries. The IEA's mandate incorporates the "Three E's" of balanced energy-policy making: energy security, economic development, and environmental protection. The organization is a major source of energy statistical data.

The Intergovernmental Panel on Climate Change

The Intergovernmental Panel on Climate Change (IPPC) was established by the United Nations Environment Programme and the World Meteorological Organization to provide a scientific view on the current state of climate change and its potential environmental and socioeconomic consequences. It reviews and assesses global scientific, technical, and socioeconomic information relevant to the understanding of climate change. In addition to publications and other resources, the organization has a data distribution center (http://www.ipcc-data.org/) that provides data related to climate change.

The Pew Center on Global Climate Change

The stated objectives of the Pew Center on Global Climate Change (http://www.pewclimate.org/) include producing analyses of key climate issues, working to keep policy makers informed, engaging the business community in the search for solutions, and reaching out to educate the key audiences. The organization works internationally and in the USA with government and business with the goal of reducing GHG emissions. It provides many publications and analyses related to this issue.

The United Nations Industrial Development Organization

The stated purpose of the Energy and Climate Change Branch of the United Nations Industrial Development Organization (UNIDO) (http://www.unido.org/) is to "promote access to energy for productive uses while at the same time supporting patterns of energy use by industry that mitigate climate change and are environmentally sustainable." The organization focuses on energy-system optimization and energy-management standards and offers support for policymakers. There are three publications related to industrial energy efficiency [28][29][30].

The World Economic Forum

The World Economic Forum (WEF) is an international organization based in Switzerland, with the objective of engaging leaders in partnerships to shape global, regional, and industry agendas. The WEF's Centre for Global Industries (http://www.weforum.org/en/about/CentreforGlobalIndustries/index.htm) has developed the Industry Partnership program, which provides networks and activities, identifies issues for collaboration, and brings partners together to initiate needed change.

The World Resources Institute

The World Resources Institute (WRI) is an environmental think tank centered on policy research and analysis addressed to global resource and environmental issues. The organization focuses on four goals: climate protection, governance, markets and enterprise, and people and ecosystems. One of the research topics on their EarthTrends site (http://earthtrends.wri.org/) is Energy and Resources, which offers a searchable database, maps, country profiles, articles, and data tables.

38.6.2 The USA

The Alliance to Save Energy

The Alliance to Save Energy (ASE) (http://ase.org/) is a nonprofit coalition of business, government, environmental, and consumer leaders that was established in 1977. The organization supports energy efficiency as a cost-effective energy resource under existing market conditions and advocates energy-efficiency policies that minimize costs to society and individual consumers and that lessen GHG emissions and their impact on the global climate. In addition to programs in the USA, the ASE has partnerships in or with 20 other countries. The organization sponsors "Star of Energy Efficiency" awards for various categories and sizes of entities.

The American Chemistry Council

The American Chemistry Council (ACC) introduced the Responsible Care initiative [31] in 1988 with the objective of significantly improving the environmental, health, safety, and security performance of member companies. This is a global initiative that is now active in 53 countries. Member companies submit to the Council annual performance data, which are made publicly available on their website. Accountability for member companies includes energy as well as other issues such as safety and product stewardship. Member companies are recognized through certification.

The American Council for an Energy-Efficient Economy

The American Council for an Energy-Efficient Economy (ACEEE) is a nonprofit organization founded in 1980 to advance energy efficiency as a means of promoting economic prosperity, energy security, and environmental protection. The organization's programs involve energy policy (primarily federal and state), research (in industry and agriculture among others), and communications (including conferences, publications, and development). The industry and agriculture section of the ACEEE (http://www.aceee.org/industry/index.htm) states as its purpose to "analyze and promote technologies, process innovations, and policies for increasing the energy efficiency and competitiveness of manufacturing industries and agriculture." The ALLY program invites organizations to help shape the nation's energy-efficiency research and policy agenda and gain access to expertise

Figure 38.3. The business and site energy leaders' network.

	EE&C Global Leader	Business EE&C Teams				
		Business A EE Leader	Business B EE Leader	Business C EE Leader	Business D EE Leader	Business E EE Leader
Site EE&C Teams Site 1 EE Leader		Plant A-1	Plant B-1	Plant C-1	Plant D-1	Plant E-1
Site 2 EE Leader		Plant A-2	Plant B-2	Plant C-2	Plant D-2	Plant E-2
Site 3 EE Leader		Plant A-3	Plant B-3	Plant C-3	Plant D-3	Plant E-3
Site 4 EE Leader		Plant A-4	Plant B-4	Plant C-4	Plant D-4	Plant E-4
Site 5 EE Leader		Plant A-5	Plant B-5	Plant C-5	Plant D-5	Plant E-5

by 20% by the year 2005 from base year 1994. The result by 2005 was a 22% EI reduction. Dow subsequently set aggressive sustainability goals to reduce the company's EI by 25% by 2015 from base year 2005, GHG emissions intensity by 2.5% annually through 2015, and absolute emissions within the company by 2025.

The Sustainability Program Management Office (PMO) helps drive the implementation of the sustainability goals by

- facilitating implementation across the EE&C system and through all regions/functions,
- tracking and reporting of performance against the goals for internal and external stakeholders,
- identifying and addressing barriers that block implementation,
- clarifying the corporate and goal-specific business rule,
- helping ensure that Dow earns respect for its sustainability commitment,
- providing strategic corporate philanthropic support for sustainability
- benchmarking of sustainability performance,
- coordinating interaction with external stakeholders.

38.8.2 The EE&C organizational structure

With more than 53 diverse business groups and hundreds of manufacturing plants, Dow relies heavily on a well-defined, institutionalized, organizational structure dedicated to driving the implementation of its energy-efficiency and -conservation objectives globally. This structure includes implementation leaders and teams and networks of energy teams whose responsibilities include the following:

- develop management systems,
- establish an implementation model,
- develop specific plans to achieve goals,
- identify energy-saving opportunities,
- implement EE&C projects,
- monitor and report progress,
- promote EE&C culture locally,
- leverage success.

A key position in this organization is the Global Energy Efficiency and Conservation Leader, who establishes and maintains the global program's key elements, including its management systems, work processes, energy-accounting rules, performance tracking and reporting systems, overall results analysis, and internal program-effectiveness-assessment program. This person also leads the Business and Site Energy Leaders Network and provides program-implementation strategy and coordination.

The Business and Site Energy Leaders form a network that spans each business across numerous global sites (vertical) and large sites across numerous businesses (horizontal) (Figure 38.3).

The Business Group Energy Efficiency and Conservation Management Program Leaders are accountable for implementing global program requirements and managing the energy programs within their respective business group. Their responsibilities include

- leading their business's energy team,
- determining potential energy savings at each of the business's manufacturing plants,
- developing and maintaining a 10-year energy improvement plan aligned to the overall business strategy and capital funding program,

Figure 38.4. Historical energy-intensity performance for The Dow Chemical Company, 1994–2009. 1 Btu/lb is 2.326 kJ/kg.

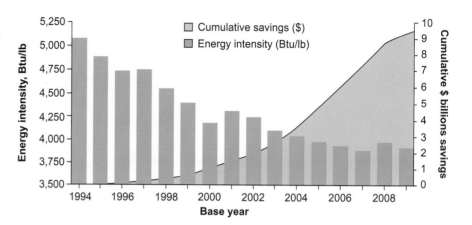

- setting annual improvement targets for each of their plants and tracking results,
- benchmarking performance,
- reporting plant and business progress against short- and long-range improvement goals and targets.

The Site Energy Teams, each led by a Site Energy Efficiency and Conservation Leader, further drive energy improvement at Dow's large multi-plant, integrated manufacturing sites. At these large sites, there exist numerous opportunities to improve energy flow and use within and between the manufacturing plants. These local teams actively engage employees, subject-matter experts, engineering and technology resources, and external experts, including energy service providers, in energy-efficiency-improvement projects. This collaboration drives an energy-efficiency mindset and culture at the local level. Similarly to the Business Energy Leader, these teams are accountable for establishing site goals, formulating 10-year energy-improvement plans, and tracking and reporting energy-intensity performance and progress against annual and long-term goals.

38.8.3 Energy management and reporting

Energy-intensity data are collected, tracked, and reported for all 170 Dow sites and 700 plants/facilities. Dow uses a highly automated and robust energy-accounting system to measure the total energy consumption of every manufacturing plant within Dow. Steam, electricity, compressed air, fuel gas, and all other forms of energy consumed in the production of Dow products are measured and converted to a total Btu unit, then an energy intensity (1 Btu/lb is 2.326 kJ/kg) footprint for each plant is automatically calculated and reported.

Dow uses data to drive performance and direct decision-making. In setting long-range performance targets, historical energy-performance data are used to establish a baseline performance for each plant, each site, and the company as a whole. Energy-accounting rules are

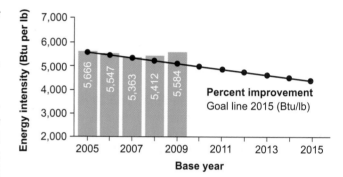

Figure 38.5. The energy-intensity performance goal (25% reduction by 2015 from 2005 baseline) of The Dow Chemical Company, 2005–2015.

established and used to consistently track and report performance on an established frequency. Energy intensity (Btu per lb) is the primary energy-performance indicator (EPI) used by Dow to set its performance goals and gauge improvement. From 1994 through 2009, the company has shown an EI reduction of 23%, with energy savings of over 1,700 trillion Btu having an estimated saved value of $9.2 billion (Figure 38.4). Figure 38.5 reflects the company goal of a 25% reduction in energy intensity by 2015 from base year 2005. The graph represents the aggregate annual average energy intensity (Btu per lb) of The Dow Chemical Company. Energy was calculated on an equivalent Btu of source natural-gas basis. Energy intensity was not run rate-adjusted, as can be seen – the economic slowdown is reflected in the 2008 and 2009 EI.

The online EI report is available to anyone within Dow. With a click of the mouse, employees can view historical performance, progression trends, current-quarter energy consumption, production, and EI. This information can be generated for a single plant, an integrated multi-plant site, a business group, any Dow geographical region, or the entire company. Each quarter, the energy performance of each plant is reviewed and analyzed. Reports are posted for each business, for integrated manufacturing sites, and for the company as a whole.

The reports compare current performance with baseline data against annual targets and progress made toward the 10-year corporate goals. The online tracking system allows any plant's footprint to be compared with that of any other plant within Dow. If an external plant's energy footprint is known, Dow plants can be compared. Plants can benchmark themselves against Dow's aggregate EI and rank plants within a business or technology.

Businesses benchmark their plants and rank them on the basis of their EI to estimate potential energy savings and develop long-range improvement plans. Specific areas of energy efficiency and conservation are identified and EI goals are set at the site and plant levels.

Traditionally, Dow's energy-reduction efforts have not focused on occupied buildings, since these represent less than 1% of its total consumption. However, in 2007, Dow retained the services of a vendor to help develop and implement a program to upgrade major buildings and obtain certification by Energy Star standards.

38.8.4 Improvement methodology

Dow's energy-improvement methodology makes extensive use of well-established global work processes in addition to some unique strategies to ensure successful performance.

Key elements of the improvement methodology for existing assets include the following.

- Initial identification of opportunities for energy savings through business and technology reviews, benchmarking studies, and internal and external energy assessments.
- A 10-year energy-improvement plan, which lists identified energy-saving opportunities/projects organized by business and plant with estimated energy savings and implementation timeframe.
- Linkage of the 10-year energy improvement plan to the Business's Capital Funding Program and the Business Strategic Plan. Opportunities are prioritized, evaluated, and funded accordingly.
- An annual review and update of the plan, in which annual improvement targets are developed.
- The Energy Measurement and Reporting System, which tracks and reports on performance and target attainment.
- Established capital-project work processes mandate formal energy reviews to ensure that the designs of new manufacturing facilities/plants are optimized.

The improvement methodology used by Dow continuously works to correct inefficiencies, find optimum solutions to defects, and establish control plans to sustain gains in energy efficiency. Dow's use of Six Sigma in concert with its established improvement methodology has been especially successful in improving energy efficiency, reducing energy costs, and managing new projects and technologies. Over the past five-plus years, hundreds of Six Sigma projects have been implemented, yielding cost savings in the hundreds of millions of dollars.

In addition to improvements to existing assets, the company works toward energy optimization of capacity-expansion projects; major retrofits of existing plants, facilities, or systems; new product plants; replacement projects; and rearrangement projects to comply with regulatory or compliance issues.

38.8.5 Goals, compensation, and metrics

Dow's goals and compensation programs drive accountability throughout the organization through the establishment of specific annual EI goals on multiple levels throughout the organization. Pre-agreed compensation elements are linked to energy-goal attainment and performance.

In addition to its comprehensive goals and compensation program, Dow also recognizes employee achievement relating to energy efficiency and conservation through its Waste Reduction Always Pays (WRAP) award. Dow's WRAP awards provide employees with an opportunity to develop ways to reprocess waste into raw materials, thus enabling Dow to capture value while reducing waste.

Since the inception of WRAP in 1986, Dow has recognized 395 projects and their sponsors through internal and external communications tools.[1] Globally, the projects have accounted for the reduction of 230,000 tons of waste, 13 million tons of wastewater, and 8 trillion Btu of energy. This energy saving would be equivalent to the energy used in more than 87,000 typical homes annually according to the US DOE. The estimated value of all of these projects is about $1 billion.

Finally, the Recognition at Dow program allows all Dow leaders and employees to recognize their colleagues for their good works and contributions to the company across a variety of disciplines. The recognition employees receive through the program can vary from an electronic thank-you card to a monetary award, depending on its impact on the company.

38.8.6 Promoting and communicating success – internally

Dow's employee-communication component of its Energy Efficiency and Conservation Management

---

[1] WRAP winner for example was waste reduction and recovery/recycling for polyvinyl chloride (PVC) and polypropylene plastic at DowBrands.

system facilitates the sharing of improvement ideas, performance reporting, energy best practices, performance benchmark information, and important messages from the management team. Monthly and quarterly progress reports can be communicated on a global, site-specific, or plant-specific basis, as needed. External communications such as press releases are communicated to employees as internal news items, as are external awards and milestones.

Dow's intranet website

Most internal communications are driven through Dow's Energy Efficiency and Conservation intranet site. The site, which is available to all Dow employees, provides a multitude of tools that allow access to information necessary to continue to meet Dow's long-term energy-efficiency and -conservation goals. Features of the Energy Efficiency and Conservation intranet site include

- current energy-efficiency and -conservation goals,
- specific web pages detailing energy-related information and performance reporting on Dow's large integrated energy-consuming sites,
- business-specific web pages detailing energy-related information and performance reporting,
- best practices that are applicable across businesses, plants, and sites around the world,
- important messages from the leadership team,
- mechanisms to encourage employee involvement and reward successes,
- a listing of external energy-conservation links such as to Energy Star, the Alliance to Save Energy, the US DOE Industrial Technology Program, SEP Energy Quick Start, and the American Council for an Energy Efficient Economy,
- industrial energy-efficiency resources for energy managers such as energy-savings-assessment tools, benchmark information, and training opportunities,
- energy-savings calculation and accounting tools such as Global Asset Utilization Reporting (GAUR) energy-consumption spreadsheets, energy-conversion calculators for utility savings projects, heat-loss calculators, and Btu accounting tools,
- global energy-efficiency success stories.

Site events

Site energy and conservation events highlight energy issues and conservation opportunities. In 2009, over 250 employees at Dow Central Germany (DCG) in Schkopau gathered during the site's first-ever Energy Day, with the goal of raising energy awareness. The event featured presentations from external energy suppliers on topics such as the feedstock/energy-market situation, heating energy savings, energy savings in houses, and energy savings by efficient water-conditioning. Throughout the day, employees also visited the gallery walk, which featured 14 posters on Dow energy policies, energy projects, and DCG goals and results. During the event, employees were also encouraged to share their energy-savings ideas.

38.8.7 Promoting and communicating success – externally

General external communication is done through press releases distributed via PR Newsline, with additional distribution to energy-specific recipients. Follow-up to distribution may include media interviews and expansion of information in congressional hearings and meetings.

One of the most important forms of external communication is through partnerships and participation with external organizations and other companies involved with energy-efficiency and -conservation issues. Dow maintains corporate involvement in the ANSI and ISO efforts toward energy standards as well as active participation in the Energy Star and Save Energy Now programs. More details on the efforts of these organizations are covered elsewhere in this chapter.

Externally recognized success stories

Dow's St. Charles Operations in Hahnville, Louisiana was featured in a Save Energy Now case study where an assessment identified opportunities for natural-gas savings in the plant's steam system. The actions taken resulted in savings of $1.9 million annually with annual natural-gas savings of 272,000 million Btu.

In 2008, Dow was recognized by the Energy Star program as a Partner of the Year [38]. Key accomplishments listed were

- reducing the average EI of its US operations by nearly 2.5% per year from the base year 2004 through 2007, saving the equivalent of nearly 47 trillion Btu and $325 million;
- instituting a world-class, comprehensive corporate energy-management program that has the personal support of senior leadership and all business units.

Dow is recognized in the Save Energy Now program as a LEADER organization [39]. These companies have committed to the Save Energy Now LEADER initiative of reducing EI by 25% or more in 10 years.

Dow was featured in a recent report prepared for the Pew Center on Climate Change on corporate energy-efficiency strategies that work and how corporations can apply them [40]. The report includes case studies covering strategies employed by Dow, United Technologies, IBM, Toyota, PepsiCo, and Best Buy. The report also summarizes best practices and provides resources for corporations seeking to improve their energy efficiency.

38.9 Summary

Improvement in industrial energy efficiency and conservation is good for business. It saves money, enhances global competitiveness, preserves jobs, and creates prosperity for shareholders. It is good for the environment by resulting in smaller GHG emissions and being part of the solution to global climate change. It is good for society by reducing demand, lowering energy bills, and promoting a stable energy supply.

Nonetheless, challenges remain, including societal issues as well as technical issues [41]. According to the International Energy Agency, the current rate of improvement in energy efficiency is not enough to overcome the other factors driving up energy consumption, including the rise in energy use in the transportation and service sectors [1], p. 15. Between 1990 and 2005 global final energy use increased by 23% while CO_2 emissions rose by 25%. The IEA has concluded that the world is headed for an unsustainable energy future unless new ways to separate energy use and emissions from economic growth are found.

As the major energy user across the globe, industry is a key player in the drive to conserve energy and reduce GHG emissions. To address the economic, security, and environmental issues raised earlier in this chapter, industry, governments, and nongovernmental organizations, such as the ISO and the IEA, need to work together to establish energy-efficiency priorities and standards, implement change, and strive for continual improvement.

38.10 Questions for discussion

1. Are there aspects of approaches to energy efficiency that could affect an industry's productivity or profitability either in the short term or in the long term? Explain whether and how gains achieved do or do not override the losses.
2. Which three manufacturing groups have historically been the largest energy consumers within the chemical industry in the USA? (Look this up.) Discuss possible reasons for this historically high consumption and research the strides in energy efficiency made by these groups in recent years.
3. Research how government standards regarding energy efficiency have changed in the last 20 years and discuss the effects on both industry and sustainability.
4. Propose a strategy for improving energy conservation at a new corporation. How might it differ from that of an established corporation?
5. Propose methods of encouraging industrial employees to participate in corporate energy-efficiency efforts.

38.11 Further reading

- **A. McKane**, **P. Scheihing**, and **R. Williams**, 2007, "Certifying industrial energy efficiency performance: aligning management, measurement, and practice to create market value," in *Proceedings of ACEEE Summer Study for Industry*, White Plains, NY.
- **International Energy Agency (IEA)**, 2008, *Energy Technology Perspectives 2008 – Scenarios and Strategies to 2050*, Paris, IEA.
- **International Energy Agency (IEA)**, 2009, *Energy Technology Transitions for Industry Strategies for the Next Industrial Revolution*, Paris, IEA.
- Lawrence Berkeley National Laboratory, Industrial Energy Analysis, http://industrial-energy.lbl.gov/publications.
- US Department of Energy, Energy Efficiency and Renewable Energy, Industrial Technologies Program, Technical Publications, http://www1.eere.energy.gov/industry/bestpractices/technical.html.

38.12 References

[1] **International Energy Agency (IEA)**, *World Wide Trends in Energy and Energy Efficiency*, http://www.iea.org/Papers/2008/Indicators_2008.pdf.

[2] **US Department of Energy (US DOE)**, **US Energy Information Administration (USEIA)**, 2008, *Independent Statistics and Analysis, International Energy Statistics*, http://tonto.eia.doe.gov/cfapps/ipdbproject/iedindex3.cfm?tid=44&pid=44&aid=2&cid=&syid=1995&eyid=2005&unit=QBTU.

[3] **US Department of Energy (US DOE)**, US Energy Information Administration (EIA), 2008, *Annual Energy Review, 2.1a Energy Consumption by Sector*, http://www.eia.doe.gov/aer/pdf/pages/sec2_6.pdf.

[4] **US Department of Energy (US DOE)**, US Energy Information Administration (EIA), 2002, *Annual Energy Review, 2.2 Manufacturing Energy Consumption for All Purposes*, http://www.eia.doe.gov/aer/pdf/pages/sec2_13.pdf.

[5] **E. Worrell**, **D. Phylipsen**, **D. Einstein**, and **N. Martin**, 2000, *Energy Use and Energy Intensity of the U.S. Chemical Industry*, Ernest Orlando Lawrence Berkeley National Laboratory, http://www.energystar.gov/ia/business/industry/industrial_LBNL-44314.pdf.

[6] **International Energy Agency (IEA)**, 2007, *Tracking Industrial Energy Efficiency and CO_2 Emissions*, http://www.iea.org/textbase/nppdf/free/2007/tracking_emissions.pdf.

[7] **E. Piñero**, 2009, "ISO 50001: setting the standard for industrial energy management," *Green Manufacturing News*, http://www.greenmfgnews.com/magazine/summer09/iso.htm.

[8] **International Standards Organization (ISO)**, 2010, *Standards Development*, http://www.iso.org/iso/iso_technical_committee.html?commid=558632.

[9] **A. McKane**, **D. Desai**, **M. Matteini** *et al.*, 2009, "Thinking globally: how ISO 50001 – Energy Management can make industrial energy efficiency standard practice," in *Proceedings of 2009 American Council for an Energy-Efficient Economy (ACEEE) Summer Study on Energy Efficiency in Industry*, Washington, DC.

[10] **American National Standards Institute (ANSI)**, 2010, *NSSN Search Engine for Standards*, http://www.nssn.org.

[11] **American National Standards Institute (ANSI)**, 2010, *International Policy Papers and Charts*, http://www.ansi.org/standards_activities/international_programs/critical_issues.aspx?menuid=3.

[12] **L. Price**, 2008, "Technologies and policies to improve energy efficiency in industry," in *Physics of Sustainable Energy, Using Energy Efficiently and Producing It Renewably*, eds. **D. Hafemeister**, **B. Levi**, **M. Levine**, and **P. Schwartz**, New York, AIP, p. 163.

[13] **D. Gielen**, **J. Newman**, and **M. K. Patel**, 2008, "Reducing industrial energy use and CO_2 emissions: the role of materials science," *MRS Bull.*, **33**, 471.

[14] **J. K. Kissock** and **C. Eger**, 2008, "Measuring industrial energy savings," *Appl. Energy*, **85**, 347–361.

[15] **V. S. Stepanov** and **S. V. Stepanov**, 1998, "Energy efficiencies and environmental impacts of complex industrial technologies," *Energy*, **23**(12), 1083–1088.

[16] **European Commission (EC)**, 2010, *Intelligent Energy eLibrary, Energy Efficiency in Industry*, http://www.iee-library.eu/index.php?option=com_jombib&limit=10&limitstart=0&Itemid=30.

[17] **Energy Star**, 2010, *Guidelines for Energy Management*, http://www.energystar.gov/ia/business/guidelines/assessment_matrix.xls.

[18] **US Department of Energy**, 2010, *Energy Efficiency and Renewable Energy, Industrial Technologies Program, Best Practices, Software*, http://www1.eere.energy.gov/industry/bestpractices/software.html.

[19] **M. Khrushch**, **E. Worrell**, **L. Price**, **N. Martin**, and **D. Einstein**, 1999, "Carbon emissions reduction potential in the US chemicals and pulp and paper industries by applying CHP technologies," in *Proceedings of 2009 American Council for an Energy-Efficient Economy (ACEEE) Summer Study on Energy Efficiency in Industry*, Washington, DC, p. 1.

[20] **US Environmental Protection Agency (USEPA)**, 2010, *Combined Heat and Power Partnership*, http://www.epa.gov/chp/.

[21] **US Department of Energy (US DOE)**, 2008, *The Smart Grid: An Introduction*, available from http://www.oe.energy.gov/SmartGridIntroduction.htm.

[22] 2010, *SuperSmart Grid*, http://www.supersmartgrid.net/.

[23] **A. McKane**, **I. Rhyne**, **A. Lekov**, **L. Thompson**, and **M. A. Piette**, 2009, "Automated demand response: the missing link in the electricity value chain," in *Proceedings of 2009 American Council for an Energy-Efficient Economy Summer Study on Energy Efficiency in Industry*, Washington, DC.

[24] **C. Valsecchi E. Watkins**, **J. Chiavari** *et al.*, 2008, *Intelligent Energy e-Library of Tools and Guidebooks on Energy Efficiency in Industry*, available from http://www.iee-library.eu/index.php?option=com_jombib&limit=10&limitstart=0&Itemid=30, 2008.

[25] **European Bank for Reconstruction and Development (EBRD)**, 2010, *Industrial Energy Efficiency*, http://www.ebrd.com/country/sector/energyef/indust.htm.

[26] **Bundesministerium für Wirtschaft und Technologie (Federal Ministry of Economics and Technology)**, 2010, *Energy*, http://www.bmwi.de/BMWi/Navigation/Energie/energieeinsparung.html.

[27] **IHS, Inc.**, 2010, *Energy*, http://www.ihs.com/energy-solutions/index.htm.

[28] **L. Price** and **A. McKann**, for the United Nations Industrial Development Organization, 2009, *Policies and Measures to Realise Industrial Energy Efficiency and Mitigate Climate Change*, available from http://www.unido.org/index.php?id=1000596.

[29] **United Nations Industrial Development Organization**, 2009, *UNIDO and Energy Efficiency: A Low-Carbon Path for Industry*, available from http://www.unido.org/index.php?id=1000596.

[30] **A. McKane**, **L. Price**, and **S. de la Rue du Can**, for the United Nations Industrial Development Organization, 2008, *Policies for Promoting Industrial Energy Efficiency in Developing Countries and Transition Economies*, available from http://www.unido.org/index.php?id=1000596.

[31] **American Chemistry Council (ACC)**, 2010, *Responsible Care*, http://www.americanchemistry.com/s_responsiblecare/sec.asp?CID=1298&DID=4841.

[32] **Industrial Energy Management Information Center**, 2010, *Energy Star*, http://www.energystar.gov/index.cfm?c=industry.bus_industry_info_center.

[33] **US Department of Energy (US DOE)**, 2010, *Energy Efficiency and Renewable Energy, Save Energy Now*, http://www1.eere.energy.gov/industry/saveenergynow/.

[34] **US Department of Energy (US DOE)**, 2010, *Energy Efficiency and Renewable Energy, Save Energy Now*, http://www1.eere.energy.gov/industry/saveenergynow/pdfs/42460.pdf.

[35] **Energy Star**, 2009, http://www.energystar.gov/ia/partners/pt_awards/2009_profiles_in_leadership.pdf.

[36] **Texas Industries of the Future**, 2007, http://texasiofces.utexas.edu/PDF/Documents_Presentations/Case_Studies/NOx%20Case%20Study%20No2%20Bayer%20Final.pdf.

[37] **Alliance to Save Energy (ASE)**, 2009, *2009 Stars of Energy Efficiency Award Winners*, http://ase.org/content/article/detail/5687.

[38] **Energy Star**, 2008, http://www.energystar.gov/ia/partners/pt_awards/2008_profiles_in_leadership.pdf.

[39] **US Department of Energy (US DOE)**, 2010, *Energy Efficiency and Renewable Energy, Save Energy Now*,

Leader Company Listings, http://www1.eere.energy. gov/industry/saveenergynow/leader_companies.html.

[40] **W. R. Prindle**, 2010, *From Shop Floor to Top Floor: Best Business Practices in Energy Efficiency*, prepared for the Pew Center on Global Climate Change and available

from http://www.pewclimate.org/energy-efficiency/ corporate-energy-efficiency-report, pp. 72–84.

[41] **M. Huesemann**, 2004, "The failure of eco-efficiency to guarantee sustainability: future challenges for industrial ecology," *Environmental Prog.*, **23**(4), 264–270.

39 Green processing: catalysis

Ronny Neumann

The Weizmann Institute of Science, Rehovot, Israel

39.1 Focus

The chemical industry is intimately linked with the realities of transformation of fossil fuels to useful compounds and materials, energy consumption, and environmental sustainability. The questions that concern us are the following: how can one minimize energy consumption in the chemical industry and reduce waste formation in chemical reactions, and can we transform this industry from one based primarily on petroleum to one that utilizes renewable feedstocks? In this chapter we will see how this "greening" of this major industrial and energy sector will require the use of catalysis and development of new catalysts.

39.2 Synopsis

Thomas Jefferson stated that "the earth belongs in usufruct to the living." This premise is the basis of present recognition that, if the natural capacity of planet Earth to deal with pollution and waste is exceeded, then our lifestyle will become unsustainable. In this context, it is difficult to imagine life in the twenty-first century without accounting for the role the chemical industry plays in our modern society. For example, in the transportation sector, the most obvious aspect is the production of efficient fuels from petroleum, but one also should consider catalytic converters that reduce toxic emissions, and engineering polymers and plastics that reduce vehicle weight and therefore reduce fuel consumption. In daily consumer life we use chemicals in products such as paints, DVDs, synthetic carpets, refrigerants, packaging, inks and toners, liquid-crystal displays, and synthetic fibers. Pesticides and fertilizers are needed in order to increase agricultural productivity and pharmaceuticals to keep us healthy. In this chapter we will try to understand the aspects involved in the sustainability of the chemical industry. We will describe how to measure and control environmental performance and then define what we mean by green processing in the chemical industry, specifying green process metrics, introducing key concepts such as atom economy, E-factors, and effective mass yields. With these concepts together with the evaluation raw material costs, waste treatment, and unit processes needed for production of a chemical one can appreciate the "greenness" of a chemical process. The second part of this chapter will describe the role catalysis has in chemical transformations and how catalysis is used to reduce energy consumption and waste formation, together leading to increased sustainability. Examples will be given for a spectrum of applications ranging over oil refining, ammonia production, manufacture of important chemical intermediates and materials, and use of catalysis for the production of commodity chemicals. Finally, we will discuss the role catalysis may play in the replacement of fossil-fuel feedstocks with renewable ones, and how catalysis may contribute to our search for solar fuels.

39.3 Historical perspective

Organic chemistry has its roots as a discipline in the early nineteenth century. In those times, compounds were considered too complicated to be synthesized, and, according to the theory of vitalism, such matter had a "vital force" and was considered "organic" in nature, thus the term organic chemistry. Chevreul in c. 1816 was probably the first to synthesize organic compounds, making soaps by reacting fats and bases or alkalis, but the first real organic synthesis marking the turning point is commonly credited to Wöhler, who in 1828 prepared urea from ammonium cyanate and essentially debunked vitalism. The industrial revolution in Europe yielded the first industrial-scale organic synthesis in 1856, when Perkin, wishing to prepare quinine, an anti-malarial natural compound, prepared instead a dye called Perkin's mauve, whose structure rather ironically was deciphered only in 1994. Such syntheses and others revolutionized the clothing industry.

Industrial manufacture of organic compounds jumped significantly starting from the mid twentieth century upon the exploitation of petroleum, coal, and natural gas, first as a source of energy and then through the application of "catalysis" developed into the petrochemical industry. The availability of petrochemicals led to an almost infinite tree of compounds and materials, leading to the development of the polymer, fabrics, and plastic industries, the agrotech industry, the pharmaceutical industry, and other fine chemicals important for our high-tech information society and high standard of living (Figure 39.1).

Thus, by 1980 organic synthesis, together with catalysis, a term coined by Jöns Jacob Berzelius, also in the early nineteenth century, had led to a mature chemical industry of enormous economic and societal importance. In this process, as was common throughout the industrial revolution, environmental dangers were either not recognized or not considered to be a problem. This turned out to be a rueful mistake, but eventually also led to a significant change in chemistry and the chemical industry. A prime example was the use of freons or chorofluorocarbons as practically the only viable refrigerants and propellants, which turned out, however, to cause catastrophic depletion of the Earth's ozone layer. Quite miraculously, the Montreal Protocol, which was signed in 1987, led to the relatively fast phasing out of freons on the one hand and development of environmentally friendly alternatives on the other. Seemingly the perfect solution had been found – quality of life had been sustained without additional environmental or economic damage.

This example and many others have led to a continuing environmental "clean up," first in the developed nations (the USA, Western Europe, and Japan) and later in the rest of the world. Abatement, however, is not a root solution but rather a correction, and the objective of modern "green processing" is to avoid environmental problems at their core by the production of environmentally friendly products and/or avoidance of the formation of waste, which is often achieved using catalysis.

39.4 Green processing

39.4.1 The need to measure greenness

One of the more difficult problems in the development of green processes is how to measure greenness. In some

Figure 39.1. Mega-industrial sites such as the Ludwigshafen compound of BASF are a physical, tangible manifestation of the importance of the chemical industry.

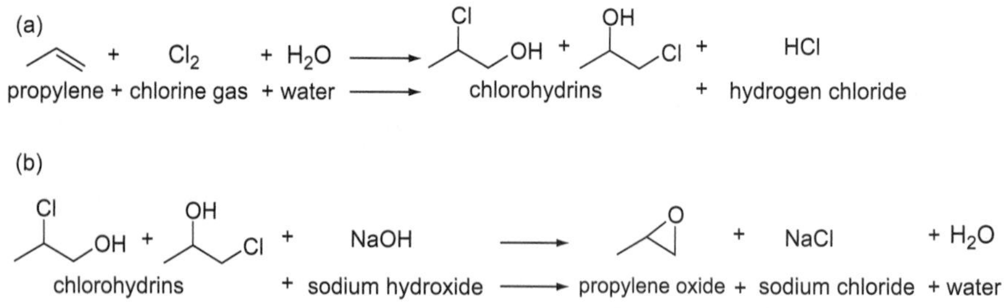

Figure 39.2. The chlorohydrin process for production of propylene oxide.

instances it may be obvious that an improvement has been made *vis-à-vis* a predecessor technology, but in other cases the situation is more complicated. Let us take, for example, the case of the manufacture of propylene oxide, an important commodity chemical. The classic approach, which, as of 2005, accounted for almost half of the worldwide production, is synthesis through a chlorohydrin process as shown in Figure 39.2 [1]. As can be seen from the scheme, the preparation of propylene oxide by the chlorohydrin pathway involves the use of toxic chlorine gas (Cl_2), which eventually is not included in the product itself. If a base such as calcium hydroxide (lime) is used, then calcium chloride is a waste product of this process. On the other hand, if sodium hydroxide (caustic soda) is used as base, as shown in Figure 39.2, then, albeit making the process more expensive, the waste product sodium chloride can be recycled by electrolysis back to chlorine gas and sodium hydroxide. The process in itself is not energy-intensive and can be carried out at ambient temperatures, although the electrolytic production of chlorine and caustic soda from sodium chloride is very energy-intensive. Thus, by integrating chlorine, caustic soda, and propylene oxide production, one can mitigate the apparently non-productive use of chlorine in the process. However, what is hidden from the eye in Figure 39.2 is the fact that the reaction of propylene with chlorinated water, besides yielding the required chlorohydrins, also produces very significant amounts of byproducts, namely 1,2-dichloropropane (6%–9%) and bischlorodi-*iso*-propyl ether (1%–3%), which have hardly any commercial value and thus amount to a major quantity of organic waste that needs to be treated and abated, typically by incineration.

In order to avoid the use of chlorine and the formation of 1,2-chloropropane waste, an inherently obvious alternative is to oxidize propylene with oxygen, O_2. Such direct procedures are not yet available, so indirect ones are applied, as described in Figure 39.3.

As can be seen from Figure 39.3, the pathway here to propylene oxide involves the essential co-production of methyl-*tert*-butyl ether (MTBE) without which the process is not economical or sustainable [1]. Methyl-*tert*-

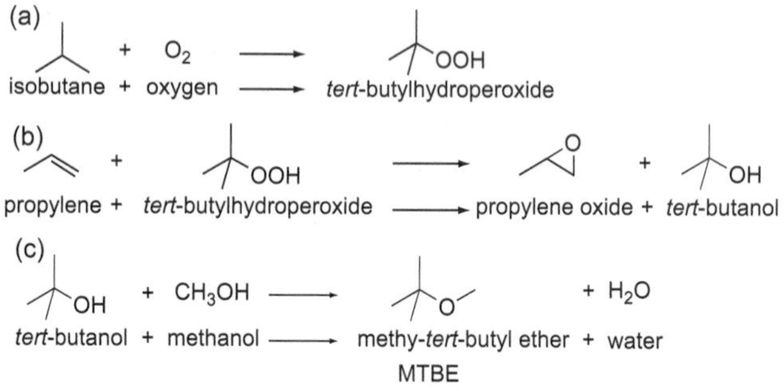

Figure 39.3. The oxidation pathway for propylene oxide production with methyl-*tert*-butyl ether (MTBE) as co-product.

butyl ether was introduced into gasoline in the 1970s in the USA to replace the noxious tetraethyl lead which had been banned and increase the octane rating of gasoline. Apparently, a greener alternative to the chlorohydrin process had been obtained. As it turned out 20 or so years later, however, MTBE, began turning up in drinking water and its use in gasoline is now being phased out [2]. A perceived green solution turned out to be not so green, and new avenues for propylene oxide production were required. Much of the plant capacity for MTBE production can be refitted for a similar process using ethylbenzene instead of isobutene as co-substrate, forming styrene, an important monomer for polymers, as co-product. The ethylbenzene hydroperoxide (EBH) yield is low and large amounts of ethylbenzene need to be recycled. Schematic flow charts for the chlorohydrin process (simple, but waste-forming) and the ethylbenzene process (complicated, but cleaner) are shown in Figure 39.4. The above narrative, although representative of problems in defining green processing, is qualitative by nature, and, in order to quantify green processes, several green metrics have been introduced.

39.4.2 Atom economy

Atom economy is a measure of how many atoms of the reactants end up in the product; the higher the atom economy the less by products and waste formed in the reaction [3]. In this way we can compare the atom economy for different reactions, for example the two

Figure 39.4. Flow charts for the chlorohydrin process (top) and ethylbenzene co-substrate process (bottom) for propylene oxide manufacture.

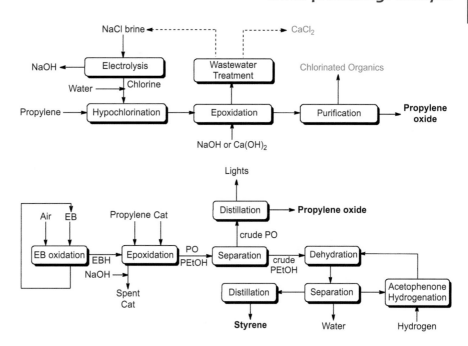

synteses of propylene oxide only as shown in Figure 39.2 and Figure 39.3:

$$C_3H_6 + Cl_2 + NaOH \longrightarrow C_3H_6O \qquad C_3H_6 + C_4H_{10} + O_2 \longrightarrow C_3H_6O$$
$$\quad 42 \quad 70.9 \quad 40 \qquad\qquad 58 \qquad\quad 42 \quad 58 \quad 32 \qquad\qquad 58$$
$$\qquad\quad 58/152.9 = 37.9\% \qquad\qquad\qquad 58/132 = 43.9\%$$

$$(39.1)$$

It should be noted that, if MTBE is included as a desirable co-product to propylene oxide, the atom efficiency for two products is much higher, $132/164 = 80.5\%$.

39.4.3 The *E*-factor, environmental quotient, and effective mass yield

An *E*-factor can be defined as the weight ratio of waste byproduct to product [4]. As an industry, oil refining has a low *E*-factor of about 0.1; however, because the scale of oil refining is about 10^8 tons of final products per year, it also produces 10^6 tons of waste annually. At the other end of the spectrum, the productivity of the highly specialized pharmaceutical industry is much lower, only about 1,000 tons per year of final products. However, since the *E*-factor for this industry is high, estimated to be up to 100, the amount of waste produced is significant, 10^5 tons per year. Waste generation, of course, has economic, environmental, and social consequences. Economically, waste generation effects direct production costs in terms of loss of raw materials and wasted energy. Indirect costs include waste treatment and disposal. The need for environmental responsibility, with or without governmental legislation, also affects the sustainability of an industry. For example, the ever-increasing legislation to decrease the amount of sulfur

in fuel that forms sulfur oxides, SO_x, leading to acid rain and smog upon combustion requires more and more complicated and expensive processes for the removal of sulfur-containing compounds from fuel [5]. Often, however, as we shall see later on in the chapter, environmental pressures can lead to the development of improved and more profitable industrial processes, to the advantage of both industry and society.

The use of the idea of an *E*-factor alone has the disadvantage that all waste is considered to be equally problematic, while it is obvious that this is not the case. Waste sodium chloride is much less toxic than, for example, its heavy-metal analog, mercuric chloride. Thus, the concept of an environmental quotient has been suggested: this is the product of the *E*-factor and the "unfriendliness" of the byproduct, *Q*, which can range from 1 (sodium choride) to 100 (mercuric chloride) [6]. There is, however, little consensus on the values of *Q* for various byproducts; the concept is viewed as being too subjective and has not been widely adopted.

Since the *E*-factor takes into account all waste, the production of benign waste such as water, ethanol, small amounts of nontoxic inorganic salts, and benign gases can make the environmental impact as expressed by the *E*-factor appear to be higher than is justifiable. In order to compensate for this "benign-waste effect" the concept of effective mass yield (EMY) has been introduced [7]. The EMY approximates to the reciprocal of the *E*-factor, but does not take into account benign waste. The application of the concepts of the *E*-factor and EMY requires more details concerning a process such as the chlorohydrin pathway to propylene oxide described in Figure 39.2. Thus, an optimal process (100% propylene

Table 39.1. A summary of metrics for "greenness"

	Green	Non-green	Chlorohydrin (lime)	Ethylbenzene (byproduct)
Atom economy	High	Low	Low	High
E-factor	Low	High	Medium–high	Low
EMY	High	Low	Low	High

conversion, 90% selectivity) will involve first the reaction of 1 mol (42 g) propylene with 1 mol Cl_2 (70.9 g) in 1 l of water to yield 0.9 mol (85 g) chlorohydrin and 0.1 mol (11.3 g) 1,2-dichloropropane. The solution is treated with 1.8 mol (72 g) sodium hydroxide to yield 0.9 mol (52.2 g) propylene oxide and 1.8 mol (105.2 g) sodium chloride. Calculation of the E-factor (g waste/g product = 116.5/52.2) yields a value of 2.2, which is rather high for a bulk chemical. The environmental quotient would be rather high due to the somewhat toxic 1,2-dichloropropane and the high concentration of sodium chloride. The EMY would be 42/184.9 or 22.7%, which is also low, thus being indicative of a "non-green" process. If, however, one takes into account recycling of the sodium chloride waste then the recovery-adjusted E-factor is much lower, 0.22. Considering the significant scale of production of propylene oxide, there is the formation of a non-trivial amount of about 500–1,000 tons per year of 1,2-dichloropropane.

See Table 39.1 for a summary of these metrics for the example considered.

39.4.4 Energy and capital cost

Finally, it should be emphasized that what has been presented above aids in the measure of the greenness of a process from the pure-chemistry point of view, but this approach is still far from being a complete description of all factors involved. The first major factor to consider is the energy costs of processes including those needed for preparation of substrates. For example, the chlorohydrin process can be carried out under ambient conditions, that is, there is no net energy consumption, but the electrolysis of sodium chloride brine to make chlorine is energy intensive, 2,500–3,000 kWh t^{-1}. The ethylbenzene process is run at higher temperature, is a multi-step process that also requires high-temperature distillation, and thus is much more heat-intensive. Another factor to take into account is that chemical plants are expensive to build, with high capital costs. They are also difficult to refit, often being technology-specific. Thus, transfer to new greener technologies is rarely a "drop-in solution" – new plants need to be built, making the economic barrier to the introduction of new

green technologies often very high, despite possible green benefits.

39.5 Catalysis in fossil-fuel-based chemistry

Catalysis is one of the most important tools with which to achieve environmental sustainability in the chemical industry. In this section we will limit ourselves to questions more related to basic chemicals and use of non-renewable fossil fuels. Catalysis in the synthesis of finer chemicals and pharmaceuticals is beyond the scope of this chapter, although, broadly speaking, the use of catalysis in this area is still relatively minor due to problems of catalyst expense and recovery. The textbook definition of catalysis is the increase in rate of a chemical reaction due to the participation of a substance called a catalyst that, unlike the other substrates that participate in the chemical reaction, is not consumed by the reaction itself. There are two very important corollaries to this definition, which are in many respects more important than the ability of catalysts to increase rate. The first consequence is that catalysis can be used to decrease the temperature needed to carry out a reaction, which translates into the saving of energy. The second result is that often catalysts can be used to control and improve the selectivity of a chemical reaction, irrespective of whether the catalyst significantly increases the rate or not. The increased selectivity translates into reactions with lower E-factors and EMYs. Often the combination of these two corollaries will allow alternative transformations with benign reagents to yield desired products, which, although thermodynamically possible, are in fact not feasible without catalysis. These additional properties of catalysis are eminently obvious in the catalytic oxidation of hydrocarbons with molecular oxygen, O_2. When hydrocarbons react with oxygen the thermodynamically preferred result is total combustion to carbon dioxide and water. In fact, at high temperatures it is difficult to avoid combustion of hydrocarbons. Therefore, an important use of catalysts in this area is the ability, by virtue of the reduction of reaction temperature and through control of the reaction pathway, to direct reactions to oxygenated hydrocarbons that are less stable than carbon dioxide. The societal importance of catalytic processes cannot be understated, and therefore there are very numerous examples of how catalysis is used in our present economy. It has been estimated that at least 90% of the chemicals used for any purpose have at some stage of their manufacture been involved in a catalytic transformation.

39.5.1 Petroleum oil refining

A primary example of the importance of catalysis is the refining of crude oil or petroleum [8]. This is a complex process whereby oil having a broad range of hydrocarbons with a broad range of boiling points is divided into a range of products such as low-pressure gases (LPG) and butanes, gasoline, jet fuel and kerosene, diesel oil, fuel oil, asphalt, petroleum coke, and sulfur for sulfuric acid. This is achieved by a combination of distillation and chemical processes. After initial removal of salts and distillation, light and heavy naphtha are treated with hydrogen (hydrotreated) in a hydrodesulfurization reaction to remove sulfur as hydrogen sulfide gas. The heavier desulfurized fractions, termed heavy naphtha or heavy gas oil, with an average boiling point of 340 °C, are then reacted in fluid catalytic cracker (FCC) units that split heavier-relative-molecular-mass fractions into lighter, more valuable fractions. Fluid catalytic cracking is needed to correct the imbalance between the demand for gasoline and the excess of heavy, high-boiling-range products resulting from the initial distillation process. There exist upwards of 400 FCC units worldwide and about one-third of all crude oil passes through such a unit. In numbers this translates into more than 10 million barrels or nearly 20 billion liters per day. A modern FCC catalyst has three major components: (1) a crystalline zeolite, faujasite or type Y, which is the active catalytic component and acts as a strong Bronsted and Lewis acid to affect the cracking reaction; (2) an alumina matrix that is needed to "dilute" the activity of the zeolite catalyst; and (3) the binder and filler components, which provide the physical strength and integrity of the catalyst. The binder is usually a silica sol and the filler is usually a clay such as kaolin. An FCC unit has both reactor and regenerator parts, since the catalyst is easily fouled by the formation of petroleum coke, which needs to be burned off in the regenerator. The burning of this residue has its energy advantages since it is an exothermic reaction, whose excess heat is used to drive the endothermic cracking reaction itself. Altogether about 5 kg of catalyst per kg feedstock is needed in order to carry out catalytic cracking. This amounts to a staggering amount of catalyst, namely 100 billion tons of catalyst used worldwide in this process alone. Fluid catalytic cracker units typically produce gasoline with low octane ratings. Octane ratings are raised by a process called catalytic re-forming by restructuring the hydrocarbons

Steam Re-forming $CH_4 + H_2O \xrightarrow[700-1000\,°C]{Ni} CO + 3H_2$

Water-Gas Shift $CO + H_2O \xrightarrow[200\,°C]{Cu/ZnO} CO_2 + H_2$

Ammonia Synthesis $N_2 + 3H_2 \xrightarrow[500\,°C]{Fe} 2NH_3$

Figure 39.5. Reactions involved in ammonia synthesis.

through cyclization, isomerization from linear to branched alkanes, and dehydrogenation to aromatic components. Hydrogen is a significant byproduct of these catalytic reactions and is used in other units such as for hydrodesulfurization and hydrocracking, a form of catalytic cracking in the presence of hydrogen. Typical catalysts consist of platinum or rhenium on an alumina–silica matrix. The metals enable dehydrogenation reactions while the acidic matrix also catalyzes cracking and isomerization reactions.

39.5.2 Ammonia synthesis

The well-known Haber–Bosch process employed to produce ammonia from nitrogen and hydrogen was developed by Fritz Haber and Carl Bosch in 1909–1910 [9]. Originally, ammonia was needed almost exclusively for the production of explosives through its oxidation to nitric acid, although today about 80%–85% of the approximately 150 million tons produced annually is used as fertilizers, mostly ammonia, but also ammonium nitrate and other ammonium salts, and urea. The rest is used to make other nitrogen-based compounds, notably nitric acid. A modern ammonia-synthesis plant can be divided into two parts, hydrogen production and formation of ammonia by reaction of nitrogen with hydrogen (the Haber–Bosch process). In principle, hydrogen can be produced by electrolysis of water, although, in the current economic climate, this is too expensive. Instead, hydrogen is generally produced by steam re-forming of methane at 700–1,000 °C over a nickel catalyst, Figure 39.5, to yield carbon monoxide, CO, and hydrogen, H_2 [10]. Other fossil fuels such as LPG, coal, and coke are also possible feedstocks in processes that are similar and called gasification. The carbon monoxide is further reacted in a water-gas shift reaction over an iron oxide catalyst, Figure 39.5, to yield additional hydrogen and carbon dioxide. Since the ammonia synthesis is extremely sensitive to carbon monoxide and carbon dioxide, they must be stringently removed; the latter, which is a major component, through adsorption or reaction with triethanol amine. The reaction of nitrogen, N_2, and hydrogen, $3H_2$, to give ammonia, $2NH_3$, is an equilibrium reaction that lies to the left at low pressure. According to Le Chatelier's

Figure 39.6. Immediate derivatives, or the ethylene tree, formed by further catalytic reactions of ethylene.

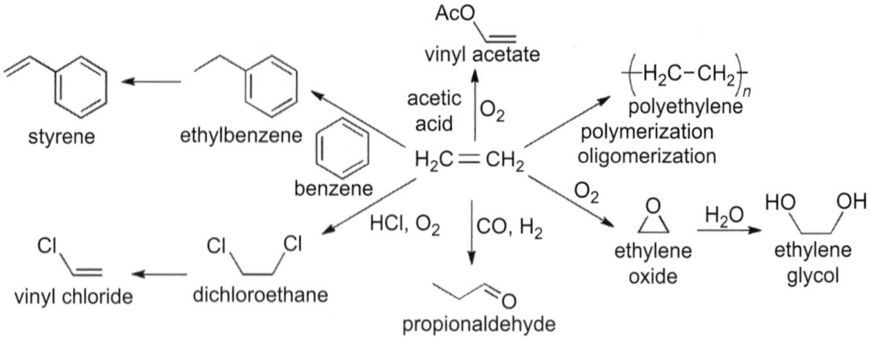

Figure 39.7. Immediate derivatives, or the propylene tree, formed by further catalytic reaction of propylene.

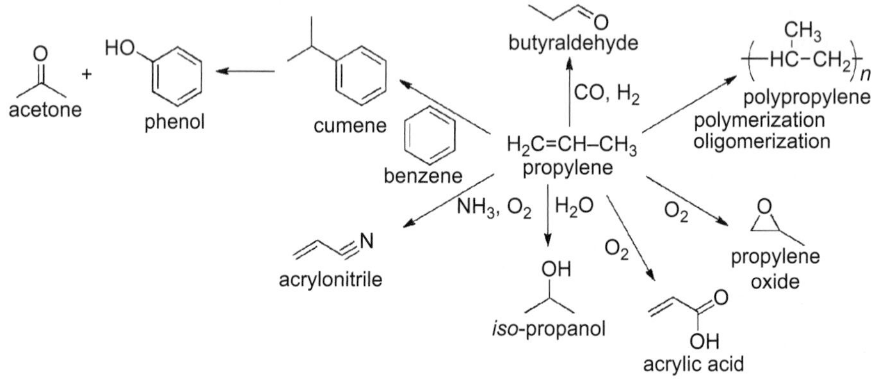

principle, the equilibrium, $K = P_{NH_3}^2 / P_{N_2} P_{H_2}^3$, can be shifted to the right under high nitrogen and hydrogen pressures of 200 bar, and the maximum yield under typical feasible catalytic conditions at 500 °C, over a porous iron catalyst, can reach approximately 15%. The process is run over several catalyst beds in series and can be engineered to obtain high ammonia yields. Importantly in the context of energy consumption, the production of ammonia is energy-intensive. It has been estimated that 1%–2% of the world's energy supply and 2%–3% of the world supply of methane goes into manufacturing ammonia. The latter may drop in relative, but not absolute, terms due to increased use of methane in power plants. Notably, while the Haber cycle itself can be considered green, with an atom economy of 100%, and a near-zero *E*-factor, the production of hydrogen by steam re-forming with formation of significant amounts of CO_2 is very un-green, with an atom economy of only 15.4% and an optimal *E*-factor of 5.5, which is very high for a basic chemical. The use of water as a benign reagent leads to a slightly more respectable EMY of 50%, which is, however, quite low for such large-scale synthesis.

39.5.3 Alkenes and their derivatives products

Alkenes represent a significant portion of the basic petrochemicals, together with aromatics such as benzene, toluene, and xylene that are used to make a myriad of further derivative compounds and materials. As will become clear from what is presented below, much of our material world is based on alkenes and these aromatic compounds. Therefore, it is evident that the production of our everyday products and materials, which we practically take for granted, is very dependent on fossil fuels. The organic compound which is synthetically prepared in the largest volumes is ethylene, $CH_2=CH_2$. Annual production probably exceeds 120 million tons. Propylene, $CH_2=CHCH_3$, is also produced in significant amounts, about 50% that of the ethylene production. Both are produced in crackers, more specifically from light naphtha and LPG by steam cracking of the feed to 750–900 °C followed by a fast temperature quench and separation of other hydrocarbons and hydrogen and gas compression [11]. The production of ethylene is net energy-intensive, requiring an estimated 10 GW on a world scale.

The various further reactions of ethylene and propylene are presented in Figures 39.6 and Figure 39.7, respectively. Polyethylene is the largest-volume synthetic polymer and plastic, and the major use of ethylene is in polymerization. Depending on how ethylene is polymerized, its properties can vary very significantly. Thus, oligomers or α-linear alkenes are used as precursors in the preparation of detergents, plasticizers, and synthetic lubricants [12]. Low-density polyethylene

(LDPE), so called because the polymer is branched and of lower density, is made by energy-intensive (very high pressure and high temperature), non-catalytic, free-radical polymerization [13]. It is used for both rigid containers and plastic film applications such as plastic bags and film wrap. High-density polyethylene (HDPE) is produced by using chromium/silica catalysts, Ziegler–Natta catalysts, or metallocene catalysts [14]. The lack of branching is ensured by an appropriate choice of catalyst and reaction conditions. Global HDPE consumption has reached a volume of more than 30 million tons and is similar to that of LDPE. Oxidation of ethylene in the presence of silver catalysts leads to the formation of ethylene oxide [15], which, upon oligomerization or polymerization, is used in surfactants and detergents, while its hydrolysis yields ethylene glycol, which is mostly used as automotive antifreeze. Oxidative chlorination of ethylene with hydrogen chloride gas and oxygen, followed by hydrodechlorination to vinyl chloride monomer, and polymerization yields polyvinyl chloride (PVC) [16]. Polyvinyl chloride is the third-most-common synthetic polymer, and is widely used in construction. In the presence of plasticizers, typically phthalate esters, it becomes soft and malleable, and is used in upholstery, flexible hoses and electrical cable insulation, and floors. The presence of chlorine reduces the greenness of PVC, and the safety of phthalate plasticizers has also been questioned. Oxidation of ethylene in the presence of palladium catalysts and acetic acid leads to the formation of vinyl acetates [17]. Polyvinyl acetate is mostly used as an adhesive, especially of porous materials, while the hydrolysis yields polyvinyl alcohol, which is used as a fiber, and, after reaction with butanal, yields an adhesive resin with high toughness and flexibility that is used mostly in laminated safety glass, most notably in cars. Catalytic alkylation of benzene with ethylene yields ethylbenzene, which, after high-temperature dehydrogenation, yields styrene, which is used to make styrene polymers and copolymers [18]. Uses of these materials include synthetic rubber, plastic insulators, pipes, car parts, and food containers. As for the products made from ethylene, materials derived from propylene such as polypropylene, phenol, and acrylic acid, among many more, are pervasively present in many materials.

39.6 Chemicals from non-petroleum and renewable resources

Considering that oil is almost exclusively used for the synthesis of transportation fuels (gasoline, kerosene, and diesel), and petrochemicals and commodity chemicals it is important at this stage to consider alternatives to a petroleum-based chemical society after its depletion after a yet-to-be-determined time. It is common to

Combustion	$C + x\,O_2 \longrightarrow CO + CO_2$
Gasification	$C + H_2O \longrightarrow CO + H_2$
Water-Gas Shift	$CO + H_2O \longrightarrow CO_2 + H_2$
Methanol Synthesis	$CO + CO_2 + x\,H_2 \longrightarrow CH_3OH$
Fischer–Tropsch	$x\,CO + (2x+1)\,H_2 \longrightarrow C_xH_{2x+2} + x\,H_2O$

Figure 39.8. Pathways from carbon sources to methanol and gasoline.

separate the problems associated with alternative sources of transportation fuels and petrochemicals due to their different scale and a different sense of urgency, but in part the boundaries of these areas may be artificially blurred. One approach to this question has been to propose the development of a coal/methanol-based fuel and chemical society. The reasoning behind this approach is that such a society is in fact immediately available through the use of proven technologies. The basis of these technologies is a series of catalytic reactions, Figure 39.8. First gasification of coal, with large proven reserves, at elevated temperatures can be used to form a mixture of carbon monoxide and hydrogen, termed also synthesis gas, or syn gas [19]. Gasification is, however, a combination of three main reactions: pyrolysis for the removal of volatiles adsorbed in coal, whereby about 70% by weight of the coal is lost; partial combustion of carbon to carbon monoxide and carbon dioxide; and the actual gasification that forms synthesis gas. Synthesis gas can be then used to make methanol using a catalyst of copper and zinc oxide on alumina at 50–100 bar at 250 °C. If the synthesis gas is rich in hydrogen, the addition of carbon dioxide can used also to form methanol from carbon dioxide. Methanol can be directly used as a fuel, or alternatively can be used to make gasoline in what is essentially a reverse cracking procedure using acid catalysis [20][21]. Another option is to form transportation fuel via Fischer–Tropsch reactions directly from synthesis gas using cobalt- or iron-based catalysts at up to 300 °C [22]. This technology was used extensively by Germany during World War II and later on in South Africa. Nowadays Fischer–Tropsch chemistry is usually applied using cheaper synthesis gas derived from natural gas. Supplementing Fischer–Tropsch-chemistry processes with additional reactions to yield alkenes and aromatics, although less developed, is certainly a viable option to allow the production of materials needed for a modern society. The major deterrent to this approach is that this coal/methanol-based chemistry is energy-intensive and a net producer of a considerable amount of carbon dioxide. In this context it should be noted that gasification of biomass and organic waste is similar to that of coal and is also possible.

Although at first glance this might appear to be an attractive proposal for a use of a renewable resource, it should be emphasized that the actual energy content of even dried biomass is only 14–17 GJ per ton. This is about one-third that of oil. In addition, the significantly higher oxygen content of biomass versus coal will lead to even greater carbon dioxide formation.

It should also be noted that biomass and certain organic wastes can be treated by anaerobic digestion, a complex series of biological (enzymatic) reactions that leads to the formation of methane and carbon dioxide [23]. This is an interesting option for use as a source for electricity, especially in smaller-scale applications, for example in remote areas. As noted above, methane in non-power-plant applications is generally used to form synthesis gas by steam re-forming, which can be followed by methanol synthesis. In this area a low-temperature direct selective oxidation of methane to methanol has been a long-standing goal in catalysis in order to avoid the high energy costs of methanol synthesis via synthesis gas [24]. However, this problem has been shown to be rather intractable and has eluded a solution. Surely new ideas and more research are still needed here.

39.6.1 Chemicals from bioethanol

As noted in Chapters 25 and 26, there is much interest in the possible use of ethanol from renewable resources as a replacement for gasoline. What has been less discussed is the idea of a bioethanol as a green resource for the production of petrochemicals and intermediates. Consideration of known catalytic processes reveals that such a bioethanol chemistry tree is certainly a viable entrance into the present basic petrochemical mixture, Figure 39.9. Thus, ethanol can be dehydrated to yield

ethylene by acid catalysis [25]. Beyond the immediate availability of ethylene, dimerization to yield butene [26], followed by metathesis, will yield propylene [27]. Benzene can be made available by dehydrogenation of ethylene to yield acetylene, which can be cyclotrimerized to yield benzene [28]. Ethanol can also be oxidized to acetic acid [29], an important compound made at present by carbonylation of methanol [30], which is available, as stated, from natural gas transformed to synthesis gas. It should be noted that such a chemical tree would be attainable through low-temperature reactions, and the atom economy and the EMY are high since no detrimental waste is formed.

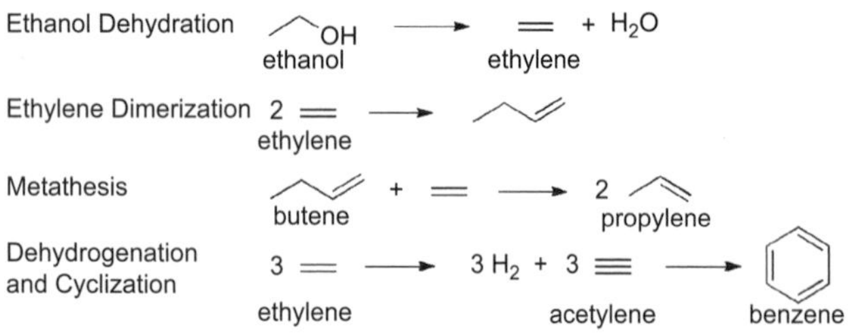

39.6.2 α-Hydroxycarboxylic acids and their derivatives

The most important α-hydroxycarboxylic acid that is produced industrially is lactic acid, $CH_3CH(OH)COOH$. At present there are both petroleum and renewable resources. The former is produced by hydrocyanation of acetaldehyde, while the later is produced by fermentation of cornstarch using *Lactobacillus acidophilus* bacteria. It would appear that recent process improvements have made the cornstarch route highly favorable and the major pathway used [31]. There are two major uses of lactic acid, one is as the ethyl ester, which is gaining considerable interest as a biodegradable industrial solvent [32]. The second use is to form polylactic acid [33]. Typically, lactic acid is cyclodehydrated by tin oxide to form L-lactide, D-lactide, and *meso*-lactide from the racemic starting lactic acid. The isomer ratio can be controlled by distillation; the unwanted lactides are hydrolyzed back to racemic lactic acid. The specific isomer ratio of the lactide determines the properties of the polylactic acid produced, Figure 39.10, which then can be applied for various applications ranging from packaging materials to bottles and fibers, all of which are biodegradable. Although polylactic acid is at present more expensive than petroleum-based polymers, perhaps in the future it will be able to compete better with polyalkenes and others.

There is also significant interest in other polyhydroxy acids due to the fact that some such polymers have properties similar to those of polyethylene and polypropylene. To date there are no large-scale, commercially viable sources of the required α-hydroxycarboxylic acids; however, it can be expected that future developments in genetic engineering may lead to such sources.

Figure 39.9. A route from ethanol to basic petrochemicals such as ethylene, propylene, and benzene.

Figure 39.10. Synthesis of polylactic acid via lactide isomers.

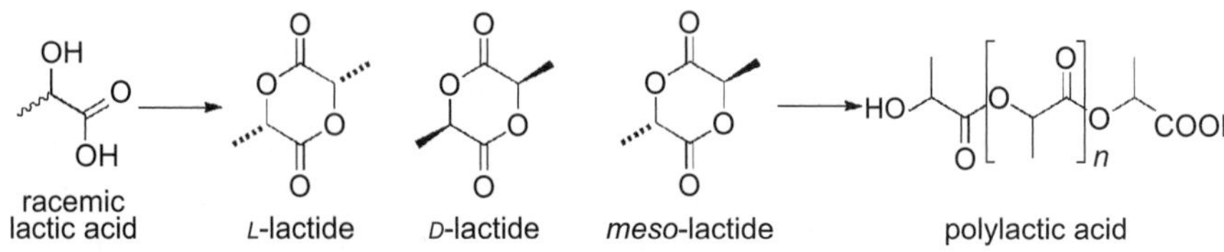

Figure 39.11. Transformations of fatty acids to synthetic products from renewable resources.

$$C_nH_{2n+1}COOR \longrightarrow C_nH_{2n+1}CH_2OH$$

$$C_nH_{2n+1}CH=CHC_mH_{2m+1}COOH \longrightarrow C_nH_{2n+1}COOH + HOOCC_mH_{2m+1}COOH$$

$$C_nH_{2n+1}CH=CHC_mH_{2m+1}COOH + CH_2=CH_2 \longrightarrow C_nH_{2n+1}CH+CH_2 + H_2C=CHC_mH_{2m+1}COOH$$

$$H_2C=CHC_mH_{2m+1}COOH \longrightarrow BrH_2CCH_2C_mH_{2m+1}COOH \longrightarrow H_2NH_2CCH_2C_mH_{2m+1}COOH$$

39.6.3 Chemicals from triglycerides and fatty acids

Currently more than 100 million tons of oils and fats are produced annually, and around 85% of this production goes into animal feeds and human foods. The remainder is available for use for the non-food chemistry industry. Modern genetic engineering methods also allow one to bring about a considerable variation of the fatty-acid components of triglycerides. Fatty acids are prepared mostly by acid-catalyzed hydrolysis of triglycerides, which, after separation from glycerol, are separated into fractions by vacuum distillation [34]. Unsaturated acids can be separated from saturated ones, for example oleic acid from stearic acid, by solvent or supercritical fluid extraction [35]. The major transformation of fatty acids, especially the saturated ones, is reduction to the fatty alcohol; further functionalization leads to formation as surfactants [38]. Unsaturated fatty acids may also be functionalized in other ways, Figure 39.11. For example, cleavage of the double bond by ozone or preferably hydrogen peroxide in the presence of catalyst yields a shorter-chain fatty acid [37], typically with an odd number of carbon atoms, such as a nonanoic acid, which can be used as a lubricant. The diacids formed, such as azaleic acid (C_{11}) and brassylic acid (C_{13}), are good materials for nylon 11,11 and nylon 13,13. Metathesis reactions of fatty acids with ethylene lead to short alkenes and fatty acids with terminal double bonds [27]. The latter may be functionalized by bromination, followed by amine substitution to yield compounds such as 11-aminoundecanoic acid, which is used to make nylon 11, which is useful for its chemical and shock-resistant properties. Some highly unsaturated acids such as linolenic acid are used directly without transformation, for example as drying agents in paints and inks.

39.7 Unanswered challenges in catalysis for greening chemistry and new chemistry of renewable resources

39.7.1 Cellulose

Cellulose is used for paper and paper products, and certain modifications such as acetylation lead to cellulose acetate, which can be used as a fiber. The abundance of unused cellulose would be sufficient to replace petroleum as a feedstock for chemicals and materials. There are, however, several obstacles to this utilization of cellulose, beyond the need for its separation from lignin. One major problem is the lack of an industrially practical method for the depolymerization of cellulose to glucose or possibly to the reduced sugar alcohol sorbitol [38]. This problem is exacerbated by the insolubility of cellulose in water, which would be the preferred solvent for such an industry. The additional problem is that, from the synthesis point of view, the chemical transformations developed and fine-tuned for many decades have been oxidative and additive in nature. That is, compounds originating from petroleum containing only carbon and hydrogen are modified specifically with oxygen, nitrogen, and other functional units. On the other hand, a chemical tree from glucose would require a reductive reaction sequence, which is not yet in hand and would need to be developed almost from scratch.

39.7.2 Methanol directly from methane and biomass

Methanol could be part of a methanol-based economy as detailed in Section 39.5, yet its synthesis to date is via very-high-temperature formation of synthesis gas via steam re-forming or gasification followed by methanol synthesis, also at high temperature. It seems intuitively obvious that catalysis should be developed for the formation of methanol directly from carbon sources. Concerning reduced carbon sources such as methane and coal, selective catalytic oxidation reactions using molecular oxygen need to be developed and invented. For oxidized carbon resources, notably biomass in general and cellulose more specifically, reductive procedures utilizing hydrogen coupled with carbon–carbon-bond cleavage need to be developed. In such a situation, cellulose, for example, with a carbon:hydrogen:oxygen ratio of 1:2:1 will need an equivalent of molecular hydrogen per carbon to attain methanol with 100% atom economy.

39.7.3 Hydrogen and oxygen by photocatalytic reduction/oxidation of water

The splitting of water to oxygen and hydrogen has long captured the imagination of scientists wanting to provide a cheap source of hydrogen fuel. Much of the interest has been directed toward hydrogen as a fuel for automobiles. In this energy sector hydrogen storage is an additional important problem since it can as yet not be stored safely at the high weight density required [39]. However, hydrogen from the splitting of water may be valuable in other contexts where it need not be stored. Considering, for example, the significant amount of energy needed for hydrogen production in ammonia synthesis, an alternative to steam re-forming of methane would be of significant value. The same argument could be made for the use of hydrogen for the reduction of carbon dioxide to methanol. In a general sense, there are two main approaches to obtaining hydrogen from water. The first is to use photo-electrolysis as opposed to electrolysis based on an energy source derived from power plants [40]. The second, less developed, idea it to use photocatalysts that would split water directly into hydrogen and oxygen [41]. Approaches range from preparing artificial photosynthetic systems to semiconductor catalysts to organometallic homogeneous catalysts that would need to be rendered heterogeneous.

39.7.4 Reduction of carbon dioxide

The reduction of carbon dioxide to a reduced form such as methanol or even carbon monoxide would be a desirable goal, and would amount to a recycling of carbon dioxide emitted into the atmosphere by the burning of fossil fuels. Most attention has been paid to the possibility of reduction with hydrogen, the typical product of these reactions being formic acid or, even more commonly, sodium formate [42]. One can also consider the reaction of hydrogen with carbon dioxide to yield carbon monoxide, in what is essentially the reverse of the water-gas-shift reaction. At room temperature, this reaction is thermodynamically unfavorable; the free energy of reaction is above 6 kcal mol^{-1}. At higher temperature, for example 300 °C, this reaction becomes slightly more favorable, and a low yield of carbon monoxide may be obtained. However, there remains a significant interest in the reduction of carbon dioxide using solar energy at room temperature [43]. Success in this area will be linked to the development of new photocatalysts on the one hand and non-fossil-fuel sources of molecular hydrogen on the other. Another option for the reduction of carbon dioxide is to use water directly as a reducing agent [44]. The most studied photocatalysts in this area of research have been titanium dioxide-based semiconductors that, upon photooxidation, yield holes and electrons that reduce carbon dioxide by one electron to yield an anion radical and then carbon monoxide and two hydroxide radicals; the latter recombine to form molecular oxygen. The one-electron oxidation of carbon dioxide is thermodynamically very unfavorable, and

results so far indicate that ultraviolet light will be needed in order to expedite this reaction. Two-electron oxidations of carbon dioxide are more thermodynamically favorable, and new photocatalysts to facilitate this reaction will need to be discovered.

39.8 Summary

In this chapter we have covered various topics related to green processing and catalysis. First, we have defined metrics such as atom economy, E-factors, and effective mass yields that aid us in quantitatively describing the "greenness" of a chemical reaction. This is a valuable tool for initial analyses; however, more complete analysis of the greenness of a chemical reaction process requires, in addition, a more complete assessment taking into account questions of energy costs, toxicity, and the integration of a specific process within the framework of various processes operating in concert. Other questions that arise deal with the "greenness" of the product itself *vis-à-vis* biodegradation versus persistence in the environment, which may lead to a change in the preferred product itself. Second, we have described the present importance of catalysis in the transformation of petroleum/crude oil to usable chemicals such as transportation fuels, ammonia, and basic compounds for important materials. This, of course, is by no means a comprehensive description, since it must be realized that catalysis is used in thousands of additional chemical transformations on various scales. Further, we have considered what role catalysis has and will have in a non-petroleum-based industrial setting. While transformations from coal are known, it's use in green processing is very limited because coal gasification leading eventually to methanol, while technologically feasible, is inherently not green. Use of biomass for methanol synthesis also is not green, despite the perception that we would be using a renewable resource. The use of specific compounds such as fatty acids and lactic acid from renewable resources as industrial chemicals, although still of minor scale, has been described as having the potential of bioethanol as a platform for industrial chemicals. Finally, we have described various unsolved challenges for desired reactions facing the catalysis research community. These challenges are difficult and complicated indeed, and will require much scientific effort throughout the twenty-first century.

39.9 Questions for discusssion

1. What are the different metrics used to measure the "greenness" of processes? What are the advantages ad disadvantages of each metric?
2. How is catalysis related to the utilization of fossil fuels, and why is it needed?

3. What are the important petrochemical products at present being produced from fossil fuels, and how are they made?

4. Which compounds can be, or are, made from renewable resources, and how is this done?

5. Explain four unsolved challenges related to the development of green processes from renewable resources.

39.10 Further reading

- **M. Lancaster**, 2002, *Green Chemistry: An Introductory Text*, Cambridge, Royal Society of Chemistry.
- **A. Lapkin** and **D. J. C. Constable**, 2009, *Green Chemistry Metrics: Measuring and Monitoring Sustainable Processes*, New York, Wiley.
- **R. H. Crabtree**, 2009, *Handbook of Green Chemistry, Volume 1: Homogeneous Catalysis*, Weinheim, VCH-Wiley.
- **R. H. Crabtree**, 2009, *Handbook of Green Chemistry, Volume 2: Heterogeneous Catalysis*, Weinheim, VCH-Wiley.
- **P. Anastas** and **J. Warner**, 1998, *Green Chemistry: Theory and Practice*, Oxford, Oxford University Press.
- **R. A. Sheldon**, **I. W. C. E. Arends**, and **U. Hanefeld**, 2007, *Green Chemistry and Catalysis*, Weinheim, Wiley-VCH.

39.11 References

[1] **T. A. Nijhuis**, **M. Makkee**, **J. A. Moulijn**, and **B. M. Weckhuysen**, 2006, "The production of propene oxide: catalytic processes and recent developments," *Indust. Eng. Chem. Res.*, **45**(10), 3447–3459.

[2] **D. McGregor**, 2006, "Methyl *tertiary*-butyl ether: studies for potential human health hazards," *Crit. Rev. Toxicol.*, **36**(4), 319–358.

[3] **B. M. Trost**, 1995, "Atom economy – a challenge for organic synthesis: homogeneous catalysis leads the way," *Angew. Chem. Int. Edn.*, **34**(3), 259–281.

[4] **R. A. Sheldon**, 2007, "The E factor: fifteen years on," *Green Chem.*, **9**(12), 1273–1283.

[5] **K. G. Knudsen**, **B. H. Cooper**, and **H. Topsøe**, 1999, "Catalyst and process technologies for ultra low sulfur diesel," *Appl. Catal. A*, **189**(2), 205–215.

[6] **R. A. Sheldon**, 1994, "Consider the environmental quotient," *Chemtech*, **24**(3), 38–47.

[7] **T. Hudlicky**, **D. A. Frey**, **L. Koroniak**, **C. D. Claeboe**, and **L. E. Brammer Jr.**, 1999, "Toward a 'reagent-free' synthesis," *Green Chem.*, **1**(2), 57–59.

[8] **W. L. Leffler**, 1985, *Petroleum Refining for the Nontechnical Person*, 2nd edn., Tulsa, OK, PennWell Books.

[9] **R. Noyes**, 1967, *Ammonia and Synthesis Gas*, Park Ridge, NJ, Noyes Data Corp.

[10] **G. W. Crabtree**, **M. S. Dresselhaus**, and **M. V. Buchanan**, 2004, "The hydrogen economy," *Phys. Today*, **57**(12), 39.

[11] **J. H. Gary**, and **G. E. Handwerk**, 2001, *Petroleum Refining: Technology and Economics*, 4th edn., Boca Raton, FL, CRC Press.

[12] **E. F. Lutz**, 1986, "Shell higher olefins process," *J. Chem. Ed.*, **63**(3), 202–203.

[13] **P. Ehrlich** and **G. A. Mortimer**, 1970, "Fundamentals of the free-radical polymerization of ethylene," *Fortschritte Hochpolymeren-Forschung*, **7**(3), 386–448.

[14] **J. Boor Jr.**, 1979, *Ziegler–Natta Catalysts and Polymerizations*, New York, Academic Press.

[15] **P. A. Kilty** and **W. M. H. Sachtler**, 1974, "The mechanism of the selective oxidation of ethylene to ethylene oxide," *Catal. Rev.*, **10**(1), 1–16.

[16] **T. Reis**, 1966, "Compare vinyl acetate processes," *Hydrocarbon Processing*, **45**(11), 171.

[17] **L. F. Albright**, 1967, "Manufacture of vinyl chloride," *Chem. Eng. Chem. Metall. Eng.*, **74**(8), 219.

[18] **J. N. Hornibrook**, 1962, "Manufacture of styrene," *Chem. Indust.*, 872–877.

[19] **C. D. Frohning** and **B. Cornils**, 1974, "Chemical feedstocks from coal," *Hydrocarbon Processing*, **53**(11), 143–146.

[20] **M. Stocker**, 1999, "Methanol-to-hydrocarbons: catalytic materials and their behavior," *Microporous Mesoporous Mater.*, **29**(1–2), 3–48.

[21] **F. J. Keil**, 1999, "Methanol-to-hydrocarbons: process technology," *Microporous Mesoporous Mater.*, **29**(1–2), 49–66.

[22] **R. B. Anderson**, 1984, *The Fischer–Tropsch Synthesis*, New York, Academic Press.

[23] **D. P. Chynoweth** and **R. Isaacson**, 1987, *Anaerobic Digestion of Biomass*, Amsterdam, Elsevier Applied Science.

[24] **J. A. Labinger** and **J. E. Bercaw**, 2002, "Understanding and exploiting C—H bond activation," *Nature*, **417**, 507–514.

[25] **A. K. Frolkova** and **V. M. Raeva**, 2010, "Bioethanol dehydration: state of the art," *Theor. Found. Chem. Eng.*, **44**(4), 545–556.

[26] **S. M. Pillai**, **M. Ravindranathan**, and **S. Sivaram**, 1986, "Dimerization of ethylene and propylene catalyzed by transition-metal complexes," *Chem. Rev.*, **86**(2), 353–399.

[27] **R. H. Grubbs**, 1978, "The olefin metathesis reaction," *Prog. Inorg. Chem.*, **24**, 1–50.

[28] **G. Sankar**, **R. Raja**, **J. M. Thomas**, and **D. Gleeson**, 2001, "Advances in the determination of the architecture of active sites in solid catalysts," in *Catalysis by Unique Metal Ion Structures in Solid Matrices: From Science to Application*, ed. **G. Centi**, Dordrecht, Springer, p. 95.

[29] **V. Kenkre**, **D. Mukesh**, and **C. S. Narasimhan**, 1985, "Oxidation of ethanol on V_2O_5 based catalyst," *Adv. Catal.*, 455–465.

[30] **G. J. Sunley** and **D. J. Watson**, 2000, "High productivity methanol carbonylation catalysis using iridium: the Cativa™ process for the manufacture of acetic acid," *Catal. Today*, **58**(4), 293–307.

[31] **R. Datta** and **M. Henry**, 2006, "Lactic acid: recent advances in products, processes and technologies – a review," *J. Chem. Technol. Biotechnol.* **81**(7), 1119–1129.

[32] **E. W. Flick**, 1998, *Industrial Solvents Handbook*, 5th edn., Norwich, NY, William Andrew Inc.

[33] **P. Gruber** and **M. O'Brien**, 2005, "Polylactides," in *NatureWorks® PLA*, Weinheim, Wiley-VCH, doi:10.1002/3527600035.bpol4008.

[34] **K. Ngaosuwan**, **E. Lotero**, **K. Suwannakarn**, **J. G. Goodwin Jr.**, and **P. Praserthdam**, 2009, "Hydrolysis of triglycerides using solid acid catalysts," *Indust. Eng. Chem. Res.*, **48**(10), 4757–4767.

[35] **S. Saito**, 1995, "Research activities on supercritical fluid science and technology in Japan – a review," *J. Supercrit. Fluids*, **8**(3), 177–204.

[36] **M. Kjellin** and **I. Johansson**, 2010, *Surfactants from Renewable Resources*, New York, John Wiley & Sons.

[37] **A. Koeckritz** and **A. Martin**, 2008, "Oxidation of unsaturated fatty acid derivatives and vegetable oils," *Eur. J. Lipid Sci. Technol.*, **110**(9), 812–824.

[38] **L. T. Fan**, **M. M. Gharpuray**, and **Y. H. Lee**, 1987, *Cellulose Hydrolysis*, Berlin, Springer-Verlag.

[39] **K. L. Lim**, **H. Kazemian**, **Z. Yaakob** and **W. R. W. Daud**, 2010, "Solid-state materials and methods for hydrogen storage: a critical review," *Chem. Eng. Technol.*, **33**(2), 213–226.

[40] **M. Woodhouse** and **B. A. Parkinson**, 2009, "Combinatorial approaches for the identification and optimization of oxide semiconductors for efficient solar photoelectrolysis," *Chem. Soc. Rev.*, **38**(1), 197–210.

[41] **D. Gust**, **T. A. Moore**, and **A. L. Moore**, 2009, "Solar fuels via artificial photosynthesis," *Acc. Chem. Res.*, **42**(12), 1890–1898.

[42] **W. Leitner**, 1995, "Carbon dioxide as a raw material: the synthesis of formic acid and its derivatives from CO_2," *Angew. Chem. Int. Edn.*, **34**(20), 2207–2226.

[43] **A. J. Morris**, **G. J. Meyer**, and **E. Fujita**, 2009, "Molecular approaches to the photocatalytic reduction of carbon dioxide for solar fuels," *Acc. Chem. Res.*, **42**(12), 1983–1994.

[44] **W. Lin**, **H. Han**, and **H. Frei**, 2004, "CO_2 splitting by H_2O to CO and O_2 under UV light in TiMCM-41 silicate sieve," *J. Phys. Chem. B*, **108**(47), 18269–18273.

40 Materials availability and recycling

Randolph Kirchain and Elisa Alonso

Materials Systems Laboratory, Engineering Systems Division, Massachusetts Institute of Technology, Cambridge, MA, USA

40.1 Focus

The financial future of firms that depend on materials can be permanently compromised if the availability of those materials is constrained. This chapter examines how limited materials availability can affect a firm; how a firm can know whether it is using materials that are at risk of becoming of limited availability; and what can be done to mitigate that risk. One mitigation strategy is to foster an effective recycling system. The chapter concludes by exploring the benefits of expanded recycling and some of the remaining challenges to making that happen.

40.2 Synopsis

Resource scarcity is a topic that has challenged scientists, engineers, and economists for centuries. Current interest in this topic stems from the central role of natural resources in our economy, the inherently finite supply of those resources, and the unprecedented rate of resource consumption.

Will limited resource availability limit economic growth? The answer to this question is complex: resource consumption or use occurs in a complex and dynamic system in which the actions of governments, firms, and consumers, as well as the geological endowment of the resources, play a role. While there has never been a documented case of a material "running out," the finite nature of the resources has implications for global material markets. For example, materials are geologically distributed unevenly, their supply may be controlled by only a few institutional entities, and they require varying amounts of effort (and cost) to extract.

On the demand side, materials-selection decisions are often made early in the product development life cycle and are not easily changed. Thus, supply disruptions can significantly damage manufacturers' profits and even result in permanent changes to a material's market and the firms that depend on that material. Strategies to mitigate such risks to firms need to be considered when materials-selection decisions are being made. Since many firms use dozens of different materials and are limited in resources, any strategy for dealing with the risk of scarcity should begin with screening for strategic materials before taking a more involved approach.

One example of a strategy to mitigate resource-availability risk is recycling. Recycling reduces primary-resource depletion and, generally, reduces the environmental impact of production. Additionally, recycling can reduce material costs and diversify the supply chain, reducing the impact of limited availability. However, technological, societal, and economic considerations vary for different products and materials, and recycling's benefits must be carefully evaluated and fostered for each situation.

40.3 Historical perspective

Throughout time, the use of materials has been a defining feature of humankind's effort to survive and thrive. As we have become more ingenious in the ways in which we manipulate those materials, we have also dramatically increased the magnitude of our use of materials. As you learned in Chapter 7, industrialization has led to a more than ten-fold growth in the per-capita usage of material resources. Within that trend, society has also transitioned from reliance on renewable biomass to dependence on nonrenewable minerals. As an example, the United States Geological Survey estimates that, only 100 years ago, about 40% of domestic raw-materials use came from renewables. Today, that figure is closer to 5%.

Although the use of renewables is not without its own challenges, this reliance on nonrenewable resources has raised concerns about the sustainability of modern society. Can economic growth continue indefinitely? Will we run out of some critical resource? Interestingly, these questions are not new. Formal treatment of these issues dates back at least to the eighteenth-century economist Thomas Malthus, who raised concern about whether available farming resources would ultimately limit society's growth (possibly in the form of catastrophe). Interestingly, such concerns have haunted reflective societies for millennia. In fact, Tertullian, a second-century theologian, remarked on a "world, which can hardly supply us from its natural elements" due to our "teeming population." Since the times of Tertullian and Malthus, we have learned a lot about resource availability and how it affects our world, but we have also dramatically changed the rate at which we tap those resources.

We have learned that economics and technology temper the impacts of increasing resource use and declining resource stocks. As the cost to acquire resources grows, so does the impetus to develop technologies to use that resource more efficiently and to switch to alternatives. Such mechanisms are often insufficient when the resource is a common good (e.g., a fisheries stock) or is not marketed (e.g., clean air). Nevertheless, even for more conventional cases, those who use materials should realize that limited materials availability, even if only temporary, can still present real economic risk. Interestingly, many of the strategies to mitigate that risk serve the dual purpose of reducing the environmental burden of resource use.

One such strategy that is familiar to most is recycling. Clearly recycling reduces primary-resource consumption, but also can provide attendant reductions in energy consumption and emissions. Contrary to some current perceptions, recycling is itself not a new phenomenon. In fact, recycling surely dates as far back as the use of materials by man. When resources are costly to acquire, they are discarded reluctantly and reused assiduously and ingeniously. Thus, valuable resources such as metals and even stones were reused or recycled wherever possible. This was mostly an informal process that depended heavily on individual manual segmentation of valuable resources. We still see this today in some parts of the world where "rag-pickers" make a subsistence living by manually sorting out recyclables from municipal solid waste (MSW). Generally, however, with affluence most societies transition away from careful segmentation and reuse, instead shipping most of their waste resources to landfills or incinerators. The past few decades have witnessed a departure from this trend. Since at least the 1970s, organized recycling programs have been expanding not on the basis of their private economics alone, but to address some of the social costs associated with removing resources from the economy. As we will discuss in more detail later, despite this development, many developed countries still "throw a lot away." Using the USA as an example, while we now recycle more than 60 million tons of MSW, we still landfill more than 130 million tons.

40.4 Introduction[1]

> The challenges facing engineering today are not those of isolated locales, but of the planet as a whole and all the planet's people. Meeting all those challenges must make the world not only a more technologically advanced and connected place, but also ... more sustainable, safe, healthy, and joyous ...
>
> *National Academy of Engineering Grand Challenges Committee*

In fact, some studies estimate that in the USA raw-materials usage is approaching 100 kg per person per day [2]. In the face of a still growing and increasingly affluent world population, such estimates have raised important questions. Will resource availability limit economic growth?

In this context, few debate the need to move toward a more cyclic materials system. However, the debate over the need to intervene in materials use is hotly contended. One side of the debate contends that economic mechanisms inherently allocate scarce resources and most efficiently cope with decreasing resource availability: as resources become scarcer, prices rise, and users find ways to use less or to shift to substitutes. Others contend that markets fail to fully reflect the long-term (not to mention broader social) value of resources and, therefore, cannot drive efficient resource decisions.

---

[1] Sections 40.6–40.8 of this chapter are based substantially on [1].

Figure 40.1. The rate of modern-day raw-materials use: *A Sunday Afternoon on the Island of La Grande Jatte* was originally completed in 1886 by neo-impressionist painter Georges Seurat. This version *Cans Seurat*, by American artist Chris Jordan, is composed of images of 100,000 aluminum cans – roughly the number used in the USA every 30 s. Unfortunately, only about half of those are recycled. (2007 © Chris Jordan, Courtesy of Kopeikin Gallery.)

In general, these debates[2] have been framed in terms of aggregate social welfare (see Figure 40.1). This chapter does not attempt to pick either side of this debate, but rather takes another perspective: it examines the question of materials vulnerability, or conversely materials availability, from the perspective of an individual firm.

Materials availability is critical to all firms, not just those that extract, refine, and process materials into products. If raw materials become difficult to acquire, market forces may shift demand to other goods. The economic impact of such shifts means that some firms would not survive. Unfortunately, reacting after the fact, after a shift has begun, might not work because materials influence so much about the way a firm does business.

This raises some difficult questions, which this chapter will attempt to answer. (1) How does limited materials availability affect a firm? 2) How can a firm know whether it is using materials that are at risk of becoming of limited availability? (3) What can that firm do to mitigate that risk? One of the answers to question (3) is to foster an effective recycling system for at-risk materials. The chapter concludes by exploring the benefits of expanded recycling and some of the remaining challenges to making that happen.

Before attempting to answer these questions, the chapter begins with a number of key definitions.

40.5 Definitions and terminology

40.5.1 Geological definitions for primary supply

This section will quickly review some of the geological terms used when discussing nonrenewable materials. The allocation of nonrenewable materials in Earth's

---

[2] Across both sides of the debate, it is broadly recognized that the limited capacity of the natural environment to absorb the effluents of production might in fact be the most eminently limiting resource. Pricing that resource and internalizing that price within market actions are inherently challenging [3].

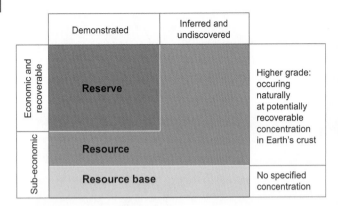

Figure 40.2. Fundamental characteristics of materials reserves and resources [4].

crust is determined by geological processes, and exploration is the means of obtaining information on the resource distribution and concentration in Earth's crust and on the ocean floor.

Geologists subdivide primary sources of metals by how well they have been identified and measured and by how economic and technically viable it is to extract them today (Figure 40.2).

The materials that are mined today form part of the global *reserves*. The reserves incorporate all ore bodies at a given *ore grade*, ore type, and location such that it is technically and economically or marginally sub-economically viable to extract. The ore grade is the concentration of the desired material in the ore body that is being extracted and is generally given as a percentage by weight. The size of global reserves changes as prices change, as new ore bodies are discovered, as extraction technology improves, and as minerals are extracted. The accuracy of the reported data on reserve size depends on how recently the numbers have been updated, how responsive individual mines are to reporting, and how statistically accurate the geological methods for identifying ore-body characteristics are.

In geological terms, the word "resource" has a very specific meaning. The resources include the reserve base and the rest of the sub-economic deposits as well as estimates of the quantity of material that has not yet been properly characterized. There is much more uncertainty in the size of the resources than in the size of the reserves.

The largest number that defines primary material sources is the estimated size of the *resource base*, which includes all material content in Earth's crust (to a certain depth) and oceans, at all concentrations. The size of the resource base can be several orders of magnitude greater than the size of the resource for a given metal, but is generally not a number that is used in analyses of depletion.

40.5.2 Secondary supply: recycling

When products reach the end of their life, they may be disposed of or collected for recycling. End-of-life products effectively become a resource. *Secondary resources*, a term used for all recycled materials, including industrial scrap (*new scrap*) and post-consumer scrap (*old scrap*), can be categorized so that economical ones can be identified and exploited through recycling. The concentration and distribution of the materials that form the secondary resources depends on where the products are discarded, how many products reach their end-of-life state, and the amount of material contained in the product. In general, many of the world's secondary resources are found in industrialized countries, which account for most of the world's consumption.

The *recycling rate* will here be defined as the amount of secondary material used as a fraction of the total amount of material used. The recycling rate has sometimes been referred to as the *static recycling rate* or *recycled content*. The *recovery rate* will be defined as the amount of old scrap collected and recycled as a fraction of the amount of total material disposed of and recycled, landfilled, or otherwise dispersed without recovery. A high recovery rate means that very few end-of-life products are reaching landfills. The recovery rate can also be called the *recycling efficiency* or *dynamic recycling rate*. An efficient recycling system is one in which the recovery rate is high, which means that most of the material in products reaching their end-of-life state is being reused. A high recovery rate could still result in a low recycling rate if the product lifetime is long and demand is growing rapidly.

40.6 A historical case study: global outcomes from decreased availability of cobalt in the 1970s

No historic case of *global* materials depletion has been documented. Nevertheless, firms have been impacted by specific examples of limited materials availability during the 20th century. This section examines one such case: the use of cobalt. This case will help us to answer our first question: How does limited materials availability affect a firm? It will also provide some suggestions for how we might judge vulnerability risk and respond to mitigate vulnerability.

40.6.1 Overview of the case

Information on historical events and data for this section were taken from [5][6][7][8].

For millennia, cobalt compounds have been used because of the blue color they impart to ceramics and glasses. In the industrial era, cobalt has also become important in many applications, including aircraft

Figure 40.3.
Historical usage of cobalt, before and after the price shock in the late 1970s. [6]

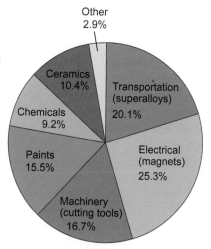

Average 1975–1977

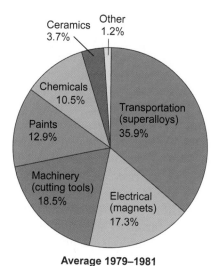

Average 1979–1981

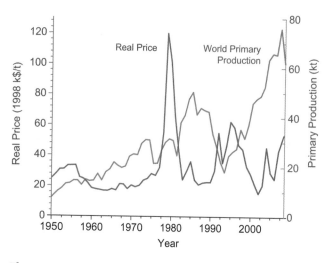

Figure 40.4. Cobalt average yearly prices and primary production 1950–2009. (After [1], data from [8].)

engines, turbines, magnets, and cutting tools. (See Figure 40.3 for a historical perspective.) More recently, cobalt has also become critical to many electronic devices and even the electrification of transportation because of its use in lithium-ion batteries.

For at least the last 50 years, the price of cobalt has always been volatile. However, between 1977 and 1979, cobalt prices experienced an unprecedented shock, rising by 380%. In response to this price swing, many aspects of the production and use of cobalt changed; in some cases, permanently.

The price spike occurred following a rebellion in the Republic of Zaire (now known as the Democratic Republic of the Congo), which at the time was home to only 0.6% of the world's population, but controlled more than 40% of world production of cobalt. In May 1978, insurgents from Angola took over parts of the Shaba province in Zaire. They cut power to most major mining facilities and killed many workers. Overall, the insurgents were in

Zaire for only about two weeks, but, because of flooding and the evacuation of most expert contractors, the mines were slow to restore operation.

During this period, a surging world economy pushed up demand for many materials, including cobalt. The concern for supply shortages in the midst of the upturn, together with real delays in transporting cobalt out of Zaire, led to speculation. In February 1979, the price of cobalt soared and prices remained high until 1982 (see Figure 40.4).

These events and the resulting price spike led both firms and governments to act to reduce the ultimate economic impact. In the short term, "upstream" firms, that is firms in the mining and refining sector, changed their operations (e.g., using air rather than land transport) to speed available material to market. As part of a longer-term response, firms in Zaire, Zambia, and Australia expanded production capacity through changes in operational practices and new technology. Because of these changes, by 2004, Zaire accounted for only 31% of the world's mined cobalt.

"Downstream" firms, such as component and product manufacturers, also reevaluated their production options in light of the price increases. The specific changes in use pattern for cobalt are shown in Figure 40.3.

Substitution to lower-cobalt-content alloys occurred quickly in the magnet industry in applications with limitations on weight, size, and energy [8]. Reducing cobalt use in superalloys was difficult because of the limited availability of substitutes and an increased demand for jet engines. In the short term, cobalt use in the transportation industry increased, with only some substitution by nickel-based alloys. A key change in cobalt use occurred with the development of a recycling process for scrap superalloy.

Figure 40.5. Pertinent stocks and flows in a materials economy. (After [1].)

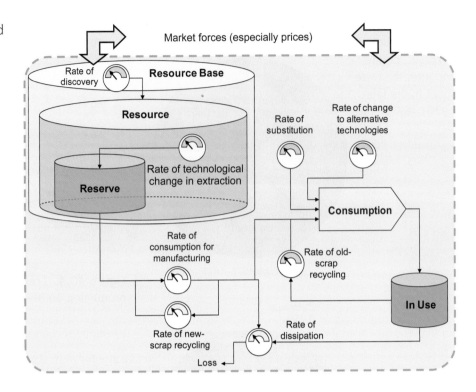

Some substitution by iron-based and nickel-based alloys also occurred in cutting tools; however, net end use of cobalt for machinery increased slightly. Cobalt use in ceramics and paints also dropped because substitution in these applications was straightforward.

Overall, as prices rose, the cobalt supply chain responded through

- materials substitution and development of new technology,
- source relocation,
- hoarding and rationing,
- supply-mode changes, and
- recycling.

Looking back at the events of 1978 in the cobalt market and how they affected firms in many industrial sectors, it is clear that limited materials availability can have real economic impacts. It is also clear that many of the responses that can reduce that impact take time to implement (e.g., developing new technology or implementing a recycling system) or are very expensive to implement once prices are high (e.g., hoarding). Clearly, it would be beneficial to recognize the risk early and develop strategies ahead of time. The next section explores some metrics of risk.

40.7 Identifying and measuring vulnerability to availability

The problem for those attempting to ascertain the risk of limited resource availability is the complexity of a materials economy. Identifying risk metrics – that is, quantities we could measure to test for risk – for such a system has been an ongoing effort. Unfortunately, no single metric retains the generality necessary to reflect all possible presentations of "scarcity." The history of mercury serves as an example of this shortcoming. In 1972, mercury was identified as becoming critically scarce [9]. However, changes in the market for mercury have meant that now, nearly 40 years later, the world has identified mercury resources sufficient to last more than 200 years [8]. Clearly, identifying resources at risk is rife with ambiguity. Nevertheless, despite this ambiguity, firms that use materials must make business decisions.

Figure 40.5 depicts many of the fundamental elements of a materials economy, particularly those that are relevant to questions of materials availability. A materials economy consists of a network of resource flows (e.g., bauxite ore or aluminum) that move among resource stocks (e.g., resource-base or in-use stocks). The magnitude of these flows (depicted by the valve gauges schematic in Figure 40.5) is driven by the demand for applications that use the resource (e.g., aluminum cans) and moderated by the availability of substitutes (e.g., PET bottles) and recycling. The rates of these flows are ultimately dictated by the economy. As such, they are dynamic and constantly shifting.

As Figure 40.5 suggests, there is a host of competing rates within a materials economy, which can also be viewed as *"drivers of availability."* Metrics must somehow assess the evolution of these rates against the amount of extracted and as-yet-unextracted resources.

Table 40.1. Measures of institutional inefficiency

Metrics and indicators	Description
Geographical structure based on supply (%)	Distribution of reserve size in top countries *Assumption: supply diversity increases efficiency*
Geographical structure based on production (%)	Distribution of production in top producing countries *Assumption: supply diversity increases efficiency*
Institutional structure based on production (%)	Distribution of control by most important company *Assumption: supply diversity increases efficiency*
Institutional structure based on use (%)	Distribution of applications and companies that use a given material, identification of new uses of the material *Assumption: demand diversity increases efficiency*
Recycling rate (%)	Scrap use divided by total use at a given time *Assumption: reliance upon recycled resource increases efficiency; greater confidence in supply*
Recycling efficiency rate (%)	Old scrap used divided by total material at end of life *Assumption: reliance upon recycled resource increases efficiency; greater confidence in supply*
Economic metric: market price ($)	Time trend and volatility of price *Assumption: efficient markets*
After [1].	

Most discussions of scarcity focus on physical constraints and raise the following question: when will we run out of, or exhaust, a resource? This perspective dates back at least to the eighteenth-century economist Thomas Malthus. In the early nineteenth century, the economist David Ricardo refined this notion of physical constraints by making the observation that resources exist in different levels of quality. Hence scarcity is not a consequence of exhaustion, but instead derives from the increasing difficulty and cost of access. This realization is critical to our understanding of scarcity today. We now realize that firms and individuals react to these economic signals to use less or shift to other resources (substitutes). However, as the cobalt case demonstrates, the economic nature of scarcity means that even temporary limitations in availability can significantly affect how firms produce and use materials.

Given that context, there are clearly two mechanisms that must be considered in evaluating the risk of scarcity.

- *Institutional inefficiency*: failures by markets, firms, and governments can result in transitory resource unavailability. Such events temporarily close or tighten key valves in the system.
- *Physical constraints*: the amount and quality of a resource are physically determined and ultimately limit resource availability. Physical reality means that eventually the stocks will decline or the valves constrict.

These perspectives on the mechanisms of scarcity provide a useful scheme to categorize metrics that have emerged over time in the literature. Sources of, and a more in depth discussion of, these metrics can be found in [1].

40.7.1 Institutional inefficiency metrics

Table 40.1 lists some useful metrics to test for risk of institutional inefficiency within a materials economy. The most broadly cited measures of vulnerability to institutional inefficiency focus on concentration within the supply chain, at either the national or firm level.

Generally, the geographical distribution of reserves depends on geophysics, past depletion, and present exploration. In the face of uncertain external factors (such as political events and natural disaster), a higher level of concentration makes a resource more susceptible to institutional inefficiency and supply disruptions. Likewise, oligopsonistic markets are more vulnerable to fluctuations in demand, leading to market volatility.

Secondary production (i.e., recycling) can reduce institutional inefficiency risk for two reasons: (1) secondary production provides an additional source of supply (diversification); and (2) secondary stocks are often located and processed in different locations and by different institutions from those for primary stocks. The recycling rate is a broadly available indicator of the importance of recycled material as a resource. Thus,

Table 40.2. Measures of physical constraint, including static and dynamic Malthusian metrics and Ricardian metrics

	Metrics	Description
STATIC	Static index of depletion (d_s, years)	Time to use supplies at constant use rate: $d_s = R/C_{present}$, where R is some estimate of the available primary supply (i.e., reserves, resource, or resource base) and $C_{present}$ is the current rate of use or primary resources.
		Assumptions: use rate constant; discovery rate negligible
DYNAMIC	Dynamic (exponential) index of depletion (d_e, years)	Time to use supplies at constant exponential growth of use rate: supply/ (projected use), where future use can be modeled as having exponential growth; $d_e = \frac{1}{r} \ln\left(r\frac{R}{C_{present}} + 1\right)$, where r is the rate of growth.
		Assumptions: use rate exponential; discovery rate negligible
	Relative rates of discovery and extraction (unitless)	Ratio of rate of discovery to rate of use
		Assumption: recycling/reuse and substitution negligible; improvement in extraction technologies negligible
	Time to peak production (years)	Time until this forecast peak is reached: this metric is based on models of future rates.
		Assumption: rate of net use (demand minus substitution) will grow faster than rates of discovery, technological improvement, and recycling/reuse
RICARDIAN	Average ore grade (%)	Concentration of metal in a given ore body
		Assumptions: efficient markets in factors and capital; technological efficiency; accessibility effects negligible
	Cost ($)	Sum of technical costs (machines, fuel, labor, etc.), environmental costs, political costs, commercial costs (marketing, insurance, stock dividends)
		Assumptions: efficient markets in factors and capital; technological efficiency
	Economic metric: market price ($)	Relative price, time trend.
		Assumption: efficient markets

After [1].

higher recycling rates can be an indicator of lower vulnerability. The ability of a supply chain to modify availability through secondary sources is ultimately limited by access to such materials. The recycling efficiency rate (RER) is a metric that provides insight into this issue. Unfortunately, the RER is difficult to measure, and data must be derived from prospective modeling.

The final metric of institutional inefficiency listed in Table 40.1 is the market price of the commodity of interest. Price is one of the best measures of scarcity insofar as the market embeds many of the issues outlined above into pricing. However, price is often not a leading indicator; while price will ultimately be the trigger that initiates supply-chain change, effective response strategies must already be in place before prices go up.

40.7.2 Physical-constraint metrics

The outcomes arising from institutional inefficiency in the cobalt case could also have occurred from physical constraints. In this section, metrics of physical-constraint risk will be briefly defined, but their interpretation will be made through the case study which follows.

Malthusian metrics
The direct approach to measuring risk to geophysical limits is to compare how much of a resource is known to be available with how fast it is being used. Table 40.2 includes Malthusian-inspired metrics. The metrics are divided into two broad categories (static or dynamic)

depending on the degree to which they treat the varying nature of the many interrelated rates (see Figure 40.5).

The static index of depletion is an estimate of the years to exhaust a material supply on the basis of present use rates and one of the four estimates of available supply: reserve, reserve base, resource, or resource base [8][10]. The dynamic index of depletion is a simple extension that includes changing use rate, for which expected use is derived from historical data. A material is considered more vulnerable if it has a low index of depletion.

The simplicity of Malthusian metrics is a major advantage: the depletion risk is related to how fast a nonrenewable resource is used. Moreover, the required data are generally readily available.

These metrics all assume a decreasing supply base for nonrenewable resources. However, new discoveries, improved technology, increased recycling, and changes in resource economics have contributed to increases in supply in the past. Taking this into account, one can classify materials with a rate of supply growth less than the rate of increasing use as vulnerable [11]. One criticism of these metrics is that, since many of the parameters are based on historical data, the effect of new technologies that may increase demand or improve efficiency is not considered.

Ricardian metrics

From a Ricardian viewpoint, scarcity should occur long before physical exhaustion since the difficulty and therefore cost of extraction increases [12]. The ore grade is a physical measure of the quality of supply [13]. In general, the lower the ore grade, the more earth will be displaced, energy will be expended, and waste will be generated to extract the resource. Ore grade alone, however, does not capture entirely the accessibility of the supply; an ore body at the surface is more accessible than one underground. Moreover, extraction from certain minerals is more difficult than extraction from others (e.g., oxide minerals vs. sulfide minerals).

A more informative measure of quality is the cost of extraction. Increasing extraction costs would be expected to correlate with increasing vulnerability. The primary barrier to using extraction cost as a metric of risk is the public availability of data. Owing to the difficulty in obtaining cost data, many rely on incomplete proxies such as energy use, or labor or capital costs.

Market price is often viewed as the ultimate Ricardian metric. As was noted previously, variations in price might not provide enough time for a firm to react to changes in availability of materials.

40.7 Exploring the utility of metrics: the case of copper

This section demonstrates the use of metrics in the context of copper. Copper is a key building block of

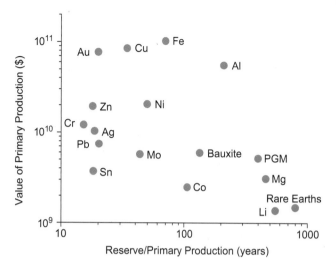

Figure 40.6. Positioning of metals: static index of depletion base on reserves and global value of primary production in 2009. Metals of increasing concern have high value and low depletion index. (After [1], updated data from [8].)

modern life and technology. As the y axis of Figure 40.6 shows, the value of global copper use is second only to those of aluminum and iron. This high value and the high rates of use that drive it contribute to apprehension about copper's long-term availability.

Specifically, we will examine (1) a simple, imperfect screening metric, (2) the criteria for action on that metric, and (3) the use of more detailed measures for additional insight on risk. (Data for this section largely derive from [14][15].)

40.7.1 Screening for risk

The simplest way to screen for scarcity-based material vulnerability is to calculate the time to deplete the current supply – a static depletion index. The smallest measure of available supply is referred to as reserves, a measure of economically and technically available primary metal. Looking back at the x axis in Figure 40.6, we can see that the static depletion index based on reserves of copper is small, only 34 years.

Of course, this metric does not reflect the dynamics in the market, the changes in consumption, or the emergence of economically viable sources. This most conservative static estimate is indicative of the time frame within which new technologies for extraction or new sources must be found in order to continue the present yearly use under present economic conditions.

The first question that arises concerning the static depletion indices is as follows: what index values indicate a need for action? Unfortunately, there is no single or simple answer to that question. Several authors have examined this question and all suggest a figure of about

30 years for the magnitude of the managed reserve life relative to current use. Figure 40.7 shows that, while the static depletion index for copper reserves has oscillated over the last century, it has frequently returned to around 30 years. Thus, 30 years may serve as a threshold indicator for concern; greater values would represent conditions under which the primary industry is unmotivated to address geophysical scarcity, while values around or below 30 years would indicate a need for further evaluation. In terms of this criterion, copper sits on the border of concern.

Metrics of supply-chain concentration are useful indicators of vulnerability due to institutional inefficiency and can be readily calculated (e.g., see the US Geological Survey's minerals commodity summaries published by the US Department of the Interior). This metric is plotted in Figure 40.8 for a range of commodities for 2009 guidelines applied by the US Department of Justice concerning supply concentration within a market. It suggests that moderate levels of concern exist when individual suppliers reach market shares around 30%, and high levels of concern exist when market shares approach 40% [16][17][18]. Using those guidelines, copper merits attention.

While recycling can provide an alternative source for materials, into the foreseeable future, for most materials, primary production will remain the dominant source. This appears to be the case for copper. Recycling of old scrap accounts for less than one-fifth of global copper use. This suggests that at present recycling does not dramatically reduce the vulnerability of the copper supply chain. This represents an opportunity for firms to reduce their risk by fostering more copper recycling.

Further investigation into supply-chain risk: Ricardian measures

Throughout history, technological improvements have made it economically feasible to exploit lower and lower grades of ore. Between 1970 and 1993, when US copper grade remained a relatively constant 0.5%, the costs of western-world copper mining decreased. Although the impact of technological improvement is impressive, the overall trend in ore grade would indicate that copper supplies are shifting into a regime of increasing vulnerability. Technological improvements will have to keep pace even at lower ore grades in order to keep costs in

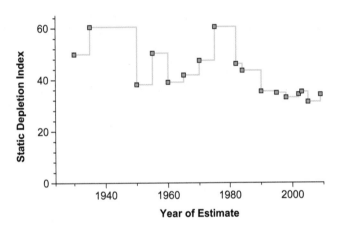

Figure 40.7. An estimate over time of the static depletion index based on reserves (in years) for copper. (After [1], updated data from [8].)

Figure 40.8. Geographical distributions of primary production for various metals, showing the top three producing countries for each metal in 2009. (After [1], updated data from [8].)

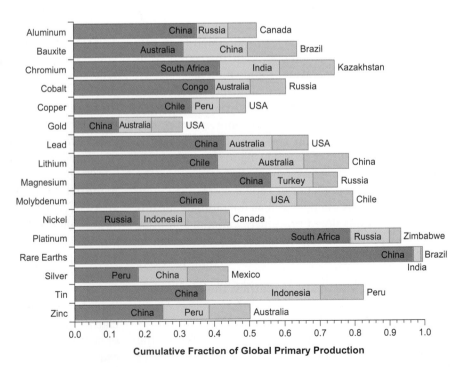

check. Unfortunately, data on ore grade are not publically available for all commodities.

Substitutes can come in many forms. The simplest source of a substitute is a recycled raw material. In addition, materials can sometimes be substituted for one another. In the case of copper, aluminum (also a good conductor of heat or power) and optical fibers (good conductors of information) are frequently considered substitutes. Finally, substitution may occur at the application level, with new products removing demand for those products that contain copper. A thorough discussion of this topic is beyond the scope of this chapter, but would be critical in any assessment of the risk exposure of a given firm.

40.7.2 Understanding copper vulnerability

What businesses should take from the above analysis is that, while the complete depletion of copper is *not* imminent, most of the metrics indicate that the risk of disruption to copper supplies is significantly greater than for other major metals (e.g., iron and aluminum), and is at or near to a historical high. A proactive business that depends upon copper materials will understand that there are actions that should be considered in order to mitigate these risks.

40.8 Responses to risk: preparing for limited availability

Because materials are so fundamental to the way in which a company operates, changing materials takes time and resources. Waiting until availability is limited probably means leaving it until it is too late to take that action. Managers need to assess risks to materials availability and, when appropriate, prepare for possible future problems.

Dealing with risk and uncertainty within the supply chain is a topic addressed by a growing literature [19]. First of all, this literature suggests that firms must "know" their supply. In the case of materials availability, this includes not only monitoring metrics of risk, but also fostering the exchange of information to ensure the accuracy of those metrics.

Additionally, firms should act to add flexibility (resilience). Conventionally this is achieved by having multiple suppliers and/or keeping inventory. In the case of materials, flexibility can be added by keeping inventory (which is expensive), developing alternative sources of supply, developing substitutes, and recycling. The latter two options take time to develop and must be in place before a crisis occurs.

Finally, firms should act to increase robustness with respect to materials-availability events by slowing primary use. This can be accomplished by developing processes that are more efficient or, as with flexibility, by ensuring that an effective recycling infrastructure exists.

Knowledge, flexibility, and robustness are broad measures for reducing supply-chain risks. Although motivated solely by private concerns, the actions that support these strategies could (1) decrease use of primary stocks, (2) facilitate transition to more sustainable substitutes, and (3) ensure viable recycling. Together, these actions drive toward a more sustainable materials system. The next section will explore one modification – increased recycling – in more detail. Increasing recycling could make a supply chain more flexible and robust, and might provide environmental benefits.

40.9 Recycling

Few discussions of strategies to address modern-day environmental challenges are complete without praise for recycling. Therefore, recycling is well established in the consciousness of citizens throughout the developed world. Nevertheless, few are familiar with the specific benefits and barriers associated with recycling or, more precisely, increasing recycling. Before discussing those in detail, it is important to point out that, like all activities, recycling is not universally beneficial. While most recycling produces real environmental benefits and all regions of the world could improve aspects of their recycling systems, there are and always will be cases in which the environmental cost of recycling outweighs any demonstrable environmental benefit [20]. This issue will be revisited later, but the next section will focus on the bulk of cases in which benefit outweighs cost and much room for improvement exists.

Before getting to those issues with recycling, it is useful to pause and consider the following question: how effectively do we recycle? As with many things, the answer is that it depends on various factors. First, let's consider how much waste we generate. In 2010, residents of OECD countries collectively generated well over 600 million tons of municipal solid waste. Rates of generation, however, vary widely, with countries such as Norway and the USA discarding more than 2 kg per person per day while Mexico, Poland, and others discard less than 1 kg [21]. Culture, income, and policy also drive vastly different practices regarding that waste. For example, in Switzerland in 2009, 51% of MSW was recovered for recycling or composting [22]. In the USA in 2008, a little over 33% of MSW was recycled or composted [23]. The effectiveness of recycling also varies dramatically by product. As Figure 40.9 shows, recycling rates in the USA can vary from nearly 100% for automotive batteries to below 30% for plastic and glass bottles. As these figures indicate, there is still a lot of room to increase the recycling of many products and materials.

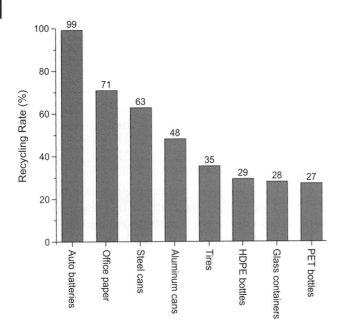

Figure 40.9. Recycling rates of various products in municipal solid waste in the USA [23].

40.9.1 Benefits of recycling

The societal benefits of recycling when compared with other forms of waste management (i.e., landfilling or incineration) can be grouped broadly into five categories. These can include

- avoiding resource depletion,
- reducing energy consumption and emissions,
- economic returns,
- diversified institutional risks,
- market stabilization.

Avoiding resource depletion
The first benefit of recycling is definitional in nature. When we use secondary raw materials we avoid the use of primary raw materials. In doing so, we leave an additional unit of primary resource undisturbed for future use. Global statistics on recycling are not uniformly available. Nevertheless, from what information we have, it seems clear that recycling is growing for most major commodities. Considering aluminum as a representative example, recovery of aluminum from secondary sources has more than tripled in the last 30 years. Unfortunately, overall demand for aluminum has matched that growth over the same period, such that today recycled aluminum accounts for roughly the same fraction of total production as it did in 1980. See Figure 40.10.

Reducing energy consumption and emissions
It is well documented that for many materials, and particularly for metals, the substitution of primary with

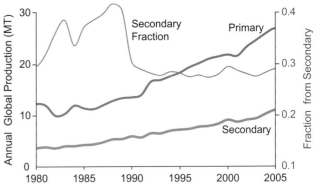

Figure 40.10. Global production of aluminum including primary production, secondary production, and fraction of production from secondary over the period 1980–2005. (Data from and regions defined in *Aluminum Statistical Review for 2006*, Arlington, VA, The Aluminum Association, Inc., 2007, p. 18.)

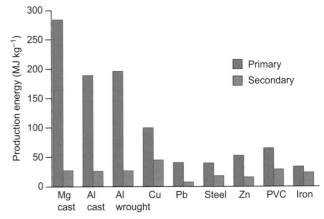

Figure 40.11. The average energy required to produce 1 kg of various materials from primary and secondary (recycled) resources [24].

secondary resources, i.e., those recovered from manufacturing waste or end-of-life products, decreases energy consumption and the attendant environmental burden of production (Figure 40.11). Aluminum, the material selected as a case for this chapter, serves as an excellent example due to the large energy differences between primary and secondary production: 175 MJ kg^{-1} for primary compared with 10–20 MJ kg^{-1} for secondary [24]. However, aluminum is by far not the most extreme case. For some precious metals, such as gold and platinum, the energy difference can exceed two orders of magnitude.

Economic returns
Recycling is an activity that has gone on for as long as humankind has used materials. This is true because, for many materials, recycling saves money. On looking at Figure 40.11 it is clear that the self-same energy benefit

that provides an environmental driver to recycle can provide a (sometimes dramatic) economic driver as well. To get a sense of the magnitude of the economic benefit, consider these facts. In 2009, the Aluminum Association (the industry association for aluminum producers) estimated that in the USA approximately 41 billion aluminum cans were landfilled. If that material alone were recovered, it could have displaced nearly 1 billion dollars' worth of primary aluminum production.

Unfortunately, for many products, the economic incentive by itself is not sufficiently large to fund an effective recycling system. For these commodities, government policies may be implemented to increase recycling rates. Generally, such programs come at a cost to taxpayers. However, authors of a range of studies have demonstrated that there are generally secondary economic returns from recycling programs. These may come from additional employment, savings for downstream processors, or the avoidance of social costs associated with pollution. Whether the benefits outweigh the costs depends upon the commodity, the characteristics of the collection system, and the scope of effects included [20]. For organic commodities, it is always important to consider whether energy recovery rather than strictly defined recycling may be the more fiscally prudent alternative [25].

Diversified institutional risks

As was discussed earlier, one characteristic of a material system that makes it more vulnerable to economically significant availability-related crises is the diversity of sources that provide that material. More diversity reduces the risk of a major crisis. The nature of geography, geology, and consumption means that, for many materials, where primary processing happens is not where secondary processing happens. Because primary production is generally more energy-intensive, primary processing facilities tend to be located where energy is least expensive. This is moderated by all the normal considerations that go into the geography of supply chains, but seems to capture a general trend. These primary production sites are often far from the wealthy countries where most consumption and, therefore, retirement of products occurs.

Considering our example system of aluminum, Figure 40.12 shows how global secondary production is more distributed, and distributed differently, than primary production.

Market stabilization

Materials markets are subject to all of the vagaries that confront any business, but are especially exposed to price variation due to their strong tie to (unpredictable) economic cycles, their position at the end of the supply chain [26], and the difficulty and expense associated with adding or removing capacity. Recent empirical

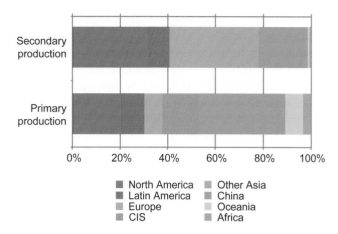

Figure 40.12. Distribution (mass fraction) of primary and secondary (recycling) production by region in 2006. (Data from and regions defined in *Aluminum Statistical Review for 2006*, Arlington, VA, The Aluminum Association, Inc., 2007, p. 18.)

evidence suggests that recycling can mute this effect in a materials market. Specifically, by more nimbly expanding and contracting capacity, secondary production allows materials markets to be more resilient with respect to natural oscillations in demand [27].

40.9.2 Issues to consider

While recycling can provide real economic and environmental benefits, realizing those benefits requires overcoming the challenges and barriers that hold down recycling rates.

Technological issues

Over the last two centuries, the minerals and mining industries have been developing the infrastructure to economically win useful materials from primary resources. Although these resources can be compositionally heterogeneous, the technologies in place have been tailored to those specific compositional challenges. Unfortunately, the forms of heterogeneity associated with secondary raw materials are very different from that of mineral ores, so heterogeneity remains a fundamental challenge to any form of recycling.

Additionally, some authors, e.g., [28] have raised concerns that the repeated recycling of a resource compounds this heterogeneity, degrading the resource and making it increasingly difficult to reuse. This is due to the *accumulation* of certain elements, compounds, or microstructural changes in the material stream. The mechanisms for accumulation are varied. Many materials, particularly metals, contain alloying elements that are added purposefully in order to achieve desired properties in the final product. Joining that occurs during product manufacturing can mix materials due to welds, rivets, nails, and adhesives as well as impairing separation at the product's end of life. Elements from cutting machinery, typically iron from steel cutting instruments, can be introduced during fabrication. Many materials become mixed during collection at their end of life as well. Other sources include end-of-life processing such as crushing, shredding, and paint removal. Fiber length, an important property for paper, decreases with increased recycling. Similarly, organics suffer degradation of chain length (or relative molecular mass) as well as introduction of impurities during their service life or through the rigors of recycling.

Regardless of the mechanism, there is a variety of operational and technological solutions that firms can employ to mitigate the impact of heterogeneity and accumulation. These include dismantling of end-of-life products, sorting of scrap, and "filtration" technologies that remove elements in the melt, such as fractional crystallization and vacuum distillation. Although it is clear that such strategies provide technical benefit, they are not always economically viable.

Consumer participation

For many materials, the largest barrier to increased recycling is one that is rarely mentioned by engineers and scientists. That is consumer participation. Given the current price of resources, if a consumer does not place a product into the recycling stream, the resources in that product are not recovered. Unfortunately, for many materials consumer recycling remains limited [29].

Limited consumer participation partly occurs because of the high cost of collection [30][31]. This cost limits the availability and scope of curbside programs and the availability of alternative collection points for recyclables. To expand recycling, it is necessary to remove or reduce these disincentives to return and collect secondary material. Finally, consumer participation is often limited by knowledge or attitudes about recycling. Although it is not a cure-all, the engineering community must work to ensure that accurate information about the benefits to recycling is available. One approach that has proven to be effective in encouraging consumer recycling is creating a financial incentive. As an example, in the USA, states that have deposit systems for beverage containers average a recycling rate over 74% (fraction of containers recovered). In contrast, those states without deposit systems have an average recycling rate of only 24%. As noted previously, the costs and benefits of recycling must always be evaluated for each case. Nevertheless, this clearly shows that much higher levels of recovery are achievable.

Economic and energetic issues

Despite all of the benefits of recycling, it is still true that some effort must be expended to separate particular waste streams and get them to appropriate processing systems so that resources can be recovered and recycled. In most cases, this requires additional energy and comes at an added cost compared with conventional waste disposal (which generally must occur whether recycling happens or not). These energy and financial costs derive from the cost of separate collection infrastructure (e.g., recycling centers and trucks), transport of recyclables to generally more geographically distributed recycling facilities, pre-processing of recyclables to segment them and remove contamination, and, finally, from the recycling process itself. Although in many cases these additional costs are more than offset by the benefits of recycling, they should always be carefully considered.

40.10 Summary

Recent price swings have placed a renewed spotlight on the business implications of raw materials. This raises three key questions for firms. (1) How does limited materials availability affect a firm? (2) How can a firm know whether it is using materials that are at risk of becoming of limited availability? (3) What can that firm do to mitigate that risk?

When a limited-availability event occurs, the supply chain could experience significant changes, specifically:

technological: materials substitution and process efficiency;

geographical: mining exploration and source relocation; and

operational: in transportation modes, increased inventory, and development of a recycling infrastructure.

In some cases, even if availability is limited for only a short period of time, these changes can be permanent.

It is important to note that the availability of materials can be limited by *institutional inefficiency*, in addition to the classically discussed mechanism of global *physical constraint*.

The materials-scarcity literature suggests a number of metrics that indicate an increased risk of limited availability. Unfortunately, due to the complexity of any materials market, no one metric captures all aspects of risk. Nevertheless, careful application of metrics offers insights that should help guide a firm's strategy.

One strategy that would benefit most firms that depend on materials is recycling. Recycling provides environmental and economic benefits. Nevertheless, increasing recycling can be challenging, given the heterogeneity of materials in most modern products, lagging customer participation, and challenging economics that militates against investment in technology.

As technology advances, we tap into a broader swath of the periodic table. In many cases, this means relying

41 Life-cycle assessment

Corrie E. Clark

Environmental Science Division, Argonne National Laboratory, Washington, DC, USA

41.1 Focus

Life-cycle assessment (LCA) evaluates the energy and material requirements and resulting environmental impacts of a product or process over its entire life cycle from raw-material extraction to disposal. This examination across the life cycle provides a systems perspective that can aid decision making for product optimization, product selection, and supply-chain management.

41.2 Synopsis

Life-cycle assessment evaluates the environmental impacts of a product or process over its entire life cycle. It can provide an environmental profile of a system or process through the evaluation of inputs, outputs, and potential environmental impacts. There are generally two approaches to conducting an LCA, namely, a process-oriented approach and an economic input–output approach. In a process-oriented assessment, the inputs and outputs are itemized for each step in the process. Specifically, the five steps considered are raw-material acquisition or extraction, material processing, product manufacturing, use, and recovery and retirement. An optional transportation stage can also be added. In contrast, the latter type of assessment considers the required materials and energy resources (inputs) of a process to estimate the resulting environmental emissions (outputs).

When completing an LCA, there are typically four steps. These include goal definition, life-cycle inventory, life-cycle impact assessment, and interpretation of results. The goal definition identifies the types of information and data that are needed and the appropriate level of accuracy. The inventory identifies and quantifies the relevant inputs and outputs. The impact assessment characterizes and assesses the environmental burdens identified in the inventory. The interpretation improves understanding of results to aid decision making.

Life-cycle approaches to problem solving have informed other decision-making tools, including life-cycle cost analysis and life-cycle management.

41.3 Historical perspective

The practice of life-cycle assessment developed in the 1960s, when concerns over resource and energy limitations led to an accounting of inputs to assist in projections of future resource and energy use [1], p. 4. At the World Energy Conference in 1963, Harold Smith presented calculations on the cumulative energy requirements for the production of chemical intermediates and products. Researchers at The Coca-Cola Company conducted the first multi-criteria study, which formed the foundation for current LCA methods, in 1969. In the 1970s, LCAs were also combined with economic input–output models, which analyze the process by which inputs from one industry or a series of industries produce outputs for consumption. The combined model could estimate environmental emissions where process-specific emission data were limited or unavailable, in an approach developed by the Russian-American economist Wassily Leontief [2]. In the early 1990s, a lack of transparency in the approaches, assumptions, data, and valuations within LCA research and its expanded use in marketing and policy development revealed a need for standardization. Standardization of LCA methods and applications occurred during the 1990s with the development of a Code of Practice in 1993 by the Society of Environmental Toxicology and Chemistry (SETAC) and a series of LCA standards by the International Organization for Standardization (ISO) between 1997 and 2000. In 2006, these standards were streamlined and revised to improve approaches to life-cycle management, inventory, and assessment [3][4]. In particular, ISO standard 14040 describes the principles and framework for LCA, whereas requirements and guidelines are described in ISO standard 14044. The standards provide an overview of the practice, applications, and limitations of LCA without going into detail on the technique or methodologies. According to the ISO standards, there are four steps to an LCA: goal definition, life-cycle inventory, life-cycle impact assessment, and interpretation of results. Through this approach, LCA evaluates the energy and material requirements and resulting impacts of a product or process from raw-material extraction to disposal (Figure 41.1). In 2002, SETAC and the United Nations Environment Program (UNEP) launched the Life Cycle Initiative, a joint international partnership that shares information about LCA and life-cycle thinking and implements life-cycle approaches around the world. As the field of LCA has developed, the concept of life-cycle thinking has extended into government regulatory programs, since it is increasingly being recognized as a useful decision-making tool.

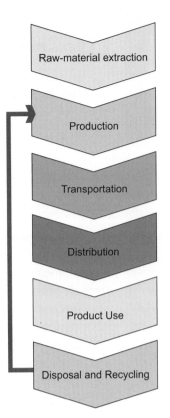

Figure 41.1. A typical life cycle for a product.

41.4 The practice of life-cycle assessment

41.4.1 Goal definition and introduction to conducting life-cycle assessments

Life-cycle assessments are used to evaluate environmental impacts for a variety of purposes. They can assist in improving the environmental aspects of products or processes through identification of impacts at various stages of the life cycle. They can aid decision making and strategic planning for industry, government, or nongovernmental organizations. They also enable the comparison of products that perform the same function (e.g., paper versus plastic versus canvas bags) and can evaluate design alternatives for the same product (e.g., plastic versus aluminum foil for yogurt-container lids). They can assist in the selection of relevant indicators for environmental performance of products or processes and can identify which impacts are the most significant across the life cycle. Such examinations of the life cycle provide a systems perspective, considering each analyzed product or process in the context of external factors with which it interacts, rather than in isolation.

At the start of an assessment, the purpose and method of the LCA must be defined. This first step, known as goal definition, includes the identification of the types of information and data that are needed in order to add value to the decision-making process and the level of accuracy that the results need to have. This step also identifies the selected method for the LCA, any assumptions, and any limitations of the study.

Practitioners define how data should be organized through a **functional unit**, which describes the purpose of the product or process being evaluated. For LCAs conducted to compare products, the functional unit for each product is typically equivalent to its use. For example, when comparing paper plates with ceramic plates, one must consider that paper plates are typically used once and discarded, whereas ceramic plates can be washed and used repeatedly. The functional unit in this example could therefore be defined as the total number of plates used over a set number of meals.

Once the goals and the functional unit of the LCA have been defined, the **life-cycle inventory (LCI)** can be conducted. The LCI identifies and quantifies the relevant inputs and outputs for a given product or system. Typically, an analyst develops a flow diagram of the product or process and a plan for data collection and then collects the data. Conducting and completing an LCI requires analysts to make a number of assumptions in order to describe a complex system. Although hybrid approaches exist, there are generally two approaches to conducting an LCI: the process-oriented approach and the economic input–output approach.

Process-based LCA

In a **process-based life-cycle assessment**, the inputs (materials and energy resources) and the outputs (emissions and wastes to the environment) are itemized for each (or any) given step in producing a product or process. For a product such as a pencil, a list of material inputs would include wood, graphite with clay binder, rubber, and crimped metal; additional inputs might include electricity or natural gas for operating the machinery to create the pencil. A list of waste outputs might include material scraps of wood, graphite, rubber, and metal, and pencils that did not meet specifications. In addition, the process-based life cycle would account for the activities upstream and downstream of the manufacturing of the pencils to include material and energy inputs in resource extraction and processing, distribution of the pencils, use of the pencils, and ultimate disposal of the pencils (Figure 41.2).

For each of these inputs, energy would be consumed to extract the raw materials and to process the materials into a usable input for manufacturing of the pencil. As depicted in Figure 41.2, the product life cycle is typically a linear progression and consists of five stages.

- The first stage is **raw-material acquisition or extraction**. Examples of raw materials include ore, wood, water, and natural gas.

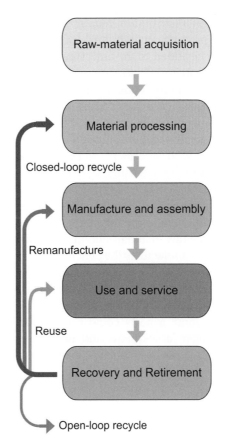

Figure 41.2. Typical life-cycle stages for a process-oriented life-cycle assessment.

- **Material processing** converts the raw materials extracted in the first stage into finished materials that are ready for use in manufacturing. Examples of material processing include the smelting of iron ore to manufacture steel or the removal of bark from trees in pulp and paper production.
- **Product manufacturing** is the assembly of processed materials into a final product. In complex systems, this stage can be subdivided into two parts: the manufacturing of components and the assembly of the separate components into the final product. For example, in automobile manufacturing, the first step would include the manufacturing of aluminum or carbon-fiber sheets into automotive body panels and the assembly of parts such as the engine and chassis. The second manufacturing step would involve the assembly of all of these separate parts into the automobile.
- The fourth stage is the **use phase** and refers to the period during which a consumer has control of the product.
- The final stage is referred to as **recovery and retirement**. At the end of its useful life, a product might be treated and discarded or broken down into component materials for reuse, remanufacturing, or recycling.

If there are significant distribution requirements for materials or the final product, an additional **transportation** stage might also be added to the assessment.

Each stage requires material and energy inputs, and each generates wastes and emissions. Waste generation can include both solid and hazardous wastes. Emissions can be released to the air or water. A process-based LCA will account for these inputs and outputs at each stage.

The main challenge with process-based LCAs is that a system boundary must be determined for each process. That is, a process-based LCA requires the analyst to define what the assessment will include and exclude. In the pencil example, an analyst might exclude the impacts of procuring the raw materials and manufacturing the equipment that manufactures the pencils as being too far removed from the actual pencil-manufacturing process. Although it is necessary to establish a boundary to keep the LCA manageable, it is important to recognize that the exclusion of some aspects could result in an inaccurate or incomplete assessment and an underestimation of the true life-cycle impacts. Nevertheless, when a thorough or detailed inventory is performed, process-based LCA is an appropriate, data-intensive approach.

Economic input–output LCA

The **economic input–output LCA (EIO-LCA)** method estimates the required materials and energy resources and the resulting environmental emissions from activities in the economy. This theory was developed by the economist Wassily Leontief in the 1970s and expanded by researchers at Carnegie Mellon University in the mid 1990s [5]. The EIO-LCA method uses information about industry transactions, such as purchases of materials by one industry from other industries, and information about industries' direct environmental emissions to estimate the total emissions throughout the supply chain. In the USA, the data available for this assessment are extensive. In other economies, the data might be more limited. Such an approach permits a broad and inclusive boundary, since the interdependent transactions and emissions occurring within the economy are captured within all industry sectors. In addition, transactions within a sector are also included. In this manner, impacts from the power-generation and -supply sector account for average impacts from power-generation activities, which would include emissions from fossil-fuel power plants, nuclear power plants, and alternative-energy power plants (e.g., geothermal). The breadth of this approach is also limiting, however, in that attempting to compare impacts within the power-generation and -supply sector itself is challenging because the sector-scale approach can lead to inaccurate assumptions for allocation of inputs and outputs when comparing subsectors.

The EIO-LCA is accomplished by representing monetary transactions in mathematical form. Typical EIO models indicate what materials or services are consumed by other industries and can be used to understand potential changes in demand or in the structure of the economy. For EIO-LCA, the traditional EIO matrix models that indicate transactions between industries are supplemented by information on emissions to the environment. This additional matrix column could be interpreted as an environment sector, where the value in each row represents the emissions output from one industry sector that is then input into the environment.

Consider the building-construction industry. Inputs into building construction include the outputs from industry sectors that produce steel or wood beams, glass windows, foam insulation, carpeting or flooring, and the fuel to operate construction equipment, among others. In turn, these industry sectors require inputs for their operations that are outputs of other sectors. Each of these requirements for goods or services between industry sectors is identified in an EIO model.

In EIO-LCA models one uses as a basis only data that are publicly available. Whereas industry-specific data for a number of environmental effects are publicly available, data that track or measure impacts on such elements as habitat destruction, disposal of nonhazardous solid wastes, and release of nontoxic pollutants to water are not publicly available. Other sources and LCA methods will need to be consulted to account for a full range of environmental impacts.

The results from the EIO-LCA method inform the inventory stage of the overall process of LCA. The results are an inventory estimate of the emissions released or the associated resources consumed within the life cycle of an industry sector. This method does not estimate the environmental impacts or impacts on human health associated with the emissions and consumption. The EIO-LCA inventory approach includes all inputs and outputs within a sector, which can lead to an overestimate of impacts by including "background noise" from the economy. For situations in which a process-based approach is unsuitable because of data or time limitations, EIO-LCA is an appropriate approach.

41.4.2 Considerations in conducting a life-cycle inventory

For both approaches to LCI, an analyst typically develops a flow diagram and a data-collection plan for the product or process, collects the data, performs the evaluation, and reports the results. Whereas both methods consider the contributions of upstream energy and transportation uses, each handles the consideration of these processes differently. For EIO-LCA one considers energy and transportation uses at a sector level.

The following sections describe how energy and transportation uses are taken into account in a process-oriented LCA. In addition to selecting the approach to use for an LCI, there are several steps to complete and aspects to consider when conducting an LCA.

Energy determination in process-based LCA

Energy consumption from a life-cycle perspective includes more than the energy consumed from the electricity grid to manufacture a product or complete a process. Energy is lost in the delivery of energy by the electricity grid and must be taken into account. In addition, energy is required to generate the electricity that is supplied to the grid. Finally, energy is required to extract the feedstock fuel that is used to generate electricity. All of these inputs are taken into account in what is referred to as the **total fuel cycle** or **total energy cycle**.

The energy found in natural resources (e.g., fossil fuels, sunlight, wind) that has not been transformed or converted into any other energy forms is known as **primary energy**. Primary energy is converted into a more convenient form of energy, such as electricity, that is known as **secondary energy**. Energy is lost in the conversion process. The energy associated with extracting raw materials, leading to production of a processed, refined primary fuel (e.g., coal, natural gas, fuel oil) is referred to as **pre-combustion energy**. The energy associated with the combustion of these primary fuels to produce electricity, to generate heat, or to provide energy for transportation is known as **combustion energy**. The sum of the pre-combustion energy and the combustion energy is the total energy.

For example, consider the total amount of energy required to produce 1 kW·h of electricity at a coal-fired power plant. According to Spath *et al.* [6], p. 135, an average of 12.55 MJ is consumed by the system per 1 kW·h of electricity delivered to the grid. The vast majority of this energy is consumed during electricity generation, when the primary fuel, coal, is combusted. This combustion stage accounts for 96.89% of the total (both pre-combustion and combustion) energy. The pre-combustion energy consumed to extract, process, and deliver the coal amounts to only 3.11% of the total energy, or 0.39 MJ.

When determining the energy consumed for a product or process, analysts also must account for the energy required in order to deliver the energy from the generator (e.g., power plant, refinery) to the end user. In the case of power plants, this amount can be found quickly by using a multiplier to account for the transmission-line losses that occur between the manufacturer and the power plant. According to the US Energy Information Agency, in 2008, the average transmission loss in the USA was 6.1% [7]. Multiplying the energy demanded by

1.061 provides the amount of energy that must be delivered to the grid. In the previous coal example, this calculation makes the total energy required to deliver 1 kW·h of electricity to the end user equal to 13.32 MJ.

The next section summarize the steps employed to account for transportation burdens.

Transportation determination in LCA

When considering contributions attributable to transportation burdens, the upstream process must be included as described in the previous section. In addition, the total energy consumed must be allocated according to the extent of vehicle use.

In product-oriented LCA, analysts are often interested in accounting for the energy required to convey a mass of product a certain distance according to a transportation method. For example, when comparing the energy and emissions requirements to convey a bottle of wine to New York City from Napa Valley, CA, or from Bordeaux, France, analysts would first account for the energy and emissions of the different transportation methods, assuming use of a heavy truck to deliver the California wine and a cargo ship for the French wine. Because cargo ships have a much larger carrying capacity than trucks, the energy or fuel burden of a single bottle of wine is much smaller when it is transported by ship than when it is transported by truck. In this example, without accounting for the inputs and outputs of other life-cycle stages, the CO_2-equivalent emissions burden of transporting a French wine has been reported to be one-third that of transporting a California wine [8], p. 22. To determine the energy burden for transporting a product a certain distance, one must know the mass of the product, the distance traveled, and the method of transport. With this information and the energy requirements to convey a tonne of cargo a unit distance as described in Table 41.1, the associated transportation energy burden can be determined.

Alternatively, one might be interested in comparing the amounts of energy required for various transit technologies. In this case, the assessment could be performed according to the functional unit of a person-trip. Table 41.2 lists selected transit methods, the average number of people conveyed per trip, and the energy consumed to convey one person a distance of 1 km.

Material acquisition and processing

At the beginning of the life cycle, raw materials are acquired through mining, drilling, or harvesting. As mentioned above, energy is required in order to extract these materials, and emissions are generated in the transporting of these materials to the processing site. Material processing is different from the manufacturing phase in that the focus of processing is to convert the raw

Table 41.1. Selected transport methods and the energy requirements for conveying 1 t of cargo 1 km in selected years

Freight-transportation mode	Energy required per tonne-kilometer (MJ)	
	2006[a]	2008[b]
Railroad	0.23	0.3
Waterborne (domestic)	0.37	0.3
Heavy truck	–	2.7
Air	–	10.0

[a] From [9].
[b] From [10], p. 3509.

Table 41.2. Selected transit methods and the energy requirements for conveying one person a distance of 1 km

Vehicle	Load factor (persons per vehicle)	Energy required per passenger-kilometer (kJ)
Cars	1.57	2,304
Personal trucks	1.72	2,587
Motorcycles	1.2	1,215
Transit buses	9.1	2,829
Commercial air	97.2	2,034
Rail (average)	26.3	1,695
Intercity rail	21.7	1,649
Transit rail	24.4	1,689
Commuter rail	34.2	1,729

From [11], p. 63.

materials into a usable form (much like the conversion of primary energy into secondary energy). This phase often includes such processes as oil refining, polymerization, or smelting of ore to produce steel or aluminum. Although each of these processes could be considered in an isolated life cycle, materials such as steel, aluminum, and polymers typically serve as material inputs into the manufacturing of more complex products, such as automobiles and electronics. It is worth noting that traditional system boundaries typically do not include the energy and materials required to produce the equipment that is used to acquire and process materials.

The manufacturing phase

In considering the manufacturing phase, analysts include the energy and materials that are input into the manufacturing process or processes, as well as any emissions generated during this phase. Processes typically taken into account in this phase include stamping, extruding, and molding. As for the system boundaries for material acquisition and processing, the energy and materials required in order to produce the equipment that manufactures a product typically are not considered. Although this assumption might lead to an underestimation of the inventory, the contribution of the materials and energy to produce the equipment to any one product is small because the inputs and outputs are allocated to all products produced by the equipment throughout its use phase.

In manufacturing processes, multiple products might be generated. However, attempts to allocate the energy, materials, and emissions according to each product can present challenges. Instead, unit processes should be broken down into subprocesses to collect the input and output data of these subprocesses separately. Alternatively, the process could be expanded to include related co-products.

When allocation is necessary, there are generally two recommended methods. The preferred allocation method is to partition the inputs and outputs according to the physical relationships. Selecting this method does not necessarily obligate analysts to allocate inputs and outputs according to the mass flow of co-products. When the physical relationship cannot be used as the basis for allocation, one alternative method is to use another relationship, such as the economic value of the products.

If the separate LCI of the co-product is known, a third method, known as displacement, can be considered. In this situation, the estimated inputs and outputs from the separate process inventory can be attributed to the co-product, with the remaining burdens allocated to the primary product.

The use phase

There is typically a transportation burden to deliver the product from the manufacturer to the user. The inputs and outputs for transportation can be assessed as described previously. Typical inputs during the use phase include electricity, fuel, and water. Routine service for maintenance or repair of parts (e.g., electronics, appliances, vehicles) might also be included.

The end-of-life management phase

After the product or process is no longer useful or needed, it enters the final phase of the life cycle, namely end-of-life management. At this stage, a product can be treated, if necessary, for disposal, or it can reenter the same life cycle or enter another life cycle. If the product reenters the life cycle, there are three possible pathways, as demonstrated in Figure 41.2. The first possibility is through **reuse**, during which the product could serve another consumer (e.g., through donation, as is often practiced with cellular telephones and computers). **Remanufacture** requires that a product be disassembled into component parts that can be directed back into the manufacturing phase and incorporated into a new product. It is important to note that not all components will necessarily be acceptable for remanufacturing. Some parts could be worn out, in which case landfilling or incineration might be the only appropriate management options. Products that are suitable for remanufacture include computers, toner cartridges, and automobiles. **Closed-loop recycling** is when materials within products or component parts are reprocessed prior to reentering the same manufacturing process. The recycling of aluminum cans is a good example, since the aluminum must be melted and reprocessed prior to reentering the manufacturing phase of the life cycle of a soft-drink can. **Open-loop recycling** refers to instances in which a material from one life cycle is directed into the processing phase of another product life cycle. Open-loop recycling is common for plastic materials, since the integrity of the plastic declines each time it is reprocessed and, therefore, the plastic is **downcycled** into the life cycle of another product of reduced functionality or quality.

When a product or parts of a product cannot be diverted for recycling, remanufacturing, or reuse, they typically are treated and disposed. For liquids, this might involve industrial or wastewater treatment plants or landfilling for toxic and hazardous materials. For solid wastes, disposal options include landfilling of the waste in the appropriate type of landfill (e.g., municipal waste, construction and demolition debris, hazardous waste) or incineration, with subsequent management of emissions to the air and smaller volumes of solids (e.g., ash). Although these treatment and disposal options are described in this context of LCA, they exist for outputs at all phases of the life cycle.

41.4.3 Life-cycle impact assessment

Once the LCI is complete, the information must be interpreted in order to understand the potential environmental impacts of a product or process. This interpretation is carried out through a **life-cycle impact assessment (LCIA)**, which characterizes and assesses – either quantitatively or qualitatively – the environmental burdens identified in the LCI.

Methodology

There are several steps to completing an LCIA: (1) identification and classification of the impacts, (2) characterization of the impacts, and (3) normalization and valuation of the results. Once the inventory is complete, analysts classify the results or assign them to different impact categories. The outputs in each category are modeled according to an indicator of different kinds of potential (e.g., global warming potential). Normalization, valuation, and interpretation are more subjective steps that are unique to each LCIA, depending on the location of the product or process under study and the goal of the LCA.

Identification and classification

Impacts of the inputs and outputs as documented from the LCI that are of interest to an impact assessment typically focus on human health, ecosystem health, and availability of natural resources. For the LCIA, the inputs and outputs of the LCI are grouped according to impact categories. For example, carbon dioxide (CO_2) emissions would be placed in the global-warming impact category. For results that have multiple impacts such as nitrogen oxides (e.g., acidification, photochemical ozone creation, and eutrophication), there are typically two methods of assigning and allocating the impact categories. For those impacts that are dependent on each other, the LCI results can be partitioned according to their contribution(s) to impact categories as would be the case for nitrogen oxides. For independent effects, however, the inventory results can be applied in their entirety to each category to which they contribute. The method of categorization of results should be clearly documented.

Characterization of impacts

Impact characterization converts LCI results into representative indicators of impacts. Different inventory inputs can be compared directly within an inventory category.

Global-warming potential (GWP) estimates a gas's contribution to global warming by using a normalized scale relative to the potential of CO_2 (i.e., CO_2 provides the baseline unit) such that CO_2 has a GWP of 1. The GWP values are calculated for a specific time scale; most often, regulators use a 100-year time period (Table 41.3). This means that, over a period of 100 years, methane will contribute to global warming at a rate that is 21 times that of carbon dioxide. The GWP for any particular gas depends on the following factors: the gas's absorption of infrared radiation, the wavelengths that are absorbed, and the lifetime of the gas in the atmosphere. A gas with a shorter lifetime might have a large GWP in a short time frame (e.g., 20 years) but less of an impact in longer time frames (e.g., 100 years).

The GWPs are estimates of the relative potential. The values will change, and have changed, as the science improves. Note that GWPs can be calculated for a single type of emission only, not for the combined effects of multiple emissions.

Table 41.3. Estimated GWP contributions of selected greenhouse gases over a 100-year period

Greenhouse gas	GWP (100 years)
Carbon dioxide (CO_2)	1
Methane (CH_4)	21
Nitrous oxide (N_2O)	310
Sulfur hexafluoride (SF_6)	23,900
Hydrofluorocarbons (HFCs)	140–11,700
Perfluorocarbons (PFCs)	6,500–17,700

Source: EPA Mandatory Reporting Rule, Federal Register, October 30, 2009, Table A-1 to Subpart A of Part 98.

The **ozone-depletion potential (ODP)** of a gaseous emission is the amount of degradation the gas can cause to Earth's stratospheric ozone relative to the degradation potential of trichlorofluoromethane (CCl_3F), which is also referred to as CFC-11, R-11, or Freon-11. Thus, trichlorofluoromethane has an ODP of 1 (Table 41.4). The ODP is defined as the ratio of the global loss of ozone attributable to a substance to the global loss of ozone attributable to the same mass of CCl_3F.

The ODPs are estimates of the chemicals' relative potentials. The values will change, and have changed, as the science improves. The ODPs can be calculated for a single type of emission only, not for the combined effects of multiple emissions.

There are several additional measures of a substance's potential for environmental impact, including acidification, photochemical ozone creation, and eutrophication (excessive fertilization of bodies of water that stimulates excessive plant growth). Certain gaseous emissions can react in the atmosphere and form acidic compounds that are removed from the atmosphere during rain events. This "acid rain" is absorbed by plants, soil, and surface waters, all of which can lead to leaf damage and the acidification of the soil. Acidification, which is defined as an increase in the concentration of positive hydrogen (H^+) ions, can lead to increased uptake of

Table 41.4. Estimated ODP contributions of selected chemicals

Chemical name	ODP (Montreal Protocol)
Dichlorofluoromethane ($CHFCl_2$, HCFC-21)	0.04
Monochlorodifluoromethane (CHF_2Cl, HCFC-22)	0.055
Monochlorofluoromethane (CH_2FCl, HCFC-31)	0.02
Tetrachlorofluoroethane (C_2HFCl_4, HCFC-121)	0.01–0.04
Trichlorodifluoroethane ($C_2HF_2Cl_3$, HCFC-122)	0.02–0.08
Dichlorotrifluoroethane ($C_2HF_3Cl_2$, HCFC-123)	0.02
Monochlorotetrafluoroethane (C_2HF_4Cl, HCFC-124)	0.022
Trichlorofluoromethane (CCl_3F, CFC-11)	1
Dichlorodifluoromethane (CCl_2F_2, CFC-12)	1
1,1,2-Trichlorotrifluoroethane ($C_2F_3Cl_3$, CFC-113)	0.8
Monochloropentafluoroethane (C_2F_5Cl, CFC-115)	0.6
Bromochlorodifluoromethane (CF_2ClBr, Halon 1211)	3
Bromotrifluoromethane (CF_3Br, Halon 1301)	10
Dibromotetrafluoroethane ($C_2F_4Br_2$, Halon 2402)	6

http://www.epa.gov/ozone/title6/phaseout/classone.html and http://www.epa.gov/ozone/title6/phaseout/classtwo.html.

heavy metals or reduced uptake of some nutrients and thereby negatively affect plant growth. Excessive acidification of soils affects the solubility of chemical compounds in the soil and changes the availability of nutrients and trace elements for plants. A measure of the ability of a chemical substance to release H^+ ions is known as the **acidification potential**. The acidification potential is normalized relative to that of sulfur dioxide (SO_2).

Tropospheric ozone formation through photochemical oxidation contributes to smog and poor air quality. This complex series of chemical reactions is dependent on ultraviolet radiation, nitrogen oxides (NO_x), and volatile organic compounds (VOCs). The **photochemical ozone-creation potential** of VOCs expresses the potential for ozone formation normalized relative to that of ethylene (C_2H_4).

Overuse of fertilizers can lead to excessive concentrations of nutrients in waterways from runoff. The presence of excessive levels of nutrients can lead to algal blooms, which can reduce the concentration of oxygen in water. Reduced oxygen concentration levels can lead to fish kills and other imbalances in aquatic ecosystems. Determination of the **eutrophication potential** of chemical emissions is limited to substances that contain either nitrogen or phosphorus and is normalized with respect to that of phosphate (PO_4^{3-}).

Calculating the impact potentials – such as those occurring as a result of acidification, photochemical ozone creation, and eutrophication – enables the characterization of different types and quantities of chemical emissions on an equivalent scale to determine the extent of impacts. In addition to the impact potentials discussed, there are several models available that assist in the characterization of impacts on human health, the environment, and natural resources. Models, including Eco-indicator 99, ReCiPE, Life-Cycle Impact Assessment Method Based on Endpoint Modeling (LIME), and USEtox, assist LCA practitioners in conducting impact assessments and aid in the comparison of products or processes.

Normalization and weighting

Although it is an optional step under ISO 14044, normalization enables the results of an LCIA to be considered relative to the total impact of this category to a system. For example, the eutrophication potential of a wastewater-treatment plant can be normalized relative to the collective eutrophication potential of all wastewater-treatment plants, agricultural sources, and atmospheric contributions within a watershed. In a large watershed such as the Chesapeake Bay, the contribution of a single wastewater-treatment plant to eutrophication might be small.

Weighting of the various impact potentials can also aid comparison. In weighting, quantitative weights are assigned to all impact categories to compare them with each other. For example, two processes might be compared where one has a greater impact on land use and the other has a greater impact on water quality. Weighting would determine the relative importance of the two impacts. For water-scarce regions, impacts on water quality might be of greater concern than those on land use; therefore, stakeholders in that region might place greater value or weight on water quality in choosing between the two processes.

41.4.4 Interpretation of results

Although there is a variety of approaches to interpreting LCI and LCIA results, all methods aid decision makers through improved understanding of the results. Analytical tools to assess the robustness of the results include sensitivity, uncertainty, and variation analyses. Understanding the sensitivity can indicate where improved data are needed and identify where small improvements could lead to significant reductions in impacts. When data are limited, combining the sensitivity and uncertainty analyses can inform those conducting an LCA where to focus future efforts by identifying where uncertainties significantly affect the results. Variation analysis or scenario assessments can interpret results under different conditions, components, or scales to assist in comparing alternatives. Table 41.5 lists additional useful analyses for interpreting results.

41.4.5 Life cycle beyond the assessment

In addition to LCA, there are several tools and approaches that incorporate life-cycle thinking into decision making. **Life-cycle cost (LCC)** analysis calculates costs over the entire life cycle of a product or process. These costs include conventional costs such as initial investment, capital, operating, maintenance, and closing costs, as well as less tangible costs. For example, when comparing the LCC of a vegetated or "green" roof with that of a conventional roof, the LCC would include stormwater management fees allocated to the impervious surface of the conventional roof and the contribution of a building's energy consumption resulting from heat loss and gain through both roof alternatives. Normally, LCC analysis does not include external costs that are not seen by the user (e.g., the building's owner in the roof example). Other single-issue LCAs include the currently popular carbon-footprint and water-footprint calculations. **Life-cycle management (LCM)** is a management approach wherein decision making at all organizational levels is conducted in a coordinated fashion and seeks to recognize operational interdependences. Such a management approach addresses an organization's improvements to its technological, economic, environmental, and social components or systems to improve the sustainability of the products and services it provides. By enabling

Table 41.5. Various analysis approaches for interpreting results of life-cycle assessments

Analysis	Purpose	Example
Contribution	Identify the extent to which an environmental load from a unit process is responsible for a potential environmental impact	Determine the contribution of CO_2 emissions resulting from the consumption of energy required to pump water in the life cycle of bottled water
Dominance	Identify the life-cycle stage that causes the greatest potential environmental impact	Determine the stage (e.g., extraction, processing, fuel use) of coal-fired power production that generates the greatest amount of emissions (e.g., criteria for air pollutants, CO_2)
Breakeven	Investigate trade-offs between products	Determine the number of times a glass milk container must be reused before the energy consumed in its production process and use phases is equal to the energy consumed to produce and use an equivalent number of disposable milk containers
Comparative	List LCA results for different alternatives	Determine the CO_2 emissions per kilowatt-hour of electricity produced for a variety of electricity-generation technologies

From [12], pp. 33–34.

a systems approach to problem solving, such life-cycle tools can help an organization address the environmental, economic, and social concerns relating to its product(s) and/or process(es) in order to support efforts to develop more sustainable systems.

41.5 Summary

In LCA one evaluates the environmental impacts of a product or process over its entire life cycle through the accounting of inputs, outputs, and potential environmental impacts. Common approaches include process-based LCA and EIO-LCA.

The steps involved in conducting an LCA include defining the scope, conducting an inventory, assessing the impacts, and interpreting the results.

- Defining a goal, scope, and definition for the LCA establishes system boundaries and the functional unit for comparison of the product or processes.
- The LCI creates an inventory of input and output data for the system.
- The LCIA evaluates the significance of the potential environmental impacts associated with the inventory-assessment results.
- Interpretation of the results provides both a summary and recommendations according to the inventory and impact assessments to assist decision makers.

41.6 Questions for discussion

1. What are the benefits and limitations of process-oriented LCA and EIO-LCA?
2. What are the differences among pre-combustion energy, combustion energy, primary energy, and secondary energy?
3. Sulfur hexafluoride (SF_6) under standard conditions is a gaseous insulator that is often used in electrical equipment, including transmission substations, current transformers, and circuit breakers. The US Environmental Protection Agency (EPA) established the SF_6 Emission Reduction Partnership for Electric Power Systems in 1999. In 1999, the partnership members measured emissions at their respective facilities of 692,652 lb (314,182 kg) of SF_6. In 2006, the partnership had reduced annual emissions to 377,140 lb (171,068 kg), a 45.6% reduction. What is the GWP of the emissions avoided in 2006?
4. When accounting for the transporting of fuels or other products, allocation of transportation energy and emission burdens is necessary. For example, the distance to transport natural gas from source to power plant is 3,000 miles (4,828 km). The main pipeline is a 30-in. (76-cm) standard steel pipe [with a wall thickness of 0.375 in. (1 cm) and a mass of 118.6 lb per linear foot (176 kg m^{-1})] that conveys the gas 80% of the distance. This pipeline is shared

by many users, and the power plant utilizes 19% of the flow delivered in this pipeline. The fuel is conveyed the remaining distance to the power plant through a steel pipeline with a diameter of 18 in. (46 cm) [wall thickness 0.375 in. (1 cm), 70.59 lb per ft (105 kg m^{-1})]. What is the total amount of steel from the pipelines allocated to the power plant?

5. Select an energy-generation technology described in earlier chapters. How do the processes employed to produce energy using that technology fit into the various life-cycle stages described in this chapter? If you conducted an LCA, what would be a goal of the assessment? What would be the data inputs and outputs needed in order to complete a life-cycle inventory?

41.7 Further reading

- Green Design Institute, 2010, *Economic Input–Output Life Cycle Assessment*, http://www.eiolca.net/.
  This resource explains EIO-LCA in further detail and provides access to models.
- **H. Bauman** and **A. Tillman**, 2004, *The Hitch Hiker's Guide to LCA*, Lund, Studentlitteratur AB.
- **T. E. Graedel** and **B. R. Allenby**, 1995, *Industrial Ecology*. Englewood Cliffs, NJ, Prentice Hall.
- **J. B. Guinée**, 2002, *Handbook on Life Cycle Assessment: Operational Guide to the ISO Standards (Eco-Efficiency in Industry and Science)*, Amsterdam, Springer Netherlands.
  These titles provide additional information on process-based LCA and LCIA and describe various types of environmental impact potentials (acidification, eutrophication, etc.).
- **W. Klöpffer**, 1996, "Allocation rule for open-loop recycling in life cycle assessment," *Int. J. Life Cycle Assessment*, **1**(1), 27–31.
  Klöpffer explains approaches to integrating recycling and allocation into LCA.
- UNEP/SETAC Life Cycle Initiative, 2009, *Guidelines for Social Life Cycle Assessment of Products*, available at http://www.uneptie.org/shared/publications/pdf/DTIx1164xPA-guidelines_sLCA.pdf.
  In addition to impacts assessed by the approaches presented here, there are social and socioeconomic impacts of human activities. Guidelines for conducting research in these areas using LCA are presented in this report.
- *The International Journal of Life Cycle Assessment* is the first journal devoted to LCA and includes articles on LCA methodology, governmental activities associated with LCA, case studies, and examples from industrial applications.

Acknowledgments

This chapter has been created by the UChicago Argonne, LLC, Operator of Argonne National Laboratory ("Argonne"). Argonne, a US Department of Energy Office of Science laboratory, is operated under Contract No. DE-AC02–06CH11357. The US Government retains for itself, and others acting on its behalf, a paid-up, non-exclusive, irrevocable worldwide license in said article to reproduce, prepare derivative works, distribute copies to the public, and perform publicly and display publicly, by or on behalf of the Government.

41.8 References

[1] Scientific Applications International Corporation (SAIC), 2006, *Life Cycle Assessment: Principles and Practice*, EPA/600/R-06/060, Washington, DC, US Environmental Protection Agency, National Risk Management Research Laboratory, Office of Research and Development, http://www.epa.gov/NRMRL/lcaccess/pdfs/600r06060.pdf.

[2] **T. Hawkins, C. Hendrickson, C. Higgins**, and **H. S. Matthews**, 2007, "A mixed-unit input–output model for environmental life-cycle assessment and material flow analysis," *Environmental Sci. Technol.*, **41**(3), 1024–1031.

[3] International Standards Organization, 2006, *Environmental Management – Life Cycle Assessment – Principles and Framework ISO 14040*.

[4] International Standards Organization, 2006, *Environmental Management – Life Cycle Assessment – Requirements and Guidelines ISO 14044*.

[5] **C. T. Hendrickson, L. Lave**, and **H. S. Matthews**, 2006, *Environmental Life Cycle Assessment of Goods and Services: An Input–Output Approach*. Washington, DC, Resources for the Future.

[6] **P. L. Spath, M. K. Mann**, and **D. R. Kerr**, 1999, *Life Cycle Assessment of Coal-Fired Power Production*, NREL/-TP-570–25119, Boulder, CO, National Renewable Energy Laboratory.

[7] Energy Information Administration (EIA), 2010, "Table 10. Supply and disposition of electricity, 1990 through 2008 (million kilowatthours)," in *United States Electricity Profile*. DOE/EIA-0348(01)/2, US Department of Energy, http://www.eia.doe.gov/cneaf/electricity/st_profiles/sept10us.xls.

[8] **A. R. Williams**, 2009, "Environment: the toll of wine," *National Geographic Magazine*, May, 29.

[9] US Department of Transportation, Bureau of Transportation, 2008, *Transportation Statistics Annual Report*, http://www.bts.gov/publications/national_transportation_statistics/2008/index.html.

[10] **C. L. Weber** and **H. S. Matthews**, 2008, "Food-miles and the relative climate impacts of food choices in the United States," *Environmental Sci. Technol.*, **42**(10), 3508–3513.

[11] **S. C. Davis, S. W. Diegel**, and **R. G. Boundy**, 2009, *Transportation Energy Data Book: Edition 28*, US Department of Energy's Energy Efficiency and Renewable Energy and Oak Ridge National Laboratory, ORNL-6984, http://cta.ornl.gov/data/tedb28/Edition28_Full_Doc.pdf.

[12] **D. Elcock**, 2007, *Life-Cycle Thinking for the Oil and Gas Exploration and Production Industry*, ANL/EVS/R-07/5, Argonne National Laboratory.

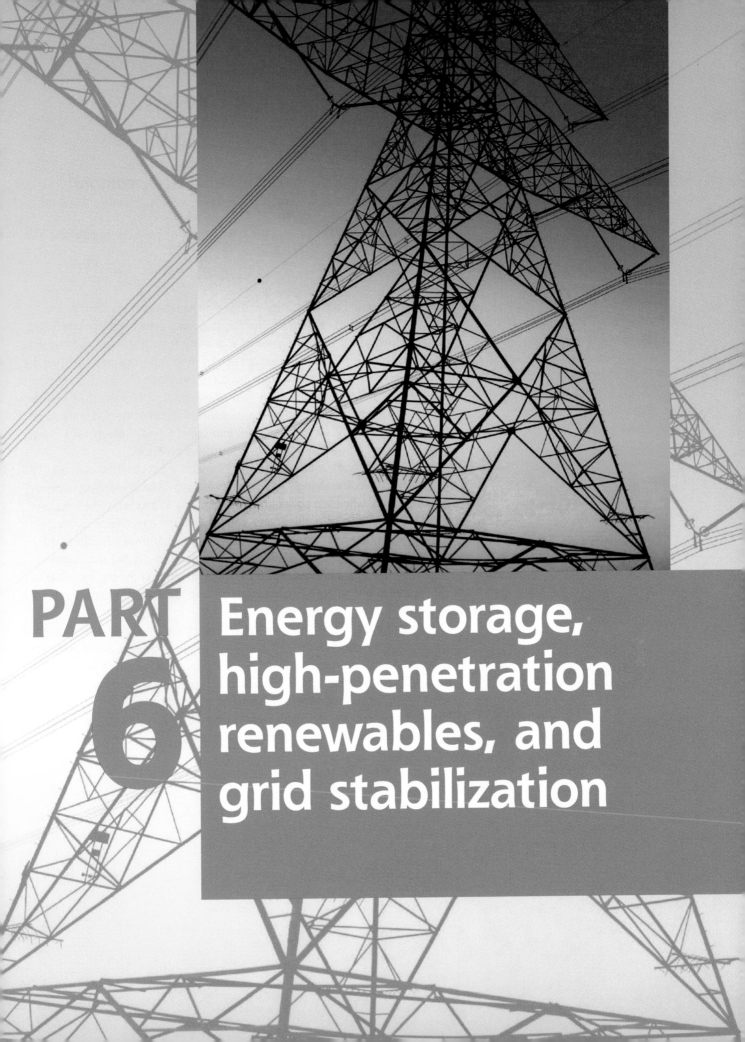

PART 6

Energy storage, high-penetration renewables, and grid stabilization

42 Toward the smart grid: the US as a case study

S. Massoud Amin[1] and Anthony M. Giacomoni[2]

[1]Technological Leadership Institute, College of Science and Engineering, University of Minnesota, Minneapolis, MN, USA

[2]Electrical and Computer Engineering Department, University of Minnesota, Minneapolis, MN, USA

42.1 Focus

Electric power systems constitute the fundamental infrastructure of modern society. Electric power grids and distribution networks, often continental in scale, reach virtually every home, office, factory, and institution in developed countries and have made remarkable, albeit remarkably insufficient, penetration in developing countries such as China and India.

42.2 Synopsis

The electric power grid can be defined as the entire apparatus of wires and machines that connects the sources of electricity, namely the power plants, with customers and their myriad needs. Power plants convert a primary form of energy, such as the chemical energy stored in coal, the radiant energy in sunlight, the pressure of wind, or the energy stored at the core of uranium atoms, into electricity, which is no more than a temporary, flexible, and portable form of energy. It is important to remember that electricity is not a fuel: it is an energy carrier. At the end of the grid, at factories and homes, electricity is transformed back into useful forms of energy or activity, such as heat, light, torque for motors, or information processing.

Electric power grids which once were "loosely" interconnected networks of largely local systems, increasingly host large-scale, long-distance wheeling of power (the movement of wholesale power from one company to another, sometimes over the transmission lines of a third-party company) from one region to another. Likewise, the connection of distributed resources, primarily small generators at the moment, is growing rapidly. The extent of interconnectedness, like the number of sources, controls, and loads, has grown with time. In terms of the sheer number of nodes, as well as the variety of sources, controls, and loads, electric power grids are among the most complex networks ever made.

This chapter focuses on the technical aspects of the challenges posed by this rapid growth: improving existing technology through engineering and inventing new technologies requiring new materials. Some materials advances will improve present technology (e.g., stronger, higher-current overhead lines), some will enable emerging technology (e.g., superconducting cables, fault-current limiters, and transformers), and some will anticipate technologies that are still conceptual (e.g., storage for extensive solar or wind energy generation).

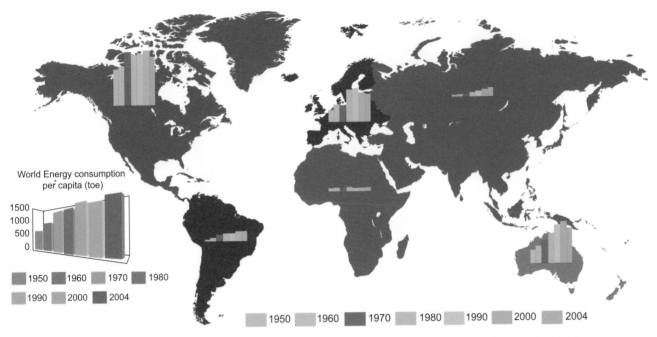

Figure 42.1. World energy consumption per capita (1950–2004). See Chapters 2 and 6.

42.3 Historical Perspective

From a historical perspective, the electric power system in the USA evolved in the first half of the twentieth century without a clear awareness and analysis of the system-wide implications of its evolution. In 1940, 10% of the energy consumption in the USA was used to produce electricity. By 1970, this had risen to 25%, and by 2002 it had risen to 40%. The grid now underlies every aspect of our economy and society, and it has been hailed by the National Academy of Engineering as the twentieth century's engineering innovation most beneficial to our civilization.

The role of electric power has grown steadily in both scope and importance during this time, and electricity is increasingly recognized as a key to societal progress throughout the world, driving economic prosperity and security, and improving the quality of life. Still, it is noteworthy that, at the time of writing, there are about 1.4 billion people in the world with no access to electricity, and another 1.2 billion people have inadequate access to electricity (meaning that they experience outages of 4 h or longer per day).

Since the industrial revolution, worldwide energy consumption has been growing steadily. In 1890, the consumption of fossil fuels roughly equaled the amount of biomass fuel burned by households and industry. In 1900, global energy consumption equaled 0.7 TW (1 TW $= 10^{12}$ W) [1]. The twentieth century saw a rapid 20-fold increase in the use of fossil fuels. Between 1980 and 2004, the worldwide annual growth rate was 2% [1]. According to the US

Energy Information Administration's (EIA's) 2006 estimate, of the estimated 15 TW of total energy consumption in 2004, fossil fuels supplied 86%. North America consumed annually about 4,610 terawatt hours (TWh) of electricity in 2007–8, or about 30% of the estimated global electricity demand. Canada consumed 536 TWh; Mexico consumed 200 TWh; and the USA consumed 3,875 TWh.

The EIA predicts that the world's electricity consumption will double in the next 25 years. Worldwide, current production is near 15,000 billion kilowatt-hours (kWh) per year. For 2030, projections reach more than 30,000 billion kWh per year, which will require a rigorous 2% increase in electricity-generating capacity each year between now and then.

In the coming decades, electricity's share of total energy is expected to continue to grow, as more efficient and intelligent processes are introduced into this network. Electric power is expected to be the fastest-growing source of end-use energy supply throughout the world. To meet global power projections, it is estimated by the US DOE/EIA that over $1 trillion will have to be spent during the next 10 years. The electric-power industry has undergone a substantial degree of privatization in a number of countries over the past few years. Power-generation growth is expected to be particularly strong in the rapidly growing economies of Asia, with China leading the way (Figures 42.1 and 42.2).[1]

---

[1] Refer to http://www.answers.com/topic/world-energy-resources-and-consumption.

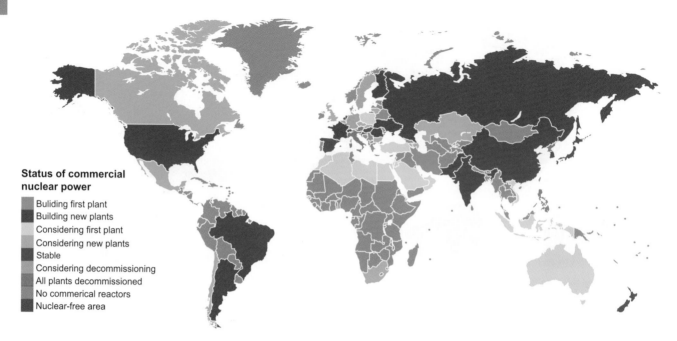

Figure 42.2. The status of commercial nuclear power plants. See Chapter 13.

42.4 The "grid"

When most people talk about the "grid" they are usually referring to the electrical transmission system, which moves the electricity from the power plants to the sub-stations located close to large groups of users. However, it also encompasses the distribution facilities that move the electricity from the substations to the individual users.

In the USA alone, the electrical network includes some 15,000 generators, with an average thermal efficiency of approximately 33% at 10,000 of these power plants. These generators send power through 240,000 miles of high-voltage (HV) transmission lines. In addition, there are about 5,600 distribution facilities. In 2002, the installed generating capacity in the USA was 981,000 MW. If the power plants ran full time, the net annual generation would be $8,590 \times 10^6$ kWh; the actual net generation was $3,840 \times 10^6$ kWh, representing a "capacity factor" of 44.7%.

While electricity demand increased by about 25% since 1990, construction of transmission facilities decreased by about 30% [2]. The planned transmission lines (230 kV or greater) for the period from 2004–2013 total approximately 7,000 miles [3]. According to the EIA, 281 GW of new generating capacity will be needed by 2025 to meet the growing demand for electricity; on the basis of current needs, this implies a need for approximately 50,000 miles of new HV transmission lines.

As currently configured, the continental-scale grid is a multiscale, multilevel hybrid system consisting of vertically integrated hierarchical networks including the generation layer and the following three basic levels.

- **Transmission**: meshed networks combining extra-high-voltage (above 300 kV) and HV (100–300 kV) lines, connected to large generation units, very large customers, and, via tie-lines, to neighboring transmission networks and to the sub-transmission level.
- **Sub-transmission**: a radial or weakly coupled network including some HV (100–300 kV) lines but typically 5–15-kV lines connected to large customers and medium-sized generators.
- **Distribution**: typically a tree network including low-voltage (110–115 or 220–240 V) and medium-voltage (1–100 kV) lines, connected to small generators, medium-sized customers, and local low-voltage networks for small customers.

In its adaptation to disturbances, a power system can be characterized as having multiple states, or "modes," during which specific operational and control actions and reactions are taking place.

- **Normal**: economic dispatch, load frequency control, maintenance, forecasting, etc.
- **Disturbance**: faults, instability, load shedding, etc.
- **Restorative**: re-scheduling, re-synchronization, load restoration, etc.

Besides these spatial, energy, and operational levels, power systems are also multiscaled in the time domain, from nanoseconds to decades, as shown in Table 42.1.

Why the need for a system of such daunting complexity? In principle, it might seem possible to satisfy a small user group – for example, a small city – with one or two generator plants. However, the electricity supply system has a general objective of very high reliability,

Table 42.1. The multiscale time hierarchy of power systems

Action/operation	Time frame
Wave effects (fast dynamics, lightning-caused overvoltages)	Microseconds to milliseconds
Switching overvoltages	Milliseconds
Fault protection	100 ms or a few cycles
Electromagnetic effects in machine windings	Milliseconds to seconds
Stability	60 Cycles or 1 s
Stability augmentation	Seconds
Electromechanical effects of oscillations in motors and generators	Milliseconds to minutes
Tie-line load frequency control	1–10 s; ongoing
Economic load dispatch	10 s to 1 h; ongoing
Thermodynamic changes from boiler control action (slow dynamics)	Seconds to hours
System structure monitoring (what is energized and what is not)	Steady state; ongoing
System state measurement and estimation	Steady state; ongoing
System security monitoring	Steady state; ongoing
Load management, load forecasting, generation scheduling	1 h to 1 day or longer; ongoing
Maintenance scheduling	Months to 1 year; ongoing
Expansion planning	Years; ongoing
Power-plant site selection, design, construction, environmental impact, etc.	2–10 years or longer

and that is not possible with a small number of generators. This is what led the industry to the existing system in North America, where there are just three "interconnects." Within each of these interconnects, all generators are tightly synchronized, and any failure in one generator immediately is covered by other parts of the system. The interconnects are the Eastern, covering the eastern two-thirds of the USA and Canada; the Western, covering the rest of the two countries; and the Electric Reliability Council of Texas (ERCOT) covering most of Texas. The interconnects have limited DC links between them.

One of the important issues about the use of electricity is that storage is very difficult, and thus generation and use must be matched continuously. This means that generators must be dispatched as needed. The US power grids have approximately 150 Control Area Operators using computerized control centers to meet this need. Generally, generators are classified as baseload generators, which are run all the time to supply the minimum demand level; peaking generators, which are run only to meet power needs at maximum load; and intermediate generators, which deal with the rest. Actually, the dispatch order is much more complicated than this, because of the variation in customer demand from day to night and season to season.

42.5 Electrical transmission lines: challenges and materials solutions

Transmitting electric power over large distances can result in losses. The major loss is heat, and this can be reduced by increasing the voltage and decreasing the current. Most of the power in the USA is alternating current (AC), and this allows the voltage from the generator to be stepped up. This is done using transformers; the unit in which this is done is called a step-up transmission substation. Long-distance transmission is typically done with overhead lines of voltages of 110–765 kV. The capacity of an overhead line varies with the voltage and distance; thus, a 765-kV line for a 100-mile length has a maximum capacity of 3.8 GW; for a 400-mile length, the capacity is 2.0 GW. To avoid system failures, the amount of power flowing over each transmission line must remain below the line's capacity.

The principal limitation on the capacity of a line is its temperature. As it gets warmer, it sags; and, in the worst cases, it may touch trees or the ground. Another factor is the mechanical strength of the support structure. Conductors with higher strength-to-weight ratios for a given current-carrying capacity may increase the overall capacity of the right-of-way. Typically, the right-of-way for a 230-kV transmission line is 75–150 feet or more.

The standard for overhead conductors in transmission systems is Aluminum Conductor Steel Reinforced (ACSR) [4]. This consists of fibers of aluminum twisted around a core of steel fibers. The steel core provides the mechanical strength and the aluminum provides the electrical conductivity. Some alternative composite cable materials have been developed over the last several years [5]; the basic candidate composite materials for the substitution of core support members include 1350 H19 aluminum, stainless steel, S-2 glass fibers, E-glass fibers, epoxy resins, T-300 carbon fibers, and Kevlar 49 fibers. For example, a composite formed of a polyester with 54 vol% of an E-glass fiber as a unidirectional satin cloth has been tested.

More recently, 3M is developing and ORNL is testing designs of advanced overhead cables, using a composite core in place of the steel; this is an aluminum metal matrix containing Nextel fibers. The conductor wires are made of an aluminum–zirconium alloy; the zirconium precipitates provide a dispersed strengthening to an essentially pure (and thus high-conductivity) aluminum. Kirby [4] points out that the improved composite conductor substituted for the traditional ACSR in an existing transmission line could carry up to three times the current without the need for tower modification or additional rights-of-way. The current objective is to develop a conductor to increase the capacity of existing corridors by five times that of ACSR at current cost by 2010. The ultimate stretch goal is to achieve transmission-corridor power densities for cables and conductors of 50 times that of ACSR by 2025.

42.6 Chief grid problems

It is inevitable that an electrical grid built on such a huge scale in a patchwork manner over 100 years will have reliability issues [6][7][8]. Several cascading failures during the past 40 years have spotlighted the need to understand the complex phenomena associated with power systems and the development of emergency controls and restoration.

In addition to the mechanical failures, overloading a line can create power-supply instabilities such as phase or voltage fluctuations. For an AC power grid to remain stable, the frequency and phase of all power-generation units must remain synchronous within narrow limits.

A generator that drops 2 Hz below 60 Hz will rapidly build up enough heat in its bearings to destroy itself. As a result, circuit breakers trip a generator out of the system when the frequency varies too much. However, much smaller frequency changes can indicate instability in the grid: in the Eastern Interconnect, a 30-mHz drop in frequency reduces the power delivered by 1 GW [9]. Moreover, power outages and power-quality disturbances cost the US economy over $80 billion annually.

The electricity grid faces (at least) three looming challenges: its organization, its technical ability to meet 25-year and 50-year electricity needs, and its ability to increase its efficiency without diminishing its reliability and security.[2] Starting in 1995, the amortization/depreciation rate exceeded utility-construction expenditures. Since that crossover point in 1995, utility-construction expenditures have lagged behind asset depreciation. This has resulted in a mode of operation of the system that is analogous to harvesting more rapidly than planting replacement seeds. As a result of these diminished "shock absorbers," the electric grid is becoming increasingly stressed, and whether the carrying capacity or safety margin to support predicted demand will exist is in question.

To assess the impacts of power outages, actual electric-power-outage data for the USA, which are available from several sources, including the EIA and the North American Electric Reliability Corporation (NERC), were analyzed. Both databases are extremely valuable sources of information and insight. However, in both databases, a report of a single event may be missing certain data elements such as the amount of load dropped or the number of customers affected. In general, the EIA database contains more events, and the NERC database gives more information about each listed event.

In both sets of data, each five-year period was worse than the preceding one (Figure 42.3). An analysis of the NERC data collected revealed that during the period 1991–2000 there were 76 outages of 100 MW or more in the second half of the decade, compared with 66 such occurrences in the first half. Furthermore, there were 41% more outages affecting 50,000 or more consumers in the second half of the 1990s than in the first half (58 outages in 1996–2000 versus 41 outages in 1991–1995), and, between 1996 and 2000, outages affected 15% more consumers than they did between 1991 and 1995 (the average size per event was 409,854 customers affected in

---

[2] Over the last decade, the August 2003 Blackout (over 50 million consumers affected and more than $6 billion in losses), Hurricane Katrina (more than 1,800 deaths and over $150 billion in economic losses), and the August 1, 2007 collapse of the I-35W bridge in Minneapolis (killing 13 and disrupting traffic and the local economy for a year), in addition to hundreds of blackouts, water-main breaks, and daily traffic gridlocks, have stimulated growing public awareness of the necessity for accelerated programs of replacement, rehabilitation, and new investment in the US infrastructure.

Figure 42.3. Power outages have
steadily increased [10].

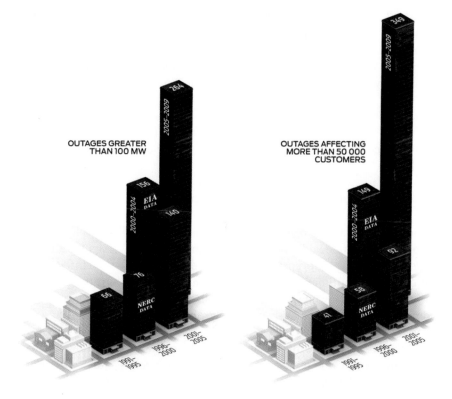

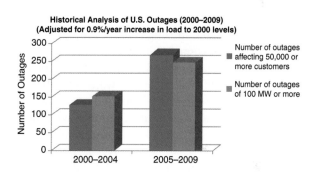

Figure 42.4. Electric power outages over 100 MW and
affecting over 50,000 consumers during 2000–2009 in
the USA (EIA data), adjusted for 0.9% annual load
increase.

the second half of the decade versus 355,204 customers
in the first half of the decade). Similar results were
determined for a multitude of additional statistics such
as the kilowatt magnitude of the outage, average load
lost, etc. These trends have persisted in this decade. The
NERC data show that during 2001–2005 we had 140
occurrences of outages of 100 MW or more, and 92
occurrences in which 50,000 or more customers were
affected.

According to data assembled by the EIA for most
of the past decade, there were 156 outages of 100 MW
or more during 2000–2004; the number of such outages
increased to 264 during 2005–2009. The number of US
power outages affecting 50,000 or more consumers
increased from 149 during 2000–2004 to 349 during
2005–2009. In 2003 EIA changed their reporting form
from EIA-417R to OE-417. In all, the reporting requirements are very similar, with OE-417 being a little
more stringent. The main change in the requirement
affecting the above figures is that all outages greater
than 50,000 customers for 1 hour or more be reported
in OE-417, where it was only required for 3 hours or
more in EIA-417R prior to 2003. Adjusting for the
change in reporting in 2003 (using all the data from
2000–2009 and only counting the outages that met
the less stringent requirements of the EIA-417R form
used from 2000–2002) and also adjusting for a 0.9% per
year increase in load to 2000 levels, a similar trend
results: There were 152 outages of 100 megawatts
or more during 2000–2004; such outages increased to

248 during 2005–2009. The number of U.S. power
outages affecting 50,000 or more consumers increased
from 130 during 2000–2004 to 272 during 2005–2009
(Figure 42.4).

Furthermore, the US electrical grid has been
plagued by ever more and ever worse blackouts over
the past 15 years. In an average year, outages total 92
minutes per year in the Midwest and 214 minutes per
year in the Northeast. Japan, by contrast, averages only
4 minutes of interrupted service each year.

As energy professionals and electrical engineers,
we find it hard to imagine how anyone could believe
that in the USA we should learn to "*cope*" with these
increasing blackouts – and that we do not have the

technical expertise, the political will, or the money to bring our power grid up to twenty-first-century standards. Coping as a primary strategy is ultimately defeatist. We absolutely can meet the needs of a pervasively digital society that relies on microprocessor-based devices in vehicles, homes, offices, and industrial facilities. Furthermore, it is not just a matter of "*can.*" We *must* – if the USA is to continue to be an economic power. However, it will not be easy or cheap.

42.7 A smart self-healing grid to the rescue?

How to control a heterogeneous, widely dispersed, yet globally interconnected system is a serious technological problem. It is even more complex and difficult to control it for optimal efficiency and maximum benefit to the consumers while still allowing all its business components to compete fairly and freely. Above, a brief overview of some of the key issues and the context in which the electricity infrastructure is being operated was provided; in this part, a strategic vision of a self-healing smart grid, an "electrinet," extending to a decade, or longer, that would enable more secure and robust systems operation, security monitoring, and efficient energy markets is presented.

During 1998–2002, through the Complex Interactive Networks/Systems Initiative (CIN/SI), the mathematical foundations and simulations for enabling the smart self-healing grid were established showed that the grid can be operated close to the limit of stability, given adequate situational awareness combined with better secure communication and controls [11].

As part of enabling a self-healing grid, adaptive protection and coordination methods that minimize impact on the whole system performance (load dropped as well as robust rapid restoration) were developed. Note that, while computation is now heavily used in all levels of the power network (e.g., for planning and optimization, fast local control of equipment, and processing of field data), coordination across the network happens on a slower time scale, being based on a system of operation that was developed in the 1960s. Some coordination occurs under computer control, but much of it is still based on telephone calls between system operators at the utility control centers, especially *during* emergencies.

In any situation subject to rapid changes, completely centralized control requires multiple, high-data-rate, two-way communication links, a powerful central computing facility, and an elaborate operations control center. But all of these are liable to disruption at the very time when they are most needed (i.e., when the system is stressed by natural disasters, purposeful attack, or unusually high demand).

When failures occur at various locations in such a network, the whole system breaks into isolated "islands," each of which must then fend for itself. However, in the proposed smart grid, in which the intelligence is distributed throughout the components in the system acting as independent agents, components in each island have the ability to reorganize themselves and make efficient use of available local resources until they are able to rejoin the network. A network of local controllers can act as a parallel, distributed computer, communicating via microwaves, optical cables, or the power lines themselves, and intelligently limiting their messages to only that information necessary to achieve global optimization and facilitate recovery after failure.

Integrated sensing, communications, and control of the power grid and pertinent issues of why/how to develop controllers for centralized vs. decentralized control, together with adaptive operation and robustness against disturbances that included various types of failures, were demonstrated. As expressed in the July 2001 issue of *Wired* magazine, "The best minds in electricity R&D have a plan: Every node in the power network of the future will be awake, responsive, adaptive, price-smart, eco-sensitive, real-time, flexible, humming – and interconnected with everything else" [12].

Since then, there have been several convergent definitions of "smart grids," including within the 2007 Energy Bill, in addition to informative reports by the EPRI (1998–present), NIST, and US DOE (2007–2010), and definitions by the IEEE, FERC, GE, and Wikipedia. Many define "smart grid" in terms of its functionalities and performance objectives (e.g., two-way communications, interconnectivity, renewable integration, demand response, efficiency, reliability, self-healing).

In a recent nationwide survey, most of the consumers in the USA (~68%) did not know what "smart grid" meant. Thus, we must assess and clearly articulate answers to the following questions.

1 What is the "smart grid"?
2 What are the costs/benefits and range of new consumer-centered services enabled by smart grids?
3 What is the smart grid's potential to drive economic growth?

42.7.1 Regarding the first question . . .

The following are definitions for the smart "self-healing" grid.

● The term "smart grid" refers to the use of computer, communication, sensing, and control technology, which operates in parallel with an electric power grid for the purpose of enhancing the reliability of electric power delivery, minimizing the cost of electric energy to consumers, improving security, quality, resilience, and robustness, and facilitating the interconnection of new generating sources to the grid.

- It is a system that uses information, sensing, control, and communication technologies to allow it to deal with unforeseen events and to minimize their adverse impact. It is a secure "architected" sensing, communications, automation/control, and energy overlaid infrastructure acting as an integrated, reconfigurable, and electronically controlled system that will offer unprecedented flexibility and functionality, and improve system availability, security, quality, resilience, and robustness.

There are many definitions, but there is one vision of a highly instrumented overlaid system with advanced sensors and computing with the use of enabling platforms and technologies for secure sensing, communications, automation, and controls as keys to (1) engage consumers, (2) enhance efficiency, (3) ensure reliability, and (4) enable integration of renewables and electric transportation.

Recent policies in the USA, China, India, the EU, and other nations, combined with the potential for technological innovations and business opportunities, have attracted a high level of interest in smart grids. Smart grids are seen as a fundamentally transformative, global imperative for helping the planet deal with its energy and environmental challenges. The ultimate goal is for an end-to-end electric power system (from fuel source to generation, transmission, distribution, and end use) that will

- allow secure and real-time two-way power and information flows,
- enable integration of intermittent renewable energy sources and help decarbonize power systems,
- enable energy efficiency, effective demand management, and customer choice,
- enable the secure collection and communication of detailed data regarding energy usage to help reduce demand and increase efficiency.

In 2007, the United States Congress passed the Energy Independence and Security Act outlining specific goals for the development of the nation's smart grid. Section 1301 of this Act [13] states that

It is the policy of the United States to support the modernization of the Nation's electricity transmission and distribution system to maintain a reliable and secure electricity infrastructure that can meet future demand growth and to achieve each of the following, which together characterize a Smart Grid:

1 Increased use of digital information and controls technology to improve reliability, security, and efficiency of the electric grid.

2 Dynamic optimization of grid operations and resources, with full cyber-security . . .

This smart grid is a concept and a range of functionalities. It is designed to be inherently flexible, accommodating a variety of energy-production sources and adapting to and incorporating new technologies as they are developed. It allows variable rates to be charged for energy, on the basis of upon supply and demand at the time. In theory, this will incentivize consumers to shift their heavy uses of electricity (such as for heavy-duty appliances or processes that are less time-sensitive) to times of the day when demand is low (called peak shaving or load leveling). As an example of the range of functionalities, in 2008, the US Department of Energy (DOE) defined the functions of a smart grid as

- "self-healing" from power-disturbance events,
- enabling active participation by consumers in demand response,
- operating resiliently against physical and cyber attacks,
- providing power quality for twenty-first-century needs,
- accommodating all generation and storage options,
- enabling new products, services, and markets,
- optimizing assets and operating efficiently.

Smart-grid conceptualization and development is occurring internationally. Some information regarding activities in the EU and in China, for example, is available at http://www.smartgrids.eu/ and http://www.juccce.com/program_events/juccce_china_smart_grid_cooperative, respectively.

42.7.2 Regarding the second set of questions . . .

To begin addressing these questions, the costs of full implementation of a nationwide smart grid in the US over a 20-year period (2010–2030).

- A study by the EPRI published in January 2010 estimates that the actual costs will be approximately $165 billion over the course of 20 years.
- According to the energy consulting firm Brattle Group, the necessary investment to achieve a smart grid is $1.5 trillion spread over 20 years (~$75 billion per year), including new generators and power-delivery systems.
- According to our estimate 1998–present, it will cost $10–13 billion per year for 10 years or longer; about $150–170 billion over this period.

Despite the costs of implementation, the following points pertain to integration of the smart grid.

1 Costs of outages will be reduced by about $49 billion per year.

2 Efficiency will be increased and emissions reduced by 12%–18% per year [14].

3 A greater than 4% reduction in energy use will be achieved by 2030, translating into $20.4 billion in savings.

4 It is more efficient to move electrical power through the transmission system than to ship fuels the same distance. From an overall system's perspective, with goals of increased efficiency, sustainability, reliability, security and resilience, we need both

- develop local microgrids (that can be as self-sufficient as possible and island rapidly during emergencies and
- develop an interconnected, smarter, and stronger power-grid backbone that can efficiently integrate intermittent sources, and provide power for end-to-end electrification of transportation.

5 There will be a reduction in the cost of grid infrastructure expansion and overhaul in response to annual peaks if demand response and smart-grid applications are employed.

6 Cyber/IT security, and overall energy security, will be increased if security is built into the design as part of a layered defense system architecture.

7 Electricity's unique capability to be produced from a wide variety of local energy sources, together with its precision, cleanliness, and efficiency, make it the ideal energy carrier for economic and social development.

From a broader perspective, in a single century, electricity became the foundation and prime mover of our modern society. Not just as a clean and convenient form of energy, but as the toolmaker's dream. Electricity opened the doors of invention to new technologies of incredible precision, intelligence, and communication, and to new forms of instrumentation and innovation.

On options and pathways forward, the question is often asked "should we have a high-voltage power grid or go for totally distributed generation, for example with microgrids?" We need both, since the "*choice*" in the question poses a false dichotomy. It is not a matter of "*this OR that*" – it is a case of "*AND.*" To elaborate briefly, from an overall energy system's perspective (with goals of efficiency, ecofriendliness, reliability, security and resilience) we need both (1) microgrids *AND* (2) a stronger and smarter power grid.

Simply replicating the existing system through expansion or replacement will not only be technically inadequate to meet the changing demands for power, but will produce a significantly higher price tag. Through the transformative technologies outlined here, the nation can put in place a twenty-first-century power

system capable of eliminating critical vulnerabilities while meeting intensified consumer demands, and, in the process, save society considerable expense.

A phased approach to system implementation will allow utilities to capture many cost synergies. Equipment purchases, for example, should emphasize switchgear, regulators, transformers, controls, and monitoring equipment that can be easily integrated with automated transmission and distribution systems. Long-term plans for equipment upgrades should also address system-integration considerations.

Energy policy and technology development and innovation require long-term commitments as well as sustained and patient investments in innovation, technology creation, and development of human capital. Given economic, societal, and quality-of-life issues and the pivotal role of the electricity infrastructure, a self-healing grid is essential.

Considering the impact of regulatory agencies, they should be able to induce the electricity producers to plan and fund the process. That may be the most efficient way to get it into operation. The current absence of a coordinated national decision-making entity is a major obstacle. States' rights and state Public Utilities Commission (PUC) regulations have removed the individual state utility's motivation for a national plan. Investor utilities face either collaboration on a national level or a forced nationalization of the industry.

Revolutionary developments in information technology, materials science, and engineering promise significant improvement in the security, reliability, efficiency, and cost-effectiveness of all critical infrastructures. Steps taken now can ensure that this critical infrastructure continues to support population growth and economic growth without environmental harm.

42.8 Enabling technologies

We have investigated whether there are leading applications of science and technology outside the traditional electric energy industry that may apply in meeting and shaping consumer needs. These applications may include entirely new technologies, not part of the portfolio of traditional electricity solutions and not identified in other tasks, which could be potentially available as well. Some technology areas include

- materials and devices – including nanotechnology, microfabrication, advanced materials, and smart devices;
- mesoscale and microscale devices, sensors, and networks;
- advances in information science: algorithms, artificial intelligence, systems dynamics, network theory, and complexity theory;

- bioinformatics, biomimetics, biomechatronics, and systems biology;
- enviromatics: development and use of new methodologies and the use of state-of-the-art information technology for improved environmental applications;
- other industries – moving to a wireless world – transportation, telecommunications, digital technologies, sensing, and control;
- markets, economics, policy, and the environment;
- end-to-end infrastructure – from fuel supply to end use.

In what follows, we briefly highlight pertinent examples of some of the six technology platforms:

- sensors,
- biotechnology,
- smart materials,
- nanotechnology,
- fullerenes,
- information technology.

These platforms have been selected on the basis of past technology road-mapping efforts to identify key underlying technologies, and the emphasis is primarily on long-term, limit-breaking developments. Higher-temperature alloys for turbines and steam-generator components, for example, are certainly important, but their development is likely to follow from conventional, near-term refinement work and is not discussed here; on the other hand, more-innovative solutions to heat-based turbine problems – and much larger improvements – may result from longer-term research on biomimetic ceramics or fullerene composite materials. The outlook and future possibilities for some of these technology platforms are summarized below.

42.8.1 Sensors

Industry has always been dependent on measurement instruments to ensure safe, efficient processes and operations, and today almost every engineering system incorporates sophisticated sensor technology to achieve these goals. However, an increased focus on cost and efficiency, together with the growing complexity of industrial processes and systems, has placed new demands on measurement and monitoring technology: operators are asking for more-accurate data on more variables from more system locations in real time.

The power industry, with its large capital investment in expensive machinery and its complicated, extremely dynamic delivery system, has an especially pressing need for advanced sensors that are small enough to be used in distributed applications throughout the power system. Continued development of digital control systems to replace far-less-accurate analog and pneumatic controls is a key research focus. Sensors that can accurately detect and measure a wide range of chemical species (e.g., emissions of CO_2, NO_x, and SO_2, and for the health-monitoring and maintenance of devices, including testing of oil in transformers) are needed, as are sensors and gauges robust enough to withstand the harsh temperatures and chemical environments characteristic of power plants. Advanced fiber-optic sensors – devices based on sapphire fibers or fiber Bragg gratings, for instance – are especially important because of their versatility, small size, and freedom from magnetic interference. Overcoming today's limitations on temperature, robustness, versatility, and size will facilitate attainment of a number of long-standing power-system needs, including real-time characterization of plant emissions and waste streams, distributed measurement of transformer winding temperatures, and online monitoring of pH in steam-plant circulation water.

42.8.2 Smart materials

Smart materials will be necessary in the future power grid in order to give it the ability to self-recover with a fast response, in milliseconds, under outage events or terrorist attacks. To accomplish this level of self-recovery, it is necessary to make each component intelligent. Such local, autonomous control will make the system much more resilient with respect to multiple contingencies. Control components must

- be reconfigurable power electronic devices with a distributed controller to function as high-speed switches to re-direct power flow,
- have an overall architecture appropriate to the new control/communication paradigm,
- possess a fault-tolerant, agent-based collaborative intelligence at the lowest level of the system to tremendously expedite its response to contingency events.

One class of materials known as "smart materials and structures" (SMSs) has the unique capability to sense and physically respond to changes in the environment – to changes in temperature, pH, or magnetic field, for example. SMS devices, generally consisting of a sensor, an actuator, and a processor, and based on such materials as piezoelectric polymers, shape-memory alloys, hydrogels, and fiber optics, can function autonomously in an almost biological manner. Smart materials have already shown up in a number of consumer products, and they are being studied extensively for aircraft, aerospace, automotive, electronics, and medical applications. In the electric power field, SMSs hold promise for real-time condition assessment of critical power-plant components, allowing continuous monitoring of remaining life and timely maintenance and component replacement. Control of power-plant-cycle chemistry could be done rapidly and automatically with smart systems to inject chemicals that counter pollutants or chemical imbalances. Control of NO_x creation in boilers could be accomplished by adjusting the combustion process with sensor and activation devices distributed at different locations within the boiler.

On the wires side of the business, smart materials could be utilized to monitor the condition of conductors, breakers, and transformers to avoid outages. Smart materials could also be used to avoid potentially catastrophic subsynchronous resonance in generating units and to adjust transmission-line loads according to real-time thermal measurements. Critical capability gaps relate to integrating smart materials into sensors, actuators, and processors; embedding the SMS components into the structure to be controlled; and facilitating communication between smart-structure components and the external world.

The growing list of smart materials encompasses a number of different physical forms that respond to a wide variety of stimuli.

- **Piezoelectric ceramics and polymers**: materials, such as lead zirconate titanate ceramics and polyvinylidene fluoride polymers, that react to physical pressure. They can be used as either sensors or actuators, depending on their polarity.
- **Shape-memory alloys**: metal alloys, such as nitinol, that can serve as actuators by undergoing a phase transition at a specific temperature and reverting to their original, undeformed shape.
- **Shape-memory polymers**: elastomers, such as polyurethane, that actuate by relaxing to their undeformed shape when heated above their glass-transition temperatures.
- **Conductive polymers**: polymers that undergo dimensional changes on exposure to an electric field. These versatile materials can be used not only as sensors and actuators but also as conductors, insulators, and shields against electromagnetic interference.
- **Electrorheological fluids**: actuator materials containing polarized particles in a nonconducting fluid that stiffens when exposed to an electric field.
- **Magnetorestrictive materials**: molecular ferromagnetic materials and other metallic alloys that change dimensions when exposed to a magnetic field.
- **Polymeric biomaterials**: Synthetic, muscle-like fibers, such as polypeptides, that contract and expand in response to temperature or chemical changes in their environment.
- **Hydrogels**: cross-linked polymer networks that change shape in response to electric fields, light, electromagnetic radiation, temperature, or pH.
- **Fiber optics**: fine glass fibers that signal environmental change through analysis of light transmitted through them. Optical fibers, which are perhaps the most versatile sensor materials, can indicate changes in force, pressure, density, temperature, radiation, magnetic field, and electric current.

These materials, when matched to an appropriate application, provide the base functionality for simple as well as higher-level smart structures and systems. Sensory structures, such as optical fibers embedded in concrete bridge support pillars, merely furnish information about system states; with no actuator, they are able to monitor the health of the structure but cannot physically respond to improve the situation. Adaptive structures contain actuators that enable controlled alteration of system states or characteristics; electrorheological materials, for example, can damp out vibrations in rotating mechanical systems when an electric field is applied. Controlled structures provide feedback between sensors and actuators, allowing the structure to be fine-tuned continuously and in real time; for example, aircraft wings instrumented with piezoelectric sensors and actuators can be programmed to subtly change shape to avoid flutter under problematic wind conditions.

Examples of higher-level smart structures and systems that can be built from smart materials and utilized in the grid include the following.

1 **Flexible AC transmission (FACTS)**. FACTS devices are a family of solid-state power-control devices that provide enhanced power-control capabilities to HVAC grid operators (see Box 42.1). FACTS controllers act like integrated circuits – but scaled up by a factor of 500 million in power. By applying FACTS devices, utilities can increase the capacity of individual transmission lines by up to 50% and improve system stability by responding quickly to power disturbances. There is a need to reduce the costs of FACTS technology to provide for broader use. One method for reducing the costs is to replace the silicon-based power electronics with wide-bandgap semiconductors such as silicon carbide (SiC), gallium nitride (GaN), and diamond.

2 **High-voltage direct-current (HVDC) transmission systems**. These transmission systems are based on the rectification of the generated AC and then inversion back to AC at the other end of the transmission line. Modern systems are based on thyristor valves (solid-state power-control devices) to perform the AC/DC/AC conversions. Conventional HVDC transmission systems have been built with power-transfer capacities of 3,000 MW and ±600 kV. A new class of HVDC converter technology has been introduced in the last few years, with devices referred to as voltage source converters (VSCs). This technology is based on gate turn-off switching technology or

Box 42.1. FACTS devices

FACTS devices are used for the dynamic control of the voltage, impedance, and phase angle of HV AC transmission lines. The main types of FACTS devices are as follows.

(1) **Static VAR compensators (SVCs)**, the most important FACTS devices, have been used for a number of years to improve transmission-line economics by resolving dynamic voltage problems. Their accuracy, availability, and fast response enable SVCs to provide high-performance steady-state and transient voltage control compared with classical shunt compensation. They are also used to dampen power swings, improve transient stability, and reduce system losses by optimized reactive power control. (VAR is Volt-Ampere-Reactive)

(2) **Thyristor-controlled series compensators (TCSCs)** are an extension of conventional series capacitors created by adding a thyristor-controlled reactor. Placing a controlled reactor in parallel with a series capacitor enables a continuous and rapidly variable series compensation system. The main benefits of TCSCs are increased energy transfer, dampening of

power oscillations, dampening of subsynchronous resonances, and control of line power flow.

(3) **STATCOMs** are GTO (gate turn-off-type thyristor)-based SVCs. Compared with conventional SVCs (see above) they do not require large inductive and capacitive components to provide inductive or capacitive reactive power to HV transmission systems. This results in smaller land requirements. An additional advantage is the higher reactive output at low system voltages, at which a STATCOM can be considered as a current source independent from the system voltage. STATCOMs have been in operation for over 10 years.

(4) **Unified power flow controllers (UPFCs)**: connecting a STATCOM, which is a shunt-connected device, with a series branch in the transmission line via its DC circuit results in a UPFC. This device is comparable to a phase-shifting transformer but can apply a series voltage of the required phase angle instead of a voltage with a fixed phase angle. The UPFC combines the benefits of a STATCOM and a TCSC.

insulated gate bi-polar transistors, IGBTs. These devices have higher switching frequency capability, and HVDC transmission is used in long-distance bulk power transmission over land, or for long submarine cable crossings. Altogether, there are more than 35 HVDC systems operating or under construction in the world today. The longest HVDC submarine cable system in operation today is the 250-km Baltic Cable between Sweden and Germany.

3 **Dynamic line rating**. The maximum power that can be carried by a transmission line is ultimately determined by how much the line heats up and expands. The "thermal rating" of a line specifies the maximum amount of power it can safely carry under specific conditions without drooping too much. Most thermal ratings today are static in the sense that they are not changed through the year. For such ratings to be reliable, they must be based on worst-case weather conditions, including both temperature and wind velocity. Dynamic line ratings use real-time knowledge about weather or line sag to determine how much power can be transmitted safely. Typically, a dynamically monitored line can increase its allowable power flow (ampacity) by 10%–15% over that with static ratings.

In the future, smart materials and structures are expected to show up in applications that span the entire electric power system, from power plant to end user.

Smart materials, in all their versatility, could be used to monitor the integrity of overhead conductor splices, suppress noise from transformers and large power-plant cooling fans, reduce cavitation erosion in pumps and hydroturbines, or allow nuclear plants to better handle structural loads during earthquakes.

42.8.3 Advanced hardware

Incorporating smart materials and higher-level smart structures and systems into the grid will require the development of advanced hardware components. These include advanced meters; advanced sensors and monitors; advanced motors (this includes superconducting motors); advanced transformers (this includes the concept of a universal transformer that would be a standardized and portable design, FACTS phase-shifting transformers capable of controlling power flow, and next-generation transformers using solid-state devices and high-temperature superconductors); power electronics (this includes FACTS, solid-state breakers, switchgear, and fault-current limiters); computers and networks; mobile devices; and smart equipment and appliances.

Advanced hardware includes the following.

- Advanced cables
  - Gas-insulated lines for underground cables
  - Advanced composite conductors, lighter and able to carry more current than current ACSR conductors

- High-temperature superconductors (it is pointed out that these could also revolutionize generators, transformers, and fault-current limiters)
- Electric storage
  - Superconducting magnetic energy storage (SMES)
  - Advanced flywheels using composites and/or superconductors for higher efficiency and capacity
  - Flow batteries that charge and discharge fluid between tanks
  - Liquid molten-sulfur batteries built to utility scale

42.8.4 Nanotechnology

The theme of the development of nanotechnology in energy application technology is geared toward two main directions, "nanomaterials for energy storage" and "nanotechnology for energy saving." Owing to the advantages of high reactivity, large surface area (200–2000 m^2 g^{-1}), self-assembly (~1–3 nm of active catalyst), super crystal characteristics (~10–30-nm nanostructures), and special opto-electronic effects of nanomaterials for energy saving, advanced countries are heavily engaged in the development of energy-related nanomaterials.

There is an expectation that with nanotechnologies we will be able to develop power-storage systems with energy densities at least several times higher than those of current batteries. Owing to the small dimensions (5–20 nm) and high specific surface area and special optical properties of nanomaterials, nanotechnology for energy saving is expected to increase the contact area of the medium. This will shorten the response time and improve the thermal conductivity by a factor of two. Nanotechnology applications for energy storage include using nanoparticles and nano-tubes for batteries and fuel cells. Nanotechnology is being used to better the performance of rechargeable batteries through the study of molecular electrochemical behavior. Newly patented lithium-ion batteries, which use nanosized lithium titanate, can provide 10–100 times greater charging/discharging rates than those of the current conventional batteries. Other new batteries that apply nanotechnology could provide added power and storage capabilities by applying a concept based on mechanical resonance using a single micro-electromechanical-system (MEMS) device. Micro-electromechanical systems are devices that use the combined technology of computers and mechanical devices. This combined device improves power density, offering significant benefits for portable equipment. (See Chapters 44 and 46).

42.8.5 Fullerenes and nanotubes

The soccer-ball-patterned C_{60} fullerene and the cylindrical carbon nano-tube have been considered by many to be the ultimate materials, and, even though only small amounts of fullerenes have as yet been produced, researchers have suggested many potential applications. Most of these involve the carbon nano-tube, a long, hollow string with tremendous tensile strength that could be wound into the strongest structural cable ever made. Use of shorter nano-tube strings in metal, ceramic, or polymer composites would create stronger, lighter, more versatile materials than are currently available in any form. Structural beams, struts, and cables for airplanes, bridges, and buildings are envisioned, in addition to uses in rocket nozzles, body armor, and rotating machinery such as flywheels and generators. Electrical applications range from highly conductive (and perhaps superconductive) wires and cables to electron emitters in flat-panel displays and magnetic recording media for data storage. Because nano-tubes are incredibly thin and have such versatile electrical properties, they are seen as ideal building blocks for nanoscale electronic devices. Realization of such possibilities is highly dependent on developing processes for producing high-quality fullerenes in industrial quantities at reasonable cost and finding ways to manipulate and orient nano-tubes into regular arrays. Cost will almost certainly determine whether fullerenes will become true universally used materials or an esoteric, high-cost/high-value option for specialized applications.

42.9 Future opportunities and challenges

To highlight further opportunities where science and technology from other industries could possibly be identified to fill these gaps we must address

- ubiquitous, hierarchical computational ability with perfect software integrated into the power system that enables dynamic control through fast simulation and modeling with complete system visualization,
- low-cost, practical electric and thermal energy storage,
- microgrids – AC and DC, both self-contained, cellular and universal energy systems, and larger building- or campus-sized systems,
- advanced (post-silicon) power electronics devices (valves) to be embedded into flexible AC and DC transmission and distribution circuit breakers, short-circuit current limiters and power-electronics-based transformers,
- power-electronics-based distribution-network devices with integrated sensors and communications,
- fail-safe communications that are transparent and integrated into the power system,
- cost-competitive fuel cells,

- low-cost sensors to monitor system components and to provide the basis for state estimation in real time,
- cost-effective integrated thermal storage (heating and cooling) devices,
- thermal appliances that provide "plug-and-play" capability with distributed generation devices,
- high-efficiency lighting, refrigerators, motors, and cooling,
- enhanced portability by improving storage devices and power-conversion devices,
- efficient, reliable, cost-effective plug-in hybrid vehicles,
- technologies and systems that enable "hardened" end-use devices,
- conductors that enable greatly increased power-flow capability,
- thermoelectric devices that convert heat directly into electricity.

However, these technologies will require sustained funding and commitment to R&D; given the state of the art in electricity infrastructure security and control as indicated in this chapter, creating a smart grid with self-healing capabilities is no longer a distant dream; we have made considerable progress. The cost of a self-healing smart grid will not be cheap – some estimates are that as much as $10–13 billion per year will be needed for a period of 10 years or more for real-world testing and installation. However, the price of electrical failure, estimated to be over $80 billion per year, is not cheap either.

There are signs too that Congress and the government recognize the need for action. In 2005, the White House Office of Science and Technology Policy and the US Department of Homeland Security declared the Self-healing Infrastructure to be one of three strategic thrust areas for the National Plan for R&D in Support of Critical Infrastructure Protection. Furthermore, in 2007, the US Congress passed the Energy Independence and Security Act outlining specific goals for the development of the nation's smart grid.

Nevertheless, considerable technical challenges as well as several economic and policy issues remain to be addressed. At the core of the power-infrastructure investment problem lie two paradoxes of restructuring, one technical and one economic. Technically, the fact that electricity supply and demand must be in instantaneous balance at all times must be reconciled with the fact that new power infrastructure is extraordinarily complex, time-consuming, and expensive to construct. Economically, the theory of deregulation aims to achieve the lowest possible price through increased competition. However, the market reality of electricity deregulation has often resulted in a business-focused drive for maximum efficiency to achieve the highest profit from existing assets and does not result in lower prices or improved reliability. Both the technical paradox and the economic paradox could be resolved by knowledge and technology.

Given economic, societal, and quality-of-life issues and the ever-increasing interdependences among infrastructures, a key challenge before us is whether the electricity infrastructure will evolve to become the primary support for the twenty-first century's digital society – a smart grid with self-healing capabilities – or be left behind as a twentieth-century industrial relic.

42.10 Summary

The electric power grid is one of the largest, most complex systems ever built. Ensuring the security and reliability of this infrastructure is imperative to maintaining our economy and quality of life. Currently, the grid faces numerous technological challenges as it is becoming increasingly stressed as worldwide electricity demand continues to increase and infrastructure investment remains stagnant. A smart self-healing grid promises the ability to overcome these technological challenges by enabling advanced sensing, communications, and control technology to allow the grid to increase energy efficiency, enable real-time two-way communication and power flows, allow the integration of intermittent renewable energy sources, and enable the secure collection of energy data. To achieve this goal, numerous leading science and technology applications outside of the traditional electric energy industry must be utilized. Several leading candidates and platform technologies have been selected and summarized on the basis of their current promise for long-term limit-breaking developments. These platforms include sensors, smart materials, advanced hardware, nanotechnology, and fullerenes. However, to achieve these long-term developments requires sustained funding and commitment to R&D. While such investments will not be cheap, the costs of not doing so are even greater.

42.11 Questions for discussion

1. There are many unanswered questions relating to the smart grid. Who will build the smart grid? Should electric companies build the smart grid when in fact it may reduce their income? If consumers are to pay for the consumer end of the smart grid, how is that to be financed? What is the overall cost–benefit analysis of a smart grid? Where are the greatest savings to be had? Where will the greatest costs be incurred?

2. *Expenditures and investments:* given the persisting lack of R&D funding, combined with reduced infrastructure expenditure during the past 20–25 years,

how can we reverse this trend with the goals of increased reliability, security, efficiency, and sustainability (both at device and at microgrid level as well as macro-level end-to-end system-wide, from energy sources to end use)?

3. *Science and technology:* what are the most efficient pathways forward to redesign, retrofit, and upgrade the North American electric power grid, consisting of over 450,000 miles (100 kV and higher) of electromechanically controlled system into a smart self-healing grid that is driven by a well-designed market approach?

4. *Infrastructure security:* cyber/physical threats are dynamic, evolving quickly, and often combined with lack of training and awareness (forgetting/ignoring the human in the equation). Installing modern communications and control equipment (elements of the smart grid) can help. How can we include security as a design criterion from the start (such as secure sensing, "defense in depth," fast reconfiguration, and self-healing), built into the infrastructure?

5. *Security versus efficiency:* the specter of future sophisticated terrorist attacks raises a profound dilemma for the electric power industry, which must make the electricity infrastructure more secure, while being careful not to compromise productivity. Resolving this dilemma will require both short-term and long-term technology development and deployment that will affect fundamental power-system characteristics. What are risk-managed, pivotal, and strategic science and technology areas that provide non-intrusive (or low-intrusive) yet high-confidence levels of security? What level of threat is the industry responsible for, and what does government need to address? Will market-based priorities support a strategically secure power system?

6. *Centralization versus decentralization of control:* for several years, there has been a trend toward centralizing control of electric power systems. The emergence of regional transmission organizations, for example, promises to greatly increase efficiency and improve customer service. But we also know that terrorists can exploit the weaknesses of centralized control; therefore, smaller and local would seem to be the system configuration of choice. In fact, strength and resilience in the face of attack will increasingly require the ability to bridge simultaneous top-down and bottom-up decision-making in real time – fast-acting and totally distributed at the local level, coordinated at the middle level. What system architecture is most conducive to assuring security while increasing or at least maintaining efficiency?

7. *Wider grid integration and increasing complexity:* system integration helps move power more efficiently over long distances and provides redundancy to ensure reliable service, but it also makes the system more complex and harder to operate. How can we develop fast mathematical approaches to simplify the operation of complex overlaid energy, power, and communications systems and make them more robust in the face of natural or man-made interruptions?

8. *Leadership:* the energy, electric power, computer, communication, and electronics industries need to begin to develop standards for communication, computer message structure, and sensing and control device interfaces before the potential for the smart grid can be realized. What are the most efficient pathways forward to enable this? Over 85% of the infrastructure in the USA is privately owned, so how do we enable public–private partnerships to make the transformation for twenty-first-century lifeline infrastructures happen?

42.12 Further reading

- **M. Amin** and **B. F. Wollenberg**, 2005, "Toward a smart grid: power delivery for the 21st century," *IEEE Power Energy Mag.*, **3**(5), 34–41. This paper provides a vision and approach to enable smart, self-healing electric power systems that can respond to a broad array of destabilizers.

- **M. Amin** (ed.), 2005, special issue on *Energy Infrastructure Defense Systems*, *Proc. IEEE*, **93**(5). This issue is devoted to the defense of energy infrastructure, including topics such as software, applications, and algorithmic developments, the use of sensors and telecommunications to increase situational awareness of operators/security monitors, signals and precursors to failures, infrastructure defense plans, wide-area protection against rare events and extreme contingencies, and rapid/robust restoration.

- **M. Amin**, 2002, "Modeling and control of complex interactive networks," *IEEE Control Systems Mag.*, **22**(1), 22–27. This paper provides a description of the issues dealing with the modeling of critical national infrastructures as complex interactive networks.

- **P. Kundur**, 1994, *Power System Stability and Control*, New York, McGraw-Hill, Inc. This book provides an in-depth explanation of voltage stability, covering both transient and longer-term phenomena and presenting proven solutions to instability problems.

Acknowledgment

This work is supported by the National Science Foundation under grant number 0831059, and grants from EPRI, Honeywell Labs, and the Sandia National Laboratories' Grand Challenge LDRD 11–0268.

42.13 References

[1] **M. Amin**, 2003, "North America's electricity infrastructure: are we ready for more perfect storms?," *IEEE Security Privacy*, **1**(5), 19.

[2] **US Department of Energy**, 2005, *GridWorks: Overview of the Electric Grid*, Washington, DC, US Department of Energy.

[3] **Edison Electric Institute**, 2005, *Meeting U.S. Transmission Needs*.

[4] **B. Kirby**, 2002, *Reliability Management and Oversight. DOE National Transmission Grid Study*.

[5] **G. Newaz**, **D. Bigg**, and **R. Eiber**, 1987, *Structural Composite Cores for Overhead Transmission Conductors*, EPRI Report EM-5110.

[6] **M. Amin** and **P. F. Schewe**, 2007, "Preventing blackouts," *Scientific American*, 60–67.

[7] **P. F. Schewe**, 2007, *"The Grid: A Journey Through the Heart of Our Electrified World*, Washington, DC, Joseph Henry Press.

[8] **S. M. Amin** and **J. Stringer**, 2008, "The electric power grid: today and tomorrow," *MRS Bull.*, **33**(4), 399–407.

[9] **E. J. Lerner**, 2003, "What's wrong with the electric grid?," *Indust. Physicist*, **9**(5), 8–13.

[10] **M. Amin**, 2011, "U.S. electrical grid gets less reliable," *IEEE Spectrum*, **48**(1), 80.

[11] **Electric Power Research Institute**, 2002, *Complex Interactive Networks/Systems Initiative: Final Summary Report*.

[12] **S. Silberman**, 2001, "The energy web," *Wired*, **9**(7), 8 pp.

[13] The Smart Grid Interoperability Panel-Cyber Security Working Group, National Institute of Standards and Technology, 2010, *Smart Grid Cyber Security Strategy and Requirements*, DRAFT NISTIR 7628.

[14] **R. Pratt**, **P. J. Balducci**, **C. Gerkensmeyer** *et al.*, 2010, *The Smart Grid: An Estimation of the Energy and CO_2 Benefits*, PNNL-19112, Pacific Northwest National Laboratory.

43 Consequences of high-penetration renewables

Paul Denholm

Strategic Energy Analysis Center, National Renewable Energy Laboratory, Boulder, CO, USA

43.1 Focus

The use of wind and solar electricity generation has grown tremendously during the last decade. This raises the important question of how these variable and uncertain resources can be effectively used while maintaining reliable electricity generation.

43.2 Synopsis

The large-scale deployment of wind and solar energy creates challenges for grid operators to maintain reliable service. Wind and solar output are variable, uncertain, and often not correlated with normal demand patterns for electricity. At low penetrations in an energy system (up to about 20% on an energy basis) these energy sources act to reduce the fuel use and emissions from conventional power plants used to meet normal variations in electricity demand. These sources can also add varying levels of "firm capacity" to the system, depending on technology and location. Studies have found that current utility systems can accommodate these levels of variable generation sources with a combination of changes in operational practices, but without massive deployment of "enabling" technologies such as energy storage. However, the variability and uncertainty impose modest cost penalties, since utilities require increased operating reserves in order to maintain reliable service. At higher penetrations (beyond 20%) new methods of integrating renewables into the grid are required, including transmitting power over long distances to take advantage of spatial diversity and new generation technologies that can ramp rapidly to respond to variations in demand. At these penetrations variable generation sources also begin to affect the operation of baseload power plants, which creates more challenges for system operators, and may lead to curtailed wind and solar generation. This will begin to decrease the environmental benefits of these renewable sources. At very high penetrations (beyond 30%) the simple coincidence of energy supply and demand limits the useful contributions of wind and solar energy, with wind potentially exceeding the demand for electricity on occasions. This will require deployment of a variety of enabling technologies, including greater use of long-distance transmission, shiftable load, new demands for electricity, such as electric vehicles, and energy storage.

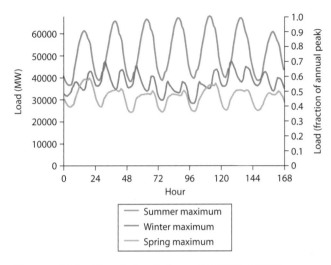

Figure 43.1. Hourly loads from ERCOT in 2005.

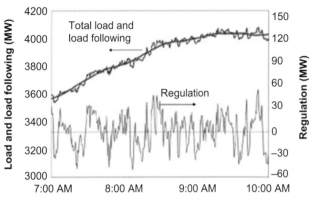

Figure 43.2. System load following and regulation. Regulation (red) is the rapidly fluctuating component of total load (green), while load following (blue) is the slower trend.

43.3 Historical perspective

The power grids in many countries have evolved over the past 100 years and are based on "dispatchable" energy sources such as fossil fuels, nuclear, and hydro-electricity. The operation of electric power systems involves a complex process of forecasting the demand for electricity, and scheduling and operating a large number of power plants to meet that varying demand. The instantaneous supply of electricity must always meet the constantly changing demand, as indicated in Figure 43.1. It shows the electricity demand patterns for three weeks for the Texas power grid during 2005. The USA operates three independent power grids – the Eastern Interconnect, with about 73% of demand, the Western Interconnect, at about 18% of demand, and the Electric Reliability Council of Texas (ERCOT), with about 9% of demand [1]. The seasonal and daily patterns are driven by factors such as the need for heating, cooling, lighting, etc. The trends in Figure 43.1 are typical for those in the USA, where demand is generally highest in the summer. In Europe, demand often is highest during the winter. To meet this demand, utilities build and operate a variety of types of power plant. Baseload plants are used to meet the large constant demand for electricity. These are often nuclear and coal-fired plants, and utilities try to run these plants at full output as much as possible. Variation in load is typically met with load-following or "cycling" plants. These units are typically hydroelectric generators or plants fueled with natural gas or oil. These "load-following" units are further categorized as intermediate-load plants, which are used to meet most of the day-to-day variable demand, and peaking units, which meet the peak demand and often run for less than a few hundred hours per year [2].

In addition to meeting the predictable daily, weekly, and seasonal variation in demand, utilities must keep additional plants available to meet unforeseen increases in demand, losses of conventional plants and transmission lines, and other contingencies. This class of responsive reserves is often referred to as operating reserves and includes meeting frequency regulation (the ability to respond to small, random fluctuations around normal load), load-forecasting errors (the ability to respond to a greater or less than predicted change in demand), and contingencies (the ability to respond to a major contingency such as an unscheduled power-plant or transmission-line outage) [3]. Both frequency regulation and contingency reserves are among a larger class of services often referred to as ancillary services, which require units that can rapidly change output. Figure 43.2 illustrates the need for rapidly responding frequency regulation (red) in addition to the longer-term ramping requirements (blue). In this utility system, the morning load increases smoothly by about 400 megawatts (MW) in two hours. During this period, however, there are rapid short-term ramps of ±50 MW within a few minutes [4].

Because of the rapid response needed by both regulation and contingency reserves, a large fraction of these reserves is provided by plants that are online and "spinning" (as a result, operating reserves met by spinning units are sometimes referred to as spinning reserves) [5]. Spinning reserves are provided by a mix of partially loaded power plants and responsive loads. Power plants providing spinning reserves are providing energy, and, although their efficiency is somewhat less than that of plants operating at rated output, they are not "idling" (burning fuel without delivering useful energy). The need for reserves increases the costs and decreases the efficiency of an electric power system compared with a system that is perfectly predictable and does not experience unforeseen contingencies. These costs result from several factors. First, the need for fast-responding units results in uneconomic dispatch – because plants providing spinning reserve must be operated at part load, they potentially displace more economic units [4]. (Flexible load-following units are often either less efficient or burn more expensive fuel than "baseload" coal or nuclear units.) Second, partial loading can reduce the efficiency of individual power plants. Finally, the reserve requirements increase the number of plants that are online at any time, which increases the capital and operation and maintenance (O&M) costs.

During the past decade, renewable electricity sources such as wind and solar have increasingly been added to the electric grid. These sources are unlike the dispatchable plants currently used to meet the baseload demand, or cycled to meet variations in demand and provide reserves. This raises questions about the reliability and cost-effectiveness of these sources, and their potential need for enabling technologies such as energy storage.

43.4 Consequences of high renewables penetration

43.4.1 Characteristics of renewable resources

There are two general types of renewable electricity generators, in terms of on their dependence on short-term weather conditions. "Dispatchable" renewables include geothermal, hydro, and biomass, and are controllable in nature due to the fact that their "fuel" is storable or controllable (hydro has stored energy behind a dam, biomass fuel can be stockpiled next to the plant, etc.). A system comprised primarily of these sources would be relatively easy to integrate into an electric power system since they are similar to the fossil plants in current use.

Variable-generation (VG) renewables such as wind and solar create greater challenges due to the fact that their output can vary over relatively short time periods and is not completely predictable. (Historically, these sources of generation have been referred to as intermittent; however, a recent trend is to label wind and solar as "variable generation" or "variable and uncertain" [3][6].)

The use of VG renewable sources such as wind and solar will change how the existing power-plant mix is operated, because their output is unlike that of conventional dispatchable generators. It is easiest to understand the impact of VG technologies on the grid by considering them as a source of demand reduction with unique temporal characteristics. Instead of considering wind or solar as a source of generation, they can be considered a reduction in load, with conventional generators meeting the "residual load" of normal demand minus the electricity produced by renewable generators.

Figure 43.3 illustrates this framework for understanding the impacts of variable renewables, in this case

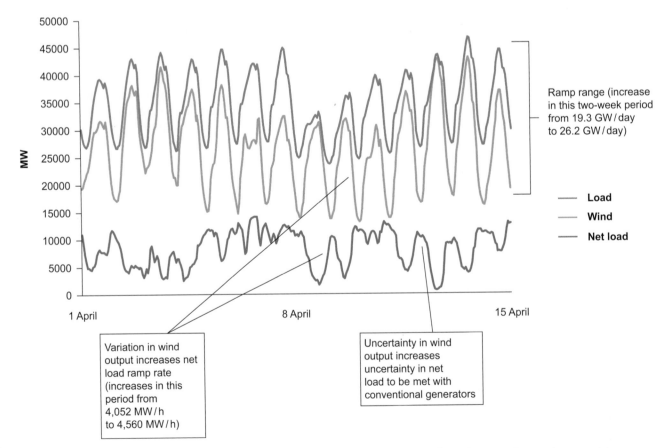

Figure 43.3. The impact of net load from increased use of renewable energy.

on the Texas grid. In this figure, 15 gigawatts (GW) of simulated wind generation is subtracted from the normal load, showing the "residual" or net load that the utility would need to meet with conventional sources. The benefits to the utility include a decrease in fuel use (and associated emissions) and a somewhat reduced need for overall system capacity (this is relatively small for wind, but can be significant for solar, given its coincidence with load). There are also four significant impacts that change how the system must be operated and affect costs. The first is the increased need for frequency regulation, because wind can increase the short-term variability of the net load (not illustrated on the chart). The second is the increase in the ramping rate, or the speed at which load-following units must increase and decrease output. The third impact is the uncertainty in the wind resource and resulting net load. The final impact is the increase in overall ramping range – the difference between the daily minimum and maximum demand – and the associated reduction in minimum load, which can force baseload generators to reduce output, and, in extreme cases, force the units to cycle off during periods of high wind output. Together, the increased variability of the net load requires a greater amount of flexibility and operating reserves in the system, with more ramping capability to meet both predicted and unpredicted variability. The use of these variable and uncertain resources will require changes in the operation of the remaining system, and this will incur additional costs, typically referred to as integration costs [2].

43.4.2 Costs of wind and solar integration

Concerns about grid reliability and the cost impacts of wind have driven a large number of wind-integration studies. These studies use utility-simulation tools and statistical analysis to model systems with and without wind and calculate the integration costs of wind. The basic methodology behind these studies is to compare a base case without wind with a case with wind, evaluating technical impacts and costs. The studies calculate the additional costs of adding operating reserves as well as the other system changes needed to reliably address the increased uncertainty and variability associated with wind generation. The cost impacts evaluated are typically the first three of the four impacts discussed previously.

- **Regulation and load following** – the increased costs that result from providing the ramping requirements (seconds to minutes for regulation and minutes to hours for load following) resulting from wind deployment. To provide regulation and load following, the additional variability requires that utilities run more flexible generators (such as gas-fired units instead of coal units, or simple-cycle turbines instead of combined-cycle turbines) to ensure that additional ramping requirements can be met. Additional costs arise from keeping units at part-load, ready to respond to the variability, or from more frequent unit starts.

- **Wind uncertainty** – the increased costs that result from having a sub-optimal mix of units online because of errors in the wind forecast. This is typically called unit commitment or scheduling cost because it involves costs associated with committing (turning on) too few or too many slow-starting, but lower operational-cost units than would have been committed if the wind forecast been more accurate.

Most large thermal generators must be scheduled several hours (or even days) in advance in order for them to be ready when needed. Ideally, utilities schedule and operate only as many plants as needed to meet energy and reserve requirements at each moment in time. If utilities over-schedule (turn on too many plants), they will have many plants running at part-load, and will have incurred higher than needed startup costs. So, if they underpredict the wind, they will commit and start up too many plants, which incurs greater fuel (and other) costs than would have been needed if the wind and corresponding net load had been forecasted accurately. Conversely, if the wind forecast is too high and the net load is higher than expected, insufficient thermal generation may be committed to cover the unexpected shortfall in capacity. A worst-case scenario would be a partial blackout; but the likely result is the use of high-cost "quick-start" units in real time, purchasing expensive energy from neighboring utilities (if available), or paying customers a premium to curtail load – all while lower-cost units are sitting idle. An example of this is the ERCOT event of February 26, 2008 [7]. On this date, a combination of events – including a greater than predicted demand for energy, a forced outage of a conventional unit, the wind forecast not having been given to the system operators, and a lower than expected wind output – resulted in too little capacity online to meet load. As a result, the ERCOT system needed to deploy high-cost quick-start units, and had to pay customers to curtail load through its "load acting as a resource" program. All of this occurred while lower-cost units were idle because the combination of events was unanticipated. It is worth noting that wind forecasting is an area of active research, and these errors are expected to decrease in time, which could potentially lead to a corresponding decrease in unit-commitment errors and associated costs.

Integration studies in the USA and Europe have found that the costs of integrating wind at penetrations

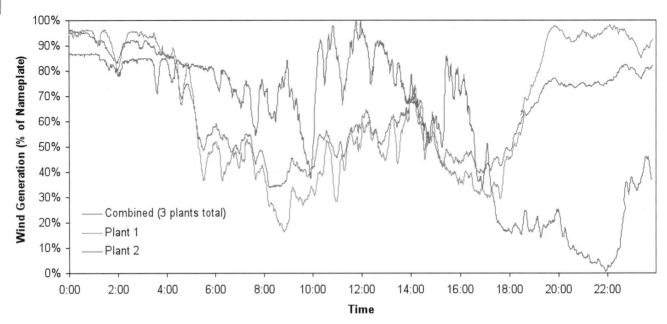

Figure 43.4. The smoothing effect of aggregating wind in the midwestern USA over one day for three wind generators.

of up to 20% on an energy basis are fairly modest, typically less than 0.5 cents per kilowatt-hour, adding less than 10% to the cost of wind energy [8]. The explanation for this relatively modest impact on costs is largely based on the already significant variation in normal load. The large amount of flexible generation already available to meet the variability in demand has the ability to respond to the greater variability caused by the large-scale deployment of wind. Furthermore, these studies have found significant benefits of spatial diversity – just because the wind isn't blowing in one location, it may be in another. The combination of multiple wind sites tends to smooth out the aggregated wind generation in a system, which reduces the per-unit size of ramps and mitigates the range of flexibility required [9]. This is illustrated in Figure 43.4 showing the effect of spatial aggregation on the minute-to-minute time scale for the midwestern USA, which substantially reduces both the rate and the total variation in wind output, reducing the need for load-following reserves.

Spatial aggregation also substantially decreases the sub-hourly ramping, illustrated in Figure 43.5. This figure shows the 10 variability (change in wind generation output as a fraction of plant size) as a function of production in the eastern USA. Aggregation decreases the need for frequency-regulation reserves [10].

Far less work has been performed on the operational impacts of large-scale solar generation, due largely to its lower deployment rate compared with wind. Preliminary studies have found the integration costs of photovoltaics (PV) to be similar to those of wind [11]; however, there are limited data sets from which to completely

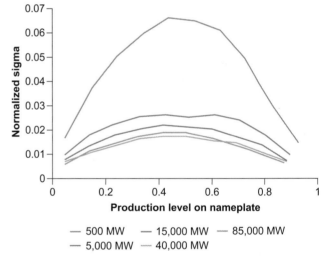

Figure 43.5. The smoothing effect of aggregating wind in the eastern USA in the hourly time-frame.

evaluate the impacts of sub-hourly ramping events due to passing clouds and the effects of spatial diversity on mitigating these effects [12]. It is expected that the impact of concentrating solar power (CSP) on short time scales will be significantly less than PV, because CSP has significant "thermal inertia" in the system that will minimize high-frequency ramping events. Furthermore, CSP has the potential advantage of utilizing high-efficiency thermal storage. Additional studies are being performed, but it will be some time until solar's impact on the grid and associated costs are understood as well as are those of wind.

Authors of wind-integration studies almost universally come to the conclusion that, at the penetrations studied to date (up to about 30% on an energy basis), the systems analyzed can accommodate wind's variability and maintain reliable service. In the studied systems, no new generation technologies are required; but some potentially significant operational changes are needed in order to maintain the present level of reliability, in addition to significant transmission additions in order to exchange resources over larger areas. This means exchanging power across most of the eastern and western interconnects in the USA [13], and much greater integration of the grid will be required in Europe. Beyond these levels of penetration, however, the coincidence of VG supply and demand, together with additional system constraints, will impose increasing economic and technical challenges, and may require additional enabling technologies such as energy storage.

43.4.3 Limits to integrating wind and solar in the existing grid

To date, integration studies in the USA and Europe have found that VG sources can be incorporated into the grid by changing operational practices to address the increased ramping requirements on various time scales. At higher penetrations (beyond those already studied), the required ramp ranges will increase, which adds additional costs and the need for fast-responding generation resources. However, there are additional constraints on the system that will present new challenges. These constraints are based on the simple coincidence of the renewable-energy supply and demand for electricity, combined with the operational limits on generators providing baseload power and operating reserves.

As discussed in Section 43.3, in current electric power systems, electricity is generated by two general types of generators: baseload generators, which run at nearly constant output; and load-following units (including both intermediate-load and peaking plants), which meet the variation in demand as well as provide operating reserves. At current penetrations of wind and solar in the USA, and at the levels studied in most integration studies, wind and solar generation primarily displaces flexible load-following generators. Figures 43.6 and 43.7 illustrate this by providing the (simulated) impacts of increasing amounts of wind and solar generation in the western USA in a week in July. Figure 43.6 provides a dispatch stack, showing the mix of generators providing energy without wind and solar, showing the baseload demand met by nuclear and coal generators, and the variation in demand met by combined-cycle and simple-cycle gas turbines. When wind and solar are added, the gas turbines are "backed off," saving fuel and associated emissions, and VG is easily accommodated. Because these generators are designed to vary output, they can do this with modest cost penalties, as analyzed in the wind-integration studies discussed previously.

In contrast, during the spring, increased wind output combined with lower load makes integration more challenging. Figure 43.8 illustrates a period during April when wind forces reduced output from baseload coal plants, which traditionally are not cycled, and the ability of these thermal generators to reduce output may have become constrained. If the baseload generators cannot reduce output, then the use of wind energy will need to be curtailed [14].

Utilities have expressed concern about their systems "bottoming out" due to the minimum-generation requirements during overnight hours, and being unable

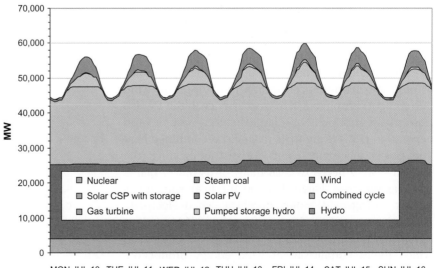

Figure 43.6. Normal system dispatch in the western USA, for the week of July 10, with no wind or solar in this and the following 2 figures the color coding runs from Nuclear bottom to hydro on the top, not all sources are in all figures.

Figure 43.7. System dispatch in the western USA for the week of July 10, 30% of annual energy from wind and solar.

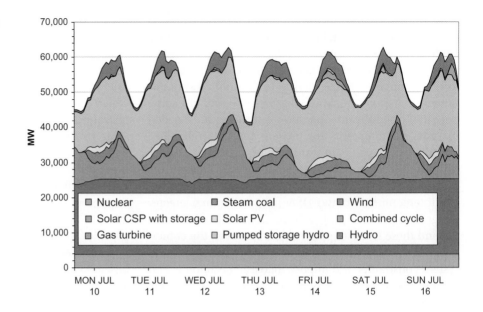

Figure 43.8. System dispatch in the western USA for the week of April 10, 30% of annual energy from wind.

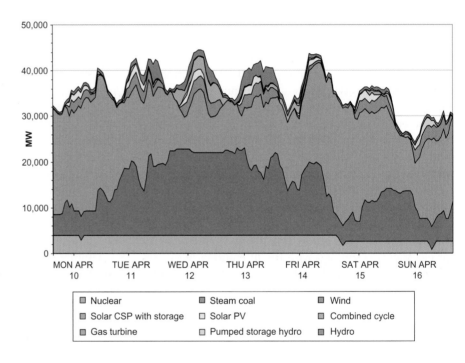

to accommodate more VG during these periods. Minimum-generation constraints (and resulting wind curtailment) are already a real occurrence in the Danish power system, which has a large installed base of wind generation [15]. Owing to its reliance on combined heat and power electricity plants for district heating, the Danish system needs to keep many of its power plants running for heat. Large demand for heat sometimes occurs during cold, windy evenings, when electricity demand is low and wind generation is high. This combination sometimes results in an oversupply of generation, which forces curtailment of wind energy production. Figure 43.9 provides an example of historical wind and

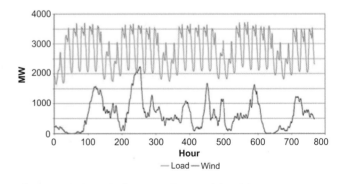

Figure 43.9. Wind generation and load in west Denmark, January 2006.

load patterns in west Denmark, showing that the wind actually exceeds the total demand for electricity on a few occasions. Currently, Denmark has access to export markets for much of this energy, but there are limits to how much energy can be exported due to transmission constraints and the use of wind by its neighbors.

Wind energy is occasionally curtailed in the USA due to transmission constraints, when the output of wind generation exceeds the local transmission capacity [16]. Modern wind turbines can reliably curtail output, but this is largely undesirable because curtailment throws away cost-free and emissions-free energy. When curtailing output, wind or other VG can supply operating reserves. In some cases, the value of the curtailed energy may actually exceed the value of the energy, but the primary value of VG is in displacing conventional generation.

The constraint on wind integration due to minimum load is a function of several factors, including the mix of conventional generation as well as the amount of reserves and the types of generators providing those reserves. The ability to cycle conventional units is both a technical and an economic issue – there are technical limits to how much power plants of all types can be turned down. The limits on cycling are uncertain in part because utilities often have limited experience with cycling large coal plants, and have expressed concern about potentially costly (but difficult to quantify) maintenance impacts [17]. In a number of markets in the USA, energy prices have dropped below the actual variable (fuel) cost of producing electricity on a number of occasions. This indicates that power-plant operators are willing to sell energy at a loss to avoid further reducing output and to avoid a forced shutdown and very expensive restart of a coal generator. With the exception of certain peaking plants such as aeroderivative turbines and fast-starting reciprocating engines, most plants have minimum up-and-down times, and require several hours to restart (at considerable cost). It should be noted that because cycling costs are not universally captured in operational models, and are difficult to quantify, they may be ignored or underestimated in studies of wind and solar integration studies [12]. The impact of VG on power-plant cycling is an active area of research, especially considering the evolving grid and the introduction of more-responsive generation [18].

Overall, the ability to accommodate a variable and uncertain net load has been described as a system's flexibility. System flexibility varies by system, with some systems using generators that can ramp rapidly and can reduce output to relatively low levels. Flexibility is also greater where plant operators have experience with cycling individual units. System flexibility will also change over time as new technologies are developed and power plants are retired. In addition, VG deployment in the USA is still relatively small, and utilities have yet to

evaluate the true cycling limits on conventional generators and their associated costs. Also, the additional reserve requirements due to VG at high penetration are still uncertain. As a result, it is not possible to precisely estimate the costs of VG integration or the amount of curtailment at very high penetration; it is also difficult to define with certainty the value of energy storage or other enabling technologies. It is clear, however, that a substantial increase in the penetration of wind energy without storage will require changes in grid operation in order to reduce curtailment.

The impact of system flexibility on curtailment and resulting usable VG can be observed in Figures 43.10 and 43.11. Figure 43.10 superimposes load data in ERCOT from 2005 with a spatially diverse set of simulated wind and solar data from the same year; however, the mix of generators (flexible and inflexible) is hypothetical and used only to illustrate the impact of VG. In this scenario the mix of wind and solar provides about 25% of the system's annual energy, with 3% of the annual renewable generation curtailed.

Increasing or decreasing the minimum-load point will change the amount of curtailment and the corresponding contribution of VG. Adding additional VG will increase its contribution, but also the amount of curtailment. This curtailment increases costs to the system, and, depending on the correlation of the wind and solar resources, the rate of curtailment can be highly nonlinear.

These results can be evaluated more generally by examining the amount of curtailment that will result (without energy storage or enabling technologies) as a function of system flexibility. Figure 43.11 shows example results, with two different minimum-load levels. The minimum-load level can also be generalized in terms of the system's "flexibility factor," which is defined as the fraction below the annual peak to which conventional generators can cycle [19]. The first chart in Figure 43.11 illustrates the average curtailment that would result if the system could not cycle below 12 GW (corresponding to a flexibility factor of 80%) [20] and 6 GW (corresponding to a flexibility factor of 90%, which largely eliminates baseload units from the generation mix). These results indicate that, without storage or other enabling technologies, deriving 50% of the energy in this system would require "throwing away" about 10% and 30%, respectively, of the renewable generation for the two flexibility cases.

Curtailment increases the cost of the VG that is actually used, because curtailment reduces the net capacity factor of the wind and solar generators. The results in Figure 43.12 concern only a specific mix of VG – and a single system – so they cannot be applied generally. However, they do illustrate the trends that may limit the contribution of VG without enabling technologies. This example can also be compared with

Figure 43.10. The effect of the minimum-load point on curtailment.

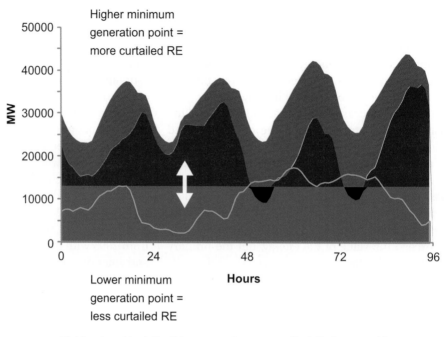

Net load met by inflexible generation

Net load met by flexible generation

Renewable energy

Curtailed renewable energy

Load met renewable energy

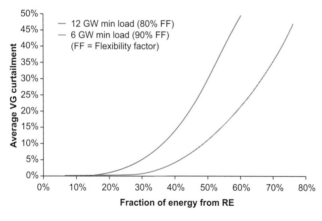

Figure 43.11. The average curtailment rate as a function of VG penetration for different flexibilities in ERCOT.

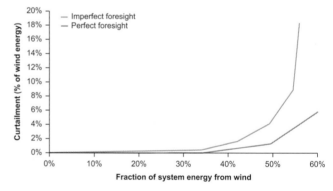

Figure 43.12. Wind curtailment as a function of penetration in the Irish grid.

the results of the previous US wind-integration studies, in which little curtailment was found up to about 30% on an energy basis, due to the existing grid flexibility, assumed low-cost operational changes, and the availability of new transmission [21]. These results can also be compared with European studies, such as an examination of curtailment in the Irish grid, illustrated in Figure 43.12, in which it was found that meeting 50% of the systems demand can produce curtailment rates below 4% of wind generation. (This depends largely on the accuracy of the forecast – imperfect foresight regarding wind values can increase reserve

requirements, thereby increasing curtailment as discussed previously.) The large difference in results indicates the importance of examining the coincidence of wind and loads across regions. In this case, greater coincidence of wind with local demand patterns, as well as a more flexible system, allows greater penetration of renewable energy [2].

Figure 43.12 also shows how the use of energy storage can reduce wind curtailment, and increase the use of wind and other renewable energy sources. While energy storage is often suggested as a means to enable deployment of renewable energy, there are also enabling technologies that can increase the coincidence of energy supply with normal electricity demand patterns.

Figure 43.13. Options for increasing the use of VG and decreasing curtailment. The figure shows the effect of load shifting and on potential use of storage and increases the flexibility of generation.

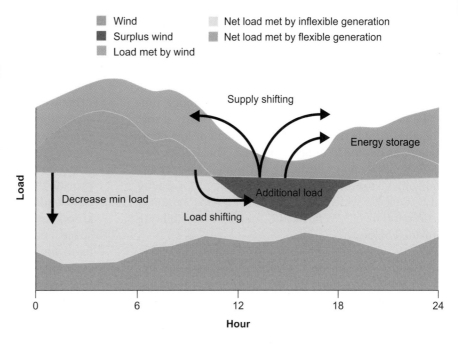

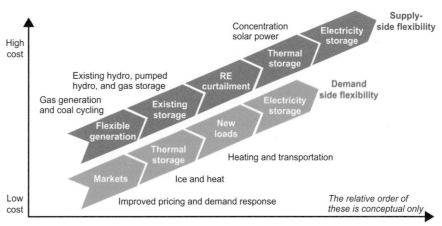

Figure 43.14. The flexibility supply curve.

43.4.4 Enabling greater penetration of wind and solar

The previous section indicates that, at high penetration of VG, fundamental changes to the grid may be required in order to accommodate the increased variability of net load and the limited coincidence of VG supply and normal electricity demand. Various techniques and technologies – described as flexibility resources – to accommodate the impacts of VG and ensure that the generation mix matches the net-load requirement have been proposed.

Figure 43.13 illustrates several options for decreasing curtailment and increasing the usefulness of VG [22][23].

Overall, these options can be expressed in terms of a flexibility supply curve that can provide responsive energy on various time scales. The flexibility supply curve is conceptually similar to other resource supply curves where (ideally) the lowest-cost resources are used until they are exhausted, then the next (higher-cost) resource is deployed. The curtailment curve in Figure 43.11, for example, was restricted to ERCOT, and did not consider the opportunity to exchange wind and solar energy with surrounding areas by building interconnections with the other US grids. It also did not consider the ability to shift load by incentivizing customers to use less electricity when VG output is low. Overall, utilities have many "flexibility" options for incorporating greater amounts of VG into the grid, many of which may cost less than more capital-intensive technologies such as energy storage. Figure 43.14 provides a conceptual supply curve including some of these options. While the figure hypothesizes an order of the supply curve, the actual costs and availability of the individual components are conjectural.

Overall, there are two general "types" of flexibility required by variable sources and offered by technologies in this curve. The first can be described as ramping flexibility, or the ability to follow the variation in net load (on the second-to-minute time scale needed for frequency regulation, or the minutes-to-hours time scale needed for load following and forecast error). This flexibility is the primary requirement at low penetration, as discussed in Section 43.4.2. The second type of flexibility is energy flexibility, or the ability to increase the coincidence of VG supply with demand for electricity services, which is described in Section 43.4.3. A description of several of the sources of flexibility is provided below.

(a) **Supply and reserve sharing**. This includes the sharing of renewable and conventional supply, operating reserves, and net loads through markets or other mechanisms that effectively increase the area over which supply and demand is balanced. Greater aggregation of loads and reserves has historically been one of the cheapest methods of dealing with demand variability, especially because it often requires operational changes and relatively little new physical infrastructure. This includes introducing sub-hourly markets that allow systems to respond faster to variability. Large independent system operator (ISOs) with 5-minute markets typically have substantially lower wind-integration costs [24]. It may also require transmission development to increase the spatial diversity of VG resources.

(b) **Flexible generation**. This includes deploying new, more flexible conventional generators as well as increasing the flexibility of existing generators. This can be accomplished by modifying equipment and operational practices to increase the load-following, ramping rate, and ramping range of the grid. This also includes introducing new generators that can be brought online quickly to respond to forecast errors [25]. This may also require increased use of natural-gas storage to increase use of flexible gas turbines and decrease contractual penalties for forecast errors in natural-gas use [26]. Another source of flexible generation is improved use of existing storage and hydro assets.

(c) **Demand flexibility**. This includes introducing market or other mechanisms to allow a greater fraction of the load to respond to price variations and provide ancillary services. Responsive demand can provide flexibility on multiple time scales by curtailing demand for short periods or shifting load over several hours. Many of these technologies and processes have been described in terms of a "smart grid" (Chapter 42) and will require regulatory and policy changes in addition to new technologies.

In many locations in the USA, demand is increasingly used as a source of grid services.

(d) **Curtailment of VG**. Overbuilding VG may result in curtailment of low-value springtime generation, but would allow a greater overall VG contribution. (This is functionally equivalent to cycling baseload generators in the spring.) Furthermore, curtailed VG provides additional benefits, because it provides a source of operating reserves and potentially allows de-commitment of thermal units that typically provide these services.

(e) **New loads**. New controllable loads can be added to absorb otherwise unusable VG. Examples include space and process heating, which currently use fossil fuels. Another possibility is fuel production such as of hydrogen via electrolysis or shale-oil heating. Electrification of transportation using electric vehicles or plug-in hybrid vehicles is also a potential large-scale application. This may also include electric vehicles providing regulation and contingency reserves with or without the use of vehicle to grid (V2G).

(f) **Electricity storage**. Electricity storage encompasses a large number of technologies. Storage can provide both ramping flexibility, by rapidly changing output, and energy flexibility, by absorbing potentially curtailed generation during periods of low load and moving it to times of high net system load (where net load is defined as the normal load minus VG). The additional flexibility storage provides is also important. Storage can provide operating reserves, which reduces the need for partially loaded thermal generators that may restrict the contribution of VG. Finally, by providing firm capacity and energy derived from VG sources, storage can effectively replace baseload generation, which reduces the minimum-loading limitations.

The general classes of flexibility resources listed above represent dozens or even hundreds of individual technologies, each with a potential contribution to increasing grid flexibility. The cost and availability of many flexibility resources has yet to be quantified, and there are regulatory barriers to completely deploying many flexibility options such as demand response. The large number of options available for increasing the penetration of VG is one of the reasons why it is not possible to precisely determine when certain technologies such as energy storage become "necessary." There is less of a technical limit than an economic one that depends on a large number of factors, such as the cost of storage compared with a vast array of alternatives. Overall, the grid of the future is likely to be characterized as a more interactive or "smart grid," where the balance of supply and demand is met by a combination of techniques,

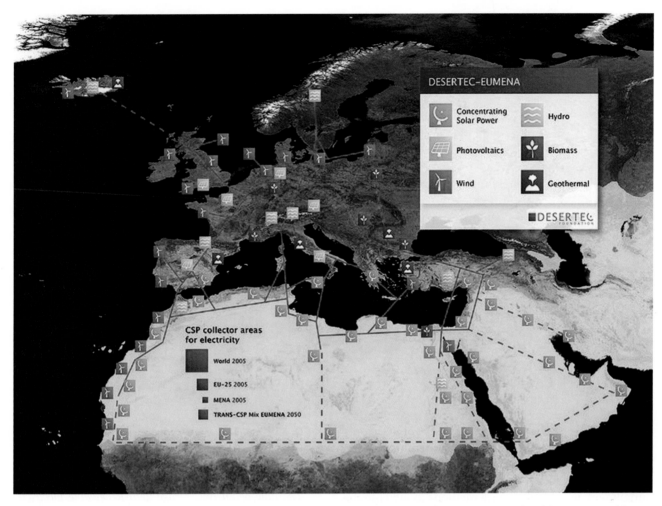

Figure 43.15. DESERTEC's conceptual European/North African supergrid connecting major renewable-resource regions with demand centers.

including responsive demand, energy storage, and long-distance transmission of new renewable resources. Figures 43.15 and 43.16 illustrate two important components of this concept. The first (Figure 43.15), illustrates the possible development of a larger transmission network (in this case in Europe and North Africa), allowing sharing of supply and demand resources over a large geographical area [27]. This will act to smooth the variability of wind and solar resources, and increase access to a greater mix of resources, ranging from concentrating solar power in North Africa, to wind in the North Sea, and even new hydroelectric resources in Iceland.

Figure 43.16 shows how the aggregated supply of bulk renewables can be more intelligently controlled and distributed within load centers. Many of these "smart-grid" concepts rely on real-time sensors, pricing, and control that allows demand patterns to respond to variations in supply and price of electricity generation, in addition to incorporating local generation and storage.

43.5 Summary

The increasing role of variable renewable sources (such as wind and solar) in the grid has prompted concerns about grid reliability and raised the question of how much these resources can contribute to reducing dependence on fossil fuels and decreasing carbon emissions in the electric sector. To date, studies of wind of integration up to about 20% on an energy basis have found that the grid can accommodate a substantial increase in VG without the need for large amounts of new technologies such as energy storage, but this will require changes in operational practices, such as sharing of generation resources and loads over larger areas. Beyond this level, the impacts and costs are less clear, but 30% or more appears feasible if additional transmission can be built and a variety of flexibility options, such as greater use of demand response, can be deployed However, there are technical and economic limits to how much of a system's energy can be provided by VG without additional enabling technologies. These limits

SMART GRID

A vision for the future — a network of integrated microgrids that can monitor and heal itself.

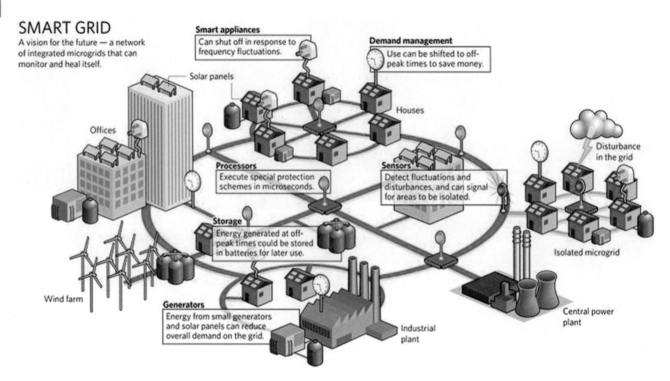

Smart appliances
Can shut off in response to frequency fluctuations.

Demand management
Use can be shifted to off-peak times to save money.

Solar panels

Offices

Houses

Processors
Execute special protection schemes in microseconds.

Sensors
Detect fluctuations and disturbances, and can signal for areas to be isolated.

Disturbance in the grid

Storage
Energy generated at off-peak times could be stored in batteries for later use.

Isolated microgrid

Wind farm

Generators
Energy from small generators and solar panels can reduce overall demand on the grid.

Industrial plant

Central power plant

Figure 43.16. The smart grid.

arise from at least two factors: coincidence of VG supply and demand; and the ability to reduce output from conventional generators. At extremely high penetration of VG, these factors may cause excessive (and costly) curtailment, which will require methods to increase the useful contribution of VG. However, the question of how much VG can be used before storage becomes the most economic option for further integration currently has no simple answer, primarily because the availability and cost of grid flexibility options are not well understood and vary by region.

43.6 Questions for discussion

1. Name several drivers behind the hourly and seasonal variation in demand for electricity.
2. Describe the difference between variability and uncertainty.
3. What is spatial diversity? How does it improve the performance of renewable energy sources?
4. Why does wind curtailment occur?
5. How would energy storage allow greater use of renewable energy sources?

43.7 Further reading

- General Electric (GE) Energy (2010). *Western Wind and Solar Integration Study*, NREL report no. SR-550–47434.

- **D. Corbus**, **D. Lew**, **G. Jordan** *et al.*, 2009, "Up with wind," *IEEE Power Energy Mag.*, **7**(6), 36–46; also available as NREL report no. JA-550–47363.
- **P. Denholm**, **E. Ela**, **B. Kirby**, and **M. Milligan**, 2010, *The Role of Energy Storage with Renewable Electricity Generation*, NREL/TP-6A2–47187.
- TradeWind (principal author **F. Van Hulle**), 2009, *Integrating Wind: Developing Europe's Power Market for the Large-Scale Integration of Wind Power*, Brussels, European Wind Energy Association.

43.8 References

[1] Energy Information Administration (EIA), 2011, *Electric Power Annual*, DIE/EIA-0348.
[2] **P. Denholm**, **E. Ela**, **B. Kirby**, and **M. Milligan**, 2010, *The Role of Energy Storage with Renewable Electricity Generation*, NREL/TP-6A2–47187.
[3] North American Electric Reliability Corporation (NERC), 2009, *Accommodating High Levels of Variable Generation*, http://www.nerc.com/docs/pc/ivgtf/IVGTF_Outline_Report_040708.pdf.
[4] **B. Kirby**, 2004, *Frequency Regulation Basics and Trends*, ORNL/TM 2004/291, Oak Ridge National Laboratory.
[5] **Y. Rebours** and **D. S. Kirschen**, 2005, *A Survey of Definitions and Specifications of Reserve Services*, Manchester, University of Manchester Press.
[6] **J. C. Smith** and **B. Parsons**, 2007, "What does 20% look like?," *IEEE Power & Energy*, November/December, 22–33.

[7] **E. Ela** and **B. Kirby**, 2008, *ERCOT Event on February 26, 2008: Lessons Learned*, NREL/TP-500–43373, National Renewable Energy Laboratory.

[8] **J. DeCesaro**, **K. Porter**, and **M. Milligan**, 2009, "Wind energy and power system operations: a review of wind integration studies to date," *The Electricity J.*, **22**(10), 34–43.

[9] **B. Palmintier**, **L. Hansen**, and **J. Levine**, 2008, "Spatial and temporal interactions of solar and wind resources in the next generation utility," in *Solar 2008*, San Diego, CA.

[10] **D. Corbus**, 2010, *Eastern Wind Integration and Transmission Study (EWITS)*, NREL report no. SR-550–47078.

[11] EnerNex Corporation, 2009, *Solar Integration Study for Public Service Company of Colorado*, for Xcel Energy, http://www.xcelenergy.com/SiteCollectionDocuments/docs/PSCo_SolarIntegration_020909.pdf.

[12] **D. Lew**, **M. Milligan**, **G. Jordan** *et al.*, 2009, *How Do Wind and Solar Power Affect Grid Operations: The Western Wind and Solar Integration Study*, NREL/CP-550–46517.

[13] **M. Milligan**, **D. Lew**, **D. Corbus** *et al.*, 2009, *Large-Scale Wind Integration Studies in the United States: Preliminary Results*, NREL/CP-550–46527.

[14] General Electric (GE) Energy, 2010, *Western Wind and Solar Integration Study*, NREL report no. SR-550–47434.

[15] **T. Ackermann**, **G. Ancell**, **L. D. Borup** *et al.*, 2009, "Where the wind blows," *IEEE Power Energy Mag.*, **7**(6) 65–75.

[16] **S. Fink**, **C. Mudd**, **K. Porter**, and **B. Morgenstern**, 2009, *Wind Energy Curtailment Case Studies*, NREL/SR-550–46716.

[17] **S. A. Lefton** and **P. Besuner**, 2006, "The cost of cycling coal fired power plants," *Coal Power Mag.*, Winter, 16–20.

[18] **N. Troy**, **E. Denny**, and **M. O'Malley**, 2010, "Baseload cycling on a system with significant wind penetration," *IEEE Trans. Power Systems*, **25**(2), 1088–1097.

[19] **P. Denholm** and **R. M. Margolis**, 2007, "Evaluating the limits of solar photovoltaics (PV) in traditional electric power systems," *Energy Policy*, **35**, 2852–2861.

[20] General Electric (GE) Energy, 2008, *Analysis of Wind Generation Impact on ERCOT Ancillary Services Requirements*, prepared for the Electric Reliability Council of Texas (ERCOT), http://www.uwig.org/AttchA-ERCOT_A-S_Study_Exec_Sum.pdf.

[21] **D. Corbus**, **M. Milligan**, **E. Ela**, **M. Schuerger**, and **B. Zavadil**, 2009, *Eastern Wind Integration and Transmission Study – Preliminary Findings*, preprint, 9 pp., NREL report no. CP-550–46505.

[22] **A. Tuohy** and **M. O'Malley**, 2011, "Pumped storage in systems with very high wind penetration," *Energy Policy*, **39**(4), 1965–1974.

[23] http://www.nrel.gov/wind/systemsintegration/energy_storage.html.

[24] **M. Milligan**, **B. Kirby**, **R. Gramlich**, and **M. Goggin**, 2009, *The Impact of Electric Industry Structure on High Wind Penetration Potential*, NREL/TP-550–46273.

[25] California Independent System Operator (CAISO), 2007, *Integration of Renewable Resources*, Folsom, CA, CAISO.

[26] **R. Zavadil**, 2006, *Wind Integration Study for Public Service Company of Colorado*, Enernex Corporation for Xcel Energy.

[27] DESERTEC Foundation, 2009, "Clean power from deserts, in *The DESERTEC Concept for Energy, Water and Climate Security*, 4th edn., DESERTEC Foundation.

44 Electrochemical energy storage: batteries and capacitors

M. Stanley Whittingham

Institute for Materials Research, SUNY at Binghamton, Binghamton, NY, USA

44.1 Focus

This chapter explains and discusses present issues and future prospects of batteries and supercapacitors for electrical energy storage. Materials aspects are the central focus of a consideration of the basic science behind these devices, the principal types of devices, and their major components (electrodes, electrolyte, separator). Both experimental and modeling issues are discussed, in the context of needed advancements in battery and supercapacitor technology.

44.2 Synopsis

Electrical energy storage is needed on many scales: from milliwatts for electronic devices to multi-megawatts for large grid based, load-leveling stations today and for the future effective commercialization of renewable resources such as solar and wind energy. Consider the example of hybrid electric vehicles (HEVs) (Chapter 31). In HEVs, batteries and/or capacitors are used to capture the energy evolved in braking, and HEV buses use an all-electric drive, which allows them to get up to traffic speed much faster than regular buses, pollute less while moving, and generate zero pollution when standing. The next generation of electric vehicles will be be plug-in hybrids (PHEVs), which require larger batteries. In addition, an all-electric vehicles (EVs) might find niche markets such as city buses, postal delivery vans, and utility-repair vehicles that stop and start frequently and have limited daily ranges; high-cost hot-rod sports cars; and small commuter cars. In all of these transportation applications, low cost and long life are essential for commercial success. Judging by these criteria, most, if not all, vehicles in a decade could be HEVs. However, while present battery and capacitor storage systems, have increased market penetration of PHEVs and EVs further technical improvements are needed to make them fully cost competitive.

Load-leveling, power-smoothing, and central backup applications demand even lower costs and higher reliability. In a typical daily electrical usage cycle, there is considerable hourly variation, and the load could be leveled, for example, by storing energy during the early-morning low-usage hours and using that energy during the peak hours of 4–6 pm. In addition to load leveling, sub-second to minute smoothing is needed, and this market can sustain a more costly system. Lithium-ion batteries recently started filling this niche, with batteries as powerful as 20 MW being installed.

To place the economic desire for uninterruptable power in context, some $US 80 billion is lost by US industry [1] each year, mainly because of short power interruptions. To ameliorate this problem, high-technology, high-cost industries such as chip-fabrication plants have large power-storage backups and frequency smoothing, often still with lead–acid batteries. Although other options are available, the most flexible energy-storage schemes are batteries and capacitors, since they can be located almost everywhere and often are maintenance-free, readily scalable, and portable.

44.3 Historical perspective

Batteries are, by far, the most common form of storing electrical energy and range in size from the button cells used in watches to megawatt load-leveling applications. They are efficient storage devices, with output energies typically exceeding 90% of input energy, except at the highest power densities. The Italian physicist Alessandro Volta is credited with the development (in 1800) of the first battery, an electrochemical cell comprising alternating disks of zinc and copper separated by cardboard with a brine solution as the electrolyte. This evolved into the two-electrolyte Daniell cell (invented by the British chemist and meteorologist John Frederic Daniell in 1836), consisting of a copper pot filled with a copper sulfate solution, in which was immersed an unglazed earthenware container filled with sulfuric acid and a zinc electrode. In this setup, the zinc acts as the anode (negative terminal), and the copper pot acts as the cathode (positive terminal). The advantage of the porous earthenware pot is that it allows ions to pass without allowing the solutions to mix, thus preserving battery life. The next major milestone was the development of the Leclanché cell (invented and patented by the French electrical engineer Georges Leclanché in 1866), which used a zinc anode and a carbon cathode. The present small dry cell using an alkaline electrolyte, zinc anode, and manganese oxide cathode was not invented until 1949, by Lew Urry at the Eveready Battery Company Laboratory in Parma, OH. Alkaline batteries proved superior to Leclanché batteries because they can supply more total energy at higher currents, and further improvements in the decades since have increased the energy storage within a battery of a given size.

The field of rechargeable, also known as secondary, batteries began with the invention of the lead–acid battery by the French physicist Gaston Planté in 1859. Next came the nickel–cadmium battery, which was invented by Waldemar Jungner of Sweden in 1899. He experimented with substituting iron for some or all of the cadmium, but found the results to be unsatisfactory because of poorer efficiency and greater production of hydrogen. Nevertheless, Thomas Edison developed the nickel–lead battery in the early 1900s for use as the energy source for electric vehicles. These batteries then evolved through nickel–metal hydride (NiMH) batteries to lithium-ion batteries (Figure 44.1). The NiMH batteries were the initial workhorse for electronic devices such as computers and cell phones. They have been almost completely displaced from that market by lithium-ion batteries because of the latter's higher volumetric energy storage capability and lower cost. The NiMH technology is used today in some hybrid electric cars, but the higher-energy and lower-cost lithium batteries are already being used in hybrid electric buses, in plug-in

Figure 44.1 One of the earliest lithium rechargeable batteries, built by Exxon for the 1975 Electric Vehicle in Chicago, comprised three 2-V cells connected in series.

electric cars and in fully electric vehicles. Of the advanced batteries, Li-ion batteries, so called because the Li ions shuttle back and forth between two intercalation electrodes, are the dominant power source for most rechargeable electronic devices.

44.4 Present status and future directions

44.4.1 What is electrochemical energy storage?

There are two predominant types of **electrochemical energy storage**, namely batteries and capacitors. In a **battery**, electrical energy is stored as chemical energy, whereas in a **capacitor** electrical energy is stored as surface charge. In a battery, the reactions occur throughout the bulk of the solid, whereas in capacitors the charge is stored on the surface of the electrode materials. Thus, the two devices involve quite different behaviors of materials. Pure capacitors can be charged and discharged millions of times without any significant degradation of the materials, whereas in batteries the chemical reactions are not always readily reversed because structural changes of the materials occur. Supercapacitors are a hybrid between the two, involving both surface charge and some heterogeneous charge-transfer reactions in the bulk of the material. These electrochemical storage systems have a close relationship with fuel cells. A **fuel cell** is essentially a primary battery in which a fuel is supplied to one electrode and the oxidant to the other. Closely related are metal–air batteries, such as zinc–air cells, which are

Table 44.1. Energy outputs of electrochemical devices and petroleum

Device or fuel	Energy output (kW·h kg^{-1})
Supercapacitors	0.01
Lithium battery	0.8
Hydrogen fuel cell	1.1
Gasoline	6.0[a]

[a] Assuming 30% combustion efficiency.

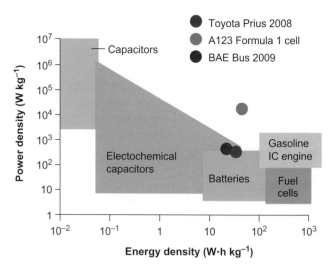

Figure 44.2. A comparison of power density and energy density for batteries, capacitors, and fuel cells. (Energy is the capacity to do work; power is the rate at which work is done.) Also shown are the figures for the nickel–metal hydride battery pack used in a Toyota Prius, the lithium–iron phosphate battery pack used in a hybrid electric bus, and that of a single cell of a lithium–iron phosphate battery used as a power booster in a Formula 1 racing car. Kinetic Energy Recovery Systems (KERS).

used in hearing aids as a primary (not rechargeable) battery. Extensive efforts are being made to develop a rechargeable Zn–air cell, as well as to overcome the immense materials hurdles to achieve a rechargeable Li–air cell. These metal–air cells are discussed in Section 44.4.7.

To place in perspective the energy-storage capabilities of the above-mentioned devices, a comparison is made in Table 44.1, from which it can be seen that the systems being considered have energy-storage capabilities that are far inferior to that of gasoline. A pure-hydrogen fuel cell has a theoretical capacity comparable to that of a lithium battery, because hydrogen containers hold only 5 wt% hydrogen using the best storage media and under high pressure and because fuel cells put out just 0.8 V (see Table 44.1).

The energy obtained from any storage device depends strongly not only on the storage capacity but also on the power output, as shown in Figure 44.2. Batteries can store higher amounts of energy (J) than capacitors, whereas capacitors are high-power (W) devices with limited energy-storage capability. In addition, the quality of the energy from capacitors is generally poor (that is, the voltage delivered varies strongly with the state of discharge), whereas batteries tend to have a fairly constant output voltage. Fuel cells, operating on liquid fuels such as methanol, can have high energy storage, but their power output is limited. Moreover, their efficiency is optimum only at constant output, and their poor response time demands that they be coupled to another storage medium such as batteries.

Batteries, capacitors, and fuel cells have many features in common, all being based on electrochemistry. Materials have always played a critical role in energy production, conversion, and storage [2], and will continue to do so. This chapter discusses the materials challenges facing electrical energy storage using batteries and capacitors. Fuel cells, which have many similar electrochemistry characteristics, are discussed in Chapter 46.

44.4.2 The fundamental science behind energy storage

In batteries and fuel cells, chemical energy is converted into electrical energy through **redox (reduction–oxidation) reactions.** These redox reactions occur at the two electrodes of an electrochemical cell: the **anode** and the **cathode.** By convention, the anode is the electrode containing the more electropositive material (for example, lithium in a lithium-ion cell), irrespective of whether the cell is being discharged or charged. Box 44.1 describes the components of a battery.

Thermodynamics
Thermodynamics describes the energy of systems including chemical reactions. The Gibbs free energy of an electrochemical reaction, or the maximum amount of electrical energy available from the reaction, is given by

$$\Delta G = -nEF,$$

where n is the number of electrons transferred during the reaction of one mole of reactant, E is the electromotive force (emf) of the reaction (i.e., the open-circuit voltage, which is measured when no current is flowing), and F is Faraday's constant (96,485 C mol^{-1}). F is the amount of charge transferred during the reaction of one

Box 44.1. Battery components

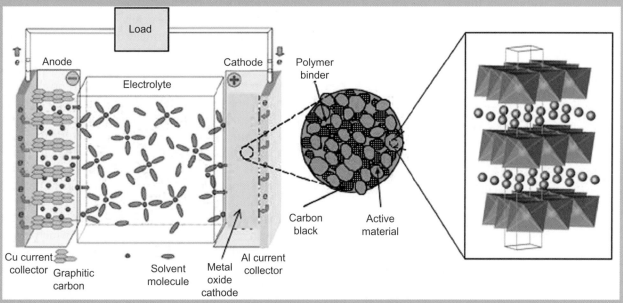

Figure 44.3. A schematic representation of a lithium-ion battery based on intercalation reactions.

Figure 44.3 shows a schematic diagram of a typical lithium cell; a battery consists of a number of such cells connected in series and/or in parallel to provide the desired voltage and power. The individual components of such a cell are as follows.

The **anode** is the electropositive electrode from which electrons are generated to do external work. In a lithium cell, the anode contains lithium, which is commonly held within graphite in the well-known lithium-ion batteries.

The **cathode** is the electronegative electrode to which, on discharge, positive ions migrate inside the cell and electrons migrate through the external electrical circuit; on charging, the positive ions and electrons flow in the opposite directions.

The **electrolyte** allows the flow of the positive ions (e.g., lithium ions) from one electrode to the other. It allows the flow only of ions, not electrons. The electrolyte is commonly a liquid solution containing a salt dissolved in a solvent. In addition, the electrolyte must be stable in the presence of both electrodes.

The **separator** (not shown) is typically a polymer that prevents the anode and cathode from making contact with one another. In cases in which a solid electrolyte is used, the electrolyte also performs the function of the separator.

The **current collectors** allow the transport of electrons to and from the electrodes. They are usually metals, and must not react with the electrode materials. Typically, copper is used for the anode and aluminum for the cathode. (However, the lighter-weight aluminum reacts with lithium and therefore cannot be used with lithium-based anodes.)

The **cell voltage** is determined by the energy of the chemical reaction occurring in the cell. The **open-circuit voltage** is the voltage when no current is flowing, whereas the **closed-circuit voltage** is the voltage when the cell is being used and is a function of the current drain rate.

The anode and cathode are, in practice, complex composites, containing, in addition to the active material, polymeric binders to hold the powder structure together and conductive diluents such as carbon black and/or carbon nano-tubes to give the whole structure electronic conductivity so that electrons can be transported to the active material. In addition, these components are combined in such a way as to leave sufficient porosity to allow the liquid electrolyte to penetrate the powder structure and the ions to reach the reacting sites.

equivalent weight of reactants, and, given the definition of amperes as $1\ A = 1\ C\ s^{-1}$, it can be expressed as 26.8 A·h for ease of calculations. The energy stored in chemical reactions is typically reported on a gravimetric or volumetric basis, which can be obtained simply by dividing nEF by the weight or volume, respectively, of the reactants. As an example, consider the reaction of lithium with iron phosphate.

- *At the anode*: each Li atom gives one Li^+ ion that migrates through the electrolyte and one electron that does work goes through the external circuit (i.e., $n = 1$).
- *At the cathode*: the Li^+ ion reacts with $FePO_4$ (and an electron), giving $LiFePO_4$.

Since the total molar mass of the reactants, Li (6.94 g mol^{-1}) and $FePO_4$ (150.82 g mol^{-1}), is approximately 158 g, the gravimetric capacity of this reaction is $(1 \times 26.8\ A\cdot h)/(158\ g) = 169\ A\cdot h\ kg^{-1}$.

Since the measured emf of the cell reaction (equal to the open-circuit voltage) is 3.45 V, the maximum energy stored on a gravimetric basis is $1 \times (3.45\ V) \times (26.8\ A\cdot h)/(158\ g) = (3.45\ V) \times (169\ A\cdot h\ kg^{-1}) = 586\ W\cdot h\ kg^{-1}$. Since the density of $LiFePO_4$ is 3.6 kg l^{-1}, the maximum energy stored on a volumetric basis is 2.11 kW·h l^{-1}.

Thermodynamics versus kinetics

Thermodynamics describes reactions occurring under equilibrium conditions and thus describes the maximum possible energy released; this is frequently referred to as the theoretical energy density. The equilibrium voltage, known as the open-circuit voltage, is reduced when a current flows because of the kinetics of the reaction. **Kinetics** controls what power can be obtained from the chemical reaction, and, even if thermodynamics indicates that a large amount of energy is stored, only kinetics can reveal whether that energy can be harvested at a reasonable rate. The kinetics of the release of energy from electrochemical storage devices is controlled by a number of factors, including charge transfer at the electrode interfaces; cell resistance; diffusion of the lithium (or other cation) into the cathode material; electron diffusion into the electrode; and, at very high rates, concentration gradients within the electrolyte. Since a driving force is needed to overcome the kinetics, the higher the power output, the lower the amount of energy that can be recovered. As explained in the preceding section, the energy stored in a battery or capacitor is normally described in terms of just one unit: kilowatt-hours per kilogram (kW·h kg^{-1}). In contrast, the kinetics of an electrochemical reaction is often described using several different units, and it is not a simple matter to convert between them. The three most common units used are amperes per square centimeter of electrode area (A cm^{-2}), amperes per gram of electrode-active material (A g^{-1}), and C rate. A 1C battery can be discharged in 1 h, a 2C battery in 30 min, and a C/10 battery in 10 h. In reading the literature, one must also be careful, since authors can choose the unit that gives the most impressive-appearing rate. Thus, it is essential in interpreting kinetic data to know the amount of material used and how much carbon diluent and polymeric binder were used. Realistic electrode loadings for the cathode in a lithium battery are around 20 mg cm^{-2}, with a combined level of carbon plus binder of around 10%; lower loadings and higher carbon levels can lead to unrealistically high rates. In complete batteries, the power is normally quoted in watts per kilogram or watts per liter, where the weight and volume refer to the entire battery. The use of large amounts of carbon results in lower energy storage, particularly on a volumetric basis, because of the low density of carbon blacks.

The structure of the electrode materials directly controls the kinetics of reaction and the lifetime of the cell. In a lithium cell, the lithium is incorporated into the crystalline lattice of the cathode material, ideally without any significant change in the crystalline lattice. This is commonly referred to as an **intercalation reaction** and is the basis of all rechargeable lithium batteries and even of the common alkaline cell, in which protons are inserted into MnO_2 to form $MnOOH$. Upon cell discharge, the lithium is shuttled from one intercalation compound, a layered carbon of approximate composition C_6Li, to another intercalation compound, such as $LiCoO_2$. The less the structure of these materials changes upon insertion and removal of the ions, the less likely it is that the structure will be damaged over the desired thousands of cycles of the reaction. Any damage to the structure will result in a loss of storage capability over multiple cycles; this is known as **capacity fade**. Another reason to minimize structural change is that the stability of the electrodes in contact with the electrolyte relies on the formation of the so-called solid electrolyte interface (SEI) layer, or interphase, which will be broken if the lattice flexes too much, and must then be rebuilt on each cycle. That rebuilding will use up the electrolyte and lithium and, at the same time, increase the internal resistance of the cell. The kinetics of the reaction will depend on the diffusion coefficient of the lithium within the electrode material assuming diffusion in the electrolyte is much faster, which is an important parameter to be measured. Desired chemical diffusion coefficients are higher than 10^{-10} cm^2 s^{-1}.

To complete the chemical reaction, electrons must flow to the reaction site, so it is essential that the electrodes be electronically conductive. This is normally accomplished by mixing the active cathode material with a conductor such as carbon black; however, this approach reduces the energy-storage capacity because of the dead weight added. No conductive diluent is needed only in the very few cases in which the electrode material itself is a metallic conductor, such as for titanium disulfide. A binder, typically made of fluorinated polymers, is also usually added to hold the electrode components together.

The structural changes occurring during the electrochemical reaction in a battery also control the nature of the voltage. For the previous example of lithium reacting

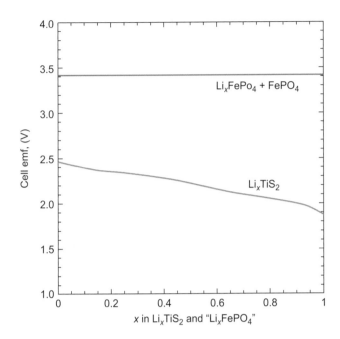

Figure 44.4. The cell emf of the single-phase system Li_xTiS_2 compared with that for the two-phase system $LiFePO_4/FePO_4$.

with iron phosphate, only two solid phases are present during the reaction, namely $LiFePO_4$ and $FePO_4$. Since these two phases are in equilibrium, the free energy of the reaction and thus the cell voltage remain constant for all degrees of reaction. In contrast, when lithium reacts with titanium disulfide, the continuous solid solution Li_xTiS_2 is formed for all values of x from 0 to 1. As the composition changes continuously during this reaction, the cell voltage also varies continuously. These cell-voltage behaviors are exemplified by the data shown in Figure 44.4 for these two cathode materials.

So far, only intercalation reactions have been described. A second group of reactions under study for chemical energy storage is known as **conversion reactions**. In these reactions, the structure of the electrode material is destroyed upon reaction and must be rebuilt upon recharge. An example is the reaction between lithium and FeF_2 to form LiF and Fe metal, which shows a constant cell emf. These electrode couples have a very high capacity, but the structure destruction/rebuilding that occurs might limit their lifetimes and rates of reaction (see Section 44.4.7).

44.4.3 The present status of advanced batteries

The lithium-ion cell, which was initially developed in the early 1970s, contains no lithium metal because of the difficulty of plating out dendrite-free lithium upon recharge; the formation of dendrites results in short-circuiting of the cell. Initially, lithium was alloyed with metals such as aluminum, which forms the one-to-one

alloy LiAl with a very large energy-storage capability (780 A·h kg^{-1}). However, the volume of the Al doubles in size to accommodate the lithium, which results in electrode crumbling, loss of electrical contact between particles, and rapid capacity loss. Carbon, however, can react with lithium to form the intercalation compound LiC_6 very readily at room temperature, with only a 5% increase in volume, and has been used in essentially all lithium batteries since 1990 [3]; most recently, purified natural graphites have replaced the expensive synthetic carbons.

Although carbon has been the workhorse of lithium-ion batteries, its potential is only 300 mV from that of lithium, which leads to possible problems upon high-rate charging, such as in regenerative breaking in HEVs, if the rate of intercalation of lithium into the carbon is too low. In such circumstances, the lithium ions can plate out as metallic lithium, which is commonly in the form of dendrites, which can penetrate the separator, short the battery, lead to thermal runaway, and eventually cause a fire. These carbon–lithium anodes also have low gravimetric and volumetric energy densities – around 340 A·h kg^{-1} and 740 A·h l^{-1}, respectively, in comparison with 3,800 A·h kg^{-1} for pure lithium. Thus, much effort has been expended in searches for new lithium-containing anode materials, particularly for alloy-like compounds with tin and silicon, both of which form lithium-rich materials ($Li_{4.4}Sn$ and Li_4Si). However, very large volume changes have, as in the case of aluminum, severely limited their extended deep cycling.[1] In a breakthrough, Sony showed that amorphous nanostructured tin anodes can be readily recharged and have a 30% higher volumetric energy density than carbon anodes. These anodes have been shown to be an almost equimolar alloy with cobalt and to contain some other critical elements such as titanium, nanosized particles of which are embedded in carbon [4]. This amorphous nanosized material is not the answer to the anode materials challenge for large markets because of the cost and scarcity of cobalt, but it provides key clues as to a possible materials approach. The challenge for the materials scientist is to fully understand why these nanosized materials work well and how to substitute an abundant low-cost material in place of the cobalt. Of potentially even greater interest would be a similar silicon-based material, with its greater energy-storage capability. Initial results from magnetic studies suggest

---

[1] Batteries, in use, can be either deep cycled or shallow cycled and are designed specifically for one or the other. A deep-cycled battery is one that is typically discharged to >50% of its capacity before being recharged; applications are golf carts, computers, and cameras. Shallow-cycled batteries are discharged to only 10%–20% of their capacity; applications include the common automotive starting, lighting, and ignition (SLI) lead–acid battery and the NiMH battery (used in the Prius HEV).

that lithium is first intercalated into the compound, forming Li$_x$SnCo, and then a conversion reaction occurs, forming Li$_y$Sn and superparamagnetic cobalt. The term **cross-over reaction** is used to describe these reactions that first intercalate lithium and then convert structure. FeF$_3$ is another such example, in which case LiFeF$_3$ is formed first.

A key materials challenge to making any anode work is the formation and retention of what is called the SEI layer. This film, which is formed on the surface of the anode during the first electrochemical cycle, stops the reaction between the electrolyte solvent and lithium and is critical to the safe and efficient operation of any lithium battery. It is believed to be a mix of organic and inorganic components, but is at present ill-characterized – another challenge for the materials scientist. This film is one of the failure mechanisms of anode materials that expand and contract upon reaction with lithium because the film will be broken and have to be re-formed. The re-formation step consumes more electrolyte and creates resistive films between the anode material particles.

The area of potentially greatest opportunity for the materials researcher is the cathode. Here, lithium batteries will be used as an example. Several materials have been used, starting with titanium disulfide, TiS$_2$ (commercialized in 1977), which has a simple layered structure, followed by lithium cobalt oxide, LiCoO$_2$ (1991), and variants with nickel and manganese, Li(NiMnCo)O$_2$, both of which have structures similar to that of LiTiS$_2$. More recently, lithium manganese spinel (LiMn$_2$O$_4$) has found use, as has LiFePO$_4$ (2006). The spinel has a close-packed oxygen lattice like the layered oxides, but with a different distribution of the manganese and lithium ions among the available cation sites. In contrast to all of the prior materials, the olivine LiFePO$_4$ has a tunnel structure. Each of these battery systems has its advantages and drawbacks, but none to date have much exceeded the 150 A·h kg^{-1} theoretical capacity [5], leading to practical capacities of no more than 75 A·h kg^{-1}. All of these batteries are based on intercalation reactions. Major materials breakthroughs are needed in order to significantly increase the energy-storage capability, while at the same time increasing lifetime and safety.

Any cathode material for a lithium battery must satisfy the following requirements.

(a) The material must contain a readily reducible/oxidizable ion, for example, a transition metal.
(b) The material must react with lithium reversibly and without major structural change to give a rechargeable cell.
(c) The material must react with lithium with a high free energy of reaction for high voltage.

(d) The material should react with Li very rapidly upon both insertion and removal of Li in order to give high power.
(e) The material should ideally be a good electronic conductor, to alleviate the need for a conductive additive.
(f) The material must be cheap and environmentally benign; its elements must be plentiful in nature.

It would be preferable if an individual redox-active material satisfied many of these criteria, but several are opposite in character. For example, high-voltage cathodes generally have low electronic conductivities, so that a conductive diluent must be admixed; carbon black is most commonly used, but it can lead to side reactions, particularly at high potentials, and is almost certainly not usable with the present electrolytes at 5 V. Figure 44.3 shows the complexity of the electrode. The electrochemical reactions occur predominantly at the point where electrolyte, cathode-active material, and electronic conductor are in contact. Thus, there would be many advantages in finding a material that does not require any conductive additive, such as metallically conducting TiS$_2$ [6][7], but that operates at 3.5–4.5 V.

All of the rechargeable cathode materials commercialized to date basically belong to two classes: close-packed structures, such as those of layered TiS$_2$, LiCoO$_2$, and cubic LiMn$_2$O$_4$; and tunnel structures such as in LiFePO$_4$. None of these have demonstrated reversible storage capacities much exceeding 500 W·h kg^{-1} under normal operating conditions; in actual battery packages, the storage capacities range from around 80 W·h kg^{-1} to under 200 W·h kg^{-1}.

The three cathode materials LiTiS$_2$, LiCoO$_2$, and LiFePO$_4$ are compared in Table 44.2 to illustrate the six key criteria discussed above.

The three cathode materials show the full range of properties expected for a cathode material. In the Table 44.2, only reaction with one lithium ion is considered.

(a) All three have a transition-metal ion that can be oxidized on removal of lithium.
(b) Both LiTiS$_2$ and LiFePO$_4$ retain their crystal structure upon lithium removal, but LiCoO$_2$ undergoes a number of structural changes, and, if most of the lithium is removed, the CoO$_2$ slabs rearrange relative to each other from a cubic close-packed lattice to a hexagonal close-packed lattice. The resulting disorder causes capacity loss on cycling, so the cycling is restricted to around 0.5–0.6 Li.
(c) All three materials react with lithium with a relatively high free energy of reaction, but with LiCoO$_2$ having the highest value. Having the highest value is not necessarily an advantage, as will be discussed below.

Table 44.2. Properties of the three cathode materials LiTiS$_2$, LiCoO$_2$, and LiFePO$_4$

Property	LiTiS$_2$	LiCoO$_2$[a]	LiFePO$_4$
Number of phases	1	Multiple	Two
Δx in reaction[c]	1	0.5	1
Capacity (A·h kg^{-1})	213	137	169
Cell voltage (V)	2.2	3.8	3.45
Energy (W·h kg^{-1})	469	520	586
Electronic conductivity	Metallic conductor	Semiconductor	Insulator
Energy (kW·h l^{-1})	1.5[b]	2.5	2.0

[a] LiNi$_{0.33}$Mn$_{0.33}$Co$_{0.33}$O$_2$ has about a 20% higher capacity.
[b] LiTiS$_2$ does not require any carbon conductive additive, so the comparative energy is about 20% higher.
[c] Δx is the maximum amount of lithium that reacts per atom of transition metal, as in the formula Li$_x$CoO$_2$.

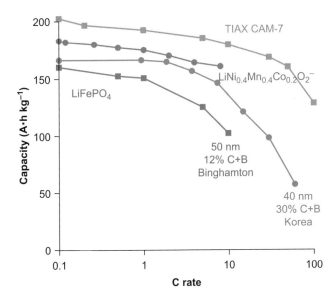

Figure 44.5. A plot showing how the capacity of cells declines as the power demand is increased. TIAX CAM-7 (orange squares) is a high-nickel-content phase, LiNi$_{1-y}$M$_y$O$_2$. LiNi$_{0.4}$Mn$_{0.4}$CoO$_2$ (green circles) is a preferred composition [5] tested on a carbon nano-tube mesh. The olivine LiFePO$_4$ materials were formed with 5% vanadium substitution (red squares) [8] and·a solvothermal material (blue circles) [9] (note: that B = binder and C = carbon).

(d) All three materials react with lithium very rapidly, with perhaps the LiCoO$_2$ having the slowest lithium diffusion in the lattice. A comparison of the rate capabilities of mixed-metal oxides [5] and LiFePO$_4$ [8][9] is shown in Figure 44.5. It should be emphasized that the highest power densities are obtained by partial substitution and by using lower materials loadings on the cathode.

(e) Only LiTiS$_2$ is a good electronic conductor, and it is sufficiently so that a conductive diluent is not needed; significant amounts of diluent (e.g., carbon black) must be added to the others. In the case of LiFePO$_4$, special carbon coatings on nanosized materials must be used.

(f) LiTiS$_2$ and LiFePO$_4$ are inherently cheap and environmentally benign, in contrast to cobalt, which is in short supply and with which there might be environmental issues. As a result, LiCoO$_2$ has been displaced from most applications by the mixed-metal oxide Li[NiMnCo]O$_2$, in which the cobalt content is ≤ 0.33; by LiFePO$_4$, with which volume is not an issue; and by the LiMn$_2$O$_4$ spinel for some applications. However, the fact that the components in the materials are cheap does not mean that the material will be cheap, unless a low-cost process is available. More effort is required in order to devise low-cost high-throughput manufacturing processes. Such efforts must determine the appropriate starting material from a search of the manufacturing tree for the desired elements.

44.4.4 Ultimate limitations to intercalation reactions

All lithium rechargeable batteries to date have used intercalation reactions. Such reactions, with their inherent structure retention, allow high rates of reaction and capacity retention over thousands of cycles. However, there is a continuing demand for higher storage levels, lower costs, and better lifetimes. As already mentioned, the olivine LiFePO$_4$ [10] has a number of advantages when there are no volume constraints, such as for utility load leveling and HEV buses. It has a lower potential than many other cathode materials, around 3.5 V versus lithium, which is well within the stability limits of many electrolytes. Moreover, the phosphate group is extremely stable, not readily releasing oxygen and not susceptible to being broken up on overcharge or discharge. Only under extreme discharge is the olivine structure destroyed, and even then the phosphate group is maintained as lithium phosphate. This material, despite being an electronic insulator, cycles exceptionally well even at high rates when a conductive film is coated onto the particles by grinding with carbon or by formation *in situ* from carbonaceous compounds. The outstanding characteristics of this material have led to its commercialization in some high-power applications, such as professional power tools [11], HEV buses, and utility energy smoothing. Its energy storage is somewhat low, a theoretical 590 W·h kg^{-1} at 3.5 V. The energy-storage capacity can be improved by switching to LiMnPO$_4$ with its 4 V potential, but this compound has been notoriously difficult to cycle. However, some recent work is showing high capacities, comparable to that of LiFePO$_4$, when a small amount of the Mn is substituted by other elements such as Fe [12]. Another approach to raising the energy stored is to synthesize compounds with a higher ratio of redox ion to phosphorus, thus reducing the amount of dead-weight phosphate. Compounds such as lipscombite, Fe$_{1.3}$PO$_4$O(H,Li), show some promise [13], but because they have Fe—O—Fe bonds, the discharge potential falls to around 3 V as a result of the lowering of the inductive effect found in the olivine structure. One way around this difficulty might be to use manganese in place of the iron, which might again give a 3.5-V cell, but with increased capacity. There are many opportunities for crystal engineering in this class of material both on the cation site and on the anion site.

The layered mixed-metal oxides exemplified by Li[NiMnCo]O$_2$, LiNi$_{0.80}$Co$_{0.15}$Al$_{0.05}$O$_2$, and lithium-rich Li$_{1+y}$[NiMnCo]$_{1-y}$O$_2$ are all, in principle, capable of capacities exceeding 250 A·h kg^{-1}. However, their capacities are mostly limited to around 180 A·h kg^{-1} for Li[Ni$_y$Mn$_y$Co$_{1-2y}$]O$_2$ when the charging voltage is limited to 4.3 or 4.4 V, as indicated in Figure 44.6. Using a higher charging voltage appears to lead to faster capacity loss on cycling. An increased nickel content can improve the capacity, but at the cost of safety, since Ni^{4+} can lead to oxygen release. This oxygen is a ready combustion agent for the oxidizable components of the cell, leading to thermal runaway, and can result in fiery explosions. LiNiO$_2$ is the worst compound in this regard, so nickel is always mixed with some other metal, for example, as in LiNi$_{0.95-y}$Co$_y$Al$_{0.05}$O$_2$. The presence of the redox-inactive

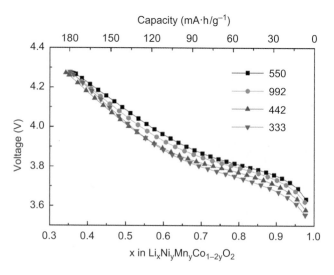

Figure 44.6. Open-circuit potential/composition curves of LiNi$_y$Mn$_y$Co$_{1-2y}$O$_2$; the compositions are represented by the nomenclature such that 442 represents 40% Ni, 40% Mn, and 20% Co. The reaction of 1 lithium is equivalent to around 270 Ah/kg, so the 0.7 lithium here is the equivalent of 180 Ah/kg.

aluminum prevents the complete charging of all the nickel to the 4+ state. The lithium- and manganese-rich compounds, which can be represented by the formula yLi$_2$MnO$_3$·(1 − y)Li[NiMnCo]O$_2$, release oxygen during their charging cycle at 4.6 V and, on subsequent cycling, they have capacities approaching 250 A·h kg^{-1} at low rates. However, the capacity falls rapidly at 2C rates, the rate needed for PHEV and many other applications. This is probably a result of disruption of the structure upon oxygen loss. This oxygen loss leads to pressure buildup in cells [14] and possible reaction with the electrolyte; lower-valence vanadium oxides have been proposed as a sink for this extra oxygen [14].

Materials showing two electrons per redox center

One way to increase the storage capacity in lithium batteries is to get more than one electron per transition-metal redox center. Several materials, including VSe$_2$ [15], VOPO$_4$ [16], and Li$_x$Ni$_{0.5}$Mn$_{0.5}$O$_2$ [17], can achieve close to two electrons, as shown in Figure 44.7. However, in each of these cases, there is a large difference between the potential of the first reduction and that of the second reduction, which could prevent both steps from being used in a practical battery. Nevertheless, VOPO$_4$ remains a possibility. Vanadium seems to be particularly attractive because of the availability of the V^{3+}/V^{5+} couple; V$_2$O$_5$ readily reacts with three lithium ions, but the structure becomes cation-disordered, giving Li$_3$V$_2$O$_5$ with a capacitor-like sloping discharge from 3.5 to 2.5 V. Some of the δ-V$_4$O$_{10}$ phases also react

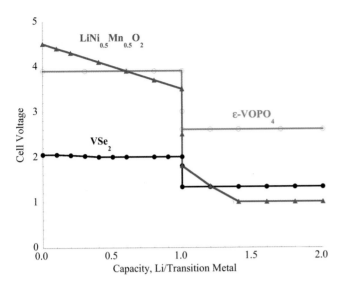

Figure 44.7. Examples of materials that show two-electron reductions per redox center: VSe_2, $VOPO_4$, and $LiMn_{0.5}Ni_{0.5}O_2$.

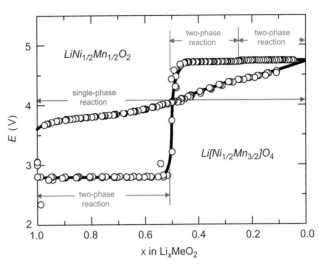

Figure 44.8. Lithium cell potentials of the two lithium nickel manganese dioxides with layered and spinel structures.

with lithium/vanadium ratios approaching 2. Similarly, $Li_3V_2(PO_4)_3$-like phases also cycle more than two Li ions. An alternative to inserting two lithium or sodium ions is to use a higher-valent cation such as magnesium or aluminum, which would also considerably reduce the weight and volume of the anode. However, the diffusion coefficients and, thus, ionic conductivities of these higher-valent cations tend to be much reduced in solids.

Higher-voltage systems

A second way to increase the storage capacity is to go to higher-potential cathodes, such as $LiMn_{3/2}Ni_{1/2}O_4$, that operate at close to 5 V. Figure 44.8 compares the voltage of two Ni–Mn compositions, both with close-packed oxygen lattices, namely the layered structure $LiMn_{0.5}Ni_{0.5}O_2$ and the spinel $LiMn_{3/2}Ni_{1/2}O_4$, demonstrating the favorable high-voltage behavior of the latter [5]. This example also demonstrates just how large an impact the chemical composition can have. However, there would be major stability issues with the electrolyte if such a cathode were used. Today's systems are not stable much beyond 4.2 V, with decomposition reactions occurring at higher voltages. In addition, the conductive carbon additives used in cathodes can react with the electrolyte salt at these potentials, forming intercalation compounds. The discovery of a high-voltage electrolyte would thus open up a number of opportunities for increased energy storage.

The electrolyte

A major cost factor and limitation to the next generation of batteries and capacitors is the electrolyte. At this time, there is a major lack of understanding of the molecular interactions occurring within the electrolyte and at the solid interfaces with which the electrolyte is in contact. Moreover, these interfaces can be on the external surface of the electrodes or in the complex internal porous structure. An advanced electrolyte, whether for a battery or an electrochemical capacitor, will need the following characteristics, compared with today's systems:

- higher ionic conductivity, to give high cycling rates over a wide range of temperature, ideally from -30 °C to 100 °C;
- higher stability, both chemical and electrochemical, to allow for higher-voltage systems and to increase safety;
- higher compatibility with other cell components, for both better wettability and lower corrosion/reaction; and
- lower cost and greater environmental friendliness.

Today, lithium batteries use electrolytes containing the salt $LiPF_6$ dissolved in a mixed carbonate solvent, whereas high-voltage capacitors use quaternary ammonium salts dissolved in acetonitrile or an organic carbonate. There is a desire to get away from the $LiPF_6$ salt, which can produce HF in even traces of moisture. This HF can cause dissolution of the cathode metals, which then migrate to and react with the lithium–graphite anode, causing significant loss of capacity. Boron-based salts are of interest because of their higher stability, but, in some cases, the SEI layers they form are too resistive. LiBOB and its fluorinated analogs are of particular interest and might lead to totally new systems over the next decade. Another research opportunity lies in ionic liquids. These are salts that are liquid under ambient conditions and do not need any solvent for operation. They also tend to have low vapor pressures and to be nonflammable, but they might be too reactive to be used

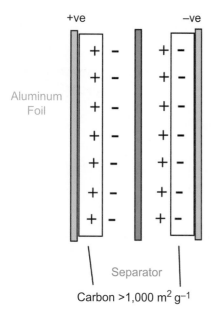

+ve   −ve

Aluminum
Foil

Separator

Carbon >1,000 m² g⁻¹

Figure 44.9. A schematic representation of energy storage in an electrochemical capacitor using high-surface-area carbon. The typical power density is 2.3 kW kg⁻¹, and the typical energy density is 1.7 W·h kg⁻¹.

with lithium and some cathode materials with which they can form complexes. However, the present materials might find application in high-power batteries, such as the dual-spinel $LiTi_yO_4/LiM_2O_4$ system, or in electrochemical capacitors. This is just the beginning of new opportunities for the electrolyte chemist, and major breakthroughs can be expected.

44.4.5 The present status of supercapacitors

Traditionally, capacitors differ from batteries by storing energy as charge on the surface of the electrodes, rather than by chemical reaction of the bulk material. As a result, the electrodes do not have to undergo structural change, and they have much longer lifetimes – essentially unlimited under perfect conditions. They thus also tend to have much higher rate capabilities than batteries, being almost instantaneously charged or discharged, and are therefore quite suitable for repetitive fast applications such as for regenerative breaking and subsequent acceleration applications. However, because capacitors use only the surface of the material for charge storage, as shown in Figure 44.9, they are very limited in energy-storage capability, with typical values of 0.1–1 W·h kg⁻¹, compared with >100 W·h kg⁻¹ for Li batteries. In addition, they provide a rather low quality of energy; typically the voltage varies in a continuous manner from the charging voltage to 0 V, rather than remaining at a relatively constant potential.

Capacitors are of various types. Those used in power and consumer electronic circuits are of the dielectric and electrolytic types and are mostly solid-state devices. They have extremely fast response times and essentially unlimited lifetimes, but they store very little energy, typically less than 0.1 W·h kg⁻¹. They thus have no place in applications for which significant amounts of energy need to be stored. The second class of capacitors consists of **electrochemical capacitors (ECs)** [18], often using high-surface-area carbon for the electrodes and sulfuric acid or acetonitrile as the electrolyte. Electrochemical capacitors can be split into two groups: **supercapacitors** and **pseudocapacitors**. Supercapacitors, also known as ultracapacitors, are the electric double-layer capacitors, in which the energy is stored at the surface of the material in the double layer. They have improved energy storage over solid-state capacitors but still less than 10 W·h kg⁻¹, and the cell voltages are limited to prevent the decomposition of the liquid electrolytes: less than 1 V for water-based and around 3 V for non-aqueous electrolytes. The ideal applications for ECs are those demanding energy for short periods, from 10^{-2} to 10^2 s, and for thousands of cycles reliably and at low cost, such as fork-lift trucks and dock-cranes.

The second class of ECs consists of pseudocapacitors, which are a hybrid between double-layer capacitors and batteries, with both the bulk and the surface of the material playing key roles. Pseudocapacitors can thus store much more energy than surface capacitors, >10 W·h kg⁻¹, but they face many of the same reliability and scientific challenges as advanced batteries. Many of the materials used in batteries also find application in pseudocapacitors, particularly transition-metal oxides such as vanadium and manganese oxides. In fact, upon reaction with lithium, some crystalline compounds with flat discharge plateaus, such as vanadium pentoxide, convert to amorphous materials, as the lithium and vanadium ions become randomized in the lattice, and show sloping discharge profiles typical of a capacitor. A compound that has generated much scientific interest is hydrated ruthenium oxide, which can store around 200 W·h kg⁻¹ (750 F g⁻¹), the highest of any capacitor. A fundamental understanding of this currently unique material is essential in order to determine its key physical and chemical behavior. This understanding might lead to a commercially viable pseudocapacitor material not containing any noble metal.

As is also the case for batteries, a goal for the materials scientist must be to design capacitive storage materials in such a way that each component preferably performs more than one function. Such a design would incorporate both chemical and surface charge storage, would determine the optimum mix of each, and would determine the optimum pore-size distribution required in order to maximize the capability of the electrolyte to

Figure 44.10. The variation of the storage capability of vanadium oxide as a function of its morphology. From [19].

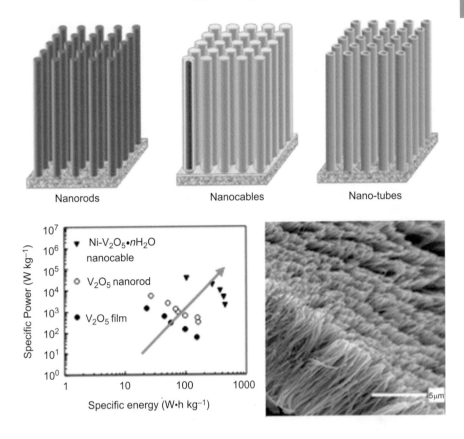

Nanorods   Nanocables   Nano-tubes

provide or remove the ions at a very high rate. It would also incorporate how the solvation of the charging ion changes with pore size from macropores to the extreme at which an intercalation reaction occurs between neighboring sheets of the material where only a minimal or zero solvation of the ion occurs. Such a design approach would also help determine the optimum electrolyte and salt combination for high conductivity and charge transfer and address the ever-present cost issue. It is expected that new computational tools will play a key role in allowing the theory, modeling, and simulation of the complex structures, including interfaces, demanded for the next generation of capacitors. As noted above for batteries, a key challenge for any electrical energy-storage device is to get a continuous electronic conductive pathway in the electrode so that the electrons can be inserted or removed extremely rapidly. This is even more important for pseudocapacitors, which are designed to be high-power devices. A natural extension of the porous materials researched today is to go to nanosized materials, with their inherently extreme surface areas and potentially enhanced chemical redox behavior.

Vanadium oxides are a class of materials that have been used in batteries, e.g., silver oxide plus vanadium pentoxide in pacemakers, and are beginning to gain increased attention because of their broad range of morphologies. For example, upon reaction with Li,

crystalline V_2O_5 rapidly becomes disordered and exhibits typical capacitive-like discharge curves. In addition, the electrochemical characteristics of nanometer-sized particulate vanadium oxides have been found to be strongly dependent on the morphology, as shown in Figure 44.10. These materials show behavior between that of a battery and a capacitor, and it is likely that a number of future applications will use such hybrid devices.

44.4.6 Materials challenges related to the inactive components

Most batteries and capacitors do not approach even 50% of their theoretical capacities. What is really limiting their specific energy storage? As described in Section 44.4.2, batteries and electrochemical capacitors contain many components that are critical to performance, but do not contribute to the energy or power density and instead reduce both the gravimetric and the volumetric capacity. Such components include the conductive additive and binders that are used in both electrodes, as well as the critically important electronically insulating separator. A challenge to the materials scientist is to design materials that are multipurpose, for example, a conductive binder or a separator that can also serve as a cell cutoff in the event of thermal runaway. Another area for exploration is the use of current collectors that could be used in a bipolar cell configuration, where one side is the anode and the other the cathode. In such batteries/capacitors, the cells are simply stacked to deliver the desired battery voltage, with no external connections between the individual cells.

A key recent finding that carbon nano-tubes, with their extremely high aspect ratio, reach the conductivity percolation threshold in mixtures at levels of only 0.1 wt% could dramatically reduce the weight and volume of the conductive diluent that needs be added to cathode mixtures (A. Windle, personal communication, 2008). This could explain the improved electrochemical behavior reported for LiFePO4 formed hydrothermally in the presence of carbon nano-tubes [20]. Included in Figure 44.5 are data corresponding to a cathode material supported on a carbon nano-tube mesh, showing the effect of using the mesh to improve the rate capability of the layered oxide material (C. Ban, personal communication, 2010).

44.4.7 Other battery systems

Lithium-ion cells have received the most attention over the past three decades, but, if transformational changes are to be made, then other systems must also be researched. For example, sodium–sulfur cells operating at around 300 °C are available commercially. The volumetric capacity of lithium cells could be significantly improved if pure lithium could be used as the anode or if a pure chalcogen could be used as the cathode. Zinc–air cells have been used as primary batteries for powering hearing aids, but not yet as secondary batteries. If a lithium–air cell could be made operable, the energy density could be twice that of present-day lithium-ion cells. However, such cells have the problems of both batteries and fuel cells: the need for non-dendritic replating of lithium, a reliable efficient electrocatalyst for the cathode, and a non-flooding electrolyte. See Box 44.2. Aside from lithium and zinc, a number of other metal–oxygen couples might be viable, including those

---

Box 44.2. The challenges facing a lithium–air battery

Consider the materials challenges facing a metal–air battery, as an illustration of the significant challenges facing any new system.

(a) Control of the reaction to prevent undesired side reactions. What is desired is a two-electron electrochemical reaction forming Li_2O_2. However, the very reactive species formed, such as LiO_2, might easily undergo chemical reaction with the electrolyte solvent. This does not result in any energy conversion to electricity. Moreover, it will use up the electrolyte, leading to rapid capacity loss. The electrolyte will have to be chosen carefully in order to avoid side reactions, and the electrocatalyst must be chosen to catalyze the desired electrochemical reaction, probably a two-electron reaction to the peroxide, and prevent undesired chemical reactions.

(b) Electrodeposition of lithium without the formation of dendrites.

(c) Protection of lithium from the air electrode, including moisture. This will almost certainly require a solid electrolyte, which is impermeable to oxygen and water. A solid electrolyte would also solve the lithium-dendrite problem. The company Polyplus has developed such an electrolyte for primary lithium–water cells.

(d) Liquid electrolytes to make contact with the solid electrolyte and with the cathode.

(e) Containment of the electrolyte, if the air electrode is open to the environment.

(f) Development of an electrocatalyst system that can assist in both the reduction and the oxidation at the cathode. The electrocatalyst must be supported on a conducting structure, typically a porous carbon; this carbon must not be oxidized during operation of the cell.

(g) Improvement of the overall efficiency, because, today, the discharge voltage is below 3 V and the charge voltage is over 4 V.

(h) Improvement of the kinetics of the electrochemical reactions to attain rates useful for real applications.

(i) A means of preventing the lithium peroxide formed from coating the electrocatalyst and negating its effectiveness. Researchers must determine whether it needs to be solubilized and, if so, what penalty that imposes in terms of weight and volume storage. A fallback position is to use an aqueous electrolyte, combined with a solid electrolyte to protect the lithium anode.

(j) A scrubbing system for the air electrode to remove all undesirables such as carbon dioxide.

None of these challenges will be overcome quickly, but the gains, if successful, are sufficiently great to warrant an intensive materials and electrochemistry effort. Moreover, the opportunity in such a study for other discoveries that might advance today's battery technology is enormous.

with aluminum and magnesium as the metal, but researchers have not developed a feasible scientific approach to date for a rechargeable metal–air battery.

Most of the electrode reactions discussed so far have been based on intercalation reactions. Conversion reactions, in which the structure is destroyed and a new one formed, can lead to much higher capacities. For example, in the case of FeF_3 forming LiF and Fe upon reaction with lithium, the capacity is a high 601 A·h kg^{-1}. However, the rate capability is lower than for intercalation reactions, and there can be a large voltage hysteresis upon reaction, which lowers the full-cycle efficiency. Research is being carried out in order to improve the rate capability and reduce this hysteresis.

Moving away from lithium altogether, sodium- or even magnesium-based systems might be considered, but the low melting point of the former leads to safety concerns for consumer applications. Sodium has been found to reversibly intercalate into the chevrel phases, $M_xMo_6S_8$; sodium also intercalates into most of the layered oxides and chalcogenides, giving a range of phases. Since the sodium can reside in different sites depending on the composition, the MX_2 layers can shift with sodium content [6]. The sodium–sulfur couple has been commercialized using sodium–β-alumina as the electrolyte at around 300 °C, and there are possibilities of reducing the temperature of operation if suitable sulfur solvents and electrolytes can be found.

In the 1970s, various flow-redox batteries were considered, such as Zn–Br$_2$. In these batteries, which resemble fuel cells in many respects, the cathode and anode reactants are held in large tanks so that the energy-storage capability can be quite high. Vanadium redox batteries are in commercial use now, as discussed earlier. The use of stacked bi-polar plates allows the battery pack to provide high power in a small volume. Such configurations have not been considered in lithium cells because of the concern of cell imbalance leading to safety issues, but they probably deserve more attention. Another potential low-cost flow system involves the aqueous Fe/Cr couple.

44.4.8 Key materials challenges and opportunities facing batteries and capacitors

A number of scientific challenges and opportunities face both batteries and capacitors, whether configured for high-power/low-energy or lower-power/high-energy applications. New lower-cost materials that readily react and store charge are obvious candidates for the materials scientist to study. It is essential to study all forms of materials, bulk and nano, and to fully understand their reaction mechanisms by performing *in situ* measurements. Such efforts can be enhanced by using modern computational tools. These are briefly discussed below.

The role of nanomaterials for energy storage

At the nanoscale, there is likely to be little that separates pseudocapacitive from storage reactions conceptually. The ability to control the structure of materials at the nanoscale adds size as a functional variable, in addition to composition and structure. These materials exhibit new phenomena beyond those associated with just the larger surface areas, such as change in their phase diagram and crystalline cell parameters. The smaller dimensions enhance the rate of reaction, both chemical and charge storage, but they can also enhance the rate of undesired side reactions, such as with the electrolyte. Some materials, such as the LiFePO$_4$ discussed earlier, work effectively only when the particles are less than 100 nm in size. The challenge to the theorist is to predict the chemical composition, structure and physical behavior of the next generation of materials, from nanoscale to bulk, to guide the experimentalist.

Materials characterization challenges

Although much research has been done on the behavior of batteries, little is really known about the chemical reactions and materials structural changes under real electrochemistry conditions. Studying the materials after removal from a battery or capacitor, which takes time and allows the materials to relax to an equilibrium state, is very different from seeing exactly what is occurring in the cell itself. A battery electrode is a living device with ions, not just the lithium ions, moving as charge is inserted and removed. Therefore, much effort to develop techniques that allow the study of materials in reacting cells is under way. These *in situ* techniques often require the use of a high-powered X-ray source, such as is available at a synchrotron, and the associated software in order to understand data relating to often amorphous or poorly crystalline materials. Advanced laboratory techniques that allow *in situ* TEM, NMR, and magnetic measurements of cycling cells are beginning to be used today.

Opportunities for modeling

Modeling can not only help in the rational design of novel electrical energy storage systems with high energy and power density, but also provide a better understanding of the design of cells to optimize the structure of the complex electrodes to allow maximum use of the active materials and the minimization of inactive materials such as the current collectors and separators. Modeling could also help us to understand the interfaces between the reacting components, in particular the critical SEI layer. The SEI layer formed between the electrodes and the electrolyte is critical to the performance, life, and safety of electrochemical devices. This complex, continually changing interface is very poorly understood and not well described in mathematical performance

Table 44.3. Some projections

Technology	Current	In 10 years	In 25 years
Key materials	Pb–acid, Li-ion	Li-ion + redox	Sodium, zinc/lithium–air
Cost of technology ($ per kW·h)	800	300	100–300
Efficiency of system, cost of energy used	80%, 20%	90%, 10%	>90%, 10%
Limitations of current technology	Cost of materials and manufacturing; lifetime	Increasing efficiency of power plants will reduce pollution (load leveling)	Electric vehicles will reduce the dispersed pollution
Environmental impact			
Key roadblocks to technology	Materials lifetime	Availability of materials	

models. It will be absolutely essential to fully understand this living layer in order to design the long-lived cells demanded by electric vehicles. This will demand a joint effort between the theorist and novel *in situ* characterization tools.

44.5 Summary

Batteries and capacitors have made remarkable advances over the last 20 years, enabling the revolution in portable electronic devices. Their continued improvement is to be expected, particularly with their need to support an electric economy based on renewable energy sources such as wind and solar, and the rapidly oncoming demand for electric vehicles. Breakthroughs in advanced materials are to be expected, with novel chemical redox couples being discovered, and new geometric structures being found for capacitors. These breakthroughs are best accomplished through a combination of experimental and theoretical studies. Efforts along several lines, such as increased-capacity anode and cathode materials, higher-potential cathodes, metal–air or metal–sulfur batteries, and use of nanomaterials appear to be promising approaches. Some projections are given in Table 44.3.

44.6 Questions for discussion

1. MnO_2 reacts with one Li to form $LiMnO_2$, with an average voltage of 4.0 V. How much energy can be stored in a Li/MnO_2 battery assuming 100% efficiency and only the masses of MnO_2 and Li?
2. What are the major challenges to the development of a room-temperature lithium-anode/sulfur-cathode battery?

3. There is much interest in flexible solar photovoltaic systems that could, for example, be used for the covers of tents or be worn as clothing. Some believe it would be ideal if a battery could be built into the flexible system back-to-back with the solar collector. Using an average solar flux, how thick would the battery have to be to store 8 h of solar energy. (Remember to give the efficiency of the solar cell.)
4. Discuss the advantages of a battery having solid electrodes and a liquid electrolyte versus liquid electrodes and a solid electrolyte.
5. Does an electric vehicle cause less pollution than a gasoline-powered vehicle if you consider the entire energy-production process? Discuss the various scenarios provided by the source of electricity.
6. Using the US Geological Survey website, determine which transition metals are most abundant in the Earth's crust. If 10 million electric vehicles are built each year for 10 years and each has 20 kW·h of battery, then which transition metals could you consider for use?

44.7 Further reading

- **M. S. Whittingham**, **R. F. Savinell**, and **T. Zawodzinski** (eds.), 2004, *Chemical Reviews*, **104**, 4243–4533. Multiple articles from an issue on batteries and fuel cells.
- **M. S. Whittingham**, 1978, "Chemistry of intercalation compounds: metal guests in chalcogenide hosts," *Prog. Solid State Chem.*, **12**, 41–99.
- **US DOE**, 2007, *Basic Research Needs for Electrical Energy Storage*, available at http://www.er.doe.gov/bes/reports/files/EES_rpt.pdf. Provides a wide-ranging discussion, with extensive background material, on the progress required to satisfy future electrical energy-storage needs.

- **R. Kötz** and **M. Carlen**, 2000, "Principles and applications of electrochemical capacitors," *Electrochim. Acta*, **45**, 2483–2498.
- **Dinorwig Power Station**, 1984, http://www.fhc.co.uk/dinorwig.htm contains excellent reviews on electrolytes and current collectors.
- **D. Linden** and **T. B. Reddy**, 2011, *Handbook of Batteries*, 4th edn., New York, McGraw-Hill.
- Several reviews cover this area – e.g., batteries and fuel cells [6] and electrochemical capacitors [18].
- **Seth Fletcher**, 2011, *Bottled Lightning*, New York, Hill and Wang, is an enlightening description of the history of the lithium battery.

44.8 References

[1] **K. H. LaCommare** and **J. H. Eto**, 2004, http://certs.lbl.gov/pdf/55718.pdf.

[2] **G. G. Libowitz** and **M. S. Whittingham**, 1979, *Materials Science in Energy Technology*, New York, Academic Press.

[3] **S. Flandois** and **B. Simon**, 1999, "Carbon materials for lithium-ion rechargeable batteries," *Carbon*, **37**, 165–180.

[4] **Q. Fan**, **P. J. Chupas** and **M. S. Whittingham**, 2007, "Characterization of amorphous and crystalline tin–cobalt anodes," *Electrochem. Solid-State Lett.*, **10**, A274–A278.

[5] **M. S. Whittingham**, 2004, "Lithium batteries and cathode materials," *Chem. Rev.*, **104**, 4271–4301.

[6] **M. S. Whittingham**, 1978, *Prog. Solid State Chem.*, **12**, 41.

[7] **M. S. Whittingham**, 1976, "Electrical energy storage and intercalation chemistry," *Science*, **192**, 1126–1127.

[8] **J. Hong**, **C. S. Wang**, **X. Chen**, **S. Upreti**, and **M. S. Whittingham**, 2009, "Vanadium modified LiFePO$_4$ cathode for Li-ion batteries," *Electrochem. Solid-State Lett.*, **12**, A33–A38.

[9] **D.-H. Kim** and **J. Kim**, 2006, "Synthesis of LiFePO$_4$ nanoparticles in polyol medium and their electrochemical properties," *Electrochem. Solid-State Lett.*, **9**, A439–A442.

[10] **A. K. Padhi**, **K. S. Nanjundaswamy**, **C. Masquelier**, **S. Okada** and **J. B. Goodenough**, 1997, "Effect of structure on the Fe^{3+}/Fe^{2+} redox couple in iron phosphates," *J. Electrochem. Soc.*, **144**, 1609–1613.

[11] A123, 2007, http://www.a123systems.com/newsite/index.php#/applications/cordless/.

[12] **T. Drezen**, **N.-H. Kwon**, **P. Bowen** *et al.*, 2007, "Effect of particle size on LiMnPO$_4$ cathodes," *J. Power Sources*, **174**, 949–953.

[13] **Y. Song**, **P. Y. Zavalij**, **N. A. Chernova**, and **M. S. Whittingham**, 2005, "Synthesis, crystal structure, and electrochemical and magnetic study of new iron (III) hydroxyl-phosphates, isostructural with lipscombite," *Chem. Mater.*, **17**, 1139–1147.

[14] **K.-S. Park**, **A. Benayad**, **M.-S. Park**, **W. Choi**, and **D. Im**, 2010, "Suppression of O$_2$ evolution from oxide cathode for lithium-ion batteries VO$_x$-impregnated 0.5Li$_2$MnO$_3$–0.5LiNi$_{0.4}$Co$_{0.2}$Mn$_{0.4}$O$_2$ cathode," *Chem. Commun.*, **46**, 4190–4192.

[15] **M. S. Whittingham**, 1978, "The electrochemical characteristics of VSe$_2$ in lithium cells," *Mater. Res. Bull.*, **13**, 959–965.

[16] **Y. Song**, **P. Y. Zavalij**, and **M. S. Whittingham**, 2005, "ε-VOPO$_4$: electrochemical synthesis and enhanced cathode behavior," *J. Electrochem. Soc.*, **152**, A721–A728.

[17] **C. S. Johnson**, **J.-S. Kim**, **A. J. Kropf** *et al.*, 2003, "Structural characterization of layered Li$_x$Ni$_{0.5}$Mn$_{0.5}$O$_2$ ($0 < x \leq 2$) oxide electrodes for Li batteries," *Chem. Mater.*, **15**, 2313–2322.

[18] **R. Kötz** and **M. Carlen**, 2000, "Principles and applications of electrochemical capacitors," *Electrochim. Acta*, **45**, 2483–2498.

[19] **Y. Wang**, **K. Takahashi**, **K. H. Lee**, and **G. Z. Cao**, 2006, "Nanostructured vanadium oxide electrodes for enhanced lithium-ion intercalation," *Advanced Functional Mater.*, **16**, 1133–1144.

[20] **J. Chen** and **M. S. Whittingham**, 2006, "Hydrothermal synthesis of lithium iron phosphate," *Electrochem. Commun.*, **8**, 855–858.

45 Mechanical energy storage: pumped hydro, CAES, flywheels

Troy McBride, Benjamin Bollinger, and Dax Kepshire

SustainX, Inc., West Lebanon, NH, USA

45.1 Focus

We are all familiar with small-scale electrical energy storage in chemical batteries, from cars to cell phones. Batteries offer near-instant response time, but cost tends to scale linearly with size, making very large batteries or systems of batteries prohibitively expensive. Mechanical energy storage, in contrast, tends to be inexpensive at large scales due to the use of relatively low-cost materials (e.g., concrete and steel) and low-cost storage media (e.g., water, air), and due to long device lifetimes. The levelized cost of energy (LCOE), which is essentially the break-even selling price per kilowatt-hour (kWh) including all lifetime costs, for pumped-hydroelectric and compressed-air storage can be much less than for smaller-scale technologies such as batteries.

45.2 Synopsis

Electrical energy can be converted into any of the three forms of mechanical energy: gravitational potential, elastic potential, or kinetic. Each of these can, in turn, be converted back to electricity. In principle, though never in practice, interconversion can be 100% efficient. The most common mechanical energy-storage technologies are pumped-hydroelectric energy storage (PHES), which uses gravitational potential energy; compressed-air energy storage (CAES), which uses the elastic potential energy of pressurized air; and flywheels, which use rotational kinetic energy.

In PHES, water is pumped uphill to a reservoir, then run downhill through turbine-generators when electric output is desired. It is by far the most common form of large-scale energy storage. However, classical PHES is limited in its future additional capacity, especially in the developed world, because the most cost-effective, environmentally suitable locations for storage reservoirs have already been developed. Nonetheless, PHES with underground storage and/or at a smaller scale may provide additional storage resources in the future.

Electric-in, electric-out CAES, although studied and widely discussed for decades, has not yet been widely deployed. The only two utility-scale CAES facilities built so far store energy as compressed air in underground caverns. This energy is recovered by combusting natural gas with the compressed air in turbines. New CAES technologies are being actively pursued, including no-fuel options such as advanced adiabatic CAES (AA-CAES) and isothermal CAES. Several companies are developing modular CAES systems storing compressed air in aboveground pressure vessels, and commercial versions of these technologies will be available in the near future.

Flywheels compete directly with chemical batteries and supercapacitors to provide rapid, high-power bursts of electricity to smooth ("frequency regulate") the power supplied by the grid. They store much smaller total amounts of energy than do PHES and CAES systems, making them appropriate for powering outage-intolerant loads (e.g., a data center) for short periods or for responding to rapid power fluctuations on the grid. They are not typically appropriate for bulk energy storage. Commercial flywheel products are available as uninterruptible power supplies, and a utility-scale (20 MW, 5 MWh) installation for frequency regulation, began operation in 2011 in New York State. Flywheels are increasingly competitive for frequency regulation and offer much longer device lifetimes than do chemical batteries.

PHES and CAES technologies largely use conventional materials (e.g., steel), are built of industrially proven components with known maintenance schedules, and offer long system lifetimes and moderate to low costs at large storage scales.

45.3 Historical perspective

45.3.1 Pumped-hydroelectricity energy storage

Modern, electrical PHES – also termed pumped-storage hydroelectricity – was made possible by basic technological advances of the nineteenth century. The Francis turbine, which was developed in the 1840s, was the first turbine design with efficiencies over 90% and remains the turbine style of choice in PHES. Electric motors/generators, the other essential mechanical component of PHES, were developed at about the same time. Fully reversible turbine-generators were not feasible, however, until the 1930s.

The first electric-in, electric-out PHES installation was commissioned in 1909, in Schaffhausen, Switzerland, and produced 1 MW of electricity; the Rocky River Plant at New Milford, Connecticut, which was commissioned in 1929, is sometimes cited as the first "major" PHES system (originally 24 MW, upped to 31 MW in 1951). Today PHES is the world's dominant grid-scale energy-storage technology, with an installed global capacity of approximately 90 GW (~5% of world generating capacity) as of 2010. PHES round-trip efficiency (the ratio of electric power in to electric power out) is typically around 65%–75%.

45.3.2 Compressed-air energy storage

In the mid to late 1800s, compressed air competed with electricity as a means of energy distribution. One reason why compressed air lost out is its thermodynamic inefficiency: as gas is compressed, it tends to heat. After compression, leakage or deliberate shedding of heat from the hot gas to its environment, if not later recovered, results in energy loss and low efficiency. Heat management remains an issue for contemporary CAES schemes.

In the 1970s, utility-scale CAES schemes for load following (adjustment of grid power output to match fluctuating demand) became attractive as oil prices increased and it was widely believed that nuclear power would soon make baseload power very cheap. Storing inexpensive power at night and using it during the day, when demand is higher and peaker plants burn expensive fuel (at the time, primarily oil), made sense. Very inexpensive baseload power did not materialize, however, and interest in CAES declined in the 1980s and beyond along with prices for the oil and natural gas used in peaker plants. In the twenty-first century, with fuel prices again high and with non-hydroelectric renewables – with their relatively high variability – penetrating the grid in large quantities for the first time, there is renewed interest in CAES.

In classical CAES, off-peak electricity is used to compress air into underground caverns or aquifers. During higher-demand periods, the stored compressed air is combusted with natural gas in turbines that drive generators. Such a system hybridizes storage with fueled power generation.

Compressed air can also be used to generate electricity directly, rather than being hybridized with gas combustion. In systems of this type, turbines or pistons recover the elastic potential energy (and, in some versions, thermal energy) of the compressed gas and direct it to electric generators. Modular no-fuel CAES systems storing compressed air in aboveground vessels, which are now on the technological horizon, will be suitable anywhere, have adjustable storage capacity, and offer faster response than classical CAES.

45.3.3 Flywheels

Flywheels storing kilowatt-hours of energy have been experimentally used in buses and other vehicles since the 1950s, but have not yet found a commercial niche for energy storage in transportation. Modern flywheels specialized for electric-in, electric-out storage can spin at tens of thousands of revolutions per minute (rpm), may be suspended magnetically in vacuum, and can store up to tens of kilowatt-hours per unit. Energy is delivered to the energy-storing rotor and extracted from it using electromagnetic fields: the rotor may not be in mechanical contact with its container.

Since the early 2000s, lower-speed flywheels (e.g., steel rotors spinning at thousands of rpm) have competed with chemical batteries in the provision of uninterruptable electric power to data centers (where the costs of even a brief power loss can be catastrophic), with power levels reaching megawatts and storage durations in the tens of seconds. Higher-speed flywheels (e.g., carbon-composite rotors spinning at tens of thousands of rpm) with storage capacities an order of magnitude larger than those of the flywheels used in data centers are being developed, as of 2010, with storage durations in the tens of minutes. The purpose of the higher-speed units under development is not to store the large amounts of energy needed for load shifting, but to supply rapid, high-power bursts of electricity that compensate for temporary mismatches between supply and demand on the grid – a function termed "frequency regulation."

45.4 Mechanical energy-storage technologies

45.4.1 Pumped hydroelectric energy storage

Principles

The most common and cost-effective means of large-scale energy storage (excluding fuels) is PHES. Electrical energy is used to pump water to an elevated reservoir, usually artificial (see Figure 45.1): most of this energy

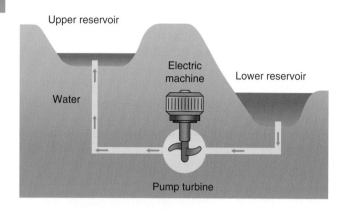

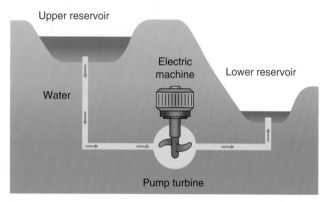

Figure 45.1. A typical PHES scheme. top: at night, when demand is low, excess generating capacity in the grid is used to run an electric machine that pumps water uphill from a lower reservoir to an upper reservoir. In effect, the gravitational potential energy of the elevated water stores the electric energy used to pump it uphill. bottom: during the day, when demand for power is high, water is run downhill from the upper reservoir through the pump turbine to the lower reservoir, converting most of the water's gravitational potential energy back to electricity, which is transmitted to the grid.

can be recovered as electricity when the water is allowed to flow downhill through turbines to a lower-altitude body of water, either a dedicated reservoir or a river or ocean. Reversible turbine-generator combination units are used both to pump the water uphill and to recover energy from the water as it runs downhill. It uses technologies that have long lifetime and low maintenance, and PHES is the least expensive per lifetime kWh large-scale storage technology yet deployed.

The type of installation just described, in which the upper reservoir is fed solely by the water pushed uphill by the turbine-generator, is sometimes referred to as "simple" or "pure" PHES. If the upper reservoir is also fed by a natural water source such as a river, then the scheme combines conventional hydroelectric generation with pumped storage and is termed "combination" or "mixed" PHES (Figure 45.2).

The amount of energy that can be stored by a pure PHES installation depends on the volume capacity of its upper reservoir and the altitude difference or "head" between its upper and lower reservoirs. (The lower reservoir must be at least large enough to fill the upper reservoir.) Energy is stored in the upper reservoir as gravitational potential energy, which is proportional to water volume and altitude: $U = \rho VgH$, where U is the potential energy, ρ is the density of water (1,000 kg m^{-3}), V is the volume of water, g is the acceleration due to gravity, and H is the height difference between the two reservoirs. The water volume V is limited by the capacity of the upper reservoir. The head H – here assumed constant, though in fact it varies slightly as the upper reservoir empties and the lower reservoir fills, or vice versa – is limited by local geography, which must allow the construction of an upper reservoir at a significantly higher altitude (e.g., 100 m or more) than the lower body

Figure 45.2. The Kinzua Dam and 400-MW Seneca Pumped Storage system in Alleghany State Park, Pennsylvania. Water is pumped to the circular artificial reservoir during low-demand periods and used to generate power during peak-demand periods. This is a type of "mixed" PHES system, with the upper reservoir providing energy storage and non-storage generating capacity available from the Kinzua Dam (lower left) [1]. When full, the upper reservoir contains 7.4 million cubic meters of water with a potential energy of 4,200 MWh. Photo: US Army Corps of Engineers.

of water. The reservoirs must be nearby so that they can be connected by a relatively short run of tunnel: a long tunnel or pipe run would be expensive to build and would waste excessive energy through pipe friction. The usual requirement for a classical PHES installation is therefore a site where one can build a large reservoir that is separated by a steep slope over 100 m high from a natural or constructed lower body of water. Water must be available in abundance and the site must not impinge on protected lands. These strict siting requirements are a major constraint on the building of ever-larger numbers of classical PHES installations.

The total energy stored in the upper reservoir is given by $U = \rho V g H$, but also relevant is the maximum power output of the installation. The available power is given by

$$P = \eta_d \rho \dot{V} g (H - \Delta H_t),$$

where η_d is the net discharge efficiency (the product of all efficiencies in the turbine-to-busbar conversion chain), ρ the density of water, $\dot{V}$ the volumetric flow rate, g the acceleration due to gravity, H the head, and ΔH_t the loss of head during generation (the pressure loss seen at the turbine due to water friction in pipes). Most energy losses arise from inefficiencies in the pump/turbine and motor/generator and from friction inside pipes and other devices. Most PHES installations worldwide can produce between 100 MW and 2 GW of power when they are emptying their upper reservoirs at full speed.

Because the total stored energy $U = \rho V g H$, more energy is stored per unit volume of water for high-head PHES (larger H): that is, for a given system size, the energy density is higher. This yields lower cost per kWh stored, since reservoir and pump-house costs do not increase as quickly for high-head systems as does the energy density. Consequently, there has been a decades-long trend in new PHES construction toward higher and higher heads: from 1960 to 2000, the average energy density of new PHES approximately quadrupled [2]. However, very high-head PHES sites are harder to find than lower-head sites.

The round-trip storage efficiency for PHES is typically 65%–75%, although projected plants with high head and improved pump/turbines may achieve 76% [3]. The round-trip efficiency of a PHES system is

$$\eta = \eta_c \cdot \eta_d,$$

where η_c is the charging efficiency (the efficiency with which electrical energy received by the PHES system is converted to potential energy in the upper reservoir) and η_d is the discharging efficiency.

Two existing installations

The PHES site at Raccoon Mountain near Chattanooga, Tennessee, was completed in 1978. Starting with a full upper reservoir, Raccoon Mountain can deliver up to 1.53 GW of electric output for up to 22 h. To produce this much energy (33.7 GWh) from storage, assuming 85% one-way efficient conversion from gravitational potential energy to electricity, requires a lake 2.14 km^2 (528 acres or almost 1 square mile) in area with a mean depth of at least 23 m (nearly 50 million cubic meters of water) and a head of approximately 300 m (1,000 ft) above the lower reservoir.

The Tianhuangping pumped-storage project in China, 175 km from Shanghai, is the largest PHES project in Asia. Its two reservoirs are separated by a head of 590 m and each stores 8 million cubic meters of water. The installation, which was completed in 2001, can produce 1.8 GW and operates at 70% round-trip efficiency. It is used to stabilize the East China power grid.

Scores of broadly similar installations exist around the world.

Current research and advances in PHES

A number of wrinkles on this century-old technology are now being investigated, all with an eye to achieving lower cost, lower environmental impact, and better grid integration of intermittent inputs from wind power and other renewables [2].

(1) **Upgrading**. Upgrading and modernizing existing PHES installations by replacing decades-old turbine-generators with higher-capacity, more-efficient modern units is a major area of PHES activity. For example, at the Blenheim–Gilboa PHES site in New York State, turbine-generators almost 40 years old were replaced with new units. The project, which was finished in 2010, increased the power capacity by 11.5%, from 1.04 GW to 1.16 GW.

(2) **Renewables penetration and variable speed**. In most PHES, the turbine-generators can operate only at a constant speed: the electrical input or output of such a PHES installation can only be incremented or decremented by the power rating of a single turbine-generator (typically there are four to eight per site). An alternative is to use variable-speed turbine-generators. These have been employed in some Japanese PHES installations since the 1990s, with a total adjustable-speed capacity of about 3 GW [2].

(3) **Underground PHES**. Schemes for underground PHES (UPHES) have been described for decades, though no such system has yet been built. In UPHES, the upper reservoir is located on the surface and the lower reservoir is an aquifer or a body of water in a large cavern, either natural or artificial [4]. As viable sites for conventional PHES are used up, UPHES may become an increasingly attractive option [5].

Status and future

The more than 200 PHES systems operating worldwide at present constitute over 99% of all grid-connected storage capacity. Since about 1970, global PHES power capacity has grown by about 3 GW per year. Today about 290 sites are in operation worldwide, with 40 under construction, and as many as 550 planned [6].

45.4.2 Compressed-air energy storage

Principles

When a quantity of gas is compressed adiabatically, that is, without exchanging heat with its environment, its pressure and temperature increase is described by the isentropic relation

$$\frac{p_2}{p_1} = \left(\frac{T_2}{T_1}\right)^{\frac{\gamma}{\gamma-1}} = \left(\frac{V_1}{V_2}\right)^{\gamma},$$

where p_1 and p_2 are the absolute pressures prior to and following compression, T_1 and T_2 are the absolute temperatures prior to and following compression, V_1 and V_2 are the gas volumes prior to and following compression, and $\gamma = C_p/C_v$ is the heat-capacity ratio of the gas, i.e., the heat capacity at constant pressure C_p divided by the heat capacity at constant volume C_v ($\gamma \approx 1.4$ for adiabatic compression of air).

Work must be performed on gas to compress it. As the air is compressed, the input work is converted into elastic (pressure) potential energy and heat. In an adiabatic compression, this heat remains in the gas, causing the gas to increase in temperature as it is compressed. Thus, for an adiabatic compression, the work input is stored in the gas as pressure potential energy and thermal energy.

Adiabatic expansion of the high-pressure, high-temperature air converts both the pressure potential energy and the thermal potential energy of the air back into work. In theory, all the work done on a quantity of gas to compress it can be retrieved as work when it re-expands to its original volume and temperature if both compression and expansion are carried out adiabatically.

When gas is compressed isothermally (at constant temperature), the pressure increases according to the ideal gas law,

$$p_1 V_1 = p_2 V_2 = nRT,$$

where n is the number of moles of gas, R the gas constant (~ 8.314 J K^{-1} mol^{-1}), and T the absolute temperature at which the gas is compressed. In an isothermal compression, all of the heat generated by compression leaves the gas as it is generated, allowing the gas to remain at constant temperature. Thus, for an isothermal compression, the work input resides partially in the gas as pressure potential energy and partially outside the gas as thermal energy at the temperature of the gas. In classical CAES, the heat of air compression is neither stored within the air

itself, as in a purely adiabatic process, nor recovered during expansion as in a purely isothermal process.

Again, in theory, all the work done on a quantity of gas to compress it can be retrieved when it re-expands if both compression and expansion are carried out isothermally. For an isothermal compression/expansion process, the high-pressure air is expanded at the constant temperature at which it was compressed, drawing thermal energy back into the air during the expansion process. For a perfect isothermal compression/expansion process, the thermal energy that flows back into the gas during expansion is equal to the thermal energy that flowed out of the gas during compression, and thus the generated work output is – as in the ideal adiabatic case – equal to the original work input.

In practice, as with all energy-storage schemes, 100% energy recovery is not possible due to component friction and efficiencies. For CAES technologies, there are thermodynamic considerations as well. These are discussed below for each of the three classes of CAES: classical, advanced adiabatic, and isothermal.

Classical CAES

In classical CAES, neither is the heat of air compression stored within the air itself as in a purely adiabatic process, nor is the heat of compression recovered during expansion as in a purely isothermal process. Instead, the compressed air storage is typically combined with a combustion-turbine [e.g., compressed natural gas (CNG)] power plant. Part of the heat of combustion replaces recovery of the heat of compression, preventing sub-freezing temperatures as the air expands. In addition, pre-compressing air enables the power plant to shift the energy cost of air compression from peak to off-peak, allowing it to increase electrical output during the peak day hours by buying inexpensive off-peak electricity for air compression at night.

Figure 45.3 shows a schematic illustration of a classical CAES process. The compression process is diabatic: that is, an electric motor drives multiple adiabatic air compressors with intercoolers (heat exchangers) between each compressor to remove heat. This reduces the maximum air temperature. The more compression/intercooler stages are used, the more the diabatic process approximates an isothermal process and the less energy is required for air compression. This is critical, since the portion of the input compression energy that is converted into heat is not recovered in a classical CAES facility during expansion. Following compression, the air is stored at ambient temperature in underground caverns.

To expand the gas to recover the stored pressure potential energy, the compressed air is first preheated by combustion gases in a recuperator at the air outlet, then combined with compressed natural gas, combusted, and expanded adiabatically through a turbine.

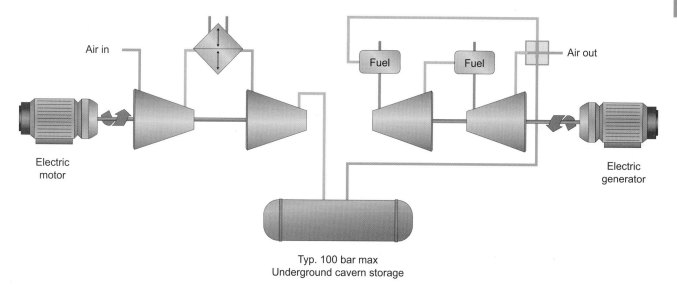

Figure 45.3. A schematic illustration of the classical CAES process, with underground storage and natural-gas turbine recovery. During energy storage (on the left), an electric motor runs a compressor with an intercooler. The compressed air is pumped into a cavern. During recovery of energy from storage (on the right), compressed air is released from the cavern and combusted with a fuel – typically natural gas being burned in a multistage turbine. The turbines drive an electric generator. (Source: SustainX, Inc.)

The mechanical work thus recovered is then used to drive an electric generator [7][8]. Multiple combustion and expander stages are used to limit the maximum temperature and increase efficiency.

Energy analysis of classical CAES systems can be complex because it is hybridized with combustion of natural gas and there is no unique efficiency metric. A total energy-in, energy-out analysis yields an efficiency of

$$\eta = \frac{E_T}{E_M + E_F},$$

where E_T is the electrical energy output generated by the CAES plant, E_M is the electrical energy consumed by the electric motor to compress air into storage, and E_F is the thermal energy released by burning the natural-gas fuel. This metric – typically about 54%, for existing CAES [7] – is physically valid but equates electrical energy to thermal energy. It also lumps into a single process the CAES and natural-gas energy generation. There are several different methods to separate these two effects in order to analyze the efficiency of the energy-storage component alone, each with its own merits and assumptions; a thorough discussion of these methods is beyond the scope of this chapter but is given by Succar and Williams [7].

Advanced adiabatic CAES

Advanced adiabatic CAES addresses a key disadvantage of classical CAES by storing the heat generated during air compression in a high-temperature thermal well. This allows the recovery of that thermal energy.

In AA-CAES, air is compressed adiabatically within one or more turbine-based compressors, without intercoolers between each stage as in classical CAES (Figure 45.4). The result is high-pressure, high-temperature compressed air. As this air passes through the high-temperature thermal well (e.g., a volume of non-fuel oil), the well absorbs thermal energy from the air, cooling the air to near-ambient temperature before it is stored in caverns. In effect, the heat generated during compression is segregated from the air undergoing compression and stored in the thermal reservoir [9][10].

To expand the air and generate electricity, compressed air drawn from the cavern storage is sent back through the high-temperature thermal well, heating it to near its post-compression temperature prior to adiabatic expansion through a turbine driving an electric generator.

Because there is no fuel input, the round-trip efficiency of an AA-CAES system is the electrical energy out divided by the electrical energy in. Since the AA-CAES process approximates an ideal adiabatic process, the round-trip efficiency is governed by the individual component efficiencies and any parasitic thermal losses. Component efficiencies are impacted by the high air temperatures that are reached, which, due to materials compatibility issues, also limit the compression ratio (and thus the energy density) of the system. Thermal losses are also a concern: the higher the temperature of the thermal well, the greater the rate of heat loss through its walls. Furthermore, the inside of the thermal well has striated layers of temperature from the hot upper side to the cool lower side, and heat transfer within the well itself reduces the temperature to which the stored compressed air can be reheated; this reduces the heat energy that can be recovered from the well.

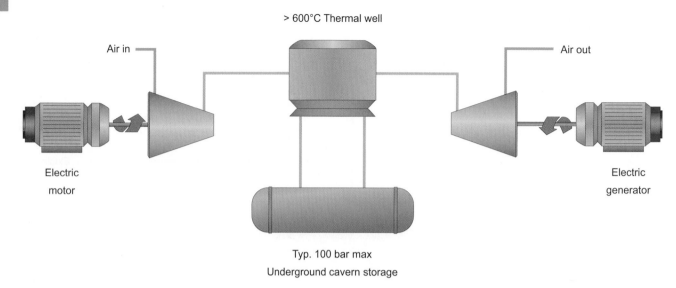

> 600°C Thermal well

Air in

Air out

Electric
motor

Electric
generator

Typ. 100 bar max
Underground cavern storage

Figure 45.4. A schematic illustration of an advanced adiabatic CAES process, with underground storage and a high-temperature thermal well. During storage (on the left), an electric motor runs a compressor. Thermal energy from the adiabatically compressed air is stored in a thermal well (e.g., a tank of hot oil) and the cooled air is stored in a cavern. During recovery of energy from storage (on the right), air is released from storage, heated with thermal energy recovered from the thermal well, and used to drive a turbine or equivalent device, which in turn drives an electric generator. A reversible compressor/expander and motor/generator may be used rather than four discrete devices as shown here for clarity. (Source: SustainX, Inc.)

Isothermal CAES

Isothermal CAES addresses the key disadvantage of classical CAES by purposely dissipating the heat generated during air compression to the ambient environment at low temperature differential, allowing almost the same amount of heat to be recovered again from the ambient environment, again at low temperature differential. The transfer of large amounts of heat at low temperature differential allows high thermal efficiencies to be achieved.

In an isothermal CAES process (Figure 45.5), the heat-exchange process is integrated into the compression device, allowing heat transfer to occur continuously throughout the compression process. If the rate of heat transfer approximates the rate of compression work, the compression approximates an isothermal process. Similar continuous heat exchange during an expansion process allows air to be near-isothermally expanded, recovering the heat energy dissipated during compression.

As with AA-CAES, there is no fuel input for isothermal CAES: the round-trip efficiency is the electrical energy out divided by the electrical energy in. The round-trip efficiency of an isothermal CAES process is governed by the individual component efficiencies as well as by deviation in the isothermal temperatures of the compression and expansion processes.

While a perfect isothermal compression and expansion process is theoretically 100% thermally efficient, no actual gas compression or expansion process can be perfectly isothermal at ambient temperature because the gas temperature must deviate from ambient by some amount in order for heat transfer to and from the environment to be possible. Near-isothermal processes achieve heat transfer at low temperature differentials, typically using large surface areas for expediting heat transfer.

Because a near-isothermal compression and expansion process can occur at ambient temperature, any heat dissipated to the ambient environment is recovered from the ambient environment during expansion, and total thermal losses, including during air storage, are minimized.

Existing installations

Only two large CAES facilities are operating, namely the 290-MW Huntorf plant in Bremen, Germany and the 110-MW plant in McIntosh, Alabama. Both are hybridized with natural-gas combustion (classical CAES).

The Huntorf plant, which was completed in 1978, stores 310,000 m^3 of air in caverns at up to 70 bar (about 70 times ordinary atmospheric pressure), and is used primarily for cyclic duty, ramping duty, and as a hot spinning reserve for industrial loads. The plant has also been used recently to level power supply from wind-turbine generators (6.5% of German supply as of 2009), demonstrating the potential of storage to support renewables [8][11].

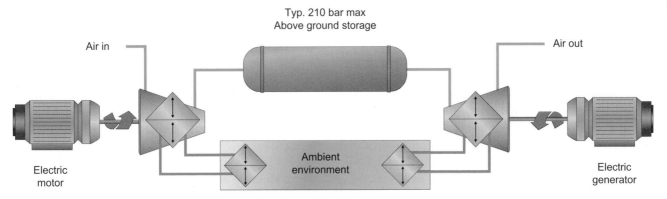

Figure 45.5. A schematic illustration of an isothermal CAES process. During energy storage (on the left), an electric motor drives a compressor integrated with a heat exchanger that continuously rejects heat from gas undergoing compression to the environment. Heat rejection takes place at a rate that keeps the air at constant temperature (an isothermal process). The cooled air is stored in aboveground storage vessels. During recovery of stored energy (on the right), air is released from storage through an expander. The expander is integrated with a heat exchanger that recovers heat from the environment and shifts it to the expanding, cooling gas. The expander transfers work to an electric generator. A reversible compressor/expander and motor/generator may be used rather than four discrete devices as shown here for clarity. (Source: SustainX Inc.)

The McIntosh plant, completed 1991, stores 540,000 m³ of air in a salt cavern at up to 74.5 bar and is used primarily for load leveling and spinning reserve. The McIntosh design includes a recuperator, which recovers waste heat from the low-pressure turbine to preheat the stored air before it goes to the dual-fuel (i.e., natural gas or fuel oil) high-pressure combustor, reducing the fuel consumption by about 25% compared with that of a Huntorf-type plant [8][12].

Current research

Improvements to classical CAES

Several companies and organizations are pursuing next-generation classical CAES installations. Compressor and turbine technologies for such installations are mature, but different implementations of said technology that provide potential reductions in system cost and/or improvements in performance are being proposed. Some of these proposed improvements are detailed in [8], pp. 15–23.

Advanced CAES

Attention is increasingly being turned to forms of no-fuel CAES (not hybridized with natural gas or other fuel combustion). As mentioned earlier, adiabatic and isothermal storage systems have the potential for high efficiencies through management of the heat of compression.

The biggest technical challenge to advanced adiabatic CAES is the large temperature change that occurs on compressing the air. Air compressed adiabatically from atmospheric pressure and temperature to 74.5 bar (as in the Alabama classical CAES plant) would reach a temperature of 750 °C (1,380 °F). The higher energy densities being investigated would require higher

pressure ratios and therefore still higher temperatures. Efforts to develop high-temperature compressors and expanders with sufficient efficiencies and lifetimes are under way. In 2010, large scale AA-CAES systems were under development in Europe and the USA. [9]. A small-scale battery-replacement CAES system (10 kW, 20 kWh) using thermal storage at ~750 °C and electric heating was also under development in the USA.

Isothermal CAES

Isothermal CAES is rapidly developing as an alternative CAES technology. This type of CAES keeps compressed air at near-ambient temperature throughout the energy-storage–recovery cycle, thus reducing component cost and potential wear. Rapid, continuous heat exchange between the environment and pressurized air at low temperature differential is key to isothermal CAES; this technology's biggest technical challenge – now being addressed by researchers – is to economically achieve these high rates of heat transfer. Low-cost, long-lifetime, modular isothermal CAES systems will likely be on the market within a few years.

Status and prospects

The next few decades will probably see greatly increased deployment of CAES as the demand for diverse, affordable forms of energy storage in the grid grows with ongoing deployment of renewables [10].

The largest CAES facility planned to date is a 2.7-GW plant in Norton, Ohio. This plant would compress air to 100 bar in a defunct limestone mine some 670 m underground. The plant has been projected for at least a decade, and it is not known whether it will be built.

Another much discussed but still unbuilt project is the Iowa Stored Energy Park, which would combine aquifer storage of high-pressure air, off-peak generation from a 100-MW$_{ac}$ (capacity) wind farm with gas-combustion turbines to produce 268 MW and supply load following, spinning reserve, and black-start services, [13]. The Iowa Stored Energy Park has also been delayed, perhaps indefinitely.

In 2009, the US Department of Energy granted funding to support two classical CAES demonstration projects. One is a 300-MW CAES plant planned by Pacific Gas & Electric that would store its air in a saline aquifer in Kern County, California. The other is a plan by New York State Electric & Gas to build a 150-MW plant using an abandoned salt mine in the Finger Lakes region as its storage cavern.

In 2010, AA-CAES is under development by a partnership of several corporations and a research center under the acronym ADELE with the goal of developing a prototype 200-MW, 1,000-MWh system for 2013.

In 2010, isothermal CAES is under development by several corporations. The DOE has funded a Smart Grid initiative with SustainX to develop a 1-MW prototype system for 2012.

45.4.3 Flywheels

Principles

The energy stored by a rotating wheel, whatever its shape, is proportional to the square of the rotational velocity of the wheel: that is, doubling the number of rotations per unit time quadruples the kinetic energy of the wheel. Fast rotation is therefore desirable in order to store more energy. However, the centripetal force experienced at every point throughout a flywheel is also proportional to the square of the rotational velocity. The energy stored in a flywheel and the mechanical stress on the wheel therefore increase together, and any attempt to increase the energy-storage density of a flywheel by speeding it up must reckon with an increased probability of breakup – that is, catastrophic failure [14].

Flywheels are being developed for vehicular storage as well as for stationary storage. The 2010 Porsche 911 GT3 R hybrid racecar, for example, uses a flywheel-based "kinetic-energy-recovery system" to store energy from braking and deliver it to boost acceleration, a process termed "regenerative braking." The Toyota Prius hybrid automobile also uses small composite-material flywheels in its regenerative braking system.

Metallic (e.g., steel) flywheels are used for flywheel energy storage with rotation speeds in the thousands of rpm. However, steel flywheels are severely constrained by low specific strength and specific modulus (stiffness-to-weight ratio), and, in failure, produce dangerous shrapnel unless expensively containerized. In the 1970s, fiber-reinforced composite materials based on carbon, arimid (aromatic polyamide), and glass that offered far greater strength per weight than steel were devised: Kevlar, for example, has about five times the tensile strength of steel on an equal-weight basis. Using such materials, it has been possible to build flywheels that rotate at up to 50,000 rpm, suspended in vacuum and levitated on magnetic fields, with their outer rims moving at two or more times the atmospheric speed of sound [15]. In failure, such materials dissolve into lightweight, nonrigid fibers rather than penetrating shrapnel. Some commercially available flywheel systems use composite fibers solely as a passive outer rim for kinetic-energy storage, combined with steel hubs, shafts, and motor/generators. Flywheels that blend powdered magnetic materials with composite fibers in a three-dimensional pattern that allows the flywheel to be magnetized as the rotor of a brushless DC motor/generator or as the rotating element of a passive magnetic bearing have also been under development since the 1990s. Some currently available utility-scale flywheel systems offer a round-trip storage efficiency of 85% [16].

Unlike PHES and CAES, flywheels store kinetic energy. From elementary mechanics, the kinetic energy E of a particle of mass m attached to an axis of rotation by a massless string so that it moves in a circular path of radius r at rotational velocity ω is given by

$$E = \frac{1}{2}mr^2\omega^2.$$

The centripetal force F experienced by this particle (i.e., the force that keeps it rotating in a circle) is

$$F = mr\omega^2.$$

Any rigid flywheel, whatever its shape, can be considered as a collection of such point masses all having the same ω. Thus, independently of shape, the energy stored by a flywheel increases as ω^2, and doubling the angular speed of a flywheel quadruples the energy it stores. Doubling the speed also quadruples the centripetal force F acting upon each particle comprising the flywheel. The tendency of a flywheel to break up thus grows in direct proportion to the energy stored.

The total energy stored by a particular flywheel at any given ω depends on its shape and density distribution. The total energy is found by appropriate integration of E for each infinitesimal volume throughout the whole flywheel volume. Alternatively, the quantity mr^2 can be integrated throughout the flywheel volume to derive a quantity J, the moment of inertia [17]. For example, one can show that a right cylinder of uniform density, total mass m, and radius r has moment of inertia $J = \frac{1}{2}mr^2$. Then, for any mass distribution (flywheel shape),

$$E = \frac{1}{2}J\omega^2.$$

The energy that can be practically stored by a flywheel having moment of inertia J can be written as a function

Figure 45.6. The NASA G2 experimental flywheel (a modular testbed for system integration and demonstration of components) was designed to operate at up to 60,000 rpm. The flywheel is levitated with magnetic bearings and operated in a near-vacuum housing (~0.001 mTorr) with a titanium hub and a carbon-fiber rim. Photo and schematic diagram: NASA.

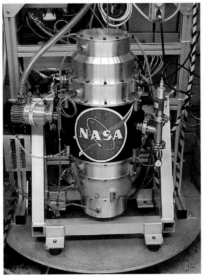

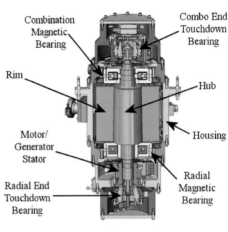

of the difference between its maximum rotational speed ω_{max} and its minimum speed ω_{min}:

$$E = \frac{1}{2}J\omega_{max}^2 - \frac{1}{2}J\omega_{min}^2 = \frac{1}{2}J\left(\omega_{max}^2 - \omega_{min}^2\right).$$

If the ratio of the maximum to minimum velocity is written $k = \omega_{max}/\omega_{min}$, then

$$E = \frac{1}{2}J\omega_{max}^2\left(1 - \frac{1}{k^2}\right).$$

This is the energy capacity of a flywheel having moment of inertia J and operating speed range ω_{min} to ω_{max}. The ratio k is related to a depth of discharge for a flywheel, where a k of 2 corresponds to a ~75% depth of discharge (i.e., $1 - 1/k^2 = 0.75$). Typically k is between 2 and 4; for a k of 3, the depth of discharge is 89%.

Losses occur due to conversion inefficiencies during charging (acceleration) of the wheel, dissipation during storage, and inefficiencies during discharging. The actual energy deliverable to an electrical load is given by $E\eta_D$, where η_D is the discharge efficiency. The round-trip efficiency can be written $\eta_{RT} = \eta_C\eta_D$, where η_C is the charging efficiency and storage losses are assumed to be zero. However, storage losses can be significant for flywheel systems – indeed, much higher, for a given time period, than for PHES or CAES, at least 3%–5% per hour, although superconducting magnetic suspensions under development may allow losses less than 0.1% per hour. Losses during storage are minimized and component reliability increased in modern flywheel systems by using magnetic bearings and evacuated containers, but this high storage-loss rate means that flywheel systems are best suited for short-term storage where a very rapid response time (milliseconds) is desired.

Existing installations

Flywheel systems are commercially available today for uninterruptible power supplies: over 600 MW of such flywheel storage has been installed as of 2010, with system power levels up to several megawatts and storage capacities in the tens of seconds.

Carbon-composite flywheel systems, primarily for frequency regulation, are under development and have been deployed in prototype applications. These systems may be ganged together to provide multi-megawatt power levels and minutes of storage. A 20-MW, 5-MWh installation (200 flywheels) to supply about 10% of New York State's frequency-regulation needs began operation in 2011. A smaller (100-kW, 25-kWh), experimental installation was already operating as part of a demonstration of wind-power firming using flywheel storage at Tehachapi, California, being carried out for the California Energy Commission. The system is part of a wind-power/flywheel demonstration project being carried out for the California Energy Commission.

Status and prospects

Flywheel systems are increasingly favored for the provision of uninterruptible power to data centers, often combined with onsite diesel generators. Such systems take up less floorspace and have far longer cycle lifetimes than batteries, and can supply large pulses of current for starting generators [18]. Larger flywheel installations for frequency regulation of grid power are being planned or built, with installation sites as of 2010 limited to the USA.

An experimental testbed is shown in Figure 45.6.

45.5 Comparing storage technologies

Technologies for energy storage are developing rapidly, which tends to render any comparison of efficiencies, LCOE, energy capacities, and response times out of date before it sees print. Storage technologies are problematic to rank or compare. The LCOE for all storage modalities, chemical and mechanical, depends on capital costs, efficiency, lifetime, operating expenses, financing mechanisms, local regulations, subsidies, siting constraints, and other variables. Moreover, the LCOE is not the only metric of value for an energy-storage system. For example, in some applications, a fast response time is crucial and is therefore worth a large per-kWh premium. Nor are raw physical measures such as energy density (kWh stored per unit mass) helpful when comparing large, stationary applications.

In general, flywheels, like chemical batteries and ultracapacitors, address market applications that require very fast response times but relatively small total energy storage, PHES addresses applications requiring very large storage, and CAES tends to address applications requiring intermediate and large quantities of energy storage. However, there are exceptions to all these generalizations and these exceptions may become more common as the storage market matures. Comparisons between storage technologies are best made on an application-by-application, site-specific basis.

45.6 Summary

Both demand and supply for grid-scale energy storage are in a state of flux, as new technologies emerge, old technologies are refined, awareness of application opportunities grows, and grids become more "smart" and more dependent on relatively variable solar and wind inputs. In this context, the three types of mechanical energy storage tend to address distinct market needs, albeit with significant overlap. See Table 45.1.

45.6.1 Pumped-hydroelectric energy storage

Pumped-hydroelectric technology has been used for over a century. It has a relatively low LCOE and can store gigawatt-hours of energy, but is viable only at very large scale – unless small-scale aquifer PHES schemes, which so far are merely speculative, someday become practical. Classical PHES has high environmental impact due to its large landscape footprint, though this may be mitigated by schemes that propose placing one reservoir underground.

45.6.2 Compressed-air energy storage

Classical, underground-storage CAES systems are still few in number and are constrained to use with combustion turbines. They have so far been built or proposed to operate at similar scale to early PHES (hundreds of megawatts). Classical CAES also has geographical siting constraints – suitable aquifers, abandoned mines, or salt formations are required. Both PHES and classical CAES projects, due to their inherently large scale and to siting constraints, suffer from long lead times and complex permitting and require large single-project capital commitments, to which managers are sometimes averse.

Advanced adiabatic CAES, which is currently being researched, proposes a different means of underground

CAES, offering similar power levels but replacing combustion turbines with air-only turbines on generation. In AA-CAES one retains the thermal energy produced by gas compression using high-temperature oil, steel, or some other medium as a thermal battery, stores the air at ambient temperature, and then reheats the air to the high temperatures of compression (e.g., 600 °C) prior to expansion through air turbines.

Alternative CAES methods, especially isothermal CAES, tend to address a smaller storage regime than classical CAES, namely from 1 MW up to tens of megawatts, though there is no theoretical upper limit. Isothermal CAES exchanges low-temperature thermal energy with its environment in order to keep the air temperature approximately constant while the pressure varies widely (e.g., from atmospheric to 200 bar). The isothermal CAES approach eliminates temperature extremes and concerns about heat leakage during storage, and can be modified to include cogeneration, that is, the harvesting of low-grade waste heat from thermal power plants or other sources. Commercially viable, no-fuel CAES systems have the potential to evade the scale, siting, and environmental constraints of classical CAES and PHES and are nearing market readiness.

45.6.3 Flywheels

Flywheels are contending for those market niches that demand near-instantaneous response and short-term storage. As of 2010, several hundred megawatts of flywheel systems had been installed as uninterruptible power supplies with near-instantaneous response times and durations in the tens of seconds. Prototype flywheel systems were being applied to frequency regulation with storage durations in the tens of minutes.

Major considerations for potential buyers of stationary energy storage include those for energy generation (e.g., durability, environmental impact, convenience, responsiveness, cost) but also round-trip efficiency. Mechanical storage modalities – PHES, CAES, and flywheels – particularly stand out for their very long lifetimes and low lifetime costs. Lifetimes for PHES, CAES, and flywheels are all estimated at over 20 years, and the LCOE for PHES and CAES can be much less than for smaller-scale technologies such as batteries [19].

45.7 Questions for discussion

1. What are the different advantages and drawbacks offered by PHES, various forms of CAES, and flywheel storage? For which applications are they suitable? What are the non-mechanical storage competitors for each application?

2. Is demand for mechanical storage technologies increasing or likely to increase? Why or why not?

Table 45.1. Summary of mechanical energy-storage technologies

Pumped–hydroelectric energy storage (PHES)

- Long lifetime
- Based on mature technology
- Low lifetime cost of energy
- Non-exotic materials (steel, concrete, copper, etc.)
- Geographically constrained (large elevation change between upper and lower reservoirs)
- >90 GW installed worldwide as of 2010
- 65%–75% round-trip efficiency

Compressed-air energy storage (CAES)

- Long lifetime
- Based on mature technology
- Low lifetime cost of energy
- Non-exotic materials (steel, concrete, copper, etc.)

Classical CAES[a]

- o Two installations worldwide (<1 GW) as of 2010
- o Combined with combustion power plant (e.g., natural-gas turbine)
- o Geographically constrained (storage via underground reservoir such as a cavern)

Advanced adiabatic CAES

- o Under development as of 2010
- o High-temperature heat storage (e.g., ceramics, oils)

Isothermal CAES

- o Under development as of 2010
- o Broad site options with aboveground air storage[b]

Flywheel energy storage

- Long lifetime
- Based on mature technology
- High-power, short-duration storage[c]
- Steel or advanced composite flywheel
- Broad site options
- <1 GW installed worldwide as of 2010
- 80%–95% efficiency[d]

[a] Energy analysis of classical CAES systems can be complex because it is hybridized with combustion of natural gas and there is no unique efficiency metric (see [7]).
[b] Higher pressures and heat recovery allow low-cost above ground storage compared with classical CAES.
[c] Flywheel typically addresses high-power, short-duration applications, not bulk energy storage, so cost per unit power is more important than cost per unit energy.
[d] Flywheel has a parasitic loss associated with friction (bearing and windage losses) that is on the order of 1% of the nominal power.

3. Why has it taken so long for mechanical storage technologies other than PHES to approach widespread commercial use?

4. What are the approximate volumetric dimensions of a storage reservoir capable of storing 100 MWh of energy for a 100-m-elevation PHES water reservoir, 100 bar of 20-°C compressed air expanded adiabatically, and 100 bar of 20-°C compressed air expanded isothermally? State your assumptions.

5. What are the current preferred material options for flywheel construction for flywheel energy storage? What are the major considerations for performance?

6. If increased reliance on relatively high-variability renewables such as wind and solar will increase the need for grid-connected energy storage, how did the US grid accommodate the addition of ~7 GW of wind-generated electricity (actual, not capacity) in 1999–2009, a more than three-fold increase, without adding significant new storage? (Hint: what is the scale of other sources of intermittent mismatch between power supply and demand – e.g., load variations, unexpected unavailability of installed conventional generators – and how does the existing grid cope?)

45.8 Further reading

- **EPRI-DOE**, 2003, *EPRI-DOE Handbook of Energy Storage for Transmission and Distribution Applications*, available for free download from http://www.epri.com.
- **P. Denholm** and **G. L. Kulcinski**, 2004, "Life cycle energy requirements and greenhouse gas emissions from large scale energy storage systems," *Energy Conversion Management*, **45**, 2153–2172.
- **R. A. Huggins**, 2010, *Energy Storage*, New York, Springer.
- **S.-i. Inage**, 2009, *Prospects for Large-Scale Energy Storage in Decarbonised Power Grids*, working paper, International Energy Agency.
- **M. G. Jog**, 1989, *Hydro-Electric and Pumped Storage Plants*, New York, John Wiley & Sons.
- **S. Succar** and **R. H. Williams**, 2008, *Compressed Air Energy Storage: Theory, Resources, and Applications For Wind Power*, Princeton, MA, Energy Systems Analysis Group, Princeton Environmental Institute, Princeton University.

45.9 References

[1] **J. P. Fitzgerald**, **G. W. Groscup**, **E. A. Cooper**, **R. T. Byerly**, and **E. C. Whitney**, 1969, "Synchronous starting of Seneca pumped storage plant," *IEEE Trans. Power Apparatus Systems*, **88**(4), 307–315.

[2] **S.-i. Inage**, 2009, *Prospects for Large-Scale Energy Storage in Decarbonised Power Grids*, working paper, International Energy Agency.

[3] **J. M. Bemtgen**, **A. Charalambous**, **M. Dionisio** *et al.*, 2008, *Report of the SETIS Workshop on Electricity Storage in Stationary Applications*, Petten, Netherlands, European Commission Directorate-General, Joint Research Centre.

[4] **G. D. Martin**, 2007, *Aquifer Underground Pumped Hydroelectric Energy Storage*, Boulder, CO, University of Colorado at Boulder.

[5] 2010, *Investigating Aquabank*, available from http://www.waterpowermagazine.com/story.asp?storyCode=2055067.

[6] **US Department of the Interior BoR**, *Hydropower: A Key to Prosperity in the Growing World*, US Department of the Interior BoR.

[7] **S. Succar** and **R. H. Williams**, 2008, *Compressed Air Energy Storage: Theory, Resources, and Applications For Wind Power: Energy Systems Analysis Group*, Princeton, MA, Princeton Environmental Institute, Princeton University.

[8] **Electric Power Research Institute/Department of Energy (EPRI-DOE)**, 2003, *EPRI-DOE Handbook of Energy Storage for Transmission and Distribution Applications*, EPRI-DOE.

[9] **W. F. Pickard**, **N. J. Hansing**, **A. Q. Shen**, 2009, "Can large-scale advanced-adiabatic compressed air energy storage be justified economically in an age of sustainable energy?," *J. Renewable Sustainable Energy*, **1**, 033102:1–10.

[10] **C. Bullough**, **C. Gatzen C. Jakiel**, *et al.* (eds.), 2004, *Proceedings of the European Wind Energy Conference*, London.

[11] **F. Crotogino**, **K.-U. Mohmeyer**, and **R. Scharf**, 2001, "Huntorf CAES: more than 20 years of successful operation," in Solution Mining Research Institute Spring Meeting, Orlando, FL.

[12] **R. Pollak**, 1994, *History of First U.S. Compressed Air Energy Storage (CAES) Plant (110-MW–26 h): Volume 2: Construction*, Palo Alto, CA, Electric Power Research Institute.

[13] **R. Haug**, 2006, "The Iowa stored energy plant: progress report," in *DOE Energy Storage Systems Annual Peer Review*.

[14] **J. A. Kirk**, 1977, "Flywheel energy storage – I: basic concepts," *Int. J. Mechanical Sci.*, **19**(4), 223–331.

[15] **D. Castelvecchi**, "High-tech reincarnations of an ancient way of storing energy," *Science News*, 171(20), 312–331.

[16] **R. Walawalkar**, and **J. Apt**, 2008, *Market Analysis of Emerging Electric Energy Storage Systems*, DOE/NETL-2008/1330, Washington, DC, National Energy Technology Laboratory, Department of Energy.

[17] **ActivePower Inc.**, 2010, *Understanding Flywheel Energy Storage: Does High-Speed Really Imply a Better Design?*, Austin, TX, ActivePower, Inc., available from http://www.activepower.com/fileadmin/documents/white_papers/WP112_FlywheelEnergyStorage.pdf.

[18] **R. Miller**, 2007, *Flywheels Gain as Alternative to Batteries*, available from http://www.datacenterknowledge.com/archives/2007/06/26/flywheels-gain-as-alternative-to-batteries/.

[19] **Electricity Storage Association**, 2009, *Technology Comparisons*, available from http://www.electricitystorage.org/ESA/technologies/technology_comparisons/.

46

Fuel cells

Shyam Kocha, Bryan Pivovar, and Thomas Gennett

National Renewable Energy Laboratory, Golden, CO, USA

46.1 Focus

Over the last two decades the problem of limited oil and gas supply versus new emerging-nation demands has created an immediate need for new technologies to alleviate global dependence on a hydrocarbon-based energy policy. This requires significant changes in the way global energy-system policy is managed and the rapid adoption/introduction of an array of new technologies that produce and use energy more efficiently and more cleanly than in the past. Specifically, these energy policy changes have directly led to more focus on commercial applications of alternative, sustainable energy policy. This chapter is centered on the establishment of a hydrogen-based economy, specifically as it relates to fuel cells. The net result is the concept of hydrogen and fuel cells as a practical foundation for implementing public policies responding to growing uncertainties about the security and long-term price of oil and environmental concerns. With the expectation that fuel cells and hydrogen can play a significant role in the global energy economy, governments are committing funds for research, development, and demonstration of hydrogen and fuel cells as they strive to create programs for viable infrastructure to support their use.[1]

---

[1] Within this chapter the balance of plant (BOP) will not be dealt with, but it consists of the following essential components: (i) an air blower or compressor, (ii) water-recovery systems and humidifiers, (iii) recirculation pumps for anode hydrogen, and (iv) filters, regulators, valves, sensors, etc. Although these are common components found in other devices, they need to be modified to meet the special needs of a fuel-cell system. This interesting subject is beyond the scope of this chapter; needless to say, improving the components of the fuel cell in terms of performance and durability (for example, improved oxygen-reduction-reaction catalysts, corrosion-resistant catalyst supports, membranes that can withstand low humidity and higher temperature) permits system simplifications and lowers the overall cost of the system.

46.2 Synopsis

Fuel cells generate energy from controlled, spontaneous oxidation–reduction (redox) reactions. A fuel cell is a multi-component device with two electrodes, separated by an ionic conductive membrane, the positive anode where the oxidation reaction (loss of electrons LEO) occurs and the negative cathode where the reduction reaction (gain of electrons GER) occurs (LEO goes GER, lose electrons oxidation, gain electrons reduction). As with battery systems, there are several kinds of fuel cells, and each operates a bit differently, but in general the fuel cell uses hydrogen (or hydrogen-rich fuel) at the anode and oxygen (air) at the cathode, to create electricity. As mentioned, the typical fuel cell consists of two electrodes: the negative electrode (or anode) and a positive electrode (or cathode), separated by an ion (charge)-conducting electrolyte. In a model system, the hydrogen is fed to the anode, and oxygen is fed to the cathode. Through the utilization of the catalyst, the activation energy barrier for the separation of hydrogen atoms into protons (H^+) and electrons (e^-) is decreased substantially, making it kinetically viable at <80 °C, i.e., the catalyst lowers the activation barrier for the chemical reactions and increases the rate at which the reactions occur. The electrons generated are forced to go through an external circuit, which creates creating a flow of electrons (electricity). In order to complete the balance of charge required in redox reactions, the protons migrate through the electrolyte to the cathode, where they react with oxygen and the electrons to produce water and heat. So, as long as fuel (hydrogen) and air are supplied, the fuel cell will generate electricity. A single fuel cell generates a small amount of electricity so, in practice, similarly to what is done every day with loading multiple batteries to operate an electronic device, fuel cells are usually assembled into a stack of multiple cells to meet the specific power and energy requirements of the particular application.

In summary, within a fuel cell (1) there is no combustion, redox reactions generate energy, with only water-vapor emissions; (2) there are no moving parts, therefore, fuel cells are quiet and reliable; (3) electricity is created electrochemically, rather than by combustion (burning in air), so thermodynamic laws that limit a conventional power plant are not applicable, therefore, fuel cells are more efficient in extracting energy; and (4) the fuel, hydrogen, and the product, water, are considered renewable and green, therefore, fuel cells will be essential to a sustainable energy program.

This chapter will give an overview of multiple types of fuel cells and their incorporation into automotive and stationary applications (such as generating electricity or heating buildings, remote grid applications).

46.3 Historical perspective

In the early 1800s it was well known that one could split water into hydrogen and oxygen with electricity, but it wasn't until 1839 that a lawyer turned scientist, Sir William Grove, was able to demonstrate in practice the veracity of his belief that the reverse of the electrolysis process, i.e., generating electricity from the reaction of gaseous oxygen with hydrogen, should also be possible. Experimentally the proof came, as the story is told, in an evening lecture during which he enclosed two platinum strips in separate sealed bottles, one containing hydrogen and the other containing oxygen, and immersed them in dilute acid solution. The current that began to flow between the two electrodes was used to illuminate himself with a carbon-filament light bulb. Grove also can be thought of the inventor of the fuel-cell stack (Figure 46.1) because he was able to increase the voltage produced by linking several of these devices in series even though he referred to it as a "gas battery." It wasn't until 1889 that two chemists, Ludwig Mond and Charles Langer, used the term "fuel cell" as they attempted to build an industrially applicable device using air and industrial coal gas.

It often happens that some development in science is forgotten not because of a lack of importance, but rather due to circumstance. At the end of the nineteenth century, the internal combustion engine and the widespread exploitation of fossil fuels sent the fuel cell the way of many other inventions, and it was labeled a mere curiosity.

There isn't really much to say about fuel cells and their applications to anything more than fundamental science for another 50 years. The press still didn't get much better, even though in 1932 Dr. Francis Thomas Bacon, of Cambridge University in England, resurrected the fuel cell developed in 1889, by the implementation of several modifications to the original design. It still took another 27 years until Bacon could produce a truly workable fuel-cell stack when, in 1959, he demonstrated a machine capable of producing 5 kW of power. Subsequently industrial interest was piqued once again, and Harry Karl Ihrig of Allis-Chalmers, a farm equipment manufacturer in the USA, demonstrated the first fuel-cell-powered vehicle. The stack contained >1,000 individual cells, and was capable of powering a 20-hp tractor.

Then on October 4, 1957 a game changer occurred, the 'Sput-Nik' – heralded the advent of the space age and, in the late 1950s and early 1960s, NASA was looking for a way to power a series of upcoming manned space flights, and decided that fuel cells were the only reasonable choice. So a benefit of the Cold War, and the race to have a manned flight reach the moon, is that NASA sponsored efforts to develop practical working

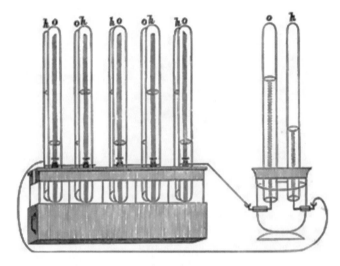

Figure 46.1. William Grove's drawing of his "gas battery" experiment, and the first fuel-cell stack. (Source: Smithsonian Institute.)

fuel cells that could be used in space flights. These efforts led to the development of the first proton-exchange membrane fuel cell (PEMFC).

While NASA was conducting research, a scientist working at General Electric (GE), Willard Thomas Grubb, used a sulfonated polystyrene ion-exchange membrane as the electrolyte in the fuel cell. Then, within three years other GE scientists devised a way of depositing platinum onto this membrane, which ultimately became known as the "Grubb–Niedrach fuel cell." General Electric and NASA developed this technology together, and it was used exclusively on the Gemini space project. This was the first commercial manufacture and use of a fuel cell.

In parallel, Pratt & Whitney focused their efforts on the alkaline fuel cell (AFC). The goal was to reduce the weight, an important consideration in space flight, and develop a longer-lasting fuel cell. The net result of this work was that Pratt & Whitney developed for NASA AFCs for the Apollo spacecraft. Derivatives of these alkali cells have since been used on most subsequent manned US space missions, including those of the Space Shuttle and the International Space Station, Figure 46.2.

Since necessity is the mother of invention, fuel-cell research was also re-invigorated in the USA and Europe during the oil embargos of the 1970s. Throughout the 1970s and 1980s, the research effort was dedicated to developing the materials needed, identifying the optimum fuel source, and drastically reducing the cost of this technology. Technical breakthroughs during the 1980s and early 1990s led to the development of the first marketable fuel-cell-powered vehicle in 1993 by the Canadian company, Ballard, which was dubbed the Ballard Fuel Cell.

Now, in the new millennium, with the ever-increasing cost of fossil fuels and the demand expected from

Table 46.1. Types of fuel cells

Type	Electrolyte	Fuel	Operating temperature (°C)	Power level (kW)	Typical applications
Solid-oxide fuel cell (SOFC)	Ceramic, solid oxide, zirconia	Hydrogen or methane	500–1,000	100–100,000	Combined heat and power CHP, power generation, transportation
Molten-carbonate fuel cell (MCFC)	Molten lithium carbonate	Hydrogen	630–650	1,000–100,000	Large stationary power
Phosphoric acid fuel cell (PAFC)	Phosphoric acid	Hydrogen	150–210	100–5,000	CHP, power generation
Proton-exchange membrane (PEMFC)	Sulfonic acid membranes	Hydrogen	50–90	0.01–1,000	Transportation, power supplies, CHP, distributed power
Alkaline fuel cell (AFC)	Potassium hydroxide	Hydrogen	50–200	10–100	Space, power generation
Direct methanol fuel cell (DMFC)	Sulfuric acid, Sulfonic acid membrane	Methanol	50–110	0.001–100	Portable power, consumer electronics

Figure 46.2. The Shuttle orbiter's alkaline fuel cell (Nasa.gov).

emerging nations as they develop a more industrially based economy, fuel cells are again being considered as an option. Climate change, pollution, and overall cost all play important roles as we try to move toward a hydrogen-based economy. It is now quite common to see stationary fuel cells that have been installed at various locations such as public-safety buildings, hospitals, and schools. The redeveloped World Trade Center will house one of the largest fuel-cell installations in the world. UTC Power of South Windsor, Connecticut, Willing to install four 1.2-MW fuel-cell systems, totaling 4.8 MW of generating capacity. In addition, most of the major automotive companies have unveiled prototypical fuel-cell-powered cars. Currently fuel-cell-powered buses have taken their place, even though most are in niche tourist applications, in most major cities in North America and Europe. The next major breakthroughs in hydrogen storage and inexpensive catalysts are needed, and it will be then that the full implementation of fuel cells in all applications will be more feasible.

46.4 Fuel cells basics

46.4.1 Introduction to fuel cells

Fuel cells generate power by electrochemically oxidizing a fuel on one electrode (anode) and reducing an oxidant such as oxygen at the cathode, producing water and heat as byproducts. As shown in Table 46.1, depending on the type of fuel, the temperature of operation, and the type of electrolyte used, fuel cells may be broadly classified as (i) proton-exchange membrane fuel cells (PEMFCs), (ii) direct methanol fuel cells (DMFCs), (iii) phosphoric acid fuel cells (PAFCs), (iii) molten-carbonate fuel cells (MCFCs), (iv) solid-oxide fuel cells (SOFCs), (v) alkaline fuel cells (AFCs), and (v) biological or microbial fuel cells (BioFCs/MFCs). Figure 46.3 is a schematic illustration of the various fuel-cell systems' operation.

Alternatively, according to their application, fuel cells may also be classified as (i) automotive fuel cells, (ii) stationary fuel cells, (iii) residential fuel cells,

Figure 46.3. An illustration of the various fuel cells outlined in Table 46.1, showing the various fuels and products (the DMFC has been omitted).

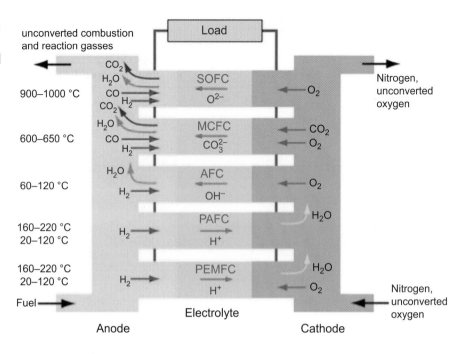

(iv) back-up power fuel cells, and (v) portable-power fuel cells. Automotive fuel cells producing 80–130 kW (of peak power) are intended for use in cars and buses. Applications for fork lifts and golf carts use similar fuel-cell power plants producing much lower power, 20–30 kW. Fuel cells for buses and larger vehicles are designed to generate about 200 kW. Stationary fuel cells are a category of larger fuel-cell power plants that may produce 100–250 kW or more of power using re-formate derived from various fossil fuels. Figure 46.4 is a collage of some the various stationary fuel-cell stacks currently in operation across the USA and Canada. Residential fuel cells are intended to power a home or a small store, produce power in the range 2–10 kW, and are fueled by re-formed methane or propane. An intermediate class of fuel cells generating power in the range 1–5 kW has application as back-up power. Portable-power fuel cells are devices having power ratings of <100 W for use in small electronic appliances such as laptop computers, with the most promising ones being DMFCs, which typically use methanol stored in a removable cartridge.

As shown in Table 46.1, PEMFCs typically operate in the temperature range from freezing to about 95 °C. The upper temperature is limited mainly by the lack of a proton-exchange membrane (PEM) that can durably withstand higher temperatures. Typically, hydrogen, methanol, or a re-formed fossil fuel containing very small quantities of CO (less than 10 ppm) and CO_2 (a few percent) are the choices for fuel. The PEMFCs employ a perfluorosulfonic acid (PFSA) or hydrocarbon (HC) material in the form of a membrane that functions as the proton-conducting electrolyte/separator and avoids system complications involved when dealing with a liquid acid or alkaline electrolyte.

46.4.2 Basics of fuel-cell operation

A PEMFC, specifically the hydrogen/oxygen fuel-cell system, is shown in Figure 46.5. The basic operation of all fuel cells is the same, with slight changes in fuel and whether it operates in an acidic or alkaline medium, with the ion moving across the electrolyte from a proton (H^+) to a hydroxyl (OH^-).

The reaction on the anode of hydrogen-fueled PEMFCs is the oxidation of hydrogen [hydrogen oxidation reaction (HOR)] to form protons and electrons,

$$2H_2 \rightarrow 4H^+ + 4e^-. \tag{46.1}$$

The reaction on the cathode is a slow $4e^-$ reduction of oxygen [oxidation reduction reaction (ORR)]. Oxygen reacts with protons that were generated on the anode side and diffused to the cathode and electrons to form product water,

$$O_2 + 4H^+ + 4e^- \rightarrow 2H_2O. \tag{46.2}$$

Today PEMFCs are rapidly approaching the initial stages of commercialization in automotive and residential applications; therefore, the core of this chapter will deal with details on the technological challenges involved in lowering their cost and increasing the durability of their components.

46.4.3 Other fuel-cell systems

Phosphoric acid fuel cells

The PAFCs are designated as intermediate-temperature fuel cells and are designed to operate at about 160–200 °C. The PACF may be considered one of the first fuel-cell technologies that has already been commercialized

Figure 46.4. Shown are installed. Fuel-cell stacks at universities, hospitals, and schools at a number of installations. Sizes range to 300 KW and in the bottom example the fuel cells use the biogas from waste as a fuel source.

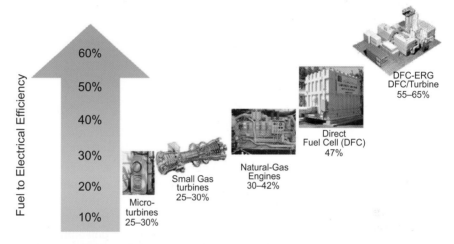

[by United Technologies Corp. Power (UTC Power), in 1992] for use as stationary power plants. With hydrogen as a fuel, the reactions at the anode and cathode are similar to those in PEMFCs. The electrolyte is concentrated phosphoric acid soaked into a SiC paste, and the anode and cathode are Teflon (PTFE)-bonded gas-diffusion electrodes (GDEs) employing Pt or Pt alloys as catalysts and graphitized carbon black as the catalyst support. Hydrogen is extracted from various hydrocarbon fuels that are re-formed externally. An advantage of the higher temperature is that the anodes can tolerate about 1% of CO in the re-formate simply using a Pt/C catalyst of moderate loading. The efficiency of PAFCs falls within the range of 40%–50% and increases to 80%

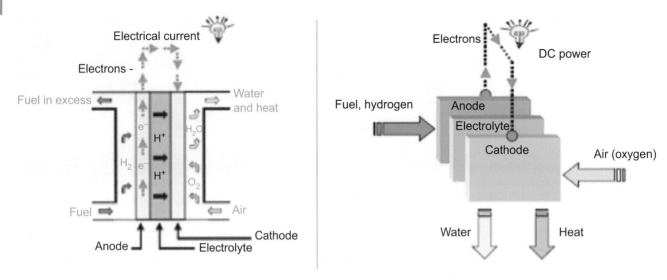

Figure 46.5. Two schematic representations of a single cell in a PEMFC illustrating the anode, electrolyte, and cathode, and the flow of molecular and ionic species and electrons.

if the waste heat is used in a co-generation scheme. Commercially (PureCell Model 400) available PAFC power plants from UTC Power can produce 400 kW of electricity and heat and are used for combined cooling, heating, and power (CCHP) applications in a variety of facilities such as supermarkets, hospitals, hotels, schools, hospitality centers, and data centers.

Molten-carbonate fuel cells

The MCFCs operate in the range 600–700 °C, in order to obtain reasonable conductivity of the carbonate electrolyte, and permit the use of non-precious-metal catalysts such as Ni–Cr and Ni–Al alloys for anodes and NiO for cathodes. The electrolyte is composed of a mixture of lithium and either potassium or sodium carbonate salt that is suspended in a ceramic matrix of $LiAlO_2$.

The reaction that occurs on the anode of MCFCs can be written as

$$H_2 + CO_3^{2-} \rightarrow H_2O + CO_2 + 2e^-. \qquad (46.3)$$

The reaction that occurs on the cathode of MCFCs can be written as below. Carbon dioxide produced at the anode is recycled and used at the cathode; we note that carbon dioxide and oxygen are necessary at the cathode to form carbonate ions:

$$\frac{1}{2}O_2 + CO_2 + 2e^- \rightarrow CO_3^{2-}. \qquad (46.4)$$

The overall reaction that occurs in MCFCs is thus

$$H_2 + \frac{1}{2}O_2 + CO_2(\text{cathode}) \rightarrow H_2O + CO_2(\text{anode}). \qquad (46.5)$$

The MCFCs have efficiencies as high as 60%, approaching 85% when the waste heat is collected and used in a co-generation cycle. The MCFCs do not need a re-former since the fuel is converted into hydrogen within the fuel cell itself (internal re-forming). The high temperature also renders the anode resistant to poisoning from carbon monoxide, and, therefore, such systems are being developed to work with fuels such as natural gas and coal gas. The high temperature and contact with the molten-carbonate electrolyte, which in inorganic chemistry is known as one of the best solvents, limits the choice of material that can be used, especially in the electrodes. For example, PTFE is introduced in PAFC electrodes to establish an electrolyte/reactant interface; in MCFCs, however, a controlled electrode structure (pore size and pore-size distribution) is the only available option. The high operating temperature also accelerates the corrosion and degradation of cathode and cell hardware and lowers the life of the stack.

Solid-oxide fuel cells

The SOFCs are the only class of fuel cells that employs a solid electrolyte that avoids the complexity involved with electrolyte management that plagues other types of fuel cells. The high operating temperature allows the use of a variety of fossil fuels without penalty; since water is generated at the anode, re-forming of hydrocarbon fuels may be carried out within the fuel cell. The SOFCs employ an electrolyte consisting of a non-porous metal oxide or ceramic such as Y_2O_3-stabilized Zr_2O_3 (YSZ), Sc_2O_3-stabilized Zr_2O_3 (SCZ), or gadolinea-doped ceria (GDC) and operate at 600–1,000 °C, at which temperature the ceramic is a good conductor of oxide anions. Ohmic losses occur due to the conductivity of the electrolyte, and can be decreased by lowering the thickness, doping the material, or raising the temperature. The cathode material is typically lanthanum strontium manganite (LSM), while the anode materials are cermets constituted of Ni combined with the electrolyte material, for example, Ni–ZrO_2 (cermet is a composite ceramic metal material). The electrochemical oxidation of hydrogen or carbon monoxide with oxygen ions takes place at the anode. The Kröger–Vink notation (devised by Freddy Kröger and H. J. Vink) is a notation used to describe the lattice position and electric charge for point-defect species in crystals. In the general format, M_s^c, M represents an atom, vacancy (V), electron (e), or hole (h). The subscript S represents the lattice site that the species occupies, including an interstitial site. The superscript C represents the electronic charge of the species relative to the site that it occupies: "×" is used to represent a null charge, "·" to indicate a single positive charge, "··" to represent two positive charges, and "′" to represent a single negative charge. The overall reaction can then be written using the Kroger–Vink notation as

$$\frac{1}{2}O_2 + 2e' + V_O^{··} \rightarrow O_O^{×}, \tag{46.6}$$

where the subscript "O" stands for an oxygen site.

The SOFCs have high efficiencies (60%) and fuel flexibility (almost any hydrocarbon can be used), and, although the problems of liquid-electrolyte management are eliminated, cheap materials and construction of stacks that can withstand the elevated temperatures and longer startup times are technical challenges remaining to be met. The applications of SOFCs range from stationary power generation (up to 2 MW) to their use for auxiliary power in automobiles. The extremely high operating temperature allows the SOFC to be used in combined heat and power systems, and results in further increased efficiencies.

Alkaline fuel cells

The AFCs typically operate on hydrogen and oxygen with 35%–50% potassium hydroxide (KOH) as the electrolyte (in a porous matrix) and gas-diffusion electrodes as anodes and cathodes. The gases have to be extremely pure and free of carbon dioxide in order to avoid the formation of solid carbonates (K_2CO_3). The carbonates lower the ionic conductivity of the electrolyte and blocks the electrode pores, impeding the cell's operation. Alkaline fuel cells have advantages that include lower overpotential (kinetic) losses at the cathode for oxygen reduction than for acid fuel cells and the possibility of using non-precious metals as catalysts.

The formation of carbonates when carbon dioxide reacts with air is as follows:

$$2KOH + CO_2 \rightarrow K_2CO_3 + H_2O. \tag{46.7}$$

The HOR at the anode can be written as

$$2H_2 + 4OH^- \rightarrow 4H_2O + 4e^-. \tag{46.8}$$

The ORR at the cathode can be written as

$$O_2 + 2H_2O + 4e^- \rightarrow 4OH^-. \tag{46.9}$$

The net overall reaction is summarized as

$$2H_2 + O_2 \rightarrow 2H_2O. \tag{46.10}$$

As mentioned in the historical section, the first application of alkaline fuel cells was in spacecraft such as the Apollo-series missions; in the Space Shuttle, static-electrolyte alkaline fuel cells (using high loadings of precious metal in the electrodes) provided power and generated potable water. However, it is important to note that in this NASA application it is mass and volume, not cost, that are the primary issues. The AFCs are highly efficient and can have efficiencies reaching 70%. They are further classified into flowing (circulating)-electrolyte cells and static-electrolyte cells; the type used in current space applications is the static type.

A version of AFC under development utilizes an anion-exchange membrane that conducts anions and is impermeable to reactant gases. Non-precious-metal catalysts can be coated onto the membrane in a manner similar to what is done for PEMFCs, these are expected to have an impact on the 2020 and 2030 future-generation devices.

Microbial fuel-cell systems

It is well known that micro-organisms generate water and carbon dioxide when they consume carbohydrates and sugar. In the absence of oxygen, i.e., under anaerobic conditions, they are capable of generating protons, electrons, and carbon dioxide. Microbial or biological fuel cells can be described as bio-electrochemical systems that produce electrical energy by utilizing the catalytic reactions of micro-organisms. Fuel is oxidized by micro-organisms on the anode (anaerobic), generating protons and electrons; the protons move through a cation-exchange membrane and combine with electrons and oxygen to form water. The MFCs or BioFCs are divided into mediator MFCs and mediator-free MFCs. In mediator MFCs, thionine, methyl blue, humic acid, methyl viologen, neutral red, etc. facilitate electron transfer from the microbial cells to the electrodes. In mediator-free MFCs, electrochemically active bacteria such as *Shewanella putrefaciens* and *Aeromonas hydrophilia* are used for electron transfer. In principle, a variety of organic materials can be used as fuel, including waste water from sewage treatment plants; clean water and a small amount of power can be produced, thereby making the plant more efficient.

In the rest of this chapter we focus primarily on the intensively researched PEMFCs fueled by pure hydrogen

Figure 46.6. A schematic diagram of components of a single cell repeat unit showing the membrane, anode and cathode catalyst layers, and diffusion layers.

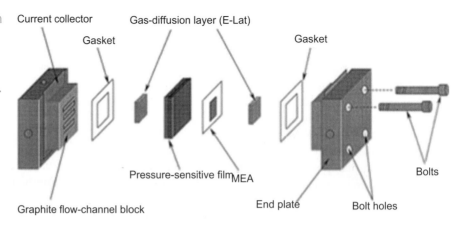

Current collector
Gasket
Gas-diffusion layer (E-Lat)
Gasket
Pressure-sensitive film MEA
Graphite flow-channel block
Bolts
End plate
Bolt holes

and ambient air, and the requirements and targets for their commercialization in automobiles.

Direct methanol fuel cells

As has probably become quite clear, most fuel cells are powered by hydrogen, which can be either introduced directly into anode or generated by re-forming hydrogen-rich fuels such as alcohols and hydrocarbons. There is another series of direct methanol fuel cells that are instead powered by the oxidation of pure methanol. Therefore, DMFCs will not have the significant level of fuel-storage problems typically associated with PEM fuel cells since methanol has a higher energy volume density than hydrogen and, since it is a liquid, it will also be easier to transport and supply to the public using the current infrastructure used for gasoline.

The optimal overall reaction for a DMFC system involves anodic oxidation of methanol,

$$CH_3OH + H_2O \rightarrow CO_2 + 6H^+ + 6e^- \qquad (46.11)$$

and Cathodic reduction of oxygen,

$$1.5O_2 + 6H^+ + 6e^- \rightarrow 3H_2O, \qquad (46.12)$$

with the net total reaction

$$CH_3OH + 1.5O_2 \rightarrow CO_2 + 2H_2O. \qquad (46.13)$$

Direct methanol fuel-cell technology is relatively new compared with that of fuel cells powered by pure hydrogen, and research and development are roughly 5–10 years behind that relating to types of other fuel cell, mostly because of research funding levels. Nonetheless, the DMFC appears to be a possible battery replacement for portable applications, with a number of manufacturers are already introducing commercial versions of these applications.

46.4.4 Basic functional requirements of PEM fuel-cell components

As we move forward in our explanation of fuel cells, we will focus on the PEM system which lies at the heart of the most important automotive applications at this time. The "guts" of the fuel cell is the membrane electrode assembly (MEA). The MEA is defined as the five-layer assembly of (i) a membrane, (ii) the anode electrode, (iii) the cathode electrode, (iv) the anode diffusion medium (DM) or gas-diffusion layer (GDL), and (v) the cathode DM or GDL. The MEA is sandwiched by two bipolar plates that consist of channels or flow fields for reactant-gas flow. A set of repeating units of MEAs and bi-polar plates held together by a set of end plates is referred to as a fuel-cell stack. Figure 46.6 shows schematically a single cell repeat unit. Figures 46.7(a) and (b) show fuel-cell vehicles.

Membrane electrode assemblies

Membrane electrode assemblies are prepared by one of two basic approaches: (i) coating the catalyst onto the membrane to form a three-layer catalyst-coated membrane (CCM), followed by placing the diffusion media on either side of the CCM to form the five-layer MEA; or (ii) coating the catalyst onto the diffusion medium and hot pressing them to the membrane. The catalyst is typically a platinum-decorated carbon (Pt/C) powder that is made into a slurry or ink (using alcohol, water, glycerol, etc.) in a consistency suitable for the coating method employed. The catalyst is coated onto the membrane or gas-diffusion layer (GDL) by methods such as slot coating, blade coating, spray coating, or screen printing. The detailed principles of MEA structure and preparation are discussed elsewhere [1][2].

Electrocatalysts

The electrocatalyst used in both the anode and the cathode today is Pt nanoparticles (or a Pt alloy) supported on carbon black. The precious-metal particle size usually falls within the range 2–4 nm, depending on the carbon support and any alloying or heat treatment that it may have been subjected to. Table 46.2 outlines the targets for automotive PEMFC catalysts and Figure 46.8 shows TEMs of a typical commercial catalyst [3].

Table 46.2. Targets (2015) for automotive PEMFC electrocatalyst activity and costs

Parameter	Target
Electrocatalyst activity	~1000 mA per mg Pt at 0.9 V
Non-pt electrocatalyst activity	>300 A cm^{-3} at 0.8 V
Durability/ECA loss with cycling	<40% after 5,000 h
Cost	~0.1g Pt per kW

The kinetics of the oxidation of hydrogen is extremely fast and thus requires a Pt loading of only about 0.05 mg cm^{-2}. This is true when the fuel is pure hydrogen; in the presence of trace (ppm) impurities of CO (that may be present if hydrogen is produced from re-forming of a fossil fuel), some reversible degradation is observed. Other impurities, such as H_2S, may lead to irreversible poisoning of the Pt surface.

The reaction on the cathode is a slow 4e$^-$ reduction of oxygen. Oxygen reacts with protons that were generated on the anode side and electrons to form water. Most of the losses (~400 mV) in the fuel cell occur due to the sluggish kinetics of the ORR, and much effort has been expended on understanding the mechanism and finding materials that exhibit lower overpotentials. The loading on cathodes that employ Pt supported on

(a)

(b)

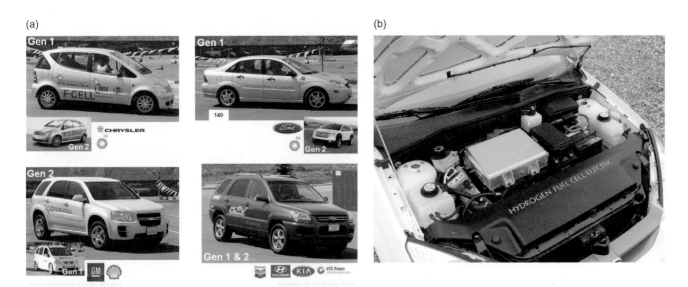

Figure 46.7. Photographs of operational fuel-cell vehicles. (Courtesy of the NREL.)

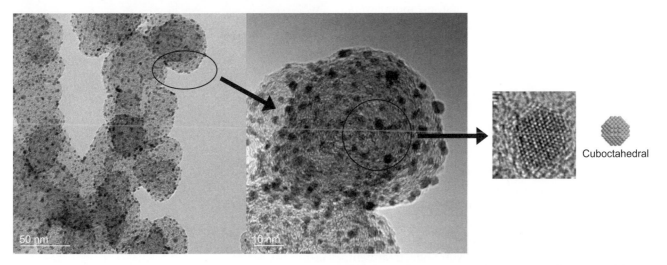

Figure 46.8. Low- and high-resolution TEMs of conventional commercial Pt nanoparticles dispersed on carbon black. (Courtesy of the NREL.)

Figure 46.9. Core–shell catalysts have been synthesized. These materials have a non-Pt core and a shell that is predominantly Pt, resulting in a significant enhancement of the mass activity (*A* per g Pt) [4].

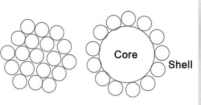

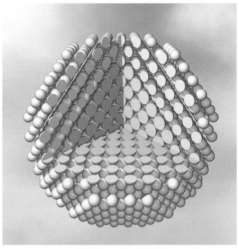

Figure 46.10. A nanostructured thin-film catalyst consisting of Pt deposited on a nonconducting support. From http://www.hydrogen.energy.gov/pdfs/review09/fc_17_debe.pdf.

carbon as the catalyst is about 0.35 mg cm^{-2}, while the use of Pt-alloy/C systems (for example, PtCo, PtNi, PtCu, etc.) can reduce the loading of Pt by ~50%.

Pathways to enhance the ORR kinetics include the use of Pt alloys, core–shell catalysts supported on carbon (Figure 46.9), and continuous thin films of catalyst (Figure 46.10) on nonconducting supports as well as continuous Pt films on structured conducting supports (Figure 46.11). The core–shell pathway involves the use of precious metal only in the first few monolayers, with the core being a less precious material to enhance the mass activity [4][5][6]. The thin-film pathway is based on the fact that bulk Pt and Pt alloys exhibit extremely high specific activity compared with nanoparticles; utilizing this property while depositing the catalyst as a very thin film would result in a high-mass-activity catalyst [7][8].

The gas-diffusion layer

The gas-diffusion layer, or diffusion medium is composed of a carbon paper (cast from a slurry of carbon fibers and particles) or carbon cloth (woven graphite fibers) that has a microporous layer, MPL constituted of carbon powders and PTFE coated onto it. The GDL may be subjected to a hydrophobic or hydrophilic treatment, depending on the stack design. An appropriate amount of hydrophobicity or hydrophilicity to maximize the

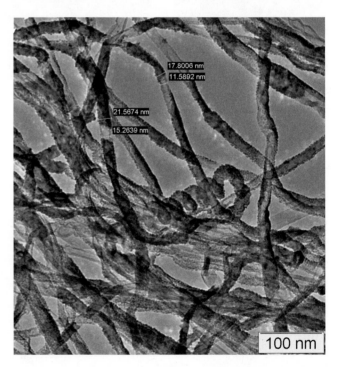

Figure 46.11. Continuous thin films of Pt on single-walled carbon nanotubes synthesized at the NREL [8]. Extended thin-film catalysts show promise of much higher specific activity than that of nanoparticles. Areas of light contrast show regions without platinum coatings. (Courtesy of the NREL.)

limiting current by minimizing the mass-transport resistance is targeted. The DM provides electronic conductivity, pathways for gas channels that allow reactants into the catalyst layer CL, and also pathways for the water generated to exit into the flow fields or channels of the bi-polar plates. The MPL is usually hydrophobic due to the PTFE used in preparing the layer, and plays a key role in providing channels for gas flow to the catalyst layer.

The ion-conductive membrane

Today's PEMFC systems are limited to a maximum temperature in the range ~80 °C, with short excursions to 90 °C. The constraint is imposed by the low proton conductivity of membrane materials at low relative humidity (RH) (below 50%); operating at high temperatures while maintaining proton conductivity would require high levels of humidification and high pressures, stressing the system and the BOP complexity and cost. Ideally, a membrane that operates at 120 °C would promote system simplification whereby heat rejection during short spurts of maximum power is easily achieved. Limited amounts of liquid water at high temperatures and low RH would lower the degradation of the catalyst and carbon and make water management facile. The properties expected of good candidates are (i) high protonic conductivity, (ii) high electronic resistance, (iii) low hydrogen and oxygen crossover, (iv) low cost, (v) high mechanical durability under RH cycling, and (vi) high chemical durability. Properties of various membranes have been reviewed by Savogado [9]. Permeability and hydrogen crossover of membranes have also been well characterized [10].

It should be noted that there are trade-offs in selecting a membrane material with adequate performance and durability. A low-EW PFSA material possesses a larger number of sulfonate sites, increasing water retention and proton conductivity. These materials, though, swell to a greater extent, and therefore are susceptible to degradation during RH cycles encountered during fuel-cell operation. Introduction of additives to membranes enables the membrane to hold water at lower RHs and higher temperatures; the additives may leach out over time and sometimes cause the membrane to become brittle and fail more easily. Thinner membranes have been used over the last decade to lower the areal resistance and enhance cell performance, but often a backbone or support of non-proton-conducting material (e.g., PTFE) is required for mechanical strength, which in turn increases the resistance.

The durability targets for membranes are similar to those for the complete fuel cell and include <10% performance loss over 5,000 h (10 years) of operation under automotive conditions that include startup/shutdown, load cycling, freezing (–40 °C), and idling. The target cost

Table 46.3. Technical targets for membranes for PEMFCs for automotive applications (target properties are at operating temperature unless specified otherwise)

Parameter	2015 Targets
Cost	$5 per kW; $20 per m^2
Conductivity/resistance at operating T, RH	70 mS cm^{-1}; 20 mΩ cm^2
Conductivity/resistance at –20 °C	10 mS cm^{-1}; 200 mΩ cm^2
Minimum electrical resistance	1.0 kΩ cm^2
Maximum O$_2$ crossover	<1 mA cm^{-2}
Maximum H$_2$ crossover	<2 mA cm^{-2}
Durability	>5,000 h
Survivability	–40 °C to 120 °C
RH cycles	20,000
Open Circuit Voltage Lifetime	500 h

of the membrane is ~$10–20 per m^2, corresponding to production levels of ~10^7 m^2 per year. More detailed membrane-performance targets from the US Department of Energy (DOE) and other sources are detailed in Table 46.3 [11].

In PEMFCs, the proton-conducting electrolyte is in the form of a membrane having a thickness of 15–30 μm. The catalyst layer is either coated directly onto the membrane or decal transferred by hot-pressing. The most commonly used membranes are PFSA-based, such as Nafion® from Dupont, Gore, Asahi Glass, and 3M. Hydrocarbon membranes are also being studied, and have been incorporated into some stacks by automakers such as Honda [12].

Membrane degradation is an important issue since it occurs in the form of gradual loss in performance (fluoride and sulfate elution) as well as by catastrophic failure (membrane thinning and pinhole formation) due to reactant crossover. Membranes based on PFSA are chemically degraded by peroxide radical attacks on non-fluorinated main-chain terminals and ensuing unzipping reactions. The peroxyl radicals are thought to be generated during the operation of the fuel cell on the surface of Pt in the presence of both hydrogen and oxygen. The Open Circuit Voltage holds of fuel cells and low RH are known to accelerate membrane degradation. The US DOE has published accelerated test protocols for evaluating both chemical and mechanical degradation of membranes.

Bi-polar plates

Bi-polar plates are the outermost component of each individual cell. They incorporate flow fields or channels for distributing the reactants to the entire cell area, and also play the role of current collectors. Bi-polar plates also incorporate internal flow fields for the flow of coolant that collect the heat rejected by each cell. Requirements for the properties of bipolar plates are shown in Table 46.4.

The materials used for bi-polar plates include graphite foils (or grafoil) constituted of expanded graphite flakes, porous graphite plates, graphite polymer composites, and metal sheets. Flexible graphite plates are processed by a continuous rolling of natural graphite flakes; the flow fields are embossed onto the flexible plates. Plates are typically composed of stainless steel or aluminum and are coated with a corrosion-resistant layer. Composite plates are composed of 80%–90% graphite, with the remainder being a thermosetting resin. The bi-polar plate is assembled (bonded together) from two half-plates that have flow fields embossed on them. Since composite plates are produced by injection or compression molding in a batch manufacturing process (due to the heating/cooling/setting times), they are not well suited to high-volume manufacturing. Metal bi-polar plates are composed of various stainless steels, or aluminum, Figure 46.12. The flow fields are stamped, pressed or etched onto the plates, and two half-plates are then welded together. Metal plates have the advantage of being very thin and suited to high-volume manufacturing. An exhaustive review of the status of bi-polar plates can be found in the work of Wang and Turner [13].

The disadvantage is the low corrosion resistance of most metals; coatings that are both conductive and economical are required in order to provide the plates with the properties of low contact resistance and durability. Stainless steels, Al-, Ni-, and Ti- based alloys have been studied extensively as possible candidates for bi-polar plates. One of the most well-studied materials for bi-polar plates is SS 316/316L (16%–18% Cr, 10%–14% Ni, 2% Mo, the rest Fe); other candidates are 310,904L, 446, and 2205. Bare stainless-steel plates form a passive 2–4-nm chromium oxide surface layer under PEMFC conditions that leads to unacceptably high interfacial contact resistance (ICR). A similar trend is observed for the other alloys, and therefore surface modification or surface coating on selected substrate material has to be considered as a pathway to meet the technical targets of low ICR and high corrosion resistance. The corrosion-resistant coating minimizes the oxidation of the metal and maintains good electronic contact.

Table 46.4. Technical targets for bi-polar plates for automotive PEMFCs

Parameter	2015 US DOE target
Cost	$3.00 per kW
Weight	0.4 kg per kW^{-1}
Hydrogen permeation at 80 °C	2×10^{-6} cm^3 per cm^2 s^{-1} at 80 °C, 3 bar
Corrosion	1 μA cm^{-2}
Resistivity	10 mΩ cm
Flexural strength	25 MPa
Flexibility	3%–5% deflection at midspan

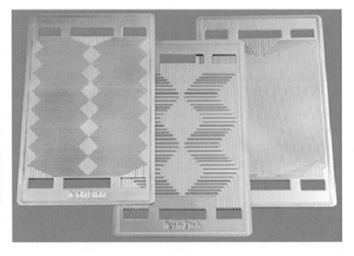

Figure 46.12. A photograph of the MEA roll to roll process and an image of bi-polar plates used in automotive PEMFC stacks [14]. (Courtesy of the NREL.)

46.4.5 Fuel-cell electrochemical performance diagnostics

The most fundamental and essential diagnostics for fuel cells are electrochemical in nature and involve measuring the performance and durability of a component, a single cell, or an entire fuel-cell stack. The diagnostics carried out in assembled subscale cells are additionally intended to isolate the contribution of individual components of the cell assembly and also separate the losses or overpotentials in terms of activity, resistance, and mass-transport losses [1]. The most preliminary and critical diagnostic measurement carried out on a fuel cell is that of that of the hydrogen crossover current and shorting resistance; these tests confirm the health of the MEA and eliminate cells that may potentially fail catastrophically. Values for hydrogen crossover currents typically lie below $1\ mA\,cm^{-2}$ (being a function of the RH, temperature, and membrane thickness) and acceptable values of electronic shorting resistance (caused by pinholes) are $>1,000\ \Omega$.

All electrochemical reactions take place at the surface and therefore an estimate of the surface area, especially that of the precious-metal catalyst, is of prime importance. One of the most critical measurements employed to estimate the utilization of the Pt-based catalyst as well as changes that takes place over the life of a fuel cell is that of the in-situ electrochemical area (ECA). The ECA is measured by recording a cyclic voltammogram (CV) under H2|N2 [reference, counter electrode (RE,CE)|working electrode (WE)] and calculating the charge under the hydrogen-adsorption peaks (Hads) using a potentiostat. Assuming that one H atom is adsorbed on each Pt atom, and knowing the number of Pt atoms per unit area, the surface area of Pt can be determined. This measurement can be carried out intermittently over the life of a fuel cell and the loss in area related in part to the total performance loss.

Figure 46.13 depicts the in-situ measurement of a CV (shown only in the 0.0–0.4 V regime) and the ECA for Pt nanoparticles on a supported carbon catalyst. The figure illustrates the impact of N_2 gas flow rates on the WE and fuel cell temperature on the hydrogen adsorption peaks as well as the onset of hydrogen evolution.

For each curve, we observe two HUPD/HAD (hydrogen under-potential deposition/hydrogen adsorption: $H^+ + e^- = Hads$) peaks that are attributed to hydrogen adsorption on specific crystal faces of the 2–3 nm Pt nanoparticles in the catalyst layer. As we sweep cathodically towards negative potentials and approach zero, we see large reducing cathodic currents corresponding to hydrogen evolution. After hydrogen evolution, (anodic scan), some of the evolved hydrogen near the electrode surface is oxidized and we observe large hydrogen oxidation currents. The five curves in the figure

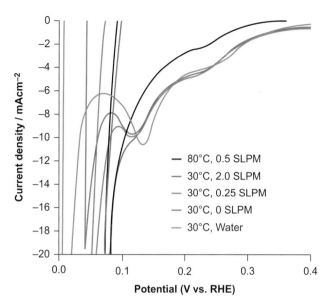

Figure 46.13. *In situ* electrochemical area, ECA, measurements in a PEMFC conducted under different conditions. Using a low temperature and low flow of N_2 (or water) allows accurate measurement of the ECA. SLPM, standard liters per minute. RHE: reversible hydrogen electrode. For electrodes of the same geometric area the current density can be used for ECA.

essentially show why it is important to use a lower temperature and lower flows of N_2 on the WE in order to get well-resolved peaks that are critical for accurate measurement of the charge (C/cm^2) and hence the ECA $(cm^2Pt/cm^2geo$ or $m^2/gPt)$. Maintaining a low N_2 flow or using water instead, traps some H_2 (originating from H_2 cross-over through the membrane and H_2 evolution at the WE) at the electrode/catalyst surface and prevents the early onset of hydrogen evolution at positive potentials and smearing of the HUPD peaks, thus producing well-resolved HUPD peaks and accurate measurements of the Pt surface area [15].

The ORR that takes place on the cathode is slow and requires the use of a large amount of expensive Pt-based catalyst. Most of the research work on catalysts is focused on improving the activity of the catalyst in an attempt to lower the Pt loadings. To accurately measure the kinetics in a fuel cell without interference from mass-transport phenomena, polarization curves need to be acquired under H2|O2. The mass flow rate (S) of O_2 is maintained at a high value $(S \sim 9)$ to ensure that the concentration of oxygen is uniform over the entire active area. The curves are corrected for the resistance by measuring the high-frequency resistance (HFR). These curves plotted on a semi-logarithmic scale can be analyzed for ORR activity as well as Tafel slopes[2]. Figure 46.14 depicts a H2|O2 Tafel

---

[2] The Tafel equation relates the rate of an electrochemical reaction to the overpotential.

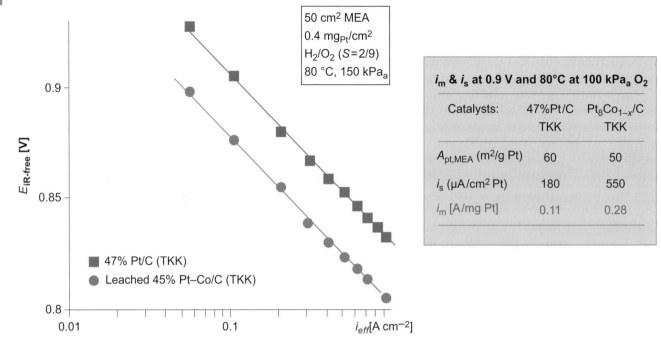

| 50 cm² MEA |
| 0.4 mg_Pt/cm² |
| H₂/O₂ (S=2/9) |
| 80 °C, 150 kPa_a |

i_m & i_s at 0.9 V and 80°C at 100 kPa_a O₂		
Catalysts:	47%Pt/C TKK	Pt₈Co₁₋ₓ/C TKK
$A_{pt,MEA}$ (m²/g Pt)	60	50
i_s (µA/cm² Pt)	180	550
i_m [A/mg Pt]	0.11	0.28

- ■ 47% Pt/C (TKK)
- ● Leached 45% Pt–Co/C (TKK)

Figure 46.14. Measurement of ORR activity in a PEMFC under H₂|O₂ at 80 °C and 100% RH [18]. In the table, i_m and i_s represent the mass and specific activity (A per g Pt and A per cm² Pt) of the catalyst under ORR. TKK refers to commercially available Pt-based catalysts from the company Tanaka Kikinzoku Kyogyo, Japan. kPa is kilopascals absolute.

plot for a Pt/C and a Pt-alloy/C catalyst and the extraction of the specific and mass activities from it. Not discussed are the issues of oxide species on the Pt surface that can significantly influence the measured current and ORR activity. The pre-conditioning and direction of sweep affect the measured current since the oxide coverage is a function of the two parameters [16][17].

Lastly, the H₂|air polarization curve (Figure 46.15) provides us with information on the actual performance of the fuel cell under realistic conditions of reactants, reactant stoichiometries, RH, and temperature. Such curves taken in subscale cells will be offset from data taken in full planform cells or short stacks due to the current and temperature distribution and design of flow fields not being the same.

We note that, for small quantities of newly synthesized catalysts, evaluation is often performed *ex situ* in rotating-disk electrodes for surface area and ORR activity [19][20]. Other standard electrochemical diagnostics not discussed here include dilute H₂ and O₂ polarization curves to study mass transport and the use of electrochemical impedance spectroscopy (EIS) to obtain the catalyst layer resistance, membrane conductivity using a four-electrode setup, corrosion currents for catalyst supports and bi-polar plates. In addition, effluent water from the anode and cathode is often collected to measure the elution of F⁻ and SO₄²⁻ from degradation of PFSA membranes. Specific tests to measure the corrosion of

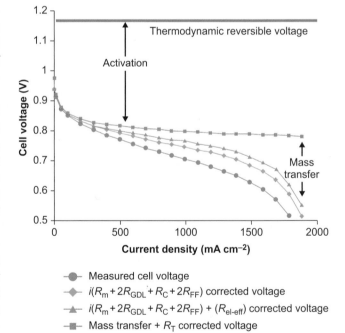

- ● Measured cell voltage
- ◆ $i(R_m + 2R_{GDL} + R_C + 2R_{FF})$ corrected voltage
- ▲ $i(R_m + 2R_{GDL} + R_C + 2R_{FF}) + (R_{el-eff})$ corrected voltage
- ■ Mass transfer + R_T corrected voltage

Figure 46.15. H₂|air polarization curves together with corrections for component resistances (m, membrane; GDL, gas-diffusion layer; FF, flow field, C, contact, el-eff, effective catalyst layer) and mass transport. The thermodynamic reversible potential is also depicted [1].

Table 46.5. Basic modes of automotive vehicle operation that result in PEMFC degradation

Operating mode	Root cause of degradation	Degradation pathway
(1) Idling/low load	Formation of peroxy radicals	Membrane degradation, Pt ECA loss
(2) Acceleration/ deceleration	Cathode-potential cycling	Pt dissolution, particle growth, Pt ECA loss
(3) Startup/shutdown	Cathode potential spikes up to 1.5 V	Support corrosion/Pt particle agglomeration
(4) Ambient air contaminant	Adsorption of air contaminants	Poisoning of cathode catalyst sites
(5) Freezing temperatures	Freezing water/fuel starvation	Support corrosion, electrode degradation

carbon-black supports include using a gas chromato-graph (GC) to measure effluent gases while the cathode is subjected to high or cycling potentials.

46.4.6 Modes of operation in automobiles

Unlike a fuel cell intended for stationary use, a fuel cell in an automobile undergoes changes in operating conditions and load identical to that experienced by a conventional internal combustion engine, (ICE). A variable load is imposed by acceleration, braking, idling conditions, and driving style, as well as by the effects of road conditions and environmental conditions on a fuel cell. The fuel cell has to respond to the stimulus by providing the power that is demanded at any point in time. In addition, the vehicle may be started and stopped arbitrarily and repeatedly to exposed sub-freezing conditions for long periods of time. The fuel-cell vehicle may also be driven in areas that have a high level of pollutants generated by ICE vehicles that might contaminate the fuel cell. All of these conditions imposed on a fuel cell result in a certain amount of degradation or decay in performance that may, or may not, be reversible or recoverable. Automotive drive cycles such as the US06, FUD, etc., have been developed by engineers to simulate the load experienced by the engine or fuel cell and are often used to evaluate the durability of fuel cells [2].

Table 46.5 briefly categorizes the modes of operation of vehicles that cause fuel-cell degradation. We will describe operating modes (2) and (3) briefly. Operating mode (3) refers to startup/shutdown degradation with the mechanism illustrated in Figure 46.16. A fuel cell that has been shut down for a period of time will be filled in both the anode and cathode compartments with air that has leaked in from the environment. On starting up the vehicle and stack, hydrogen is injected into the anode compartment. This results in a hydrogen–air front passing through the anode, leading to extremely high potentials at the cathode. The potentials are high

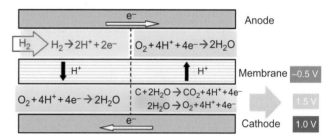

Figure 46.16. A schematic diagram of the startup/shutdown degradation mechanism. A high potential of 1.5 V at the cathode catalyst layer leads to catastrophic degradation in a short period of time.

enough to cause decomposition of the carbon support and cause severe irreversible degradation. The degradation due to startup/shutdown has been significantly lowered mainly by controlling the operating conditions rather than by implementing a materials change. Work on new corrosion-resistant support materials such as WO_3, TiO_2, etc., that can replace carbon blacks and tolerate 1.5 V to provide a solution that will minimize system complexity is under way. Issues with some of the new corrosion-resistant materials are lower electronic conductivity and surface area.

The degradation mechanism involved in operating mode (2) is the dissolution of Pt during normal potential cycling operations referred to as "acceleration/deceleration" in Table 46.5. The thermodynamics of Pt indicates that Pt dissolves at ~1.18 V at 80 °C; this potential is lowered for small nanoparticles that have a high surface energy. The repeated formation and removal of Pt oxide species on the catalyst surface has been implicated in accelerating Pt dissolution. For fast ramp rates, the potential on the oxide-free surface of Pt can approach high potentials for short periods of time during which the dissolution rate is raised. The dissolution process of Pt often leads to a band of Pt being deposited into the membrane (Figure 46.16) and has not yet been mitigated [2]. Certain Pt alloys or heat-treated Pt suppress the dissolution partially; it is also expected that the use of newer catalysts in the form of continuous extended films might lower these losses significantly once they are implemented in practical fuel-cell electrodes.

In terms of contamination, the PEMFC acts as a filter and is susceptible to trace impurities in the air caused by pollution and exhaust emissions of ICE vehicles. Dust, SO_2, NO_x, chloride, etc., all degrade the performance of fuel cells. Some of the impurities adsorbed on the catalyst are oxidized and flushed out during fuel-cell operation.

46.4.7 Recent global progress in PEMFCs

Tremendous progress has been made over the last decade in the advancement of hydrogen-fueled PEMFC technology, especially for automotive applications. Major automobile companies have invested tens of millions of dollars a year in the research and development of PEMFCs. At the beginning of this decade, re-formed gasoline was considered as a fuel for automobiles, but was abandoned when it became clear that placing an entire chemical re-forming plant inside a vehicle posed serious problems of size and cost. The inability of low-temperature (60–80 °C) PEMFCs to tolerate even ppm levels of CO and other contaminants found in re-formates exacerbated the problems. The ensuing research on PEMFCs carried out in automobile companies as well as research carried out in universities and national laboratories funded in part by government agencies such as the US DOE brought forth scientific advances that have led to significant improvements in performance, durability, and cost.

According to a US DOE report, the status of fuel cells today is as follows: 52,000 fuel cells have been shipped worldwide, with a 50% increase from 2007 to 2008. In the transportation area, there are about 200 fuel-cell vehicles, 20 fuel-cell buses, and 60 fueling stations. about 10 million tons of hydrogen are being produced annually in the USA (some 70 million tonnes worldwide) and 1,200 miles of hydrogen pipelines have been put down [22].

Figure 46.17 depicts the cost reduction in PEMFCs over the last 10 years. The cost distribution of the various components of a fuel-cell system is illustrated in Figure 46.18. Figure 46.19 illustrates the detailed

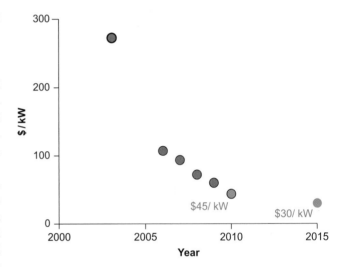

Figure 46.17. Cost reduction of automotive PEMFCs over the last 10 years. The current ICE cost has been marked at $30 per. Replotted from [21].

Figure 46.18. The distribution of automotive PEMFC system cost on the basis of US DOE projections. Replotted from [21].

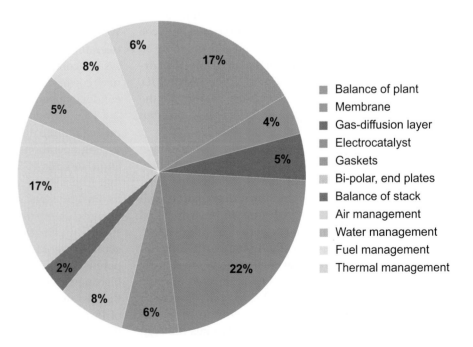

breakdown of fuel-cell system costs on the basis of estimates for 2008 and projected targets for 2010 and 2015.

The 2008 status for automotive PEMFCs was about 2,000–3,000 h and the 2015 target life is 5,000 h (150,000 miles). For stationary PEMFCs, the 2003 durability was 15,000 h and the 2011 target is 40,000 h.

Table 46.6. Overall status and targets for PEMFC systems, hydrogen storage, production, and delivery [12]

	Status	Target
Fuel-cell system cost ($ per kW)	61	30
Fuel-cell system durability (h)	2,000–3,000	5,000
Hydrogen production ($ per gge)	3–12	2–3
Hydrogen delivery ($ per gge)	2.30–3.30	1
Hydrogen storage, gravimetric (wt%)	3.0–6.5	7.5
Hydrogen storage volumetric (g L^{-1})	15–50	70
Hydrogen storage cost ($ per kWh)	15–23	2

If the world is to move to a hydrogen economy, a hydrogen infrastructure that includes a combination of distributed and centralized production needs to evolve. Hydrogen-pipeline networks already exist in some regions, often to provide hydrogen to the refining and food-processing industry; transport by trucks is also prevalent. In the USA, there are about 60 hydrogen fueling stations (there are ~350 worldwide), 1,200 miles of hydrogen pipelines, and ~10 million tonne of hydrogen are produced every year. Hydrogen storage is often categorized as physical (or molecular) and chemical (or dissociative) storage. On-board physical storage methods include compressed gas, liquid hydrogen, and cryo-adsorbed hydrogen; chemical storage includes metal hydrides and liquid organic carriers. Compressed hydrogen (35–70 MPa) in one or two tanks (~4–8 kg H$_2$ depending on the target range) is stored on board fuel-cell vehicles today. Fuel-cell vehicles already meet the driving range of conventional ICE vehicles at this time. The safety of the hydrogen fuel tank has been demonstrated to be a non-issue; fuel-cell vehicles successfully pass the front, rear, and side impact tests that are typically applied to ICE vehicles. The overall status and targets for PEMFC systems for automotives are summarized in Table 46.6.

46.5 Summary

Significant progress has been made in the advancement of H$_2$|air PEMFC technology for use in automobiles, with commercialization expected to commence around 2015. The main thrust and focus of research for PEMFCs is

Figure 46.19. A detailed breakdown of automotive PEMFC system costs on the basis of estimates for 2008 and projected targets for 2010 and 2015 [21].

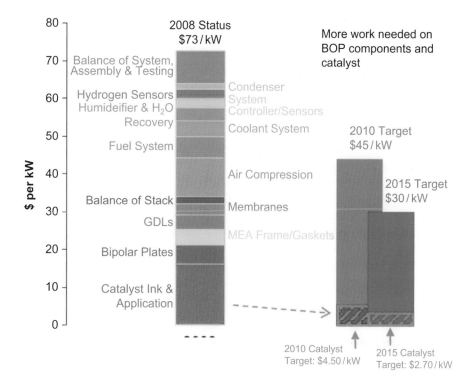

in the area of new highly active and durable electro-catalysts and membrane materials with adequate conductivity that survive low RH levels and generally in attempts to lower the cost of the fuel-cell stack and system. For electrocatalysts, the amount of Pt used in a stack is targeted to be 0.1 g per kW or 10 g in a 100-kW stack to ensure acceptable costs as well as adequate availability of Pt. Current fuel-cell vehicles have a driving range comparable to that of ICE vehicles and a life of 70% of the target value. For a hydrogen economy to come into being, the fuel-cell stack must first become commercially viable, following which a hydrogen infrastructure for distribution will naturally follow.

46.6 Questions for discussion

1. What are the 2015 DOE targets for performance and durability of PEMFC stacks for automotives?
2. What percentage of the total cost of the stack is attributed to the electrocatalyst?
3. How many grams of Pt or precious metal are used in catalytic convertors of ICE vehicles?
4. Outline the main pathways for ongoing and future electrocatalyst research.
5. What is the main breakthrough in membrane material property that would be significant for PEMFCs?
6. What are the desirable properties of a bi-polar plate for PEMFCs?
7. What is the range of power densities ($kW L^{-1}$) of current-generation automotive fuel cells?
8. What are the main modes of operation of fuel-cell vehicles that result in degradation of the fuel-cell stack?
9. What are the sources from which hydrogen can be obtained?
10. What is the range of current fuel-cell vehicles? How many kg of hydrogen do they carry? What is the pressure of hydrogen in compressed tanks used in fuel-cell vehicles?
11. How many miles of hydrogen pipelines exist in the USA at this time?

46.7 Further reading

- R. O'Hayre, S.-W. Cha, W. Colella, and F. B. Prinz, 2009, *Fuel Cell Fundamentals*, New York, John Wiley.
- J. L. and A. Dicks, 2003, *Fuel Cell Systems Explained*, 2nd edn., New York, John Wiley.
- US DOE, *Fuel Cell Handbook*, now in its seventh edition and available freely on the web.

46.8 References

[1] S. S. Kocha, 2003, "Principles of MEA Preparation," in *Handbook of Fuel Cells – Fundamentals, Technology and Applications*, eds. W. Vielstich, A. Lamm, and H. Gasteiger, New York, John Wiley & Sons, Ltd., pp. 538–565.

[2] M. Uchimura, S. Sugawara, Y. Suzuki, J. Zhang, and S. S. Kocha, 2008, "Electrocatalyst durability under simulated automotive drive cycles," *ECS Trans.*, 16(2), 225–234.

[3] P. N. Ross, 2005, *DOE Annual Merit Review*, http://www.hydrogen.energy.gov/pdfs/review05/fc10_ross.pdf.

[4] R. R. Adzic, 2010, "Contiguous platinum monolayer oxygen reduction electrocatalysts on high-stability-low-cost supports," in *2010 DOE Annual Merit Review*, available from http://www.hydrogen.energy.gov/pdfs/review10/fc009_adzic_2010_o_web.pdf.

[5] J. Zhang, H. B. Lima, M. H. Shao *et al.*, 2005, "Pt mono-layer on noble metal–noble metal core core–shell nano-particle electocatalysts for O_2 reduction," *J. Phys. Chem. B*, 109(48), 22701–22704.

[6] J. Zhang, Y. Mo, M. B. Vukmirovic *et al.*, 2004, "Pt–Pd core–shell," *J. Phys. Chem. B*, 108, 10955–10964.

[7] M. K. Debe, 2003, "NSTF Catalysts," in *Handbook of Fuel Cells – Fundamentals, Technology and Applications*, eds. W. Vielstich, A. Lamm, and H. Gasteiger, New York, John Wiley & Sons, Ltd.

[8] B. Pivovar, 2010, "Extended, continuous Pt nanostructures in thick, dispersed electrodes," in *2010 DOE Annual Review Meeting*, available from http://www.hydrogen.energy.gov/pdfs/review10/fc007_pivovar_2010_o_web.pdf.

[9] O. Savadogo, 1998, "Emerging membranes for the electrochemical systems: (I) solid polymer electrolyte membranes for fuel cell systems," *J. New Mater. Electrochem. Systems*, 47–66.

[10] S. S. Kocha, D. J. Yang, and J. S. Yi, 2006, "Characterization of gas crossover and its implications in PEM fuel cells," *AIChE J.*, 52(5), 1916–1925.

[11] US DOE, 2007, *Fuel Cell Targets*, 2007, available from http://www1.eere.energy.gov/hydrogenandfuelcells/mypp.

[12] Honda, 2010, *Honda: Fuel Cell Electric Vehicle*, available from http://world.honda.com/FuelCell/.

[13] H. Wang and J. A. Turner, 2010, "Reviewing metallic PEMFC bipolar plates," *Fuel Cells*, 10(4), 510–519.

[14] GreenCarCongress, 2009, *GM Highlights Engineering Advances with Second Generation Fuel Cell System and Fifth Generation Stack; Poised for Production Around 2015*, available from http://www.greencarcongress.com/2009/09/gm-2gen-20090928.html.

[15] M. Uchimura and S. S. Kocha, 2008, "The influence of Pt-oxide coverage on the ORR reaction order in PEMFCs," in *ECS Transactions 2008*, http://www.electrochem.org/meetings/scheduler/abstracts/214/0914.pdf.

[16] M. Uchimura and S. Kocha, 2007, "The impact of cycle profile on PEMFC durability," *ECS Trans.*, 11(1), 1215–1226.

[17] M. Uchimura and S. S. Kocha, 2007, "The impact of oxides on activity and durability of PEMFCs," in *Annual AIChE Meeting*, Utah.

[18] S. S. Kocha and H. A. Gasteiger, 2004, "The use of Pt-alloy catalyst for cathodes of PEMFCs to enhance

performance and achieve automotive cost targets," in *2004 Fuel Cell Seminar*, San Antonio, TX.

[19] **Y. Garsany**, **O. A. Baturina**, **K. E. Swider-Lyons**, and **S. S. Kocha**, 2010, "Experimental methods for quantifying the activity of platinum electrocatalyst for the oxygen reduction reaction," *Anal. Chem.*, **82**, 6321–6328.

[20] **I. Takahashi** and **S. S. Kocha**, 2010, "Examination of the activity and durability of PEMFC catalysts in liquid electrolytes," *J. Power Sources*, **195**(19), 6312–6322.

[21] **S. Satyapal**, 2009, *Hydrogen Program Overview*, 2009 DOE Hydrogen Program and Vehicle Technologies Program 2009, available from http://www.hydrogen.energy.gov/pdfs/review09/program_overview_2009_amr.pdf.

[22] **S. Satyapal**, 2009, *Fuel Cell Project Kickoff*, US DOE, EERE Fuel Cell Technologies Program 2009, available from http://www1.eere.energy.gov/hydrogenandfuelcells/pdfs/satyapal_doe_kickoff.pdf.

47 Solar fuels

Christian Jooss[1] and Helmut Tributsch[2]

[1]Institute of Materials Physics, Georg August University Göttingen, Germany
[2]Free University Berlin and Helmholtz Center Berlin for Materials and Energy

47.1 Focus

Solar fuels are substances that store solar energy in the form of usable chemical energy. For them to be appropriate substances have to meet various requirements, which include the ability for their efficient production, sufficient energy density, and flexible conversion into heat, electrical, or mechanical energy. An essential requisite is environmental friendliness in order to sustainably incorporate conversion products into the global circulation of matter of the biosphere. This can be fulfilled now only by H_2 and to some extent by some carbohydrates directly or indirectly produced by a solar energy source. Direct conversion of solar energy into free chemical energy – either by hydrogen or by hydrocarbon production – requires the development of efficient catalysts for the oxidation of water. This represents a huge materials design challenge, because multiple requirements for catalyst materials must be addressed simultaneously. After an introduction concerning the materials requirements for solar fuel production, storage, transport, and consumption, this chapter focuses on the topic of water-oxidation catalysis: what we can we learn from evolution for the development of an artificial oxygen-evolution center?

47.2 Synopsis

Solar fuels are substances that store solar energy as usable chemical energy at rates that allow sustainable conversion into other forms of energy. Oxidation of water driven by solar light, in order to gain a proton source for either hydrogen or hydrocarbon formation, seems to be the key issue. There are three fundamental steps in the conversion of solar energy into chemical energy (the first two of which also apply to the conversion of solar energy into electrical energy in photovoltaic cells; see Chapter 18). The first is light capture – absorbing the sunlight and transforming it into chemical energy of excited electron–hole pairs. The second is electron and hole transfer – separating and transporting sunlight-excited electrons and holes from their original sites in order to use them. The third is catalysis – the efficient generation and breaking of chemical bonds using the electrons and holes produced in this process to reduce and oxidize compounds, respectively. Although much progress has been made in catalytic control of reaction paths for the production of hydrocarbon and ammonia compounds, detailed understanding of the atomic processes occurring at catalysts and materials design of highly active and specific catalysts are still in their early stages. Here, the outstanding problem is the control of multielectron-transfer reactions, such as the evolution of oxygen from water, which requires four electrons for the liberation of one molecule of oxygen. Thermodynamics promise an energetically most favorable possibility for the oxidation of water when all four electrons are extracted in a correlated way with a minimum identical energy input (1.23 eV). Multielectron transfer requires a well-balanced coordinated chemical reaction involving multiple electron transfers through a sufficiently complex reacting chemical catalyst.

Each fundamental step of solar energy conversion, including the multielectron-transfer mechanisms, builds on the preceding ones, so if all could be done in a single or multi-component material, this would be most efficient. The development of such catalysts is a huge challenge, because of multiple requirements for the catalyst material that must be simultaneously addressed (e.g., semiconductor energy band-gap, band-edge position at the surface, stability against corrosion, low overpotential, catalytic activity for multielectron transfer).

47.3 Historical perspective

The main form of fuel that historically has been used by humans is carbohydrates, which are formed by photosynthesis. In the early history of humans this was primarily in the form of biomass. Exhaustive and non-sustainable use of wood for firing and wood charcoal for metal production produced severe environmental damage, e.g., in Roman times in the Mediterranean area and also during the sixteenth to eighteenth centuries in central Europe, the UK, and, later, in areas of the USA, due to deforestation (Figure 47.1). The industrialization of northern countries and their expanded energy demand was then related to an increasing use of fossil fuels in the form of coal carbon and hydrocarbons, which had been produced by photosynthetic energy conversion in the distant past.

The origin of technologies that are relevant for the production and use of sustainable solar fuels goes back more than 200 years. Since solar fuels depend on a large variety of different technologies, it is not the goal of this section to give an overview of this history but, rather, to show that some of these technologies lay dormant not due to scientific reasons but more because of political and economical causes.

Figure 47.1. Sustainable (top) and environmentally damaging (bottom) use of biomass. The top photo shows the fabrication of palm oil after deforestation of natural rain forest in Indonesia. The bottom photo shows the biomass reactor of the German village Jühnde, which sustainably converts biomass to heat and electricity using about 10% of the agricultural area for energy farming. (Credits: top *by Nick Lyon*, www.films4.org; bottom: Institut für Bioenergiedörfer Göttingen e. V., Benedikt Sauer.)

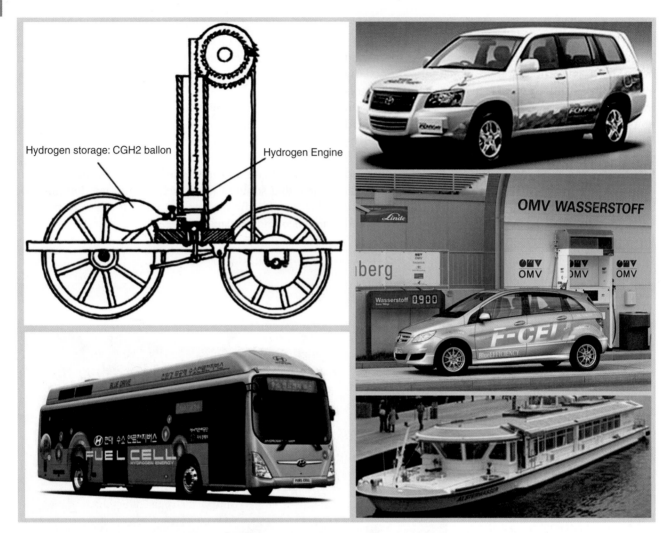

Hydrogen storage: CGH2 ballon

Hydrogen Engine

OMV WASSERSTOFF

Figure 47.2. Historical and actual examples for the development of hydrogen vehicles. Top left: the first vehicle driven by a combustion engine, constructed by Francois Isaac de Rivaz, in France in 1807. Top right: the Toyota FCHV vehicle (2008), fueled with 156 l compressed H_2; it has a range of 830 km and a maximum speed of 155 kph. Middle right: the Mercedes-Benz B-Class F-Cell (2009), which was produced in a small series. It is fueled by compressed H_2, and has a range of 400 km and a top speed of 170 kph. Bottom left: the Hyundai second-generation Fuel Cell Bus (2009) with a capacity of 28 persons, a range of 360 km, and a top speed of 100 kph. Bottom right: Eco-boat Alsterwasser, Hamburg, Germany (2008), a fuel-cell-driven electro-boat with a compressed-H_2 tank. Credits: middle left, Daimler AG; all others, www.H2mobility.org.

The production of H_2 by electrolysis of water was demonstrated in 1800 by William Nicholson and Anthony Carlisle, who decomposed water into H_2 and O_2 by electrolysis with a voltaic pile.

Indeed, the first vehicle driven by a combustion engine was fueled by H_2. It was constructed in 1807 by Francois Isaac de Rivaz in France and was tested in 1813 by powering it with H_2 gas (Figure 47.2). Rudolf Diesel, who invented the diesel compressor engine in 1897,

used peanut oil for fuel, and he wanted to prove that this was the only fuel source that would ever be required. In fact, biodiesel was used in diesel engines until the 1920s; it was only then that manufacturers decided to make use of petroleum as fuel. Even Henry Ford believed in the concept propounded by Rudolf Diesel, and he built a factory that was equipped for mass production of biofuels. In the 1940s fossil-fuel companies that manufactured petroleum-based products started mass marketing and advertising campaigns, and, since the petroleum products were cheaper than bio-fuels, this soon stopped the production of alternatives.

As early as 1939 Hans Gaffron discovered that certain algae can switch between producing O_2 and H_2, in an environment that starves them of sulfur. This is a trait that is nowadays being further explored for the mass production of H_2.

In 1967 a fuel-cell car was developed by GM, using stored liquid H_2 and a 5-kW fuel cell.

The scientific history of direct conversion of solar light into the chemical energy of hydrogen started in 1967 when Akira Fujishima and Kenichi Honda discovered the photocatalytic properties of TiO_2 for water oxidation (this work was published in 1972 [1]). This material, although very successful for photocatalytic self-cleaning technology, later in fact also distracted researchers from water-splitting research. It involves oxidizing water only by one-electron extraction processes, which is facilitated by TiO_2's large energy gap (3.03 eV for rutile and 3.18 eV for anatase). It absorbs, however, <5% of the sunlight in the ultraviolet (UV) spectral region.

After some initial enthusiasm for research into the field of photocatalysts, mainly in Japan, progress came to a halt. It is being revitalized now due to the urgent need for functioning materials for solar-to-chemical energy conversion. The idea, however, is that TiO_2 should serve as a stable matrix for incorporated catalytic water-splitting centers, which support the utilization of visible light [2]. Such centers are expected to require less energy for water splitting than the TiO_2-only process, because they can operate near the four-electron oxidation potential for water (see Section 47.4.9).

47.4 Materials basics for solar fuels

47.4.1 The definition of a solar fuel

A solar fuel is a renewable substance that releases usable chemical energy through an oxidation–reduction reaction with an oxidizer. In a combustion (burning) reaction the fuel is burnt in oxygen and chemical energy is transformed into heat energy. All combustion reactions are exothermic. Explosions are forms of combustion. In an explosive combustion reaction, the fuel is exploded (as in a car engine), releasing mechanical energy. The efficiency is given by the Carnot efficiency (see Chapter 48) and can theoretically be as high as 60% in ideal systems [3]; however, practically achievable efficiencies in piston engines are only about 30%.

In a fuel-cell reaction the chemical energy is transformed in a more controlled way into electrical energy (see Chapter 46). Since fuel cells do not operate on a thermal cycle, the conversion efficiency can be much

higher and is proportional to the cell voltage divided by the Gibbs free energy, ΔG, of the redox reaction. The maximum theoretical efficiency can be as high as 83% for an ideal fuel cell operating reversibly on pure H_2 and O_2 under standard (pressure and temperature) conditions in the case of an H_2–O_2 reaction. The practical efficiency of a fuel cell depends on the amount of power drawn from it and the purity of the fuel and oxidizer. An H_2 cell operated at $U = 0.7$ V has an efficiency $\eta = Ue/\Delta G$ of typically 50%–60%, releasing the remaining energy as heat.

47.4.2 Contrasting solar fuels to nonrenewable fuels

Fuels can be divided into renewable and nonrenewable fuels. Fossil fuels such as coal, oil, and natural gas mainly consist of black carbon and hydrocarbons in the form of C_nH_{2n+2}. They were formed from the products of natural photosynthesis, as over periods on the order of hundreds of millions of years these products were modified by geological activity such as high temperature and pressure under anaerobic conditions. In addition to fossil fuels, nuclear fuels such as ^{235}U and ^{239}Pu belong to the class of nonrenewable fuels (see Chapters 13 and 15). The application of both types of fuels is related to severe long-term environmental damage such as accumulation of CO_2 in the atmosphere (see Chapter 1) and pollution, e.g., penetration of radioactive and toxic nuclei into the biosphere. Apart from the environmental problems, there is another drawback, namely the limited reserves of oil, natural gas, and coal (~10^{23} J [4]), and the fact that the current world annual need for energy is high (~5×10^{20} J per year in 2007) and rising.

As discussed in Chapters 6 and 17, solar energy provides 3.8×10^{24} J per year at the Earth's surface, constituting an immense renewable energy resource. Solar fuels resulting from the conversion of photon energy into chemical energy may become the only future renewable type of fuel. This may include solar H_2, carbohydrates $(C_m(H_2O)_n)$ formed by photosynthesis and converted by plants into sustainably produced biomass, followed by their conversion into more usable forms such as biogas or alcohols (see Chapters 25 and 26), and various artificial solar fuels such as methane (CH_4), alcohols ($C_nH_{2n+1}OH$), other C-based compounds, and ammonia (NH_3) (see Section 47.4.5 in this chapter and Chapter 39).

47.4.3 Biomass as a solar fuel

Historically, and still also today, the main form of solar fuel is biomass. The annual solar energy stored via photosynthesis in biomass is ~3×10^{21} J per year,

Table 47.1. An overview of formation processes for various possible solar fuels based on H_2 or hydrogen compounds, using solar light or atmospheric N_2 or CO_2 from the atmosphere or biomass, respectively. The Gibbs free-energy change, $G°$ per mole, under standard (pressure and temperature) conditions the number of electrons passed during the conversion reaction, n, and the average energy per electron for the uphill reaction are given. The wavelength, λ_{max}, indicates the lowest photon energy required for the given number n of electrons passed during the conversion reaction. In photosynthesis the reaction splitting water to give $\frac{1}{2} O_2$ is performed with four, rather than with two, electrons, allowing the use of red and infrared photons

Fuel	Reaction	ΔG (kJ mol^{-1})	n	ΔE (eV)	λ_{max} (nm)
Hydrogen	$H_2O \rightarrow H_2 + \frac{1}{2}O_2$	237	2	1.23	611
Formic acid	$CO_2 + H_2O \rightarrow HCOOH + \frac{1}{2}O_2$	270	2	1.40	564
Formaldeyhde	$CO_2 + H_2O \rightarrow HCHO + O_2$	519	4	1.34	579
Ethanol	$2CO_2 + 3H_2O \rightarrow C_2H_5OH + 3O_2$	1,371	6	2.36	515
Methanol	$CO_2 + 2H_2O \rightarrow CH_3OH + \frac{3}{2}O_2$	702	6	1.21	617
Methane	$CO_2 + 2H_2O \rightarrow CH_4 + 2O_2$	818	8	1.06	667
Glucose	$CO_2 + H_2O \rightarrow \frac{1}{6}C_6H_{12}O_{6(8)} + O_2$	480	4	1.24	608
Ammonia	$N_2 + 3H_2O \rightarrow 2NH_3 + \frac{3}{2}O_2$	679	6	1.17	629

Source: A. J. Nozik, National Renewable Energy Laboratory, USA and for ethanol.

$<10^{-3}$ of the available solar energy [5]. In other words, the low efficiency of the complete photosynthetic process combined with the relatively low efficiency of biomass-to-fuel conversion leads to a solar-to-fuel conversion efficiency of ~0.1%.

How much of the 3×10^{21} J of chemical energy stored in photosynthetic biomass per year can be used in a sustainable way depends not only on factors such as energy balance, emissions, and efficiencies of conversion of biomass into usable gas and liquids, but also on economical implications and social aspects. In a world where still about one-sixth of all human beings are undernourished, competition between food and biofuel is a big issue (see Chapter 25). The goal of ecological sustainability is even more imperative if we consider the problem of disparity. The mobility of humans, for example, depends enormously on the availability of *economically affordable* fuels. Thus, the average US resident consumes twice as much energy as the average Western European, eight times more than the Chinese, 15 times more than the Indian, and a staggering 30 times more than the average African person. Disparity in affordability of fuel has a deep impact on the infrastructure of societies, e.g., the organization of the relation between the workplace and living areas.

Humans already use 30%–40% of net produced biomass as food, feed, fiber, and fuel, which corresponds to between 12% and 18% of the actual world's primary energy supply [6]. There is a gap of solar fuel production of ~4 × 10^{20} J per year for the transition to a renewable-fuel basis. Even with enhanced conversion efficiencies for biofuel production and large-scale use of biomass waste, clearing sludge, animal wastes, and wood it will be hard to provide ALL the required amounts of renewable chemical energy necessary to fuel an industrializing world economy and to cover the needs of today's world population and that of the future. The actual contribution of artificially produced solar fuels such as H_2 and carbohydrates is, however, still vanishingly small, and strenuous efforts will be needed in order to overcome this bottleneck hindering passage to a sustainable energy basis.

47.4.4 Definition of basic properties – overview

There are several important material properties that characterize fuels in general and solar fuels in particular. The most important for a chemical to be called a fuel is the free energy, ΔF, per mole of the oxidation reaction. Table 47.1 gives an overview of the Gibbs free energy per mole for various possible solar fuels based on H_2, hydrocarbon, and ammonia compounds. Uphill reactions for direct production from H_2O, CO_2, and N_2, together with the numbers of charge carriers transferred, are given. The combustion of fuels is usually related to the formation of gases, which have to displace other gases by pushing them out against a pressure. The appropriate quantity for describing the energy content of a fuel with respect to a specific oxidation reaction is the change in free enthalpy (also called Gibbs free energy) of the process, $\Delta G = \Delta F + pV = \Delta H - T\Delta S$, where pV gives the work due to volume change under constant pressure, ΔH is the enthalpy and ΔS the entropy change due to release of heat (see Appendix A).

For mobile or stationary energy-storage applications three different types of specific energies are very relevant, the enthalpy per mole h, the enthalpy mass density $h_m = \Delta H / m$, and the enthalpy volume density $g_v = \Delta H / V$. For chemical compounds such as metal hydrides and metal amides, which are decomposed to liberate their H_2 or NH_3 content, the compound is considered as an integral part of the fuel. Corresponding h_m and g_v values are shown in Table 47.2.

There are also external parts that enter the mass and volume energy densities of a fuel. The ΔH of a fuel cannot be defined without selecting the oxidizer, which is usually atmospheric O_2. Since the oxidizer is not transported as part of the fuel, the mass or volume contribution of the oxidizer is not included in the specific free energy. However, other external components for fuel storage (such as vessels or chemical carriers) and components making it accessible for the reaction (such as heaters or pressure transducers) must be considered. This is the so-called effective enthalpy density, h_{eff}, of a fuel. The values given in Table 47.2 are only rough estimates, since the exact values strongly depend on the details of the storage technology used.

The reaction of a fuel in a combustion engine or fuel cell requires sufficient mobility of the chemical species and, thus, a liquid or gaseous state of matter. With respect to small g_v values, the liquid or compressed gaseous states are favorable. For fuel selection, the vapor pressure depends on the temperature, and the boiling temperature T_B is a key parameter. For fuels that are stored in a compound, the desorption temperature T_{Des} must also be considered. Other quantities, which are mainly important for daily usage of a fuel and safety requirements, are summarized for the examples of H_2, CH_4, and gasoline in Table 47.3.

47.4.5 Properties of solar fuels and synthesis routes

In the following sections, properties, application conditions, and fabrication routes of various solar fuels are discussed by way of examples, to give the student some insights into the background for contemporary challenges and developments.

Hydrogen

Hydrogen, H_2, has a long tradition as an energy and chemical raw material. Its free-enthalpy density per mass of 141.5 MJ kg^{-1} is much higher than that of, e.g., gasoline (45.7 MJ kg^{-1}) and thus was it singled out early

on as the ideal fuel where the weight plays a dominant role, e.g., in automotive transport and spacecraft. The free-enthalpy density per volume strongly depends on the state of the H_2. Liquid hydrogen has a much smaller volume free-enthalpy density, 9.6 GJ m^{-3}, than does gasoline (~33 GJ m^{-3}). It is cryogenic and boils at 20.27 K. It can also be stored in gaseous form, compressed gaseous form, as a chemical hydride, and in some other hydrogen-containing compounds such as hydrocarbons and ammonia. Hydrogen gas forms explosive mixtures with air in the concentration range 4%–74% (volume percentage of H_2 in air). The mixtures are spontaneously detonated by a spark, heat, or sunlight. Pure H_2–O_2 flames emit UV light and are nearly invisible to the naked eye. The detection of a burning H_2 leak may thus be an issue, but, because H_2 is buoyant in air, H_2 flames tend to ascend rapidly and thus cause much less damage than do hydrocarbon fires. With respect to emission of undesired byproducts, H_2 is the most environmentally friendly fuel. In a combustion machine, only small amounts of NO_x, which are formed from the molecular N_2 in the air, are emitted. H_2 can be converted to water in almost all currently available types of fuel cells. The controlled catalytic conversion of H_2 and O_2 into water is an emission-free clean technology.

At present, almost all of the ~70 million metric tonnes of H_2 per year used in industrial processes stems from fossil sources. The current dominant technology for direct production is steam re-forming of methane or natural gas, whereby at high temperatures (700–1,100 °C) steam (H_2O) is reacted with methane (CH_4) to yield syngas ($CH_4 + H_2O \rightarrow CO + 3H_2 + 191.7$ kJ mol^{-1}). In a second stage, further H_2 is generated through the lower-temperature water-gas-shift reaction, which is performed at about 130 °C: $CO + H_2O \rightarrow CO_2 + H_2 - 40.4$ kJ mol^{-1}. An intermediate step toward fully renewable H_2 production may be solar upgrading of fuels for the generation of electricity [10] or solar-assisted steam methane re-forming for H_2 production. The various pathways for H_2 production using solar light are described later in this chapter.

Ammonia

Ammonia, NH_3, has been proposed as a potential fuel [7][11] that can be cheaply produced from pure hydrogen or syngas (a mixture of CO and H_2) via the well-established Haber–Bosch process. It can be considered as a chemical hydrogen-storage compound. By adding nitrogen, NH_3 is formed over catalysts and then separated, with no energy penalty, via condensation. Similarly to H_2, NH_3 can be used as an energy carrier and storage medium because NH_3 can potentially be combusted in an environmentally benign way, exhausting H_2O, N_2, and NO_x emissions, which must be further reduced to N_2. NH_3 can be used directly as a fuel in alkaline [12] and

Table 47.2. A comparison of the pressure p, mass density ρ, molar mass M, molar enthalpy density h of the oxidation reaction, mass free-enthalpy density h_v and estimated effective free-enthalpy density h_{eff} for various solar and fossil fuels. If not noted otherwise, quantities are given at standard conditions (1 bar pressure, 298.15 K)

Fuel	Storage system	P (bar)	ρ (kg m⁻³)	M (kg mol⁻¹)	g (kJ mol⁻¹)	g_m (MJ kg⁻¹)	g_v (GJ m⁻³)	g_{eff} (GJ m⁻³)	T_B (K)	T_{Des} (K)
Octane[a] C_8H_{18}	Liquid tank	1	703	0.114	5430	47.5	33.4	37.6	399.0	
Methane CH_4	Gas-storage system[b]	250	188	0.016	891.8	55.6	10.5	27.0	109.0	
Methanol CH_3OH	Liquid tank	1	790	0.032	637.6	19.9	15.7	16.0	338.0	
Hydrogen H_2	Cryogenic-liquid tank	>1	67.8	0.00202	285.8	141.5	9.6	30.0	20.3	
Hydrogen H_2	Gas standard condenser	1	0.085	0.00202	285.8	141.5	0.01	112.0	20.3	
Hydrogen H_2	Gas pressurized	300	21.5	0.00202	285.8	141.5	3.04	35.0	$-$[c]	
Hydrogen H_2	Metal hydrides[d,e]	10–200	700–5,000	0.017–0.053	85.0–1,500	5–65	20.0–45.0	3.7–50		300–600
Hydrogen H_2	Metal hydride[d] $LiBH_4$	155	669	0.022	1,413	64.8	43.4	52.0		500.0
Ammonia NH_3	Gas pressurized	10	603	0.017	679	39.8	24.0	32.0	240.0	
Ammonia NH_3	Metal ammine[d] $Mg(NH_3)_6Cl_2$	1	610	0.197	4,204	21.3	13.0	18.0		620.0
Formic acid HCOOH	Liquid tank	1	1,220	0.046	270	5.9	7.2	5.7	374.0	
Ethanol C_2H_5OH	Liquid tank	1	789	0.046	1,371	29.8	23.5	18.8	352.0	

[a] A major component of gasoline, which is a mixture of C_nH_m (n = 4–8) compounds, with average $h_m = 44$ MJ kg⁻¹.
[b] Compressed natural gas or biogas mainly contains CH_4; at 1 bar h_v is only 0.0378 GJ m⁻³.
[c] State is above critical point at 33 K and 12.98 bar.
[d] Properties include the metal compound because it is considered an integral part of the fuel.
[e] Range of values for various metal hydrides.
Sources: [7], plus [8] for ammonia and [9] for the metal amide.

Table 47.3. Properties of solar and fossil fuels

Property	Hydrogen	Methane	Gasoline
Self-ignition temperature (°C)	585	540	228–501
Flame temperature (°C)	2,045	1,875	2,200
Ignition limits in air (vol%)	4–74	5.3–15	1.0–7.6
Minimal ignition energy (mJ)	0.02	0.29	0.24
Flame propagation in air (m s^{-1})	2.65	0.4	0.4
Detonation limits (vol%)	13–65	6.3–13.5	1.1–3.3
Explosion energy (kg TNT m^{-3})	2.02	7.03	44.22
Diffusion coefficient in air (cm^2 s^{-1})	0.61	0.16	0.05

solid-oxide fuel cells (SOFCs) [13] to produce steam and NO_x as exhausts (see Chapter 46); the NO_x can be at least partially reduced by Pt catalysts. To have NH_3 as a fuel source for proton-exchange-membrane (PEM) fuel-cell (see Chapter 46) vehicles, it is first cracked catalytically into N_2 and H_2; this reaction is thermally driven by a 350–400 °C heat source [14] fed by a small part of the generated H_2. Internal combustion engines fueled directly with NH_3 must have special technical features because the ammonia's ignition behavior is different from that of conventional fuels. Recent developments include homogeneous charge compression ignition technology and decomposition routes, whereby the mixture of hydrogen, nitrogen, ammonia, and air can be adjusted to have combustion characteristics comparable to those of gasoline.

Until today the primary material used to produce NH_3 has been methane from fossil sources. For application as a solar fuel, a solar hydrogen source or biomass-to-methane conversion is required (see Chapter 39). Furthermore, NH_3 can be produced by special micro-organisms. All three routes avoid CO_2 emissions from fossil carbon sources and the problem of PEM electrode poisoning with CO produced by, e.g., the methanol re-forming process is completely eliminated. As an alternative option, hydrogen can also be recaptured via decomposition of NH_3 using electrolysis [15]. A main advantage of NH_3 is that it can be stored in the liquid state at 8 kbar at room temperature or by absorption in porous metal ammine complexes. The ammine can be shaped in the desired form and has reasonable mass and volume enthalpy densities (it can store 9 kg H_2 per 100 kg and 100 kg H_2 per m^3). This technique has been mentioned as a way to store NH_3 on board for vehicular applications so as to avoid any danger related to a crash. A main drawback with using liquid NH_3 storage in passenger vehicles is the toxicity problem. Leakage may lead to harmful effects to living species.

Methane

Methane, CH_4, is a natural gas fuel. Today, CH_4 provides approximately 30% of the US energy requirement. Methane is mostly used in homes, as gas applied for cooking or for heating systems. Its greenhouse warming potential is 72 times that of CO_2 and, thus, if it were to be used as a fuel on a larger scale, maintenance to keep it from leaking would become an extremely important issue. Methane can be produced from syngas (a mixture of CO_2, CO, and H_2), which can be obtained from coal gasification, from pure CO_2, or by biomass conversion. Apart from the biomass-to-methane conversion via micro-organisms, all routes require a hydrogen source. Solar fuel routes that avoid the liberation of fossil carbon need a solar hydrogen source and can use C obtained either from sustainable biomass or by capturing atmospheric CO_2.

Direct capture of CO_2 from the air to form carbohydrates represents a CO_2-emission-neutral substitute for natural gas or other hydrocarbons. Various technologies exist for extraction of CO_2 from the atmosphere or from exhausts (see Chapter 8 and [16][17]).

Present-day absorption technologies include adsorption on solid sorbents, cryogenic condensation, and membrane separation. They use much energy, and membranes for separation are still not stable. Very promising, but still under development, is absorption in solutions such as, e.g., caustic NaOH scrubbing solution, to form Na_2CO_3. Here, absorption rates of typically 45% can be achieved at an energy demand of 2.7 GJ per t CO_2. CO_2 is recovered from the carbonate produced by acidifying the solution with sulfuric acid. The caustic scrubber solution and the sulfuric acid are regenerated in an electrodialytical unit with bi-polar membranes,

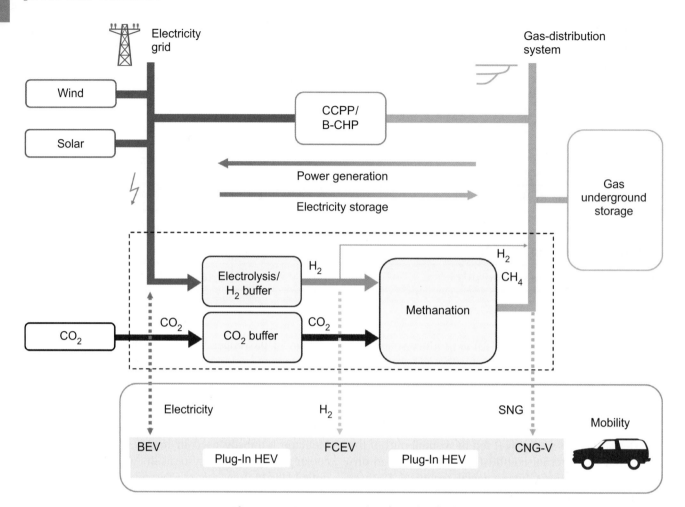

Figure 47.3. An example of a solar fuel application based on renewable electricity, which is produced by fluctuating wind and photovoltaic sources. Power that is not directly transmitted into the electric grid can be used for H_2 production that can then be either directly used as a fuel in fuel cells or converted into hydrocarbons such as methane using atmospheric CO_2. CCPP, combined-cycle power plant; B-CHP, block-type combined heat and power station; EV electric vehicle; BEV, battery electric vehicle; FCEV, fuel-cell electric vehicle; SNG substitute natural gas, CNG-V, compressed-natural-gas vehicle; Plug-In HEV, plug-in hybrid electric vehicle. Credit: M. Specht, Center for Solar and Hydrogen Research, Stuttgart, Germany.

consuming 7 GJ per t CO_2. This two-stage neutralization process enables full CO_2 recovery, and a modular setup of electrodialysis units can be employed [18].

CO_2 methanation can be performed in several ways [19][20], and one of the most important contemporary materials challenges is the development of suitable catalysts. A possible direct route is

$$4H_2 + CO_2 \leftrightarrow CH_4 + 2H_2O \quad (\Delta h = -164.9 \, \text{kJ} \, \text{mol}^{-1}).$$

It is composed of a reversed endothermal water-gas-shift reaction, $H_2 + CO_2 \leftrightarrow CO + H_2O$ ($\Delta h = 41.5$ kJ mol^{-1}) and an exothermal CO methanation, $3H_2 + CO \leftrightarrow CH_4 + 2H_2O$ ($\Delta h = -206.4$ kJ mol^{-1}). This reaction is called the Sabatier process and was discovered in 1913.

It is still being studied, but laboratory tests show that CO_2 methanation rates of up to 95% at a pressure of 5–10 bar at 250–500 °C can be achieved.

Combining the efficiencies of electrolysis and methanation results in a renewable-power methane-generation efficiency in the range of 46%–86%, on average 63%, which is further reduced by about 15% when using atmospheric CO_2 in highly efficient absorption technologies. A recent study shows routes for methane production, using wind-power electricity and atmospheric CO_2, with an efficiency of 61.6% and using another 11.7% as process heat [21]. An interesting alternative currently under development represents conversion of CO_2 dissolved in seawater. Since ocean water contains about 140 times the atmospheric CO_2 concentration, a direct conversion using electrolysis combined with well-suited catalysts is very promising.

CO_2 reduction

CO_2 reduction, either to CO for syngas production or directly to hydrocarbons other than methane, would be an important step in recycling emitted CO_2 and creating a basis for solar fuels. Since the thermal route by reversing the water-gas-shift reaction yields only a small output (see Chapter 39), efficient photocatalysts using solar light are highly desirable. This requires a well-catalyzed chemical process, since the first electron-transfer step to CO_2 is highly inhibited and a full multielectron transfer is difficult to achieve. Since the early work of Honda and co-workers using semiconductors such as TiO_2, GaP, and CdS [22], some progress has been achieved recently by using nanostructured and heterostructured catalysts, e.g., Cu/Pt-doped nano-tube arrays [23] or $CdSe/Pt/TiO_2$ heterostructures [24].

Formic acid

Formic acid, HCOOH, is the simplest carboxylic acid and an important intermediate in chemical synthesis of hydrocarbons [25]. With respect to currently installed technologies for methane production from atmospheric CO_2 it may play an important role as an intermediate to build up a thermodynamically and kinetically favorable process route

$$CO_2 \rightarrow HCOOH \rightarrow CH_2O \rightarrow CH_3OH \rightarrow CH_4$$

for production of methanol and methane. Formic acid can be stored in liquid form under standard conditions; however, it is a skin- and eye-irritant in concentrated form. In fuel applications, it can be directly decomposed into CO_2 via direct formic acid fuel cells. Alternatively, H_2 can be generated from formic acid–amine adducts at room temperature in fuel cells using ruthenium phosphine systems as catalysts in this transformation [26].

Ethanol

Ethanol, C_2H_5OH, is a volatile, colorless liquid. It burns with a smokeless blue flame. Complete combustion of ethanol forms carbon dioxide, water, and NO_x as a waste product. It is used as a transport fuel, mainly as a biofuel additive for gasoline. World ethanol production for transport fuel increased between 2000 and 2009 from 17 billion to 73.9 billion liters, increasing the share of ethanol in global gasoline-type fuel use to more than 5%. This development created serious food-related arguments due to mass production of corn and sugar-cane in agricultural areas, as addressed above. Ethanol is produced by microbial fermentation of the sugar,

followed by distillation and further dehydration steps to decrease the water content to a level below ~4%. Fermentation of glucose by micro-organisms consists of several steps, including the breaking up of the glucose molecule (glycolysis), and can be summed up by the formal reaction

$$C_6H_{12}O_6 \rightarrow 2C_2H_5OH + 2CO_2.$$

There has been a wide debate about how much greenhouse-gas emission is avoided by replacing gasoline by ethanol. On the basis of a total-life-cycle assessment, we conclude that this depends on many details such as the type of biomass feedstock, transport processes, and the process used for biomass-to-ethanol conversion. Typically reductions of greenhouse-gas emissions using corn ethanol instead of gasoline are only between 18% and 28%; in some cases the actual CO_2 emissions are even increased. An attractive approach for sustainable ethanol production is based on hydrogenation of carbon dioxide. This requires catalysts such as, e.g., an Rh-based catalyst, an Fe-based modified Fischer–Tropsch catalyst, or Cu-based modified methanol-synthesis catalysts [27].

Other reduced-C materials

Saturated hydrocarbons (see Box 47.1) are the basis of petroleum fuels and are found as either linear or branched species. Hydrocarbons in the form of gasoline or diesel can chemically be generated from carbohydrates via pyrolysis and additions of H_2. Alternative, but well-established, procedures for forming alkanes from solid forms of carbon or biomass are based on the **Fischer–Tropsch process** (see Chapter 39). It

---

Box 47.1. Carbohydrates and hydrocarbons

Carbohydrates, $C_m(H_2O)_n$ (formed, e.g., in photosynthesis), **hydrocarbons**, C_nH_{2n+m} (the main naturally occurring compounds in crude oil and natural gas), and their interconversion may be part of a solar fuel route whereby reduced C is used to chemically store hydrogen. Carbohydrates are synonymous with saccharides, which can be divided into different groups (monosaccharides, disaccharides, oligosaccharides, and polysaccharides). In general, the monosaccharides and disaccharides, which are smaller (lower-relative-molecular-mass) carbohydrates, are commonly referred to as sugars [28]. Hydrocarbons are organic compounds consisting entirely of hydrogen and carbon, such as methane, CH_4, and ethylene, C_2H_4. Saturated hydrocarbons (C_nH_{2n+2} alkanes) are the simplest hydrocarbon species, possessing only single C—C bonds.

involves a series of chemical reactions that lead to a variety of hydrocarbons such as alkanes:

$$(2n + 1)H_2 + nCO \rightarrow C_nH_{2n+2} + nH_2O \quad (n > 1).$$

Historically, the major H_2 source, which was simultaneously used to adjust the H_2:CO ratio necessary for Fischer–Tropsch catalysis, was the water-gas-shift reaction (see above).

Fischer–Tropsch-based conversion using a solar hydrogen source combined with a sustainable carbon process, which can be incorporated into the global carbon cycle without producing imbalances and CO_2 accumulation, may turn out to be an important element in the necessary rapid global of a construction means of solar fuel delivery.

47.4.6 The solar fuel life cycle

All solar fuels that we have discussed are based either on pure H_2 or on hydrogenation of C or N atoms. All these fuel molecules have in common that they possess a large enough Gibbs free energy for performing exothermic reactions in combustion so that atmospheric O_2 can be used as the oxidizer and that the reaction products can be controlled sufficiently and can become part of the naturally existing global C or N cycle. They have further in common that their production requires the reduction of C or its oxides, or N or its oxides and, thus, the availability of hydrogen sources.

The challenge of solar fuel production can then be outlined in the simplified way given in Figure 47.4. An essential factor is the use of solar radiation to generate primary electron flow in the form of electron and proton activity by consuming H_2O and liberating O_2. Such energy technology, indicated as photovoltaics and photocatalysis in Figure 47.4, is an extension of primary natural photosynthesis. Hydrogen can now be attached to CO_2, to form carbohydrates or hydrocarbons. However, some bacteria activate electrons and protons to directly generate H_2, albeit with a modest energy efficiency. At present it seems much more efficient to produce H_2 sustainably via solar or other renewable energy sources, such as wind energy or ocean power. If hydrogen is added to thermolyzed biomass, then gasoline, diesel, or any other chemical can be synthesized. If hydrogen is catalytically added to CO_2, as can be done via some bacteria or via photocatalytic of reduction CO_2, fuels and chemicals can be generated. H_2 itself can conveniently be used in fuel cells for electricity generation and electricity can directly be generated from primary and secondary solar energy for H_2 generation via electrolysis. Solar heat as well as waste heat can also assist diverse technological processes, e.g., thermolysis and heat-assisted electrolysis of water.

Most of these energy-conversion processes have, in principle, been demonstrated, but require significant further improvement. A key challenge remains catalysis, especially multielectron-transfer catalysis for splitting water (oxidation of water to O_2) and the combination of CO_2 with H_2 for fuel production. Another challenge is, of course, cost reduction, especially with the solar generation of electricity, the production of hydrogen fuel cells, and the generation of fuels from hydrogenated biomass.

Summarizing the challenge for thermodynamics, we conclude that, for the uphill reaction for the generation of H_2 and reduced C, suitable energy must come from light or electrical energy generated in a sustainable way.

The kinetics is usually extremely slow, and very efficient catalysts are necessary in order to avoid the need for high reaction temperatures [29]. Nature has evolved ways to operate all essential catalytic reactions for energy conversion at ambient or body temperature, which shows that such energy-conversion catalysts are possible.

47.4.7 Catalysis – controlling reaction paths

Catalysis, be it homogeneous catalysis, heterogeneous catalysis, enzyme-based catalysis, photocatalysis, or electrocatalysis, will be the basic science enabling a breakthrough in large-scale production of solar fuels. Leaving aside for a moment all of the detailed technical, environmental, and economic requirements which were outlined above and coming back to the fundamental mechanisms involved in the production, storage, and consumption of solar fuel (see also Chapter 39), the suitability of a specific substance serving as a solar fuel is determined by both of the following.

- Thermodynamics, i.e., the free energy needed for formation and consumption of solar fuel. This basically determines the general potential of a redox couple for energy storage.
- Kinetics, i.e., control of the reaction paths, avoiding undesired side products and energy losses and minimizing activation barriers during production and consumption of solar fuel. In principle, this can be achieved by developing suitable catalysts with high selectivity. Control of the reaction kinetics mainly determines the practical efficiencies in production and consumption processes and how solar fuels become part of the global material cycles involved in the biosphere and the chemical industry.

Thermodynamic calculations and arguments focus only on the initial and final states of a system (see Appendix A). The path by which a change takes place is not considered. Intuitively, for exothermic processes one might expect a spontaneous reaction. Usually, due to the presence of activation barriers, this is not true.

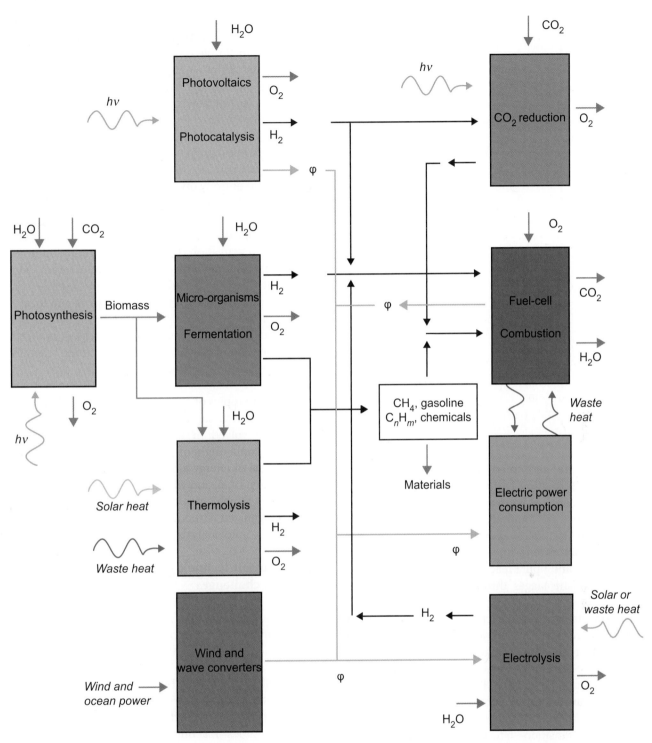

Figure 47.4. An energy and process scheme explaining the generation, transformation, and consumption of solar fuels in a sustainable energy economy. Electricity production is denoted by φ and solar radiation by $h\nu$.

For example, combustion of CH_4 or H_2 does not proceed spontaneously, but needs to be spark- or flame-initiated. Once the process has started, the heat produced by the combustion serves to maintain the reaction until one or both of the reactants have been completely consumed. In fuel cells, well-designed catalysts can reduce activation barriers and may

enable controlled conversion of chemical energy into other useful energy with a minimum of dissipation (see Chapter 46).

This also applies to endothermic processes such as the production of solar fuels. The underlying uphill reactions can take place only with sufficient energy input; however, the reaction path determines whether the

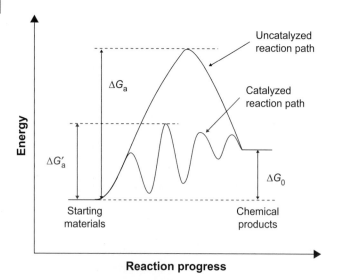

Figure 47.5. Reaction paths with and without catalyst for the example of an uphill reaction, where ΔG_0 represents the difference in Gibbs free energy between the starting materials and the formed products in thermodynamic equilibrium under standard conditions; ΔG_a and $\Delta G'_a$ are the activation Gibbs free energy without and with catalyst, respectively. In solar fuel production, ΔG_0 may be delivered by solar photons, electric energy, or solar heat.

process can be performed near the thermodynamic value of the formation enthalpy. In endothermic reactions, the presence of barriers determines whether large excess energies are required, e.g., the presence of over potentials in H_2O splitting and H_2 production by electrolysis or photocatalysis.

Downhill and uphill reactions have in common that the intermediate stages through which reacting molecules pass on the way to products are described by transition states. Every reaction in which bonds are broken will necessarily have a higher-energy transition state on the reaction path that must be traversed before products can form. This is true both for exothermic and for endothermic reactions. To allow the reactants to reach this transition state, energy must be supplied from the surroundings. The energy needed to raise the reactants to the transition-state energy level is called the activation Gibbs free energy, $\Delta G_a = \Delta H_a - T \Delta S_a$, where ΔH_a denotes the activation enthalpy and ΔS_a the activation entropy. In transition-state theory, the reaction rate r is given by [30]

$$r = \kappa \, \frac{k_B T}{h} \, e^{-\frac{\Delta G_a}{RT}} = \kappa \, \frac{k_B T}{h} \, e^{-\frac{\Delta H_a}{RT}} \, e^{\frac{\Delta S_a}{RT}}.$$

Here $k_B T/h$ is a frequency factor for crossing the transition state (equal to about 6 ps^{-1} at 300 K in solution as well as in the gas phase), k_B is Boltzmann's constant, T is the absolute temperature, h is Planck's constant, R is the

ideal gas constant, and κ is the generalized transmission coefficient that relates the actual reaction rate to the rate obtained from simple transition-state theory ($\kappa = 1$).

Catalysts are providing alternative reaction routes and thus a pathway for the reduction of ΔG_a (see, e.g., [31]). Since the rate at which chemical reactions proceed is exponentially dependent on their activation enthalpies and entropies, catalysts can drastically enhance the rate of a chemical reaction. Various microscopic mechanisms contribute to the change of transition-state enthalpy and entropy due to catalysts. Examples involve the change of molecular orientations or bond strain, manipulation of the transition state by proton or electron acceptors/donors, quantum-mechanical tunneling through an activation barrier, and manipulation of charged transition states by electrostatic forces, e.g., due to the presence of ionic bonds at the catalyst surface. It is also possible to take advantage of the formation of covalent bonds between the active site of the catalyst and one or several of the species involved. The covalent bond must, at a later stage in the reaction, be broken in order to regenerate the active area of the catalyst.

The main topics of actual materials research in catalysis are a microscopic understanding of the atomic processes involved in alternative reaction pathways, the microscopic origin of the entropy term ΔS_a through the dispersal of energy among a large number of translational, rotational, and vibrational quantum states, and approaches by which to overcome entropy production by self-organized structures related to multielectron transfer and autocatalysis [32].

47.4.8 Current technology and materials challenges

Which main technologies for generation of H_2 and its conversion into various solar fuels exist, and what are their efficiencies, challenges, and key problems?

Electrolysis – still mainly based on noble-metal catalysts

The key problem is electrode stability of non-noble materials. Holes reacting at the interface are often equivalent to missing chemical bonds (e.g., taking away electrons from O^{2-} means elimination of a polar bond in an oxide electrode). However, it was shown that holes in d-states do not initially break essential chemical bonds, but can induce coordination-chemical reactions. Generation of a hole at the interface means that the oxidation state of the transition metal concerned is increased, which is compensated for by addition of suitable chemical ligands from the electrolyte. If this ligand is water and if the transition metal can reach a sufficiently high oxidation state, oxygen may be liberated from water.

This is the case for RuO_2 electrodes. In presence of chloride Cl_2 is evolved from water. RuO_2 is technically the best electrode for production of oxygen and chlorine in the dark.

Nature avoids the use of noble metals, and instead performs electrolysis reactions with clusters of abundant transition metals such as Fe, Cu, Mo, Co, and Mn. Enzymes with only Fe as metal ions are applied by certain bacterial hydrogenases for efficient evolution of H_2 (see Chapter 25). An Mn-containing cluster is, as has previously been mentioned, applied for evolution of O_2 (see Chapters 24 and 28). Transition-metal clusters of non-noble elements can replace noble metals because the clusters can accommodate corresponding oxidizing or reducing equivalents needed for the reactions.

Photoelectrolysis and photocatalysis – competition between catalysis and corrosion

Photoelectrolysis requires well-designed photoactive semiconducting materials (see Chapter 49). For the above-mentioned reasons (photogenerated holes are equivalent to missing chemical bonds) most semiconductors (e.g., Si, CdS, GaAs, and GaP) corrode during the process. Holes accumulating at the semiconductor interface decompose the material. If, however, a material with a valence band derived from d-states, such as RuS_2, is selected, then stability is observed with photoevolution of O_2 from H_2O. However, in this case the forbidden energy gap ($E_g = 1.3$ eV) is too small for purely photo-induced splitting of H_2O and an external electric potential is needed in order to drive the reaction.

Photocatalysts still exhibit poor conversion efficiencies. For example, thin TiO_2 surface layers are intensively used for solar cleaning of · interfaces. The UV-light-induced photoreaction with water, which generates radicals that oxidize organic pollutants, is of limited benefit for energy-conversion purposes, because it involves one-electron-transfer reactions from water that require an unfavorably high potential ($E = 2.8$ V) compared with $E = 1.23$ V for a multiple-electron-transfer process.

Solar thermal water splitting – the need for catalysts stable at high temperature

Thermal decomposition, which is also called thermolysis (see Chapter 48), breaks up H_2O into various combinations of H and O atoms, mostly H, H_2, O, O_2, and OH at sufficiently high temperatures. At $p_{H_2O} = 1$ bar, the potential for splitting water into H_2 and O_2 decreases from $\Delta E = 1.23$ V at 25 °C to 0.919 V at 1,000 °C, and 0 V at 4,013 °C [33]. According to the law of mass action, this translates in a shift of the concentrations in the reaction mixture. For example, at 2,200 °C about 3% of all H_2O molecules are dissociated, and at 3,000 °C more than half of the water molecules are decomposed.

At higher water pressure, the decrease of ΔE with increasing temperature is even stronger, e.g., $\Delta E = 0.58$ V at $p_{H_2O} = 1$ bar and 1,000 °C. A major challenge for efficient H_2 generation is the manipulation of the recombination rates of split gaseous species; thus development of catalysts that are stable at high temperatures will decrease the process temperatures.

Thermochemical splitting of water

Solar-powered thermochemical production of H_2 from water uses an oxidation–reduction cycle of a working substance to generate H_2. Typically such a cycle involves a two-step process whereby, e.g., a metal is oxidized in water, liberating H_2. The second step consists of the regeneration of the metal from the oxide compound by a high-temperature metal-oxide-decomposition process. According to a US Department of Energy review more than 352 thermochemical cycles have been demonstrated, including metal oxides, metal sulfides, and metal chlorides. For examples, see Chapter 48. A recently started 100-kW pilot project using concentrated solar power is based on a thermochemical two-step water-splitting process, whereby Fe_3O_4/FeO, ZnO/Zn, and Mn_3O_4/MnO redox couples are used as oxidants, with a calculated efficiency of 43% [34]. Main challenges are the improvement of reaction kinetics, e.g., via the use of porous materials and increasing the number of process cycles that can be implemented while avoiding aging of the material under the applied extreme conditions.

47.4.9 Materials science needs for catalytic O_2 evolution

The main challenge for solar H_2 production is the solar-generated electrochemical O_2 evolution during the water-splitting reaction. While there has been substantial progress in finding efficient catalysts for reduction of the produced H^+ ions to H_2 [e.g., Pt, Ru–chromophore complexes, colloidal metal (Pt, Au) catalysts, metal oxide (e.g., Fe_2O_3, TiO_2) or core–shell-type nanoparticle dye-sensitized oxides such as chromophores on platinized TiO_2 particles, and H_2-evolving enzymes (hydrogenases)], development of an efficient catalyst for O_2 evolution constitutes the bottleneck for direct splitting of water by sunlight, although recent progress in using Mn and Co compounds indicates that a breakthrough may soon occur.

Electrolysis of water always involves both cathodic H_2 evolution and anodic O_2 evolution in series. Both electrochemical steps generate electrochemical potential losses (see Appendix B). They are much more critical and significant for O_2 evolution, which requires the extraction of four electrons from water (see Figure 47.6). As has already been mentioned, a minimum

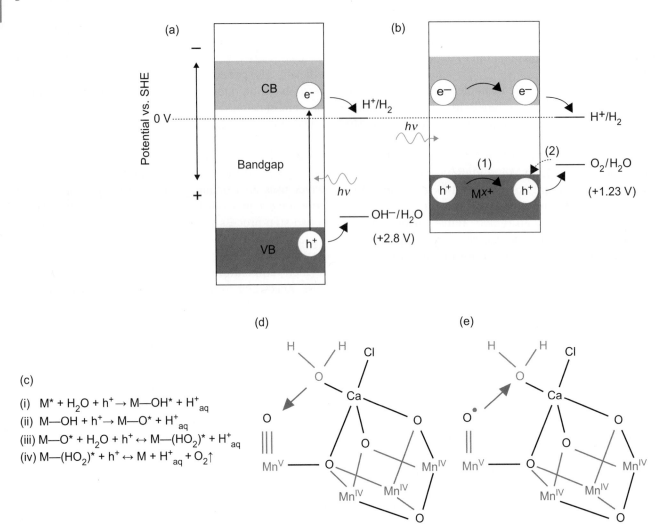

(c)

(i) $M^* + H_2O + h^+ \rightarrow M\text{—}OH^* + H^+_{aq}$
(ii) $M\text{—}OH + h^+ \rightarrow M\text{—}O^* + H^+_{aq}$
(iii) $M\text{—}O^* + H_2O + h^+ \leftrightarrow M\text{—}(HO_2)^* + H^+_{aq}$
(iv) $M\text{—}(HO_2)^* + h^+ \leftrightarrow M + H^+_{aq} + O_2\uparrow$

Figure 47.6. Photocatalytic H_2O splitting within semiconductor Excitation Schemes (a,b). In an appropriate photocatalyst material solar light excites electron–hole pairs, where holes with sufficiently positive potential oxidize water, leading to O_2 evolution, and electrons with sufficiently negative potential reduce protons, leading to H_2 evolution. (a) A process with individual hole transfer involving surface adsorption of unfavorable OH^- intermediates which requires a minimum potential of 2.8 V. (b) A cooperative hole-transfer process at the optimum thermodynamic potential. It must avoid the formation of undesired intermediates and requires a feedback loop (2) for hole transfer and a flexible charge reservoir (1) using different valence states x of a metal M. (c) Mechanistic model for a possible reaction path for the stepwise water oxidation at an active metal site M^*. (d,e) (d) and (e) alternative mechanisms for O=O formation within step (iii) of the reaction (c) (see text).

energy for O_2 evolution (negligible energy loss) would be required if the thermodynamic free energy for O_2 evolution were divided by four, so that each electron would have to overcome only a potential of 1.23 V (Figure 47.6 (b)). This would correspond to an ideal four-electron-transfer reaction, as suggested in Figure 47.6(c). Typically, however, significantly more energy is required. The reason is that extraction of individual electrons from

water generates radicals with unfavorable electronic structure. If only one electron is extracted in a first step, yielding an $OH^·$ radical, approximately 2.8 eV is needed, generating an effective energy loss of 1.57 eV [32] (Figure 47.6(a)). The extraction of the following three electrons will produce additional losses. In terms of free-energy curves versus the reaction coordinate this simply means that unfavorable curves are involved, leading to significant losses of entropic and enthalpic energy. A sophisticated catalyst that is able to bind water and to undergo subsequent structural changes during electron transfer is needed in order to minimize overall energy losses. In this respect, Ru complexes have received by far the most attention [35].

Since the water-oxidizing TiO_2 photoanode was demonstrated by Fujishima and Honda [1], various various metal-oxide (TiO_2, $SrTiO_3$, Fe_2O_3, ...) semiconductors have been studied as photo-electrodes [36]. But all these oxides engage only in subsequent one-electron-transfer processes, which are energetically unfavorable. Depending on their band structure and on their large bandgaps, several of these oxides are stable against photocorrosion, but their practical conversion efficiencies remain <2%. Because of their role in photosynthetic O_2 evolution, many Mn compounds have been tested in order to evaluate their catalytic properties for oxidation of water. For example, Mn clusters covalently linked to Ru-tris(bipyridyl) sensitizers were investigated to see whether they constitute valuable synthetic models for key electron-transfer steps in natural systems [37][38]. The results up to now have not been very encouraging. However, dimer- and tetramer-type Mn–oxo complexes and Co phosphate compounds that have the ability to acquire different/mixed valence states in common show continuous O_2 evolution under light [39]. Thus, in the final paragraphs, the basic mechanisms and principles of materials design for designing water-oxidation catalysts employing inorganic Mn and other flexible-valence transition-metal compounds that may mimic the natural oxygen-evolution center will be discussed.

Transition-metal oxide compounds with flexible valence state

The basic question behind the materials development challenge is how to couple the optical absorption process of individual photons to a coupled charge-transfer process of four electrons all working at, or at least nearby, a potential of 1.23 $\Delta G°$. This needs materials that have adequate bandgaps and a high flexibility in valence state in order to coordinate and transform water to O_2 with energy losses as small as possible (see Figure 47.6). On the basis of experience from the electrochemical and catalytic literature and studies of the natural O_2 evolution center $CaMn_4O_x$ (see Chapters 24 and 28), some general postulates may be advanced.

- Transition-metal compounds are a necessary and successful ingredient of water-oxidation catalysts because they enable coordination with H_2O as a precondition for the redox interaction. Without transition metals water cannot easily be involved in energy conversion. The underlying reason is the presence of partially filled d-shells.

- Why did nature select a Mn complex? Mn has very flexible valence states and thus has various possibilities for coordination geometries coupled to charge transfer (process (1) in Figure 47.6b).

- Why is O required as a ligand? One important issue is that it cannot be further oxidized during oxidation of water and thus enables to some extent corrosion stability of the catalyst. However, the crucial point is the chemistry of interaction with Mn. There is a mixture of ionic and covalent contributions to chemical bonding, the ratio of which depends on the Mn valence state and the Mn—O bond length. The presence of the Mn—O bond with strongly covalent character is a crucial prerequisite for charge transfer and charge balance during the increase of the oxidation state of the $CaMn_4O_x$ cluster during the S-cycle in the natural O_2 evolution process (see Figure 24.7 and the text in Chapter 24). This is particularly valid for the formation of the O=O double bond and subsequent release of O_2 in the $S_4 \rightarrow S_0$ transition (see Chapters 24 and 28). This dominant covalency is the basis for two alternative models in Figure 47.6(d) and (e), suggesting that either the high oxidation state (Mn^V) or the formation of a radical at the Mn^{IV} outside the Mn_3CaO_4 cubane gives rise to the formation of O_2 [40]. Depending on the valence state, the electronic levels and splitting are strongly coupled to the charge state and ligand geometry in Mn_xO_y. This couples sequential charge-transfer process of four electrons to the energy position of the state involved in process (2) in Figure 47.6(b) and thus enables a high degree of selectivity of the stepwise reactions of two H_2O molecules to produce O_2. It is clear that the theoretical description of such a process needs to transcend the single-particle picture and the related "rigid-band" approximation, in which the electronic structure is considered to be independent of the occupation of the states.

- The role of bound Ca is not well understood; it may contribute to the regulation of the oxidation/charge state of Mn and may constitute a flexible adsorption site for H_2O as suggested in Figures 47.6(d) and (e). However, since the overlap of Ca 4s states with Mn 3d orbitals is very weak, it also may function as a structurally stabilizing cation, which forms an ionic bond to the MnO anions. In artificial Mn clusters other dopants may play a similar role.

- The key question for the development of an artificial O_2-evolution center is to what extent the coupled structural and valence flexibility of Mn (or other transition metals, such as Co) is important for the dynamics of valence states and for a cooperative charge-transfer process. During oxidation of H_2O the valence states of Mn will necessarily change. This will require a transition through structurally different stages of the Mn complex. In addition, efficient H^+ transport will be required. An artificial H_2O-oxidation catalyst should therefore have a loose (nanomaterial) structure to allow the corresponding chemical dynamics to occur.

- These considerations limit the choice to Mn and some other transition-metal compounds, which should be considered. To temporarily store four electronic charge carriers in a Mn complex, there would have to be Mn clusters. Clusters of related transition metals (Cr, Co, Fe) could also be used. Only certain noble metals that provide high oxidation states (e.g., Ru in RuO_2) can oxidize water without the presence of transition-metal clusters.

Numerous Mn compounds that have been studied in the literature have turned out not to be suited for oxidation of water. The critical point appears to be the chemical nature of the bonding, which typically combines polar and covalent contributions (e.g., via hybridization of specific Mn 3d orbitals with O 2p orbitals with appropriate Mn valence and bond length). Polar oxygen bonds to manganese ions are simply broken and lead to corrosion when electrons are extracted. However, if electrons are extracted from Mn 3d states with the right symmetry they can attach water as an electron-donating ligand. The oxidation state of Mn will be relevant. Mn^{3+} might not be sufficient for an effective water-oxidation process, but Mn^{4+} appears to be adequate (Figure 47.6 (e)). It is possible that, during oxidation of the complex, Mn^V states have to be reached temporarily (Figure 47.6 (d)). It appears, therefore, to be essential that the most weakly bound electrons in the Mn cluster are situated in specific Mn 3d states. It turns out to be a challenging problem for theoretical chemistry to search for corresponding Mn clusters; however, the successful artificial synthesis of dimer and tetramer Mn–oxo complexes with mixed-valence Mn^{III}, Mn^{IV}, and Mn^V states seems to be a promising route [39]. It is interesting to note that, in simple perovskite-type $CaMnO_3$, which contains the same elements (Mn, Ca, O) as the photosynthetic reaction center, the partially covalent nature of the O 2p–Mn^{IV} 3d t_{2g} states provides reasonably high catalytic water-oxidation efficiency.

Before deciding which inorganic materials can mimic the photosynthetic $CaMn_4O_y$ complex we must better understand the complex's electronic structure, the detailed reaction path for four-step oxidation of water, and how the electron-transfer process is coupled to structural reorganization and proton release.

So it appears to be necessary to learn how to tailor transition-metal clusters with flexible or even mixed valence states in an oxide environment for oxidation of water. Important aspects that can be also studied in

doped $CaMnO_3$ perovskites are the optimization of electrical conductivity by doping, by tailoring electron–lattice interaction and the resulting nonlinear transport, by varying the chemical composition, and by shortening the electronic and protonic transport pathways by using a nano-composite structure. Such nanomaterials may also guarantee the structural flexibility necessary for significant Mn coordination changes during oxidation of water.

47.5 Summary

The transition from a fossil to a solar energy basis requires sustainable mass production of solar fuels. The storage of light and of electric energy in chemical bonds of molecules that can be easily stored and transported and the energy of which can be converted to other useful forms of energy in redox reactions is, from the viewpoint of thermodynamics, the most efficient form of energy storage. This includes the energetics of the conversion process as well as the achievable energy densities for storage. However, the interconversion of such simple molecules as H_2, H_2O, CH_4, and CO_2 still represents a huge challenge for chemistry. The main bottleneck is the development of durable, available, and cheap catalysts for the photo-induced and/or thermally assisted mutual conversion of various redox pairs, and, within this bottleneck, the catalyst for O_2 evolution is the central issue.

A total life-cycle assessment of energy efficiency for each solar-fuel route must include also the energy and material costs for mass production of catalysts. This severely limits the application of noble metals and points toward the material design of new catalyst materials in the three sub-disciplines of heterogeneous, homogeneous, and biological catalysis. The conversion challenges of small and simple molecules probably cannot be solved by one simple approach. Complex process chains in which catalysts and reactants may mutually interchange their roles may be developed. Since all solar fuels that have been considered depend on the development of a sustainable solar H source, the synthesis of effective and cheap catalysts for solar water splitting could trigger an accelerated development in this area.

We focused in this chapter on the possible role of transition-metal oxides for O_2 evolution and arrived at conclusions about necessary material and electronic properties. Transition metals are frequently used in enzymes, which are biological catalysts. Transition-metal oxides are found in large concentrations on Earth, and most of them are environmentally friendly and can be easily recycled under ambient conditions. Materials design of complex oxides on the atomic scale and on the nanoscale, including nano-composite heterosystems, may thus be a promising pathway toward the development of catalysts with well-adjusted band structure that enable cooperative charge-transfer processes and, thus, mimic natural catalysts in biological and natural systems.

47.6 Questions for discussion

1. What chemical properties must a photocatalyst have in order for it to be suitable for splitting water molecules, which do not themselves absorb light?
2. Some energy experts claim that with currently available sustainable energy technologies (photovoltaics, wind turbines, heat collectors, …) we can overcome our dependence on fossil fuel. Do you agree, and how would the chemical industry respond?
3. For electrolytic water splitting one needs an O_2-evolving and an H_2-evolving electrode. Why is the development of an efficient O_2-evolving electrode much more demanding?
4. Discuss the possible role of direct photoconversion of solar energy into chemical energy for a solar energy strategy. What are the requirements for the photocatalysts?
5. What can we learn from the natural O_2-evolution center for the development of artificial catalysts? Compare single-electron-transfer processes with cooperative multiple charge transfer and discuss microscopic mechanisms for cooperative behaviour.
6. It has been argued that electricity generated from primary and secondary solar energy for domestic use, industry, and traffic would be the most intelligent basis for a sustainable energy future. What would our world look like if biological evolution had followed a similar strategy?
7. Early *Homo sapiens* with all his activities required biological energy of approximately 100 W, whereas an average car requires 100 kW. Why are the applied technologies so different? Hint: consider the different optimization strategies which have been applied in natural evolution and in human production.
8. Methanol, hydrazine, and ammonia have been proposed as chemical energy carriers. Why would you hesitate to use them on a large scale?

47.7 Further reading

- H. Arakawa, M. Aresta, J. N. Armor *et al.*, 2001, "Catalysis research of relevance to carbon management: progress, challenges, and opportunities," *Chem. Rev.*, **101**, 953–996. Gives an overview on catalysis involving conversion of carbonhydrates.
- R. Eisenberg and G. D. Nocera (eds.), 2005, special issue *Forum on Solar and Renewable Energy, Inorg.* *Chem.*, 44, 6799–7260. A collection of articles in this issue describe various fundamental aspects of solar-to-chemical energy conversion.
- S. L. Lewis and D. G. Nocera, 2006, "Powering the planet: chemical challenges in solar energy utilization," *Proc. Natl. Acad. Sci.*, **103**, 15729–15735.
- R. Schloegl, 2010, "The role of chemistry in the energy challenge," *ChemSusChem* 3, 209–222. Describes the role that chemistry will have to play in the era of renewable energy systems, where the storage of solar energy in chemical carriers and batteries is a key requirement.
- H. Tributsch, 2008, "Photovoltaic hydrogen generation," *Int. J. Hydrogen Energy*, **33**, 5911–5930, doi: 10.1016/j.ijhydene.2008.08.017. Explains general problems hindering the move toward a hydrogen economy.
- H. Tributsch, 2007, "Multi-electron transfer catalysis for energy conversion based on abundant transition metals," *Electrochim. Acta*, **52** (6), 2302–2316. Focuses on challenges for multielectron catalysis.
- H. Tributsch, T. Bak, J. Nowotny, and M. K. Nowotny, and L. R. Sheppard, 2008, "Photo-reactivity models for titanium dioxide with water," *Energy Mater.: Mater. Sci. Eng. Energy Systems*, **3** 158–168. Focuses on water-oxidation catalysts incorporated into TiO_2.

47.8 References

[1] A. Fujishima and K. Honda, 1972, "Electrochemical photolysis of water at a semiconductor electrode," *Nature*, **238**, 37–38.

[2] H. Tributsch, T. Bak, J. Novotny, M. K. Novotny, and L. R. Sheppard, 2008, "Photo-reactivity models for titanium dioxide with water," *Energy Mater.: Mater. Sci. Eng. Energy Systems*, **3**, 158–168.

[3] R. Ebrahimi, 2010, "Theoretical study of combustion efficiency in an Otto engine," *J. Am. Sci.* **6**, 113.

[4] A. Lovins, 2005, *Winning the Oil Endgame: Innovation for Profit, Jobs and Security*, Snowmass, CO, Rocky Mountain Institute.

[5] K. Miyamoto (ed.), 1997, *Renewable Biological Systems for Alternative Sustainable Energy Production*, FAO Agricultural Services Bulletin.

[6] S. Rojstaczer, S. M. Sterling, and N. J. Moore, 2001, "Human appropriation of photosynthesis products," *Science*, **294**, 2549–2551.

[7] G. Aylward and T. Findlay, 1999, *SI Chemical Data Book*, 4th edn., Richmond, VIC, Jacaranda Wiley.

[8] C. Zamfirescu and I. Dincer, 2008, "Using ammonia as a sustainable fuel," *J. Power Sources*, **185**, 459–465.

[9] C. H. Christensen, R. Z. Sørensen, T. Johannessen, *et al.*, 2005, "Metal ammine complexes for hydrogen storage," *J. Mater. Chem.*, **15**, 4106–4108.

[10] R. Tamme, R. Buck, M. Epstein, U. Fisher, and CH. Sugarmen, 2001, "Solar upgrading of fuels for generation of electricity," *J. Solar Energy Eng.*, **123**, 160–163.

[11] **R. Schloegl**, 2010, "The role of chemistry in the energy challenge," *ChemSusChem*, **3**, 209–222.

[12] **K. Kordesh**, **R. R. Aronsson**, **P. Kalal**, **V. Hacker**, and **G. Faleschini**, 2007, "Hydrogen from cracked ammonia for alkaline fuel cell–rechargeable battery hybrids and ICE vehicles," in *Proceedings of the Ammonia Conference*, San Francisco, CA.

[13] **K. Xie**, **Q. Ma**, **Y. Jiang** et al., 2007, "An ammonia fuelled SOFC with a $BeCe_{0.9}Nd_{0.1}O_{3-\delta}$ thin electrolyte prepared with a suspension spray," *J. Power Sources*, **170**, 38–41.

[14] **S. F. Yin**, **Q. H. Zhang**, **B. Q. Xu** et al., 2004, "Investigation on catalysis of CO_x-free hydrogen generation from ammonia," *J. Catal.*, **224**, 384–396.

[15] **F. Vitse**, **M. Cooper**, and **G. G. Botte**, 2005, "On the use of ammonia electrolysis for hydrogen production," *J. Power Sources*, **142**, 18–26.

[16] **M. Sterner**, 2009, *Bioenergy and Renewable Power Methane in Integrated 100% Renewable Energy Systems*, Kassel, Kassel University Press.

[17] **J. K. Stolaroff**, **D. W. Keith**, and **G. V. Lowry**, 2008, "Carbon dioxide capture from atmospheric air using sodium hydroxide spray," *Environmental Sci. Technol.*, **42**, 2728–2735.

[18] **A. Bandi**, 1995, "CO_2 recycling for hydrogen storage and transportation. Electrochemical CO_2 removal and fixation," *Energy Conversion Management*, **36**, 89.

[19] **G. A. Mills** and **F. W. Steffgen**, 1974, "Catalytic methanation," *Catal. Rev.*, **8**, 159–210.

[20] **V. Subramani** and **S. K. Gangwal**, 2008, "A review of recent literature to search for an efficient catalytic process for the conversion of syngas to ethanol," *Energy Fuels*, **22**, 814–839.

[21] **M. Specht**, **J. Brellochs**, **V. Frick** et al., 2010, "Speicherung von Bioenergie und eneuerbarem Strom im Erdgasnetz," in *DGMK-Fachbereichstagung "Konversion von Biomassen,"* pp. 181–190.

[22] **T. Inoue**, **A. Fujishima**, **S. Konishi**, and **K. Honda**, 1979, "Photoelectrocatalytic reduction of carbon dioxide in aqueous suspensions of semiconductor powders," *Nature*, **277**, 637–638.

[23] **O. K. Varghese**, **M. Paulose**, **Th. J. LaTempa**, and **C. A. Grimes**, 2009, "High-rate solar photocatalytic conversion of CO_2 and water vapour to hydrocarbon fuels," *Nano Lett.*, **9**, 731–737.

[24] **C. Wang**, **R. L. Thompson**, **J. Baltrus**, and **Ch. Matranga**, 2010, "Visible light photoreduction of CO_2 using $CdSe/Pt/TiO_2$ heterostructured catalysts," *Phys. Chem. Lett.*, **1**, 48–53.

[25] **M. Isaacs**, **J. C. Canales**, **A. Riquelme** et al., 2003, "Contribution of the ligand to the electroreduction of CO_2 catalyzed by a cobalt (II) macrocyclic complex," *J. Coordination Chem.*, **56**, 1193–1201.

[26] **B. Loges**, **A. Boddien**, **H. Junge**, and **M. Beller**, 2008, "Controlled generation of hydrogen from formic acid amine adducts at room temperature and application in H_2/O_2 Fuel Cells," *Fuel Cells*, **47**, 3962–3965.

[27] **T. Inui** and **T. Yamamoto**, 1998, "Effective synthesis of ethanol from CO_2 on polyfunctional composite catalysts," *Catal. Today*, **45**, 209–214.

[28] **S. L. Flitsch** and **R. V. Ulijn**, 2003, "Sugars tied to the spot," *Nature*, **421**, 219–220.

[29] **N. Sutin**, **C. Creutz**, and **E. Fujita**, 1997, "Photoinduced generation of dihydrogen and reduction of carbon dioxide using transition metal complexes," *Comments Inorg. Chem.*, **19**, 67–92.

[30] **H. Eyring** and **A. E. Stern**, 1939, "The application of the theory of absolute reaction rates to proteins," *Chem. Rev.*, **24**, 253–270.

[31] **G. Rothenberg**, 2008, *Catalysis, Concepts and Green Applications*, Weinheim, Wiley-VCH.

[32] **H. Tributsch**, 2007, "Multi-electron transfer catalysis for energy conversion based on abundant transition metals," *Electrochim. Acta*, **52**(6), 2302–2316.

[33] **S. Licht**, 2003, "Solar water splitting to generate hydrogen fuel: photothermal electrochemical analysis," *J. Phys. Chem. B*, **107**, 4253–4260

[34] **M. Roeb**, **C. Sattler**, **R. Klüser** et al., 2005, "Solar hydrogen production by a two-step cycle based on mixed iron oxides," in *Proceedings of ISEC 2005: ASME International Solar Energy Conference*.

[35] **F. E. Osterloh**, 2008, "Inorganic materials as catalysts for photochemical splitting of water," *Chem. Mater.*, **20**, 35–54.

[36] **A. Kudo** and **Y. Miseki**, 2009, "Heterogeneous photocatalyst materials for water splitting," *Chem. Soc. Rev.*, **38**, 253–278.

[37] **R. Lomoth**, **A. Magnuson**, **M. Sjoedin** et al., 2006, "Mimicking the electron donor side of photosystem II in artificial photosynthesis," *Photosynth. Res.*, **87**, 25–40.

[38] **M. Borgstrom**, **N. Shaikh**, **O. Johansson** et al., 2005, "Light induced manganese oxidation and long-lived charge separation in a $Mn_2^{II,II}$–$Ru^{II}(bpy)_3$–acceptor triad,", *J. Am. Chem. Soc.*, **127**, 17504–17515.

[39] **H. J. M. Hou**, 2010, "Structural and mechanistic aspects of Mn-oxo and Co-based compounds in water oxidation catalysis and potential applications in solar fuel production," *J. Integrative Plant Biol.*, **52**, 704–711.

[40] **J. Messinger**, 2004, "Evaluation of different mechanistic proposals for water oxidation in photosynthesis on the basis of Mn_4O_xCa structures for the catalytic site and spectroscopic data," *Phys. Chem. Chem. Phys.*, **6**, 4764–4771.

48 Solar thermal routes to fuel

Michael Epstein

Solar Research Facilities Unit, Weizmann Institute of Science, Rehovot, Israel

48.1 Focus

The conversion of solar energy to alternative fuels is becoming a vital need in view of the current oil prices, the possible ecological damage associated with oil drills, especially off-shore, and the global distribution of oil reserves. There are several routes by which to convert solar energy to fuels, such as electrochemical, photochemical, photobiological, and the thermochemical route, the last of which is the focus of this chapter. This route involves using solar heat at high temperatures to operate endothermic thermochemical processes. It offers some intriguing thermodynamic advantages, with direct economic implications. It is also an attractive method of storage for solar energy in chemical form. An important vector of this route is the production of hydrogen, a potentially clean alternative to fossil fuels, especially for use in the transportation sector.

48.2 Synopsis

There is a pressing need to develop advanced energy technologies to address the global challenges of clean energy, climate change, and sustainable development. The conversion of solar energy to fuels can basically be done through three routes, separately or in combination: the electrochemical route, which uses solar electricity; the photochemical/photobiological route, which makes direct use of solar photons; and the thermochemical route, which utilizes solar heat, usually at high temperatures, for endothermic processes.

Concentrated solar (CS) energy used to execute endothermic reactions at high temperatures to produce energy carriers (fuels, hydrogen) is a promising route that differs from photovoltaics (PV) and wind, which produce electricity directly from the Sun's rays as the energy vector.

Producing synthetic fuels always involves hydrogen. The benchmark for producing solar hydrogen is electrolysis of water using solar electricity by PV or CS power. The overall efficiency of solar to hydrogen is only about 10%, with current electrolysis technologies. The efficiency of thermal processes using CS energy to produce hydrogen is estimated at 30% from Sun to hydrogen.

Full-scale solar-assisted natural-gas reforming plants can be built in good solar sites (e.g., the southwestern USA, central Asia, North Africa). Solar hydrogen can be blended to 10% volume with natural gas in the existing gas pipelines and distribution networks as a near-future step, without further investment in this infrastructure.

When fuel-cell vehicles are introduced into the market, solar hydrogen can be produced to fill them with zero-carbon fuel. Once a solar synthesis-gas mixture has been produced via reforming or gasification, its conversion to synthetic liquid fuels for transportation and industrial application is possible with commercially available technology.

Concentrating solar radiation requires a clear sky, which is usually found in semi-arid and desert regions, in a range of latitude between 15° and 40°. The sunlight strikes the Earth's surface both directly and indirectly after numerous reflections and absorption in the atmosphere. On a clear day, the direct irradiance which can be used for its concentration represents 80%–90% of the total amount of solar radiation reaching the ground.

The direct component of the solar irradiance measured as the direct normal irradiance (DNI) can be concentrated on a small target area using various appropriate reflecting geometries (mirrors). Good annual DNI resources are between 1,900 and 2,300 kWh m^{-2} per year. The DNI resources are accurately known in places where the DNI has been monitored on the ground, but today good estimates can be obtained from satellite data correlated with ground measurements for sufficient accuracy.

48.3 Historical perspective

See Chapter 21 for basic historical development of solar thermal conversion.

48.3.1 The thermodynamics of solar thermochemical conversion (cf. Appendix A)

A solar thermochemical process converts solar concentrated energy into chemical energy in an endothermic reaction. The solar energy is used as a source of high-temperature process heat. In particular, this conversion is aimed at production of fuels such as hydrogen and synthesis gas (a mixture of H_2 and CO). The energy value of the products is higher than the energy content of the reactants.

The thermodynamics of such a process comprises the chemical part, the concentration of the solar radiation, and the energy input and loss of the chemical reactor.

The total internal energy change, DELTA U (cf. Appendix A) of a chemical system (the final internal energy minus the initial internal energy) is a state function that does not depend on how the change was brought about, in other words, the path. The initial energy, however, includes work and heat that do depend on the path. The sum of these two path-dependent energies is thus a state function that is independent of path. We define a quantity called enthalpy, H, which is the total internal energy of the chemical process under controlled pressure. The change of enthalpy of a chemical system, ΔH, under constant pressure is the heat (gain or loss) in the process:

$$\Delta H_p = H_{products} - H_{reactants} = q_p \left[kJ\,mol^{-1} \right]. \quad (48.1)$$

If $\Delta H_p < 0$ we say that the reaction is exothermic. That is, the system releases heat to the surroundings. When $\Delta H_p > 0$, we say that the reaction is endothermic – the system accepts heat from the surroundings.

In processes in which only liquids and solids are involved, the change in pressure and volume, $\Delta(pV)$, is small. When gases are involved $\Delta(pV)$ is not necessarily small, and

$$\Delta(pV)_{gas} = RT\,\Delta n_{gas}, \quad (48.2)$$

where Δn_{gas} is the difference in the number of moles of gaseous products and reactants. In this case the heat of a chemical reaction that we measure at constant pressure can be written as

$$q_p = q_v + RT\,\Delta n_{gas}, \quad (48.3)$$

where q_v is the heat we measure if the reaction runs at constant volume.

The second law of thermodynamics can be introduced in the following verbal statements:

- heat cannot move spontaneously from a cold to a hot body with no other effect, and

- heat cannot be converted quantitatively into work with no other effect.

This can alternatively be expressed by stating that the heat change divided by the temperature in a reversible process dq_{rev}/T is independent of the path. This state function is called entropy, S:

$$\Delta S = \frac{dq_{rev}}{T} \quad \text{or} \quad \Delta q_{rev} = T\,\Delta S. \quad (48.4)$$

When a process is irreversible,

$$\Delta S > \frac{dq_{rev}}{T} \left[kJ\,mol^{-1}\,K^{-1} \right]. \quad (48.5)$$

The second law can provide a thermodynamic definition of equilibrium in chemical system. In a closed isolated system at equilibrium any spontaneous change must leave the entropy unchanged:

$$\Delta S = 0. \quad (48.6)$$

Incorporating the second law into the definition of enthalpy results in

$$\Delta H = T\,\Delta S + V\,\Delta p. \quad (48.7)$$

We can now define the Gibbs free energy of a chemical system as

$$\Delta G = \Delta H - T\,\Delta S \left[kJ\,mol^{-1} \right] \quad (48.8)$$

for a constant temperature.

For a process at constant temperature and pressure the change in Gibbs free energy gives all the reversible work except for the pV work. This work can include electrical work and so on.

This can be written

$$\Delta G_{T,P} = \Delta q - T\,\Delta S. \quad (48.9)$$

Any reaction for which ΔG is negative is spontaneous and the products have lower Gibbs free energy than that of the reactants.

Any reaction for which ΔG is positive is unfavorable:

- favorable or spontaneous reaction: $\Delta G < 0$
- unfavorable or non-spontaneous reaction: $\Delta G > 0$
- equilibrium: $\Delta G = 0$.

The free energy of a reaction at standard conditions of 0.1 MPa and 298 K (25 °C) is written as

$$\Delta G^{\circ} = \Delta H^{\circ} - T\,\Delta S^{\circ}. \quad (48.10)$$

48.3.2 Concentration of solar radiation

Concentration of solar radiation is achieved by reflecting the solar flux incident on an aperture area, A_c, of a concentrator onto a smaller receiver/reactor area A_r. The optical concentration ratio, C_o, is defined as the ratio of solar flux, I_r, on the receiver to the flux, I_c, on the collector aperture, or

$$C_o = \frac{I_r}{I_c}, \qquad (48.11)$$

while the geometric concentration ratio C_g is based on the ratio of the areas,

$$C_g = \frac{A_c}{A_r}. \qquad (48.12)$$

If we define a view angle, 2θ (which is sometimes called the acceptance angle), between the receiver and edges of the collector aperture, the maximum optical concentration in air can be written as

$$C_{o,max} = \frac{1}{\sin^2\theta}. \qquad (48.13)$$

Since the Sun's disk is not a point source but rather has a size that can be viewed at $1/4°$ from Earth, the upper limit of concentration is

$$C_{o,max} = \frac{1}{\sin^2(1/4°)} \cong 46{,}000 \qquad (48.14)$$

in air. In practice, this level of concentration is not achievable because of tracking errors and surface imperfections of the concentrating devices. Concentrations of 1,000–10,000 have been demonstrated in practice. The second law treats not only the geometric limit of concentration but also the operating temperature limit of a solar concentrating device.

Assuming that convection and conduction losses could be eliminated, all the heat loss q_L is by radiation according to

$$q_L = \varepsilon A_r \sigma T_r^4, \qquad (48.15)$$

where ε is the effective emissivity and σ the Stefan–Boltzmann constant ($\sigma = 5.6705 \times 10^{-8}$ W m^{-2} K^{-4}).

The highest temperature an ideal solar receiver can achieve is defined as the stagnation temperature, at which the yield of the system is zero (the heat absorbed equals that lost by radiation). It can be calculated from

$$T_{stagnation} = \left(\frac{I_c C_o}{\sigma}\right)^{0.25}. \qquad (48.16)$$

Indeed, without concentration ($C_o = 1$) the stagnation temperature on a bright-sky day ($I_c = 1{,}000$ W m^{-2}) can reach only 364 K.

However, as the temperature increases, the thermal losses from the reactor, mainly by re-radiation, also increase. To reach high temperatures and to reduce these losses, the solar radiation must be concentrated (cf. Chapters 20 & 21). For example, in order to run a chemical process at an optimal working temperature of 1,500 K, it is desirable to concentrate the solar radiation by a factor of 5,000.

Two concentrating technologies – dish and tower – are shown in Figure 48.1. The absorption efficiency of a solar reactor is defined by the net rate at which the solar energy is absorbed by the chemical process, divided by the solar energy input. Assuming only radiation losses, this efficiency can be formulated as [1]

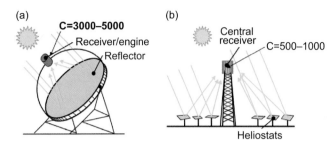

Figure 48.1. Solar concentrating technologies: (a) parabolic dish and (b) central receiver or solar tower (from the SolarPACES home page, http://www.solarpaces. org/CSP_Technology/csp_technology.htm).

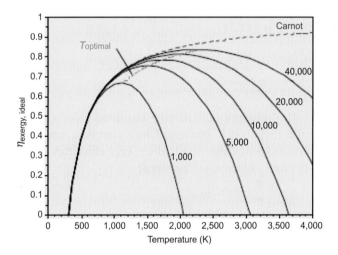

Figure 48.2. The exergy efficiency (cf. Chapter 6) as a function of the operating temperature for a blackbody reactor converting concentrated solar radiation into chemical energy. The Carnot efficiency and the locus of the optimum reactor temperature are shown at different concentration levels.

$$\eta_{absorption} = \alpha_s - \frac{\varepsilon \sigma T^4}{IC}, \qquad (48.17)$$

where I is the intensity of the radiation; T is the operating temperature of the reactor; C is the concentration ratio; α_s and ε are the absorbance of the solar radiation and the emittance of the reactor, respectively; and σ is the Stefan–Boltzmann constant.

The conversion of the solar heat to chemical free energy of the reaction products is limited by the Carnot efficiency (eq. 21.5 in Ch. 21) and the overall system efficiency can be defined as

$$\eta_{system} = \eta_{absorption}\left(1 - \frac{T_L}{T_H}\right), \qquad (48.18)$$

where T_L and T_H are the temperatures of the cold and hot states, respectively. The maximum system efficiencies (assuming blackbody properties: $\alpha_s = \varepsilon = 1$) that can be obtained for operation temperatures of 1,100 and 1,700 K are 65% and 80% at concentration ratios of 1,000 and 10,000, respectively [2], see Figure 48.2 [3].

Table 48.1. Thermodynamic parameters and temperatures for which $\Delta G = 0$ for several reactions described in this chapter

Reaction	ΔH°_{298K} (kJ)	ΔG°_{298K} (kJ)	T at $\Delta G = 0$ (K)	S at T* (J K^{-1})
(1) $CH_4 + H_2O \rightarrow CO + 3H_2$	250	150.5	753	314
(2) $CH_4 + CO_2 \rightarrow 2CO + 2H_2$	247	170.5	923	283.3
(3) $ZnO \rightarrow Zn + 0.5O_2$	350.5	320.4	2,235	201.3
(4) $ZnO + C \rightarrow Zn + CO$	240	183	1,243	194
(5) $Fe_3O_4 \rightarrow 3FeO + 0.5O_2$	313.7	275.2	3,073	160.5
(6) $CH_4 \rightarrow C + 2H_2$	74.6	50.5	823	106.0
(7) $C + H_2O \rightarrow CO + H_2$	175.3	100.0	723	213.0
(8) $H_2O \rightarrow H_2 + 0.5O_2$	285.8	237.1	4,300	66
* for which $\Delta G = 0$				

The measure of how well absorbed solar concentrated radiation is converted into chemical energy in an endothermic reaction is the exergy efficiency, which is defined as (see also Chapter 6)

$$\eta_{\text{exergy}} = \frac{-n\Delta G^\circ_{298K}}{Q_{\text{solar}}}, \quad (48.19)$$

where ΔG is the maximum amount of work that can be obtained from the reaction products at their standard conditions (1 bar; 298 K). As an example, the thermodynamic parameters of the major reactions analyzed in this chapter are summarized in Table 48.1.

48.4 Solar thermal fuels

48.4.1 General

Kinetics and catalysts

Thermodynamics is a very powerful tool for determining the maximum possible yield from a given chemical reaction assuming that equilibrium is established. However, the actual yield will always be smaller and will be determined by the kinetics of the particular reaction.

The rate of a reaction can be given in terms of the concentration of any one of the reactants or products. The way in which the rate varies with concentration is defined in terms of the order of the reaction.

The rate of a chemical reaction can be greatly increased by the use of a suitable catalyst. Catalysts are also used to direct a reaction in one particular way or to inhibit undesired side reactions. Most industrial catalytic processes, e.g., methane reforming, involve a heterogeneous gas-phase catalyst consisting of, for instance, a porous bed of solid pellets, honeycomb, foam, or other fixed configuration through which the reacting gases are passed.

Since the catalyst is unchanged during the course of the chemical reaction, it can have no effect on the point of equilibrium. However, the catalyst does enter into the reaction, presumably by forming intermediate compounds with the reactants, which then break down to give the products of the reaction and simultaneously regenerate the catalyst.

In the case of solid catalysts, the reaction is assumed to take place in the following steps: (1) diffusion of the reactants to the surface of the catalyst; (2) adsorption of the reactants on the surface; (3) reaction on the surface; (4) desorption of the products; and (5) diffusion of the products away from the surface. The overall rate of reaction depends on the rate of each individual step, and it can happen that one or more of the steps may control the rate. Many of the catalysts used for reforming have micropores, and the diffusion is usually important. The amount of catalyst required in a given process can be calculated if the reaction rates are known as a function of temperature and pressure. However, in practice, the design calculations are usually based on the space velocity, namely the rate of gas flow measured at standard conditions which can be used for each unit volume of catalyst. This is equal to the reciprocal of the catalyst contact time. The method is more or less empirical, since space velocities vary with temperature, pressure, and the concentrations of the reactants. However, the method is widely used because of its simplicity and the ease of scaling up laboratory experiments to actual commercial designs.

Solar reactors

The thermochemical route offers thermodynamic advantage. Irrespective of the type of fuel produced, higher reaction temperatures yield higher energy-conversion efficiencies. Solar reactors use concentrated solar radiation as an external source of heat. Since most of the reactions described in this chapter (see Table 48.1) are conducted at temperatures higher than 1,000 K, the reactors are placed in a cavity enclosure to minimize energy losses mainly as a result of re-radiation. Since these losses are proportional to the re-radiating area, they can be minimized using cavity-type solar receivers. These receivers are well-insulated enclosures equipped with a small aperture that allows the entry of the concentrated radiation. The aperture is positioned exactly at the focal plane of the concentrator. The size of the aperture is optimized to be small enough to minimize radiation losses and large enough to intercept and receive the CS radiation with minimal spillage loss around the opening. The incoming solar radiation undergoes multiple internal reflections by the internal walls and is finally absorbed by the reactor tubes, usually arranged along the walls. The larger the ratio of the cavity depth to the aperture diameter, the closer the receiver approaches a blackbody cavity.

The materials used to build the reactor tubes must be able to withstand high temperatures, severe thermal shocks, and transitory conditions. For reactions at temperatures below 1,100 K the reactors can be made of metal tubes. Above these temperatures some other solution must be implemented. Additionally, designing a good solar reactor means matching the reaction kinetics to the rate of heat transfer. The reforming reactions (reactions (1) and (2) in Table 48.1) are good examples. The kinetics is fast but the heat conduction through the reactors metal walls is the bottleneck causing high surface temperatures, which result in higher radiation losses and shorten the lifetime of the tube's material.

For higher temperatures direct illumination of the reactants is a preferable method. In this case the solar reactor is equipped with a quartz window through which the concentrated radiation is accessed. These reactors are also called "volumetric" reactors. In the case of the re-former, the radiation directly heats the catalyst bed through which the reactants pass. Since the kinetics is fast, much larger heat fluxes can be absorbed than in a tubular re-former, which means a more compact and efficient reactor. The directly irradiated reactor can also heat gas particle suspensions or fluidized beds, as in the example of methane cracking (reaction (6) in Table 48.1). This reaction is a non-catalytic one, and is conducted usually at 1,700–1,800 K. The concentrated radiation is absorbed by a cloud of tiny carbon particles produced in the reaction and floating inside the reactor. Since methane is practically transparent to the solar spectrum, the particles, which have a very large heat-transfer area, absorb the solar radiation and heat the methane mainly by convection.

In some examples, such as the dissociation of $CaCO_3$ or Fe_3O_4 (reaction (5) in Table 48.1), the reactor can be open (no window) and the radiation directly strikes the solid reactants.

Worldwide scope

The main products in the solar thermochemical routes to produce fuels are either hydrogen or synthesis gas (syngas, a mixture of hydrogen and carbon monoxide). These products can be used directly as gaseous fuels in electricity generation, preferably via Brayton cycles using gas turbines (GT) or combined cycles (CC) [topping with GT and bottoming with a condensing steam turbine (ST)]. These products can also be used as raw material for synthetic liquid fuels such as methanol or in Fischer–Tropsch synthesis (FTS) (Chapter 39), which is in principle a carbon-chain-building process, where by CH_2 groups are attached to the carbon chain. The resulting overall reaction can be presented as follows:

$$nCO + \left(n+\frac{m}{2}\right)H_2 \rightarrow C_nH_m + nH_2O, \quad (48.20)$$

$$CO + 2H_2 \rightarrow \text{—}CH_2\text{—} + H_2O$$
$$\Delta H^\circ_{300\,K} = -165\,\text{kJ mol}^{-1}. \quad (48.21)$$

Depending on the temperature and the catalyst, different hydrocarbons can be synthesized. Other families of materials produced via solar thermochemistry are metals such as Zn, Al, Mg, Sn, B, and Cd. These metals have been produced from their oxides either by direct thermal decomposition [4] or by a carboreduction process [5]. Such metals can be used as commodities, can be hydrolyzed to generate hydrogen, and can be used to store solar energy in chemical form. Zinc can also be used in zinc–air batteries. When the carboreduction is executed with biomass charcoal as the reducing agent, the system can be considered benign from an ecological point of view.

48.4.2 Solar reforming of methane with steam or CO_2

The reforming of methane with steam or CO_2 is an endothermic catalytic reaction with as its product a gaseous mixture of primarily H_2 and CO:

$$CH_4 + H_2O \rightarrow CO + 3H_2 \quad \Delta H^\circ_{298K} = 250\,\text{kJ mol}^{-1}, \quad (48.22)$$

$$CH_4 + CO_2 \rightarrow 2CO + 2H_2 \quad \Delta H^\circ_{298K} = 247\,\text{kJ mol}^{-1}. \quad (48.23)$$

Steam reforming of methane is accompanied mainly by the side reaction of the water-gas shift (WGS), which is slightly exothermic:

$$CO + H_2O = CO_2 + H_2 \quad \Delta H^o = -36\,\text{kJ mol}^{-1}. \quad (48.24)$$

One of the less favorable possible side reactions is the cracking of methane, resulting in deposition of carbon onto the catalyst:

$$CH_4 = C + 2H_2. \quad (48.25)$$

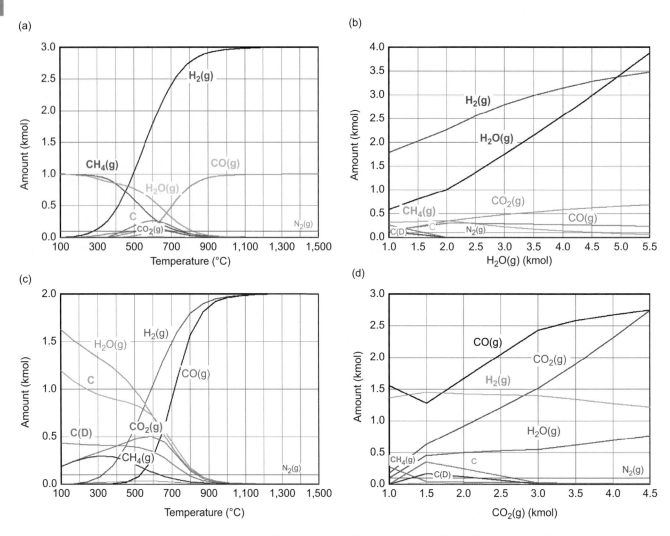

Figure 48.3. The equilibrium composition of methane reforming with steam (a) at 1 bar pressure at various temperatures and (b) at 600 °C with different steam-to-CH_4 molar ratios; and with CO_2 (c) at 1 bar at various temperatures and (d) at 700 °C with different CO_2-to-CH_4 molar ratios C(D) = diamond carbon, also some negligible amounts of amorphous carbon are formed in the process (not shown).

From Figure 48.3(b) one can see that an excess of steam can be used to reduce the tendency for carbon formation via reactions (5) and (6) in Table 48.1. However, it can be seen from Figure 48.3(a) that working at lower temperature increases the probability of carbon deposition. These temperature fluctuations are typical of solar operation, for which the process is subjected to frequent startups, clouds, etc. Therefore the catalyst plays an extremely important role in this process by selectively slowing down the methane cracking relative to the reforming reaction. Figure 48.3(a) shows the equilibrium composition of the steam reforming reaction under stoichiometric composition conditions (H_2O:CH_4 = 1:1). It can be seen that, at temperatures above 1,000 °C, there is almost full conversion of the CH_4 to H_2 and CO. At around 550 °C the CH_4 conversion is about 50%, but the concentration of CO_2 is high and therefore

solid carbon is deposited. Figure 48.3(b) shows that, in order to avoid the deposition of carbon at around 600 °C, one needs at least H_2O:CH_4 = 2:1. In the case of CO_2 reforming of methane under stoichiometric composition conditions (CO_2:CH_4 = 1:1), Figure 48.3(c), solid carbon is obtained already below 900 °C and one is required to start with a molar ratio of CO_2:CH_4 = 3:1 in order to avoid carbon deposition, e.g., at 700 °C (see Figure 48.3(d)).

These reactions ideally upgrade the calorific value of the methane feed by more than 25%. This upgrade can be contributed to by CS radiation, stored and transported with the syngas, and used as fuel in a GT or CC. When combusted in a modern CC plant the conversion of heat to electricity can reach an efficiency of 55%–60%. Since these reactions require high temperatures for full conversion of the methane (1,100 K and 1,200 K at 1 bar, respectively), the real contribution of the solar energy is higher than the 25% input to the chemistry because a substantial amount of energy is required to raise the temperature of the reactants (their sensible heat) to the operating temperatures. The solar-produced syngas can also be converted into transportable liquid fuels such as methanol.

Unlike the steam reforming of methane, reaction (5), which is commonly used in industry to produce hydrogen, the CO_2 reforming reaction (6) has been extensively developed in connection with the closed chemical heat pipe (CHP) for transport and storage of solar energy in chemical form [6][7]. The concept of the CHP is to carry out the highly endothermic reversible reaction (6) at the solar site and transport the product syngas at ambient temperature by pipeline to the customer's site, where the reaction is reversed. The reversed reaction is exothermic and can be used to supply process heat for industrial and commercial customers. The original products are recovered and recycled back to the solar site. The end result is that solar energy enters the closed system and process heat is released at the other end, with no emission of CO_2 to the atmosphere. The concept suffered economically because of the large investment required for the pipeline infrastructure to connect the solar site with the customer and back to close the loop. Steam reforming of methane and combustion of the product syngas in a micro-GT has successfully been demonstrated [8]. The CO_2 reforming can be useful to process biogas (a mixture of CH_4 and CO_2) and landfill gas and to fix CO_2 from gas streams rich in this compound, such as contaminated natural gas and flue gas from coal power plants (13%–15% CO_2) or from CO_2 sequestration sources, for reuse following solar conversion to syngas.

While commercial steam reforming is carried out on a nickel-based catalyst, it was essential to develop new catalysts for the CO_2 reforming. Catalysts based on rhodium [9][10] and ruthenium were developed and later adapted to re-form also CH_4 with steam, CO_2, and a mixture of the two [11][12]. These catalysts show superior performance compared with nickel at high temperatures, which can be developed in a directly illuminated solar reactor and with a much lower steam-to-methane molar ratio, without the undesirable phenomenon of carbon deposition. The syngas product of the solar reformer can also be further processed through the WGS reaction:

$$CO + H_2 + H_2O \rightarrow CO_2 + 2H_2. \qquad (48.26)$$

The CO_2 is separated for disposal or reuse and the hydrogen can be used in fuel cells for generating electricity at high efficiency or on board a vehicle.

Solar reformers and catalysts

Different concepts of solar reformers have been developed during the last three decades. They can be divided into two approaches – indirect heating and direct heating of the catalytic system.

In the indirect concept the catalyst is introduced into a tubular re-former. The re-former tubes are heated by a secondary fluid such as air [13]. The hot air flows in an annular channel around the re-former tube. Other designs used a heat pipe to introduce the energy into the catalyst bed [14] and sodium vapor condensed on the surface of the re-former tubes. A pool of liquid sodium was boiled in a solar furnace and isothermal operation was achieved [15]. In another design the re-former tubes were installed in a solar-cavity receiver and indirectly heated by internal reflection inside the cavity. Experiments with this design were first done in a 20-kW solar furnace at the Weizmann Institute of Science, Israel (WIS) [16] on a scale of a few kilowatts, which was later scaled up to 8.5 kW with a solar furnace and up to 480 kW with a solar tower. Methane conversion of 94% was demonstrated.[1] In the solar tower approach the cavity receiver was equipped with a secondary concentrator placed next to its aperture to enhance the concentration and increase the thermal efficiency of the receiver to 84%.

The second approach to solar re-formers is direct illumination of the catalyst with CS radiation through a transparent window made of quartz or fused silica. The main rationale behind this approach is the enhanced heat transfer to the catalyst. Methane reforming at 1,000–1200 K has fast kinetics, and the rate-limiting mechanism in the tubular re-former is heat transfer via conduction through the tube's wall and into the catalyst bed. The resistance to heat transfer causes a high surface temperature of the tube, and one is therefore limited by the tube's material of construction. The low heat-transfer rate through the tubular re-former (20–100 kW m^{-2}) dictates, therefore, a larger surface area and dilution of the CS radiation. For all these reasons, illuminating directly the catalyst system and the reaction sites could result in more efficient solar-flux absorption (a peak incident solar flux of 2.5 MW m^{-2}) [19]. Two main designs for directly irradiated absorber/catalyst configurations (known as volumetric receivers) have been developed and demonstrated during the last two decades. One was based on

---

[1] The first prototype was constructed from an inconel reactor tube (of internal diameter 1.3 cm) filled with rhodium on an alumina catalyst, inserted into a tubular cavity (of internal diameter 0.16 m and length 0.45 m) equipped with an aperture of diameter 0.08 m. The power absorbed into the tube was 27 kW at a space velocity of 37,000 h-1. Methane conversion of 86 % was obtained at a temperature of 1,176 K of the gases at the outlet. This concept was scaled up to 8.5 kW with the solar furnace [17] and further to 480 kW at the solar tower of the WIS [18]. This re-former used a ruthenium catalyst. The receiver cavity was shaped like a pentagon, and contained eight re-former tubes made of inconel. Each tube was 0.05 m in diameter and 3 m long.

(a)

(b)

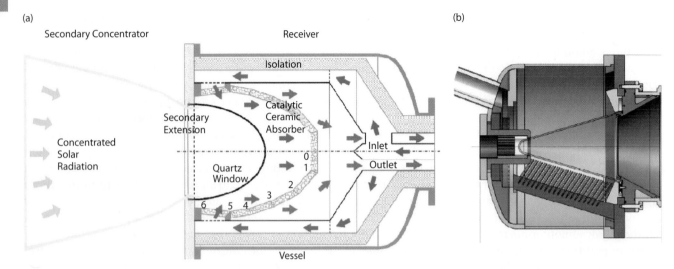

Figure 48.4. Solar reactors for methane reforming (reaction with H_2O or CO_2, or both): (a) catalyst on foam absorber (DLR) and (b) catalyst on pin-fins absorber (WIS).

ceramic foam made of α-Al_2O_3 or SiC on which the catalyst was applied (this approach was developed by the DLR, Germany [20][21]) and the other one was based on ceramic pin fins made of alumina (developed by the WIS, Israel [22]) as illustrated in Figure 48.4. In contrast to the reforming catalysts used in conventional tubular heated reactors, a solar-irradiated volumetric receiver requires a new, solar-specific catalyst. Owing to the direct irradiation, it functions both as a solar absorber and as a heat-transfer unit. Therefore it should have low reflectivity and high resistance to thermal shock, since the absorber can often be heated to high temperatures ($>$1,000 K) in a very short time due to the high-flux irradiation.

A design[2], with SiC foam coated with Alumina and Rh as a catalyst, was a very efficient heat transfer medium and was highly resistant to thermal shock, so that high-temperature operation was achieved, enabling CH_4 conversion of over 95% to be achieved at 1,300 K. Other catalysts based on metal-foam absorbers,

such as Ni–Cr–Al and WO_3/W, have been developed and reported [24][25].

48.4.3 Solar thermochemical splitting of water

Direct thermolysis of water to H_2 and O_2 requires very high temperatures of above 2,500 K in order to achieve a reasonable degree of dissociation. Thereafter an effective technique to separate the H_2 very fast at this high temperature and finally to recover a substantial part of the sensible heat is needed in order to obtain an efficient process. For the separation, ceramic membranes based on stabilized zirconia have been tested [26], but failed to withstand the thermal shocks and exhibited limited longevity due to the sintering process. Rapid quenching by diluting with cold product gas and expansion through a nozzle were tested successfully, but the drop in the energy efficiency and the explosive gas mixture produced caused the abandonment of this approach.

Water-splitting thermochemical cycles alleviate most of the difficulties encountered in direct splitting. They operate at lower temperatures and bypass the H_2–O_2-separation problem.

Numerous cycles have been considered and analyzed in the last two decades [27]. They have been screened according to criteria including the maximum process temperature, number of intermediate reactions and separation steps, number of elements, technical feasibility, expected exergy efficiency, materials, and environmental issues [28]. This chapter details the metal oxides family only. These are two-step cycles using relatively simple metal oxide redox reactions [29]: a reduction step

---

[2] The alumina-foam absorber (porosity 92%, surface area 990 m_2 per m_3) was coated with 30 wt% of g-Al_2O_3 as a washcoat. After sintering, 11 wt% of Rh catalyst, with respect to the mass of the washcoat, was loaded. The alumina foam was susceptible to thermal shocks and suffered from cracks after a relatively short period of operation under real solar conditions. Therefore it was replaced by SiC foam (porosity 94% and a surface area of 550 m_2 per m_3), which was coated with a 10% g-Al_2O_3 washcoat and then 11% Rh. These absorbers were tested at a solar input of about 300 kW and produced above 90% methane conversion [10]. The pin-fins absorber used a high-temperature catalytic system (1,200–1,400 K). The pins were made of sintered alumina (outer diameter 3.0 mm, length 40 mm). The Ru catalyst (2 wt%) was supported on an alumina washcoat promoted with Mn oxides [23].

$$M_xO_y \rightarrow xM + \frac{y}{2}O_2 \qquad (48.27)$$

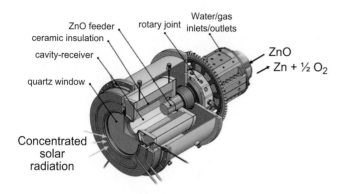

Figure 48.5. The scheme of a "rotating-cavity" (PSI) solar reactor for thermal dissociation of zinc oxide (SolarPACES solar fuels document).

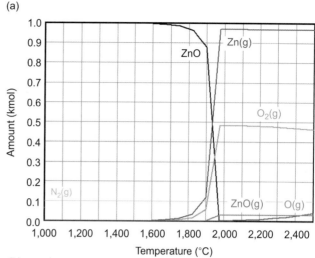

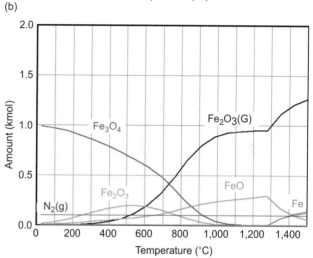

Figure 48.6. Equilibrium compositions for thermolysis of ZnO and Fe_3O_4 at 1 bar pressure.

and an oxidation step

$$xM + yH_2O \rightarrow M_xO_y + yH_2, \qquad (48.28)$$

where M is the metal and M_xO_y is the corresponding oxide. The reduction step is endothermic and, with the aid of CS radiation, the metal oxide is thermally dissociated to the elemental metal or to a lower-valence oxide. In the second step the metal is exothermically hydrolyzed, and the water is split to form H_2 and to give recovery of the oxide, which is recycled to the reduction step. One example of such a system is the Zn/ZnO cycle which has been studied extensively during the last decade. At 2,340 K, $\Delta G^\circ_{298\,K} = 0$ and $\Delta H^\circ_{298\,K} = 395\,kJ\,mol^{-1}$. The products of the ZnO decomposition are Zn and O_2 in the gaseous phase, and they readily recombine. Therefore a quenching step is necessary. The exergy efficiency of this cycle is 29% without any heat recovery [30], and the theoretical upper limit of the cycle with complete heat recovery during the quenching and the hydrolysis step could reach 82% [31]. A high-temperature solar-rotary-reactor concept has been developed at the Paul Scherrer Institute (PSI), Switzerland and tested on a 10-kW scale. This concept features a windowed rotating cavity lined with ZnO particles that are held by centrifugal force. The ZnO is directly exposed to high solar flux and serves simultaneously the function of radiation absorber, thermal insulator, and chemical reactant [32], as depicted in Figure 48.5 [33].

Another example of a two-step redox pair is the Fe_3O_4/FeO pair [34], for which the process proceeds as follows:

$$Fe_3O_4 \rightarrow 3FeO + \frac{1}{2}O_2, \qquad (48.29)$$

$$H_2O + 3FeO \rightarrow Fe_3O_4 + H_2. \qquad (48.30)$$

The solar step, reaction (10) in Table 48.1, is performed at 1,300 K at 1 bar while the hydrolysis step proceeds below 1,000 K [35]. It was found necessary to quench the products of in order to avoid the back reaction. Quenching introduces into the process an energy penalty of up to 80% of the solar input. Equilibrium compositions of the system of zinc and iron oxides are presented in Figure 48.6.

Other redox pairs of metal oxides such as Mn_3O_4/MnO (thermal reduction at 1,700 K) and Co_3O_4/CoO have been considered [33], but the H_2 yield (9) was low for practical purposes. Partial substitution of the iron in Fe_3O_4 by other metals such as Mn and Ni that form mixed oxides of the type $(Fe_{1-x}M_x)_3O_4$ gave lower reduction temperatures. The reduced phase $(Fe_{1-x}Mx)_{1-y}O$ is still capable of hydrolysis reaction. The Ni–Mn ferrite system was reduced at around 1,000 K in a solar furnace. It was found that the two-step process consisted of formation of a non-stoichiometric ferrite phase with cation excess at temperatures above 1,073 K and water splitting with cation-excess ferrite at <1,073 K [36]. The hydrogen yield was low and the cyclic repeatability of the process was not satisfactory. The mixed iron oxide was tested on a 100-kW scale in a dual-chamber monolithic solar reactor. The reactor was constructed from ceramic multi-channeled monoliths coated with mixed nanostructured doped iron oxide, which are activated by heating to around 1,500 K. After releasing the O_2, it is capable of splitting water vapor passing through the reactor and releasing the H_2 at around 1,100 K. Quasi-continuous H_2 production by alternate operation of two or more reactor chambers has been demonstrated [37]. The doped iron oxide had the form of a single-phase spinel, $A_xB_{1-x}Fe_2O_4$. Mixed ferrites such as $ZnFe_2O_4$, $Mn_{0.5}Zn_{0.5}Fe_2O_4$, and $Ni_{0.2}Zn_{0.8}Fe_2O_4$ have been evaluated for their water-splitting activity at 1,100 K [38].

48.4.4 Carbothermal reduction of metal oxides

The solar carbothermal reduction of metal oxides using carbonaceous materials as reducing agents can be performed at much lower temperatures. The general reactions are

$$M_xO_y + yC(s) \rightarrow xM + yCO \qquad (48.31)$$

with charcoal and

$$M_xO_y + yCH_4 \rightarrow xM + y(2H_2 + CO) \qquad (48.32)$$

with methane. In addition to the metal, CO or syngas is produced, which can be further processed to produce more H_2, reaction (7) in Table 48.1, or liquid fuels, reaction (3).

The most advanced system is ZnO/C/Zn: 300-kW solar pilot-scale production of zinc by carbothermal reduction of ZnO has been demonstrated (see Figure 48.7). The batch reactor was operated at 1,300–1,500 K, producing 50 kg per hour of zinc powder ($\Delta H^{\circ}_{1500\,K} = 350\,kJ\,mol^{-1}$). Zinc vapor and CO were quenched immediately after exiting the reactor to avoid the back reaction [39]. Beech charcoal was used as the reducing agent. The conceptual design of an industrial production plant is presented in [40]. The hydrolysis of zinc to produce hydrogen by bubbling steam through liquid zinc and via solid–gas-phase processes has been studied and demonstrated [41][42]. It was shown that the hydrogen yield was much higher in the case of the reaction between steam and micrometer-sized zinc particles. Preheating of the zinc particles to about 400 °C was necessary in order to attain a spontaneous reaction. However, since the hydrolysis is an exothermal reaction, close control of its temperature is imperative in order to avoid zinc losses via evaporation and back reaction between the hydrogen and the zinc oxide product. The zinc powder was also experimentally

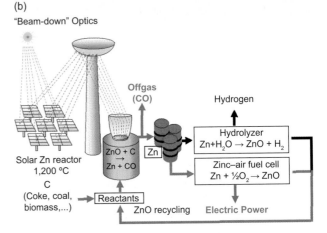

Figure 48.7. (a) The process scheme of carboreduction of ZnO and (b) a solar reactor for 350 kW solar input and a nominal production rate of 50 kg per hour of zinc (SOLZINC project, WIS).

tested for use in zinc–air batteries. The use of solar energy offers a sustainable option for recharging this type of battery.

Other similar systems such as SnO_2/Sn have been studied. The advantage of this system is the lower temperature (1,100–1,200 K) and the fact that the metal remains in the reactor and does not evaporate [43]. Other metal oxides of interest are Fe_2O_3 and MgO, which show significant free-metal formation via the carboreduction reaction. Carboreduction of oxides of Al, Ca, Si, and Ti will result in their carbides, which are stable. CaC_2 is a valuable material and known as the feedstock for the production of acetylene. The hydrolysis or acidolysis of Fe_3C yields liquid hydrocarbons, and the hydrolysis of various carbides of manganese yields H_2 and hydrocarbons in different proportions. Carboreduction of MgO to magnesium is useful also to produce secondary materials, such as boron from its oxide B_2O_3. Solid boron can react with steam and efficiently produce hydrogen, which could be used in future vehicles operated with fuel cells [44].

48.4.5 Producing H_2 by decarbonization of fossil fuels

Currently the most widely used process to produce hydrogen is catalytic steam reforming. This involves primarily natural gas (CH_4) as feedstock, reaction (5) in Table 48.1, followed by the WGS (water gas shift) reaction to produce additional hydrogen, reaction (7). An alternative method for producing H_2 is the thermal dissociation of CH_4. This reaction is endothermic and can be written as follows:

$$CH_4 \rightarrow 2H_2 + C \qquad \Delta H^\circ_{298\,K} = 75\,kJ\,mol^{-1}. \quad (48.33)$$

At thermodynamic equilibrium the dissociation of the methane can be completed at around 1,500 K [45]. However, at this temperature, side products such as C_2H_6, C_2H_4, and C_2H_2 are formed [46]. Therefore temperatures in the range 1,800–2,100 K are preferable for the non-catalytic reaction.

(a)

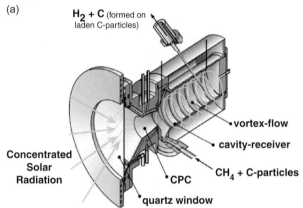

(b)

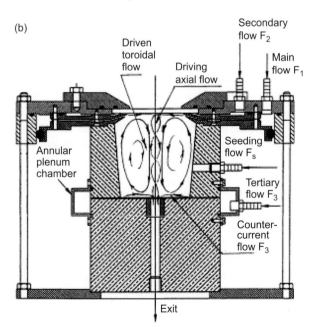

Two approaches to solar-reactor design have been investigated during the last few years: direct heating of the methane and indirect heating. In the direct concept the CS radiation enters the reactor through a transparent window, and the heating of the CH_4 is achieved by seeding the feed gas with a cloud of fine absorbing particles (see Figure 48.8). Highly effective absorption can be attained if the linear dimension of the seed particles is smaller than the characteristic absorption length for the light passing through the suspension. It was reported that, for sub-micrometer particles, a seeding-mass loading on the order of 0.1 g of carbon-black powder per cubic meter of CH_4 is sufficient to obtain adequate absorption of solar radiation [47]. A major hurdle in the direct approach is the protection of the window from carbon deposition during the CH_4 decomposition process. This is done by controlling the fluid dynamics inside the reactor. In one approach a confined-tornado flow configuration has been developed as a means for protection of the window of the reactor from contact with incandescent solid particles in the gas suspension [48].

Another method developed to protect the window used streaming of inert gas (argon) that was injected through a ring manifold around the aperture, forming an aerodynamic gas curtain that prevented particle flow from reaching the quartz window [49]. Almost full CH_4 conversion (99%) and over 99% H_2 yield were reported. The solid carbon product obtained contained substantial amounts of nano-fibers and agglomerates of multi-wall-carbon-nano-tubes (MWCNT) as can be seen in Figure 48.9 [50].

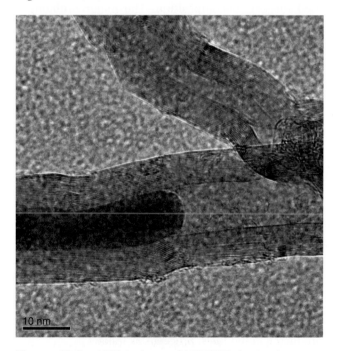

Figure 48.8. Solar reactors for thermal decomposition of CH_4: (a) a "vortex" reactor (PSI) and (b) a "tornado" reactor (WIS).

Figure 48.9. A TEM photo of MWCNT obtained by solar cracking of CH_4 at the WIS.

The indirect-heating approach involves an opaque tubular reactor made of graphite. The tube wall absorbs the solar radiation and then heats the gas by gas–solid convection. The reactor tube is placed inside a graphite cavity closed with a hemispherical quartz window that covers its aperture. The graphite cavity is housed inside stainless steel and filled with an inert atmosphere to prevent oxidation of the graphite. Methane conversion of 95% at 2,073 K was reported [51].

A different solution to the indirect arrangement is the fluid-wall aerosol flow reactor. The reactor is composed of three concentric vertical tubes. The innermost is made of porous graphite, the central tube is solid graphite, and both are contained within a quartz outer tube. The sunlight heats the central solid tube, which then radiates to the porous tube. The radiation heat from the porous tube is absorbed by carbon particles formed in the gas stream as a result of the methane-splitting reaction. Argon is fed into the annular region between the two graphite tubes and the gas forced through the porous tube wall. It serves to protect the inner tube wall from carbon deposition and possible clogging. Argon was fed also into the annular space between the quartz tube and the solid graphite tube. Approximately 90% conversion of methane to H_2 was obtained at a reactor-wall temperature of 2,133 K with an average residence time of about 0.01 s [52]. The average size of the carbon-black particles was in the range 58–77 nm, depending on the flow speed.

Catalytic cracking of methane was also studied. Metal catalysts including Ni, Fe, Co, and Pd have been used for CH_4 decomposition. The major problem is their deactivation associated with the build-up of carbon on the catalyst surface. The carbon produced can be burned off with oxygen or air, or gasified with steam to regenerate the original activity of the metal catalyst. Carbon-based catalysts have also been studied. They offer the advantage of avoiding both the need for regeneration and the contamination of the product H_2 with carbon oxides. Methane decomposition rates on different activated carbons and carbon blacks have been reported [53]. Different types of reactor such as tubular, fluid-wall, spouted-bed, and fluidized-bed reactors have been analyzed [50]. A similar method can be applied to decarbonization of other hydrocarbons to generate H_2 with decreased CO_2 emissions [54]. Dehydrogenative coupling of methane is also a compelling endothermic reaction in solar thermochemistry.

The conversion of methane to ethylene, which is an important raw material in the industrial process of producing liquid fuels, proceeds as follows:

$$2CH_4 \rightarrow C_2H_4 + 2H_2 \qquad \Delta H^\circ_{298\,K} = 202\,kJ\,mol^{-1}. \quad (48.34)$$

Hydrogen is an additional product, and, when the reaction is carried out using solar energy, the calorific value of the methane is upgraded by 11%. The reaction proceeds catalytically. A catalyst based on $SnO_2/MgFe_2O_4/SiO_2$ was reported to achieve a high conversion rate and yield, with 92% C_2 selectivity at 1,223 K [55].

48.4.6 Solar pyrolysis/gasification of carbonaceous materials

Research has demonstrated the feasibility of solar-assisted gasification of carbonaceous materials to form primarily synthesis gas (syngas). The potential feedstocks include coal, oil-shale, residual oil, coke, and biomass. The advantages of such processing are yields of syngas with calorific value above those of the carbonaceous feedstocks, syngas quality suited to production of hydrogen, methanol, or bulk Fischer–Tropsch fuels, and the ability to process low-grade and waste materials with essentially no emission of CO_2 to the atmosphere.

Coal gasification involves two basic chemical steps: pyrolysis and char gasification. Pyrolysis proceeds when coal is heated in the absence of air and is essentially completed at 1,200 K according to [56].

The solar-energy requirement for the char gasification with steam of CO_2 is as follows:

$$C + H_2O(l) \rightarrow CO + H_2 \qquad \Delta H^\circ_{298\,K} = 175\,kJ\,mol^{-1}, \quad (48.35)$$

$$C + CO_2 \rightarrow 2CO \qquad \Delta H^\circ_{298\,K} = 172\,kJ\,mol^{-1}. \quad (48.36)$$

Because of these highly endothermic reactions the calorific values of the product syngas are ideally higher than that of the initial coal by 44%–45% [57]. Basic design options for solar coal-gasification reactors have been analyzed, such as moving-bed and fluidized-bed reactors. The concentrated sunlight enters the reactor through a window and is absorbed directly on a mixed layer of coal and ash. The reactor window is physically separated from the coal and maintained clean and relatively cool by a steam or CO_2 feed around the window [58]. In the steam gasification a typical dry composition of 54% H_2, 25% CO, 16% CO_2, and 4% CH_4 was obtained. Gasification of charcoal with steam and CO_2 in a packed-bed windowed reactor using a vertical solar furnace was also demonstrated [59]. Steam was generated by spraying water directly onto the surface of the charcoal while simultaneously heating the charcoal at the focus of the solar furnace. At an optimum rate of steam flow, about 50% of the steam reacted with the char.

Catalytic solar gasification with alkaline-earth metals and some transition-metal compounds has been studied extensively. Improvement of the kinetics and reduction of temperatures with improved conversion have been demonstrated [60][61]. Also the carboreduction of metal oxides, reaction (12) in Table 48.1, with charcoal can be considered as gasification of the carbon and its conversion to CO. At the same time the metal product is used also to store solar energy and release it as hydrogen through hydrolysis reaction (9) detached from the solar hours.

48.4.7 Thermochemical conversion of biomass

Biomass thermochemical conversion processes are, in general, endothermic, and the heat required can be supplied by CS energy in such a way that the total energy evolved from the products ideally represents the sum of the energy stored during the photosynthesis and the thermal processes.

The biomass sources can be wood (including residues like bark, sawdust, etc.), agricultural waste (like straw and husks), urban waste, industrial organic waste, and aquatic biomass. The main constituent of the biomass is cellulose, which is a linear polymer with a degree of polymerization of up to 10,000-carbon-atom anhydrous-glucose sugar units. The cellulose fibers are held together in a matrix of lignin and hemicellulose (cf. Ch. 26). The elementary analysis of wood, for example, leads to an average mean formula of $CH_{1.44}O_{0.66}$, which is valid for a wide variety of sources [62]. The oxygen content of this biomass is therefore 40–45 wt% on a dry and ash-free basis.

The main solar thermal routes for biomass conversion (apart from combustion) are gasification and pyrolysis. Biomass is gasified at high temperatures (above 850 °C) and the conversion process can be expressed as

$$\text{biomass} + \text{heat} + \text{steam} \rightarrow H_2 + CO + CH_4 + \text{tars} + \text{char.} \quad (48.37)$$

In a conventional gasification process the heat is supplied by partial oxidation of the biomass either with air (low-heating-value gas is produced, containing up to 60% N_2 and having a typical heating value of 4–6 MJ $N^{-1}m^{-3}$) or with oxygen (which yields a better quality of gas with a heating value of 10–15 MJ N^{-1} m^{-3}, but requires an O_2 supply from an air-separation plant). These drawbacks clearly explain why solar energy has been considered to provide the heat: it provides a high calorific value of the products, less CO_2, and no N_2.

Most solar gasifiers have used a fluidized-bed reactor [63] or free-falling particles with direct illumination. Steam gasification of bio-char was also performed via indirect heating, whereby particles of beech-wood charcoal were used as the biomass feedstock in a continuous steam–particle flow through a 3-kW tubular absorber [64]. A previous heat-transfer model [65] was applied to couple the heat-transfer mechanisms to the reaction kinetics and was validated experimentally.

A scale-up of the reactor to 1 MW is planned, with an expected solar-to-chemical energy-conversion efficiency of 50%.

Pyrolysis of biomass is heating it in the absence of air to decompose it into liquid oil, solid charcoal, and gaseous compounds. This pyrolysis can be further classified into slow and fast processes. In the slow process, charcoal yields are as high as 42%–62% [66], depending on the moisture content of the feed and the pressure in the reactor. Fast (flash) pyrolysis requires rapid heating, which can be achieved with direct heating exploiting the CS radiation. These conditions favor gas production with high fractions of CO and H_2 and a calorific value of 18 MJ $N^{-1}m^{-3}$.

The major problem is the poor heat- and mass-transfer characteristics of the biomass feedstocks which screen the incoming solar radiation, causing two consequences: the internal layers of virgin biomass receive less heating and the radiation intensity is attenuated. Therefore the flash pyrolysis process deteriorates with increasing formation of charcoal. Another constraint associated with direct solar heating is connected to the use of a transparent window. The product liquids (tar, oil) and dust (charcoal, ash) which are produced in the form of aerosols have to be prevented from reaching the window.

Flash pyrolysis was mainly considered for production of bio-oils. The oils are complex mixtures of a large number of species derived from almost complete depolymerization of the biomass components.

Although fraction of oils higher than 70% can be attained, the crude bio-oils need further upgrading in order to match them to transportation applications. Another approach is to maximize the gaseous products, primarily H_2, CO, CO_2, and CH_4, which can be processed via FTS to liquid fuels [67]. The methane can be steam re-formed to produce more hydrogen. The liquid products can be catalytically converted into gases. Since tar is difficult to gasify, extensive studies on the catalytic effect of dolomite (CaO plus K_2CO_3) and various metal oxides (ZrO_2, TiO_2, and Cr_2O_3) have been published [68]. A novel approach to overcome the heat-transfer limitation in fluidized- or moving-bed solar reactors and to catalytically shift the flash pyrolysis to gaseous products, minimizing the tar and char fractions, using a molten-salt mixture as the medium into which the biomass particles are fed has been demonstrated (see the schematic layout in Figure 48.10). Eutectic mixtures of sodium (56 wt%) and potassium carbonates were used (melting point 983 K). The reactor was heated to 1,123 K, and above 95% conversion to gaseous products was reported [69]. Tar and other condensates were totally eliminated. A large-scale solar reactor is conceptually described in [69]. The molten-salt medium can also be used as a storage medium to smooth operation in case of the passage of clouds.

Figure 48.10. A solar biomass gasifier in a molten salt medium.

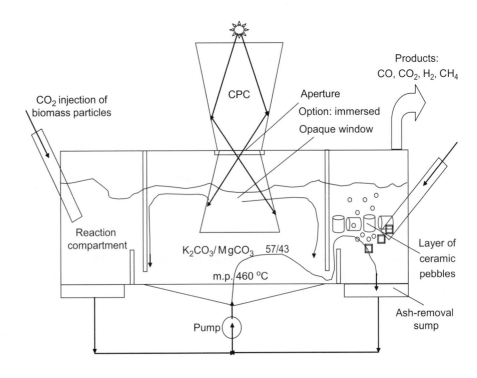

48.4.8 Future outlook

Both the US DOE and the European Commission (EC) have set clear targets for H_2 production costs. The US targets for 2012 and 2017 were $6 and $3 per gasoline gallon equivalent (gge; 1 gge is about 1 kg of H_2), respectively, and the EC target for 2020 is €3.5 per kg of H_2. Predicted solar thermal hydrogen-production costs are $2–4 per kg of H_2 for selected solar thermochemical cycles. The economics of massive solar H_2 production have been assessed in numerous studies, which have indicated that solar H_2 produced thermochemically can be competitive compared with other methods.

In a comparison study, both the hybrid-sulfur cycle and a metal-oxide-based cycle were operated by solar-tower technology for multistage water splitting [70]. The electricity required for the alkaline electrolysis was produced by a parabolic-trough power plant. For each process, the investment, operating, and hydrogen-production costs were calculated on a 50-MW$_{th}$ scale. The authors point out the economic potential of sustainable hydrogen production using solar energy and thermochemical cycles compared with commercial electrolysis. A sensitivity analysis was done for three different cost scenarios. Hydrogen-production costs ranged from €3.9 to €5.6 per kg for the hybrid-sulfur cycle, from €3.5 to €12.8 per kg for the metal-oxide-based cycle, and from €2.1 to €6.8 per kg for alkaline electrolysis. The weaknesses of these economic assessments are primarily related to the uncertainties in the viable efficiencies and investment costs of the various solar components due to their early stage of development and their economy of scale [71].

Solar-produced synthesis gas, a mixture of predominantly CO and H_2 obtained either from reforming of natural gas or biogas or via gasification of biomass or coal, can be converted into a multi-component mixture of hydrocarbons in a commercial Fischer–Tropsch process. Fuels produced with the Fischer–Tropsch (Chapter 39) process are of high quality due to their very low aromaticity and the absence of sulfur. These fuels can also be used as blending stocks for transportation fuels derived from crude oil. Other valuable products besides fuels can be tailor-made with the Fischer–Tropsch process in combination with upgrade processes: for example, ethane, propene, α-olefins, ketones, solvents, alcohols, and waxes.

The development of large-scale use of solar hydrogen as an energy carrier requires the development of end-use technologies, construction of appropriate infrastructure, and significant cost reduction in order to compete with alternative technologies.

Hydrogen can be a complement fuel for future hybrid vehicles in addition to biofuels and renewable electricity. Safety factors involved in its distribution and storage on board vehicles are considerable challenges. The option of producing the hydrogen on board using intermediate compounds and their recycling using CS energy is an interesting approach that could solve the safety and storage issues but requires systematic R&D efforts.

Numerous challenges regarding components, materials, and operational issues have to be resolved in connection with solar reactors, catalysts, and the intermittent nature of the Sun as an energy resource.

In the near future (the next 10–15 years) solar reforming, gasification of biomass, and carboreduction of metal oxides could be the main technologies that can be scaled up to pilot demonstration. To minimize CO_2 emission, the carbon used as a reducing agent should come from biomass resources. Carbon dioxide can also be separated from various flue-gas streams (e.g., coal power plants, industrial process) and recycled to reform natural gas to produce synthesis gas followed by synthesis of liquid fuels. In parallel to the scale-up efforts there is huge scope for R&D efforts in developing efficient catalysts and high-temperature materials for construction of solar reactors that can reliably withstand the severe operating conditions.

In the next phase, 15–25 years into the future, the most promising CO_2-free thermochemical processes should reach pilot-scale demonstration. It is vital to know the limitations of materials and manufacturing technologies required in order to structure the solar reactors. As a result of this, the size of a single module of a solar reactor should be optimized in close connection with the relevant solar concentrating technology.

Solar energy is free, abundant, and inexhaustible, but the two phases described above will be necessary for successful market penetration of solar fuels and hydrogen.

48.5 Summary

This chapter describes several high-temperature thermochemical processes to convert carbon-containing materials to hydrogen and synthesis gas with the aid of CS energy. The carbonaceous materials can be biomass, organic wastes, biogas, CO_2, and natural gas. The products can be further processed to yield synthetic liquid fuels for transportation and industry.

The processes are reforming of methane with steam, CO_2, or a mixture thereof, methane cracking, and gasification. Since the upgrading of synthesis gas to liquid fuels involves continuous industrial processes, the storage of solar energy in chemical form becomes critical. For that purpose the carboreduction of metal oxides is performed during the sunny hours. The product metals can be easily stored and react later, detached from the solar operational hours, with steam, CO_2, or a mixture of the two, to continuously produce the syngas and recover the metal oxide for processing the next day.

These thermochemical processes have the potential of achieving solar-to-fuel efficiencies exceeding 40% and therefore of becoming viable options for the efficient, cost-effective, clean, and industrial-scale production of hydrogen and liquid fuels.

48.6 Questions for discussion

1. Develop the correlation among receiver efficiency, solar concentration ratio, and working temperatures for non-blackbody ($\alpha = \varepsilon \neq 1.0$) and for selective-body ($\alpha \neq \varepsilon$) systems.
2. Compare direct water splitting vs. two-step thermochemical splitting using the redox pair Fe_3O_4/FeO from the energetic point of view.
3. Analyze the option of CO_2 capture from coal-power-plant flue gases and its fixation by solar reforming of methane and find out the final CO_2 emission per unit energy output and the solar contribution.
4. Find the theoretical yield of H_2 as a function of the O_2 content in various biomass feeds and compare biological and thermochemical pathways from biomass to hydrogen (e.g., anaerobic digestion, fermentation, gasification, and pyrolysis).
5. Compare life-cycle analysis for the production of ethanol as an alternative fuel via biological and thermochemical routes from the energetic point of view.

48.8 References

[1] **E. Fletcher**, 1984, "On the thermodynamics of solar energy use," *J. Minnesota Acad. Sci.*, **49**(2), 30–34.
[2] **A. Steinfeld** and **M. Schubnell**, 1993, "Optimum aperture size and operation temperature for a solar cavity-receiver," *Solar Energy*, **50**(1), 19–25.
[3] **E. Fletcher** and **R. Moen**, 1977, "Hydrogen and oxygen from water," *Science*, **197**(4308), 1050–1056.
[4] **R. Palumbo**, **J. Lédé**, **O. Boutin** *et al.*, 1998, "The production of Zn from ZnO in a high-temperature solar decomposition quench process, I. The scientific framework for the process," *Chem. Eng. Sci.*, **53**(14), 2503–2517.
[5] **I. Vishnevetsky**, **A. Berman**, and **M. Epstein**, 2009, "Boron, zinc, tin and cadmium as candidates for thermal chemical redox cycles for solar hydrogen production," in *Proceedings of SolarPACES 2009 Conference*, Berlin, paper no. 11168.
[6] **D. Fraenkel**, **R. Levian**, and **M. Levy**, 1986, "A solar thermochemical pipe based on the CO_2–CH_4 (1:1) system," *Int. J. Hydrogen Energy*, **11**(4), 267–277.
[7] **M. Levy**, **R. Levitan**, **H. Rosin**, and **R. Rubin**, 1993, "Solar energy storage via a closed-loop chemical heat pipe," *Solar Energy*, **50**(2), 179–189.
[8] **U. Fisher**, **C. Sugerman**, **R. Tamme**, **R. Buck**, and **M. Epstein**, 2000, "Solar upgrading of fuels for generation of electricity," in *Proceedings of the 10th SolarPACES Symposium*, Sydney, pp. 19–20.
[9] **J. Richardson** and **S. Paripatyadar**, 1990, "Carbon dioxide reforming of methane with supported rhodium," *Appl. Catalysis*, **61**, 293–309.
[10] **A. Wörner** and **R. Tamme**, 1998, "CO_2 reforming of methane in a solar-driven volumetric receiver–reactor," *Catalysis Today*, **46**, 165–174.

[11] A. Berman, R. K. Karn, and M. Epstein, 2005, "Kinetics of steam reforming of methane on Ru/Al₂O₃ catalyst promoted with Mn oxides," *Appl. Catalysis A: General*, **282**(1–2), 73–83.

[12] A. Berman, R. K. Karn, and M. Epstein, 2006, "A new catalyst system for high temperature solar reforming of methane," *Energy Fuels*, **20**(2), 455–462.

[13] M. Böhmer, U. Langnickel, and M. Sanchez, 1991, "Solar steam reforming of methane," *Solar Energy Mater.*, **24**, 441–448.

[14] J. T. Richardson, S. A. Paripatyadar, and J. C. Shen, 1988, "Dynamics of sodium heat pipe reforming reactor," *AICHE J.*, **34**, 743.

[15] R. B. Diver, J. D. Fish, R. Levitan *et al.*, 1992, "Solar test of an integrated sodium reflux heat pipe receiver/reactor for thermochemical energy transport," *Solar Energy*, **48**(1), 21–30.

[16] R. Levitan, H. Rosin, and M. Levy, 1989, "Chemical reactions in a solar furnace – direct heating of the reactor in a tubular receiver," *Solar Energy*, **42**(3), 267–272.

[17] M. Levy, R. Levitan, H. Rosin, and R. Rubin, 1993, "Solar energy storage via a closed-loop chemical heat pipe," *Solar Energy*, **50**(2), 179–189.

[18] M. Epstein, I. Spiewak, A. Segal *et al.*, 1996, "Solar experiments with tubular reformer," in *Proceedings of the 8th International Symposium on Solar Thermal Concentrating Technology*, Cologne, vol. 3, pp. 1209–1229.

[19] R. Scocyper Jr., R. Hogan, and J. Muir, 1994, "Solar reforming of methane in a direct absorption catalytic reactor on a parabolic dish II. Modeling and analysis," *Solar Energy*, **52**(6), 479–490.

[20] R. Buck, M. Abele, H. Bauer, A. Seitz, and R. Tamme, 1994, "Development of a volumetric receiver–reactor for solar methane reforming," *ASME J. Solar Energy Eng.*, **116**, 73–78.

[21] M. Abele, H. Bauer, R. Buck, R. Tamme, and A. Wörner, 1996, "Design and test results of a receiver–reactor for solar methane reforming," *ASME J. Solar Energy Eng.*, **118**, 339–346.

[22] R. Ben-Zvi and J. Karni, 2007, "Simulation of a volumetric solar reformer," *ASME J. Solar Energy Eng.*, **129**, 197–204.

[23] A. Berman, K. K. Rakesh, and M. Epstein, 2007, "Steam reforming of methane on a Ru/Al₂O₃ catalyst promoted with Mn oxides for solar hydrogen production," *Green Chem.*, **9**, 626–631.

[24] A. Kiyama, Y. Kondoh, T. Yokohama, K. I. Shimizu, and T. Kodama, 2002, "New catalytically-activated metal/ceramic foam absorber for solar reforming receiver–reactor," in *Proceedings of the 11th Solar-PACES Symposium*, Zurich, pp. 337–343.

[25] T. Kodama, H. Ohtake, S. Matsumoto, *et al.*, 2000, "Thermochemical methane reforming using a reactive WO₃/W redox system," *Energy*, **20**, 411–425.

[26] A. Kogan, 1998, "Direct solar thermal splitting of water and on-site separation of the products, II. Experimental feasibility study," *Int. J. Hydrogen Energy*, **23**(2), 89–98.

[27] J. E. Funk, 2001, "Thermochemical hydrogen production: past and present," *Int. J. Hydrogen Energy*, **16**, 185–190.

[28] S. Abenades, P. Charvin, G. Flamant, and P. Neven, 2006, "Screening of water-splitting thermochemical cycles potentially attractive for hydrogen production by concentrated solar energy," *Energy*, **31**, 2805–2822.

[29] A. Steinfeld, P. Kuhn, A. Reller, *et al.*, 1998, "Solar processed metals as clean energy carriers and water-splitters," *Int. J. Hydrogen Energy*, **23**, 767–774.

[30] A. Steinfeld, 2002, "Solar hydrogen production via a 2-step water-splitting thermochemical cycle based on Zn/ZnO redox reactions," *Int. J. Hydrogen Energy*, **27**, 611–619.

[31] A. Weidenkaff, A. Reller, F. Sibiende, A. Wokaun, and A. Steinfeld, 2000, "Experimental investigations on the crystallization of zinc by direct irradiation of zinc oxide in a solar furnace," *Chem. Mater.*, **12**, 2175–2181.

[32] L. O. Schunk, W. Lipinski, and A. Steinfeld, 2009, "Heat transfer model of a solar receiver–reactor for the thermal dissociation of ZnO – experimental validation at 10 kW and scale-up to 1 MW," *Chem. Eng. J.*, **150**, 502–508.

[33] L. Schunk, P. Haeberling, S. Wepf *et al.*, 2008, "A solar receiver–reactor for the thermal dissociation of zinc oxide," *ASME J. Solar Energy Eng.*, **130**, 021009.

[34] T. Nakamura, 1997, "Hydrogen production from water utilizing solar heat at high temperatures," *Solar Energy*, **19**, 467–475.

[35] F. Sibiende, M. Ducavior, A. Tofighi, and J. Ambriz, 1982, "High-temperature experiments with solar furnace: the decomposition of Fe₃O₄, Mn₃O₄, CdO," *J. Hydrogen Energy*, **7**, 79–88.

[36] T. Tamaura, A. Steinfeld, P. Kuhn, and K. Ehrensberger, 1995, "Production of solar hydrogen by a novel, 2-step, water-splitting thermochemical cycle," *Energy*, **20**(4), 325–330.

[37] C. Agrafiotis, M. Roeb, A. G. Konstandopoulos *et al.*, 2005, "Solar water splitting for hydrogen production with monolithic reactors," *Solar Energy*, **79**, 409–421.

[38] S. Lorentzou, A. Zygogianni, K. Tousimi, C. Agrafiotis, and A. F. Konstandopoulis, 2009, "Advanced systhesis of nanostructured materials for environmental applications," *J. Alloys Compounds*, **483**, 302–305.

[39] C. Wieckert, U. Frommherz, S. Kraul *et al.*, 2007, "A 300 kW solar chemical pilot plant for the carbothermic production of zinc," *ASME J. Solar Energy*, **129**, 191–196.

[40] M. Epstein, G. Olalde, S. Santen, A. Steinfeld, and C. Wieckert, 2008, "Towards the industrial solar carbothermal production of zinc," *ASME J. Solar Energy*, **130**, 104505:1–4.

[41] A. Berman and M. Epstein, 2000, "The kinetics of hydrogen production in the oxidation of liquid zinc with water vapour," *Int. J. Hydrogen Energy*, **25**, 957–967.

[42] I. Vishnevetsky and M. Epstein, 2007, "Production of hydrogen from solar zinc in steam atmosphere," *Int. J. Hydrogen Energy*, **32**, 2791–2802.

[43] **I. Vishnevetsky** and **M. Epstein**, 2009, "Tin as a possible candidate for solar thermochemical redox process for hydrogen production," *ASME J. Solar Energy Eng.*, **131**, 021007:1–8.

[44] **I. Vishnevetsky**, **M. Epstein**, **T. Abu-Hamed**, and **J. Karni**, 2008, "Boron hydrolysis at moderate temperatures: first step to solar fuel cycle for transportation," *ASME J. Solar Energy Eng.*, **130**, 014506:1–5.

[45] **D. Hirsch**, **M. Epstein**, and **A. Steinfeld**, 2001, "The solar thermal decarbonization of natural gas," *Int. J. Hydrogen Energy*, **26**(10), 1023–1033.

[46] **A. Holman**, **O. Olsvik**, and **O. A. Rokstad**, 1995, "Pyrolysis of natural gas: chemistry and process concepts," *Fuel Process Technol.*, **42**(2–3), 249–267.

[47] **A. Kogan**, **M. Kogan**, and **S. Barak**, 2005, "Production of hydrogen and carbon by solar thermal methane splitting. III. Fluidization, entrainment and seeding particles into a volumetric solar receiver," *Int. J. Hydrogen Energy*, **30**, 35–43.

[48] **A. Kogan**, **M. Israeli**, and **E. Alcobi**, 2007, "Production of hydrogen and carbon by solar thermal methane splitting. IV. Preliminary simulation of a confined tornado flow configuration by computational fluid dynamics," *Int. J. Hydrogen Energy*, **32**, 4800–4810.

[49] **G. Maag**, **G. Zanganeh**, and **A. Steinfeld**, 2009, "Solar thermal cracking of methane in a particle-flow reactor for the co-production of hydrogen and carbon," *Int. J. Hydrogen Energy*, **34**, 7676–7685.

[50] **D. Hirsch** and **A. Steinfeld**, 2004, "Solar hydrogen production by thermal decomposition of natural gas using a vortex-flow reactor," *Int. J. Hydrogen Energy*, **29**, 47–55.

[51] **S. Rodat**, **S. Abanades**, and **G. Flamant**, 2009, "High temperature solar methane dissociation in a multitubular cavity-type reactor in the temperature range 1823–2073 K," *Energy Fuels*, **23**, 2666–2674.

[52] **J. K. Dahl**, **K. J. Buechler**, **A. W. Weimer**, **A. Lewandowski**, and **C. Bingham**, 2004, "Solar-thermal dissociation of methane in a fluid-wall aerosol flow reactor," *Int. J. Hydrogen Energy*, **29**, 725–773.

[53] **N. Z. Muradov**, 2001, "Hydrogen via methane decomposition: and application for decarbonisation of fossil fuels," *Int. J. Hydrogen Energy*, **26**, 1165–1175.

[54] **N. Z. Muradov** and **T. N. Veziroğlu**, 2005, "From hydrocarbon to hydrogen–carbon to hydrogen economy," *Int. J. Hydrogen Energy*, **30**, 225–237.

[55] **T. Himizu**, **Y. Kitayama**, and **T. Kodama**, 2001, "Thermochemical conversion of CH_4 to C_2-hydrocarbons and H_2 over $SnO_2/FeO_4/SiO_2$ in methane–water-co-feed system," *Energy Fuels*, **15**, 463–469.

[56] **D. W. Gregg**, **R. W. Taylor**, and **J. H. Campbell**, 1980, "Solar gasification of coal, activated carbon, coke and coal and biomass mixtures," *Solar Energy*, **25**. 353–364.

[57] **T. Kodama**, 2003, "High-temperature solar chemistry for converting solar heat to chemical fuels," *Prog. Energy Combustion Sci.*, **29**, 567–597.

[58] **D. W. Gregg**, **W. R. Aiman**, **H. H Otsuki**, and **C. B. Thorsness**, 1980, "Solar coal gasification," *Solar Energy*, **24**, 313–321.

[59] **R. W. Taylor**, **R. Berjoan**, and **J. P. Coutures**, 1983, "Solar gasification of carbonaceous materials," *Solar Energy*, **30**(6), 513–525.

[60] **F. Kapteijn**, **H. Porre**, and **J. Moulijn**, 1986, "CO_2 gasification of activated carbon catalyzed by alkaline earth elements," *AIChE J.*, **32**(4), 691–695.

[61] **H. Ohme** and **T. Suzuki**, 1996, "Mechanisms of CO_2 gasification of carbon catalyzed with group VIII metals, I. Iron catalyzed CO_2 gasification," *Energy Fuels*, **10**, 987–987.

[62] **X. Deglise** and **J. Lede**, 1982, "The upgrading of the energy of biomass by thermal methods," *Int. Chem. Eng.*, **22**, 631–646.

[63] **J. P. Murray** and **E. A. Fletcher**, 1994, "Reaction of steam with cellulose in a fluidized bed using concentrated sunlight," *Energy*, **19**, 1083–1098.

[64] **T. Melchior**, **C. Perkins**, **P. Lichty**, **A. W. Weimer**, and **A. Steinfeld**, 2009, "Solar-driven biochar gasification in particle-flow reactor," *Chem. Eng. Processing*, **48**, 1279–1287.

[65] **T. Melchior**, **C. Perkind**, **A. W. Weiner**, and **A. Steinfeld**, 2008, "A cavity-receiver containing tubular absorber for high-temperature thermodynamical processing using concentrated solar energy," *Int. J. Thermal Sci.*, **47**, 1496–1503.

[66] **M. J. Antal**, **E. Croiset**, **X. Dai** *et al.*, 1996, "High-yield biomass charcoal," *Energy Fuels*, **10**, 652–658.

[67] **C. Perkins** and **A. W. Weiner**, 2009, "Solar thermal production of renewable hydrogen," *AIChE J.*, **55**, 286–293.

[68] **D. Sutton**, **B. Kelleher**, and **J. R. H. Ross**, 2001, "Review of literature on catalysts for biomass gasification," *Fuel Processing Technol.*, **73**, 155–173.

[69] **R. Adinberg**, **M. Epstein**, and **J. Karni**, 2004, "Solar gasification of biomass: a molten salt pyrolysis study," *J. Solar Energy Eng.*, **126**, 851–857.

[70] **T. Pregger**, **D. Graf**, **W. Krewitt** *et al.*, 2009, "Prospects of solar thermal hydrogen production processes," *Int. J. Hydrogen Energy*, **34**, 4256–4267.

[71] **A. Meier** and **A. Steinfeld**, 2010, "Solar thermochemical production of fuels," *Adv. Sci. Technol.*, **74**, 303–312.

49 Photoelectrochemistry and hybrid solar conversion

Stuart Licht

Department of Chemistry, George Washington University, Washington, DC, USA

49.1 Focus

Photoelectrochemistry studies photo-driven electrochemical processes (light-driven processes which inter-convert electrical and chemical energy). As with photovoltaics, photoelectrochemical processes can directly convert sunlight into electricity, but have the additional capabilities of being able to store energy, as in solar batteries, or to directly convert solar energy to chemical energy, as in the production of hydrogen fuel or disinfectants. The challenges involved, which have impeded the development of photoelectrochemical devices, can include corrosion, lower solar-energy-conversion efficiency, and packaging vulnerabilities of liquid systems. (i) Dye-sensitized solar cells, (ii) STEP energetic chemical generation and (iii) photoelectrochemical waste treatment are technologies that address many of these challenges.

49.2 Synopsis

Society's electrical needs are largely continuous. However, clouds and darkness dictate that photovoltaic (PV) solar cells have an intermittent output. A photoelectrochemical solar cell (PEC) can generate not only electrical but also electrochemical energy, thereby providing the basis for a system with an energy-storage component. Sufficiently energetic insolation incident on semiconductors can drive electrochemical oxidation/reduction and generate chemical, electrical, or electrochemical energy. Aspects include efficient dye-sensitized or direct solar-to-electrical energy conversion, solar electrochemical synthesis (electrolysis), including the splitting of water to form hydrogen, the generation of solar fuels, environmental cleanup, and solar-energy-storage cells. The PEC utilizes light to carry out an electrochemical reaction, converting light to both chemical and electrical energy. This fundamental difference between the PV solar cell's solid/solid interface and the PEC's solid/liquid interface has several ramifications in cell function and application. Energetic constraints imposed by single-bandgap semiconductors have limited the demonstrated values of photoelectrochemical solar-to-electrical energy-conversion efficiency for and using multiple-bandgap tandem cells can lead to significantly higher conversion efficiencies. Photoelectrochemical systems not only may facilitate solar-to-electrical energy conversion, but also have led to investigations into the solar photoelectrochemical production of fuels, photoelectrochemical detoxification of pollutants, and efficient solar thermal electrochemical production (STEP) of metals, fuels, and bleach, and carbon capture.

49.3 Historical perspective

Photoelectrochemistry and PV share a common history, dating back to the nineteenth century when in 1839 Becquerel discovered the PV effect while experimenting with two metal electrodes in an aqueous solution [1]. Today, Becquerel's configuration would be called a photoelectrochemical PV cell. Up until the middle of the twentieth century photoelectrochemistry was not an area of special interest, and was studied more through interest in related fields. The pioneering work of Gerischer on the fundamental energetics of photoelectrochemical systems formalized the science of photoelectrochemistry after the mid twentieth century [2][3]. In 1972, Fujishima and Honda demonstrated that a wide-bandgap, E_g, semiconductor such as titanium dioxide, TiO_2 ($E_g = 3.4$ eV), can use near-UV radiation to split water and form hydrogen [4]. This study has also been the basis for photoelectrochemical-driven disinfectants and waste treaments [5].

With the discovery of more efficient photoelectrochemical solar cells in the mid 1970s by research groups at the Weizmann Institute of Science [6], Bell Labs [7], and MIT [8], the field of photoelectrochemistry blossomed. These studies provided a basis for the wide exploration of a large variety of semiconductors and redox couples, as summarized in early reviews [9][10][11] and more recent monographs on photoelectrochemistry [12][13] and solar-driven hydrogen fuel [14][15].

A photoelectrochemical solar cell with *in situ* storage was demonstrated in the 1970s [6][16]. During this period, photoelectrochemical etching as a useful surface treatment for the semiconductor industry was developed [17][18], and the understanding of solution-effect limitations on photoelectrochemical charge transfer led to semiconductor-photoabsorber photoelectrochemical solar cells with increasing solar-to-electrical energy conversion efficiency [19][20][21][22].

Rather than a semiconductor as the photoabsorber, Tributsch demonstrated that a coating of dye on a semiconductor film (zinc oxide) can drive photoelectrochemical charge transfer [23][24]. Gratzel's discovery in 1991 of a low-cost, high-efficiency solar cell based on dye-sensitized colloidal TiO_2 films led to intense interest in these systems [25][26]. The solar-to-electrical efficiency of dye-sensitized solar cells remains lower than those of solid-state devices such as thin film, silicon, and concentrator PV.

In 2002, the issue of light-driven endothermic electrochemical formation of energetic molecules was brought to the forefront [27][28], demonstrating that a small-bandgap, E_g, semiconductor such as Si ($E_g = 1.12$ eV) could drive water splitting ($E°(25 °C) = 1.23$ V), when excess solar heat was used to decrease the effective potential needed for electrolysis [29]. As summarized in Section 49.4.5, this STEP process was generalized in 2009 from water splitting to all endothermic electrolytic processes, including solar efficient carbon capture and the formation of carbon fuels, bleach, and metals [30].

49.4.1 Semiconductor/electrolyte electrical energy conversion

Radiation incident on semiconductors can drive electrochemical oxidation/reduction and generate chemical, electrical, or electrochemical energy. Light is absorbed by the semiconductor and drives charge generation. If charge recombination occurs, heat, rather than useful work, is the result. Useful work requires charge separation.

Photo-driven semiconductor/electrolyte processes maintain similarities with solid-state solar cells (Chapter 18). In traditional (PV, rather than photoelectrochemical) solar cells, light is absorbed, which induces formation of an electron–hole pair. Electron–hole separation occurs across a space-charge field gradient formed at a p–n junction (the intimate contact of p-doped and n-doped semiconductors). The maximum photopotential of a single p–n junction, VOC, is constrained by the photo-current through the junction and dark current I_D (Chapter 18.4.3).

Rather than a p–n junction, the space-charge field may be formed by a single p- or n-type semiconductor in contact with another material, such as a metal or an electrolyte. In illuminated semiconductor systems the absorption of photons generates excited electronic states. These excited states have lifetimes of limited duration. Without a mechanism of charge separation their intrinsic energy would be lost through relaxation (recombination). Several distinct mechanisms of charge separation have been considered in designing efficient photoelectrochemical systems. These include the effect of the band bending in the semiconductor and the charge in the double layer.

Depending on the relative rates of charge transfer, I_D may be constrained by either solid-state or electrochemical limitations, and is respectively termed the saturation current, or the equilibrium exchange current [11].

This concept of carrier generation is illustrated in Figure 49.1 (for an n-type PEC), and has been the theoretical basis for several efficient semiconductor/redox-couple PEC cells. Illumination of the electrode surface with light whose photon energy is greater than the bandgap promotes electrons into the conduction band, leaving holes in the valence band. In the case of a photoanode, band bending in the depletion region

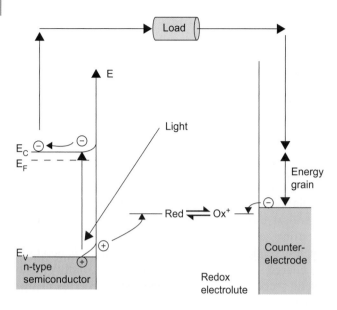

Figure 49.1. Carrier generation under illumination arising at the semiconductor/liquid interface.

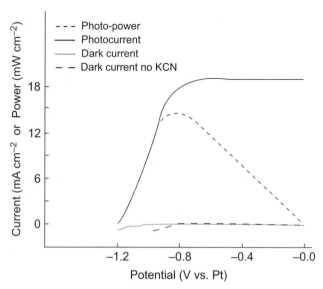

Figure 49.2. Outdoor PEC current–voltage characteristics for an illuminated, or dark, n-CdSe electrode in a solar cell with 0.25 M $K_4Fe(CN)_6$, 0.0125 M $K_3Fe(CN)_6$, 0.1 M KCN, and 0.5 M KOH. The measured solar-to-electrical conversion efficiency was 16(±0.4)%.

drives electrons in the conduction band into the interior of the semiconductor and eventually to the contact, and drives holes in the valence band toward the electrolyte, where they participate in an oxidation reaction. The electrons removed through the contact drive an external load and eventually reach the counter-electrode or storage electrode, where they participate in a reduction process. Under illumination under open-circuit conditions, a negative potential is created in a photoanode, and as a result the Fermi level for the photoanode shifts in the negative direction, thus reducing the band bending. Under illumination with increasing intensity, the semiconductor Fermi level shifts continually toward negative potentials until the band bending effectively is reduced to zero, which corresponds to the flat-band condition. At this point a photoanode exhibits its maximum photovoltage, which is equal to the barrier height.

Electrolyte modification, and an understanding of the distribution (speciation) and role of chemical species in the solution, can substantially impact photoelectrochemical charge transfer and lead to improved solar-to-electrical conversion efficiency. Fundamental understanding of the photo-oxidized species, the nature of the counter-ion, and competing reactions can overcome limits to photoelectrochemical charge transfer. Such effects have been shown in sulfide, selenide, iodide, and ferrocyanide electrolytes [31]. For example, additions to a ferri/ferrocyanide electrolyte can increase by 0.4 V the photopotential of an immersed, illuminated n-CdSe photoelectrode, and the addition of cesium cations can also improve photoelectrochemical charge transfer [19].

49.4.2 Semiconductor/electrolyte electrochemical energy storage

Photoelectrochemical solar cells can generate not only electrical but also electrochemical energy. Figure 49.3. presents one configuration of a PEC combining *in situ* electrochemical storage and solar conversion capabilities, providing continuous output that is insensitive to daily variations in illumination. A cell configuration of this type with high solar-to-electric conversion efficiency was demonstrated in 1987 and utilized a Cd(Se, Te)/S_x^{2-} conversion half-cell and an Sn/SnS storage system resulting in a solar cell with a continuous output [16]. Under illumination, as seen in Figure 49.3(a), the photocurrent drives an external load. Simultaneously, a portion of the photocurrent is used in the direct electrochemical reduction of metal cations in the device's storage half-cell. In darkness or below a certain level of light, the storage compartment spontaneously delivers power, by metal oxidation, as shown in Figure 49.3(b).

A variety of two-electrode configurations have been investigated as PEC storage systems. Important variations of these photoelectrochemical conversion and storage configurations are summarized in Table 49.1. In each case, and as summarized in Figure 49.3 for the simplest configurations, exposure to light drives separate redox couples and a current through the external load. There is a net chemical change in the system, with an overall increase in free energy. In the absence of

Table 49.1. Two-electrode photoelectrochemical conversion and storage configurations. Components of these systems include a semiconductor photoelectrode (SPE) and counter-electrode (CE). At the electrode/electrolyte interface, redox couples A and B are in solution (|Redox|), counter-electrode-confined (|Redox$_{CE}$-CE), or confined to the semiconductor photo-electrode (SPE-Redox A$_{SPE}$|).

Scheme	Electrode 1	Electrolyte(s)	Electrode 2
I	SPE	| Redox A Redox B	| CE
II	SPE	| Redox A–membrane–Redox B	| CE
III	SPE	| Redox A	| Redox B$_{CE}$-CE
IV	SPE-Redox A$_{SPE}$	| Redox B	| CE
V	SPE	| Redox A–membrane–Redox B	| SPE

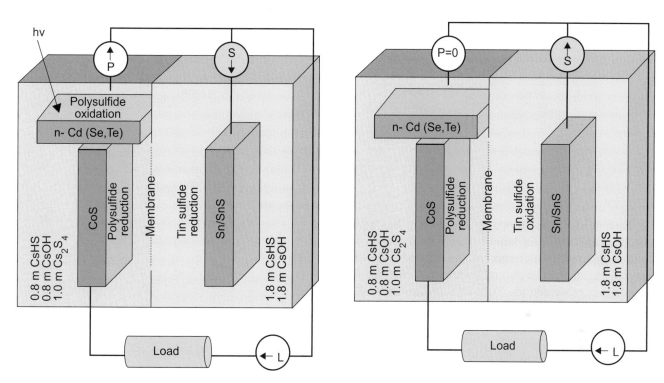

Figure 49.3. Schematic representations of a photoelectrochemical solar cell combining both solar conversion and storage capabilities: (a) under illumination and (b) in the dark.

illumination, the generated chemical change drives a spontaneous discharge reaction. The electrochemical discharge induces a reverse current in the storage electrode, but the same direction through the load as during light-induced charging. In each case in Table 49.1, exposure to light drives separate redox couples, and drives current through the external load.

Consistently with Figure 49.1, in a regenerative PEC, illumination drives work through an external load, without inducing a net change in the chemical composition of the system. This should be compared with the two-electrode PEC storage configurations shown in Figures

49.4(a) and (b). Unlike for a regenerative system, there is a net chemical change in the system, with an overall increase in free energy. In the absence of illumination, the generated chemical change drives a spontaneous discharge reaction. The electrochemical discharge induces a reverse current. Changes taking place in the system during illumination can be reversed in the dark utilizing two quasi-reversible chemical processes. Similarly to a secondary battery, the system discharges, producing an electric flow in the opposite direction and the system gradually returns to the same original chemical state.

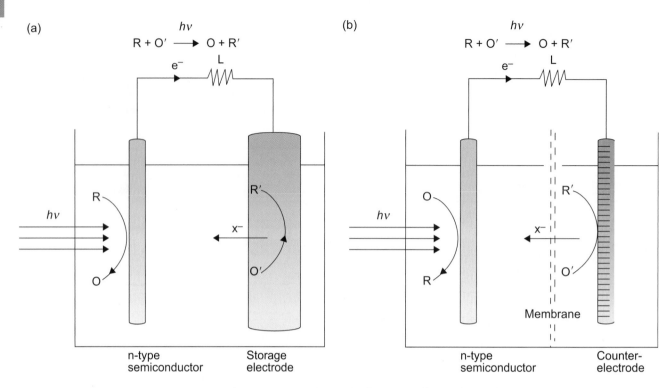

Figure 49.4. Schematic diagrams of two-electrode storage cells: (a) the storage electrode is an insoluble redox couple; and (b) the storage electrode has a soluble redox couple.

Each of the cells shown in Figure 49.4. has some disadvantages. Both for bound (Figure 49.4(a)) and for soluble (Figure 49.4(b)) redox couples, the redox species may chemically react with, and impair, the active materials of the photo-electrode. Furthermore, during the discharge process, the photo-electrode is kinetically unsuited to perform as a counter-electrode. In the absence of illumination, the, photo-electrode P, in this case a photoanode, now assumes the role of a counter-electrode by supporting a reduction process. For the photoanode to perform efficiently during illumination (charging), the very same reduction process should be inhibited, to minimize photo-oxidation back-reaction losses. Hence, the same photo-electrode cannot efficiently fulfill the dual role of being kinetically sluggish with respect to reduction during illumination and yet being kinetically facile with respect to the same reduction during dark discharge. The configuration represented in Figure 49.4 has another disadvantage, namely the disparity between the small surface area needed to minimize photocurrent dark-current losses, and the large surface area necessary to minimize storage polarization losses in order to maximize storage capacity.

Several of the disadvantages of the two-electrode configuration can be overcome by considering a three-electrode storage-cell configuration as shown in Figure 49.5. In Figure 49.5, the switches E and F are generally

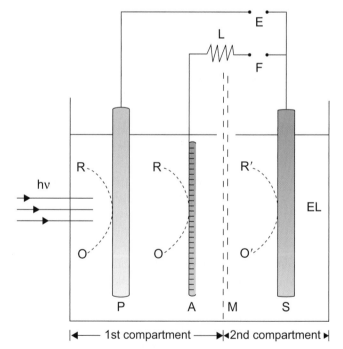

Figure 49.5. A schematic diagram for a storage system with a third electrode (counter-electrode) in the photo-electrode compartment. P, photo-electrode; A, counter-electrode; M, membrane; S, storage electrode; EL, electrolyte; E and F, electrical switches; and L, load.

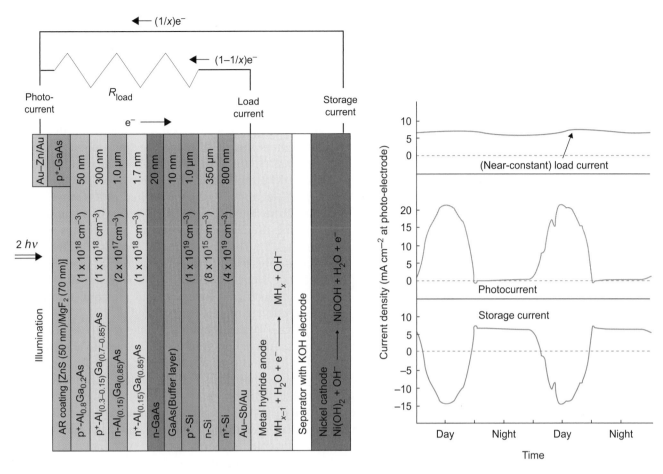

Figure 49.6. A solar conversion and storage cell with high solar-to-electrical efficiency (18%) [33][34].

alternated during charge and discharge. During charging, just switch E may be closed, facilitating the storage process; and during discharge E is kept open while F is closed. In this case chemical changes that took place during the storage phase are reversed, and a current flow is maintained from the storage electrode to a third (counter-)electrode, which is kept in the first compartment. To minimize polarization losses during the discharge, this third electrode should be kinetically fast with respect to the redox couple used in the first compartment.

A still further improved situation would be to have both switches closed all the time. In this case electric current flows from the photo-electrode both to the counter-electrode and to the storage electrode. The system is energetically tuned such that when insolation is available a significant fraction of the converted energy flows to the storage electrode. In the dark, or under diminished insolation, the storage electrode begins to discharge, driving continued current through the load. In this system a proper balance should be maintained between the potential of the solar energy conversion process and the electrochemical potential

of the storage process. There may be residual electric flow through the photo-electrode during dark cell discharge, since the photo-electrode is sluggish, but not entirely passive, with respect to a reduction process. This can be corrected by inserting a diode between the photo-electrode and the outer circuit.

Systems for photoelectrochemical charge storage using schemes I–V have recently been reviewed [32]. A logical evolution of scheme I in Table 49.1 is the case in which a PV cell drives a battery. This prevents any electrolytic corrosion of the semiconductor, since the semiconductor is only in electronic (wire) contact with the battery and never contacts the battery electrolyte. A high-efficiency version of this case is shown in Figure 49.6 and delivers a nearly constant power output, with little variation at night or under cloudy conditions.

The single cell in Figure 49.6 contains both multiple-bandgap and electrochemical storage, which, unlike conventional PV, provides a nearly constant energetic output in illuminated *and* in dark conditions [33], and exhibits excellent long-term stability [34]. The cell combines bipolar AlGaAs ($E_g = 1.6$ eV) and Si ($E_g = 1.0$ eV) and AB$_5$ metal hydride/NiOOH storage.

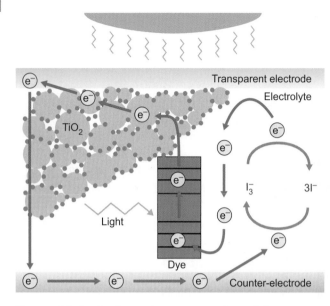

Figure 49.7. A schematic representation of the operation of a dye-sensitized solar cell.

Appropriate lattice matching between AlGaAs and Si is critical in order to minimize the dark current, provide ohmic contact without absorption loss, and maximize cell efficiency. The metal hydride/NiOOH storage process is near ideal for the AlGaAs/Si due to the excellent match of the storage and photocharging potentials. The electrochemical storage processes utilizes oxidation of the metal hydride (MH) and reduction of nickel oxyhydroxide:

$$\mathrm{MH + OH^- \rightarrow M + H_2O + e^-} \quad E_{\mathrm{M/MH}} = -0.8\,\mathrm{V\ vs.\ SHE},$$

$$\mathrm{NiOOH + H_2O + e^- \rightarrow Ni(OH)_2 + OH^-} \quad E_{\mathrm{NiOOH/Ni(OH)_2}} = 0.4\,\mathrm{V\ vs.\ SHE}.$$

49.4.3 Dye-sensitized solar cells

Excitation can also occur in molecules directly adsorbed and acting as mediators at the semiconductor interface. In this dye-sensitization mode, the function of light absorption is separated from charge-carrier transport. Photoexcitation occurs at the dye and photogenerated charge is then injected into a wide-bandgap semiconductor. This alternative carrier-generation mode also can lead to effective charge separation as illustrated in Figure 49.7.

The first example of a dye-sensitized solar cell with high solar-to-electric conversion efficiency was presented in 1991 [25]. Its high efficiency was achieved through the use of high-surface-area n-TiO$_2$ (nanostructured thin film), coated with a well-matched trimeric ruthenium-complex dye immersed in an aqueous poly-iodide electrolyte. The unusually high surface area of the transparent semiconductor coupled to the well-matched

spectral characteristics of the dye leads to a device that harvests a high proportion of insolation.

The functional components of a dye-sensitized solar cell are shown in Figure 49.7. An electrically conducting transparent electrode is generally made of fluorine-doped tin dioxide, SnO$_2$, which is deposited onto glass. This is coated with nanoporous, very-high-surface-area TiO$_2$ onto which is adsorbed a monolayer of dye. Sunlight enters the cell through the transparent electrode and strikes the dye. Photogenerated electrons are injected into the TiO$_2$, where they diffuse to the transparent electrode. The photoexcited dye removes electrons from iodide ions (injects holes into the iodide) to form triiodide (composed of iodine plus iodide) ions, I$_3^-$/I$^-$. The reverse redox process (the reduction of triiodide back to iodide) occurs at the counter-electrode. Hence the redox couple is regenerated during solar-to-electrical conversion and the concentration of the electrolyte remains unchanged.

This solar cell is made possible by the use of a high-surface-area nanoporous layer. The need to absorb more of the incident light was the driving force for the development of mesoscopic semiconductor materials with a very-high-surface-area morphology. A single monolayer of the dye on the semiconductor surface was sufficient to absorb essentially all the incident light within a reasonable thickness (several micrometers) of the nanoporous semiconductor film. TiO$_2$ became the semiconductor of choice since it is inexpensive, abundant, and nontoxic. The original dye was tris(2,2′-bipyridyl-4,4′-carboxylate)ruthenium(II). The function of the carboxylate group in the dye is to attach the semiconductor oxide substrate by chemisorption. The dye must carry attachment groups such as carboxylate or phosphonate to firmly graft onto the TiO$_2$ surface. The attachment group of the dye ensures that it spontaneously assembles as a molecular layer upon exposing the oxide film to a dye solution, and increases the probability that, once a photon has been absorbed, the excited state of the dye molecule will relax by electron injection to the semiconductor conduction band. An alternative is the "black dye," with the structure shown in Figure 49.8, tri(cyanato-2,2′,2″-terpyridyl-4,4′,4″-tricarboxylate)Ru(II), whose response extends 100 nm further into the IR.

As described by Wei [25], because of the encapsulation problem posed by the use of a liquid electrolyte, research on how to form an all-solid-state dye-sensitized solar cell with enhanced stability is under way. In principle, a solid p-type conductor can replace the liquid electrolyte. The redox levels of the dye and p-type materials have to be adapted carefully, as Figure 49.9 shows, to inject an electron into the conduction band of n-type semiconductors (e.g., TiO$_2$) and a hole into the valence band of the p-type conductor.

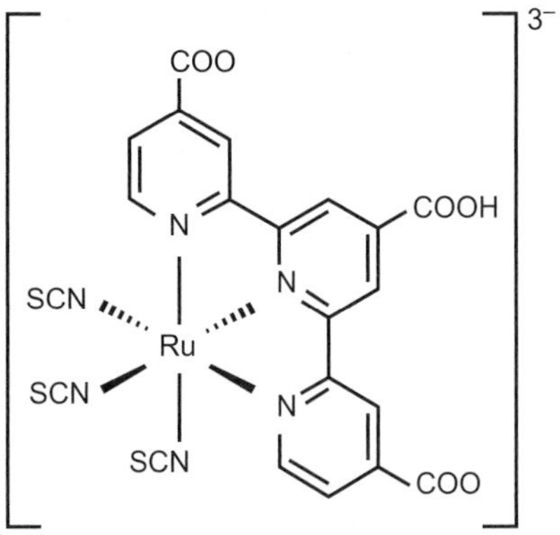

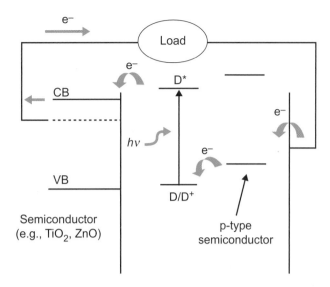

Figure 49.8. The "black dye" has a wider spectral response than that of the original ruthenium bypyridine used in dye-sensitized solar cells. (Figure modified from [35].)

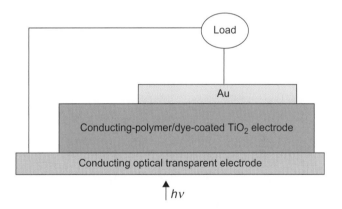

Figure 49.9. The mechanism of a dye-sensitized solar cell with a p-type semiconductor to replace the liquid electrolyte. D, dye sensitizer; D*, electronically excited dye; D+, oxidized dye; CB, semiconductor conduction band, VB, semiconductor valence band. (Figure modified from [26].)

Wei has also reviewed recent efforts to form a conducting-polymer dye-sensitized solar cell [26]. A conducting polymer, pyrrole, was electrochemically polymerized on a dye-sensitized porous nanocrystalline TiO_2 electrode, and functions as a hole-transport layer connecting dye molecules anchored on TiO_2 to the counter-electrode. Conducting polyaniline has also been used in solid-state solar cells sensitized with methylene blue with conducting polyaniline-coated electrodes sandwiched with a solid polymer electrolyte, poly(vinyl alcohol) with phosphoric acid. The prototype of this kind of conducting-polymer dye-sensitized solar cell is shown in Figure 49.10, and to date these cells offer only ~40% the output of comparable liquid-electrolyte dye-sensitized solar cells [35]. The solar-to-electrical efficiency of dye-sensitized solar cells generally remains lower than those of solid-state devices such as thin film, silicon, and concentrator PV. Dye-sensitized solar cells have been widely studied during the period from 1991 through 2010, and reviews are available [26][36][37][38][39][40].

49.4.4 Multi-bandgap semiconductor/electrolyte electrical energy conversion

Energetic constraints imposed by single-bandgap semiconductors have limited values of photoelectrochemical solar-to-electrical energy-conversion efficiency to date to 12%–16% [20][41]. Multiple-bandgap devices can provide efficient matching of the solar spectra [42][43][44][45][46] (Chapter 20). A configuration with two or more bandgaps will lead, per unit surface area, to more efficient solar energy conversion, and in solid-state multiple-bandgap solar cells a conversion efficiency of more than 40% has been achieved [47][48].

A limited fraction of incident solar photons will have sufficient energy (greater than the bandgap) to initiate

Figure 49.10. The conducting-polymer dye-sensitized solar cell. (Figure modified from [26].)

charge excitation within a semiconductor. Owing to the low fraction of short-wavelength solar light, wide-bandgap solar cells generate a high photovoltage, but have a low photocurrent. Smaller-bandgap cells can utilize a larger fraction of the incident photons, but generate a lower photovoltage. As shown in Figure 49.11, multiple-bandgap devices can overcome these limitations. In stacked multijunction systems, the topmost cell absorbs (and converts) energetic photons, but is transparent to lower-energy photons. Subsequent layer(s) absorb the lower-energy photons, and conversion efficiencies can be enhanced.

Several distinct types of multiple-bandgap PEC (MPEC) configurations are possible, each with its advantages and disadvantages [49]. The simplest MPEC configurations contain two different bandgaps, which can be aligned in the cell either in a bi-polar (or "tandem")

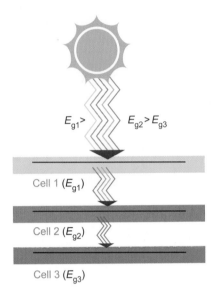

Cell 1 (E_{g1})

Cell 2 (E_{g2})

Cell 3 (E_{g3})

Figure 49.11. In stacked multijunction solar cells, the top cell converts higher-energy photons and transmits the remainder into layers, each of smaller bandgap than the layer above, for more effective utilization of the solar spectrum.

Figure 49.12. The energy diagram for a bi-polar-bandgap Schottky regenerative MPEC.

configuration or in a less conventional inverted manner. In either configuration, the PEC solid/electrode interface can consist of either an ohmic or a Schottky interface. The ohmic interface can consist of either direct (semiconductor/electrolyte) or indirect (semiconductor/metal and/or electrocatalyst/electrolyte) interfaces. The bipolar arrangement provides a conceptually simpler PEC, and generates a large open-circuit photopotential, V_{oc}. The bi-polar photovoltage, V_{photo}, generated is the sum of the potentials of the individual bandgap layers, minus cathodic and anodic polarization losses incurred in driving a regenerative redox couple:

$$V_{photo} = V_w + V_s - (h_{cathode} + h_{anode}).$$

The energy diagram of a bi-polar-bandgap photocathodic electrochemical Schottky configuration is presented in Figure 49.12. The scheme comprises a two-photon, one-electron photoelectrochemical process ($2h\nu \rightarrow e^-$), which may be generalized, for an n-bandgap configuration, to an n-photon process ($nh\nu \rightarrow e^-$). Light shown incident from the left side of the configuration first enters the wide-bandgap layer(s) in which more energetic photons are absorbed; less energetic photons are transmitted through this upper layer, and are absorbed by the small-bandgap layer. The resultant combined potential of the photo-driven charge sustains reduction at the photocathode interface, and drives extractable work through the external load, R_{load}. For the wide-bandgap (w) and small-bandgap (s) layers with

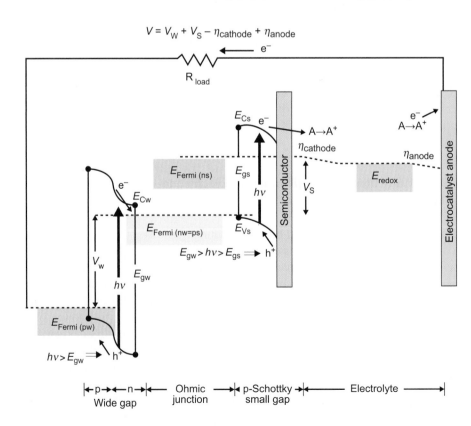

respective valence and conduction bands, E_V and E_C, and bandgap E_g:

$$E_{gw} = E_{Cw} - E_{Vw}; \qquad E_{gs} = E_{Cs} - E_{Vs}.$$

Wide-bandgap-layer charge separation occurs across a p–n-junction space-charge field gradient, while charge separation in the small-bandgap layer is maintained with a field formed by the Schottky semiconductor/electrolyte interface. In the bi-polar Schottky MPEC configuration, generated charge flows through all layers of the cell, providing the additional constraint. The current through each of cells have to be matched (Chapter 20).

A bi-polar-gap *direct ohmic* photoelectrochemical system comprises either a bi-polar-bandgap p–n–p–n/ electrolyte ohmic photoelectrochemical cell, with reduction occurring at the photoelectrode/electrolyte interface and regenerative oxidation occurring at the electrolyte/counter-electrode (anode) interface, or, alternatively, an n–p–n–p/electrolyte bi-polar-bandgap cell with oxidation occurring at the semiconductor/ electrolyte interface and regenerative reduction occurring at the electrolyte/counter-electrode interface.

In an alternative bi-polar regenerative configuration, the bi-polar (or multiple) bandgap configurations may contain consecutive space-charge field gradients generated solely via solid-state phenomena. For example, in the case of two consecutive bi-polar p–n junctions, the lowest semiconductor layer (the small-bandgap n-type layer in Figure 49.12) may remain in direct contact with the electrolyte, but the contact is ohmic, and is not the source of the small-bandgap space-charge field. In this case, the lowest semiconductor layer is restricted to electronic, not ionic, contact with the electrolyte, through use of an intermediate (bridging) ohmic electrocatalytic surface layer. This can facilitate charge transfer to the solution-phase redox couple, and prevents any chemical attack of the semiconductor. In the bi-polar cases (including Schottky, direct, and indirect ohmic configurations), the photo-power generated by a bi-polar regenerative MPEC is given by the product [11]

$$P_{\text{bipolar regenerative}} = j_{\text{ph}}[V_W + V_s - (\eta_{\text{cathode}} + \eta_{\text{anode}})].$$

The example presented here combines multijunction solid-state layers consisting of a bi-polar AlGaAs (E_{gw} = 1.6 eV) wide bandgap, overlaid on an Si (E_{gs} = 1.0 eV) small bandgap, and used in an electrolytic cell [49]. Absorption of light by the electrolyte can interfere with the cell and should be avoided. Figure 49.13 overlays the optical characteristics of the solid and solution phases of the AlGaAs/Si solid-state and $V^{3+/2+}$ electrolyte optimized components within a bi-polar-gap photoelectrochemical cell. Solution transmission is measured through a pathlength (1 mm) typical of many

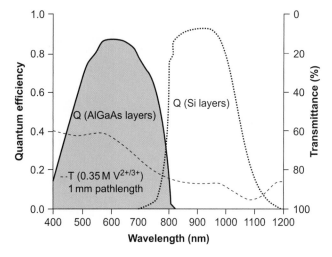

Figure 49.13. Overlay of the optical characteristics of the solid and solution phases of the AlGaAs/Si solid state and $V^{3+/2+}$ electrolyte constituents within a bi-polar-gap photoelectrochemical cell. Transmission of the $V^{3+/2+}$ electrolyte is measured through a pathlength of 1 mm. As described in the text, the Si bottom cell consists of a p^+-Si, n-Si, and n^+-Si multijunction. The $Al_{(0.3-0.15)}Ga_{(0.7-0.85)}$ As top cell utilizes a graded-band emitter.

experimental front-wall photoelectrochemical cells. As is evident, light-transmission interference will occur for the top AlGaAs layers and bottom Si layers through this or substantially shorter electrolyte pathlengths. The solid-state component includes a graded-band emitter, varying in the range $Al_{(0.3-0.15)}Ga_{(0.7-0.85)}$ As, with overlayers of p^+-Al_xGa_{x-1} As on n-Al_xGa_{x-1}As. The growth sequence and graded-band emitter layer improve collection efficiency. The Si bottom cell consists of a p^+-Si, n-Si, and n^+-Si multijunction. The band edges observed in Figure 49.13 at approximately 800 nm and 1100 nm are consistent with the respective AlGaAs and Si bandgaps.

For efficient electron–hole-pair charge generation, incident photons need to be localized within the multiple-bandgap semiconductor small- and wide-bandgap regions, rather than lost through competitive electrolyte light absorption. As can be seen in Figure 49.13, the vanadium electrolyte can significantly block light, over a wide range of visible and near-IR wavelengths, preventing it from entering the wide- and small-bandgap layers of the multiple-bandgap photoelectrochemical cell. This deleterious effect is prevented by use of the back-wall multiple-bandgap photoelectrochemical cell presented in Figure 49.14. Light does not pass through the solution. As shown, illumination enters directly through antireflection films of ZnS of thickness 50 nm on 70 nm of MgF$_2$. An evaporated Au–Zn/Au grid provides electrical contact to the wide-gap AlGaAs layers

Figure 49.14. A schematic description of the components in the bi-polar-gap direct ohmic AlGaAs/Si–V$^{3+/2+}$ photoelectrochemical solar cell.

	Illumination	Au–Zn/Au
		p$^+$-GaAs
AR coating [ZnS(50 nm)/MgF$_2$(70 nm)]		
p$^+$-Al$_{0.8}$Ga$_{0.2}$As	(1×10^{18}cm^{-3})	50 nm
p$^+$-Al$_{(0.3-0.15)}$Ga$_{(0.7-0.85)}$As	(1×10^{18}cm^{-3})	300 nm
n-Al$_{(0.15)}$Ga$_{(0.85)}$As	(2×10^{17}cm^{-3})	1.0 μm
n$^+$-Al$_{(0.15)}$Ga$_{(0.85)}$As	(1×10^{18}cm^{-3})	1.7 nm
n-GaAs		20 nm
GaAs(Buffer layer)		10 nm
p$^+$-Si	(1×10^{19}cm^{-3})	1.0 μm
n-Si	(8×10^{15}cm^{-3})	350 μm
n$^+$-Si	(4×10^{19}cm^{-3})	800 nm

Au

e$^-$

$$V^{3+} + e^- \rightarrow V^{2+}$$

$$E°, V^{2+/3+} \text{ Electrolyte } -0.3\text{V vs H}_2$$

$$V^{2+} \rightarrow V^{3+} + e^-$$

Carbon

through a bridging p$^+$-GaAs layer. Internally, a bridging GaAs buffer layer provides an ohmic contact between the wide-bandgap AlGaAs junctions and the lower Si layers. An intermediate contact layer, labeled "Au," is used only for probing separated characteristics of the wide- and small-bandgap junctions; it is not utilized in the complete cell. Photogenerated charge at the indicated silicon–electrolyte interface induces solution-phase reduction of vanadium, and a carbon counter-electrode provides an effective (low-polarization) electrocatalytic surface for the reverse process in a regenerative cell, in accord with

$$V^{3+}(+h\nu) \rightarrow V^{2+} + h^+; \qquad V^{2+} \rightarrow V^{3+} + e^-.$$

Maximization of the photopower necessitates minimization of the anodic and cathodic polarization losses, η_{anode} and $\eta_{cathode}$ during charge transfer through the photo-electrode and counter-electrode interfaces. In the current domain investigated, polarization losses are linear for both anodic and cathodic processes at 2.5–3.5 mV mA^{-1} cm^{-2}, and can create small, but significant, losses on the order of 10–100 mV in the MPEC.

Figure 49.15 presents the outdoor characteristics of the bi-polar direct ohmic AlGaAs/Si–V$^{2+/3+}$ photoelectrochemical cell under solar illumination. The system comprises the individual components illustrated in Figure 49.14, and uses an electrolyte of aqueous HF containing vanadium to improve photocurrent stability (0.35 M V(II) + V(III), 4 M HCl, 0.2 M HF). The electrolyte cathode is 0.2 cm n$^+$-Si, and the electrolyte anode is 0.2 cm^2 C. As shown, under 75 mW cm^{-2} insolation, the AlGaAs/Si–V$^{2+/3+}$ electrolyte PEC exhibits an open-circuit potential of V_{oc} = 1.4 V, a short-circuit

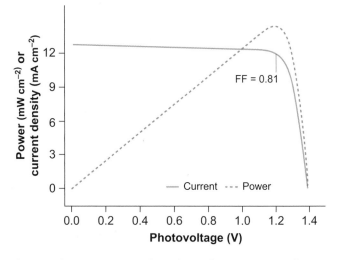

Figure 49.15. Measured outdoor photocurrent–voltage characteristics of the bi-polar-gap direct ohmic AlGaAs/Si–V$^{3+/2+}$ PEC [49]. FF, fill factor.

photocurrent of J_{sc} = 12.7 mA cm^{-2}, and a fill factor of 0.81, determined from the fraction of the maximum power, P_{max}, compared with the product of the open-circuit potential and the short-circuit current. The multiple-bandgap solar-to-electrical conversion efficiency of 19.1% compares favorably with the maximum solar-to-electrical energy conversion of 15%–16%.

A common disadvantage of photoelectrochemical systems is photo-induced corrosion of the semiconductor, which originates at the semiconductor/solution interface. The corroded surface dissolves the semiconductor or inhibits charge transfer, which diminishes the photocurrent. A stable solid/solution interface that both facilitates charge transfer and impedes semiconductor photocorrosion is provided by an electrocatalyst placed between the semiconductor and the electrolyte. The multiple-bandgap photoelectrochemical cell can utilize this electrocatalyst interface, as well as a bi-polar series arrangement of wide- and small-bandgap semiconductors, to enhance energy conversion. In this indirect ohmic photoelectrochemistry, such as with a modified GaAs/Si–$V^{3+/2+}$ MPEC, electrolyte-induced photocorrosion of the silicon is entirely inhibited by utilization of an electrocatalyst (a second carbon electrode) bridging charge transfer between the semiconductor and the electrolyte, while retaining a high solar-to-electrical conversion efficiency in.

45.4.5 Solar thermal electrochemical production of energetic molecules

Rather than using solar generation to form electricity as a product, solar energy can be used to form chemicals (Chapter 45). This section explores an electrochemical approach to efficiently generate useful chemicals using solar energy. The new process is synergetic, using both the visible and the thermal components of sunlight, to achieve higher solar energy conversion efficiencies than with processes that use only one of these components. The captured solar energy is stored within the generated energetically rich chemical products, including iron and hydrogen fuel, and used to proactively convert anthropogenic CO_2 generated in burning fossil fuels [30][50][51][52].

With PV solar cells, charge transfer occurs only up to the semiconductor band edge, excluding the use of long-wavelength (thermal) radiation. STEP uses a combination of the PV effect and the energy not used by the PV cell to generate heat which can help drive endothermic reactions. Some examples of electrolysis reactions that are endothermic are presented in Figure 49.16.

One example of the STEP process is the reduction of the greenhouse gas carbon dioxide to carbon, or to carbon monoxide. The latter product is a useful starting point for the synthesis of a range of useful hydrocarbon products. CO_2 is a stable gas, and normally difficult to activate and utilize. However, as can be seen in Figure 49.17(a), CO_2 is readily electrically reduced to carbon in high-temperature molten carbonates. As summarized in the figure, the principal product switches from C to CO at elevated temperature, the rates of electrolysis are high (amps per square centimeter), and the electrochemical potential decreases with increasing temperature. Solar energy drives the process at from 30% to over 50% conversion efficiency, depending on the efficiency of the solar heating used to increase the temperature of the CO_2 reactant [50]. Similarly, iron oxide is readily reduced to iron metal and oxygen. In this example of

a STEP iron-production process, visible light drives conventional PV electronic charge transfer, and all of the excess heat is used to increase the electrolysis temperature. Iron is produced at a low electrolysis potential (less than 1 V), and without the CO_2 emissions associated with the industrial carbothermic process used to form iron [51][52].

Light-driven water splitting to generate H_2 was originally demonstrated with TiO_2 ($E_g \sim 3.0$ eV) [4] because, only a small fraction of sunlight has sufficient energy to photoexcite of TiO_2, researchers sought to lower the semiconductor bandgap to provide a better match to the electrolysis potential [53], or to use multiple photons to drive the multiple-bandgap electrolysis of water [54]. With STEP a small-bandgap semiconductor, such as silicon ($E_g = 1.12$ V) was used to directly spit water producing hydrogen [29][55]. The key is that, in contrast to previous attempts to tune the bandgap of the semiconductor to make it more compatible with the water-splitting energetics, STEP tunes the water-splitting potential to match the semiconductor bandgap. A single small-bandgap material, such as silicon, cannot generate the minimum potential required to drive many room-temperature redox couples as illustrated in Figure 49.18(a). Rather than tuning the bandgap to provide a better energetic match to the electrolysis potential, the STEP process instead tunes the redox potential to match the bandgap. Figure 49.18(b) presents the energy diagram for a STEP process.

At any electrolysis temperature, T_{STEP}, and at unit activity, the reaction has electrochemical potential, E°_T. This may be calculated from consistent, compiled unit-activity thermochemical data sets, such as the NIST condensed-phase and fluid properties data sets [13], as

$$E^{\circ}_T = -\Delta G^{\circ}(T = T_{STEP})/(nF); \quad E^{\circ}_{ambient} \equiv E^{\circ}_T(T_{ambient}),$$

where $T_{ambient} = 298.15K = 25°C$.

Given a stable high-temperature electrolysis environment, the experimental STEP solar conversion efficiency is the product of the electrolysis efficiency and the electronic solar efficiency:

$$\eta_{STEP} = \eta_{PV}\eta_{electrolysis}.$$

STEP's high-temperature pathway decreases the thermodynamic energy requirements for processes whose electrolysis potential decreases with increasing temperature. The extent of the decrease in the electrolysis potential, E_{redox}, may be tuned (will vary) with the temperature.

49.5 Current and emerging technologies

The simplest mode of photoelectrochemical energy conversion occurs with illumination of a semiconductor immersed in an electrolyte. In this mode solar energy

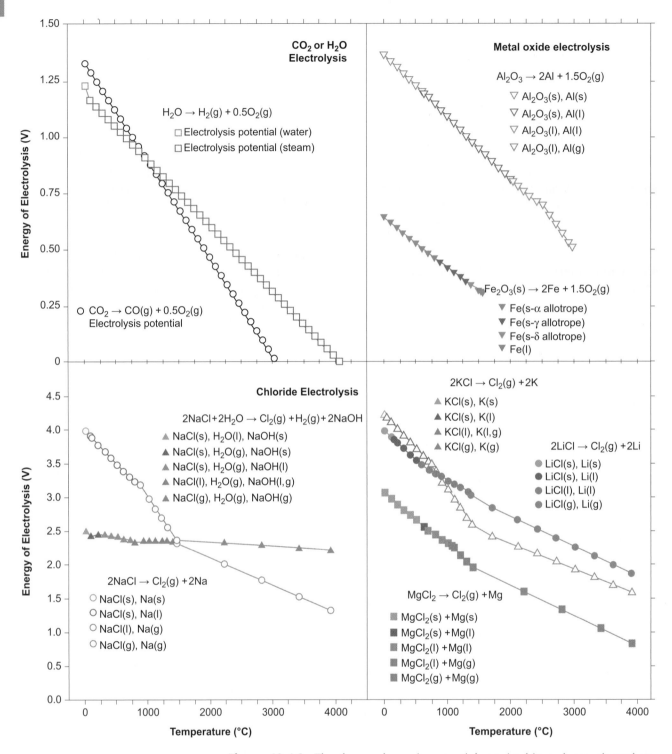

Figure 49.16. The thermodynamic potential required in order to electrolyze selected oxides (top) and chlorides (bottom). The decrease in electrolysis energy with increasing temperature provides energy savings in the STEP process in which high temperature is provided by excess solar heat. From [30].

conversion efficiencies of up to 16% have been achieved. However, these systems remain a technological challenge due to corrosion and photocorrosion at the semiconductor/electrolyte interface, which decreases the output over time. Despite their not having been commercialized as solar cells in this mode, related photoelectrochemistry methodologies have commercial potential including photoelectrochemical etching as a useful surface treatment for the semiconductor industry [17][18].

(a)

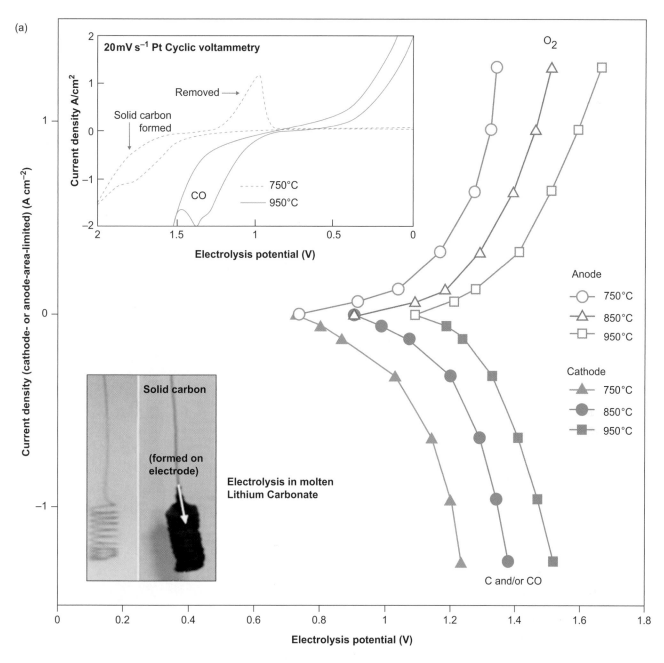

Figure 49.17. (a) Photograph inset: electrode before (left) and after (right) carbon capture at 750 °C in molten carbonate. Carbon dioxide fed into the electrolysis chamber is converted into solid carbon in a single step. Main figure: the electrolysis full-cell potential, in molten Li_2CO_3. Figure inset: cyclic voltammetry in molten Li_2CO_3. From [51]. (b) In STEP iron production, solar thermal can be separated from visible sunlight using a cold mirror, to provide one thermal component, Q, to heat a molten carbonate electrolysis chamber. Iron ore, such as hematite (Fe_2O_3), is added to the carbonate and is electrolyzed to iron metal using power from the solar cell driven by visible sunlight. From [51].

(b)

Li$_2$CO$_3$ heated by
Q$_{solar-PV}$, Q$_{impedance}$, Q$_{product}$, Q$_{solar-IR (below)}$

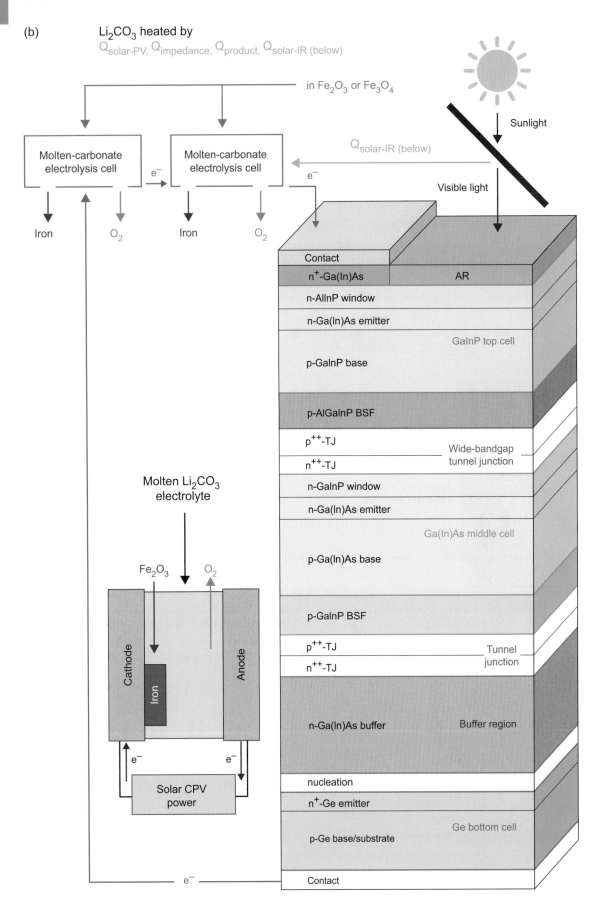

Figure 49.17. (cont.)

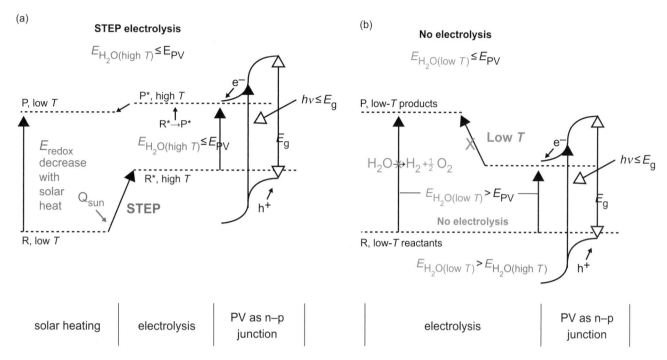

Figure 49.18. A comparison of the STEP and ambient-temperature solar-driven water-electrolysis energy diagrams. (a) A conventional efficient PV system generates a voltage too low to drive the indicated generic redox transfer. (b) The STEP process uses both visible and thermal solar energy for higher efficiency. The process uses this thermal energy for the indicated step decrease of the necessary electrolysis energy and forms an energetically allowed pathway to drive electrochemical charge transfer.

Another approach is to use TiO_2 as a highly stable photoabsorber/electrolyte interface, even though its wide bandgap ($E_g = 3.2$ eV) can access only the near-UV, not the visible and IR, portions of sunlight. Commercial disinfectant products, which use the oxidizing power available at the illuminated TiO_2 aqueous surface, have been developed, including self-cleaning tiles and headlights, air purifiers, and materials for drinking-water sterilization [5][56].

Another alternative means to avoid electrolyte contact and the corrosion at the interface is to use a back-wall cell configuration, in which the semiconductor-as-photoabsorber layer does not come into contact with the electrolyte [57]. The alternative which has received the most attention is to replace the semiconductor-as-photoabsorber with a dye-as-photoabsorber. While the solar efficiencies of these dye-sensitized solar cells have improved only modestly in nearly two decades [25], the production of cells with higher surface area is improving and several companies claim to have made low cost, effective cells, and are promising the availability of dye-sensitized solar cells in the near future, including SolarPrint, G24innovations, Hydrogen Solay, Dyesol, Konarka, Aisin Seiki, 3G Solar and Sony [35].

One-third of the global industrial sector's annual emission of 10^{10} metric tons of the greenhouse gas CO_2 is released in the production of metals and chlorine.

The photoelectrochemical generation of these staple chemicals without CO_2 emission is viable [30]. This, together with the additional CO_2 emissions for electrical generation, heating, and transportation, comprises the majority of anthropogenic CO_2 emissions. The STEP process proactively converts anthropogenic CO_2 generated in burning fossil fuels, and also eliminates CO_2 emissions associated with the generation of hydrogen fuel, bleach, iron, aluminum, chlorine, magnesium, lithium, sodium, carbon monoxide, and synthetic diesel and jet fuels. This process captures sunlight more efficiently than does PV. STEP's synergistic coupling of endothermic electrolysis with PV is promising, and is intrinsically more efficient than PV alone. Photovoltaic efficiencies above 40% have been reported [47], which will provide impetus to develop new STEP technologies. STEP hydrogen generators are under development by Lynntech, Inc., College Station, TX for the US Airforce [58], and new STEP processes continue to be introduced [59].

49.6 Summary

Photoelectrochemistry is the study of photo-driven electrochemical processes (light-driven processes that interconvert electrical and chemical energy). A photoelectrochemical solar cell (PEC) can generate not only electrical but also electrochemical energy, and can thereby provide the basis for a system with an energy-storage component Sufficiently energetic insolation incident on semiconductors can drive electrochemical oxidation/reduction and generate chemical, electrical, or electrochemical energy. Aspects include efficient dye-sensitized or direct solar-to-electrical energy conversion, solar electrochemical synthesis (electrolysis), including splitting of water to form hydrogen, environmental cleanup, and solar energy-storage cells. The PEC utilizes light to carry out an electrochemical reaction, converting light into both chemical and electrical energy. This fundamental difference between the photovoltaic (PV) solar cell's solid/solid interface and the PEC's solid/liquid interface has several ramifications for cell function and application. The energetic constraints imposed by single-bandgap semiconductors have limited the demonstrated values of photoelectrochemical solar-to-electrical energy conversion efficiency to 16%, and multiple-bandgap cells can lead to significantly higher conversion efficiencies. Photoelectrochemical systems not only may facilitate solar-to-electrical energy conversion, but also have led to investigations into solar photoelectrochemical production of fuels and photoelectrochemical detoxification of pollutants, and efficient solar thermal electrochemical production (STEP) of metals, fuels, and bleach and carbon capture.

49.7 Questions for discussion

1. What are the differences among conventional photoelectrochemical (without dye), dye-sensitized, and STEP solar energy conversion processes?
2. What are the advantages of photoelectrochemical energy conversion when *in situ* storage is included?
3. What are the similarities and differences between photovoltaic and photoelectrochemical energy conversion?
4. Why have photoelectrochemical technologies been deployed less to date than photovoltaic technologies?
5. In Figure 49.16, one of the electrolysis reactions displayed describes the conventional alkali-chlor industrial process in which bleach is made. Which reaction is this? Is this reaction suitable in the STEP process, and, if so, why? What is the energy effect of removing water from this electrolysis, and is the resultant reaction suitable for the STEP process?

49.8 Further reading

- **S. Licht** (ed.), 2002, *Semiconductor Electrodes and Photoelectrochemistry*, Weinheim, Wiley-VCH.
- **M. Archer** and **A. Nozik**, 2008, *Nanostructured and Photochemical and Photoelectrochemical Approaches to Solar Energy Conversion*, London, World Scientific.
- **K. Rajeshwar**, **S. Licht**, and **R. McConnell**, 2008, *The Solar Generation of Hydrogen: Towards a Renewable Energy Future*, New York, Wiley Press.
- **L. Vayssieres** (ed.), 2010, *Solar Hydrogen and Nanotechnology*, New York, Wiley.

49.9 References

[1] **E. Becquerel**, 1839, "Mémoires sur les effets électriques produits sous l'influence des rayons," *Comptes Rendues*, **9**, 561–567.
[2] **H. Gerischer**, 1961, *Advances in Electrochemistry and Electrochemical Engineering*, New York, Interscience, p. 139.
[3] **H. Gerischer**, 1970, *Physical Chemistry: An Advanced Treatise*, vol. **9**A, New York, Academic Press.
[4] **A. Fujishima** and **K. Honda**, 1972, "Electrochemical photolysis of water at a semiconductor electrode," *Nature*, **238**, 37–38.
[5] **T. Rao**, **D. A. Tryk**, and **A. Fujishima**, 2002, "Applications of TiO_2 photocatalysis," in *Semiconductor Electrodes and Photoelectrochemistry*, ed. **S. Licht**, Weinheim, Wiley-VCH.
[6] **G. Hodes**, **J. Manassen**, and **D. Cahen**, 1976, "Photoelectrochemical energy conversion and storage using polycrystalline chalcogenide electrodes," *Nature*, **261**, 402–404.
[7] **A. B. Ellis**, **S. W. Kaiser**, and **M. S. Wrighton**, 1976, "Visible light to electrical energy conversion. Stable cadmium sulfide and cadmium selenide photoelectrodes in aqueous electrolytes," *J. Am. Chem. Soc.*, **98**, 1635–1637.
[8] **B. Miller** and **A. Heller**, 1976, "Semiconductor liquid junction solar cells based on anodic sulphide films," *Nature*, **262**, 680–681.
[9] **A. J. Nozik**, 1978, "Photoelectrochemistry: applications to solar energy conversion," *Ann. Rev. Phys. Chem.*, **29**, 189–222.
[10] **M. A. Butler** and **D. S. Ginley**, 1980, "Principles of photoelechemical solar energy conversion," *J. Mates. Sci.*, **15**, 1–91.
[11] **R. Memming**, 1991, "Improvements in solar energy conversion," in *Photochemical Conversion and Storage of Solar Energy*, eds. **E. Pelizzetti** and **M. Schiavello**, Dordrecht, Kluwer, p. 193.
[12] **S. Licht** (ed.), 2002, *Semiconductor Electrodes and Photoelectrochemistry*, Weinheim, Wiley-VCH.
[13] **M. Archer** and **A. Nozik** (eds.), 2008, *Nanostructured and Photochemical and Photoelectrochemical Approaches to Solar Energy Conversion*, London, World Scientific.
[14] **K. Rajeshwar**, **S. Licht**, and **R. McConnell** (eds.), 2008, *The Solar Generation of Hydrogen: Towards a Renewable Energy Future*, New York, Wiley.
[15] **L. Vayssieres** (ed.), 2010, *Solar Hydrogen and Nanotechnology*, New York, Wiley.
[16] **S. Licht**, **G. Hodes**, **R. Tenne**, and **J. Manassen**, 1987, "A light variation insensitive high efficiency solar cell," *Nature*, **326**, 863–864.

[17] **R. Tenne** and **G. Hodes**, 1980, "Improved efficiency of CdSe photanodes by photoelectrochemical etching," *Appl. Phys. Lett.*, **37**, 428–430.

[18] **R. Tenne** and **G. Hodes**, 1983, "Selective photoelectrochemical etching of semiconductor surfaces," *Surf. Sci.*, **135**, 453–478.

[19] **S. Licht**, 1987, "A description of energy conversion in photoelectrochemical solar cells," *Nature*, **330**, 148–151.

[20] **S. Licht** and **D. Peramunage**, 1990, "Efficient photoelectrochemical solar cells from electrolyte modification," *Nature*, **345**, 330–333.

[21] **S. Licht** and **D. Peramunage**, 1992, "Rational electrolyte modification of n-CdSe/(KFe(CN)$_6$)$^{3-/2-}$ photoelectrochemistry," *J. Electrochem. Soc.*, **139**, L23–L26.

[22] **S. Licht**, 2001, "Multiple bandgap semiconductor/electrolyte solar energy conversion," *J. Phys. Chem. B*, **105**, 6281–6294.

[23] **H. Tributsch**, 1972, "Reaction of excited chloroohyll molecules at electrodes and in photsynthesis," *Photochem. Photobiol.*, **16**, 261–269.

[24] **H. Tsubomura**, **M. Matsumura**, **Y. Nomura**, and **T. Amamiya**, 1976, "Dye sensitised zinc oxide: aqueous electrolyte: platinum photocell," *Nature*, **261**, 402–403.

[25] **B. O'Regan** and **M. Grätzel**, 1991, "A low-cost, high-efficiency solar cell based on dye-sensitized colloidal TiO$_2$ films," *Nature*, **353**, 737–740.

[26] **D. Wei**, 2010, "Dye sensitized solar cells," *Int. J. Molec. Sci.*, **11**, 1103–1113.

[27] **S. Licht**, 2002, "Efficient solar generation of hydrogen fuel – a fundamental analysis," *Electrochem. Commun.*, **4**, 789–794.

[28] **S. Licht**, 2003, "Solar water splitting to generate hydrogen fuel: photothermal electrochemical analysis," *J. Phys. Chem. B*, **107**, 4253–4260.

[29] **S. Licht**, 2003, "Electrochemical potential tuned solar water splitting," *Chem. Commun.*, 3006–3007.

[30] **S. Licht**, 2009, "STEP (solar thermal electrochemical photo) generation of energetic molecules: a solar chemical process to end anthropogenic global warming," *J. Phys. Chem. C*, **113**, 16283–16292.

[31] **S. Licht**, 2002, "Optimizing photoelectrochemical solar energy conversion: multiple bandgap and solution phase phenomenon," in *Semiconductor Electrodes and Photoelectrochemistry*, ed. **S. Licht**, Weinheim, Wiley-VCH.

[32] **S. Licht** and **G. Hodes**, 2008, "Photoelectrochemical storage cells," in *Nanostructured and Photochemical and Photoelectrochemical Approaches to Solar Energy Conversion*, ed. **M. Archer** and **A. Nozik**, London, World Scientific.

[33] **S. Licht**, **B. Wang**, **T. Soga**, and **M. Umeno**, 1999, "Light invariant, efficient, multiple bandgap AlGaAs/Si/metal hydride solar cell," *Appl. Phys. Lett.*, **74**, 4055–4057.

[34] **B. Wang**, **S. Licht**, **T. Soga**, and **M. Umeno**, 2000, "Stable cycling behavior of the light invariant AlGaAs/Si/metal hydride solar cell," *Solar Energy Mater. Solar Cells*, **64**, 311–320.

[35] **H. Snaith**, **A. Moule**, **C. Klein** *et al.*, 2007, "Efficiency enhancements in solid-state hybrid solar cells via reduced charge recombination and increased light capture," *Nano Lett.*, **7**, 3372–3376.

[36] **M. K. Naseeruddin** and **M. Grätzel**, 2002, "Dye-sensitized regenerative solar cells," in *Semiconductor Electrodes and Photoelectrochemistry*, ed. **S. Licht**, Weinheim, Wiley-VCH.

[37] **J. Nelson**, "Charge transport in dye-sensitized systems," in *Semiconductor Electrodes and Photoelectrochemistry*, ed. **S. Licht**, Weinheim, Wiley-VCH.

[38] **K. Uzaki**, **T. Nishimura**, **J. Usagawa** *et al.*, 2010, "Dye-sensitized solar cells consisting of 3D-electrodes – a review: aiming at high efficiency from the view point of light harvesting and charge collection," *J. Solar Energy Eng. Trans. ASME*, **132**, 021204.

[39] **J. H. Wu**, **Z. Lan**, **S. C. Hao**, *et al.*, 2008, "Progress on the electrolytes for dye-sensitized solar cells," *Pure Appl. Chem.*, **80**, 2241–2258.

[40] **T. W. Hamann**, **R. A. Jensen**, **A. B. F. Martinson**, *et al.*, 2008, "Advancing beyond current generation dye-sensitized solar cells," *Energy Environmental Sci.*, **1**, 66–78.

[41] **B. Miller**, **S. Licht**, **M. E. Orazem**, and **P. C. Searson**, 1994, "Photoelectrochemical systems," *Crit. Rev. Surf. Chem.*, **3**, 29.

[42] **C. H. Henry**, 1980, "Limiting efficiencies of ideal single and multiple energy gap terrestrial solar cells," *J. Appl. Phys.*, **51**, 4494–4500.

[43] **D. J. Friedman**, **S. R. Kurtz**, **K. Bertness** *et al.*, 1995, "30.2% Efficient GaInP/GaAs monolithic two-terminal Ptandem concentrator cell," *Progr. Photovolt.*, **3**, 47–50.

[44] **J. P. Benner**, **J. M. Olson**, and **T. J. Coutts**, 1992, "Recent advances in high-efficiency solar cells," in *Advances in Solar Energy*, ed. **K. W. Boer**, Boulder, CO, American Solar Energy Society, Inc., pp. 125–165.

[45] **M. A. Green**, **K. Emery**, **K. Bücher**, **D. L. King**, and **S. Igari**, 1996, "Solar cell efficiency tables (version 8)," *Progr. Photovolt.*, **4**, 321–325.

[46] **T. Soga**, **T. Kato**, **M. Yang**, **M. Umeno**, and **T. Jimbo**, 1995, *J. Appl. Phys.*, **78**, 4196.

[47] **R. R. King**, **D. C. Law**, **K. M. Edmondson** *et al.*, 2007, "40% Efficient metamorphic GaInP/GaInAs/Ge multijunction solar cells," *Appl. Phys. Lett.*, 183516–183518.

[48] 2010, "Spire pushes solar cell record to 42.3%," optics.org/news/15/5.

[49] **S. Licht**, 2001, "Multiple bandgap semiconductor/electrolyte solar energy conversion," *J. Phys. Chem. B*, **105**, 6281–6294.

[50] **S. Licht**, **B. Wang**, **S. Ghosh** *et al.*, 2010, "A new solar carbon capture process: solar thermal electrochemical photo (STEP) free production of iron," *J. Phys. Chem. Lett.*, **1**, 2363–2368.

[51] **S. Licht** and **B. Wang**, 2010, "High solubility pathway to the carbon dioxide free production of iron," *Chem. Commun.*, **46**, 7004.

[52] **S. Licht**, **H. Wu**, **Z. Zhang**, and **H. Ayub**, 2011, "Chemical mechanism of the high solubility pathway for the

carbon dioxide free production of iron," *Chem. Commun.*, **47**, 3081–3083.

[53] **Z. Zou**, **J. Ye**, **K. Sayama**, and **H. Arakawa**, 2001, "Direct splitting of water under visible light irradiation with an oxide semiconductor photocatalyst," *Nature*, **414**, 625–627.

[54] **S. Licht**, **B. Wang**, **S. Mukerji** *et al.*, 1998, "Over 18% solar energy conversion to generation of hydrogen fuel; theory and experiment for efficient solar water splitting," *Int. J. Hydrogen Energy*, **280**, 425–659.

[55] **S. Licht**, **O. Chitayat**, **H. Bergmann** *et al.*, 2010, "Efficient STEP (solar thermal electrochemical photo) production of hydrogen – an economic assessment," *Int. J. Hydrogen Energy* **35**, 10867–10882.

[56] **J. Ng**, **X. Zhang**, **T. Zhang**, *et al.*, "Construction of self-organized free-standing TiO_2 nanotube arrays for effective disinfection of drinking water," *J. Chem. Technol. Biotechnol.*, **85**, 1061–1066.

[57] **S. Licht** and **F. Forouzan**, 1995, "Solution modified n-GaAs/aqueous polyselenide photoelectrochemistry," *J. Electrochem. Soc.* **142**, 1539–1545.

[58] **C. P. Rhodes**, **A. Cisar**, **H. Lee** *et al.*, 2008, "Effect of temperature on the electrolysis of water in concentrated alkali hydroxide solutions," in *215th Electrochemical Society Meeting*, San Francisco.

[59] **S. Licht**, **B. Wang**, **H. Wu**, 2011, "STEP – a solar chemical process to end anthropogenic global warming II: experimental results." *J. Phys. Chem. C*, **115**, 11 803–11 821.

Summary

David S. Ginley and David Cahen

Likely only the very brave will reach this point after having studied all of the "materials" (no pun intended) presented in this book. Most students will have used this book a bit as one chooses courses, focusing on what is needed, interesting, and/or challenging. For the reader, we would now like to raise some crosscutting concepts that any student of this information should consider.

To a large extent this book presents sets of chapters with insight into specific areas. In many cases the chapters present technology-specific views and understanding of the current state of the art or state of affairs for the specific area. The more technology-specific chapters provide an idea of what is needed in order to advance specific technologies, and present a picture of how that technology or the situation brought about by the technology may evolve.

The exceptions are Part I and two chapters in Part V, which look at some of the crosscutting areas of environment, energy flows, and materials availability. These parts help to establish the complexity of the interrelationship of all of the other chapters. Overall this can be viewed as a series of "grand challenges," technical, economic, and social, all of which affect both energy and the environment. Solutions to these challenges may be be found only by combining two or more of the approaches in the following sections. It should be clear that the ultimate solution must lie in a matrix of new energy sources and energy-efficiency measures, coupled with an evolution of the way in which we live (if possible without decreasing our standard of living), while meeting the goals of cost, efficiency, and minimizing impact.

We know very little about the interactions between technological approaches when they are implemented globally. For example, recently there has been much discussion of the application of bio-char (charcoal created by the pyrolysis of biomass) for the purpose of sequestering carbon. (J. Burges, 2010, *The Biochar Debate*, Chelsea Green Publishing). The result of the process is charcoal, which is then put back in the soil. This has the potential to restore fertility to over-cultivated land. However, the process uses energy and has other environmental impacts, the bio-impact of large amounts of charcoal dispersed in the soil is unknown, and the process must be compared with alternative approaches. The interplay of the environment, economics, social priorities, and human impact is complex and not understood regarding the application of bio-char. Understanding that interplay of all the diverse alternatives presented in the book adds multiple dimensions that make the issue fascinating but also frustratingly hard. In fact, this clearly illustrates that one size does not fit all, e.g., an approach that may work well for fields in India or areas with vast amounts of downed timber in the USA might not be usable at all elsewhere, i.e., in many parts of Europe or Africa. As is discussed below, there are many non-technical aspects to consider, an issue that we, as scientists and engineers, mostly prefer not to acknowledge.

Clearly, being able to meet the challenges posed by energy and environmental needs is a very complex problem with no "one" unique solution.

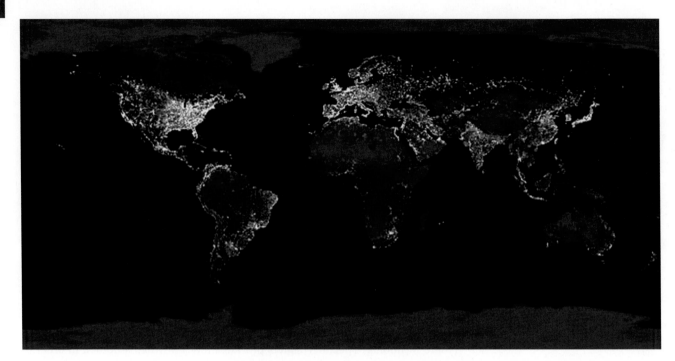

Figure 1. A composite photograph of Earth from space at night, showing not only the impact of humans but also the significant inequity of energy use, by way of our use of artificial lighting (see Chapter 35). Credit: C. Mayhew & R. Simmon (NASA/GSFC), NOAA/NGDC, DMSP Digital Archive.

The primary way to understand this may be to apply the new science of complex systems. Interestingly, to date there is not a clear definition of that science; however, a reasonable attempt goes as follows: "a system composed of interconnected parts that as a whole exhibits one or more properties (behavior among the possible properties) not obvious from the properties of the individual parts." This is certainly the case for the environment alone, but, when coupled to our diverse uses of, and needs for energy, this is even more the case.

A clear ultimate goal of humankind is to develop a sustainable and equitable way to live worldwide. However, even the definition of sustainability is not clear:

- the ability to be sustained for an indefinite period without damaging the environment and without depleting a resource;
- a means of configuring civilization and human activity so that society, its members, and its economies are able to meet their needs and express their greatest potential in the present, while preserving biodiversity and natural ecosystems, and while planning and acting for the ability to maintain these ideals for future generations;

and, from the Brundtland Commission,

- "meeting the needs of the present without compromising the ability of future generations to meet their own needs."

These definitions raise the issue of environmental vs. technological sustainability. While there is overlap between the two, it is the technical aspects of sustainability that tend to be emphasized, as this book has done. Nonetheless, there are many connections in the chapters, and certainly in reality, between technology and environmental sustainability, and between these and the need globally to maintain biodiversity.

Achieving sustainability requires a global scale. The world has flattened considerably, in large part due to the increase in international commerce and the ease of travel worldwide. These developments encourage global conversations about the kind of issues presented here. However, the considerable diversity of cultures, languages, per-capita income, and political priorities are all obstacles to being able to implement a global strategy. An apparent solution for North America or Western Europe will be very different from that in India and China, which in turn will be very different from that in sub-Saharan Africa or the Maghreb region. But true sustainability must entail a coupling of different solutions, emphasizing again that there is no "one" solution. A number of factors must be taken into account, including the following.

- *Complexity* – we have already noted that both the existing state (situation) and the new final sustainable state will be complex systems. The change from one state to the other must occur on a large number of length scales from global to individual.

Figure 2. Solutions to complex problems will be complex. They will require unique and novel integrations of what we know and a willingness to adopt things outside our current comfort/belief zone. (Credit: Stéphane Foulon, Venturi Automobiles.)

Historically, lack of global sustainability is in part due to emergent behavior whereby use patterns in the developed world impact all systems globally. Consequently it may be that emergent behavior can also catalyze global change – i.e., economic incentives may change global use and generation patterns.

- *Socioeconomic factors* – it is increasingly clear that the global economy is constrained by social factors and that these social factors are constrained also by environmental factors. Engines of change must therefore address all three, the economy, the social factors, and the environment. In addition, security is embedded in all three, and, while this is predominantly a national issue, it must be addressed also globally with respect to the stability and vulnerability of any solution. The complexity of trying to achieve a solution in which all the conditions are optimized does not exist at present for any technology. Thus, for example, we cannot jump to an all-renewable energy economy directly, but need to wean ourselves from conventional fuels by gradually implementing a range of solutions employing diverse technologies when they become ready for large-scale deployment. Just as we must be willing to adapt to a diversity of solutions, we must also begin to develop a more realistic set of expectations. Much of the world has wanted to live like North Americans, Western Europeans, or the Japanese, but these countries can themselves not afford to live that way anymore. It is thus unrealistic for the rest of the world or ourselves to expect that we can consume resources at such a non-sustainable level. Rather we must assess what is realistic vs. time and begin to change our lifestyles.

- *Life cycle* – many of the systems and resources discussed in this book have multi-generational time constants. Coal supplies can last for hundreds of years, we expect a photovoltaic system to run for at least 25 years and nuclear power plants for 50–75 years, and the sustainability of biomass must be viewed on a time scale of 5–20 years. Clearly this is another complex system, but it also shows the importance of life-cycle analysis for each new system. Thus, the scarcity of certain critical rare-earths elements, In and Te for example for the scale of anticipated technology deployment have come to the forefront. Thus we must begin to look at where rare elements can be best deployed, how they can be recycled, and how we do this on a global scale.

- *Interconnectivity* – while nearly all approaches presented in this text can be implemented locally, they will really only have the desired impact if implemented globally. This means that, as above, we need start viewing energy/environment as a complex system with a global length scale requiring solutions on that length scale. That will require an increased level of economic connectivity, global data sharing, and long-term support for the various approaches in order to give them a chance to reach deployment. Integration on this length scale will cause social barriers to be removed and change our global interdependence, which may have significant ramifications for global politics and explains the need to develop binding ways to implement change on a global scale. The Kyoto Accord (and the Montreal one, before that) is an example of the beginnings of this kind of approach.

- *Equity* – currently there are tremendous global inequities in food, energy, income, and standards of living. To develop an ultimately sustainable world, we need look at achieving equity in these areas not only as a moral imperative but also as a practical necessity. The latter will be especially clear to those students versed in thermodynamics (Appendix A), which teaches us about the instability of systems with widely different potential energies ("social thermodynamics").

Key to you, the reader, is that the perspective you have gained from studying parts or all of the book will help you to advocate actions/decisions that help to drive us toward a sustainable future. This is a key role for everyone, i.e., no one can afford to be a spectator, whether in terms of making appropriate life decisions, advocating suitable political decisions, or as a technologist/researcher to develop solutions. We hope that the information in this book will help you to set out, or continue, on your journey.

Appendix A

Thermodynamics

Jennifer A. Nekuda Malik

Department of Materials and Centre for Plastic Electronics, Imperial College London, London, UK

A.1 General concepts of thermodynamics

Thermodynamics is the branch of science that deals with the conversion of one form of energy into another. The laws of thermodynamics govern all materials processes and reactions and can be used to predict how a system and its surroundings will behave under various defined conditions. Changes in temperature (T), pressure (p), and volume (V) all play a role in the overall thermodynamic behavior of a system and can be related to one another through thermodynamic principles. The conditions for equilibrium are established by thermodynamics, but the rate at which equilibrium is achieved cannot be determined by thermodynamics [1].

To better understand thermodynamics, it is important first to define some important thermodynamic concepts. A thermodynamic system can be defined as the material and volume of interest. The system is separated from its surroundings by boundaries (which can be real or imaginary) and can be defined as either closed or open. A closed system contains a fixed mass while for an open system mass (typically in the form of a fluid) flows through the boundaries of the system. A system can also be defined as adiabatic if it is insulated and does not exchange heat with its surroundings or as isolated if the system cannot exchange either energy or matter with its surroundings.

A thermodynamic process can be either spontaneous – meaning that once it is started it will proceed to completion without any further stimulation, or non-spontaneous – meaning that energy (in some form) must be supplied to maintain or complete the process. Spontaneous reactions are typically exothermic (heat energy is released upon reaction), and a process that is spontaneous in one direction is non-spontaneous in the opposite direction.

In addition to being spontaneous or non-spontaneous, a process is also either reversible or irreversible. A reversible process is one for which, upon an exact reversal of the process, the original states both of the system and of the surroundings are achieved. An irreversible process cannot be restored to the initial conditions simply by reversal of the process, but rather requires external stimulation in the form of work or heat. By this definition, a spontaneous process is also an irreversible process.

The *first law of thermodynamics* deals with the conservation of energy. It states that the total energy of the Universe is constant and energy cannot be created or destroyed, but is rather converted from one form to another or transferred from one system to another. This simple law can be expressed mathematically thus:

$$\Delta U_{\text{system}} + \Delta U_{\text{surroundings}} = 0. \tag{A.1}$$

The internal energy of a system, ΔU, is the sum of the kinetic energy and potential energy of the system. The internal energy can be expressed as the sum of the heat (added to or taken away from the system), Q, and the work done by or on the system, W:

$$\Delta U = Q - W. \tag{A.2}$$

Heat and work are exchanged between the system and the surroundings, but the total amount of energy is conserved. It is also important to note that, while heat and work are path-dependent processes, ΔU is a state function and is therefore path-independent.

The first law can be further defined by defining both work and heat. Work can be mechanical (a force applied over a distance), electrical (movement of charge over a potential difference), or volumetric (a change in volume caused by a constant pressure). Thermodynamic work is commonly expressed as a function of pressure and volume, and this definition of work will be used throughout this section:

$$W = p\,\Delta V. \tag{A.3}$$

It is important to note that for systems with only solid or liquid phases the change in volume is very small, and therefore the work term is negligible. For a system that contains gases, it is useful to utilize the ideal-gas law to determine the value for pressure and volume change. The ideal-gas law defines the pressure and volume as a function of the number of moles of the gas (n), the gas constant ($R = 8.314\,\mathrm{J\,mol^{-1}\,K^{-1}}$) and the temperature ($T$):

$$pV = nRT. \tag{A.4}$$

In a system with a constant volume the work term is zero and the change in internal energy, ΔU, represents the change in heat content in the system. The derivative of internal energy with respect to temperature therefore yields the heat capacity (C_v) of the system at constant volume:

$$\left(\frac{\partial U}{\partial T}\right)_v = \left(\frac{\partial Q}{\partial T}\right)_v = C_v. \tag{A.5}$$

On solving Equation (A.5) for Q it is easy to see that any change in the heat of the system at a constant volume is directly related to the change in internal energy and can be calculated thus:

$$Q = \Delta U = \int_{T_1}^{T_2} C_v\,dT. \tag{A.6}$$

In most thermodynamic systems it is likely that volume will change but pressure will remain constant. In the constant-pressure case, the change in heat content with respect to temperature can be determined first by differentiating and then by rearranging Equation (A.2) as follows:

$$\left(\frac{\partial U}{\partial T}\right)_p = \left(\frac{\partial Q}{\partial T}\right)_p - P\left(\frac{\partial V}{\partial T}\right)_p \Rightarrow \left(\frac{\partial Q}{\partial T}\right)_p$$

$$= \left(\frac{\partial U}{\partial T}\right)_p + P\left(\frac{\partial V}{\partial T}\right)_p \Rightarrow \left(\frac{\partial Q}{\partial T}\right)_p \tag{A.7}$$

$$= \left(\frac{\partial (U + PV)}{\partial T}\right)_p.$$

From Equation (A.7), the quantity $U + PV$ has been defined as the enthalpy or heat content and assigned the variable H such that

$$H = U + pV. \tag{A.8}$$

Therefore, at constant pressure, Q is equal to H. It is important to note that, while heat (Q) is dependent on path, enthalpy (H) assumes a constant pressure and is therefore a state function and independent of path. The heat capacity of a system at constant pressure can be defined as

$$\left(\frac{\partial q}{\partial T}\right)_p = \left(\frac{\partial H}{\partial T}\right)_v = C_p. \tag{A.9}$$

Again Equation (A.9) can be solved for Q, yielding

$$Q = \Delta H = \int_{T_1}^{T_2} C_p\,dT. \tag{A.10}$$

Furthermore, at a constant pressure, Equation (A.2) can be rewritten as

$$\Delta U_p = \Delta H - p\,\Delta V. \tag{A.11}$$

The change in enthalpy for a chemical reaction carried out under constant-pressure conditions can be calculated from the enthalpies of the products and reactants as follows:

$$\Delta H_p = \sum_{products} \Delta H - \sum_{reactants} \Delta H. \tag{A.12}$$

When ΔH_p is positive, this means that the reaction is endothermic and requires heat from the surroundings. Conversely, when ΔH_p is negative the reaction is exothermic and releases heat to the surroundings.

Entropy, S, is a measure of the disorder or randomness present in a system. Entropy is related to the various distinguishable states that are possible in a materials system, and increases with increasing number of degrees of freedom. Water can be used to illustrate an increase in entropy: in its solid form the ice molecules can only vibrate; in the liquid form the water molecules are free to vibrate and rotate, and they have a limited ability to translate; in the gaseous form water vapor or steam molecules can vibrate, rotate, and translate. Therefore, entropy increases with increasing disorder: $S(\mathrm{s}) < S(\mathrm{l}) < S(\mathrm{g})$. If one recalls that changes of state occur at a constant temperature (ice $\Leftrightarrow$ water, $T = 0\,^\circ\mathrm{C}$; water $\Leftrightarrow$ steam, $T = 100\,^\circ\mathrm{C}$) then the relationship among entropy (S), heat (Q), and temperature (T) can be defined as

$$\Delta S = \int \frac{dQ}{T} = \frac{Q}{T}. \tag{A.13}$$

The second and third laws of thermodynamics address entropy. The *second law of thermodynamics* states that, for a reversible process, entropy does not change ($\Delta S = 0$); but, for an irreversible process, the entropy of the universe always increases ($\Delta S > 0$). Therefore, entropy can be produced but not destroyed, and the change in entropy with any process is always greater than or equal to zero:

$$\Delta S_{universe} = \Delta S_{system} + \Delta S_{surroundings} \geq 0. \tag{A.14}$$

Similarly to internal energy (ΔU) and enthalpy (ΔH), entropy is a thermodynamic property (state function), and therefore is independent of path. Also like with enthalpy, the change in entropy for a chemical reaction can be calculated from the entropies of the products and reactants as follows:

$$\Delta S = \sum_{\text{products}} \Delta S - \sum_{\text{reactants}} \Delta S. \tag{A.15}$$

The first and second laws of thermodynamics can be manipulated and combined to generate fundamental relationships among internal energy, enthalpy, entropy, temperature, pressure, and volume. In a system with only volume work, Equations (A.3) and (A.13) can be substituted into Equation (A.2), yielding

$$\Delta U = T\,\Delta S - p\,\Delta V. \tag{A.16}$$

To determine the relationship to enthalpy (ΔH), Equation (A.8) must be differentiated and then the left-hand side of Equation (A.16) must be substituted for ΔU as follows:

$$\Delta H = \Delta U + p\,\Delta V + V\,\Delta p \Rightarrow$$
$$\Delta H = T\,\Delta S - p\,\Delta V + p\,\Delta V + V\,\Delta p \Rightarrow \tag{A.17}$$
$$\Delta H = T\,\Delta S + V\,\Delta p.$$

Two thermodynamic free energies can now be defined, the Helmholtz (A) and Gibbs (G) free energies. The Helmholtz free energy is a measure of the work obtained from a closed thermodynamic system at a constant temperature and volume. The Gibbs free energy is a thermodynamic potential that measures the work obtained from a closed system under constant temperature and pressure. The Helmholtz and Gibbs free energies can be expressed as follows:

$$\text{Helmholtz}: \quad A = U - TS, \tag{A.18}$$

$$\text{Gibbs}: \quad G = H - TS. \tag{A.19}$$

At standard conditions of 1 atm pressure and 298 K (25 °C) the Gibbs free energy can be rewritten as

$$\Delta G^\circ = \Delta H^\circ - T\,\Delta S^\circ. \tag{A.20}$$

Similarly to enthalpy and entropy, the Gibbs free energy of a chemical reaction can also be calculated from the free energies of the products and the reactants. A reaction with a negative ΔG value is spontaneous, whereas a reaction with a positive ΔG value is unfavorable:

$$\Delta G = \sum_{\text{products}} \Delta G - \sum_{\text{reactants}} \Delta G. \tag{A.21}$$

Equilibrium of a thermodynamic system can be predicted using either the Gibbs free energy or entropy. When pressure and temperature are constant, equilibrium is achieved when ΔG is equal to zero. When a spontaneous chemical process occurs, entropy increases until the reaction reaches equilibrium. When the reaction has reached equilibrium, the volume (V) and internal energy (U) are constants, the entropy of the system is a maximum, and the change in entropy (ΔS) is zero.

There are two further laws that govern all thermodynamic systems. The *third law of thermodynamics* states that the entropy of a pure crystalline substance at absolute zero (0 K) is zero. The *zeroth law of thermodynamics*, implicitly assumed in the development of the other three laws of thermodynamics but only later recognized and named [2], states that two systems in thermal equilibrium with a third system must also be in equilibrium with each other.

For a more comprehensive study of thermodynamics please refer to standard thermodynamics textbooks such as [1].

A.2 Use of thermodynamics in energy and environmental science

Thermodynamics is fundamental to all materials systems and processes. The laws of thermodynamics can be applied across a broad range of fields to predict chemical reactions, changes of state, and the conditions for equilibrium within a given system. The relationships developed through thermodynamic principles are discussed throughout this book and indeed provide a very useful tool by means of which to examine, study, and predict advances in energy and environmental science.

A.3 References

[1] **O. F. Devereux**, 1989, *Topics in Metallurgical Thermodynamics*, reprint edn., Malabar, FL, Krieger Publishing Company.

[2] **J. P. Pickett**, **S. Kleinedler**, and **S. Spitz** (eds.), 2005, "Thermodynamics," in *The American Heritage Science Dictionary*, Boston, MA, Houghton Mifflin Company, p. 626.

Electrochemistry

Jennifer A. Nekuda Malik

Department of Materials and Centre for Plastic Electronics, Imperial College London, London, UK

B.1 General concepts of electrochemistry

Electrochemistry is the branch of chemistry that examines the relation between coupled electrical processes and chemical reactions. Most commonly, electrochemistry studies the chemical changes produced by an electrical current, at a given electric potential difference (with a reference), to cause a chemical reaction or, conversely, the electrical power produced by a chemical reaction. The general concepts of electrochemistry that will be discussed here can also be applied to a broad range of applications and technologies including devices (e.g., fuel cells, batteries, and sensors), methods of electroplating metals or preventing corrosion, and industrial production of various materials (most notably Al and Cl_2) [1].

The most basic electrochemical cell consists of two conductive electrodes (the anode and the cathode) and an ionic conductor known as an electrolyte. Figure B.1 shows a galvanic Zn/Cu electrochemical cell. A galvanic cell is essentially two half-cells consisting of a metal and a solution of the metal salt separated by a semi-porous membrane that prevents the solutions from mixing but allows charge transport between the solutions. An electrode reaction, characterized by the transfer of electrons across the interface between the electrode and the electrolyte, is anodic if the electrolyte is oxidized and loses electrons to the electrode (anode) and is cathodic if the electrolyte is reduced and gains electrons from the electrode (cathode). It is important to note that conventional current flow is for positive charge and is thus opposite to the direction of electron flow; therefore current flows from the cathode to the anode.

Chemical cells have their own specific symbolic notation whereby the cell reactants are expressed starting at the left with the anode material. In this notation, vertical bars between reactants represent phase boundaries, double vertical lines are used to indicate a phase boundary that has a negligible potential, and a comma is used to separate two components in the same phase [1][2][3]. In the electrochemical cell shown in Figure B.1 the phase boundary between the solutions at the semi-porous membrane has a negligible potential and the cell can be expressed as follows:

$$Zn(s)|Zn^{2+}(aq)||Cu^{2+}(aq)|Cu(s). \tag{B.1}$$

An electrochemical cell can be divided into two half-cells, where the cathode and anode are joined by an external circuit (through which electrons flow) and the electrolyte (where ions flow). By breaking the cell into half-cells, the overall electrochemical reaction can also be broken into two reactions: oxidation and reduction. The oxidation and reduction reactions that occur for the cell in Figure B.1 are

$$\text{Anode}: \quad Zn(s) \rightarrow Zn^{2+} + 2e^- \quad \text{oxidation}, \tag{B.2}$$

$$\text{Cathode}: \quad Cu^{2+} + 2e^- \rightarrow Cu(s) \quad \text{reduction}. \tag{B.3}$$

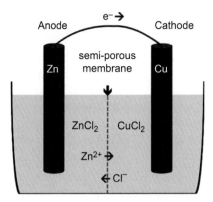

Figure B.1. A Zn/Cu galvanic electrochemical cell, showing the Zn anode, the Cu cathode, electrolyte solutions, the semi-porous membrane, and electron- and ion-transport directions.

The combination of the two half-cell reactions gives the net cell reaction:

$$Zn(s) + Cu^{2+} \rightarrow Zn^{2+} + Cu(s). \tag{B.4}$$

It is possible to determine whether the reaction will proceed in the indicated direction by calculating the reaction quotient (Q). The reaction quotient is a function of the activities or partial pressures of the species involved in the reaction. Mathematically, Q is calculated by dividing the activities of the products ($a_{products}$) by the activities of the reactants ($a_{reactants}$). Q takes into account the amount of each product and reactant in the reaction because each activity is raised to the power of the stoichiometric coefficient of the species in the reaction. Q for the reaction in Equation (B.4) can be written as follows:

$$Q = \frac{[a_{Zn^{2+}}]^1 [a_{Cu}]^1}{[a_{Zn}]^1 [a_{Cu^{2+}}]^1}. \tag{B.5}$$

It is important to note that solids are considered to have an activity of unity and the activity of a gaseous species is typically expressed as its partial pressure. As a reaction occurs the concentrations of the reactants will change, which in turn changes the activities of the aqueous species (or partial pressures of any gaseous species), thus changing the value of Q. When the reaction reaches equilibrium, the reaction quotient will be equal to the equilibrium constant, K.

In addition to the metal–metal-ion electrode reaction shown in Figure B.1, there are various other types of electrodes, including ion–ion electrodes, gas electrodes, and insoluble-salt electrodes [3]. Several different reactions are possible at the interfaces of these electrodes, ranging from simple electron transfer to metal deposition, corrosion, or oxidation to gas evolution [2]. In all cases, there are both oxidation and reduction reactions taking place. For more in-depth information on the various different types of electrode reactions refer to [1][2][3].

An electrochemical reaction can occur only in a connected cell where electrons are passing through the external electric circuit and ions are passing through the electrolyte between the anode and cathode. The rate of the reaction is governed both by the rates of the reactions at the electrodes and by the mobility of the ions moving through the electrolyte. Furthermore, to maintain charge balance (electroneutrality), the amount of reduction occurring at the cathode must match the amount of oxidation occurring at the anode. For the reaction to be spontaneous, the net energy change associated with the cell reaction must be negative. Non-spontaneous reactions can be made to occur by applying a potential (voltage) between the electrodes. The net cell potential can be defined by the following equation:

$$E_{cell} = \Delta V = E_{cathode} - E_{anode}. \tag{B.6}$$

The half-cell potentials cannot be measured directly, but tables of standard electrode-potential values have been tabulated using the internationally accepted standard hydrogen electrode (SHE) as a reference. The SHE half-cell consists of a platinum electrode with a specially treated surface over which hydrogen gas is bubbled. The SHE is operated at the standard conditions of 1 atm H_2 pressure, 25 °C temperature, and a pH of 0. *By definition*, the half-cell potential of the SHE is zero [3]. If a SHE were added to the cell in Figure B.1 and the copper electrode selected as the working electrode, then the cell could be represented as follows:

$$Pt(s)|H_2(g)|H^+(aq) \,\|\, Cu^{2+}(aq)\,|Cu(s) \tag{B.7}$$

with a net reaction of

$$H_2(g) + Cu^{2+} \rightarrow 2H^+ + Cu(s), \tag{B.8}$$

and the corresponding cell potential would yield the Cu half-cell potential thus:

$$\begin{aligned} E_{cell} = \Delta V &= E_{cathode} - E_{anode} \\ &= E_{Cu(s)} - E_{Cu^{2+}(aq)} - E_{Pt(s)} + E_{H^+(aq)} \\ &= E_{Cu(s)} - E_{Cu^{2+}(aq)}. \end{aligned} \tag{B.9}$$

A reference electrode is often very useful in electrochemistry because electrochemists are very often concerned with only the processes occurring at one electrode [1][2][3] Because all measurements must be made on a complete cell, a reference electrode is commonly used to make up one half of the cell, allowing electrochemists to focus solely on the working electrode. In addition to the SHE, other electrodes can be used as a reference, provided that they display a stable potential. Two other common reference electrodes are the silver–silver chloride electrode and the saturated calomel (mercury(I) chloride) electrode (SCE) [3].

Electrical work is done by an electrochemical cell when an electric charge, q, moves through a potential difference. The potential difference between the electrodes of a cell is a measure of the change in Gibbs free energy associated with the cell reaction. This relationship can be expressed by the following equation:

$$\Delta G° = -nFE° \quad \text{or similarly} \quad E° = \frac{-\Delta G°}{nF}. \tag{B.10}$$

In the above equation $\Delta G°$ is the free energy (in units of J mol^{-1}), n is the number of moles of charge active in the reaction, F is the Faraday constant (96,485 C mol^{-1}, which is the charge that one mole of electrons would have), and $E°$ is the cell potential measured in volts (V). A positive cell potential yields a negative change in the Gibbs free energy, and indicates that the reaction will proceed spontaneously. It is important to note that $\Delta G°$ is a measure of the maximum amount of useful work that can be electrically extracted from the cell, and the equation above is for the complete conversion of products to reactants [1][2][3].

All of the electrochemical cells that have been considered up to this point have been assumed to be standard. This means that all dissolved species are at a concentration of 1 M (1 mol l^{-1}) and all gases are at a pressure of 1 atm. Often non-standard conditions are employed in electrochemistry, and the change in cell potential can be predicted by modifying the free-energy equation given above. Consider the example Zn/Cu cell given in Figure B.1. If the concentration of the Zn^{2+} ions in the solution were reduced from 1 M to 0.001 M, the cell would be represented thus:

$$Zn(s)|Zn^{2+}(aq, 0.001\,M)\,\|\,Cu^{2+}(aq)|Cu(s). \qquad (B.11)$$

The decrease of the concentration of zinc ions in the solution changes the activity of Zn^{2+} ions, thus reducing the value of the reaction quotient (Q) in the cell and driving the reaction to the right, making it more spontaneous. From the free-energy equation above we know the relationship between the free energy and cell potential, and by analogy we can also write the more general equation (which does not assume full conversion of products to reactants and is valid for any value of Q)

$$\Delta G = -nFE. \qquad (B.12)$$

The thermodynamic equation for the Gibbs free energy $\Delta G = \Delta G° + RT\ln(Q)$, can be applied, and rewritten as follows:

$$-nFE = -nFE° + RT\ln(Q). \qquad (B.13)$$

This can be rearranged to yield the Nernst equation:

$$E = E° - \frac{RT}{nF}\ln(Q). \qquad (B.14)$$

The Nernst equation is important because it takes into account the activities of the reactive species and predicts the cell potential on the basis of these activities and the standard potential. It is important to note that the Nernst equation is valid only for relatively dilute solutions (total ionic concentration $<10^{-3}$ M) because, in more concentrated solutions, ions of opposite charge become bound in ion pairs, therefore reducing the total number of ions that are free to participate in the reaction [1][3]. The Nernst equation is also often simplified ($R = 8.3145$ J K^{-1} mol^{-1}) and expressed for a reaction at 25 °C and in the log$_{10}$ form shown here:

$$E = E° - \frac{0.059}{n}\log(Q). \qquad (B.15)$$

Often the pH of a solution plays a significant role in the potential of an electrochemical cell because many electron-transfer reactions involve either H^+ or OH$^-$ ions. The Nernst equation can be used to calculate how the cell potential changes with changes in pH, and the results can be graphed in E vs. pH plots known as Pourbaix diagrams. These diagrams are an excellent way to visualize what species are stable under various conditions [3].

For a comprehensive study of electrochemistry please refer to references [1][2][3].

B.2 Use of electrochemistry in energy and environmental science

One of the most common applications of electrochemistry is to drive a chemical reaction by applying an external voltage. This process, known as *electrolysis*, is commonly used to split water into hydrogen and oxygen as discussed in Chapters 27 and 47. In addition, Chapters 27, 28, 47, and 49 also discuss *photoelectrochemical cells* that convert light either to electrical energy or directly to various chemical products. Thus, the electrical power generated in the photoelectrochemical cell can be used directly or can be used to drive a reaction such as the splitting of water or CO_2 reduction. One of the main reasons why scientists are interested in splitting water is to produce hydrogen, an alternative fuel source, which is discussed in Chapters 28, 46, 47, and 49.

Electrochemical storage systems, including capacitors and batteries, are a second common area derived from electrochemistry. In a *battery*, electrical energy is stored as chemical energy and an electrochemical reaction is used to produce a voltage. In a *capacitor* the electrical energy is stored as surface charge on the electrodes. Batteries and capacitors are discussed in detail in Chapter 44.

The *fuel cell* is another common electrochemistry-based energy-supply system. A fuel cell is basically a battery in which one electrode is supplied with a fuel and the other electrode is supplied with an oxidizer to maintain a continuous electrochemical reaction and produce electricity. Fuel cells are also discussed in further detail in Chapters 44, 46, and 47.

B.3 References

[1] **A. J. Bard** and **L. R. Faulkner**, 2001, *Electrochemical Methods. Fundamentals and Applications*, 2nd edn., Hoboken, NJ, John Wiley & Sons, Inc.

[2] Southampton Electrochemistry Group, University of Southampton, *et al.*, 2001, *Instrumental Methods in Electrochemistry*, Chichester, Horwood Publishing Limited.

[3] **S. K. Lower**, 2004, "Chemical reactions at an electrode, galvanic and electrolytic cells," in *A Chem 1 Reference Text*, Burnaby, BC, Canada.

Appendix C | Units

Throughout this book an effort has been made to use Si and SI-derived units, but the diversity of the field and the, often independent development of areas that comprise it, have led to well-established use of non-SI units in some cases.

Where that was the case, we provide conversion in the chapter text or figure captions, if the figures that are suitable for the text, use non-SI units. Here we summarize the SI, SI-derived and other units, to facilitate use of the various parts of the book.

What is measured	SI (symbol) SI-derived unit (symbol)	Other units (symbol)
Distance	meter (m)	mile (m) $\approx$ 1609 m foot (f) $\approx$ 0.3048 m inch $\approx$ 0.0254 m
Area	m^2 Hectare (ha) $= 10^4$ $m^2 km^2$	acre $= 4046.85642$ m^2
Volume	cubic meter (m^3) liter (l) $= 0.001$ m^3	gallon [US] $= 3.7854$ l Barrel [of oil] $\approx$ 159 l Cubic feet (cf) $\approx$ 28. 32 l
Weight	kilogram (kg)	tonne (t) $=$ metric ton $= 1000$ kg pound (lb) $=$
Pressure	Pascal (Pa)	bar $= 100,000$ Pa Psi $= 6,895$ Pa
Energy	J ($=$ W.sec)	britsh thermal unit (btu) $= 0.293$ Wh
	Watt-hour (Wh) $= 3600$ J	Quad $= 10^{15}$ btu $= 1.055$ EJ
	Watt-year (Wyr) $= 8760$ Wh	barrel of oil equivalent (boe) $= 1,628$ kWh toe (ton of oil equivalent) $=$ 11,630 kWh
Power	Watt (W)	W_e : electrical power W_p : power at peak
Radioactivity	Becquerel (Bq)	Curie (Ci) $= 37$ GBq
Disintegrations/sec		
Exposure	Coulomb/kg (C/kg)	Roentgen (R) $= 258$ μC/kg
Magnetic Field Strength		
Velocity	km/h (kph)	miles/h (mph) $\approx$ 1.61 km/h

Prefixes and their symbols for orders of $10 > 0$, for use in the SI system

Yotta	Peta	Zetta	Exa	Tera	Giga	Mega	Kilo
Y	P	Z	E	T	G	M	k
10^{24}	10^{21}	10^{18}	10^{15}	10^{12}	10^{9}	10^{6}	10^{3}

Miscellaneous

Fuel consumption	m/l	miles/gal $\approx$ 425 m/l
	J/l	Btu/gal $\approx$ 279 m/l

Electrochemical potential abbreviations/symbol used in the book

Normal Hydrogen Electrode   NHE
Standard Hydrogen Electrode   SHE

Reversible Hydrogen Electrode   RHE
E_m mid-potential (used mostly in biology)

Insolation

It is often given in terms of Air Mass (AM) the mass of air through which the sunlight passes, where Air Mass 1 is the situation with the sun at its zenith in midsummer and midwinter at the equator, with completely clear sky.

Mostly AM1.5 is used where the solar zenith is at an angle of 48.2° from the normal to the earth's surface.

AM1.5G (where G stands for global, i.e., including direct and diffuse, scattered, radiation), corresponds to an insolation of 964 W/m² , with a given spectral distribution.

AM1.5D stands for the part of the radiation that is Direct, non-scattered radiation (~ 80%).

Index